Advances in Global Change Research

Volume 78

This book series is indexed in SCOPUS.

Advances in Global Change Research

Aims and Scopes

This book series is aimed at addressing a range of environmental issues through state-of-the-art and/or interdisciplinary approaches. The books in the series can either be monographs or edited volumes based, for example, on the outcomes of conferences and workshops, or by invitation of experts. The topics that the series can consider publishing include, but are not limited to:

Physical and biological elements of earth system science, in particular

- Climate change
- Biodiversity
- Sea-level rise
- Paleo-climates and paleo-environments

Social aspects of global change

- Environmentally-triggered migrations
- Environmental change and health
- Food security and water availability
- Access to essential resources in a changing world

Economic and policy aspects of global change

- Economic impacts
- Cost-benefit analyses
- Environmental governance
- Energy transition

Methodologies for addressing environmental issues

- Planetary data analysis
- Earth observations from space
- Proxy data analyses
- Numerical modeling
- Statistical analyses

Solutions to global environmental problems

- Sustainability
- Ecosystem services
- Climate services
- Technological/engineering solutions

Books in the series should be at least 200 pages in length, and include a table of contents.

Images supplied in colour will be reproduced in both the print and electronic versions of the book at no cost to the author/editor.

Manuscripts should be provided as a Word document and will be professionally typeset at no cost to the authors/editors.

All contributions to an edited volume should undergo standard peer review to ensure high scientific quality, while monographs should also be reviewed by at least two experts in the field.

Manuscripts that have undergone successful review should then be prepared according to the Publisher's guidelines manuscripts: https://www.springer.com/gp/authors-editors/book-authors-editors/book-manuscript-guidelines.

Himanshu Pathak · Dibyendu Chatterjee ·
Saurav Saha · Bappa Das
Editors

Climate Change Impacts on Soil-Plant-Atmosphere Continuum

Springer

Editors
Himanshu Pathak
ICAR-National Institute of Abiotic Stress Management
Baramati, Maharashtra, India

Saurav Saha
ICAR-Research Complex for NEH Region, Sikkim Center
Tadong, Gangtok, India

Dibyendu Chatterjee
ICAR-National Rice Research Institute
Cuttack, Odisha, India

Bappa Das
ICAR-Central Coastal Agricultural Research Institute
Old Goa, Goa, India

ISSN 1574-0919 ISSN 2215-1621 (electronic)
Advances in Global Change Research
ISBN 978-981-99-7934-9 ISBN 978-981-99-7935-6 (eBook)
https://doi.org/10.1007/978-981-99-7935-6

This Springer imprint is published by the registered company Springer Nature Singapore Pte Ltd.
The registered company address is: 152 Beach Road, #21-01/04 Gateway East, Singapore 189721, Singapore

Paper in this product is recyclable.

Preface

Anthropogenic GHG-emission-induced climate change became one of the most pressing global problems for our modern civilization, which has significant adverse effects on ecosystem functioning. The net impact of climate change on soil, plants and atmosphere continuum relies on the complex interaction among the soil, plants and atmosphere functions became obvious in terms of far-reaching detrimental consequences on ecosystem productivity, agricultural growth and human health. This book provides a comprehensive overview of the current understanding of the impact of climate change on the soil-plant-atmosphere continuum and improved our present understanding on mass and energy transitions in soil, plant and atmospheric processes in order to develop effective strategies to adapt and mitigate the negative effects of climate change.

The book primarily focused on examining the fundamental soil processes in the soil-plant-atmosphere continuum and their significant role in sustaining soil environment under modified climate. It then delves into the various processes actively involved in nutrient cycling, soil quality, resource use efficiency, soil carbon sequestration, land degradation, plant growth and water availability, and discusses how they are impacted by the climate change. It examines the impact of climate change on soil microorganisms and the potential implications for soil fertility and food security. The book also explores the impact on plant ecophysiology with respect to evaluation of dry matter allocation, rhizosphere interactions, management of other biotic and abiotic stress factors in modifying the net agro-ecosystem productivity towards anthropogenic climate change; followed by effective adaptation strategies for agronomic crops, weed management, manipulation of seeds physiology, selection of breeding ideotype and crop simulation modelling. The impact on atmosphere will account for the greenhouse gas measurements, mitigation options, eddy covariance measurement of GHGs, satellite-based monitoring, abiotic stress management options, air pollution and atmospheric modelling. The book also considers the potential for mitigating climate change impacts on the soil-plant-atmosphere continuum through various strategies, including changes in land use and agricultural practices, carbon sequestration and ecosystem restoration.

Throughout the book, the authors provide insights from their own research and draw on the latest scientific literature to present a comprehensive and up-to-date overview of the impacts of climate change on the soil-plant-atmosphere continuum. This book will help researchers, students and policymakers in understanding climate change impacts on the interaction processes among the atmosphere, soil and plant community from local to regional scale.

Baramati, India — Himanshu Pathak
Cuttack, India — Dibyendu Chatterjee
Tadong, India — Saurav Saha
Old Goa, India — Bappa Das

Contents

Contributors

Sneha Aggarwal School of Environmental Sciences, Jawaharlal Nehru University, New Delhi, India;
Department of Environmental Science, Stockholm University, Stockholm, Sweden

Javed Akhter Department of Atmospheric Sciences, University of Calcutta, Kolkata, India

Annamalai Anandan Indian Council of Agricultural Research, Indian Institute of Seed Science, Regional Station, Bangalore, India

Jayakishore Ankireddypalli Central Laboratory for Soil and Plant Analysis, Division of Soil Science and Agricultural Chemistry, ICAR-Indian Agricultural Research Institute, New Delhi, India

Koushik Bag ICAR-Indian Agricultural Research Institute, New Delhi, India

Santanu Kumar Bal ICAR-Central Research Institute for Dryland Agriculture (CRIDA), Santoshnagar, Hyderabad, India

Koushik Banerjee ICAR - Mahatma Gandhi Integrated Farming Research Institute, Piprakothi, Motihari, Bihar, India

Tanumoy Bera Texas A&M AgriLife Research, Beaumont, TX, USA

Basabduttaa Bhabai Central Seed Research Station for Jute and Allied Fibre, ICAR-CRIJAF, Budbud, West Bengal, India

Arti Bhatia Division of Environment Science, ICAR-Indian Agricultural Research Institute, New Delhi, India

Sayantan Bhattacharjee West Bengal University of Animal and Fishery Sciences, Kolkata, West Bengal, India

Niraj Biswakarma ICAR-Indian Agricultural Research Institute, New Delhi, India

Sunanda Biswas ICAR-Indian Agricultural Research Institute, New Delhi, India

Bidisha Chakrabarti Division of Environment Science, ICAR-Indian Agricultural Research Institute, New Delhi, India

Debasish Chakraborty ICAR-Research Complex for NEH Region, Umiam, Meghalaya, India

Ratul Chakraborty Department of Agricultural Meteorology and Physics, Mohanpur, Nadia, West Bengal, India

Somsubhra Chakraborty Agricultural and Food Engineering Department, IIT Kharagpur, Kharagpur, West Bengal, India

M. A. Sarath Chandran ICAR-Central Research Institute for Dryland Agriculture (CRIDA), Santoshnagar, Hyderabad, India

Dibyendu Chatterjee ICAR National Rice Research Institute, Cuttack, Odisha, India

B. U. Chaudhury ICAR Research Complex for NEH Region, Umiam, Meghalaya, India

Aaradhana Chilwal Punjab Agricultural University, Ludhiana, Punjab, India

Ashok Choudhury Uttar Banga Krishi Viswavidyalaya, Cooch behar, West Bengal, India

Ratul Roy Choudhury Department of Atmospheric Sciences, University of Calcutta, Kolkata, India

Saikat Chowdhury ICAR-Indian Agricultural Research Institute, New Delhi, India

Bappa Das ICAR-Central Coastal Agricultural Research Institute, Old Goa, India

Debarup Das Division of Soil Science and Agricultural Chemistry, ICAR-Indian Agricultural Research Institute, New Delhi, India

Lalu Das Department of Agricultural Meteorology and Physics, Mohanpur, Nadia, West Bengal, India

Saikat Ranjan Das ICAR-National Rice Research Institute, Cuttack, Odisha, India

Sanjib Kumar Das Department of Agronomy, Bidhan Chandra Krishi Viswavidyalaya, Mohanpur, Nadia, West Bengal, India

Shubhadip Dasgupta Department of Agricultural Chemistry and Soil Science, Faculty of Agriculture, Bidhan Chandra Krishi Viswavidyalaya, Nadia, West Bengal, India;
Agricultural and Food Engineering Department, IIT Kharagpur, Kharagpur, West Bengal, India

S. P. Datta Division of Soil Science and Agricultural Chemistry, ICAR-Indian Agricultural Research Institute, New Delhi, India;
ICAR-Indian Institute of Soil Science, Bhopal, India

Konjengbam Sarda Devi ICAR Research Complex for NEH Region, Manipur Centre, Imphal, Manipur, India

Abir Dey Central Laboratory for Soil and Plant Analysis, Division of Soil Science and Agricultural Chemistry, ICAR-Indian Agricultural Research Institute, New Delhi, India

Souvik Dey Mohanpur, Nadia, West Bengal, India

M. S. Dhanya Central University of Punjab, Bhatinda, Punjab, India

Gopal Dutta Bidhan Chandra Krishi Viswavidyalaya, Nadia, West Bengal, India

Susanta Dutta Birsa Agricultural University, West Singhbhum, Jharkhand, India

Tushar K. Dutta Division of Nematology, ICAR-Indian Agricultural Research Institute, New Delhi, India

Tridiv Ghosh ICAR-Indian Agricultural Research Institute, New Delhi, India

Nitika Gupta Advanced Centre for Plant Virology, Division of Plant Pathology, ICAR-Indian Agricultural Research Institute, New Delhi, India

Jaydeep Halder Division of Crop Protection, ICAR-Indian Institute of Vegetable Research, Varanasi, Uttar Pradesh, India

B. Hari Krishna ICAR-Indian Agricultural Research Institute, New Delhi, India

Lingaraj Huggi CoEWM, UAS, GKVK, Bangalore, India

Niveta Jain Division of Environment Science, ICAR-Indian Agricultural Research Institute, New Delhi, India

Krishna Kumar Jangid ICAR-National Institute of Abiotic Stress Management, Baramati, Maharashtra, India

Deepak Jhajharia Department of Soil and Water Conservation Engineering, College of Agricultural Engineering and Post Harvest Technology, Central Agricultural University (Imphal), Ranipool, Gangtok, Sikkim, India

Sudha Kanojia Division of Environment Science, ICAR-Indian Agricultural Research Institute, New Delhi, India

Megha Kaviraj ICAR National Rice Research Institute, Cuttack, Odisha, India

Shivani Kothiyal Punjab Agricultural University, Ludhiana, Punjab, India

Umesh Chandra Kulshrestha School of Environmental Sciences, Jawaharlal Nehru University, New Delhi, India

Dhiraj Kumar ICAR-Indian Institute of Soil Science, Bhopal, India

Sandeep Kumar Division of Environment Science, ICAR-Indian Agricultural Research Institute, New Delhi, India

Upendra Kumar ICAR National Rice Research Institute, Cuttack, Odisha, India

V. Lavanya Agricultural and Food Engineering Department, IIT Kharagpur, Kharagpur, West Bengal, India

Jayanta Layek ICAR-Indian Institute of Agricultural Biotechnology, Ranchi, Jharkhand, India

Anumalla Mahender Rice Breeding Innovation Platform, International Rice Research Institute (IRRI), Los Baños, Philippines

Debjyoti Majumder Krishi Vigyan Kendra - Malda, Ratua, West Bengal, India

Debashis Mandal ICAR-Indian Institute of Soil and Water Conservation, Dehradun, Uttarakhand, India;
ICAR-Indian Agricultural Research Institute, New Delhi, India

Naorem Loya Mangang ICAR Research Complex for NEH Region, Manipur Centre, Imphal, Manipur, India

M. H. Manjunatha AICRP on Agrometeorology, UAS, GKVK, Bangalore, India

Jitendra Kumar Meher Kreate Energy (I) Pvt. Ltd. Ansal Plaza, Sector-1, Vaishali, Ghaziabad, India

Kiran Kumar Mohapatra ICAR-National Rice Research Institute, Cuttack, India

Khitish Chandra Moharana ICAR-National Rice Research Institute, Cuttack, India

Surajit Mondal ICAR Research Complex for Eastern Region, Patna, Bihar, India;
ICAR Central Citrus Research Institute, Nagpur, Maharashtra, India

Nilimesh Mridha ICAR-National Institute of Natural Fibre Engineering and Technology, Kolkata, West Bengal, India

Baishnab Charan Muduli ICAR National Rice Research Institute, Cuttack, Odisha, India

Sushmita Munda ICAR National Rice Research Institute, Cuttack, Odisha, India

Varunseelan Murugaiyan Rice Breeding Innovation Platform, International Rice Research Institute (IRRI), Los Baños, Philippines

Manoj Kumar Nanda Mohanpur, West Bengal, India

Poulomi Nandy Department of Agronomy, Bidhan Chandra Krishi Viswavidyalaya, Mohanpur, Nadia, West Bengal, India

Manish Kumar Naskar Department of Agricultural Meteorology and Physics, Mohanpur, Nadia, West Bengal, India

Amaresh Kumar Nayak ICAR-National Rice Research Institute, Cuttack, India

Bitish Kumar Nayak ICAR-National Rice Research Institute, Cuttack, Odisha, India

Aditi Pahari Department of Agronomy, Bidhan Chandra Krishi Viswavidyalaya, Mohanpur, Nadia, West Bengal, India

Siddharth Panda Institute of Agricultural Sciences, SOA (DU), Bhubaneswar, Odisha, India

Prashant Pandey ICAR-Research Complex for North Eastern Hill Region, Tadong, Sikkim, India

Bappa Paramanik Dakshin Dinajpur KVK, Uttar Banga Krishi Viswavidyalaya, Dakshin Dinajpur, West Bengal, India

Subhra Parija ICAR National Rice Research Institute, Cuttack, Odisha, India

Himanshu Pathak Indian Council of Agricultural Research, New Delhi, India

Krishnayan Paul ICAR-Indian Agricultural Research Institute, New Delhi, India

Victor Phani Department of Agricultural Entomology, College of Agriculture, Dakshin Dinajpur, West Bengal, India

Abhijit Pradhan ICAR National Rice Research Institute, Cuttack, Odisha, India

Achintya Pramanik Discipline of Agricultural Entomology, Quality and Productivity Improvement Division, ICAR-Indian Institute of Natural Resins and Gums, Ranchi, Jharkhand, India

Tapan Jyoti Purakayastha ICAR-Indian Agricultural Research Institute, New Delhi, India

Feroze Hasan Rahman ICAR-National Bureau of Soil Survey and Land Use Planning, Regional Center, Kolkata, West Bengal, India

Jagadish Rane ICAR-National Institute of Abiotic Stress Management, Baramati, Maharashtra, India

Deb Prasad Ray ICAR-National Institute of Natural Fibre Engineering and Technology, Kolkata, West Bengal, India

Prasenjit Ray Division of Soil Science and Agricultural Chemistry, ICAR-Indian Agricultural Research Institute, New Delhi, India

Debasish Roy ICAR-Indian Agricultural Research Institute, New Delhi, India

Trisha Roy ICAR-Indian Institute of Soil and Water Conservation, Dehradun, Uttarakhand, India

Sonam Sah G.B. Pant, University of Agriculture and Technology, Pantnagar, Uttarakhand, India

Salil Saha AMFU, Regional Research Station (Old Alluvial Zone), Majhian, West Bengal, India

Saurav Saha ICAR-Research Complex for NEH Region, Sikkim Centre, Tadong, Sikkim, India

Sunayan Saha ICAR-Central Potato Research Institute, Regional Station Jalandhar, Punjab, India

Bhabagrahi Sahoo School of Water Resources, Indian Institute of Technology, Kharagpur, West Bengal, India

Rajalaxmi Sahoo ICAR-National Rice Research Institute, Cuttack, India

Soumyaranjan Sahoo School of Water Resources, Indian Institute of Technology, Kharagpur, West Bengal, India;
Istituto di Ricerca per la Protezione Idrogeologica (IRPI), Consiglio Nazionale delle Ricerche (CNR), Perugia, Italy;
Surface Water Hydrology Division, National Institute of Hydrology, Roorkee, Uttarakhand, India

Suman Samui Department of Agronomy, Faculty of Agriculture, Bidhan Chandra Krishi Viswavidyalaya, Mohanpur, Nadia, West Bengal, India

Disha Sandilaya Division of Environment Science, ICAR-Indian Agricultural Research Institute, New Delhi, India

Abhijit Sarkar ICAR-Indian Institute of Soil Science, Bhopal, M.P., India

Ayan Sarkar ICAR-Indian Agricultural Research Institute, New Delhi, India

Abhiliasha Sharma Division of Environment Science, ICAR-Indian Agricultural Research Institute, New Delhi, India

Aribam Priya Mahanta Sharma Department of Soil and Water Conservation Engineering, College of Agricultural Engineering and Post Harvest Technology, Central Agricultural University (Imphal), Ranipool, Gangtok, Sikkim, India

Susheel Kumar Sharma ICAR Research Complex for NEH Region, Manipur Centre, Imphal, Manipur, India;
Advanced Centre for Plant Virology, Division of Plant Pathology, ICAR-Indian Agricultural Research Institute, New Delhi, India

Nihar Ranjan Singh Ravenshaw University, Cuttack, Odisha, India

R. N. Singh ICAR-National Institute of Abiotic Stress Management, Baramati, Maharashtra, India

Raghuveer Singh ICAR Research Complex for NEH Region, Arunachal Pradesh Centre, Basar, Arunachal Pradesh, India

Shankarappa Sridhara Center for Climate Resilient Agriculture, KSNUAHS, Shivamoga, India

Rajiv Kumar Srivastava Institute of Soil Science and Land Evaluation, Biogeophysics, University of Hohenheim, Stuttgart, Germany

Chinmaya Kumar Swain ICAR-National Rice Research Institute, Cuttack, Odisha, India

M. N. Thimmegowda AICRP on Agrometeorology, UAS, GKVK, Bangalore, India

Ankita Trivedi ICAR-Indian Agricultural Research Institute, New Delhi, India

Ghanashyam Singh Yurembam Department of Soil and Water Conservation Engineering, College of Agricultural Engineering and Post Harvest Technology, Central Agricultural University (Imphal), Ranipool, Gangtok, Sikkim, India

Part I
Climate Change and Soil Environment

Chapter 1
Impacts of Climate Change on Soil Processes

Dibyendu Chatterjee, Saikat Ranjan Das, Saurav Saha, Abhijit Sarkar, and Himanshu Pathak

Abstract Understanding the interactions between climate change and various soil processes helps in understanding the impacts and developing strategies to adapt and mitigate climate change. In this chapter, we have described a brief overview of climate change and greenhouse gas concentrations, as well as different representative concentration pathways (RCP) scenarios and changes in soil processes on a global scale. We have also described the impacts of climate change on various soil processes, including soil formation, clay mineral transformation, soil erosion, structural degradation, hydrological processes, soil air, and thermal regimes, nutrient cycling, ion exchange, nutrient availability, acidification, salinization, alkalinization, microbial changes, and rhizosphere chemistry. In addition, we also described the effects of climate change on seven soil functions and soil processes such as C sequestration, aggregation, porosity, humification, podsolization, translocation of C deep into the soil profile, soil water regime, and nitrification–denitrification.

Keywords Adaptation · Climate change · Greenhouse gas · Mitigation · Soil processes · Soil properties

D. Chatterjee · S. R. Das
ICAR-National Rice Research Institute, Cuttack 753006, Odisha, India

S. Saha
ICAR-Research Complex for North Eastern Hill Region, Tadong 737102, Sikkim, India

A. Sarkar
ICAR-Indian Institute of Soil Science, Navibagh, Bhopal 462038, M.P., India

H. Pathak (✉)
Indian Council of Agricultural Research, New Delhi 110001, India
e-mail: hpathak.iari@gmail.com

H. Pathak et al. (eds.), *Climate Change Impacts on Soil-Plant-Atmosphere Continuum*, Advances in Global Change Research 78, https://doi.org/10.1007/978-981-99-7935-6_1

1.1 Introduction

Urbanization, industrialization, fossil fuel use, land use change, improper fertilizer use, and intensive agriculture are the possible causes of increasing emission of greenhouse gasses (GHGs) such as carbon dioxide (CO_2), methane (CH_4), and nitrous oxide (N_2O) into the atmosphere, leading to global climate change (Pathak, 2015; Pathak et al., 2010, 2019). In India, agriculture contributes 417.22 million tons of CO_2 equivalent per year, about 16% of India's total GHG emissions, of which methane (CH_4) accounts for 74% and nitrous oxide (N_2O) accounts for 26% (MoEFCC, 2018; Saha et al., 2018). Climate change impacts physical, biological, and ecological processes due to the direct and indirect effects of rising temperature, increase in GHG concentration in the atmosphere, changing precipitation patterns, sea level rise, soil erosion and degradation, biodiversity loss, erratic precipitation pattern, increased frequency of extreme precipitation events, flash floods, increased frequency and magnitude of floods, droughts, cyclones, heat stress, cold stress and prolonged dry spells (Chatterjee & Saha, 2015; Nayak et al., 2020). Climate change occurs slowly, with small changes in temperature, CO_2, and precipitation accumulating over a long period. Soils as a natural resource are linked to slow climate change, which directly and indirectly affects various soil properties such as soil formation, soil erosion, soil structure degradation, soil hydrology, soil air and heat regimes, nutrient cycling (C, N, P, and S), ion exchange, nutrient availability, soil acidification, soil salinization, soil alkalinization, soil microbial changes, and soil rhizosphere chemistry. All these changes in soil processes ultimately affect soil fertility and crop productivity posing a threat to food security. Nutritional security is also at stake as increasing CO_2 content in the atmosphere leads to a decrease in protein content in major food crops by about 10–15% (Taub et al., 2007). In addition, higher CO_2 levels may lead to forced closure of stomata, hindering the uptake of important micro- and macronutrients from the soil (Loladze, 2002; Saha et al., 2020). Soil type, pH, soil moisture, salinity, and organic matter content are the factors responsible for increased mineralization of organic matter. Elevated temperatures increase soil labile C and β-glucosidase activities, which increase C, N, P, and S mineralization and increase nitrate formation through enhanced nitrification and promote more N losses through nitrate leaching (Chatterjee & Saha, 2018a, 2018b; Manna et al., 2005; Pendall et al., 2004). Soil microbial processes are also affected by climate change through altered environmental factors such as GHG content, soil moisture, soil temperature, enzyme activity, and nutrient availability (Bradford, 2013; IPCC, 2007).

1.2 Climate Change Scenarios and Soil Processes

Intergovernmental Panel on Climate Change (IPCC) presented the concept of Representative Concentration Pathway (RCP) in its Fifth Assessment Report (AR5) (IPCC, 2014a, 2014b). The RCP is the greenhouse gas concentration trajectory based on the number of GHGs to be emitted in the coming years that can produce radiative forcing of 2.6, 4.5, 6.0, and 8.5 W m^{-2} in 2100 AD. Four RCPs i.e., RCP 2.6, RCP 4.5, RCP 6.0, and RCP 8.5 are used in different models to predict the possible changes in soil processes.

The RCPs 2.6, 4.5, 6.0, and 8.5 have been designed by different models such as IMAGE (PBL Netherlands Environmental Assessment Agency, Netherlands), GCAM (Pacific Northwest National Laboratory's Joint Global Change Research Institute, USA), AIM (National Institute for Environmental Studies, Japan) and MESSAGE (International Institute for Applied Systems Analysis, Austria) (van Vuuren et al., 2011, Wayne, 2014). Mean temperature is predicted to change by 0.3–4.8 °C, with the highest increase (average 3.7 °C) predicted in RCP 8.5 (Table 1.1). The increase in greenhouse gas emissions is predicted for 490–1370 ppm CO_2 equivalent. In RCP 2.6, CO_2 equivalent peaks around 2050, followed by a decline to about 400 ppm in 2100 AD, while CO_2 equivalent is expected to stabilize in RCP 4.5 and 6.0, but rapidly increasing concentrations are observed in RCP 8.5 (Wayne, 2014). Sea level rise is projected to be between 0.4 and 0.63 m at the end of the century.

Soil processes are also affected differently by the changes in GHG concentrations, temperature rise, and sea level change that occur under RCPs at different levels. In a study in the high plains of the USA, grasslands were observed to increase soil organic C sequestration under both moderate (RCP 4.5) and severe warming (RCP 8.5) even by 2100 AD, while wheat field crop rotations realized C increase only under moderate warming (RCP 4.5) by 2050 (Robertson et al., 2018). One study estimated that C sequestration of 0.68 Pg C $year^{-1}$ for 85 continuous years is capable of reducing global temperature by 0.1 °C in 2100 under RCP 2.6 (Mayer et al., 2018). Global water erosion is projected to increase by 30–66% (Borrelli et al., 2020). Land use change will have a differential impact on global water erosion, as it can either decrease by 10% under RCP2.6 or increase by 2% (RCP4.5) to 10% (RCP8.5) by 2070 (Borrelli et al., 2020). In contrast, some parts of the world, including Australia, Southern China, the eastern United States, Western Europe, South America, and Sub-Saharan Africa will face deficit moisture stress within another 40 years (2060 AD), as projected under both RCP 2.6 and 8.5.

Table 1.1 Different RCPs and projected change in GHGs (CO_2 equivalent), temperature, and sea level at the end of the twenty-first century (2081–2100)

RCP	Temperature (°C)	CO_2 equivalent (ppm)	Sea level rise (m)
2.6	+1.0 (0.3–3.8)	+490	+0.40 (0.26–0.55)
4.5	+1.8 (1.1–2.6)	+650	+0.47 (0.32–0.63)
6.0	+2.2 (1.4–3.1)	+850	+0.48 (0.33–0.63)
8.5	+3.7 (2.6–4.8)	+1370	+0.63 (0.45–0.82)

Source Wayne (2014)

1.3 Climate Change Impacts on Various Soil Processes

1.3.1 Soil Formation

Weathering of parent rock and formation of soil is a function of climate, biota, relief/topography, parent material, and time. Thus, climate is one of the five soil-forming factors that control soil formation and properties. Soil formation in hot and humid climates is faster than in cold and dry climates due to faster chemical reactions and biological activity in hot and humid climates.

1.3.2 Transformation of Clay Minerals

Soil formation also involves the transformation of minerals from primary to secondary and the neosynthesis of secondary minerals from silica and alumina-enriched solutions. Climate change can accelerate or inhibit the transformation of clay minerals (Ghosh et al., 2019; Karmakar et al., 2016). Increased atmospheric CO_2 due to climate change can enhance the assimilation of C and its allocation below ground, which in turn increases the microbial population that enhances mineral weathering (Karmakar et al., 2016). Various processes related to clay mineral transformation are also affected by global climate change, such as hydrolysis, cheluviation, and ferrolysis, which are accelerated by increased leaching rates (Karmakar et al., 2016). Sea level rise due to climate change will result in less mineral formation due to oxidation of sulfide minerals in coastal areas. Clay minerals such as montmorillonite, palygorskite, and analcime, which are formed under neutral to strongly alkaline conditions, can be observed in arid areas (Bush et al., 2010).

Clay minerals can mitigate climate change by increasing the C sequestration potential of some soils. Soils developed from volcanic ash are rich in short-range order minerals and show very high C sequestration potential (Chatterjee et al., 2021a, 2021b). The content of such minerals can also be increased by agronomic practices (Chatterjee et al., 2014a). In a study with maize, it was observed that certain types of short-range order minerals are capable of accumulating non-labile fractions of C that may have negative feedback to climate change (Chatterjee et al., 2013, 2014a, 2014b, 2015a, 2015b, 2016).

1.3.3 Soil Erosion

Soil erosion causes a loss of agriculturally important topsoil of 5.3 billion tons at a rate of 16.4 tons ha^{-1} $year^{-1}$ in India (Bhattacharyya et al., 2015). The global estimate of soil erosion is 43 Pg per year, estimated based on high spatial resolution Revised Universal Soil Loss Equation (RUSLE)-based semiempirical modeling approach (GloSEM) (Borrelli et al., 2020). Climate change has increased the number and frequency of extreme events and soil erosion is probably one of the quick visible responses to climate change (Borrelli et al., 2020). An increase in intensity of heavy rainfall increases surface runoff which is the main reason for increase in soil erosion caused by water. Apart from rainfall intensity, topography, soil erodibility, and vegetation cover influence soil erosion. Climate change may alter vegetation cover and land use. Moreover, degradation of land caused by soil erosion gradually makes it unsuitable to use for agricultural activities and growth of any other types of vegetation. Reducing the number of vegetation is also responsible for increasing greenhouse gas concentration and increasing the direct impact of rainfall, which increases soil erosion.

Problem of wind erosion is more prominent in a dry and arid regions. Climate change impacts global precipitation patterns with more extreme events. It is predicted that the wet areas will be converted wetter than before and the dry areas will become drier compared to previous situations (Feng & Zhang, 2015). A drier climate will promote accelerated wind erosion under climate change scenario. Since accelerated wind erosion (also true for water erosion) removes the top agriculturally fertile layer of soil, there will be an acute deficiency of plant-available nutrients in highly eroded soils, which again affects the vegetation cover (Ioras et al., 2014).

1.3.4 Soil Structural Degradation

Impact of climate change on soil structural changes (formation or degradation) involves complex processes as climate change influences the type, arrangement, and stability of aggregates in soil (Várallyay, 2010). Soil structure is directly related to soil porosity, aeration, and stable aggregates, which can physically protect organic matter by preventing O_2 transport through small pores and thus aerobic microbial decomposition, which in turn is a critical factor in the release of CO_2 and CH_4 from the soil, a major greenhouse gas that contributes to climate change (Tóth et al., 2018). Destruction of soil aggregate is impacted directly by the splashing of raindrops, surface runoff, and infiltration during heavy rainfall events caused by climate change (Gelybó et al., 2018; Várallyay, 2010). Climate change indirectly impacts soil structure through changes in land use and vegetation cover. The process of soil structural degradation is also influenced by increase in temperature, low water availability, and low level of soil organic matter which reduce the size and stability of soil aggregates (Carrizo et al., 2015).

1.3.5 Soil Hydrological Process

Hydrological processes are sensitive to climate change as it modifies the precipitation pattern, which exhibits a direct impact on water availability (Pandey et al.,

2017; Yadav et al., 2010). In a study using soil and water assessment tool (SWAT) model, it was reported that modifications in the level of CO_2, temperature, and precipitation levels under climate change scenarios influenced water productivity and evapotranspiration (Ficklin et al., 2009; Pandey et al., 2017).

1.3.6 Soil Air and Heat Regime

A close relationship is observed between soil and air temperature. An increase in air temperature due to climate change will also influence soil temperature positively. The incoming and outgoing solar radiation is responsible for maintaining the heat regime in the soil through the processes of conduction i.e., heat transfer through the soil profile, convection i.e., heat transfer through the water and gas molecules, and evaporation (Karmakar et al., 2016). Increased soil temperature is responsible for acceleration of various soil processes such as an increase in microbial activity, rapid decomposition of soil organic matter, higher nutrient mobilization rate, and increased nitrification rate. Climate change increases the temperature and decreases the moisture content of soil, which leads to a higher metabolic rate of microbes and a higher CO_2 concentration in soil pores (Feng et al., 2017). It is predicted that the gradual shift in the behaviour of climate may change the growing zones, making some zones unsuitable for some crops (McCarl, 2017).

1.3.7 Nutrient Cycles (C, N, P, and S)

Climate change affects nutrient supply, nutrient cycling, and soil–plant interactions in the biosphere. Soil is the storehouse of about 1550 Pg of soil organic C (Chatterjee & Saha, 2018a, 2018b). An increase in temperatures affects the soil organic C stock, which reduces soil organic C content and C:N ratio. Soil C stocks help to hold water and other nutrients in the soil and also serve as a source of energy for decomposing microbes. This C can be lost from the soil through increasing microbial respiration and subsequently accumulated into the atmosphere in the form of CO_2. Global C cycle is controlled by C exchange into and out of various pools such as geological (4130 Pg), oceanic (38,153 Pg), pedologic (2730 Pg), atmospheric (780 Pg), and biotic (500–650 Pg) (Lal, 2008a, 2008b). Exchange of CO_2 within the atmosphere, vegetation, fossil fuels, oceans, and soils are major drivers controlling global C cycle (Bhattacharyya et al., 2018). Soil processes such as soil respiration (by roots and microbes), decomposition, and mass transfer are the main causes of emission of CO_2 into the atmosphere.

The nutrient cycling of N is closely linked to the cycling of soil organic C. Climate change induced increased in temperature and precipitation strongly affect N deposition in the atmosphere and also affect the nutrient cycling of other nutrients such as P and S (Kumar & Swarup, 2012). The process of N mineralization increases with increased temperature. The available N can promote C storage in the soil by allowing higher plant growth. However, when the availability of N exceeds a certain limit, it can be lost as nitrate leaches from the soil and accumulate in rivers, leading to eutrophication. Global monitoring of N status is important to analyze how available N is affected by global climate change. The C:N ratio in soil decreases as temperature increases. In one experiment, the lowest values (9.29 and 10.45) were observed at temperatures between 36 and 41 °C and higher values (12.15–15.95) were observed in the temperature range between 27 and 35 °C (Adeyeye et al., 2018). This increase in temperature harms soil C:N ratio and plant performance is also reduced. Higher temperature activates most of the enzymes controlling the N cycle and the effect becomes more visible when temperature is limiting (winter) and during the time of higher microbial activities (spring) in soil (Beier et al., 2008). Formation of NH_4 and its concomitant uptake by plants decreases its content, while enhanced nitrification results in increased soil NO_3 content, which increases the chances of its leaching (Zhang et al., 2020). However, the effect of moisture is more prominent in controlling different processes in relation to global N cycling (Beier et al., 2008). Under elevated CO_2 in atmosphere, allocation of C to below-ground C is increased through root exudation, which may stimulate either microbial N immobilization or mineralization due to priming effect depending on the C:N ratio of a decomposing substrate (Bengtson et al., 2012).

Warmer soil temperatures significantly manipulated soil phosphorus status by altering plant phenology. When the temperature changes from 33 to 38 °C, P loss increases by 16–20%, but above the temperature of 38 °C, soil P loss was suppressed by 1% (Nord & Lynch, 2009). In a study, it was reported that the P availability decreased by 18.8% under elevated temperature (+2 °C) at the jointing stage of wheat (Zhang et al., 2014). Phosphatase enzymes that promote P mineralization produced by root and soil microbes are boosted under elevated CO_2 (Jin et al., 2015). Soil available P content was reported to be increased by 14.3% in the jointing stage of wheat under elevated CO_2 (Zhang et al., 2014).

Climate change potentially affects the S cycle in soil. Temperature controls the rates of mineralization of native organic sulphur and oxidation (to SO_4^{2-}) and transformation of elemental sulphur to soil organic S in soil (Jaggi et al., 1999). The rate of sulphur mineralization at 36 °C was higher in neutral soil, followed by alkaline and acidic soils (Jaggi et al., 1999). Under elevated CO_2, there will be dissolution of CO_2 into the ocean that cause a reduction of oceanic pH and acidify it. Due to ocean acidification led by climate change the formation of dimethyl sulphide will be increased (Dutta & Dutta, 2016). This compound acts as a nuclei for cloud and cause acid rain (sulphuric acid) through which sulphur from the oceans is transported to the soil (Glöckner et al., 2012). The impact of climate change on C, N, P, and S cycle is illustrated in Fig. 1.1.

Climate Change Impact	Nutrient Cycle: C cycle	N cycle	P cycle	S cycle
Elevated Temp	Reduce soil organic carbon content Reduce C:N ratio (Wan et al. 2011)	Nitrogen mineralization rate increases (Joshi et al. 2006)	Phosphorus loss increases (Zhang et al. 2014)	Increase the rates of mineralization of native organic S (Jaggi et al. 1999)
Elevated CO_2	Increase soil respiration (by roots and microbes) and decomposition (Carney et al. 2007)	Stimulate either microbial N immobilization (Zak et al. 2000) or mineralization due to priming effect (Blagodatskaya and Kuzyakov 2008)	P mineralization is boosted and increased available P in soil (Dijkstra et al. 2003)	Transport of oceanic sulphur to soil due to oceanic acidification (Glöckner et al. 2012)
Altered Moisture level	Reduced decomposition of soil organic matter under deficit moisture (Chatterjee and Saha 2015)	Increased leaching of NO_3 in excess moisture Limited enzymes related to mineralisation and nitrification under deficit moisture (Beier et al. 2008)	P mineralization is affected under deficit stress of moisture (Chatterjee and Saha 2015)	S mineralization is affected under deficit stress of moisture (Weil and Magdoff 2004)

Fig. 1.1 Impact of climate change on C, N, P, and S cycles

1.3.8 Ion Exchange

Cation exchange capacity (CEC) and sorption are considered to be the most important properties for the retention of important soil cations such as K^+, Ca^{2+}, Mg^{2+}, and fixation of potentially toxic cations Al^{3+} and Mn^{3+}. Due to increased global temperature, a decrease in CEC is caused by the loss of SOM from the low active clay soil and coarse-textured soils (Davidson & Janssens, 2006). This may lead to an increase in the leaching of basic cations in response to extreme rainfall, allowing alkalinity from the soil to enter the aquatic environment. It is reported that an increase in temperature from 30 to 41 °C can lead to a decrease in CEC from 7.71 to 26.75% under the experimental conditions when there is a deficiency of Mg^{2+}, K^+ and other cations (Joseph et al., 2018). Anion exchange capacity is increased with decrease in pH of the soil which may happen due to an increase in temperature and an altered precipitation pattern (Joseph et al., 2018).

1.3.9 Nutrient Availability

The availability of nutrients to plants depends on the chemical properties of the soil. Factors such as the location of the nutrient near the root surface and the length of the path an ion travels to reach the root surface also determine nutrient availability in the soil. To become available, the nutrient must be transformed biologically. The transformation between the different pools i.e., the organic and inorganic pools is highly influenced by the temperature and moisture regime of the soil, so climate change can have a strong influence on the concentration of N and S in the soil solution as well as on the changing C pool. Although increased CO_2 in the atmosphere does not directly affect the process of N mineralization in the soil, the associated warming of the soil leads to an increased rate of N mineralization, which results in the movement of a higher amount of N in solution (Pendall et al., 2004). The availability of the available potassium fraction in the soil solution is not much controlled by biological activity. With increased soil temperature, rates of adsorption/desorption reactions in the soil increase, altering potassium availability. In plants, higher temperatures affect stomatal closure by increasing the vapour pressure deficit, and nutrient transport by mass flow also decreases (Cramer et al., 2009). The combination of increasing CO_2 and warming improved the availability of several micronutrients such as Zn, Fe, and Cu in the soil (Wang et al., 2016). In addition, these environmental factors had a significant effect on micronutrient uptake and translocation.

1.3.10 Acidification

Significant rainfall increases the downward movement of water along with basic cations through the soil profile causing soil acidification. The process of soil acidification alters soil chemistry by lowering pH, which contributes to the mobilization of potentially toxic elements (mostly heavy metals such as Pb, Cr, Cd, and Hg) and can lead further to the depletion of basic cations through leaching. Experimental studies have shown a decrease in aqueous pH of 1–3 units in water present in soil pores due to exposure to a higher concentration of CO_2 gas (Vong et al., 2011; Wei et al., 2011). The presence of highly reactive carbonate clay minerals shows a high buffering capacity of the soil that can resist the change in pH. On the other hand, sandy soils generally have low buffering capacity, and a drop in pH is often observed, which can have long-term effects compared to a system with a high buffering capacity (Wilkin & DiGiulio, 2010).

1.3.11 Salinization

Soil salinization is determined by many factors, such as excessive water evaporation due to a low water table and inundation with saline water in coastal regions. The phenomenon of global warming increases the temperature in the soil, that adversely impact the soil moisture regime and triggers the process of rapid evapotranspiration (Dagar et al., 2016). The increase in temperature and evapotranspiration can lead to capillary movement of water and consequently evaporation of groundwater, which causes salt accumulation (i.e., salinization) within the soil profile (Várallyay, 2007). Another consequence of climate change is sea level rise. The rising sea level rapidly increases the number of flood incidences in the coastal areas and brings these areas under the influence of seawater intrusion. The phenomenon of soil salinization occurs in different geographical regions characterized by different climatic conditions. However, arid and semi-arid regions are more affected and face serious challenges of soil salinization (Kurylyk & MacQuarrie, 2013).

1.3.12 Alkalization

Climate change is not only triggering temperature rise in many parts of the world but is also responsible for unpredictable precipitation patterns. The change in temperature and precipitation increased evaporation, which brings alteration in land use. Increased alkalinization occurs in areas where precipitation has decreased significantly. Alkalization increases capillary rise due to increased evaporation and scarcity of rainfall which brings salts like NaCl onto the surface and forms saline-sodic soils. The occurrence of leaching during precipitation events may also be limited by surface sealing of particles, which increases subsoil drying and increases the concentration of salts within the soil solution.

1.3.13 Microbial Changes

Climate change mainly affects the function and abundance of soil microorganisms by changing the respiration rate of soil microorganisms and some soil functions mediated by them, such as nitrification, denitrification, N fixation, and methanogenesis (Bakken et al., 2012). Soil organic matter (SOM) can be considered as the most available C fraction and a good indicator of changes in various soil processes related to soil nutrient and energy dynamics (Saha & Mandal, 2009). The activity of soil microorganisms is mainly affected by abrupt climate change. Increased temperatures due to more greenhouse gas emissions change precipitation patterns, which may directly affect soil microbes (Bardgett et al., 2008). The interaction between climate and microorganisms in the rhizosphere region increases the activity of heterotrophic microorganisms (Classen et al., 2015), which will increase the emission of CO_2 into the atmosphere. It is estimated that global warming will increase the net shift of the C pool from the soil to the atmosphere due to the higher respiration rate of soil microbes.

1.3.14 Rhizosphere Chemistry

The rhizosphere is considered as a region under the influence of the plant root. It is the region surrounding the plant root where vital plant-soil interactions take place, including synergistic, antagonistic, and neutral reactions (Baetz, 2016). Microbial activity in this region exerts an important role in soil nutrient dynamics i.e., the behavior, fate, and uptake of nutrients by the plant (Kayler et al., 2017). The fluctuation in global temperature, rise in the water table, and increase in CO_2 concentration are characterized by global climate change (Dutta & Dutta, 2016). The increased CO_2 concentration due to oxidation of soil organic C may accelerate the photosynthetic rate of plants and increase the exudation of roots, thus increasing the concentration of toxic substances in the root zone. Due to the presence of higher toxic substances, microbial activity in the root zone was hindered and increased immobilization of N and denitrification occurred (Tripathi et al., 2015). However, amino acids, organic acids, and sugars abundant in root exudates are rapidly assimilated by soil microorganisms, especially bacteria, and are thought to play a key role in controlling microbial population dynamics in the root rhizosphere.

1.4 Soil Function and Climate Change

Since the climate change impact on soils is a slow process, the direct and visible impacts on soils are less common. For instance, temperature-mediated decomposition of soil organic matter or change in soil moisture due to evapotranspiration by higher plants. Quantification of changes in soil function affected due to climate change is a complicated issue. Moreover, the interplay of various soil-forming processes, more specifically the soil biological soil-forming processes, makes this quantification further complicated. In order to address such problems, the European Commission adopted some strategies in 2006 for sustainable use of soil by protecting the soil ecosystem. The Commission highlighted seven vital functions of soil, which are: (i) food and biomass production, (ii) storing, filtering, transforming, and recycling water and nutrients, (iii) providing habitat and gene pool, (iv) serving as soil organic C (SOC) pool, (v) providing raw materials, (vi) serving as a physical and cultural environment for the mankind and (vii) storing the geological and archaeological heritage. We will discuss the above functions of soils in this chapter.

1.4.1 Food and Biomass Production

Global climate change will impact the four components of food security: food accessibility, food openness, food utilization, and food framework dependability (Al et al., 2008). The impact is now being perceived in worldwide food markets and is probably going to be critical in the marginal lands (problem soils) and knowledge-poor rural areas where yield decline is very common. This will lead to the scarcity of food in rural areas and further affect the supply chains to urban and peri-urban areas that will soar food prices, loss in resources and business opportunities, reduction in purchasing power, and ultimately threaten food and nutritional security of humans and livestock. Environmental climate change can directly affect food security by disrupting the stability of the food system through increasing events of climate hazards, such as drought, flood, cyclone, heat wave, cold waves, etc. Apart from providing food to all, agricultural sector also provides jobs to about 36% of the world's all-out labour force (Pareek, 2017). Nations like India are even more vulnerable considering the tropical climate and poor capability of adaptation by small and marginal farmers. It is predicted that environmental change will fundamentally impact agriculture through immediate and divergent impacts on crops, soils, animals, and pests. However, environmental change is a slow cycle with moderately little changes in precipitation and temperature throughout an extensive period. It doesn't assess the other conceivably huge effects that environmental change could have on the worldwide food framework, especially on market costs. These effects include the cost of energy and water consumed in preparing, transporting, cold-storing, and producing food, as well as the cost of the food itself. A rise in sea level and increase in frequency of extreme weather conditions present a new threat to the resources of individuals living in marginal areas, undermining livelihoods and increasing a threat to future food security globally. The expected impact of environmental changes on food security should be seen in the bigger global arena of changing climate.

1.4.2 Storing, Filtering, Transforming, and Recycling Water and Nutrients

Environmental changes are predicted to negatively affect human health and nutrition. Accessibility of nutrients by the plant in the soil is a component of soil chemical properties as nutrient has to travel a path within the soil to reach the root adsorption site. Altered precipitation and rise in atmospheric temperature fundamentally affect temperature and moisture regime in root zone. Notably, soil temperature and moisture are essential determinants of nutrient availability, root development, and belowground C allocation due to which it is expected that change in these two variables will impact the storing, filtering, transforming, and recycling of water and nutrients interaction in rhizosphere. The degree and nature of the adjustment of these two parameters will be soil-specific and site-specific. Nutrient use efficiency is mainly influenced by direct effects on root surface area and inflow rate (Brouder & Volenec, 2008). Natural transformation among inorganic and organic pools is more likely to be impacted by temperature and moisture, and in this manner, climate change may impact N, P, and S concentrations in soil solutions. N mineralization is likely to be increased, resulting in more N in solution phase available for plants. Rates of adsorption/desorption in soil are predicted to accelerate with soil moisture and temperature changes, which may further be adjusted by modifying the ionic strength of the soil solution. Carbon dioxide (CO_2), for example, can negatively affect the nutrient content of some crops like wheat, rice, potatoes, and peas. For example, zinc and iron concentrations in wheat and rice will decrease due to increased emissions of ozone-depleting gasses (Myers et al., 2014). It is believed that higher CO_2 levels combined with environmental changes will lead to micronutrient deficiencies (Smith et al., 2017).

Environmental changes will affect soil water levels through immediate climatic influences (temperature and precipitation control on evaporation), environmental changes in soil water extraction rates by plants, plant development rates, vegetation, and the impact of increased CO_2 levels on plant events. Alteration in soil water movement may contribute to droughts by diminishing accessible moisture and modifying soil water distribution in soil profile. Increased CO_2 level decreases stomatal conductance in numerous species. Conductance of leaf was reported to decrease by 77–86% under the elevated CO_2 compared to ambient (Saha et al., 2020). Increased CO_2 levels will bring about site-specific changes in water availability, a decrease in total water use, and an increase in water use efficiency, which may alter total nutrient requirements by plants.

1.4.3 Providing Habitat and Gene Pool

Soil serves as a good biological habitat and gene pool reserve for a variety of soil-dwelling species or microorganisms. Soil is considered as a living entity and therefore can support the survival of many living organisms. In the soil ecosystem, all organisms can interact not only with each other but also with the components of the soil and the atmosphere. Global climate change interferes with the distribution of species dwelling in soil and their way of interacting with each other. Soil biota cooperates in complex networks with a variety of different species, some of which are helpful, some of which are pathogenic, and some of which have almost no influence. Climatic changes directly affect the interaction of soil microorganisms and plants, and implicitly affect them. Shifts in species associations in light of environmental change naturally lead to changes in the biodiversity and capacity of terrestrial biological systems (Langley & Hungate, 2014). Changes in the biodiversity of living organisms managing particular cycles may directly affect the rate of their transformation (for example, N mineralization). Environmental changes altered the overall abundance of soil bacterial populations and expanded the ratio of bacteria to fungi (DeAngelis et al., 2015). Microbial populations respond to warming and different other factors by offering resistance to it, which is more controlled by microbial diversity (Allison & Martiny, 2008). It is, largely, accepted that the elevated CO_2 controls the soil C accessibility to microorganisms with more preference to fungal community over bacteria, which will enhance fungal biomass compared to bacterial biomass. It is a well-known fact that fungi play an important role in organic matter decomposition, soil aggregate formation, plant nutrition, and nutrient cycling, changes in fungal populations may largely affect soil functions. As an effect of climate change, the pore space will be loaded with CO_2 as ambient CO_2 increases, which reduces the dispersion of CO_2 from soil air to the atmosphere. The higher concentration of CO_2 leads to the dissolution of CO_2 in soil moisture to form carbonic acid and lower the soil pH, leading to a boost in fungal community. The fungal population is also less likely to be impacted by lowering of soil moisture.

1.4.4 Serving as Soil Organic Carbon (SOC) Pool

Soils organic C (1550 Pg) is about twice the atmospheric pool (800 Pg) and about three times that in biotic pool (620 Pg). Even slight losses from this vast stock could fundamentally affect the emission trajectory of CO_2 into the atmosphere. The degree of SOC in a given soil is controlled by numerous factors, including climatic factors (temperature and moisture regime) and edaphic factors (cation exchange capacity, clay content, parent material) (Dawson & Smith, 2007). The rate at which organic matter decays influences the rate at which C is lost from the soil. Decomposition of soil organic matter ultimately results in losses of soil C as CO_2 and CH_4 to the environment, as particulate organic C and dissolved organic C through decomposition activities, and as losses of dissolved, particulate, and vapor C to surface waters (Dawson & Smith, 2007). The response of soil C to climate change relies upon the equilibrium between the translocation of below-ground plant biomass that build-up SOC stocks and enhanced decomposition due to warmer temperatures that reduce SOC stocks. In the event when increased C emissions from soils due to a rise in temperature surpass below-ground C allocation, soils could act as a net emitter of atmospheric CO_2 (Davidson & Janssens, 2006). Elevated CO_2 leads to a decrease in the distribution of N-rich plant metabolites, while an increase in C-enriched plant metabolites in root exudates (Tarnawski & Aragno, 2006). This leads to a boom in microbial population that generates more CO_2 in soil, which most likely exerts an adverse impact on SOC stock and C sequestration in soils. It has been seen that the priming effect has a critical negative impact on climate change mitigation due to enhanced soil microbial activity at elevated CO_2, reducing the sequestration capacity of soils. The C isotope tracer techniques have shown that the formation of CO_2 in the rhizosphere by roots and microorganisms is fundamentally promoted by growth and development of plants under elevated CO_2. The release of CO_2 through respiration in the rhizosphere (increased by 56–74%) might be a lot higher than the increase of below-ground root biomass (increased by 15–26%) (Gschwendtner et al., 2015). Microbial biomass, soil respiration, and soil organic matter decomposition are much influenced by a rise in temperature (Bradford et al., 2008). The impacts of global warming on soil depend upon the removal of labile C by increased microbial activity.

1.4.5 Providing Raw Materials

Soil is a diversified natural resource and so it can supply innumerable raw materials for human use. Soil can give development materials like coarse-grained sand, concrete, soil, stones, rocks, etc. In modern times, as well as in ancient times, the soil was used to make earthenware. Extreme climatic events, particularly flooding, pose a significant threat both in terms of ecological impact and disruption to the supply of various raw materials. The occurrence of frequent and intensive heavy rainfall, hurricanes, snowmelt, cyclones, droughts, and floods due to climate change poses a real threat to this soil function. Flooding is particularly dangerous as it not only causes widespread ecological damage but also destroys livelihoods, affects infrastructure, and even causes fatalities. Flooding can negatively impact freshwater accessibility as sea level rise can lead to intrusion of brackish water to the inland. Besides, flooding can lead to uncontrolled arrival of metal-polluted wastewater (IPCC, 2014a, 2014b). Contrarily, drought can lead to water shortages, affecting both natural aspects and supply prospects.

1.4.6 Serving as a Physical and Cultural Environment for the Mankind

Physical and cultural environment is significant for understanding climate change and its mitigation and obviously for portraying climate change as a serious phenomenon to society. Climate change affects culture by affecting individuals who are forced to leave their ancestral practice and compelled to adopt a new practice (Crate et al., 2008, Garrett, 2010). For instance, with decreasing snow and ice cover in icy areas, the man and other animals dwelling there are forced to change the target fish and other hunt species and leave traditional dwellings (Furgal & Seguin, 2006). Distressed ecosystems due to climate change influence plant and animal species and put them under threat of extinction and unnatural cultural and social expressions (Brown et al., 2011). Communities feel the pressure of various stressors (e.g. social imbalance) as climate change acts as a "hazard multiplier" and puts additional pressure on communities.

1.4.7 Storing the Geological and Cultural Heritage

Climate change presents a threat to the preservation and protection of cultural and social resources and heritage. The cultural and social resources and heritage assume a significant part in supporting monetary development and employment, promoting education and learning, influencing socio-economic capital, and contributing to conservation (Brabec & Chilton, 2015). Climate change presents dangers to the security, safeguarding, and transmission of these assets to people in the future. Climate change negatively impacted cultural and social resources and heritage through changes in atmospheric temperature, flooding, sea level rise, and ocean acidification, increasing humidity, and extreme climate events such as dry spells, storms, and typhoons (Ronco et al., 2014). Geological and cultural heritage sites in coastal regions are particularly in danger of sea level rise and storm surges, while the hills are in threat of hazards such as erosion, landslides, and water shortage which can also have devastating impacts on geological and cultural heritage.

1.5 Potential of Soil Processes in Adaptation and Mitigation of Climate Change

The global air temperature increased by 0.74 °C between 1906 and 2005 and projected that tropical countries like India may face a hike in temperature during *kharif* season (July to October) by 1.7 °C and *rabi* season (November to March) by 3.2 °C at the end of year 2070 (Pathak et al., 2012). In India, rainfed agriculture accounts for more than 60%. More than 70% of population from developing countries survives with the practice of rainfed agriculture and it is considered a "food-basket for the poor" (Bhattacharyya, 2019; UNESCO, 2009). Abnormality in rainfall intensity and distribution may significantly hamper rainfed agriculture and increasing temperature may adversely affect the staple cropping systems by exerting remarkable influence on soil processes. However, the strong buffering capacity of soil limits the changes that are expected or predicted due to climate change (Chatterjee & Saha, 2018a, 2018b). Some of the soil processes such as C sequestration, aggregation, porosity, humification, podsolization, C translocation to the depth of the soil profile, soil water regime, nitrification, and denitrification are related to the observed changes in atmospheric temperature and rainfall are subjected to climate change adaptation and mitigation. Unlike climate-smart technologies, soil processes cannot be designated as options for climate change adaptation or mitigation but are associated with decreasing vulnerability of agroecosystems in order to withstand the hostile impacts of projected seasonal climate change impact (Table 1.2).

Table 1.2 Linkage of soil processes with adaptation and mitigation benefit

Soil processes	Adaptation benefits	Mitigation benefits
Carbon sequestration	• Better soil health • Increased yield	• Reduced C emission due to inaccessibility of C source
Soil aggregation	• Improved soil health • Increased yields	• Protects C and reduces C emission
Soil porosity	• Better aeration leads to better root growth	• Ultramicro, micro, and mesopores protect C and reduce C emission
Humification	• Better soil health • Increased yield	• Reduced GHGs because of formation of almost stable humic substances
Podsolization	• Increased and stable production • Higher efficiency of inputs	• Reduced C emission to atmosphere
Translocation and burial	• Increased and stable production • Saving fertilizers	• Reduced GHGs due to shorter duration and high NUE
Soil water regime	• Saving of water • Increased water use efficiency • Increased production and income	• Reduced CH_4 emission by avoiding waterlogging
Nitrification and denitrification	• Increased nutrient use efficiency	• Slow nitrification reduces N_2O emission

1.5.1 Carbon Sequestration

Carbon sequestration involves capturing atmospheric C (usually in the form of CO_2) and storing it in a sink (usually in the soil) in such a way that the stored C cannot be readily released back to the source; mostly aim to curb the greenhouse gas emission (Lal et al., 2015). C sequestration is possible either through direct methods (immediately binding of C compounds at the source before entering the atmosphere) or through indirect methods (sequestering of C and C compounds through photosynthesis and subsequently converted to soil organic C via. afforestation, conservation agriculture, peat bogs, sequestration through the soil's biosphere as well as fertilization of water bodies or algal photosynthesis, etc.) (Eloka-Eboka & Inambao, 2017). Modern advanced methods are also used to sequester C directly by capturing CO_2 in different industrial sectors like food industry, production of carbonated brine, and production of CO_2-containing fertilizers (Lee et al., 2016; Stockmann et al., 2013). Lal et al. (2011) estimated that through photosynthesis 120 Pg C y^{-1} can be sequestered from the atmosphere to root biomass; detritus of the plants can store approximately 60 Pg C y^{-1} in the soil. There are several techniques like restoration of eroded or degraded soils, adoption of recommended management practices for agricultural soils, increasing forest cover, fire management, integrated nutrient management (INM), restoration of urban parks and polluted areas, etc. could be successfully used to improve soil organic C pool through C sequestration (Lal, 2004,

2008a, 2008b; Lal et al., 2011). Mineralogical control of soil C, particularly by short-range order minerals, occurs through protection in the mesopore (Chatterjee et al., 2013). Land use can also alter the amount of C stored in soils. In a study with six land-use systems and three elevated temperatures, it was reported that the reduction in total organic C (TOC) with a rise in temperature ranges 4.1–31.4% among different land-use systems, however, orchard-based systems are more resistant to change in C compared to an agricultural system (Chatterjee et al., 2019a, 2019b, 2019c).

The process of C sequestration can be enhanced by various agronomic practices such as conservation tillage (Chatterjee, 2016; Lal et al., 2019), crop diversification (Gautam et al., 2021), use of biochar (Munda et al., 2018), a mineralogical transformation leading to higher formation of short-range order minerals (Chatterjee et al., 2014a), and land use change (Chatterjee et al., 2019a, 2019b, 2019c). A study of zero tillage in a rice-maize cropping system quantified the reduction in C footprint at about 39% compared to conventional tillage. Zero tillage also reduces 20% of N_2O emissions, improves soil health, increases labile C status, and increases the population of soil microbes (Lal et al., 2019).

Adoption of INM (NPK fertilizers + farm yard manure) increased the C accumulation rate (0.69 Mg C ha^{-1} yr^{-1}) more than the control plots of a long-term fertilization experiment in Inceptisol (Ghosh et al., 2018). Particularly, INM increased the recalcitrant pools of SOC within macroaggregates and microaggregates. Moharana et al. (2019) reported that due to application of tree leaf compost (TLC) and TLC + 50% of recommended doses of NPK fertilizers in a four-year-old field experiment could increase the total organic C by 37 and 49%, respectively. Combined use of balanced fertilizers and native organic residues could be able to increase the SOC sequestration in the Alfisol under semi-arid tropics climatic conditions (Srinivasarao et al., 2012). Likewise, application of organic residues and N or P separately recorded higher SOC in Mollisol (Russell et al., 2005) and Vertisol (Wani et al., 2006).

Application of biochar represented the most effective approach for increasing SOC content (39%; Bai et al., 2019). Biochar is organic material produced by burning in the presence of limited oxygen. These materials are very stable, porous, and usually contain C in a non-labile form (Munda et al., 2018). Biochar plays a pivotal role in the process of C storage, fractionation, and mineralization in soil. In a study, it was observed that six variable doses ranging from 0.5 to 10.0 t ha^{-1} of rice husk biochar (RHB) applied in rice fields increased rice productivity and soil C storage in the form of total organic C (43–169%) (Munda et al., 2018). It increases soil C sequestration potential by increasing the relative proportion of passive C pools (>300%) as compared to active C pools when applied above 4 t ha^{-1} of RHB. Biochar application simultaneously enhances soil porosity, pore size distribution, and SOC content.

Crop diversification, a component of conservation agriculture, as a substitute for rice-fallow or rice-rice improves soil C fractions (Gautam et al., 2021). A more diversified cropping system with legumes could be a promising cropping practice that is able to increase SOC through higher humification rate and microbial activity (Saha et al., 2021). Rice-based cropping systems allow large amounts of organic C to be stored in the soil, which is related to incomplete anaerobic decomposition due to waterlogging (Bhaduri et al., 2017). However, such waterlogged condition is harmful for a long time due to the formation of methane gas, which has 28–36 times higher potential to trap heat in the Earth's atmosphere (Chatterjee et al., 2018; Swain et al., 2018a, 2018b, 2018c).

1.5.2 Soil Aggregation

Soil aggregates are secondary structural units formed by grouping of a number of primary particles with the association of organic and/or inorganic binding agents (Bronick & Lal, 2005). Encapsulation of C in soil aggregates by formation of organo-mineral complexes causes protection of C from microbes (Bhattacharyya et al., 2018). Of the several factors involved in soil aggregation, vegetation, soil flora and fauna, free and adsorbed cations, interaction of organic matter, clay minerals, and clay-organic matter interactions (Kumar et al., 2013). Soils contain numerous complex three-dimensional aggregates composed of clusters of primary mineral particles (sand, silt, and clay) and organic/inorganic binders (organic matter, carbonates, divalent cations) that hold together more tightly than the adjacent bulk soil and cannot be broken down even during wetting events (Wilpiszeski et al., 2019). The increase in stable micro aggregates (<250 μm) due to the formation of organo-mineral complexes encapsulates C and protects it from microbial activities (Kumar et al., 2013). In addition to protection from microbial activities, micro aggregates bind with soil organic C and provide protection from removal by soil erosion (Wilpiszeski et al., 2019). When the zeta-potential of the particles decreased either naturally or in addition to flocculating agents it forms aggregates. Polysaccharides excreted by soil microbes are responsible for macroaggregate (>250 μm) stabilization; whereas, micro aggregates (53–250 μm) are generally stabilized with help of humic compounds (Ghosh et al., 2018).

There are two conceptual models that elucidate the process of soil aggregation related to C sequestration. The conventional conceptual model explains association of soil aggregation related to C sequestration, soil biological activity, and plant growth; however, current conceptual model mainly concentrated on C sequestration in the soil aggregates (Six et al., 2002). Aggregates can sequester C and ultimately be stored as clay-humus complexes (Lal, 2004). Among the processes, physical entrapment in the macro aggregates and micro aggregates, physicochemical protection due to attraction of particles and formation of organo-mineral complexes, chemical protection through adsorption of CO_2 due to chemical reactions and biochemical stabilization through recalcitrance condensation (Kumar et al., 2013; Six et al., 2002). An increase in stable micro-aggregates through the formation of organo-mineral complexes encapsulates C and protects it against microbial activities. Liu et al. (2019) concluded that SOC stabilization could be increased through agricultural abandonment and macro aggregates are solely responsible for controlling the SOC storage recovery in abandoned cropland. Continuous cereal-cereal monocropping resulted in soil aggregate disintegration and C loss, but pigeon pea-wheat and maize-wheat-mungbean facilitated macroaggregate formation, C sequestration and resulted in a higher C management index (Hazra et al., 2019).

1.5.3 Soil Porosity

The subject of C sequestration is now becoming more and more specific to soil aggregates and porosity. Soil pores serve as transport channels for soil solution, soil nutrients, dissolved organic C, various gases, and harbor a variety of microorganisms. Basically, soil pores come in different sizes: macropores (>75 μm), mesopores (30–75 μm), micropores (5–30 μm), and ultramicropores (0.1–5 μm). Soil ultramicropores, micropores, and mesopores protect soil-C from decomposition by soil microbes or associated enzymatic activities. In a study, higher emission of carbon dioxide was observed with higher porosity caused by tillage operation in Oxisol managed with sugarcane (de Souza et al., 2019). They further interpreted that the role of increased porosity due to tillage is more important in clayey soils than in sandy soils, as the former contains more C than the latter. Kravchenko et al. (2019) reported that soil pores of 30–150 μm promote higher enzymatic activities, which mainly determine whether the addition of C is stored in the soil or lost to the atmosphere. Tillage, crop diversification, high root length density, and rapid root turnover are some factors controlling porosity (Dignac et al., 2017; Kravchenko et al., 2019). Under zero tillage cultivation, development of micropores triggered that are responsible for higher infiltration rate, water storage capacity, and lower runoff losses than tilled soils through enhancing soil aggregate stability and formation of vertical bio pores created by worms and roots; such bio pores are more continuous, less tortuous, and more stable than macropores created during ploughing (Hubbard et al., 2001). Hence, zero tillage helps to form more stable soil aggregates, soil pores, and improved soil organic C content.

1.5.4 Humification

Humification is the formation of humic substances, which are the dark-colored, decomposition-resistant final decomposition product of organic matter. Humification of organic matter results in the formation of humus, humate, humic acid, fulvic acid, and humic, which have varying degrees of resistance to further decomposition. Humification is assisted by various other soil processes, such as charring, encrustation, cementation, occlusion, sorption, coagulation, sedimentation, flocculation, complexation, encapsulation, intercalation, and entrapment of macro organic, particulate, and soluble organic matter in micropores to form stable humus (Semenov et al., 2013). Soil amendments like crop residues, organic wastes, composts, etc. have a beneficial effect on soil organic matter retention and stabilize the easily decomposable organic matter to humus (Ghosh et al., 2019; Saha et al., 2021). Alpine biomes are extremely sensitive carbon accumulators and play critical roles in ongoing global change (IPCC, 2014a, 2014b). Nikodem et al. (2013) reported that humification degrees of birch, fir, willow, cypress, larch, and azalea forest litter are correspondingly 18–45%, 15–40%, 8–30%, 14–26%, 9–25%, and 7–19%, respectively. In addition, application of humified materials, crop diversification, crop residue composition, and hydrothermal regimes of the soil greatly affect distribution of stable organic matter in the soil (Moharana et al., 2019; Saha et al., 2021). In some cases, organo-mineral complex helps to further stabilize humic substances. For example, stabilization of humic acid by Ca, Al, and Fe in alpine meadow soils (Han et al., 2019). Such stabilization reduces C emissions to the environment.

1.5.5 Podzolization

Podzolization is a soil process that usually occurs under cool temperate and humid climates under forest or heath vegetation with acidic parent materials and in soils with free drainage and unbuffered conditions. Characteristic of this process is the leaching of organic matter and iron oxides from the topsoil and their accumulation in the lower part of the soil profile (Bs and Bh horizon) (Brock et al., 2020). From a long-term experiment in Vladimir, Russia it was concluded that C stock may increase by 21% compared to the initial level if the supplemental manurial application is maintained at 10 t ha^{-1} y^{-1} in Podzols (Ilichev et al., 2021). Podzolization involved vertical movement of SOC and cations to the deeper soil layers through leaching and translocation. Translocated cations and SOM form metal–organic complexes with Fe and Al. Among several reasons, polymerization of SOC in the subsoil and higher turnover time of SOC in Podzol are prime reasons that control C sequestration. The topsoil looks like a bleached ash-grey colour, while the illuviated region is dark in colour, mostly brown or black for humus and reddish where Al and/or Fe compounds dominate (Nikodem et al., 2013). Because podzols are formed in cold climates, soil microbial activity is comparatively lower in the topsoil and the removal of organic matter from the topsoil and its accumulation in the lower part of the soil profile makes it almost inaccessible to microorganisms (Brock et al., 2020). The slower microbial decomposition of soil organic matter in podzols is due to the low pH and water stagnation above the impermeable B horizon, resulting in an accumulation of organic C in the order of 13.6 kg m^{-2} in 0–30 cm soil or 59.1 kg m^{-2} in 0–200 cm soil (Batjes, 2014). Such C stock is very important in the context of climate change mitigation in boreal areas that are likely to be affected by the increase in mean atmospheric temperature (Soja et al., 2007).

1.5.6 Translocation and Burial of Carbon Enriched Sediments

Carbon-enriched sediments can be buried in depressed lands and aquatic ecosystems, causing very slow aerobic decomposition in the presence of diffuse oxygen (forming CO_2) and mostly anaerobic decomposition (slow process) to form CH_4. In one estimate, it was found that translocation of about 0.4–0.6 Gt C per annum to the subsoil/depressed land/aquatic system, while about 0.8–1.2 Gt C per annum is emitted to the atmosphere (Lal, 2003). One study reported that continental margins store about 90% of organic C in the ocean floor and its protectiveness, mobilization, and redistribution are controlled by sediment grain size fractions and organic-mineral interactions (Ausín et al., 2021). Fine silt fraction mainly plays role in protecting organic C at continental margins (Ausín et al., 2021).

1.5.7 Soil Water Regime

Soil water regime is an important factor controlling the emission of CH_4 and N_2O from soil, which directly affect global warming and climate change. Rice is a unique crop grown under standing water (about 5–8 cm) and therefore has a different energy and water balance than commonly grown arable crops (Chatterjee et al., 2019a, 2019b, 2021a, 2021b; Swain et al., 2018a, 2018b) and produces a higher amount of CH_4 (Swain et al., 2018c). Xu et al. (2004) reported that both CH_4 emission and production increased significantly with the increasing soil moisture content. The higher the soil water content, the quicker soil redox potential (Eh) declined and the earlier CH_4 emission initiated after rice transplantation. As soil Eh decreases quickly with the increasing soil water content, hence CH_4 production and emission become faster after rice transplantation. Adhya et al. (1994) reported that soils with transplanted rice produce more CH_4 than the unplanted fields. As well as plant growth stages significantly influence total CH_4 production and emission. During reproductive stage, rice plants enhance CH_4 production due to decreased oxidation in rice rhizosphere. On the other hand, ripening, and maturity stages of rice retard the CH_4 production due to an increase in root porosity and rhizosphere oxidation (Neue et al., 1996). With the increasing temperature of soil CH_4 production significantly increased (Adhya et al., 1994; Huang et al., 2002). Available Cu is also negatively affecting CH_4 emission (Huang et al., 2002). Various water-smart technology alternate wetting and drying (AWD), tensiometer-based irrigation, and dry direct sowing help to grow this semiaquatic food crop with minimal water that reduces emission of methane to a large extent (40–70%) (Nayak et al., 2020; Tran et al., 2018), however, it may increase N_2O emission (Tarlera et al., 2016). Repeated alternate wetting and drying cycles increase formation of dissolved organic C which in turn increases microbial C consumption that leads to increased O_2 consumption which may stimulate denitrification (Fork & Heffernan, 2014).

1.5.8 Nitrification and Denitrification

Nitrification (NH_3/NH_4^+ to NO_3^-) and denitrification (NO_3^- to N_2O/N_2) control nitrous oxide emission into our atmosphere (Pathak et al., 2016). Total N content as well as NH_4^+–N content negatively correlated with the CH_4 emission in puddled soil (Chatterjee & Saha, 2018a, 2018b). These two processes are most likely to be controlled by N fertilization (right source, rate, time, and place) which is necessary for successful crop production (Chatterjee et al., 2018; Pathak et al., 2019). Besides, N application in the form of chemical or organic fertilizers provides substrates to aerobic (nitrification) and anaerobic (denitrification) microbes involved in these biochemical processes that control anthropogenic N_2O emission. Excess use of N fertilizers is becoming the main source of N_2O in the atmosphere; therefore, use of either nitrification inhibitors or slow-release N fertilizers could be an option to curb the N_2O emission. Excess nitrification is simultaneously increasing the reverse process of denitrification by increasing the nitrate concentration. Nitrifiers are classified as 'r-strategist' (fast-growing) and 'k-strategist' (slow-growing) bacteria based on their kinetic values (Yu et al., 2020). Slowing these processes by any means/climate–smart technologies can lower N_2O emission to atmosphere as 70% of global N_2O emission is contributed by microbial nitrification and denitrification (Butterbach-Bahl et al., 2013). Some chemicals are able to control the nitrification rate by retarding the activity of the nitrifiers known as nitrification inhibitors. Dicyandiamide (DCD), Pronitradine (Centuro™), Nitrapyrin (N-Serve™ and Instinct™), etc. are examples of nitrification inhibitors (Gonzaga et al., 2018; Yu et al., 2020). However, these nitrification inhibitors get adsorbed in the soil, which limits their effectiveness. Zhang et al. (2020) reported that surface adsorption of nitrapyrin in soil followed the following order black soil > planosol > chernozems; with increasing temperature adsorption also increases. Apart from this, chemo-denitrification in acidic soil and decomposition of hydroxylamine also contributed to N_2O emission (Wang et al., 2019). Sometimes another soil process, leaching of NO_3^- to lower soil profile slows down the denitrification process by lowering the concentration of substrate (Bárta et al., 2017; Giles et al., 2012).

1.5.9 Auto-reclamation

Problematic soils are defined as soils affected by low pH or salt affected. Salinity, sodicity, or acidity directly affects plant growth and indirectly affects SOC stocks, generally leading to less SOC. Generally, salt-affected soils are classified as Aridisol (regions, where precipitation is usually <500 mm per year) and usually contain carbonates, which contribute to soil inorganic C (SIC) (Lal & Kimble, 2000). In global soil, approximately 940 Pg of C is present up to 1 m as SIC pool (Eswaran et al., 2000). However, salinity and sodicity affect the SOC stocks differently. In sodic soils, increasing dispersion of aggregates hasten SOC mineralization as well as increasing bulk density that restricts access to a substrate for mineralization. Contrarily, saline conditions can increase the decomposability of soil organic matter but also restrict access to substrates due to flocculation of aggregates as a result of high concentrations of soluble salts. The application of vermicompost could improve soil physical properties (Di et al., 2019), reduce soil salinity (Ibrahim et al., 2015) and alkalinity (Gupta et al., 2016). Acid soils can also be reclaimed by applying manures, compost, and other organic residues; because organic matter could supply a good quantity of base cations. Tillage is an integral part of crop production, simultaneously tillage is also capable of reclaiming saline and sodic soils. Reduced tillage and zero tillage are more efficient than deep tillage in normal soils; however, deep tillage is more beneficial for reclaiming sodic soils (Ding et al., 2021). Because, deep tillage involves subsoiling, which increases the possibility of crop production and increased below-ground crop biomass ultimately helps to reclaim the problematic soils. *Melaleuca cajuputi* forestry in Vietnam could be used as a sodicity reclaiming plantation. Despite this, exchangeable sodium percentage of *Melaleuca* forests on sandy soil showed high sodicity, while those on clay soil varied from low to moderate sodicity (Tran et al., 2015). Carbon sequestration on reclaimed sites depends on reclamation age, plantation species composition, conditions of geo-mining (soil characteristics) as well as climate of that region. *Albizia lebbeck*, *Dalbergia sissoo*, and *Bambusa arundinacea* could be planted to reclaim coal mine soils (Das & Maiti, 2016). C stocks of *A. lebbeck*, *D. sissoo* and *B. arundincea* were correspondingly 7.8, 3.5, and 1.2 Mg C ha^{-1}.

1.6 Conclusion

Soil acts as a major sink as well as a source for C. The effects of climate change on soil processes are sometimes seen as positive, such as increased mineralization to increase soil fertility and nutrient availability, CO_2 fertilization that can increase photosynthesis, crop productivity, and water use efficiency, higher precipitation leading to better soil moisture availability which could lead to better vegetation cover and reduced soil erosion, as well as higher soil C sink, automatic reclamation of alkaline calcareous soils due to dissolution of calcium carbonate and protection of organic matter by flooding soils at lower elevations for longer periods of time. However, these positive impacts are short-lived and in the long run, climate change will negatively affect the soil including its processes and functions. Appropriate site-specific strategies need to be adopted to manage the negative effects of climate change. Understanding the effects of climate change on soil processes and vice versa could help balance profitability and environmental sustainability in the face of changing climate.

References

Adeyeye, J., Akintan, O., & Omoju, O. (2018). Climate change effect and adaptation measures on selected soil properties. *Journal of Soil Science and Environmental Management, 9*(10), 154–163.

Adhya, T. K., Rath, A. K., Gupta, P. K., Rao, V. R., Das, S. N., Parida, K. M., Parashar, D. C., & Sethunathan, N. (1994). Methane emission from flooded rice fields under irrigated conditions. *Biology and Fertility of Soils, 18*, 245–248.

Al, W., Orking, G., & Clima, O. (2008). *Climate change and food security: A framework document*. FAO Rome.

Allison, S. D., & Martiny, J. B. (2008). Resistance, resilience, and redundancy in microbial communities. *Proceedings of the National Academy of Sciences, 105*(Suppl. 1), 11512–11519.

Ausín, B., Bruni, E., Haghipour, N., Welte, C., Bernasconi, S. M., & Eglinton, T. I. (2021). Controls on the abundance, provenance and age of organic carbon buried in continental margin sediments. *Earth and Planetary Science Letters, 558*, 116759.

Baetz, U. (2016). Root exudates as integral part of belowground plant defence. *Belowground defence strategies in plants* (pp. 45–67). Springer.

Bai, X., Huang, Y., Ren, W., Coyne, M., Jacinthe, P., Tao, B., Hui, D., Yang, J., & Matocha, C. (2019). Responses of soil carbon sequestration to climate smart agriculture practices: A meta-analysis. *Global Change Biology, 25*, 2591–2606.

Bakken, L. R., Bergaust, L., Liu, B., & Frostegård, Å. (2012). Regulation of denitrification at the cellular level: A clue to the understanding of N_2O emissions from soils. *Philosophical Transactions of the Royal Society B, 367*, 1226–1234.

Bardgett, R. D., Freeman, C., & Ostle, N. J. (2008). Microbial contributions to climate change through carbon cycle feedbacks. *ISME Journal, 2*, 805–814.

Bárta, J., Tahovská, K., Šantrůčková, H., & Oulehle, F. (2017). Microbial communities with distinct denitrification potential in spruce and beech soils differing in nitrate leaching. *Scientific Reports, 7*(1), 1–15.

Batjes, N. H. (2014). Total carbon and nitrogen in the soils of the world. *European Journal of Soil Science, 65*(1), 10–21.

Beier, C., Emmett, B. A., Peñuelas, J., Schmidt, I. K., Tietema, A., Estiarte, M., Gundersen, P., Llorens, L., Riis-Nielsen, T., Sowerby, A., & Gorissen, A. (2008). Carbon and nitrogen cycles in European ecosystems respond differently to global warming. *Science of the Total Environment, 407*(1), 692–697.

Bengtson, P., Barker, J., & Grayston, S. J. (2012). Evidence of a strong coupling between root exudation, C and N availability, and stimulated SOM decomposition caused by rhizosphere priming effects. *Ecology and Evolution, 2*(8), 1843–1852.

Bhaduri, D., Mandal, A., Chakraborty, K., Chatterjee, D., & Dey, R. (2017). Interlinked chemical-biological processes in anoxic waterlogged soil—A review. *Indian Journal of Agricultural Sciences, 87*(12), 1587–1599.

Bhattacharyya, A. (2019). Water-use efficiency under changing climatic conditions. In A. Bhattacharyya (Ed.), *Changing climate and resource use efficiency in plants* (pp. 111–180). Academic Press, Elsevier.

Bhattacharyya, P., Dash, P. K., Swain, C. K., Nayak, A. K., Chatterjee, D., Padhy, S. R., Saha, R., & Barman, D. (2018). Carbon dynamics in soil-plant-environment system on climate change perspective: Special reference to rice. In *Advances in crop environment interaction* (pp. 3–23). Springer.

Bhattacharyya, R., Ghosh, B. N., Mishra, P. K., Mandal, B., Rao, C. S., Sarkar, D., Das, K., Anil, K. S., Lalitha, M., Hati, K. M., & Franzluebbers, A. J. (2015). Soil degradation in India: Challenges and potential solutions. *Sustainability, 7*(4), 3528–3570.

Blagodatskaya, E., & Kuzyakov, Y. (2008). Mechanisms of real and apparent priming effects and their dependence on soil microbial biomass and community structure: Critical review. *Biology and Fertility of Soils, 45*, 115–131.

Borrelli, P., Robinson, D. A., Panagos, P., Lugato, E., Yang, J. E., Alewell, C., Wuepper, D., Montanarella, L., & Ballabio, C. (2020). Land use and climate change impacts on global soil erosion by water (2015–2070). *Proceedings of the National Academy of Sciences, 117*(36), 21994–22001.

Brabec, E., & Chilton, E. (2015). Toward an ecology of cultural heritage. *Change Over Time, 5*(2), 266–285.

Bradford, M. A., Davies, C. A., Frey, S. D., Maddox, T. R., Melillo, J. M., Mohan, J. E., Reynolds, J. F., Treseder, K. K., & Wallenstein, M. D. (2008). Thermal adaptation of soil microbial respiration to elevated temperature. *Ecology Letters 11*(12):1316–1327. https://doi.org/10.1111/j.1461-0248.2008.01251.x

Bradford, M. A. (2013). Thermal adaptation of decomposer communities in warming soils. *Frontiers in Microbiology, 4*, 333. https://doi.org/10.3389/fmicb.2013.00333

Brock, O., Kalbitz, K., Absalah, S., & Jansen, B. (2020). Effects of development stage on organic matter transformation in Podzols. *Geoderma, 378*, 114625. https://doi.org/10.1016/j.geoderma.2020.114625

Bronick, C. J., & Lal, R. (2005). Soil structure and management: A review. *Geoderma, 124*, 3–22.

Brouder, S. M., & Volenec, J. J. (2008). Impact of climate change on crop nutrient and water use efficiencies. *Physiologia Plantarum, 133*(4), 705–724.

Brown, T., Budd, L., Bell, M., & Rendell, H. (2011). The local impact of global climate change: Reporting on landscape transformation and threatened identity in the English regional newspaper press. *Public Understanding of Science, 20*(5), 658–673.

Bush, R. T., Sullivan, L. A., Burton, E. D., Johnston, S. G., Keene, A. F., Wong, V. N., & Mosley, L. (2010). Climate change impacts on acid sulfate soil landscapes. In *Proceedings of the Earth, Energy and the Environment: 20th Annual VM Goldschmidt Conference*, June 13–18, 2010, Geochemical Society and European Society for Geochemistry, Knoxville, Tennessee.

Butterbach-Bahl, K., Baggs, E. M., Dannenmann, M., Kiese, R., & Zechmeister-Boltenstern, S. (2013). Nitrous oxide emissions from soils: How well do we understand the processes and their controls? *Philosophical Transactions of the Royal Society B, 368*, 20130122. https://doi.org/10.1098/rstb.2013.0122

Carney, K. M., Hungate, B. A., Drake, B. G., & Megonigal, J. P. (2007). Altered soil microbial community at elevated CO_2 leads to loss of soil carbon. *Proceedings of the National Academy of Sciences of the United States of America, 104*, 4990–4995.

Carrizo, M. E., Alesso, C. A., Cosentino, D., & Imhoff, S. (2015). Aggregation agents and structural stability in soils with different texture and organic carbon contents. *Scientia Agricola, 72*(1), 75–82.

Chatterjee, D. (2016). Strengths-weaknesses-opportunities-threats (SWOT) analysis of conservation agriculture. *Indian Journal of Hill Farming, 29*(1), 18–23.

Chatterjee, D., Datta, S. C., & Manjaiah, K. M. (2013). Clay carbon pools and their relationship with short-range order minerals: Avenues to mitigate climate change? *Current Science, 105*(10), 1404–1410.

Chatterjee, D., Datta, S. C., & Manjaiah, K. M. (2014a). Transformation of short-range order minerals in maize (*Zea mays* L.) rhizosphere. *Plant, Soil and Environment, 60*(6), 241–248.

Chatterjee, D., Datta, S. C., & Manjaiah, K. M. (2014b). Fractions, uptake and fixation capacity of phosphorus and potassium in three contrasting soil orders. *Journal of Soil Science and Plant Nutrition, 14*(3), 640–656. https://doi.org/10.4067/S0718-95162014005000051

Chatterjee, D., Datta, S. C., & Manjaiah, K. M. (2015a). Effect of citric acid treatment on release of phosphorus, aluminium and iron from three dissimilar soils of India. *Archives of Agronomy and Soil Science, 61*(1), 105–117. https://doi.org/10.1080/03650340.2014.919449

Chatterjee, D., Datta, S. C., & Manjaiah, K. M. (2015b). Characterization of citric acid induced transformation of short-range order minerals in Alfisol, Inceptisol and Vertisol of India. *European Journal of Mineralogy, 27*, 551–557. https://doi.org/10.1127/ejm/2015/0027-2446

Chatterjee, D., Datta, S. C., & Manjaiah, K. M. (2016). Citric acid induced potassium and silicon release in Alfisols, Vertisols and Inceptisols of India. *Proceedings of the National Academy*

of Sciences, India Section B: Biological Sciences, 86(2), 429–439. https://doi.org/10.1007/s40011-014-0464-y
Chatterjee, D., Datta, S. C., Manjaiah, K. M., & Mandal, N. (2021a). Short range order (SRO) minerals and implications in sustainable agriculture. In N. Mandal, A. Dey, & R. Rakshit (Eds.), *Soil management for sustainable agriculture new research and strategies* (p. 570). Apple Academic Press (in press).
Chatterjee, D., Kuotsu, R., Ao, M., Saha, S., Ray, S. K., & Ngachan, S. V. (2019a). Does rise in temperature effect adversely on soil fertility, carbon fractions, microbial biomass and enzymatic activities under different land use? *Current Science, 116*(12), 2044–2054.
Chatterjee, D., Mohanty, S., Guru, P. K., Swain, C. K., Tripathi, R., Shahid, M., Kumar, U., Kumar, A., Bhattacharyya, P., Gautam, P., & Lal, B. (2018). Comparative assessment of urea briquette applicators on greenhouse gas emission, nitrogen loss and soil enzymatic activities in tropical lowland rice. *Agriculture, Ecosystems & Environment, 252*, 178–190.
Chatterjee, D., Nayak, A. K., Vijayakumar, S., Debnath, M., Chatterjee, S., Swain, C. K., Bihari, P., Mohanty, S., Tripathi, R., Shahid, M., Kumar, A., & Pathak, H. (2019b). Water vapor flux in tropical lowland rice. *Environmental Monitoring and Assessment, 191*, 550. https://doi.org/10.1007/s10661-019-7709-4
Chatterjee, D., & Saha, S. (2015). Climate change and agriculture: a multidimensional perspective. In Chatterjee, D., Sangma, C. B. K., Kikon, Z. J., Ray, S. K., Chowdhury, P., Bordoloi, L. J., & Deka, B. C. (Eds.), *Resource conservation technologies in the context of climate change* (pp. 1–8). ICAR Research Complex for NEH Region, Nagaland Centre.
Chatterjee, D., & Saha, S. (2018a). Response of soil properties and soil microbial communities to the projected climate change. In *Advances in crop environment interaction* (pp. 87–136). Springer.
Chatterjee, D., & Saha, S. (2018b). Response of soil properties and soil microbial communities to the projected climate change. In S. Bal, et al. (Eds.), *Advances in crop environment interaction* (pp. 87–136). Springer.
Chatterjee, D., Swain, C. K., Chatterjee, S., Bhattacharyya, P., Tripathi, R., Lal, B., Gautam, P., Shahid, M., Dash, P. K., Dhal, B., & Nayak, A. K. (2021b). Is the energy balance in a tropical lowland rice perfectly closed? *Atmósfera, 34*(1), 59–78.
Chatterjee, D., Tripathi, R., Chatterjee, S., Debnath, M., Shahid, M., Bhattacharyya, P., Swain, C. K., Tripathy, R., Bhattacharya, B. K., & Nayak, A. K. (2019c). Characterization of land surface energy fluxes in a tropical lowland rice paddy. *Theoretical and Applied Climatology, 136*(1–2), 157–168. https://doi.org/10.1007/s00704-018-2472-y
Classen, A. T., Sundqvist, M. K., Henning, J. A., Newman, G. S., Moore, J. A., Cregger, M. A., Moorhead, L. C., & Patterson, C. M. (2015). Direct and indirect effects of climate change on soil microbial and soil microbial-plant interactions: What lies ahead? *Ecosphere, 6*(8), 1–21.
Cramer, M. D., Hawkins, H. J., & Verboom, G. A. (2009). The importance of nutritional regulation of plant water flux. *Oecologia, 161*, 15–24.
Crate, B. S., Einarsson, N., Ford, J. D., Lahsen, M., Paolisso, M., Schlee, G., & Crate, S. A. (2008). Gone the bull of winter? Grappling with the cultural implications of and anthropology's role (s) in global climate change. *Current Anthropology, 49*(4), 569–595.
Dagar, J. C., Sharma, P. C., Chaudhari, S. K., Jat, H. S., & Ahamad, S. (2016). Climate change vis-a-vis saline agriculture: Impact and adaptation strategies. In *Innovative saline agriculture* (pp. 5–53). Springer.
Das, R., & Maiti, S. K. (2016). Estimation of carbon sequestration in reclaimed coalmine degraded land dominated by Albizia lebbeck, Dalbergia sissoo and Bambusa arundinacea plantation: A case study from Jharia Coalfields, India. *International Journal of Coal Science & Technology, 3*(2), 246–266.
Davidson, E. A., & Janssens, I. A. (2006). Temperature sensitivity of soil carbon decomposition and feedbacks to climate change. *Nature, 440*, 165–173.
Dawson, J. J., & Smith, P. (2007). Carbon losses from soil and its consequences for land-use management. *Science of the Total Environment, 382*(2–3), 165–190.

de Souza, L. C., Fernandes, C., Moitinho, M. R., da Silva, B. E., & La Scala, N. (2019). Soil carbon dioxide emission associated with soil porosity after sugarcane field reform. *Mitigation and Adaptation Strategies for Global Change, 24*(1), 113–127.

DeAngelis, K. M., Pold, G., Topçuoğlu, B. D., van Diepen, L. T., Varney, R. M., Blanchard, J. L., Melillo, J., & Frey, S. D. (2015). Long-term forest soil warming alters microbial communities in temperate forest soils. *Frontiers in Microbiology, 6*, 104.

Di, W., Yanfang, F., Lihong, X., Manqiang, L., Bei, Y., Feng, H., & Linzhang, Y. (2019). Biochar combined with vermicompost increases crop production while reducing ammonia and nitrous oxide emissions from a paddy soil. *Pedosphere, 29*(1), 82–94.

Dignac, M. F., Derrien, D., Barre, P., Barot, S., Cécillon, L., Chenu, C., Chevallier, T., Freschet, G. T., Garnier, P., Guenet, B., & Hedde, M. (2017). Increasing soil carbon storage: Mechanisms, effects of agricultural practices and proxies. A review. *Agronomy for Sustainable Development, 37*(2), 14.

Dijkstra, F. A., Breemen, N. V., Jongmans, A. G., Davies, G. R., & Likens, G. E. (2003). Calcium weathering in forested soils and the effect of different tree species. *Biogeochemistry, 62*, 253–275.

Ding, Z., Kheir, A. M. S., Ali, O. A. M., Hafez, E. M., ElShamey, E. A., Zhou, Z., Wang, B., Lin, X., Ge, Y., Fahmy, A. E., & Seleiman, M. F. (2021). A vermicompost and deep tillage system to improve saline-sodic soil quality and wheat productivity. *Journal of Environmental Management, 277*, 111388.

Dutta, H., & Dutta, A. (2016). The microbial aspect of climate change. *Energy, Ecology and Environment, 1*(4), 209–232. https://doi.org/10.1007/s40974-016-0034-7

Eloka-Eboka, A. C., & Inambao, F. L. (2017). Effects of CO_2 sequestration on lipid and biomass productivity in microalgal biomass production. *Applied Energy, 195*, 1100–1111.

Eswaran, H., Beinroth, F. H., & Virmani, S. M. (2000). Resource management domains: A biophysical unit for assessing and monitoring land quality. *Agriculture, Ecosystems & Environment, 81*(2), 155–162.

Feng, H., & Zhang, M. (2015). Global land moisture trends: Drier in dry and wetter in wet over land. *Scientific Reports, 5*(1), 1–6.

Feng, J., Wang, J., Ding, L., Yao, P., Qiao, M., & Yao, S. (2017). Meta-analyses of the effects of major global change drivers on soil respiration across China. *Atmospheric Environment, 150*, 181–186.

Ficklin, D. L., Luo, Y., Luedeling, E., & Zhang, M. (2009). Climate change sensitivity assessment of a highly agricultural watershed using SWAT. *Journal of Hydrology, 374*(1), 16–29.

Fork, M. L., & Heffernan, J. B. (2014). Direct and indirect effects of dissolved organic matter source and concentration on denitrification in northern Florida rivers. *Ecosystems, 17*(1), 14–28.

Furgal, C., & Seguin, J. (2006). Climate change, health, and vulnerability in Canadian northern Aboriginal communities. *Environmental Health Perspectives, 114*(12), 1964–1970.

Garrett, B. L. (2010). Drowned memories: The submerged places of the Winnemem Wintu. *Archaeologies, 6*(2), 346–371.

Gautam, P., Lal, B., Panda, B. B., Bihari, P., Chatterjee, D., Singh, T., Nayak, P. K., & Nayak, A. K. (2021). Alteration in agronomic practices to utilize rice fallows for higher system productivity and sustainability. *Field Crops Research, 260*, 108005.

Gelybó, G., Tóth, E., Farkas, C., Horel, Á., Kása, I., & Bakacsi, Z. (2018). Potential impacts of climate change on soil properties. *Agrokémia és Talajtan, 67*(1), 121–141.

Ghosh, A., Bhattacharyya, R., Meena, M. C., Dwivedi, B. S., Singh, G., Agnihotri, R., & Sharma, C. (2018). Long-term fertilization effects on soil organic carbon sequestration in an Inceptisol. *Soil and Tillage Research, 177*, 134–144.

Ghosh, S., Mukhopadhyay, J., & Chakraborty, A. (2019). Clay mineral and geochemical proxies for intense climate change in the Permian Gondwana rock record from eastern India. *Research, 2019*, Article 8974075. https://doi.org/10.34133/2019/8974075

Giles, M. E., Morley, N. J., Baggs, E. M., & Daniell, T. J. (2012). Soil nitrate reducing processes—Drivers, mechanisms for spatial variation, and significance for nitrous oxide production. *Frontiers in Microbiology, 3*, 407. https://doi.org/10.3389/fmicb.2012.00407

Glöckner, F. O., Gasol, J. M., McDonough, N., & Calewaert, J. B. (2012). *Marine microbial diversity and its role in ecosystem functioning and environmental change*. European Science Foundation. https://hdl.handle.net/10261/65386

Gonzaga, L. C., Carvalho, J. L. N., ves de Oliveira, B. G., Soares, J. R., & Cantarella, H. (2018). Crop residue removal and nitrification inhibitor application as strategies to mitigate N_2O emissions in sugarcane fields. *Biogas Bioenergy, 119*, 206–216.

Gschwendtner, S., Leberecht, M., Engel, M., Kublik, S., Dannenmann, M., Polle, A., & Schloter, M. (2015). Effects of elevated atmospheric CO_2 on microbial community structure at the plant-soil interface of young beech trees (*Fagus sylvatica* L.) grown at two sites with contrasting climatic conditions. *Microbial Ecology, 69*(4), 867–878.

Gupta, M., Srivastava, P. K., Shikha, N. A., & Tewari, S. K. (2016). Use of a bioaugmented organic soil amendment in combination with gypsum for Withania somnifera growth on sodic soil. *Pedosphere, 26*(3), 299–309.

Han, B., Kitamura, K., Hirota, M., Shen, H., Tang, Y., Suzuki, T., & Fujitake, N. (2019). Humus composition and humification degree of humic acids of alpine meadow soils in the northeastern part of the Qinghai-Tibet Plateau. *Soil Science and Plant Nutrition, 65*(1), 11–19.

Hazra, K. K., Nath, C. P., Singh, U., Praharaj, C. S., Kumar, N., Singh, S. S., & Singh, N. P. (2019). Diversification of maize-wheat cropping system with legumes and integrated nutrient management increases soil aggregation and carbon sequestration. *Geoderma, 353*, 308–319.

Huang, Y., Jiao, Y., Zong, L. G., Zheng, X. H., Sass, R. L., & Fisher, F. M. (2002). Quantitative dependence of methane emission on soil properties. *Nutrient Cycling in Agroecosystems, 64*(1–2), 157–167.

Hubbard, R. M., Ryan, M. G., Stiller, V., & Sperry, J. S. (2001). Stomatal conductance and photosynthesis vary linearly with plant hydraulic conductance in ponderosa pine. *Plant, Cell and Environment, 24*(1), 113–121.

Ibrahim, M. M., Mahmoud, E. K., & Ibrahim, D. A. (2015). Effects of vermicompost and water treatment residuals on soil physical properties and wheat yield. *International Agrophysics, 29*, 157–164.

Ilichev, I., Romanenkov, V., Lukin, S., Pavlova, V., Siptits, S., & Krasilnikov, P. (2021). Arable Podzols are a substantial carbon sink under current and future climates: Evidence from a long-term experiment in the Vladimir Region, Russia. *Agronomy, 11*, 90.

Ioras, F., Bandara, I., & Kemp, C. (2014). Introduction to climate change and land degradation. In *Climate change and restoration of degraded land* (pp. 15–52). Colegio de Ingenieros de Montes.

IPCC. (2007). *Climate change 2007: The physical science basis, contribution of working group I to the fourth assessment report of the Intergovernmental Panel on Climate Change*. Cambridge University Press.

IPCC. (2014a). In C. B. Field, V. R. Barros, D. J. Dokken, K. J. Mach, M. D. Mastrandrea, T. E. Bilir, M. Chatterjee, K. L. Ebi, Y. O. Estrada, R. C. Genova, B. Girma, E. S. Kissel, A. N. Levy, S. MacCracken, P. R. Mastrandrea, & L. L. White (Eds.), *Climate change 2014: Impacts, adaptation, and vulnerability. Part A: Global and sectoral aspects. Contribution of working group II to the fifth assessment report of the Intergovernmental Panel on Climate Change* (p. 1132). Cambridge University Press.

IPCC. (2014b). In Core Writing Team, R. K. Pachauri, & L. A. Meyer (Eds.), *Climate change 2014: Synthesis report. Contribution of working groups I, II and III to the fifth assessment report of the Intergovernmental Panel on Climate Change*. IPCC. https://archive.ipcc.ch/pdf/assessmentreport/ar5/syr/SYR_AR5_FINAL_full_wcover.pdf

Jaggi, R. C., Aulakh, M. S., & Sharma, R. (1999). Temperature effects on soil organic sulphur mineralization and elemental sulphur oxidation in subtropical soils of varying pH. *Nutrient Cycling in Agroecosystem, 54*(2), 175–182.

Jin, J., Tang, C., & Sale, P. (2015). The impact of elevated carbon dioxide on the phosphorus nutrition of plants: A review. *Annals of Botany, 116*(6), 987–999.

Joseph, A., Oluwakemi, A., & Olanrewaju, O. (2018). Climate change effect and adaptation measures on selected soil properties. *Journal of Soil Science and Environmental Management, 9*(10), 154–163.

Joshi, A. B., Vann, D. R., & Johnson, A. H. (2006). Litter quality and climate decouple nitrogen mineralization and productivity in Chilean temperate rainforests. *Soil Science Society of America Journal, 70*(1), 153–162.

Karmakar, R., Das, I., Dutta, D., & Rakshit, A. (2016). Potential effects of climate change on soil properties: A review. *Science International, 4*, 51–73.

Kayler, Z., Keitel, C., Jansen, K., & Gessler, A. (2017). Experimental evidence of two mechanisms coupling leaf-level C assimilation to rhizosphere CO_2 release. *Environmental and Experimental Botany, 135*, 21–26.

Kravchenko, A. N., Guber, A. K., Razavi, B. S., Koestel, J., Quigley, M. Y., Robertson, G. P., & Kuzyakov, Y. (2019). Microbial spatial footprint as a driver of soil carbon stabilization. *Nature Communications, 10*(1), 1–10. https://doi.org/10.1038/s41467-019-11057-4

Kumar, M., & Swarup, A. (2012). Impact of elevated atmospheric CO_2 and temperature on plant available phosphorus in soil—Assessment using ^{32}P tracer technique. *Journal of the Indian Society of Soil Science, 60*(4), 312–316.

Kumar, R., Rawat, K. S., Singh, J., Singh, A., & Rai, A. (2013). Soil aggregation dynamics and carbon sequestration. *Journal of Applied and Natural Science, 5*(1), 250–267.

Kurylyk, B. L., & MacQuarrie, K. T. (2013). The uncertainty associated with estimating future groundwater recharge: A summary of recent research and an example from a small unconfined aquifer in a northern humid-continental climate. *Journal of Hydrology, 492*, 244–253.

Lal, B., Gautam, P., Nayak, A. K., Panda, B. B., Bihari, P., Tripathi, R., Shahid, M., Guru, P. K., Chatterjee, D., Kumar, U., & Meena, B. P. (2019). Energy and carbon budgeting of tillage for environmentally clean and resilient soil health of rice-maize cropping system. *Journal of Cleaner Production, 226*, 815–830. https://doi.org/10.1016/j.jclepro.2019.04.041

Lal, R. (2003). Soil erosion and the global carbon budget. *Environment International, 29*, 437–450.

Lal, R. (2004). Soil carbon sequestration impacts on global climate change and food security. *Science, 304*(5677), 1623–1627.

Lal, R. (2008a). Carbon sequestration. *Philosophical Transactions of the Royal Society B: Biological Sciences, 363*(1492), 815–830.

Lal, R. (2008b). Sequestration of atmospheric CO_2 in global carbon pools. *Energy and Environmental Science, 1*, 86–100. https://doi.org/10.1039/B809492F

Lal, R., Delgado, J. A., Groffman, P. M., Millar, N., Dell, C., & Rotz, A. (2011). Management to mitigate and adapt to climate change. *Journal of Soil and Water Conservation, 66*(4), 276–285.

Lal, R., & Kimble, J. M. (2000). Pedogenic carbonate and the global carbon cycle. In R. Lal, J. M. Kimble, H. Eswaran, & B. A. Stewart (Eds.), *Global climate change and pedogenic carbonates* (pp. 1–14). CRC Press.

Lal, R., Negassa, W., & Lorenz, K. (2015). Carbon sequestration in soil. *Current Opinion in Environment Sustainability, 15*, 79–86.

Langley, J. A., & Hungate, B. A. (2014). Plant community feedbacks and long-term ecosystem responses to multi-factored global change. *AoB Plants, 6*.

Lee, S., Kim, J. W., Chae, S., Bang, J. H., & Lee, S. W. (2016). CO_2 sequestration technology through mineral carbonation: An extraction and carbonation of blast slag. *Journal of CO2 Utilization, 16*, 336–345.

Liu, Z., Zhu, M., Wang, J., Liu, X., Guo, W., Zheng, J., Bian, R., Wang, G., Zhang, X., Cheng, K., Liu, X., Li, L., & Pan, G. (2019). The responses of soil organic carbon mineralization and microbial communities to fresh and aged biochar soil amendments. *GCB Bioenergy*. https://doi.org/10.1111/gcbb.12644

Loladze, I. (2002). Rising atmospheric CO_2 and human nutrition: Toward globally imbalanced plant stoichiometry? *Trends in Ecology and Evolution, 17*, 457–461.

Manna, M. C., Swarup, A., Wanjari, R. H., Ravankar, H. N., Mishra, B., Saha, M. N., Singh, Y. V., Sahi, D. K., & Sarap, P. A. (2005). Long-term effect of fertilizer and manure application on soil organic carbon storage, soil quality and yield sustainability under sub-humid and semi-arid tropical India. *Field Crops Research, 93*, 264–280.

Mayer, A., Hausfather, Z., Jones, A. D., & Silver, W. L. (2018). The potential of agricultural land management to contribute to lower global surface temperatures. *Science Advances, 4*(8), eaaq0932.

McCarl, B. A. (2017). Economics, energy, climate change, and soil security. In D. J. Field et al. (Eds.), *Global soil security, progress in soil science* (pp. 195–205). Springer. https://doi.org/10.1007/978-3-319-43394-3_17

MoEFCC. (2018). *Reports at environmental impact assessment division*. New Delhi, India: Ministry of Environment, Forests, and Climate Change, The Government of India.

Moharana, P. C., Biswas, D. R., Ghosh, A., Sarkar, A., Bhattacharyya, R., & Meena, M. D. (2019). Effects of crop residues composts on the fractions and forms of organic carbon and nitrogen in subtropical Indian conditions. *Soil Research, 58*(1), 95–108.

Munda, S., Bhaduri, D., Mohanty, S., Chatterjee, D., Tripathi, R., Shahid, M., Kumar, U., Bhattacharyya, P., Kumar, A., Adak, T., & Jangde, H. K. (2018). Dynamics of soil organic carbon mineralization and C fractions in paddy soil on application of rice husk biochar. *Biomass and Bioenergy, 115*, 1–9.

Myers, S. S., Zanobetti, A., Kloog, I., Huybers, P., Leakey, A. D., Bloom, A. J., Carlisle, E., Dietterich, L. H., Fitzgerald, G., Hasegawa, T., & Holbrook, N. M. (2014). Increasing CO_2 threatens human nutrition. *Nature, 510*(7503), 139–142.

Nayak, A. K., Chatterjee, D., Tripathi, R., Shahid, M., Vijayakumar, S., Satapathy, B. S., Kumar, A., Mohanty, S., Bhattacharyya, P., Mishra, P., Kumar, U., Mohapatra, S. D., Panda, B. B., Rajak, M., Bhaduri, D., Munda, S., Chakraborty, K., Priyadarsani, S., Swain, C. K., …Moharana, K. C. (2020). *Climate smart agricultural technologies for rice production system in Odisha* (p. 366). ICAR-National Rice Research Institute.

Neue, H. U., Wassmann, R., Lantin, R. S., Alberto Ma, C. R., Aduna, J. B., & Javellana, A. M. (1996). Factors affecting methane emission from rice fields. *Atmospheric Environment, 30*(10–11), 1751–1754.

Nikodem, A., Pavlů, L., Kodešová, R., Borůvka, L., & Drábek, O. (2013). Study of podzolization process under different vegetation cover in the Jizerskéhory Mts. region. *Soil and Water Research, 8*(1), 1–2.

Nord, E. A., & Lynch, J. P. (2009). Plant phenology: A critical controller of soil resource acquisition. *Journal of Experimental Botany, 60*, 1927–1937.

Pandey, B. K., Gosain, A. K., Paul, G., & Khare, D. (2017). Climate change impact assessment on hydrology of a small watershed using semi-distributed model. *Applied Water Science, 7*(4), 2029–2041.

Pareek, N. (2017). Climate change impact on soils: Adaptation and mitigation. *MOJ Ecology & Environmental Sciences, 26*(3).

Pathak, H. (2015). Greenhouse gas emission from Indian agriculture: Trends, drivers and mitigation strategies. *Proceedings of the Indian National Science Academy, 81*(5), 1133–1149.

Pathak, H., Aggarwal, P.K., & Singh, S.D. (2012). *Climate change impact, adaptation and mitigation in agriculture: Methodology for assessment and applications* (p. 302). Indian Agricultural Research Institute.

Pathak, H., Bhatia, A., Jain, N., & Aggarwal, P. K. (2010). Greenhouse gas emission and mitigation in Indian agriculture—A review. *ING Bulletins on Regional Assessment of Reactive Nitrogen, Bulletin, 19*, 1–34.

Pathak, H., Chattarjee, D., & Saha, S. (2019). Fertilizer and environmental pollution-from problem to solution. *Indian Journal of Fertilisers, 15*(3), 262–280.

Pathak, H., Jain, N., Bhatia, A., Kumar, A., & Chatterjee, D. (2016). Improved nitrogen management: A key to climate change adaptation and mitigation. *Indian Journal of Fertilisers, 12*(11), 151–162.

Pendall, E., Bridgham, S., Hanson, P. J., Hungate, B., Kicklighter, D. W., Johnson, D. W., Law, B. E., Luo, Y., Megonigal, J. P., Olsrud, M., Ryan, M. G., & Wan, S. (2004). Below-ground

process responses to elevated CO_2 and temperature: A discussion of observations, measurement methods, and models. *New Phytologist, 162*, 311–322.
Robertson, A. D., Zhang, Y., Sherrod, L. A., Rosenzweig, S. T., Ma, L., Ahuja, L., & Schipanski, M. E. (2018). Climate change impacts on yields and soil carbon in row crop dryland agriculture. *Journal of Environmental Quality, 47*(4), 684–694.
Ronco, P., Gallina, V., Torresan, S., Zabeo, A., Semenzin, E., Critto, A., & Marcomini, A. (2014). The KULTURisk regional risk assessment methodology for water-related natural hazards—Part 1: Physical–environmental assessment. *Hydrology and Earth System Sciences, 18*, 5399–5414.
Russell, A. E., Laird, D. A., Parkin, T. B., & Mallarino, A. P. (2005). Impact of nitrogen fertilization and cropping system on carbon sequestration in midwestern Mollisols. *Soil Science Society of America Journal, 69*(2), 413.
Saha, M., Das, M., & Sarkar, A. (2021). Distinct nature of soil organic carbon pools and indices under nineteen years of rice-based crop diversification switched over from uncultivated land in eastern plateau region of India. *Soil and Tillage Research, 207*, 104856.
Saha, N., & Mandal, B. (2009). Soil health—A precondition for crop production. In *Microbial strategies for crop improvement* (pp. 161–168). Springer.
Saha, S., Chatterjee, D., Swain, C. K., & Nayak, A. K. (2018). Methane emission from wetland rice agriculture-biogeochemistry and environmental controls in projected changing environment. In *Advances in crop environment interaction* (pp. 51–85). Springer.
Saha, S., Das, B., Chatterjee, D., Sehgal, V. K., Chakraborty, D., & Pal, M. (2020). Crop growth responses towards elevated atmospheric CO_2. In M. Hasanuzzaman (Ed.), *Plant ecophysiology and adaptation under climate change: Mechanisms and perspectives I* (pp. 147–198). Springer.
Semenov, V. M., Tulina, A. S., Semenova, N. A., & Ivannikova, L. A. (2013). Humification and nonhumification pathways of the organic matter stabilization in soil: A review. *Eurasian Soil Science, 46*(4), 355–368.
Six, J., Feller, C., & Denef, K. (2002). Soil organic matter, biota, and aggregation in temperate and tropical soils-effects of no-tillage. *Agronomie, 22*, 755775.
Smith, M. R., Golden, C. D., & Myers, S. S. (2017). Potential rise in iron deficiency due to future anthropogenic carbon dioxide emissions. *GeoHealth, 1*(6), 248–257.
Soja, A. J., Tchebakova, N. M., French, N. H., Flannigan, M. D., Shugart, H. H., Stocks, B. J., Sukhinin, A. I., Parfenova, E. I., Chapin, F. S., III., & Stackhouse, P. W., Jr. (2007). Climate-induced boreal forest change: Predictions versus current observations. *Global and Planetary Change, 56*(3–4), 274–296.
Srinivasarao, Ch., Venkateswarlu, B., Lal, R., Singh, A. K., Kundu, S., Vittal, K. P. R., Sharma, S. K., Sharma, R. A., Jain, M. P., & Chary, G. R. (2012). Sustaining agronomic productivity and quality of a Vertisolic Soil (Vertisol) under soybean safflower cropping system in semi-arid central India. *Canadian Journal of Soil Science, 92*, 771785.
Stockmann, U., Adams, M. A., Crawford, J. W., Field, D. J., Henakaarchchi, N., Jenkins, M., Minasny, B., McBratney, A. B., De Courcelles, V. D., Singh, K., & Wheeler, I. (2013). The knowns, known unknowns and unknowns of sequestration of soil organic carbon. *Agriculture, Ecosystems & Environment, 164*, 80–99.
Swain, C. K., Bhattacharyya, P., Nayak, A., Singh, N., Neogi, S., & Chatterjee, P. H. (2018a). Dynamics of net ecosystem methane exchanges on temporal scale in tropical lowland rice. *Atmospheric Environment, 191*, 291–301.
Swain, C. K., Bhattacharyya, P., Nayak, A. K., Singh, N. R., Chatterjee, D., Dash, P. K., Neogi, S., & Pathak, H. (2018b). Temporal variation of energy fluxes during dry season in tropical lowland rice. *MAPAN-Journal of Metrology Society of India*. https://doi.org/10.1007/s12647-018-0260-x
Swain, C. K., Nayak, A. K., Bhattacharyya, P., Chatterjee, D., Chatterjee, S., Tripathi, R., Singh, N. R., & Dhal, B. (2018c). Greenhouse gas emissions and energy exchange in wet and dry season rice: Eddy covariance-based approach. *Environmental Monitoring and Assessment, 190*(7), 423.

Tarlera, S., Capurro, M. C., Irisarri, P., Scavino, A. F., Cantou, G., & Roel, A. (2016). Yield-scaled global warming potential of two irrigation management systems in a highly productive rice system. *Scientia Agricola, 73*(1), 43–50.

Tarnawski, S., & Aragno, M. (2006). The influence of elevated [CO_2] on diversity, activity and biogeochemical functions of rhizosphere and soil bacterial communities. In *Managed ecosystems and CO_2* (pp. 393–412). Springer.

Taub, D. R., Miller, B., & Allen, H. (2007). Effects of elevated CO_2 on the protein concentration of food crops: A meta-analysis. *Global Change Biology, 14*, 565–575.

Tóth, E., Gelybó, G., Dencső, M., Kása, I., Birkás, M., & Horel, Á. (2018). Soil CO_2 emissions in a long-term tillage treatment experiment. In *Soil management and climate change* (pp. 293–307). Academic Press.

Tran, D. B., Hoang, T. V., & Dargusch, P. (2015). An assessment of the carbon stocks and sodicity tolerance of disturbed Melaleuca forests in Southern Vietnam. *Carbon Balance and Management, 10*, 15.

Tran, D. H., Hoang, T. N., Tokida, T., Tirol-Padre, A., & Minamikawa, K. (2018). Impacts of alternate wetting and drying on greenhouse gas emission from paddy field in Central Vietnam. *Soil Science and Plant Nutrition, 64*(1), 14–22.

Tripathi, V., Fraceto, L. F., & Abhilash, P. C. (2015). Sustainable clean-up technologies for soils contaminated with multiple pollutants: Plant-microbe-pollutant and climate nexus. *Ecological Engineering, 82*, 330–335.

UNESCO. (2009). *The United Nations world water development report 3: Water in a changing world*. Paris/London: UNESCO/Earthscan.

Van Vuuren, D. P., Edmonds, J., Kainuma, M., Riahi, K., Thomson, A., Hibbard, K., Hurtt, G. C., Kram, T., Krey, V., Lamarque, J. F., & Masui, T. (2011). The representative concentration pathways: An overview. *Climatic Change, 109*(1), 5–31. https://doi.org/10.1007/s10584-011-0148-z

Várallyay, G. (2007). Potential impacts of climate change on agro-ecosystems. *Agriculturae Conspectus Scientificus, 72*, 1–8.

Várallyay, G. (2010). The impact of climate change on soils and on their water management. *Agronomy Research, 8*(Special Issue II), 385–396.

Vong, C. Q., Jacquemet, N., Picot-Colbeaux, G., Lions, J., Rohmer, J., & Bouc. (2011). Reactive transport modeling for impact assessment of a CO_2 intrusion on trace elements mobility within fresh groundwater and its natural attenuation for potential remediation. *Energy Procedia, 4*, 3171–3178.

Wan, Y., Lin, E., Xiong, W., & Guo, L. (2011). Modeling the impact of climate change on soil organic carbon stock in upland soils in the 21st century in China. *Agriculture, Ecosystems & Environment, 141*(1–2), 23–31.

Wang, J., Zhang, X., Li, L., Cheng, K., Zheng, J., Zheng, J., Shen, M., Liu, X., & Pan, G. (2016). Changes in micronutrient availability and plant uptake under simulated climate change in winter wheat field. *Journal of Soils and Sediments, 16*(12), 2666–2675.

Wang, M., Hu, R., Ruser, R., Schmidt, C., & Kappler, A. (2019). Role of chemodenitrification for N_2O emissions from nitrate reduction in rice paddy soils. *ACS Earth and Space Chemistry, 4*(1), 122–132.

Wani, S. P., Dwivedi, R. S., Ramana, K. V., Vadivelu, A., Navalgund, R. R., & Pande, A. B. (2006). Spatial distribution of rainy season fallows in Madhya Pradesh: Potential for increasing productivity and minimizing land degradation. *Journal of SAT Agricultural Research, 2*(1), 1–36.

Wayne, G. P. (2014). *The beginner's guide to representative concentration pathways. Skeptical science. Version 1.0* (p. 24).

Wei, Y., Maroto-Valer, M., & Steven, M. D. (2011). Environmental consequences of potential leaks of CO_2 in soil. *Energy Procedia, 4*, 3224–3230.

Weil, R. R., & Magdoff, F. (2004). Significance of soil organic matter to soil quality and health. In R. R. Weil & F. Magdoff (Eds.), *Soil organic matter in sustainable agriculture* (pp. 1–43). CRC Press.

Wilkin, R. T., & DiGiulio, D. C. (2010). Geochemical impacts to groundwater from geologic carbon sequestration: Controls on Ph and inorganic carbon concentrations from reaction path and kinetic modeling. *Environmental Science and Technology, 44*(12), 4821–4827.

Wilpiszeski, R. L., Aufrecht, J. A., Retterer, S. T., Sullivan, M. B., Graham, D. E., Pierce, E. M., Zablocki, O. D., Palumbo, A. V., & Elias, D. A. (2019). Soil aggregate microbial communities: Towards understanding microbiome interactions at biologically relevant scales. *Applied and Environmental Microbiology, 85*(14), e00324-e419. https://doi.org/10.1128/AEM.00324-19

Xu, Z., Zheng, X., Wang, Y., Han, S., Huang, Y., Zhu, J., & Butterbach-Bahl, K. (2004). Effects of elevated CO_2 and N fertilization on CH_4 emission from paddy rice fields. *Global Biogeochemical Cycles, 18*(3), GB3009.

Yadav, R. K., Rupa Kumar, K., & Rajeevan, M. (2010). Climate change scenarios for Northwest India winter season. *Quaternary International, 213*(1–2), 12–19.

Yu, L., Chen, S., Chen, W., & Wu, J. (2020). Experimental investigation and mathematical modeling of the competition among the fast-growing "r-strategists" and the slowgrowing "K-strategists" ammonium-oxidizing bacteria and nitrite-oxidizing bacteria in nitrification. *Science of the Total Environment, 702*, 135049.

Zak, D. R., Pregitzer, K. S., King, J. S., & Holmes, W. E. (2000). Elevated atmospheric CO_2, fine roots and the response of soil microorganisms: A review and hypothesis. *New Phytologist, 147*, 201–222.

Zhang, Y., Chen, X., Zhang, C., Pan, G., & Zhang, X. (2014). Availability of soil nitrogen and phosphorus under elevated [CO_2] and temperature in the Taihu Lake region. *China Journal of Plant Nutrition and Soil Science, 177*, 343–348. https://doi.org/10.1002/jpln.201200526

Zhang, Y., Zhang, N., Yin, J., Zhao, Y., Yang, F., Jiang, Z., Tao, J., Yan, X., Qiu, Y., Guo, H., & Hu, S. (2020). Simulated warming enhances the responses of microbial N transformations to reactive N input in a Tibetan alpine meadow. *Environment International, 141*, 105795. https://doi.org/10.1016/j.envint.2020.105795

Chapter 2
Nutrient Cycling and Climate Change

Debarup Das, Prasenjit Ray, and S. P. Datta

Abstract Climate change and global warming are the major challenges in the agriculture sector for nutrient use efficiency, sustainability of natural resources, and food security. Sustainability of soil resources for food security in the era of climate change usually revolves around soil carbon and its possible sequestration in more stabilized forms. Carbon sequestration in soil is a widely studied approach to mitigate climate change, and global warming since soil carbon and its stock is highly vulnerable to the changing climate. It is known that carbon cycling and transformation affects directly or indirectly the transformation and cycling of nutrients in soil. Understanding the processes and mechanisms of nutrient cycling and transformations as affected by climate change is of utmost importance as far as nutrient use efficiency and appropriate management strategies are concerned. Evidence of alterations in nutrient transformation and cycling due to the changes in climatic variables has been well documented. Nevertheless, research efforts are lacking in addressing the effect of climate change on the availability of nutrients, the transformation, and cycling of which are not directly associated with carbon cycling in soil. This chapter aims at highlighting the impact of climate change on the cycling and transformations of important plant nutrient elements and the future prospects in climate change research pertaining to nutrient transformation and availability in soil.

Keywords Climate change · Drying-wetting · Elevated temperature · Elevated CO_2 · Freeze–thaw · Nutrient cycling

D. Das · P. Ray · S. P. Datta (✉)
Division of Soil Science and Agricultural Chemistry, ICAR-Indian Agricultural Research Institute, New Delhi 10012, India
e-mail: spdatta.manish@gmail.com

D. Das
e-mail: debarup.das@icar.gov.in

P. Ray
e-mail: prasenjit.ray@icar.gov.in

S. P. Datta
ICAR-Indian Institute of Soil Science, Bhopal 462038, India

H. Pathak et al. (eds.), *Climate Change Impacts on Soil-Plant-Atmosphere Continuum*, Advances in Global Change Research 78, https://doi.org/10.1007/978-981-99-7935-6_2

2.1 Introduction

Carbon dioxide (CO_2), a greenhouse gas, is one of the major causes of global climate change. Its concentration in the atmosphere is increasing since industrial revolution. Global atmospheric CO_2 concentration was <300 ppm in 1850, which rose to ~421 ppm in 2022 (https://www.co2levels.org/, accessed on 21 April 2022). At current rate, it may increase up to 550 ppm by 2050 (Carter et al., 2007), and reach 970 ppm by 2100 (IPCC, 2007a, 2007b). Consequently, global surface temperature is also increasing and is likely to rise by 1.5–2 °C by 2100 (IPCC, 2022). Additionally, anomalies and variability in the annual precipitation as well as the potential evapotranspiration (PET) are becoming increasingly unpredictable (Follett et al., 2012). As a result, some places on the earth will become wetter (e.g., Northern Hemisphere high latitudes), whereas others will become drier (e.g., the Mediterranean region), with an overall global trend going towards more dryness (Cook et al., 2014). Average summer precipitation is estimated to decrease by ~25% in many areas by 2100 (IPCC, 2012; Stocker et al., 2014). There could be a global surge in the aridity of terrestrial ecosystems (Fu & Feng, 2014), leading to an 11–23% rise in the worldwide extent of drylands by 2100 (Huang et al., 2016). Besides, frequency and severity of droughts and intermittent heavy rainfall events have largely increased in different regions across the world (IPCC, 2013; Mpelasoka et al., 2008; Reichstein et al., 2013; Spinoni et al., 2015, 2017). Such changes in drying (drought) and wetting (rainfall) patterns associated with climate change have a strong influence on nutrient cycling in terrestrial ecosystems (Gao et al., 2020). Snow present during major/entire parts of the year in areas of high altitudes, high latitudes, and temperate regions insulates and prevents the underlying soils from freezing when the temperature is too low, thereby reducing soil temperature variability. Global warming is likely to reduce the time and thickness of snow cover, due to which freeze–thaw cycles (FTC) can also become intense and more frequent altering the soil structure and processes (Groffman et al., 2011). Such anomalies in FTCs could disintegrate soil aggregates (Chai et al., 2014; Xiao et al., 2019), rupture microbial cells (Larsen et al., 2002; Yanai et al., 2004), destroy plant litter, plant roots (Campbell et al., 2014; Pelster et al., 2013; Reinmann & Templer, 2016), etc. Disturbances by FTCs would ultimately enhance substrates' solubility and availability, and thereby affect soil organic matter (SOM) turnover and nutrient transformation (Gao et al., 2018; Song et al., 2017a, 2017b).

All these together are likely to alter the processes involved in nutrient cycling and transformation to various extents depending on the nutrient itself, existing climate, land use, management practices, and other unspecified factors.

2.2 Carbon (C)

Carbon present in soil, especially the organic pool is perhaps the most important variable for nutrient cycling, transformation, and ultimately soil fertility. Both organic and inorganic pools of C are important from a C sequestration point of view and can be affected by elevated temperature, CO_2 concentration, and/or changes in water availability. Soil C stocks are vulnerable to both rise in temperature and reduced water availability. A study conducted across three different land uses in the great plains of the United States of America (USA) showed a strong negative correlation between soil C stocks and mean annual temperature (MAT), but a positive correlation of soil C stock with the ratio of mean annual precipitation (MAP) to PET (Follett et al., 2012). They estimated that around 486 Tg ha^{-1} soil organic C (SOC) may be lost depending on MAP: PET over a period of 30 years from the uppermost 10 cm layer for an increase of 1 °C in MAT. Many previous researchers have observed decreases in SOC stocks with increased MAT (Dijkstra & Morgan, 2012; Homann et al., 2007; Jobbagy & Jackson, 2000). Chatterjee et al. (2019) incubated soils from six land-use systems distributed over two soil orders (Alfisols and Entisols) at four temperatures, viz. 27 °C (ambient temperature of the region), 36 °C (maximum temperature of the region), 39 °C (elevated by 3 °C over maximum temperature), and 42 °C (elevated by 6 °C over maximum temperature) for one month. Averaged across the land-use systems, they found variable changes in different SOC fractions due to elevation in temperature (Fig. 2.1). There was an increase in oxidizable organic C from 27 to 36 °C, which remained at par up to 39 °C, but decreased significantly at 42 °C. The labile C increased progressively up to 39 °C, then decreased significantly upon a further increase in temperature by 3 °C. On the contrary, total organic C and dissolved organic C contents gradually decreased with increasing temperature (Fig. 2.1). In general, increased temperature (up to a limit) causes more soil respiration and organic matter decomposition, hence decreasing SOC stocks (Conant et al., 2004; Del Grosso et al., 2005). The quality of SOC dictates the extent of loss to increased temperature. As per the C quality-temperature hypothesis, Q_{10} of SOM decomposition by microbes should rise as activation energy increases; hence, a recalcitrant pool of SOC would show a greater response to increased temperature than labile SOC (Craine et al., 2010; Davidson & Janssens, 2006). However, both positive (Xu et al., 2014) and negative (Fierer et al., 2006) correlations are possible between Q_{10} of SOM decomposition and its quality parameters because of complex interactions among microbes, organic C, and minerals in soil (Cotrufo et al., 2013; Fontaine et al., 2003). Ghosh et al. (2020) observed that soils in northeast India with low-quality SOM had a higher Q_{10}. In areas with low temperature (e.g., hills), that indirectly affects nutrient availability to plants by lowered decomposition rate of SOM, an increase in temperature is likely to cause greater mineralization and better release of nutrients from SOM, thereby improving plant growth (Rawes & Welch, 1969). However, temperature alone will not predict the result, as factors like MAP and PET may become limiting under changing climate scenarios. Water availability interacts with temperature to determine soil respiration (and hence SOC loss), especially during dry and warm periods of the

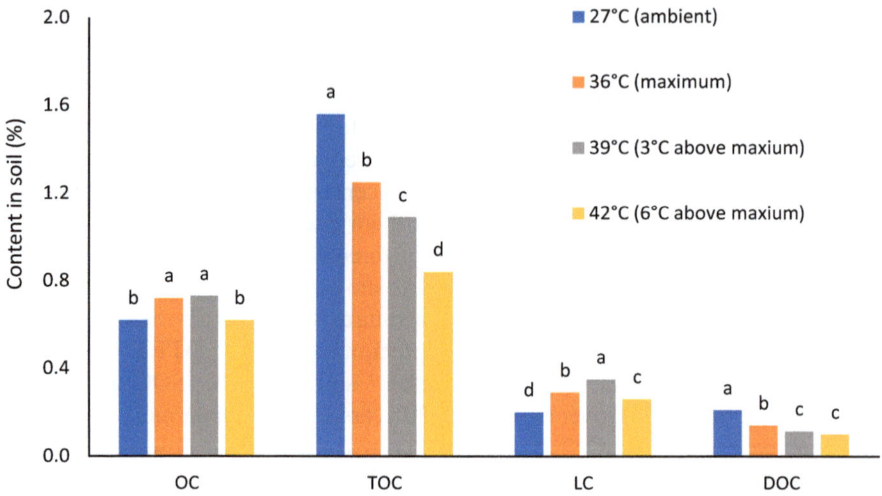

Fig. 2.1 Effect of incubation temperatures on contents of oxidizable organic C (OC), total organic C (TOC), labile C (LC), and dissolved organic C (DOC) in soils averaged over six land-use systems; the columns within a C fraction with no common letters are significantly different ($P = 0.05$) (Adapted from Chatterjee et al., 2019)

year. Direct responses of soil respiration to increased temperature ease off with a decrease in soil moisture (Conant et al., 2004). Generally, CO_2 enrichment in the atmosphere boosts the growth of C3 plants, whereas increased temperature favours C4 grasses (Morgan et al., 2011). However, such responses of plants to elevated CO_2 and increased temperature may not manifest in SOC dynamics in the long term, as decomposition will also increase with increased C input through plants (Carrillo et al., 2011). In contradiction, it is also thought that CO_2 enrichment may increase the C:N ratio of plants, which would cause decomposition of plant residues to occur at a slower rate, affecting nutrient release (Pathak, 2015).

Carbon cycling in terrestrial ecosystems can also be influenced by changes in the intensity and frequency of drying-wetting (DW) and FTC associated with climate change. Drying of soils reduces SOM decomposition through reduced microbial and enzymatic activities (Manzoni et al., 2012; Schimel, 2018; Sun et al., 2017), whereas wetting of dried soils can largely enhance mineralization and nutrient release from SOM (Borken & Matzner, 2009). Likewise, FTC significantly augmented soil dissolved organic C (DOC, by 38%), DOC leaching (by 62%), methane (CH_4) fluxes (instant flux by 41%, and cumulative flux by 59%), whereas reduced microbial biomass C (MBC, by 10%), microbial C:N ratio (by 8.1%), and cellulase activity (by 16%), as perceived with a global meta-analysis based on 75 articles with 2471 observations (Gao et al., 2021).

Drivers of climate change can affect the soil inorganic C (SIC) pool and vice versa. The SIC affects climate either by liberating CO_2 into the atmosphere or sequestering CO_3^{2-} (Groshans et al., 2018; Monger et al., 2015; Zamanian et al., 2016). The SIC

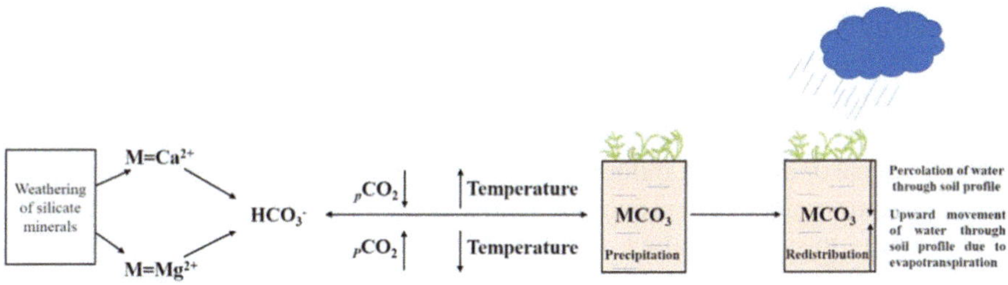

Fig. 2.2 Accumulation of pedogenic carbonates in soil under the influence of climatic variables

is present in solid form as solid carbonates (mostly $CaCO_3$), and in solution form as CO_3^{2-} (aq), HCO_3^- (aq), H_2CO_3 (aq), and CO_2 dissolved in soil water (Groshans et al., 2018; Monger, 2014; Schindlbacher et al., 2019; Zamanian et al., 2016). The CO_2-partial pressure (pCO_2) in the atmosphere equipoised with the soil solution mainly controls the dynamics of the SIC forms present in soil solution (Andrews & Schlesinger, 2001; Karberg et al., 2005; Zamanian et al., 2016). The CO_2 enrichment in atmosphere along with the CO_2 generated in the soil through respiration of soil organisms and plant roots can affect the SIC pool through acidification, mineral dissolution, shifts in cation exchange reaction rates, and buffering capacity (Oh & Richter, 2004). An increase in the atmosphere results in its increased dissolution in the soil water forming more of H_2CO_3. If it is associated with the leaching of basic cations, e.g., Ca^{2+}, Mg^{2+}, K^+, Na^+, under heavy rainfall, the net result would be acidification. Hence, elevated CO_2 under leaching environments can generate soil acidity and enhance weathering processes, which might have negative feedback on the SIC pool in long run through a dissolution of carbonate minerals and subsequent movement to waterbodies (Andrews & Schlesinger, 2001; Oh & Richter, 2004).

The accumulation of SIC in the form of carbonates is highly controlled by atmospheric temperature. It is known that formation of pedogenic carbonates (of Ca and or Mg) in soil is faster under arid/semi-arid regions, provided the concentration of Ca and/or Mg in soil is not limiting (Monger et al., 2015; Zamanian et al., 2016). With the increase in atmospheric temperature and subsequent increase in soil temperature, the solubility of CO_2 in soil solution decreases and it results in the supersaturation of carbonates in soil solution (Barker & Cox, 2011). Once formed, the redistribution of pedogenic carbonates after dissolution in soil is controlled by mean annual precipitation (MAP) and evapotranspiration (Fig. 2.2).

2.3 Nitrogen (N)

Nitrogen (N) can severely limit sequestration of CO_2 in terrestrial and marine ecosystems (Gruber & Galloway, 2008). It is estimated that limitation of N in CO_2 sequestration by terrestrial ecosystems could further bring a 2 °C temperature increase by the end of twenty-first century (Wang & Houlton, 2009). Processes involved in N transformation, e.g., mineralization, nitrification, losses through leaching or in gaseous forms (NH_3, N_2O, NO_x, etc.) are influenced by the CO_2 enrichment, increased temperature, variability in MAP and PET, altered frequency, and intensity of DW and FTC, etc. associated with climate change. However, the type (positive or negative) and extent of such influence largely depend on the existing climatic conditions, vegetation, slope, soil properties, land management, etc. As mentioned earlier, a soil's ability to store C in the form of SOM is largely limited by N; because SOM needs more N per unit C than plants (Cotrufo et al., 2019). The CO_2 enrichment in atmosphere might increase the soil C:N ratio permitting greater SOC accumulation with a lesser than expected amount of N (Hungate et al., 2003). Under such circumstances, however, the microbes involved in SOM decomposition require more N, which can decline N mineralization and reduce inorganic N supply to plants (Gill et al., 2002; Hungate et al., 2003). This is why an increase in soil C:N due to elevated CO_2 in atmosphere is not likely to be translated into a significant enlargement of SOC stocks.

Like SOC, soil organic N (SON) is also vulnerable to increased temperature and variations in MAP:PET. Follett et al. (2012) estimated that around 180 kg ha^{-1} SON can be lost from 0 to 10 cm soil of the great plains of USA over 30 years because of a 1 °C increase in MAT. The reasons are similar to that for SOC as mentioned earlier. Chatterjee et al. (2019) observed an increase in available N in soils, averaged over six land-use systems, after one month of incubation at 36 °C over that at 27 °C (Fig. 2.3). However, beyond 36 °C, increase in temperature up to 39 or 42 °C did not significantly increase soil available N. More specifically, the increase in available N due to temperature rise was confined to soils from agricultural or orchard lands, but not from a forest or pasture lands. Temperature rise makes the enzymes involved in N cycling more active (Sardans & Peñuelas, 2015; Steinweg et al., 2013), which may increase the SON mineralization causing greater soil available N under higher temperatures (Chatterjee et al., 2019). They also observed a corresponding significant increase in microbial biomass C (MBC) from 27 to 36 °C, and beyond that for each 3 °C rise up to 42 °C; and in microbial biomass N (MBN) from 27 to 36 °C and 36 to 39 °C. Temperature increases up to a non-harmful limit can increase the release of substrates from plant roots and dead soil organisms, ultimately enhancing MBC and MBN. Beyond a certain temperature, however, mesophilic microbes may undergo thermal denaturation (Joergensen et al., 1990). However, an increase in enzyme activity due to elevated temperature, as mentioned above, might only occur when soil moisture is sufficient (Sardans et al., 2008). Beier et al. (2008) even reported that SON mineralization could be non-responsive to temperature change, but responsive to alterations in soil moisture. Higher temperature and lower precipitation might also reduce snow cover at certain locations, increase the frequency of FTCs, cause a physical breakdown of soil aggregates, microorganisms, and plant root cells, and facilitate release of plant available form of nutrients in soil (Budge et al., 2011; Garcia-Pausas et al., 2017; Guan et al., 2018), and also higher mineralization of organic compounds containing plant nutrients (Song et al., 2017a, 2017b; Wang et al., 2016a, 2016b). For similar reasons, Andrade-Linares et al. (2021) observed higher concentrations of total dissolved N at a site with MAT of 13.1 °C and MAP of

546 mm compared with a site with lower MAT (11.2 °C) and higher MAP (890 mm), as observed through short-term exposure to imitated climate change in the montane region of Southern Germany. An increase in gross SON mineralization and plant available N with higher temperature was also observed in alpine ecosystems (Dawes et al., 2017; Wang et al., 2016a, 2016b). According to Wang et al. (2016a, 2016b), the turnover of soil N is sensitive to increased temperature (+2 °C MAT) and decreased precipitation (−500 mm MAP), and more frequent FTCs. Similar climate changes revealed higher gross turnover of soil N as well as a greater abundance of nitrifying microbes in frozen soils (Wang et al., 2016a, 2016b). Accelerated decomposition of SON under increased temperature though increases plant available N; in the long run, however, the same could diminish the SON content and reduce soil fertility level (Kamewada & Yoshizawa, 2018).

Loss of SON due to increased temperature means processes like mineralization and nitrification are involved in producing plant-available NH_4^+ and NO_3^-. Besides, volatilization of NH_3, leaching, and/or denitrification of NO_3^- etc. will also occur. There could be a production of N_2O too, having positive feedback on climate change (Follett et al., 2012). Both N mineralization and nitrification might increase with increased temperature. Hart (2006) conducted a study, where he transferred whole soil cores (0–15 cm) from a higher elevation forest (F_{high}) to a lower elevation forest (F_{low}, 730 m lower) and assessed the effect of climate change on N mineralization, nitrification, and N_2O emission apart from other variables. The F_{low} had a 2.5 °C higher mean annual soil temperature (MAST) than F_{high}. Over a period of 13 months,

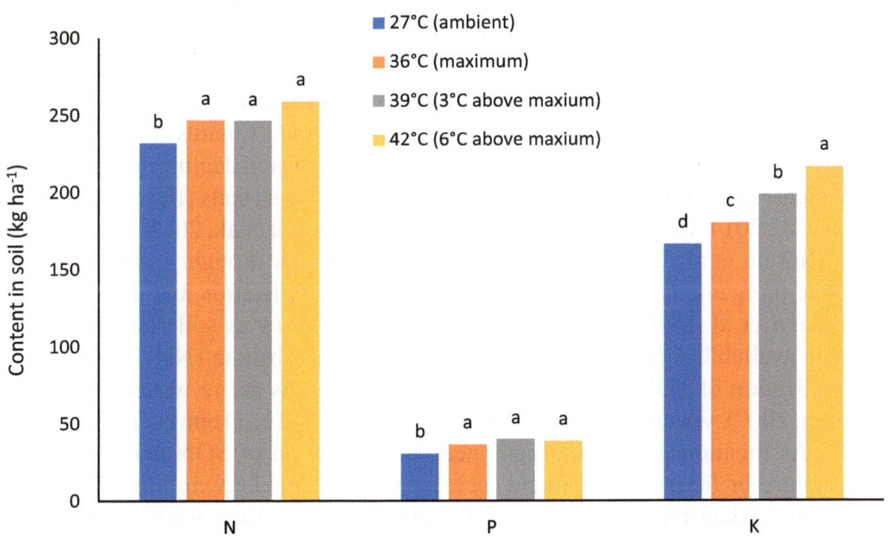

Fig. 2.3 Effect of incubation temperatures on available nitrogen (N), available phosphorus (P), and available potassium (K) in soils averaged over six land-use systems; the columns for a particular nutrient with no common letters are significantly different ($P = 0.05$) (Adapted from Chatterjee et al., 2019)

net N mineralization and nitrification were raised by around 80% when soil cores from F_{high} were incubated at F_{low} relative to soil cores of F_{high} incubated in situ, despite no difference in soil moisture. Only an increased temperature of 2.5 °C could cause such high rates of N mineralization and nitrification. However, such transfer did not generally alter the potential leaching of N (i.e., leaching when there is no plant uptake) or the N_2O emission. Hence, an increase in nitrification though leads to the production of more NO_3^- in the soil solution, does not necessarily mean that N losses by leaching or N_2O emission will be more. Moreover, when increased temperature is associated with an extension of plant growing seasons and enhanced photosynthesis causing better plant growth and more nutrient uptake (Beerling & Woodward, 1994), and sometimes more tissue N concentration, e.g., in temperate hilly regions, leaching losses may rather decrease (Ineson et al., 1998). Increased temperature can also reduce the growing season, especially in warmer regions, and might sometimes cause high-temperature sterility in crops (Kamewada & Yoshizawa, 2018; Yoon & Choi, 2020), which could result in lower N uptake. The effect of temperature rise on N leaching might be different under such situations. In addition, alterations in N leaching would depend on the variations in MAP:PET associated with climate change for a particular site. There are studies where increased N_2O formation was observed besides higher SON mineralization and nitrification with warming (Hu et al., 2017; Larsen et al., 2011; Rustad et al., 2001). Interaction between temperature and moisture affects all these processes (Auyeung et al., 2013, 2015; Dannenmann et al., 2016). Some also opined that N losses through NO_3^- leaching and gaseous emissions of N_2O might increase due to climate change over short-term periods (Kuypers et al., 2018; Stein & Klotz, 2016; Zistl-Schlingmann et al., 2019). Short-term exposure to a higher temperature (by ~1.9 °C) and lower precipitation (by ~350 mm) might also reduce the profusion of free-living N_2-fixing bacteria but increase in SON mineralizing bacteria, as observed by downslope translocation of intact plant-soil mesocosms in the montane grasslands of southern Germany (Andrade-Linares et al., 2021).

Losses of N through leaching and N_2O emission from soil decline due to drought, but increase after re-wetting compared with perpetual moist soils (Beare et al., 2009; Chen et al., 2014; Guo et al., 2014a, 2014b; Harrison-Kirk et al., 2013; Leitner et al., 2017; Norton et al., 2008). Because drying of soil under drought can significantly decrease inorganic N transformation processes like nitrification and denitrification (Hartmann et al., 2013). However, nitrification increases as soil start to dry due to more availability of O_2 but later decreases due to substrate (NH_4^+) limitations. Mineralization of SON increases sharply when dry soils are re-wetted (Borken & Matzner, 2009), which might increase NO_3^- leaching and N_2O emissions (Gao et al., 2020). As mentioned earlier, accelerated decomposition of SON under increased temperature or frequent DW might increase plant available N, in the long run the same could exhaust the SON more, declining soil fertility (Kamewada & Yoshizawa, 2018). Hence, increased SON mineralization that promotes plant productivity might cause mining of soil N especially if external N inputs fall short of N outputs by enhanced plant growth (Schlingmann et al., 2020).

2.4 Phosphorus (P)

Considering the global P limitation in terrestrial ecosystems (Augusto et al., 2017), a better understanding of P cycle as affected by the changing climate would be helpful for its inclusion into various earth system models (Reed et al., 2015), and also better management in agricultural lands in the coming days. Availability of soil P is both directly and indirectly affected by climatic variables (Hou et al., 2018), e.g., MAT, MAP, MAP:PET, etc. Temperature can directly affect the sorption–desorption of P in soil (Barrow, 1983); whereas, greater MAP might cause more P leaching (Sims et al., 1998). Indirectly, climatic variables can affect P availability to plants by altering its different forms in soil; influencing important soil properties like pH, organic C, etc.; and by affecting P uptake and immobilization by plants and microbes, respectively (Hou et al., 2018). An elevated temperature may cause a greater shift from available and secondary mineral P forms to occluded forms in soil (Barrow, 1983; Siebers et al., 2017). On the contrary, under sufficient soil moisture, an increase in temperature up to a certain limit might even increase the content of available P in soil under non-leaching conditions, as observed after one-month incubation in laboratory conditions (Fig. 2.3; Chatterjee et al., 2019). However, under field conditions with MAP > 2,000 mm, the secondary mineral P and occluded P forms might largely be lost due to heavy precipitation (Austin & Vitousek, 1998). A combination of increased temperature and precipitation hastens the processes like mineralization, immobilization, and weathering of P-containing rocks; leading to rapid changes in distribution of different P forms in soil, and availability to plants (Delgado-Baquerizo et al., 2013; Dixon et al., 2016; Siebers et al., 2017). On the other hand, more aridity causing lesser plant cover might increase the wind-erosion loss of fine soil particles (Delgado-Baquerizo et al., 2013), which have greater content and sorption capacity for P compared with coarser particles (Ohalloran et al., 1987). Based on data from around the globe (from 96 published articles) of Hedley P fractions and major related physiochemical properties of 760 seminatural/natural soils, Hou et al. (2018) assessed the effect of climatic variables on P cycle in soil and its availability in terrestrial ecosystems. Their observation indicated that soil available P or the Hedley labile inorganic P significantly declined with an increase in MAT and MAP. The negative influence of MAT on soil available P was caused by lower contents of organic P and primary mineral P, and greater sand content in soil. Earlier researchers also observed negative effect of MAT on the availability of soil P due to intense weathering and greater plant uptake (Dixon et al., 2016; Unger et al., 2010). The MAP had a declining effect on primary mineral P and hence on soil available P. On the other hand, higher MAP tends to increase organic P fraction in soil; besides higher MAP sites had greater finer soil particles, which partly offset the negative effect of MAP on soil available P. Consequently, Hou et al. (2018) observed negligible net impact of MAP on soil available P. Examples of positive (Miller et al., 2001; Siebers et al., 2017) as well as negative (Austin & Vitousek, 1998; Vitousek et al., 2010; Walker & Syers, 1976) impacts of MAP on soil P availability are present; yet negative effects have been reported more frequently. The net effect of climate change on soil P cycling will depend on all the direct and indirect effects. Because of the complexity of interactions among climatic variables, soil P forms, and vital soil properties affecting its dynamics, the net impact of climatic variables on soil P at local or regional level has often been inconsistent (Austin & Vitousek, 1998; Jiao et al., 2016; Tian et al., 2010; Vitousek & Chadwick, 2013; Zhang et al., 2005).

Elevated CO_2 levels in atmosphere may indirectly affect the P cycling through the mediation of soil biological activities (Chatterjee & Saha, 2018). Elevated CO_2 increases the production of ectoenzymes released by plant roots, mycorrhizal hyphae, and soil microbes, which are involved in reactions governing P availability in soil (Chatterjee et al., 2014, 2015; Lagomarsino et al., 2008; Read et al., 2004). Elevated CO_2 may also increase the mineralization of organic P compounds and solubilization of inorganic P compounds through increased belowground C availability (Chatterjee & Saha, 2018). A study in China showed an increase of ~14% in soil available P at the jointing stage of wheat due to elevated CO_2 (Zhang et al., 2014). Bhattacharyya et al. (2014) in India observed a significant increase (>20%) in soil available P in three *kharif* seasons under elevated CO_2 over ambient CO_2. This could be attributed to the large increase (~40%) in organic acids in root exudates under elevated CO_2 over ambient CO_2. They also observed a >30% increase in phosphatase activity and microbial biomass P (MBP) due to elevated CO_2. Kumar et al. (2012) observed that elevated CO_2 had a positive, while the rise in temperature (by 3 °C) had a negative effect on plant P uptake. However, a combination of elevated CO_2 and temperature increased the P uptake by ~17% in an Inceptisol under a subtropical climate.

Like N, alterations in DW patterns linked to climate change may disturb P dynamics in terrestrial ecosystems. The type and extent of response to DW perturbations, however, would depend on the factors previously mentioned for N. A meta-analysis based on 79 studies (1882 observations) by Gao et al. (2020) showed that experimental drying could significantly increase soil available P by 72%, whereas decreased the microbial biomass P (MBP) and phosphatase activity to the extent of 28% and 13%, respectively. Re-wetting of dried soils significantly increased extractable organic phosphorus, phosphatase activity, and leaching of dissolved inorganic P by 27%, 12%, and 116%, respectively. The decline in phosphatase activity due to drought indicates suppression of inorganic P release from SOM. Reduced mineralization of P under prolonged drought was also recorded in a heathland (Andresen et al., 2010). Not only mineralization, P immobilization by microbes is also suppressed due to drought (Andresen et al., 2010; Dijkstra et al., 2015). Hence, there would be a minor general impact of drought on net mineralization of P. Again, the mineralization and immobilization rates would be overlaid by the release of labile organic P and inorganic P from broken aggregates and ruptured microbial cells in very dry soils, and desorption of P from different minerals (Bünemann et al., 2013; Dinh et al., 2016). Together all these are likely to enhance available P in soil under drought (Gao et al., 2020). Similar to DW, alterations in FTC due to climate change also impact P cycle in terrestrial ecosystems. A meta-analysis of global data from 75 publications (2471 observations) by Gao et al. (2021) showed that experimental FTC had a significant positive effect on total salt-extractable P (27%), dissolved organic P (DOP, 9.4%), leaching of dissolved total P (312%), DOP (30%), and dissolved inorganic P (DIP, 115%); but significant negative impact on microbial biomass P (MBP, −10%).

The C:N:P ratios in soil are quite variable compared with marine ecosystems (Redfield, 1958), and are prone to alteration by climatic factors owing to the influence of temperature and precipitation on geochemical and biological cycles of these elements (Bui & Henderson, 2013; Delgado-Baquerizo et al., 2013; Tian et al., 2010; Yu et al., 2018). The geochemical and biological processes involved in balancing C, N, and P in natural ecosystems are stronger and quicker under wetter and warmer environments (Zhang et al., 2008; Li et al., 2012). Even in agricultural lands, climatic variables either strengthen or weaken the effect of crops and management practices on C, N, and P cycling (Deng et al., 2017; Fuchslueger et al., 2019; Li et al., 2012). Increased frequency of DW cycles can induce an imbalance between N and P, and more so in forest soils than agricultural lands. The anticipated increase in the frequency and severity of droughts has potential to decouple the N and P cycles, having unprecedented impacts on nutrient leaching, and availability to plants and microbes (Gao et al., 2020).

2.5 Potassium (K)

Climate models predict an upsurge in the degree of drought across large tracts of land worldwide that are already dry, e.g., the Mediterranean Basin and the Sahel (IPCC, 2007a, 2007b). Increase in aridity increases the importance of K in plant nutrition manyfold due to its role in water-use efficiency (Römheld & Kirkby, 2010; Ruan et al., 1997). K has often been considered the most crucial nutrient limiting crop production in the Mediterranean region (Ryan & Sommer, 2012). However, plants' resorption efficiency for K is generally higher in drier areas, which compensates to some extent for the lower K uptake therein (compared with plants in relatively moist regions) (Fayyaz et al., 2013; Tomlinson et al., 2013). Resorption is the process of re-translocation of a nutrient from leaves, before their abscission, to other living plant parts (Killingbeck, 1996). Besides, increased aridity is likely to reduce K leaching, thereby increasing plants' ability to retain it (Singer, 1989).

Prolonged droughts tend to increase total K, but decrease available K in soil (Sardans & Peñuelas, 2007). Droughts reduce K releasing ability of soils (Kaya et al., 2005; Oliveira et al., 2004; Ruan et al., 1997). Lesser weathering, changes in humus quality due to lower microbial activity, and perhaps the poor activity of enzymes could be the reasons behind lower soluble and available K with greater unavailable K accumulation in soils under drought (Sardans & Peñuelas, 2005, 2007). Consequently, greater K accumulation in soils of arid than in mesic ecosystems has been found (Singer, 1989). Higher water deficit and weakened ability of plants to resist drought due to poor K availability and uptake might become a serious problem in arid regions threatened by even drier circumstances (Sardans & Peñuelas, 2015). An increase in torrential rains and a decrease in MAP, as anticipated for the Mediterranean Basin and the Sahel, would further aggravate the problem (IPCC, 2007a,

2007b; Sardans & Peñuelas, 2015). On the other hand, K tends to concentrate more on biomass with an increase in MAP (Sardans et al., 2011). An elevated temperature may also increase soil available K under non-leaching conditions. Chatterjee et al. (2019), in a laboratory experiment, observed that available K in soils, averaged across different land-use systems, increased significantly from 27 to 36 °C, and beyond for every 3 °C rise up to 42 °C after one-month incubation (Fig. 2.3).

As far as we know, existing models for the prediction of ecosystem's responses to projected climate change do not include K. Moreover, studies on climate change's effect on K cycling are rare. Whatever studies are present mostly show indirect effects. The lack of interest of researchers in studying the climate change impacts on K apparently could be due to the fact that neither is there any organic form of K present in soil or plants that could show changes in mineralization rates under increased temperature like N or P nor is there any gaseous loss of K from the ecosystem. However, given the uncertainty in the upcoming variations in MAP:PET associated with climate change bringing water-use efficiency of crops into the picture, K surely would become a point of keen interest for environmental and agricultural scientists in future.

2.6 Calcium (Ca), Magnesium (Mg), and Sulphur (S)

In general, a temperature rise is associated with the increased dissolution of silicate rocks releasing Si and Na; but the same may not cause much dissolution and leaching of Ca and Mg (White et al., 1999). However, in some cases increases in Ca and Mg concentrations or at least a slower decrease rate compared with that of sulphate concentration were observed in lakes having Si-rich bedrock while they were recovering from acidification (Houle et al., 2010; Mast et al., 2011; Rogora et al., 2013). Such phenomena were thought to be due to faster weathering caused by a warmer environment associated with climate change. An indirect effect of climate change on soil Ca and Mg can be seen through the changes in CEC. In soils with low-activity clays or with a coarse texture, SOM is a major contributor to CEC (Weil & Magdoff, 2004). Enhanced decomposition of SOM due to increased temperature under changing climate might reduce the CEC of these soils by SOM loss (Allen et al., 2011; Davidson & Janssens, 2006). Lowering of CEC might cause enhanced leaching of base cations like Ca^{2+}, Mg^{2+}, etc., especially under frequent and intense rainfall, causing alkalinity to move from soil to various waterways (Allen et al., 2011). Also, salinization, i.e., accumulation of salts of Ca^{2+}, Mg^{2+}, Na^{+}, etc. due to climate warming can degrade the quality and availability of freshwater (IPCC, 2007b). Climate warming is likely to expand the requirement for irrigation in semi-arid regions, e.g., Indo-Gangetic plains of northern India, Pakistan, and parts of Africa (Yeo, 1999). Increased irrigation with higher surface water evaporation caused by elevated temperature may cause more agricultural salinization (St.Clair & Lynch, 2010), through the accumulation of soluble salts of Ca^{2+}, Mg^{2+}, etc. On the contrary, there is a theoretical possibility that increased CO_2 levels might counteract the accumulation of soluble salts of Ca^{2+}, Mg^{2+}, etc. (Nicolas et al., 1993; Yeo, 1999). Moreover, rising sea levels caused by climate change may also cause intrusion of seawater in coastal aquifers (Antonellini et al., 2008; Don et al., 2006), causing soluble salts of Ca^{2+}, Mg^{2+}, etc. to accumulate.

As discussed earlier, SOM is largely affected by changes in climatic variables like MAT and MAP. A significant portion of S is part of the SOM, hence, elevated temperature and large fluctuations in MAP associated with climate change are likely to alter organic S forms in soil. Climatic variables like MAT and MAP were reported to affect soil organic S (SOS) content in the cultivated and native grassland soils of the Great Plains of North America (Wang et al., 2006). There, an increase in MAT was found to decrease the contents of ester-SO_4 S, C-bonded S, and total organic S. Accelerated decomposition of SOM under elevated temperature leads to a such decrease in SOS forms. The impact of MAP on SOS forms was less obvious compared with MAT in the native grasslands, but a major loss of SOS could be seen in the cultivated soils with an apparent increase in MAP (Wang et al., 2006). Hence, the same climatic factor can have a different extent of impact on SOS dynamics depending on the land use.

2.7 Micronutrients

Micronutrient availability to plants and microorganisms in soil depends, besides their total content and different soil properties, on the climatic factors, with the latter assuming increasing importance under the changing climate scenario associated with atmospheric warming and CO_2 enrichment (Johnson, 2013). Potential increase in C input via root exudates under elevated CO_2 can alter soil pH and SOM content, and thereby increase the bioavailability of soil-derived nutrients (Ehrenfeld et al., 2005; Paul, 2007; Phillips et al., 2009; Wang et al., 2016a, 2016b). Reports are there showing higher accumulation of micronutrients by rice due to CO_2 enrichment (Johnson, 2013; Seneweera et al., 1996). In a study with winter wheat on neutral soil, atmospheric CO_2 enrichment (500 ppm) increased the availability of Fe, Mn, Zn, and Cu by ~48%, ~23%, ~114%, and ~60%, respectively, and their uptake by wheat (Wang et al., 2016a, 2016b). This was partly due to the slight drop in soil pH (by 0.35 units) induced by a greater amount of root exudates. Enhanced availability of soil Fe, Mn, and Zn under elevated CO_2 was also observed by Guo et al. (2014a, 2014b). As mentioned earlier, plants release more root exudates when grown under elevated than under ambient CO_2 levels (Bhattacharyya et al., 2013; Phillips et al., 2009; Yin et al., 2013). The root exudates are made up of amino acids, organic acid polymeric carbohydrates, saccharides, fatty acids, and polymer degrading enzymes (Doornbos et al., 2012; Fransson & Johansson, 2010), which cause the drop in soil pH in the rhizosphere. Moreover, higher root exudates and organic debris under elevated CO_2 tend to enhance microbial activity in the rhizosphere, often seen as increased MBC (Jin et al., 2014; Wang et al., 2016a, 2016b). There was a positive correlation between MBC and the availability of Fe, Zn, and Cu (Wang et al., 2016a, 2016b), due to better nutrient cycling associated with higher MBC (Allard et al., 2005). Besides, higher soil moisture under elevated CO_2 was also observed sometimes (Niklaus et al., 1998; Wang et al., 2016a, 2016b). Higher soil moisture tends to promote a reducing environment enhancing the availability of Mn (the reduced form i.e., Mn^{2+} being more soluble than the oxidized form, e.g., Mn^{4+}) (Wang et al., 2016a, 2016b). Wang et al. (2016a, 2016b) also found that warming of air under a wheat canopy by 2 °C over ambient temperature increased the availability of Fe, Zn, and Cu by ~60%, ~15%, and ~24%, respectively; but not of Mn. The combined effects of CO_2 enrichment (500 ppm) and warming also resulted in significantly increased concentrations of Fe, Mn, Zn, and Cu in soil under wheat crop at the jointing stage in comparison to that recorded with warming alone (Fig. 2.4). Such increases in availability, however, did not always render increased uptake. In an earlier study, it was found that an increase of 0.56 °C in soil temperature under dragon spruce (*Picea asperata*) increased the Mehlich-3 extractable Mn, Zn, and Cu in soil, but slightly decreased the available Fe (Li et al., 2014). In that study, a 0.55 °C rise in soil temperature decreased the Mehlich-3 extractable Fe, Mn, Zn, and Cu in the soils under natural forest. Hence, the effect of warming on micronutrient availability in soil varies with plant species (Wang et al., 2016a, 2016b).

Anticipated upsurge in global aridity caused by climate change may have a robust impact on the bioavailability of micronutrient cations via indirect effects on various soil properties (Moreno-Jiménez et al., 2019). For instance, aridity declines the rate of weathering, leaching of base cations, hindering the processes of reactive clay minerals formation, soil acidification, and preservation and build-up of SOM (Maestre et al.,

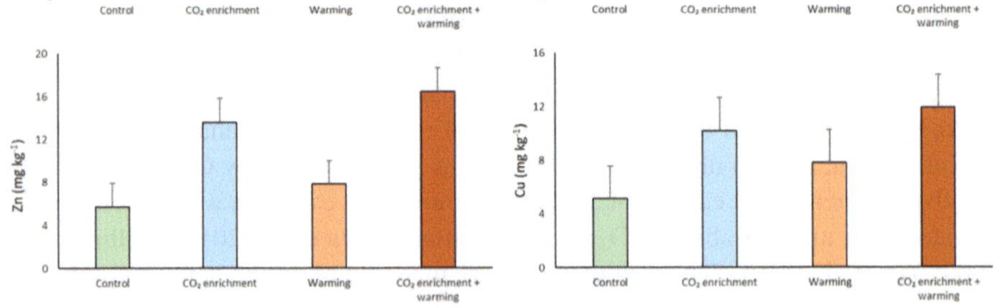

Fig. 2.4 Effect of CO_2 enrichment and warming on the concentrations of micronutrients in soil under wheat crop (Adapted from Wang et al., 2016a, 2016b)

2016; Plaza et al., 2018), which in turn would affect solubility and sorption–desorption of micronutrient cations. After analyzing 143 dryland soils all over the world (except Antarctica), Moreno-Jiménez et al. (2019) observed lower concentrations of total and available Fe, Mn, Zn, and Cu in dryland soils compared with soils of natural and agricultural ecosystems. Aridity resulted in increased soil pH and lower SOM content; consequently, showed a negative effect on bioavailability of those micronutrients. Available Zn:Fe ratio increased exponentially with increasing aridity, indicating that aridity may cause stoichiometric alterations in micronutrient cations. Hence, increased global aridity caused by climate change might decline the availability of micronutrient cations, especially Fe and Zn, to plants and soil microbes. This along with other adversities of climate change may jeopardize major ecological functions and services (Moreno-Jiménez et al., 2019).

Changes in rainfall and temperature regimes associated with climate change can also alter redox dynamics and hence, nutrient cycling. Laboratory studies under controlled conditions showed an exponential increase in Fe-reduction with temperature in mine pit lake sediments having high acidity (Meier et al., 2005). Whereas, dissimilatory microbial reduction of Fe^{3+} was more in sub-alpine wetlands at higher temperatures (Schilling et al., 2019). In an incubation experiment, Fe^{3+} reduction in soils showed higher temperature sensitivity at a higher temperature range (26–35 °C) than lower temperature range (8–26 °C) under anoxic conditions (Chari et al., 2021).

Overall, climate change is caused and/or manifested by changes in variables like temperature, CO_2 concentration, rainfall, etc. These, in turn, alter the processes involved in nutrient cycling, and either enhances or reduces the mobility and availability of soil nutrients and consequently their uptake by plants (Fig. 2.5).

2.8 Conclusion

Nutrient cycling and transformations *vis-à-vis* climate change are mostly associated with factors, namely increase in atmospheric CO_2, temperature, and changes in precipitation pattern along with the water availability in soil. The differential impacts of such factors associated with climate change on the availability and transformation of nutrients are usually attributed to the changes in soil reaction, mineralization pattern of soil organic matter, microbial and associated enzymatic activities, weathering rates of nutrient-containing rocks and minerals, and redox dynamics in soil. Moreover, changes in several nutrient-specific processes involved in nutrient dynamics (e.g., nitrification in case of N and sorption–desorption in case of P) also govern the availability of nutrients in soil under climate change scenarios. The available information suggests that an increase in mean annual temperature and soil aridity may result in a short-term increase in plant available N with the consequent decrease in organic N pool of soil. The organic pools of P and S are also expected to decline due to an increase in soil temperature. Such findings indicate the vulnerability of SOC stock to the changing climate, which needs immediate attention of the natural resource managers and policymakers for the implementation of appropriate

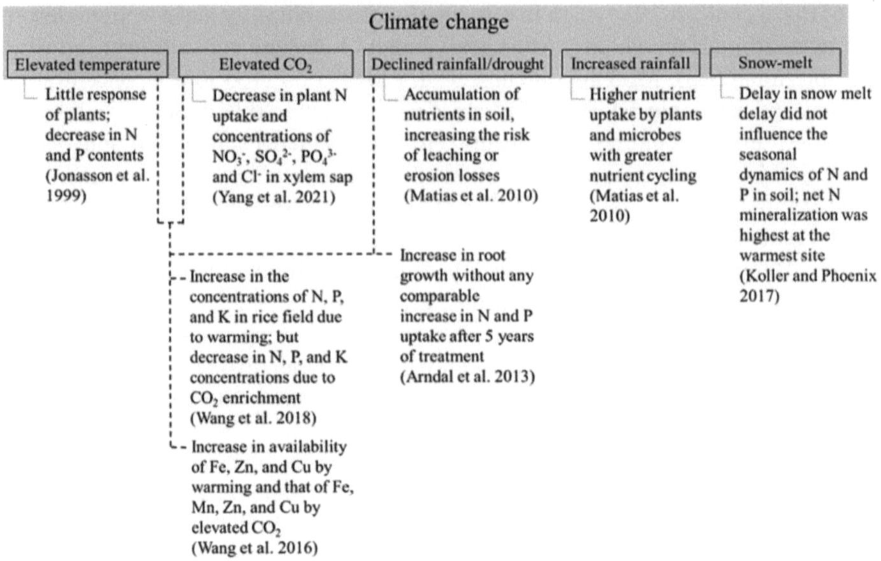

Fig. 2.5 Effect of climate change on plant availability and uptake of nutrients; solid blue line indicates sole effect; dotted black line indicates combined effect

management practices. Very limited studies about the impact of climate change on the availability of K and micronutrients underline the need for further research in this direction. Because of the possible increase in aridity and formation of pedogenic carbonates of Ca and/or Mg in soil, formulation of appropriate soil-cum-nutrient management strategies is also an important challenge in changing climate.

References

Allard, V., Newton, P., Lieffering, M., Soussana, J., Carran, R., & Matthew, C. (2005). Increased quantity and quality of coarse soil organic matter fraction at elevated CO_2 in a grazed grassland are a consequence of enhanced root growth rate and turnover. *Plant and Soil, 276*, 49–60.

Allen, D. E., Singh, B. P., & Dalal, R. C. (2011). Soil health indicators under climate change: A review of current knowledge. In B. Singh, A. Cowie, & K. Chan (Eds.), *Soil health and climate change. Soil biology* (Vol. 29). Springer. https://doi.org/10.1007/978-3-642-20256-8_2

Andrade-Linares, D. R., Zistl-Schlingmann, M., Foesel, B., Dannenmann, M., Schulz, S., & Schloter, M. (2021). Short term effects of climate change and intensification of management on the abundance of microbes driving nitrogen turnover in montane grassland soils. *Science of the Total Environment, 780*, 146672. https://doi.org/10.1016/j.scitotenv.2021.146672

Andresen, L. C., Michelsen, A., Jonasson, S., Schmidt, I. K., Mikkelsen, T. N., Ambus, P., & Beier, C. (2010). Plant nutrient mobilization in temperate heathland responds to elevated CO_2, temperature and drought. *Plant and Soil, 328*, 381–396.

Andrews, J. A., & Schlesinger, W. H. (2001). Soil CO_2 dynamics, acidification, and chemical weathering in a temperate forest with experimental CO_2 enrichment. *Global Biogeochemical Cycles, 15*(1), 149–162.

Antonellini, M., Mollema, P., Giambastiani, B., Bishop, K., Caruso, L., Minchio, A., Pellegrini, L., Sabia, M., Ulazzi, E., & Gabbianelli, G. (2008). Salt water intrusion in the coastal aquifer of the southern Po Plain, Italy. *Hydrogeology Journal, 16*, 1541–1556.

Arndal, M. F., Merrild, M. P., Michelsen, A., Schmidt, I. K., Mikkelsen, T. N., & Beier, C. (2013). Net root growth and nutrient acquisition in response to predicted climate change in two contrasting heathland species. *Plant and Soil, 369*, 615–629.

Augusto, L., Achat, D. L., Jonard, M., Vidal, D., & Ringeval, B. (2017). Soil parent material—A major driver of plant nutrient limitations in terrestrial ecosystems. *Global Change Biology, 23*, 3808–3824. https://doi.org/10.1111/gcb.13691

Austin, A. T., & Vitousek, P. M. (1998). Nutrient dynamics on a precipitation gradient in Hawai'i. *Oecologia, 113*, 519–529. https://doi.org/10.1007/s004420050405

Auyeung, D. S. N., Martiny, J. B. H., & Dukes, J. S. (2015). Nitrification kinetics and ammonia oxidizing community respond to warming and altered precipitation. *Ecosphere, 6*, 83. https://doi.org/10.1890/es14-00481.1

Auyeung, D. S. N., Suseela, V., & Dukes, J. S. (2013). Warming and drought reduce temperature sensitivity of nitrogen transformations. *Global Change Biology, 19*, 662–676. https://doi.org/10.1111/gcb.12063

Barker, S. L., & Cox, S. F. (2011). Oscillatory zoning and trace element incorporation in hydrothermal minerals: Insights from calcite growth experiments. *Geofluids, 11*, 48–56.

Barrow, N. (1983). A mechanistic model for describing the sorption and desorption of phosphate by soil. *Journal of Soil Science, 34*, 733–750. https://doi.org/10.1111/j.1365-2389.1983.tb01068.x

Beare, M. H., Gregorich, E. G., & St-Georges, P. (2009). Compaction effects on CO_2 and N_2O production during drying and rewetting of soil. *Soil Biology & Biochemistry, 41*, 611–621.

Beerling, D. J., & Woodward, F. I. (1994). Climate change and the British scene. *Journal of Ecology, 82*, 391–397.

Beier, C., Emmett, B. A., Penuelas, J., Schmidtd, I. K., Tietemae, A., Estiartec, M., Gundersend, P., Llorensc, L., Riis-Nielsend, T., Sowerbyb, A., & Gorissenf, A. (2008). Carbon and nitrogen cycles in European ecosystems respond differently to global warming. *Science of the Total Environment, 407*, 692–697.

Bhattacharyya, P., Roy, K. S., Neogi, S., Dash, P. K., Nayak, A. K., Mohanty, S., Baig, M. J., Sarkar, R. K., & Rao, K. S. (2013). Impact of elevated CO_2 and temperature on soil C and N dynamics in relation to CH_4 and N_2O emissions from tropical flooded rice (*Oryza sativa* L.). *Science of the Total Environment, 461*, 601–611.

Bhattacharyya, P., Roy, K. S., Dash, P. K., Neogi, S., Shahid, M., Nayak, A. K., Raja, R., Karthikeyan, S., Balachandar, D., & Rao, K. S. (2014). Effect of elevated carbon dioxide and temperature on phosphorus uptake in tropical flooded rice (*Oryza sativa* L.). *European Journal of Agronomy, 53*, 28–37.

Borken, W., & Matzner, E. (2009). Reappraisal of drying and wetting effects on C and N mineralization and fluxes in soils. *Global Change Biology, 15*, 808–824.

Budge, K., Leifeld, J., Hiltbrunner, E., & Fuhrer, J. (2011). Alpine grassland soils contain large proportion of labile carbon but indicate long turnover times. *Biogeosciences, 8*, 1911–1923. https://doi.org/10.5194/bg-8-1911-2011

Bui, E. N., & Henderson, B. L. (2013). C:N: P stoichiometry in Australian soils with respect to vegetation and environmental factors. *Plant and Soil, 373*, 553–568.

Bünemann, E. K., Keller, B., Hoop, D., Jud, K., Boivin, P., & Frossard, E. (2013). Increased availability of phosphorus after drying and rewetting of a grassland soil: Processes and plant use. *Plant and Soil, 370*, 511–526.

Campbell, J. L., Socci, A. M., & Templer, P. H. (2014). Increased nitrogen leaching following soil freezing is due to decreased root uptake in a northern hardwood forest. *Global Change Biology, 20*, 2663–2673.

Carrillo, Y., Pendall, E., Dijkstra, F. A., Morgan, J. A., & Newcomb, J. M. (2011). Response of soil organic matter pools to elevated CO_2 and warming in a semi-arid grassland. *Plant and Soil, 347*, 339–350.

Carter, T. R., Jones, R. N., Lu, X., Bhadwal, S., Conde, C., Mearns, L. O., O'Neill, B. C., Rounsevell, M., & Zurek, M. B. (2007). New assessment methods and the characterisation of future conditions. In M. L. Parry, O. F. Canziani, J. P. Palutikof, P. J. Van der Linden, & C. E. Hanson (Eds.), *Climate change 2007: Impacts, adaptation and vulnerability. Contribution of working group II to the fourth assessment report of the Intergovernmental Panel on Climate Change* (pp. 133–171). Cambridge University Press.

Chai, Y. J., Zeng, X. B., E, S. Z., Bai, L. Y., Su, S. M., & Huang, T. (2014). Effects of freeze-thaw on aggregate stability and the organic carbon and nitrogen enrichment ratios in aggregate fractions. *Soil Use and Management, 30*, 507–516.

Chari, N. R., Lin, Y., Lin, Y. S., & Silver, W. L. (2021). Interactive effects of temperature and redox on soil carbon and iron cycling. *Soil Biology & Biochemistry, 157*, 108235. https://doi.org/10.1016/j.soilbio.2021.108235

Chatterjee, D., Datta, S. C., & Manjaiah, K. M. (2014). Fractions, uptake and fixation capacity of phosphorus and potassium in three contrasting soil orders. *Journal of Soil Science and Plant Nutrition, 14*, 640–656.

Chatterjee, D., Datta, S. C., & Manjaiah, K. M. (2015). Effect of citric acid treatment on release of phosphorus, aluminium and iron from three dissimilar soils of India. *Archives of Agronomy and Soil Science, 61*, 105–117.

Chatterjee, D., Kuotsu, R., Ao, M., Saha, S., Ray, S. K., & Ngachan, S. V. (2019). Does rise in temperature adversely affect soil fertility, carbon fractions, microbial biomass and enzyme activities under different land uses? *Current Science*, 116, 2044–2054. https://doi.org/10.18520/cs/v116/i12/2044-2054

Chatterjee, D., & Saha, S. (2018). Response of soil properties and soil microbial communities to the projected climate change. In *Advances in crop environment interaction* (pp. 87–136). Springer. https://doi.org/10.1007/978-981-13-1861-0_4

Chen, C., Whalen, J. K., & Guo, X. (2014). Earthworms reduce soil nitrous oxide emissions during drying and rewetting cycles. *Soil Biology & Biochemistry, 68*, 117–124.

Conant, R. T., Dalla-Betta, P., Klopatek, C. C., & Klopatek, J. M. (2004). Controls on soil respiration in semiarid soils. *Soil Biology & Biochemistry, 36*, 945–951.

Cook, B. I., Smerdon, J. E., Seager, R., & Coats, S. (2014). Global warming and 21st century drying. *Climate Dynamics, 43*, 2607–2627. https://doi.org/10.1007/s00382-014-2075-y

Cotrufo, M. F., Ranalli, M. G., Haddix, M. L., Six, J., & Lugato, E. (2019). Soil carbon storage informed by particulate and mineral-associated organic matter. *Nature Geoscience, 12*, 989–994. https://doi.org/10.1038/s41561-019-0484-6

Cotrufo, M. F., Wallenstein, M. D., Boot, C. M., Denef, K., & Paul, E. (2013). The Microbial efficiency-matrix stabilization (MEMS) framework integrates plant litter decomposition with soil organic matter stabilization: Do labile plant inputs form stable soil organic matter? *Global Change Biology Bioenergy, 19*, 988–995. https://doi.org/10.1111/gcb.12113

Craine, J. M., Fierer, N., & McLauchlan, K. K. (2010). Widespread coupling between the rate and temperature sensitivity of organic matter decay. *Nature Geoscience, 3*, 854. https://doi.org/10.1038/ngeo1009

Dannenmann, M., Bimüller, C., Gschwendtner, S., Leberecht, M., Tejedor, J., Bilela, S., Gasche, R., Hanewinkel, M., Baltensweiler, A., Kögel-Knabner, I., Polle, A., Schloter, M., Simon, J., & Rennenberg, H. (2016). Climate change impairs nitrogen cycling in European beech forests. *PLoS ONE, 11*, e0158823. https://doi.org/10.1371/journal.pone.0158823

Davidson, E. A., & Janssens, I. A. (2006). Temperature sensitivity of soil carbon decomposition and feedbacks to climate change. *Nature, 440*, 165–173.

Dawes, M. A., Schleppi, P., Hättenschwiler, S., Rixen, C., & Hagedorn, F. (2017). Soil warming opens the nitrogen cycle at the alpine treeline. *Global Change Biology, 23*, 421–434. https://doi.org/10.1111/gcb.13365

Del Grosso, S. J., Parton, W. J., Mosier, A. R., Holland, E. A., Pendall, E., Schimel, D. S., & Ojima, D. S. (2005). Modelling soil CO_2 emissions from ecosystems. *Biogeochemistry, 73*, 71–91.

Delgado-Baquerizo, M., Maestre, F. T., Gallardo, A., Bowker, M. A., Wallenstein, M. D., Quero, J. L., Ochoa, V., Gozalo, B., García-Gómez, M., Soliveres, S., García-Palacios, P., Berdugo, M., Valencia, E., Escolar, C., Arredondo, T., Barraza-Zepeda, C., Bran, D., Carreira, J. A., Chaieb, M., … Zaady, E. (2013). Decoupling of soil nutrient cycles as a function of aridity in global drylands. *Nature, 502*, 672–676. https://doi.org/10.1038/nature12670

Deng, Q., McMahon, D. E., Xiang, Y., Yu, C. L., Jackson, R. B., & Hui, D. (2017). A global meta-analysis of soil phosphorus dynamics after afforestation. *New Phytologist, 213*, 181–192.

Dijkstra, F. A., He, M., Johansen, M. P., Harrison, J. J., & Keitel, C. (2015). Plant and microbial uptake of nitrogen and phosphorus affected by drought using ^{15}N and ^{32}P tracers. *Soil Biology & Biochemistry, 82*, 135–142.

Dijkstra, F. A., & Morgan, J. A. (2012). Elevated CO_2 and warming effects on soil carbon sequestration and greenhouse gas exchange in agroecosystems: A review. In M. A. Liebig, A. J. Franzluebbers, & R. F. Follett (Eds.), *Managing agricultural greenhouse gases: Coordinated agricultural research through GRACEnet to address our changing climate* (Chap. 27). Academic Press (in press).

Dinh, M. V., Schramm, T., Spohn, M., & Matzner, E. (2016). Drying-rewetting cycles release phosphorus from forest soils. *Journal of Plant Nutrition and Soil Science, 179*, 670–678.

Dixon, J. L., Chadwick, O. A., & Vitousek, P. M. (2016). Climate-driven thresholds for chemical weathering in postglacial soils of New Zealand. *Journal of Geophysical Research: Earth Surface, 121*, 1619–1634.

Don, N. C., Hang, N. T. M., Araki, H., Yamanishi, H., & Koga, K. (2006). Salinization processes in an alluvial coastal lowland plain and effect of sea water level rise. *Environmental Geology*, 49, 743–751. https://doi.org/10.1007/s00254-005-0119-7

Doornbos, R. F., van Loon, L. C., & Bakker, P. A. (2012). Impact of root exudates and plant defence signalling on bacterial communities in the rhizosphere. A review. *Agronomy for Sustainable Development, 32*, 227–243.

Ehrenfeld, J. G., Ravit, B., & Elgersma, K. (2005). Feedback in the plant-soil system. *Annual Review of Environment and Resources, 30*, 75–115.

Fayyaz, P., Etemadi, E., Julaiee-Manesh, N., & Zolfaghari, R. (2013). Sodium and potassium allocation under drought stress in Atlas mastic tree (*Pistacia atlantica* subsp. *mutica*). *Forest-Biogeosciences and Forestry, 6*, 90–94.

Fierer, N., Colman, B. P., Schimel, J. P., & Jackson, R. B. (2006). Predicting the temperature dependence of microbial respiration in soil: A continental scale analysis. *Global Biogeochemical Cycles, 20*(3). https://doi.org/10.1029/2005GB002644

Follett, R. F., Stewart, C. E., Pruessner, E. G., & Kimble, J. M. (2012). Effects of climate change on soil carbon and nitrogen storage in the US Great Plains. *Journal of Soil and Water Conservation, 67*, 331–342. https://doi.org/10.2489/jswc.67.5.331

Fontaine, S., Mariotti, A., & Abbadie, L. (2003). The priming effect of organic matter: A question of microbial competition? *Soil Biology & Biochemistry, 35*, 837–843. https://doi.org/10.1016/S0038-0717(03)00123-8

Fransson, P., & Johansson, E. M. (2010). Elevated CO_2 and nitrogen influence exudation of soluble organic compounds by ectomycorrhizal root systems. *FEMS Microbiology Ecology, 71*, 186–196.

Fu, Q., & Feng, S. (2014). Responses of terrestrial aridity to global warming. *Journal of Geophysical Research: Atmospheres, 119*, 7863–7875.

Fuchslueger, L., Wild, B., Mooshammer, M., Takriti, M., Kienzl, S., Knoltsch, A., Hofhansl, F., Bahn, M., & Richter, A. (2019). Microbial carbon and nitrogen cycling responses to drought and temperature in differently managed mountain grasslands. *Soil Biology & Biochemistry, 135*, 144–153.

Gao, D., Bai, E., Li, M., Zhao, C., Yu, K., & Hagedorn, F. (2020). Responses of soil nitrogen and phosphorus cycling to drying and rewetting cycles: A meta-analysis. *Soil Biology & Biochemistry, 148*, 107896. https://doi.org/10.1016/j.soilbio.2020.107896

Gao, D., Bai, E., Yang, Y., Zong, S., & Hagedorn, F. (2021). A global meta-analysis on freeze-thaw effects on soil carbon and phosphorus cycling. *Soil Biology & Biochemistry, 159*, 108283. https://doi.org/10.1016/j.soilbio.2021.108283

Gao, D., Zhang, L., Liu, J., Peng, B., Fan, Z., Dai, W., Jiang, P., & Bai, E. (2018). Responses of terrestrial nitrogen pools and dynamics to different patterns of freeze-thaw cycle: A meta-analysis. *Global Change Biology, 24*, 2377–2389.

Garcia-Pausas, J., Romanyà, J., Montané, F., Rios, A. I., Taull, M., Rovira, P., & Casals, P. (2017). Are soil carbon stocks in mountain grasslands compromised by land-use changes? In J. Catalan, J. M. Ninot, & M. M. Aniz (Eds.), *High mountain conservation in a changing world* (pp. 207–230). Springer International Publishing.

Ghosh, A., Das, A., Das, D., Ray, P., Bhattacharyya, R., Biswas, D. R., & Biswas, S. S. (2020). Contrasting land use systems and soil organic matter quality and temperature sensitivity in North Eastern India. *Soil and Tillage Research, 199*, 104573.

Gill, R. A., Polley, H. W., Johnson, H. B., Anderson, L. J., Maherali, H., & Jackson, R. B. (2002). Nonlinear grassland responses to past and future atmospheric CO(2). *Nature, 417*(6886), 279–282.

Groffman, P. M., Hardy, J. P., Fashu-Kanu, S., Driscoll, C. T., Cleavitt, N. L., Fahey, T. J., & Fisk, M. C. (2011). Snow depth, soil freezing and nitrogen cycling in a northern hardwood forest landscape. *Biogeochemistry, 102*, 223–238.

Groshans, G. R., Mikhailova, E. A., Post, C. J., & Schlautman, M. A. (2018). Accounting for soil inorganic carbon in the ecosystem services framework for United Nations sustainable development goals. *Geoderma, 324*, 37–46.

Gruber, N., & Galloway, J. N. (2008). An Earth-system perspective of the global nitrogen cycle. *Nature, 451*, 293–296.

Guan, S., An, N., Zong, N., He, Y., Shi, P., Zhang, J., & He, N. (2018). Climate warming impacts on soil organic carbon fractions and aggregate stability in a Tibetan alpine meadow. *Soil Biology & Biochemistry, 116*, 224–236.

Guo, J., Zhang, M. Q., Wang, X. W., & Zhang, W. J. (2014a). A possible mechanism of mineral responses to elevated atmospheric CO_2 in rice grains. *Journal of Integrative Agriculture, 14*, 1–14.

Guo, X., Drury, C. F., Yang, X., Daniel-Reynolds, W., & Fan, R. (2014b). The extent of soil drying and rewetting affects nitrous oxide emissions, denitrification, and nitrogen mineralization. *Soil Science Society of America Journal, 78*, 194.

Harrison-Kirk, T., Beare, M. H., Meenken, E. D., & Condron, L. M. (2013). Soil organic matter and texture affect responses to dry/wet cycles: Effects on carbon dioxide and nitrous oxide emissions. *Soil Biology & Biochemistry, 57*, 43–55.

Hart, S. C. (2006). Potential impacts of climate change on nitrogen transformations and greenhouse gas fluxes in forests: A soil transfer study. *Global Change Biology, 12*, 1032–1046.

Hartmann, A. A., Barnard, R. L., Marhan, S., & Niklaus, P. A. (2013). Effects of drought and N fertilization on N cycling in two grassland soils. *Oecologia, 171*, 705–717.

Homann, P. S., Kapchinske, J. S., & Boyce, A. (2007). Relations of mineral-soil C and N to climate and texture: Regional differences within the conterminous USA. *Biogeochemistry, 85*, 303–316.

Hou, E., Chen, C., Luo, Y., Zhou, G., Kuang, Y., Zhang, Y., Heenan, M., Lu, X., & Wen, D. (2018). Effects of climate on soil phosphorus cycle and availability in natural terrestrial ecosystems. *Global Change Biology, 24*, 3344–3356.

Houle, D., Couture, S., & Gagnon, C. (2010). Relative role of decreasing precipitation sulfate and climate on recent lake recovery. *Global Biogeochemical Cycles, 24*, GB4029.

Hu, Y., Zhang, Z., Wang, Q., Wang, S., Zhang, Z., Wang, Z., Xu, G., Du, M., & Dijkstra, F. A. (2017). Variations of N_2O fluxes in response to warming and cooling in an alpine meadow on the Tibetan Plateau. *Climate Change, 143*, 129–142.

Huang, J., Yu, H., Guan, X., Wang, G., & Guo, R. (2016). Accelerated dryland expansion under climate change. *Nature Climate Change, 6*, 166–171.

Hungate, B. A., Dukes, J. S., Shaw, M. R., Luo, Y., & Field, C. B. (2003). Nitrogen and climate change. *Science, 302*, 1512.

Ineson, P., Taylor, K., Harrison, A. F., Poskitt, J., Benham, D. G., Tipping, E., & Woof, C. (1998). Effects of climate change on nitrogen dynamics in upland soils. 1. A transplant approach. *Global Change Biology, 4*, 143–152.

IPCC. (2007a). In S. Solomon, et al. (Eds.), *Climate change 2007: The physical science basis. Contribution of working group I. Fourth assessment report of the Intergovernmental Panel on Climate Change* (pp. 1–32). Cambridge University Press.

IPCC. (2007b). *Working group II fourth assessment report.* Cambridge University Press.

IPCC. (2012). Managing the risks of extreme events and disasters to advance climate change adaptation. In C. B. Field, V. Barros, T. F. Stocker, Q. Dahe, D. J. Dokken, K. L. Ebi, et al. (Eds.), *A special report of working groups I and II of the Intergovernmental Panel on Climate Change.* Cambridge University Press.

IPCC. (2013). Climate change 2013: The physical science basis. In T. F. Stocker, D. Qin, G. K. Plattner, M. Tignor, S. K. Allen, J. Boschung, A. Nauels, Y. Xia, V. Bex, & P. M. Midgley (Eds.), *Contribution of working group I to the fifth assessment report of the Intergovernmental Panel on Climate Change* (pp. 1–1535). Cambridge University Press.

IPCC. (2022). Summary for policymakers [H.-O. Pörtner, D.C. Roberts, E.S. Poloczanska, K. Mintenbeck, M. Tignor, A. Alegría, M. Craig, S. Langsdorf, S. Löschke, V. Möller, A. Okem (eds.)]. In: Climate Change 2022: Impacts, Adaptation, and Vulnerability. Contribution of Working Group II to the Sixth Assessment Report of the Intergovernmental Panel on Climate Change [H.-O. Pörtner, D.C. Roberts, M. Tignor, E.S. Poloczanska, K. Mintenbeck, A. Alegría, M. Craig, S. Langsdorf, S. Löschke, V. Möller, A. Okem, B. Rama (eds.)]. Cambridge University Press.

Jiao, F., Shi, X., Han, F., & Yuan, Z. (2016). Increasing aridity, temperature and soil pH induce soil CNP imbalance in grasslands. *Scientific Reports, 6*, 19601.

Jin, J., Tang, C., Robertson, A., Franks, A. E., Armstrong, R., & Sale, P. (2014). Increased microbial activity contributes to phosphorus immobilization in the rhizosphere of wheat under elevated CO_2. *Soil Biology & Biochemistry, 75*, 292–299.

Jobbagy, E. G., & Jackson, R. B. (2000). The vertical distribution of soil organic carbon and its relation to climate and vegetation. *Ecological Applications, 10*, 423–436.

Joergensen, R. G., Brookes, P. C., & Jenkinson, D. S. (1990). Survival of the soil microbial biomass at elevated temperatures. *Soil Biology & Biochemistry, 22*, 1129–1136.

Johnson, A. A. (2013). Enhancing the chelation capacity of rice to maximise iron and zinc concentrations under elevated atmospheric carbon dioxide. *Functional Plant Biology, 40*, 101–108.

Jonasson, S., Michelsen, A., Schmidt, I. K., & Nielsen, E. V. (1999). Responses in microbes and plants to changed temperature, nutrient, and light regimes in the arctic. *Ecology, 80*, 1828–1843.

Kamewada, K., & Yoshizawa, C. (2018). Impacts of climate change on soil nitrogen kinetics and rice production in Andisol paddy fields. *Soil Science and Plant Nutrition, 64*, 752–766.

Karberg, N. J., Pregitzer, K. S., King, J. S., Friend, A. L., & Wood, J. R. (2005). Soil carbon dioxide partial pressure and dissolved inorganic carbonate chemistry under elevated carbon dioxide and ozone. *Oecologia, 142*, 296–306.

Kaya, C., Higgs, D., & Kirnak, H. (2005). Influence of polyethylene mulch, irrigation regime, and potassium rates on field cucumber yield and related traits. *Journal of Plant Nutrition, 28*, 1739–1753.

Killingbeck, K. T. (1996). Nutrients in senesced leaves: Keys to the search for potential resorption and resorption proficiency. *Ecology, 77*, 1716–1727.

Koller, E. K., & Phoenix, G. K. (2017). Seasonal dynamics of soil and plant nutrients at three environmentally contrasting sites along a sub-Arctic catchment sequence. *Polar Biology, 40*, 1821–1834.

Kumar, M., Swarup, A., Patra, A. K., Chandrakala, J. U., & Manjaiah, K. M. (2012). Effect of elevated CO_2 and temperature on phosphorus efficiency of wheat grown in an Inceptisol of subtropical India. *Plant, Soil and Environment, 58*, 230–235.

Kuypers, M. M. M., Marchant, H. K., & Kartal, B. (2018). The microbial nitrogen-cycling network. *Nature Reviews Microbiology, 16*, 263.

Lagomarsino, A., Moscatelli, M. C., Hoosbeek, M. R., Angelis, P. D., & Grego, S. (2008). Assessment of soil nitrogen and phosphorous availability under elevated CO_2 and N-fertilization in a short rotation poplar plantation. *Plant and Soil, 308*, 131–147.

Larsen, K. S., Andresen, L. C., Beier, C., Jonasson, S., Albert, K. R., Ambus, P., Arndal, M. F., Carter, M. S., Christensen, S., Holmstrup, M., Ibrom, A., Kongstad, J., Linden, L. V. D., Maraldo, K., Michelsen, A., Mikkelsen, T. N., Pilegaard, K., Priemé, A., Ro-Poulsen, H., … Stevnbak, K. (2011). Reduced N cycling in response to elevated CO_2, warming, and drought in a Danish heathland: Synthesizing results of the CLIMAITE project after two years of treatments. *Global Change Biology, 17*, 1884–1899.

Larsen, K. S., Jonasson, S., & Michelsen, A. (2002). Repeated freeze-thaw cycles and their effects on biological processes in two arctic ecosystem types. *Applied Soil Ecology, 21*, 187–195.

Leitner, S., Homyak, P. M., Blankinship, J. C., Eberwein, J., Jenerette, G. D., Zechmeister-Boltenstern, S., & Schimel, J. P. (2017). Linking NO and N_2O emission pulses with the mobilization of mineral and organic N upon rewetting dry soils. *Soil Biology & Biochemistry, 115*, 461–466.

Li, D., Niu, S., & Luo, Y. (2012). Global patterns of the dynamics of soil carbon and nitrogen stocks following afforestation: A meta-analysis. *New Phytologist, 195*, 172–181.

Li, D.-D., Li, Y.-J., Liang, J., Zhao, C.-Z., Yin, H.-J., Yin, C.-Y., Cheng, X.-Y., & Liu, Q. (2014). Responses of soil micronutrient availability to experimental warming in two contrasting forest ecosystems in the eastern Tibetan Plateau, China. *Journal of Soils and Sediments, 14*, 1050–1060.

Maestre, F. T., Eldridge, D. J., Soliveres, S., Kéfi, S., Delgado-Baquerizo, M., Bowker, M. A., García-Palacios, P., Gaitán, J., Gallardo, A., Lázaro, R., & Berdugo, M. (2016). Structure and functioning of dryland ecosystems in a changing world. *Annual Review of Ecology Evolution and Systematics, 47*, 215–237.

Manzoni, S., Schimel, J. P., & Porporato, A. (2012). Responses of soil microbial communities to water stress: Results from a meta-analysis. *Ecology, 93*, 930–938.

Mast, M. A., Turk, J. T., Clow, D. W., & Campbell, D. H. (2011). Response of lake chemistry to changes in atmospheric deposition and climate in three high-elevation wilderness areas of Colorado. *Biogeochemistry, 103*, 27–43.

Matias, L., Castro, J., & Zamora, R. (2010). Soil-nutrient availability under a global-change scenario in a Mediterranean mountain ecosystem. *Global Change Biology, 17*, 1646–1657.

Megan, S. J., Dukes, J. S., Paul, E. A., & Wallenstein, M. D. (2013). Microbial responses to multi-factor climate change: Effects on soil enzymes. *Frontiers in Microbiology, 4*, 1–11.

Meier, J., Costa, R., Smalla, K., Boehrer, B., & Wendt-Potthoff, K. (2005). Temperature dependence of Fe(III) and sulfate reduction rates and its effect on growth and composition of bacterial enrichments from an acidic pit lake neutralization experiment. *Geobiology, 3*, 261–274.

Miller, A. J., EaG, S., & Chadwick, O. A. (2001). Redox control of phosphorus pools in Hawaiian montane forest soils. *Geoderma, 102*, 219–237.

Monger, H. C. (2014). Soils as generators and sinks of inorganic carbon in geologic time. *Soil Carbon* (pp. 27–36). Springer.

Monger, H. C., Kraimer, R. A., Khresat, S., Cole, D. R., Wang, X., & Wang, J. (2015). Sequestration of inorganic carbon in soil and groundwater. *Geology, 43*, 375–378.

Moreno-Jiménez, E., Plaza, C., Saiz, H., Manzano, R., Flagmeier, M., & Maestre, F. T. (2019). Aridity and reduced soil micronutrient availability in global drylands. *Nature Sustainability, 2*, 371–377.

Morgan, J. A., LeCain, D. R., Pendall, E., Blumenthal, D. M., Kimball, B. A., Carrillo, Y., & Williams, D. G. (2011). C4 grasses prosper as carbon dioxide eliminates desiccation in warmed semi-arid grassland. *Nature, 476*, 202–205.

Mpelasoka, F., Hennessy, K., Jones, R., & Bates, B. (2008). Comparison of suitable drought indices for climate change impacts assessment over Australia towards resource management. *International Journal of Climatology, 28*, 1283–1292.

Niklaus, P. A., Spinnler, D., & Körner, C. (1998). Soil moisture dynamics of calcareous grassland under elevated CO_2. *Oecologia, 117*, 201–208.

Nicolas, M. E., Munns, R., Samarokoon, A. B., & Gifford, R. M. (1993). Elevated CO_2 improves the growth of wheat under salinity. *Australian Journal of Plant Physiology, 20*, 349–360.

Norton, U., Mosier, A. R., Morgan, J. A., Derner, J. D., Ingram, L. J., & Stahl, P. D. (2008). Moisture pulses, trace gas emissions and soil C and N in cheatgrass and native grass dominated sagebrush-steppe in Wyoming, USA. *Soil Biology & Biochemistry, 40*, 1421–1431.

Oh, N. H., & Richter, D. D., Jr. (2004). Soil acidification induced by elevated atmospheric CO_2. *Global Change Biology, 10*, 1936–1946.

Ohalloran, I. P., Stewart, J. W. B., & Kachanoski, R. G. (1987). Influence of texture and management practices on the forms and distribution of soil phosphorus. *Canadian Journal of Soil Science, 67*, 147–163.

Oliveira, R. H., Rosolem, C. A., & Trigueiro, R. M. (2004). Importance of mass flow and diffusion on the potassium supply to cotton plants as affected by soil water and potassium. *Revista Brasileira Ciencia Solo, 28*, 439–445.

Pathak, H. (2015). Enhancing resilience of Indian agriculture to climate change. In H. Pathak & B. Chakrabarti (Eds.), *Climate change and agriculture: Technologies for enhancing resilience* (pp. 1–18). CESCRA, ICAR-IARI.

Paul, E. A. (2007). *Soil microbiology, ecology and biochemistry* (3rd ed., pp. 389–430). Academic Press.

Pelster, D. E., Chantigny, M. H., Rochette, P., Angers, D. A., Laganiere, J., Zebarth, B., & Goyer, C. (2013). Crop residue incorporation alters soil nitrous oxide emissions during freeze-thaw cycles. *Canadian Journal of Soil Science, 93*, 415–425.

Phillips, R. P., Bernhardt, E. S., & Schlesinger, W. H. (2009). Elevated CO_2 increases root exudation from loblolly pine (*Pinus taeda*) seedlings as an N-mediated response. *Tree Physiology, 29*, 1513–1523.

Plaza, C., Zaccone, C., Sawicka, K., Méndez, A. M., Tarquis, A., Gascó, G., Heuvelink, G. B. M., Schuur, E. A. G., & Maestre, F. T. (2018). Soil resources and element stocks in drylands to face global issues. *Science and Reports, 8*, 13788.

Rawes, M., & Welch, D. (1969). *Upland productivity of vegetation and sheep at Moor House National Nature Reserve, Westmorland, England, Oikos* (Suppl. 11).

Read, D. J., Leake, J. R., & Perez-Moreno, J. (2004). Mycorrhizal fungi as drivers of ecosystem processes in heathland and boreal forest biomes. *Canadian Journal of Botany, 82*, 1243–1263.

Redfield, A. C. (1958). The biological control of chemical factors in the environment. *American Scientist, 46*, 230–221.

Reed, S. C., Yang, X., & Thornton, P. E. (2015). Incorporating phosphorus cycling into global modeling efforts: A worthwhile, tractable endeavor. *New Phytologist, 208*, 324–329.

Reichstein, M., Bahn, M., Ciais, P., Frank, D., Mahecha, M. D., Seneviratne, S. I., Zscheischler, J., Beer, C., Buchmann, N., Frank, D. C., Papale, D., Rammig, A., Smith, P., Thonicke, K., van der Velde, M., Vicca, S., Walz, A., & Wattenbach, M. (2013). Climate extremes and the carbon cycle. *Nature, 500*, 287–295.

Reinmann, A. B., & Templer, P. H. (2016). Reduced winter snowpack and greater soil frost reduce live root biomass and stimulate radial growth and stem respiration of red maple (*Acer rubrum*) trees in a mixed hardwood forest. *Ecosystems, 19*, 129–141.

Rogora, M., Colombo, L., Lepori, F., Marchetto, A., Steingruber, S., & Tornimbeni, O. (2013). Thirty years of chemical changes in alpine acid-sensitive lakes in the Alps. *Water, Air, and Soil Pollution, 224*, 1746–1765.

Römheld, V., & Kirkby, E. S. (2010). Research on potassium in agriculture: Needs and prospects. *Plant and Soil, 335*, 155–180.

Ruan, J., Wu, X., & Härdter, R. (1997). The interaction between soil water regime and potassium availability on the growth of tea. *Communications in Soil Science and Plant Analysis, 28*, 89–98.

Rustad, L., Campbell, J., Marion, G., Norby, R., Mitchell, M., Hartley, A., Cornelissen, J., Gurevitch, J., & GCTE-NEWS. (2001). A meta analysis of the response of soil respiration, net nitrogen mineralization, and aboveground plant growth to experimental ecosystem warming. *Oecologia, 126*, 543–562.

Ryan, J., & Sommer, R. (2012). Soil fertility and crop nutrition research at an international center in the Mediterranean region: Achievements and future perspective. *Archives of Agronomy and Soil Science, 58*, S41–S54.

Sardans, J., & Peñuelas, J. (2015). Potassium: A neglected nutrient in global change. *Global Ecology and Biogeography, 24*, 261–275.

Sardans, J., & Peñuelas, J. (2005). Drought decreases soil enzyme activity in a Mediterranean holm oak forest. *Soil Biology & Biochemistry, 37*, 455–461.

Sardans, J., & Peñuelas, J. (2007). Drought changes phosphorus and potassium accumulation patterns in an evergreen Mediterranean forest. *Functional Ecology, 21*, 191–201.

Sardans, J., Penuelas, J., & Estiarte, M. (2008). Changes in soil enzymes related to C and N cycle and in soil C and N content under prolonged warming and drought in a Mediterranean shrubland. *Applied Soil Ecology, 39*, 223–235.

Sardans, J., Rivas-Ubach, A., & Peñuelas, J. (2011). Factors affecting nutrient concentration and stoichiometry of forest trees in Catalonia (NE Spain). *Forest Ecology and Management, 262*, 2024–2034.

Schilling, K., Borch, T., Rhoades, C. C., & Pallud, C. E. (2019). Temperature sensitivity of microbial Fe(III) reduction kinetics in subalpine wetland soils. *Biogeochemistry, 142*, 19–35.

Schimel, J. P. (2018). Life in dry soils: Effects of drought on soil microbial communities and processes. *Annual Review of Ecology, Evolution, and Systematics, 49*, 409–432.

Schindlbacher, A., Beck, K., Holzheu, S., & Borken, W. (2019). Inorganic carbon leaching from a warmed and irrigated carbonate forest soil. *Frontiers in Forests and Global Change, 2*, 40.

Schlingmann, M., Tobler, U., Berauer, B., Garcia-Franco, N., Wilfahrt, P., Wiesmeier, M., Jentsch, A., Wolf, B., Kiese, R., & Dannenmann, M. (2020). Intensive slurry management and climate change promote nitrogen mining from organic matter-rich montane grassland soils. *Plant and Soil, 456*, 81–98.

Seneweera, S., Blakeney, A., & Milham, P. (1996). Influence of rising atmospheric CO_2 and phosphorus nutrition on the grain yield and quality of rice (*Oryza sativa* cv. Jarrah). *Cereal Chemistry, 73*, 239–243.

Siebers, N., Sumann, M., Kaiser, K., & Amelung, W. (2017). Climatic effects on phosphorus fractions of native and cultivated north American grassland soils. *Soil Science Society of America Journal, 81*, 299–309.

Sims, J. T., Simard, R. R., & Joern, B. C. (1998). Phosphorus loss in agricultural drainage: Historical perspective and current research. *Journal of Environmental Quality, 27*, 277–293.

Singer, A. (1989). Illite in the hot-aridic soil environment. *Soil Science, 147*, 126–133.

Song, Y., Zou, Y., Wang, G., & Yu, X. (2017a). Stimulation of nitrogen turnover due to nutrients release from aggregates affected by freeze-thaw in wetland soils. *Physics and Chemistry of the Earth, 97*, 3–11.

Song, Y., Zou, Y. C., Wang, G. P., & Yu, X. F. (2017b). Altered soil carbon and nitrogen cycles due to the freeze-thaw effect: A meta-analysis. *Soil Biology & Biochemistry, 109*, 35–49.

Spinoni, J., Naumann, G., Vogt, J., & Barbosa, P. (2015). European drought climatologies and trends based on a multi-indicator approach. *Global and Planetary Change, 127*, 50–57.

Spinoni, J., Naumann, G., & Vogt, J. V. (2017). Pan-European seasonal trends and recent changes of drought frequency and severity. *Global and Planetary Change, 148*, 113–130.

St.Clair, S. B., & Lynch, J. P. (2010). The opening of Pandora's Box: Climate change impacts on soil fertility and crop nutrition in developing countries. *Plant and Soil, 335*, 101–115.

Stein, L. Y., & Klotz, M. G. (2016). The nitrogen cycle. *Current Biology, 26*, R94–R98.

Steinweg, M. J., Dukes, J. S., Paul, E. A., & Wallenstein, M. D. (2013). Microbial responses to multi-factor climate change: Effects on soil enzymes. *Frontiers in Microbiology, 4*, 1–11.

Stocker, T. F., Qin, D., Plattner, G. K., Tignor, M. M. B., Allen, S. K., Boschung, J., Nauels, A., Xia, Y., Bex, V., & Midgley, P. M. (2014). *Climate change 2013: The physical science basis: Working group I contribution to the fifth assessment report of the Intergovernmental Panel on Climate Change*. Cambridge University Press.

Sun, D. S., Li, K. J., Bi, Q. F., Zhu, J., Zhang, Q. C., Jin, C. W., Lu, L. L., & Lin, X. Y. (2017). Effects of organic amendment on soil aggregation and microbial community composition during drying-rewetting alternation. *Science of the Total Environment, 574*, 735–743.

Tian, H., Chen, G., Zhang, C., Melillo, J. M., & Hall, C. A. (2010). Pattern and variation of C: N: P ratios in China's soils: A synthesis of observational data. *Biogeochemistry, 98*, 139–151.

Tomlinson, K. W., Pooter, L., Sterck, F. J., Borghetti, F., Ward, D., de Bie, S., & van Langevelde, F. (2013). Leaf adaptations of evergreen and deciduous trees of semi-arid and humid savannas on three continents. *Journal of Ecology, 101*, 430–440.

Unger, M., Leuschner, C., & Homeier, J. (2010). Variability of indices of macronutrient availability in soils at different spatial scales along an elevation transect in tropical moist forests (NE Ecuador). *Plant and Soil, 336*, 443–458. https://doi.org/10.1007/s11104-010-0494-z

Vitousek, P. M., & Chadwick, O. A. (2013). Pedogenic thresholds and soil process domains in basalt-derived soils. *Ecosystems, 16*, 1379–1395. https://doi.org/10.1007/s10021-013-9690-z

Vitousek, P. M., Porder, S., Houlton, B. Z., & Chadwick, O. A. (2010). Terrestrial phosphorus limitation: Mechanisms, implications, and nitrogen-phosphorus interactions. *Ecological Applications, 20*, 5–15. https://doi.org/10.1890/08-0127.1

Walker, T. W., & Syers, J. K. (1976). The fate of phosphorus during pedogenesis. *Geoderma, 15*, 1–19. https://doi.org/10.1016/0016-7061(76)90066-5

Wang, C., Chen, Z., Unteregelsbacher, S., Lu, H., Gschwendtner, S., Gasche, R., Kolar, A., Schloter, M., Kiese, R., Butterbach-Bahl, K., & Dannenmann, M. (2016a). Climate change amplifies gross nitrogen turnover in montane grasslands of Central Europe in both summer and winter seasons. *Global Change Biology, 22*, 2963–2978. https://doi.org/10.1111/gcb.13353

Wang, J., Li, L., Lam, S. K., Zhang, X., Liu, X., & Pan, G. (2018). Changes in nutrient uptake and utilization by rice under simulated climate change conditions: A 2-year experiment in a paddy field. *Agricultural and Forest Meteorology, 250*, 202–208.

Wang, J., Solomon, D., Lehmann, J., Zhang, X., & Amelung, W. (2006). Soil organic sulfur forms and dynamics in the Great Plains of North America as influenced by long-term cultivation and climate. *Geoderma, 133*, 160–172.

Wang, J., Zhang, X., Li, L., Cheng, K., Zheng, J., Zheng, J., Shen, M., Liu, X., & Pan, G. (2016b). Changes in micronutrient availability and plant uptake under simulated climate change in winter wheat field. *Journal of Soils and Sediments*. https://doi.org/10.1007/s11368-016-1464-8

Wang, Y. P., & Houlton, B. Z. (2009). Nitrogen constraints on terrestrial carbon uptake: Implications for the global carbon-climate feedback. *Geophysical Research Letters, 36*, L24403. https://doi.org/10.1029/2009GL041009

Weil, R. R., & Magdoff, F. (2004). Significance of soil organic matter to soil quality and health. In R. R. Weil & F. Magdoff (Eds.), *Soil organic matter in sustainable agriculture* (pp. 1–43). CRC Press.

White, A. F., Blum, A. E., Bullen, T. D., Vivit, D. V., Schulz, M., & Fitzpatrick, J. (1999). The effect of temperature on experimental and natural weathering rates of granitoid rocks. *Geochimica et Cosmochimica Acta, 63*, 3277–3291.

Xiao, L., Zhang, Y., Li, P., Xu, G. C., Shi, P., & Zhang, Y. (2019). Effects of freeze-thaw cycles on aggregate-associated organic carbon and glomalin-related soil protein in natural succession grassland and Chinese pine forest on the Loess Plateau. *Geoderma, 334*, 1–8.

Xu, Z. Z., Shimizu, H., Ito, S., Yagasaki, Y., Zou, C. J., & Zhou, G. S. (2014). Effects of elevated CO_2, warming and precipitation change on plant growth, photosynthesis and peroxidation in dominant species from North China grassland. *Planta, 239*, 421–435. https://doi.org/10.1007/s00425-013-1987-9

Yanai, Y., Toyota, K., & Okazaki, M. (2004). Effects of successive soil freeze-thaw cycles on soil microbial biomass and organic matter decomposition potential of soils. *Soil Science and Plant Nutrition, 50*, 821–829.

Yang, X., Zhang, P., Wei, Z., Liu, J., Hu, X., & Liu, F. (2021). Effects of elevated CO_2 and nitrogen supply on leaf gas exchange, plant water relations and nutrient uptake of tomato plants exposed to progressive soil drying. *Scientia Horticulturae, 292*, 110643.

Yeo, A. (1999). Predicting the interaction between the effects of salinity and climate change on crop plants. *Scientia Horticulturae, 78*, 159–174.

Yin, H., Li, Y., Xiao, J., Xu, Z., Cheng, X., & Liu, Q. (2013). Enhanced root exudation stimulates soil nitrogen transformations in a subalpine coniferous forest under experimental warming. *Global Change Biology, 19*, 2158–2167.

Yoon, P. R., & Choi, J. Y. (2020). Effects of shift in growing season due to climate change on rice yield and crop water requirements. *Paddy and Water Environment, 18*, 291–307. https://doi.org/10.1007/s10333-019-00782-7

Yu, Z., Wang, M., Huang, Z., Lin, T. C., Vadeboncoeur, M. A., Searle, E. B., & Chen, H. Y. H. (2018). Temporal changes in soil C-N-P stoichiometry over the past 60 years across subtropical China. *Global Change Biology, 24*, 1308–1320.

Zamanian, K., Pustovoytov, K., & Kuzyakov, Y. (2016). Pedogenic carbonates: Forms and formation processes. *Earth-Science Reviews, 157*, 1–17.

Zhang, C., Tian, H. Q., Liu, J. Y., Wang, S. Q., Liu, M. L., Pan, S. F., & Shi, X. Z. (2005). Pools and distributions of soil phosphorus in China. *Global Biogeochemical Cycles, 19*, GB1020.

Zhang, D., Hui, D., Luo, Y., & Zhou, G. (2008). Rates of litter decomposition in terrestrial ecosystems: Global patterns and controlling factors. *Journal of Plant Ecology, 1*, 85–93.

Zhang, Y., Chen, X., Zhang, C., Pan, G., & Zhang, X. (2014). Availability of soil nitrogen and phosphorus under elevated [CO_2] and temperature in the Taihu Lake region, China. *Journal of Plant Nutrition and Soil Science, 177*, 343–348.

Zistl-Schlingmann, M., Feng, J., Kiese, R., Stephan, R., Zuazo, P., Willibald, G., Wang, C., Butterbach-Bahl, K., & Dannenmann, M. (2019). Dinitrogen emissions: An overlooked key component of the N balance of montane grasslands. *Biogeochemistry, 143*, 15–30. https://doi.org/10.1007/s10533-019-00547-8

Chapter 3
Soil Carbon Sequestration in the Context of Climate Change

Tanumoy Bera, Suman Samui, Abir Dey, and Jayakishore Ankireddypalli

Abstract Soil carbon (C) sequestration is one of the major pathways to reverse climate change and reduce its deleterious impact on human civilization. Removal of atmospheric CO_2 into other stable pools with long residence time is called C sequestration. To sequester C in soil of agricultural, grazing, forest, and wetlands, adaptation of established and proven technologies is necessary. In this chapter, details of those processes are described with their advantages and disadvantages. Inclusion of leguminous species, use of inputs including fertilizer and irrigation water, and residue retention are important practices in sequestering more C in agricultural, forest, and grazing lands. Managing wetland C is complex and needs to be dealt with caution. Engineering and chemical technologies are new concepts and are mostly at the research stage. The economics and practical adaptability of these practices are yet to be established.

Keywords Carbon · Carbon sequestration · Agriculture · Grassland · Wasteland

3.1 Introduction

Soil organic carbon (SOC) is a component of the global carbon (C) cycle, which involves C cycling through the soil, plants, ocean, and atmosphere. Besides C buried deep into earth's crust as a rock (10^8 Pg C), there are five major C pools: (i) oceanic C pool (39×10^3 Pg) that is growing at 2.3 Pg C per year; (ii) fossil fuel C pool ($5–10 \times 10^3$ Pg) that is mined and burned at 8 Pg C per year; (iii) pedologic C pool

T. Bera (✉)
Texas A&M AgriLife Research, Beaumont, TX 77713, USA
e-mail: tanumoy.soil@gmail.com

S. Samui
Department of Agronomy, Faculty of Agriculture, Bidhan Chandra Krishi Viswavidyalaya, P.O. Krishi Viswavidyalaya, Mohanpur, Nadia 741252, West Bengal, India

A. Dey · J. Ankireddypalli
Central Laboratory for Soil and Plant Analysis, Division of Soil Science and Agricultural Chemistry, ICAR-Indian Agricultural Research Institute, New Delhi 110012, India

H. Pathak et al. (eds.), *Climate Change Impacts on Soil-Plant-Atmosphere Continuum*, Advances in Global Change Research 78, https://doi.org/10.1007/978-981-99-7935-6_3

to 1-m soil depth (2.5 × 10^3 Pg) composed of 1.55 × 10^3 Pg of SOC and 950 Pg of soil inorganic C (SIC); (iv) atmospheric C pool (780 Pg) that is increasing at 4 Pg C per year; and (v) the biotic C pool (560 Pg) comprised of live biomass and 60 Pg of detritus material (Lal, 2010). The terrestrial C pool, which is made up of both the pedologic and biotic pools, has a mass of 3120 Pg, or roughly four times that of the atmospheric pool. C is not stabilized in the pools, but rather flows among pools due to natural and anthropogenic activities. In contrast to the fossil fuel pool, which has been a significant source of carbon dioxide (CO_2) since around 1750, terrestrial C pools, depending on land use and management, can either be a source or a sink for atmospheric C. About 39% increase in CO_2 concentration since pre-industrial era to till date along with other greenhouse gases (such as nitrous oxide [N_2O] and methane [CH_4]) in the atmosphere may make radiative forcing stronger and change the Earth's average temperature and rainfall (IPCC, 2007). Because of the considerable effect of radiative forcing on global warming and climate change, there is an increasing need to identify measures that will reduce atmospheric CO_2 enrichment by at least anthropogenic emissions. Anthropogenic sources related to agricultural activities includes burning biomass, deforestation, improper land management, and land-use conversion including drainage of peatlands, animal husbandry, etc., besides fossil fuel burning and industrial activities. Therefore, the focus of this article is to delineate the strategies or methods to improve C sequestrations in territorial pools by emphasizing agriculture, forest, grass, and wetland ecosystems.

3.2 Carbon Sequestration and Climate Change Mitigation

Between 1750 and 2003, an estimate suggested anthropogenic emissions were 292 Pg from fossil fuel combustion (Holdren, 2008) and 136 ± 30 Pg from land-use changes, deforestation, and soil cultivation (IPCC, 2001). At present total anthropogenic emission is 9.9 Pg C yr^{-1} comprising 8.3 Pg C yr^{-1} from fossil fuel combustion and 1.6 Pg C yr^{-1} from deforestation, land-use change, and soil cultivation (IPCC, 2007). Of the total anthropogenic emission, atmosphere and ocean absorbed 4.2 and 2.3 Pg C yr^{-1}, respectively, the remainder being absorbed by unidentified terrestrial sinks. Thus, atmospheric enrichment is needed to be reversed to offset global warming and global climate change and limit atmospheric CO_2 concentration below 441 ppm by 2010 (UNFCCC, 2009). To mitigate the weather risk of increasing atmospheric CO_2 concentration, removal of atmospheric CO_2 into other stable pools with long residence time; a process called C sequestration is necessary. In 2008, Lal proposed three major types of C sequestration process: (i) sequestering C in biomass, soil organic matter, and other component of terrestrial biosphere; (ii) conversion of C involving engineering techniques; and (iii) chemical transformations of C.

There is a strong interaction between terrestrial and atmospheric C pools through photosynthesis and respiration by plants and animals comprising biotic C pools. Annually, 120 Pg C is fixed by photosynthesis. However, because the photosynthesized C is returned to atmosphere by respiration, conversion from natural to managed ecosystems, extractive farming practices based on low external input, and soil degrading land usage, terrestrial C poll is continuously depleted. Most of the present agricultural soils contain much less C compared to their counterparts under natural ecosystems due to depletion of SOC pools. Oxidation, mineralization, leaching, and erosion are the major processes in soil to deplete the SOC. Thus, soil C sequestration involves increasing the SOC in agricultural, forestry, and wetland systems. Table 3.1 describes the extent and additional potentials of several natural pathways including forest, grassland, agricultural, and wetland management in soil C sequestration along with identified co-benefits.

3.3 Best Management Practices for Agricultural Soils

3.3.1 Agroforestry

Agroforestry systems (AFS) have gained increased recognition in recent decades for their productivity and sustainability. One of the more recent environmental advantages of this traditional method of land management is the C sequestration potential. The United Nations Framework Convention on Climate Change (UNFCCC) recognized C sequestration through afforestation and reforestation (A&R) as a way to offset greenhouse gas (GHG) emissions. Since most agroforestry is carried out by subsistence farmers in developing nations, there is a considerable chance that C trade with developed nations will result in financial rewards. CO_2 levels in the atmosphere could be reduced by using agroforestry biomass instead of fossil fuels for bio-power and biofuels.

Bio-sequestration is the process by which CO_2 is absorbed by living organisms and turned into inert, long-lasting C-containing materials (Lorenz & Lal, 2014b). The agroforestry system is an example of bio-sequestration as some of the C can last for thousands of years. In agroforestry systems, C is stored in the stem, branch, and leaves above the ground, in the roots below the ground, and the soil. Bastin et al. (2017) reported that forests cover between 4.0 and 5.0 billion hectares of land around the world and stores ~860 Pg C. More than 40% of all the organic C in land ecosystems is stored in forest soils around the world (Wei et al., 2014). In 2009 Nair et al. estimated C sequestration rates of AFC ranging from 0.29 to 15.2 Mg C ha^{-1} year^{-1} aboveground and from 30 to 300 Mg C ha^{-1} year^{-1} in soils down to 1 m depth. The C sequestration capacity of various agroforestry systems (Table 3.2) has been reported to range from 12 to 228 Mg ha^{-1}, with a typical value of 95 Mg ha^{-1} (Nair et al., 2009; Rai et al., 2013). However, recent research indicates that (sub)tropical and temperate agroforestry systems sequester 725 $\pm$ 100 kg C ha^{-1} year^{-1} on average (Wiesmeier et al., 2020). Agroforestry systems are more efficient than monocultures in capturing site resources for biomass growth, which results in larger C inputs to the soil through specific processes. These include: (a) returning woody species pruning to the soil as mulch and allowing plenty of tree litter to decompose on site; (b) letting

Table 3.1 Maximum potential extent and additional mitigation potential of natural pathways along with co-benefits by 2030

Pathway	Maximum potential extent	Maximum additional mitigation potential (Tg CO_2e yr^{-1})	Co-benefits
Avoided forest conversion	Not applicable	3,603	Biodiversity conservation, greater water availability and regulation, soil water flow regulation, soil health benefits, and air quality improvement
Reforestation	678 Mha	10,124	New wild habitat creation, drought mitigation and avoidance, water retention, and flow regulation
Natural forest management	1,914 Mha	1,470	Species richness of invertebrates, amphibians, and mammals decreases, soil health, and productivity reduce
Improved plantations	257 Mha	443	Enhances diversity and promotes community plantation involving native species
Fire management	Not quantified	212	Improve forest diversity, contain more organic matter, better water retention and flow regulation, occasionally decreased air quality
Avoid woodfuel harvest	2,800 M people	367	Increase food and habitat for forest organisms reduces soil compaction, and promote efficient cook stoves in household
Avoided grassland conversion	Not applicable	116	Important habitat for nesting and foraging birds, biological flood control, ecosystem water balance, check on breeding wading birds by soil macroinvertebrates
Biochar	1,670 Tg yr^{-1} crop residue	1,102	Enhances soil quality and fertility
Cropland nutrient management	44 Tg N yr^{-1} used	706	Increases fish species richness, drinking water quality, recreation opportunities and health benefits, soil fertility and reduces greenhouse gas and ammonia emission
Conservation agriculture	352 Mha	413	Reduce water demand through cover crop, soil erosion, and disturbances
Trees in croplands	608 Mha	1,040	Provided habitat, erosion control, water retention, capture airborne particles pollutant gases
Grazing—optimal intensity	712 Mha	148	Reduces overall disturbance to plant–insect interactions, reduces water use, improves soil health

(continued)

Table 3.1 (continued)

Pathway	Maximum potential extent	Maximum additional mitigation potential (Tg CO_2e yr^{-1})	Co-benefits
Grazing—legumes in pastures	72 Mha	147	Higher insect herbivore and insect predator diversity, improved soil structure, erosion protection, and biodiversity
Grazing—improved feed	1,400 M head cattle	680	
Grazing—animal management	1,400 M head cattle	200	
Improved rice cultivation	163 Mha	265	Improved water management decreases overall water consumption by reducing gray water
Avoided coastal wetland impacts	Not applicable	304	Maintain nutrients and primary productivity of seafood, provide economic value, cross-system nutrient transfer, and improve air quality
Avoided peatland impacts	Not applicable	754	Provide specific insect habitat, flooding avoidance, and reduce possibility of peat subsidence or loss and peat fire risk
Coastal wetland restoration	29 Mha	841	Maintain nutrients and primary productivity of seafood, flood control, and water filtration, cross-system nutrient transfer, and improve air quality
Peatland restoration	46 Mha	815	Establish diverse community, water treatment, and stormwater remediation, high productivity systems with increased risk of a peat fire

Table adapted from tables S2, S4, and S5 in Griscom et al. (2017)

livestock graze and add manure to the soil; (c) allowing woody species grow and add surface and belowground litter during crop fallow seasons; (d) integrating trees and their litter input into animal production systems; (e) allowing litter inputs from shade-tolerant plants growing under trees; and (f) benefiting from the soil C inputs of agricultural crops grown during early stages of the forestry establishment.

The quantity of the forest soil C stock is influenced by the factors (temperature, vegetation, terrain, soil chemical, physical and biological qualities, and parent material) that interact to determine the site's fertility and productivity, as well as how the land is used and managed. Changes in the soil's organic C pool, which is the largest in most forest ecosystems, can have an impact on the ecosystem as a whole. A global

Table 3.2 Effect of different agroforestry systems on soil C sequestration

Systems	Soil carbon (Mg ha^{-1})	Age (yr)	Soil depth (cm)
Tree based pastures (*Gliricidia sepium* + maize (*Zea mays*))	123	10	0–200
Tree based pastures (slash pine (*Pinus elliottii*) + bahia grass (*Paspalum notatum*))	6.9–24.2	8–40	0–125
Fodder bank (*Pterocarpus erinaceus, P. lucens* and *Gliricidia sepium*)	33.4	6–9	0–100
Silvopastoral system (*Guazuma ulmifolia* + *Coradia alliodora* + *Brachiaria brizantha*)	132	10–16	0–100
Silvopastoral system (*Arachis pintoi* + *Acacia mangium*)	173	10–16	0–100
Agrisilviculture (*Gmelina arborea* + field crops)	27.4	5	0–60
Agro forestry (*Trifolium subterraneum* + *Pseudotsuga menzeiscii*)	95.89	11	0–45
Alley cropping system (*Erythrina poeppigiana* + bean (*Phaseolus vulgaris*) and maize (*Zea mays*))	1.62	19	0–40
Agrisilviculture (Indo-Gangetic region of India)	2.24–19.9	3–22	–
Plantation crop-based agroforestry (Eastern and Northeastern India)	0.08–76.16	3–22	–
Agrihorticulture (Northern Himalayas of India)	8.05–81.68	3–22	–
Block plantation (Southern India)	14.75–152.16	3–22	–
Tree-based intercropping (hybrid poplar + barley)	78.5	13	0–20
Sivopasture: slash pine (Pinus elliottii Engelm.) + bahiagrass (Paspalum notatum Flugge)	6.9–24.2	8–40	0–125

Adapted from: Haile et al. (2008); Nair et al. (2009, 2010); Rai et al. (2013); Panwar et al. (2022)

perspective needs, on the one hand, to measure the synergies and trade-offs between forest and ecosystem C storage, including the soil C sequestration, greenhouse gas emission, and, on the other hand, estimate the substitution and storage impacts of using the forest to create lumber, fiber, or electricity. Additionally, it is important to consider how variations in the quantity and mobility of the soil's C pool will affect the cycling of nutrients. To maintain consistent stand and nutrition, soil organic matter C to nutrient ratios must remain constant, especially if soil C stocks rise as a result of forest management.

3.3.2 Zero Tillage

Zero tillage (ZT) is effective in minimizing soil disintegration, enhancing soil fertility and SOC sequestration. Adaptation of ZT increases the proportion of macro-aggregates (0.25–2 mm) and decreases the fraction of micro-aggregates (0.05–0.25 mm) relative to conventional tillage techniques while increasing the aggregates' mean weight diameter (Six et al., 2002). Thus, ZT minimizes erosional losses and increases C sequestration. In general, C encapsulated within the soil micro-aggregates is characterized as humified organic matter with a low C/N ratio. However, micro-aggregates in ZT soil were found with a higher C/N ratio suggesting SOC accompanied with that fraction comprises less decay material and fundamentally consists of fungal hyphae and plant roots (Six et al., 2002). Under the ZT system, soil microbial biomass C is usually higher and basal respiration (CO_2 evolution) is lower. Increased biomass C addition and decreased labile C losses suggest better C sequestration in the soil (Hussain et al., 2021). Zero-till agriculture improves the efficiency of the farming system by making it easier for water to get into the soil, stopping soil from washing away, making aggregates more stable, encouraging biological tillage, lowering soil temperatures, and reducing the number of weeds.

3.3.3 Crop Residue Retention

Crop growth has an impact on SOC sequestration rates in agricultural systems, but crop residue applied as mulch has an impact on SOC retention. To achieve long-term advantages and stop soil deterioration, C storage, and climate change, agricultural residues on the soil surface must be retained and managed sustainably. Crop residue removal decreased soil C by reducing input and nutrient availability, and increasing microbial activity (Ansari et al., 2022). Crop residue retention using the ZT approach adds C to the soil, prevents erosion and aggregate disturbance, encourages infiltration and water storage, and establishes/maintains a favorable microbial habitat, which increases SOC turnover and may even increase SOC stocks over time (Sekaran & Kumar, 2021). The C:N ratio of agricultural residue appears to determine whether C is deposited in SOC or respired (Finn et al., 2016). Compared to residue removal, residue retention increased soil C pools, C-lability index, microbial biomass C, and C stocks by more than 2.2 Mg ha^{-1} (Ansari et al., 2022). The amount of organic matter generated and returned to the soil, as well as the quality of these C sources, which impacts microbial activity and SOM decomposition, are factors that influence the increase in SOC. According to Finn et al. (2016), the addition of N to wheat crop residue increased C retention, resulting in a steady SOC proportion (humus). Therefore, crop residue management aids in the soil's ability to store C. At least, it can sustain the SOM level if the C (energy source) for microbial development is external like stubble or extra organic matter.

3.3.4 Crop Diversification

The amount of C that is sequestered in the soil increases as a result of crop rotations that contribute to high C input through crop residue and increased root biomass. By using legumes or pulses that shed their leaves, SOC sequestration in intensive cereal-cereal farming systems can be improved. Narrow C:N ratios in pulse crop residues accelerate decomposition, which increases SOM (Yadav et al., 2017). According to Ghosh et al. (2019), chickpea inclusion in rice–wheat and maize-wheat systems increased total SOC in the surface soil by 7% and 13%, respectively. Legumes can improve both labile and recalcitrant SOC pools, although labile pools do so more quickly, enhancing SOC lability (Jat et al., 2022). Crop diversification strategies may have varying effects on the amount of C in the soil because some strategies, such as agroforestry, intercropping, and cover crops, produce more above- and below-ground litter, which facilitates in the formation and accumulation of organic C in the soil (Beillouin et al., 2021). These practices must not only increase biomass and C uptake but also (i) not reduce yields (which would result in displacement effects), (ii) not increase greenhouse gases like nitrous oxide (N_2O) or methane, (iii) not increase energy use for processes like making fertilizer, and (iv) last for a long time (Guenet et al., 2021).

3.3.5 Integrated Nutrient Management (INM)

The greatest strategy to maintain healthy, sustainable soil systems while enhancing crop yield is integrated nutrient management (INM), which uses both organic manures and chemical fertilizers concurrently. The SOC pool is affected by a wide range of agricultural management practices, such as tilling, managing residues, applying fertilizer and manure, managing water, and draining the soil. One method for recovering soil organic C in depleted and damaged soils while also enhancing crop yields is integrated nutrient management. When organic manure is added to soils, it has a variety of effects on SOC and crop yields. Correctly combining chemical fertilizer with organic manures aids in increasing C in the soil.

The soil C fractions saw significant changes because of various integrated nutrient management schemes. Chemical fertilizers increased agricultural yields, but they only slightly improved soil organic C storage. It appears to be a good idea to combine organic manures and inorganic fertilizers to increase the soil's capacity to store C (Ghosh et al., 2019). However, the type and quality of organic manure that is added are likely to influence the type of C fraction that is stored (passive or labile) and the pace of C storage in soils. To increase the soil's labile C fractions, which in turn enhances the soil's quality, vermicompost (VC) can be used in place of using inorganic fertilizers alone. The division of the organic C in the soil into several pools is also greatly influenced by the C to N ratio in organic manures. An example is using rice straw in conjunction with inorganic fertilizer may be preferable in terms of storing

C in more stable pools. Animal dung is more effective than straw at increasing soil C, according to Liu et al. (2013). This may be due to straw containing less humified and resistant C forms than animal dung. Manure application leads to more C storage than plant residues for the same C intake because it is more resistant to microbial destruction than plant residues (Feng and Li, 2002). The effect of soil microbial communities on biogeochemical processes is influenced by the addition of various organic and inorganic fertilizers to soils. The addition of organic manure encourages the growth and activity of soil microorganisms, demonstrating a direct relationship between microbial activity and the rise in biomass C. As a result, the soil's C content increased above its initial value. These C inputs finally lead to SOC sequestration in profiles with balanced fertilization. For example, SOC accumulation rates in the 0–90 cm soil profile were 529 kg ha^{-1} $year^{-1}$ greater in balanced fertilized plots (compared to control plots) (Ghosh et al., 2019). The combined use of organic and inorganic fertilizers had a substantial effect on the C sequestration in post-harvest soil. Integrated nutrient management is therefore essential for C sequestration and sustained crop production.

3.3.6 Green Manure

It is critical to highlight the benefits of C sequestration and crop yield through continuous green manure use for sustainable agricultural development. Green manure is important because it incorporates organic materials into the soil. Green manure is a method of improving the soil that involves ploughing or incorporating green manure crops while they are still green or shortly after they begin flowering. Leguminous green manure crops including sunhemp, dhaincha, green gram, and black gram are highly desirable because of their rapid growth, high biomass production, ability to fix atmospheric nitrogen (N) through root nodules and sequester C. Green manure biomass incorporation into cereal-based intensive cropping systems contributes to the accumulation of organic matter, recycling of nutrients, and the improvement of soil hydro-physical properties. Green manures promote faster N mineralization because they have a lower C:N ratio than grains (Ansari et al., 2022). In mountainous and erodible ecosystems, green manures are also used as cover crops to decrease soil erosion and the loss of nutrients and organic C. Brown manuring, a form of green manuring with brown foliage, is a technique for minimizing weed pressure because it acts as a cover crop, effectively limiting weed growth at the germination stage (Maity & Mukharjee, 2011). Brown manure refers to the loss of chlorophyll in the leaves as a result of post-emergence herbicide treatment on green manure leaves (Tanwar et al., 2010). It is possible to achieve brown manuring by growing green manure crops as an intercrop, such as *Sesbania* (dhaincha), sunnhemp, and others and then spraying them with post-emergence herbicides. This is an alternative technique for increasing biomass and C sequestration in the soil. Green and brown manuring (preferably legumes) is known to improve soil organic C accumulation and is an important C sequestration method in tropical and subtropical soils (Nandan et al.,

2019). Ansari et al. (2022) discovered that using Sesbania as green manure led to the highest total biomass accumulation (>30 Mg ha^{-1}) and the highest C supply to the soil (>14.0 Mg ha^{-1}). *Sesbania aculeata* has been shown to have higher shoot and root biomass due to its rapid and deterministic growth, resulting in a significant rise in soil C content (Pooniya et al., 2012). Incorporating green and brown manure is another effective method of increasing and sustaining soil organic C, which improves soil physical characteristics, and C stock, resulting in C sequestration.

3.3.7 Water Management

Irrigation increases the plant's available moisture in soils, and thus increases net primary production in arid environments. Increased plant growth increases organic C inputs into agricultural soils, however rapid C and N mineralization may reduce irrigation's benefits for SOC (Trost et al., 2013). A global meta-analysis found that the C sequestration capacity of irrigated soil ranged between 50 and 100 kg C ha^{-1} $year^{-1}$ (Meena et al., 2020). Irrigation, on the other hand, reduced SOC pool in the top 25–30 cm of soil by 144 kg C ha^{-1} $year^{-1}$, according to field studies. Irrigation has a C sequestration potential of 1378 kg C ha^{-1} $year^{-1}$ (Francaviglia et al., 2019). The impact of irrigation and fluctuations in the water table on SOC stocks and decomposition should be studied in detail. In this instance, the newly developed subsoil soil irrigation system—a device implanted at a depth of 25 cm in the soil—may provide substantial advantages (better water use efficiency) over conventional irrigation systems (Chenu et al., 2019). Deficit irrigation systems that use precise irrigation technologies can sustain crop productivity while minimizing negative externalities, increasing the capacity of this practice to improve increasing SOC stock. Water consumption efficiency, which is the connection between plant productivity (biomass or crop output) and water use, is used to evaluate sustainable irrigation (Hatfield & Dold, 2019). Because the C costs of pumping water can exceed the benefits of increased SOC stock, the impact of irrigation on SOC sequestration can be insignificant or negative.

3.3.8 Biochar

Biochar is produced by thermally converting (pyrolyzing) organic compounds derived from plants or animals at temperatures higher than 350 °C while limiting oxygen supply. The majority of the C-rich, porous, fine-grained biochar is resistant to microbial decomposition in soil (Bera et al., 2018; Lehmann, 2007). Because of its concentrated aromatic character and high recalcitrance, biochar stays in soil for a very long time in soils (Bera et al., 2019) being a durable C sink. Only a small portion of biochar is labile, and when applied to soil, it can cause "priming," which accelerates the microbial decomposition of soil organic matter in a short period of time

(Kuzyakov, 2010). Additionally, there is conflicting evidence about the influence of priming, with favorable, neutral, and adverse impacts on SOC stocks (Purakayastha et al., 2019; van Zwieten et al., 2017). In addition to these immediate effects, biochar application appears to increase primary output, refractory fractions of SOC, and SOC pools in the subsoil, all of which increase SOC stocks in agricultural areas (Schmidt et al., 2014). Additionally, it might increase soil biota activity, aggregate stability, soil water retention, and soil erosion avoidance (Bera et al., 2016; Lorenz & Lal, 2014a). Several studies suggest that applying biochar has an average C sequestration potential of between 1.6 and 5.14 Mg C ha^{-1} $year^{-1}$ (Majumder et al., 2019). Although studies on biochar's stability are presently being conducted, in general, biochar's recalcitrance is the most significant factor in determining its C sequestration capacity (Leng et al., 2019).

3.3.9 Organic Agriculture

Organic agricultural systems increase SOC stocks by (1) utilizing organic fertilizers inherently greater in C content, (2) diversifying crop rotation with legumes, and (3) growing cover crops (Müller et al., 2016). Organic farming techniques averaged 28.7–102 kg C ha^{-1} $year^{-1}$ greater C storage than conventional farming techniques (Autret et al., 2019). In organic farming systems, organic fertilization with bovine dung and green manure (alfalfa) sequestered 160 kg C ha^{-1} $year^{-1}$ and 180 kg C ha^{-1} $year^{-1}$, respectively (Blanco-Canqui et al., 2017). Organic fertilization is the most influential component in organic farming systems for creating SOC stocks, according to a global meta-analysis of 68 datasets from 32 peer-reviewed studies (Leifeld & Fuhrer, 2010). In addition, turning all cropland in the European Union to organic farming would immediately sequester 30 Mt of C, which is equivalent to one-fourth of the present agricultural emissions. Cover crops, legumes in crop rotation, and organic fertilizers can all be employed in traditional agriculture. For instance, if organic fertilizers are added at the same rate to conventionally managed crops as they are in organic farming systems, the SOC stocks in both farming systems will be comparable (Leifeld & Fuhrer, 2010).

Organic agriculture utilizes fertilizers, cover crops, and crop rotations that are, on average, more diversified than conventional agriculture. In contrast, tillage, a management method that tends to deplete soil organic matter, is commonly applied to suppress weeds in organic agriculture (Müller et al., 2016). In addition to increasing biodiversity and decreasing the risk of eutrophication and water contamination, organic farming has the ability to sequester C. Low artificial external inputs are also typical of organic management techniques (omission of chemically synthesized products such as pesticides, herbicides, or easily soluble mineral fertilizers). Consequently, these techniques require less primary energy than standard agricultural practices (Agrimonti et al., 2021). Compared to conventional agriculture, organic agriculture produces lower emissions per hectare. Given that organic farming may result in lower crop yields, the organic farming emissions per crop yield (product) remain debatable.

3.3.10 *Eco-Restoration of Degraded Lands*

In recent years, land degradation has been a major factor in diminishing soil quality. It has been challenging to restore deteriorated land. It calls for land-use planning. Sustainable and livelihood-secured land use options are desired by society for increasing food, fodder, and firewood as well as reducing and adapting to climate change (Kumar et al., 2019). The process of recovering a degraded ecosystem is called ecological restoration (Martin, 2017). Conditions favorable to tree-grass-based eco-restoration prevail in tropical regions characterized by abundant precipitation (Qi et al., 2019). Through C sequestration, enzyme activity, and other mechanisms, tree-grass interactions will alter the efficacy and scope of eco-restoration. All trees contributed equally to the sequestration of organic C in the surface soil. All grasses have the same effect on surface soil organic C uptake and sequestration. Due to the addition of leaf litter, fine roots, and residues, as well as an increase in microbial activity, eco-restored land accumulated and retained more C than barren ground (Zhao et al., 2015). In rehabilitated regions, C sequestration is a significant ecological service. The relationship between SOC accumulation and sequestration and the efficacy of trees to restore ecosystems is important. The effectiveness of grass eco-restoration was positively correlated with SOC accumulation and sequestration. Most rehabilitated land has the capacity to sequester C. Massive amounts of C stored in mature trees will also contribute to the global C balance, although the C sink may diminish and reach saturation as the trees continue to grow (Xu et al., 2017).

3.4 Best Management Practices for Grassland Soil

Globally, grasslands are commonly used for fodder and food production due to their naturally fertile soils. Over 20% of the world's croplands and 20–25% of its grasslands are already degraded (Bai et al., 2008). Several causes, including decreasing C inflow to the soil, are responsible for the soil C losses. Grassland C storage is prone to soil C losses since they are sensitive to management. Carbon inputs, soil texture, climate, and soil organic matter (SOM) degradation rate influence C storage in soils (Johnston et al., 2009). To improve C storage, management practices should either, (i) increase the C input to the soil or (ii) decrease the rate of C decomposition in the soil. Under rigorous management, higher C sequestration is reported because C stocks respond differentially to management approaches, while C losses are commonly linked to excessive grassland management (Ammann et al., 2007). On grasslands in general, the following management practices may affect soil C stocks: intensive grazing management, production inputs (e.g., fertilizer, irrigation), fire management, grassland disturbance management, species selection of favorable forage grasses and legumes, restoring degraded lands, including grass in the rotation cycle, and adoption of agroforestry practices. In general, soils that have already lost C due to poor management or annual crop production have the highest potential for replenishing C stores (Bruce et al., 1999).

3.4.1 Intensive Grazing Management

It is widely accepted that excessive grazing is detrimental to the ecosystem and soil C stocks (Conant & Paustian, 2002; Milchunas & Lauenroth, 1993). Carbon stocks in grassland ecosystems can be restored by reversing management methods that deplete soil C levels, thereby sequestering atmospheric CO_2 (Follett et al., 2001). It is possible to alter grazing patterns that regularly deplete a considerable proportion of above-ground biomass. Well-managed grazing in grassland ecosystems enhances the growth of herbaceous plants and increases N cycling. Grazing has been shown to increase aboveground productivity, tillering, and rhizome production (McNaughton et al., 1996; Schuman et al., 1990). Improved grazing management (production enhancement management) increases soil C stores by 0.35 MgC ha^{-1} $year^{-1}$ on average (Conant et al., 2001). Additionally, grazing can improve root respiration and exudation rates (Dyer & Bokhari, 1976). The faeces and urination of grazing animals have a significant impact on N cycle and movement in grazing habitats. All of these variables are reported to contribute to the observed increases in soil C stores. It was reported that animal traffic enhances the physical disintegration, soil incorporation, and breakdown rate of aboveground plant biomass (Schuman et al., 2002). When grazing intensity is reduced or grazing is delayed, forage species may flourish, leading to a decrease in C removal. Therefore, sustainably managed grazing may increase C fluxes and soil C stores without diminishing forage yield. Additionally, these management methods may be used to restore high-yielding forage species to further increase soil C levels. Conant et al. (2016) reported mean positive C-level changes for better grazing practices (0.28 t C ha^{-1} $year^{-1}$), fertilization (0.57 t C ha^{-1} $year^{-1}$), and legume planting (0.66 t C ha^{-1} $year^{-1}$) when looking at individual techniques.

The effects of grazing on soil C can be influenced by interactions among soils, plant species, and climate. Grazing often increases soil C on sandy, coarse-textured soils in regions with greater precipitation, whereas clay soils respond with modest gains to considerable losses. Compared to sandy soils, fine-textured clay soils display the greatest increase in soil C due to grazing in desert grasslands (McSherry & Ritchie, 2013). These increases in response to effective grazing management contribute to a generally increasing trend, which a global analysis sets as a 2.9% increase in C stocks (Conant et al., 2001). In dry conditions, the higher water-holding capacity of clay soils may enhance this benefit (Steffens et al., 2008), whereas, in wet environments where soil moisture is not a limiting factor, compaction of clay soils can inhibit root growth and deplete soil C over time (Sigua & Coleman, 2010).

3.4.2 Production Input Uses

In several instances, grasslands are N-deficient, and N fertilizer has resulted in increased yield and water use efficiency. Adding organic and inorganic fertilizers to soil can increase SOC levels while directly and indirectly increasing grassland plant productivity (Jokubauskaite et al., 2016; Sykes et al., 2020). According to Smith et al. (2008), application of municipal biosolids or manure to the soil sequestered C at a rate of 0.42–0.76 t C ha^{-1} yr^{-1}, depending on the location (in wet areas, sequestration rates are higher than in dry areas). Application of N fertilizer enhanced tall grass prairie plant productivity and led to a 1.6 Mg ha^{-1} increase in SOC (Rice, 2000). In natural pastures, the application of 40 kg ha^{-1} N increased dry forage production and soil C significantly (Rai et al., 2013). The rate of soil C accumulation was 1.5 times that of natural grassland (0.74 g kg^{-1} yr^{-1}). Application of other nutrients was also reported to increase organic C storage in regions where it was deficient. Following seven long-term mineral fertilizer trials on temperate grasslands, Poeplau et al. (2018) found a considerable positive effect of phosphorus and N fertilization on SOC stocks in the 0–30 cm soil depth. To evaluate the overall impact on the environment, however, the benefits of increased soil C stocks must be compared against the C released for fertilizer production (Schuman et al., 2002). Enhanced net primary productivity (NPP) and consequently larger C inputs to the soil are frequently given as reasons for the good impacts of mineral fertilizer on croplands (Christopher & Lal, 2007; Kätterer et al., 2012).

Several chemical and biological processes involved in SOC production and stability are affected by the soil's pH (Madigan et al., 2022). When more calcium carbonate is made available by liming, more organocalcium aggregates are generated, which increases the potential of soil organic C sequestration (Tu et al., 2018). Increased pH, on the other hand, can reduce the stability of some organoaluminium aggregation compounds and increase the C respiration rates of soil bacteria (Miyazawa et al., 2013; Paradelo et al., 2015). Therefore, several soil parameters must be evaluated before lime application to ensure long-term favorable effects on soil C reserves and enhanced C sequestration.

3.4.3 Suitable Forage Grasses and Legumes

Increased biomass production can be accomplished by introducing or promoting improved and favored forage species that are better adapted to the local climate, more graze-resistant, drought-resistant, and capable of enhancing soil fertility. Increased biomass production increases C inputs and improves C sequestration. According to Rai et al. (2013), *Macroptilium lathyroides* led to a 1.29-fold increase in SOC compared to natural grassland. Other legumes produced more diverse biomass and increased soil TOC (Fig. 3.1). Among them, *Micropitilum lathyroildes* had the highest SOC accumulation rate (42%). Except for *Alysicarpus ragouts* and *Clitoris tenanted*, the potential of legume-incorporated grassland to sequester C is greater than that of natural grassland. The silvipasture system can contribute significantly to rehabilitating damaged land. Hazra (1995) examined the relative effectiveness of eight different tree species as components of *Cenchrus ciliaris*-based silvipasture compared to *Cenchrus ciliaris* as pasture (Fig. 3.2). Variations in litterfall and biomass production beneath diverse plant species dictated SOC content of the surface soil. Within the same time frame, *Leucaena leucocephala* induced a 3.4-fold increase in SOC above the initial values (Hazra, 1995). In grasslands with diverse species of swords, SOC levels are often higher (Fornara & Tilman, 2008; Nobilly et al., 2013).

3.4.4 Grassland Disturbance and Management

Globally, approximately 20% of grasslands have been converted to farmland (Ramankutty et al., 2008). In annual row crop production, soil disturbance (i.e., conventional tillage) is a common management technique that results in large organic matter losses. Due to conversion to agricultural land, many Midwestern soils in the US have lost 30–50% of their C (25–40 metric tons of C per hectare) (Lal, 2002). Soil disturbance disrupts soil structure, exposes organic C to decomposition by soil microbes and invertebrates, and increases decomposer activity by raising soil warmth and aeration. Disturbance accelerates soil erosion, leading to deposits elsewhere on the landscape and additional C losses as CO_2 in the atmosphere (Kimble et al., 1998).

Potential C sinks include agricultural fields converted to permanent pastures or rangelands that are well-managed (Conant et al., 2001). They reported a 5% average annual increase in the soil C content of converted agricultural areas to pasture. In addition, grazing land converted from native land has the lowest annual increase in C content and concentration, at 1.9% and 1.6%, respectively. Therefore, soil C can be improved in established pastures by eliminating soil disturbances and enhancing primary production.

It is a recognized fact that the 0–15 cm depth of soil has the greatest proportion of grass roots and soil organic C. Therefore, at the time of pasture renewal, inversion tillage provides a chance to raise C sequestration by inverting the C-rich organic layer into the subsoil and bringing up subsoil with a lower C content to the surface (Lawrence-Smith et al., 2021). Additionally, the top layer of soil also has the highest rates of SOC decomposition and microbiological activity. Soil C sequestration is therefore not proportional to soil depth always. As subsoils have a low concentration of SOC, they are in an unsaturated state. Thus, subsurface soil layers have a high ability to effectively stabilize C in the zone, hence they continue to hold substantial potential for long-term SOC sequestration (Paustian et al., 2016). Environmental factors, such as low temperature, less availability of organic materials, and insufficiency of nutrients and oxygen, are averse to microbial growth, leading to decreased subsoil C losses by respiration (Don et al., 2013; Whitehead et al., 2018). Recent research by Pereira et al. (2017) found that pasture restoration by full inversion tillage over four years led to an increase of roughly 14 t C ha^{-1} in New Zealand grasslands.

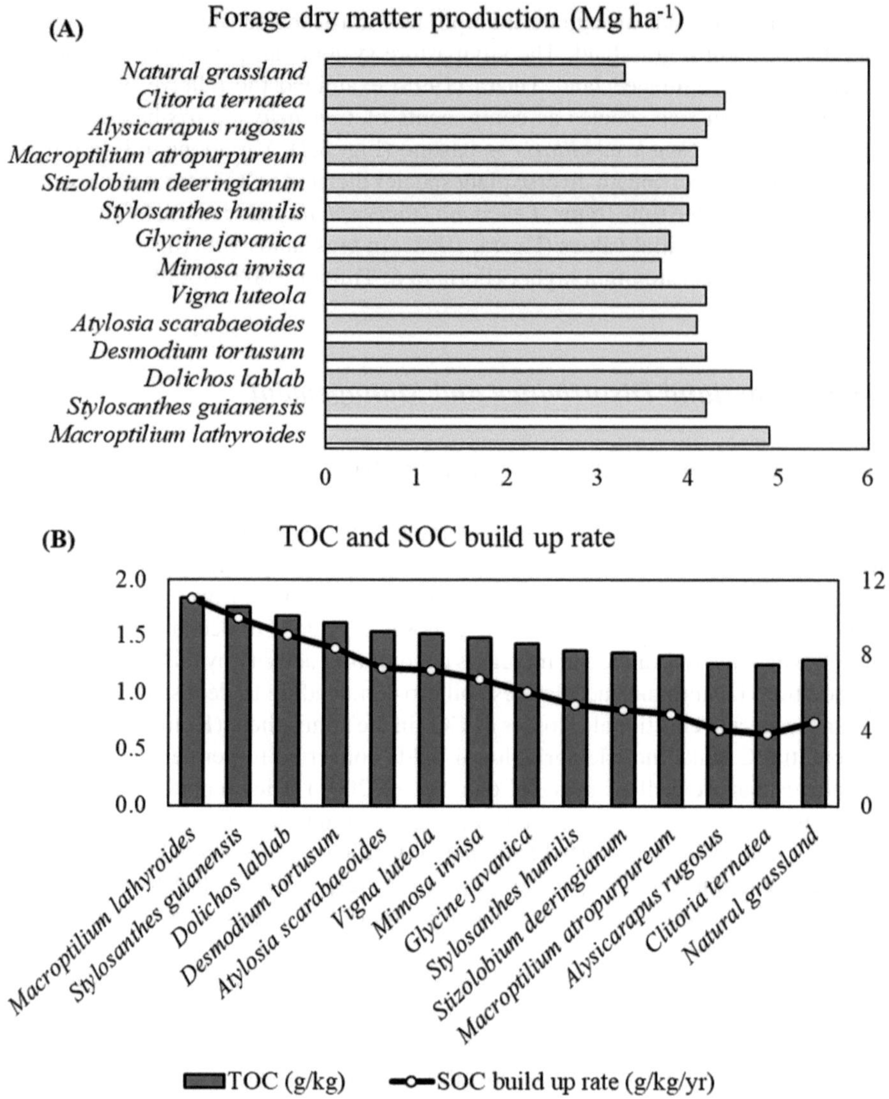

Fig. 3.1 Effect of different legumes on **a** forage dry matter production (Mg ha^{-1}) and **b** soil organic carbon (SOC) build up rate (g kg^{-1} yr^{-1}) and total organic carbon (TOC) (g kg^{-1}) (Rai et al., 2013)

Commonly, pasture reseeding is necessary every five to ten years as fodder yields and grassland quality deteriorates with age due to sward degradation caused by a range of causes including severe weed infestation, soil compaction, and inadequate drainage (Carolan & Fornara, 2016).

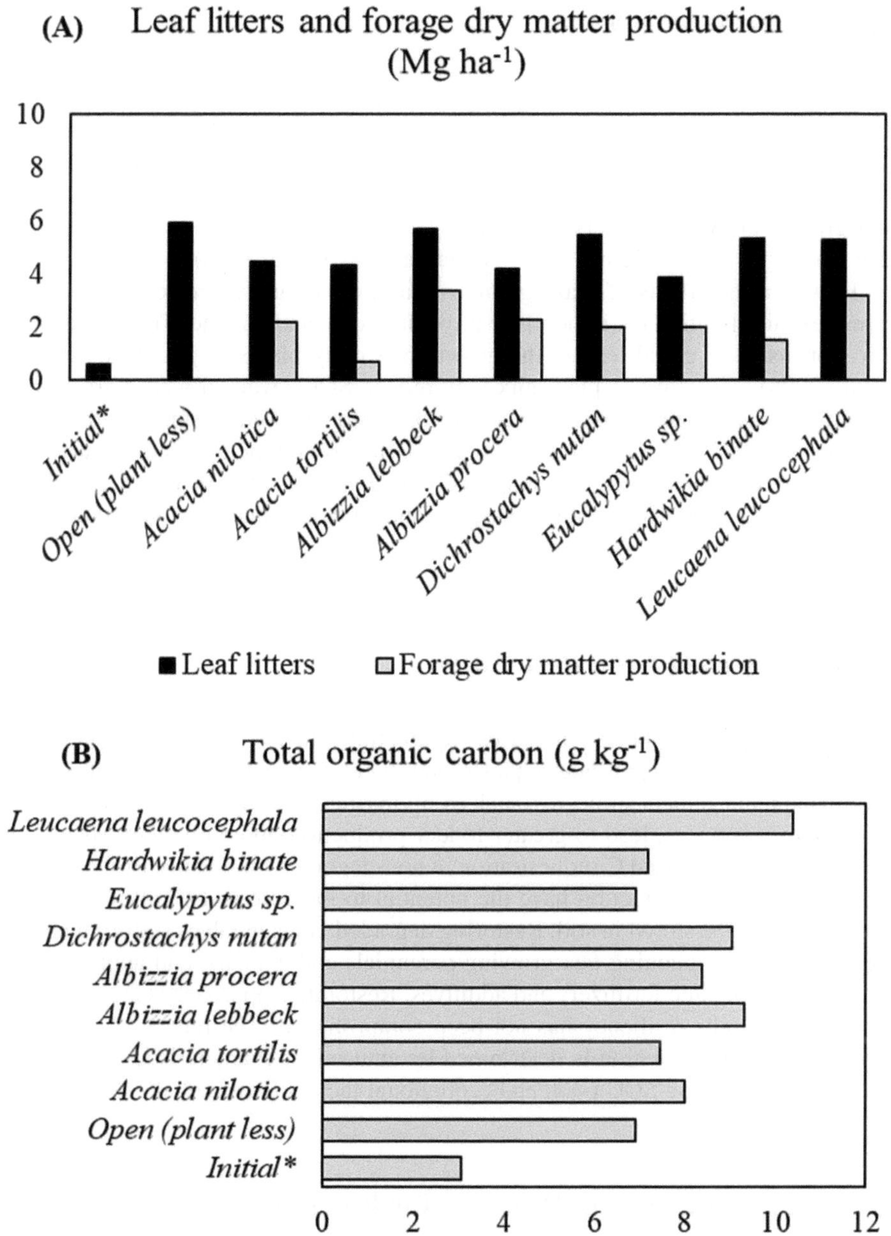

Fig. 3.2 Effect of different tree species in *Cenchrus ciliaris* silvipasture system on **a** Leaf litters and forage dry matter production (Mg ha^{-1}) and **b** total organic carbon (g kg^{-1}) (Hazra, 1995). **Heteropogon contortus*, *Aristida sps.* and *Eromopogon faveolatus* were the initial vegetation

3.4.5 Fire Management

In many types of grassland, fire suppression can alter the soil C storage by influencing the invasion of woody plant species. When biomass is burned, charcoal is generated, which is an extremely refractory form of C that may account for a significant portion of the C stored in the grassland. After 10 years of yearly fire and grazing in the tall grass prairie, researchers detected a 2.2 Mg ha^{-1} increase in soil C storage, accounting for a 0.22 Mg ha^{-1} yr^{-1} annual increase (Rice, 2000).

However, the effects of grazing intensity and fire on grassland soil C stocks are complex, highlighting the importance of plants, soils, and climate. The exclusion of fire, for instance, could lead to the growth of woody plants in grasslands (Knapp et al., 2008). This shift in plant mix may reduce SOC levels in wetter grasslands; nevertheless, encroachment may increase soil C stocks in increasingly arid habitats (Jackson et al., 2002). Other researchers have discovered the opposite impact, with woody encroachment from fire suppression increasing soil C in wetter grasslands (Tilman et al., 2000), indicating the significance of additional variables such as soil textures.

3.4.6 Degraded Land Restoration

Restoring degraded lands increases production in low-herbivore-productivity regions, as well as C inputs and sequestration. It has been demonstrated that changes in grazing management strategies, such as silvipastures, hortipastures, and enhanced grazing management, lead to greater fodder production, more efficient use of land resources, and increased C sequestration in non-degraded grasslands. Strategies that encourage forage production have the potential to increase soil C, allowing atmospheric CO_2 to be stored in soil. Restoring degraded lands involves restoring natural vegetation cover, planting fast-growing perennials and annuals, and applying both artificial and organic fertilizers and additives. Restorative approaches increase SOC levels if the soil's resiliency has not been compromised. The rate of C sequestration during soil restoration is determined by antecedent characteristics, restorative methods, the starting SOC pool, and ecoregional factors (Bruce et al., 1999).

3.5 Best Management Practices for Forest Soil

The world's forests cover 4–5 billion hectares and contain around 860 Pg C (Bastin et al., 2017). More than forty percent of total SOC is stored in forests (Wei et al., 2014). The total C stocks in boreal, temperate, and tropical forests were estimated as 202, 69, and 155 Pg, respectively (Pan et al., 2011). By altering inflow or efflux rates of C from soils, forest management strategies can have a substantial impact on

SOC stocks in forests. Above- and below-ground plant litter is the primary source of organic matter. Rhizodeposition, which is secreted by plant roots and their symbionts, as well as feces and soil biota bodies, are also essential organic inputs. This section examines 13 typical ways for managing forests effectively for soil C sequestration, all of which are essential and valuable to forest professionals.

3.5.1 Afforestation or Reforestation

Afforestation, reforestation, or replanting of agricultural or degraded land is suggested as a practice for mitigating climate change by sequestering C in situ in soil and plant biomass (Bastin et al., 2019). Several co-variables must be considered when evaluating the influence of afforestation on soil C reserves. After replanting, the rate of change in the soil C stock is influenced by soil type, historical land use, environment, forest age, tree species, and care. Even while an increase in the rate of C sequestration can be observed within decades following the conversion of cropland to a forest, it may take millennia for soil C stocks to settle. According to some recent research works, the conversion of former cropland to forest can increase C levels in soil over the coming millennia (Bárcena et al., 2014; Li et al., 2012; Nave et al., 2013; Poeplau et al., 2011). However, Poeplau et al. (2011) showed that 75% of all conversions of forest land to grassland have suffered C losses from the soil even after 100 years. Following forestation, the rate of soil C sequestration increases in direct proportion to temperature and precipitation. Positive C-level changes are observed regardless of soil type or land use history in the organic or forest surface soil layer (Mayer et al., 2020). However, negative soil C changes have been found in numerous studies when deep peat and shallow soils are replanted (Chapman et al., 2013; Simola et al., 2012; Vanguelova et al., 2019).

Variation in tree species (such as conifer–broadleaf) influences soil C sequestration rates. This is mostly owing to the higher initial rate of C sequestration in the forest floor of conifers. According to a site-level study, the distribution of C stocks sequestered in the soil profile changes with time. Since the afforestation of cropland in Denmark, resampling of oak and Norway spruce has indicated that in the first three decades, C sequestration in soil occurs mostly on the surface horizon (spruce > oak), with small losses in soil C stocks in the mineral soil (0–5 cm) (Vesterdal et al., 2002).

The rate of soil C accumulation can also be affected by soil type. After replanting, Bárcena et al. (2014) reported that coarse-textured soils had a greater capacity to collect C than fine-textured soils. Before afforestation, Finland's drained peatlands saw a 1.5 Mg ha^{-1} yr^{-1} drop in soil C sequestration rate (Simola et al., 2012). Numerous investigations conducted in the United Kingdom about the afforestation of Sitka spruce (Picea sitchensis (Bong.) Carr.) in peaty gley soils on the moorland revealed soil C losses in the peat layer during the first three decades (Vanguelova et al., 2019). They also discovered that after a century of following this replanting,

C building on forest floor compensated for this loss, and after two forest cycles, total C stock was restored to its prior levels (Vanguelova et al., 2019).

3.5.2 *Conversion of Primary Forest*

According to the FAO, a primary forest is a "naturally regenerated forest of native species with no clearly obvious indications of human activity and unaffected biological processes" (Bernier et al., 2017). Although regeneration in damaged forests may aid in the recovery of soil C stocks, primary forest conversion is only viable if such forests continue to exist. The benefit depends on the type of forest conversion; when converted to a controlled forest, even replanting with a single species has fewer negative effects than farmland conversion. The conversion of forest to farming results in a large decrease in SOC in the surface soil layer (0–30 cm) (e.g., 52%, 41%, and 31%, respectively, in temperate, tropical, and boreal climates) (Wei et al., 2014). The effect on SOC will be greater when a major forest is initially converted to agriculture following the conversion of a secondary forest. Nave et al. (2019) reported that soil C losses were greater on afforested sites than on secondary forest sites.

Secondary forests with sound management methods provide greater yields than primary forests and hence have a greater ability to absorb CO_2 from the atmosphere (Turner et al., 2017). Depending on their eventual use, well-managed forests can provide biofuels and various wood products that can operate as long-term C sinks and replace C-rich fossil fuels (Laganière et al., 2017; Soimakallio et al., 2016; Timmons et al., 2016). Compared to converted agricultural or commercial forest land, a primary forest may provide less immediate economic benefit. Protecting surviving primary forests, on the other hand, can provide additional ecosystem services, such as biodiversity conservation, biological heritage, recreational and social benefits, water supply and quality, and the potential for an ecotourism enterprise.

3.5.3 *Forest Harvesting*

All across the world, forest harvesting is practiced, but different nations, regions, and landowners employ different methods. Soil C stocks are influenced by the level and severity of soil disturbance associated with certain harvesting procedures, soil type, soil moisture present during harvesting, and tree species. Clear-cut harvesting is the most prevalent harvesting practice worldwide, and it has a negative influence on soil C stocks. Clearcutting reduces soil C by ten percent throughout the entire soil profile, with the forest floor suffering the most (Achat et al., 2015). According to Nave et al. (2010), forest harvesting reduced SOC by a mean of 8% in temperate forests; the amount of C stored on the forest floor declined by 30%, while the changes in mineral horizons were negligible. Carbon losses from the surface horizon of hardwood forests are greater than those of mixed species or coniferous forests (Nave et al., 2010).

Longer harvesting periods, retaining harvest debris on-site, and minimizing soil disturbance will help offset the consequences of harvesting cycles (Achat et al., 2015; Jandl et al., 2007; Law & Waring, 2015; Noormets & Nouvellon, 2015). This negative impact can be mitigated by employing harvesting techniques that disturb the soil the least (e.g., winter harvesting from snow-covered or frozen soils, forwarding hoe) (Mayer et al., 2020). Forest harvesting techniques such as continuous cover, maintaining living trees, cautious logging, selective cutting, and variable retention may aid in reducing soil C losses. According to Strukelj et al. (2015), after 9 years of harvesting, clearcut areas of boreal aspen forests acted as a net C source, whereas partially harvested regions acted as C sink.

3.5.4 *Collection of Harvest Residues*

Harvest residues are the low-value plant components that remain on-site following forest harvesting. Examples of harvest residue include twigs, leaves, bark, branches, dry stalks, and stumps. The impacts of residue removal on SOC are affected by soil type, climate, and forest management techniques. Whole tree harvesting (WTH), stem wood-only harvesting (SwOH), stem-only harvesting (SOH), and stump harvesting are common harvesting techniques. The sorts of residue removal after harvest vary between various approaches, as do the necessary pieces of equipment. Stump harvesting causes greater soil disturbance and minimizes the amount of root debris in the soil. Elimination of forest harvest residue has the biggest effect on the forest floor (organic horizon). According to Johnson and Curtis (2001), WTH results in a 6% decrease in soil C stocks, while SOH results in an 18% increase. According to Achat et al. (2015), harvest residue clearance led to a significant depletion of SOC in the forest surface soil horizon (10–45%) as well as in the subsoil (10%) down to a depth of 20 cm. This study indicates that the export of harvest leftovers from the site has a greater impact on temperate forests than boreal forests. Exporting harvest leftovers resulted in an 11% loss of SOC throughout the entire soil profile. Negative effects appeared to be more pronounced in sandy soils than in soils with a higher clay component (Oliveira et al., 2018). In medium and coarse-textured soils, SOH increased ~7% more than WTH, however in fine-textured soils, there was no significant difference between the two residue management strategies (Wan et al., 2018). On organic soils, stump harvesting appears to have a greater effect on soil C; for example, 50% SOC was removed from the 0–80 cm soil layer four years after stump harvesting (Vanguelova et al., 2017).

3.5.5 *Minimize Soil Disturbance*

Deliberate soil disturbance is a popular technique for preparing sites in the forests to enhance restoration effectiveness. It improves the development and survival of tree seedlings by eliminating competing vegetation, increasing nutrient availability, raising soil moisture and temperature, and decreasing the potential negative effects of small animals and various insects (Löf et al., 2012; Mjöfors et al., 2017; Van Miegroet & Olsson, 2011). Mechanical site preparation procedures that cause varied degrees of soil disturbance include tilling, disc trenching or blade scarification, pit mounding, and bedding. Scarification includes removing both the pioneer plants and the organic soil's top layer. The organic layer is turned over by disc-trenching/harrowing to create ridges that form continuous furrows (Piirainen et al., 2015). With soil that is mixed or inverted, mounding creates deeper furrows that can be made for continuous plowing and bedding, or it can create higher mounds and pits. Major reductions in total soil C content have been observed after 6–7 decades of site preparation by mechanical techniques (Örlander et al., 1996), but the SOC changes were minimal upto 1 m depth of sub-soils, so the underlying organic matter would have been included (Prescott et al., 2017). In Canada, disc trenching with herbicide treatment reduces SOC by two to three times when controlling understory vegetation in coniferous forests (Burgess et al., 1995). In the Ponderosa pine forest of the United States, scarification caused a 50% decrease in soil C after 24 years of continuous pesticide application (Jiménez Esquilín et al., 2008).

3.5.6 Nitrogen Application

Increasing soil C stocks are prevalent in forests that have been fertilized with nitrogenous fertilizers or have N deposition by some species (Janssens et al., 2010). According to Nave et al. (2009), N additions increased SOC reserves by 7.7% (including mineral soil and forest floor). Mineral soils can increase their SOC reserves by 12.2%. Increased litter inputs and decelerated decomposition are two mechanisms by which N fertilization can increase forest soil SOC levels. Higher litter and biomass generation increases C inflow to forest soils where primary productivity is limited by a lack of N (Huang et al., 2011); nevertheless, increased SOC stock has also been observed in the absence of increased return (production) (Cusack et al., 2011; Frey et al., 2014).

The significance of soil microbial products, biomass, and residue materials in the synthesis of SOM provides an additional reason for the increase in soil C following N application (Kallenbach et al., 2016; Paul, 2016; Wiesmeier et al., 2013). Nitrogen inputs can increase N concentrations in soil microorganisms (i.e., decrease C:N), resulting in increased production of nitrogenous microbial products that create robust SOM (Li et al., 2014).

However, in N-rich soil profiles as a result of atmospheric N deposition or on agricultural land, the addition of N is not suggested because it may increase nitrate losses through the denitrification process (Gao et al., 2015). Low organic matter content soil will benefit from the addition of N in organic forms, such as manure, municipal biosolids, etc., as this will increase N retention (Larney & Angers, 2012). As the production, shipping, and application of synthetic fertilizers, all involve the burning of fossil fuels and the release of CO_2, the final benefits of N fertilization should be weighed against the associated environmental damage. In many instances, the introduction of N-fixing species or legumes may be preferred to the use of synthetic fertilizers, since similar gains in soil C can be attained through these methods (Nave et al., 2010).

3.5.7 N-Fixing Species Selection

Nitrogen-fixing plant species modify soil qualities rapidly compared to other plant species, as well as enhance SOC levels continuously. With N-fixing plants, Nave et al. (2009) observed that mineral soil C stock increased by 12%. Binkley (2005) reported that N-fixing tree species *Ceanothus* and *Alnus* in temperate forests and *Leucaena*, *Casuarina*, and *Falcataria* in tropical forests accumulate an average of 870 kg C ha^{-1} $year^{-1}$ in forest soil. On eroded and damaged soils, N-fixing plants can replace soil C reserves considerably sooner than other species. Thus, in degraded areas, N-fixing plantations can be used to increase fertility to some degree. Studies comparing N-fixing plantations to nearby non-fixing trees that have been regularly fertilized have shown that N-fixing plants accumulate more SOC in the mineral layer (Forrester et al., 2013; Nave et al., 2010), and additional processes may increase SOC accumulation in forests containing N-fixing plants. On N-deficient soils, the enhanced biomass production of N-fixing plants presumably results in increased soil organic C reserves (Epron et al., 2013; Forrester et al., 2013).

It is well-established that N-fixing plants have a positive effect on SOC levels, meaning that adding N-fixing species into forests could aid in C sequestration. As tropical and subtropical climates are favorable for N-fixing species, the combination of N-fixing and non-fixing plant species is prevalent in the tropics and subtropics. However, climate change may increase the number of N-fixers at mid-and high latitudes in the coming decades (Liao et al., 2017; Steidinger et al., 2019). However, the dangers associated with introducing N-fixing plants outside of their normal habitats are comparable to those associated with introducing any non-native species. It has been discovered that exotic N-fixing plants have deleterious effects on tropical and subtropical grasslands worldwide (Richardson et al., 2011). Because the competitive advantage of N-fixing plants may make them invasive, native species are typically favored over exotics.

3.5.8 Selection of Appropriate Tree Species

Choosing acceptable species of forest tree is a common planning strategy for management and conservation, and it has also been studied in recent decades as a method for sequestering C in soils. There is a clear distinction between broadleaved and conifer species in terms of the acquisition, distribution, and stability of soil C stores.

Since conifers grow faster than other broad-leaved trees in harsh or poor fertility conditions, it is possible to establish a relationship between plant productivity and SOC levels (Augusto et al., 2014; Bellassen & Luyssaert, 2014; Chapin et al., 1993). Under optimal conditions (fertile soil, temperate climate), broad-leaved trees are more effective for sustained C sequestration in forests, however, under less favorable conditions, conifer species are projected to accumulate greater total C. According to Laganiere et al. (2010), broadleaf tree species resulted in a 25% increase in soil C stores 2–3 decades after afforestation, but conifers only resulted in a 2% increase. Coniferous plants have greater soil C levels than broadleaf angiosperms, whereas broadleaf angiosperms have greater mineral SOC levels (Vesterdal et al., 2013). The species with the highest N concentrations in their roots and the lowest levels of acid-insoluble compounds have the most C-stable soil, whereas those with the highest calcium tissue concentrations have the least C-stable soil. Angst et al. (2019) discovered that SOC stability varies greatly from plant to plant, with soil C having no effect on this heterogeneity. Consequently, the abiotic and biotic qualities of a site may impact the types of plantations chosen to maximize the SOC pool.

3.5.9 Tree Species Diversity Management

Species mixture has a smaller and less uniform effect on soil C than species identity (broadleaves vs. conifers). Species identity, whether it be a single species or a functional group such as conifers or N-fixers, may have a bigger impact on C sequestration than species diversity (Dawud et al., 2017; Wang et al., 2013). In large-scale comparisons of broadleaf, coniferous, and mixed forests, this difference may be minor (Brunel et al., 2017). The outcomes are also affected by numerous variables, suggesting that they may change depending on the temperature, species traits, or soil types (Dawud et al., 2017; Ratcliffe et al., 2017). Carbon output fluxes might be impacted despite the fact that greater organic material inputs to soils are anticipated as a result of above- and below-ground over-yielding.

3.5.10 Forest Stands Density

The basal area of dense forests is greater than that of low-density forests. Consequently, plantations with a greater population should have greater C inputs to the soil than those with a lower population, potentially raising their soil C reserves. These expected effects on soil C levels in densely established stands are identical to those in the less thinned plantation.

Laganiere et al. (2010) reported no difference in soil C levels (in 0–26 cm soil depth) between intensively planted forests (>1600 stems ha^{-1}) and lesser density plantations (1600 stems ha^{-1}). However, new research indicates contradictory results on the effect of plant density on soil C stores. Recent studies conducted 10–17 years after planting in Canada (with Populus cultivars; Truax et al., 2018) and China (with Fraxinus mandshurica cultivars; Sun et al., 2019) indicate that dens areas have more soil C storage than sparsely forested areas. In Quercus stands in Spain, another study indicated that tree density has a positive effect on SOC levels (González et al., 2012).

It is projected that forest thinning, in which some of the established stems are chopped down, will decrease soil C levels due to lesser plant litter inflow and a high rate of degradation in forest soils due to abundant moisture and higher temperature (Vesterdal et al., 1995). If thinning is significant, such as a 50% drop in a basal area relative to an unthinned control, organic C reserves can be depleted on the forest floor (Achat et al., 2015; Bravo-Oviedo et al., 2015; Powers et al., 2012). Zhang et al. (2018) reported that light thinning with a 30% volume reduction of plant basal area/stems increases soil C by 17%, whereas heavy thinning with >65% reduced soil C by 8%, and medium thinning did not affect soil C stores.

3.5.11 Herbivory Regulation

Reducing herbivory grazing pressure will increase SOC levels in ungulate herbivores, which are crucial forest ecosystem regulators (Tanentzap & Coomes, 2012). Numerous studies have indicated that ungulate herbivores have a substantial negative effect on tree regeneration, abundance, growth, structure, density, and richness (Ramirez et al., 2018), as well as a substantial negative influence on litter production and SOC in tropical, boreal, and temperate forests (Tanentzap & Coomes, 2012). Herbivores favor faster-growing species over slower-growing ones, which mostly influences litter or biomass production or decomposability (Pröll et al., 2015). Herbivores can also affect soil properties and decomposition mechanisms by excavating and trampling, as well as by depositing dung and urine (Andriuzzi & Wall, 2017; Valdés-Correcher et al., 2019). SOC levels increased by 107% in moderately grazed and ungrazed plots of tropical dry forest topsoil compared to severely grazed plots (Schulz et al., 2016). This study also revealed that when herbivore grazing was minimal, litter inflow rose, and SOC losses decreased. When red deer were eradicated from broad-leaved woodlands in Scotland, dissolved SOC increased by 36% due to higher deposition of soluble soil C components from root exudates (Harrison & Bardgett, 2004). High-density afforestation under controlled browsing and grazing provides greater soil protection against landslides, erosion, rock fall, and avalanches.

3.5.12 Regulation on Removal of Forest Litter and Biomass

In many regions of the world, rural inhabitants collect fuel wood, cut trees or branches for their fodder stocks, as well as collect litter for their animals' bedding and, when augmented with urine and feces, use them as manure for farmlands. Continuous C and N extraction with these techniques have deleterious effects on forest soils.

Ebermayer (1876) identified the negative effects of trash scraping on woodland soil, particularly in mountainous regions, including differences in water penetration and storage, variances in organic matter contents, decreased fertility, increased erosion, and changes in tree species composition. In addition, he discovered 36% less soil organic matter and 42% less N in a litter-raked stand compared to a nearby conserved stand (forest floor soil with 47 cm mineral soil). Still, popular subsistence farming operations in the Himalayas include trash raking, forest grazing, forest tree chopping, and manure harvesting (Baral & Katzensteiner, 2015). When compared to a neighboring restricted forest area, they observed 23% less SOC in heavily utilized forest sections (where 80% of annual terrestrial litter fall was cleared by the community).

Forest management policies may restrict the harvesting of biomass, enabling vegetative biomass and increasing the C stores of the soil. Exclosures can enhance plant density, composition, and diversity, as well as rehabilitate degraded soils and increase SOC supplies (Birhane et al., 2017; Yayneshet et al., 2009). This type of restriction, on the other hand, maybe socially unacceptable in areas where small farmers rely on these resources for a living, and it may exacerbate pressure in other regions. The most effective technique for restoring soil C levels may be forest management by the local people, in which forest is maintained for several purposes, including soil and plantation conservation. This negative cycle of forest degradation can be reversed if individuals assume responsibility for managing the forest's resources using sustainable practices (MFSC, 2013). Nepal has pioneered community-based forest management since the 1970s, providing vast tracts of land to the forest user community for long-term forest management (Baral et al., 2018a, 2018b; MFSC, 2013). In addition to extending forest area and increasing C stocks, community-based forestry has led to the development of revenue and has played a crucial role in enhancing rural people's livelihoods (Baral et al., 2019a, 2019b; Zewdie, 2008).

3.5.13 Management of Forest Fire

Each year, about 350 million ha of land, primarily in savannas, is charred by forest fires, which affect forest biological systems worldwide. Burned forest areas of 62%, 48%, and 36% are predominant in tropical Asia, boreal Northern America, and boreal Asia, respectively (Giglio et al., 2013). The quantity of C consumed and its effect on SOC stock are determined by the intensity and duration of the fire. The effects of forest fires on SOC contents vary considerably. Fire decreased total soil C by 26%, especially at the forest floor organic horizon (average SOC levels in the forest floor plummeted by 59%) (Nave et al., 2011). They also stated that it takes at least 128 years for the SOC level in burned forest soil to recover. The accumulation of C on the forest floor was reduced by 46% due to prescribed fires, as opposed to 67% due to uncontrolled fires. The effect of prescribed fires on C concentrations in mineral soil was minimal, whereas the effect of wildfires was substantial.

Several factors, such as land use changes and human habitation, forest segmentation, and climate change, affects fire weather (Liu et al., 2010), fuels for wildfire

(Flannigan et al., 2016), and possibly enhanced the sparks of lightning (Veraverbeke et al., 2017), thus contributing to the increase in wildfire risk. Wildfires can only be delayed but not prevented in the long run by fire suppression. Transformations of fire-prone areas to adapt wildfire in a specific ecosystem are largely focused on regulating wildfire fuels through controlled burning as well as prescribed practices of different silvicultural systems. As a result of these interventions, particularly managed fire, soil C is reduced and takes a very long time to recover; nonetheless, compared to wildfire soil C losses, it may be preferable to implement these interventions (James et al., 2018; Nave et al., 2011). Conversions of plantations to fewer flammable species, such as broadleaves as opposed to conifers, are another option for fire management (Schelhaas et al., 2010). Open terrain, agricultural croplands, and species that are relatively less combustible can operate as fire breaks, slowing the spread of fire and conserving terrestrial ecosystems (Khabarov et al., 2016; Lloret et al., 2002; Syphard et al., 2014).

Carbon sequestration in forest soils has the potential to reduce the rate of CO_2 enrichment in the atmosphere. Various forest management practices, which include afforestation or reforestation, site preparation, fire control, fertilizer application, grazing control, suitable species selection, and forest harvest management, can increase the C stocks in forest soils. Due to CO_2 fertilization, which has a positive effect on net primary productivity (NPP), and increased availability of N as a result of mineralization, it is expected that climate change will have a positive effect on soil C stocks; however, increased soil temperature and associated oxidation result in a decrease in soil C stocks. While there are numerous methods for monitoring soil C stocks and fluxes and significant strides have been made in measuring and scaling activities, quantifying C dynamics in forest soils remains challenging.

3.6 Best Management Practices for Wetlands

Except for Antarctica, wetlands can be found in every climate zone, from the tropics to the tundra. Wetlands are dynamic, natural ecosystems that are characterized by having standing water for at least a portion of the year. They make up around 5% of the land surface on Earth. According to numerous studies (Adhikari et al., 2009; Lal, 2008; Mitra et al., 2005), wetlands are the most significant, affordable, and effective options among natural pathways for sequestering atmospheric CO_2. In comparison to other ecosystems like forests and grass/shrublands, wetlands are the most productive and have the highest soil C density (Kayranli et al., 2010; Villa & Bernal, 2018). However, estimating the C stored in wetlands remains a challenge. Though there are several estimates of wetland C stock, a broad consensus among the researchers indicated wetland C ranged between 350 and 535 Gt C corresponding to 20–25% world's organic soil C (Gorham, 1995; Lal, 2008; Mitsch & Wu, 2018). To mitigate the predicted climate change, managing the wetlands soil to retain and improve C is essential (Griscom et al., 2017; Were et al., 2019).

3.6.1 Wetland Protection

Natural wetland preservation has been advocated in numerous studies as an essential step to improve C sequestration (Villa & Bernal, 2018). Villa and Bernal (2018) demonstrated that natural wetlands store more C than any other biome, at a global average of 400 Pg C in the top 1 m of soil. Carbon sequestration in wetlands is very sensitive to land use change. Thus, protecting wetlands in their natural state against anthropogenic disturbances including changes in plant, hydrology, and soil regimes. In a comparative study, Nahlik and Fennessy (2016) showed natural wetlands stored more C (2.25 ± 0.28 Pg C) in only 5.5×10^6 ha compared to that in disturbed wetlands (1.63 ± 0.33 Pg C), despite covering a larger area (7.0×10^6 ha) in topsoil horizon up to 1.2 m. Similar findings corroborated by a study in China in which it was demonstrated that wetland conversion resulted in the largest C loss (113 Tg) between 1995 and 2010 when compared to other natural ecosystems (Zhang et al., 2015). However, protecting wetlands in their natural form is not always beneficial in increasing C sequestration. In a study on Finnish wetlands, Minkkinen et al. (2002) showed C sequestration had increased from 2.2 Tg year^{-1} in 1900 to 3.6 Tg year^{-1} in 2002 when about 60% of natural wetlands had been disturbed by drainage.

Wetland disturbances can also change the direction of C sequestration either by being a sink or source depending on the methane (CH_4) emission. Natural wetlands, in certain instances, have increased CH_4 emission than artificially created ones due to the history of greater primary production that provides ample substrate for C reduction and production of CH_4. Sha et al. (2011) reported increased CH_4 emission (215 g CH_4 ha^{-1} h^{-1}) from Ohio's natural wetland. Additionally, a comparison study between natural and disturbed freshwater cypress (Taxodium distichum) wetlands in Florida, USA Pereyra and Mitsch (2018) reported natural wetlands emitted 156–495 g CH_4 ha^{-1} d^{-1} that were significantly greater than disturbed wetland emission. They even reported disturbed wetlands to serve as a sink for CH_4.

Dissolved organic C (DOC) is a major pool of soil organic C that regulates global C cycles ultimately influencing C sequestration. Wetlands are frequently reported to serve as a source of DOC that ultimately converted to CO_2 by degassing reducing potential of C sequestration. Sharifi et al. (2013) noted that about 15% of terrestrial organic C flux to surface water bodies originates from wetlands. In a study of the central Amazon river and flood basins, wetland C export contributed to 0.21 Pg C yr^{-1} emission as CO_2 (Abril et al., 2014; Richey et al., 2002).

The above findings shades doubt on the policy of natural wetland protection as an effective means of increasing C sequestration but do not confirm it as a net CO_2 emitter. Thus, a comprehensive understanding and assessment of C flux and pools of natural wetlands along with regional and global scale accounting are also necessary to make a final decision. Globally, natural wetlands are also facing challenges specifically in Asian countries where more land is required to produce food for a growing population. Rice being the favored diet for Asian population, draining and managing wetlands for rice cultivation became a natural choice. Though there is contradictory evidence, studies in China (Ti et al., 2012), Japan (Nishimura et al., 2008), and India

(Nath et al., 2016) confirmed greater C sequestration in converted wetlands used for rice cultivation.

3.6.2 *Wetland Restoration*

Restoring degraded wetlands to their initial stages provides the lost ecological services including C sequestration. The view was also supported by Ramsar Convention's Scientific and Technical Review Panel (STRP) when they highlighted that degraded wetlands are already a significant source of atmospheric CO_2 and their restoration is 100 times more economical than other options of C mitigations (Were et al., 2019). In a recent study, Lamers et al. (2015) showed that restoration of drained European wetlands sequesters 4 Mg C ha^{-1} yr^{-1} on average that is greater than earlier studies. In another study in Illinois, USA, Chen et al. (2017) showed that C stock of the restored wetland was 25–46% greater than that of natural counterparts just after 3 years. Globally, if 230 million ha of degraded wetlands can be restored, a net C sequestration rate of 4 Mt C $year^{-1}$ can be achieved by 2010 (IPCC, 2000).

Even though restoration aims to bring back lost ecological services of wetlands, restoring degraded wetlands is faced with several challenges. Beginning with enormous resources required to restore wetlands is not always considered economical in developing counties. This challenge is compounded by the fact there is no detailed inventory of natural wetlands and their degradation stages. Lack of knowledge on pristine wetlands restricts the success of restoration steps as there are no reference points to consider for evaluating the ecological services of restored wetlands.

3.6.3 *Wetland Creation*

Globally wetlands are created (known as constructed wetlands) to treat wastewater or stormwater. There are many constructed wetlands in the US and Europe but are very fewer in Asia and developing counties. Chen et al. (2017) reported constructed wetlands to be net C sequester. The reasoning is that constructed wetlands have shallower water columns and large sediments that can hold greater C per unit area. In Northern America, more wetlands need to be created to increase current net C sequestration (Mitsch et al., 2013). This is corroborated by the fact that constructed riverine wetland in Ohio sequesters more C (1.4 ± 0.016 kg C ha^{-1} $year^{-1}$) 15 years after creation compared to its natural counterpart (Mitsch et al., 2014).

Like wetland protection and restoration, creation is also faced with economic challenges compounded by a lack of enough arable land to produce food for a growing population in developing countries. Besides, when constructed wetlands are used to extract pollutants for stormwater, then after a certain period mature vegetation needs to be removed to effectively maintain the remediation function of the wetlands. This includes physical disturbances that may lead to a disturbed steady state of C flux and result in positive emission and reduced stock. Thus, wetlands are an excellent natural pathway to sequester C, often limited by socio-economic and biogeochemical considerations.

3.7 Engineering and Chemical Techniques for Enhanced C Sequestration

3.7.1 Enhanced Weathering of Rocks

The increased weathering (EW) of rocks is a potentially significant but still barely understood CO_2 removal (CDR) option. The purpose of accelerated weathering is to increase the natural geological process of C sequestration. In this method, crushed calcium (Ca) and magnesium (Mg)-bearing rocks are used to intensively maintain agricultural farmlands (Schuiling & Krijgsman, 2006). Enhanced weathering accelerates the interactions of atmospheric CO_2 with these rocks' mineral components. With the release of base cations (Ca^{2+} and Mg^{2+}) and the delivery of bicarbonate (HCO_3^-) and, to a lesser extent, carbonate (CO_3^{2-}) anions, the chemical breakdown of carbonate and silicate rocks in soils is accelerated during aqueous reactions within the elevated CO_2 environment of the soil because of root and microbial respiration (Beerling et al., 2018). Through runoff, these weathered chemicals reach the surface waters, and ultimately the ocean (Beerling et al., 2018). The EW utilizes the ocean as a C sink. The stable dissolved inorganic alkaline forms of C (carbonates and bicarbonates) are stored for a very long time in the oceanic pool, where they may precipitate as insoluble $CaCO_3$ (Fig. 3.3). The potential of the world's ocean to store CO_2 is 45 times the present amount of CO_2 in the atmosphere. On human timescales, the residence time of dissolved inorganic C in the oceans is around 10^5–10^6 years, making the oceans effectively a permanent C storage reservoir (Renforth & Henderson, 2017). The EW basically taps into the limitless future C storage capacity of the world's oceans.

If soil pores water chemistry results in the precipitation of secondary carbonate minerals from base cation release, silicate weathering on land can also absorb atmospheric CO_2 without affecting the oceans. In this instance, precipitated carbonate rather than ocean becomes the CO_2 sink. Assuming a sufficiently small particle size for successful dissolution on decadal timeframes, it has been reported that basalt has a C capture capability of up to 0.3 t CO_2 t^{-1} (Beerling et al., 2020). Beerling et al. (2018) also projected that application of basalt rock dust at rates of 10–30 t ha^{-1} yr^{-1} on two-thirds of the fertile agricultural soils (9×10^8 ha) may sequester 0.5–4 Pg CO_2 yr^{-1} by 2100. Alternatively, cation release from basalt weathering may boost soil cation exchange capacity and consequently nutrient availability (Gillman et al., 2001). The United States, China, and India, which are the largest fossil fuel CO_2 polluters in the world, have a substantial EW-based C sequestration

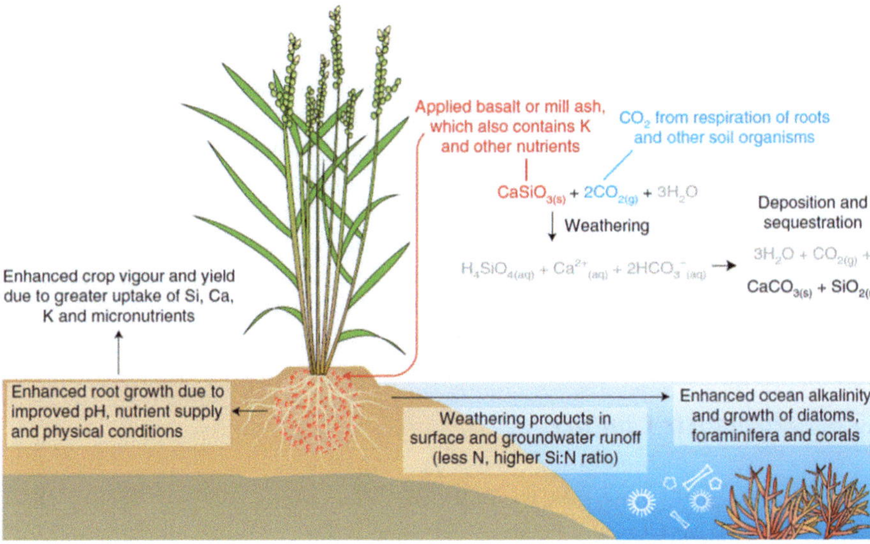

Fig. 3.3 Summary of the potential effects of weathering of crushed basalt or silicate-rich wastes applied to croplands. *Source* Beerling et al. (2018)

potential (Beerling et al., 2020). The potential of EW as a CDR is relatively undeveloped and requires additional research, development, and demonstration across a variety of crops, soil types, climates, and spatial scales.

3.7.2 Mineral Carbonation

Mineral carbonation is the process of artificially bringing high concentrations of CO_2 into contact with metal oxide-bearing minerals and fixing them as carbonates (Lackner et al., 1995). For this process, naturally occurring abundant silicate rocks containing serpentine or olivine minerals can be used. For silicate rocks, carbonation can be accomplished either ex-situ in a chemical processing facility following mining and pretreatment of the silicates, or in-situ by injecting CO_2 into silicate-rich geological formations or alkaline aquifers. In contrast, industrial leftovers can be carbonated in the same facility where they are produced like slag from steel industries. In the process, the CO_2 generated by industry is channeled and preheated between 100 and 150 °C. The interaction between CO_2 and dry metal oxides is extremely sluggish and almost impossible. Alternately, solid materials are suspended in an aqueous solution before dissolving and releasing metal ions. These metal ions then react with carbonic acid to produce carbonates.

In mineral carbonation, it takes between 1.6 and 3.7 tons of rock to fix one ton of CO_2. The reaction is thermodynamically exothermic and yields energy (IPCC, 2005). However, implantation procedures are still in the developmental phase and require a substantial amount of energy to reach economic conversion rates. For successful C sequestration, the resultant products that store atmospheric C should have a long turnover period. Currently, industrial mineral carbonation does not produce products with a long geological residence time. Therefore, more research must be conducted on the technology. The metal oxide dissolving is the rate-limiting stage, and the majority of research efforts have focused on accelerating metal extraction from solid input materials. This could be accomplished by either activating the mineral to make it more labile and reactive or by increasing the metal oxide extraction by adding additives or catalysts to the solution. Currently, activation is exceedingly energy-intensive, rendering the procedure extremely expensive and ineffective.

3.7.3 Direct Air Capture

To attain net zero C emissions by 2050, the International Energy Agency (IEA) predicts that 980 Mt CO_2 will be absorbed using industrial direct air capture (DAC) by 2050 (IEA, 2022). DAC technologies currently in use include solid and liquid approaches. The solid DAC (S-DAC) is based on solid adsorbents that cycle between adsorption and desorption. While adsorption occurs at ambient temperature and pressure, CO_2 is released at low pressure and moderate temperature (80–100 °C) using a temperature–vacuum swing process. A single adsorption/desorption unit can capture several tens of tons of CO_2 per year (for example, 50 tCO_2/year) and can be used to extract water from the atmosphere when local conditions permit (early prototypes were able to remove around 1 ton of water per ton of CO_2). An S-DAC plant is designed to be modular and can include as many units as needed. For instance, the largest operating S-DAC plant currently captures 4000 tons of CO_2 annually. Liquid DAC (L-DAC) is composed of two chemically closed loops. The first circuit occurs in an apparatus known as the contactor, which puts ambient air into contact with an aqueous basic solution (such as potassium hydroxide) capable of trapping CO_2. In a sequence of units working at high temperatures (between 300 and 900 °C), the second loop releases the CO_2 taken from the solution. A large-scale L-DAC facility can remove around 1 Mt CO_2 per year from the environment. Depending on the local climate, water replenishment may be required. As it does not compete for land resources with the food and energy industries, DAC technology is a very efficient technology that contributes to zero C emission targets.

3.8 Conclusions

Soil C sequestration is a process to not only increase C in a soil profile but also improve soil health and crop production simultaneously. The major pathways of soil C sequestration are the inclusion of leguminous species, the use of inputs including fertilizer and irrigation water, and residue retention in agricultural, grazing, and forest land. Though these major processes are not free of any detrimental effects on ecosystem function. One such effect is the introduction of leguminous tree species in areas where it acts as invasive species and disturbs the ecosystem balance. Then again, c sequestration in wetland systems is complex because of its dynamics between being the source as well as the sink of C. The economic considerations of wetlands are less important in Asian countries making them more vulnerable. Engineering and chemical technologies are new concepts and are mostly at the research stage. The economics and practical adaptability of these practices are yet to be established.

References

Abril, G., Martinez, J. M., Artigas, L. F., Moreira-Turcq, P., Benedetti, M. F., Vidal, L., et al. (2014). Amazon River carbon dioxide outgassing fuelled by wetlands. *Nature, 505*, 395–398.

Achat, D. L., Fortin, M., Landmann, G., Ringeval, B., & Augusto, L. (2015). Forest soil carbon is threatened by intensive biomass harvesting. *Scientific Reports, 5*(1), 1–10.

Adhikari, S., Bajracharaya, R. M., & Sitaula, B. K. (2009). A review of carbon dynamics and sequestration in wetlands. *Journal of Wetlands Ecology, 21*, 42–46.

Agrimonti, C., Lauro, M., & Visioli, G. (2021). Smart agriculture for food quality: Facing climate change in the 21st century. *Critical Reviews in Food Science and Nutrition, 61*(6), 971–981.

Ammann, C., Flechard, C. R., Leifeld, J., Neftel, A., & Fuhrer, J. (2007). The carbon budget of newly established temperate grassland depends on management intensity. *Agriculture, Ecosystems & Environment, 121*(1–2), 5–20.

Andriuzzi, W. S., & Wall, D. H. (2017). Responses of belowground communities to large aboveground herbivores: Meta-analysis reveals biome-dependent patterns and critical research gaps. *Global Change Biology, 23*(9), 3857–3868.

Angst, G., Mueller, K. E., Eissenstat, D. M., Trumbore, S., Freeman, K. H., Hobbie, S. E., Chorover, J., Oleksyn, J., Reich, P. B., & Mueller, C. W. (2019). Soil organic carbon stability in forests: Distinct effects of tree species identity and traits. *Global Change Biology, 25*(4), 1529–1546.

Ansari, M. A., Choudhury, B. U., Layek, J., Das, A., Lal, R., & Mishra, V. K. (2022). Green manuring and crop residue management: Effect on soil organic carbon stock, aggregation, and system productivity in the foothills of Eastern Himalaya (India). *Soil and Tillage Research, 218*, 105318.

Augusto, L., Davies, T. J., Delzon, S., & De Schrijver, A. (2014). The enigma of the rise of angiosperms: Can we untie the knot? *Ecology Letters, 17*(10), 1326–1338.

Autret, B., Beaudoin, N., Rakotovololona, L., Bertrand, M., Grandeau, G., Gréhan, E., Ferchaud, F., & Mary, B. (2019). Can alternative cropping systems mitigate nitrogen losses and improve GHG balance? Results from a 19-yr experiment in Northern France. *Geoderma, 342*, 20–33.

Bai, Z. G., Dent, D. L., Olsson, L., & Schaepman, M. E. (2008). Proxy global assessment of land degradation. *Soil Use and Management, 24*(3), 223–234.

Baral, S., Chhetri, B. B., Baral, H., & Vacik, H. (2019a). Investments in different taxonomies of goods: What should Nepal's community forest user groups prioritize? *Forest Policy and Economics, 1*(100), 24–32.

Baral, S., Chhetri, B. B. K., Baral, H., & Vacik, H. (2019b). Investments in different taxonomies of goods: What should Nepal's community forest user groups prioritize? *Forest Policy and Economics, 100*, 24–32.

Baral, S., Gautam, A. P., & Vacik, H. (2018a). Ecological and economical sustainability assessment of community forest management in Nepal: A reality check. *Journal of Sustainable Forestry, 37*(8), 820–841.

Baral, S. K., & Katzensteiner, K. (2015). Impact of biomass extraction on soil properties and foliar nitrogen content in a community forest and a semi-protected natural forest in the central mid-hills of Nepal. *Tropical Ecology, 56*, 323–333.

Baral, S., Gautam, A. P., & Vacik, H. (2018b). Ecological and economical sustainability assessment of community forest management in Nepal: A reality check. *Journal of Sustainable Forestry, 37*, 820–841.

Bárcena, T. G., Kiær, L. P., Vesterdal, L., Stefánsdóttir, H. M., Gundersen, P., & Sigurdsson, B. D. (2014). Soil carbon stock change following afforestation in Northern Europe: A meta-analysis. *Global Change Biology, 20*(8), 2393–2405.

Bastin, J. F., Berrahmouni, N., Grainger, A., Maniatis, D., Mollicone, D., Moore, R., Patriarca, C., Picard, N., Sparrow, B., Abraham, E. M., & Aloui, K. (2017). The extent of forest in dryland biomes. *Science, 356*(6338), 635–638.

Bastin, J. F., Finegold, Y., Garcia, C., Mollicone, D., Rezende, M., Routh, D., Zohner, C. M., & Crowther, T. W. (2019). The global tree restoration potential. *Science, 365*(6448), 76–79.

Beerling, D. J., Leake, J. R., Long, S. P., Scholes, J. D., Ton, J., Nelson, P. N., Bird, M., Kantzas, E., Taylor, L. L., Sarkar, B., & Kelland, M. (2018). Farming with crops and rocks to address global climate, food and soil security. *Nature Plants, 4*(3), 138–147.

Beillouin, D., Ben-Ari, T., Malézieux, E., Seufert, V., & Makowski, D. (2021). Positive but variable effects of crop diversification on biodiversity and ecosystem services. *Global Change Biology, 27*(19), 4697–4710.

Beerling, D. J., Kantzas, E. P., Lomas, M. R., Wade, P., Eufrasio, R. M., Renforth, P., ... Banwart, S. A. (2020). Potential for large-scale CO_2 removal via enhanced rock weathering with croplands. *Nature, 583*(7815), 242–248.

Bellassen, V., & Luyssaert, S. (2014). Carbon sequestration: Managing forests in uncertain times. *Nature, 506*(7487), 153–155.

Bera, T., Collins, H. P., Alva, A. K., Purakayastha, T. J., & Patra, A. K. (2016). Biochar and manure effluent effects on soil biochemical properties under corn production. *Applied Soil Ecology, 107*, 360–367.

Bera, T., Purakayastha, T. J., Patra, A. K., & Datta, S. C. (2018). Comparative analysis of physicochemical, nutrient, and spectral properties of agricultural residue biochars as influenced by pyrolysis temperatures. *Journal of Material Cycles and Waste Management, 20*(2), 1115–1127.

Bera, T., Vardanyan, L., Inglett, K. S., Reddy, K. R., O'Connor, G. A., Erickson, J. E., & Wilkie, A. C. (2019). Influence of select bioenergy by-products on soil carbon and microbial activity: A laboratory study. *Science of the Total Environment, 653*, 1354–1363.

Bernier, P. Y., Paré, D., Stinson, G., Bridge, S. R., Kishchuk, B. E., Lemprière, T. C., Thiffault, E., Titus, B. D., & Vasbinder, W. (2017). Moving beyond the concept of "primary forest" as a metric of forest environment quality. *Ecological Applications, 27*(2), 349–354.

Binkley, D. (2005). How nitrogen-fixing trees change soil carbon. In *Tree species effects on soils: Implications for global change* (pp. 155–164). Springer.

Birhane, E., Treydte, A. C., Eshete, A., Solomon, N., & Hailemariam, M. (2017). Can rangelands gain from bush encroachment? Carbon stocks of communal grazing lands invaded by Prosopis juliflora. *Journal of Arid Environments, 141*, 60–67.

Blanco-Canqui, H., Francis, C. A., & Galusha, T. D. (2017). Does organic farming accumulate carbon in deeper soil profiles in the long term? *Geoderma, 5*(288), 213–221.

Bravo-Oviedo, A., Ruiz-Peinado, R., Modrego, P., Alonso, R., & Montero, G. (2015). Forest thinning impact on carbon stock and soil condition in Southern European populations of P. sylvestris L. *Forest Ecology and Management, 357*, 259–267.

Bruce, J. P., Frome, M., Haites, E., Janzen, H., Lal, R., & Paustian, K. (1999). Carbon sequestration in soils. *Journal of Soil and Water Conservation, 54*(1), 382–389.

Brunel, C., Gros, R., Ziarelli, F., & Da Silva, A. M. F. (2017). Additive or non-additive effect of mixing oak in pine stands on soil properties depends on the tree species in Mediterranean forests. *Science of the Total Environment, 590*, 676–685.

Burgess, D., Baldock, J. A., Wetzell, S., & Brand, D. G. (1995). Scarification, fertilization and herbicide treatment effects on planted conifers and soil fertility. *Plant and Soil, 168*, 513–522.

Carolan, R., & Fornara, D. A. (2016). Soil carbon cycling and storage along a chronosequence of re-seeded grasslands: Do soil carbon stocks increase with grassland age? *Agriculture, Ecosystems & Environment, 218*, 126–132.

Chapin, F. S., III., Autumn, K., & Pugnaire, F. (1993). Evolution of suites of traits in response to environmental stress. *The American Naturalist, 142*, S78–S92.

Chapman, S. J., Bell, J. S., Campbell, C. D., Hudson, G., Lilly, A., Nolan, A. J., Robertson, A. H., Potts, J. M., & Towers, W. (2013). Comparison of soil carbon stocks in S cottish soils between 1978 and 2009. *European Journal of Soil Science, 64*(4), 455–465.

Chen, H., Popovich, S., McEuen, A., & Briddell, B. (2017). Carbon and nitrogen storage of a restored wetland at Illinois' Emiquon preserve: Potential for carbon sequestration. *Hydrobiologia, 804*, 139–150.

Chenu, C., Angers, D. A., Barré, P., Derrien, D., Arrouays, D., & Balesdent, J. (2019). Increasing organic stocks in agricultural soils: Knowledge gaps and potential innovations. *Soil and Tillage Research, 188*, 41–52.

Christopher, S. F., & Lal, R. (2007). Nitrogen management affects carbon sequestration in North American cropland soils. *Critical Reviews in Plant Sciences, 26*(1), 45–64.

Conant, R. T., Cerri, C. E. P., Osborne, B. B., & Paustian, K. (2016). Grassland management impacts on soil carbon stocks: A new synthesis. *Ecological Applications, 27*, 662–668.

Conant, R. T., Paustian, K., & Elliott, E. T. (2001). Grassland management and conversion into grassland: Effects on soil carbon. *Ecological Applications, 11*(2), 343–355.

Conant, R. T., & Paustian, K. (2002). Potential soil carbon sequestration in overgrazed grassland ecosystems. *Global Biogeochemical Cycles, 16*(4), 90–91.

Cusack, D. F., Silver, W. L., Torn, M. S., & McDowell, W. H. (2011). Effects of nitrogen additions on above-and belowground carbon dynamics in two tropical forests. *Biogeochemistry, 104*(1), 203–225.

Dawud, S. M., Raulund-Rasmussen, K., Ratcliffe, S., Domisch, T., Finer, L., Joly, F. X., Hättenschwiler, S., & Vesterdal, L. (2017). Tree species functional group is a more important driver of soil properties than tree species diversity across major European forest types. *Functional Ecology, 31*(5), 1153–1162.

Don, A., Rödenbeck, C., & Gleixner, G. (2013). Unexpected control of soil carbon turnover by soil carbon concentration. *Environmental Chemistry Letters, 11*(4), 407–413.

Dyer, M. I., & Bokhari, U. G. (1976). Plant-animal interactions: Studies of the effects of grasshopper grazing on blue grama grass. *Ecology, 57*(4), 762–772.

Ebermayer, E. (1876). *The entire doctrine of forest litter with regard to the chemical statics of silviculture: On the basis of the in the Königl. investigations carried out by the state forests of Bavaria. jumper*.

Epron, D., Nouvellon, Y., Mareschal, L., e Moreira, R. M., Koutika, L. S., Geneste, B., Delgado-Rojas, J. S., Laclau, J. P., Sola, G., de Moraes Goncalves, J. L., & Bouillet, J. P. (2013). Partitioning of net primary production in Eucalyptus and Acacia stands and in mixed-species plantations: Two case-studies in contrasting tropical environments. *Forest Ecology and Management, 301*, 102–111.

Feng, Y., & Li, X. (2002). A tool to determine long-term sustainable manure application rate for Alberta soils. In *Report to Canadian-Alberta beef industry development fund* (p. 120).

Finn, D., Page, K., Catton, K., Kienzle, M., Robertson, F., Armstrong, R., & Dalal, R. (2016). Ecological stoichiometry controls the transformation and retention of plant-derived organic matter to humus in response to nitrogen fertilisation. *Soil Biology and Biochemistry, 99*, 117–127.

Flannigan, M. D., Wotton, B. M., Marshall, G. A., De Groot, W. J., Johnston, J., Jurko, N., & Cantin, A. S. (2016). Fuel moisture sensitivity to temperature and precipitation: Climate change implications. *Climatic Change, 134*(1), 59–71.

Follett, R. F., Kimble, J. M., & Lal, R. (Eds.). (2001). *The potential of US grazing lands to sequester carbon and mitigate the greenhouse effect*. CRC Press LLC.

Fornara, D. A., & Tilman, D. (2008). Plant functional composition influences rates of soil carbon and nitrogen accumulation. *Journal of Ecology, 96*(2), 314–322.

Forrester, D. I., Pares, A., O'hara, C., Khanna, P. K., & Bauhus, J. (2013). Soil organic carbon is increased in mixed-species plantations of Eucalyptus and nitrogen-fixing Acacia. *Ecosystems, 16*(1), 123–132.

Francaviglia, R., Di Bene, C., Farina, R., Salvati, L., & Vicente-Vicente, J. L. (2019). Assessing "4 per 1000" soil organic carbon storage rates under Mediterranean climate: A comprehensive data analysis. *Mitigation and Adaptation Strategies for Global Change, 24*(5), 795–818.

Frey, S. D., Ollinger, S., Nadelhoffer, K. E., Bowden, R., Brzostek, E., Burton, A., Caldwell, B. A., Crow, S., Goodale, C. L., Grandy, A. S., & Finzi, A. (2014). Chronic nitrogen additions suppress decomposition and sequester soil carbon in temperate forests. *Biogeochemistry, 121*(2), 305–316.

Gao, W., Yang, H., Kou, L., & Li, S. (2015). Effects of nitrogen deposition and fertilization on N transformations in forest soils: A review. *Journal of Soils and Sediments, 15*(4), 863–879.

Ghosh, P. K., Hazra, K. K., Venkatesh, M. S., Nath, C. P., Singh, J., & Nadarajan, N. (2019). Increasing soil organic carbon through crop diversification in cereal–cereal rotations of Indo-Gangetic plain. *Proceedings of the National Academy of Sciences, India Section B: Biological Sciences, 89*(2), 429–440.

Giglio, L., Randerson, J. T., & Van Der Werf, G. R. (2013). Analysis of daily, monthly, and annual burned area using the fourth-generation global fire emissions database (GFED4). *Journal of Geophysical Research: Biogeosciences, 118*(1), 317–328.

Gillman, G. P., Burkett, D. C., & Coventry, R. J. (2001). A laboratory study of application of basalt dust to highly weathered soils: Effect on soil cation chemistry. *Soil Research, 39*(4), 799–811.

González, I., Grau Corbí, J. M., Fernández Cancio, A., Jiménez Ballesta, R., & González Cascón, M. R. (2012). Soil carbon stocks and soil solution chemistry in Quercus ilex stands in Mainland Spain. *European Journal of Forest Research, 131*(6), 1653–1667.

Gorham, E. (1995). In G. M. Woodwell & F. T. MacKenzie (Eds.), *Biotic feedbacks in the global climatic system* (pp. 169–187). Oxford University Press.

Griscom, B. W., Adams, J., Ellis, P. W., Houghton, R. A., Lomax, G., Miteva, D. A., Schlesinger, W. H., Shoch, D., Siikamäki, J. V., Smith, P., & Woodbury, P. (2017). Natural climate solutions. *Proceedings of the National Academy of Sciences, 114*(44), 11645–11650.

Guenet, B., Gabrielle, B., Chenu, C., Arrouays, D., Balesdent, J., Bernoux, M., Bruni, E., Caliman, J. P., Cardinael, R., Chen, S., & Ciais, P. (2021). Can N_2O emissions offset the benefits from soil organic carbon storage? *Global Change Biology, 27*(2), 237–256.

Haile, S. G., Nair, P. R., & Nair, V. D. (2008). Carbon storage of different soil-size fractions in Florida silvopastoral systems. *Journal of Environmental Quality, 37*(5), 1789–1797.

Harrison, K. A., & Bardgett, R. D. (2004). Browsing by red deer negatively impacts on soil nitrogen availability in regenerating native forest. *Soil Biology and Biochemistry, 36*(1), 115–126.

Hatfield, J. L., & Dold, C. (2019). Water-use efficiency: Advances and challenges in a changing climate. *Frontiers in Plant Science, 10*, 103.

Hazra, C. R. (1995). Soil and water conservation in relation to land use and biomass production. In *Forage production and utilization*. Indian Grassland and Fodder Research Institute.

Holdren, J. P. (2008). Science and technology for sustainable well-being. *Science, 319*(5862), 424–434.

Huang, Z., Clinton, P. W., Baisden, W. T., & Davis, M. R. (2011). Long-term nitrogen additions increased surface soil carbon concentration in a forest plantation despite elevated decomposition. *Soil Biology and Biochemistry, 43*(2), 302–307.

Hussain, S., Hussain, S., Guo, R., Sarwar, M., Ren, X., Krstic, D., Aslam, Z., Zulifqar, U., Rauf, A., Hano, C., & El-Esawi, M. A. (2021). Carbon sequestration to avoid soil degradation: A review on the role of conservation tillage. *Plants, 10*(10), 2001.

IEA. (2022). *Direct air capture: A key technology for net zero*. International Energy Agency.

IPCC. (2000). Land use, land use change, and forestry. Cambridge: Cambridge University Press. http://98.131.92.124/sites/default/files/2000%20Watson%20IPCC.pdf.

IPCC. (2007). *Climate change-the physical science basis: Working group I contribution to the fourth assessment report of the IPCC* (Vol. 4). Cambridge University Press.

IPCC. (2001). *Climate change 2001: The scientific basis. Contribution of working group 1 to the third assessment report of the IPCC*. Cambridge University Press.

IPCC. (2005). In B. Metz, O. Davidson, H. C. de Coninck, M. Loos, & L. A. Meyer (Eds.), *IPCC special report on carbon dioxide capture and storage. Prepared by working group III of the Intergovernmental Panel on Climate Change* (p. 442). Cambridge University Press.

Jat, M. L., Chakraborty, D., Ladha, J. K., Parihar, C. M., Datta, A., Mandal, B., … & Gerard, B. (2022). Carbon sequestration potential, challenges, and strategies towards climate action in smallholder agricultural systems of South Asia. *Crop and Environment, 1*(1), 86–101.

Jackson, R. B., Banner, J. L., Jobbágy, E. G., Pockman, W. T., & Wall, D. H. (2002). Ecosystem carbon loss with woody plant invasion of grasslands. *Nature, 418*(6898), 623–626.

James, J. N., Kates, N., Kuhn, C. D., Littlefield, C. E., Miller, C. W., Bakker, J. D., Butman, D. E., & Haugo, R. D. (2018). The effects of forest restoration on ecosystem carbon in western North America: A systematic review. *Forest Ecology and Management, 429*, 625–641.

Jandl, R., Lindner, M., Vesterdal, L., Bauwens, B., Baritz, R., Hagedorn, F., Johnson, D. W., Minkkinen, K., & Byrne, K. A. (2007). How strongly can forest management influence soil carbon sequestration? *Geoderma, 137*(3–4), 253–268.

Janssens, I. A., Dieleman, W., Luyssaert, S., Subke, J. A., Reichstein, M., Ceulemans, R., Ciais, P., Dolman, A. J., Grace, J., Matteucci, G., & Papale, D. (2010). Reduction of forest soil respiration in response to nitrogen deposition. *Nature Geoscience, 3*(5), 315–322.

Jiménez Esquilín, A. E., Stromberger, M. E., & Shepperd, W. D. (2008). Soil scarification and wildfire interactions and effects on microbial communities and carbon. *Soil Science Society of America Journal, 72*(1), 111–118.

Johnson, D. W., & Curtis, P. S. (2001). Effects of forest management on soil C and N storage: Meta analysis. *Forest Ecology and Management, 140*(2–3), 227–238.

Johnston, A. E., Poulton, P. R., & Coleman, K. (2009). Soil organic matter: Its importance in sustainable agriculture and carbon dioxide fluxes. *Advances in Agronomy, 101*, 1–57.

Jokubauskaite, I., Karčauskienė, D., Slepetiene, A., Repsiene, R., & Amaleviciute, K. (2016). Effect of different fertilization modes on soil organic carbon sequestration in acid soils. *Acta Agriculturae Scandinavica, Section B—Soil & Plant Science, 66*(8), 647–652.

Kallenbach, C. M., Frey, S. D., & Grandy, A. S. (2016). Direct evidence for microbial-derived soil organic matter formation and its ecophysiological controls. *Nature Communications, 7*(1), 1–10.

Kätterer, T., Bolinder, M. A., Berglund, K., & Kirchmann, H. J. (2012). Strategies for carbon sequestration in agricultural soils in northern Europe. *Acta Agriculturae Scandinavica, Section A–Animal Science, 62*(4), 181–198.

Kayranli, B., Scholz, M., Mustafa, A., & Hedmark, Å. (2010). Carbon storage and fluxes within freshwater wetlands: A critical review. *Wetlands, 30*, 111–124.

Khabarov, N., Krasovskii, A., Obersteiner, M., Swart, R., Dosio, A., San-Miguel-Ayanz, J., Durrant, T., Camia, A., & Migliavacca, M. (2016). Forest fires and adaptation options in Europe. *Regional Environmental Change, 16*(1), 21–30.

Kimble, J. M., Follett, R. F., & Cole, C. V. (1998). *The potential of US cropland to sequester carbon and mitigate the greenhouse effect*. CRC Press.

Knapp, A. K., Briggs, J. M., Collins, S. L., Archer, S. R., Bret-Harte, M. S., Ewers, B. E., Peters, D. P., Young, D. R., Shaver, G. R., Pendall, E., & Cleary, M. B. (2008). Shrub encroachment in North American grasslands: Shifts in growth form dominance rapidly alters control of ecosystem carbon inputs. *Global Change Biology, 14*(3), 615–623.

Kumar, S., Singh, A. K., Singh, R., Ghosh, A., Chaudhary, M., Shukla, A. K., Kumar, S., Singh, H. V., Ahmed, A., & Kumar, R. V. (2019). Degraded land restoration ecological way through horti-pasture systems and soil moisture conservation to sustain productive economic viability. *Land Degradation & Development, 30*(12), 1516–1529.

Kuzyakov, Y. (2010). Priming effects: Interactions between living and dead organic matter. *Soil Biology and Biochemistry, 42*, 1363–1371.
Lackner, K. S., Wendt, C. H., Butt, D. P., Joyce, E. L., Jr., & Sharp, D. H. (1995). Carbon dioxide disposal in carbonate minerals. *Energy, 20*(11), 1153–1170.
Laganiere, J., Angers, D. A., & Pare, D. (2010). Carbon accumulation in agricultural soils after afforestation: A meta-analysis. *Global Change Biology, 16*(1), 439–453.
Laganière, J., Paré, D., Thiffault, E., & Bernier, P. Y. (2017). Range and uncertainties in estimating delays in greenhouse gas mitigation potential of forest bioenergy sourced from Canadian forests. *GCB Bioenergy, 9*(2), 358–369.
Lal, R. (2002). Soil carbon dynamics in cropland and rangeland. *Environmental Pollution, 116*(3), 353–362.
Lal, R. (2008). Sequestration of atmospheric CO_2 in global carbon pools. *Energy & Environmental Science, 1*, 86–100.
Lal, R. (2010). Managing soils and ecosystems for mitigating anthropogenic carbon emissions and advancing global food security. *BioScience, 60*(9), 708–721.
Lamers, L. P., Vile, M. A., Grootjans, A. P., Acreman, M. C., van Diggelen, R., Evans, M. G., Richardson, C. J., Rochefort, L., Kooijman, A. M., Roelofs, J. G., & Smolders, A. J. (2015). Ecological restoration of rich fens in Europe and North America: From trial and error to an evidence-based approach. *Biological Reviews, 90*(1), 182–203.
Larney, F. J., & Angers, D. A. (2012). The role of organic amendments in soil reclamation: A review. *Canadian Journal of Soil Science, 92*(1), 19–38.
Law, B. E., & Waring, R. H. (2015). Carbon implications of current and future effects of drought, fire and management on Pacific Northwest forests. *Forest Ecology and Management, 355*, 4–14.
Lawrence-Smith, E. J., Curtin, D., Beare, M. H., McNally, S. R., Kelliher, F. M., Calvelo Pereira, R., & Hedley, M. J. (2021). Full inversion tillage during pasture renewal to increase soil carbon storage: New Zealand as a case study. *Global Change Biology, 27*(10), 1998–2010.
Lehmann, J. (2007). A handful of carbon. *Nature, 447*, 143–144.
Leifeld, J., & Fuhrer, J. (2010). Organic farming and soil carbon sequestration: What do we really know about the benefits? *Ambio, 39*(8), 585–599.
Leng, L., Huang, H., Li, H., Li, J., & Zhou, W. (2019). Biochar stability assessment methods: A review. *Science of the Total Environment, 647*, 210–222.
Li, D., Niu, S., & Luo, Y. (2012). Global patterns of the dynamics of soil carbon and nitrogen stocks following afforestation: A meta-analysis. *New Phytologist, 195*(1), 172–181.
Li, P., Yang, Y., Han, W., & Fang, J. (2014). Global patterns of soil microbial nitrogen and phosphorus stoichiometry in forest ecosystems. *Global Ecology and Biogeography, 23*(9), 979–987.
Liao, W., Menge, D. N., Lichstein, J. W., & Ángeles-Pérez, G. (2017). Global climate change will increase the abundance of symbiotic nitrogen-fixing trees in much of North America. *Global Change Biology, 23*(11), 4777–4787.
Liu, E., Yan, C., Mei, X., Zhang, Y., & Fan, T. (2013). Long-term effect of manure and fertilizer on soil organic carbon pools in dryland farming in northwest China. *PLoS ONE, 8*(2), e56536.
Liu, Y., Stanturf, J., & Goodrick, S. (2010). Trends in global wildfire potential in a changing climate. *Forest Ecology and Management, 259*(4).
Lloret, F., Calvo, E., Pons, X., & Díaz-Delgado, R. (2002). Wildfires and landscape patterns in the Eastern Iberian Peninsula. *Landscape Ecology, 17*(8), 745–759.
Löf, M., Dey, D. C., Navarro, R. M., & Jacobs, D. F. (2012). Mechanical site preparation for forest restoration. *New Forests, 43*(5), 825–848.
Lorenz, K., & Lal, R. (2014a). Biochar application to soil for climate change mitigation by soil organic carbon sequestration. *Journal of Plant Nutrition and Soil Science, 177*(5), 651–670.
Lorenz, K., & Lal, R. (2014b). Soil organic carbon sequestration in agroforestry systems. A review. *Agronomy for Sustainable Development, 34*(2), 443–454.
Madigan, A. P., Zimmermann, J., Krol, D. J., Williams, M., & Jones, M. B. (2022). Full Inversion Tillage (FIT) during pasture renewal as a potential management strategy for enhanced carbon sequestration and storage in Irish grassland soils. *Science of the Total Environment, 805*, 150342.

Maity, S. K., & Mukherjee, P. K. (2011). Effect of brown manuring on grain yield and nutrient use efficiency in dry direct seeded Kharif rice (Oryza sativa L.). *Indian Journal of Weed Science, 43*(1–2), 61–66.
Majumder, S., Neogi, S., Dutta, T., Powel, M. A., & Banik, P. (2019). The impact of biochar on soil carbon sequestration: Meta-analytical approach to evaluating environmental and economic advantages. *Journal of Environmental Management, 250*, 109466.
Martin, D. M. (2017). Ecological restoration should be redefined for the twenty-first century. *Restoration Ecology, 25*, 668–673.
Mayer, M., Prescott, C. E., Abaker, W. E., Augusto, L., Cécillon, L., Ferreira, G. W., James, J., Jandl, R., Katzensteiner, K., Laclau, J. P., & Laganière, J. (2020). Tamm Review: Influence of forest management activities on soil organic carbon stocks: A knowledge synthesis. *Forest Ecology and Management, 466*, 118127.
McNaughton, S. J., Milchunas, D. G., & Frank, D. A. (1996). How can net primary productivity be measured in grazing ecosystems? *Ecology, 77*(3), 974–977.
McSherry, M. E., & Ritchie, M. E. (2013). Effects of grazing on grassland soil carbon: A global review. *Global Change Biology, 19*(5), 1347–1357.
Meena, R. S., Kumar, S., & Yadav, G. S. (2020). Soil carbon sequestration in crop production. In *Nutrient dynamics for sustainable crop production* (pp. 1–39). Springer.
MFSC. (2013). *Persistence and change—Review of 30 years of community forestry in Nepal.* Ministry of Forests and Soil Conservation, Government of Nepal.
Milchunas, D. G., & Lauenroth, W. K. (1993). Quantitative effects of grazing on vegetation and soils over a global range of environments: Ecological archives M063–001. *Ecological Monographs, 63*(4), 327–366.
Minkkinen, K., Korhonen, R., Savolainen, I., & Laine, J. (2002). Carbon balance and radiative forcing of Finnish peatlands 1900–2100—The impact of forestry drainage. *Global Change Biology, 8*, 785–799.
Mitra, S., Wassmann, R., & Vlek, P. L. (2005). An appraisal of global wetland area and its organic carbon stock. *Current Science, 88*, 25–35.
Mitsch, W. J., Bernal, B., Nahlik, A. M., Mander, Ü., Zhang, L., Anderson, C. J., Jørgensen, S. E., & Brix, H. (2013). Wetlands, carbon, and climate change. *Landscape Ecology, 28*(4), 583–597.
Mitsch, W. J., & Wu, X. (2018). Wetlands and global change. *Soil Management and Greenhouse Effect, 6*, 205–230.
Mitsch, W. J., Zhang, L., Waletzko, E., & Bernal, B. (2014). Validation of the ecosystem services of created wetlands: Two decades of plant succession, nutrient retention, and carbon sequestration in experimental riverine marshes. *Ecological Engineering, 72*, 11–24.
Miyazawa, M., Takahashi, T., Sato, T., Kanno, H., & Nanzyo, M. (2013). Factors controlling accumulation and decomposition of organic carbon in humus horizons of Andosols. *Biology and Fertility of Soils, 49*(7), 929–938.
Mjöfors, K., Strömgren, M., Nohrstedt, H. Ö., Johansson, M. B., & Gärdenäs, A. I. (2017). Indications that site preparation increases forest ecosystem carbon stocks in the long term. *Scandinavian Journal of Forest Research, 32*(8), 717–725.
Müller, A., Bautze, L., Meier, M., Gattinger, A., Gall, E., Chatzinikolaou, E., Meredith, S., Ukas, T., & Ullmann, L. (2016). *Organic farming, climate change mitigation and beyond. Reducing the environmental impacts of EU agriculture.*
Nahlik, A. M., & Fennessy, M. S. (2016). Carbon storage in US wetlands. *Nature Communication, 7*, 13835.
Nair, P. R., Nair, V. D., Kumar, B. M., & Haile, S. G. (2009). Soil carbon sequestration in tropical agroforestry systems: A feasibility appraisal. *Environmental Science & Policy, 12*(8), 1099–1111.
Nair, P. R., Nair, V. D., Kumar, B. M., & Showalter, J. M. (2010). Carbon sequestration in agroforestry systems. *Advances in Agronomy, 108*, 237–307.

Nandan, R., Singh, V., Singh, S. S., Kumar, V., Hazra, K. K., Nath, C. P., Poonia, S., Malik, R. K., Bhattacharyya, R., & McDonald, A. (2019). Impact of conservation tillage in rice-based cropping systems on soil aggregation, carbon pools and nutrients. *Geoderma, 340*, 104–114.

Nath, A. J., Brahma, B., Pathak, K., Das, A. K. (2016). Why should we preserve wetlands? *Current Science, 110*, 1619–1620.

Nave, L. E., DeLyser, K., Butler-Leopold, P. R., Sprague, E., Daley, J., & Swanston, C. W. (2019). Effects of land use and forest management on soil carbon in the ecoregions of Maryland and adjacent eastern United States. *Forest Ecology and Management, 448*, 34–47.

Nave, L. E., Swanston, C. W., Mishra, U., & Nadelhoffer, K. J. (2013). Afforestation effects on soil carbon storage in the United States: A synthesis. *Soil Science Society of America Journal, 77*(3), 1035–1047.

Nave, L. E., Vance, E. D., Swanston, C. W., & Curtis, P. S. (2009). Impacts of elevated N inputs on north temperate forest soil C storage, C/N, and net N-mineralization. *Geoderma, 153*(1–2), 231–240.

Nave, L. E., Vance, E. D., Swanston, C. W., & Curtis, P. S. (2010). Harvest impacts on soil carbon storage in temperate forests. *Forest Ecology and Management, 259*(5), 857–866.

Nave, L. E., Vance, E. D., Swanston, C. W., & Curtis, P. S. (2011). Fire effects on temperate forest soil C and N storage. *Ecological Applications, 21*(4), 1189–1201.

Nishimura, S., Yonemura, S., Sawamoto, T., Shirato, Y., Akiyama, H., Sudo, S., & Yagi, K. (2008). Effect of land use change from paddy rice cultivation to upland crop cultivation on soil carbon budget of a cropland in Japan. *Agriculture, Ecosystems & Environment, 125*(1–4), 9–20.

Nobilly, F., Bryant, R. H., McKenzie, B. A., & Edwards, G. R. (2013). Productivity of rotationally grazed simple and diverse pasture mixtures under irrigation in Canterbury. In *Proceedings of the New Zealand Grassland Association* (pp. 165–172).

Noormets, A., & Nouvellon, Y. (2015). Introduction for special issue: Carbon, water and nutrient cycling in managed forests. *Forest Ecology and Management, 355*, 1–3.

Oliveira, F. C. C., Silva, I. R., Ferreira, G. W. D., Soares, E. M. B., Silva, S. R., & Silva, E. F. (2018). Contribution of Eucalyptus harvest residues and nitrogen fertilization to carbon stabilization in Ultisols of southern Bahia. *Revista Brasileira de Ciência do Solo, 42*.

Örlander, G., Egnell, G., & Albrektson, A. (1996). Long-term effects of site preparation on growth in Scots pine. *Forest Ecology and Management, 86*(1–3), 27–37.

Pan, Y., Birdsey, R. A., Fang, J., Houghton, R., Kauppi, P. E., Kurz, W. A., Phillips, O. L., Shvidenko, A., Lewis, S. L., Canadell, J. G., & Ciais, P. (2011). A large and persistent carbon sink in the world's forests. *Science, 333*(6045), 988–993.

Panwar, P., Mahalingappa, D. G., Kaushal, R., Bhardwaj, D. R., Chakravarty, S., Shukla, G., Thakur, N. S., Chavan, S. B., Pal, S., Nayak, B. G., et al. (2022). Biomass production and carbon sequestration potential of different agroforestry systems in India: A critical review. *Forests, 13*, 1274. https://doi.org/10.3390/f13081274

Paradelo, R., Virto, I., & Chenu, C. (2015). Net effect of liming on soil organic carbon stocks: A review. *Agriculture, Ecosystems & Environment, 202*, 98–107.

Paul, E. A. (2016). The nature and dynamics of soil organic matter: Plant inputs, microbial transformations, and organic matter stabilization. *Soil Biology and Biochemistry, 98*, 109–126.

Paustian, K., Lehmann, J., Ogle, S., Reay, D., Robertson, G. P., & Smith, P. (2016). Climate-smart soils. *Nature, 532*(7597), 49–57.

Pereira, R. C., Hedley, M. J., Arbestain, M. C., Bishop, P., Enongene, K. E., & Otene, I. J. J. (2017). Evidence for soil carbon enhancement through deeper mouldboard ploughing at pasture renovation on a Typic Fragiaqualf. *Soil Research, 56*(2), 182–191.

Pereyra, A. S., & Mitsch, W. J. (2018). Methane emissions from freshwater cypress (Taxodium distichum) swamp soils with natural and impacted hydroperiods in Southwest Florida. *Ecological Engineering, 114*, 46–56.

Piirainen, S., Finér, L., & Starr, M. (2015). Changes in forest floor and mineral soil carbon and nitrogen stocks in a boreal forest after clear-cutting and mechanical site preparation. *European Journal of Soil Science, 66*(4), 735–743.

Poeplau, C., Don, A., Vesterdal, L., Leifeld, J., Van Wesemael, B. A. S., Schumacher, J., & Gensior, A. (2011). Temporal dynamics of soil organic carbon after land-use change in the temperate zone–carbon response functions as a model approach. *Global Change Biology, 17*(7), 2415–2427.

Poeplau, C., Zopf, D., Greiner, B., Geerts, R., Korvaar, H., Thumm, U., Don, A., Heidkamp, A., & Flessa, H. (2018). Why does mineral fertilization increase soil carbon stocks in temperate grasslands? *Agriculture, Ecosystems & Environment, 265*, 144–155.

Pooniya, V., Shivay, Y. S., Rana, A., Nain, L., & Prasanna, R. (2012). Enhancing soil nutrient dynamics and productivity of Basmati rice through residue incorporation and zinc fertilization. *European Journal of Agronomy, 41*, 28–37.

Powers, M. D., Kolka, R. K., Bradford, J. B., Palik, B. J., Fraver, S., & Jurgensen, M. F. (2012). Carbon stocks across a chronosequence of thinned and unmanaged red pine (Pinus resinosa) stands. *Ecological Applications, 22*(4), 1297–1307.

Prescott, C. E., Reid, A., Wu, S. Y., & Nilsson, M. C. (2017). Decomposition rates of surface and buried forest-floor material. *Canadian Journal of Forest Research, 47*(8), 1140–1144.

Pröll, G., Darabant, A., Gratzer, G., & Katzensteiner, K. (2015). Unfavourable microsites, competing vegetation and browsing restrict post-disturbance tree regeneration on extreme sites in the Northern Calcareous Alps. *European Journal of Forest Research, 134*(2), 293–308.

Purakayastha, T. J., Bera, T., Bhaduri, D., Sarkar, B., Mandal, S., Wade, P., Kumari, S., Biswas, S., Menon, M., Pathak, H., & Tsang, D. C. (2019). A review on biochar modulated soil condition improvements and nutrient dynamics concerning crop yields: Pathways to climate change mitigation and global food security. *Chemosphere, 227*, 345–365.

Qi, W., Li, H., Zhang, Q., & Zhang, K. (2019). Forest restoration efforts drive changes in land-use/land-cover and water-related ecosystem services in China's Han River basin. *Ecological Engineering, 126*, 64–73.

Rai, A. K., Ghosh, P. K., Ram, S. N., Singh, S., Kumar, S., Mahanta, S. K., Maity, S. B., Singh, J. P., Dixit, A. K., Tiwari, S. K., & Roy, A. K. (2013). *Carbon sequestration in forage based land use systems* (p. 52). Indian Grassland and Fodder Research Institute.

Ramankutty, N., Evan, A. T., Monfreda, C., & Foley, J. A. (2008). Farming the planet: 1. Geographic distribution of global agricultural lands in the year 2000. *Global Biogeochemical Cycles, 22*(1).

Ramirez, J. I., Jansen, P. A., & Poorter, L. (2018). Effects of wild ungulates on the regeneration, structure and functioning of temperate forests: A semi-quantitative review. *Forest Ecology and Management, 424*, 406–419.

Ratcliffe, S., Wirth, C., Jucker, T., van der Plas, F., Scherer-Lorenzen, M., Verheyen, K., Allan, E., Benavides, R., Bruelheide, H., Ohse, B., & Paquette, A. (2017). Biodiversity and ecosystem functioning relations in European forests depend on environmental context. *Ecology Letters, 20*(11), 1414–1426.

Renforth, P., & Henderson, G. (2017). Assessing ocean alkalinity for carbon sequestration. *Reviews of Geophysics, 55*(3), 636–674.

Rice, C. W. (2000). Soil organic C and N in rangeland soils under elevation CO_2 and land management. In: *Advances in Terrestrial Ecosystem Carbon Inventory, Measurements and Monitoring Conference in Raleigh, North Carolina* (Vol. 5, pp. 15–24).

Richardson, D. M., Carruthers, J., Hui, C., Impson, F. A., Miller, J. T., Robertson, M. P., Rouget, M., Le Roux, J. J., & Wilson, J. R. (2011). Human-mediated introductions of Australian acacias—A global experiment in biogeography. *Diversity and Distributions, 17*(5), 771–787.

Richey, J. E., Melack, J. M., Aufdenkampe, A. K., Ballester, V. M., & Hess, L. L. (2002). Outgassing from Amazonian rivers and wetlands as a large tropical source of atmospheric CO_2. *Nature, 416*, 617.

Schelhaas, M. J., Hengeveld, G., Moriondo, M., Reinds, G. J., Kundzewicz, Z. W., Ter Maat, H., & Bindi, M. (2010). Assessing risk and adaptation options to fires and windstorms in European forestry. *Mitigation and Adaptation Strategies for Global Change, 15*(7), 681–701.

Schmidt, H. P., Kammann, C., Niggli, C., Evangelou, M. W., Mackie, K. A., & Abiven, S. (2014). Biochar and biochar-compost as soil amendments to a vineyard soil: Influences on plant growth, nutrient uptake, plant health and grape quality. *Agriculture, Ecosystems & Environment, 191*, 117–123.

Schuiling, R. D., & Krijgsman, P. (2006). Enhanced weathering: An effective and cheap tool to sequester CO_2. *Climatic Change, 74*(1), 349–354.

Schulz, K., Voigt, K., Beusch, C., Almeida-Cortez, J. S., Kowarik, I., Walz, A., & Cierjacks, A. (2016). Grazing deteriorates the soil carbon stocks of Caatinga forest ecosystems in Brazil. *Forest Ecology and Management, 367*, 62–70.

Schuman, G. E., Booth, D. T., & Waggoner, J. W. (1990). Grazing reclaimed mined land seeded to native grasses in Wyoming. *Journal of Soil and Water Conservation, 45*(6), 653–657.

Schuman, G. E., Janzen, H. H., & Herrick, J. E. (2002). Soil carbon dynamics and potential carbon sequestration by rangelands. *Environmental Pollution, 116*(3), 391–396.

Sekaran, U., & Kumar, S. (2021). Responses of soil carbon storage, compaction, and biological properties under no-till and conventional-till systems. In *Conservation agriculture: A sustainable approach for soil health and food security* (pp. 359–378). Springer.

Sha, C., Mitsch, W. J., Mander, Ü., Lu, J., Batson, J., Zhang, L., & He, W. (2011). Methane emissions from freshwater riverine wetlands. *Ecological Engineering, 37*(1), 16–24.

Sharifi, A., Kalin, L., Hantush, M. M., Isik, S., & Jordan, T. E. (2013). Carbon dynamics and export from flooded wetlands: A modeling approach. *Ecological Modelling, 263*, 196–210.

Sigua, G. C., & Coleman, S. W. (2010). Spatial distribution of soil carbon in pastures with cow-calf operation: Effects of slope aspect and slope position. *Journal of Soils and Sediments, 10*(2), 240–247.

Simola, H., Pitkänen, A., & Turunen, J. (2012). Carbon loss in drained forestry peatlands in Finland, estimated by re-sampling peatlands surveyed in the 1980s. *European Journal of Soil Science, 63*(6), 798–807.

Six, J., Feller, C., Denef, K., Ogle, S., de Moraes Sa, J. C., & Albrecht, A. (2002). Soil organic matter, biota and aggregation in temperate and tropical soils-Effects of no tillage. *Agronomie, 22*(7–8), 755–775.

Smith, P., Martino, D., Cai, Z., Gwary, D., Janzen, H., Kumar, P., McCarl, B., Ogle, S., O'Mara, F., Rice, C., & Scholes, B. (2008). Greenhouse gas mitigation in agriculture. *Philosophical Transactions of the Royal Society B: Biological Sciences, 363*(1492), 789–813.

Soimakallio, S., Saikku, L., Valsta, L., & Pingoud, K. (2016). Climate change mitigation challenge for wood utilization—The case of Finland. *Environmental Science & Technology, 50*(10), 5127–5134.

Steffens, M., Kölbl, A., Totsche, K. U., & Kögel-Knabner, I. (2008). Grazing effects on soil chemical and physical properties in a semiarid steppe of Inner Mongolia (PR China). *Geoderma, 143*(1–2), 63–72.

Steidinger, B. S., Crowther, T. W., Liang, J., Van Nuland, M. E., Werner, G. D., Reich, P. B., Nabuurs, G. J., de-Miguel, S., Zhou, M., Picard, N., & Herault, B. (2019). Climatic controls of decomposition drive the global biogeography of forest-tree symbioses. *Nature, 569*(7756), 404–408.

Strukelj, M., Brais, S., & Paré, D. (2015). Nine-year changes in carbon dynamics following different intensities of harvesting in boreal aspen stands. *European Journal of Forest Research, 134*(5), 737–754.

Sun, X., Wang, W., Razaq, M., & Sun, H. (2019). Effects of stand density on soil organic carbon storage in the top and deep soil layers of Fraxinus mandshurica plantations. *Austrian Journal of Forest Science, 136*(1), 27–44.

Sykes, A. J., Macleod, M., Eory, V., Rees, R. M., Payen, F., Myrgiotis, V., Williams, M., Sohi, S., Hillier, J., Moran, D., & Manning, D. A. (2020). Characterising the biophysical, economic and social impacts of soil carbon sequestration as a greenhouse gas removal technology. *Global Change Biology, 26*(3), 1085–1108.

Syphard, A. D., Brennan, T. J., & Keeley, J. E. (2014). The role of defensible space for residential structure protection during wildfires. *International Journal of Wildland Fire, 23*(8), 1165–1175.

Tanentzap, A. J., & Coomes, D. A. (2012). Carbon storage in terrestrial ecosystems: Do browsing and grazing herbivores matter? *Biological Reviews, 87*(1), 72–94.

Tanwar, S. P., Singh, A. K., & Joshi, N. (2010). Changing environment and sustained crop production: A challenge for agronomy. *Journal of Arid Legumes, 7*(2), 91–100.

Ti, C., Pan, J., Xia, Y., & Yan, X. (2012). A nitrogen budget of mainland China with spatial and temporal variation. *Biogeochemistry, 108*, 381–394.

Tilman, D., Reich, P., Phillips, H., Menton, M., Patel, A., Vos, E., Peterson, D., & Knops, J. (2000). Fire suppression and ecosystem carbon storage. *Ecology, 81*(10), 2680–2685.

Timmons, D. S., Buchholz, T., & Veeneman, C. H. (2016). Forest biomass energy: Assessing atmospheric carbon impacts by discounting future carbon flows. *GCB Bioenergy, 8*(3), 631–643.

Trost, B., Prochnow, A., Drastig, K., Meyer-Aurich, A., Ellmer, F., & Baumecker, M. (2013). Irrigation, soil organic carbon and N_2O emissions. A review. *Agronomy for Sustainable Development, 33*(4), 733–749.

Truax, B., Fortier, J., Gagnon, D., & Lambert, F. (2018). Planting density and site effects on stem dimensions, stand productivity, biomass partitioning, carbon stocks and soil nutrient supply in hybrid poplar plantations. *Forests, 9*(6), 293.

Tu, C., He, T., Lu, X., Luo, Y., & Smith, P. (2018). Extent to which pH and topographic factors control soil organic carbon level in dry farming cropland soils of the mountainous region of Southwest China. *CATENA, 163*, 204–209.

Turner, J., Lambert, M., & Turner, S. (2017). Long term carbon and nutrient dynamics within two small radiata pine catchments. *Forest Ecology and Management, 389*, 1–14.

UNFCCC UN. (2009). *Kyoto protocol reference manual on accounting of emissions and assigned amount*. eSocialSciences.

Valdés-Correcher, E., Sitters, J., Wassen, M., Brion, N., & Olde, V. H. (2019). Herbivore dung quality affects plant community diversity. *Scientific Reports, 9*(1), 1–6.

Van Miegroet, H., & Olsson, M. (2011). Ecosystem disturbance and soil organic carbon—A review. In *Soil carbon in sensitive European ecosystems: From science to land management* (pp. 85–117).

Van Zwieten, L., Singh, B. P., Tavakkoli, E., Joseph, S., Macdonald, L. M., Rose, T. J., Rose, M. T., Kimber, S. W., Morris, S., Cozzolino, D., & Araujo, J. R. (2017). Biochar built soil carbon over a decade by stabilizing rhizodeposits. *Nature Climate Change, 7*(5), 371–376.

Vanguelova, E. I., Crow, P., Benham, S., Pitman, R., Forster, J., Eaton, E. L., & Morison, J. I. L. (2019). Impact of Sitka spruce (Picea sitchensis (Bong.) Carr.) afforestation on the carbon stocks of peaty gley soils—A chronosequence study in the north of England. *Forestry: An International Journal of Forest Research, 92*(3), 242–252.

Vanguelova, E. I., Pitman, R., Benham, S., Perks, M., & Morison, J. I. (2017). Impact of tree stump harvesting on soil carbon and nutrients and second rotation tree growth in Mid-Wales, UK. *Open Journal of Forestry, 7*(01), 58.

Veraverbeke, S., Rogers, B. M., Goulden, M. L., Jandt, R. R., Miller, C. E., Wiggins, E. B., & Randerson, J. T. (2017). Lightning as a major driver of recent large fire years in North American boreal forests. *Nature Climate Change, 7*(7), 529–534.

Vesterdal, L., Clarke, N., Sigurdsson, B. D., & Gundersen, P. (2013). Do tree species influence soil carbon stocks in temperate and boreal forests? *Forest Ecology and Management, 309*, 4–18.

Vesterdal, L., Dalsgaard, M., Felby, C., Raulund-Rasmussen, K., & Jørgensen, B. B. (1995). Effects of thinning and soil properties on accumulation of carbon, nitrogen and phosphorus in the forest floor of Norway spruce stands. *Forest Ecology and Management, 77*(1–3), 1–10.

Vesterdal, L., Ritter, E., & Gundersen, P. (2002). Change in soil organic carbon following afforestation of former arable land. *Forest Ecology and Management, 169*(1–2), 137–147.

Villa, J. A., & Bernal, B. (2018). Carbon sequestration in wetlands, from science to practice: An overview of the biogeochemical process, measurement methods, and policy framework. *Ecological Engineering, 114*, 115–128.

Wan, X., Xiao, L., Vadeboncoeur, M. A., Johnson, C. E., & Huang, Z. (2018). Response of mineral soil carbon storage to harvest residue retention depends on soil texture: A meta-analysis. *Forest Ecology and Management, 408*, 9–15.

Wang, H., Liu, S., Wang, J., Shi, Z., Lu, L., Zeng, J., Ming, A., Tang, J., & Yu, H. (2013). Effects of tree species mixture on soil organic carbon stocks and greenhouse gas fluxes in subtropical plantations in China. *Forest Ecology and Management, 300*, 4–13.

Wei, X., Shao, M., Gale, W., & Li, L. (2014). Global pattern of soil carbon losses due to the conversion of forests to agricultural land. *Scientific Reports, 4*(1), 1–6.

Were, D., Kansiime, F., Fetahi, T., Cooper, A., & Jjuuko, C. (2019). Carbon sequestration by wetlands: A critical review of enhancement measures for climate change mitigation. *Earth Systems and Environment, 3*, 327–340.

Whitehead, D., Schipper, L. A., Pronger, J., Moinet, G. Y., Mudge, P. L., Pereira, R. C., Kirschbaum, M. U., McNally, S. R., Beare, M. H., & Camps-Arbestain, M. (2018). Management practices to reduce losses or increase soil carbon stocks in temperate grazed grasslands: New Zealand as a case study. *Agriculture, Ecosystems & Environment, 265*, 432–443.

Wiesmeier, M., Mayer, S., Burmeister, J., Hübner, R., & Kögel-Knabner, I. (2020). Feasibility of the 4 per 1000 initiative in Bavaria: A reality check of agricultural soil management and carbon sequestration scenarios. *Geoderma, 369*, 114333.

Wiesmeier, M., Prietzel, J., Barthold, F., Spörlein, P., Geuß, U., Hangen, E., Reischl, A., Schilling, B., von Lützow, M., & Kögel-Knabner, I. (2013). Storage and drivers of organic carbon in forest soils of southeast Germany (Bavaria)—Implications for carbon sequestration. *Forest Ecology and Management, 295*, 162–172.

Xu, L., Wen, D., Zhu, J., & He, N. (2017). Regional variation in carbon sequestration potential of forest ecosystems in China. *Chinese Geographical Science, 27*(3), 337–350.

Yadav, M. R., Parihar, C. M., Kumar, R., Yadav, R. K., Jat, S. L., Singh, A. K., Ram, H., Meena, R. K., Singh, M., Meena, V. K., & Yadav, N. (2017). Conservation agriculture and soil quality—An overview. *International Journal of Current Microbiology and Applied Sciences, 6*, 1–28.

Yayneshet, T., Eik, L. O., & Moe, S. R. (2009). Seasonal variations in the chemical composition and dry matter degradability of exclosure forages in the semi-arid region of northern Ethiopia. *Animal Feed Science and Technology, 148*(1), 12–33.

Zewdie, M. (2008). *Temporal changes of biomass production, soil properties and ground flora in Eucalyptus globulus plantations in the central highlands of Ethiopia* (Vol. 2008, No. 2008, p. 18).

Zhang, M., Huang, X., Chuai, X., Yang, H., Lai, L., & Tan, J. (2015). Impact of land use type conversion on carbon storage in terrestrial ecosystems of China: A spatial-temporal perspective. *Scientific Reports, 5*(1), 1–3.

Zhang, X., Guan, D., Li, W., Sun, D., Jin, C., Yuan, F., Wang, A., & Wu, J. (2018). The effects of forest thinning on soil carbon stocks and dynamics: A meta-analysis. *Forest Ecology and Management, 429*, 36–43.

Zhao, F., Kang, D., Han, X., Yang, G., Feng, Y., & Ren, G. (2015). Soil stoichiometry and carbon storage in long-term afforestation soil affected by understory vegetation diversity. *Ecological Engineering, 74*, 415–422.

Chapter 4
Soil Health and Climate Change

Tapan Jyoti Purakayastha, Ankita Trivedi, Sunanda Biswas, and Dhiraj Kumar

Abstract Soil's capacity to perform diversified functions, such as sustainability in the production of plants and animals and at the same time preserving and/or improving the environment, is referred to as soil health. Soil organic carbon, biodiversity, infiltration of water, susceptibility to drought, drainage, and soil resistance and resilience are key indices of soil health. One health idea has recently received a lot of attention by emphasizing the linkage between human and animal health with soil health. Numerous soil functions and planned land use are key factors in managing soil health. As part of a framework for soil management assessment, productivity, and environmental protection goals are mainly employed to assess soil health, improve water quality, and mitigate and adapt to changing climate. Anticipated climate change is likely to decrease organic matter content in soil, weaken soil structure, and make the soil susceptible to erosion and other degrading processes, thereby negatively impacting soil health. As a result, in the projected scenario of climate change, increasing carbon sequestration potential through the use of various resource conservation technologies is a practical and affordable method of reducing human-induced greenhouse gas emissions and simultaneously enhancing environmental quality and promoting food security.

Keywords Carbon sequestration · Climate change · Conservation agriculture · Soil Health · Soil health indicators

T. J. Purakayastha (✉) · A. Trivedi · S. Biswas
ICAR-Indian Agricultural Research Institute, New Delhi, India
e-mail: tpurakayastha@gmail.com

D. Kumar
ICAR-Indian Institute of Soil Science, Bhopal, India

H. Pathak et al. (eds.), *Climate Change Impacts on Soil-Plant-Atmosphere Continuum*, Advances in Global Change Research 78, https://doi.org/10.1007/978-981-99-7935-6_4

4.1 Introduction

'The ability of soil to continue functioning as an essential living system within ecological and land-use boundaries, sustain biological productivity, maintain or improve the quality of air and water, and support plant, animal, and human health' is known as soil health (Doran & Zeiss, 2000). However, "soil quality" as well as "soil health" have been mostly used synonymously or interchangeably. But a second school of thought denies the synonymous use of both terms. According to Gregorich and Carter (1997), soil health only includes dynamic quality of soil whereas soil quality includes both dynamic as well as inherent qualities of soil. Therefore, depending upon the management practices, soil has good health instead of having poor inherent soil quality. Similar to human health, soil health is determined by keeping important soil qualities at their optimum levels (Lal, 2011; Magdoff, 2001). Assessment of soil health is a tedious task; it can be measured indirectly from soil properties as a direct measurement is not possible. The important properties representing soil's physicochemical and biological properties might be sensitive to change due to management practices that regulate soil health. A reliable soil health index is necessary to evaluate data from various soil measures and determine if a given management and land use strategy have the desired impact on system productivity, environmental quality as well as soil health. A proper assessment of soil health is essential for taking necessary rehabilitation measures. The comprehensive assessment is extremely essential for identification of sensitive soil health indicators as well as soil health assessment across various soil types, climates, cropping systems, and management. The important properties include optimum soil organic matter (SOM), soil texture, structure, optimum aeration in soil, water and nutrient retention capacity, soil pH, presence of optimum soil biodiversity, good resistance, resilience against perturbation, less susceptibility to erosion and less pest and pathogen in soil. Soil health integrates soil physicochemical and biological properties and is evaluated by certain indicators. It is to be noted that all those properties are interrelated and their interaction is governed by a more complex phenomenon. Therefore, soil health is a new concept where all the physical, chemical, and/or biological properties important for sustainable production and environmental quality are integrated into soil health index. Soil health components and representative properties of those components are depicted in Fig. 4.1.

Human health analogy is applied to soil health (Magdoff, 2001). Key soil properties (indicators) and processes need to be maintained at their optimum levels to determine soil health. Favourable structure and texture of soil or tilth, efficient internal drainage, ideal nutrient, and water holding capacities, and soil reaction are key soil qualities that are crucial to sustain good soil health. Soil organic carbon is considered as a global soil health indicator thus maintenance of its optimal level is essential to drive almost all the soil functions and the key soil properties that describe the particular soil function. Maintenance of soil health is crucial for agricultural sustainability, which drives the productivity of agro-ecosystem and prevents soil degradation. However, maintenance of soil health has increasingly become a herculean

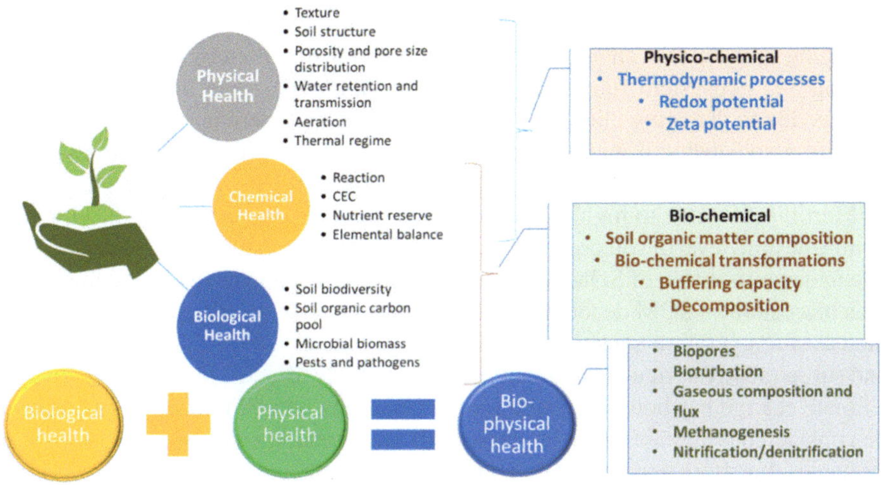

Fig. 4.1 Components of soil health

task because of deteriorating soil health due to desertification, deforestation, non-judicious application of agrochemicals, salinization, dumping of pollutants directly on soil, and climate change. Among these, climate change is a factor that affects soil health either directly and/or indirectly to a worrisome level. Every aspect of human life has been profoundly altered by industrialization, which is associated with population explosion and urbanization, which have triggered global climate changes. These events have also marked a significant turning point in ecology and the relationship between man and environment (IPCC, 2007). Rise of temperature is more on land than over the oceans, associated with increased frequency of high-temperature events on land, global average temperature is projected to rise by 0.7 from 0.3 °C from 2016 to 2035 as compared to the period between 1986 and 2005 (Kirtman et al., 2013). The Intergovernmental Panel on Climate Change (IPCC) has projected a rise in global temperatures by 1.6–6.4 °C, an increase in atmospheric CO_2 concentrations up to 550 ppm, and an alteration in precipitation at least by 20%. According to Meehl et al. (2007), cultivated lands will experience larger rises in global warming than uncultivated lands during the next two to three decades, at a rate of 0.2 °C every 10 years. Consequently, agricultural lands which are under irrigated and non-irrigated conditions around the world will suffer a greater loss of soil moisture (FAO, 2013). Spatio-temporal fluctuations in rainfall patterns, rising temperature, and elevated levels of greenhouse gases (GHGs) associated with climate change will have a definite adverse impact on various soil processes and parameters and might pose a serious threat to agriculture. There are varying scenarios that may arise due to changing climate at different locations which will affect the soil properties differently. Different regions of the world will observe either increased or decreased rainfall along with higher temperatures due to a variable and changing climate. Leaching of bases from soil will occur when precipitation increases. Higher biomass production caused by increased temperature and elevated CO_2 will also lead to acidification of soil (Rengel, 2011). Changing climate also alters soil microbial composition thereby affecting nutrient cycling and ecosystem functioning (Dubey et al., 2019). The migration of plant species due to changing climate is well established and there is evidence of a shift of plant-associated soil microbial communities (Pearson, 2006). Increasing global temperature is associated with rapid decomposition of SOM which is very central to soil fertility. But due to high temperature and rapid decomposition of SOM, maintaining desired level of soil organic carbon is a challenging issue. The CO_2 is liberated due to decomposition of SOM by microbes. Climate change raises soil temperature, which accelerates SOM decomposition (Singh et al., 2017), influences priming, and increases temperature sensitivity of SOM (Fang et al., 2017). The quality of soil and subsequently agricultural output gets significantly affected by changing climate. Climate change has both direct and indirect bearings on soil functions, which must be considered while assessing its impact. To implement the necessary management strategies to restore and/or sustain soil health, a critical understanding of direct and indirect effects of climate change on key soil attributes is needed.

4.2 Soil Health: Concept, Definition, Progress Made, Soil Health Indicators, Sensitive Indicators

In order to produce food and fibre as well as to maintain environmental sustainability, soil health remains one of the most essential components. Soil is one of the dynamic and living natural system. The quality of the soil can have a noteworthy impact on plant, animal, and human health apart from its role to maintain productivity and sustainability of the land (Acton & Gregorich, 1995). The basic idea of soil health first came into light in the early 2000s, and now it is associated with the idea of "One Health", which links the well-being of people, animals, and the environment (Keith et al., 2016; Lehmann et al., 2020). Management practices of soil significantly influence the ecosystem and the inhabiting organisms. Soil health is difficult to assess directly, the inference must be derived from changes in soil characteristics because of the adoption of different management techniques. To assess sustainability of management approaches and land use, soil health indicators are also employed (Shukla et al., 2006). Different scientists around the globe have opined varying perspectives on soil quality. Soil quality is an idea that helps in quantification of integrated physical, chemical, and biological responses of soil towards crop management practices. The concept of ecological sustainability becomes inextricably linked to soil quality. From production perspective, concept of soil quality has been broadened (Larson & Pierce, 1991) to incorporate ecological factors and thus soil quality is defined as "the capacity of a soil to function within its ecosystem boundaries and interact positively with the environment external to that ecosystem". Soil quality and its environmental concerns were highlighted by Acton and Gregorich (1995), who defined it as "the soil's capacity or fitness to support crop growth without resulting in soil degradation or otherwise, harming the environment". The Soil Science Society of America developed a more elaborated definition of soil quality as "the capacity of a specific kind of soil to function, within natural or managed ecosystem boundaries, to sustain plant and animal productivity, maintain environmental quality and promote plant, animal and human health" (Allan et al., 1995). Considering a strong reliance on the biological component of the soil ecosystem, soil health is the outcome of ongoing conservation and degradation processes. It has an impact on plant and environmental health, food quality and safety, and many more. The terms soil quality and soil health are used interchangeably to mean "the soil's fitness to support crop growth without becoming degraded or otherwise harming the environment" (Acton & Gregorich, 1995). Soil health often refers to the overall conditions of soil and basically to its vital functions (Gil-Sotres et al., 2005). In a national context, soil health could serve as a strong foundation for national policy to protect the environment. The concept of soil health has been introduced in the late 1980s. Soil health can be defined as "the continued capacity of soil to function as a living system, within ecosystem and land use boundaries, to sustain biological productivity, maintain the quality of air and water environment and promote plant, animal, and human health" (Doran et al., 1996). The soil health concept integrates soil biological, chemical, and physical attributes to assess a soil's capacity to function (Karlen et al., 1997). Since it is very difficult to measure soil health directly, indicators and attributes are utilized instead. These attributes can be described as quantifiable soil characteristics that affect the soil's ability to perform a certain function (Acton & Padbury, 1993). The following criteria should be met by indicator: (i) It should encompass ecosystem process (ii) It should respond to changing climate and management strategies (iii) It should be easy to measure and reproduce (iv) It should be a part of an existing soil database (v) It should be easy to access and should have applicability in field conditions; and (vi) It should integrate physical chemical, and biological properties and processes of soil (Doran & Perkin, 1994). A minimum dataset (MDS) was proposed by Larson and Pierce (1991) based on these characteristics, and Doran and Parkin (1994) added other biological characteristics to it. To preserve soil health, physicochemical characteristics of the soil are very crucial. Some of the physical indicators are bulk density, soil aggregate stability, water retention capacity, and hydraulic conductivity. The chemical factors influencing soil health include soil reaction (pH), organic carbon content, cation exchange capacity, clay content, and soil nutrient status. Microbial biomass C and N, mineralizable C and N, dehydrogenase, acid and alkaline phosphatase, urease, aryl sulphatase activities, soil microbial diversity, etc. are a few soil biological indicators that primarily govern soil health (Doran & Parkin, 1994).

According to USDA, soil health is the continued capacity of soil to operate as a vital living ecosystem that supports all life forms on the planet. Nevertheless, the idea of soil health is still in evolving phase. However, there is some confusion between "soil health" and "soil quality" terms. Both phrases are distinct from one another because soil quality refers to soil's effect on health of plants, animals, and entire ecosystems. The scope of soil quality typically focuses on ecosystem services with regard to humans, but the scope of soil health extends beyond human health to even include sustainability goals that encompass planetary health (Lehman et al., 2020).

There has been a history of the emergence of soil health concept. Its history ways back to the term "soil quality" which can be identified as the narrowest form of soil health which just relates to its role in crop production. It focuses on production of crops for meeting human demand for food, fiber, and fuel. The attempt to define soil's diversified role continued and the soil quality concept came into place followed by the concept of soil health. Now progress has already been made on the concept of soil health as well and a new term was introduced in 2012 namely "Soil security". It is the most recent and broadest term which relates to the need for access to soil ecosystem services to be on the same level as other human rights (Koch et al., 2013). Soil security is a broader term than soil fertility and soil quality where the soil is a common good like air or water than only private property and therefore, it is used in a policy context and legal aspect of management. More like water and air, the term 'soil security' advocates fruitful conversation about soil as a common good, rather than considering it as private property. The progress made in emergence of soil health as a concept has been modified by Lehman et al. (2020) and given in Fig. 4.2.

The following are the four key aspects for choosing reliable soil health indicators and methods (Stott, 2019):

I. Effectiveness of soil health indicators.

   A. Sensitive to management: Rapid response of indicator to changing agricultural and land management practices.
   B. Short-term Sensitivity: When significant management changes are implemented, the indicator can typically identify changes in sub-humid to humid

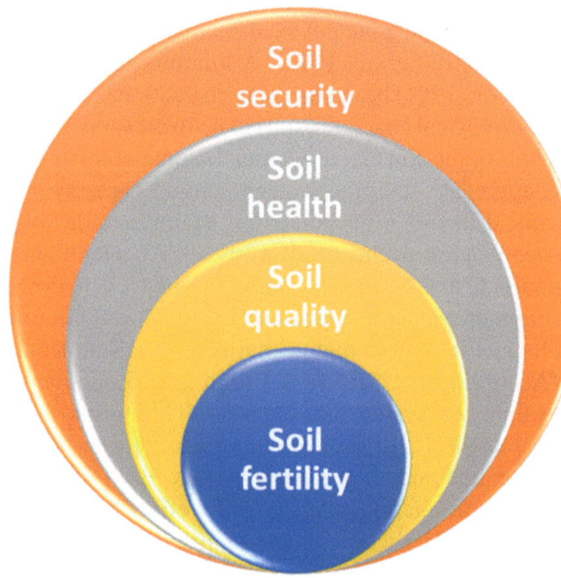

Fig. 4.2 Emergence and progress made on the concept of soil health

regions within 1 to 3 years. In conditions ranging from semiarid to arid, or with little management adjustments, changes are likely to take longer.

C. Interpretable
   1. The indicator (by itself) describes particular soil processes or conditions that are physical, chemical, or biological and have an impact on agricultural output as well as environmental outcomes.
   2. Interpretation in conjunction with additional tests: If a single test is not enough to represent a particular process/condition, it should be understandable by measuring it with one or more test(s).

D. Useful: The indicator offers information that can be used to evaluate soil health status of regions and to address particular resource concerns.

II. Production Readiness: Readiness for use in commercial production laboratories in terms of:

A. Ease of use:
   1. Sampling (for field conservation planners, consultants, other agricultural service providers, and producers).
   2. Sample submission.
   3. Sample preparation for laboratories.
   4. Laboratory measurements.

B. Cost-effectiveness of the sample:
   1. Labor and material costs.
   2. The price of specialized equipment
   3. Space and time needs for the lab, as well as overhead.

III. Repeatability of measurements. The method's degree of precision falls within allowable boundaries.

IV. A decision-making tool for agricultural management.

A. Values that are "directionally understood" (more is better/less is better/optimum).
B. Management practices that improve the measure are known.
C. Regional potential ranges to define relative poor/good functioning are known.
D. Outcome-based (yield, resilience, risk, environmental) thresholds are known.

Other details of recommended indicators and the methods for assessment of soil health have been given by Stott (2019). The classification of soil health indicators does not have a clear-cut boundary, rather the indicators are interrelated.

Physical soil health indicators: Simple, quick, and inexpensive techniques are typically used as physical soil health indicators. Important physical soil health indicators in relation to changing climate include soil aggregate stability, root depth, infiltration of water, bulk density, and soil surface cover.

Chemical soil health indicators: Soil's ability to supply plant nutrients and/or retain chemicals that might harm plant growth or the environment can be categorized as chemical soil health indicators. These include the cation exchange capacity, pH, electrical conductivity, and nutrient availability of the soil.

Biological soil health indicators: Biological monitoring fundamentally involves measuring how living things react to changes in their environment (Pandolfini et al., 1997). Microbial biomass, basal and/or substrate-induced respiration, mineralizable nitrogen, enzyme activity, microflora and fauna richness, soil biodiversity, plant biodiversity, and plant growth are just a few of the many soil biological indicators that have been evaluated. However, all those indicators have their own set of vices and virtues. Soil enzymatic activity has been proposed to be a suitable soil health indicator because of its prominent role in nutrient cycling of soil, degradation of organic pollutants, immobilization of heavy metals, and the positive or negative effect on plant growth (Alkorta et al., 2003). Besides, the procedure for determining soil enzyme activity is comparatively inexpensive, rapid, and easy which can be done in routine analysis. Management practices prove sensitive to soil enzymatic activities and those changes are detected early, much earlier than the changes observed in soil organic matter. Thus, these can be used as a potential indicator to detect soil health through determination of enzyme activity comes with its own share of limitations. Those enzymatic indicators can be used in integration with other physio-chemical and biological indicators of soil health, but not alone.

4.3 Climate Change: Brief History, Projections

The IPCC defines climate change as "any change in climate over time, whether as a result of natural variability or human activity". Contrary to the Framework Convention on Climate Change (FCCC), which defines climate change as a "change in climate that is added to natural climate variability observed over comparable time periods and that is attributed either directly or indirectly to human activity that alters the composition of the global atmosphere". According to IPCC10, from pre-industrial times to 2005, the amount of CO_2, CH_4, and N_2O emissions into the atmosphere has increased globally (Fig. 4.3). Last ten years have witnessed larger annual CO_2 concentrations (1995–2005, 1.9 ppm year^{-1}) than it has been since the beginning of continuous direct atmospheric measurements (1960–2005, average 1.4 ppm year^{-1}). The sea level rises 1.8 mm per year due to dissolution of glaciers and ice caps. Over the majority of geographical areas, the frequency of heavy precipitation occurrences has increased. The climatic systems' energy balance is affected by variations in solar radiation, the quantity of GHGs and aerosols present in the atmosphere, and the characteristics of the land surface. Continental drift, volcanoes, ocean currents, the earth's tilt, comets, and meteorites are the prominent natural factors that significantly influence climate change. Water vapour, CO_2, CH_4, and N_2O concentrations in the atmosphere, all of which are GHGs that contribute to global warming, increased significantly as a result of anthropogenic activities such as industrialization, deforestation, and pollution. Addition of CO_2 into the atmosphere as a result of anthropogenic activities supersedes ocean and plant absorption rate of

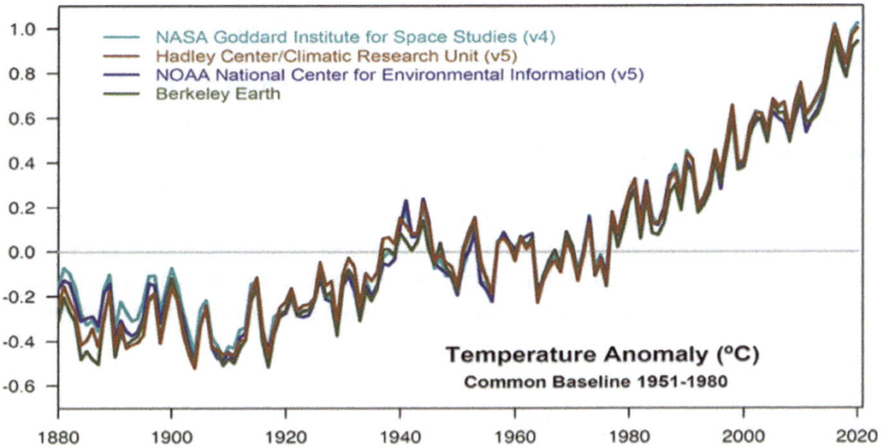

Fig. 4.3 Global warming over the decades (*Source* NASA)

CO_2. Since GHGs have a very long lifetime in the atmosphere, even if all of their emissions were stopped today, global warming would not end right away. The earth radiates long wave radiations that get absorbed by the GHGs which raises global mean temperature to 15 °C; otherwise, in absence of an atmosphere, the planet would have experienced an estimated temperature of -18 °C. This phenomenon is commonly referred to as the "Greenhouse" effect. Therefore, the greenhouse effect is called a necessary evil. Water vapour, by far is considered as the most important GHG, but its contribution is not considered as the concentration of this in the atmosphere is beyond our control. However, carbon dioxide makes up a significant portion of the atmosphere's greenhouse gas emissions, whereas methane, nitrous oxide, ozone, and chlorofluorocarbons (CFCs) make up a smaller portion.. The global warming potential of CFCs, nitrous oxide, and methane are estimated to be 23,900–140, 310, and 21 times of carbon dioxide in a time horizon of 100 years (UNFCC). Higher concentrations of these gases since 1860 might have increased the average global surface temperature by about 0.5 °C, and it is anticipated that increased concentrations of those gases could result in a warming of 1.5 °C over the next 40 years.

4.4 Consequences of Changing Climate on Soil Health

Across the globe, changing climate has impacted our daily lives in so many ways, either subtly or severely. Some key soil properties are highly sensitive to climate change and climate variability (Thornton et al., 2014). Changing climate has raised global concern because of its adverse effect on various soil processes influencing soil health. Climate variability brings heavy impacts on soil and agricultural production by accelerating the decomposition of soil organic matter (Fig. 4.4). As SOM is key to agricultural productivity and therefore, loss of it in response to increased atmospheric temperature will accelerate soil erosion, decrease soil fertility and other ecosystem services and ultimately deteriorate soil health which is directly linked to animal and human health. The capacity to change the current management and the characteristics of the soils both influence the ability of the agricultural sector to respond and react. When assessing the implications of projected climate change on soil functions, both direct and indirect impacts on soil properties and processes must be taken into account. Soils come under stress as a consequence of changing climate; therefore it becomes imperative to understand climate and management-driven changes on soil physical, chemical, and biological health and its functions.

4.4.1 Soil Physical Health

Soil physical health gets affected by climate and management-driven factors and the major determinants of soil physical health include soil aggregate stability, bulk density, SOM, soil aeration, and water infiltration through the soil profile (Luo et al., 2017). "Soil structural stability" is considered one of the most essential components of healthy soil because it influences flow and management of water, oxygen, and nutrients (Ussiri et al., 2009). Anthropogenic activities directly and climate change indirectly influence soil structural stability. The increased risk associated with soil salinization, receding water table, loss of soil biodiversity, alteration in soil C and N dynamics as well as nutrient storage are anticipated to result from elevated temperature and CO_2 concentrations, uneven rainfall distribution associated with high-intensity rainfall events and their interactions because of changing climate (Benbi & Kaur, 2009). By affecting water flow through the soil, root penetration, and water congestion, soil physical processes have impacted soil health. The following are significant soil physical characteristics impacting soil health due to climate change:

4.4.1.1 Soil Texture

Proportionate distribution of sand, silt, and clay in soil determine its texture, which directly affects climate changes. Arid, semi-arid, sub-humid, and humid climates are the four possible climatic scenarios that govern textural differentiation in the soil profile (Brinkman & Brammer, 1990).

4.4.1.2 Soil Structure and Aggregate Stability

The soil structure which is an arrangement and orientation of primary particles controls the distribution of soil air and water. Soil aggregate stability provides resistance to heavy rainfall. Soil structure along with some chemical and biological characteristics and management techniques influence cultivation practices (Dalal &

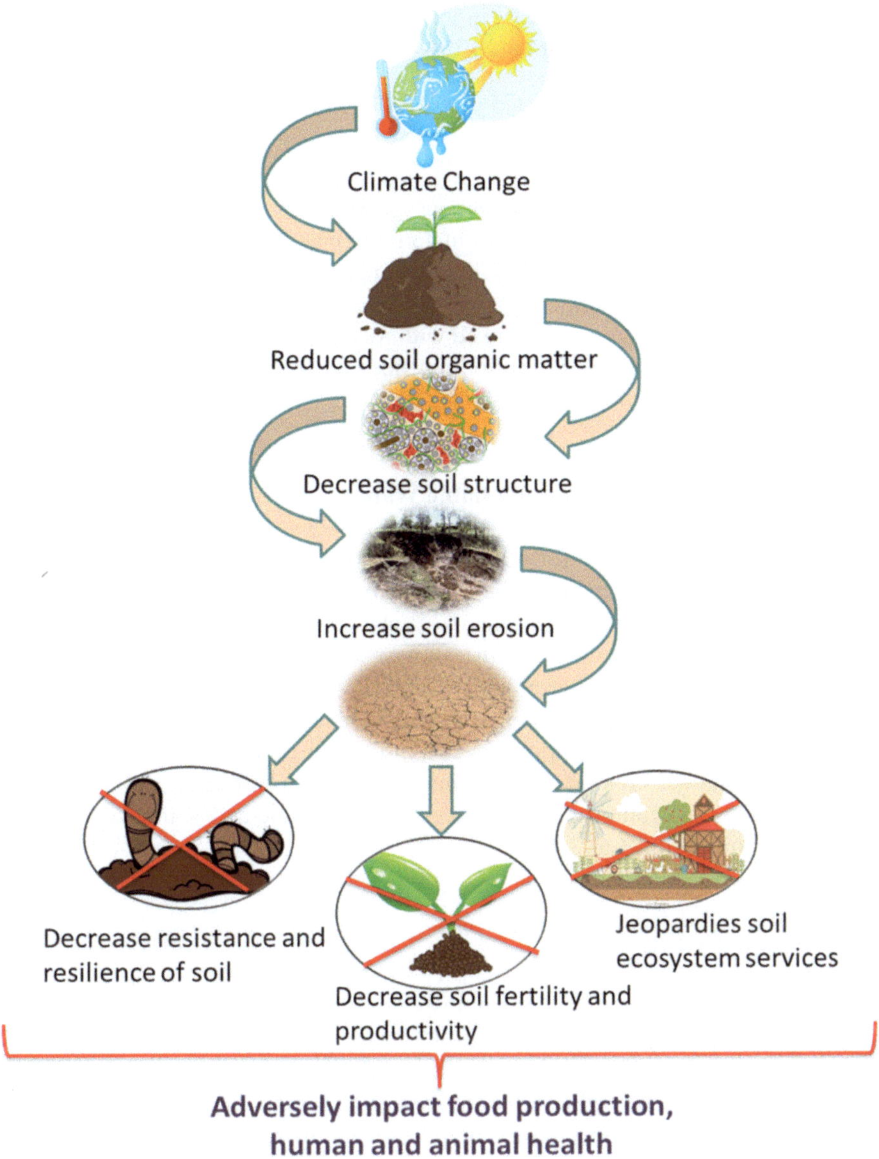

Fig. 4.4 Climate change and its impact on soil health

Moloney, 2000). It is regarded as one of the key indicators of soil health due to its significant contribution to the preservation and improvement of important ecosystem functions in soil, such as accumulation of organic carbon, the ability of the soil to move and store water, microbial and root activity. Future climate change is likely to cause a decrease in SOM that could lead to a decrease in soil aggregate stability, reduction in infiltration rates, increase in soil compaction, and increased susceptibility to runoff and erosion (Karmakar et al., 2016).

4.4.1.3 Porosity

The optimum air–water proportion for root growth is provided by porosity, which is a measure of total void spaces in a material expressed as a fraction of its total volume (Reynolds et al., 2002). Physical features of the soil such as bulk density, microporosity, and pore volume functions all have a significant impact on pore characteristics. Soil porosity is closely linked with root development and its distribution. Besides, soil enzyme activities are also influenced by void spaces present in soil that in turn influence soil biological health.

4.4.1.4 Infiltration and Plant Available Water

The amount of water needed for plant growth and soil processes depends on a few of the soil characteristics that include soil texture, field capacity, porosity, and lower limit of plant available water (Reynolds et al., 2002). To evaluate management-induced changes, integrative soil health tests have taken plant available water capacity into account. Including cover crop in the cropping system, adoption of conservation tillage strategies and incorporation of organic matter helps to improve and/or maintain available water in soil and improve water infiltration rate that can reduce severe rainfall, drought, and erosion events. Thus, increased availability and distribution of water in soil can promptly respond to changing climate and will have a positive response on soil health (Lal, 1995).

4.4.1.5 Bulk Density

It represents the compactness of soil in response to management practices. Usually, a negative correlation exists between bulk density and soil organic carbon (SOC) content. According to Davidson and Janssens (2006), the loss of SOC from accelerated decomposition brought on by the increasing temperature may increase bulk density and make soil compact. Soil compaction and increased soil bulk density are also related to land use planning and climate change stresses from variable and high-intensity rainfall and drought events (Birkas et al., 2009). Compact soil restricts the rooting depth of plants and subsequently, availability of water, subsoil salinity, and SOC content get affected by changes in rooting depth.

4.4.1.6 Soil Surface Cover

Improving water and nutrient holding capacity, C and N fixation, and supporting native seed germination are some of the crucial ecological functions that soil surface cover provides (Box & Bruce, 1996). Under the scenario of climate change, soil health is described by structural features of the soil, such as the development of surface crusts and seals, which have an impact on soil activities like water infiltration, surface runoff, and water evaporation from the soil.

4.4.1.7 Soil Temperature

The evaporation process, heat conduction through the soil profile, incoming and outgoing solar radiation at/from the soil's surface, conduction of heat, and convective movement of air and water affect soil temperature (Karmakar et al., 2016). Soil temperature in association with soil moisture regulates the majority of soil processes. Soil temperature increases the rate of decomposition of SOM, microbial activity, and chemical weathering. Under climate change conditions, soil temperature changes the type of vegetation grown.

4.4.2 Soil Chemical Health

4.4.2.1 Soil pH

Parent material, weathering time, vegetation, and climate influence soil pH. Besides, soil pH being sensitive to agricultural management practices is considered as an important indicator of soil health. Changes in pH occur in most of the soil because of an increase in temperature, elevated CO_2 level, variable precipitation, and atmospheric N deposition because of changing climate. The status of SOM, nutrient cycling, C cycling, plant available water, and system productivity change because of climate change drivers, which ultimately impact soil pH (Reth et al., 2005).

4.4.2.2 Electrical Conductivity

In soil, salt concentration is measured by soil electrical conductivity (EC) which represents soil salinity, nutrient cycling, crop performance, and soil enzymatic activity. It can serve as a substitute for measuring soil structural decline along with pH, especially in sodic soils (Arnold et al., 2005). As a chemical indicator, electrical conductivity has been employed to assess soil biological quality in response to crop management techniques (Gil et al., 2009). An increase in temperature and decrease in rainfall because of changing climate causes an increase in electrical conductivity.

4.4.2.3 Sorption and Cation Exchange Capacity

For retention of essential plant nutrients like Ca^{2+}, Mg^{2+}, K^{+}, and immobilization of potentially toxic cations like Al^{3+} and Mn^{3+}, sorption and cation exchange capacity (CEC) play a vital role. The CEC is an important chemical indicator of soil health as it influences the capacity of soil to absorb nutrients, pesticides, and other chemical (Ross et al., 2008). Since SOM is accountable for the CEC of low-activity clay and coarse-textured soils, accelerated decomposition of SOM because of elevated temperature causes the loss of CEC of these soils (Davidson & Janssens, 2006). Alkalinity is transported from soil to the waterways as high rainfall causes leaching loss of bases.

4.4.2.4 Plant Available Nutrients

The soil extractable nutrients contribute to plant growth and development. Those nutrients also help in identification of the critical or threshold values for environmental hazard assessment (Dalal & Moloney, 2000). The cycling of soil organic carbon is strongly linked with soil N cycling (Weil & Magdoff, 2004). As a result, climate change drivers including elevated temperatures, variable precipitation, and atmospheric N deposition are likely to affect the cycling of N as well as possible to other plant-available nutrients like phosphorus and sulfur.

4.4.3 *Soil Biological Health*

Under changing climate scenarios, soil microorganisms act as an important indicator of soil health assessment as soil biota is adaptive to climate change. Important biological indicators such as SOM and its components, soil respiration, soil C, and microbial biomass of soil are more important.

4.4.3.1 Soil Organic Matter

The SOM is the most heterogeneous and complex system having varying properties, functions, and turnover rates (Weil & Magdoff, 2004). It provides essential nutrients for plant growth and development, serves as a source and sink for carbon and nitrogen, and also controls the cycling of phosphorus and sulphur. It can combine with both organic molecules and multivalent ions to form complexes. SOM serves both as a source and a sink for nutrients, and its decline causes the loss of soil structure, a decrease in water holding capacity of soil, and an increase in bulk density. According to Haynes (2008), it also affects soil aggregate stability, water, and nutrient retention as well as hydraulic conductivity.

4.4.3.2 Land Use and Management Practices

For the mitigation of global warming, adoption of the management practices concerning land use that builds up SOM, is quite necessary. The SOM can play a significant role in reducing the effects of flooding at the time of high rainfall events by boosting water storage and through water storage in the event of droughts. SOM also makes the soil resilient.

4.4.3.3 Light Fraction and Macro Organic Matter

Mineral-free particulate plant and animal residues make up the majority of the light fraction and macro organic components of SOM, which act as a labile nutrient reservoir and quickly decomposable substrate for soil microorganisms (Wagai et al., 2009). Those fractions are very sensitive to management practices under the scenario of changing climate (Gregorich et al., 1994).

4.4.3.4 Soil Carbon and C:N Ratio

Under climate change, an increase in temperature and erratic rainfall events lead to microbial activity decline which results in lower biomass accumulation and depletion of soil C thereby decreasing C:N ratio (Lal, 2004). A rise in atmospheric CO_2 concentration increases plant water use efficiency (WUE). Under conditions of increased water shortage, a decomposition rate exceeds net primary production. This process creates a dry environment that is ideal for reducing organic carbon. Adoption of conservation tillage practices, crop residue management, green manuring, and intercropping are part of the management strategies that enhances carbon storage in soil.

4.4.3.5 Potentially Mineralizable C and N

During the process of nutrient cycling, mineralizable organic matter serves as a link between plants and heterotrophic microorganisms (Gregorich et al., 1994). However, because climate change impacts nutrient dynamics within specified growing seasons, the mineralizable organic matter might be relevant to evaluate soil health under these scenarios.

4.4.3.6 Soil Respiration

Due to its positive correlation with SOM concentration, soil respiration is utilized as a biological measure of soil health. Because of its sensitivity to temperature soil respiration is a critical link between climate change and global C cycle (Wixon & Balser, 2009).

4.4.3.7 Soil Microbial Biomass

Microbial biomass carbon is the only living pool of carbon and as it is the most labile pool thus acts as a very sensitive indicator to soil processes especially nutrients and energy dynamics (Saha & Mandal, 2009). However, soil microbial biomass has been demonstrated, which is comparable to labile C, to be susceptible to short-term environmental changes (Haynes, 2008).

4.4.3.8 Enzyme Activity

Biogeochemical cycles of all major nutrients are dictated by enzymatic activities in soil and are easily measurable, and respond quickly to changes in soil management practices (García-Ruiz et al., 2009). Climate change alters the composition of vegetation which modifies the organic input available for microorganisms in soil. Also, elevated CO_2 is supposed to increase plant biomass, and stimulate microorganisms thereby altering C turnover and modifying the above and below-ground biomass. However, the role of soil microbial enzymes in organic C turnover, nutrient cycling, and GHG emissions is presently unclear.

4.4.3.9 Soil Fauna and Flora

All soil organisms are crucial for the retention, decomposition, and integration of plant remnants, and nutrient cycling, to maintain soil structure and porosity. Because soil fauna and flora have a pretty wide temperature range as their optimum, elevated temperature because of climate change might not have a significant impact on their ecological structure. However, through increased temperature and rainfall fluctuations, changing ecosystems and the migration of plant zones might have a negative impact on relatively sedentary soil flora and fauna. Under climate change at enhanced CO_2 levels plant growth is accelerated and carbon accumulation at below ground level increases microbial population which accelerates N_2 fixation rates, N immobilization, and denitrification, enhanced mycorrhizal associations, soil aggregation improvement, and lastly increased mineral weathering. However, as was already mentioned, a lot will depend on what kind of equilibrium develops as a result of climate change between higher plants on one side and increased SOM decomposition on the other.

4.4.3.10 Functional and Genetic Biodiversity of Soils

Under climate change scenario due to an increase in temperature community structure of soil heterotrophic microorganisms. Due to potential evolutionary changes that could facilitate the spread of virulence factors and genes that help organisms survive in their environment, changing climatic conditions are also a cause for concern (French et al., 2009). Recently community structure and microbial diversity have also become part of indicators of soil health due to advanced molecular techniques (Arias et al., 2005), specifically within food web and nutrient cycling models, to better improve climate change predictions on soil health.

4.4.4 Overall Climate Change Impact on Soil Health

While "Soil health" implies a dynamic condition in which human impact results in a shift in quality, soil quality could be partially considered as a static (qualitative) indicator of the capabilities of soil. There are many potential signs of healthy or good soil. They can be broadly divided into four groups: (i) visual (includes runoff loss, response of plant and abundance of weeds), (ii) physical (includes depth of topsoil, bulk density, soil aggregation, soil crusting, and compaction), (iii) chemical (includes soil reaction, soil salinity and/or alkalinity, SOM, CEC, and contaminant concentrations), and (iv) biological (includes enzymatic activity, soil biodiversity, microbial biomass carbon, nitrogen, etc.). Among the potential indicators soil organic carbon is the most important one to infer health status of soil. The SOM acts both as a source and sink of all nutrients in soil and it serves as the source of carbon and energy for soil microorganisms. Besides that, SOM performs various ecosystem services. Within the terrestrial biosphere, soil C can behave either as a source or sink for atmospheric CO_2 which depends on soil management practices adopted, thus playing a key role in either accelerating or mitigating the greenhouse effect. Change in temperature and precipitation cycle due to global warming strongly influence soil organic carbon and accelerate GHG emissions. Overall, the impact of climate change on soil health if ignored, will make us pay a heavy price.

4.5 Selection of Soil Health Key Indicators Under a Scenario of Climate Change

Research is needed to verify existing indicator frameworks as it helps in opting for the best management practice, rather than simply comparing with the status quo (Stockdale & Watson, 2009). However, recent reviews emphasize on measurement of individual indicators or a suite of indicators to assess changes in soil health in the context of soil's physical, chemical, and/or biological properties (Zagal et al.,

2009). Because of their connection and complexity, soil health indicators have a direct impact on primary productivity of plants, soil flora, and fauna diversity and associated activities, enzymatic activities along with C and N cycling in soil (Kibblewhite et al., 2008). This section summarises key soil health indicators, their roles, and processes under envisioned climate change, their importance to the effects of climate change, and how frequently they are included in a minimal data set for soil health evaluation. Management practices led to affect sensitive key indicators, climatic variation integrates all physical, chemical and biological properties, applicable under field conditions and user friendly (Karlen et al., 2003). Under climate change, the key indicators should reflect the adaptive and mitigation capacity over short and long-term periods (Table 4.1).

4.6 Management of Soil Health

To improve/restore soil health, various soil management strategies can be taken up which include diversification of crops, minimizing mechanical disturbance to soil, and addition of various organic amendments to soil (Kibblewhite et al., 2008). The findings by Williams et al. (2020) suggest that to achieve good soil health, soil management is the key; suggesting to go for crop diversification, reducing soil disturbance by opting for minimum tillage practices and addition of SOM. Turmel et al. (2015) emphasized the role of residue retention in improving soil's physical, chemical, and biological properties. The soil management practices that lead to improved soil health are briefly discussed (Fig. 4.5) as follows:

4.6.1 Soil Organic Carbon

Besides improving soil tilth, infiltration of water, water holding capacity, cation exchange capacity, and soil microbial activity, SOM also enhances the availability of essential soil nutrients. Quantity, quality, and dynamics/turnover of SOC are critical to soil health (Lal, 2014). In the root zone of soil, SOC has a threshold level of 1.5–2.0%. It is essential to maintain SOC pool above this threshold level to: (1) have good soil tilth and aeration by soil aggregate formation (2) improve water use efficiency and WHC thereby improving drought and heat wave tolerance and to fight the adversity of changing climate (3) enhance nutrient use efficiency and nutrient holding capacity which decreases potential loss of nutrient, eutrophication and soil pollution(4) promote various rhizospheric processes leading to a transformation of elements and suppression of disease-causing pathogens (5) regulate emissions of various GHGs like CO_2, CH_4, N_2O which have a direct bearing on climate change. Various technological options can be explored to have a positive C budget which can

Table 4.1 Soil health indicators under changing climate

Indicators of soil health	Soil processes influenced	Relevance to assess climate change impacts	Inclusion in a minimum data set
Physical			
Soil structure	Aggregate stability, retention of SOM	Medium	Frequently included
Porosity	Plant available water, field capacity	High	May be occasional or frequent inclusion
Infiltration	Availability and movement of soil water	High	Occasionally included
Bulk density	Soil structure and compaction	Low	Frequently included
Soil depth	Water holding capacity of soil and subsoil salinity	Medium	Occasionally included
Soil/plant available water and distribution	Permanent wilting point and field capacity, macropore flow, soil texture	High	Frequently included
Chemical			
pH	Soil chemical properties and biological activities	Medium	Frequently included
EC	Soil salinity influencing nutrient availability, microbial activities, plant activities	Medium	Frequently included
Plant available N, P, and K	Soil nutrient status and potential losses	Medium	Frequently included
Biological			
Organic matter content of soil	Quality and quantity of soil organic matter	High	Frequently included
Mineralisable C and N	Soil N mineralization, immobilization, and net flux, catabolic and anabolic activities of soil organisms	High	Occasionally included
Soil respiration	Microbial activity in soil	High	May be occasional or frequent inclusion
Microbial biomass C and N	Microbial activity in soil,	High	Occasionally included
Microbial quotients	Efficiency of substrate use	High	Occasionally included

(continued)

Table 4.1 (continued)

Indicators of soil health	Soil processes influenced	Relevance to assess climate change impacts	Inclusion in a minimum data set
Microbial diversity	Nutrient cycling, ecosystem function	High	Occasionally included
Other microbiological indicators, enzyme activity	Soil structural stability, lability of soil carbon, K_m, V_{max}, K_i, Q_{10}	High	Occasionally included

(Modified from Allen et al., 2011)

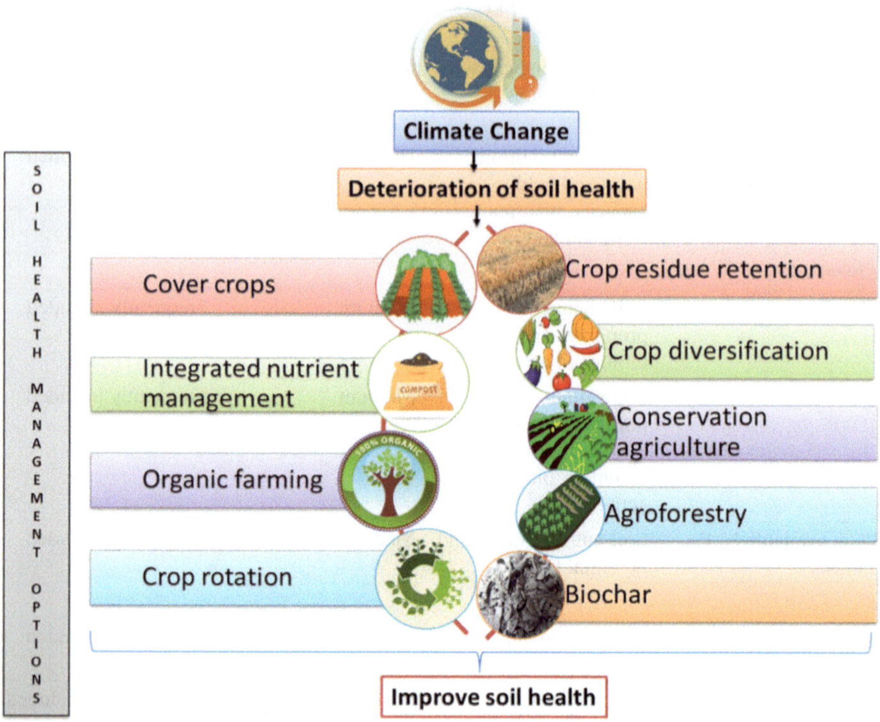

Fig. 4.5 Soil management strategies to enhance soil health under a scenario of climate change

increase SOC pool. Some of the important technological interventions are conservation agriculture (CA), integrated nutrient management strategies, adoption of diversified cropping/farming practices, using good quality organic inputs, and other options that can restore soil and ecosystem functions. Managing SOC is a global issue and its political significance can be realized by the recommendation "4 per Thousand" program at COP21 held in Paris. Lal (2016) opined that food and nutritional security, mitigation and adoption of climate change, reduction of pollution from nonpoint sources and eutrophication, enhancing soil biodiversity, increasing nutrient use efficiency, and sustainable intensification of agro-ecosystem can be achieved through managing SOC.

4.6.2 Conservation Agriculture

It is one of the crop production systems that emphasizes on conservation of resources. The three basic principles of this agro-ecological approach are: (1) Minimizing mechanical disturbance to soil (No tillage, Minimum tillage, direct seeding) (2) Keeping the soil permanently covered with cover crops and/or crop residues (3) Adopting crop rotation practices to promote species diversification (Corsi et al., 2012). Slowing down physical and biotic processes in conservation agriculture caused significant improvement in soil health (Thierfelder & Wall, 2009). Long-term study under conservation tillage and residue management resulted in an improvement of soil properties and nutrient availability (N, P, K, Cu, Mn, and Zn) in soil surface when compared with conventional farming (Zahid et al., 2020). Incorporation of crop residues in conservation tillage has the potential of improving organic carbon concentration in soil, increasing availability of major nutrients in soil, stabilization of soil aggregates, improving water retention capacity of soil, aeration, and promoting enzymatic activities in soil. Even after knowing the importance of conservation agriculture in sustaining and/or improving soil health, adoption of the technology becomes difficult because of the obstacles like inadequate knowledge, non-availability of implements specific to conservation agriculture unavailability, and paucity of research on conservation tillage. Besides, long-term agroecosystem sustainability can be ensured by adoption of conservation tillage practices as it can reverse the soil degradation induced by conventional agriculture. This technology also plays a significant role in carbon sequestration thereby offsetting GHG emissions.

4.6.3 Organic Farming

Many researchers have reported that well-managed organic farms lead to enhancement in soil quality when compared with nearby farms that adopted conventional farming practices (Edmeades, 2003; Gomiero et al., 2008; Mäder et al., 2002; Pimentel et al., 2005; Reganold et al., 2001). The organic system has demonstrated improvement in microbial diversity, activity, and microbial biomass C and N which ultimately affect nutrient cycling and have a direct bearing on soil health. Besides organically managed farms show greater biological properties and processes, which help in suppressing pathogen activity, mainly due to antagonism and competition.

4.6.4 Integrated Nutrient Management (INM)

The INM is defined as "maintenance and/or adjustment of soil fertility and supply of plant nutrients to an optimum level for sustaining the desired crop productivity through optimization of benefit from all possible sources of plant nutrients in an integrated manner". The main ingredient of INM includes inorganic fertilizers, organic manures, legumes, crop residues, industrial by-products, and bio-fertilizers. Integrated application of organic manures and inorganic fertilizers reduces soil bulk density when compared with chemical fertilizer alone due to build-up of SOM. INM also improves soil structure. In long-term fertilization experiments, application of manure either alone or in conjunction with NPK fertilizers increased large macroaggregate formation while reducing the percentage of microaggregates in red soil of China under a subtropical climate (Huang et al., 2010). The INM is a sustainable strategy that is agronomically feasible, economically viable, and environmentally friendly which enhances soil fertility and facilitates soil C sequestration, reducing N losses and greenhouse gas emissions (Aulakh, 2010). Meena et al. (2019) concluded that organic manures like FYM and poultry manure along with STCR-based fertilizers should be advised for improving crop productivity and maintaining soil health for long term. Javaria and Khan (2010) analyzed several studies with tomatoes and concluded that INM improved crop yield and quality besides improving soil fertility.

4.6.5 Biochar

It is different from charcoal in terms of its applicability as a soil amendment that is produced by degradation of organic matter at a defined temperature without oxygen supply (pyrolysis) (Lehmann & Joseph, 2009). Biochar sequesters C and simultaneously improves soil health (Lehman et al., 2006). Biochar use has several benefits which include C sequestration potential, generation of bioenergy, reduction of N_2O gas emission from agricultural fields, increasing soil microbial activity, adsorption of pesticides and various nutrient ions, soil structural improvement, and water holding capacity and suppression of soil-borne diseases (Najar et al., 2015). Biochar addition in soil increases the negative charges on soil surface thereby decreasing the zeta potential which leads to an increase in cation exchange capacity of soil. Remediating heavy metal-polluted soil through the use of biochar has also been well established. Biochar also decreases nutrient leaching in soil because of the reactions involving cation and anion exchange which help in nutrient retention. Biochar also acts as a soil conditioner which improves soil's physical properties (Sohi et al., 2010). However, its effect on soil enzymatic activity is variable and depends on the type of enzymatic activity (Najar et al., 2015). The multiple benefits incurred by use of biochar make it a suitable tool for managing and/or improving soil health.

4.6.6 Crop Residue Retention

Retention of crop residues leads to the addition of organic matter in soil. Minimum tillage along with retention of crop residues (complete/partial), either incorporated in soil or left as such on-soil surface is having more advantages than complete removal or incorporation of the residue in intensively tilled soils. Retention of crop residue is having many benefits which include improvement in soil aggregation and reduction of soil erosion; however, soil might become compacted and soil loss can increase when crop residue is retained rather than incorporated. In regions that are prone to erosion and humid tropical regions favoring SOM decomposition, residue retention on soil surface should be promoted. Mobilization of N is the most common negative impact which reduces its availability to crops, specifically in regions where N is limiting (Turmel et al., 2015). However, benefits harnessed due to crop residue retention are site-specific and depend not only on agroclimatic conditions but also on farming systems followed and socio-economic factors. However, a lot of farmers use crop residue as livestock feed that provides a short-term economic benefit, which otherwise is not obtained when the residue is retained to maintain/improve soil fertility.

4.6.7 Crop Diversification

In order to provide food security to ever-increasing global population, crop intensification is needed to produce more food per unit area per unit of time. However, crop intensification causes the removal of large amounts of soil available nutrients thereby altering soil fertility and causing immense damage to soil health. So, we need to take up the crops that can maintain soil fertility by adding nutrients back to the soil along with maintaining crop diversity. The methods like incorporation of leguminous crops in cropping cycle and keeping the crop residue in the field itself are easy and convenient to adopt. Diversified cropping practice benefit soil health by supplying organic matter by growing crops that adds a lot of organic matter through their roots, root exudates, leaves, and stems into the soil and also reduce nutrient loss through erosion and leaching besides providing economic benefits to the farmers. Kumar et al. (2021) concluded that diversification of rice-rice or rice-jute cropping system by including vegetable pea or mustard-mung bean supplied with 100% recommended dose of fertilizer along with crop residue can sustain the cropping system besides improving soil quality.

4.6.8 Resource Conservation Technologies (RCTs)

Overexploitation of water and soil resources over the years in conventional agriculture reduced the efficiency of fertilizer, irrigation water, and energy in a dramatic manner. Besides, it has caused soil health problems like loss of SOC, salinization of soil, depletion of groundwater, soil compaction, and acceleration of soil erosion thereby removing a top fertile layer of soil. Crop residue burning to facilitate crop intensification has caused air pollution, SOM loss, loss of soil nutrients, and GHG emissions. In developing countries, maintaining soil health under intensively cultivated commercial agricultural practices and a faster pace of economic growth sets serious challenges for sustaining the use of agricultural resources. The practices or technologies that enhance resource or input use efficiency such as use of certain varieties that take up N in a more efficient way, zero or minimum tillage practices that save energy and water, etc., may be regarded as RCTs. Conservation agriculture is one of the RCTs that is also discussed in another section. Other popular RCTs also include:

4.6.8.1 Conservation Tillage

Conservation tillage refers to any tillage and planting system where crop residues cover at least 30% of the soil surface post-planting to arrest water erosion. In areas that are susceptible to wind erosion, it is recommended to maintain at least 1000 kg/ha of small grain residues equivalent on soil surface during a critical wind erosion period (Ghosh et al., 2017). Techniques such as no-tillage, minimum tillage, ridge tillage, mulch tillage, chisel plough, etc. are included under conservation tillage. Those methods of tillage help in the decline of soil erosion, increasing the use efficiency of inputs, improving SOM, decreasing production cost and environmental degradation.

4.6.8.2 Maintenance of Soil Cover

Maintenance of permanent or semi-permanent soil cover is possible by inclusion of cover crops, management of crop residue, and mulching which protect soil from direct exposure to the sun, air, and wind. Incorporation of legumes as a cover crop decreases the risk of soil erosion and increases soil organic matter input, increases soil N content through biological nitrogen fixation besides suppressing disease and increasing system productivity as a whole. Organic or inorganic mulching protects soil from erosion, increases SOM content, and maintains soil moisture and temperature. Management of crop residue includes retention of crop residues and incorporation which facilitates build-up of organic matter, suppresses weed infestation, protects soil from direct abiotic stresses, and prevents soil erosion.

4.6.9 Crop Rotation

It refers to the incorporation of crops that differ in their rooting patterns and nature in a rotational manner combined with least soil disturbance systems generally promoting an extensive network of roots, and macropores in the soil. Legume crop inclusion in the crop rotation has the advantages of adding additional N through biological N_2 fixation, nutrient recycling from different depths of soil, reducing soil compaction, input of SOM, breaking the life cycles of weeds and pests, and minimizing the potential allelopathic effects.

4.6.9.1 Agro-Forestry

This is a management system that is self-sustaining and ecologically sound. This system involves long-term conservation and integration of trees, livestock, and crops in a profitable and eco-friendly manner. Scientists and land use planners recommend the system for sustainability in agricultural production, nutrient recycling, natural resource conservation, problem soil management, and mitigation of climate change by sequestrating SOC (Ghosh et al., 2017).

4.7 Sensitive Indicator of Soil Health Under Changing Climate

In regions where agriculture experiences climate extremes, detecting small but incremental changes in the health of these fragile soils is critical to support productivity. Testing similar parameters along with the tests that examine different forms of nitrogen in soil should be included that can provide information about potential GHG emissions including N_2O to manage soil health for climate change mitigation. An indicator that monitors a management practice, which improves soil health must be sensitive to changes in management practices. Under changing climate conditions, it is recommended to include soil aggregation, infiltration capacity of soil, an abundance of earthworms, soil organic C and N fractions, and mineralizing enzymes in soil health testing. Besides, new indicators that are more directed to measure non-agricultural soil services including human health and quality water should be included in routine soil analysis. Soil pathogens, parasites, biodiversity, toxins that are bioavailable and mobile, and diversity of pore size are among those indicators that should be included. To measure emissions of GHGs in the field, information about processes that are having high temporal dynamics is needed which requires new generation sensors that are based on autonomous gas and solute detection powered by bioreactors and a variety of energy-harvesting technologies. Molecular and soil structural diversity should be explored and included for knowing soil organic carbon persistence and sequestration. Microorganisms can also serve as a sensitive indicator of soil health in changing climate. If a microbial parameter is representative of the environmental properties (state variables) and impacts the environment in a manner that can be interpreted beyond the information that the measured or observed parameter represents by itself is defined as a microbial indicator (Christensen et al.,

1992). Microorganisms are capable of giving an integrated measure of soil health, which otherwise is difficult to obtain from any physical or chemical measures and/ or analyses of diversity of higher organisms. Microorganisms rapidly adapt to the environmental conditions around them because of their ability to respond quickly to changes. Therefore, changes in the population of soil microbes and activities can serve as an excellent indicator to assess changes in soil health (Kennedy & Papendick, 1995; Nielsen et al., 2002). However, the limitation is the large variations of microbial properties on spatial and temporal scales that needs to be considered while selecting those indicators for the assessment of soil health.

4.8 Pitfalls in Management of Soil Health

Quantification of SOM, pH and plant available P, K are part of >60% soil health testing framework while >50% of those frameworks also include water storage and bulk density (Bünemann et al., 2018). Inclusion of microbial biomass N, soil respiration, and N mineralization to address soil biological properties and stability of soil structure is recommended in around 33% of soil testing frameworks. At least 40% of chemical indicators are involved in 90% of soil health assessment schemes which underscores the involvement of chemical indicators in assessing soil health. Biological indicators typically constitute less than 20% of the indicators (Lehmann et al., 2020). This triggers misinterpretation of soil health under changing climate scenarios. Incorporation of a new set of indicators to assess soil health also needs implementation of a new instrument, soil test methods, and infrastructure which is also a challenging step. Creating a soil health index is also a difficult task due to the availability of a relatively low number of indices. To determine soil health score, often the indicators are given equal importance which is practically not true. Adoption of management practices that promote soil health is also not encouraging. Making people understand the importance of soil health and sustainability of agriculture requires a well-framed approach that is in progress. Establishing a common understanding and bridge between scientists and stakeholders is also a challenging task and the research should be oriented in such a fashion that it takes into consideration the problem and issues of stakeholders along with improving/restoring soil health.

4.9 Conclusion

It is imperative to monitor the climate change impacts on soil health for evaluation of the severity and magnitude of the problem, to follow the trends, and to devise necessary crop management strategies to combat its effect on soil health and so as agriculture. To understand soil health and the effect on it due to climate change, the indicators should be used in such a way that addresses various physicochemical and biological properties that can be related to ecological functions and subsequently can be monitored in the context of changing climate and sustainable land management. The indicators should be sensitive to changing climate and land use management and those indicators should also have the capacity to integrate and relate with other soil functions, they should be easy to use, economic to handle and should have good repeatability. Indicators having all those desirable traits should be selected in a minimum data set devised to monitor soil health in the context of climate change. The issues of climate change and land degradation are interconnected to each other and conservation farming is a promising technique to minimize land degradation. Monitoring climate change's impact on soil systems should be done regularly in different agroecological regions. Harmonization of database related to land degradation is highly needed keeping in view the losses in productivity and economics vis-à-vis effects of climate change. The impact of climate change on soil properties can be permanent or temporary, but the factors that affect adoption rates for sustainable practices in the croplands are very much specific to the techniques, skills of knowledge, and growers' to reduce the impact of climate change. Therefore, it becomes imperative to fortify the nexus between soil health and climate change by tailoring new strategic and climate-resilient approaches while considering the practicality of soil health knowledge at the farmers' paddocks. Nevertheless, devising novel management strategies that can set new directions to address the soil productivity constraints under the ever-changing climatic and land use scenarios. There should be a common understanding between stakeholders and scientists to make soil health concept live up to its potential as a unifying concept that encompasses soil functions. Holistic soil health indices should include multi-criteria decision analysis. There is an urgent need to develop a comprehensive soil health index that will be acceptable globally. Government policy intervention is very much essential for widespread adoption of management practices that are oriented towards soil health improvement considering the sensitivity of agriculture under the scenario of changing climate. Adoption of conservation tillage leaving behind the conventional one is a Herculean task that can only become possible through interventions at national and political scales. Further, there has to be a strong linkage between farmers, researchers, and extension workers. Appropriate training to the farmers at ground level should be given to bring significant behavioral changes in the farming community to adopt conservation agriculture practices so that the real benefit of the technique can be harnessed.

References

Acton, D. F., & Gregorich, L. J. (eds.). (1995). The health of our soils. Towards sustainable agriculture in Canada. Agriculture and Agri-Food Canada, Ottawa, Canada. Soil Sci Soc Am. Statement on soil quality. Agronomy News. 7 p.

Acton, D. F., & Padbury, G. A. (1993). A conceptual framework for soil quality assessment and monitoring. In D. F. Acton (ed.) *A Program to Assess and Monitor Soil Quality in Canada: Soil*

Quality Evaluation Program Summary (Interim). Centre for Land and Biological Resources Research Contribution No. 93–94. Research Branch, Agriculture Canada, Ottawa, 2–7.

Alkorta, I., Aizpurua, A., Riga, P., Albizu, I., Amézaga, I., & Garbisu, C. (2003). Soil enzyme activities as biological indicators of soil health. *Reviews on Environmental Health, 18*(1), 65–73.

Allan, D. L., Adriano, D. C., Bezdicek, D. F., Cline, R. G., Coleman, D. C., Doran, J. W., Haberen, J., Harris, R. G., Juo, A. S. R., Mausbach, M. J., & Peterson, G. A. (1995). SSSA statement on soil quality. Agronomy News, 7p.

Allen, D. E., Singh, B. P., & Dalal, R. C. (2011). Soil health indicators under climate change: A review of current knowledge. In B. Singh, A. Cowie, & K. Chan (eds) *Soil Health and Climate Change.* Soil Biology, vol. 29. Springer, Berlin, Heidelberg.

Arias, M. E., González-Pérez, F. J., & González-Vila, A. S. B. (2005). Soil health—a new challenge for microbiologists and chemists. *International Microbiology, 8*, 13–21.

Arnold, S. L., Doran, J. W., Schepers, J., & Wienhold, B. (2005). Portable probes to measure electrical conductivity and soil quality in the field. *Communications in Soil Science and Plant Analysis, 36*, 2271–2287.

Aulakh, M. S. (2010). August. Integrated nutrient management for sustainable crop production, improving crop quality and soil health, and minimizing environmental pollution. In *19th world congress of soil science, soil solutions for a changing world*, pp. 1–6.

Benbi, D. K., & Kaur, R. (2009). Modeling soil processes in relation to climate change. *Journal of Indian Society Soil Science, 57*, 433–444.

Birkas, M., Dexter, A., & Szemök, A. (2009). Tillage-induced soil compaction, as a climate threat increasing stressor. *Cereal Research Communication, 37*, 379–382.

Box, J. E., & Bruce, R. R. (1996). The effect of surface cover on infiltration and soil erosion. In M. Agassi (ed) *Soil erosion, conservation and rehabilitation.* Marcel Dekker Inc, New York.

Brinkman, R., & Brammer, H. (1990). The influence of a changing climate on soil properties. In *Transactions 14th International Congress of Soil Science*, Kyoto, Japan, Volume V. pp. 283–288.

Bünemann, E. K., Bongiorno, G., Bai, Z., Creamer, R. E., De Deyn, G., de Goede, R., Fleskens, L., Geissen, V., Kuyper, T. W., Mäder, P., & Pulleman, M. (2018). Soil quality–a critical review. *Soil Biology & Biochemistry, 120*, 105–125.

Christensen, H., Griffiths, B., & Christensen, S. (1992). Bacterial incorporation of tritiated thymidine and populations of bacteriophagous fauna in the rhizosphere of wheat. *Soil Biology & Biochemistry, 24*(7), 703–709.

Corsi, S., Friedrich, T., Kassam, A., Pisante, M., & Sà, J. D. M. (2012). Soil organic carbon accumulation and greenhouse gas emission reductions from conservation agriculture: a literature review. *Food and Agriculture Organization of the United Nations (FAO).*

Dalal, R. C., & Moloney, D. (2000). Sustainability indicators of soil health and biodiversity. In P. Hale, A. Petrie, D. Moloney, & P. Sattler (eds.) *Management for sustainable ecosystems.* Centre for Conservation Biology, Brisbane, pp. 101–108.

Davidson, E. A., & Janssens, I. A. (2006). Temperature sensitivity of soil carbon decomposition and feedbacks to climate change. *Nature, 440*, 165–173.

Doran, J. W., & Parkin, T. B. (1994). Defining and assessing soil quality. In J. W. Doran, D. C. Coleman, D. F. Bezdicek, & B. A. Stewart (eds). *Defining Soil Quality for a Sustainable Enviroment.* SSSA Spec. Publ. No. 35, Madison, Wisconsin, USA, pp. 3–21.

Doran, J. W., Sarrantonio, M., & Liebig, M. A. (1996). *Soil Health and Sustainability. Advances of Agronomy, 56*, 1–54.

Doran, J. W., & Zeiss, M. R. (2000). Soil health and sustainability: Managing the biotic component of soil quality. *Applied Soil Ecology, 15*, 3–11.

Dubey, A., Malla, M. A., Khan, F., Chowdhary, K., Yadav, S., Kumar, A., Sharma, S., Khare, P. K., & Khan, M. L. (2019). Soil microbiome: A key player for conservation of soil health under changing climate. *Biodiversity and Conservation, 28*(8), 2405–2429.

Edmeades, D. C. (2003). The long-term effects of manures and fertilisers on soil productivity and quality: A review. *Natural Cycling of Agroecosystems, 66*(2), 165–180.

Fang, Y., Singh, B. P., Matta, P., Cowie, A. L., & Van Zwieten, L. (2017). Temperature sensitivity and priming of organic matter with different stabilities in a vertisol with aged biochar. *Soil Biology & Biochemistry, 115*, 346–356.

FAO (2013) Report of the First Meeting of the Plenary Assembly of the Global Soil Partnership. (Rome, Italy; 11–12 June 2013). GSPPA-I/13/Report. Food and Agriculture Organization of the United Nations, Rome, Italy. 22 pp.

French, S., Levy-Booth, D., Samarajeewa, A., Shannon, K. E., Smith, J., & Trevors, J. T. (2009). Elevated temperatures and carbon dioxide concentrations: Effects on selected microbial activities in temperate agricultural soils. *World Journal of Microbiology & Biotechnology, 25*, 1887–1900.

García-Ruiz, R., Ochoa, V., Vinegla, B., Hinojosa, M. B., Pena-Santiago, R., Liébanas, G., Linares, J. C., & Carreira, J. A. (2009). Soil enzymes, nematode community and selected physico-chemical properties as soil quality indicators in organic and conventional olive oil farming: Influence of seasonality and site features. *Applied Soil Ecology, 41*(3), 305–314.

Gil, S. V., Meriles, J., Conforto, C., Figoni, G., Basanta, M., Lovera, E., & March, G. J. (2009). Field assessment of soil biological and chemical quality in response to crop management practices. *World Journal of Microbiology & Biotechnology, 25*, 439–448.

Gil-Sotres, F., Trasar-Cepeda, C., Leirós, M. C., & Seoane, S. (2005). Different approaches to evaluating soil quality using biochemical properties. *Soil Biology & Biochemistry, 37*, 77–87.

Ghosh, P. K., Palsaniya, D. R., & Kumar, T. K. (2017). Resource conservation technologies for sustainable soil health management. *Adaptive Soil Management: From Theory to Practices* (pp. 161–187). Springer.

Gomiero, T., Paoletti, M. G., & Pimentel, D. (2008). Energy and environmental issues in organic and conventional agriculture. *Critical Reviews in Plant Sciences, 27*(4), 239–254.

Gregorich, E. G., Carter, M. R., Angers, D. A., Monreal, C. M., & Ellert, B. H. (1994). Towards a minimum data set to assess soil organic matter quality in agricultural soils. *Canadian Journal of Soil Science, 74*, 367–385.

Gregorich, E. G., & Carter, M. R. (Eds.) (1997). *Soil quality for crop production and ecosystem health*. Elsevier.

Haynes, R. J. (2008). Soil organic matter quality and the size and activity of the microbial biomass: Their significance to the quality of agricultural soils. In Q. Huang, P. M. Huang, & A. Violante (Eds.), *Soil mineral-microbe-organic interactions: Theories and applications* (pp. 201–230). Springer.

Huang, S., Yan-Ni, S. U. N., Wen-Yi, R. U. I., Wu-Ren, L. I. U., & Zhang, W. J. (2010). Long-term effect of no-tillage on soil organic carbon fractions in a continuous maize cropping system of Northeast China. *Pedosphere, 20*(3), 285–292.

Javaria, S., & Khan, M. Q. (2010). Impact of integrated nutrient management on tomato yield quality and soil environment. *Journal Plant Natural, 34*(1), 140–149.

IPCC. (2007). Climate Change 2007: Mitigation. Contribution of Working Group III to the Fourth Assessment Report of the Intergovernmental Panel on Climate Change. Cambridge University Press, Cambridge, UK., New York.

Karlen, D. L., Doran, J. W., Andrews, S. S., & Wienhold, B. J. (2003). Soil Quality—Humankinds' foundation for survival. *Journal of Soil and Water Conservation, 58*, 171–179.

Karmakar, R., Das, I., Dutta, D., & Rakshit, A. (2016). Potential effects of climate change on soil properties: A review. *Science Int, 4*(2), 51–73.

Keith, A. M., Schmidt, O., & McMahon, B. J. (2016). Soil stewardship as a nexus between ecosystem services and One Health. *Ecosystem Services, 17*, 40–42.

Kennedy, A. C., & Papendick, R. I. (1995). Microbial characteristics of soil quality. *Journal of Soil and Water Conservation, 50*(3), 243–248.

Kibblewhite, M. G., Ritz, K., & Swift, M. J. (2008). Soil health in agricultural systems. *Philosophical Transactions of the Royal Society B: Biol Sci, 363*(1492), 685–701.

Kirtman, B., Power, S. B., Adedoyin, A. J., Boer, G. J., Bojariu, R., Camilloni, I., Doblas-Reyes, F., Fiore, A. M., Kimoto, M., Meehl, G., & Prather, M. (2013). *Near-term climate change: projections and predictability*.

Koch, A., McBratney, A., Adams, M., Field, D., Hill, R., Crawford, J., Minasny, B., Lal, R., Abbott, L., O'Donnell, A., & Angers, D. (2013). Soil security: Solving the global soil crisis. *Global Policy, 4*(4), 434–441.

Kumar, M., Mitra, S., Mazumdar, S. P., Majumdar, B., Saha, A. R., Singh, S. R., Pramanick, B., Gaber, A., Alsanie, W. F., & Hossain, A. (2021). Improvement of soil health and system productivity through crop diversification and residue incorporation under jute-based different cropping systems. *Agronomy, 11*(8), 1622.

Lal, R. (2011). Soil health and climate change: an overview. *Soil Health and Climate Change,* 3–2.

Lal, R. (2014). Societal value of soil carbon. *Journal of Soil Water Conservation, 69*(6), 186A-192A.

Lal, R. (2016). Soil health and carbon management. *Food Energy Secur, 5*(4), 212–222.

Lal, R. (1995). Tillage systems in the tropics: management options and sustainability implications. Food and Agriculture Organization of the United Nations Soils Bulletin 71, FAO, Rome.

Lal, R. (2004). Soil carbon sequestration to mitigate climate change. *Geoderma, 123,* 1–22.

Larson, W. E., & Pierce, F. J. (1991). Conservation and enhancement of soil quality. In J. Dumanski, E. Pushparajah, M. Latham, R. Myers (eds.). *Evaluation for Sustainable Land Management in the Developing World*, Volume 2: Technical Papers, IBSRAM Proceedings No 12 (2), Bangkok, pp. 175–203.

Luo, Z., Feng, W., Luo, Y., Baldock, J., & Wang, E. (2017). Soil organic carbon dynamics jointly controlled by climate, carbon inputs, soil properties and soil carbon fractions. *Global Change Biology, 23*(10), 4430–4439.

Lehmann, J., & Joseph, S. (2009). Biochar for environmental management: An introduction. In J. Lehmann & S. Joseph (Eds.), *Biochar for Environmental Management: Science and Technology* (pp. 1–12). Earthscan.

Lehmann, J., Bossio, D. A., Kögel-Knabner, I., & Rillig, M. C. (2020). The concept and future prospects of soil health. *Nature Reviews Earth and Environment, 1*(10), 544–553.

Mäder, P., Fliessbach, A., Dubois, D., Gunst, L., Fried, P., & Niggli, U. (2002). Soil fertility and biodiversity in organic farming. *Science, 296*(5573), 1694–1697.

Magdoff, F. (2001). Concept, components, and strategies of soil health in agroecosystems. *Journal of Nematology, 33*(4), 169.

Meehl, G. A., Stocker, T. F., Collins, W. D., Friedlingstein, P., Gaye, A. T., Gregory, J. M., Kitoh, A., Knutti, R., Murphy, J. M., Noda, A., & Raper, S. C. (2007). Global climate projections. Chapter 10.

Meena, B. P., Biswas, A. K., Singh, M., Chaudhary, R. S., Singh, A. B., Das, H., & Patra, A. K. (2019). Long-term sustaining crop productivity and soil health in maize–chickpea system through integrated nutrient management practices in Vertisols of central India. *Field Crops Research, 232*, 62–76.

Najar, G. R., Ganie, M. A., & Tahir, A. L. I. (2015). Biochar for sustainable soil health: A review of prospects and concerns. *Pedosphere, 25*(5), 639–653.

Nielsen, M. N., Winding, A., & Binnerup, S. (2002). *Microorganisms as indicators of soil health.*

Pandolfini, T., Gremigni, P., & Gabbrielli, R. (1997). Biomonitoring of soil health by plants. *Biological Indicators of Soil Health,* 325–347.

Pearson, R. G. (2006). Climate change and the migration capacity of species. *Trends in Ecology & Evolution, 21*(3), 111–113.

Pimentel, D., Hepperly, P., Hanson, J., Seidel, R., & Douds, D. (2005). Organic and conventional farming systems: *Environmental and Economic Issues.*

Reganold, J. P., Glover, J. D., Andrews, P. K., & Hinman, H. R. (2001). Sustainability of three apple production systems. *Nature, 410*(6831), 926–930.

Rengel, Z. (2011). *Soil pH, soil health and climate change. In Soil health and climate change.* Springer, Berlin, Heidelberg, pp. 69–85.

Reth, S., Reichstein, M., & Falge, E. (2005). The effect of soil water content, soil temperature, soil pHvalue and root mass on soil CO_2 efflux. *Plant and Soil, 268*, 21–33.

Reynolds, W. D., Bowman, B. T., Drury, C. F., Tan, C. S., & Lu, X. (2002). Indicators of good soil physical quality: Density and storage parameters. *Geoderma, 110*, 131–146.

Ross, D. S., Matschonat, G., & Skyllberg, U. (2008). Cation exchange in forest soils: The need for a new perspective. *European Journal of Soil Science, 59*, 1141–1159.
Saha, N., & Mandal, B. (2009). Soil health—a precondition for crop production. In M. S. Khan, A. Zaidi, & J. Musarrat (Eds.), *Microbial strategies for crop improvement* (pp. 161–216). Springer.
Shukla, M. K., Lal, R., & Ebinger, M. (2006). Determining soil quality indicators by factor analysis. *Soil till Research, 87*, 194–204.
Singh, M., Sarkar, B., Biswas, B., Bolan, N., & Churchman, G. J. (2017). Relationship between soil clay mineralogy and carbon protection capacity as influenced by temperature and moisture. *Soil Biology & Biochemistry, 109*, 95–106.
Sohi, S. P., Krull, E., Lopez-Capel, E., & Bol, R. (2010). A review of biochar and its use and function in soil. *Advances in Agronomy, 105*, 47–82.
Stockdale, E. A., & Watson, C. A. (2009). Biological indicators of soil quality in organic farming systems. *Renewable Agriculture Food System, 24*, 308–318.
Stott, D. E. (2019). Recommended Soil Health Indicators and Associated Laboratory Procedures. Soil Health Technical Note No. 450–03. U.S. Department of Agriculture, Natural Resources Conservation Service.
Thierfelder, C., & Wall, P. C. (2009). Effects of conservation agriculture techniques on infiltration and soil water content in Zambia and Zimbabwe. *Soil till Research, 105*(2), 217–227.
Thornton, P. K., Ericksen, P. J., Herrero, M., & Challinor, A. J. (2014). Climate variability and vulnerability to climate change: A review. *Global Change Biology, 20*(11), 3313–3328.
Turmel, M. S., Speratti, A., Baudron, F., Verhulst, N., & Govaerts, B. (2015). Crop residue management and soil health: A systems analysis. *Agricultural Systems, 134*, 6–16.
Ussiri, D. A. N., Lal, R., & Jarecki, M. K. (2009). Nitrous oxide and methane emissions from longterm tillage under a continuous corn cropping system in Ohio. *Soil till Research, 104*(2), 247–255.
Wagai, R., Mayer, L. M., & Kitayama, K. (2009). Nature of the "occluded" low-density fraction in soil organic matter studies: A critical review. *Soil Science Plant Nutrition, 55*, 13–25.
Weil, R. R., & Magdoff, F. (2004). Significance of soil organic matter to soil quality and health. In R. R. Weil & F. Magdoff (Eds.), *Soil organic matter in sustainable agriculture* (pp. 1–43). CRC Press.
Williams, H., Colombi, T., & Keller, T. (2020). The influence of soil management on soil health: An on-farm study in southern Sweden. *Geoderma, 360*, 114010.
Wixon, D. L., & Balser, T. C. (2009). Complexity, climate change and soil carbon: A systems approach to microbial temperature response. *Systems Research and Behavioral Science, 26*, 601–620.
Zagal, E., Munoz, C., Quiroz, M., & Córdova, C. (2009). Sensitivity of early indicators for evaluating quality changes in soil organic matter. *Geoderma, 151*, 191–198.
Zahid, A., Ali, S., Ahmed, M., & Iqbal, N. (2020). Improvement of soil health through residue management and conservation tillage in rice-wheat cropping system of Punjab, Pakistan. *Agronomy, 10*(12), 1844.

Chapter 5
Climate Change Impact on Soil Erosion and Land Degradation

Debashis Mandal and Trisha Roy

Abstract The climate change process is intricately linked with soil erosion which is a major land degradation issue, decreasing soil productivity and causing soil carbon losses, which in turn accelerates various climate change processes. Climate change will give rise to extreme climatic events and rainfall erosivity which is a major driving force of soil erosion by water is likely to increase manifolds across different regions in the globe. This increase in rainfall erosivity will increase soil loss through erosion. Soil erosion from cropland is also going to increase if intensive agricultural practices are being followed without the adoption of proper soil conservation measures. Soil carbon loss due to the displacement of eroded soil and its subsequent mineralization is another phenomenon of concern that is going to increase under climate change scenario. Almost 30% of the displaced C is likely to be mineralized depleting the soils of C, paralyzing their capacity to provide various ecosystem services, and also increasing the atmospheric C concentration. The present chapter summarizes the various impacts of climate change on soil erosion with a concluding note of how the present world is striving to achieve sustainable soil management through various programs, designed at international and national levels. The need of the hour is to formulate soil legislation at the country and state levels to better address the problem and protect our soils from further degradation and achieve land degradation neutrality by 2030.

Keywords Soil degradation · Soil erosion · Climate change · Soil carbon · Rainfall erosivity · Sustainable development goals (SDG)

D. Mandal (✉) · T. Roy
ICAR-Indian Institute of Soil and Water Conservation, 218, Kaulagarh Road, Dehradun, Uttarakhand 248 195, India
e-mail: debashisiiswc@gmail.com

D. Mandal
ICAR-Indian Agricultural Research Institute, New Delhi, India

H. Pathak et al. (eds.), *Climate Change Impacts on Soil-Plant-Atmosphere Continuum*, Advances in Global Change Research 78, https://doi.org/10.1007/978-981-99-7935-6_5

5.1 Introduction

Land degradation is a pervasive and systemic phenomenon occurring across the globe that declines the capacity of the land and soil to function. With the exponential rate of increase in world population, the pressure on land and all natural resources is increasing. Climate change which involves the global rise in atmospheric greenhouse gas concentration leading to increased temperature, precipitation, and erratic climatic phenomenon puts land resources at further risk.

Soil erosion, a major form of land degradation, refers to the accelerated removal of soil from the land's surface by various erosive agents such as water and wind, and it is a major risk to soil resources worldwide (FAO, 2015). The soil under influence of various physical, chemical, and even biological weathering forces gets formed by the disintegration of rocks and minerals in due course of time. Soil erosion is a three-step process comprising of disintegration and detachment of the soil particles followed by their transportation and finally deposition by different causal agents. The key erosive agents that play in different parts of the world's are water, wind, gravity, and glaciers. The nature of erosive agent dominant in a specific region is governed by the climatic conditions, thus, water is the major erosive agent in humid areas, while in aridic conditions, the force of wind is the dominant erosive agent.

At equilibrium conditions, the rate of soil erosion is neutralized by the natural rate of soil formation, maintaining natural balance in the ecosystem. However, with an intensification of agriculture and more pressure on land resources globally, the soil erosion rate has accelerated and is a major threat to the path of sustainable development (Mandal et al., 2021). Soil equivalent of one soccer pitch (approximately 0.74 ha) is eroded every five seconds (FAO & ITPS, 2015) which might help us to fathom the intensity of the problem. Rates of soil erosion well in excess of rates of soil loss tolerance limit are a recipe for disaster and thus, it is necessary to have a crystal clear understanding of soil loss tolerance for planning and implementation of robust and site-specific soil conservation strategies (Mandal & Sharda, 2011, 2013).

Land degradation causes a decline in biomass productivity of land and affects economy, resulting from various biophysical and social factors including climate aberration and anthropogenic dominance, leading to loss of ecosystem services. The term "desertification" is a division of land degradation and referred to the occurrence of land degradation in dryland regions (UNEP, 1992). Degradation may affect a range of land use systems counting forested land, grassland, pasture land, and cropland. Land degradation can be caused by a variety of factors, such as changes in the climate or a series of processes brought on by anthropogenic activity. Concerns about land degradation and desertification are growing on a global level as it endangers water quality, biodiversity, ecology, economic growth and living status of the people (UNCCD, 2017).

Nearly 2 billion ha of once productive land is now affected by numerous forms of degradation across the globe. Annually, near about 6 Mha cropland is being lost worldwide due to land degradation (Lal & Stewart, 1992). Globally, food security of every 1 in 6 people is affected by land degradation in three-quarters of drylands (about 3.6 billion ha) and one-quarter of the total global land. Continent wise Asia has the highest degraded land (37%) followed by Africa (25%), Latin America (14%), Europe (11%), North America (4%), and Central America (3%). Land degradation statistics in India revealed that about 73.3 Mha cropland and 9.30 Mha forested (<40% canopy) are primarily affected by soil erosion due to water where the erosion rate is at least 10 t ha^{-1} yr^{-1} (NASS, 2010). Additionally, about 1.30 Mha of salt-affected land is also affected by water erosion. Similarly, about 5.72 Mha of cropland and about 7.13 Mha of forested is affected by dual problem of soil acidity and water erosion. Thus, a total of 96.7 Mha (80.2 Mha of cropland and 16.5 Mha of forested land) area is affected either exclusively by soil erosion or in conjunction with salinity/acidity. In India, northern states (Punjab, Haryana, Uttarakhand, Uttar Pradesh) and all the states in the north-eastern states, Central plateau (Madhya Pradesh, Chattisgarh, and others), and other states like Tamil Nadu, Karnataka, Orissa, Andhra Pradesh, and Kerala showed higher soil erosion (Mandal et al., 2020). Land resource region-wise, the Black soil region has the highest erosion rate (23.7–112.5 t ha^{-1}) followed by Shivalik soil region (80 t ha^{-1}), shifting cultivated land of North-Eastern region (27–40 t ha^{-1}) and the least (2.1 t ha^{-1}) in forest soil of North Himalayan region. Considering the urgency of the problem, efforts at national and international levels are made to combat desertification and land degradation. Sustainable management of the land to make it neutral from further degradation is now worldwide agenda of many countries.

5.2 Factors Affecting Soil Erosion

There are various factors that impact the rate of soil erosion. Factors such as intensity and amount of rainfall, surface runoff, land steepness (length and gradient), land use and land cover, and presence or absence of conservation practices are responsible for soil erosion. All these factors interplay to cause loss of soil by erosion from any particular location. However, all these factors can be classified broadly into three main categories.

5.2.1 Energy

The energy term indicates the power or potential of natural eroding agents like wind and water to cause soil erosion. Water is the most dominant erosive agent across the globe contributing to more than 80% of soil loss due to erosion. Both the intensity and duration of rainfall are key determining factors for soil erosion. Soil erosion process incepts the phenomenon of the raindrop impact on soil particles/aggregates which disintegrates and detaches the soil particles, the fine particles, in turn, clogs the soil pore spaces and reduces infiltration, thus forming a surface crust. These increase the runoff and soil loss from surface. These various rainfall characteristics have been considered and assimilated to form a single factor called rainfall kinetic energy. The rainfall kinetic energy is used for estimating the rainfall's potential ability to cause erosion, defined as rainfall erosivity.

5.2.2 *Resistance*

The soil has the potential to resist the erosive forces of a raindrop or wind by its inherent or managed characteristics. Of these, an important parameter is the soil aggregate stability which plays a crucial role in determining the soil's ability to resist detachment. Aggregate stability can be enhanced by various management practices and this resistive power of soil is termed soil erodibility. Thus, erodibility is a measure of how vulnerable soil is towards erosion.

5.2.3 *Protection*

After the interaction between the energy of the eroding agent and resistance of soil, the third layer allowing or resisting erosion is soil protection. The different soil conservation measures and protection of the ground surface limit the interface between the erosive agent and soil, thus, preventing soil loss by erosion. The rates of erosion are significantly influenced by vegetation, both natural and cultivated, as well as its residues. The vegetation primarily affects erosion rate through a variety of interventions, including canopy interception of rainfall, physical obstruction of runoff, protection from higher temperatures, surface coverage by litter and residues, improved porosity, improvement of soil strength and structure, soil moisture retention, etc.

5.3 Soil Erosion and Climate Change

Soil loss due to erosion is a significant environmental issue across the globe (Mondal et al., 2016). It leads to the degradation of fertile soils and negatively impacts the environment and agricultural productivity (Mandal et al., 2020). Of the various types of erosion water erosion is the dominant class affecting almost 55% of the eroded land surfaces (Bridges & Oldeman, 1999). According to Garca-Ruiz et al. (2015), rate of soil erosion is very high in semi-arid and sub-humid climates. Thus, it is plausible to assert that soil erosion in the United States, Australia, China, India, and portions of Europe, Africa, and South America is already severe by now and may become worse as a consequence of climate change (Li & Fang, 2016). Climate change is likely to increase the intensity of hydrologic cycles making the precipitation more intense (IPCC, 2007). The most important factor which is likely to be influenced is the erosive power of rainfall thereby increasing soil erosion by water (Borrelli et al., 2020). Climate change affects soil erosion in both direct and indirect ways. The direct impact on soil erosion is by rainfall amount, intensity, and spatiotemporal distribution of rainfall. The indirect consequences are related to increasing global warming potential. Climate change affects the soil moisture status and vegetation cover which will impact soil erosion (Nearing et al., 2004). Ziadat and Taimeh (2013) reported that in cultivated landscapes soil moisture conditions mostly influenced soil erosion while in non-cultivated landscapes it is primarily controlled by the steepness of the slope. Soil erosion is also induced by differences in land use and any change in land use impacted by climate change will alter the soil erosion. Cultivated lands due to lower aggregate stability and higher soil erodibility are more prone to soil erosion compared to scrub land where aggregate stability is better due to lesser soil disturbances (Creda, 2000).

One major impact of climate change is the rise in atmospheric temperature which directly causes an enhancement in the evapotranspiration losses from soil and vegetation. Contrasting to the negative impact of soil erosion lower antecedent soil moisture may result in increased infiltration rates, thereby reducing runoff and soil loss (Xu, 2003). The increasing CO_2 concentration results in higher photosynthetic activity and more biomass production by crops (Rosenzweig & Hillel, 1998). This aids canopy interception and reduces runoff and soil erosion. However, such a positive scenario of climate change is less likely to occur because increased temperature will reduce the moisture availability to crops and negatively impact crop growth. Soil erosion, in particular, may contribute to the intensification of the effects of climate change by drastically lowering soil quality and fertility. Erosion disrupts the soil and degrades the aggregates, resulting in the partial mineralization of soil organic carbon and the release of CO_2 into the atmosphere (Lal, 2004; Lal & Pimentel, 2008). Carbon (C) interactions among land ecosystems and the atmosphere are being altered by global processes such as an increase in temperature and atmospheric CO_2 concentrations (Bond-Lamberty et al., 2018). Warming stimulates biological activities and breakdowns at first, resulting in a reduction in SOC stocks (Li et al., 2016a, 2016b; Melillo et al., 2017).

5.4 Change in Rainfall Erosivity (R-factor) Due to Climate Change

The rainfall erosivity or R factor is the energy component of rainfall which is proportional to the capacity of rain to cause soil erosion. The rainfall erosivity bears a non-linear relationship and is largely dependent on factors like the size of raindrops, intensity, and quantity of effective rain (Salles et al., 2002). Different projections on the influence of climate change on the erosive energy of rainfall in different locations across the globe have been worked out by several researchers. A compilation of selected few is presented in Table 5.1.

Table 5.1 Variation in rainfall erosivity in different areas under different rainfall scenarios (1961–2090)

Sl. No	Area/ Location	Climate Models used	Baseline period	Projected Years	Impact on rainfall erosivity	References
1	Yellow River Basin, Loess Plateau, China	HadCM (Hadley Centre for Climate Prediction and Research) (HadCM3 for A2 and B2** scenario) (IPCC, 2001)	1961–1990	Scenario A2 & B2 2020s 2050s 2080s	+8.5% (A2); +18.2% (B2) +22.1% (A2); + 11.5% (B2) +34.5% (A2); + 19.3% (B2)	Zhang et al. (2005)
2	North-eastern part of the Republic of Uzbekistan (41°10′00″N, 69°45′00″E)	ACCESS1-3 (4.5)	1975–2005	2030s 2050s 2070s	+36% +46% +47%	Gafforov et al. (2020)
		ACCESS1-3 (8.5)_		2030s 2050s 2070s	+43% +49% +55%	
		Average of various climate models		2030s 2050s 2070s	+11.8% +14.1% +16.3%	
3	Huai Luang Watershed, Thailand	Multivariate climate model (IPCC AR5 consisting of CCSM4, CSIRO-MK3.6.0 and others)	1981–2010	2030s 2050s 2070s 2090s	+12% +24% +43% +41%	Pherawat and Udmale (2017)
4	Nepal	Ten GCMs	1961–1990	2070s	+27.8% in high mountain +17.76% in upper hill, + 17.7% in mid hill, + 19.68% in siwalik, + 16.5% in Tarai plain	Talchabhadel et al. (2020)

(continued)

Table 5.1 (continued)

Sl. No	Area/ Location	Climate Models used	Baseline period	Projected Years	Impact on rainfall erosivity	References
5	Brazil	HADGeM 2-ES (RCP 8.5)	1961–2005	2040s	South (−6%) South east (–34%), North (−8%), North east (−17%), Central-west (−15%)	Almagro et al. (2017)
				2070s	South (+5%) South east (−30%), North (−7%), North east (−18%), Central-west (−12%)	
				2100s	South (+11%) South east (−41%), North (−20%), North east (−25%), Central-west (−18%)	
6	Narmada River Basin, India	HADCM3 (Hadley GCM 3 HADCM3(A2) HADCM3(B2) and CGCM3 (Canadian Center for Global Climate Model) CGCM3(A1B) CGCM3(A2)	1961–2001	2020s	−32.91 to + 24.12%	Mondal et al. (2016)
				2050s	−18.82 to + 75.48%	
				2080s	+20.95 to + 202.40%	

** Climate Scenario A2 specifies a very mixed world economy. The underlying theme is self-reliance and safeguarding local identities

The B2 scenario designates a global situation in which more emphasis is given to local solutions to both economic, and environmental sustainability. The population growth here is lower than A2

The precipitation elasticity of rainfall erosivity is generally observed to be more than 1 indicating that any change in total rainfall is amplified when the corresponding rainfall erosivity is considered (Zhang et al., 2005). As per the Universal Soil Loss Equation (USLE), a one percent increase in rainfall erosivity directly increases the soil erosion rates by 1% while other factors are kept constant. Therefore, any increment in rainfall erosivity in different geographical locations across the globe will increase soil erosion manifold. Thus, future projections under different climate change scenarios and how spatially and temporally the rainfall erosivity variation is likely to occur are of great significance to manage the adverse consequences of climate change. For example, the effect of future climate change on rainfall erosivity values in Nepal (Table 5.1) shows a maximum increase in the hilly terrains which are already susceptible to higher soil erosion. Thus, climate-proofing and conservation measures should be convergent with future projections to sustain ecosystems and landforms. Projected climate change scenarios in Brazil show a decline in rainfall erosivity over certain areas which indicates a reduction in the soil erosion rates. Production in the agricultural sector is likely to be enhanced in the North (Amazon) and Central-West (Cerrado) driving land use changes that favours conversion of natural forests into croplands (Morton et al., 2016; Oliveira et al., 2014).

However, the global circulation models (GCMs) which are used for estimating future climate change are not very well equipped to calculate the predicted R-factor values which are necessary to evaluate the GCM output relative to changes in rainfall erosivity (Nearing, 2001).

A modified R factor has been given by Renard and Freimund (1994) which is derived based on strong statistical relationships between the R-factor and both total annual precipitation of any given location and a modified Fournier coefficient (Arnoldus, 1977; Fournier, 1960), F, calculated from monthly rainfall amounts as:

$$F = \frac{\sum_{i=1}^{12} p_i^2}{P} \tag{5.1}$$

where, pi(mm) = average monthly precipitation and P(mm) = average annual precipitation.

The derived relationships between R-factor and P developed by Renard and Freimund (1994) were:

$$\mathrm{R-factor} = 0.04830\mathrm{P}1.610\left(\mathrm{r}^2 = 0.81\right) \tag{5.2}$$

$$\mathrm{R-factor} = 587.8 - 1.219\mathrm{P} + 0.004105\mathrm{P}2\ \left(\mathrm{r}^2 = 0.73\right) \tag{5.3}$$

and the relationships between R-factor and F were

$$\mathrm{R-factor} = 0.7397\mathrm{F}1.847\ \left(\mathrm{r}^2 = 0.81\right) \tag{5.4}$$

$$\mathrm{R-factor} = 95.77 - 6.081\mathrm{F} + 0.04770\mathrm{F}2\ \left(\mathrm{r}^2 = 0.75\right) \tag{5.5}$$

where, the R-factor is in units of MJ mm ha^{-1} $year^{-1}$.

These relationships can be used directly for understanding the trends in R-factor changes as influenced by the precipitation outputs of several GCMs. The second and fourth Equations are better fit for data pertaining to lower end while the third and fifth equations fit better on the upper end. Based on the fit, Eq. 5.2 is recommended for areas where P is <850 mm and Eq. 5.3 where P is greater than 850 mm. Similarly, where F was less than 55 mm Eq. 5.4 is recommended and Eq. 5.5 for locations with F greater than 55 mm is used (Renard & Freimund, 1994) These relationships have been successfully used by different researchers for calculating the precipitation output from GCMs and analyzing the trends in R-factor changes (Nearing, 2001).

5.5 Impact of Climate Change on Soil Erosion from Croplands

Climate change is likely to impact the rates at which soil gets eroded from different land parcels, through changes in different climatic parameters. Of them, the changes in rainfall erosivity are one of the prime causes which will have a direct bearing on soil erosion (Nearing, 2001; Pruski & Nearing, 2002). The variation in the intensity and duration of rainfall is a major consequence of climate change. This variation can have a tremendous impact on crop growth and productivity and it will enhance the risk of extreme natural events like flood and drought, thus directly magnifying the quantum of water erosion-induced soil loss from arable lands. These will also have a major impact on agricultural production, increasing the risk of hunger, water scarcity, changing the behaviour of stream flow and impact the availability of fresh water in different river basins across the country (Gosain et al, 2006). The United Nations Office for Disaster Risk Reduction has estimated that India is likely to suffer direct economic losses to the tune of USD 79.5 billion between 1998 to 2017, most of which are due to climate-related events like flood and drought (Wallemacq et al., UNISDR, and CRED, 2018). A 9% increase in soil erosion rates by 2090 is projected as an influence of climate change (Yang et al., 2003).

Deficit rainfall is predicted to occur in several pockets over eastern Madhya Pradesh, Chhattisgarh, and the Northeast region in Central and Eastern India, as observed in the trend of long-term changes in rainfall patterns across Indian states (Rao et al, 2003). The west coast, northern Andhra Pradesh, and sections of northwest India, on the other hand, exhibit increased rainfall activity of +10 to 12% (National Action Plan on Climate Change, 2008). Changes in the monthly patterns of rainfall are also seen in the southern peninsular region. A comparison of the mean annual rainfall from 1991 onwards with the earlier normal (1961–90) rainfall in Bengaluru showed that the peak monthly rainfall had shifted from September to October. In general, the shift seems to be of about 20–25 days. Similar patterns are also observed at Anantapur and Bijapur. The moisture condition before rainfall largely controls hydrological parameters including infiltration and runoff generation. However, the surface condition and surface cover are also very important controlling factors of infiltration. A runoff would predominantly be regulated by rainfall amount in relatively wetter areas, such as the west coast, northern Andhra Pradesh, and parts of NW India, which is consistent with the saturation or storage control model of runoff generation. In these regions quantity of runoff generated would be directly proportional to the amount of precipitation. Data from all 36 subdivisions of the country show an increase in June and August rainfall contributions, whereas July rainfall contributions have significantly declined trends (Guhathakurta, 2006; Guhathakurata et al., 2015). Of the 36 subdivisions, 19 showed an increase in June rainfall, while the remaining 17 showed a declining trend. Interestingly the areas with reduced July rainfall activity were the ones that showed increased rainfall during August. This is prominent in Central and west Peninsular India.

Rainfall trend analysis for winter rains; pre and post-monsoon activities have also shown significant variations than normal. Eighteen subdivisions witnessed declining rainfall during winter season while Central India has shown a declining trend in the pre-monsoon rainfall. The pre-monsoon showers are mostly caused by convective activity which is declining in the Central part of the country as evidenced by lower rainfall during this period. Contrastingly in the post-monsoon season, a significant increase in rainfall is observed in these sub-divisions in Gujarat viz., Saurashtra and Kutch, Marathwada, and Rayalseema.

In the case of rainfall, the observed southwest monsoon seasonal rainfall at all India levels does not show any significant trend (Rupa Kumar et al, 1992) though regional variations are witnessed. Analysis of rainfall events (between 1951–2000) with intensities of 10, 100, and above 100 mm in Central part of the country (Goswami et al, 2006) indicates that events with increased intensities are of higher frequency while there is a decrease in moderate events. Climate change phenomenon triggers the occurrence of these extreme events which can lead to enhanced soil erosion thereby reducing the productive capacity of the soils (Mandal et al., 2010).

A 10% increase in both rainfall amount and rainfall intensity is projected to occur over the Indian subcontinent in 2050 under a climate change scenario, with a concomitant increase in rainfall erosivity. The link between intensity and rainfall erosivity is not linear, and a 1% increase in intensity could result in a 2.0% rise in rainfall erosivity (Sharda & Ojasvi, 2016). According to Ojasvi et al. (2006), a 1% increase in rainfall intensity may result in a 1.5% increase in soil loss from croplands. Also, this will change the erosion rates and by 2050, the water erosion class with the erosion rates of 5–10 t ha^{-1} yr^{-1} is likely to increase to more than 10 t ha^{-1} yr^{-1}, which is presently designated as the erosion rate causing land degradation. This will bring almost 66 Mha of cropland in our country under degraded category due to climate-induced changes in rainfall unless strong ameliorative measures are employed. These projections suggest that the country should be proactive in developing and updating a comprehensive knowledge base on land degradation scenarios as driven by various degrading agents, for the benefit of various stakeholders. In drier regions, the alteration in runoff production due to changes in soil water regime is more likely to be influenced indirectly through vegetation cover and soil faunal activity than directly. So, any increase in precipitation should result in enhanced protective vegetative cover which in turn produces low runoff. Under natural conditions, the erosion of the surface crust is much better controlled by faunal activity, the vegetative cover, and litterfall which completely neutralize the effect of annual rainfall on runoff coefficient (runoff depth divided by runoff depth stated in percentage terms) at a regional scale. As a result, the runoff coefficient increases with increasing climate aridity.

The influence of climate change on vegetation cover and biomass, and consequently its impact on changes in soil erosion is a complex process (Pruski & Nearing, 2002). Levels and rates of biomass production are affected by temperature alterations in complicated ways. For example, the maize biomass may increase due to increasing temperature and extension in growing season but may decline if temperature is too high (Rosenzweig & Hillel, 1998). The biomass and vegetation impact the soil surface cover which greatly impacts erosion. The alteration in a number of days of snowfall and rainfall changes the erosion rates, as decreased snowfall corresponding to increased rainfall activities will enhance the erosion rates. Soil surface conditions, like surface roughness, sealing, and crusting, may also change with changing climatic conditions, hence impacting erosion rates.

Except for the regions of Rajasthan, Gujarat, and Bihar, where large negative trends are recorded, the geographic patterns of mean annual temperature trends reveal a major positive trend over most of the country. Seasonal temperature trends showed that the post-monsoon season (0.7 °C) saw the largest increase in temperature, followed by the winter season (0.67 °C), the pre-monsoon season (0.5 °C), and the monsoon season (0.3 °C). Rainfall, evapotranspiration rates as well as air temperature determine the soil moisture content at any particular time. Infiltration of water at the soil-air interface is influenced by soil consolidation and surface cover. The canopy and ground cover are dynamic variables that influence infiltration, runoff, and erosion in different interactive ways and this complex interaction can be studied using models like (WEPP) for assessing climate change impacts on erosion.

Borrelli et al. (2020) worked out the estimated erosion rates from different land-use systems across the globe under different Shared Socioeconomic Pathways and Representative Concentration Models (SSP-RCP) (2.6, 4.5, and 8.5). Considering 2015 as the base year, global soil erosion was predicted to be 43 Pg yr^{-1}, out of which 54% of soil erosion is contributed by agricultural land across the globe (~23.4 Pg yr-1). The future scenarios suggest that with SSP-RCP 2.6 if global temperature is allowed to rise by a maximum of 2 °C by 2100, the share of erosion from agricultural land is likely to reduce to 48% from 54% in 2070. However, with SSP-RCP 4.5 and SSP-RCP 8.5 in place, the projections suggested an increase in erosion from arable lands by 56% and 59%, respectively for 2070 (Borrelli et al., 2020). The increase is primarily associated with the increase in agricultural areas in Brazil, Myanmar, Sub-Saharan Africa, India, and some parts of China. However, if average soil erosion rates from all land use are considered, under the three SSP-RCPs, +30% (SSP1-RCP2.6), +51% (SSP2-RCP4.5) and +66% (SSP5-RCP5) increase in soil erosion rates in 2070 compared to base year 2015 is observed. Therefore, it is evident that climate change scenarios will in all probability increase the risk of water-induced soil erosion reducing the productivity of croplands and inducing food and nutritional insecurity.

5.6 Land Degradation in Drylands

Drylands are the regions where the potential evapotranspiration (PET) greatly exceeds precipitation (P) on annual basis. In technical terms, drylands are the region where the aridity index i.e. the ratio of P/PET is less than 0.65 (Feng & Fu, 2013). Based on aridity index, the drylands include hyper arid, arid, semiarid, and dry subhumid areas. Globally, dry lands occupy about 41% of land surfaces (White & Nackoney, 2003) producing nearly 40% of the biomass productivity (Wang et al., 2012). More than 38% of the world's population resides in this area (UNDP, 2014a, 2014b), and 90% of these people reside in under-developing economies. The low soil fertility level of dryland makes this area sensitive to degradation in conjunction with a temperature rising and increased human activities (Li et al., 2016a, 2016b; Zhou et al., 2016). Developing countries suffer the most due to climate change (Huang et al., 2017; Fu and Mao, 2017).

Many researchers have reported that global warming has a significant role in dryland expansion. Altogether, worldwide about 280 M ha has been expanded to dryland with the largest expansion in semi-arid regions since 1960. Based on the converted figures from semiarid to arid and from arid to semiarid it is noted that more land in semiarid regions becomes drier (Huang et al., 2017). This result also emphasized that about 73% of the newly developed semiarid regions are largely converted from sub-humid and humid climatic regions.,

According to the RCP 8.5, it is reported that by 2100, 11% larger area compared to the temporal expansion between 1961–1990 periods, 5Feng and Fu (2013). A major shift of hyperarid area is expected over northern and southern Africa, parts of Saudi Arabia, and Iraq while the main expansions of arid regions will occur over southwest North America, the northern part of Africa, southern Africa, the Middle East, and Australia. The results of two different temporal scales (by the end of 21st-century relative to 1961–1990 showed that under RCP8.5 the world's drylands will expand by 23% compared to RCP4.5 (11%) which means around 56% of the total land surface will come under drylands by the end of this century. Around 78% of the extension will happen in developing economies, making the dryland exposure rate in these countries to the extent of 61% by 2100 under RCP8.5.

Desertification and land degradation data of India is being periodically monitored by SAC, Ahmedabad. The remote sensing-based analysis revealed that a total 97.85 million ha (29.77%) area is degraded by different forms of land degradation during 2018–19 (Fig. 5.1). However, during timeframe 2011–13 and 2003–05 the total degraded land was about 96.40 million ha (29.32% of the TGA) and 94.53 million ha (28.76% of the TGA), respectively (SAC, 2021). A cumulative increase of 1.45 million ha of degraded area (0.44% of the TGA) was observed in 2018–2019 compared to timeframe 2011–13. Whereas, between the timeframe 2003–05

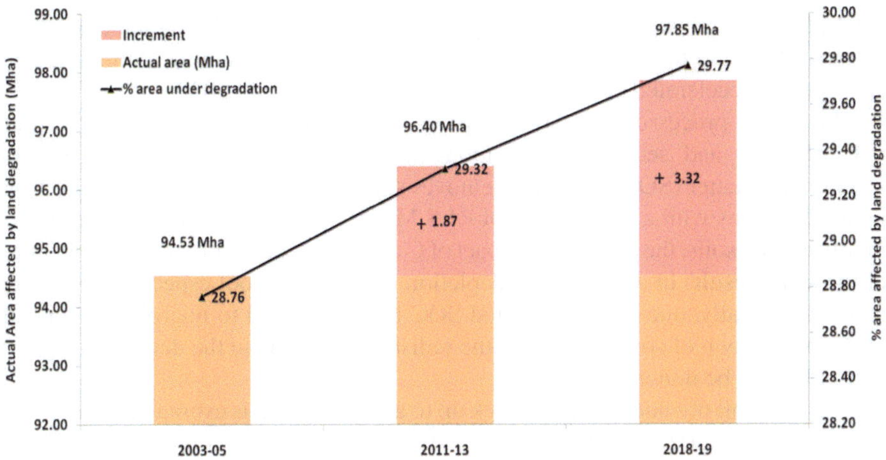

Fig. 5.1 Land degradation status of India in different time scales as per data from Land Degradation Atlas adapted from SAC (2021)

to 2011–13, a cumulative increase of 1.87 million ha area (0.57% of the TGA) was observed. Land degradation in the dryland regions occupy 83.69 million ha area in 2018–19 timeframe. Whereas, the degraded area under dry land during 2011–13 and 2003–05 occupied 82.64 million ha and 81.48 million ha, respectively. Desertification data revealed that there is a cumulative increase of 1.05 million ha area under dryland from timeframe 2011–13 to 2018–19. The increase in area under desertification from 2003–05 to 2011–13 is 1.16 million ha. Wind erosion is the most significant land degradation process in arid regions whereas, water erosion and vegetation degradation processes are dominant processes in semi-arid and sub-humid climatic regions.

5.7 Impact of Climate Change on Soil C Vis-A-Vis Soil Erosion

Soils are the largest reservoir of reactive carbon fractions and terrestrial C pools (Lal, 2013). Soil organic C is one of the major components linked to proper functioning of soil and maintaining soil health. One of the main factors contributing to land degradation is the loss of soil organic C, which has to be controlled to improve soil health and minimize different kinds of degradation. The cycle of soil C loss, climate change, and land degradation is a vicious one, where all three components are intricately linked with each other.

The emission of C from soil during the past two centuries has been a major contributor to the increased atmospheric CO_2 concentration and consequently global warming. Sanderman et al. (2017) have estimated that 116 Pg of C (up to 2 m soil depth) have been lost from agricultural land use (particularly during the conversion of native land use to agricultural land uses). The loss of C is more from areas that are prone to accelerated erosion due to wind and water (Lal., 2020).

The four processes of soil erosion—detachment, breakdown, transport/redistribution, and sediment deposition—all have a significant impact on the soil organic matter (SOM). Erosion removes soil organic matter preferentially from lighter fractions with a low density of <1.8 Mg m^{-3}. In comparison to uneroded or mildly eroded soils, the combined impact of C export and concomitant mineralization of the same results in a substantial depletion of the SOC pool especially in eroded soil. Additionally, due to the protected SOC being exposed to microbes as a result of the breakdown of soil aggregates, the redistributed SOC in the deposition sites is more likely to be mineralized.

The total soil organic C (SOC) stock up to a depth of 1 m is estimated to be 1500–2400 Pg C (Lal., 2020). However, the stability of the SOC stocks is intricately linked with climatic variables like temperature and precipitation. Under changing climate scenarios, the SOC stock may get destabilized, particularly in the temperate areas/permafrost region. This destabilization will not only emit CO_2 into the atmosphere but also increase the prospects of erosion-induced land degradation. The debate of whether soil erosion increases the potential of the soil to sequester more C by making it a sink or aggravates emissions by acting as a source is yet to receive any conclusive climax (Lal, 2003; Lal, 2018; VanOost et al., 2007). The major rift is with the various assumptions about the extent of SOC oxidation caused by erosion. Some researchers are of opinion that the eroded C is protected and very minimum oxidation takes place (Smith et al., 2001) while others believed that eroded SOC is more susceptible to oxidation (Schlesinger, 1995; Van Oost et al, 2007). The fate of the eroded C may be varied depending on various factors:

(a) The composition of the organic matter
(b) Protection of organic matter through aggregation
(c) Environmental covariates like temperature, rainfall
(d) Sediment enrichment ratio

It has been noticed that roughly 20–30% of the eroded SOC is mineralized while being transported with the eroded deposits (Jacinthe & Lal, 2001; Lal, 1995, 2005; Polyakov & Lal, 2008). Contrarily, some studies claim that 1/3 of all carbon emissions from land covers have been offset by the process of erosion since the inception of agriculture (Wang et al., 2017). If a delivery ratio (fraction of sediment delivered to the river system) of 10% is considered, rivers transfer between 15–20 billion Mg of sediment to aquatic habitats, and then ultimately to oceans. If the sediment's SOC content is between 2–6%, this adds up to a loss of 4.0–6.0 Pg C every year as river deposits (Lal, 2003). The actual loss due to mineralization is estimated to be 0.8–1.2 Pg C yr^{-1} if rate of mineralization is 20%. Soil erosion, therefore, has a significant impact on the global C cycle, and this factor must be taken into account when evaluating the global C budget. Implementing conservation-effective strategies may decrease the likelihood of C emission and sequester C in soil and biota.

One-half and one-fourth of all soil degradation are caused by accelerated erosion brought on by water and wind, respectively (Pimentel et al., 1995). Every year, high rates of soil erosion affect 1.1×10^9 hectares of land around the world (Berc et al., 2003; Jacinthe & Lal, 2001). This causes sediment to be transported and redistributed, which causes reservoirs to silt up and lakes to become eutrophicated. Cropland-related soil erosion is projected to result in \$400 billion in annual losses (Pimentel et al., 1995). The mineralization rate of the translocated SOC as a result of erosion is still up for debate (Starr et al., 2000) and justification for either its enhanced mineralization or tempered rates has been quoted by several researchers. The scientific community has focused heavily on the fate of the erosion-induced C, and under the climate change scenarios where erosion will be greatly accelerated; this is of the utmost concern (Berc et al., 2003).

Assessment of erosion-induced C losses is generally based on a comparison of C emissions from disturbed and undisturbed sites (Harden et al., 1999). However, the observations are difficult to be generalize and cannot be extrapolated beyond the study area. One way to overcome this problem is modeling the SOC losses due to water or wind erosion (Starr et al., 2000). For predicting and assessing the erosion-induced release of CO_2 into the atmosphere, integrating field results at various scales under different climatic and soil conditions with current C dynamics is required.

Lugato et al, (2018) expressed the net C exchange between soil and atmosphere using the equation:

$$C_{net} = C_I - C_H - C_{Tm} - C_{Bm} \tag{5.6}$$

where C_I = atmospheric C fixed in litter inputs

C_H = vertical fluxes representing heterotrophic respiration in soil

C_{Tm} = C mineralized during transportation of soil particles after detachment

C_{Bm} = C mineralized in the buried deposits.

The C emission under accelerated erosion scenario under RCP 4.5 using the climate scenario of HADGEM2 climate model between 2016–2100 was 10.1 Tg C yr^{-1} under conditions when erosion induced C sink effect is minimized. Changes in rainfall erosivity led to 35% higher C displacement from the entire European Union and the net loss of C due to accelerated erosion scenario was 2.34 Tg C yr^{-1}.

The erosion-induced C loss based on existing databases is 115.36 Tg C yr^{-1} in the Indian scenario (Mandal et al., 2020). State-wise contribution (top five states) towards erosion displaced C was as follows: Madhya Pradesh (20.33 Tg C yr^{-1}) > Chhattisgarh (9.98 $\pm$ 2.32 Tg C yr^{-1}) > Maharashtra (9.18 $\pm$ 0.91 Tg C yr^{-1}) > Uttar Pradesh (9.11 $\pm$ 2Tg C yr^{-1}) > Andhra Pradesh (8.70 $\pm$ 2 Tg C yr^{-1}) (Mandal et al., 2020). Assuming that 30% of the displaced C is mineralized, the C emitted per year is 34.61 Tg C yr^{-1} which is a significant amount. Also, the amount of C lost from different states was proportional directly to the amount of soil lost by erosion (Mandal et al., 2020). This study provides a strong background to believe that under changing climate scenarios with increasing rainfall erosivity, increased soil erosion will accelerate the loss of soil C and reduce the productivity of already marginal landscapes. Changes in rainfall intensity under different climate change projections

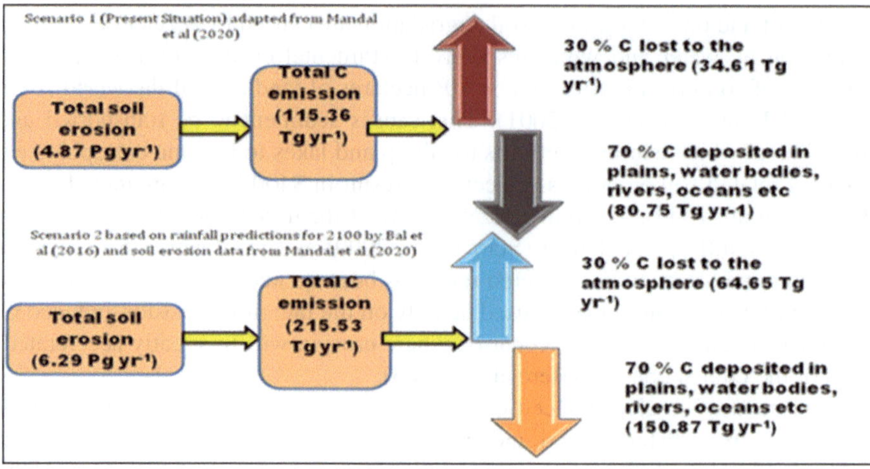

Fig. 5.2 Projected erosion-induced soil Carbon loss during 2020 and 2100

have been worked out by several researchers at different time scales. Bal et al (2016) reported a general change in annual rainfall intensity of India by 15–24% by the end of the century (2100). Assuming a 1.5% increase in soil erosion due to a 1% change in rainfall intensity (Sharda & Ojasvi, 2016), the mean change in soil erosion due to an increase in annual rainfall intensity is calculated to be 29.25%. The projected soil loss by the end of the century (2100) amounts to 6.29 Pg yr^{-1} and the erosion-induced soil C loss concomitantly is 215.53 Tg ha^{-1} yr^{-1}. Figure 5.2 depicts the erosion-induced C loss in the two-time scales.

5.8 Climate Change and Soil Biodiversity: How Does It Impact Erosion?

Agriculture forms the heart of the development of all contemporary societies where cultivated area accounts for almost 38% of the land surface (Foley et al., 2011). Human beings are entirely dependent on the capacity of the soil to uphold production in the agriculture and livestock sector which essentially contributes 95% of global food production (FAO, 2015). Interestingly this very agriculture sector contributing to the sustainability of the population through food production is also one of the major drivers of land degradation. Loss of habitat, changes in species abundance, and biodiversity are all significantly impacted by land degradation. It causes ecological services and biodiversity to deteriorate. India loses 15.7% of its total production of cereal, oilseed, and pulse (COP) crops which is equivalent to ₹292.03 billion in monetary terms as per 2015–16 price indexes. Of the different land degradation processes soil erosion by wind and water is one of the prime degrading agents and is a major threat worldwide (FAO, ITPS, 2015). The major causes of soil erosion are land use changes and climate change-induced extreme weather phenomenon which harms the soil (O'Neal et al, 2005). If we analyze it critically, changes in land use hold a major share of the carbon dioxide emitted into the atmosphere in the post-industrial era which has contributed to global warming and climate change. One major aspect of climate change's impact on soil erosion is the loss of soil organic carbon (SOC) (Lal, 2004; Lal & Pimentel, 2008). SOC is the key to maintaining soil health and essentially allow the soil to perform the different ecosystem services. It is also important for nutrient cycling, fostering microbial growth, and maintaining soil biodiversity. However, the impact of climate change induced on erosion and subsequent linking to soil biodiversity is understudied with a major focus on C sequestration and the C cycle (Nielsen et al., 2015).

The relationship between soil biodiversity and soil erosion losses is multifaceted and non-unidirectional (Orgiazzi & Panagos, 2018). Climate change scenario can lead to two possibilities in terms of soil biodiversity and its impact on soil erosion:

- In the recessive process, increased CO_2 concentration may result in increased photosynthesis, higher above ground biomass, and plant growth, which in turn accelerates the growth of microbes. Under such conditions, the increased microbial activity will lead to better soil aggregation, improved soil structure, and reduced soil erosion.
- The dominant process with a more likely chance of occurrence is increased CO_2, increased temperature, and less precipitation which will have a drastic negative impact on soil moisture and above and below-ground biodiversity of soil. The positive impact of below-ground soil diversity in soil binding and soil aggregate formation will be nullified and result in increased soil erosion.

According to a global study by Guerra et al. (2020), soil erosion will not only affect the soil's vulnerability but also its biodiversity, with 6.4% of these susceptible areas for soil macrofauna and 7.6% of them for soil fungus coexisting with areas with high soil biodiversity. Thus, soil erosion risks the loss or depletion of soil biodiversity from biodiversity-rich areas which is accelerated under the influence of climate change scenarios.

5.9 Policies and Programs in the Promenade of Mitigating CC

The phenomenon of land degradation is pervasive, systemic in occurrence, and negatively influences all life forms on this planet. It is a barrier to harvesting ecosystem services from natural resources to their full potential and results in enormous loss of both above and belowground diversity (IPBES, Assessment Report on Land Degradation, 2018). In the face of climate change, land degradation impacts are likely to be exacerbated and the damages caused may be irreversible, limiting the options to address the degradation problem. Tackling both challenges are also important to achieve the objectives of Sustainable Development Goals (SDG) laid down by the United Nations. All the SDGs are in tandem intimately linked to addressing the problems of land degradation and climate change with SDG 13 specifically addressing climate change. Though SDG13 specifically addresses climate action, however, achievements of all SDGs are intimately linked to addressing problems of land degradation and climate change.

The climate change concerns began in the first Earth summit organized in Stockholm, Sweden between 5 to 16th June 1972. Sixteen years later in 1988, the phenomenon of global warming and ozone layer depletion became more prominent and the Intergovernmental Panel on Climate Change (IPCC) was established by the United Nations Environment Program (UNEP). The year 1989 is often designated as a watershed year for climate change as the two most significant declarations, namely, the "Male' Declaration" (Global warming and Rising Sea Level) and the "Helsinki declaration" (Protection of Ozone Layer) was assumed on May 2, 1989. Side by side, the "Montreal Protocol" on substances that diminishes the ozone layer also entered into force. The climate change problem was revisited time and again by the global community. In the second Earth Summit, held in Rio de Janiro, Brazil, in 1992 the United Nations Framework Convention on Climate Change (UNFCC) was formed to control and stabilize the atmospheric concentrations of greenhouse gases. Later in 1997, the Kyoto Protocol was signed in Japan under the UNFCC which was by far the most influential climate change action. The Kyoto Protocol was promising as it targeted to reduce emissions from industrialized nations by at least 5% below the 1990 level from 2008 to 2012.

Parallel along with climate change, desertification and loss of biodiversity were also identified as major challenges in the path of humanity during the 1992 Rio Earth Summit. Consequently, in 1994 the UN General Assembly created the UNCCD (United Nations Convention to Combat Desertification), later a legal agreement linking the environment and development to sustainable land management which came into force in 1996 (UNEP Story, 2020).

As people understood climate change and land degradation as serious problems, quality of human life and protecting different life forms from anthropogenic harmful chemicals, improving the living conditions as well as improving the quality of the natural resources, gained focus. In 2002, the world summit on Sustainable Development took place in Johannesburg, just a decade before the Rio + 20, the United Nations Conference on Sustainable Development, which was held in 2012.

The positive impacts of various climate actions at the global level were visible as the ozone layer started healing with all due credit to the Montreal Protocol. In 2015, a landmark climate agreement was embarked upon which led to the adoption of 17 Sustainable Development Goals (SDGs) which focussed on life forms, soil, water, air, clean energy, food security, and overall sustainability of the planet. Under the 17 SDGs, achievement of Land Degradation Neutrality placed under SDG 15.3 is one major goal which aims to restore, enhance or stabilize land resources that are essential to support various ecosystem services. All signatory countries have set targets for various SDGs by year 2030. India is committed to the achievement of LDN under which it aims to restore and stabilize 26 Mha of degraded land. In 2018, in a regional workshop of UNCCD, India committed itself to combat land degradation, desertification, and drought. In this regard, the present Government has launched various schemes which target stakeholders at multiple levels to capacitate them in achieving LDN. The several schemes launched are the Pradhan Mantri Fasal Bima Yojna, National food security Mission, Soil Health Card, Pradhan Mantri Krishi Sinchayi Yojna, Per Drop More Crop, Har Khet ko Paani, Swachh Bharat Yojna and National Rural drinking water program (https://pib.gov.in/Pressreleaseshare.aspx?PRID=1530019). All these programs are being implemented and executed at grass root levels to achieve sustainable management of resources.

5.10 Conclusions

The degradation of soil fertility brought on by soil erosion has a detrimental effect on the ecosystem, threatens agriculture, and degrades soil fertility. The accelerating pace of soil erosion is influenced by a variety of anthropogenic processes, including changes in land use and the impact of a changing climate. The United Nation General Assembly has declared the decade 2021–2030 as the United Nations Decade on Ecosystem Restoration. The challenges of climate change and accelerated soil erosion are pressing the world and the survival of the global community. There are numerous projections and highlights, which emphasize the magnitude of the problems of land degradation, and climate change, the planet is going to face in the coming years. Presently, diseases like COVID-19 are threatening the survival of humanity and researchers have linked several important issues like land use changes, degradation of natural resources, declining forest areas, enhanced animal-human conflicts, and climate change to the occurrence of such diseases. If things continue as business as usual, surfacing of such deadly diseases will be more likely. Though action-oriented policies and programs have been launched to combat the problems, the execution of the programs at ground level has to be done more meticulously. Despite the launch of different awareness programs, the visible impacts and their quantification is necessary to make a common man believe in the strength of these programs. This task will also ensure the participation of common mass in management of natural resources. Boer and Hannam (2019, 2021) have argued much on "soil legislation" and making it operational worldwide to protect the soil and manage the problems of land degradation. To achieve LDN in future and ensure the sustainable management of natural resources the countries can have specific legislations, which will help to manage the natural resources more efficiently.

References

Almagro, A., Oliveira, P. T. S., Nearing, M. A., & Hagemann, S. (2017). Projected climate change impacts in rainfall erosivity over Brazil. *Scientific Reports, 7*(1), 1–12.

Arnoldus, H. M. J. (1977). Methodology used to determine the maximum potential average annual soil loss due to sheet and rill erosion in Morocco. https://agris.fao.org/agris-search/search.do?recordID=XF8001961.

Bal, P. K., Ramachandran, A., Palanivelu, K., et al. (2016). Climate change projections over India by a downscaling approach using PRECIS. *Asia-Pacific Journal of Atmospheric Sciences, 52*(4), 353–369.

Berc, J. et al. (2003). *Conservation implications of climate change: Soil erosion and runoff from croplands: A report from the soil and water conservation society*. Ankeny (IA), Soil and Water Conservation Society.

Boer and Hannam. (2019). B. Boer, I. Hannam Land degradation E Lees, J. Viñuales (Eds.), *The Oxford Handbook of Comparative Environmental Law*, Oxford University Press, pp. 438–459.

Boer and Hannam. (2021). B. Boer, I. Hannam Restoration of ecosystems and land degradation neutrality N. Kakar, N. Robinson, V. Popovski (Eds.), *Fulfilling the Sustainable Development Goals*, Routledge, Abingdon, pp. 392–404.

Bond-Lamberty, B., Bailey, V. L., Chen, M., et al. (2018). Globally rising soil heterotrophic respiration over recent decades. *Nature, 560*(7716), 80–83.

Borrelli, P., Robinson, D. A., Panagos, P., et al. (2020). Land use and climate change impacts on global soil erosion by water (2015–2070). *Proceedings of the National Academy of Sciences, 117*(36), 21994–22001.

Bridges, E. M., & Oldeman, L. R. (1999). Global assessment of human-induced soil degradation. *Arid Soil Research and Rehabilitation, 13*(4), 319–325.

Cerda, A. (2000). Aggregate stability against water forces under different climates on agriculture land and scrubland in southern Bolivia. *Soil and Tillage Research, 36*, 1–8.

FAO, Healthy Soils Are the Basis for Healthy Food Production (Food and Agriculture Organization of the United Nations [FAO], 2015), pp. 1–4. http://www.fao.org/documents/card/en/c/645883cd-ba28-4b16-a7b8-34babbb3c505/.

FAO, ITPS. (2015). *The Status of the World's Soil Resources (Main Report).*

Feng, S., & Fu, Q. (2013). Expansion of global drylands under a warming climate. *Atmospheric Chemistry and Physics, 13*(19), 10081–10094.

Foley, J. A., Ramankutty, N., Brauman, K. A., et al. (2011). Solutions for a cultivated planet. *Nature, 478*(7369), 337–342.

Fournier, F. (1960). *Climat et érosion* (p. 201). Presses universitaires de France.

Fu, C., & Mao, H. (2017). *Aridity Trend in Northern China*, World Sci., Singapore.

Gafforov, K. S., Bao, A., Rakhimov, S., et al. (2020). The assessment of climate change on rainfall-runoff erosivity in the Chirchik-Akhangaran Basin. *Uzbekistan. Sustainability, 12*(8), 3369.

García-Ruiz, J. M., Beguería, S., Nadal-Romero, E., et al. (2015). A meta-analysis of soil erosion rates across the world. *Geomorphology, 239*, 160–173.

Gosain, A. K., Rao, S., & Basuray, D. (2006). Climate change impact assessment on hydrology of Indian river basins. *Current Science*, 346–353.

Goswami, B. N., Venugopal, V., Sengupta, D., et al. (2006). Increasing trend of extreme rain events over India in a warming environment. *Science, 314*(5804), 1442–1445.

Guerra, C. A., Rosa, I. M., Valentini, E., et al. (2020). Global vulnerability of soil ecosystems to erosion. *Landscape Ecology, 35*, 823.

Guhathakurta, P. (2006). Long-range monsoon rainfall prediction of 2005 for the districts and sub-division Kerala with artificial neural network. *Current Science, 90*(6), 773–779.

Guhathakurta, P., Rajeevan, M., Sikka, D. R., & Tyagi, A. (2015). Observed changes in southwest monsoon rainfall over India during 1901–2011. *International Journal of Climatology, 35*(8), 1881–1898.

Harden, J. W., Sharpe, J. M., Parton, W. J., et al. (1999). Dynamic replacement and loss of soil carbon on eroding cropland. *Global Biogeochemical Cycles, 13*(4), 885–901.

https://pib.gov.in/Pressreleaseshare.aspx?PRID=1530019. Accessed 27 January, 2021.

https://www.iucn.org/resources/issues-briefs/conserving-healthy-soils. Accessed 5 February, 2021.

https://www.unep.org/news-and-stories/story/environmental-moments-un75-timeline. Accessed 2 February, 2021.
Huang, J., Xie, Y., Guan, X., et al. (2017). The dynamics of the warming hiatus over the Northern Hemisphere. *Climate Dynamics, 48*(1–2), 429–446.
IPBES. (2018). The IPBES assessment report on land degradation and restoration. L. Montanarella, R. Scholes, & A. Brainich (eds.). *Secretariat of the Intergovernmental Science-Policy Platform on Biodiversity and Ecosystem Services*, Bonn, Germany. 744 pp. https://doi.org/10.5281/zenodo.3237392.
IPCC CWT. (2007). Climate change 2007: synthesis report (p. 104). Geneva, Switzerland: IPCC.
Jacinthe, P. A., & Lal, R. (2001). A mass balance approach to assess carbon dioxide evolution during erosional events. *Land Degradation & Development, 12*(4), 329–339.
Lal, R. (1995). Erosion-crop productivity relationships for soils of Africa. *Soil Science Society of America Journal, 59*(3), 661–667.
Lal, R. (2003). Soil erosion and the global carbon budget. *Environment International, 29*, 437–450.
Lal, R. (2004). Soil carbon sequestration impacts on global climate change and food security. *Science, 304*(5677), 1623–1627.
Lal, R. (2005). Forest soils and carbon sequestration. *Forest Ecology and Management, 220*(1–3), 242–258.
Lal, R. (2013). Soil carbon management and climate change. *Carbon Management, 4*(4), 439–462.
Lal, R. (2018). Accelerated soil erosion as a source of atmospheric CO_2. *Soil Tillage Research*. https://doi.org/10.1016/j.still.2018.02.001
Lal, R. (2020). Soil erosion and gaseous emissions. *Applied Sciences, 10*(8), 2784.
Lal, R., & Pimentel, D. (2008). Soil erosion: A carbon sink or source? *Science, 319*(5866), 1040–1042.
Li, J., Liu, Z., He, C., et al. (2016a). Are the drylands in northern China sustainable? A perspective from ecological footprint dynamics from 1990 to 2010. *Science of the Total Environment, 553*, 223–231.
Li, Y., Zhou, G., Huang, W., et al. (2016b). Potential effects of warming on soil respiration and carbon sequestration in a subtropical forest. *Plant and Soil, 409*, 247–257. https://doi.org/10.1007/s11104-016-2966-2
Li, Z., & Fang, H. (2016). Impacts of climate change on water erosion: A review. *Earth-Science Reviews, 163*, 94–117.
Lugato, E., Smith, P., Borrelli, P., et al. (2018). Soil erosion is unlikely to drive a future carbon sink in Europe. *Science Advances, 4*(11), 3523.
Mandal, D., Giri, N., & Srivastava, P. (2020). The magnitude of erosion-induced carbon (C) flux and C-sequestration potential of eroded lands in India. *European Journal of Soil Science, 71*(2), 151–168.
Mandal, D., & Sharda, V. N. (2011). Assessment of permissible soil loss in India employing a quantitative bio-physical model. *Current Science, 100*(3), 383–390.
Mandal, D., & Sharda, V. N. (2013). Appraisal of soil erosion risk in the eastern Himalayan region of India for soil conservation planning. *Land Degradation and Development, 24*, 430–437.
Mandal, D., Chandrakala, M., Alam, N. M., et al. (2021). Assessment of soil quality and productivity in different phases of soil erosion with the focus on land degradation neutrality in tropical humid region of India. *CATENA, 204*, 105440. https://doi.org/10.1016/j.catena.2021.105440
Mandal, D., Sharda, V. N., & Tripathi, K. P. (2010). Relative efficacy of two biophysical approaches to assess soil loss tolerance for Doon Valley soils of India. *Journal of Soil and Water Conservation, 65*, 42–49.
Melillo, J. M., Frey, S. D., DeAngelis, K. M., et al. (2017). Long-term pattern and magnitude of soil carbon feedback to the climate system in a warming world. *Science, 358*(6359), 101–105.
Mondal, A., Khare, D., & Kundu, S. (2016). Impact assessment of climate change on future soil erosion and SOC loss. *Natural Hazards, 82*(3), 1515–1539.

Morton, D. C., Noojipady, P., Macedo, M. M., et al. (2016). Reevaluating suitability estimates based on dynamics of cropland expansion in the Brazilian Amazon. *Global Environmental Change, 37*, 92–101.

Nearing, M. A. (2001). Potential changes in rainfall erosivity in the US with climate change during the 21st century. *Journal of Soil and Water Conservation 56*(3), 229–232.

Nearing, M. A., Pruski, F. F., & O'neal, M. R. (2004). Expected climate change impacts on soil erosion rates: a review. *Journal of Soil and Water Conservation, 59*(1), 43–50.

Nielsen, U. N., Wall, D. H., & Six, J. (2015). Soil biodiversity and the environment. *Annual Review of Environmental Resources, 40*, 63–90.

O'Neal, M. R., Nearing, M. A., Vining, R. C., et al. (2005). Climate change impacts on soil erosion in Midwest United States with changes in crop management. *CATENA, 61*, 165–184.

Ojasvi, P. R., Sharda, V. N., & Prakash, O. (2006). Evaluation of interrill erodibility parameter for soil erosion estimation in a sub-humid climate, Part I. *Development of model using simulated rainfall data. Indian Journal of Soil Conservation, 34*(3), 178–182.

Oliveira, P. T. S., Nearing, M. A., Moran, M. S., et al. (2014). Trends in water balance components across the Brazilian Cerrado. *Water Resources Research, 50*(9), 7100–7114.

Van Oost, K., Quine, T. A., Govers, G., et al. (2007). The impact of agricultural soil erosion on the global carbon cycle. *Science, 318*(5850), 626–629.

Orgiazzi, A., & Panagos, P. (2018). Soil biodiversity and soil erosion: It is time to get married: Adding an earthworm factor to soil erosion modelling. *Global Ecology and Biogeography, 27*(10), 1155–1167.

Pheerawat, P., & Udmale, P. (2017). Impacts of Climate change on Rainfall Erosivity in the Huai Luang Watershed. *Thailand. Atmosphere, 8*(8), 143.

Pimentel, D., Harvey, C., Resosudarmo, P., et al. (1995). Environmental and economic costs of soil erosion and conservation benefits. *Science, 267*(5201), 1117–1123.

Polyakov, V. O., & Lal, R. (2008). Soil organic matter and CO_2 emission as affected by water erosion on field runoff plots. *Geoderma, 143*(1–2), 216–222.

Pruski, F. F., & Nearing, M. A. (2002). Climate-induced changes in erosion during the 21st century for eight US locations. *Water Resources Research, 38*(12), 34–41.

Rao, A. R., Hamed, K. H., & Chen, H. L. (2003). *Nonstationarities in hydrologic and environmental time series* (Vol. 45). Springer Science & Business Media.

Renard, K. G., & Freimund, J. R. (1994). Using monthly precipitation data to estimate the R-factor in the revised USLE. *Journal of Hydrology, 157*(1–4), 287–306.

Rosenzweig, C., & Hillel, D. (1998). *Climate change and the global harvest: Potential impacts of the greenhouse effect on agriculture.* Oxford University Press.

Rupa Kumar, K., Pant, G. B., Parthasarathy, B., & Sontakke, N. A. (1992). Spatial and sub-seasonal patterns of the long-term trends of Indian summer monsoon rainfall. *International Journal of Climatology, 12*, 257–268.

SAC. (2021). Desertification and Land Degradation Atlas of India. Space Applications Centre Indian Space Research Organisation Department of Space, Government of India, Ahmedabad.

Salles, C., Poesen, J., & Sempere-Torres, D. (2002). Kinetic energy of rain and its functional relationship with intensity. *Journal of Hydrology, 257*, 256–270.

Sanderman, J., Hengl, T., & Fiske, G. J. (2017). Soil carbon debt of 12,000 years of human land use. *Proceedings of the National Academy of Sciences, 114*(36), 9575–9580.

Schlesinger, W. H. (1995). Soil respiration and changes in soil carbonstocks. In G. M. Woodwell & F. T. Mackenzie (Eds.), *Biotic Feedbacks in the Global Climate System: Will the Warming Feed the Warming?* (pp. 159–168). Press, New York.

Sharda, V. N., & Ojasvi, P. R. (2016). A revised soil erosion budget for India: Role of reservoir sedimentation and land-use protection measures. *Earth Surface Processes and Landforms, 41*(14), 2007–2023.

Smith, S. V., Renwick, W. H., Buddenmeier, R. W., & Crossland, C. J. (2001). Budgets of soil erosion and deposition for sediments and sedimentary organic carbon across the conterminous United States. *Global Biogeochemical Cycles, 15*, 697–707.

Starr, G. C., Lal, R., Malone, R., et al. (2000). Modeling soil carbon transported by water erosion processes. *Land Degradation & Development, 11*(1), 83–91.

Talchabhadel, R., Nakagawa, H., Kawaike, K., & Prajapati, R. (2020). Evaluating the rainfall erosivity (R-factor) from daily rainfall data: An application for assessing climate change impact on soil loss in Westrapti River basin. *Nepal. Modeling Earth Systems and Environment, 6*(3), 1741–1762.

UNCCD. (2017). Global Land Outlook. First. UNCCD, Bonn, Germany, 336 pp. United Nations, Geneva, Switzerland.

UNEP. (1992). World Atlas of Desertification [Middleton, N. and D.S.G. Thomas (eds.)]. UNEP.Edward Arnold, London, UK, 69 pp.

United Nations Development Programme (UNDP). (2014a). Environment and energy, drylands development centre, where we work. Accessed 19 Jan 2014.

United Nations Development Programme (UNDP). (2014b). Environment and energy, drylands development centre, where we work, Accessed 19 Jan 2014.

Wallemacq, P., Below, R., & McLean, D. (2018). UNISDR and CRED report: Economic Losses, Poverty & Disasters (1998–2017); Published in: CRED.

Wang, L., D'Odorico, P. J. P., Evans, D. J., et al. (2012). Dryland ecohydrology and climate change: Critical issues and technical advances. *Hydrology and Earth System Sciences, 16*(8), 2585–2603.

Wang, Z., Hoffmann, T., Six, J., et al. (2017). Human induced erosion has offset one-third of carbon emissions from land cover change. *Nature Climate Change, 7*, 345–349.

White, R. P., & Nackoney, J. (2003). *Drylands, People and Ecosystem Goods and Services*, World Resources Institute, Washington.

Xu, J. (2003). Sedimentation rates in the lower Yellow River over the past 2300 years as influenced by human activities and climate change. *Hydrological Processes, 17*(16), 3359–3371.

Yang, D., Kanae, S., Koike, T., & Musiake, K. (2003). Global potential soil erosion with reference to land use and climate changes. *Hydrological Processes, 17*, 2913–2928.

Zhang, G. H., Nearing, M. A., & Liu, B. Y. (2005). Potential effects of climate change on rainfall erosivity in the Yellow River basin of China. *Transactions of the ASAE, 48*(2), 511–517.

Zhou, L., Chen, H., Hua, W., et al. (2016). Mechanisms for stronger warming over drier ecoregions observed since 1979. *Climate Dynamics, 47*(9), 2955–2974.

Ziadat, F. M., & Taimeh, A. Y. (2013). Effect of rainfall intensity, slope, land use and antecedent soil moisture on soil erosion in an arid environment. *Land Degradation & Development, 24*(6), 582–590.

Chapter 6
Response and Behavior of Paddy Soil Microbiota Towards Environmental Change

Upendra Kumar, Subhra Parija, and Megha Kaviraj

Abstract Climate change as a global issue is characterized by alterations in assessable parameters of climate and other environmental factors, at an alarming rate. These factors are potent enough in modifying the existing dynamics of soil microbiomes in crop fields. Recent studies highlight that the formation of microbial communities associated with roots is selectively influenced by the host plants. Advancements in molecular technologies like high throughput sequencing and ecological network analysis have proved to be revolutionary tools in revealing the microbial diversity, structural community, and their interactions along with determining their habitat affinities with the changed environmental factors. Altered environmental factors like elevated CO_2, high temperature, imbalanced fertilizer application, change in moisture regimes, pesticide pollution, etc. have been discussed in this chapter along with their influence on the paddy soil microbiome. A shift in the establishment patterns, diversities, community structure, and functioning is observed significantly owing to these climatic and environmental variations. Crops gradually tend to acclimatize to different adverse external stress conditions generated by environmental change factors using their inbuilt biological mechanisms. The plant and its microbiome interact with each other through different metabolic pathways and form stress tolerance strategies. This chapter, therefore, aims to outline the major environmental change factors and how they have affected the soil microbes, their community composition, and functioning abilities in paddy fields, along with addressing some of the mitigation strategies mediated by microbes in soil nutrient cycle.

Keywords Climate change · Elevated CO_2 · High temperature · Greenhouse gas emissions · Soil microbiome · Microbial diversity · Microbial community structure

U. Kumar (✉) · S. Parija · M. Kaviraj
ICAR National Rice Research Institute, Cuttack 753 006, Odisha, India
e-mail: ukumarmb@gmail.com; upendra.kumar1@icar.gov.in

H. Pathak et al. (eds.), *Climate Change Impacts on Soil-Plant-Atmosphere Continuum*, Advances in Global Change Research 78, https://doi.org/10.1007/978-981-99-7935-6_6

6.1 Introduction

Currently, one of the world's most pressing issues is climate change and this major environmental issue is mostly linked to a raise in greenhouse gas (GHG) emissions (Ritchie & Roser, 2020). China currently leads the list of top five GHG generating countries, emitting over twice as much as the second-highest emitter, the United States followed by India, Indonesia, and Russia. According to the 5th Intergovernmental Panel on Climate Change (IPCC), global average land temperatures have increased roughly twice as much as the ocean temperatures. Up to one-third of all human GHG emissions are attributed to fertilizer production, application, storage, and their packaging, according to the most recent data from the Consultative Group on International Agricultural Research (CGIAR) (Gilbert, 2012). Thus, GHG emissions in the atmosphere are rising and ultimately resulting in increasing the global warming potential day by day.

The topmost global soil layer stores around 1500 Pg of total C and around 136 (92–140) Pg of total N, constituting the greatest terrestrial carbon and nitrogen reserves in the world (Padbhushan et al., 2021, 2022). Important climate-relevant trace gases such as CH_4, N_2O, and CO_2 are unevenly distributed (by 3–4 orders of magnitude for carbon) in the atmosphere (Kutsch et al., 2009; Nieder & Benbi, 2008; Schaufler et al., 2010). Previous case studies have shown that the lowest (night) temperature has been expanding quicker than the outside temperature during the last 100 years. Thus, the concentrations of atmospheric GHGs have increased (Easterling et al., 1997).

Among the GHGs, CO_2 is mainly responsible for global warming (Lacis et al., 2010). The cumulative warming effect of the greenhouse gasses produced in the Earth's atmosphere from 1990 to 2019 has increased by 45%, whereas that due to CO_2 alone has increased by 36% (NOAA, 2020). Agriculture, forest management, and other land use account for around 24% of direct CO_2 emissions, while industrial emissions account for 21%. CO_2 is also responsible for 20% of thermal adsorption, and its natural sources are organic degradation and respiration (Cassia et al., 2018). Therefore, CO_2 is considered the most significant greenhouse gas emitted by human activities, and it also raises global temperatures along with altering the precipitation patterns.

Agriculture is expected to account for 10–12% of total GHG emissions anthropogenically, with 60% of global N_2O and 50% of global CH_4 emissions, respectively (Smith et al., 2007). Based on agricultural management, animal feeds and production of rice were identified as the key sources of emissions of GHGs (Mohanty et al., 2020), while other cereals, fruits, and vegetables emit significantly less GHG emissions (Vetter et al., 2017). Climate change, as a global issue characterized by rising temperatures, is having an impact on agricultural productivity and quality (DaMatta et al., 2010) and could have major positive or negative effects on paddy cultivation and straw decomposition, depending on its geographical locations.

Climate change is not only confined to increasing levels of atmospheric carbon or increasing global temperature but also is associated with numerous other factors like water stress conditions, chemical pollution, changes in land use and soil reactions, etc. These factors are equally effective in modifying the existing dynamics of soil microbiomes. Hence, this chapter aims to outline the major environmental change factors and how they have affected the structure and function of soil microorganisms in paddy fields, along with addressing some of the mitigation strategies mediated by microbes in soil nutrient cycle.

6.2 Climate Change and Soil Microbiome: A Priority Research

Soil microbes are an essential component of the ecosystem contributing to the sustainability of agricultural production. The rhizosphere being the hot spot for microbial activity in soil harbors numerous plant–microbe interactions thereby resulting in the most complex ecosystem (Mendes et al., 2013). Scientific research has recently begun concentrating on the assembly and variation of plant and soil microbiomes in various rhizospheric compartments, including bulk soils, rhizosphere, rhizoplane, and endosphere. These studies have also suggested that host plants may have a selective influence on the emergence of root-associated microbial communities (Edwards et al., 2015; Kumar et al., 2018a, 2018b, 2019, 2021a, 2021b; Santos-Medellín et al., 2017). Microbes inhabiting the rhizosphere might prove beneficial in the acquisition of plant nutrients and providing resistance to various types of stress either directly or synergistically, as a result of their functional traits (Lu et al., 2018). Numerous factors influencing the community assembly of soil microbes have been identified by in-depth characterizations of root-associated microbiomes in a variety of plant systems (Wagner et al., 2016; Banik et al., 2019).

Soil microbial biomass (SMB) is considered to be a source as well as a sink for plant nutrients and plays an active part in nutrient cycling. Since it is affected by changes in the physical and chemical soil environments, although SMB is an indicator of basic soil fertility (Shibahara & Inubushi, 1997). Advancements in molecular technologies like high-throughput sequencing and ecological network analysis have proved to be revolutionary tools in revealing the microbial diversity, structural community, and their interactions along with determining their habitat affinities about the changed environmental factors (Collavino et al., 2014).

To better understand how the network of interactions between soil microbiomes is impacted by difficult environmental conditions, it is imperative to examine the impact of extreme environments on soil microbiome diversity and community composition. Finding the root-associated microorganisms that flourish in unfavourable conditions and discovering the positive interplay of microbial characteristics providing stress resistance can both be aided by research in this direction (Rodriguez et al., 2016). The various climate change factors exhibiting a strong effect on rice soil microbiota are presented in Fig. 6.1.

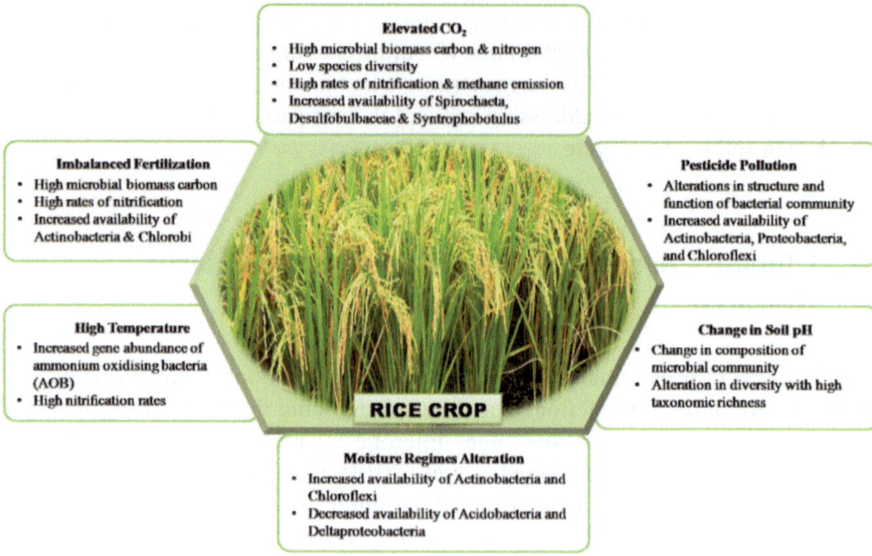

Fig. 6.1 Schematic diagram of different environmental factors and their effect on rice soil microbiome

6.3 Impact of Drivers of Climate Change on Paddy Soil Microbiota

6.3.1 Elevation in Carbon Dioxide (eCO_2) Levels

Changes in atmospheric CO_2 concentration have a significant impact on microbial activity because the majority of soil organisms are heterotrophic and dependent on organic C intake (Drigo et al., 2008; Mohanty et al., 2017). But as the amount of CO_2 in the soil is greater than the atmospheric CO_2 content, it is not so likely that elevated levels of CO_2 in the atmosphere will have a direct influence on the rhizospheric microbiome. However, it might so happen that increased CO_2 levels in the atmosphere might impose a direct effect on soil–plant-microbe association through rhizodeposition and root growth. According to Bhattacharyya et al. (2014), increased rates of photosynthesis under rising CO_2 lead to increased partitioning of photosynthates in the underground plant parts and considerable changes in the functional and structural diversity of soil microorganisms (Kumar et al., 2020a, 2020b; Panneerselvam et al., 2019; Sahoo et al., 2017). By enhancing the metabolism in plants and secretion of root exudates in C_3 plants, elevated atmospheric CO_2 indirectly affects soil microorganisms (Willey et al., 2009). These compounds and secretions encourage the growth of bacteria, fungi, and mycorrhizae in the rhizosphere. The increased amount of rhizo-deposited materials not only alters the microbial population but also transforms the structure and function of their community (Sadowsky & Schortemeyer, 1997).

Increased concentration of atmospheric CO_2 results in higher microbial biomass carbon and nitrogen. A considerable rise in the labile N fractions like microbial biomass-N is observed under elevated CO_2 conditions (Table 6.1). Rhizospheric denitrifier population was reported to be increased by 39% while there was a decrease in nitrifier population by 41% under elevated CO_2 as compared to ambient CO_2 conditions. The population of both nitrifier and denitrifier communities is found to be high at panicle initiation (PI) in comparison to maximum tillering (MT) stage of rice (Kumar et al., 2021a, 2021b).

The biological nitrogen fixation ability of the microbes is enhanced due to increased abundance of *nifH* gene under elevated CO_2 levels, also leading to higher acetylene reductase activity (Inubushi et al., 2002). Elevated CO_2 increases dissolved methane in soil, resulting in intensified methane flux and emission from rice canopy (Ziska et al., 1998). According to Xu et al. (2020), increased CO_2 caused significant losses of ammonium-nitrogen due to anaerobic ammonium oxidation coupled with iron reduction. They discovered a brand-new autotrophic *Burkholderiales* bacterium that used soil CO_2 to anaerobically connect the ammonium oxidation and iron reduction processes.

Microbial biomass, species richness, and fungal abundance were increased under elevated CO_2 (de Vries & Bardgett, 2014; Hu et al., 2001; Lipson et al., 2014; Procter et al., 2014). Amplification in the number of bacterial groups like Acidobacteria, Chloroflexi, Proteobacteria, and Nitrospirae, Planctomycetes is also observed higher under elevated CO_2 (Castro et al., 2010). Yu et al. (2018) reported that in highly responsive paddy cultivars, there is a marked reduction in α-diversity. There is a sharp rise in Spirochaeta, Desulfobulbaceae, and Syntrophobotulus phylotypes whereas the phylotypes Azospira, Thiorhodospira, Accumulibacter, and Cupriavidus suffer a reduction in their richness. According to Luo et al. (2021), oligotrophic bacteria such as Acidobacteriaceae and Bryobacteraceae grew well at elevated CO_2 levels rather than copiotrophic bacteria, resulting in community recovery and improved carbon utilization efficiency. Another study found that, except a 3.5-fold drop in the relative abundance of Acidobacteria, the bacterial biomass, richness, and community composition had all changed with rising CO_2 (Dunbar et al., 2012). Changes in community shift was also observed for Pseudomonas (Marilley et al., 1999) and Rhizobium species (Montealegre et al., 2000) under influence of elevated CO_2. When the plants were grown in high CO_2 conditions, the proportion of Pseudomonas that produced HCN likewise reduced (Tarnawski & Aragno, 2006).

Table 6.1 Responses of rice soil microbiota towards environmental change factors

Study Site	Microbiome	Methodology	Factors	Effects	References
Shizukuishi & Tsukuba, Japan	Cyanobacteria & Methanogens	FACE (free-air CO_2 enrichment), Growth Chamber Experiment	Elevated CO_2(eCO_2)	• Increase in acetylene reductase activity • Increase in microbial biomass carbon (MBC) and nitrogen (MBN) • Increase in total methane emission	Inubushi et al. (2002)
IRRI, Philippines (121°15′ E, 14°13′ N)	Methanogens	Open Top Chamber (OTC) Experiment	Elevated CO_2(eCO_2)	• Increase in dissolved soil methane • Increase in methane flux from rice canopy	Ziska et al. (1998)
Yanghzou City, China (119°42′ E, 32°35″5′ N)	Diazotrophs	FACE (free-air CO_2 enrichment)	Elevated CO_2(eCO_2)	• Decrease in α-diversity for strong responsive cultivars • Increase in abundance of nifH gene • Increase in Spirochaeta, Desulfobulbaceae, & Syntrophobotulus and decrease in Azospira, Thiorhodospira, Accumulibacter & Cupriavidus phylotypes	Yu et al. (2018)
NRRI, Cuttack, Odisha (80°85′ E, 20°25′ N)	Diazotrophs	OTC Experiments, Illumina MiSeq Sequencing	Elevated CO_2 (eCO_2), Higher dose of N (HDN) & Water Deficit Stress (WDS)	• Increase in MBC under eCO_2 and HDN • Higher abundance of *nifH* gene under HDN • Abundance of Cyanobacteria & Proteobacteria and suppression of Actinobacteria & Chlorobi under $eCO_2$and the reverse under HDN	Kumar et al. (2020a, 2020b)

(continued)

Table 6.1 (continued)

Study Site	Microbiome	Methodology	Factors	Effects	References
Jingzhou, Hubei Province, China (112°09′ E, 30°21′ N)	Ammonia oxidizing archaea (AOA) & Ammonia oxidizing bacteria (AOB)	OTC Experiments, qPCR, High Throughput Sequencing	Elevated CO_2 & High temperature	• Low sensitivity of AOA community than AOB • Enhancement in AOB gene abundance and nitrification rates	Waqas et al. (2021)
California, USA	Endospheric & Rhizospheric Bacteria	High Throughput Sequencing	Drought stress	• Enrichment of Actinobacteria and Chloroflexi • Depletion of Acidobacteria and Deltaproteobacteria	Medellín et al. (2017)
Sanghai, China	Rhizospheric Bacteria (Cyanobacteria, Acidobacteria, Actinobacteria, Proteobacteria, etc.) & Fungi (Zygomycota, Basidiomycota, Glomeromycota, Ascomycota, etc.)	qPCR, Illumina MiSeq Sequencing	Cultivation of drought tolerant & insect resistant cultivars	• Lower abundance of bacterial population • Structural alterations in bacterial communities • Insignificant effect on structure and abundance of fungal communities	Li et al. (2018)
Maritsa River Basin, Bulgaria	Proteobacteria, Actinobacteria, & Acidobacteria	High Throughput Sequencing	Soil pH and Hydrological regime	• Variation in α-diversity with highest taxonomic richness	Ilieve et al. (2019)
Davis, California, USA	Rhizospheric microbes	Meta-analysis	Imbalanced Fertilization	• Increase in MBC • Change in community composition through changes in soil properties	Geisseler et al. (2017)
Jiaxing, Zhejiang, China	Rhizospheric bacteria & fungi	Green House Experiment Nano drop Spectroscopy PCR	Organochlorine (Lindane) pesticide pollution	• Significant alterations in bacterial community as compared to fungal taxa • Stable community structure of Actinobacteria, Proteobacteria, and Chloroflexi for hybrid cultivars under Lindane stress	Feng et al. (2019)

6.3.2 Rise in Atmospheric Temperature

Temperature rise leads to a shift in microbial abundance from highly sensitive communities to less sensitive ones (Table 6.1). The annual variation in the temperature of the environment has an impact on the microbial activity in the soil. Lower than 5 °C temperatures decrease soil microbial activity (Singurindy et al., 2009), whereas higher soil temperatures (at least up to 37 °C) increase microbial activity and soil respiration, which reduces soil oxygen levels and increases diffusion in soil medium under the availability of ideal moisture conditions. An increase in temperature encourages the growth of bacteria and fungi (Castro et al., 2010; de Vries & Bardgett, 2014). However, because there are fewer labile substrates in environments with elevated temperatures on their own, which favors fungal community members over bacteria, the effect of the relationship between CO_2 and temperature is more sensitive to bacterial abundance (Castro et al., 2010).

A drastic change in climatic conditions with a consistent rise in global temperature suggests that the oxidation of ammonia in paddy soil must be largely regulated by the ammonia-oxidizing bacterial communities. There is an increased diversity of ammonia-oxidizing archaea (AOA) as compared to the ammonia-oxidizing bacteria (AOB) in paddy soils owing to a rise in the atmospheric temperature. High temperatures result in high ammonification and nitrification rates by increasing the copy numbers of related genes of the associated microbes involved in nitrogen mineralization (Waqas et al., 2021). As the soil temperature drops, the N_2O:N_2 ratio during denitrification increases (Maag and Vinther 1996). On account of the improved activity of thermophilic nitrate respirers, chemo-denitrification of NO_2^- predominates at higher temperatures (>50 °C) (Dalal et al., 2003).

6.3.3 Change in Moisture Regime

It's possible that variations in soil moisture won't always lead to noticeable or expected changes in the number of bacteria and fungi (Williams, 2007). The abundance of bacterial phyla is influenced by precipitation (Table 6.1). Additionally, an alteration in the relative abundance of acidobacteria and proteobacteria has been seen. Relative to Acidobacteria, Proteobacteria are more prevalent in wet treatments than in dry treatments, and vice versa (Castro et al., 2010).

Compared to bacteria, the prevalence of fungi rises during drought periods (de Vries & Bardgett, 2014). Drought conditions alter the overall bacterial and fungal compositions in the endospheric and rhizospheric zone significantly. Lack of proper moisture availability in paddy soil leads to the enrichment of Actinobacteria and Chloroflexi communities while there is depletion in the abundance of Acidobacteria and Deltaproteobacteria groups (Medellín et al., 2017). While mycorrhizal colonization is typically connected with rice crops cultivated in highland areas, it is less common in lowland areas or those that are flooded because of the anoxic environment

(Kumar et al., 2018b). However, recent research revealed that some AMF species might colonize successfully in rice wetland conditions (Sahoo et al., 2017).

Certain groups of microbes living in the plant rhizosheath are generally formed in drier soils. Rhizosheaths are cylinder-shaped structures that surround the roots of plants under stress and are made of mucilage deposits from crops and supporting soil particles. When compared to microbes that are isolated from the rhizospheric zone, those in rhizosheaths have a greater capacity to withstand drought stress (Etesami, 2021).

6.3.4 Imbalanced Fertilizer Application

For meeting the demand of the expanding population, there has been a trend of applying high doses of inorganic fertilizers in crop fields to increase productivity. The adverse effect of such human behavior leads to increased emission of harmful greenhouse gases like CH_4, N_2O, etc., aggravating the ill effects of climate change. Applications of fertilizers (mostly N) in high doses than required pose a significant impact on soil microbial communities (Table 6.1). The response of fertilizer applications on soil microorganisms largely depends on the environmental factors and can differ with locations specifically (Kumar et al., 2017a, 2018a). Through changes in soil properties due to heavy fertilization, there is a change in microbial community composition in paddy soil (Geisseler et al., 2017). Apart from increased microbial biomass carbon, there is an abundance of *nifH* genes and nitrogen fixation rates (Kumar et al., 2017b).

It has been observed that under high fertilizer application, there is an increased number of communities in Actinobacteria and Chlorobi while there is a suppression of cyanobacteria and Proteobacteria community abundance (Kumar et al., 2020a, 2020b). As per the findings of Wang et al. (2017), application of chemical fertilizers enhanced the profusion of oligotrophic bacteria such as Acidobacteria, Bacteroidetes, and Zygomycota phylum of fungi. Application of organic manures increased the number of copiotrophic bacteria like Agaricomycetes and Proteobacteria, and Orbiliomycetes classes of fungi. As per the findings of Lin et al. (2016), application of primary nutrients resulted in a negligible impact on soil antibiotic-resistant gene abundances in comparison to the untreated plot. Application of manure in combination with primary nutrients from inorganic sources increased the soil antibiotic-resistant gene abundance. Manure application must have changed the soil resistance without changing the microbial community, but its excessive usage might result in reduction of soil microbial activity.

Ikeda et al. (2014) stated the shifts in bacterial communities in paddy rice ecosystems with varied levels of N fertilizer application. The abundance of *Bradyrhizobium, Burkholderia,* and *Methylosinus* was greatly increased while that of methanogenic archaea was reduced under the lower dose of N-fertilizer relative to optimum doses. The addition of nitrogenous fertilizer marked a reduction in the complexity and connectivity of soil bacterial networks. Most of the key modules from non-dominant taxa were lost in upland soils that were fertilized with nitrogen, thereby indicating a reduction in the stability of the soil micro-ecosystem. In comparison to organic N sources, the stimulatory effect of chemical N fertilizers proves detrimental to the stability and sustainability of the agroecosystem (Zhang et al., 2021).

6.3.5 Changes in Soil Reaction

The sudden, drastic, and prolonged effects of climate change on environmental factors not only affect the above-ground biota but also have significant effects on below-ground microbial communities. Soil being the abode of a wide number of microorganisms; suffers huge alterations in its physico-chemical and biological properties due to several anthropogenic factors, which in turn lead to changes in the structure, function, composition, and richness of microbial associations. The most significant soil parameter regulating these microbial variations is the soil pH. Certain microbially driven soil reactions like denitrification, mineralization, and immobilization are directly influenced by soil pH. Between pH 3.9 and pH 9.0, denitrification occurs over a varied range of soil conditions (Wlodarczyk, 2000). Acidic soil slows down denitrification rates compared to slightly alkaline conditions (Ussiri & Lal, 2013). However, the nitrification process primarily occurs in soils with pH values between 3.8 and 8.5 (Subbarao et al., 2006). The ideal pH range for *Nitrobacter* ranges from 7.5 to 8.0, while *Nitrosomonas* needs a pH between around 7.0 and 8.0. Although it happens much more slowly, nitrification can take place in acidic soils having a pH as low as 3.8 (Tisdale & Nelson, 1970). A change in soil pH also results in variation in the α-diversity of the microbial communities subsequently bringing about a high abundance in taxonomic richness (Ilieve et al., 2019).

The rates of nitrification and denitrification reactions frequently decrease with rising soil acidity when there is an excess supply of NO_3^- in the soil (Dalal et al., 2003; Wlodarczyk, 2000). Since the conversion of N_2O to N_2 is more responsive to acidic soil conditions than the conversion of NO_3^- to N_2O. Moreover, N_2O reductase activity is highly sensitive to proton activity than NO_3^- reductase (Lesschen et al., 2011). The slower rate of N_2O reduction in acidic soils may be the cause of the larger proportion of N_2O in the total N products produced during denitrification. This encourages soil N_2O emission, which peaks at pH 5.0, and enables the soil environment to accumulate net N_2O (Van den Heuvel et al., 2011). When soil pH increases, the denitrification products tend to tilt more or fully toward N_2 production with a noticeable decrease in the N_2O:N_2 ratio because the net N_2 production rates, frequently exceed the net NO_3^- reduction rates (Focht & Verstraete, 1977). With the alterations in these soil reactions, the microorganisms involved in these processes are also altered in their abundance and community composition.

6.3.6 Imbalanced Pesticide Application

By interfering with critical functions like photosynthesis, respiration, and biosynthetic reactions, along with cell growth and division and molecular composition, pesticides harm non-target microbes in soil (DeLorenzo et al., 2001; Kumar et al., 2017a, 2017b, 2017c, 2020b; Sahoo et al., 2016). The rise in application of high amounts of pesticides (mostly organochlorines) to soil has led to wide alterations not only in soil properties but also affected several microbial communities (Table 6.1). There have been significant changes in bacterial communities while the fungal taxa have been comparatively stable, but they might have some synergistic effects with bacteria in response to such pesticide pollution in paddy soil. Pesticide use initially reduces microbial activity and population, but as the chemical is used repeatedly, bacteria begin to adapt and recolonize themselves (Kalia & Gosal, 2011). The microbial community colonization is promoted by selective cultivars, promoting the structural stability in Actinobacteria, Proteobacteria, and Chloroflexi for hybrid paddy varieties under such stress conditions (Feng et al., 2019).

According to Al-Ani et al. (2019), the three pesticides (Glyphosate, Malathion, and Alphacypermethrin) dramatically reduced the activity of microbes and the counts of soil bacteria, fungi, and actinomycetes. According to reports, the amount of pesticides added to the soil has an inverse relationship with the number and activity of microbes. Gupta et al. (2000) demonstrated a negative impact of pesticide application on all soil microorganisms with a drop in the average population of all groups evaluated in soil samples taken from locations with a rice–wheat farming system. Increasing pesticide concentrations and toxicity in the order insecticide > fungicide > herbicide resulted in a decrease in soil dehydrogenase activity, according to a laboratory study to investigate the effects of pesticide (insecticide, herbicide, and fungicide) application on the health of paddy field soil (Subhani et al., 2002).

6.3.7 Interaction Effect of Environmental Change Factors

Depending on the emissions forecast for the ensuing few decades, the atmospheric carbon dioxide (CO_2) level is predicted to reach roughly 570 ppm by 2050, causing the average worldwide air temperature to rise by 1.8–6.4 °C by 2100 AD (Chaterjee & Saha, 2018). According to Houghton et al. (2001), the expected increase in atmospheric CO_2 concentration is expected to elevate the average global air temperature by 1.5–8.8 °C, which will likely disrupt the balance of precipitation and could lead to more frequent occurrence of abiotic stresses like drought and flood (Bates et al., 2008; Ray et al., 2020). Strong drivers of soil microbial responses are the interaction (e.g., additive, subtractive, or multiplicative) impacts of various circumstances (Slaughter, 2012). Variations in shift of the microbial community are primarily explained by interactions between soil moisture and high CO_2 (Guenet et al., 2012). Increased warmth is likely to prevent microbial activity when there is insufficient soil moisture (Bontti et al., 2009; Butenschoen et al., 2011). Bacterial abundance increased in situations with high atmospheric CO_2, but decreased in environments with low atmospheric CO_2, according to a study by Castro et al. (2010). Certain bacterial groups' relative abundance changed significantly in response to the precipitation. As opposed to either treatment alone, the combined effect of temperature and precipitation reduces the seasonal increase in amoebae abundance (Darby et al., 2011). The microbial decomposition of soil organic matter (SOM) is highly sensitive to environmental changes, such as temperature rise, which have the potential to change the enzyme kinetics and associated nutrient availability in the soil system by altering the way resources are allocated and the community structure of the soil biota (Steinweg et al., 2012; Chaterjee et al. 2019). The number of nitrifying bacteria greatly increased with an increase in nitrogen dose, but it significantly decreased under conditions of water stress as compared to well-watered conditions. Higher N dose and well-watered conditions were associated with significantly higher denitrifying bacterial populations (Kumar et al., 2021a, 2021b). When compared to water stress and lesser doses of nitrogenous fertilizer, the N_2O emission was much higher when the soil was well-watered and fertilized (Kumar et al., 2021a, 2021b).

6.4 Mitigation Strategies

Crops need to acclimatize to different adverse external stress conditions generated by environmental change factors using their inbuilt biological mechanisms. Microbiomes associated with specific crops provide fundamental support to the plants in uptaking mineral nutrients, providing resistance against diseases, and tolerating different abiotic stresses. The plant and its microbiome interact with each other through different metabolic pathways and form stress tolerance strategies. Microbes generally produce a wide variety of metabolites that act as signals during stress conditions.

Although elevated atmospheric CO_2 levels frequently promote plant development, they can also cause soil nitrogen levels to become limited. Growing high-yielding cultivars with higher CO_2, which has been discovered to alter the structure and function of the soil ecosystem, can help offset this (Hu et al., 2017). The findings of Luo et al. (2021) suggested that elevated CO_2 mitigates the toxicity of nano-Cr_2O_3 and the degree of mitigation is soil dependent, which is due to the difference in clay and organic matter content, flexibility of the resistant bacterial taxa, and the complexity of distinct microbial network in soils. Jalil and Ansari (2018) reported that the bacteria, *Trichoderma harzianum* is seen to respond against stress in rice by activating stress-related genes like malondialdehyde, dehydrin, aquaporin, etc. to prepare the crop to tolerate the stress condition. By increasing drought stress tolerance in a variety of crop plants like wheat, rice, barley, legumes, etc., rhizosheath development has been recognized as a prospective characteristic for improving future agricultural sustainability and promoting food security against the negative effects of climate change (Etesami, 2021). According to Ikeda et al. (2014), lower N fertilizer doses resulted in higher abundances of the functional genes for the metabolism of N, S, Fe, and aromatic compounds in the paddy root microbiome, indicating that low-N fertilizer management plays a significant role in determining the composition of the microbial communities that are important for plant associations and biogeochemical processes in paddy rice ecosystems.

6.5 Conclusion

The majority of GHGs emitted or fluxed are sourced from soil C-N pools. The soil microbial population is involved in the decomposition, mineralization, and assimilation of GHGs. Agriculture alone is responsible for 10–12% of total global anthropogenic GHG emissions, accounting for around 60% of global N_2O and 50% of global CH_4 emissions (Smith et al., 2007). Due to a concomitant surge in methanogenic archaea, the combined impacts of eCO_2 and temperature increased CH_4 emissions, which are largely measured by the availability of C. Similarly, in N-fertilized fields, eCO_2 treatments result in a considerable rise in N_2O emissions. Though the rate of N_2O emission is mostly determined by the range of O_2 concentrations in the soil, the interchange of regulatory variables complicates the nitrification and denitrification process. Because soil microbes play such an important part in biogeochemical cycling, their reactions modify the microbial community's abundance, composition, and functional features. Microbial degradation is influenced by interactions between soil C and SOM storage capacity. Sudden changes in interaction flooding frequency, pollutants concentration, temperature variations, and eCO_2 can result in the loss of C supplies, as well as changes in the microbial community, which must be investigated more intensively. Furthermore, eCO_2-induced changes in soil micro-environments may have long-term implications for terrestrial biogeochemical cycles and climate change. The implications of various related mitigation measures such as adjusting sowing and harvesting dates, pulse cropping, inter-cropping, fertilizer managemnt on the agricultural C and N cycle require more research.

References

AL-Ani, M. A., Hmoshi, R. M., Kanaan, I. A., & Thanoon, A. A. (2019). Effect of pesticides on soil microorganisms. *Journal of Physics: Conference Series,1294*(7), 072007.

Banik, A., Dash, G. K., Swain, P., Kumar, U., Mukhopadhyay, S. K., & Dangar, T. K. (2019). Application of rice (*Oryza sativa* L.) root endophytic diazotrophic *Azotobacter* sp. strain Avi2 (MCC 3432) can increase rice yield under green house and field condition. *Microbiological Research, 219*, 56–65.

Bates, B. C., Kundzewicz, Z. W., Wu, S., & Palutik, J. P. (2008). Climate Change and Water. Technical Paper of the Intergovernmental Panel on Climate Change IPCC Secretariat Geneva.

Bhattacharyya, P., Roy, K. S., Dash, P. K., Neogi, S., Shahid, M., Nayak, A. K., Raja, R., Karthikeyan, S., Balachandar, D., & Rao, K. S. (2014). Effect of elevated carbon dioxide and temperature on phosphorus uptake in tropical flooded rice (Oryza sativa L.). *European Journal of Agronomy, 53*, 28–37.

Bontti, E. A., Decant, J. P., Munson, S. M., Gathany, M. A., Przeszlowska, A., Haddix, M. L., Owens, S., Burke, I. C., Parton, W. J., & Harmon, M. E. (2009). Litter decomposition in grasslands of Central North America (US Great Plains). *Global Change Biology, 15*, 1356–1363.

Butenschoen, O., Scheu, S., & Eisenhauer, N. (2011). Interactive effects of warming, soil humidity and plant diversity on litter decomposition and microbial activity. *Soil Biology and Biochemistry, 43*, 1902–1907.

Cassia, R., Nocioni, M., Correa-Aragunde, N., & Lamattina, L. (2018). Climate change and the impact of greenhouse gasses: CO_2 and NO, friends and foes of plant oxidative stress. *Frontiers in Plant Science, 9*, 273.

Castro, H. F., Classen, A. T., Austin, E. E., Norby, R. J., & Schadt, C. W. (2010). Soil microbial community responses to multiple experimental climate change drivers. *Applied Environmental Microbiology, 76*, 999–1007.

Chatterjee, D., & Saha, S. (2018). Response of soil properties and soil microbial communities to the projected climate change. In *Advances in Crop Environment Interaction* (eds Bal, S. et al.) Springer Singapore: 87–136.

Chatterjee, D., Kuotsu, R., Ao, M., Saha, S., Ray, S. K., & Ngachan, S. V. (2019). Does rise in temperature adversely affect soil fertility, carbon fractions, microbial biomass and enzyme activities under different land uses? *Current Science, 116*(12), 2044–2054.

Dalal, R. C., Wang, W., Robertson, G. P., & Parton, W. J. (2003). Nitrous oxide emission from Australian agricultural lands and mitigation options: A review. *Australian Journal of Soil Research, 41*(2), 165–195.

DaMatta, F. M., Grandis, A., Arenque, B. C., & Buckeridge, M. S. (2010). Impacts of climate changes on crop physiology and food quality. *Food Research International, 43*(7), 1814–1823.

Darby, B. J., Neher, D. A., Housman, D. C., & Belnap, J. (2011). Few apparent short-term effects of elevated soil temperature and increased frequency of summer precipitation on the abundance and taxonomic diversity of desert soil micro- and meso-fauna. *Soil Biology and Biochemistry, 43*, 1474–1481.

De Vries, F. T., & Bardgett, R. D. (2014). Climate change effects on soil biota in the UK. Biodiversity Report Card 2014–15, The University of Manchester, Manchester, UK.

DeLorenzo, M. E., Scott, G. I., & Ross, P. E. (2001). Toxicity of pesticides to aquatic microorganisms: A review. *Environmental Toxicology and Chemistry, 20*, 84–98.

Drigo, B., Kowalchuk, G. A., & van Veen, J. A. (2008). Climate change goes underground: Effects of elevated atmospheric CO_2 on microbial community structure and activities in the rhizosphere. *Biology and Fertility of Soils, 44*, 667–679.

Dunbar, J., Eichorst, S. A., Gallegos-Graves, L., Silva, S., Xie, G., Hengartner, N. W., Evans, R. D., Hungate, B. A., Jackson, R. B., Megonigal, J. P., Schadt, C. W., Vilgalys, R., Zak, D. R., & Kuske, C. R. (2012). Common bacterial responses in six ecosystems exposed to 10 years of elevated atmospheric carbon dioxide. *Environmental Microbiology, 14*, 1145–1158.

Easterling, D. R., Horton, B., Jones, P. D., Peterson, T. C., Karl, T. R., Parker, D. E., Salinger, M. J., Razuvayev, V., Plummer, N., Jamason, P., & Folland, C. K. (1997). Maximum and minimum temperature trends for the globe. *Science, 277*(5324), 364–367.

Edwards, J., Johnson, C., Santos-Medellin, C., Lurie, E., Podishetty, N. K., Bhatnagar, S., Eisen, J. A., & Sundaresan, V. (2015). Structure, variation and assembly of the root-associated microbiomes of rice. *Proceedings of the National Academy of Sciences, 112*(8), E911–E920.

Etesami, H. (2021). Potential advantage of rhizosheath microbiome, in contrast to rhizosphere microbiome, to improve drought tolerance in crops. *Rhizosphere, 20*, 100439.

Feng, J., Xua, Y., Maa, B., Tanga, C., Brookesa, P. C., Yan Hea, Y., & Xua, J. (2019). Assembly of root-associated microbiomes of typical rice cultivars in response to lindane pollution. *Environment International, 131*, 104975.

Focht, D. D., & Verstraete, W. (1977). Biochemical ecology of nitrification and denitrification. In M. Alexander (ed). *Advances in Microbial Ecology Plenum* Press, New York: 135–214.

Geisseler, D., Linquist, B. A., & Lazicki, P. A. (2017). Effect of fertilization on soil microorganisms in paddy rice systems: A meta-analysis. *Soil Biology and Biochemistry, 115*, 452–460.

Gilbert, N. (2012). *One-third of our greenhouse gas emissions come from agriculture.* Nature News.

Guenet, B., Lenhart, K., Leloup, J., Giusti-Miller, S., Pouteau, V., Mora, P., Nunan, N., & Abbadie, L. (2012). The impact of long-term CO_2 enrichment and moisture levels on soil microbial community structure and enzyme activities. *Geoderma, 170*, 331–336.

Gupta, R. P., Singh, J., Sultan, M. S., Hujan, R. K., Gosal, S. K., Sahota, H., & Sharma, S. (2000). Impact of pesticides on soil biota and non-target organisms in rice wheat cropping system (Abstract). 41st annual conference of AMI, Birla Research Institute, Jaipur. India.

Houghton, J. T., Ding, Y. D. J. G., Griggs, D., Noguer, M., van der Linden, P. J., Dai, X., Maskell, K., & Johnson, C. A. (2001). *Climate change 2001: The scientific basis.* The Press Syndicate of the University of Cambridge.

Hu, S., Chapin, F. S., Firestone, M. K., Field, C. B., & Chiariello, N. R. (2001). Nitrogen limitation of microbial decomposition in a grassland under elevated CO_2. *Nature, 409*, 188–191.

Hu, Z., Zhu, C., Chen, X., Bonkowski, M., Griffiths, B., Chen, F., Zhu, J., Hu, S., Hu, F., & Liu, M. (2017). Responses of rice paddy micro-food webs to elevated CO_2 are modulated by nitrogen fertilization and crop cultivars. *Soil Biology and Biochemistry, 114*, 104–113.

Ikedal, S., Sasaki, K., Okubo, T., Yamashita, A., Terasawa, K., Bao, Z., Liu, D., Watanabe, T., Murase, J., Asakawa, S., Eda, S., Mitsui, H., Sato, T., & Minamisawa, K. (2014). Low nitrogen fertilization adapts rice root microbiome to low nutrient environment by changing biogeochemical functions. *Microbes and Environments, 29*(1), 50–59.

Ilieve, I., Marhova, M., Kostadinova, S., Gochev, V., Tsankova, M., Ivanova, A., Yahubyan, G., & Baev, V. (2019). Metagenomic analysis of the microbial community structure in protected wetlands in the Maritza River Basin. *Biotechnology and Biotechnological Equipment, 33*(1), 1721–1732.

Inubushi, K., Hoque, M. M., Cheng, W., Kobayashi, K., Yagi, K., Sakai, H., Kim, Y., & Okada, M. (2002). Elevated carbon dioxide effects on microbial processes in submerged soil. In *Proceedings on the 17th World Conferences of Soil Sciences.*

Jalil, S. U., & Ansari, M. I. (2018). Plant microbiome and its functional mechanism in response to environmental stress. *International Journal of Green Pharmacy, 12*(1), S81–S92.

Kalia, A., & Gosal, S. K. (2011). Effect of pesticide application on soil microorganisms. *Archives of Agronomy and Soil Science, 57*(6), 569–596.

Kumar, A., Padhy, S. R., Das, R. R., Shahid, M., Dash, P. K., Senapati, A., Panneerselvam, P., Kumar, U., Chatterjee, D., Adak, T., Tripathi, R., Nayak, P. K., & Nayak, A. K. (2021a). Elucidating relationship between nitrous oxide emission and functional soil microbes from tropical lowland rice soil exposed to elevated CO_2: A path modelling approach. *Agriculture, Ecosystems & Environment, 308*, 107268.

Kumar, U., Shahid, M., Tripathi, R., Mohanty, S., Kumar, A., Bhattacharyya, P., Lal, B., Gautam, P., Raja, R., Panda, B. B., Jambhulakar, N. N., Shukla, A. K., & Nayak, A. (2017a). Variation of functional diversity of soil microbial community in sub-humid tropical rice-rice cropping system under long-term organic and inorganic fertilization. *Ecological Indicators, 73*, 536–543.

Kumar, U., Panneerselvam, P., Govindasamy, V., Vithalkumar, L., Senthilkumar, M., Banik, A., & Annapurna, K. (2017b). Long-term aromatic rice cultivation effect on frequency and diversity of diazotrophs in its rhizosphere. *Ecological Engineering, 101*, 227–236.

Kumar, U., Berliner, J., Adak, T., Rath, P. C., Dey, A., Pokhare, S. S., Jambhulkar, N. N., Panneerselvam, P., Kumar, A., & Mohapatra, S. D. (2017c). Non-target effect of continuous application of chlorpyrifos on soil microbes, nematodes and its persistence under sub-humid tropical rice-rice cropping system. *Ecotoxicology and Environmental Safety, 135*, 225–235.

Kumar, U., Nayak, A. K., Shahid, M., Gupta, V. V. S. R., Panneerselvam, P., Mohanty, S., Kaviraj, M., KumarA, C. D., Lal, B., Gautam, P., & TripathiR, P. B. B. (2018a). Continuous application of inorganic and organic fertilizers over 47 years in paddy soil alters the bacterial community structure and its influence on rice production. *Agriculture, Ecosystems and Environment, 262*, 65–75.

Kumar, U., Panneerselvam, P., Dangar, T. K., Kumar, A., Chatterjee, D., Parmeswaran, C., Mohapatra, S. D., Prasanthi, G., Chakraborty, K., Swain, P., & Nayak, A. K. (2018b). Microbial Resources for Alleviating Abiotic and Biotic Stresses and Improving Soil Health in Rice Ecology. In *Rice research for enhancing productivity, profitability and climate resilience.*

Kumar, U., Kaviraj, M., Panneerselvam, P., Priya, H., Chakraborty, K., Swain, P., Chatterjee, S. N., Sharma, S. G., Nayak, P. K., & Nayak, A. K. (2019). Ascorbic acid formulation for survivability and diazotrophic efficacy of *Azotobacter chroococcum* Avi2 (MCC 3432) under hydrogen peroxide stress and its role in plant-growth promotion in rice (*Oryza sativa* L.). *Plant Physiology and Biochemistry, 139*, 419–427.

Kumar, U., Nayak, A. K., Sahoo, S., Kumar, A., Kaviraj, M., & Shahid, M. (2020a). Combined effects of elevated CO_2, N fertilizer and water deficit stress on diazotrophic community in sub-humid tropical paddy soil. *Applied Soil Ecology, 155*, 103682.

Kumar, U., Behera, S., Saha, S., Das, D., Guru, P. K., Kaviraj, M., Munda, S., Adak, T., & Nayak, A. K. (2020b). Non-target effect of bispyribac sodium on soil microbial community in paddy soil. *Ecotoxicology and Environmental Safety, 189*, 110019.

Kumar, U., Kaviraj, M., Rout, S., Chakraborty, K., Swain, P., Nayak, P. K., & Nayak, A. K. (2021b). Combined application of ascorbic acid and endophytic N-fixing *Azotobacter chroococcum* Avi2 modulates photosynthetic efficacy, antioxidants and growth-promotion in rice under moisture deficit stress. *Microbiological Research, 250*, 126808.

Kutsch, W. L., Bahn, M., & Heinemeyer, A. (eds.) (2009). *Soil carbon dynamics: an integrated methodology*. Cambridge University Press.

Lacis, A. A., Schmidt, G. A., Rind, D., & Ruedy, R. A. (2010). Atmospheric CO_2: Principal control knob governing Earth's temperature. *Science, 330*(6002), 356–359.

Lesschen, J. P., Velthof, G. L., de Vries, W., & Kros, J. (2011). Differentiation of nitrous oxide emission factors for agricultural soils. *Environmental Pollution, 159*(11), 3215–3222.

Li, P., Ye, S., Liu, H., Pan, A., Ming, F., & Tang, X. (2018). Cultivation of drought-tolerant and insect-resistant rice affects soil bacterial, but not fungal, abundances and community structures. *Frontiers in Microbiology, 9*(1390).

Lin, H., Sun, W., Zhang, Z., Chapman, S. J., Freitag, T. E., Fu, J., Zhang, X., & Ma, J. (2016). Effects of manure and mineral fertilization strategies on soil antibiotic resistance gene levels and microbial community in a paddy upland rotation system. *Environmental Pollution, 211*, 332–337.

Lipson, D. A., Kuske, C. R., Gallegos-Graves, L. V., & Oechel, W. C. (2014). Elevated atmospheric CO_2 stimulates soil fungal diversity through increased fine root production in a semiarid shrub land ecosystem. *Global Change Biology, 20*, 2555–2565.

Lu, T., Ke, M., Peijnenburg, W., Zhu, Y., Zhang, M., Sun, L., Fu, Z., & Qian, H. (2018). Investigation of rhizospheric microbial communities in wheat, barley and two rice varieties at the seedling stage. *Journal of Agriculture and Food Chemistry, 66*, 2645–2653.

Luo, J., Guo, X., Liang, J., Song, Y., Liu, Y., Li, J., Dua, Y., Mua, Q., Jiang, Y., Zhaoa, H., & Li, T. (2021). The influence of elevated CO_2 on bacterial community structure and its co-occurrence network in soils polluted with Cr_2O_3 nanoparticles. *Science of the Total Environment, 779*, 146430.

Marilley, L., Hartwig, U. A., & Aragno, M. (1999). Influence of an elevated atmospheric CO_2 content on soil and rhizosphere bacterial communities beneath *Lolium perenne* and *Trifolium repens* under field conditions. *Microbial Ecology, 38*, 39–49.

Medellín, C. S., Edwards, J., Liechty, Z., Nguyen, B., & Sundaresana, V. (2017). Drought stress results in a compartment-specific restructuring of the rice root-associated microbiomes. *American Society for Microbiology, 8*(4), e00764-e817.

Mendes, R., Garbeva, P., & Raaijmakars, J. M. (2013). The rhizosphere microbiome: Significance of plant beneficial, plant pathogenic, and human pathogenic microorganisms. *FEMS Microbiology Reviews, 37*(5), 634–663.

Mohanty, S., Nayak, A. K., Swain, C. K., Dhal, B. R., Kumar, A., Kumar, U., Tripathi, R., Shahid, M., & Behera, K. K. (2020). Impact of integrated nutrient management options on GHG emission, N loss and N use efficiency of low land rice. *Soil and Tillage Research, 200*, 104616.

Mohanty, S., Swain, S. K., Sethi, S. K., Dalai, P. C., Bhattachrayya, P., Kumar, A., Tripathi, R., Shahid, M., Panda, M. M., Kumar, U., Lal, B., Gautam, P., Munda, S., & Nayak, A. K. (2017). Crop establishment and nitrogen management affect greenhouse gas emission and biological activity in tropical rice production. *Ecological Engineering, 104*, 80–98.

Montealegre, C. M., Van Kessel, C., Blumenthal, J. M., Hur, H. G., Hartwig, U. A., & Sadowsky, M. J. (2000). Elevated atmospheric CO_2 alters microbial population structure in a pasture ecosystem. *Global Change Biology, 6*, 475–482.

Nieder, R., & Benbi, D. K. (2008). Carbon and nitrogen in the terrestrial environment. Springer Science & Business Media.

NOAA. (2020). Climate Change Indicators: Climate Forcing. https://www.epa.gov/climate-indicators/climate-change-indicators-climate-forcing.

Padbhushan, R., Kumar, U., Sharma, S., Rana, D. S., Kohli, A., Kaviraj, M., Kumari, P., Parmar, B., Kumar, R., Annapurna, K., Sinha, A. K., & Gupta, V. V. S. R. (2022). Impact of land-use changes on soil properties and carbon pools in India: A meta-analysis. *Frontiers in Environmental Science, 9*, 794866.

Padbhushan, R., Sharma, S., Kumar, U., Rana, D. S., Kohli, A., Kaviraj, M., Parmar, B., Kumar, R., Annapurna, K., Sinha, A. K., & Gupta, V. V. S. R. (2021). Meta-analysis approach to measure effect of integrated nutrient management on crop performance, microbial activity and carbon stocks in Indian soils. *Frontiers in Environmental Science, 9*, 724702.

Panneerselvam, P., Sahoo, S., Senapati, A., Kumar, U., Mitra, D., Parameswaran, C., Anandan, A., Kumar, A., Jahan, A., & Nayak, A. K. (2019). Understanding interaction effect of arbuscular mycorrhizal fungi in rice under elevated carbon dioxide conditions. *Journal of Basic Microbiology, 59*, 1217–1228.

Procter, A. C., Ellis, J. C., Fay, P. A., Polley, H. W., & Jackson, R. B. (2014). Fungal community responses to past and future atmospheric CO_2 differ by soil type. *Applied Environmental Microbiology, 80*, 7364–7377.

Ray, K., Attri, S. D., Pathak, H., Kumar, A., & Chaterjee, D. (2020). Climate. In Mishra, B.B. (Ed.), *In The Soils of India. Springer Nature Switzerland AG 2020*, 41–55.

Rodriguez, R. J., Henson, J., Van Volkenburgh, E., Hoy, M., Wright, L., Beckwith, F., Kim, Y. O., & Redman, R. S. (2008). Stress tolerance in plants via habitat-adapted symbiosis. *International Society for Microial Ecology Journal, 2*, 404–416.

Sadowsky, M. J., & Schortemeyer, M. (1997). Soil microbial responses to increased concentration of atmospheric CO_2. *Global Change Biology, 3*, 217–224.

Sahoo, S., Bagchi, T. B., Kumar, U., Berliner, J., Mishra, B. B., Saha, S., Munda, S., & Adak, T. (2016). Non-target effects of pretilachlor on microbial properties in tropical rice soil. *Environmental Science and Pollution Research, 23*, 7595.

Sahoo, S., Panneerselvam, P., Chowdhury, T., Kumar, A., Kumar, U., Jahan, A., Senapati, A., & Anandan, A. (2017). Understanding the AM fungal association in flooded rice under elevated CO_2 condition. *ORYZA-an International Journal of Rice, 54*(3), 290–297.

Santos-Medellín, C., Edwards, J., Liechty, Z., Bao, N., & Sundaresan, V. (2017). Drought stress results in a compartment-specific restructuring of the rice root-associated microbiomes. *mBio, 8*(4).

Schaufler, G., Kitzler, B., Schindlbacher, A., Skiba, U., Sutton, M. A., & Zechmeister-Boltenstern, S. (2010). Greenhouse gas emissions from European soils under different land use: Effects of soil moisture and temperature. *European Journal of Soil Science, 61*, 683–696.

Shibahara, F., & Inuushi, K. (1997). Effects of organic matter application on microbial biomass and available nutrients in various types of paddy soils. *Soil Science and Plant Nutrition, 43*(1), 191–203.

Singurindy, O., Molodovskaya, M., Richards, B. K., & Steenhuis, T. S. (2009). Nitrous oxide emission at low temperatures from manure-amended soils under corn (*Zea mays L.*). *Agriculture, Ecosystems & Environment, 132*(1–2), 74–81.

Slaughter, L. C. (2012). *Soil microbial community response to climate change: Results from a temperate kentucky pasture*. Theses and dissertations-Plant soil sciences.

Smith, P., Martino, Z., & Cai, D. (2007). Agriculture. In *Climate change 2007*: mitigation.

Steinweg, J. M., Dukes, J. S., & Wallenstein, M. D. (2012). Modelling the effects of temperature and moisture on soil enzyme activity: Linking laboratory assays to continuous field data. *Soil Biology and Biochemistry, 55*, 85–92.

Subbarao, G. V., Ito, O., Sahrawat, K. L., Berry, W. L., Nakahara, K., Ishikawa, T., Watanabe, T., Suenaga, K., Rondon, M., & Rao, I. M. (2006). Scope and strategies for regulation of nitrification in agricultural systems-challenges and opportunities. *Critical Review in Plant Sciences, 25*, 303–335.

Subhani, A., Min, L., Yong, H. C., & Miao, X. Z. (2002). Alteration of certain soil microbiological indices of a paddy soil under anthropogenic stress. *J Zhejiang Univ. Science Advances, 3*(4), 467–474.

Tarnawski, S., & Aragno, M. (2006). The influence of elevated CO_2 on diversity, activity and biogeochemical function of rhizosphere and soil bacterial communities. In: Nösberger, J, Long SP, Norby RJ, et al. (eds) *Managed ecosystems and CO_2-case studies, processes and perspectives. Ecological studies series Springer* 187, 393–409.

Tisdale, S. L., & Nelson, W. L. (1970). *Soil fertility and fertilizers* (2nd ed.). Macmillan.

Ussiri, D., & Lal, R. (2013). *Formation and release of nitrous oxide from terrestrial and aquatic ecosystems, soil emission of nitrous oxide and its mitigation* (pp. 63–96). Springer.

van den Heuvel, R. N., Bakker, S. E., Jetten, M. S. M., & Hefting, M. M. (2011). Decreased N_2O reduction by low soil pH causes high N_2O emissions in a riparian ecosystem. *Geobiology, 9*(3), 294–300.

Vetter, S. H., Sapkota, T. B., Hillier, J., Stirling, C. M., Macdiarmid, J. I., Aleksandrowicz, L., Green, R., Joy, E. J., Dangour, A. D. & Smith, P. (2017). Greenhouse gas emissions from agricultural food production to supply Indian diets: Implications for climate change mitigation. *Agriculture, Ecosystems and Environment, 237*, 234–241.

Wagner, M. R., Lundberg, D. S., Del Rio, T. G., Tringe, S. G., Dangl, J. L., & Mitchell-Olds, T. (2016). Host genotype and age shape the leaf and root microbiomes of a wild perennial plant. *Nature Communications, 7*(1), 1–15.

Wang, J., Song, Y., Ma, T., Raza, W., Li, J., Howland, J. G., Huang, O., & Shen, Q. (2017). Impacts of inorganic and organic fertilization treatments on bacterial and fungal communities in a paddy soil. *Applied Soil Ecology, 112*, 42–50.

Waqas, M. A., Li, Y., Ashraf, M. N., Ahmed, W., Wang, B., Sardar, M. F., Ma, P., Li, R., Wan, Y., & Kuzyakov, Y. (2021). Long-term warming and elevated CO_2 increase ammonia-oxidizing microbial communities and accelerate nitrification in paddy soil. *Applied Soil Ecology, 166*, 104063.

Willey, J. M., Sherwood, L. M., & Woolverton, C. J. (2009). *Prescott's principles of microbiology*. McGraw-Hill.

Williams, M. A. (2007). Response of microbial communities to water stress in irrigated and drought prone tall grass prairie soils. *Soil Biology and Biochemistry, 39*, 2750–2757.

Wlodarczyk, T. (2000). N_2O emission and absorption against a background of CO_2 in Eutric Cambisol under different oxidation- reduction conditions. *Acta Agrophysica, 28*, 39–89.

Xu, C., Zhang, K., Zhu, W., Xiao, J., Zhu, C., Zhang, N., Yu, F., Li, S., Zhu, C., Tu, Q., Chen, X., Zhu, J., Hu, S., Koide, R. T., Firestone, M. K., & Cheng, L. (2020) Large losses of ammonium-nitrogen from a rice ecosystem under elevated CO_2. *Science Advances, 6*, eabb7433.

Yu, Y., Zhang, J., Petropoulos, E., Baluja, M. Q., Zhu, C., Zhu, J., Lin, X., & Feng, Y. (2018). Divergent responses of diazotrophic microbiome to elevated CO_2 in two rice cultivars. *Frontiers in Microbiology, 9*, 1139.

Zhang, M., Zhang, X., Zhang, L., Zeng, L., Liu, Y., Wang, X., He, P., Li, S., Liang, G., Zhou, W., & Ai, C. (2021). The stronger impact of inorganic nitrogen fertilization on soil bacterial community than organic fertilization in short-term condition. *Geoderma, 382*, 114752.

Ziska, L. H., Moya, T. B., Wassmann, R., Namuco, O. S., Lantin, R. S., Aduna, J. B., Abao, E., Bronson, K. F., Neue, H. U., & Olszyk, D. (1998). Long-term growth at elevated carbon dioxide stimulates methane emission in tropical paddy rice. *Global Change Biology, 4*(6), 657–665.

Chapter 7
Contemporary Use of Sensors for Soil Qualitative and Quantitative Assessment in the Context of Climate Change

Shubhadip Dasgupta, V. Lavanya, Somsubhra Chakraborty, and Deb Prasad Ray

Abstract There are ample opportunities to use newly introduced several low-cost digital sensing devices and technologies in soil testing and resource management services. These advanced sensors may help judicious fertilizer recommendations for nutrient-exhaustive cropping systems in different countries. It would be more convenient for the analyst, users, and government to capture the process and disseminate the required data or decisions in a time-saving manner. Farmers can get soil test reports from the laboratory in a short time. These efforts would encourage farmers of remote areas to test soil, particularly in tropical countries with more prevalent nutrient loss. This chapter discusses some low-cost proximal soil sensing devices like portable X-ray fluorescence spectrometer, color sensors, smartphone-based sensors, their operational theory, and contemporary uses with artificial intelligence (AI) and digital soil mapping techniques in soil characterization and mapping.

Keywords Soil fertility · Soil sensors · Soil testing · Portable X-ray fluorescence spectrometer · Color sensors · Smartphone · Digital soil mapping

S. Dasgupta
Department of Agricultural Chemistry and Soil Science, Faculty of Agriculture, Bidhan Chandra Krishi Viswavidyalaya, Nadia, West Bengal, India

S. Dasgupta · V. Lavanya · S. Chakraborty (✉)
Agricultural and Food Engineering Department, IIT Kharagpur, Kharagpur, West Bengal, India
e-mail: somsubhra@agfe.iitkgp.ac.in

D. P. Ray
ICAR-National Institute of Natural Fibre Engineering and Technology, Kolkata, West Bengal, India

H. Pathak et al. (eds.), *Climate Change Impacts on Soil-Plant-Atmosphere Continuum*, Advances in Global Change Research 78, https://doi.org/10.1007/978-981-99-7935-6_7

7.1 Introduction

Soil is one of the most important but neglected components of the ecosystem. Given soil is the second largest C sink, healthy and well-managed soils can diminish ill effects of climate change. Soil is a critical and fragile natural variable resource, and different environmental, biotic, and pedogenic factors are involved as soil-forming factors. These dependent factors continually interact in a complex, orderly manner, which causes variations in soil properties. However, anthropogenic activity in present-day technology-driven intense agriculture and civilization practices and their interrelated chain of changes drastically affect the soil body (Yaalon & Yaron, 1966). This influence has become prominent and extensive under the purview of escalated population pressure and the global climate-changing scenario. Therefore, the rising level of atmospheric temperature and carbon dioxide, erratic rainfall pattern, high rates of emission of greenhouse gases and other local activities, and their consequential, potential effects on physic-chemical and biological functions of soil needs to be addressed. Therefore, spatial variability characterization and classification of the soil physicochemical properties and their complex relations are of prime concern. Figure 7.1 shows the framework of sustainable soil management.

The spatial distribution pattern of essential soil elements for plants is also relatively heterogeneous. Due to more temperatures and precipitation, tropical soils need more fertilizer than temperate regions. The accelerated weathering-leaching phenomenon accounts for the higher rate of nutrient release from weatherable minerals, raising fertilizer requirements (Benedet et al., 2021; Resende et al., 2014). Furthermore, this widespread chemical application accelerates several nutrient imbalances, loss, and deficiencies in soil (Dasgupta et al., 2021). The indiscriminate NPK fertilizer

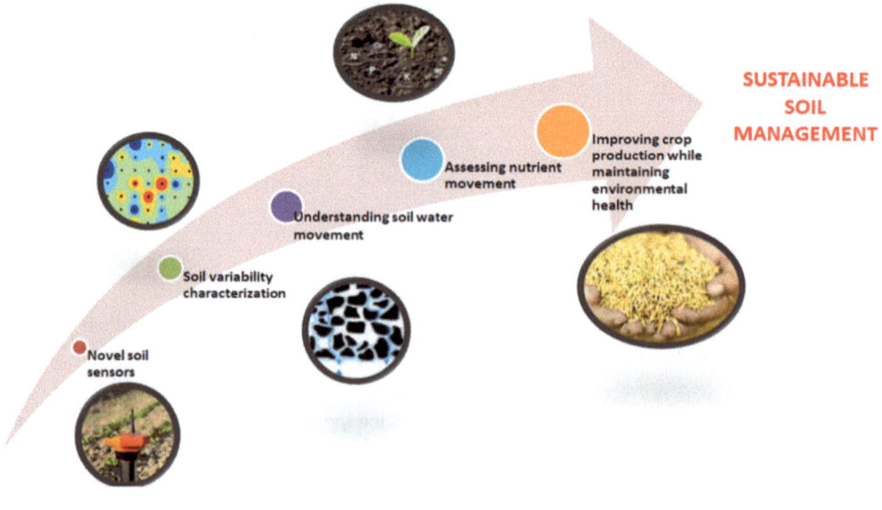

Fig. 7.1 Steps and basic framework of sustainable soil management

application, especially nitrogenous, is disrupting the desired fertilizer consumption ratio. Sharp rise in fertilizer cost and elevated fertilizer use are reflected in high cultivation cost, declined soil fertility, and productivity in the long run. On the flip side, manure and organic fertilizers alone are insufficient for optimal plant nutrient supply (Dasgupta et al., 2017). This overall complex phenomenon is flattening nutrient bio-availability and agricultural progress side by side (Dasgupta et al., 2021).

Global food crisis and nutritional insecurity are driving a greater extent of shrinking soil fertility and mismanagement. Moreover, it is impossible to go beyond a certain threshold level of cultivable land area expansion. This situation highlights the vehement demand to assess soil fertility and classify the land as per land capability classification for crop suitability, pasture afforestation, etc. (Sengupta et al., 2022). Quantitative information on soil thus is significant for ensuring sustainable agricultural land-use planning, natural resource management, and economic farmer's service. For efficient soil management and sustainability, it is essential to recognize soil property distribution patterns for better understanding and intervention in soil fertility management. To ensure this, a rapid, convenient soil testing facility may provide an index of the plant-available nutrients in soils to determine the number of supplemental nutrients to be applied for a particular crop under the prevailing agro-climatic conditions. Figure 7.2 represents the relationship between smart soil sensing networks, site-specific nutrient management, and sustainable agriculture.

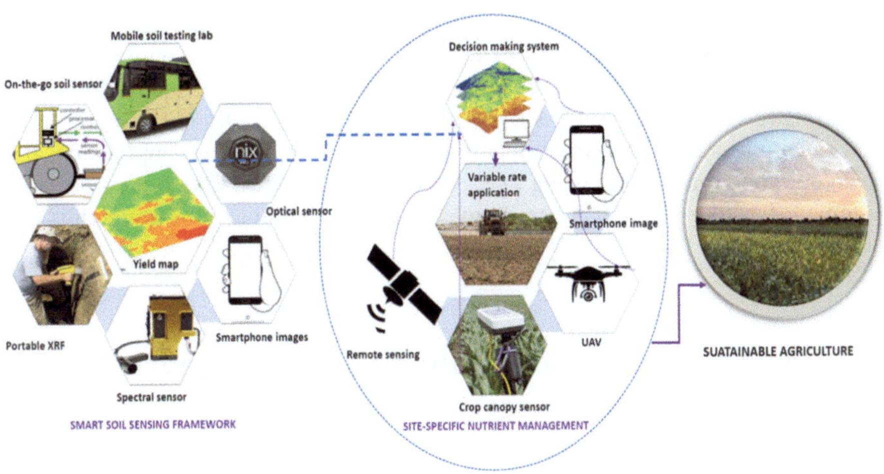

Fig. 7.2 Interconnection between smart soil sensing network, site-specific nutrient management, and sustainable agriculture

7.2 Laboratory-Based Analytical Methods for Assessing Soil Fertility

In the last few decades, tremendous progress in soil research and planning has been up-scaling by using advanced technologies in soil survey, mapping, digital and spatial soil information, and their systematic interpretations. Even qualitative improvements in pedological, geological, and micromorphological studies are being used as tools to address soil fertility comprehensively. Although soil spatial information varies widely, analytical infrastructure is still inadequate and lacks adequate financial and technical support. The appropriate nutrient extractant and process selection require soil-crop specificity, rapidity, accuracy, and user-friendliness. Unfortunately, there is no conclusive evidence of single and established soil analytical protocols suitable for precisely representing plant nutrient uptake from soil. Complex, expensive, laborious, and time-consuming protocols are still a concern in routine soil testing services everywhere.

The traditional and existing soil nutrient analytical methods in Asia and Africa, including India, effectively estimate plant nutrient uptake patterns or total soil elemental composition. But this involves separate nutrient extraction procedures that are time-consuming and costly and require a specially designed laboratory set up with glassware, chemicals, and are responsible for the unsafe disposal of chemical agents to the environment (Wang et al., 2015). These pre-requisite setup infrastructure and cost involvement have made this overall soil testing service facility not accessible well enough to the large extent of farmers. Besides that, minimum efforts are provided for monitoring soil physical properties or soil water holding attributes in existing routine soil testing. As a result, the results are usually not available at the right time, particularly before crop cultivation or before fertilizer application. It is also required to repeat soil testing in a specific interval for existing or new cropping systems for every soil type. A paradigm shift in soil nutrient status research is thus inevitable.

7.3 Problems Faced by Conventional Soil Testing Service in India

Approximately 800 and 429 newly proposed static soil testing laboratories are operating throughout India to cater to the farmers' needs (PIB, 2020). Soil Health Card Scheme was launched in 2015–2019 to issue soil health cards (SHC) to farmers on soil nutrient status and fertilizer recommendations as per their cropping pattern in all states in India for improving soil fertility and nutrient use efficiency. It was evident from the farmers' feedback survey study on 'Soil Testing Infrastructure for Faster Delivery of Soil Health Card in India' carried out by the National Productivity Council (NPC) that the use of reasonable fertilizer recommendations through this soil-test-based soil health card reporting has saved nitrogen fertilizers like urea which reduces the overall cost of cultivation. The overall crop yield increase was reported to be 5–6%, and chemical fertilizer application has declined in the range of 8–10% among the farmers. The prime concern is how well the soil health monitoring system can be linked with the distribution system of inorganic fertilizers for up-scaling fertilizer use efficiency side-by-side reduces expenditure on fertilizer subsidy. Also, it would offer an improved nationwide scenario to introduce soil-crop-climate-specific customized fertilizers and other innovative fertilizer technology.

Although the physical and financial impact of the SHC scheme is very encouraging, huge working and analytical pressure have been created over the soil testing laboratories to achieve the target. This also revealed several shortcomings in the execution of the scheme, like colossal demand and supply gap, inadequate laboratory infrastructure, shortage in the workforce, high-cost involvement, and unavailability of other logistics and support services (NITI AYOG, 2018; PIB, 2020). At the same time, digital soil mapping (DSM) at a national scale, especially in large countries like India, is a severe issue due to the unavailability of continuous and legacy data at dense sampling intervals and complex attempts to assemble with a geostatistical approach. So, these shortcomings need to be revised for further alternative considerations for efficient and convenient universal solutions. Non-destructive, systematic, synoptic fast and repetitive spatial coverage of various soil nutrient status assessments must be introduced. The adoption of precision agriculture and Internet of Things (IoT) technologies has opened a new paragon in crop production operations (Agriculture 4.0) (Benedet et al., 2021). Therefore, advanced sensing techniques may play a vital role in overcoming these shortcomings. A time and labor-saving, cost-effective way for analyzing different soil elements is increasingly required, which can be attained by relevant proximal soil sensors (Nawar et al., 2019).

7.4 Proximal Soil Sensing Approaches

In general, grid sampling is followed by sample preprocessing through grinding, sieving, and packaging before conventional laboratory analysis. Moreover, laboratory soil analysis requires extensive instrumentation with various chemicals. These practices also include costly and sophisticated instrumentations like an Atomic Absorption spectrometer (AAS), Inductively Coupled Plasma or ICP analytical setup, Kjeldahl wet digestion setup, etc. In addition, different colorimeters, Ion-selective electrodes, and Ion-selective field-effect transistors are also used for soil elemental analysis. These cumbersome, complex, time-consuming setups require a furnished laboratory with an excellent analytical hand and a steady budget flow.

On the contrary, different cost-effective portable sensors such as portable X-Ray fluorescence spectrometer (PXRF), Vis–NIR diffuse reflectance spectroscopy (DRS), NIR spectrometer, colour sensors like Nix, optical sensors like smartphones, Digital Camera, USB microscope, and colorimeters are free from these mentioned limitations. Moreover, most of these instruments are non-destructive and may be used in situ field conditions or even laboratory conditions with minimal sample preprocessing (Andrade et al., 2020).

Various scientific literature has experimented with using these sensors worldwide for many applications, especially in soil sciences (Weiser et al., 2007; Sarathjith et al., 2014; Stiglitz et al., 2016; Teixeira et al., 2018; Cozzolino, 2016; Benedet et al., 2021; Swetha & Chakraborty, 2021; Gorthi et al., 2021; Swetha et al., 2022; Dasgupta et al., 2022). These are precise and reproducible for rapidly identifying and quantifying different elemental compositions. Furthermore, these applications have brought the opportunity to explore natural resources extensively in a Spatio-temporal manner without a laboratory setup. The following sections have summarized a detailed literature survey on the above-mentioned proximal sensors used for soil fertility characterizations.

7.5 Usage of PXRF as a Sensor for the Advancement of Soil Science

Several authors have compiled PXRF use in advanced soil science globally to predict soil properties in tropical and temperate climate situations. Recently, PXRF has proved to be reliable, cheap, and fast for soil characterization to quantify several elements in seconds (Silva et al., 2021; Weindorf et al., 2014; Wang et al., 2015; Benedet et al., 2021; Dasgupta et al., 2022).

7.5.1 Working Principle of X-Ray Fluorescence Spectrometer (XRF)

Although X-ray fluorescence spectrometry (XRF) has been used in different scientific fields since 1950, the techniques got attention mainly after 1970 due to the development of solid-state detectors. Bench-top and PXRF instruments can be widely used for direct, rapid, and simultaneous multiple soil element measurements from solid soil samples with minimal or no sample preparation and without generating any chemical residue (Fig. 7.3). The working principle of XRF is related to some basis of fundamental physics that describes atom–radiation interaction. The effectiveness of the PXRF techniques largely depends on atomic number (Z). The high energy source X-ray emits from XRF excites and ionizes the soil elements. The incoming X-ray energy causes inner shell electron excitement. Inner shell electron is ejected from its orbital by absorbing incoming energy. Then, an outer shell electron jumps into the inner shell to fulfill the void. But doing so, they must emit a discrete amount of energy, which is called fluorescence. Each atom releases fluorescent X-ray photons from these external energy sources. Typically, electrons move from L to K (K_α transition), M to K (K_β transition), or M to L (L_α transition) orbital. K_α has a higher

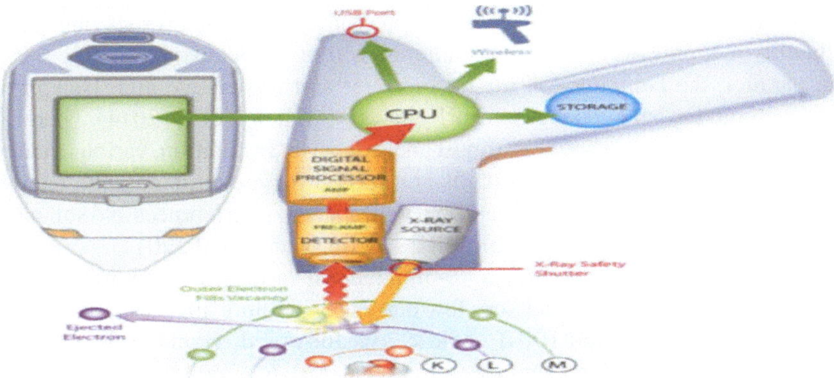

Fig. 7.3 Functional schematic of a portable pXRF instrument (López-Núñez, 2022)

intensity than K_{β}. Each element has its characteristic peak fluorescence in various transition lines, enabling us to identify and quantify it (Weindorf et al., 2014). This is characteristic of each element, which is used for its identification. The element identification is based on atomic number, emission wavelength, and their relationship in-between. A particular element can be quantified from the line intensity (number of photons for each wavelength) (Kaniu et al., 2012). Therefore, the elemental measurement (identification and quantification) is based on the characteristic energy at which fluorescence occurs, unique for each element, and the fluorescence emissions intensity (counts/second). Together, identification and quantification of each component can be achieved at a lower cost (Andrade et al., 2020; Weindorf et al., 2014).

7.5.2 Use of PXRF for Soil Fertility Characterization

PXRF can measure various elements from Mg to U of the periodic table. However, due to low energy response, all periodic table elements cannot be effectively quantified, and there is a limit to detection depending on the elemental content of the sample (Hou et al., 2004; Silva et al., 2016). The multi-element detection obtained through this process represents total elemental content. This includes elements present in soil solution, adsorbed and absorbed in soil exchangeable, non-exchangeable, mineral structure, fertilizers, and ameliorants. Using mathematical/statistical techniques, PXRF elemental results may also provide helpful information to predict bio-available, exchangeable soil nutrient status and other essential soil properties, primarily when total and available nutrient contents are related (Benedet et al., 2021; Rawal et al., 2019). Researchers are now exploring more modern data analytics than the simple linear models incorporated in the instrumentation software (Kenna et al., 2011). The use of chemometrics or machine learning tools for XRF spectra analysis is usually called re-calibration of the instrument over the manufacturer calibration models. It also efficiently gathers more soil information; widens the scope for spatial nutrient availability mapping across an area of interest with more details.

In parallel, the complex soil matrix behavior and associative high background noise levels put further challenges in XRF analysis (Weindorf et al., 2014). Therefore, the re-calibration model should be robust and extensive. In addition, various modern-age machine learning-based algorithms are needed for rapid quantitative analysis, which can minimize spectral noise and deal with spectral interferences and soil matrix effects when using the multi-variable XRF spectra (Kaniu et al., 2012).

PXRF application in tropical agro-climatic scenarios for land use planning remains mainly unexplored, and very few pieces of literature are documented. Teixeira et al. (2018) reported that PXRF could successfully predict soil fertility properties in Inceptisols under different land uses. Silva et al. (2016) aimed to examine the efficiency of digital terrain models (DTM) and proximal sensors, such as magnetometers and PXRF, to improve prediction models for soil classes and properties in tropical conditions and mapping. They suggested the idea of using PXRF as field support for soil mapping. Nawar et al. (2019) experimented on a wide range of soils with 105 soil samples originating from 4 different land uses (urban, forest, agriculture, and landfill) collected from 10 different countries on two different continents; Europe, and Africa, and then scanned with an XRF spectrometer. They found the best random forest (RF) model predicted result for soil K followed by Ca, Mg, and P with a coefficient of determination R^2 values of 0.83, 0.76, 0.69, and 0.47, and root mean squared error of prediction (RMSEP) of 2283.8, 6818.7, 1511.8, and 538.08 mg kg^{-1}, respectively. Their findings concluded that PXRF and its spectral data analysis followed by RF modeling can produce high soil K and Ca prediction accuracy. However, they found moderate performance for soil P and Mg, requiring more research.

Rawal et al. (2019) experimented in subtropical soils with a PXRF by using 300 soil samples and reported that random forest (RF) and regression tree (RT) resulted in moderate predictive performance and surpassed multiple linear regression (MLR) in predicting soil base saturation percentage (BSP) through measuring Mg, Ca, and K and cation exchange capacity (CEC).

Andrade et al. (2020) evaluated the efficiency of PXRF and machine learning algorithms [stepwise generalized linear models (SGLM), and RF] for the prediction of exchangeable Ca^{2+}, Mg^{2+}, Al^{3+}, and available P and K^+ contents alone and associated with soil texture in Brazilian Coastal Plain soils. Their prediction model resulted in moderate to fair accuracy for available K, exchangeable Ca^{2+}, and Mg^{2+} with R^2 0.67, 0.67, and 0.60 respectively. Nevertheless, models did not generate a reliable prediction for available P (0.14) and exchangeable Al^{3+} (0.47). Tavares et al. (2020) worked on modeling approaches consisting of Compton normalization (CN) and multivariate regressions (e.g., MLR and partial least squares regression (PLSR)) to overcome soil matrix effects along with more accuracy for soil fertility detection. They found improved performance for all attributes which are under study using both MLR as well as PLSR. By combining use of multivariate regressions (MLR or PLSR) along with CN showed higher performance for ex-K ($R^2 = 0.94$), CEC ($R^2 = 0.87$), ex-Ca ($R^2 = 0.96$), clay ($R^2 = 0.92$).

Benedet et al. (2021) collected 1975 soils from seven Brazilian states under varied land use and wider conditions to experiment with PXRF for exchangeable or available Al^{3+}, K^+, Ca^{2+}, Mg^{2+}, and P. They used stepwise SGLM and RF algorithms to predict relevant soil properties. The predictive power ranged from 0.49 to 0.83, in the following order: exchangeable Al < available K < available P < exchangeable Mg < P-remaining < exchangeable Ca while RF models showed more accuracy than the GLM algorithm. Pelegrino et al. (2022) applied four different models, viz. Linear regression (LR), 2nd-degree polynomial regression (PR), power regression (PwR), and Stepwise Multiple Linear Regression (SMLR) for available soil nutrient prediction in Brazilian soils. ArcGIS software spatially represented the nutrient contents in a 10 m × 10 m resolution. Also, the PXRF results were spatialized throughout the study area using Multilevel B-splines interpolation with SAGA GIS software for continuous data set generation. Subsequently, the best model for each soil nutrient was used with this spatial information to generate nutrient-wise spatial variable maps.

Dasgupta et al. (2022) experimented with PXRF in combination with agro-climatic, physiographic, physicochemical attributes (pH, EC, loss on ignition organic matter, and organic carbon) and parent-material as auxiliary inputs to predict various soil fertility parameters[available Ca, Fe, K, Zn, B, Mn, Mg, Cu, K/Mg ratio, sulfur availability index (SAI), total exchangeable bases (TEB), via four different machine-learning algorithms (support vector regression, stepwise multiple linear regression, random forest, and their averaged model). Although no universal model was suited for all 11 soil parameters, their averaged model produced a reasonable model performance with an $R^2 > 0.50$ for most of the cases. Site-wise major findings involving X-ray fluorescence (PXRF) spectrometry are shown in Table 7.1.

7.6 Use of DRS for Rapid Soil Characterization

Vis–NIR diffuse reflectance spectroscopy (DRS) operates with the reflectance spectroscopy principle, particularly in the electromagnetic spectrum of Vis–NIR-MIR regions. For decades, reflectance spectroscopy has been used to identify and predict various soil nutrients. Spectral signature indicates spectral absorption or reflection of a substance as a function of wavelength. The energy reflected on particular wavelengths depends on the substance. These sensors are active or passive, depending on their light source. Near-Infrared (NIR) spectral features arise due to molecular absorptions of overtones and combinations caused by the stretching and bending of structural groups of atoms. Mid-Infrared (MIR) spectral features occur due to fundamental vibrations, while the ultraviolet (UV) and visible spectrum cause electronic transitions in molecules. The sensor detects the amount and intensity of light absorbed, and it is plotted as a function of wavelength to yield spectral response curves. DRS applications have increased many folds with the development of portable DRS spectrometers. The DRS spectral library approach was proposed earlier to model and manage soil risk evaluations in various streams (Shepherd & Walsh, 2002). It was found that creating a spectral library was very useful for modeling multiple

Table 7.1 Site-wise major findings involving X-ray fluorescence (PXRF) spectrometry

Site	Findings	References
Southern Louisiana, USA	PXRF was used to document the spatial distribution of As, Cd, and Cr in the peri-urban agricultural fields coupled with GPS	Weindrof et al. (2012)
West Texas, USA	PXRF, VISNIR-DRS, and RS were used to predict soil salinity	Aladabaa et al. (2015)
Seward County, Nebraska, Kern County, California, and Lubbock County, Texas, USA	Total carbon and total nitrogen were predicted by fusing PXRF and VISNIR of 675 soil samples having diverse physicochemical properties	Wang et al. (2015)
Romania	Soil samples were collected from a contaminated place and the concentration of different heavy metals was analyzed by PXRF. The heavy metals contaminated hotspots were identified by non-parametric indicator kriging. The hotspots were found around the smelting plant and dispersal stack and the concentration of the heavy metals exceeded the government-mandated limits	Chakraborty et al. (2017)
Transylvanian Plain, Romania near the towns of Cojocna (Cluj County), Caianu (Cluj County), and Zau de Câmpie (Mureș County)	Different soil properties like total carbon, total nitrogen, and soil texture were predicted more accurately when PXRF and DRS were used together	Duda et al. (2017)
Italy and United States of America (Texas and New Mexico)	PXRF + VisNIR DRS was used to predict the total carbon and total nitrogen in 136 organic laden soil samples collected from the upland and non-saturated 'O' horizon	Cardelli et al. (2017)
Compost processing laboratories in both the USA and Canada	PXRF can be used to predict the CEC of compost	Li et al. (2018)
West Bengal, India	The PXRF can be used to classify soil, based on different land use and land cover system. To check the accuracy, different machine learning algorithms were used like random forest, linear and non-linear support vector machine, and classification and regression tree	Chakraborty et al. (2019)
Dhapa landfill site of Kolkata, India	PXRF and NixPRO™ sensors were applied for landfill agricultural soil assessment. Elevated soil organic carbon and heavy metal levels were quantified	Mukhopadhyay et al. (2020)

soil properties using existing soil samples and simultaneously predicting new soil properties by including them in the library. DRS has been used primarily to predict clay (Waiser et al., 2007) due to the characteristic absorption features of various clay functional groups of clay minerals, organic matter (Rossel et al., 2006; Vasques et al., 2009), soil texture (Sarathjith et al., 2014), soil color (Rossel et al., 2009), heavy elements (Cozzolino, 2016), etc.

7.7 Use of Soil Color for Assessing Soil Fertility

The soil color is an excellent indicator of soils chemical compositions and physical features. Strong interrelationships have been reported with many other soil properties, fertility, compositions, water-holding capacity, texture, organic matter, and many others. The analysis of soil color can efficiently achieve a considerable amount of soil data. Soil color acts as an indicator of productivity and mineralogy (Evans & Franzmeier, 1988). Soils have been historically grouped into various groups based on their color. Soil color was therefore used to predict these properties and achieved reasonable accuracy. Therefore, proper identification of soil color is necessary. Thus, the Munsell Soil Color Chart has been used for defining soil color qualitatively. The Munsell color system describes color using three attributes: Hue, Value, and Chroma in an irregular 3-dimensional sphere. It was developed to explain the visual color differences and is perceptually uniform. However, though it is widely used, it has the limitations of subjective vision, limited color chips, and is difficult to perform statistical analyses as it is based on visual perceptions rather than mathematical models (Blavet et al., 2000).

The Munsell soil color chart (MSCC) has been used in soil science since the 1930s (Shields et al., 1966). Testing in the 1940s (Gardner, 1944) directed to a form publicizing the Munsell soil color charts during the updated Soil Survey Manual (Soil Survey Staff, 1951). The design of the chart started where the arrangement of the chips was restricted to a single hue per page (10R, 2.5YR, 5YR, 7.5YR, 10YR, 2.5Y, and 5Y), and every single page comprised chips with chroma in additions of two units (0, 2, 4, 6, etc.). Numerous studies have recommended that it is suitable for trained scientists to interpose between chips to assess colors to the closest unit of hue and closest half unit of both value and chroma (Shields et al., 1966). However, this card divides the color space into small sections, mainly on personal experiences and under other illumination conditions. These personal interferences are prone to lead to erroneous results. In modern decades, digital colorimeters have become useful to researchers and progressively applied in soil color determination (Wutscher & McCollum, 1993) by reducing the need for subjective visual description using color chips. Swetha et al. (2022) experimented with the relationship between Nix color sensor and MSCC values for classifying three contrasting soil types (red and laterite zone, coastal saline zone, and Gangetic alluvial zone) along with soil organic carbon. Here, the potential accuracy with the Nix sensor resulted in soil classification with both classification tree and linear support vector machine algorithms. In addition, a combined dataset of $Nix_{NON\text{-}STD}$ datasets and MSCC showed the highest OC prediction accuracy ($R^2 = 0.66$).

Though for many reasons like suitability, rapidity, and small-scale diversity like redoximorphic characteristics and ped layers, the necessity will persist for field color evaluation with the help of standardized charts. The characterization of soil color uses a 'Cartesian system' that can carry out strictly or exchange already available Munsell data into another method using statistical regression (Barouchas & Moustakas, 2004).

With recent and gradual progress in digital applications of reflectance spectra, various technologies are being used to capture soil images in the visible and near-infrared spectra to identify soil types, fertility, and even soil characterizations. Different inexpensive, portable options can be used as alternative platforms on these aspects.

7.8 New Age Image and Color Sensors for Soil Characterization

Lab standard colorimeters are highly accurate in describing soil color. However, they are expensive and cannot identify soil color as in portable devices. Therefore, mobile phone cameras were used to capture soil images and identify soil colors in RGB (Gómez-Robledo et al., 2013) format. Current technology has permitted us to attain digital soil color data and use the same for describing soil color more quickly, reliably, and precisely. The era of digital color-based soil description can be reached either through digital photography at visible wavelengths (Das et al., 2022) or visible or near-infrared spectrometry (Mouazen et al., 2007). Instead of Munsell soil color charts, handheld, portable colorimetry (Campos & Demattê, 2004), soil spectrometry (Shields et al., 1966), and digital photography using smartphone cameras (Moonrungsee et al., 2015) have been used. An instrument like the Nix color sensor may offer field-based color estimation. It successfully predicted the soil color and was comparable to laboratory standard colorimeters. However, the vast differences in imaging capability across different mobile devices pose a problem in obtaining consistent results. Nix is a portable, battery-operated, wireless color sensor with inbuilt calibration. It expresses color in various color models and Color results can be recorded in various color systems such as RGB (red, green, and blue) CMYK (cyan, magenta, yellow and black), XYZ (red, green, and blue), and CIEL*a*b* (lightness, redness, and yellowness), It was compared with the Munsell soil color chart to measure soil color (Stiglitz et al., 2016). The results showed the potential of the Nix sensor for measuring soil color, and values were found closer to the lab standard colorimeter. The Nix color variables were also used to predict soil organic carbon, organic matter, and total nitrogen (Mukhopadhyay & Chakraborty, 2020; Mukhopadhyay et al., 2020; Stiglitz et al., 2017; Swetha & Chakraborty, 2021; Swetha et al., 2022).

7.9 Use of Android-Based Smartphone Sensors

7.9.1 Use of Smartphone as a Proximal Soil Sensor

In recent years, with smartphones' high popularity, android-based smartphone sensors are being used in many aspects of agriculture, like soil science. Smartphone detectors have advantages like convenience, good portability, cheapness, and wide availability among the farming community. There is no knowledge required about the camera's characteristics, and data capture and analysis can be done using available commercial software. Pongnumkul et al. (2015) systemically reviewed various literature on smartphone sensors for agricultural use by including 22 articles in depth. Out of that, 12 described farming applications, six on-farm management applications, three on information services, and four on extensive service applicability. Different smartphones have been tested under different natural illumination conditions and compared using Munsell color charts.

Levin et al. (2005) used 370 sandy soils to compare RGB values from digital images, which correlate highly with a field spectrometer performance. Han et al. (2016) analyzed and reported possibilities of the visible spectrum and machine vision technology for soil classification. They used complementary metal oxide semiconductor (CMOS) devices or smartphones as sensors and flashguns as light sources. The associated arrangements were an external lens, shading devices, and a color calibration card for this image acquisition technology. In many studies, standard color chips or a standard white reference were taken parallel within the scene to calibrate the RGB values properly.

The colors of soil were converted into RGB signals for rapid soil classifications. Several authors have reported computational solutions to transform digital images into standard color spaces from every digital RGB image pixel. Fan et al. (2017) reported the results of a new soil color app for mobile phones. Using a spectrophotometer, they compared the measure and estimated soil colors with "true" colors. Different intensities of natural illumination conditions highly influence the accuracy of smartphone images. For example, a higher accuracy (lower variance) was found with natural sunny conditions than in cloudy situations. A controlled artificial illumination was commonly suggested for ex-situ laboratory image acquisitions to check inconsistencies from the light source. Gómez-Robledo et al. (2013) developed an android application for soil sample picture acquisition. The mobile applications should distinguish between brownish, reddish, and yellowish hues, from dark to light, and their variable intensity. There has been significant improvement in image processing and data preprocessing. First, a database is generated from images from smartphones or digital cameras. Subsequently, the images are preprocessed, and extracted features are used for predicting soil properties. Many modeling techniques use neural networks, support vector machines, fuzzy logic, etc.

7.9.2 Colorimetric and Spectrophotometric Analysis of Nutrients via Smartphone Sensor

Colorimetry is the most important method used in physical chemistry and analytical chemistry to computing the concentration of colored solutions (Capitán-Vallvey et al., 2015). In color science, colorimetry explains and describes a specific color. Colorimeters are tools that compute soil and water phosphorus and NO_3^- concentration via absorbance calculations using spectrophotometry and Beer-Lambert law. They are accessible for both laboratory use and field use, depending on the intricacy of the analysis. Smartphone-based colorimetry and spectrophotometry are attaining considerable significance because of extensive progress of instruments with cumulative computational power, relative cost-effectiveness, and convenient designs associated with user-friendly interfaces, their compatibility with data attainment and analyzing capabilities (Fig. 7.4). Smartphone-based soil studies have been recently popular and found to be favorable in different fields like chemical sensing, biological sensing, water quality sensing, and many more (Shen et al., 2012). Smartphone spectrometers use various modules of wavelength, which provide spectral data produced by soil and water samples (Hossain et al., 2015). Then the color spectrum photograph is converted into numerous color spaces for mining quantitative information. The spectrum wavelength usually fluctuates between 400 and 700 nm due to the optical filters fixed in the forward-facing of the camera in the built-up process. The spectral data is used in numerous applications (Özdemir et al., 2017). The determination of soil As concentrations by smartphone was carried out by using a correlation ratio of G to R values obtained from the captured RGB photographs with the target sample's As concentration (Siddiqui & Chaudhry, 2018). The RGB color index and HSV (Hue, Saturation, and Value) color model are predominantly used for smartphone camera-based nutrient detection. Smartphone-based colorimetric sensors have been used for the measurement of K (García et al., 2011), Cl (Sumriddetchkajorn et al., 2013), NO_2^- (Puangpila et al., 2018), texture (Swetha et al., 2020), organic matter (Gorthi et al., 2021).

7.9.3 Mobile Applications for Smartphone-Based Colorimetric Soil and Water Analysis

Smartphone applications were technologically developed to create the whole structure user-friendly (Kim et al., 2017). The "Colorimetric Test Reader app" was developed by Yetisen et al. (2014), which estimates glucose, pH values, and protein values. Solmaz et al., (2018) developed the "ChemTrainer app," which determines the peroxide content using a machine-learning algorithm on the remote server. After taking an image of strips of the colorimetric test, the application directs average RGB values for an area of concern to the remote server as the machine learning algorithm requires only average values for the arrangement. The message line provision was

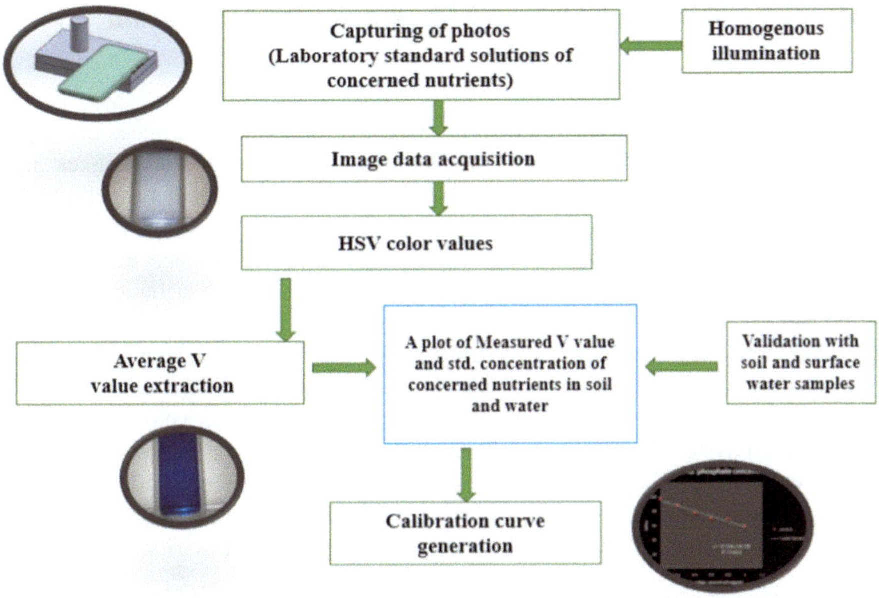

Fig. 7.4 Workflow for estimation of soil and water nutrients through colorimetric approach and smartphone-based image acquisition technique

made to allow many users to access the server instantaneously. Two available android applications with open access "The RGB color detector" and "stanXY" have been used to estimate phosphate concentrations in both soil and water samples, respectively. An application "Akvo Caddisfly" quantified mineral forms of soil N by using test strips of colorimetric technique (Golicz et al., 2019). Initial efforts were made to develop a handy and convenient colorimetric sensor to determine plant-available phosphate (Moonrungsee et al., 2015). A smartphone was attached to the device mainly to capture a series of photographs of soil extracts. Then the images were examined for RGB values through the custom-built "Phosphorus Analysis App". Photographs attained from the smartphone were successively examined by the application and found that the images were significantly correlated ($R^2 = 0.996$) with the laboratory (standard spectrophotometric) methods. The smartphone has also been used for measuring soil phosphorus in the field (Campbell et al., 2015). Other apps like "BaiKhao" was used to estimate the deficit levels of both K and N in *Oryza sativa* (Asian rice) plants (Intaravanne & Sumriddetchkajorn, 2015), and the "Smart SPAD" application was used to quantify the chlorophyll content of maize plants precisely and accurately (Vesali et al., 2015).

7.10 Sensor-Based NO_3^- and Phosphate Determination in Soil and Water

For maintaining agricultural productivity, fertilizer application is necessary at the right time, at the right place, in the right quantity, and the right method (Dinnes et al., 2002). The soil NO_3^- levels in agricultural fields, soil chemical properties, and physical properties show significant differences spatially and temporally (Heuvelink et al., 1998). Many studies increased our understanding of the impacts of soil NO_3^--on crop production systems (Almekinders et al., 1995). The distribution of NO_3^- in the soil is governed by several factors like soil formation, chemical application, microbial activity, plant progress, geometric characteristics, environmental factors, and soil management factors that impact soil crop dynamics (Wagenet et al., 1983). In the last two decades, efforts have been made to design and develop a soil NO_3^- measurement system based on mainly the following sensors types: (i) ion-selective field-effect transistor, (ii) ion-selective electrode, and (iii) spectrophotometer. The simplicity and robustness of UV absorbance spectroscopy for determining NO_3^- concentration in water samples make it possibly adoptable for *in-situ* application in soil.

In contrast, phosphate is an essential and significant nutrient responsible for supplying energy to plants that drive most biochemical functions. Phosphorus is a significant component of the plant's growth and development, and it has a vast influence on terrestrial and aquatic life (Batterman et al., 2013). However, to compensate for the low P accessibility in soil, many P-comprising fertilizers are used in many agricultural fields. However, the over-fertilization of the element may lead to the leaching of nutrients, water contamination, and permanent impairment of water life (Korostynska et al., 2012). According to the World Health Organization (WHO), the allowable concentration of phosphate level in drinking water should be 0.1 mg L^{-1}, and the permissible phosphate levels in natural and wastewater should be 0.2 and 10 mg L^{-1}, respectively (Brady et al., 2008 and WHO, 2005). Thus, managing phosphate limits in both agricultural soil and water is crucial to maintaining the fertility of the soil and the quality of water. There are numerous methods available to estimate phosphate. The methods mainly include colorimetric, spectroscopic, and electrochemical techniques. However, all these methods are generally laboratory-limited and involve more apparatus like spectrophotometers and flame photometers which are expensive, need experts to handle, and do not apply to independent task in the field.

7.11 Digital Soil Mapping Approaches for Rapid Soil Information and Resource Map Generation

Soil maps give a detailed description of the soil's spatial and temporal features. The soil mapping benefits the user in a thorough understanding of soil-forming processes and evaluates the locations of the different soil classes of varying soil types. It also helps in predicting the variation of soil properties. Soil mapping was traditionally associated with generating a theoretical understanding of soil-forming processes, which was used to forecast the spatial distribution of various soil classes (Fig. 7.5). Frequently, descriptive and analytical soil profile features were utilized to categorize the soil (Hole & Campbell, 1985). For an extended time, aerial photography and satellite pictures delivered the spatial perspective for assessing 'class-type' soil maps. Since the 1990s, the data from both spatial and attribute information have been combined into computerized databases and GIS to refine the usage and ease of access to soil information.

DSM was developed from the 'state-factor' soil-forming model (Jenny, 1930) for explaining the connection between soil development and soil distribution. The model demonstrates that the soil profile features are parent material, climate, relief, slope, microorganisms, and time. These are called soil-forming factors. In initial soil mapping actions, the empirical correlation between soil-forming factors and

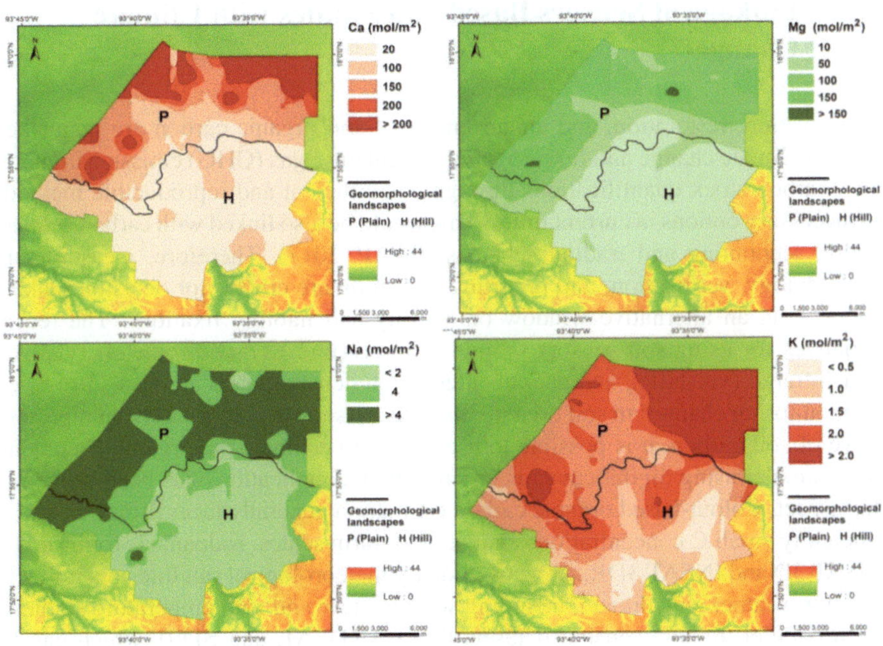

Fig. 7.5 Spatial distribution of exchangeable cations (López-Castañeda, et al., 2022)

characteristics of the soil profile was based on 'Jenny's equation.' Then, it was executed by examiners using soil-landscape-related models (Hudson, 1992).

Notably, DSM is the technical progression that pursues to refine the precision, processing, accurateness, and conveyance of soil info at several scales worldwide. Digital soil maps extensively use progressive technologies like GIS (Geographic information system), field scanners, spectral signatures, GPS, remote sensing, digital elevation models, and computational progress like machine learning, deep learning algorithms, and geostatistical interpolation. The spatial distribution of soil classes and soil assets can be mapped along with the uncertainty of prediction with DSM. The DSM can be used to modernize the data of soil surveys and generate specific thematic analyses (Carré et al., 2007). DSM is also called 'predictive soil mapping' given it constructs spatial soil data structures using statistical models that collect the spatial differences of soil characteristics and types of soil mainly from related environmental factors and soil analysis (Lagacherie et al., 2006). Furthermore, DSM assures better soil information and improved reporting of mapped regions. Digital soil mapping mainly prepares geographically addressed soil maps produced by relating spatially available environmental information and field and laboratory soil measurements (McBratney et al., 2003). Globally, DSM is accepted as the most appropriate method for getting the spatial variation of soil properties (Fig. 7.6).

7.12 Linkage of Sensors Based Soil Studies with Climate Change

Conservation, restoration, and enhancement of soil organic carbon are major key steps of climate change and for mitigating greenhouse gas (GHG) emissions. Therefore, carbon stock quantifications through cost-efficient and reproducible methods are necessary options in current times. This would be also linked with carbon sequestration accounting and trading schemes. (ICAP, 2017). Therefore, measurement methodology, verification, and reporting of carbon are crucial for decision-makers for opening an alternative window on priority and liability fixations. The recent initiatives for using soil sensors — mid-infrared (mid-IR) spectrometers, visible–near-infrared (vis–NIR), and gamma attenuation densitometers are now introduced for monitoring changes in soil organic C stocks. Also, soil colour has been representative of soil carbon, it is possible to use colour to estimate the organic C content. For example, using digital cameras or smartphone applications has been proven in this aspect. Colour-based organic carbon estimation generally faces difficulties with inherently small C concentrations, and at field–farm scales. Kalpana-1, followed by INSAT-3D and INSAT-3DR, are geostationary meteorological satellites operating to monitor weather and climate, including soil moisture and soil temperature studies.

In parallel, remote sensing datasets like LANDSAT, and SPOT which can be utilized to derive environmental variables like digital elevation model (DEM), slope, topographic wetness and position Index, aspect, and others. SRTMDEM, AsterDEM,

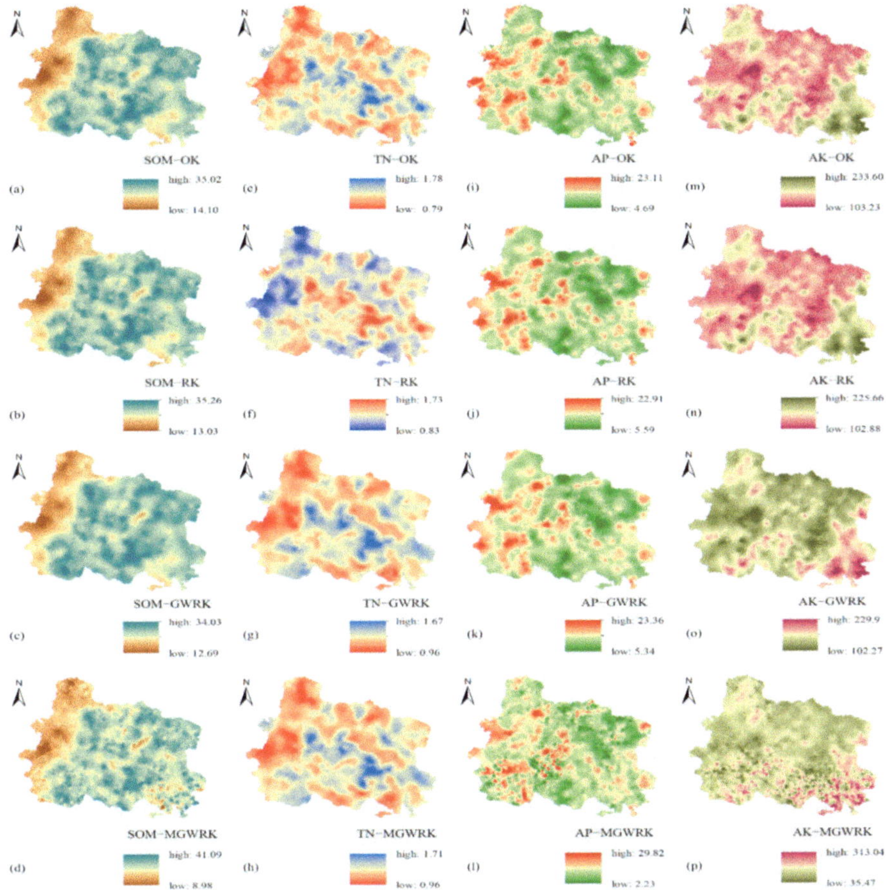

Fig. 7.6 Distribution of soil nutrients derived from different models (Gao, et al., 2021)

and Cartosat-1 DEM (10 m resolution) data can be extracted by using GIS and are available on different platforms like Land Cover Facility website (http://glcfapp.umiacs.umd.edu) or from http://srtm.csi.cgiar.org in various resolutions (Dharumarajan et al., 2019). Environment and Climate Variables may be gathered from the Environmental Information system (ENVIS), India. Long-term climate databases can be generated from Earth Observation (EO) satellites. This remote sensing-derived information's has been used for the successful prediction of climate changes.

7.13 Conclusion

Several environment-friendly low-cost methodologies have shown their potential for soil fertility assessment and characterizations and therefore modernized the soil science discipline. Proximal soil sensors-based models should be applied worldwide under similar climatic conditions, reducing cost and time compared with conventional laboratory analyses. Moreover, these models can be applied to spatial data, providing spatial variability maps of soil fertility properties at both landscape and profile scales. It is anticipated that technological advancements will soon develop more user-friendly proximal soil sensors. Also, coupling data from multiple sensors may provide a robust model to predict a wide range of physical, chemical, and morphological soil parameters.

References

Aldabaa, A. A. A., Weindorf, D. C., Chakraborty, S., & Sharma, A. (2015). Combination of proximal and remote sensing methods for rapid soil salinity quantification. *Geoderma, 239–240*, 34–46.

Almekinders, C. J., Fresco, L. O., & Struik, P. C. (1995). The need to study and manage variation in agro-ecosystems. *Netherlands Journal of Agricultural Science, 43*(2), 127–142.

Andrade, R., Faria, W. M., Silva, S. H. G., Chakraborty, S., Weindorf, D. C., Mesquita, L. F., & Curi, N. (2020). Prediction of soil fertility via portable X-ray fluorescence (PXRF) spectrometry and soil texture in the Brazilian Coastal Plains. *Geoderma, 357*, 113960.

Barouchas, P. E., & Moustakas, N. K. (2004). Soil colour and spectral analysis employing linear regression models. I. Effect of organic matter. *International Agrophysics, 18*(1), 118–124.

Batterman, S. A., Wurzburger, N., & Hedin, L. O. (2013). Nitrogen and phosphorus interact to control tropical symbiotic N 2 fixation: A test in Inga punctata. *Journal of Ecology, 101*(6), 1400–1408.

Benedet, L., Acuña-Guzman, S. F., Faria, W. M., Silva, S. H. G., Mancini, M., dos Teixeira, A. F. S., Pierangeli, L. M. P., Acerbi, F. W., Jr., Gomide, L. R., & Pádua, A. L., Jr. (2021). Rapid soil fertility prediction using X-ray fluorescence data and machine learning algorithms. *CATENA, 197*, 105003.

Blavet, D., Mathe, E., & Leprun, J. C. (2000). Relations between soil colour and waterlogging duration in a representative hillside of the West African granito-gneissic bedrock. *CATENA, 39*(3), 187–210.

Brady, N. C., Weil, R. R., & Weil, R. R. (2008). *The nature and properties of soils* (Vol. 13, pp. 662–710). Upper Saddle River, NJ: Prentice Hall.

Campbell, E. R., Warsko, K., Davidson, A. M., & Campbell, W. H. B. (2015). Determination of phosphate in soil extracts in the field: A green chemistry enzymatic method. *MethodsX, 2*, 211–218.

Campos, R. C., & Demattê, J. A. M. (2004). Soil color: Approach to a conventional assessment method in comparison to an automatization process for soil classification. *Revista Brasileira De Ciência Do Solo, 28*, 853–863.

Capitán-Vallvey, L. F., Lopez-Ruiz, N., Martinez-Olmos, A., Erenas, M. M., & Palma, A. J. (2015). Recent developments in computer vision-based analytical chemistry: A tutorial review. *Analytica Chimica Acta, 899*, 23–56.

Cardelli, V., Weindorf, D. C., Chakraborty, S., Li, B., De Feudis, M., Cocco, S., Agnelli, A., Choudhury, A., Ray, D. P., & Corti, G. (2017). Non-saturated soil organic horizon characterization via advanced proximal sensors. *Geoderma, 288*, 130–142.

Carré, F., McBratney, A. B., Mayr, T., & Montanarella, L. (2007). Digital soil assessments: Beyond DSM. *Geoderma, 142*(1–2), 69–79.

Chakraborty, S., Li, B., Weindorf, D. C., Deb, S., Acree, A., De, P., & Panda, P. (2019). Use of portable X-ray fluorescence spectrometry for classifying soils from different land use land cover systems in India. *Geoderma, 338*, 5–13.

Chakraborty, S., Weindorf, D. C., Weindorf, C. A., Das, B. S., Li, B., Duda, B., Pennington, S., & Ortiz, R. (2017). Semiquantitative evaluation of secondary carbonates via portable X-ray fluorescence spectrometry. *Soil Science Society of America Journal, 81*(4), 844–852.

Cozzolino, D. (2016). Near infrared spectroscopy as a tool to monitor contaminants in soil, sediments and water—State of the art, advantages and pitfalls. *Trends in Environmental Analytical Chemistry, 9*, 1–7.

Das, P., Chetry, B., Paul, S., Bhattacharya, S. S., & Nath, P. (2022). Detection and quantification of phosphate in water and soil using a smartphone. *Microchemical Journal, 172*, 106949.

Dasgupta, S., Chakraborty, S., Weindorf, D. C., Li, B., Silva, S. H. G., & Bhattacharyya, K. (2022). Influence of auxiliary soil variables to improve PXRF-based soil fertility evaluation in India. *Geoderma Regional, 30*, e00557.

Dasgupta, S., Sarkar, A., Chaitanya, A. K., Saha, A., Dey, A., & Mondal, R. (2017). Response of potato crop to integrated nutrient management in the Indo-Gangetic alluvial soils of West Bengal, India. Journal of Experimental Agriculture International, 1–10.

Dasgupta, S., Sengupta, S., Saha, S., Sarkar, A., & Anantha, K. C. (2021). Approaches in advanced soil elemental extractability: Catapulting future soil–plant nutrition research. In *Soil Science: Fundamentals to Recent Advances* (pp. 191–236). Springer, Singapore.

Demand and Supply Projections Towards 2033; Crops, Livestock, Fisheries And Agricultural Inputs; The Working Group Report, February, 2018, Niti Ayog.

Dharumarajan, S., Hegde, R., Janani, N., & Singh, S. (2019). The need for digital soil mapping in India. *Geoderma Regional, 16*, e00204. https://doi.org/10.1016/j.geodrs.2019.e00204

Dinnes, D. L., Karlen, D. L., Jaynes, D. B., Kaspar, T. C., Hatfield, J. L., Colvin, T. S., & Cambardella, C. A. (2002). *Review and interpretation: Nitrogen management strategies to reduce nitrate leaching in tile-drained Midwestern soils.*

Duda, B. M., Weindorf, D. C., Chakraborty, S., Li, B., Man, T., Paulette, L., & Deb, S. (2017). Soil characterization across catenas via advanced proximal sensors. *Geoderma, 298*, 78–91.

Evans, C. V., & Franzmeier, D. P. (1988). Color index values to represent wetness and aeration in some Indiana soils. *Geoderma, 41*(3–4), 353–368.

Fan, Z., Herrick, J. E., Saltzman, R., Matteis, C., Yudina, A., Nocella, N., Crawford, E., Parker, R., & Van Zee, J. (2017). Measurement of soil color: A comparison between smartphone camera and the munsell color charts. *Soil Science Society of America Journal, 81*(5), 1139–1146.

Gao, L., Huang, M., Zhang, W., Qiao, L., Wang, G., & Zhang, X. (2021). Comparative study on spatial digital mapping methods of soil nutrients based on different geospatial technologies. *Sustainability, 13*(6), 3270.

García, A., Erenas, M. M., Marinetto, E. D., Abad, C. A., de Orbe-Paya, I., Palma, A. J., & Capitán-Vallvey, L. F. (2011). Mobile phone platform as portable chemical analyzer. *Sensors and Actuators b: Chemical, 156*(1), 350–359.

Gardner, R. A. (1944). California soil color standards and their relation to the Iscc-Nbs method of designating colors. *Soil Science Society of America Journal, 8*(C), 355–357.

Golicz, K., Hallett, S. H., Sakrabani, R., & Pan, G. (2019). The potential for using smartphones as portable soil nutrient analyzers on suburban farms in central East China. *Scientific Reports, 9*(1), 1–10.

Gómez-Robledo, L., López-Ruiz, N., Melgosa, M., Palma, A. J., Capitán-Vallvey, L. F., & Sánchez-Marañón, M. (2013). Using the mobile phone as Munsell soil-colour sensor: An experiment under controlled illumination conditions. *Computers and Electronics in Agriculture, 99*, 200–208.

Gorthi, S., Swetha, R. K., Chakraborty, S., Li, B., Weindorf, D. C., Dutta, S., & Majumdar, K. (2021). Soil organic matter prediction using smartphone-captured digital images: Use of reflectance image and image perturbation. *Biosystems Engineering, 209*, 154–169.
Han, P., Dong, D., Zhao, X., Jiao, L., & Lang, Y. (2016). A smartphone-based soil color sensor: For soil type classification. *Computers and Electronics in Agriculture, 123*, 232–241.
Heuvelink, G. B. (1998). Uncertainty analysis in environmental modelling under a change of spatial scale. In *Soil and Water Quality at Different Scales* (pp. 255–264). Springer, Dordrecht.
Hole, F. D., & Campbell, J. B. (1985). *Soil landscape analysis.* Government Institutes.
Hossain, M. A., Canning, J., Ast, S., Cook, K., Rutledge, P. J., & Jamalipour, A. (2015). Combined "dual" absorption and fluorescence smartphone spectrometers. *Optics Letters, 40*(8), 1737–1740.
Hou, X., He, Y., & Jones, B. T. (2004). Recent advances in portable X-ray fluorescence spectrometry. *Applied Spectroscopy Reviews, 39*, 1–25.
Hudson, B. D. (1992). The soil survey as paradigm-based science. *Soil Science Society of America Journal, 56*(3), 836–841.
ICAP: Emissions Trading Worldwide: Status Report 2017, Report, International Carbon Action Partnership (ICAP), 2017.
Intaravanne, Y., & Sumriddetchkajorn, S. (2015). Android-based rice leaf color analyzer for estimating the needed amount of nitrogen fertilizer. *Computers and Electronics in Agriculture, 116*, 228–233.
Jenny, H. (1930). *Factors of soil formation: A system of quantitative pedology*. Dover Publications, New York.
Kaniu, M. I., Angeyo, K. H., Mwala, A. K., & Mwangi, F. K. (2012). Energy dispersive X-ray fluorescence and scattering assessment of soil quality via partial least squares and artificial neural networks analytical modeling approaches. *Talanta, 98*, 236–240.
Kenna, T. C., Nitsche, F. O., Herron, M. M., Mailloux, B. J., Peteet, D., Sritrairat, S., Sands, E., & Baumgarten, J. (2011). Evaluation and calibration of a field portable X-Ray fluorescence spectrometer for quantitative analysis of siliciclastic soils and sediments. *Journal of Analytical Atomic Spectrometry, 26*, 395–405.
Kim, H., Awofeso, O., Choi, Jung, Bae. (2017). Colorimetric analysis of saliva–alcohol test strips by smartphone-based instruments using machine-learning algorithms. *Applied Optics, 56*(1), 84–92.
Korostynska, O., Arshak, K., Velusamy, V., Arshak, A., & Vaseashta, A. (2012). Recent advances in point-of-access water quality monitoring: Water quality. *Technological Innovations in Sensing and Detection of Chemical, Biological, Radiological, Nuclear Threats and Ecological Terrorism*, 261–268.
Lagacherie, P., McBratney, A., & Voltz, M. (2006). *Digital soil mapping: An introductory perspective.* Elsevier.
Levin, N., Ben-Dor, E., & Singer, A. (2005). A digital camera as a tool to measure color indices and related properties of sandy soils in semi-arid environments. *International Journal of Remote Sensing, 26*(24), 5475–5492.
Li, B., Chakraborty, S., Sosa, M. F. G., Kusi, N. Y. O., & Weindorf, D. C. (2018). Compost cation exchange capacity via portable X-ray fluorescence (PXRF) spectrometry. *Compost Science & Utilization, 26*(4), 271–278.
López-Castañeda, A., Zavala-Cruz, J., Palma-López, D. J., Rincón-Ramírez, J. A., & Bautista, F. (2022). Digital mapping of soil profile properties for precision agriculture in developing countries. *Agronomy, 12*(2), 353.
López-Núñez, R. (2022). Portable X-ray fluorescence analysis of organic amendments: A review. *Applied Sciences, 12*, 6944. https://doi.org/10.3390/app12146944
McBratney, A. B., Santos, M. M., & Minasny, B. (2003). On digital soil mapping. *Geoderma, 117*(1–2), 3–52.
Moonrungsee, N., Pencharee, S., & Jakmunee, J. (2015). Colorimetric analyzer based on mobile phone camera for determination of available phosphorus in soil. *Talanta, 136*, 204–209.

Mouazen, A. M., Karoui, R., Deckers, J., De Baerdemaeker, J., & Ramon, H. (2007). Potential of visible and near-infrared spectroscopy to derive colour groups utilising the Munsell soil colour charts. *Biosystems Engineering, 97*(2), 131–143.

Mukhopadhyay, S., & Chakraborty, S. (2020). Use of diffuse reflectance spectroscopy and Nix pro color sensor in combination for rapid prediction of soil organic carbon. *Computers and Electronics in Agriculture, 176*, 105630.

Mukhopadhyay, S., Chakraborty, S., Bhadoria, P. B. S., Li, B., & Weindorf, D. C. (2020). Assessment of heavy metal and soil organic carbon by portable X-ray fluorescence spectrometry and NixPro™ sensor in landfill soils of India. *Geoderma Regional, 20*, e00249.

Nawar, S., Delbecque, N., Declercq, Y., De Smedt, P., Finke, P., Verdoodt, A., Meirvenne, M., & Mouazen, A. (2019). Can spectral analyses improve measurement of key soil fertility parameters with X-ray fluorescence spectrometry?. *Geoderma, 350*. 29–39.https://doi.org/10.1016/j.geoderma.2019.05.002

Özdemir, G. K., Bayram, A., Kılıç, V., Horzum, N., & Solmaz, M. E. (2017). Smartphone-based detection of dyes in water for environmental sustainability. *Analytical Methods, 9*(4), 579–585.

Pelegrino, M. H. P., Silva, S. H. G., de Faria, Á. J. G., Mancini, M., Teixeira, A. F. D. S., Chakraborty, S., Weindorf, D. C., Guilherme, L. R. G., & Curi, N. (2022). Prediction of soil nutrient content via PXRF spectrometry and its spatial variation in a highly variable tropical area. *Precision Agriculture, 23*(1), 18–34. https://doi.org/10.1007/s11119-021-09825-8

PIB. (2020). Press Information Bureau: Government of India. Ministry of Agriculture & Farmers Welfare on 17.02.2020. Retrived from https://pib.gov.in/Pressreleaseshare.aspx?PRID=1603379 on 05.12.2021.

Pongnumkul, S., Chaovalit, P., & Surasvadi, N. (2015). Applications of smartphone-based sensors in agriculture: A systematic review of research. *Journal of Sensors, 2015.*

Puangpila, C., Jakmunee, J., Pencharee, S., & Pensrisirikul, W. (2018). Mobile-phone-based colourimetric analysis for determining nitrite content in water. *Environmental Chemistry, 15*(7), 403–410.

Rakshit, A., Chakraborty, S., Parihar, M., Singh Meena, V., Kumar Mishra, P., & Bahadur Singh. H. (Eds.). *Innovation in Small-Farm Agriculture: Improving Livelihoods and Sustainability* (1st ed.). CRC Press. https://doi.org/10.1201/9781003164968

Rawal, A., Chakraborty, S., Li, B., Lewis, K., Godoy, M., & Laura, P. (2019). Determination of base saturation percentage in agricultural soils via portable X-ray fluorescence spectrometer. *Geoderma, 338*, 375–382. https://doi.org/10.1016/j.geoderma.2018.12.032

Resende, M., Curi, N., Rezende, S. B., Corr^ea, G. F., & Ker, J. C. (2014). *Pedologia: Base para Distinç˜ao de Ambientes* (6th ed.). Editora UFLA.

Rossel, R. V., Cattle, S. R., Ortega, A., & Fouad, Y. (2009). In situ measurements of soil color, mineral composition and clay content by vis–NIR spectroscopy. *Geoderma, 150*(3–4), 253–266.

Rossel, R. V., Walvoort, D. J. J., McBratney, A. B., Janik, L. J., & Skjemstad, J. O. (2006). Visible, near infrared, mid infrared or combined diffuse reflectance spectroscopy for simultaneous assessment of various soil properties. *Geoderma, 131*(1–2), 59–75.

Sarathjith, M. C., Das, B. S., Vasava, H. B., Mohanty, B., Sahadevan, A. S., Wani, S. P., & Sahrawat, K. L. (2014). Diffuse reflectance spectroscopic approach for the characterization of soil aggregate size distribution. *Soil Science Society of America Journal, 78*(2), 369–376.

Sengupta, S., Dasgupta, S., Bhattacharyya, K., Chakraborty, S., & Dey, P. (2022). *A Pandemic Resilient Framework for Sustainable Soil Health and Food Security: Response beyond Covid-19.*

Shen, L., Hagen, J. A., & Papautsky, I. (2012). Point-of-care colorimetric detection with a smartphone. *Lab on a Chip, 12*(21), 4240–4243.

Shepherd, K. D., & Walsh, M. G. (2002). Development of reflectance spectral libraries for characterization of soil properties. *Soil Science Society of America Journal, 66*(3), 988–998.

Shields, J. A., St. Arnaud, R. J., Paul, E. A., & Clayton, J. S. (1966). Measurement of soil color. *Canadian Journal of Soil Science, 46*(1), 83–90.

Siddiqui, S. I., & Chaudhry, S. A. (2018). A review on graphene oxide and its composites preparation and their use for the removal of As3+ and As5+ from water under the effect of various parameters: Application of isotherm, kinetic and thermodynamics. *Process Safety and Environmental Protection, 119*, 138–163.

Silva, S. H. G., Poggere, G. C., Menezes, M. Dd., Carvalho, G. S., Guilherme, L. R. G., & Curi, N. (2016). Proximal sensing and digital terrain models applied to digital soil mapping and modeling of Brazilian Latosols (Oxisols). *Remote Sens, 8*, 614.https://doi.org/10.3390/rs8080614

Silva, S. H. G., Ribeiro, B. T., Guerra, M. B. B., de Carvalho, H. W. P., Lopes, G., Carvalho, G. S., Guilherme, L. R. G., Resende, M., Mancini, M., Curi, N., Rafael, R. B. A., Cardelli, V., Cocco, S., Corti, G., Chakraborty, S., Li, B., & Weindorf. D. C. (2021). pXRF in tropical soils: Methodology, applications, achievements and challenges. *Advances in Agronomy, 167*, 1–62

Solmaz, M. E., Mutlu, A. Y., Alankus, G., Kılıç, V., Bayram, A., & Horzum, N. (2018). Quantifying colorimetric tests using a smartphone app based on machine learning classifiers. *Sensors and Actuators b: Chemical, 255*, 1967–1973.

Staff, S. S., & Staff, S. S. (1951). *Soil survey manual*. USDA.

Stiglitz, R., Mikhailova, E., Post, C., Schlautman, M., & Sharp, J. (2016). Evaluation of an inexpensive sensor to measure soil color. *Computers and Electronics in Agriculture, 121*, 141–148.

Stiglitz, R., Mikhailova, E., Post, C., Schlautman, M., & Sharp, J. (2017). Using an inexpensive color sensor for rapid assessment of soil organic carbon. *Geoderma, 286*, 98–103.

Sumriddetchkajorn, S., Chaitavon, K., & Intaravanne, Y. (2013). Mobile device-based self-referencing colorimeter for monitoring chlorine concentration in water. *Sensors and Actuators b: Chemical, 182*, 592–597.

Swetha, R. K., Bende, P., Singh, K., Gorthi, S., Biswas, A., Li, B., & Chakraborty, S. (2020). Predicting soil texture from smartphone-captured digital images and an application. *Geoderma, 376*, 114562.

Swetha, R. K., & Chakraborty, S. (2021). Combination of soil texture with Nix color sensor can improve soil organic carbon prediction. *Geoderma, 382*, 114775.

Swetha, R. K., Dasgupta, S., Chakraborty, S., Li, B., Weindorf, D. C., Mancini, M., Silva, S. H. G., Ribeiro, B. T., Curi, N., & Ray, D. P. (2022). Using Nix color sensor and Munsell soil color variables to classify contrasting soil types and predict soil organic carbon in Eastern India. *Computers and Electronics in Agriculture, 199*, 107192.

Tavares, T. R., Mouazen, A. M., Alves, E. E. N., dos Santos, F. R., Melquiades, F. L., Pereira de Carvalho, H. W., & Molin, J. P. (2020). Assessing soil key fertility attributes using a portable X-ray Fluorescence: A simple method to overcome matrix effect. *Agronomy, 10*, 787. https://doi.org/10.3390/agronomy10060787

Teixeira, A. F. D. S., Weindorf, D. C., Silva, S. H. G., Guilherme, L. R. G., & Curi, N. (2018). Portable X-ray fluorescence (PXRF) spectrometry applied to the prediction of chemical attributes in Inceptisols under different land uses. *Cienc. Agrotec, 42*(5), 501–512. https://doi.org/10.1590/1413-70542018425017518

Vasques, G. M., Grunwald, S., & Sickman, J. O. (2009). Modeling of soil organic carbon fractions using visible–near-infrared spectroscopy. *Soil Science Society of America Journal, 73*(1), 176–184.

Vesali, F., Omid, M., Kaleita, A., & Mobli, H. (2015). Development of an android app to estimate chlorophyll content of corn leaves based on contact imaging. *Computers and Electronics in Agriculture, 116*, 211–220.

Wagenet, R. J. (1983). Principles of salt movement in soils. *Chemical Mobility and Reactivity in Soil Systems, 11*, 123–140.

Waiser, T. H., Morgan, C. L., Brown, D. J., & Hallmark, C. T. (2007). In situ characterization of soil clay content with visible near-infrared diffuse reflectance spectroscopy. *Soil Science Society of America Journal, 71*(2), 389–396.

Wang, D., Chakraborty, S., Weindorf, D. C., Li, B., Sharma, A., Paul, S., & Ali, M. N. (2015). Synthesized use of VisNIR DRS and PXRF for soil characterization: Total carbon and total nitrogen. *Geoderma, 243–244*, 157–167.

Weindorf, D. C., Bakr, N., & Zhu, Y. (2014). Advances in portable X-ray fluorescence (PXRF) for environmental, pedological, and agronomic applications. *Advances in Agronomy, 128*, 1–45.

Weindorf, D. C., Zhu, Y., Haggard, B., Lofton, J., Chakraborty, S., Bakr, N., Zhang, W., Weindorf, W. C., & Legoria, M. (2012). Enhanced pedon horizonation using portable X-ray fluorescence spectrometry. *Soil Science Society of America Journal, 76*(2), 522–531.

World Health Organization. (2005). Nutrients in drinking water (No. WHO/SDE/WSH/05.09). World Health Organization.

Wutscher, H. K., & McCollum, T. G. (1993). Rapid, objective measurement of soil color with a tristimulus colorimeter. *Communications in Soil Science and Plant Analysis, 24*(17–18), 2165–2169.

Yaalon, D. H., & Yaron, B. (1966). Framework for man-made soil changes-an outline of metapedogenesis. *Soil Science, 102*, 272–277. https://doi.org/10.1097/00010694-196610000-00010

Yetisen, A. K., Martinez-Hurtado, J. L., Garcia-Melendrez, A., da Cruz, V. F., & Lowe, C. R. (2014). A smartphone algorithm with inter-phone repeatability for the analysis of colorimetric tests. *Sensors and Actuators b: Chemical, 196*, 156–160.

Chapter 8
Assessing Spatially-Distributed Soil Moisture Under Changing Land Uses and Climate

Soumyaranjan Sahoo and Bhabagrahi Sahoo

Abstract Soil moisture distribution across the hillslopes governs the partition of precipitation to surface runoff, infiltration, groundwater recharge, and moisture availability for evapotranspiration highlighting its ecohydrological importance. Similarly, surface soil moisture availability is essential for vegetation growth and crop production which in turn affects the agriculture-based economy. This reflects the economic importance of soil moisture. Hydrological modeling enables to assess the spatio-temporal variability of soil moisture in response to hydro-meteorological and edaphic conditions. However, in most hydrological modeling practices, it is often estimated using conceptual approaches. In such a scenario, the parameter of the conceptual model may not remain valid in response to climate and land use changes that are dynamic. Hence, the physically-based models, considering the dynamics of moisture or fluid flow by accounting for the conservation of mass and momentum seem to be a possible alternative. In this chapter, a hillslope-based model, namely the enhanced hillslope storage Boussinesq (eHSB) model is presented. The one-dimensional eHSB model traces its derivation from the governing Darcy and continuity equations. The eHSB model formulation considers the Richards equation for vadose zone modeling that simulates the spatio-temporal variation of soil moisture. Hence, the modeling approach simulates the dynamic hillslope saturated storage in the aquifer and unsaturated soil moisture storage in the overlying vadose zone. The model was tested in the Baitarani River basin in eastern India that performed with a root mean square error of $\leq$1.84 m in simulating the water table. The tested setup, when forced with future climate and recharge accounting for land use/ land cover change, gives the future soil moisture distribution and aquifer storage as output. From the predicted basin-scale

S. Sahoo · B. Sahoo (✉)
School of Water Resources, Indian Institute of Technology, Kharagpur, West Bengal 721302, India
e-mail: bsahoo2003@yahoo.com; bsahoo@swr.iitkgp.ac.in

S. Sahoo
Istituto di Ricerca per la Protezione Idrogeologica (IRPI), Consiglio Nazionale delle Ricerche (CNR), 06128 Perugia, Italy

Surface Water Hydrology Division, National Institute of Hydrology, Roorkee, Uttarakhand 247667, India
e-mail: soumyasahoo.nihr@gov.in

H. Pathak et al. (eds.), *Climate Change Impacts on Soil-Plant-Atmosphere Continuum*, Advances in Global Change Research 78, https://doi.org/10.1007/978-981-99-7935-6_8

soil moisture distribution, it could be inferred that the high-altitude (750–1000 m above mean sea level) areas that are farther from the stream network are going to experience a reduced soil moisture availability during 2061–2080.

Keyword Soil moisture · Hillslope hydrology · Climate change · Land use change · Agriculture

8.1 Introduction

The soil–plant-atmosphere (SPA) continuum describes the movement of soil water or soil moisture between soil, plant, and atmosphere (Novák, 2012). Research on SPA traces back its origin to Philip (1966). The dynamic SPA affects the land–atmosphere energy and water feedback, thereby controlling the weather and climate of a region (Deng et al., 2017). This could be inferred from the fact that soil moisture availability dictates subsequent evapotranspiration (Wang et al., 2018). The increased evapotranspiration could increase the precipitation later as inferred from a case study in Illinois (Findell & Eltahir, 1997). Hence, SPA continuum is a subject of interest in agriculture, hydrology, meteorology, and atmospheric science. Conventionally, hydrology is studied as the hydrosphere-atmosphere-lithosphere continuum that could be considered a manifestation of SPA continuum. The anthropogenic activities have resulted in climate change (Plantico et al., 1990; Poortinga et al., 2011), which is characterized by increased average global surface air temperature (Easterling & Wehner, 2009; Kaufmann et al., 2011), influencing the global precipitation anomaly (Dore, 2005; Portmann et al., 2009; Wang et al., 2019) and basin-scale water availability. Consequently, all the components of the hydrologic cycle, viz. recharge (Jyrkama & Sykes, 2007), soil moisture availability, groundwater storage, and baseflow (Cooper et al., 1995), evapotranspiration (Dinpashoh et al., 2019; Lin et al., 2018), and streamflow (Li & Jin, 2017) are affected.

Similarly, the anthropogenic activity-induced global land use/land cover (LULC) change affects the basin-scale water resources in multiple ways: (i) land use partitions the precipitation into different water balance components, such as evapotranspiration, recharge, surface runoff, soil moisture storage and groundwater flow; (ii) reduction in forest cover increases the river discharge during the rainy season and reduces the lean season flow (Easterling & Wehner, 2009) as the opportunity time for infiltration is reduced; (iii) conversion from non-agricultural (viz., forest and grassland) to agricultural land uses may increase the non-point source pollution load by enhancing the sediment, nutrient, and chemical fluxes into the nearby rivers (Bennett et al., 2001; Foley et al., 2011); (iv) urbanisation can result in reduced recharge, soil moisture storage and groundwater overdraft (Foster, 2001; Wakode et al., 2018), thereby causing declination in water table; and (v) agricultural land use causes diversion and withdrawal from watercourses to meet irrigation demand (Leng et al., 2015, 2017). Hence, for sustainable agricultural production owing to climate and LULC change, the impacts of climate and LULC change on the SPA continuum need to be assessed.

This enables the local policymakers to prioritize potential threats to the agro-hydro-ecosystem. Further, as climate change is a global phenomenon, at the local scale, only land can be managed as an immediate measure.

Generally, there is an indirect impact of climate and LULC change on the spatio-temporal variation of soil moisture as the moisture storage gets influenced by soil recharge (Jyrkama & Sykes, 2007). The water movement in the root-soil matrix is usually modeled by employing a moisture stress function as in the case of HYDRUS (e.g., Feddes et al., 1978) or using Darcy's law as in LEACHAM (Hutson, 2003). However, parametrization of these two models is difficult and requires spatiotemporally distributed soil moisture data which is usually not available unless the area under study is an instrumented one. Consequently, in most of the earlier studies, semi-conceptual surface hydrological models, such as the variable infiltration capacity (VIC) model with the simplified representation of the subsurface domain are forced with the outputs from the general circulation model (GCM) or regional climate model (RCM) (e.g., Hurkmans et al., 2009). Although this approach reduces the computational demand, it lacks a physical basis. Moreover, to the authors' knowledge, there are limited efforts in the literature for studying the integrated impact of future LULC and climate change on soil moisture storage (e.g., Chen et al., 2009). Therefore, as alterations in soil infiltration rate are affecting the basin-scale recharge dynamics and soil moisture distribution subject to future climate change scenarios and predicted LULC, modeling the LULC and climate change impact on soil moisture is essential for holistic water management.

To appropriately model the dynamism of root zone soil moisture affected by land use alterations and climate change, the physically-based frameworks are preferred over the conceptual or data-driven approaches. Conversely, computational demand and data availability for the physically-based models are the two major constraints in basin-scale climate change impact assessment. Hence, to compromise with the computational demand, most of the earlier attempts use either the conceptual or black-box models for basin-scale application (Erler et al., 2019). Further, the water management-related policy decisions are taken at the basin scale. However, many times, the parameters of the black box and conceptual models remain static with the dynamism of future land use and climate inputs, resulting in improper simulation of the future. This necessitates using simple, yet computationally efficient physically-based approaches with limited data requirements. Moreover, the use of a process-based model would help in understanding the feedback among the water balance components. The hillslope-storage Boussinesq (hsB, Troch et al., 2003) is a such typical simplified hydrodynamic model that is process-based and requires limited hydrogeological data to model the subsurface saturated and unsaturated storage and discharge (Sahoo & Sahoo, 2019; Sahoo et al., 2017), which can be applied to data-scarce basins.

Based on the previous discussion, it is evident that the conventional modeling frameworks in the literature generally do not consider the dynamic (future) LULC change and climate change while modeling the future soil moisture status. Hence, this chapter focuses on presenting a modeling framework: (i) to forecast the future LULC (2020–2099) of the study area; (ii) to downscale and bias-correct the representative concentration pathway (RCP) scenarios (2.6, 4.5, 8.5) for the study area using the observed meteorological data during the retrospective period; and iii) to model the soil moisture distribution under future LULC and multi-RCP ensemble scenarios by forcing the model with RCM-predicted future climate and predicted future LULC.

8.2 The Modeling Framework

For assessing the spatially-distributed soil moisture, the modeling framework involves (i) enhancement of the hillslope storage Boussinesq model for studying the soil moisture and groundwater dynamics; (ii) forecasting the future LULC of the study area; and (iii) deriving the future climate forcings to the hydrologic model from the climate model output.

8.2.1 Enhanced Hillslope Storage Boussinesq Model

The enhanced hillslope-storage Boussinesq (eHSB) model operates on hillslopes as elements. This model could be considered semi-distributed as the hillslope storage variables (viz., soil moisture storage in the vadose zone and groundwater storage in aquifers) is simulated as a function of the distance, x from the stream network of the basin. The assumptions/simplifications used for developing the eHSB model are (i) the recharge, R_{ch} is essentially a result of rainfall and is considered uniform for a hillslope; (ii) the validity of the Dupuit-Forchheimer assumptions hold; and (iii) both saturated and unsaturated subsurface storage across the hillslopes is presented as a function of the longitudinal distance x from the stream network, indicating the variability is ignored in the transverse direction. A schematic of the flow domain along with the different conceptual notations of a hillslope aquifer is illustrated in Fig. 8.1.

8.2.1.1 Model Derivation

As per Darcy's law, the groundwater discharge to the stream network per unit aquifer width, q_{GD} can be expressed as:

$$q_{GD} + K_{sc} h_{WT} \frac{\partial h_{WT}}{\partial x} = 0 \tag{8.1}$$

where K_{sc} = saturated hydraulic conductivity [LT^{-1}]; h_{WT} = elevation of water table measured with respect to the impermeable datum [L]; x = flow distance of a location from the stream network [L]; and $\frac{\partial h_{WT}}{\partial x}$ = hydraulic gradient of saturated subsurface storage [−].

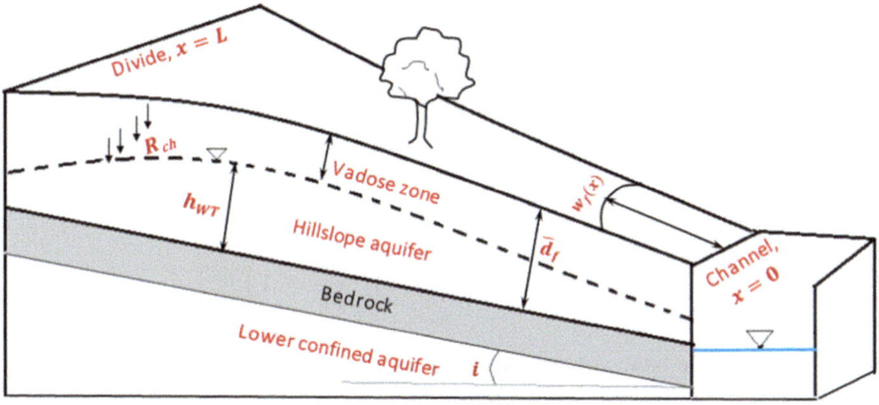

Fig. 8.1 Sketch of a sloping aquifer with a detailed cross-section of the flow domain that rests on an impermeable inclined bedrock at an angle i with horizontal, in which the hillslope length $L >> \max(\bar{d}_f)$, the total depth of the hillslope unconfined aquifer, and w_f is the hillslope width function

The continuity equation for the saturated hillslope aquifer discharge can be expressed as:

$$p\frac{\partial h_{WT}}{\partial t} + \frac{\partial h_{WT}}{\partial x}dx - R_{ch}dx = 0 \tag{8.2}$$

where p = drainable porosity [−]; t = time [T]; and R_{ch} = effective recharge rate [LT^{-1}] to hillslope aquifer resulting from precipitation only.

Combining Eqs. (8.1 and 8.2), the Boussinesq equation for horizontal aquifer can be written as:

$$\frac{\partial h_{WT}}{\partial t} + \frac{K_{sc}}{p}\frac{\partial}{\partial x}\left(h_{WT}\frac{\partial h_{WT}}{\partial x}\right) - \frac{R_{ch}}{p} = 0 \tag{8.3}$$

In the case of hillslopes, the aquifer rests on an impermeable layer and is sloping in nature (Troch et al., 2003). Hence, the modified coordinate system (x, h_{WT}) is followed as illustrated in Fig. 8.1. Such modification implies that the Dupuit-Forchheimer assumption of the streamlines being parallel to the bedrock holds. For this modified coordinate system (x, h_{WT}), Darcy's equation is expressed as (Boussinesq, 1877; Childs, 1971):

$$q_{GD} + K_{sc}h_{WT}\left(\frac{\partial h_{WT}}{\partial x}\cos i + \sin i\right) = 0 \tag{8.4}$$

where i = angle of the hillslope with horizontal.

Combining Eqs. (8.2) and (8.4), one can obtain the Boussinesq equation of subsurface flow along a unit-width hillslope with sloping bedrock, and a spatially uniform recharge rate, R_{ch} as:

$$\frac{\partial h_{WT}}{\partial t} - \frac{K_{sc}}{p}\left[\cos i \frac{\partial}{\partial x}\left(h_{WT}\frac{\partial h_{WT}}{\partial x}\right) + \sin i \frac{\partial h_{WT}}{\partial x}\right] - \frac{R_{ch}}{p} = 0 \tag{8.5}$$

where h_{WT} is measured perpendicular to the underlying impermeable layer in the coordinate system (x, h_{WT}).

In the eHSB formulation, the soil–water movement in the vadose zone can be modeled to estimate the R_{ch} using the Richards equation as:

$$\frac{\partial}{\partial t}\theta_{vz}(p_h) = \frac{\partial}{\partial z}\left[K_{vt}(p_h)\left(\frac{\partial p_h}{\partial z} + 1\right)\right] \tag{8.6}$$

where θ_{vz} = vadose zone volumetric soil moisture content [L^3L^{-3}]; p_h = soil water potential [L]; K_{vt} = vertical soil hydraulic conductivity [LT^{-1}] that can be estimated using the Mualem-van Genuchten model (Mualem, 1976; van Genuchten, 1980) as:

$$K_{vt}(p_h) = K_{sc} S_{re}^{\lambda}\left[\left(1 - S_{re}^{1/m}\right)^{m}\right]^2 \tag{8.7}$$

$$S_{re} = \frac{\theta_{vz} - \theta_{res}}{\theta_{sat} - \theta_{res}} = \left[\frac{1}{1 + [\mu_s p_h]^n}\right]^m \tag{8.8}$$

where S_{re} = relative soil moisture saturation [−]; θ_{res} = residual soil moisture content [L^3L^{-3}]; θ_{sat} = saturated soil moisture content [L^3L^{-3}]; μ_s = shape parameter [L^{-1}]; λ (= 0.5) = shape parameter that accounts for tortuosity (Mualem, 1976); and n = 1/(1-m) = shape parameter that accounts for pore size distribution in different soil types.

Solving for the vadose zone water balance, one gets the spatially varying dynamic recharge, R_{ch} to the aquifer as:

$$R_{ch}(x,t) = R_{WB}(x,t) + \frac{\partial}{\partial t}\left(\int_0^{D_{vz}} \left[\theta_{vz}(z,t) - \theta_{res}\right] dz\right) \tag{8.9}$$

where D_{vz} = depth of vadose zone [L]; θ_{vz} = volumetric soil moisture content in the vadose zone [L^3L^{-3}]; and R_{WB} is the deep percolation loss from the root zone through the vadose zone [LT^{-1}] estimated using the water balance approach as:

$$R_{WB}(x,t) = I(x,t) - ET_a(x,t) - D_{rz}\frac{d}{dt}\theta_{rz} \tag{8.10}$$

where D_{rz} = root zone depth [L]; θ_{rz} = root zone volumetric soil moisture content [L^3L^{-3}]; and ET_a = actual evapotranspiration that depletes the soil water storage [LT^{-1}], estimated using the FAO-56 Penman–Monteith equation (Allen et al., 1998) for different land use classes (e.g., Srivastava et al., 2017; Swain & Sahoo, 2015).

Equation (8.5) estimates the hillslope toe-end discharge to the river network along unit aquifer width. However, applying this equation for estimation of the saturated subsurface flow in real-world hillslopes has the limitation of not accounting for the 3D soil mantle, which is the real flow domain. Further, to fully capture the hillslope subsurface storage-discharge dynamics of naturally occurring hillslope forms, consideration of the flow domain geometry is highly essential.

Incidentally, Fan and Bras (1998) pioneered a method to account for the topographic control of the hillslope hydrologic processes. This approach, considering that the soil mantle is made up of pore space, reduces a natural 3D soil mantle to a 1D soil profile. Considering an equivalent 1D profile of a 3D hillslope on top of an impermeable bedrock (Fig. 8.1), the maximum possible soil moisture storage, i.e., soil moisture storage capacity function, $S_{mc}(x)$ can be expressed as (Fan & Bras, 1998):

$$S_{mc}(x) = w_f(x)\overline{d}_f(x)p \tag{8.11}$$

where $w_f(x)$ = width of hillslope [L] at a flow distance of x (the hillslope width function) and $\overline{d}_f(x)$ = average soil depth [L] at any flow distance x (the hillslope soil depth function).

Equation (8.11) presents the soil mantle as a consolidated formation including the pore space along the hillslope. The topographical feature is accounted for by the plan shape as the hillslope width function, and the profile curvature by the soil depth function. Similar to the concept of S_{mc}, the saturated aquifer storage, S_{GW} can be presented as $S_{GW} = w_f \times h_{WT} \times p$, where $h_{WT}(x)$ is the average elevation of the water table above the bedrock at a flow distance x.

As the hillslope's storage discharge dynamics are predominately controlled by the plan shape and profile curvature, as per the naturally occurring plan shape forms, the hillslopes can be broadly categorized into three types, viz., convergent, characterized by $\left(\partial w_f/\partial x\right) > 0$; divergent, characterized by $\left(\partial w_f/\partial x\right) < 0$; and uniform $\left(\left(\partial w_f/\partial x\right) \approx 0\right)$, where the width of the hillslope remains constant with distance from the channel. Similarly, in nature, there are three kinds of profile curvature encountered, viz., convergent, divergent, and planar.

Incorporating the HWF and S_{mc} in the continuity form of Eq. (8.5) and combining it with Darcy's equation for sloping aquifers, i.e., Eq. (8.4), the 1D eHSB can be rewritten as (Sahoo et al., 2018):

$$p\frac{\partial S_{GW}}{\partial t} - \frac{K_{sc}\cos i}{p}\frac{\partial}{\partial x}\left[\frac{S_{GW}}{w}\left(\frac{\partial S_{GW}}{\partial x} - \frac{S_{GW}}{w}\cdot\frac{\partial w_f}{\partial x}\right)\right] + K_{sc}\sin i\frac{\partial S_{GW}}{\partial x} = pR_{ch}w_f \tag{8.12}$$

8.2.2 *Forecasting Approach for Future Land Use*

The Dyna-CLUE model (Verburg & Overmars, 2009) is a widely acclaimed, freely available land use prediction framework that can predict future scenario-specific land use in a river basin accounting for a wide range of management conditions.

The land use management conditions are handled by the 'Demand' (non-spatial) and 'Allocation' (spatial) modules (Fig. 8.2). The 'Demand module' in Dyna-CLUE estimates the demands possessed by each land use type for each year which is passed to the 'Allocation module' iteratively for allocating the land use to each grid. The spatial data inputs for the 'Allocation module' are comprised of: (a) land use restriction policies on its future expansion, (b) land use type-specific conversion settings, encompassing land use conversion elasticity and its transition sequence, and (c) locational suitability of land use. The restriction policy restricts the desired raster grids of LULC from allocation in the future, which is protected by specific laws or policies by the enforcing agency (e.g., reserved forests). The conversion elasticity is a measure of the reversible nature of a land use change once allocated. The land use transition sequence keeps information on the number of time steps an allocated land use needs to remain unchanged for its further conversion to another land use, e.g., a time duration is required for forest growth. The locational suitability of land use is decided by three factors, such as (i) physical drivers of elevation, slope, and aspect; (ii) spatial neighbourhood as the distance to road network, built-up, water bodies, and forest; and (iii) socioeconomic and demographic factors as population.

The probability of allotment of a land use class to a grid can be expressed by logistic regression (Giri et al., 2019):

$$P(LC) = \frac{e^{\mu_0+\mu_1\rho_1+\mu_2\rho_2+\cdots+\mu_n\rho_n}}{1+e^{\mu_0+\mu_1\rho_1+\mu_2\rho_2+\cdots+\mu_n\rho_n}} \tag{8.13}$$

where $P(LC)$ = probability that a land use class, LC will occur at a given raster grid; $\mu_0, \mu_1, \mu_2, \ldots, \mu_n$ are the (constant) coefficients; and $\rho_1, \rho_2, \ldots, \rho_n$ are the influencing factors, such as distance to the existing road network, built up, water bodies and forest.

With the historical LULCs, the total probability for allocating given land use, *lu* to a raster grid, *i* at a future time, *t* is expressed as:

$$P_{total}(i,t,lu) = P_{ls}(i,t,lu) + P_{nbr}(i,t,lu) + CA(t,lu) + CE(lu) \tag{8.14}$$

where P_{ls} = locational suitability (−); P_{nbr} = neighbourhood suitability (−); CA = competitive advantage (−); and CE = conversion elasticity (−).

Both the neighbourhood suitability and the locational suitability factors in Eq. (8.14) can be determined using past modeling experiences in the basin (e.g., Sahoo et al., 2018), and from the expert knowledge in the land use change modeling studies in the adjacent river basins (e.g., Behera et al., 2018). The Dyna-CLUE model iteratively determines the competitive advantage until the allocated area by the model matches the demand (observed area) for each land use. This implies that if the allocated area by the model is less than the demand for particular land use, then the value of *CA* in Eq. (8.14) is increased by 1% and vice versa. The conversion elasticity denotes the cost of conversion (penalty factor) of a land use class to another. The higher the *CE*, the greater cost involved in changing land use, resulting in a high probability of the land use remaining unaltered (Fig. 8.2).

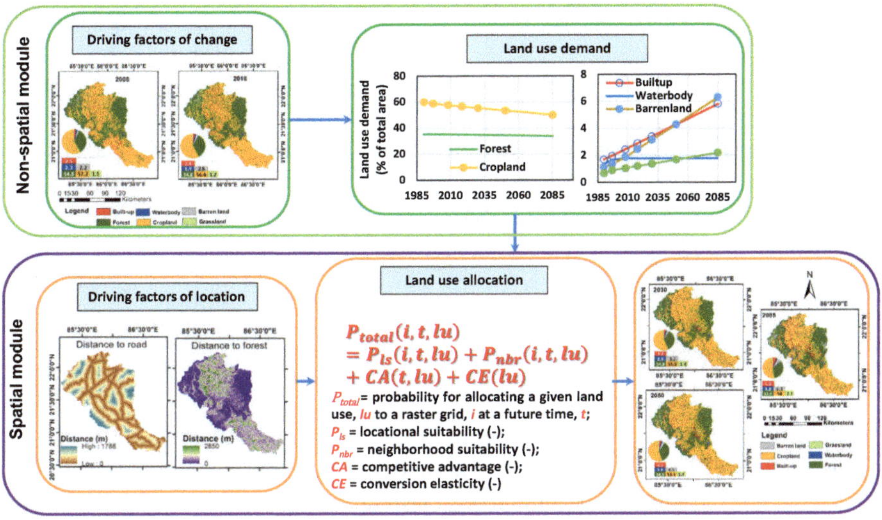

Fig. 8.2 Step-by-step procedure to forecast land use land cover using the Dyna-CLUE model

8.2.3 *Forecasting Approach for Future Climate*

The general circulation models (GCMs) provide a reasonable assessment of climate change at a coarser resolution, whereas the regional circulation models (RCMs) nested with the GCM outputs, simulate the future climate at a much finer resolution (Eden et al., 2014). Generally, the simulation grid size of any land surface model is less than the spatial scale of the downloaded RCM data. Hence, the downloaded RCM data has to be bias-corrected for the observed data collected at any ground weather station. Out of the different methods available for bias correction, the CDF-transformation method (Michelangeli et al., 2009) was used to bias correct the RCM output for the gridded India Meteorological Department (IMD) dataset.

For bias correction, three datasets and their corresponding cumulative distribution functions (CDFs) were prepared at a monthly scale using the available in situ (IMD-observed) and RCM-simulated datasets, namely: (i) the observed gridded IMD dataset for the retrospective baseline period (1970–2005) with the fitted CDF, $F_{gd,ret}(m_v)$; (ii) the concurrent RCM-simulated retrospective dataset with the fitted CDF, $F_{RCM,ret}(m_v)$; and (iii) the RCM-simulated dataset for a future period (2021–2099) with the fitted CDF, $F_{RCM,fut}(m_v)$; where m_v = meteorological variable to be bias-corrected. The transfer function, T_{fun} expressing the bias between the datasets (i) and (ii) can be expressed as (Michelangeli et al., 2009):

$$T_{fun}\big(F_{RCM,ret}(m_v)\big) = F_{gd,ret}(m_v) \tag{8.15}$$

Assuming the same transfer function, T_{fun} holds in the future, its identical expression to obtain the required CDF of bias-corrected future data can be expressed as:

$$F_{gd,fut}(m_v) = T_{fun}\left(F_{RCM,fut}(m_v)\right) \tag{8.16}$$

Combining Eqs. (8.15) and (8.16), one obtains (Michelangeli et al., 2009):

$$F_{gd,fut}(m_v) = F_{gd,ret}\left(F^{-1}_{RCM,ret}\left(F_{RCM,fut}(m_v)\right)\right) \tag{8.17}$$

8.3 Study Area and Datasets

For this study, the Baitarani River basin (12,900 km^2) in eastern India is chosen, which lies within 20.67–22.28°N latitudes and 85.16–86.89°E longitudes (Fig. 8.3). The entire river basin can be effectively assumed as groundwater model domain boundary as for integrated water resources management involving both the surface and groundwater resources, it has been emphasized for river basin-scale modeling studies (Sahoo and Sahoo, 2020). Moreover, the eHSB hydrologic model considered herein divides the entire river basin into several constituent hillslopes further based on the local topography. Hence, the entire basin ranging from low-lying delta areas to high-altitude headwater catchments can be effectively considered as the groundwater model domain. The Baitarani River originates from the Gonasika Hills towards the northern part at an altitude of 900 m above the mean sea level (msl) and drains into the Bay of Bengal in the East traversing a distance of about 350 km. The mean annual precipitation of the study area over the years 1985–2015 is 1456 mm/year. As per Köpen climate classification system, tropical savannah climate prevails over the study area (Jena et al., 2021) with 80% of precipitation occurring during June–September with the annual minimum–maximum temperature range of 4–47 °C (Sahoo et al., 2018). The geology of the selected basin is constituted by granites and granite gneisses in high lands and alluvium in the nearby stream areas with the major soil types of sandy-clay-loam, sandy-loam, and clay (Fig. 8.3c).

The observed meteorological data for the retrospective period of 1970–2005 was obtained from the India Meteorological Department, Bhubaneswar. Similarly, the future RCP scenarios from the RCMs were downloaded from the Indian Institute of Tropical Meteorology <http://cccr.tropmet.res.in/home/ftp_data.jsp>. The RCMs considered for this study were the Conformal Cubic Atmospheric Model (CCAM), the Rossby Centre regional Atmosphere model version 4 (RCA4; Samuelsson et al., 2011), and the MPI Regional Model (REMO, Teichmann et al., 2013). The meteorological outputs from different CORDEX experiments involving eight different combinations of RCMs and the driving GCMs (mentioned in parenthesis) as used in this study were: CCAM(ACCESS1.0), CCAM(CCSM4), CCAM(NorESM1-M), CCAM(GFDL-CM3), CCAM(CNRM-CM5), CCAM(MPI-ESM-LR), REMO(MPI-ESM-LR), and RCA4(ICHEC). The RCM simulated data

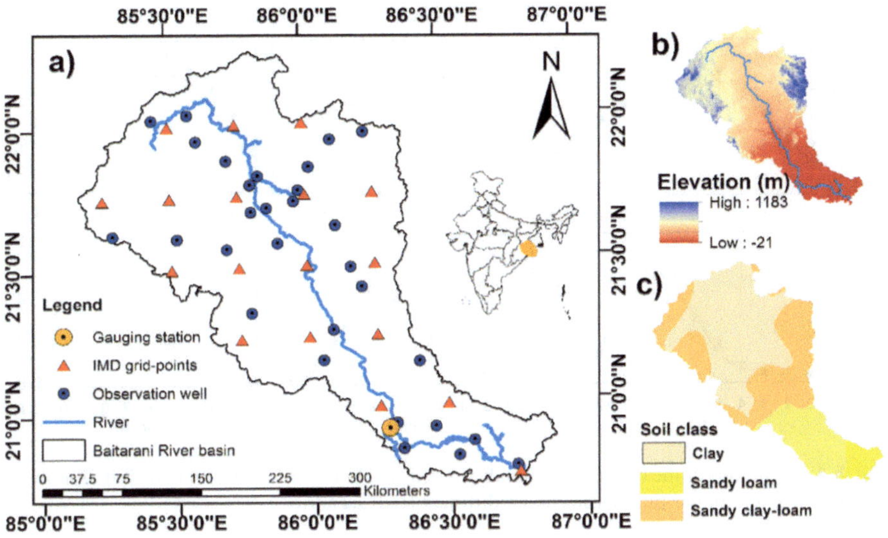

Fig. 8.3 **a** Index map with the streamflow gauging station, groundwater observation wells, and observed meteorological grid points; **b** Elevation; and **c** Soil maps of the Baitarani River basin (*Source* Odisha Remote Sensing Application Centre)

was downloaded both for the retrospective period (1970–2005) and the future period (2021–2099).

To prepare the 30 m resolution LULC maps (for 1990, 1997, and 2008) using supervised classification in the ERDAS Imagine software environment, satellite imageries of the thematic mapper (TM) sensor onboard LANDSAT-5 were downloaded from the NASA earth explorer website through <https://earthexplorer.usgs.gov>. Similarly, for the year 2018, the imageries were from the operational land imager (OLI) onboard LANDSAT-8. Further, for the elevation dataset, a 30 m SRTM digital elevation model (DEM) was downloaded from the earth explorer and was geoprocessed. The observed daily-scale river stage and flow data for the study period were obtained from the Central Water Commission (CWC), Bhubaneswar, and the India-Water Resources Information System (India-WRIS) portal <http://indiawris.gov.in/wris/#/dataAvailability>. The three-monthly depth-to-water table data from 30 observation wells, distributed throughout the study basin, was obtained from the Central Ground-Water Board (CGWB), Bhubaneswar.

8.4 Basin-Scale Climate Forecasts in Three Time-Horizons

Figure 8.4 depicts the variation in future precipitation and temperature datasets over the study area during the retrospective and future periods, considering the mean of the eight CORDEX experiment outputs. The entire future period was divided into

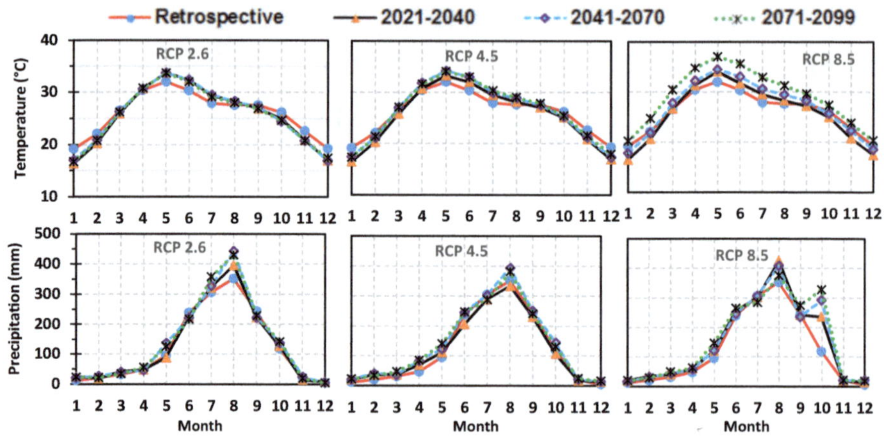

Fig. 8.4 Variations of the basin-scale monthly mean absolute temperature and precipitation for the retrospective and future time-lapses under different RCP scenarios

three time-lapses, viz., near-future (2021–2040), mid-future (2041–2070), and far-future (2071–2199) (Giri et al., 2019). As depicted in Fig. 8.4, for all the RCPs, future winter-temperature (November, December, and January) would be lower than the corresponding retrospective estimate. Conversely, summer would be warmer. For example, among all the time-lapse/RCP combinations, time-lapse 2071–2099/ RCP8.5 would be the warmest with about a 5 °C increase in the mean temperature. The monsoon season precipitation would likely to be increased under all the RCP scenarios, with the highest increase during the time-lapse 2041–2070 for both the RCP2.6 and RCP4.5 scenarios. Under the RCP8.5 scenario, during October, a sudden rise in precipitation is predicted. This could be due to increased frequency of tropical cyclones making landfall on the east coast of India (Dash & Walia, 2020).

8.5 Basin-Scale LULC Forecasts in Three Time-Horizons

Based on the future land area requirements during different years, the Dyna-CLUE model assigns different LULC classes to different grids depending on the total probability of assigning a particular land use class to a grid. Figure 8.5 illustrates the predicted LULC maps for the typical years 2030, 2050, and 2085, representing the time lapses of 2021–2040, 2041–2070, and 2071–2099, respectively. For calculation of the changes in LULC, the historical LULC map of 1997 was considered as the baseline as the classification accuracy was the highest for this year. The generated future LULC maps with a business-as-usual strategy revealed that, concerning the baseline, the cropland would decrease by 5% (366 km^2) and 12.5% (936 km^2) by the years 2030 and 2085, respectively. Similarly, in comparison to the base period, the forests are likely to decrease by 12.5%; whereas the built-up area and barren

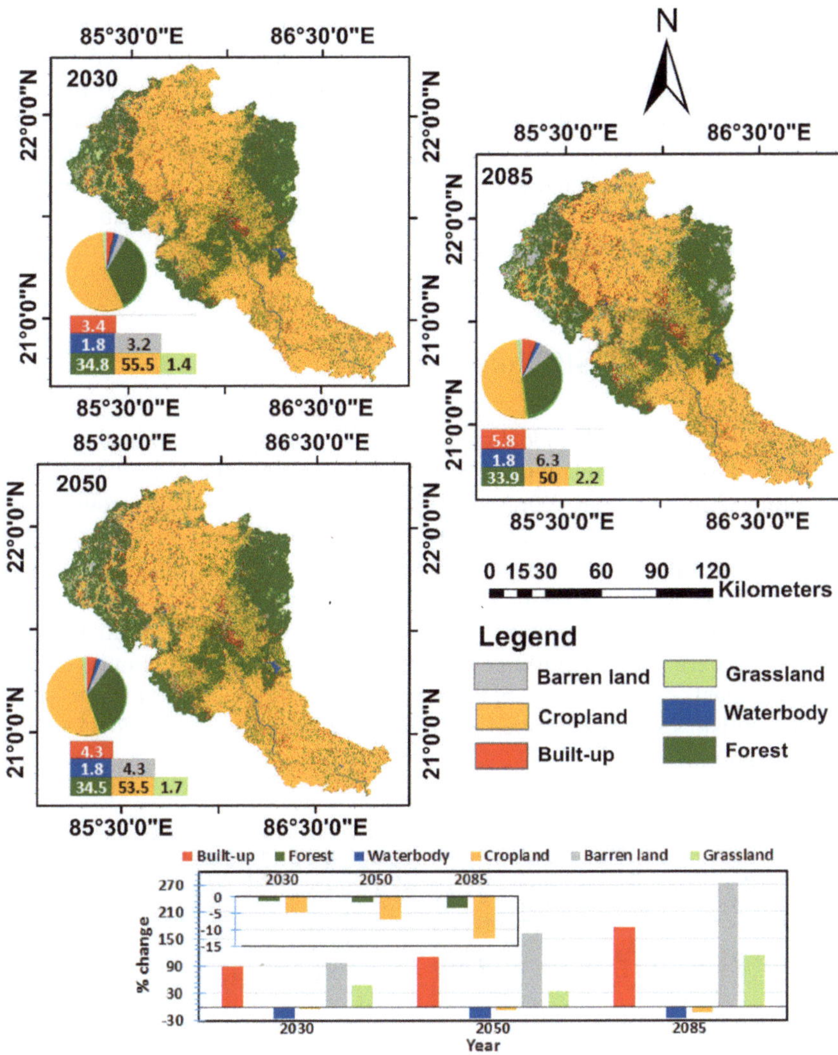

Fig. 8.5 Forecasted LUCC map of the Baitarani River basin depicting the percentage change in different land use classes

land are likely to increase by 275% and 176%, respectively, by 2099. Of the land area, which was designated as agricultural or cropland in past and is predicted to change, about 61% would be converted to built-up and 39% to barren land. As cropland ($\approx$7500 km^2) is the major land use class, its alteration over the years could significantly impact the hydrologic processes and water resources availability in the basin. The conversion of cropland to built-up area would increase the runoff as a significant amount of monsoon precipitation is usually stored in the paddy fields as ponded water. This fraction of precipitation gradually contributes to groundwater recharge and is subsequently available as delayed streamflow. Further, if the recharge in this process brings the water table within the capillary zone, the recharged groundwater contributes to non-monsoon season soil moisture through capillary rise. This happens in many locations of the delta areas in the study area. Although about 40% of the forest land in the basin is part of the 'Similipal National Park and Biosphere Reserve', a conserved forest, the non-conserved area was allowed to get converted into barren land or grassland. Similarly, croplands in the plateau and alluvial plain were allowed to get converted into built-up areas.

8.6 Setting Up of eHSB Model

Figure 8.6 presents the eHSB model simulated WTE during calibration and validation at six well locations (notations W1, W2,..., W6) within the basin. The typical well locations were selected to represent the basin-wide spatiotemporal groundwater fluctuation spanning from lower alluvial formation to high-elevation hilly terrains. The eHSB model reproduced the available seasonal depth to water table (DWT) with RMSE $\leq$ 1.84 m, MAE $\leq$ 1.47 m, and $R^2 \geq 0.68$ during both the calibration and validation. Hence, it can be surmised that the eHSB model performed satisfactorily in the selected basin, which could be further used for LULC and climate change impact assessment. After calibration, the ranges of the calibrated parameters for the constituting hillslopes of the selected basin were estimated as $K_s = 2.5$–85 m/day, and $f_e = 0.22$–0.37.

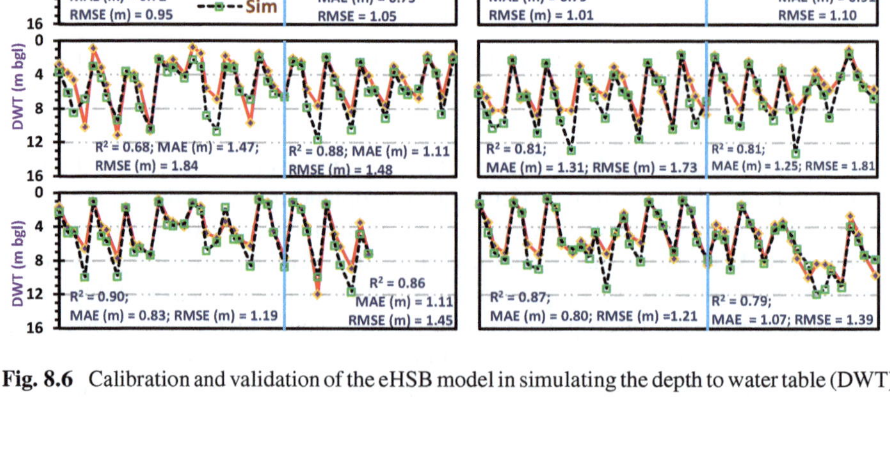

Fig. 8.6 Calibration and validation of the eHSB model in simulating the depth to water table (DWT)

8.7 Assessment of Spatially-Distributed Soil Moisture Storage

The eHSB model calibrated for estimating the water table was also used to simulate the basin-scale variation of soil moisture and for assessing future drought scenarios. Further, the setup was used to predict the future distribution of soil moisture. Figure 8.7 depicts the future distribution of soil moisture during the years 2071–2099 across the basin for the RCP 4.5 and 8.5, denoting moderate and extreme climate change scenarios, respectively. It can be inferred that the change in the root zone (assumed 45 cm) soil moisture between historical and future periods is less for the region lying nearer to the coastal region than that of the region farther from the streamline. This indicates that the high-altitude region farther from the river network may experience a decline in soil moisture storage that could amplify the possible occurring future agricultural drought. The headwater region may experience drying of the surface soil leading to loss of vegetation as also confirmed by the predicted LULC by the Duna-CLUE model (Fig. 8.5).

The predicted future distribution of soil moisture indicates that the dry zones would be agglomerated in moderate to high-elevation areas that are devoid of vegetation. The hillslope topography and the depth of the water table from the ground surface influence these future moisture stress areas. As the depth to the water table is less in the vicinity of the stream network, depending on the capillary rise supported by the overlying formation, soil moisture migrates to the vadose zone or could be available in the root zone for extraction by the vegetation. Conversely, as the depth to the water table is comparatively higher in the areas that are farther from the stream network, such soil moisture availability through the capillary transport is limited only up to a certain height above the water table, not reaching up to the root zone or topsoil layer to influence the surface soil moisture characteristics.

A suitable crop calendar is to be prepared according to the change in soil moisture in the future. Currently, in the rain-fed agricultural system, during the non-monsoon season, groundnut (*Arachis hypogaea*), green gram (*Vigna radiata*), black gram (*Vigna mungo*) and red gram (*Cajanus cajan*) are the main crops grown in the study area. With a reduced soil moisture availability in the future period, crops demanding even less soil moisture, such as finger millet (*Eleusine coracana*), sorghum (*Sorghum bicolor*), and sesamum (*Sesamum indicum*) could replace the crop calendar. Suitable drought-resistant varieties are to be developed and cultivated.

8.8 Summary and Conclusions

In this study, the impact assessment was performed strategically considering the LULC and climate change as the primary factors determining the future soil moisture status. An integrated modeling framework encompassing eHSB hillslope model and Dyna-CLUE future LULC prediction model was advocated considering the RCP

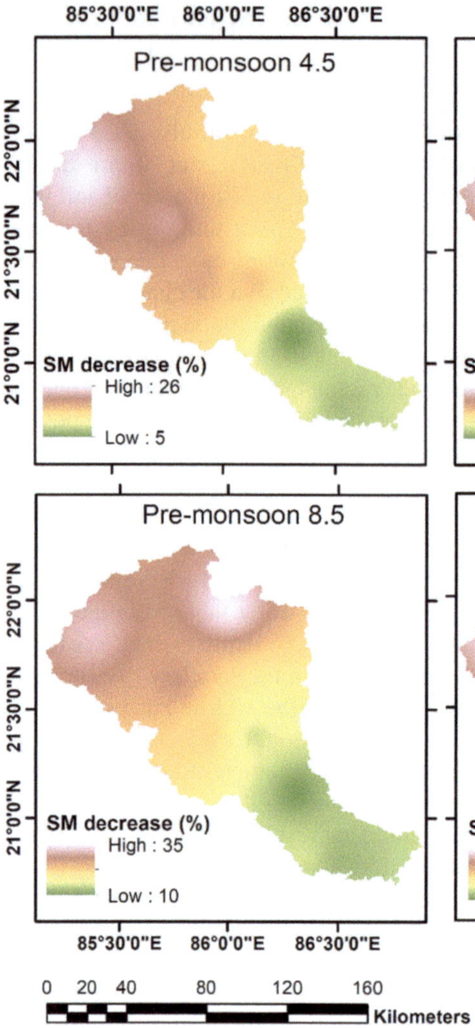

Fig. 8.7 Future distribution of soil moisture across the Baitarani basin in different Representative Concentration Pathways of 4.5 and 8.5 owing to LULC and climate change

2.6/4.5/8.5 climate scenarios during three typical future time-lapses of 2021–2040, 2041–2070, and 2071–2099.

The following specific conclusions were drawn from this study:

(i) Because of the contrasting runoff coefficients in these LULCs, both will have a cascading effect to increase the monsoon streamflow and decrease the opportunity time to induce recharge.

(ii) Compared to the baseline period, the basin may experience colder winters and warmer summers in the future. An increase in the monsoon and early post-monsoon precipitation is predicted for all the RCP scenarios for which suitable crop calendar and irrigation and drainage scheduling may be practiced to cope with resulting waterlogging.
(iii) The eHSB model for modeling the groundwater and soil moisture storage reproduced the seasonal depth to water table with RMSE $\leq$ 1.84 m, MAE $\leq$ 1.47 m, and $R^2 \geq 0.68$ during both the calibration and validation.
(iv) The locations farther from the river network could experience reduced top surface soil moisture in comparison to the locations closer to the river network. Hence, agricultural production and crop security are going to be impacted negatively in these areas.
(v) As the monsoon precipitation is predicted to increase, suitable paddy varieties tolerant for waterlogging are to be prescribed. Conversely, for the non-monsoon period, drought-tolerant crops and varieties are prescribed for the crop calendar.

Acknowledgements The authors would like to thank the Central Ground-Water Board (CGWB) and Odisha Space Application Centre (ORSAC) for providing the groundwater level and soil class map of the study area, respectively.

References

Allen, R. G., Pereira, L. S., Raes, D., & Smith, M. (1998). Chapter 8—ETc under soil water stress conditions, crop evapotranspiration-Guidelines for computing crop water requirements-FAO Irrigation and drainage paper 56, Irrigation and Drainage

Behera, M. D., Tripathi, P., Das, P., Srivastava, S. K., Roy, P. S., Joshi, C., Behera, P. R., Deka, J., Kumar, P., Khan, M. L., Tripathi, O. P., Dash, T., & Krishnamurthy, Y. V. N. (2018). Remote sensing based deforestation analysis in Mahanadi and Brahmaputra river basin in India since 1985. *Journal of Environmental Management, 206*, 1192–1203. https://doi.org/10.1016/j.jenvman.2017.10.015

Bennett, E. M., Carpenter, S. R., & Caraco, N. F. (2001). Human Impact on erodable Phosphorus and eutrophication: A global perspective. *BioScience, 51*, 227. https://doi.org/10.1641/0006-3568(2001)051[0227:hioepa]2.0.co;2

Boussinesq, J. (1877). Essai sur la théorie des eaux courantes. *Mem. Acad. Sci. Inst. Fr., 23*(1), 252–260.

Chen, X., Zhang, Z., Chen, X., & Shi, P. (2009). The impact of land use and land cover changes on soil moisture and hydraulic conductivity along the karst hillslopes of southwest China. *Environmental Earth Science 59,* 811–820 (2009). https://doi.org/10.1007/s12665-009-0077-6

Childs, E. C. (1971). Drainage of groundwater resting on a sloping bed. *Water Resources Research, 7*(5), 1256–1263. https://doi.org/10.1029/WR007i005p01256

Cooper, D. M., Wilkinson, W. B., & Arnell, N. W. (1995). The effects of climate changes on aquifer storage and river baseflow. *Hydrological Sciences Journal, 40*, 615–631. https://doi.org/10.1080/02626669509491448

Dash, B., & Walia, A. (2020). Role of multi-purpose cyclone shelters in India: Last mile or neighbourhood evacuation. *Tropical Cyclone Research and Review, 9*(4), 206–217. https://doi.org/10.1016/J.TCRR.2020.11.002

Deng, Z. H., Guan, H., Hutson, J., Forster, M. A., Wang, Y., & Simmons, C. T. (2017). A vegetation-focused soil-plant-atmospheric continuum model to study hydrodynamic soil-plant water relations. *Water Resources Research, 53*, 4965–4983. https://doi.org/10.1002/2017WR020467

Dinpashoh, Y., Jahanbakhsh-Asl, S., Rasouli, A. A., Foroughi, M., & Singh, V. P. (2019). Impact of climate change on potential evapotranspiration (case study: West and NW of Iran). *Theoretical and Applied Climatology, 136*, 185–201. https://doi.org/10.1007/s00704-018-2462-0

Dore, M. H. I. (2005). Climate change and changes in global precipitation patterns: What do we know? *Environment International*. https://doi.org/10.1016/j.envint.2005.03.004

Easterling, D. R., & Wehner, M. F. (2009). Is the climate warming or cooling? *Geophysical Research Letters, 36*, L08706. https://doi.org/10.1029/2009GL037810

Eden, J. M., Widmann, M., Maraun, D., & Vrac, M. (2014). Comparison of GCM- and RCM-simulated precipitation following stochastic postprocessing. *Journal of Geophysical Research Atmospheres, 119*, 11040–11053. https://doi.org/10.1002/2014JD021732

Erler, A. R., Frey, S. K., Khader, O., d'Orgeville, M., Park, Y., Hwang, H., Lapen, D. R., Richard, P. W., & Sudicky, E. A. (2019). Simulating climate change impacts on surface water resources within a lake-affected region using regional climate projections. *Water Resources Research, 55*, 130–155. https://doi.org/10.1029/2018WR024381

Fan, Y., & Bras, R. L. (1998). Analytical solutions to hillslope subsurface storm flow and saturation overland flow. *Water Resources Research, 34*, 921. https://doi.org/10.1029/97WR03516

Feddes, R. A., Kowalik, P. J., & Zaradny, H. (1978). *Simulation of field water use and crop yield*. Wiley.

Findell, K. L., & Eltahir, E. A. (1997). An analysis of the soil moisture-rainfall feedback, based on direct observations from Illinois. *Water Resources Research, 33*, 725–735.

Foley, J. A., Ramankutty, N., Brauman, K. A., Cassidy, E. S., Gerber, J. S., Johnston, M., Mueller, N. D., O'connell, C., Ray, D. K., West, P. C., Balzer, C., Bennett, E. M., Carpenter, S. R., Hill, J., Monfreda, C., Polasky, S., Rockström, J., Sheehan, J., Siebert, S., … Zaks, D. P. M. (2011). Solutions for a cultivated planet. *Nature*. https://doi.org/10.1038/nature10452

Foster, S. S. D. (2001). The interdependence of groundwater and urbanisation in rapidly developing cities. *Urban Water*. https://doi.org/10.1016/S1462-0758(01)00043-7

Giri, S., Arbab, N. N., & Lathrop, R. G. (2019). Assessing the potential impacts of climate and land use change on water fluxes and sediment transport in a loosely coupled system. *Journal of Hydrology, 577*, 123955. https://doi.org/10.1016/J.JHYDROL.2019.123955

Hurkmans, R. T. W. L., Terink, W., Uijlenhoet, R., Moors, E. J., Troch, P. A., & Verburg, P. H. (2009). Effects of land use changes on streamflow generation in the Rhine basin. *Water Resources Research, 45*. https://doi.org/10.1029/2008WR007574

Hutson, J. L. (2003). LEACHM—A process-based model of water and solute movement, transformations, plant uptake and chemical reactions in the unsaturated zone, version 4, Res. Ser. R03–1, Dep. of Crop and Soil Sci., Cornell Univ., Ithaca, New York

Jena, S., Panda, R. K., Ramadas, M., Mohanty, B. P., Samantaray, A. K., & Pattanaik, S. K. (2021). Characterization of groundwater variability using hydrological, geological, and climatic factors in data-scarce tropical savanna region of India. *Journal of Hydrology: Regional Studies, 37*, 100887. https://doi.org/10.1016/J.EJRH.2021.100887

Jyrkama, M. I., & Sykes, J. F. (2007). The impact of climate change on spatially varying groundwater recharge in the grand river watershed (Ontario). *Journal of Hydrology, 338*, 237–250. https://doi.org/10.1016/j.jhydrol.2007.02.036

Kaufmann, R. K., Kauppi, H., Mann, M. L., & Stock, J. H. (2011). Reconciling anthropogenic climate change with observed temperature 1998–2008. *Proceedings of the National Academy of Sciences of the U. S. A., 108*, 11790–11793. https://doi.org/10.1073/pnas.1102467108

Leng, G., Huang, M., Tang, Q., & Leung, L. R. (2015). A modeling study of irrigation effects on global surface water and groundwater resources under a changing climate. *Journal of Advances in Modeling Earth Systems*. https://doi.org/10.1002/2015MS000437

Leng, G., Leung, L. R., & Huang, M. (2017). Significant impacts of irrigation water sources and methods on modeling irrigation effects in the ACME land model. *Journal of Advances in Modeling Earth Systems*. https://doi.org/10.1002/2016MS000885

Li, Z., & Jin, J. (2017). Evaluating climate change impacts on streamflow variability based on a multisite multivariate GCM downscaling method in the Jing River of China. *Hydrology and Earth System Sciences, 21*, 5531–5546. https://doi.org/10.5194/hess-21-5531-2017

Lin, P., He, Z., Du, J., Chen, L., Zhu, X., & Li, J. (2018). Impacts of climate change on reference evapotranspiration in the Qilian mountains of China: Historical trends and projected changes. *International Journal of Climatology, 38*, 2980–2993. https://doi.org/10.1002/joc.5477

Michelangeli, P. A., Vrac, M., & Loukos, H. (2009). Probabilistic downscaling approaches: Application to wind cumulative distribution functions. *Geophysical Research Letters, 36*. https://doi.org/10.1029/2009GL038401

Mualem, Y. (1976). A new model for predicting the hydraulic conductivity of unsaturated porous media. *Water Resources Research, 12*, 513–522. https://doi.org/10.1029/WR012i003p00513

Novák, V. (2012). Movement of water in soil during evaporation. In: Evapotranspiration in the Soil-Plant-Atmosphere System. Springer, pp. 59–83

Philip, J. R. (1966). Plant water relations: Some physical aspects. *Annual Review of Plant Physiology, 17*, 245–268. https://doi.org/10.1146/annurev.pp.17.060166.001333

Plantico, M. S., Karl, T. R., Kukla, G., & Gavin, J. (1990). Is recent climate change across the United States related to rising levels of anthropogenic greenhouse gases? *Journal of Geophysical Research, 95*. https://doi.org/10.1029/jd095id10p16617

Poortinga, W., Spence, A., Whitmarsh, L., Capstick, S., & Pidgeon, N. F. (2011). Uncertain climate: An investigation into public scepticism about anthropogenic climate change. *Global Environmental Change, 21*, 1015–1024. https://doi.org/10.1016/j.gloenvcha.2011.03.001

Portmann, R. W., Solomon, S., & Hegerl, G. C. (2009). Spatial and seasonal patterns in climate change, temperatures, and precipitation across the United States. *Proceedings of the National Academy of Sciences U. S. A., 106*, 7324–7329. https://doi.org/10.1073/pnas.0808533106

Sahoo, S., & Sahoo, B. (2019). A geomorphology-based integrated stream–aquifer interaction model for semi-gauged catchments. *Hydrological Processes, 33*, 1362–1377. https://doi.org/10.1002/hyp.13406

Sahoo, S., Sahoo, B., & Panda, S. N. (2018). Hillslope-storage Boussinesq model for simulating subsurface water storage dynamics in scantily-gauged catchments. *Advances in Water Resources, 121*, 219–234. https://doi.org/10.1016/j.advwatres.2018.08.016

Sahoo, B., Sahoo, S., & Swain, R. (2017). Modeling the river-aquifer flow-interaction using a coupled hsB-VPMM Approach. In World Environmental and Water Resources Congress 2017. American Society of Civil Engineers, Reston, VA, pp. 172–182. https://doi.org/10.1061/9780784480618.017

Samuelsson, P., Jones, C. G., Will´En, U., Ullerstig, A., Gollvik, S., Hansson, U., Jansson, E., Kjellstro, M. C., Nikulin, G., & Wyser, K. (2011). The Rossby Centre Regional Climate model RCA3: model description and performance. *Tellus A: Dynamic Meteorology and Oceanography, 63*, 4–23.https://doi.org/10.1111/j.1600-0870.2010.00478.x

Srivastava, A., Sahoo, B., Raghuwanshi, N. S., & Singh, R. (2017). Evaluation of variable-infiltration capacity model and MODIS-Terra satellite-derived grid-scale evapotranspiration estimates in a river basin with tropical monsoon-type climatology. *Journal of Irrigation and Drainage Engineering, 143*(8), 04017028. https://doi.org/10.1061/(ASCE)IR.1943-4774.0001199

Swain, R., & Sahoo, B. (2015). Variable parameter McCarthy–Muskingum flow transport model for compound channels accounting for distributed non-uniform lateral flow. *Journal of Hydrology, 530*, 698–715. https://doi.org/10.1016/j.jhydrol.2015.10.030

Teichmann, C., Eggert, B., Elizalde, A., Haensler, A., Jacob, D., Kumar, P., Moseley, C., Pfeifer, S., Rechid, D., Remedio, A., Ries, H., Petersen, J., Preuschmann, S., Raub, T., Saeed, F., Sieck, K., & Weber, T. (2013). How does a regional climate model modify the projected climate change signal of the driving GCM: A Study over different CORDEX regions using REMO. *Atmosphere (Basel), 4*, 214–236. https://doi.org/10.3390/atmos4020214

Troch, P. A., Paniconi, C., Emiel van Loon, E. (2003). Hillslope-storage Boussinesq model for subsurface flow and variable source areas along complex hillslopes: 1. Formulation and characteristic response. *Water Resources Research, 39*, 1316. https://doi.org/10.1029/2002wr001728

van Genuchten, M. T. (1980). A closed-form equation for predicting hydraulic conductivity of unsaturated soils. *Social Science Society of America Journal 44*(5), 892–898. https://doi.org/10.2136/sssaj1980.03615995004400050002x

Verburg, P. H., & Overmars, K. P. (2009). Combining top-down and bottom-up dynamics in land use modeling: Exploring the future of abandoned farmlands in Europe with the Dyna-CLUE model. *Landscape Ecology, 24*, 1167–1181. https://doi.org/10.1007/s10980-009-9355-7

Wakode, H. B., Baier, K., Jha, R., & Azzam, R. (2018). Impact of urbanization on groundwater recharge and urban water balance for the city of Hyderabad, India. *International Soil and Water Conservation Research, 6*, 51–62. https://doi.org/10.1016/j.iswcr.2017.10.003

Wang, Y., Yang, J., Chen, Y., De Maeyer, P., Li, Z., & Duan, W. (2018). Detecting the causal effect of soil moisture on precipitation using convergent cross mapping. *Scientific Reports, 8*(1), 12171. https://doi.org/10.1038/s41598-018-30669-2

Wang, S., Zuo, H., Yin, Y., Hu, C., Yin, J., Ma, X., & Wang, J. (2019). Interpreting rainfall anomalies using rainfall's nonnegative nature. *Geophysical Research Letters, 46*, 426–434. https://doi.org/10.1029/2018GL081190

Chapter 9
Prospect of Organic Agriculture in the Present Climate Change Scenario

Suman Samui, Saikat Ranjan Das, Aditi Pahari, Poulomi Nandy, Bitish Kumar Nayak, and Sanjib Kumar Das

Abstract In this chapter, the potential of organic agriculture for climate change mitigation and adaptation is addressed. Through soil carbon sequestration, organic agriculture has a great potential to lower atmospheric carbon. According to various studies, the potential to reduce emissions by eliminating chemical fertilizers is about 20% and the potential to sequester carbon is around 40–72% of present yearly greenhouse gas (GHG) emissions from agriculture worldwide. Organically managed systems can help to mitigate climate change through careful regulation of soil nutrients, resulting in reduced nitrous oxide emissions from soils. Organic agriculture offers many opportunities to develop sustainable systems of food production for climate change adaptation uncertainties because it diversifies farms and enriches soils with organic matter. In addition, organic farming offers an alternative to energy-intensive inputs such as synthetic fertilizers, whose use by the rural poor is likely to be further restricted as energy prices rise. Organic farming systems provide yields that are comparable to or even higher than those produced by conventional agricultural systems in underdeveloped nations, making it a potentially critical option for sustainable livelihoods and food security of the rural poor in the context of climate change. Authorized organically produced products offer farmers more income opportunities and can thus support climate-friendly farming methods worldwide.

Keywords Carbon sequestration · Climate change · Food security · Organic agriculture · Sustainable production

---

S. Samui · A. Pahari · P. Nandy · S. K. Das
Department of Agronomy, Bidhan Chandra Krishi Viswavidyalaya, Mohanpur, Nadia, West Bengal 741252, India

S. R. Das (✉) · B. K. Nayak
ICAR-National Rice Research Institute, Cuttack, Odisha 753006, India
e-mail: saikatranjandas96@gmail.com

H. Pathak et al. (eds.), *Climate Change Impacts on Soil-Plant-Atmosphere Continuum*, Advances in Global Change Research 78, https://doi.org/10.1007/978-981-99-7935-6_9

9.1 Introduction

Emissions of greenhouse gases into the atmosphere are increased by anthropogenic activities, leading to global warming and climate change (Chatterjee & Saha, 2018; Chatterjee et al., 2020; Elbasiouny et al., 2020; Hossain et al., 2019; Lotze-Campe, 2011; Saha et al., 2018, 2020). Agroecosystems are directly affected by variations in atmospheric CO_2, temperature, and rainfall, and are also responsible for over one-third of global greenhouse gas (GHG) emissions resulting from nitrogen fertilizer, livestock, and rice cultivation, and tropical deforestation (Chatterjee et al., 2019; Hossain et al., 2019; Lotze-Campe, 2011). Agriculture has a dual role in the face of climate change: on the one hand, it is a source of GHG emissions; on the other hand, it is a storehouse for various GHGs (Chatterjee et al., 2018; Swain et al., 2018a, 2018b, 2018c). Over the past century, conventional, input-intensive agriculture has been the largest contributor to climate change after the energy sector. Communities that use a lot of pesticides and other synthetic inputs are more vulnerable to climate change's consequences. Since the publication of the IPCC 4th Assessment Report (AR4) in 2007, emissions from the agriculture, forestry, and other land use (AFOLU) sector have remained stable, but this sector's share of anthropogenic emissions has declined to 24% (in 2010) as emissions from the energy sector have increased (Smith et al., 2014). By 2030, emissions of methane (CH_4) and nitrous oxide (N_2O) are predicted to increase from 35 to 60% owing to increasing use of nitrogenous fertilizers and increasing livestock production in response to rising food demand (Venkateswarlu & Shanker, 2009). Additionally, overgrazing, deforestation, and the extensive use of technologies that degrade the soil have all been caused by intensive agriculture. These types of land use impact significantly global CO_2 emissions (Niggli et al., 2008). Consequently, sustainable methods of agriculture and food supply are more important than ever. The ability of agricultural production has to be improved to accommodate adverse and intense weather conditions i.e., floods and droughts, reduce emissions of GHGs from staple food production, and reduce the loss of soil organic carbon (SOC) (Nayak et al., 2020).

Organic farming is a method of optimizing the effective utilization of community resources while avoiding the use of inorganic inputs, genetically modified organisms (GMOs), and various synthetic substances used as food additives (Chatterjee et al., 2021; Gomiero et al., 2011; IFOAM, 2008, 2010). It offers proven techniques that contribute to poverty alleviation and mitigate climate change while increasing food security (Jordan et al., 2009). Organic agriculture (OA) produces far fewer GHG emissions and sequesters carbon in the soil in a timely, cost-effective, and efficient manner. In addition, OA contributes to better climate adaptation for farms and communities because it uses water efficiently, is resilient to severe weather, and lowers the chance of crop loss (Jordan et al., 2009). Organic agriculture offers a wide range of productions, certifications, and marketing knowledge and methods that have been established over many years in various climates, topographies, and cultures throughout the world. Organic agriculture can now be considered in the development of national and international food security as well as climate change strategies. As per a recent study, the Asian Development Bank has recognized OA for its resilient farming techniques and climate-smart practices (ADB, 2009). FAO has recognized OA as a potential method for agriculture as an adaptation strategy to climate change and to minimize its effects (Niggli et al., 2009). Without specifically mentioning OA, the IPCC 4th Assessment Report endorses several agricultural emissions reduction methods currently popular in OA, *e.g.,* reusing biowaste as a source of nutrients and combining cropping and livestock in a single system of agricultural production (IPCC, 2007b).

It is well known that the addition of fresh organic matter from outside can change the rate of native soil organic matter decomposition in lowland soils through priming effects. For example, the growth of soil microbes following the addition of fresh organic matter as an energy source could lead to positive priming that causes the co-metabolism of physiologically more resilient soil organic carbon (Blagodatskaya & Kuzyakov, 2008). Both positive and negative priming effects of organic matter input on soil organic matter decomposition and its subsequent mineralization to CO_2 and CH_4 have been found for flooded rice fields (Bhaduri et al., 2017), but this time we know very little about these processes. Consequently, organic nutrient sources act as CO_2 sources in an aerobic environment and as CH_4 sources in an anaerobic environment. In contrast, many studies depicted that the growth of certified organic agriculture is positively correlated with total agricultural GHG emissions, not with reductions in GHG emissions directly caused by agricultural production (Constance et al., 2008; Guthman, 2004; McGee, 2015; Venkat, 2012; Williams, 2006). Williams (2006) found that because of lower yields, tomatoes cultivated organically emit 30% more greenhouse gases than tomatoes grown conventionally. When comparing the life cycle trends of 12 conventional and organic crops in California, Venkat (2012) found that GHG emissions from organic farming were 10.6% greater than those from conventional cultivation. Thus, this chapter's objective is to talk about how organic farming techniques affect the environment, both positively and negatively, in addition to the different organic agriculture mitigation and adaptation measures about global climate change.

9.2 Status of Organic Agriculture

Organic farming and food are becoming increasingly popular across the world, especially in Western developed nations. Since 1985, the total area of cropland used for organic farming has increased during the past three decades (Willer & Lernoud, 2019). Overall, 69.8 million hectares (Mha) of organic land were recorded globally in 2017, depicting a 20% increase of 11.7 Mha over the past year. The rate of organic agriculture expansion is now the greatest ever noted (Willer & Lernoud, 2019). With over 35.65 Mha under cultivation, Australia now produces more certified organic produce compared to any other nation worldwide (Tal, 2018).

India ranked eighth with 1.78 Mha of total organic cropland (Willer & Lernoud, 2019). Asia accounts for 40% of global organic output (globally highest), with India

Table 9.1 Top 10 nations with the most organic farmland and largest numbers of organic producers in 2017 (Willer & Lernoud, 2019)

Position	Nations with the most organic farmland		Nations with the largest numbers of producers	
	Nation	Area (M. ha)	Nation	No. of organic producers
1	Australia	35.65	India (2016)	835,000
2	Argentina	3.39	Uganda (2016)	210,352
3	China	3.02	Mexico (2016)	210,000
4	Spain	2.08	Ethiopia (2015)	203,602
5	USA	2.03	Philippines	166,001
6	Italy	1.91	Tanzania (2013)	148,610
7	Uruguay	1.88	Peru	87,460
8	India	1.78	Turkey	75,067
9	France	1.74	Italy	66,773
10	Germany	1.37	Paraguay	58,258

Table 9.2 Continent-wise distribution of organic farming land and organic producers in 2017 (Willer & Lernoud, 2019)

Sl. no	Continent	Organic farmland share (%)	Organic producers share (%)
1	Asia	9	40
2	Africa	3	28
3	South America	11	16
4	Europe	21	14
5	North America	5	1
6	Oceania	51	1

accounting for the biggest number of organic producers (835,000) (Table 9.1). 85% of the overall organic agriculture produce (14,000 tonnes) of India is exported (Chopra et al., 2013). India has the highest number of organic farmers worldwide (Tal, 2018). Sikkim, Gujarat, Kerala, Karnataka, Rajasthan, Tamil Nadu, Uttarakhand, Himachal Pradesh, Maharashtra, and Madhya Pradesh are among India's major states following organic agriculture (Chandrashekar, 2010) (Table 9.2).

9.3 Organic Agriculture and Climate Change—The Linkage

Agriculture, going from being the second greatest emitter in the world to a lesser emitter or perhaps a net sink of greenhouse gases (GHGs), has enormous climate change mitigation potential, and may reduce its GHG emissions (Greenpeace, 2008).

Agriculture is contributing to GHG emissions, biodiversity loss, pesticide contamination, and degradation of soil, even though it produces increasing amounts of food and other items (Amundson et al., 2015; Rockström et al., 2009). The majority of these negative environmental effects are brought on by arable land, which makes up around 12% of the overall land area (FAOSTAT, 2015). It is a tremendous task to feed a burgeoning population which is predicted to increase to 9–10 billion in 2050 while simultaneously protecting the environment. Adopting truly sustainable agriculture practices on a wide scale will give us the best opportunity of overcoming this significant challenge and ensure future food and environmental security. Concerns about the sustainability of conventional agriculture have led to increased interest in integrated, organic, and conservation agriculture (Anon, 2009; De Schutter, 2010). The country's environmental security as well as food production both threatened by the depletion of water supplies and soil deterioration.

Since 1940s, organic agriculture has provided an effective substitute for the industrialized techniques used in conventional agriculture, encouraging agricultural practices that are more animal-friendly, less soil-intensive, and in line with natural ecology (Guthman, 2004). Conservative tilling and crop rotations, which are more frequent in organic agriculture (OA) than in conventional agriculture (CA), have been proven to enhance SOC sequestration (Govaerts et al., 2009; Soil Association, 2012; Chatterjee, 2016). Replacing the conventional tillage with zero tillage and soil surface covers with residue mulch decreased the energy inputs, nitrous oxide emission, carbon foot prints, and improved the soil quality, fertility, and microbial health (Lal et al., 2019). Furthermore, due to lower soil respiration rates and higher biomass levels, organic agriculture was shown to have greater CO_2 sinks in soil than conventional agriculture (OECD, 2003). Environmental risks are supposedly reduced by organic farming since it avoids the use of chemical pesticides that can harm water, soil, and native aquatic and terrestrial fauna (Oquist et al., 2007). Additionally, organic farms are more adept at resource conservation than conventional farms because of crop rotation strategies (Das et al., 2020). Organic farming enhances the physico-biological qualities of soil as compared to conventional farming *i.e.,* improved soil stability, biomass, organic matter, greater enzymatic activity, enhanced water holding capacity and percolation, reduced water and wind erosion (Chatterjee et al., 2015; Fliessbach & Mäder, 2000; Fliessbach et al., 2007). Organically cultivated farmlands are having superior quality soil and more water holding capacity (WHC), causing organic farmland to produce more even in years of drought (Pimentel et al., 2005). Also, Organic farming uses less energy as well as produces less trash per unit of production or area (Hansen et al., 2001; Stolze et al., 2000).

The main four principles of organic farming are health, fairness, care, and ecology, as proposed by the International Federation of Organic Agriculture Movements (IFOAM, 2021) (Fig. 9.1). Encouragement and reinforcement of biological cycles in agricultural production, to conserve and increase soil quality, to lower all kinds of pollution, to refrain from using inorganic fertilizers and pesticides, to maintain food's genetic diversity, to take into account the enormous socio-ecological influence on food production and making healthy food materials in adequate amounts, are the basic principles and practices of organic farming system (IFOAM, 1998).

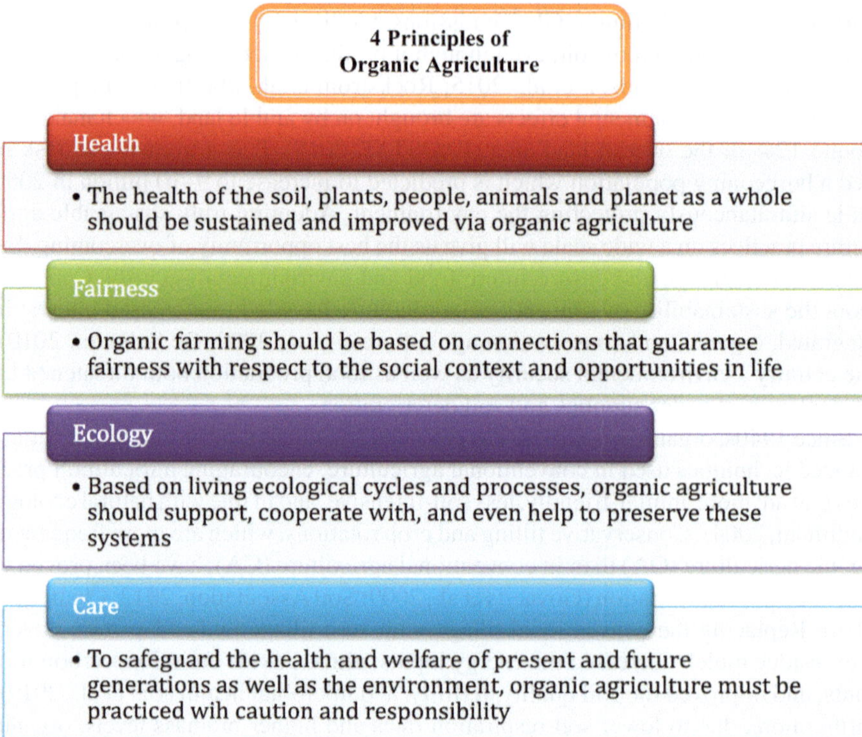

Fig. 9.1 Principles of organic agriculture

9.4 Potential of Organic Agriculture in Mitigating Climate Change

Organic agriculture (OA) is regarded as a particularly environmentally friendly method of farming (IFOAM, 2021). Using organic agricultural methods might reduce greenhouse gas (GHG) emissions, methane, and nitrous oxide emissions, boost soil carbon sequestration (Chatterjee et al., 2013) as well as strengthen the effects of organic agricultural techniques.

9.4.1 Reducing the GHG Emissions

The global warming potential (GWP) of organic farming is typically much lower compared to traditional or integrated systems. All organic crops' GWP decreased by 18%, according to two long-term trials conducted in Switzerland (Nemecek et al., 2005). Due to superior traditional yields, the difference decreases when assessed per unit of production in case of organic systems (Badgley et al., 2007). Organic farms accounted for a much lesser carbon footprint than integrated and conventional farms (Hillier et al., 2009). Crops only absorbed 17% of the 100 Mt of artificial nitrogen in conventional agriculture and the rest was discharged into the environment which leads to a higher GWP (Erisman et al., 2008).

9.4.2 Reduced Discharges of Nitrous Oxide (N_2O)

It is anticipated that the prohibition on the use of inorganic nitrogenous fertilizers and the fewer livestock units per unit area of land on organic farming will reduce the amount of readily available inorganic nitrogen in the soils, which reduces N_2O emissions. Additionally, compared to the soils with conventional farming, organically managed soils are superior in physical characteristics like soil structure due to the increased SOM contents. As a result, denitrification which releases N_2O occurs significantly less frequently in soils that have been treated organically. It was found that farms in southern Germany had N_2O emissions that were reduced by 50% when no mineral nitrogenous fertilizer was added, as well as when only a small amount of outside animal feed was provided (Zeddies, 2002). In five European nations, organically managed farms produced fewer N_2O emissions than conventional while only after taking into account emissions linked to yield out of the equation (Petersen et al., 2006).

9.4.3 Reducing Methane (CH_4) Emissions

Man-made activities account for two-thirds of global methane (CH_4) emissions, particularly from rice production systems (Smith & Conen, 2004) and from enteric ruminant fermentation (Steinfeld et al., 2006). Therefore, it is crucial to reduce methane emissions from the agricultural sector to combat global warming. Generally, there isn't much distinction between conventional and organic farming when considering the CH_4 emissions from rice production. Implementing superior rice production methods, such as adopting low CH_4 -generating cultivars (Aulakh et al., 2001), modifying the timing of the addition of organic residues (Cai & Xu, 2004), practicing mid-season drainage or eliminating constant floods (Smith & Conen, 2004) and lower C/N ratio composted manures application (Singh et al., 2003), had been demonstrated to reduce CH_4 emissions. The advantage of draining wetland rice, according to Akiyama et al. (2005), could be counterbalanced by increased N_2O emissions (Table 9.3).

Table 9.3 Potential of organic agriculture to mitigate climate change

GHG emission sources	Effect of organic agriculture
CH_4 produced by enteric fermentation of ruminants	Boosted by lesser performance and reduced dietary energy content, but decreased by lesser multi-use breeds and replacement rate
N_2O released from soils	Improved efficiency of nitrogen usage
Rice cultivation	Decreased by aquatic weeds and drainage but raised by organic amendments additions
Burning of biomass	According to organic standards, burning should be avoided
Manure management	No change in N_2O emissions but a decrease in CH_4 emissions
Direct emissions caused by the cutting down of forests for agriculture	Primary ecosystem disturbance is prohibited
Commercial fertilizers	Use of commercial fertilizers is restricted
Grassland	Greater organic matter content in the soil
Crop land	Greater organic matter content in the soil

Source Scialabba and Muller-Lindenlauf (2010)

9.4.4 Organic Farming Sequestering Carbon in the Soil

In the agricultural sector, soil carbon sequestration is recognized as the method with the greatest potential for mitigating emissions, accounting for an estimated 90% of what is theoretically possible (Smith et al., 2008). Soil carbon sequestration from atmospheric CO_2 would be a major co-benefit through using the increased quantity of soil organic matter due to organic farming. Green manuring, intercropping rotations, cover crops, composts, and animal dung are all used in organic farming, which increases soil nutrient status and the environmental factors that favor plant development, as well as increase carbon storage in the soil and minimize the stored carbon losses through soil erosion (Goh, 2004). Organic farmland soil has greater SOM content compared to those conventional systems, even if soil carbon sequestration varies widely in long-term field studies (Goh, 2011). Conventional farming had lower worldwide soil carbon reserves, even though this tendency is still present now (Lal, 2004a). To improve soil carbon storage or minimize carbon losses, modern organic agricultural methods are greatly preferred. Therefore, it is anticipated that converting to organic farming may decrease soil carbon losses, or perhaps even enhance soil net C sequestration and C concentrations over time (Niggli et al., 2009).

9.4.5 Reducing the Use of Fossil Fuels and Energy Inputs

Fossil fuel and solar energy are used in both organic and traditional farming to produce food. Any decrease in fossil fuel usage reduces climate change because the second-largest source of global greenhouse gas emissions comes from the use of fossil fuels in agriculture sector (Smith et al., 2014). Pimentel et al. (2005) asserts that switching to organic practices can lessen farmers' reliance on energy and improve energy consumption efficiency per unit of output. Because organic agriculture uses less or no synthetic pesticides and fertilizers, it was discovered that organic farming generally requires 50% fewer energy inputs per unit area than conventional farming (Topp et al., 2007).

9.5 Mitigative and Adaptive Approaches for the Organic Agriculture System to Reduce Greenhouse Gas (GHG) Emissions

To address climate change, adaptation (responding to climatic consequences) and mitigation (lowering GHG emissions) are important complements. Neither mitigation nor adaptation measures alone may avert substantial climate change consequences, when performed jointly; they can greatly lower risks (IPCC, 2007a). Mitigation is essential to slow the rate and intensity of climate change, whereas adaptation is required to minimize the unavoidable consequences of climate change.

9.5.1 Mitigation Approaches

In this mitigation method, the principles of organic agriculture are discussed in relation to lowering GHG emissions and sequestration of CO_2 in biomass. The strategies of lowering emissions and increasing sequestration are very cost-effective and easy to implement. Many of the following practices satisfy the needs mentioned by the IPCC to mitigate climate change in agriculture (Klein et al., 2007).

9.5.1.1 Building Soil Organic Carbon (SOC) and Fertility

The best way to mitigate climate change in agroecosystems is to maintain and increase soil organic carbon content. This is the process of collecting CO_2 from the atmosphere with the help of plants and storing fixed carbon as soil organic matter (SOM) (Lal, 2004b). Soil organic carbon is enhanced by biomass return in the soil in organic agriculture (Drinkwater et al., 1998; Pimentel et al., 1995). With this strategy, organic agriculture has enormous potential by adopting several practices such as use of

Table 9.4 Rates of carbon sequestration in different organic trials

Experiments/trials	Result
German Bavarian farms[a]	Sequestration rates are 0–396 kg carbon per ha per year and Integrated Pest Management lands wasted 249 and 55 kg of carbon
DOK trial data for 1978–1998, Switzerland[b]	a. Biodynamic with composted FYM: SOM level held steady b. Conventional (mineral fertilizer only): 191 kg per ha less SOM than the biodynamic type
Farming trials of USA[c]	Compared to the conventional systems; organic systems have a 14% greater SOC content
Rodale experiments[d]	a. Organic system with manure application: 981 kg per ha per year more SOC b. Organic system with leguminous crop: 574 kg per ha per year more SOC
Meta studies of improved fertilizer management on SOC stock[e]	a. SOC changes in organic fertilization relative to no fertilization (OF-NF) with 0.49 t C ha^{-1} $yr^{-}1$ b. In case of organic relative to inorganic fertilization (OF-IF) SOC changes recorded 0.33 t C ha^{-1} yr^{-1}

[a] Kiistermann et al. (2007); [b] Fliessbach et al. (2007); [c] Marriott and Wander (2006); [d] Pimentel et al. (2005); [e] Lessmann et al. (2022)

organic manures in substitute of inorganic fertilizers, cultivation of green legumes, cover crops, catch crops, using crop residues, etc. that can convert reductions of soil organic carbon to gain (Goh, 2001, 2004) (Table 9.4) resulting in improved soil structure with higher infiltration capacity (Goh, 2001; Lotter et al., 2003). These practices not only maintain a good quality soil structure with higher porosity, good aeration, and greater SOM content that results in more carbon sequestration but also restrict soil and carbon loss through erosion. Compost, farm yard manure, and legumes have the highest capacity to covert carbon into soil carbon as compared to arable crop residues which mineralized rapidly (Jordan et al., 2009).

9.5.1.2 Maintaining Soil Cover and Avoiding Crop Residue Burning

Maintaining soil cover with vegetation is important to prevent the soil from exposing it to actions which increase greenhouse gas emissions from carbon stored in the soil. The use of green manures and cultivation of cover crops keep away emissions and store nutrients for future utilization. Cover crops also provide root exudates continuously into the soil that helps in the growth of various soil microorganisms. Practice of growing cover crops is one of the ways to cover the bare soil throughout the year in the absence of arable crops (Poeplau & Don, 2015). Growing cover crops to increase soil organic carbon is one of its advantages which has been reported in the U.K. (Fullen & Auerswald, 1998; Johnston, 1973), Netherlands (Van Dijk,

1982), Sweden (Nilsson, 1986), Hungary (Berzseny and Gyrffy, 1997) and Europe (Smith et al., 1998). In most developed countries practice of burning crop residue and biomass waste in agricultural fields is restricted but it is not followed in conventional agricultural systems in developing countries. In organic agriculture to avoid burning, crop biomass is recycled which helps to enhance soil fertility status as well as to bring down CH_4 and N_2O emissions more than conventional agriculture.

9.5.1.3 Integration of Livestock and Crop Production

Conventional agriculture system is largely depending on use of synthetic nitrogenous fertilizers whereas stored manure and slurry on livestock farms are responsible for gas emissions and create environmental problems to some extent. This difficulty can be solved in organic farming where FYM is applied in an integrated or cooperative approach between agricultural and livestock activities. By adopting this practice N use efficiency will be increased by reducing nitrogen loss through emissions (Pathak et al., 2016). Greenhouse gas emission is lower in organic livestock production systems compared to the conventional system because animals are raised in pastures in organic agriculture which reduces the requirement of feed and their manures are used for building up soil organic carbon. But in case of the conventional system animals are reared in an industrial manner which promotes production of animal manure in large quantities and they mainly depend upon grain-based feed.

9.5.1.4 Avoid Chemical Fertilization

During manufacturing process of chemical products CO_2 and N_2O are released which pollutes the environment. Agriculture sector accounts for almost 13.5–30% of gaseous emissions globally (Foereid et al., 2008) during the application of inorganic fertilizers and these emissions have been pointed out as fastest growing which have increased 45% since 2001 (FAO, 2016). Continuous application of inorganic fertilizers for about five decades in paddy soil reported to alter the bacterial community structure and impact yield (Kumar et al., 2018). In organic agriculture, use of manure and compost or cultivation of nitrogen-fixing leguminous plants are some sources of nitrogen input as no chemical nitrogenous fertilizers are used. Emission of N_2O does not occur due to biological nitrogen fixation through leguminous plants but soil incorporation of nitrogen-rich plant residues results in high emission of N_2O. Concentration of soil available mineral N is less in case of organic farming due to manure and compost application, and biological N fixation by legumes which promotes a slow and steady release of organic N reducing the N_2O emission. Besides this, recycling manures and other organic materials has a huge capacity to improve soil organic carbon levels (Bernal, 2017; Cordovil, 2004). Biochar can also be used in conjunction with organic amendments to boost soil fertility. Using less labile sources of carbon like rice husk biochar (RHB) not only improved the physical quality of soil but also reduce the emission of greenhouse gases (Munda et al., 2018). Besides. RHB increased rice yield by 1–24% and enhance C-storage (43–169%) in soil (Munda et al., 2018). In an annual cropping system, soil CO_2 inflow is significantly enriched by manure incorporation (Gregorich et al., 1998; Rochette et al., 2000, 2004). Soil carbon content is reduced with the higher rate of application of chemical inputs because stability of CO_2 as soil fraction is less. This may be a probable reason for CO_2 loss in conventional agriculture (Jordan et al., 2009). A higher proportion of grassland in organic farming avoids bare fallow land and a large quantity of organic matter is added to the soil in organic farming practices (Khan et al., 2007).

9.5.1.5 Limited Use of Fossil Energy

Organic agriculture requires less energy as it is not using chemically synthesized products as industrially manufactured fertilizers, pesticides and herbicides require energy input for their production (Lampkin, 2007). According to reports, organic agricultural systems utilize 20–50% less energy than conventional farming systems (Muller et al., 2012; Pimentel et al., 2005; Schader et al., 2011). According to Aher et al. (2012) despite of 45% less energy used in organic farming; it yields equal to conventional yields proving more efficient in organic agriculture than a conventional system. Pretty and Ball (2001) reported that a similar yield is achieved in USA despite 22–120% less energy consumption in organic systems than input in an intensive industrialized system. It was estimated that organic tillage farming and organic no-till corn production save 47 and 61% fuel ha^{-1} than conventional tillage practice and conventional no-till corn production respectively (Pimentel et al., 2005). Mitigation by using organic materials in place of fossil fuels is attainable when crop cultivation does not emit GHGs due to requirement of chemical inputs (Lal, 2002).

9.5.2 Adaptive Approaches

The actions made to lessen the negative consequences of climate change while simultaneously using new opportunities are referred to as adaptation to climate change. In reaction to real or expected climate changes, it involves changing policies and actions. Adaptation can be responsive, acting in reaction to climatic impacts, or proactive, occurring before climate change consequences become apparent. Proactive adaptation will, in most cases, save money in the long run and be more successful than responsive adaptations. Climate change consequences necessitate both regional and sector-specific creative measures (Laukkonen et al., 2009).

9.5.2.1 Maintaining Soil Health and Preventing Soil Erosion

Healthy biological activity of soil got deteriorated due to faulty agricultural practices like use of an excessive dose of chemical fertilizer or sometimes may be due to lack of nutrients. Biologically active soils are highly desired for agriculture's adaptability to counteract climate change, which can provide plants enough nourishment so that application of external inputs can be minimized. Increased biological diversification of various soil microbes, macro-organisms like insects, and earthworms, thus resulting in improved soil quality have been reported by Mäder et al. (2006). In organic farming use of manure, compost, green manuring, cultivation of cover crops, short-duration legumes as catch crops, rotation of crops reduce dependency on chemical fertilizers as well as enrich soil health. Erosion-prone soil can be obstructed by adopting mixed farming, alley cropping, and reduced or conservation tillage practices in organic agriculture. High organic matter content and pronounced microbial activity ameliorate soil structural stability. According to Tuomisto et al. (2012) organically maintained farms have better soil organic matter with lower nutrient loss properties than conventional agriculture. Improved aggregate stability, restricted soil erosion (Siegrist et al., 1998), and higher microbial activities (Lampkin, 2007) have been found by applying organic farming practices.

9.5.2.2 Increased Water Usage Efficiency, Resistance to Drought and Flooding

Effect of climate change is more pronounced on irregular and erratic rainfall patterns. For obtaining sufficient yield to feed the increasing population storing water for future use is one important measure. Water retention power of soils under organic management is better than conventional cultivation practices because of superior soil structure as well as the presence of humic substances and other organic matter compounds in soil that significantly prevents chances of drought and flood (ITC & FiBL, 2007; Niggli et al., 2009). Thus, organic farming is more resilient to adverse weather circumstances like drought, floods, or water stagnation than conventional systems (Bescansa et al., 2006; Scialabba and Muller-Lindenlauf 2010) that are expected to become periodic because of climatic changes (Ahmed et al., 2009). Yield loss because of water logging or surface water flooding situation is prevented by SOM which enhances drainage in soil. Use of crop residues and mulching practices improve the contents of SOM thus helping to retain soil moisture, prevent erosion, and enrich natural biodiversity which is linked to improve drought and flood resistance power consequently to adapt climate change (Agriculture Atlas, 2019; FAO, 2015).

9.5.2.3 Crop Biodiversity

In organic agriculture greater level of crop biodiversity is found through crop rotations and various cultivation practices. This biodiversity improves ecological and economic stability. Organic agriculture promotes the usage of traditional regional varieties as well as in situ breeding programs of varieties are also conducted for future needs. Maintenance of agro-diversity is going to be an efficient adaptation to combat continuous climate change. According to Mader et al. (2002) soil fertility and biodiversity has been increased with application of organic agriculture. Severity of pest outbreaks, plant and animal diseases reduce as a beneficial effect of biodiversity as well as soil water and nutrients are used effectively simultaneously (Altieri et al., 2005; Smith et al., 2011). Farmers cultivating interspecific diversified crops have more potential to combat adverse weather phenomena than monoculture by conventional farmers (Altieri & Koohafkan, 2008; Niggli et al., 2009).

9.6 Limitation of Organic Agriculture in Context of Climate Change

For many years, the scientific community has debated the environmental consequences of organic agriculture. Organic agriculture's contribution to environmental and resource issues is still debatable, as is really whether its adoption is an acceptable policy strategy to solve current socio-ecological challenges. As climate change and global food production have become much greater public problems in recent years, the discussion over the benefits as well as drawbacks of organic farming has grown in popularity (Alim, 2013).

9.6.1 Land-Use Efficiency

Agricultural production occupies around 40% of the ice-free territory of the world (Foley et al., 2011). Constant shifting of land use (e.g., deforestation) is linked to a variety of environmental concerns, including biodiversity loss and soil carbon emissions into the environment. As a result, matching agricultural production and environmental objectives will ultimately necessitate more effective utilization of land and other environmental resources. In comparison to traditional approaches, organic systems use the land less effectively. According to various research findings, organic system yields are comparatively lower than conventional (de Ponti et al., 2012; Ponisio et al., 2015; Seufert et al., 2012). De Ponti et al. (2012) reported 20, 21 and 26% reduction in vegetables, cereals and oil seed crops yield respectively, whereas Seufert et al. (2012) reported this reduction to a magnitude of 33, 26 and 11%, respectively. Ponisio et al. (2015) also depicted 19% reduction in all crop yield in organic farming compare to inorganic farming. Furthermore, crop rotations in organic systems frequently contain such crops, which are unfit for consumption purposes of humans.

Additional negative impacts of organic and conventional agricultural practices on the environment are also influenced by land use demands. Environmental consequences are often described per unit of production and per unit of land area to take into consideration the less efficient use of land by biological systems. With the increased demand for agricultural produce and food, assessing environmental consequences per unit of production appears to be more significant from a global viewpoint. Even this technique, however, does not account for the environmental consequences of large-scale organic agriculture transformation. Only a small percentage of global farmlands are certified organic now. Organic conversion on a large scale would almost certainly need to incorporate more natural areas into cropland. Environmental costs connected with such extra land-use changes would not be fully taken into account by merely stating per unit of production (Leifeld, 2016).

9.6.2 Emissions from Manure

The treatment and storage of manure may significantly affect GHG emissions. Intensive livestock production systems, where additional bedding material is often intermingled with the feces, tend to build liquid manure more often than organic livestock systems, where it (intensive livestock production system) tends to cause greater emissions. Composting manure might result in a 50–120% increase in the indirect emission of N_2O gas (Pardo et al., 2015). Generally, due to their composition, liquid-organic fertilizers provide greater N_2O emissions than solid-organic fertilizers (Cayuela et al., 2017; Walling et al., 2020). Organic fertilizers can increase the soil's microbial activity, biodiversity as well as nutrient availability. Alternatively, as SOC increases, so does microbial activity, which consequently accelerates CO_2 emissions through soil respiration (Oertel et al., 2016). Recent literature reports a direct impact of organic fertilization, with compost and digestate, on CH_4 emission increases (Severin et al., 2015; Verdi et al., 2018).

9.6.3 Energy Usage and Greenhouse Gas Emissions

Organic agriculture, based on the research, compared to traditional agriculture, consumes a lesser amount of energy per unit of area and to a smaller degree, per unit of yield (Table 9.5). The absence of synthetic fertilizers and pesticides in organic farming accounts for the majority of this variance. Farming practices consume equal amounts of fuel throughout systems. However, in organic cultivation, additional fuel is often needed for certain crops (particularly vegetables) when frequent thermal or mechanical weed control is required (Scialabba & Muller-Lindenlauf, 2010; Lynch et al., 2011; Smith et al., 2015). In some cases, these types of field operations may result in increased total energy usage in organic systems (Lee et al., 2015).

Table 9.5 Effects on the environment of organic over conventional farming

Category	Per unit of output	Per unit of area
Land use	Organic diets land use 40% higher[d] 20–110% higher land use[a,b,c]	Not applicable
Greenhouse gas (GHG) emissions	GHG emissions were 0–10% lesser[a,b,c] N_2O emissions 8% higher[a] Organic diet equivalent to conventional[d]	39% lesser GHG emissions[c] Emission of N_2O 14–31% lower[a,c]
Energy use	15–21% less energy uses[a,b]	Energy use 10–70% lower[e]
Soil quality	Not applicable	Greater and much more active microbial populations in the soil [f] 6–7% more organic matter in the soil[a,c,g]
Nutrient leaching/ eutrophication potential	Ammonia emissions have increased by 11%[a] Increased by 13% in terms of acidification potential (in sulfur dioxide equivalents)[b] In the EU[a], nitrate leaching increased by 49%, whereas it decreased by 5% globally[c] 36% increased eutrophication potential (in terms of Phosphate Equivalents)[b]	18% less ammonia emissions[a] Nitrate leaching is 30–31% less than conventional[a,c] Phosphorus losses are 1% less[a]
Biodiversity	The biodiversity loss because of indirect land use change has not been reported	Organism abundance 50% higher[h] 30–34% greater species diversity[h,i] Species evenness higher[j]
Pesticide use to arable crops	100–72% less pesticide use[k]	Reduced by 100–81%[j]
Productivity of arable crops		Relative difference between organic/conventional on per area unit −68 to +32%[k]

[a] Tuomisto et al. (2012); [b] Clark and Tilman (2017); [c] Mondelaers et al. (2009); [d] Treu et al. (2017); [e] Gomiero et al. (2011); [f] Lori et al. (2017); [g] Mader et al. (2002); [h] Bengtsson et al. (2005); [i] Tuck et al. (2014); [j] Crowder et al. (2010); [k] Meier et al. (2015)

9.6.4 Leaching of Nutrients and Water Quality

Nutrient leaching is a concern because it contributes to the eutrophication of aquatic and marine environments, particularly in areas with intensive agriculture (Halberg, 2012). Nitrate (NO^{3-}) leaching per unit of land decreased as a result of organic management, but not per unit of yield. The average potential of eutrophication of organic systems is generally greater per unit of output when evaluated in PO_4 equivalents as well as the average potential of acidification is also higher when measured in SO_2 equivalents (Clark & Tilman, 2017). Avoiding chemical fertilizers is typically related to reducing leaching potential, however eliminating imbalances between plant demand and nutrient flow in organic systems may be challenging, which could result in a greater loss of nutrients (Gomiero et al., 2011; Halberg, 2012; Tuomisto et al., 2012). Pesticide contamination of water bodies is reduced since chemical pesticides are prohibited in organic farming (Reganold & Wachter, 2016). On the other hand, several non-chemical pesticides use for organic farming, can be harmful to aquatic life. For example, in organic horticulture, copper (Cu) based treatments are commonly used to cure fungal infections. (Niggli, 2015).

In general, organic agriculture's environmental advantages are less apparent than commonly assumed. Organic systems often outperform conventional systems in terms of environmental performance when calculated per unit of area; however, the advantage disappears when assessed per unit of production. As a result, organic agriculture appears to be better suited to resolving specific environmental concerns (such as soil deterioration) compared to other global issues (such as climate change, and land-use change) (Mondelaers et al., 2009). Some of the data on environmental consequences should be noted as premature. Present findings may not be internationally representative due to a lack of data from developing nations (Clark & Tilman, 2017; Gattinger et al., 2012; Lee et al., 2015; Skinner et al., 2014).

9.7 Conclusion

Organic agriculture can be a viable solution to minimize the impacts of global warming and climate change by sequestering atmospheric CO_2 into the soil or lowering GHG emissions. Organic agriculture is less hazardous than traditional agriculture in terms of climate and environmental consequences when assessed per unit of area; however, the advantage disappears when assessed per unit of production. Considering the sizeable and growing global demand for agricultural products and food, it is more important from a broader perspective to reflect climatic and environmental effects per unit of production. New technology and production methods will be required to protect natural resources while also making farming more sustainable and less reliant on synthetic chemicals. Localized solutions must be developed, taking into consideration all prospective scientific fields. Neither organic nor conventional approaches offer ideal solutions to these problems yet. Smart integration of both forms of agriculture is expected to be required for sustainable production systems. Organic farming is well suited for improving soil quality and reducing land degradation, but in many cases, environmental efficiency as well as if these techniques were integrated with lower amounts of chemical fertilizers as well as the most current findings in plant genetic modification, agricultural yield may be increased much further. To prepare the path for the innovation and implementation of more sustainable farming practices, ideological obstacles between advocates and critics of organic agriculture must be removed.

References

ADB. (2009). International Food Policy Research Institute/Asian Development Bank, 2009: Building Climate Resilience in the Agriculture Sector of Asia and the Pacific.

Agriculture Atlas. (2019). Facts and figures on EU farming policy. In C. Chemnitz (Chief Executive Editor), Jointly Published by: Heinrich Boll Foundation; Friends of the Earth Europe; BirdLife Europe and Central Asia. https://eu.boell.org/en/2019/05/14/agriculture-atlas-2019-facts-and-figures-eu-farming-policy

Aher, S. B., Sengupta, B., & Bhaveshanada, S. (2012). Organic agriculture: Way towards sustainable development. *ARPN Journal of Science and Technology, 2*, 318–324.

Ahmed, S. A., Diffenbaugh, N. S., & Hertel, T. W. (2009). Climate volatility deepens poverty vulnerability in developing countries. *Environmental Research Letters, 4*(3), 034004.

Akiyama, H., Yagi, K., & Yan, X. (2005). Direct N_2O emissions from rice paddy fields: summary of available data. *Global Biogeochemical Cycles, 19*(1). https://doi.org/10.1029/2004GB002378

Alim. (2013). Organic Farming—Pros and Cons. Agriculture Blogs, https://alim277.wordpress.com/2013/02/18/organic-farming-pros-and-cons/ Accessed on February 18, 2013.

Altieri, M. A., Ponti, L., & Nicholls, C. I. (2005). Enhanced pest management through soil health: Toward a belowground habitat management strategy. *Biodynamics (Summer)*, 33–40.

Altieri, M. A., & Koohafkan, P. (2008). Enduring farms: Climate change, smallholders and traditional farming communities. TWN Environment and Development Series 6, Third World Network (TWN), Penang.

Amundson, R., Berhe, A. A., Hopmans, J. W., Olson, C., Sztein, A. E., & Sparks, D. L. (2015). Soil and human security in the 21st century. *Science, 348*(6235), 1261071.

Anon. (2009). International Assessment of Agricultural Science and Technology for Development Agriculture at a Crossroads: Global Report (Island, 2009).

Aulakh, M. S., Wassmann, R., & Rennenberg, H. (2001). Methane emissions from rice fields—Quantification, mechanisms, role of management, and mitigation options. *Advances in Agronomy, 70*, 193–260.

Badgley, C., Moghtader, J., Quintero, E., Zakem, E., Chappell, M. J., Aviles-Vazquez, K., Samulon, A., & Perfecto, I. (2007). Organic agriculture and the global food supply. *Renewable Agriculture and Food Systems, 22*(2), 86–108.

Bengtsson, J., Ahnstr ¨om, J., & Weibull, A. -C. (2005). The effects of organic agriculture on biodiversity and abundance: A meta-analysis. *Journal of Applied Ecology, 42*, 261–269.

Bernal, M. P. (2017). Grand challenges in waste management in agroecosystems. *Frontiers in Sustainable Food Systems, 1*(1). https://doi.org/10.3389/fsufs.2017.00001

Berzsenyi, Z., & Győrffy, B. (1997). Effect of crop rotation and fertilization on maize and wheat yields and yield stability in long-term experiments. *Agrokémia És Talajtan, 46*(1–4), 77–98.

Bescansa, P., Imaz, M. J., Virto, I., Enrique, A., & Hoogmoed, W. B. (2006). Soil water retention as affected by tillage and residue management in semiarid Spain. *Soil and Tillage Research, 87*(1), 19–27.

Bhaduri, D., Mandal, A., Chakraborty, K., Chatterjee, D., & Dey, R. (2017). Interlinked chemical-biological processes in anoxic waterlogged soil-a review. *Indian Journal of Agricultural Sciences, 87*(12), 1587–1599.

Blagodatskaya, E., & Kuzyakov, Y. (2008). Mechanisms of real and apparent priming effects and their dependence on soil microbial biomass and community structure: Critical review. *Biology and Fertility of Soils, 45*(2), 115–131.

Cai, Z. C., & Xu, H. (2004). Options for mitigating CH_4 from rice fields in China. *Journal of Geophysical Research, 105*, 17231–17232.

Cayuela, M. L., Aguilera, E., Sanz-Cobena, A., Adams, D. C., Abalos, D., Barton, L., Ryals, R., Silver, W. L., Alfaro, M. A., Pappa, V. A., & Smith, P. (2017). Direct nitrous oxide emissions in Mediterranean climate cropping systems: Emission factors based on a meta-analysis of available measurement data. *Agriculture, Ecosystems & Environment, 238*, 25–35.

Chandrashekar, H. M. (2010). Changing Scenario of organic farming in India: An overview. *International NGO Journal, 5*, 34–39.

Chatterjee, D., Kuotsu, R., & Saha, S. (2015) Nutrient recycling and resource management through organic farming. In S. K. Ray, D. Chatterjee, B. M. Tungoe, M. Meyase, B. C. Deka (Eds.), *Livelihood improvement of tribal farmers through integrated farming system (IFS) approach* (pp. 28–39). Training manual published in May, 2015 by Joint Director, ICAR Research Complex for NEH Region, Nagaland Centre, Jharnapani, Medziphema-797 106.

Chatterjee, D., Saha, S., Swain, B., Chakraborty, D., Nayak, A. K., Pathak, H., & Singh, M. P. (2020). Monitoring and impact assessment of climate change on agriculture using advanced research techniques. In A. Rakshit, S. Ghosh, S. Chakraborty, V. Philip, A. Datta (Eds.), *Soil analysis: recent trends and applications* (pp. 33–53). Springer. https://doi.org/10.1007/978-981-15-2039-6

Chatterjee, D. (2016). Strengths-weaknesses-opportunities-threats (SWOT) analysis of conservation agriculture. *Indian Journal of Hill Farming, 29*(1), 18–23.

Chatterjee, D., Datta, S. C., & Manjaiah, K. M. (2013). Clay carbon pools and their relationship with short-range order minerals: Avenues to mitigate climate change? *Current Science, 105*(10), 1404–1410.

Chatterjee, D., Kuotsu, R., Ao, M., Saha, S., Ray, S. K., & Ngachan, S. V. (2019). Does rise in temperature effect adversely on soil fertility, carbon fractions, microbial biomass and enzymatic activities under different land use? *Current Science, 116*(12), 2044–2054.

Chatterjee, D., Mohanty, S., Guru, P. K., Swain, C. K., Tripathi, R., Shahid, M., Kumar, U., Kumar, A., Bhattacharyya, P., Gautam, P., & Lal, B. (2018). Comparative assessment of urea briquette applicators on greenhouse gas emission, nitrogen loss and soil enzymatic activities in tropical lowland rice. *Agriculture, Ecosystems & Environment, 252*, 178–190.

Chatterjee, D., Nayak, A. K., Mishra, A., Swain, C. K., Kumar, U., Bhaduri, D., Panneerselvam, P., Lal, B., Gautam, P., & Pathak, H. (2021). Effect of long-term organic fertilization in flooded rice soil on phosphorus transformation and phosphate solubilizing microorganisms. *Journal of Soil Science and Plant Nutrition, 21*(2), 1368–1381. https://doi.org/10.1007/s42729-021-00446-8

Chatterjee, D., & Saha, S. (2018). Response of soil properties and soil microbial communities to the projected climate change. *Advances in crop environment interaction* (pp. 87–136). Springer.

Chopra, A., Rao, N. C., Gupta, N., & Vashisth, S. (2013). Come sunshine or rain; organic foods always on tract: A futurist perspective. *International Journal of Nutrition, Pharmacology, Neurological Diseases, 3*(3), 202–205.

Clark, M., & Tilman, D. (2017). Comparative analysis of environmental impacts of agricultural production systems, agricultural input efficiency, and food choice. *Environmental Research Letters, 12*, 64016.

Constance, D. H., Choi, J. Y., & Lyke-Ho-Gland, H. (2008). Conventionalization, bifurcation, and quality of life: Certified and non-certified organic farmers in Texas. *Journal of Rural Social Sciences, 23*(1), 208–234.

Cordovil, C. (2004). *Dinâmica do azoto na reciclagem de resíduos orgânicos aplicados ao solo (Nitrogen Dynamics from Organic Residues Application to Soil)* (p. 56). Instituto do Ambiente, Alfragide, Lisboa.

Crowder, D. W., Northfield, T. D., Strand, M. R., & Snyder, W. E. (2010). Organic agriculture promotes evenness and natural pest control. *Nature, 466*, 109–112.

Das, S., Chatterjee, A., & Pal, T. K. (2020). Organic farming in India: A vision towards a healthy nation. *Food Quality and Safety, 4*(2), 69–76.

De Ponti, T., Rijk, B., & Van Ittersum, M. K. (2012). The crop yield gap between organic and conventional agriculture. *Agricultural Systems, 108*, 1–9.
De Schutter, O. (2010). Report Submitted by the Special Rapporteur on the Right to Food: Human Rights Council Sixteenth Session: Agenda Item 3: Promotion and Protection of All Human Rights, Civil, Political, Economic, Social and Cultural Rights, Including the Right to Develop. United Nations (UN).
Drinkwater, L. E., Wagoner, P., & Sarrantonio, M. (1998). Legume-based cropping systems have reduced carbon and nitrogen losses. *Nature, 396*(6708), 262–265.
Elbasiouny, H., Elbanna, B. A., Al-Najoli, E., Alsherief, A., Negm, S., El-Nour, A., Nofal, A., & Sharabash, S. (2020). Agricultural waste management for climate change mitigation: Some implications to Egypt. *Waste management in MENA regions* (pp. 149–169). Springer.
Erisman, J. W., Sutton, M. A., Galloway, J., Klimont, Z., & Winiwarter, W. (2008). How a century of ammonia synthesis changed the world. *Nature Geoscience, 1*(10), 636–639.
FAO. (2015). Soils help to combat and adapt to climate change by playing a key role in the carbon cycle. Food and Agriculture Organization of the United Nations, http://www.fao.org/3/a-i4737e.pdf
FAO. (2016). Conservation agriculture fact sheet. Food and Agriculture Organization of the United Nations. http://www.fao.org/3/a-i6169e.pdf
FAOSTAT. (2015). Food and Agriculture Organization of the United Nations. Online Database http://faostat.fao.org/site/377/default.aspx#ancor. Accessed on August 2015
Fliessbach, A., & Mäder, P. (2000). Microbial biomass and size-density fractions differ between soils of organic and conventional agricultural systems. *Soil Biology and Biochemistry, 32*(6), 757–768.
Fliessbach, A., Oberholzer, H. R., Gunst, L., & Mäder, P. (2007). Soil organic matter and biological soil quality indicators after 21 years of organic and conventional farming. *Agriculture, Ecosystems & Environment, 118*(1–4), 273–284.
Foereid, B., Hastings, A., & Smith, P. (2008). *Cool farming: Climate impacts of agriculture and mitigation potential* (p. 28). Greenpeace International.
Foley, J. A., Ramankutty, N., Brauman, K. A., Cassidy, E. S., Gerber, J. S., Johnston, M., Mueller, N. D., O'Connell, C., Ray, D. K., West, P. C., & Balzer, C. (2011). Solutions for a cultivated planet. *Nature, 478*(7369), 337–342.
Fullen, M. A., & Auerswald, K. (1998). Effect of grass ley set-aside on runoff, erosion and organic matter levels in sandy soil in east Shropshire, UK. *Soil and Tillage Research, 46*, 41–49.
Gattinger, A., Muller, A., Haeni, M., Skinner, C., Fliessbach, A., Buchmann, N., Mäder, P., Stolze, M., Smith, P., Scialabba, N. E., & Niggli, U. (2012). Enhanced top soil carbon stocks under organic farming. *Proceedings of the National Academy of Sciences, 109*(44), 18226–18231.
Goh, K. M. (2001). Managing organic matter in soils, sediments and water. In: R. S. Swift, K.M. Sparks (Eds.), *Understanding and managing organic matter in soils, sediments and waters, Proceedings of the 9th International Conference Int. Humic Substances Soc* (pp. 269–278). Adelaide, Australia, 2001.
Goh, K. M. (2004). Carbon sequestration and stabilization in soils: Implications for soil productivity and climate change. *Soil Science and Plant Nutrition, 50*(4), 467–476.
Goh, K. M. (2011). Greater mitigation of climate change by organic than conventional agriculture: A review. *Biological Agriculture and Horticulture, 27*(2), 205–229.
Gomiero, T., Pimentel, D., Paoletti, M. G. (2011). Environmental impact of different agricultural management practices: conventional versus organic agriculture. *Critical Reviews in Plant Sciences, 30*(1–2), 95–124.
Govaerts, B., Verhulst, N., Castellanos-Navarrete, A., Sayre, K. D., Dixon, J., & Dendooven, L. (2009). Conservation agriculture and soil carbon sequestration: Between myth and farmer reality. *Critical Reviews in Plant Sciences, 28*(3), 97–122.
Greenpeace. (2008). Cool farming: Climate impacts of agriculture and mitigation potential. http://www.greenpeace.org/international/press/reports/cool-farming-full-report

Gregorich, E. G., Rochette, P., McGuire, S., Liang, B. C., & Lessard, R. (1998). Soluble organic carbon and carbon dioxide fluxes in maize fields receiving spring-applied manure. *Journal of Environmental Quality, 27*, 209–214.

Guthman, J. (2004). *Agrarian dreams: The paradox of organic farming in California.* University of California Press.

Halberg, N. (2012). Assessment of the environmental sustainability of organic farming: Definitions, indicators and the major challenges. *Canadian Journal of Plant Science, 92*, 981–996.

Hansen, B., Alrøe, H. F., & Kristensen, E. S. (2001). Approaches to assess the environmental impact of organic farming with particular regard to Denmark. *Agri. Eco. Env., 83*(1–2), 11–26.

Hillier, J., Hawes, C., Squire, G., Hilton, A., Wale, S., & Smith, P. (2009). The carbon footprints of food crop production. *International Journal of Agricultural Sustainability, 7*(2), 107–118.

Hossain, M. S., Arshad, M., Qian, L., Zhao, M., Mehmood, Y., & Kächele, H. (2019). Economic impact of climate change on crop farming in Bangladesh: An application of Ricardian method. *Ecological Economics, 164*, 106354.

IFOAM. (1998). The IFOAM basic standards for organic production and processing. General Assembly, Argentina, November, IFOAM, Germany.

IFOAM. (2008). Organic Agriculture's Role in Countering Climate Change. www.ifoam.org/growing/1_arguments_for_oalenvironmentalbenefits/climatechnage.html

IFOAM. (2010). Definition of Organic Agriculture. IFOAM http://www.ifoam.org/growingorganic/definitions/doa/index.html, and http://www.ifoam.org. Accessed on 10 March 2010.

IFOAM. (2021). Priciples of organic agriculture: Preamble. https://www.ifoam.bio/sites/default/files/2020-03/poa_english_web.pdf. Accessed 8 November 2021.

IPCC. (2007b). From Climate Change 2007: Chapter Agriculture. www.ipcc.ch/pdf/assessmentreport/ar4/wg3/ar4-wg3-chapter8.pdf

IPCC. (2007a). Climate Change 2007: The Physical Science Basis. Contribution of working group I to the fourth assessment report of the Intergovernmental Panel on Climate Change. Cambridge University Press, Cambridge, United Kingdom and New York, NY, USA.

ITC and, FiBL. (2007). *ITC (International Trade Centre) UNCTAD/WTO and Research Institute of Organic Agriculture (FiBL) 2007: Organic Farming and Climate Change.* ITC.

Johnston, A. E. (1973). Effects of ley and arable cropping systems on the amounts of soil organic matter in the Rothamsted and Woburn ley-arable experiments. Rothamsted Exp Sta Rep.for 1972. *Part, 2*, 131–159.

Jordan, R., Müller, A., & Oudes, A. (2009). High sequestration, low emission, food secure farming. organic agriculture-a guide to climate change & food security. IFOAM.

Khan, S. A., Mulvaney, R. L., Ellsworth, T. R., & Boast, C. W. (2007). The myth of nitrogen fertilization for soil carbon sequestration. *Journal of Environmental Quality, 36*(6), 1821–1832.

Kiistermann B, Wenske K, Hiilsbeergen K (2007) Modellierung betrieblicher C- und NFliisseals Grundlage einer Emissionventur. http://orgprints.org/9654/

Klein, R., Toth, F., Denton, F., Downing, T., Huq, S., Richels, R., Robinson, J., & Burch, S. (2007). Inter-relationships between adaptation and mitigation. *The Fourth Assessment Report of IPCC Working Group II, Climate Change 2007: Impacts* (pp. 745–777). Cambridge University Press, Cambridge, UK.

Kumar, U., Nayak, A. K., Shahid, M., Gupta, V. V., Panneerselvam, P., Mohanty, S., Kaviraj, M., Kumar, A., Chatterjee, D., Lal, B., & Gautam, P. (2018). Continuous application inorganic of and organic fertilizers over 47 years in paddy soil alters the bacterial community structure and its influence on rice production. *Agriculture, Ecosystems & Environment, 262*, 65–75.

Lal, B., Gautam, P., Nayak, A. K., Panda, B. B., Bihari, P., Tripathi, R., Shahid, M., Guru, P. K., Chatterjee, D., Kumar, U., & Meena, B. P. (2019). Energy and carbon budgeting of tillage for environmentally clean and resilient soil health of rice-maize cropping system. *Journal of Cleaner Products, 226*, 815–830.

Lal, R. (2002). The potential of soils of the tropics to sequester carbon and mitigate the greenhouse effect. *Advances in Agronomy, 76*, 1–30.

Lal, R. (2004a). Soil carbon sequestration impacts on global climate change and food security. *Science, 304*(5677), 1623–1627.

Lal, R. (2004b). Soil carbon sequestration to mitigate climate change. *Geoderma, 123*, 1–22.

Lampkin, N. (2007). Organic farming's contribution to climate change and agricultural sustainability. Welsh organic producer conference, 18th October 2007.

Laukkonen, J., Blanco, P. K., Lenhart, J., Keiner, M., Cavric, B., & Kinuthia-Njenga, C. (2009). Combining climate change adaptation and mitigation measures at the local level. *Habitat International, 33*(3), 287–292.

Lee, K. S., Choe, Y. C., & Park, S. H. (2015). Measuring the environmental effects of organic farming: A meta-analysis of structural variables in empirical research. *Journal of Environmental Management, 162*, 263–274.

Leifeld, J. (2016). Current approaches neglect possible agricultural cutback under large-scale organic farming. A comment to Ponisio et al. *Proceedings of the Royal Society B: Biological Sciences, 283*(1824), 20151623.

Lessmann, M., Ros, G. H., Young, M. D., & de Vries, W. (2022). Global variation in soil carbon sequestration potential through improved cropland management. *Global Change Biology, 28*(3), 1162–1177. https://doi.org/10.1111/gcb.15954

Lori, M., Symnaczik, S., M¨ader, P., de Deyn, G., & Gattinger, A. (2017). Organic farming enhances soil microbial abundance and activity—A meta-analysis and meta-regression. *PLOS ONE, 12*(7), e0180442.

Lotter, D. W., Seidel, R., & Liebhardt, W. (2003). The performance of organic and conventional cropping systems in an extreme climate year. *American Journal of Alternative Agriculture, 18*(3), 146–154.

Lotze-Campe, H. (2011). Climate change, population growth, and crop production: an overview. In S. S. Yadav, R. J. Redden, J. L. Hatfield, H. Lotze-Campen, & A. E. Hall (Eds.), *Crop adaptation to climate change* (1st ed., pp. 1–11). John Wiley & Sons.

Lynch, D. H., MacRae, R., & Martin, R. C. (2011). The carbon and global warming potential impacts of organic farming: Does it have a significant role in an energy constrained world? *Sustainability, 3*(2), 322–362.

Mäder, P., Fliessbach, A., Dubois, D., Gunst, L., Fried, P., & Niggli, U. (2002). Soil fertility and biodiversity in organic farming. *Science, 296*(5573), 1694–1697.

Mäder, P., Fliessbach, A., Dubois, D., Gunst, L., Jossi, W., Widmer, F., Oberson, A., Frossard, E., Oehl, F., Wiemken, A., Gattinger, A., & Niggli, U. (2006). The DOK experiment (Switzerland). In J. Raupp, C. Pekrun, M. Oltmanns, & U. Köpke (Eds.), *ISOFAR scientific series* (pp. 41–58). University of Bonn.

Marriott, E. E., & Wander, M. (2006). Qualitative and quantitative differences in particulate organic matter fractions in organic and conventional farming systems. *Soil Biology and Biochemistry, 38*(7), 1527–1536.

McGee, J. A. (2015). Does certified organic farming reduce greenhouse gas emissions from agricultural production? *Agriculture and Human Values, 32*(2), 255–263.

Meier, M. S., Stoessel, F., Jungbluth, N., Juraske, R., Schader, C., & Stolze, M. (2015). Environmental impacts of organic and conventional agricultural products–Are the differences captured by life cycle assessment? *Journal of Environmental Management, 149*, 193–208.

Mondelaers, K., Aertsens, J., & Van Huylenbroeck, G. (2009). A meta-analysis of the differences in environmental impacts between organic and conventional farming. *British Food Journal, 111*, 1098–1119.

Muller, A., Olesen, J. E., Davis, J., Dytrtova, K., Gattinger, A., Lampkin, N., & Niggli, U. (2012). Reducing global warming and adapting to climate change: The potential of organic agriculture. Working paper, draft version, 2012, FiBL Schweiz/Suisse, info.suisse@fibl.org, www.fibl.org

Munda, S., Bhaduri, D., Mohanty, S., Chatterjee, D., Tripathi, R., Shahid, M., Kumar, U., Bhattacharyya, P., Kumar, A., Adak, T., & Jangde, H. K. (2018). Dynamics of soil organic carbon mineralization and C fractions in paddy soil on application of rice husk biochar. *Biomass and Bioenergy, 115*, 1–9.

Nayak, A. K., Chatterjee, D., Tripathi, R., Shahid, M., Vijayakumar, S., Satapathy, B. S., Kumar, A., Mohanty, S., Bhattacharyya, P., Mishra, P., Kumar, U., Mohapatra, S. D., Panda, B. B., Rajak, M., Bhaduri, D., Munda, S., Chakraborty, K., Priyadarsani, S., Swain, C. K., ... Pathak, H. (2020). Climate smart agricultural technologies for rice production system in Odisha. ICAR-National Rice Research Institute (p. 366).

Nemecek, T., Huguenin-Elie, O., Dubois, D., & Gaillard, G. (2005). Ökobilanzierung von Anbausystemen im schweizerischen Acker-und Futterbau. Zürich, Schriftenreihe der FAL: Agroscope FAL Reckenholz (p. 155).

Niggli, U., Schmid, H., Fliessbach, A. (2008). *Organic farming and climate change.* International Trade Centre (ITC).

Niggli, U. (2015). Sustainability of organic food production: Challenges and innovations. *The Proceedings of the Nutrition Society, 74*, 83–88.

Niggli, U., Fließbach, A., Hepperly, P., & Scialabba, N. (2009). Low greenhouse gas agriculture: Mitigation and adaptation potential of sustainable farming systems. *Ökologie & Landbau, 141*, 32–33.

Nilsson, L. G. (1986). Data of yield and soil analysis in the long-term soil fertility experiments. *Journal of Royal Swedish Academy, 18*, 32–70.

OECD (Organization for Economic Co-operating and Development Organic Agriculture). (2003). *Sustainability, markets and policies.* CABI.

Oertel, C., Matschullat, J., Zurba, K., Zimmermann, F., & Erasmi, S. (2016). Greenhouse gas emissions from soils—A review. *Geochemistry, 76*(3), 327–352.

Oquist, K. A., Strock, J. S., & Mulla, D. J. (2007). Influence of alternative and conventional farming practices on subsurface drainage and water quality. *Journal of Environmental Quality, 36*(4), 1194–1204.

Pardo, G., Moral, R., Aguilera, E., & Del Prado, A. (2015). Gaseous emissions from management of solid waste: A systematic review. *Global Change Biology, 21*(3), 1313–1327.

Pathak, H., Jain, N., Bhatia, A., Kumar, A., & Chatterjee, D. (2016). Improved nitrogen management: A key to climate change adaptation and mitigation. *Indian Journal of Fertilisers, 12*(11), 151–162.

Petersen, S. O., Regina, K., Pöllinger, A., Rigler, E., Valli, L., Yamulki, S., Esala, M., Fabbri, C., Syväsalo, E., & Vinther, F. P. (2006). Nitrous oxide emissions from organic and conventional crop rotations in five European countries. *Agriculture, Ecosystems & Environment, 112*(2–3), 200–206.

Pimentel, D., Harvey, C., Resosudarmo, P., Sinclair, K., Kurz, D., McNair, M., Crist, S., Shpritz, L., Fitton, L., Saffouri, R., & Blair, R. (1995). Environmental and economic costs of soil erosion and conservation benefits. *Science, 267*(5201), 1117–1123.

Pimentel, D., Hepperly, P., Hanson, J., Douds, D., & Seidel, R. (2005). Environmental, energetic, and economic comparisons of organic and conventional farming systems. *BioScience, 55*(7), 573–582.

Poeplau, C., & Don, A. (2015). Carbon sequestration in agricultural soils via cultivation of cover crops–A meta-analysis. *Agriculture, Ecosystems & Environment, 200*, 33–41.

Ponisio, L. C., M'Gonigle, L. K., Mace, K. C., Palomino, J., De Valpine, P., & Kremen, C. (2015). Diversification practices reduce organic to conventional yield gap. *Proceedings of the Royal Society B: Biological Sciences, 282*(1799), 20141396.

Pretty, J., & Ball, A. (2001). Agricultural influences on carbon emissions and sequestration: A review of evidence and the emerging trading options. *Centre for Environment and Society Occasional Paper, 3*, 31.

Reganold, J. P., & Wachter, J. M. (2016). Organic agriculture in the twenty-first century. *Nature Plants, 2*, 15221.

Rochette, P., Angers, D. A., Co^té, D. (2000). Soil carbon and nitrogen dynamics following application of pig slurry for the 19th consecutive year I. Carbon dioxide fluxes and microbial biomass carbon. *Soil Science Society of America Journal, 64*(4), 1389–1395.

Rochette, P., Angers, D. A., Chantigny, M. H., Bertrand, N., & Côté, D. (2004). Carbon dioxide and nitrous oxide emissions following fall and spring applications of pig slurry to an agricultural soil. *Soil Science Society of America Journal, 68*(4), 1410–1420.

Rockström, J., Steffen, W., Noone, K., Persson, Å., Chapin, F. S., Lambin, E. F., Lenton, T. M., Scheffer, M., Folke, C., Schellnhuber, H. J., & Nykvist, B. (2009). A safe operating space for humanity. *Nature, 461*(7263), 472–475.

Saha, S., Chatterjee, D., Swain, C. K., & Nayak, A. K. (2018). Methane emission from wetland rice agriculture-biogeochemistry and environmental controls in projected changing environment. *Advances in crop environment interaction* (pp. 51–85). Springer.

Saha, S., Das, B., Chatterjee, D., Sehgal, V. K., Chakraborty, D., & Pal, M. (2020). Crop growth responses towards elevated atmospheric CO_2. In M. Hasanuzzaman (Ed.), *Plant ecophysiology and adaptation under climate change: mechanisms and perspectives I* (pp. 147–198). Springer.

Schader, C., Stolze, M., Gattinger, A. (2011). Environmental performance of organic agriculture. In J. Boye, Y. Arcand (Eds.), *Green technologies in food production and processing*. Springer.

Seufert, V., Ramankutty, N., & Foley, J. A. (2012). Comparing the yields of organic and conventional agriculture. *Nature, 485*(7397), 229–232.

Severin, M., Fuß, R., Well, R., Garlipp, F., & Van den Weghe, H. (2015). Soil, slurry and application effects on greenhouse gas emissions. *Plant, Soil and Environment, 61*(8), 344–351.

Siegrist, S., Schaub, D., Pfiffner, L., & Mäder, P. (1998). Does organic agriculture reduce soil erodibility? The results of a long-term field study on loess in Switzerland. *Agriculture, Ecosystems & Environment, 69*(3), 253–264.

Singh, S. N., Verma, A., & Tyagi, L. (2003). Investigating options for attenuating methane emission from Indian rice fields. *Environment International, 29*(5), 547–553.

Skinner, C., Gattinger, A., Muller, A., Mäder, P., Flieβbach, A., Stolze, M., Ruser, R., & Niggli, U. (2014). Greenhouse gas fluxes from agricultural soils under organic and non-organic management—A global meta-analysis. *Science of the Total Environment, 468–469*, 553–563.

Smith, P., Martino, D., Cai, Z., Gwary, D., Janzen, H., Kumar, P., McCarl, B., Ogle, S., O'Mara, F., Rice, C., & Scholes, B. (2008). Greenhouse gas mitigation in agriculture. *Philosophical Transactions of the Royal Society B: Biol. Sci., 363*(1492):789–813

Smith, J., Wolfe, M., Woodward, L., Pearce, B., Lampkin, N., & Marshall, H. (2011) *Organic farming and biodiversity: a review of the literature*. Organic Center Wales, Aberystwyth.

Smith, P., Bustamante, M., Ahammad, H., Clark, H., Dong, H., Elsiddig, E. A., Haberl, H., Harper, R., House, J., Jafari, M., Masera, O., Mbow, C., Ravindranath, N. H., Rice, C. W., Robledo Abad, C., Romanovskaya, A., Sperling, F., & Tubiello, F. (2014). Agriculture, Forestry and Other Land Use (AFOLU). In O. Edenhofer, R. Pichs-Madruga, Y. Sokona, E. Farahani, S. Kadner, K. Seyboth, A. Adler, I. Baum, S. Brunner, P. Eickemeier, B. Kriemann, J. Savolainen, S. Schlömer, C. von Stechow, T. Zwickel, J. C. Minx (Eds.), Contribution of Working Group III to the Fifth Assessment Report of the Intergovernmental Panel on Climate Change. Climate Change 2014: Mitigation of Climate Change. Cambridge University Press.

Smith, K. A., & Conen, F. (2004). Impacts of land management on fluxes of trace greenhouse gases. *Soil Use and Management, 20*(2), 255–263.

Smith, L. G., Williams, A. G., & Pearce, B. D. (2015). The energy efficiency of organic agriculture: A review. *Renewable Agriculture and Food Systems, 30*(3), 280–301.

Smith, P. E., Powlson, D. S., Glendining, M. J., & Smith, J. U. (1998). Preliminary estimates of the potential for carbon mitigation in European soils through no-till farming. *Global Change Biology, 4*(6), 679–685.

Steinfeld, H., Gerber, P., Wassenaar, T. D., Castel, V., Rosales, M., Rosales, M., & de Haan, C. (2006). Livestock's long shadow: Environmental issues and options. Food & Agriculture Organization.

Soil Association. (2012). http://soilassociation.org. Accessed May 26, 2012.

Stolze, M., Piorr, A., Haring, A. M., & Dabbert, S. (2000). Environmental impacts of organic farming in Europe. In: Organic Farming in Europe: Economics and Policy, vol. 6, University of Hohenheim, Stuttgart, Germany. Accessed 15 May 2011. http://orgprints.org/8400/1/Organic_Farming_in_Europe_Volume06_The_Environmental_Impacts_of_Organic_Farming_in_Europe.pdf

Swain, C. K., Bhattacharyya, P., Nayak, A. K., Singh, N. R., Chatterjee, D., Dash, P. K., Neogi, S., Pathak, H. (2018a). Temporal variation of energy fluxes during dry season in tropical lowland rice. *MAPAN-Journal Metrology Society of India.* https://doi.org/10.1007/s12647-018-0260-x

Swain, C. K., Bhattacharyya, P., Nayak, A. K., Singh, N. R., Neogi, S., Chatterjee, D., & Pathak, H. (2018b). Dynamics of net ecosystem methane exchanges on temporal scale in tropical lowland rice. *Atmospheric Environment, 191*, 291–301.

Swain, C. K., Nayak, A. K., Bhattacharyya, P., Chatterjee, D., Chatterjee, S., Tripathi, R., Singh, N. R., & Dhal, B. (2018c). Greenhouse gas emissions and energy exchange in wet and dry season rice: Eddy covariance-based approach. *Environmental Monitoring and Assessment, 190*(7), 423.

Tal, A. (2018). Making conventional agriculture environmentally friendly: Moving beyond the glorification of organic agriculture and the demonization of conventional agriculture. *Sustainability, 10*(4), 1078.

Topp, C. F., Stockdale, E. A., Watson, C. A., & Rees, R. M. (2007). Estimating resource use efficiencies in organic agriculture: A review of budgeting approaches used. *Journal of the Science of Food and Agriculture, 87*(15), 2782–2790.

Treu, H., Nordborg, M., Cederberg, C., Heuer, T., Claupein, E., Hoffmann, H., & Berndes, G. (2017). Carbon footprints and land use of conventional and organic diets in Germany. *Journal of Cleaner Production, 161*, 127–142.

Tuck, S. L., Winqvist, C., Mota, F., Ahnstr ¨om, J., Turnbull, L. A., Bengtsson, J. (2014). Land-use intensity and the effects of organic farming on biodiversity: A hierarchical meta-analysis. *Journal of Applied Ecology, 51*, 746–755.

Tuomisto, H. L., Hodge, I. D., Riordan, P., & Macdonald, D. W. (2012). Does organic farming reduce environmental impacts?–A meta-analysis of European research. *Journal of Environmental Management, 112*, 309–320.

Van Dijk, H. (1982). Survey of Dutch soil organic research with regard to humification and degradation rates in arable land. In D. D. Boels, B. Davis, & A. E. Johnston (Eds.), *Land use seminar on land degradation* (pp. 133–143). Balkema.

Venkat, K. (2012). Comparison of twelve organic and conventional farming systems: A life cycle greenhouse gas emissions perspective. *Journal of Sustainable Agriculture, 36*(6), 620–649.

Venkateswarlu, B., & Shanker, A. K. (2009). Climate change and agriculture: Adaptation and mitigation stategies. *Indian Journal of Agronomy, 54*(2), 226–230.

Verdi, L., Mancini, M., Ljubojevic, M., Orlandini, S., & Dalla Marta, A. (2018). Greenhouse gas and ammonia emissions from soil: The effect of organic matter and fertilisation method. *Italian Journal of Agronomy, 13*(3), 260–266.

Walling, E., & Vaneeckhaute, C. (2020). Greenhouse gas emissions from inorganic and organic fertilizer production and use: A review of emission factors and their variability. *Journal of Environmental Management, 76*, 111211.

Willer, H., & Lernoud, J. (2019). The world of organic agriculture. Statistics and emerging trends 2017. Research Institute of Organic Agriculture FiBL and IFOAM-Organics International. https://www.organicworld.net/yearbook/yearbook-2019.html

Williams, A. (2006). Determining the environmental burdens and resource use in the production of agricultural and horticultural commodities: Defra project report IS0205. Zu finden in: http://randd.defra.gov.uk/Default.aspx

Zeddies, J. (2002). Vermeidungspotentiale der Landwirtschaft. Ziele und Handlungsoptionen. Tagungsband zur, 34.

[illegible]

Chapter 10
Impact of Conservation Agriculture on Soil Health and Environmental Sustainability

Surajit Mondal, Saurav Saha, Saikat Ranjan Das, and Dibyendu Chatterjee

Abstract The ever-rising global population poses significant challenges to global food security, particularly in developing countries. In order to feed the growing population, natural resources are severely depleted due to unsustainable agricultural practices leading to the widespread problem of land degradation and loss of natural habitat. Therefore, agricultural research should focus on developing eco-friendly and sustainable crop production systems with higher crop productivity to achieve sustainable development goals. Conventional agriculture encourages excessive tillage and crop residue removal. Excessive tillage and crop residue removal are deteriorating soil health due to loss of soil organic carbon (SOC) content, reduction of air-filled porosity, increased compaction, nutrient imbalance, and destruction of soil fauna, etc. Therefore, alternative methods of cultivation are being developed to maintain or improve soil quality. Conservation agriculture (CA) includes several resource conservation technologies facilitating (i) minimum soil disturbance, (ii) soil surface cover in the forms of crop residue retention or cover crop and (iii) crop rotation for improving soil health. It is now widely reported by several researchers that CA improves the status of SOC, particularly in the surface soil layer, as a possible means of soil carbon sequestration. Permanent residue retention or cultivation of cover crops also protects the soil from raindrop splashing impact and enhances soil aggregate stability, thereby reducing soil erosion. In addition, the soil at CA promotes microbial growth, maintains natural diversity, and encourages the earthworm population to better ecosystem services. Researchers have argued that CA is an effective means of C-sequestration and could be an effective mitigation strategy for climate change. A global initiative of '4 per mile' has also been initiated to achieve the goal. As a

S. Mondal (✉)
ICAR Research Complex for Eastern Region, Patna, Bihar 800 014, India
e-mail: surajit.iari@gmail.com

ICAR Central Citrus Research Institute, Nagpur, Maharashtra 440 033, India

S. Saha
ICAR Research Complex for North East Hill Region, Sikkim Centre, Gangtok 737 102, India

S. R. Das · D. Chatterjee
ICAR National Rice Research Institute, Cuttack, Odisha 753 006, India

H. Pathak et al. (eds.), *Climate Change Impacts on Soil-Plant-Atmosphere Continuum*, Advances in Global Change Research 78,
https://doi.org/10.1007/978-981-99-7935-6_10

courtesy, global policy initiatives should promote CA practices to conserve natural resources and better maintain soil health, as our sustainable future strategy for crop production.

Keywords Resource conservation technology · Residue management · Soil microbial diversity · Soil quality · Agricultural sustainability

10.1 Introduction

The world's population is expected to grow to 9.8 billion by 2050 and so is the growing demand for food to feed the growing population; food production needs to be increased significantly (54% after FAO, 2011). Cultivable lands are limited and unlikely to increase due to severe competition with other sectors. Increased pressure on cultivable land leads to widespread degradation in the form of soil health and hence about 80% of the increase in production must be achieved through sustainable intensification of existing agricultural practices (FAO, 2011). Our current research needs in agriculture and allied sectors should focus on the development of eco-friendly crop management systems for the development of sustainable crop production systems. The future production system should ensure the protection of already degraded resources while ensuring more efficient use of agricultural inputs, protection of soil and water resources, and reduction in fossil-fuel consumption and greenhouse gas emission. The documentation on the degradation of soil resources and functions has long been conducted by European Commission (2006). More recently, great importance has been given to UN Sustainable Development Goals (United Nations, 2015) for better soil functions and ecosystem services (Keesstra et al., 2016). Over-reliance on soil resources has severely impacted their abilities and could be a prospective threat to achieving food security in the long term (Blum, 2013).

Multiple ecological issues, including soil erosion, GHG emissions, and declining soil fertility, are brought on by intensive tillage along with a lack of proper soil conservation practices in traditional agriculture (Shahid et al., 2018). The CA has the potential to reduce greenhouse gas emissions, contribute to climate change mitigation, and improve production sustainability. CA acts as a sink of C, thus reducing emission of CO_2 and CH_4, and helping in reducing global warming (Jat et al., 2011). Even with only temporary adoption, growing a legume crop boosted C retention in surface soils under conservation tillage (Bhattacharyya et al., 2015).

10.2 Conventional Versus Conservation Agriculture

Conventional agriculture involves extensive cultivation in which the upper layer of the soil is completely disturbed with various heavy tilling implements to obtain a fine tilth for proper crop establishment. Crop residues are either burned in situ or removed from the field for other uses like animal feed and fuel. Conservation agriculture (CA) on the other hand incorporates three principles viz. minimum disturbance of soil by adopting no- or reduced tillage, maximum surface coverage either through crop residues or canopy cover of growing crops, and diversification of existing cropping systems with other crops (FAO, 2011; Hobbs et al., 2008). Minimum or no-tillage, the central component of CA, does not turn over the soil and produced 'minimal' or 'nil' disturbance of the soil. According to the Conservation Technology Information Center (CTIC), under conservation tillage, at least 30% of the soil surface is covered by residue after tilling or planting operations. As might be expected, there could be a large difference in the impact of conservation and conventional tillage on soil's physical properties. However, the change varies greatly with climate, soil, agricultural management, and duration of the adoption of a tillage system (Mondal et al., 2020).

The conventional practices i.e., puddling during rice cultivation and intensive tillage during the winter season are deteriorating soil health and adversely affecting crop productivity. Yield plateauing is emerging as a major challenge for maintaining food security shortly in Indo-Gangetic Plains. In this context, CA could be a viable option for better resource use efficiency, better soil health, lower greenhouse gas emissions, and higher productivity of the rice–wheat cropping system (Hobbs et al., 2008). No-till is currently estimated to cover ~157 M ha, globally (AQUASTAT FAO, 2018). It involves the complete elimination of primary tillage causing minimal soil disturbance and is often considered a potential alternative to traditional intensive cultivation practice, which involves repeated plowing of the soil to obtain a desirable tilth. The residue retained on the soil surface undergoes microbial decomposition and improves the aggregate stability, nutrient-cycling microbial populations, earthworm abundance, and soil organic carbon. Additionally, the surface retained residue protects the soil from raindrop impact during rainfall, reduces runoff water and rainwater gets more time to infiltrate into the soil. The improved physical condition of the soil maintains optimum air–water balance and is congenial for root penetration and proliferation causing more efficient use of applied water and nutrients. Due to its multiple benefits, CA is actively being promoted by many international research organizations like International Maize and Wheat Improvement Center), IRRI (International Rice Research Institute (CIMMYT), Commonwealth Scientific and Industrial Research Organisation (CSIRO), etc., in India and other countries, to contain land degradation, water pollution and achieve food security without causing much damage to the environment. The UN Framework Convention on Climate Change (UNFCC, 2016) has recognized the CA as a means of achieving sustainable development goals.

10.3 Conservation of Agriculture and Soil Health

Soil health can be defined as the ability of the soil to function as a living system and support all organisms that depend on it. It is also a concept that links agriculture in general and soil science in particular to policymakers and stakeholders for framing policies and sustaining supply-chain management. Earlier, soil health was used to be determined based on crop productivity. However, the soil is linked with many interlinked parameters like water quality, environmental emissions, human health, groundwater pollution, soil biodiversity, etc., and all those aspects should be assessed to effectively determine soil health. Chemical indicators are still mostly used for soil health measurement and the use of soil biodiversity parameters is still insignificant due to limited functional knowledge and a lack of effective methods. Several factors that influence soil health under CA system are hereunder:

10.3.1 Soil Bulk Density

Bulk density (BD) which gives an idea of soil compactness, plays a dominant role in the air–water relationship of soil and thereby influences the growth and productivity of crops. Hence, BD is often taken as an important parameter during soil quality assessment. Many soil parameters like penetration resistance, soil moisture content, total porosity, air permeability, and hydraulic conductivity are directly or indirectly affected by soil BD (Doran et al., 1996).

Different agricultural management practices affect the soil BD differently (Fig. 10.1). Conventional agriculture, which encourages repetitive soil disturbance has a different effect on bulk density than conservation agriculture which promotes minimum or no-tillage. The bulk density of agricultural fields changes significantly over the crop growing season due to different agricultural management practices, rainfall, and irrigation. Immediately after tillage, the surface soil remains loose and usually has the lowest BD. But as the season progress, the soil BD increases as a result of compaction due to traffic during intercultural operations. Rainfall or irrigation also settles the aggregates in the surface layer and tends to increase the soil BD (Osunbitan et al., 2005). The surface layer exhibits maximum fluctuations from seeding to harvesting, which can be as high as 40% (Liu et al., 2014; Logsdon, 2012). Different soil physical changes like swelling-shrinkage, freezing–thawing, clay migration, and erosion can have a significant influence on soil BD (Hamza & Anderson, 2005; Logsdon, 2012).

Regarding the impact of tillage on BD, there are two opposing viewpoints. According to one set of authors, BD is higher in the NT (No-till) system than in the CT (Conventional till), however another group of authors concluded that BD is decreased by NT practice. However, the effect of BD could vary greatly based on the time of measurement and years of NT adoption before measurement. Most of the authors have agreed that the absence of tillage under NT, densifies the surface soil during the initial years of NT adoption. The omission of tillage prevents the soil from being loosen periodically, and the movement of machinery and human traffic for different intercultural operations further compacts the surface soil. However, retained crop residues protect the surface soil from external forces and make it friable. Hence, NT adoption for a sufficiently long period can maintain or even decrease the soil bulk density in the surface layer. In contrast, CT reduces aggregate stability, augments oxidation of SOC, and helps the formation of a compact layer just below the plough

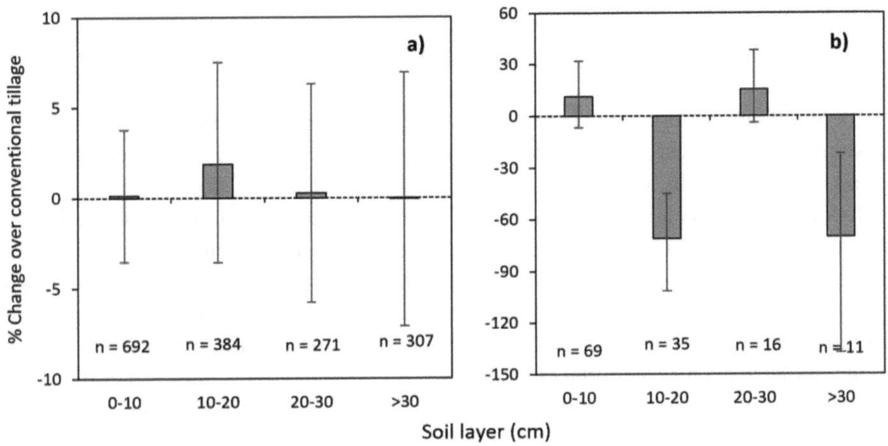

Fig. 10.1 Effect of tillage on **a** soil bulk density and **b** saturated hydraulic conductivity in different soil layers (Mondal et al., 2022a)

layer particularly due to puddling. Hardpan formation just below the surface layer has been reported by many authors (Aggarwal et al., 2006; Ahmad et al., 2018; de Moraes et al., 2016). The timing of the measurement can alter the measured BD significantly. The BD values are supposed to be lower if measured immediately after tillage than at the end of the season (Alletto & Coquet, 2009).

10.3.2 Penetration Resistance

Penetration resistance determines the soil compaction for root development. The cone penetrometer, usually used for penetration measurement, imitates the growing roots and registers hindrances offered by the soil to roots during growth and proliferation. Generally, the soil compaction or mechanical impedance happens below the soil surface and therefore, is very difficult to detect. It is considered a major problem and can affect crop productivity significantly. It is estimated that about 4% of the global land area is affected by soil compaction (Soane & van Ouwerkerk, 1994). Research has suggested the value of 2 MPa penetration resistance as the threshold for limiting root growth and causing economic yield reduction unless cracks, decayed root channels, bio-pores, and crevices are present in the soil profile (Bengough et al., 2011). The presence of compacted layers restricts root penetration to the deeper soil layers and impairs air and water balance limiting the water and nutrient use efficiency (McGarry & Sharp, 2003). The presence of a small amount of SOC can exacerbate the compaction process. Soil compaction generally happens due to mismanagement of tilling equipment or improper methods of tilling (Whitmore et al., 2011). It remains hidden below the soil surface and therefore, is long-lasting or even permanent if proper corrective measures are not taken. Soil type, soil moisture condition during tilling or intercultural operations, aggregate stability, and bulk density are the controlling factor for determining the extent of soil compaction (Horn et al., 2005; Materechera, 2009).

It has been widely reported that the puddling of rice in rice–wheat cropping systems in Southeast Asia favours the destruction of soil structure and physical breakdown of stable aggregates leading to the formation of compacted soil layer below the plough zone (Aggarwal et al., 2006; Singh et al., 2014). However, despite the numerous benefits of conservation tillage, there is no agreement that it can alleviate soil compaction. CA that avoids puddling during rice cultivation and intensive tillage during winter crops can be a viable solution for reducing soil compaction. Better soil aggregation under CA can prevent yield loss due to compaction (Drewry et al., 2008). Recently, Mondal et al. (2019a) reported a reduction in soil compaction and higher root proliferation in the subsurface soil layers under CA in comparison to CT.

10.3.3 Soil Aggregation

Soil aggregation is an important soil physical property and is the most extensively used soil health indicator. Aggregates are a combination of primary particles (sand, silt, and clay), soil organic matter, and cementing agents, and are formed through a series of processes of flocculation and cementation (Bronick & Lal, 2005). Water-stable aggregates are very much important for the protection of soil against water and wind erosion. In a well-structured soil, the soil organic carbon remains encapsulated within aggregates and is less prone to microbial oxidation.

Different agricultural management practices viz. tillage, residue incorporation, etc., have a direct bearing on soil aggregation (Madari et al., 2005). Conventional tillage greatly alters soil aggregation in two major ways. Firstly, it physically breaks down the stable aggregates and secondly, it promotes the microbial decomposition of soil organic carbon and thereby indirectly affecting the aggregation process (Bronick & Lal, 2005; Hamza & Anderson, 2005). Macro-aggregates are reported to have a lesser turnover time under CT (Six et al., 2000b). The dispersed aggregates can easily be transported by moving water or wind causing soil erosion and loss of SOC and plant-available nutrients. Soil disturbance also accelerates the oxidation of organic binding materials (Six et al., 2000a).

On the other hand, CA that favours minimum or no tillage promotes stable aggregate formation (Kumari et al., 2011; Mondal et al., 2019b) by increasing the cementing agents in the forms of different organic compounds. Residues retained on the soil surface or cover crop further protect the aggregates from the splashing impact of raindrops or running water. Organic substrate availability under CA assists microbial activity and thus macro-aggregates formation (Six et al., 2004; Zuber & Villamil, 2016).

The beneficial role of earthworms on macro-aggregate formation has also been reported by many researchers (Arai et al., 2017; Kawaguchi et al., 2011). Earthworms feed on soil organic matter along with soil particles. Therefore, earthworms cast are rich in organic matter and have higher stability. Hence, earthworm activity is taken as a health indicator and indicates better soil health. Conventional tillage is detrimental to the earthworm population by physically killing the earthworms during tilling operations or reducing the food (Briones & Schmidt, 2017).

10.3.4 Soil Hydraulic Parameters: Hydraulic Conductivity and Infiltration

Soil hydraulic parameters like saturated and unsaturated hydraulic conductivity, infiltration, and sorptivity are highly variable both spatially and temporally. Unsaturated hydraulic conductivity can change many folds with a slight change in soil water content (Strudley et al., 2008). Soil hydraulic properties do not only depend on total pore space but on pore geometry and continuity (Kutilek, 2004). The presence of a higher amount of pore space does not necessarily increase the soil infiltration rate if pores are not continuous. The macro- and micropores play a major role during saturated and unsaturated hydraulic conductivity, respectively (Richard et al., 2001).

Tillage greatly alters the soil's physical condition as well as pore spaces and thereby affects the soil water movement significantly (Fig. 10.1). Commonly, the tillage makes the soil more open, and usually, soils have higher infiltration or hydraulic conductivity immediately after tillage. However, soil particles clog the pores after irrigation or rainfall (Martínez et al., 2008) and gradually decrease the rate of water movement (Bormann & Klaassen, 2008; Schwartz et al., 2003).

Adoption of CA can promote soil water infiltration due to the presence of surface residue and more continuous soil pores. The surface retained crop residues absorbs the kinetic energy of moving water (Das et al., 2017a, 2017b; Nyamadzawo et al., 2007) and therefore, allow less destruction of stable soil aggregates. Hence, during infiltration, soils under NT remain more stable and pores are less prone to clogging causing higher infiltration (Das et al., 2018a, 2018b). In contrast, under CT, slaked soil particles clogged the soil pores and cause surface sealing (Kahlon et al., 2013). Better pore connectivity, vertical cracks (Alvarez & Steinbach, 2009; Huang et al., 2015), presence of decayed root channels and earthworm burrows (Busari et al., 2015; Castellanos-Navarrete et al., 2012) can radically alter the soil hydraulic characteristics of the soil. Greater continuity of pores, mostly the macropores and mesopores largely improved soil infiltration even though total porosity was less. The presence of macro-pores can cause preferential water flow under ponded conditions (Beven & Germann, 2013). Soil with a higher bulk density can have a higher infiltration due to the presence of more number earthworm burrows and decayed root channels (Osunbitan et al., 2005).

10.3.5 Soil Porosity

Water flows via connected pores and it is associated with the concept of structural hierarchy (Dexter et al., 2008). Tillage has a strong impact on soil porosity (Lipiec et al., 2006; Mondal et al., 2022a) (Fig. 10.2). The physical breakdown of soil aggregates caused by tillage in CT leads to the destruction of pore geometry and continuity, and gradually soil pores are produced through the rearrangement or settlement of soil particles after irrigation or rainfall. In contrast, biological activities (both floral and faunal) are responsible processes for the generation of soil pores under no-tilled conditions. The predominance of decayed root channels, earthworm or other faunal burrows, bio-pores, and, cracks favours water flows deeper into the soil profile and sometimes causes groundwater pollution by nitrates and pesticides (Gerke, 2006; Jarvis, 2007). The stability of macro-pores generated through CT is less and these are usually effective for a short period of time after tillage (Dexter, 2004). On the other hand, pore network under NT is more stable and less prone to deformation facilitating better water and air movement even in compacted soil (Schäffer et al., 2008; Wahl et al., 2004).

In general, the CT system caused reduced bulk density and increased porosity, particularly in the topsoil layer, and undergoes a wide variation over the season, while pore spaces in NT mostly remain unchanged over the crop growing period. In a global meta-analysis, Mondal et al. (2022a) observed a 3% reduction in total porosity, a large reduction of 20–32% in total porosity, and a 4–7% increase in microporosity. Air and water permeability can be used to determine the ecological functions that soil will carry out in an agroecosystem. Due to various management techniques, air permeability is more sensitive and can serve as an indicator of change in the pore system (Schjønning et al., 2013).

10.3.6 Soil Water Retention

Soil water serves as an important component for the success of agriculture. Conventional agriculture which practices fine tilth formation opens up the soil that enhances soil water evaporation leading to poor water productivity, thus influencing yield adversely. (Hou et al., 2012). Removal of crop residues from the soil surface or incorporation of residues makes the soil bare and aggravates the evaporation process causing lower efficiency of water and nutrients (Kumar et al., 2013).

Conservation tillage that has at least 30% surface coverage with crop residues or cover crops after tillage is widely accepted as an efficient management practice for soil and water conservation (Corsi et al., 2012). Crops under NT are less prone to water stress and therefore, NT is more successful in maintaining crop productivity under water scarcity regions (arid and semi-arid) (Ngigi et al., 2006). A better soil moisture condition under NT also reduces the penetration resistance of soil than in CT (Mondal et al., 2018). The surface roughness created by the crop residue offers

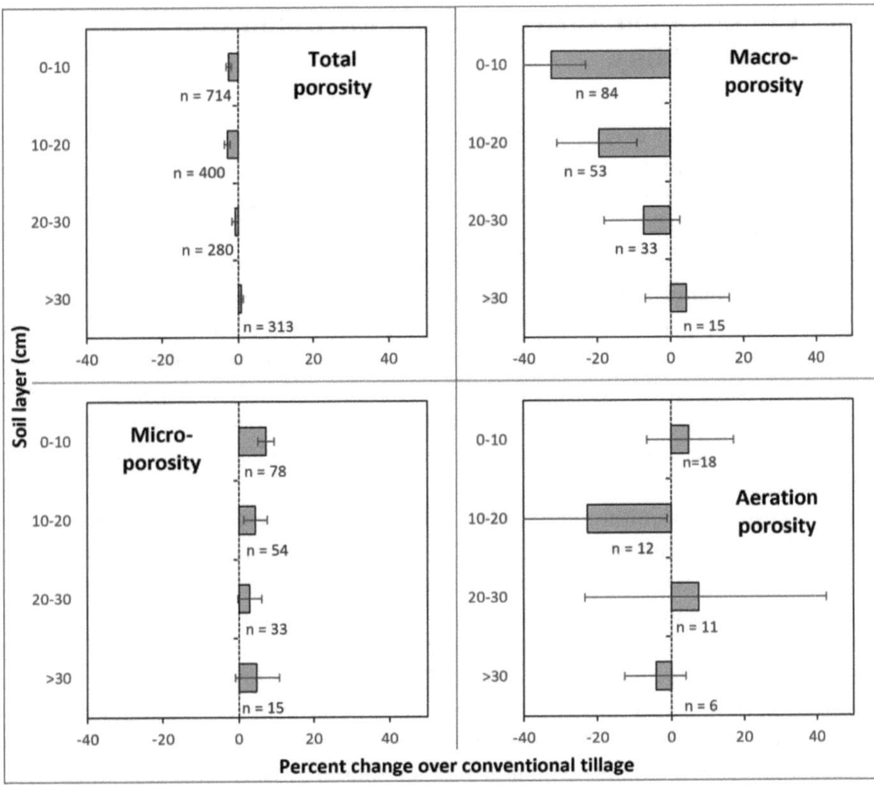

Fig. 10.2 Effect of tillage on soil porosity **a** total porosity, **b** microporosity, **c** microporosity, and **d** aeration porosity (Mondal et al., 2022a)

the water an extended time to infiltrate the soil and usually causes better soil water recharge under NT (Van Wie et al., 2013).

Crop residues on the soil surface intercept the radiation and reduce soil evaporation (Van Wie et al., 2013), and moderate temperature fluctuations (Alletto et al., 2011; Mondal et al., 2018). In contrast, higher evapotranspiration in NT has been documented by Su et al. (2007) due to wetter soil surface for a longer time allowing more water movement from deeper layers to meet the demand of the evapotranspiration. The presence of surface residue also reduces the radiation available for heating causing slow warming of soil under NT than CT (Alletto et al., 2011). A higher number of storage pores has been reported by Mondal et al. (2013) under no-tillage which subsequently increased the plant's available water in comparison to CT (Kargas et al., 2012). Due to an improvement in the soil hydro-thermal environment, conservation tillage is considered a better alternative to CT (Silburn et al., 2007).

10.4 Conservation Agriculture and Environmental Sustainability

The increased global average temperature, more occurrences of extreme weather events, glacier melting, etc. as evitable in climate change imposes a threat to environmental sustainability. The success of agriculture greatly depends on the dominant climate and hence, the development of agriculture must be in harmony with the environment. On average about 30% of the total greenhouse gas emissions are contributed by agriculture mainly due to the use of chemical fertilizers (particularly nitrogenous fertilizer), pesticides, fossil fuels for machinery, and animal wastes. Recently, researchers have emphasized the agricultural soil as a sink of carbon for capturing atmospheric CO_2 in the form of soil organic carbon for a longer period and gave the concept of '4 per 1000' (Minasny et al., 2017). The concept enumerates that a 0.4% carbon sequestration rate can effectively offset 20–35% of total anthropogenic greenhouse gas emissions.

Conservation agriculture could play a pivotal role in restricting the agricultural contribution to greenhouse gas emissions and can even be a mitigation option. Minimum soil disturbance and residue retention favours soil carbon build-up and augment carbon sequestration. Moreover, the improvement in soil organic carbon status maintains better soil health which in turn supports sustainable food production with minimal effects on the environment. The presence of organic matter causes the slow release of fertilizer nutrients and prevents the eutrophication of water bodies. A higher SOC content under CA can reduce the requirement for N-fertilizer (Mondal & Chakraborty, 2022a, 2022b) which is a major source of N_2O emission. In the rice-maize system, conservation tillage with residue retention reduces energy usage by 56%, carbon footprints by 39%, and reduced N2O emission by 20%. The population of bacteria, fungi, and actinomycetes was raised by 21.3, 51.2 and 27.6% in this system. The crop productivity, microbial biota, and enzymatic activities of soil were all greatly increased by crop residue retention as residue mulching with rice straw (Lal et al., 2019). However, there is a yield penalty of 10–15% during initial years. Incorporation of locally available legume or oilseed crops after rice in a rice-fallow system in wet season build-up of soil organic carbon, nutrients, and microbial population (Gautam et al., 2020).

Rice, a major staple food crop for Asian countries, is a tillage and water-intensive crop. Rice paddy is also reported to emit a large number of greenhouse gases (Chatterjee et al., 2018, 2021; Nayak et al., 2017; Saha et al., 2018; Swain et al., 2018). Natural lowlands which are submerged for most part of the year (marshy land) and the soils were very soft, can be utilized to practice zero-tillage or non-puddled rice transplanting (Munda et al., 2021). This causes minimal soil disturbance, reduces labour and water requirements, as well as the cost of rice establishment while retaining some of the benefits of transplanting and soil floodings, such as weed control and increased nutrient availability. Furthermore, non-puddled rice transplanting reduces greenhouse gas emissions from field, and retention of crop residues with the non-puddled rice transplanting sometimes works as a profitable alternative to existing conventional tillage operations.

10.4.1 Preserving Soil Organic Carbon

Increased concentration of atmospheric greenhouse gases and the concurrent climate change have led to a paramount interest in the sequestration of atmospheric C as soil C. SOC, generated from microbial decomposition of various organic inputs, is the chief constituent of soil organic matter (SOM). The rate of formation of SOM and its turnover depends on many factors including the rate of organic matter added to the soil, different soil physicochemical properties, and the prevalent environment in terms of temperature and humidity (Taylor et al., 2009). Better agriculture management can maintain or even augment the SOC content of soil along with other benefits such as improved physical health, superior microbial balance, improved soil fertility, and higher soil water retention (Blanco-Canqui & Lal, 2010).

Tillage operations which are primarily practiced for seedbed preparation, weed control, and residue incorporation, could be instrumental in reducing SOC levels and altering the soil's physical health (Baker et al., 2007; Victoria et al., 2012). C in the form of crop residue gets incorporated into the soil by various tillage operations, as opposed to the regular movement of soluble, particulate, or colloidal carbon. However, repeated conventional tillage breaks down the stable soil aggregates and exposes the either protected SOC to the environment which is then rapidly decomposed by aerobic microorganisms (Al-Kaisi & Yin, 2005). Removal of crop residues further accelerates SOC loss. It has been estimated that CT practices alone can remove as high as 75% of the SOC stock from the native land (Lal et al., 2007).

No-tillage can improve the SOC level, which is a mitigation option for atmospheric CO_2. The process of storing atmospheric CO_2 in soil C is called carbon sequestration, and the source-sink relationship of C is affected by different agricultural management practices (Lin et al., 2002). Turnover time of SOC can have a dominant role in C-sequestration and is determined by soil mineralogy and climatic conditions (Chatterjee et al., 2013, 2014a, 2014b, 2015a, 2015b, 2016). Root biomass and root exudates of plants increase the SOC content of fertilized soils compared to unfertilized soils (Chaudhary et al., 2012). Residue mulch or living mulch in the form of cover crop under NT improves the SOC level and thereby accelerates the C-sequestration process. Minimal soil disturbance in NT favours macro-aggregate formation and these macro-aggregates protect the intra-aggregate SOC making it less accessible to soil microbes (Six et al., 2000a). Crop residues remaining on the soil surface decompose slowly and increase the SOC level (Feng et al., 2010; Johansen et al., 2012). Literature advocates the beneficial role of crop residue extensively (Chaplot et al., 2015; Guo et al., 2015). The beneficial role of conservation agriculture (Brouder and Gomez, 2014) is often explained by the increased biological activities and/or formation of physical structures like macro-aggregates by earthworms, indicating unique SOC dynamics (Higashi et al., 2014; Villami and Nafziger, 2015). In rice-green gram cropping system of Eastern India, CA practices especially rice sown zero till-seed drills, harvested by leaving 30 cm stubble in the field and then green gram is sown by using zero-till seed drill, helps to increase SOC stabilization by 17% over conventional practice (Shahid et al., 2021). Similarly, CA practices in rice-maize help C build up over the conventional rice-green gram system (Tripathi et al., 2021).

10.4.2 Reducing Emission of Greenhouse Gas and Environmental Pollution

The beneficial role of conservation agriculture in preserving the environment has widely been accepted. Crops grown under CA requires lesser fuel due to the omission of tillage and it reduces the emission of greenhouse gases. The sequestration of carbon in the soil under CA captures the atmospheric C and acts as a mitigation strategy for climate change. The presence of surface residue or cover crops also helps to reduce soil erosion and thereby preserves soil nutrients by reducing soil erosion and runoff loss of water. Better nutrient use efficiency under CA could indirectly reduce nutrient loss and hence prevents eutrophication of water bodies and pollution of groundwater.

The main source of nitrogen for crop production is fertilizer and without that global food production will be sufficient enough to feed only half of the population (Connor, 2008). According to an estimate, about 67% of the fertilizer N is eventually transformed into the non-reactive atmospheric N_2, and the rest 33% form reactive species and have transient effects on the global ecosystem (Ladha et al., 2016). The reactive forms of N i.e., ammonia, nitrate, and nitrogen dioxide have a direct bearing on water pollution, radiative forcing, and biodiversity loss (Reay et al., 2012).

The nitrogen use efficiency, the fraction of applied N harvested as a product, has been proposed as an index for assessing the progress of achieving sustainable environmental goals. Increasing NUE in agriculture is crucial for achieving triple challenges of food security, ecosystem degradation, and climate change (Zhang et al., 2015). The average global NUE needs to be increased from ~0.4 to ~0.7 to achieve food security and environmental stewardship in 2050 (Zhang et al., 2015). Therefore, efficient N management is of paramount importance to alleviate malnutrition and improve environmental protection. Among many, no-tillage can offset the use of N-fertilizer to some extent by improving N use efficiency (Lassaletta et al., 2014; Mondal et al., 2022b) (Fig. 10.3).

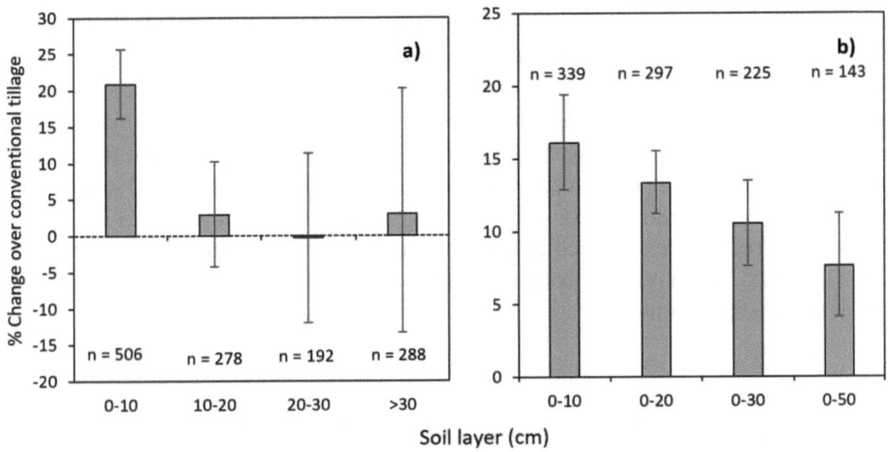

Fig. 10.3 N concentration and stock as affected by tillage practices (Mondal et al., 2022b)

10.4.3 Preserve Biodiversity

Conservation agriculture is considered to be eco-friendly and augments the diversity of soil micro- as well as macro-fauna and flora. The minimal disturbance of soil under CA preserves the habitat of soil micro- and macro-organisms whereas conservation tillage destroys the habitat or even physically harms the organisms like earthworms. Literatures are replete on the effect of tillage on an earthworm. Crop residues left on the soil surface or cover crops are grown under CA change the state of the soil's organic matter and encourage soil microbial activity and nitrogen mineralization (Samal et al., 2017). For improved soil resilience and interactions between the rhizosphere and plants, microbial diversity in the soil must be maintained. Diverse organic materials are produced by different crop rotations, maintaining a greater diversity of microbial communities in the soil (Lal et al., 2020). It is generally known that soil macrofauna may have positive effects on the physical properties of soil (Zida et al., 2011).

Because of their significant influence on the soil environment, earthworms are referred to as "ecosystem engineers" (Jones & Eggleton, 2000). Burrowing and casting by earthworms can change the structure of the soil, and both of these activities have a big impact on the soil's aeration, infiltration, and hydrology (Blanchart et al., 2004). Earthworm manipulates soil particles, and use saliva to bind them together, the soil's properties and structure change as a consequence. For the long-term health of the soil, it is preferable to provide conditions that encourage the settlement of soil macrofauna such as earthworms, arthropods, etc. The population of soil fauna is influenced by factors such as soil type, climate, and geographic location. Extreme soil conditions, such as high or low soil temperatures, low soil moisture, poor drainage, coarse or fine soil texture, low SOM content, and low pH, can be harmful to earthworms. Management techniques including manure and straw input, tillage, and cropping type have an impact on their activity as well. There are very few studies on the effects of short- and long-term crop management strategies on soil macrofauna diversity and the status of the pest-predator complex. As a result, defining and further developing sustainable conservation soil management will benefit from determining how short- and long-term management approaches affect communities of soil macrofauna. The majority of research on soil macrofauna to date has concentrated on describing and quantifying the impacts of soil invertebrates on soil processes such as the formation of soil structure (Pulleman et al., 2005), related soil physical processes, SOM decomposition (Brauman, 2000), and nutrient transformation.

10.5 Impact of Conservation Tillage on Plant Root Parameters

The roots of a plant are considered a vital component of the terrestrial ecosystem with a major role in uptaking water and nutrients from soil and supplying them to the plant (Spedding et al., 2004). In addition, shoot development and grain yield are also highly reliant on the below-ground root growth (Jeschke et al., 1997) and the water and nutrient uptake capacity of the plant is completely governed by the underground root distribution and development (Fageria, 2005). Generally, root growth characteristics are determined by the combined effects of prevalent soil characteristics and the genotype of the plant (Lynch, 2011).

The ability of a plant to uptake water along with essential nutrients depends on root growth and proliferation within the soil profile. Tillage modifies both the biotic and abiotic processes of soil by modifying soil cracks, breaking down stable aggregates, incorporating surface residue, and disrupting pore continuity (Huwe, 2003). All those changes have a great impact on the root growth of the plant and ultimately affect the overall crop growth and yield. Mosaddeghi et al. (2009) reported that the most important influence of tillage on root function and development could be due to modification in a root growth environment.

Root growth of surface, as well as subsurface soil layers, can seriously be affected by soil compaction. Lack of root proliferation restricts the utilization of water and nutrients from the deeper soil layer and thereby increases the probability of nutrient leaching particularly, nitrate in the groundwater (Thorup-Kristensen, 2011). The formation of a subsurface compact layer has widely been reported by many authors under conventional tillage (Aggarwal et al., 2006; Ahmad et al., 2018; Mondal et al., 2019a). Root proliferation is maximum in the surface soil layer due to the higher availability of water and nutrients and the presence of a compact layer impedes root penetration, proliferation, and overall root development (Batey, 2009; Guaman et al., 2016; Tracy et al., 2013).

10.6 Crop Yield Under Conservation Tillage

Food security is a high priority and an urgent issue to feed the burgeoning global population (Chen et al., 2011). According to FAO estimates, the world population will be about 9.1 billion by 2050 (FAO, 2011). In addition, climate change further threatens the challenge of food and nutritional security (Wheeler & von Braun, 2013). Increasing productivity is the only way to achieve food security, and must be done through superior production systems that use inputs more efficiently, conserve soil resources, are less dependent on fossil fuels, and ensures lower greenhouse gas emissions. CA can be a viable option to augment or maintain crop yields sustainably (Bhatt et al., 2016). Moreover, it has a higher climate adaptation capacity than conventional agricultural practices. Rising input costs, water and labour shortages, climate change, and ecosystem degradation necessitate the adoption of CA (Kassam et al., 2009). However, the yield benefit of NT is a highly debatable issue. Previous studies reported conflicting results on the relative benefits of NT. Higher crop yields have been extensively reported with NT than with CT (Bottinelli et al., 2017; Mitchell et al., 2017). Lower yields at NT were reported by many researchers (Botta et al., 2010; Tolon-Becerra et al., 2011). Crop yields were similar in some reports (Alvarez & Steinbach, 2009). Better physical health, higher SOC content, and better nutrient availability under NT led to higher yield and water use efficiency than in CT (Deubel et al., 2011; He et al., 2009). Alvarez and Steinbach (2009) and Zhang et al. (2018) found no significant yield advantages for wheat and maize in NT.

Weather conditions could be a major factor in the success of CA. A benefit of NT can better be understood in dry climates due to higher soil water retention (Triplett & Dick, 2008) whereas, in well-drained conditions, it has very little or even negative effects (Martens, 2001). Long-term tillage experiments also showed that crop productivity in NT was higher only in dry years than in wet years (Bogunovic et al., 2018; Kuhn et al., 2016). The CA may be the best option in areas that have higher seasonal rainfall variability due to its resilience (Sithole et al., 2016).

Pittelkow et al. (2015) reported a reduction in average yield under no-tillage than conventional tillage. However, crop rotation and residue retention in NT caused a significant increase in crop yield in dry climates. Overall, a 5.7% lower crop yield was observed under NT than CT, although under certain situations NT produced the same or even significantly higher yield than CT. Adoption of NT without residue retention and crop rotation always resulted in lower crop yield. Toliver et al. (2012) also reported a similar result and concluded that the average yield was lower with NT.

In a study in Odisha, conservation agriculture was reported to increase the system productivity by 55–60% in a rice-maize system and the benefit–cost ratio by 31.4% over a conventional rice-green gram (farmer's practice) (Tripathi et al., 2021). Similarly, conservation agriculture in the rice-green gram system increased yield by 2.5–5% and income by 22–25% over farmer's practices (conventional rice followed by green gram under rainfed conditions) (Shahid et al., 2021).

10.7 Conservation Agriculture in Hill Agroecosystem

The CA practiced in the unique and complex topography of the hills is quite different from the CA practiced in plains towards the development of more location-specific CA packages. Conservation measures in the hills primarily include land configuration/shaping like raised and sunken beds, raised beds (*bun*), crop planning/selection, conservation contour terracing, agroforestry, and rainwater harvesting toward the development of a micro watershed-based integrated farming system (IFS) and permanent pasture grasses on sloping lands, and intensive IFS in valleys (Das et al., 2014, 2017a, 2017b). Hilly areas are often classified in land use capability classes IV to VIII and continuous crop cultivation in these high hills through maintaining ecological sustainability is questionable in long run (Das et al., 2018a, 2018b). In high-rainfall regions of northeast India, the absence of winter rainfall often optimizes the need for *in-situ* or *ex-situ* moisture conservation measures to grow the second crop in rice or maize fallow systems (Yadav et al., 2019a). Adaptation of conservation tillage methods viz. minimum tillage/no-till and residue incorporation prove to have good potential to reduce carbon emissions and enhance soil structural stability, thus reducing erosion risk on steep slopes of eastern Himalaya (Hazarika et al., 2009). Locally adapted traditional resource conservation techniques (RCTs) viz. use of locally available plant biomass from weeds, green leaf manuring, and zero-till/establishment of successive crops minimum tillage helps to conserve soil moisture and facilitate timely sequential crop establishment in Shifting cultivation practices; thus provide higher economics benefit and enhanced resource use efficiency (Ghosh et al., 2010). Low soil moisture reserves after rice harvesting followed by the rapid decline of soil moisture with the advancement of low rainfall winter months often resulted in soil moisture deficit stress at flowering and pod-filling stage resulting in reduced crop productivity and loss of net farming profitability (Layek et al., 2014). No-till (NT) planting of a pea, lentil and leafy mustard in rice fallows enhanced resource productivity, enhance soil moisture content, improve soil aggregation, SOC stock, and soil nutrient availability (Das et al., 2019, 2021). Reduced tillage was reported to sequester 1.30 t C ha^{-1} in the rice-rice system with a 50% rise in net crop yield (Yadav et al., 2019b). Other agronomic soil and water conservation measures (2–8% slopes) such as contour farming, conservation tillage, modification of crop geometry, residue mulching, vegetative barriers, legume intercropping or strip cropping reduced 8–40% runoff; with consequent soil loss reduction by 6–35% along with rising crop productivity (~3–28%; Ghosh et al., 2012; Sharma et al., 2017).

Plantation of fast-growing multipurpose trees (MPTs) along with cultivation of site-specific pulse intercropping (soybean, rice bean, cowpea), cash crops (ginger, turmeric, large cardamom) are also economically viable, ecologically sustainable, and socially acceptable to the residents of the northeastern hills as an essential component of the CA system in hills (NITI Aayog, 2018). Moreover, hedgerow cropping of *Gliricidia sepium, Indigofera heymann, Tephrosia purpurea*, etc. across steep hill slopes reduces runoff, conserves soil, and facilitates soil fertility build-up (Layek et al., 2017). Strip intercropping of legumes such as soybean and green gram/black gram in rice and maize-based crop cultivation systems reportedly improves SOC and soil biological quality with increased availability of soil nutrients in *jhum* cultivation systems (Lungmuana et al., 2018). Adapting of conservation tillage integrated with green manuring (*Crotalaria tetragona*)/weed biomass (*Ambrosia artemisiifolia*) increased resource use efficiency by 50%, enhanced water holding capacity (WHC) and infiltration rate through SOC build-up and improvement of associated soil properties (Das et al., 2020). In hilly agroecosystems, grass-based vegetative barriers/filters/grass-hedge have proven to be low-cost and effective soil conservation measures in hill agroecosystems (Ghosh et al., 2015).

10.8 Constraints of Conservation Agriculture

Despite many benefits of conservation agriculture in terms of soil health, resilience, and climate, the rate of adoption has been slow. The main barriers of CA can be listed as follows:

- Lack of awareness among the farming community could be an important factor for its slower adoption. Widespread awareness programmes need to be organized through field day demonstrations, on-field trials, etc. to address that problem.
- Most of the meta-analyses indicate that crop yields are lower under CA as compared to conventional practices (Pittelkow et al., 2015). Therefore, current research should focus on closing the yield gap between CA and conventional systems.
- The shortage of farm machinery is also a bottleneck for the widespread adoption of CA. The development of custom-hiring centres could be a possible option to address this problem.
- The smaller size of landholdings is also a problem for mechanization under CA. Community farming could be adopted.
- Lack of advice to the farmer during the cropping system could be another reason for the lower success of CA and the resulting lower adoption rate. Weeds are a threat to the CA system and a lack of timely intervention leads to significant yield losses (Chatterjee, 2016).

10.9 Conclusion

Conservation agriculture affects soil health by improving the physical, chemical, and microbial properties of the soil. The greatest impact on soil health is probably the build up of carbon, which is mostly aggravated by retaining crop residues/cover crops and the practice of reduced tillage. The three pillars of conservation agriculture have their importance in improving soil health: crop rotations improve soil fertility (nitrogen) and microbial activity; cover crops/residues help build up carbon and conserve soil moisture, and reduced tillage improves soil structure and water-holding capacity. Even after improving soil properties, large-scale adoption of conservation agriculture is still far from reality. This is probably due to initial yield losses, shortage of farm machinery, marginal land holding, and lack of awareness among the farming community.

References

NITI Aayog. (2018). Shifting Cultivation: Towards a Transformational Approach. Report of Working Group III, NITI Aayog, Government of India, 30p.

Aggarwal, P., Choudhary, K. K., Singh, A. K., & Chakraborty, D. (2006). Variation in soil strength and rooting characteristics of wheat in relation to soil management. *Geoderma, 136*, 353–363.

Ahmad, M., Chakraborty, D., Aggarwal, P., Bhattacharyya, R., & Singh, R. (2018). Modelling soil water dynamics and crop water use in a soybean-wheat rotation under chisel tillage in a sandy clay loam soil. *Geoderma, 327*, 13–24.

Al-Kaisi, M. M., & Yin, X. (2005). Tillage and crop residue effects on soil carbon and carbon dioxide emission in corn–soybean rotations. *Journal of Environmental Quality, 34*, 437–445.

Alletto, L., & Coquet, Y. (2009). Temporal and spatial variability of soil bulk density and near-saturated hydraulic conductivity under two contrasted tillage management systems. *Geoderma, 152*, 85–94.

Alletto, L., Coquet, Y., & Justes, E. (2011). Effects of tillage and fallow period management on soil physical behaviour and maize development. *Agricultural Water Management, 102*, 74–85.

Alvarez, R., & Steinbach, H. S. (2009). A review of the effects of tillage systems on some soil physical properties, water content, nitrate availability and crops yield in the Argentine Pampas. *Soil and Tillage Research, 104*, 1–15.

AQUASTAT FAO. (2018). FAO's information system on water and agriculture. http://www.fao.org/ag/ca/6c.html. Accessed 21 May 2018.

Arai, M., Miura, T., Tsuzura, H., Minamiya, Y., & Kaneko, N. (2017). Two-year responses of earthworm abundance, soil aggregates, and soil carbon to no-tillage and fertilization. *Geoderma, 332*, 135–141.

Baker, J. M., Ochsner, T. E., Venterea, R. T., & Griffis, T. J. (2007). Tillage and soil carbon sequestration—What do we really know? *Agriculture, Ecosystems & Environment, 118*, 1–5.

Batey, T. (2009). Soil compaction and soil management–a review. *Soil Use and Management, 25*, 335–345.

Bengough, A. G., McKenzie, B. M., Hallett, P. D., & Valentine, T. A. (2011). Root elongation, water stress, and mechanical impedance: A review of limiting stresses and beneficial root tip traits. *Journal of Experimental Botany, 62*, 59–68.

Beven, K., & Germann, P. (2013). Macropores and water flow in soils revisited. *Water Resources Research, 49*(6), 3071–3092.

Bhatt, R., Kukal, S. S., Busari, M. A., Arora, S., & Yadav, M. (2016). Sustainability issues on rice–wheat cropping system. *International Soil and Water Conservation Research, 4*, 64–74.

Bhattacharyya, R., Ghosh, B., Mishra, P., Mandal, B., Rao, C., Sarkar, D., Das, K., Anil, K., Lalitha, M., Hati, K., & Franzluebbers, A. (2015). Soil degradation in India: Challenges and potential solutions. *Sustainability., 7*(4), 3528–3570.

Blanchart, E., Albrecht, A., Brown, G., Decaens, T., Duboisset, A., Lavelle, P., Mariani, L., & Roose, E. (2004). Effects of tropical endogeic earthworms on soil erosion. *Agriculture, Ecosystems & Environment, 104*, 303–315.

Blanco-Canqui, H., & Lal, R. (2010). Soil and water conservation. In *Principles of soil conservation and management*. Springer.

Blum, W. E. H. (2013). Soil and land resources for agricultural production: General trends and future scenarios-A worldwide perspective. *International Soil and Water Conservation Research, 1*, 1–14.

Bogunovic, I., Pereira, P., Kisic, I., Sajko, K., & Sraka, M. (2018). Tillage management impacts on soil compaction, erosion and crop yield in Stagnosols (Croatia). *CATENA, 160*, 376–384.

Bormann, H., & Klaassen, K. (2008). Seasonal and land use dependent variability of soil hydraulic and soil hydrological properties of two Northern German soils. *Geoderma, 145*, 295–302.

Botta, G. F., Tolon-Becerra, A., Lastra-Bravo, X., & Tourn, M. (2010). Tillage and traffic effects (planters and tractors) on soil compaction and soybean (*Glycine max* L.) yields in Argentinean pampas. *Soil Tillage and Research, 110*, 167–174.

Bottinelli, N., Angers, D. A., Hallaire, V., Michot, D., Le Guillou, C., Cluzeau, D., Heddadj, D., & Menasseri-Aubry, S. (2017). Tillage and fertilization practices affect soil aggregate stability in a Humic Cambisol of Northwest France. *Soil Tillage and Research, 170*, 14–17.

Brauman, A. (2000). Effect of gut transit and mound deposit on soil organic matter transformations in the soil feeding termite: A review. *European Journal of Soil Biology, 36*, 117–125.

Briones, M. J. I., & Schmidt, O. (2017). Conventional tillage decreases the abundance and biomass of earthworms and alters their community structure in a global meta-analysis. *Global Change Biology, 23*(10), 4396–4419.

Bronick, C. J., & Lal, R. (2005). Soil structure and management: A review. *Geoderma, 124*, 3–22.

Brouder, S. M., & Gomez-Macpherson, H. (2014). The impact of conservation agriculture on smallholder agricultural yields: A scoping review of the evidence. *Agriculture, Ecosystems & Environment, 187*, 11–32.

Busari, M. A., Kukal, S. S., Kaur, A., Bhatt, R., & Dulazi, A. A. (2015). Conservation tillage impacts on soil, crop and the environment. *International Soil and Water Conservation and Research, 3*, 119–129.

Castellanos-Navarrete, A., Rodriguez-Aragones, C., De Goede, R. G. M., Kooistra, M. J., Sayre, K. D., Brussaard, L., & Pulleman, M. M. (2012). Earthworm activity and soil structural changes under conservation agriculture in central Mexico. *Soil Tillage and Research, 123*, 61–70.

Chaplot, V., Abdalla, K., Alexis, M., Bourennane, H., Darboux, F., Dlamini, P., Everson, C., Mchunu, C., Muller-Nedebock, D., Mutema, M., Quenea, K., Thenga, H., & Chivenge, P. (2015). Surface organic carbon enrichment to explain greater CO_2 emissions from short-term no-tilled soils. *Agriculture, Ecosystems & Environment, 203*, 110–118.

Chatterjee, D., Datta, S. C., & Manjaiah, K. M. (2014a). Transformation of short-range order minerals in maize (Zea mays L.) rhizosphere. *Plant, Soil and Environment, 60*(6), 241–248.

Chatterjee, D., Datta, S. C., & Manjaiah, K. M. (2016). Citric acid induced potassium and silicon release in alfisols, vertisols and inceptisols of India. *Proceedings of the National Academy of Sciences, India, Section B Biological Sciences, 86*(2), 429–439. https://doi.org/10.1007/s40011-014-0464-y

Chatterjee, D., Saha, S., Pradhan, A., Swain, C. K., Venkatramaiah, E., Nayak, A. K., & Pathak, H. (2021). Reducing methane emission from lowland rice ecosystem. In A. Rakshit, S. K. Singh, P. C. Abhilash, A. Biswas (Eds.), *Soil science: fundamentals to recent advances*. Springer. https://doi.org/10.1007/978-981-16-0917-6_25

Chatterjee, D. (2016). Strengths-weaknesses-opportunities-threats (SWOT) analysis of conservation agriculture. *Indian Journal of Hill Farming, 29*(1), 18–23.

Chatterjee, D., Datta, S. C., & Manjaiah, K. M. (2013). Clay carbon pools and their relationship with short-range order minerals: Avenues to mitigate climate change? *Current Science, 105*(10), 1404–1410.

Chatterjee, D., Datta, S. C., & Manjaiah, K. M. (2014b). Fractions, uptake and fixation capacity of phosphorus and potassium in three contrasting soil orders. *Journal of Soil Science and Plant Nutrition, 14*(3), 640–656. https://doi.org/10.4067/S0718-95162014005000051

Chatterjee, D., Datta, S. C., & Manjaiah, K. M. (2015a). Effect of citric acid treatment on release of phosphorus, aluminium and iron from three dissimilar soils of India. *Archives of Agronomy and Soil Science, 61*(1), 105–117. https://doi.org/10.1080/03650340.2014.919449

Chatterjee, D., Datta, S. C., & Manjaiah, K. M. (2015b). Characterization of citric acid induced transformation of short range order minerals in alfisol, inceptisol and vertisol of India. *European Journal of Mineralogy, 27*, 551–557. https://doi.org/10.1127/ejm/2015/0027-2446

Chatterjee, D., Mohanty, S., Guru, P. K., Swain, C. K., Tripathi, R., Shahid, M., Kumar, U., Kumar, A., Bhattacharyya, P., Gautam, P., Lal, B., Dash, P. K., & Nayak, A. K. (2018). Comparative assessment of urea briquette applicators on greenhouse gas emission, nitrogen loss and soil enzymatic activities in tropical lowland rice. *Agriculture, Ecosystems and Environment, 252*, 178–190. https://doi.org/10.1016/j.agee.2017.10.013

Chaudhary, D. R., Saxena, J., Lorenz, N., & Dick, R. P. (2012). Distribution of recently fixed photosynthate in a switchgrass plant-soil system. *Plant, Soil and Environment, 58*(6), 249–255.

Chen, G., & Weil, R. R. (2011). Root growth and yield of maize as affected by soil compaction and cover crops. *Soil Tillage and Research, 117*, 17–27.

Commission, E. (2006). *Thematic strategy for soil protection.* European Commission.

Connor, D. J. (2008). Organic agriculture cannot feed the world. *Field Crops Research, 106*(2), 187.

Corsi, S., Friedrich, T., Kassam, A., Pisante, M., & Sà, J. D. M. (2012). Soil organic carbon accumulation and greenhouse gas emission reductions from conservation agriculture: A literature review. Food and Agriculture Organization of the United Nations (FAO).

Das, A., Devi, M. T., Babu, S., Ansari, M., Layek, J., Bhowmick, S. N., Yadav, G. S., & Singh, R. (2018). Cereal-legume cropping system in Indian Himalayan region for food and environmental sustainability. In *Legumes for soil health and sustainable management* (pp. 33–76). Springer.

Das, A., Basavaraj, S., Layek, J., Ramkrushna, G. I., Lal, R., Rangappa, K., Yadav, G. S., Babu, S., Ghosh, P. K., & Ngachan, S. (2020). Can conservation tillage and residue management enhance energy use efficiency and sustainability of rice–Pea system in the Eastern Himalayas? *Archives of Agronomy and Soil Science, 66*(6), 830–846.

Das, A., Ghosh, P. K., Lal, R., Saha, R., & Ngachan, S. (2017a). Soil quality effect of conservation practices in maize–rapeseed cropping system in Eastern Himalaya. *Land Degradation and Development, 28*, 1862–1874.

Das, A., Layek, J., Idapuganti, R. G., Basavaraj, S., Lal, R., Rangappa, K., Yadav, G. S., Babu, S., & Ngachan, S. (2021). Conservation tillage and residue management improves soil properties under a upland rice–rapeseed system in the subtropical eastern Himalayas. *Land Degradation and Development, 31*(14), 1775–1791.

Das, A., Layek, J., Ramkrushna, G. I., Patel, D. P., Choudhury, B. U., Chowdhury, S., & Ngachan, S. V. (2014). Raised and sunken bed land configuration for crop diversification and crop and water productivity enhancement in rice paddies of the north eastern region of India. *Paddy and Water Environment, 13*, 571–580.

Das, A., Layek, J., Ramkrushna, G. I., Rangappa, K., Lal, R., Ghosh, P. K., Choudhury, B. U., Mandal, S., Ngangom, B., Dey, U., & Prakash, N. (2019). Effects of tillage and rice residue management practices on lentil root architecture, productivity and soil properties in India's lower Himalayas. *Soil Tillage and Research, 194*, 104313.

Das, A., Lyngdoh, D., Ghosh, P. K., LalR, L. J., & Idapuganti, R. G. (2018b). Tillage and cropping sequence effect on physico-chemical and biological properties of soil in Eastern Himalayas, India. *Soil Tillage and Research, 180*, 182–193.

Das, A., Ramkrushna, G. I., Makdoh, B., Sarkar, D., Layek, J., Mandal, S., & Lal, R. (2017b). Managing soils of the lower Himalayas. *Encyclopedia in Soil Science, Third Edition.* https://doi.org/10.1081/E-ESS3-120053284pp.1382-1387

de Moraes, M. T., Debiasi, H., Carlesso, R., Franchini, J. C., da Silva, V. R., & da Luz, F. B. (2016). Soil physical quality on tillage and cropping systems after two decades in the subtropical region of Brazil. *Soil Tillage and Research, 155*, 351–362.

Deubel, A., Hofmann, B., & Orzessek, D. (2011). Long-term effects of tillage on stratification and plant availability of phosphate and potassium in a loess chernozem. *Soil Tillage and Research, 117*, 85–92.

Dexter, A. R. (2004). Soil physical quality: Part II. Friability, tillage, tilth and hard-setting. *Geoderma, 120*, 215–225.

Dexter, A. R., Czyż, E. A., Richard, G., & Reszkowska, A. (2008). A user-friendly water retention function that takes account of the textural and structural pore spaces in soil. *Geoderma, 143*, 243–253.

Doran, J. W., & Parkin, T. B. (1996). Quantitative indicators of soil quality: a minimum data set. Methods for Assessing Soil Quality, (methodsforasses), pp. 25–37.

Drewry, J. J., Cameron, K. C., & Buchan, G. D. (2008). Pasture yield and soil physical property responses to soil compaction from treading and grazing—A review. *Soil Research, 46*, 237–256.

Fageria, N. K. (2005). Influence of dry matter and length of roots on growth of five field crops at varying soil zinc and copper levels. *Journal of Plant Nutrition, 27*, 1517–1523.

FAO. (2011). Save and Grow: A Policymaker's Guide to the Sustainable Intensification of Smallholder Crop Production 1–37 (FAO, 2011).

Feng, F. X., Huang, G. B., Chai, Q., & Yu, A. Z. (2010). Tillage and straw management impacts on soil properties, root growth, and grain yield of winter wheat in Northwestern China. *Crop Science, 50*, 1465–1473.

Gautam, P., Lal, B., Panda, B. B., Bihari, P., Chatterjee, D., Singh, T., Nayak, P. K., & Nayak, A. K. (2020). Alteration in agronomic practices to utilize rice fallows for higher system productivity and sustainability. *Field Crops Research, 260*, 108005. https://doi.org/10.1016/j.fcr.2020.108005

Gerke, H. H. (2006). Preferential flow descriptions for structured soils. *Journal of Plant Nutrition and Soil Science, 169*, 382–400.

Ghosh, B. N., Dogra, P., Sharma, N. K., Bhattacharyya, R., & Mishra, P. K. (2015). Conservation agriculture impact for soil conservation in maize–wheat cropping system in the Indian sub-Himalayas. *International Soil Water and Conservation Research, 3*(2), 112–118.

Ghosh, B. N., Sharma, N. K., Dogra, P., & Dadhwal, K. S. (2012). Effect of integrated organic input management in maize- wheat cropping system in sloping lands of north-west Himalayas. *Indian Journal of Soil Conservation, 40*(1), 84–89.

Ghosh, P. K., Das, A., Saha, R., Kharkrang, E., Tripathi, A. K., Munda, G. C., & Ngachan, S. V. (2010). Conservation agriculture towards achieving food security in North East India. *Cursos e Congresos Da Universidade De Santiago De Compostela, 99*(7), 915–921.

Guaman, V., Båth, B., Hagman, J., Gunnarsson, A., & Persson, P. (2016). Short time effects of biological and inter-row subsoiling on yield of potatoes grown on a loamy sand, and on soil penetration resistance, root growth and nitrogen uptake. *European Journal of Agronomy, 80*, 55–65.

Guo, L. J., Zhang, Z. S., Wang, D. D., Li, C. F., & Cao, C. G. (2015). Effects of short-term conservation management practices on soil organic carbon fractions and microbial community composition under a rice-wheat rotation system. *Biology and Fertility of Soils, 51*, 65–75.

Hamza, M. A., & Anderson, W. K. (2005). Soil compaction in cropping systems: A review of the nature, causes and possible solutions. *Soil Tillage and Research, 82*, 121–145.

Hazarika, S., Parkinson, R., Bol, D. L., Russell, P., Donovan, S., & Allen, D. (2009). Effect of tillage system and straw management on organic matter dynamics. *Agronomy for Sustainable Development, 29*, 525–533.

He, J., Kuhn, N. J., Zhang, X. M., Zhang, X. R., & Li, H. W. (2009). Effects of 10 years of conservation tillage on soil properties and productivity in the farming–pastoral ecotone of Inner Mongolia, China. *Soil Use and Management, 25*, 201–209.

Higashi, T., Yunghui, M., Komatsuzaki, M., Miura, S., Hirata, T., Araki, H., Kaneko, N., & Ohta, H. (2014). Tillage and cover crop species affect soil organic carbon in Andosol, Kanto, Japan. *Soil Tillage and Research, 138*, 64–72.

Hobbs, P. R., Sayre, K., & Gupta, R. (2008). The role of conservation agriculture in sustainable agriculture. *Philosophical Transactions of the Royal Society of London B: Biological Sciences, 363*, 543–555.

Horn, R., Fleige, H., Richter, F. H., Czyz, E. A., Dexter, A., Diaz-Pereira, E., Dumitru, E., Enarche, R., Mayol, F., Rajkai, K., de la Rosa, D., & Simota, C. (2005). SIDASS project: Part 5: Prediction of mechanical strength of arable soils and its effects on physical properties at various map scales. *Soil Tillage and Research, 82*, 47–56.
Hou, X., Jia, Z., Han, Q., Sun, H., Wang, W., Nie, J., & Yang, B. (2012). Effects of different rotational tillage patterns on soil structure, infiltration and water storage characteristics in dryland. *Transactions of the Chinese Society of Agricultural Engineering, 28*, 85–94.
Huang, M., Liang, T., Wang, L., & Zhou, C. (2015). Effects of no-tillage systems on soil physical properties and carbon sequestration under long-term wheat–maize double cropping system. *CATENA, 128*, 195–202.
Huwe, B. (2003). The role of soil tillage for soil structure. In A. El Titi (Ed.), *Soil tillage in agroecosystems* (pp. 27–49). CRC Press.
Jarvis, N. J. (2007). A review of non-equilibrium water flow and solute transport in soil macropores: Principles, controlling factors and consequences for water quality. *European Journal of Soil Science, 58*, 523–546.
Jat, S. L., Parihar, C. M., Singh, A. K., Jat, M. L., Jat, R. K., Singh, D. K., & Sai Kumar, R. (2011). Conservation agriculture in maize production systems. DMR Technical Bulletin 2011/4. Directorate of Maize Research, Pusa Campus, New Delhi-110 012, p. 25.
Jeschke, W. D., Baig, A., & Hilpert, A. (1997). Sink-stimulated photosynthesis, increased transpiration and increased demand-dependent stimulation of nitrate uptake: Nitrogen and carbon relations in the parasitic association Cuscuta reflexa-Coleus blumei. *Journal of Experimental Botany, 48*, 915–925.
Johansen, C., Haque, M. E., Bell, R. W., Thierfelder, C., & Esdaile, R. J. (2012). Conservation agriculture for small holder rainfed farming: Opportunities and constraints of new mechanized seeding systems. *Field Crops Research, 132*, 18–32.
Jones, D. T., & Eggleton, P. (2000). Sampling termite assemblages in tropical forests: Testing a rapid biodiversity assessment protocol. *Journal of Applied Ecology, 37*, 191–203.
Kahlon, M. S., Lal, R., & Ann-Varughese, M. (2013). Twenty two years of tillage and mulching impacts on soil physical characteristics and carbon sequestration in Central Ohio. *Soil Tillage and Research, 126*, 151–158.
Kargas, G., Kerkides, P., & Poulovassilis, A. (2012). Infiltration of rain water in semi-arid areas under three land surface treatments. *Soil Tillage and Research, 120*, 15–24.
Kassam, A., Friedrich, T., Shaxson, F., & Pretty, J. (2009). The spread of conservation agriculture: Justification, sustainability and uptake. *International Journal of Agricultural Sustainability, 7*, 292–320.
Kawaguchi, T., Kyoshima, T., & Kaneko, N. (2011). Mineral nitrogen dynamics in the casts of epigeic earthworms (Metaphire hilgendorfi: Megascolecidae). *Soil Science and Plant Nutrition, 57*, 387–395.
Keesstra, S. D., Bouma, J., Wallinga, J., Tittonell, P., Smith, P., Cerdà, A., Montanarella, L., Quinton, J. N., Pachepsky, Y., van der Putten, W. H., Bardgett, R. D., Moolenaar, S., Mol, G., Jansen, B., & Fresco, L. O. (2016). The significance of soils and soil science towards realization of the United Nations Sustainable Development Goals. *The Soil, 2*, 111.
Kuhn, N. J., Hu, Y., Bloemertz, L., He, J., Li, H., & Greenwood, P. (2016). Conservation tillage and sustainable intensification of agriculture: Regional versus global benefit analysis. *Agriculture, Ecosystems & Environment, 216*, 155–165.
Kumar, V., Saharawat, Y. S., Gathala, M. K., Jat, A. S., Singh, S. K., Chaudhary, N., & Jat, M. L. (2013). Effect of different tillage and seeding methods on energy use efficiency and productivity of wheat in the Indo-Gangetic Plains. *Field Crops Research, 142*, 1–8.
Kumari, M., Chakraborty, D., Gathala, M. K., Pathak, H., Dwivedi, B. S., Tomar, R. K., Garg, R. N., Singh, R., & Ladha, J. K. (2011). Soil aggregation and associated organic carbon fractions as affected by tillage in a rice–wheat rotation in North India. *Soil Science Society of America Journal, 75*(2), 560–567.

Kutilek, M. (2004). Soil hydraulic properties as related to soil structure. *Soil Tillage and Research, 79*, 175–184.

Ladha, J. K., Tirol-Padre, A., Reddy, C. K., Cassman, K. G., Verma, S., Powlson, D. S., Van Kessel, C., de B Richter, D., Chakraborty, D., & Pathak, H. (2016). Global nitrogen budgets in cereals: A 50-year assessment for maize, rice and wheat production systems. *Scientific Reports, 6*(1), 1–9.

Lal, B., Gautam, P., Nayak, A. K., Panda, B. B., Bihari, P., Tripathi, R., Shahid, M., Guru, P. K., Chatterjee, D., Kumar, U., & Meena, B. P. (2019). Energy and carbon budgeting of tillage for environmentally clean and resilient soil health of rice-maize cropping system. *Journal of Cleaner Production, 226*, 815–830. https://doi.org/10.1016/j.jclepro.2019.04.041

Lal, B., Gautam, P., Panda, B. B., Tripathi, R., Shahid, M., Bihari, P., Guru, P. K., Singh, T., Meena, R. L., & Nayak, A. K. (2020). Identification of energy and carbon efficient cropping system for ecological sustainability of rice fallow. *Ecological Indicators, 115*, 106431.

Lal, R. (2007). Carbon management in agricultural soils. *Mitigation and Adaptation Strategies for Global Change, 12*, 303–322.

Lassaletta, L., Billen, G., Grizzetti, B., Anglade, J., & Garnier, J. (2014). 50 year trends in nitrogen use efficiency of world cropping systems: The relationship between yield and nitrogen input to cropland. *Environmental Research Letters, 9*(10), 105011.

Layek, J., Das, A., Ramkrushna, G. I., Panwar, A. S., Verma, B. C., & Roy, A. (2017). Improving rice production under shifting cultivation: A case study. In A. Das, K. P. Mohapatra, S. V. Ngachan, A. S. Panwar, D. J. Rajkhowa, G. I. Ramkrushna & J. Layek (Eds.), *Conservation agriculture for advancing food security in changing climate*. Today & Tomorrow's Printers and Publishers, New Delhi - 110 002, India.

Layek, J., Chowdhury, S., Ramkrushna, G. I., & Das, A. (2014). Evaluation of different lentil cultivars in lowland rice fallow under no-till system for enhancing cropping intensity and productivity. *Indian Journal of Hill Farming, 27*(2), 4–9.

Lin, E., Li, Y. E., & Guo, L. (2002). Carbon emissions and sinks in agro-ecosystems of China. Science in China. *Series C, Life Sciences, 45*, 30–39.

Lipiec, J., Kuś, J., Słowińska-Jurkiewicz, A., & Nosalewicz, A. (2006). Soil porosity and water infiltration as influenced by tillage methods. *Soil Tillage and Research, 89*(2), 210–220.

Liu, E., Teclemariam, S. G., Yan, C., Yu, J., Gu, R., Liu, S., He, W., & Liu, Q. (2014). Long-term effects of no-tillage management practice on soil organic carbon and its fractions in the northern China. *Geoderma, 213*, 379–384.

Logsdon, S. D. (2012). Temporal variability of bulk density and soil water at selected field sites. *Soil Science, 177*, 327–331.

Lungmuana, C. B. U., Saha, S., Singh, S. B., Das, A., Buragohain, J., Dayal, V., Singh, A. R., Boopathi, T., & Dutta, S. K. (2018). Impact of post-burn jhum agriculture on soil carbon pools in the north-eastern Himalayan region of India. *Soil Research, 56*, 615–622.

Lynch, J. P. (2011). Root phenes for enhanced soil exploration and phosphorus acquisition: Tools for future crops. *Plant Physiology, 156*, 1041–1049.

Madari, B., Machado, P. L., Torres, E., de Andrade, A. G., & Valencia, L. I. (2005). No tillage and crop rotation effects on soil aggregation and organic carbon in a Rhodic Ferralsol from southern Brazil. *Soil Tillage and Research, 80*, 185–200.

Martens, D. A. (2001). Nitrogen cycling under different soil management systems. *Advances in Agronomy, 70*, 143–193.

Martínez, E., Fuentes, J. P., Silva, P., Valle, S., & Acevedo, E. (2008). Soil physical properties and wheat root growth as affected by no-tillage and conventional tillage systems in a Mediterranean environment of Chile. *Soil Tillage and Research, 99*, 232–244.

Materechera, S. A. (2009). Tillage and tractor traffic effects on soil compaction in horticultural fields used for peri-urban agriculture in a semi-arid environment of the North West Province, South Africa. *Soil Tillage and Research, 103*, 11–15.

McGarry, D., & Sharp, G. (2003). A rapid, immediate, farmer-usable method of assessing soil structure condition to support conservation agriculture. In *Conservation agriculture* (pp. 375–380). Springer.

Minasny, B., Malone, B. P., McBratney, A. B., Angers, D. A., Arrouays, D., Chambers, A., Chaplot, V., Chen, Z. S., Cheng, K., Das, B. S., Field, D. J., & Winowiecki, L. (2017). Soil carbon 4 per mille. *Geoderma 292*, 59–86.

Mitchell, J. P., Shrestha, A., Mathesius, K., Scow, K. M., Southard, R. J., Haney, R. L., Schmidt, R., Munk, D. S., & Horwath, W. R. (2017). Cover cropping and no-tillage improve soil health in an arid irrigated cropping system in California's San Joaquin Valley, USA. *Soil Tillage and Research, 165*, 325–335.

Mondal, S., Chakraborty, D., Das, T. K., Shrivastava, M., Mishra, A. K., Bandyopadhyay, K. K., Aggarwal, P., & Chaudhari, S. K. (2019a). Conservation agriculture had a strong impact on the sub-surface soil strength and root growth in wheat after a 7-year transition period. *Soil Tillage and Research, 195*, 104385.

Mondal, S., & Chakraborty, D. (2022b). Soil nitrogen status can be improved through no-tillage adoption particularly in the surface soil layer: A global meta-analysis. *Journal of Cleaner Products*, 132874.

Mondal, S., & Chakraborty, D. (2022a). Global meta-analysis suggests that no-tillage favourably changes soil structure and porosity. *Geoderma, 405*, 115443.

Mondal, S., Chakraborty, D., Bandyopadhyay, K., Aggarwal, P., & Rana, D. S. (2020). A global analysis of the impact of zero-tillage on soil physical condition, organic carbon content, and plant root response. *Land Degradation and Development, 31*(5), 557–567.

Mondal, S., Chakraborty, D., Tomar, R., Singh, R., Garg, R., Aggarwal, P., Sindhu, G., & Behra, U. (2013). Tillage and residue management effect on soil hydro-physical environment under pigeonpea (*Cajanus cajan*)-wheat (*Triticum aestivum*) rotation. *Indian Journal of Agricultural Sciences, 83*, 502–507.

Mondal, S., Das, A., Pradhan, S., Tomar, R., Behera, U., Sharma, A., Paul, A., & Chakraborty, D. (2018). Impact of tillage and residue management on water and thermal regimes of a sandy loam soil under pigeonpea-wheat cropping system. *Journal of the Indian Society of Soil Science, 66*, 40–52.

Mondal, S., Das, T., Thomas, P., Mishra, A., Bandyopadhyay, K., Aggarwal, P., & Chakraborty, D. (2019b). Effect of conservation agriculture on soil hydro-physical properties, total and particulate organic carbon and root morphology in wheat (*Triticum aestivum*) under rice (*Oryza sativa*)-wheat system. *Indian Journal of Agricultural Sciences, 89*(1), 46–55.

Mosaddeghi, M. R., Mahboubi, A. A., & Safadoust, A. (2009). Short-term effects of tillage and manure on some soil physical properties and maize root growth in a sandy loam soil in western Iran. *Soil Tillage and Research, 104*, 173–179.

Munda, S., Nayak, A. K., Saha, S., Shahid, M., Panda, B. B., Guru, P. K., Tripathi, R., Khanam, R., & Chatterjee, D. (2021). Zero tillage rice transplanting: A resource conservation technology. Technology Bulletin 163. ICAR-National Rice Research Institute Cuttack India, p 6

Nayak, A. K., Sarkar, R. K., Chattopadhyay, K., Reddy, J. N., Lal, B., & Chatterjee, D. (Eds.). (2017). Enhancing climate resilience in rice: Abiotic stress tolerance and greenhouse gas mitigation. ICAR-National Rice Research Institute, Cuttack, Odisha, 753006, India. ISBN: 81–88409–03–0, pp 118+vi.

Ngigi, S. N., Rockström, J., & Savenije, H. H. (2006). Assessment of rainwater retention in agricultural land and crop yield increase due to conservation tillage in Ewaso Ng'iro river basin, Kenya. *Physics and Chemistry of the Earth, Parts A/B/C, 31*, 910–918.

Nyamadzawo, G., Chikowo, R., & Giller, K. E. (2007). Improved legume tree fallows and tillage effects on structural stability and infiltration rates of a kaolinitic sandy soil from central Zimbabwe. *Soil Tillage and Research, 96*, 182–194.

Osunbitan, J. A., Oyedele, D. J., & Adekalu, K. O. (2005). Tillage effects on bulk density, hydraulic conductivity and strength of a loamy sand soil in southwestern Nigeria. *Soil Tillage and Research, 82*, 57–64.

Pittelkow, C. M., Linquist, B. A., Lundy, M. E., Liang, X., Van Groenigen, K. J., Lee, J., Gestel, N. V., Six, J., Venterea, R. T., & Van Kessel, C. (2015). When does no-till yield more? A global meta-analysis. *Field Crops Research, 183*, 156–168.

Pulleman, M. M., Six, J., van Breeman, N., & Jongmans, A. G. (2005). Soil organic matter distribution as affected by Agricultural management and earthworm activity. *European Journal of Soil Science, 56*, 453–467.

Reay, D. S., Davidson, E. A., Smith, K. A., Smith, P., Melillo, J. M., Dentener, F., & Crutzen, P. J. (2012). Global agriculture and nitrous oxide emissions. *Nature Climate Change, 2*(6), 410–416.

Richard, G., Cousin, I., Sillon, J. F., Bruand, A., & Guérif, J. (2001). Effect of compaction on the porosity of a silty soil: Influence on unsaturated hydraulic properties. *European Journal of Soil Science, 52*, 49–58.

Saha, S., Chatterjee, D., Swain, C. K., & Nayak, A. K. (2018). Methane emission from wetland rice agriculture-biogeochemistry and environmental controls in projected changing environment. In *Advances in crop environment interaction* (pp. 51–85). Springer. https://doi.org/10.1007/978-981-13-1861-0_3

Samal, S. K., Rao, K. K., Poonia, S. P., Kumar, R., Mishra, J. S., Prakash, V., Mondal, S., Dwivedi, S. K., Bhatt, B. P., Naik, S. K., Choubey, A. K., & McDonald, A. (2017). Evaluation of long-term conservation agriculture and crop intensification in rice–wheat rotation of Indo–Gangetic Plains of South Asia: Carbon dynamics and productivity. *European Journal of Agronomy, 90*, 198–208.

Schäffer, B., Mueller, T. L., Stauber, M., Müller, R., Keller, M., & Schulin, R. (2008). Soil and macro-pores under uniaxial compression. II. Morphometric analysis of macro-pore stability in undisturbed and repacked soil. *Geoderma, 146*, 175–182.

Schjønning, P., Lamandé, M., Berisso, F. E., Simojoki, A., Alakukku, L., & Andreasen, R. R. (2013). Gas diffusion, non-Darcy air permeability, and computed tomography images of a clay subsoil affected by compaction. *Soil Science Society of America Journal, 77*, 1977–1990.

Schwartz, R. C., Evett, S. R., & Unger, P. W. (2003). Soil hydraulic properties of cropland compared with reestablished and native grassland. *Geoderma, 116*, 47–60.

Shahid, M., Nayak, A. K., Tripathi, R., Mohanty, S., Chatterjee, D., Kumar, A., Bhaduri, D., Guru, P., Munda, S., Kumar, U., Khanam, R., Mondal, B., Bhattacharyya, P., Saha, S., Panda, B. B., & Nayak, P. K. (2018). Resource conservation technologies under rice-based system in eastern India. In H. Pathak, A. K. Nayak, M. Jena, O. N. Singh, P. Samal & S. G. Sharma (Eds.), *Rice research for enhancing productivity, profitability and climate resilience* (pp. 290–306). ICAR-National Rice Research Institute, Cuttack, Odisha, India, p. x+542.

Shahid, M., Nayak, A. K., Munda, S., Tripathi, R., Panda, B. B., Guru, P., Khanam, R., Saha, S., Lal, B., Gautam, P., Chatterjee, D., & Nayak, P. K. (2021). Conservation agriculture technology for rice-green gram cropping system in Eastern India. Technology Bulletin. ICAR-National Rice Research Institute Cuttack India, 8p.

Sharma, N. K., Singh, R. J., Mandal, D., Kumar, A., Alam, N. M., & Keesstra, S. (2017). Increasing farmer's income and reducing soil erosion using intercropping in rainfed maize-wheat rotation of Himalaya, India. *Agriculture, Ecosystems & Environment, 247*, 43–53.

Silburn, D. M., Freebairn, D. M., & Rattray, D. J. (2007). Tillage and the environment in sub-tropical Australia—Tradeoffs and challenges. *Soil Tillage and Research, 97*, 306–317.

Singh, A., Phogat, V. K., Dahiya, R., & Batra, S. D. (2014). Impact of long-term zero till wheat on soil physical properties and wheat productivity under rice–wheat cropping system. *Soil Tillage and Research, 140*, 98–105.

Sithole, N. J., Magwaza, L. S., & Mafongoya, P. L. (2016). Conservation agriculture and its impact on soil quality and maize yield: A South African perspective. *Soil Tillage and Research, 162*, 55–67.

Six, J., Bossuyt, H., Degryze, S., & Denef, K. (2004). A history of research on the link between (micro) aggregates, soil biota, and soil organic matter dynamics. *Soil Tillage and Research, 79*, 7–31.

Six, J. AET, Elliott, E. T., & Paustian, K. (2000b). Soil macroaggregate turnover and microaggregate formation: A mechanism for C sequestration under no-tillage agriculture. *Soil Biology & Biochemistry, 32*, 2099–2103.

Six, J., Paustian, K., Elliott, E. T., & Combrink, C. (2000a). Soil structure and organic matter I. Distribution of aggregate-size classes and aggregate-associated carbon. *Soil Science Society of America Journal, 64*, 681–689.

Soane, B. D., & van Ouwerkerk, C. (1994). Soil compaction problems in world agriculture. *In Developments in Agricultural Engineering, 11*, 1–21.

Spedding, T. A., Hamel, C., Mehuys, G. R., & Madramootoo, C. A. (2004). Soil microbial dynamics in maize-growing soil under different tillage and residue management systems. *Soil Biology & Biochemistry, 36*, 499–512.

Strudley, M. W., Green, T. R., & Ascough, J. C., II. (2008). Tillage effects on soil hydraulic properties in space and time: State of the science. *Soil till Res, 99*, 4–48.

Su, Z., Zhang, J., Wu, W., Cai, D., Lv, J., Jiang, G., Huang, J., Gao, J., Hartmann, R., & Gabriels, D. (2007). Effects of conservation tillage practices on winter wheat water-use efficiency and crop yield on the Loess Plateau, China. *Agricultural Water Management, 87*, 307–314.

Swain, C. K., Nayak, A. K., Bhattacharyya, P., Chatterjee, D., Chatterjee, S., Tripathi, R., Singh, N. R., & Dhal, B. (2018). Greenhouse gas emissions and energy exchange in wet and dry season rice: Eddy covariance-based approach. *Environmental Monitoring and Assessment, 190*(7), 1–7. https://doi.org/10.1007/s10661-018-6805-1

Taylor, L. L., Leake, J. R., Quirk, J., Hardy, K., Banwart, S. A., & Beerling, D. J. (2009). Biological weathering and the long-term carbon cycle: Integrating mycorrhizal evolution and function into the current paradigm. *Geobiology, 7*, 171–191.

Thorup-Kristensen, K. (2011). Strick Danish regulations on nitrogen use and how understanding vegetable crop root growth may help us improve nitrogen use efficiency. In: IV Jornadas Fertilizaci6n SECH. Sociedad Española de Ciencias. Hortícolas, Castelldefels (Barcelona), Spain. pp. 4–7.

Toliver, D. K., Larson, J. A., Roberts, R. K., English, B. C., De La Torre Ugarte, D. G., & West, T. O. (2012). Effects of no-till on yields as influenced by crop and environmental factors. *Agronomy Journal, 104*, 530–541.

Tolon-Becerra, A., Tourn, M., Botta, G. F., & Lastra-Bravo, X. (2011). Effects of different tillage regimes on soil compaction, maize (*Zea mays* L.) seedling emergence and yields in the eastern Argentinean Pampas region. *Soil Tillage Research, 117*, 184–190.

Tracy, S. R., Black, C. R., Roberts, J. A., & Mooney, S. J. (2013). Exploring the interacting effect of soil texture and bulk density on root system development in tomato (*Solanum lycopersicum* L.). *Environmental and Experimental Botany, 91*, 38–47.

Tripathi, R., Nayak, A. K., Lal, B., Shahid, M., Gautam, P., Munda, S., Mohanty, S., Guru, P. K., Khanam, R., Panda, B. B., Satapathy, B. S., Chatterjee, D., Nayak, P. K., Kumar, A., Kumar, U., & Mohapatra, S. D. (2021) Conservation agriculture based on rice-maize cropping system for Eastern India. Technology Bulletin. ICAR-National Rice Research Institute Cuttack India, 6p.

Triplett, G. B., & Dick, W. A. (2008) No-tillage crop production: A revolution in agriculture!. *Agronomy Journal, 100*(Supplement_3), S–153.

UNFCC. (2016). United Nations Framework Convention on Climate Change. Views on issues relating to agriculture. Subsidiary Body for Scientific and Technological Advice Forty-fourth session Bonn, 16–26 May 2016, https://unfccc.int/resource/docs/2016/sbsta/eng/misc01.pdf. Accessed 26 July 2018.

United Nations. (2015). *Transforming our world: The 2030 agenda for sustainable development*. United Nations.

Van Wie, J. B., Adam, J. C., & Ullman, J. L. (2013). Conservation tillage in dryland agriculture impacts watershed hydrology. *Journal of Hydrology, 483*, 26–38.

Victoria, R., Banwart, S., Black, H., Ingram, J., Joosten, H., Milne, E., Nollemeyer, E., & Baskin, Y. (2012). The benefits of soil carbon. *Foresight Chapter in UNEP Yearbook, 2012*, 19–33.

Villamil, M. B., & Nafziger, E. D. (2015). Corn residue, tillage, and nitrogen rate effects on soil carbon and nutrient stocks in Illinois. *Geoderma, 253*, 61–66.

Wahl, N. A., Bens, O., Buczko, U., Hangen, E., & Hüttl, R. F. (2004). Effects of conventional and conservation tillage on soil hydraulic properties of a silty-loamy soil. *Physics and Chemistry of the Earth, Parts A/B/C, 29*, 821–829.

Wheeler, T., & Von Braun, J. (2013). Climate change impacts on global food security. *Science, 341*, 508–513.

Whitmore, A. P., Whalley, W. R., Bird, N. R., Watts, C. W., & Gregory, A. S. (2011). Estimating soil strength in the rooting zone of wheat. *Plant and Soil, 339*, 363–375.

Yadav, G. S., Lal, R., Babu, S., Datta, M., Meena, R. S., Patil, S. B., & Singh, R. (2019a). Impact of no-till and mulching on soil carbon sequestration under rice (*Oryza sativa* L.)-rapeseed (*Brassica campestris* L. var. rapeseed) cropping system in hilly agro-ecosystem of the Eastern Himalayas, India. *Agriculture, Ecosystems & Environment, 275*, 81–92.

Yadav, G. S., Lal, R., Meena, R. S., Babu, S., Das, A., Bhowmik, S. N., Datta, M., Layek, J., & Saha, P. (2019b). Conservation tillage and nutrient management effects on productivity and soil carbon sequestration under double cropping of rice in north eastern region of India. *Ecological Indicators, 105*, 303–315.

Zhang, X., Davidson, E. A., Mauzerall, D. L., Searchinger, T. D., Dumas, P., & Shen, Y. (2015). Managing nitrogen for sustainable development. *Nature, 528*(7580), 51–59.

Zhang, Y., Wang, S., Wang, H., Wang, R., Wang, X., & Li, J. (2018). Crop yield and soil properties of dryland winter wheat-spring maize rotation in response to 10-year fertilization and conservation tillage practices on the Loess Plateau. *Field Crops Research, 225*, 170–179.

Zida, Z., Ouédraogo, E., Mando, A., & Stroosnijder, L. (2011). Termite and earthworm abundance and taxonomic richness under long-term conservation soil management in Saria, Burkina Faso, West Africa. *Applied Soil Ecology, 51*, 121–129.

Zuber, S. M., & Villamil, M. B. (2016). Meta-analysis approach to assess effect of tillage on microbial biomass and enzyme activities. *Soil Biology & Biochemistry, 97*, 176–187.

Part II
Climate Change and Crop Ecophysiology

Chapter 11
Prospects of Modified Plant Micro-Climate in Global Climate Change Research

Debjyoti Majumder, Salil Saha, M. A. Sarath Chandran, and Santanu Kumar Bal

Abstract Rising temperatures, altered rainfall patterns, and the frequency of pests and diseases are just few of the indications of climate change (CC) impacting agricultural productivity, produce quality, storability etc. CC has a profound and significant impact on production risk, especially in places with limited soil and water resources. Agriculture's response to CC could be influenced not just by changing weather conditions, but also by its ability to adapt to technological and food demand changes. The climate associated risks and vagaries in agriculture, as well as food and nutritional security and the impact on farming livelihoods, are widely established. The physiological response of a plant and the energy exchange activity between the plant and its environment are regulated and influenced by microclimate or environmental conditions in the immediate area of a plant. Therefore, under the changing climate, induced alterations in microclimate will help to increase soil and crop productivity against extreme weather risks, ensuring food and nutritional security.

Keywords Climate change · Microclimate · Agriculture · Vulnerability · Adaptation

---

D. Majumder
Krishi Vigyan Kendra - Malda, Uttar Banga Krishi Viswavidyalaya, Ratua, West Bengal 732205, India

S. Saha
AMFU, Regional Research Station (Old Alluvial Zone), Uttar Banga Krishi ViswavidyalayaDakshin Dinajpur, Majhian, West Bengal 733133, India

M. A. S. Chandran · S. K. Bal (✉)
ICAR-Central Research Institute for Dryland Agriculture (CRIDA), Santoshnagar, Hyderabad 500059, India
e-mail: santanu.bal@icar.gov.in

H. Pathak et al. (eds.), *Climate Change Impacts on Soil-Plant-Atmosphere Continuum*, Advances in Global Change Research 78,
https://doi.org/10.1007/978-981-99-7935-6_11

Abbreviations

CC	Climate Change
ET	Evapotranspiration
FACE	Free Air CO_2 Enrichment
ROS	Reactive oxygen species
VPD	Vapour pressure deficit
RH	Relative Humidity
WUE	Water Use Efficiency
CIR	Crop Irrigation Requirements
IW	Irrigation Water
CPE	Cumulative Pan Evaporation
CA	Conservation Agriculture
ZT	Zero Tillage
MT	Minimum Tillage
SOC	Soil Organic Carbon

11.1 Climate Change and Agriculture

Agriculture and associated sectors are more sensitive to and prone to climate change (CC) than other industries because of its susceptibility to weather variables, resulting in severe economic implications (Mendelsohn, 2009). Crop productivity is significantly impacted by anomalies in climatic events, including temperature and precipitation (Bal et al., 2022b). In accordance with the crop type, location, and degree of climatic parameter change, the effects of rising temperatures, inconsistent precipitation, and CO_2 fertilization vary. It has been found that the effect of steadily rising temperatures on productivity and yield is diminishing, whereas the effect of rising rainfall is likely to balance or attenuate the trend in rising temperatures (Adams et al., 1998). CC has a different impact on crop production based on the agro-climatic region and modes of irrigation system used. Expanding irrigated areas can increase crop output, but it may have a negative influence on the environment where there are chances of reduced precipitation which may ultimately results in reducing the water productivity (Kang et al., 2009). The increasing temperature is anticipated to shorten the growing cycle of crops, resulting in lower yields (Mahato, 2014). Rice, wheat, and maize production are predicted to drop if both temperate and tropical regions warm by 2 °C (Challinor et al., 2014). A risk of over-prediction of CC's value exists in the absence of factors like wind speed and atmospheric water vapour (humidity), which affect agricultural output in addition to temperature and precipitation. CC is anticipated to reduce rice, maize, and wheat yields in China by 36.25%, 18.26% and 45.10% respectively, until 2100, according to Zhang et al. (2017). A significant and unfavourable effect on wheat output in the Dutch province has been caused by the rising frequency of extreme weather occurrences during the 1900s. The severity of the loss in wheat production was influenced by the week in which an extreme weather event occurred (Powell & Reinhard, 2016). CC is predicted to cause more dry periods in most parts of the world in the near future, with drought-affected areas increasing from 15.4 to 44.0% by 2100. In Africa's drought-stricken regions, it's predicted that by 2050, crop production would have decreased by more than half, and by 2100, it will have decreased by more than 90% (Li et al., 2009). In India, Bal et al. (2022c) reported that dry spells have impacted yield of cotton, groundnut, maize, pearl millet, pigeon pea and sorghum in more than 65% rainfed growing regions. Additionally, according to their research, yield can be boosted by twofold by almost 55%, 49%, and 42% of the places where pearl millet, pigeon peas, and groundnuts are grown, respectively, by reducing the combined effect of dry periods. The global food supply, on the other hand, is less affected by CC, but developing countries will suffer severe effects (Rosenzweig & Parry, 1994). Rising temperatures in India between 2.3 and 4.8 °C, an increase in CO_2 levels by twofold, and regular protracted heat waves will all have a negative impact on the agriculture economy of the nation. Three major grains (rice, maize, and wheat) are expected to experience yield deficits which are 10–25% exacerbated by a 1 °C increase in the average total surface temperature (Deutsch et al., 2018). In sub-Saharan Africa, CC is predicted to reduce harvest yields by 6–24% (Waha et al., 2013). Solomon Island's food security will be seriously compromised as per-capita consumption of fish will decline in 2050 due to an overabundance of demand (Dey et al., 2016). CC, on the other hand, has a favourable impact on agriculture production in some areas. But regardless of scale, these provincial changes, whether they are increasing or decreasing, won't lead to significant alterations, and they will be most noticeable in low-scope areas. However, if the temperature exceeds the point at which CO_2 levels multiply; major financial catastrophes may occur (Aydinalp & Cresser, 2008). CC will have a significant negative influence on tropical parts of developing countries, however this will mostly rely on the climate scenario for the area. The drier northern and eastern regions of Sri Lanka would be disastrous for agri-horticulture sectors, as contrasted to the cooler central region, where production scenarios would be stable or even on the rise due to rising temperatures. Since the rate at which CC occurs affects its effect, the cost of variation and natural strategy should be dynamic and adaptable to respond equitably to the changes (Zilberman et al., 2004).

11.2 Factors Affecting Existing Microclimate

A microclimate is a meteorological phenomenon that occurs near the earth's surface and incorporates environmental elements like temperature, humidity, wind, and light that can be used to influence biological activity and habitat choice (Naiman et al., 2005). Microclimate is linked to habitat on a microscale and is critical for species. As microclimate affects ecological processes including soil respiration, plant regeneration, the choice of a wildlife habitat, and the nutrient cycle, microclimate has grown in importance in current research (Chen et al., 1999). Air temperature, wind speed

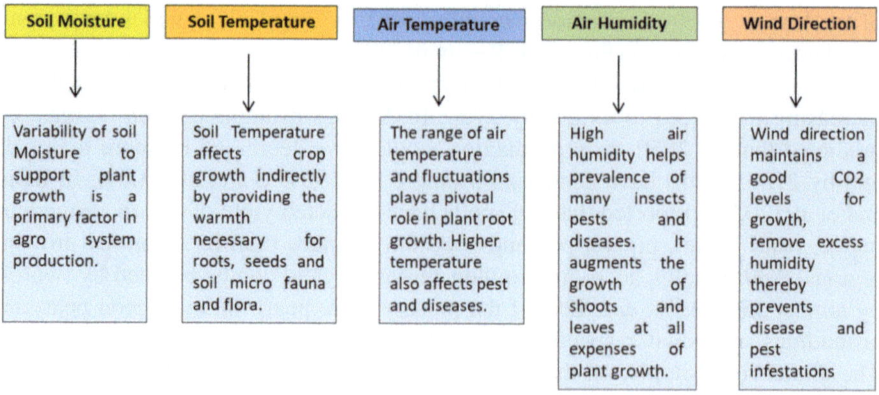

Fig. 11.1 Microclimate indicators and agriculture

and direction, soil moisture, soil temperature, radiation distribution, soil acidity, and CO_2 level are microclimate factors that have an impact on agricultural productivity. By restricting these factors, crop production quality and output can be artificially increased (Bramer et al., 2018). Microclimate isn't just restricted to the surface; it can also reach the subsurface. Organisms can live anywhere and everywhere where they create favourable microclimatic conditions and interact (Mislan & Helmuth, 2008). The effect of microclimate on plant capacities like photosynthesis, respirations, and movement eventually decides plant efficiency. We should have the option to monitor the various constituents of the microclimate in the regular habitat to comprehend how plant generally exercises respond to such activities (Fig. 11.1).

11.3 Crop Microclimate

The weather just above and inside the crop canopy to the root zone makes up the crop microclimate, which can be modified by daily management practices at different time scales (Stigter, 1994a). Crop microclimates should be the most favourable environment for plant response, i.e., crop productivity maximisation. Microclimate modification involves altering or regulating weather elements on a micro-scale, with the goal of generating a climate that is conducive to plant growth. Weather variables like temperature, humidity, wind and turbulence, dew, frost, and heat balance can all have an impact on a microclimatic scenario. Crop behavioural responses can be managed through the management of radiation budgets (Bhagat et al., 2017), heat balances and moisture balances in response to microclimate (Stigter, 1994b). Soil type has a considerable impact on microclimates. Sandy soils are prone to temperature extremes. Vegetation regulates the amount of water vapour is being released into the atmosphere (Britannica, 2020). Furthermore, plants can insulate the underlying soil and prevent temperature variations (Fig. 11.2).

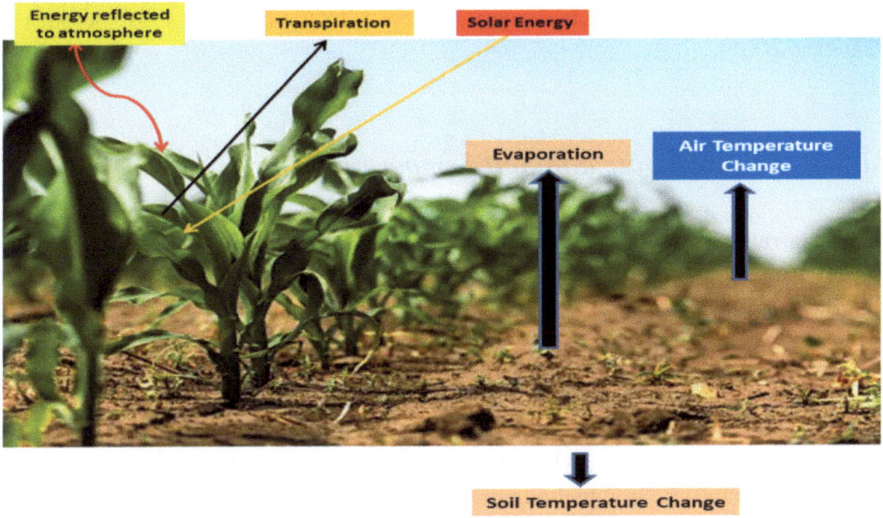

Fig. 11.2 Energy fluxes for two different landscapes: desert (left) and highly vegetated area (right)

11.3.1 Microclimate and Soil Moisture

Soil moisture is one of the most crucial microclimate elements. Thermal conductivity and heat capacity of soil are considerably increased by the presence of moisture (Bonan, 2016). Evapotranspiration uses far more energy than heating the air does. Those soils which are rich in soil moisture have a more balanced microclimate, which results in cooler air and soil temperatures. This affects local rainfall patterns, weather patterns, and plant growth in addition to plant growth. Contrary to popular opinion, adjacent surface air temperatures rise when soil moisture is limiting because there is more energy available for sensible heating. Many studies conducted in the last decades have concentrated on the connection between the absence of soil moisture and the incidence of heat waves and severe temperatures, both locally and regionally (Seneviratne et al., 2010). Because wet soils stay warmer for longer during ice events than dry soils, the adjusting effect of soil moisture also considers low temperatures. In great soil dampness conditions, biotic life can flourish. Microorganisms help in the breakdown of natural materials and the intrusions of supplements that add to soil enrichment when dampness involves around 60% of accessible soil dampness. An abundance of water keeps the stock of oxygen from being renewed, making microbial development end, stop, or turn anaerobic, influencing plant improvement (Bot and Benites 2005). As a result, in areas where flooding is common, a large amount of waste is expected to assure exceptional soil fertility. In addition, too much water can result in advancements being lost owing to filtering. Low moisture levels, on the other hand, hinder supplement conveyance processes by reducing compound impetus activity. Land levelling in slopy areas of dryland can improve the spatial distribution of rainwater, thereby improving use efficiency (Bal et al., 2015). Nitrogen cycling is particularly harmed, as it is already a limiting factor in many soils (Sardans & Peñuelas, 2005).

11.3.2 Microclimate and Soil Temperature

Incoming radiation, thermal conductivity, and the soil's heat capacity all have an impact on the temperature of the ground. The amount of incoming radiation that is absorbed or reflected is affected by the soil tone. Soil that is darker absorbs more solar energy, while lighter soils reflect it and are consequently colder. High thermal conductivity soils often release energy more quickly thereby it cools down at a faster rate. This is crucial for understanding water movement and dew production. Heat capacity manages the capacity of soil to conserve heat. This can be observed on cooler evenings at the point where heat slowly escapes and the surface air warms. Soil moisture conservation techniques are often employed in the highlands to minimize the occurrence of frost. However, frost heaving can occur when temperatures are constantly low and can be forecasted using weather-based prediction models (Bal et al., 2021). The earth swells due to the volume expansion of the ice in the soil (Bonan, 2016). Porosity, soil moisture, organic content, and soil texture or mineral structure all has an impact on heat capacity and thermal conductivity. Moisture in the soil considerably enhances heat conductivity since water is 20 times more conductive than air (Bonan, 2016). During the day, heat travels into the soil, transporting heat away from the direct surface and causing lower temperatures. The soil's heat transmission is reversed at night, when the surface is cooler, and heat is also transmitted to the surface, counteracting the limitations. Additionally, a similar exchange takes place across longer periods of time when heat is distributed during cold winter months and conserved during summertime (Bonan, 2016). The temperature of the soil affects the amount of accumulated heat needed for the development of seeds, plant roots, and small life (microorganisms) in the soil. Elevated temperatures of the soil can have the antagonistic effect on plant development, while extreme temperatures can stifle the normal cycles of tiny animals (FAO, 2016). Low soil temperatures, of course, reduce plant water uptake, restrict nitrification, and, as a result, decrease soil richness and increase drying when temperatures of the air is on higher side (Gliessman, 2015). High and low soil temperatures are likely to have an impact on plant evapotranspiration (ET). Higher soil moisture will enhance dissipation, absorbing more energy and, as a result, lowering daytime neighbourhood temperatures while raising evening surface temperatures.

11.3.3 Microclimate and Air Temperature

One of the most crucial variables of local air temperature as well as temperature inside the crop canopy is incoming and outgoing radiation. The temperature within the crop canopy, as well as the local air temperature, can be reduced by increasing transpiration. Vegetation can provide shading (somewhat) depending on canopy structure and prevent radiation from reaching to the soil surface. The contribution of energy to the critical processes of air and soil warming, photosynthesis, and ET is shown by net radiation. Albedo is an important factor in determining local air temperatures, and it varies with the weather. The orientation/angle of the slope affects the quantity of energy received as well as shadow areas, also the geographical location of a particular area has a considerable impact on incoming radiation. Soil albedo is heavily reliant on moisture content since precipitation affects albedo and supplies moisture for evaporation during rainfall temperature interactions. In general, wet soil has a lower albedo than dry soil. With a higher albedo than forest, agricultural crops reflect more radiation and emit less surface heat (Jackson et al., 2008). Temperature, humidity, and wind interact with surface processes and characteristics. Shade, or vegetation, modifies the radiation balance and acts as a wind barrier (Gliessman, 2015). The peak air temperature is found in the upper canopy, which occurs an hour or two after local noon (Foken & Napo, 2008). The temperature is lower during the day beneath the crop cover. The lower night-time temperatures in the upper canopy can be attributed to radiative cooling, which is the cooling of the earth's surface and the air adjacent to it. This is especially true when the sky is clear, the winds are mild, and the humidity is low. It was also observed in Veracruz, Mexico, that temperature affects coffee yield by shortening the reproductive phenological stage and that coffee production may not be economically sustainable in the near future, with current production down by almost 34% (Gay et al., 2006).

11.3.4 Microclimate and Humidity

ET, which is triggered by available radiation and moisture, produces humidity as an additional consequence. It is responsible for the creation of clouds, mist, and dew. Warm air can store more water vapour than cold air because it can hold more water vapour as a function of temperature. A hotter climate advances evaporation, which thus increases air humidity. In arid and semi-arid locations, atmospheric moisture can supplement the water required by plants significantly (Nicholson, 2012). The albedo from bared or cropped soils as well as from crop canopies is also influenced by dew. Dew often moistens the soil making it darker thereby lowering it's albedo, while also acting as a reflecting surface on the leaves, raising their albedo. Moist or humid air slows plant transpiration because it collects water vapour more slowly than dry air. As the humid air are being transported away from the vegetation, the local scale wind acts as an external force to mix the local air in the atmosphere (Bonan, 2016; Nicholson, 2012). Mild breezes were shown to aid dew development in unsheltered regions, but medium to high speed winds were found as factor in reducing the dew production. Saturated air, when combined with a shift in air temperature, can result in precipitation on a land surface. Light local breezes have been reported to aid with dew development, but moderate to strong winds in exposed areas have been proven to restrict dew formation (Richards, 2004). Plants act as a windbreak and provide shade, reducing wind speed and influencing the creation and duration of dew. It also reduces dew formation and duration of dew as local warm winds are not removed. Dew formation chances increase because vegetation's shade helps to reduce surface temperature. Humidity in the standing crop can be increased or decreased and it totally depends on the presence of local breezes and its nature viz. warm air, moist or dry air.

11.3.5 Microclimate and Wind

Air movement depends on pressure differences at a larger scale and at a smaller scale, it depends on temperature difference, topography and barriers. In case of local breezes landscape morphology such as mountains, water bodies and temperature differences influence wind directions and speeds (Foken & Napo, 2008). During the summer months in day time, the land surface heats up faster than the water surface, resulting a local wind system called sea breezes but water bodies lose heat slower than the land surface and at night winds blow from the reverse direction which is called land breezes. Warm air rises upslope (Valley wind) in mountains and hilly terrain during the day, and cold wind moves downslope at night (mountain wind) (Gliessman, 2015). Wind can cool plants by replacing warm air with unsaturated air around the plant. This affects the plant's water intake while simultaneously causing higher transpiration. Depending on the ambient temperature and water vapour in the atmospheric air, the prevalent wind can influence the existing temperatures by making it hotter or colder (Bonan, 2016; Gliessman, 2015). Furthermore, movement of the air within the crop canopy is critical for maintaining adequate CO_2 concentrations, removing excess humidity to maintain reduced humidity, and reducing the risk of disease for healthy growth. Soil particles, seeds, pests, and diseases can all be transported by the wind. Many cereals crops are pollinated by wind, diseases move to new hosts and pests move to new places. (Gliessman, 2015). Wind changes the vegetation potential and soil moisture storage capacity, which has a cascade effect on the microclimate (Ong et al., 2015).

11.4 Plant Response to Changes in Climatic Parameters

11.4.1 Physical/ Morphological Response

Crops are susceptible to CC, which includes changes in precipitation and temperature as well as rising atmospheric CO_2 concentration. Increases in temperature are more likely than other variables to have a detrimental effect on agricultural productivity (Ottman et al., 2012; Rosenzweig et al., 2014; Wheeler & von Braun, 2013).

Changes in developmental processes are primarily responsible for functional differences between plant genotypes and species that grow in various environments. The molecular, cellular, and morphological responses to stress vary throughout tissues and over time as a result of the diverse ways that various environmental pressures influence various organs and tissues within a plant. Success of a plant depends on its capacity to dynamically modify its developmental processes in response to its environment and on the variety of developmental processes present in different genotypes (Nicotra et al., 2010). Changes in the beginning of developmental processes, the timing of developmental events, and the final form or architecture of particular organs and full plants are only a few examples of how plants respond to their environment during the developmental process. Repressed lateral root initiation in response to water constraint is one instance of a changed developmental event's commencement (Babé et al., 2012). For instance, an early transition from vegetative to reproductive development in response to higher temperature can be used to investigate how the environment affects developmental time in *Arabidopsis thaliana* (Balasubramanian et al., 2006).

11.4.2 How Does the Plant Respond to Rising CO_2 Concentrations?

The appearance of more leaf nodes and larger leaves in response to elevated CO_2 levels can be seen in the plant's final form or architecture, both at the organ and plant levels. In the presence of the plant's complete look or architectural styles, for example, the presence of additional leaf nodes and larger leaves in response to elevated CO_2 can be observed at the level of individual organs as well as the overall plant's structure (Dermody et al., 2006). CO_2 levels in the atmosphere have risen from 280 parts per million to more than 420 parts per million during the Industrial Revolution (Unanimous, 2021). In twelve Free Air CO_2 Enrichment (FACE) studies, increasing CO_2 enhanced photosynthetic carbon absorption rates by an average of 31% across 40 species (Ainsworth and Long 2006). According to Reich et al. (2014), increased CO_2 stimulated aboveground biomass by up to 33% in grassland, albeit the degree of stimulation was dependent on water and nitrogen availability, with lower biomass stimulation recorded in drier, lower nutritional contexts. In response to higher CO_2 levels, root biomass has been observed to increase significantly in a variety of crop species (Madhu & Hatfield, 2013). Crop output in Iran is impacted by climate variables and is dependent on adaptive abilities, crop type, environmental circumstances, and the influence of CO_2 fertilization (Karimi et al., 2018).

11.4.3 Crop Morphological Response to Temperature Alterations

Different species have shown varying degrees of acclimatisation, which is partly determined by the climate they are adapted to. Animals in colder climates tend to develop quicker and produce greater photosynthetic capacity in response to increased temperatures (Gunderson et al., 2010; Way & Sage, 2008). Species that survive in warmer low-latitude settings, on the other hand, exhibited reduced tree growth and carbon uptake, as well as lower photosynthetic capabilities, according to data from research in warmer climes (Cheesman & Winter, 2013; Crous et al., 2013; Feeley et al., 2007). Tree development and a range of physiological activities are influenced by temperature. With forecasts of 2–5 °C warming during the twentieth century, many species are presently experiencing unparalleled rates of warming in evolutionary history. Plants, particularly trees, are sessile and have a longer lifespan than humans, therefore they may need to physiologically adapt to warmer temperatures. Most plants can adapt to warmer temperatures, and they usually do so in a way that preserves or enhances carbon gain. In response to warming, reduced respiration rates (Atkin & Tjoelker, 2003), increased total leaf area (Way & Oren, 2010), and occasionally greater assimilation rates at warmer growth temperatures can all be achieved, enhancing carbon uptake and growth (Way & Sage, 2008). In consequence to warming scenarios, most species should alter their photosynthetic thermal optimum requirement (Crous et al., 2013). Crop growth and development have been impacted by global climate change. Changes in agricultural phenology are one of the most important crop reactions to CC. Global warming, according to Cline (2007), is expected to diminish agricultural production in poor countries by 9 to 21%. Temperature variations result in higher yields in the mid to high latitudes, but the benefits diminish when temperature changes exceed 30 °C. You et al. (2005) showed that, rising temperatures have resulted in a 2.4% decline in wheat output in China over the last two decades. According to Schlenker and Roberts (2009), the threshold temperature for soybean is 30 °C; increasing the temperature to the optimal level boosted soybean output, while increasing the temperature beyond that level lowered yield dramatically. According to Eastburn et al. (2010), rising ozone and CO_2 concentrations in the atmosphere have already altered disease types, and disease susceptibility in soybean as global temperatures rise.

Vose et al. (2005) concluded that increase in mean global a minimum at a faster rate as compared to mean maximum temperatures, resulted in decrease in diurnal temperature range. Wheat crops have shown to be vulnerable to increases in maximum temperature, while rice crops were found to be sensitive to increases in lowest temperature, according to reports (Lal et al., 1998). According to a research reported by Lobell and Christopher (2007), between 1997 and 2002, global warming decreased crop yields generally, including those of wheat, corn, and barley, by 40 mt year. High temperature during the grain filling stages resulted in reduction in wheat yields (Wardlaw et al., 1989). Another major aspect that might have a significant impact on crop productivity is periodic temperature. Pre-anthesis phenological periods in wheat, which define prospective yield attributes, are extremely heat—sensitive (Abrol & Ingaram, 1996). Wheat yield is dramatically reduced when planting time is shifted from optimal to later. Meanwhile, meteorological variables like as rainfall and temperature influenced agronomic characters (Hayat & Ahmed 2014). According to Matthew et al. (2017), CC has caused a 3.5–12.9% yield reduction in wheat. Richards et al. (2011) studied that under stress, photosynthesis and glucose translocation into grains was disrupted, which led to a decrease in grain quantity and weight. Stress also triggered assimilates remobilization, as well as early senescence and grain filling in wheat (Yang et al., 2001). Abiotic stress is very crucial for photosynthesis, which is a yield limiting factor (water and temperature). Cornic (2000) discovered that metabolic insufficiency causes a decrease in photosynthesis. Meristematic cell divisions that result in daughter cells and the following rapid development of the young cells cause crop expansion. When there is a severe water scarcity, higher plant cell elongation can be prevented by blocking water transport from the xylem to the surrounding elongating cells (Nonami, 1998). Due to cell elongation and expansion, drought inhibited mitosis, resulting in reduced growth and yield attributes (Hussain et al., 2008). By lowering the soil's water potential, water deficits decrease the number of leaves per plant, the size of each leaf, and the vitality of each leaf. The amount of nutrients available for growth, as well as leaf turgor and canopy temperature, all-together influences leaf area expansion and canopy development.

11.4.4 Drought Impact on Crop Withstand

Drought-induced decreases in leaf area are assumed to be caused by reduced photosynthesis, which inhibits leaf growth (Rucker et al., 1995). The decline in fresh and dry biomass output is a frequent unfavourable impact of water stress on agricultural plants (Zhao et al., 2006). The maximum temperature for all crops under consideration seems to be a risk-reducing factor, whereas the minimum temperature for rice, cotton, and sorghum seems to be a risk-increasing factor. Finally, it has been determined that rainfall increases threat for rice and peanuts but decreases risk for jowar and cotton. Drought stress reduces dry matter production primarily by inhibiting leaf growth and development, which results in lower light interception (Nam et al., 1998). In sorghum, maize and barley, the combined impact of heat and drought conditions was found to have more negative consequences than individual stress (Wang & Huang, 2004). Rice fertilisation and anthesis are both affected by drought stress. The harvest index has been reduced to 60% due to a lack of water, decreasing the grain set (Garrity & O'Toole, 1995). Large drought conditions in West Africa during the 1980s El Nino years significantly reduced cocoa output (Ruf et al., 2015). In the flowering stage of the mash bean plant (Vigna mungo L.), drought stress reduced the yield from 31 to 57%, whereas in the reproductive stage, it decreased the output by 26%. (Baroowa & Gogoi, 2014). As per study conducted by Maleki et al. (2013), drought stress has had a significant impact on soybean production, with a 42% decline seen at the grain filling stage.

11.4.5 Solar Radiations and Its Role in Crop Productivity

Photosynthesis and crop productivity are both influenced by radiation (Gholipoor & Shahsavani, 2008). As per Amthor and Loomis (1999) crop growth and production are proven to be derived photosynthesis, which is reliant on receipt and solar radiation is captured. According to Richards (2006), seasonal fluctuations in solar radiation have a considerable influence on the outcome of radiation on yield. It's possible that the crop's sensitivity to sun radiation is linked to a change in the length of photosynthesis. Despite quantum irradiance varying from 200 to 1100 μmol m^{-2} s^{-1}, Meziane and Shipley (2001) found that at low nutritional supply, the mass concentration of nitrogen remained almost constant at low nutrient supply. Pathak et al. (2003) attribute declining rice and wheat potential yields in India's Indo-Gangetic plains to negative solar radiation trends and a rise in minimum temperature. Temperature and differences in sun radiation have been discovered to cause significant changes in wheat output (Ahmed et al., 2010; Li et al., 2010).

11.4.6 Biochemical Response

Global CC, particularly regional and global temperature fluctuations, has a substantial impact on plant production and vegetation. Temperature has a significant impact on plant development and output. Plants lose yield when exposed to extreme temperatures for an assortment of reasons. Temperature and water pressure can likewise altogether affect the actual microenvironment, affecting plant physiological and biochemical cycles like pressure opposition compound action and osmotic material collection (Slama et al., 2007). Heat pressure brings down the proficiency of photosystem II, the most intensity sensitive part of photosynthetic cycles, in chloroplasts (Barnabás et al., 2008). Drought-stricken plants reduce their transpirations and net photosynthetic rate; these physicochemical responses are prevalent in areas with high evaporative demand (Feng & Cao, 2005). The ability of cotton plants to withstand high temperatures has a direct impact on germination and following growth stages. This suggests that if a plant can withstand heat stress during seed germination, it will be tolerant in later phases. Assessing layer warm strength is the appropriate and beneficial technique to test heat resilience in cotton, as well as a variety of harvests. Cotton's primary adaptable components for high temperature resilience are soluble sugars, proline, and solvent proteins (Ashraf et al., 1994). Temperature has a variety of effects on plants; for example, 35 °C in tomatoes and 15 °C in watermelon produce the smallest natural product loads. As a result, each of these temperatures results in heat pressure in the plants (Rivero et al., 2001). Table 11.1 shows the response of different crops to temperature induced heat stresses. The overall anthocyanin content of rose blossoms may drop when cultivated at high temperatures, and the flower's pigment types may be altered to a higher pelargonidin/cyanidin ratio (Dela et al., 2003). Different pressures, such as heat, drought, or other stresses, have an impact on serious agricultural injuries across the world. It appears that heat stress transcription factors (HSFs) impact HSP production, which is important for heat stress resistance and thermo-tolerance in both plants and other species (Kotak et al., 2007). Under water stress, protection devices against excessive decreasing power are thus an important tactic (Chaves et al., 2009). Reactive oxygen species (ROS) can be produced in chloroplasts when photosynthesis is suppressed and light excitation energy is greater than that required for photosynthesis. This can happen when the receiving wire's photosynthetic shades are overexcited (Munne-Bosch et al., 2003). The creation of reactive oxygen species is perhaps the earliest metabolic response of eukaryotic cells to biotic and abiotic stresses (ROS). The oxidative burst, also known as the age of ROS in plants, occurs early in the plant's protective response to water pressure and serves as a precursor for the plant's later safeguard reaction. ROS, which include oxygen particles, free radicals, and peroxides, are produced as a by-product of regular oxygen digestion and play an important role in cell flagging. In any case, during adverse circumstances, such as the dry season, ROS levels rise, causing oxidative damage to proteins, DNA, and lipids (Apel & Hirt, 2004). Dry season promotes oxidative pressure in plants, according to Farooq et al. (2009), by causing the formation of reactive oxygen species (ROS). ROS such as O_2, H_2O_2, and OH^{-1} revolutionaries can directly attack film lipids, resulting in lipid peroxidation (Mittler, 2002). When the rate of CO_2 retention is slowed in a negative environment, chloroplasts are exposed to excessive excitation energy, and the age of sensitive oxygen intermediates increases, resulting in oxidative pressure (Hare & Cress, 1997). When plants, that are resistant to warm pressure are exposed to varying pressure circumstances, their cell reinforcement chemicals and cancer prevention agents change (Yin et al., 2008).

11.4.7 Physiological Response of Plants to Climatic Parameters

CC is fuelled by existing trimming frameworks with little variety and high information and fixations (Reckling et al., 2018). Dry seasons and heavy rains will become more often, as will temperature variations, salt, and pest infestations, all of which will reduce yield and increase the risk of starvation (Dhankher & Foyer, 2018). Temperature fluctuations (Fig. 11.5), as well as precipitation, have harmed crop variation (Kang et al., 2009). Expanding worldwide temperatures are likewise expected to cause expanding paces of surface dissipation and urge bigger paces of ET from plants, bringing about higher paces of evaporations and ET from plants (Dai, 2013). High temperature stress appears to be causing significant harm to plants, according to certain data. For each degree increase in temperature, estimates range up to a 17% drop in agricultural output (Lobell & Asner, 2003). Agricultural production, on the

Table1 Impacts and response Mechanism of different crops towards environmental stresses

Species	Impacts	Plant response/ mechanism	References
Cucumber (*Cucumis sativus*)	Higher concentrations of osmolytes in the leaves and root, such as proline, glycine betaine, and solvent sugars	Antioxidant enzymes (SOD, POD, CAT, APX, GPX and GTR) activities	Balal et al. (2016)
Chinese Cabbage (*Brassica campestris* L.)	The heat-tolerant cultivar demonstrated less damage to the photosynthetic apparatus and less generation of ROS than the heat-sensitive cultivar after five days of exposure to 400/300C (day/night)	The activity of the antioxidant enzymes CAT, SOD, and guaiacol peroxidase has risen	Zou et al. (2017)
Blueberry (*Vaccinum* spp.)	Heat tolerance was found greater in the Jersey cultivar than in the Diana cultivar	The tolerant cultivar showed increased levels of antioxidant enzymes and genes involved in oxidative protein repair after 18 h of exposure to H_2O_2 and superoxide radicals at 48C.	Yu et al. (2016)
Aradiopsis thaliana	When exposed to high temperatures (30, 34, and 38 °C) for one day, the lipidome changes	Heat stress increased the fraction of saturated lipids in the chloroplast membrane, resulting in a decrease in the quantity of polyunsaturated fatty acids	Higashi et al. (2015)
Soybean (*Glycine max* L.)	The roots had higher MDA readings than the hypocotyl after being exposed to 10 °C for up to 8 days	The hypocotyl produces more antioxidant enzymes, meaning that it is more resistant to the effects of ROS damage	Posmyk et al. (2005)
Populuscathayana	Both sexes showed increased H_2O_2 and APX levels after 14 days of cultivation at 40 °C, but male plants had greater SOD and GR levels, suggesting that male plants of this species are more tolerant to cold stress than females	Changes in gene expression may be linked to changes in enzyme activity in response to cold	Zhang et al. (2011)

(continued)

Tablel (continued)

Species	Impacts	Plant response/ mechanism	References
Bromeliad *A. imperialis*	At low temperatures (15 °C), sucrose was clearly replaced by trehalose, raffinose, and stachyose, but at high temperatures (30 °C), myo-inositol, glucose, fructose, and sucrose were found in greater abundance. The most starch is found in plants that were grown at high temperatures. The smallest amount of starch was produced at low temperatures	To protect plants from the cold, they are converted into soluble carbohydrates	Mollo et al. (2011)
Wheat (*Triticum aestivum*)	Because of higher photosynthetic rates under drying out and faster recovery after re-watering, the open-minded cultivar experienced less dramatic creation declines than the touchy cultivar	During vegetative stages, the plant's capacity to hold capacities throughout a dry spell and quickly recover after re-watering is critical for evaluating final creation in wheat	Abid et al. (2018)

other hand, is the outcome of a range of physical cycles that begins with CO_2 dispersion throughout the stomata and ends with diffusion into the cells from outside due to differential concentration gradient.

This has an impact on the allocation of biomass to various organs, which is affected by their anticipated course of events and extension as well as the net carbon balance (Sharkey & Schrader, 2006). Assimilates interact with their sources, and growing organs function as sinks for them. Understanding plant responses to external variables depends on this interplay. At increasing temperatures, leaf mass reduces (Poorter et al., 2009). In hotter typical conditions, some research suggests that increased CO_2 can counteract the negative effects of high temperatures on photosynthesis and horticulture development, resulting in increased productivity (Polley, 2002). However, it's worth noting that the effects of high CO_2 and high temperature on plants are frequently combined, meaning that the combined consequences of these changes can't be predicted only based on data on their individual effects (Long, 1991). Additionally, the degree and direction of crop responses to elevated CO_2 and temperature change appear to be species- and even cultivar-specific (Polley, 2002).

However, the responses of boreal and temperate plants differ significantly from those of tropical plants. Tropical species have a stronger influence on leaf mass production when the temperature changes (some alterations in plant physiological mechanisms have been found. As a result, growth temperature affects the photosynthetic rate's thermal dependency (temperature acclimation) (Hikosaka et al., 2006). Garruna-Hernandez et al. (2014) discovered that the optimal temperature for

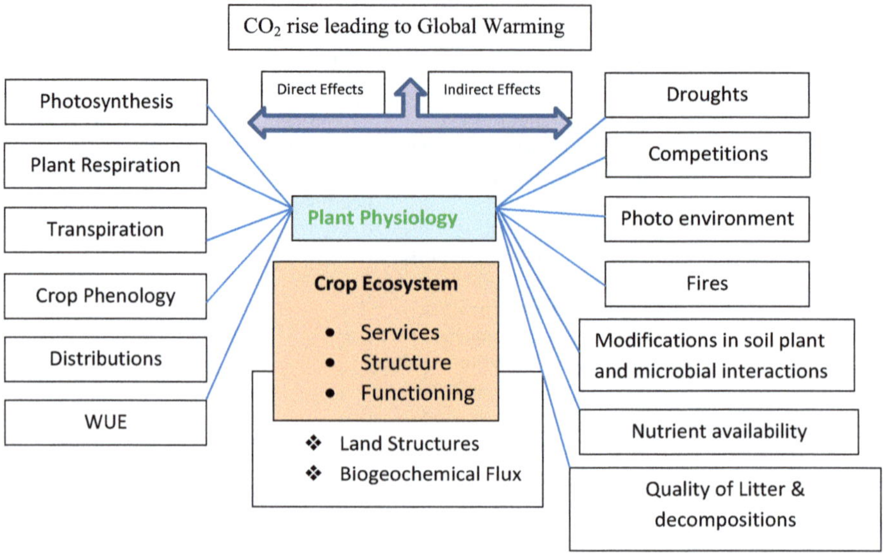

Fig. 11.5 Major driving factors affecting plant physiological responses

increasing photosynthetic rate increases when development temperature increases in a variety of species. An increase in diurnal air temperature in Habanero peppers increased both stomatal conductance and growth rate which resulted in a heat stress (Air temperature–Leaf temperature). Because of an increase in stomatal restriction, physiological boundaries shrank when leaf temperature was raised to 40 °C. In a tropical harvest like the Habanero pepper, the warm optimal reach (leaf temperature) was thought to be between 30 and 35 °C. According to some theories, higher temperatures could limit net carbon gain by encouraging plant respiration rather than photosynthesis. By promoting plant respiration over photosynthesis, warmer temperatures are expected to lower net carbon gain. However, productivity exhibits a substantial nonlinear increasing trend from 15 to 40 °C, followed by a rapid and practically vertical drop. It is true that the light-soaked photosynthetic rate of C_3 crops like wheat and rice peaks between 20 and 32 °C (Porter & Semenov, 2005). Expanded temperature could potentially reduce photosynthetic performance due to the intensity lability of Rubisco activase on the one hand, and the restriction of electron transport in the chloroplast on the other (Sage et al., 2008). Lobell and Field (2007) examined past productivity data for wheat, rice, and maize from the world's major proponents and found that a rise in the diurnal temperature range was associated with lower rice and maize yields in various agricultural locations throughout the world. This indicates the non-linear relationship of yield to temperature, which is presumably due to increased heat stress on warm days.

Enhanced plant water requirement owes to increasing transpiration rate at elevated temperatures which is an indirect impact of global warming that might possibly impair plant productivity (Peng et al., 2004). The effect of rising temperatures on vapour pressure deficit (VPD) and, as a result, barometrical ET will be insignificant if future temperature changes are anything like those observed in recent years, when global minimum temperatures have increased twice as quickly as most extreme temperatures, minimizing the diurnal temperature range (Folland et al., 2001). (Roderick & Farquhar, 2002). Rainfall distribution quantities can affect agricultural productivity in both good and negative ways. Higher rainfall, for example, might promote plant growth in (semi-) arid settings, but lesser precipitation could further limit plant growth and productivity. On the other hand, excessive rainfall in elevated zones can result in soil waterlogging, which can hinder crop growth (Jiang et al., 2008), or nutrient leaching in sandy soils (Anderson et al., 1998), and decreased rainfall on these soils could aid in reducing the negative effects of waterlogging and nutrient leaching. Precipitation during anthesis is very important for yearly crop productivity because it ensures water availability during the grain set and grain filling stages (Sadras & Connor, 1991). Crop growth and production will be affected by changes in future in-season rainfall distribution. One strategy for managing erratic rainfall coverage and insufficient irrigation water is regulating development and water usage before and during anthesis.

11.5 Modification of Bio-Climate (Microclimate) Through Interventions

Microclimates are affected by solar radiation, soil moisture, soil temperature, air temperature, air stickiness, and wind direction and speed. These microclimatic communications occur on a small scale, with a flat range of a few centimetres to two or three hundred metres and an upward range of a few centimetres to ten metres above ground level; or the covering limit layer manages them (Yoshino, 1987). Microclimates are thus the forming conditions for yields, and they are an important consideration in horticulture. This microclimatic scale is in fact unstable due to the extensive local diversity in soil types, surface slope and direction, and plant cover. Just a few metres above, the microclimate differs from the macroclimate, where air mixing processes are more active, resulting in more moderate and predictable weather (Unwin & Corbet, 1991). Microclimates, unlike macroclimates, can be controlled. The climate can be brought back to more reasonable levels by concentrating on a microclimatic scale of a single planting mound or a small area (Hadid & Toknok, 2020). Several integrated approach may be followed to adapt under the changing climatic scenarios.

11.5.1 Agronomic Interventions to Modify Crop Microclimate

11.5.1.1 Crop Spacing

The canopy temperature and net radiation are strongly influenced by crop spacing. The crop's capacity to absorb radiation decreases as rows are spaced farther apart, more radiation hits the soil surface, and soil temperature rises. However, if row spacing is decreased, radiation absorption by the crop rises and radiation transmission to the soil surface falls, resulting in a drop-in soil temperature. Crop geometry can alter relative humidity within the wheat crop canopy. 15 cm row spacing had the highest relative humidity, followed by 22.5 and 30 cm which clearly indicate high density planting results in higher RH thereby increasing chances of pest and disease incidence (Sandhu & Dhaliwal, 2016). Lower planting densities with optimum crop density are advised in humid locations to encourage faster leaf senescence and drying of fruits following rain and dew, which can be crucial as an indirect approach to prevent bacterial and fungal infections (Tamm et al., 2004). Paddy yield to be dependent on the panicles number per unit square metre and the quantity of seeds per panicle. The high correlation between the yield and the number of panicles per m^2 justifies that result, increasing the number of hills per unit is required to enhance the yield per hectare of upland rain-fed rice under changing climatic conditions (Marie-Noel et al., 2021). According to a study by Singh et al. (2005), the effect of spacing on canopy microclimate, vegetative development, and yield quality was significantly diminished. Light interception was better in guava plants at wider spacing. Due to maximal light absorption and an ideal microclimate in the orchard, wider plant spacing was proven to be desirable (Kalra et al., 1994), which led to increased plant production, higher fruiting density, and yield efficiency as demonstrated by Brar et al. (2012). At high planting densities, the canopy microclimate generated by lower temperature, higher RH, and greater fraction of light interception had a detrimental impact on yield attributes in cotton (Zhang et al., 2021). In maize plants with close spacing (6 cm plant-plant), they found decreased dry root weights, considerably reduced root reductive activity in each soil layer, and lower root biomass and plant nitrogen uptake. Close spacing increased the capacity for producing dry matter, harvest index, and nitrogen use efficiency in grains.

11.5.1.2 Adjusting Sowing Time as Adaptation Strategies for Climate Change

Temperature controls the cereal crop phenological progression from emergence to blossoming and maturity, but day duration and other physiological stresses also play a role. The responses of different species and types may differ. CC will affect cereal plant developmental time, but the precise changes will be determined by alterations in varieties as a result of plant breeding and variety selection. Temperature variations will have an impact on agricultural development and growth, as well as the unique interactions of crops with the biotic and abiotic environment at important periods of crop growth (Hakala et al., 2012). According to Bassu et al. (2009) and Turner (2004) sowing dates can help crop growing periods align with precipitation concentrations, increase water-use efficiency, and provide an escape from frost threat. Crop and cropping system adaptation are typically associated with optimising crop development stage timing to reduce stresses and maximise resource optimization. In order to help farmers and other stakeholders plan their crops while considering extreme weather events under CC scenarios across India, Bal et al. (2022c) used rainfall amount, wet spell duration, dry spell duration, and dry spell search period to define the start of crop growing season across 19 Agro-Ecological Regions in India. Forecasts of changes in cereal phenology due to CC by 2040, as recommended by Olesen et al. (2012), revealed a 1–3 week advance in the sowing date of spring cereals. Flowering and maturity dates are expected to be moved forward by 1–3 weeks. The highest alterations were seen in grain maize and the lowest in winter wheat. Plant breeding will need to change the duration of plant phenological phases in response to climate warming. In a similar study conducted in India sowing of Bt cotton on 28 July will incur higher productivity and profitable margins (Sankaranarayanan et al., 2020). As a result, the cotton sowing window should be re-standardized from July 21 to August 18. Huang et al. (2020) discovered that adjusting the sowing date by increasing growth-period precipitation contributed more (44.5–96.7%) to yield improvements across all maize planting regions than shifting cultivars (0–50.8%) or CC (53.1–23.0%), with the exception of the wet southwestern regions of China's maize belt. Adjusting sowing date (10–27 May) may incur better yield by obtaining more precipitation, potentially improving potato adaptation to changes in climate in the near future as compared to present sowing window which area affected by duration, thermal time, and precipitation (Hu et al., 2017). Klepeckas et al. (2020) suggested mid-September as the optimum sowing time for Wheat cultivars in Lithuania. Similarly, Climate variables such as temperature and precipitation, according to Wang et al. and He et al. (2015), can have a considerable impact on wheat harvest. A temperature increase may also hasten crop growth, whereas longer-growing cultivars or a later sowing date may cause growth to be delayed. Under baseline precipitation (57% at 1 °C), warming alone induced identical rates of yield losses across locales and crop kinds (Pirttioja et al., 2015; Sandhu et al., 2020). Unlikely, Ding et al. (2016) proposed that precisely deferred planting dates successfully maintained up with wheat productions, reduced ET, raised WUE, and slowed the decline in wheat yield and WUE since the 1950s due to dangerous atmospheric deviation. The wheat planting date could be deferred by 10–20 days in wet and medium years, and by 20–25 days in dry years, compared to the existing planting date, to maintain a high level of creation. As a result, farmers on the Loess Plateau and other semiarid areas throughout the globe could employ a delayed sowing date to adapt with global warming. A decision support system for taking agronomic decisions on sowing time and irrigation scheduling for rainfed regions of India was developed by Vijaya Kumar et al. (2021).

11.5.1.3 Irrigation Management

CC adaptation efforts in irrigation schemes must be considered as part of the overall irrigation modernization process. Improved water distribution methods, unambiguous transmission of water scarcity alerts to farmers, and technological and managerial modifications are all required for advanced irrigation systems to provide more flexible and reliable water delivery (FAO, 2007). In an experimental study by Deligios et al. (2019), in globe artichoke heads yield can be optimized while also resulting in a significant water reduction (−34%), thanks to the combination of evaporative cooling practice and precise irrigation technique. Furthermore, the study demonstrated that canopy-cooling techniques could be a successful tactic for reducing climatic variations and heat stress circumstances. As per a research on olive by Branquinho et al. (2021), Water deficits at higher levels might result in water savings of up to 22%. The findings shown that controlled deficit irrigation solutions can reduce the predicted crop irrigation requirement (CIR) growth for the CC scenarios. Rosa et al. (2020) found out that a soft-direction irrigation enlargement with small month-to-month water garage and deficit irrigation has the capacity to increase irrigated land with the aid of using 70 million hectares and feed three hundred million extra humans globally if global mean temperature enhances by up to 3 °C (Rockstrom et al., 2017). Majumder et al. (2016) proposed that need-based water system planning and water application in light of an IW/CPE proportion of 1.0 for Maize crop in mix with sub-surface manuring can be useful in upscaling the crop water use efficiency under Punjab conditions, considering limited water accessibility and changing climatic conditions. Subba Rao et al. (2022) reported that delayed sowing, increase in fertilizer dose, supplemental irrigation and their combinations can reduce the negative impact of climate change in future and improve yield in 16 major maize growing districts of India. Using farm-level data from three West Coast states, Olen et al. (2016) discovered that growers utilise sprinklers to reduce the risk of crop damage during extreme weather. Crop simulation models, according to Georgiou and Karpouzos (2017) and da Cunha et al. (2021), can be utilised as a decision-making tool for water system and rural region executives since they can determine the optimal arrangements for net benefit with limited water assets under future climatic fluctuation. Water stress as a technique would enhance the irrigation water supply's vulnerability and reliability. Under rising temperatures, the method of increasing irrigation efficiencies will sustain crop production in the area of the study conducted (Ashofteh et al., 2017).

11.5.1.4 Tillage

Corsi et al. (2012) defined conservation agriculture (CA) as a method for managing agro-ecosystems for increased and sustained output, greater revenues, and nutritional security while maintaining and enhancing the natural resource usage and sustainability of the environment. They went on to state that the three core concepts of CA are little mechanical soil disturbance, exceptionally long-lasting natural soil cover, and harvest assortment. Since the mid-1990s, the nation has been pushed over by protection cultivating methods such as least culturing and edit turn. Despite the fact that preservation farming (or sections of it) is increasingly being used as an environmental change transformation approach (Deressa et al., 2008; Nyanga et al., 2011), it isn't yet fully developed (Haggblade & Tembo, 2003). According to the results of a survey research conducted by Kuntashula et al. (2014), roughly 12% of farms practise minimum tillage. Minimum tillage tactics enhanced on-farm maize productivity by around 26% to 38%. Minimum tillage enhanced overall unit maize production as well. According to an FAO (2012) evaluation, the climate adaptation benefits of no-tillage can be significant. According to the study, wheat grown without tillage was more tolerant to Kazakhstan's 2012 drought and hot temperatures, yielding better yields than conventionally grown crops. According to Busari et al. (2015), of the several cultivation strategies used in environmental change relief and transformation processes, zero tillage (ZT) has been shown to be the most earth friendly. As a result, Conservation Tillage (CT), which includes ZT and minimum tillage (MT), may be able to break the surface minimised zone in soil while reducing soil upsetting influence, resulting in improved soil climate and harvest production while having a low natural effect. Studies on conservation agriculture strongly advise MT and ZT practices to reduce harmful environmental externalities when producing wheat in Bangladesh. Due to its capacity to increase Soil Organic Carbon (SOC) accumulation, lower both water loss and GHG emissions without compromising yield, and increase both CSA and sustainable wheat intensification in Bangladesh, the MT was found to be the strategy that was most effective in this study. Specifically, 15 million hectares in the United States, 24 million hectares in Brazil, 18 million hectares in Argentina, and 13 million hectares in Canada are covered by zero-tillage farming. The five countries with the highest hectarage include Australia, which has 9 million hectares under zero-tillage farming, for a total of 79 million hectares. The usage of zero-tillage is expanding globally. By 2005, it has more than doubled to 95 million hectares from its use of 45 million hectares in 1999. Using the latter data, only 17% of the total was made up of nations other than the top five. Brazil and Argentina have had the best success with conservation tillage. In these nations, conservation agriculture practices are used on 45–60% of total agricultural land (Uprety et al., 2012). Wakchaure et al. (2015) have reported significant changes in soil and canopy microclimate in laser levelled field compared to unlevelled one for taking sorghum crop.

11.5.1.5 Mulching

In order to create a better environment for plant growth and development, organic or inorganic components are applied to the soil's surface as part of an agricultural and soil protection practice known as mulching (Kamal & Singh, 2011). Mulches are notable for altering the heat/energy and water balance at the soil's surface, which benefits plant growth. Mulches maintain soil moisture by lowering evaporation (Hillel, 1982), however the composition and optical characteristics of the mulch have an impact on the mulch's effects on soil temperature (Ham et al., 1993). In contrast, organic mulches lower maximum soil temperatures while raising minimum ones, while polyethylene mulches boost both maximum and minimum soil temperatures when compared to unmulched soil (Ham et al., 1993; Teasdale & Mohler, 1993). Sarkar and Singh (2007) suggested that straw mulch can lower maximum soil temperature due to incoming solar radiation interception, high reflectivity, and low heat conductivity; however, the management of which is reliant on soil wetness, incidental radiation, mulch rate application, and season. The favourable impacts of three types of mulches, namely banana, coconut, and vetiver, were identified in a study performed in Nasola Valley, USA by Lalljee (2013). The findings revealed that runoff losses of soil and nutrients were dramatically decreased; the mechanisms causing these reductions include mechanical barriers to raindrops on soil aggregates, greater water infiltration, and stimulation of soil microbial activity. Mulches can reduce, and in some cases totally remove, the requirement to water crop plants (Kader et al., 2019; Iqbal et al., 2019). Few organic mulches often behave as sponges, retaining rainfall and irrigation water thereby preventing runoff while also providing water when crops are in need. According to Deng et al. (2019), under future climatic conditions, biodegradable mulch sheets can improve water use efficiency by an average of 9.5%. The findings also show that promoting the usage of bio-degradable mulch films in hot and arid regions could be beneficial for improving agricultural development under global climate changing scenarios. By employing straws as mulch, the runoff was reduced by 43% (Borst & Woodburn, 1942). Mulches significantly minimize the need for supplemental irrigation by retaining water and lowering soil profile drainage at the same time (Smith, 2000). Mhlanga et al. (2021) evaluated the impact of mulch as a potential enhancing and stabilising agent in an experimental investigation. Mulch combined with regular tillage produces a consistent yield on sandy soils with minimal rainfall. This is the first study to point out by the usage of mulch. Agricultural systems in southern Africa are becoming more stable and resilient, as well as their ability to adapt to environmental changes is enhancing. Organic mulches also restore many soil borne bacteria by directly competing with or degrading harmful organisms through a variety of enzymatic processes (Crohn & Bishop, 1999). The majority of soil borne bacteria that dwells inside the mulch produces cellulolytic enzymes, which destroys it (Downer et al., 2002). Utilizing materials for mulching with the intention of boosting microbial colonization can also help with the biological control of a number of illnesses (Entry et al., 2005). In a field trial conducted to evaluate the impact of tillage and mulch on cotton yield and productivity, Robert et al. (2000) found that plots blanketed with wheat stubble mulch produced 35% more lint than the control plots (no mulch). Evaporation was also lower and water use efficiency was significantly high under mulched conditions than under open field conditions.

11.5.2 Agronomic/ Crop Improvement Strategies

11.5.2.1 Crop Geometry

Planting geometry may be optimised to reduce the negative impacts of CC on phenology. Planting geometry often influences plant phenology as well (Hyles et al., 2020). Crop microclimate is influenced by planting geometry, which includes planting density, row spacing, and seed charge. Reduced planting density, wider plant or row-to-row spacing, and bypass-row designs are only a few adaptation techniques that can be utilized in dryland farming to maximize the use of the soil–water resources available (Singh et al., 2012). Sun radiation interception, canopy coverage, crop increase price, and biomass accumulation are all affected by planting geometry (Raza et al., 2019). Furthermore, because of higher photo-assimilates, optimal planting geometry increased overall dry remember production and, ultimately, grain yield growth (Mahato & Adhikary, 2017). According to Huang et al. (2021), planting practises have a significant impact on maize productivity. In dryland farming structures, soil temperature and boom period precipitation are decisive elements. As a result, they recommended the use of proper planting patterns in accordance with the limiting elements in order to boost dryland agriculture.

11.5.2.2 Crop Duration and Cropping System

Souza et al. (2019) suggested that crop cycle lengths may no longer be able to mitigate the negative effects of CC on Brazilian corn. As per a study conducted by Pires et al. (2016) soybean yields may rise during next few years, provided that it is grown as a sole crop; if double cropping is chosen, soybeans must be sown earlier, which could decline its yields by up to 25% (if planted at the start) to introduce maize in the cropping system. Likely due to the unpredictable pattern of the rainfall, 17% reduction in double cropping system is expected in the Central Brazilian region (Abrahao & Costa, 2018). Unlikely, under a harsher and more unfavourable climate for soybean-maize cultivation, the timing of soybean sowing dates (early sowing) will become more crucial in influencing the dynamics of later maize sowing dates and yields (Andrea et al., 2020). In the south of South Africa, sequential cropping (together with intercropping) has been utilised to boost crop yields in rainfed areas (Francis, 1986). Sequential cropping is the practise of producing two or more crops on the same land in succession or concurrent developmental growth stages (relay cropping). Warmer temperatures in Benin, South Africa, according to Duku et al. (2018), may significantly reduce crop production improvements obtained via the extension of sequential crop planting. Based on the CC scenario, 50 to 95% of lands currently which can promote sequential cropping pattern need to be gradually transformed under monocropping system. In northeast China, new cultivars have been developed to extend all crop growing seasons in order to better meet climatic warming effects. In the region, the growth durations of recently certified rice cultivars, corn, and soybean varieties have been extended by 14.0, 7.0, and 2.7 days, respectively, since the 1950s (Chen et al., 2012). Zhang et al. (2013) concludes that single rice sowing dates in the northeast region of China were 0.8, 0.9, and 0.3 days $°C^{-1}$ in advance in early maturing (short duration) and late maturing (Long duration) rice in the southern regions of China respectively. It was also observed that during the period 1981–2009 in northern provinces of China, sowing date, emergence date, and dormancy of wheat were delayed on by almost 1.5, 1.7, and 1.5 days, respectively once every 10 years; although the planting time of Maize progressed by a considerable delay in harvest date (Xiao et al., 2013).

11.5.2.3 Plant Composition

Hatfield and Prueger (2011) give a brief overview of the CC's agro-ecological effects on plant responses, with a focus on growth, yield, and quality, as well as the importance of agro-ecological research. Interactions between elements such as rising CO2, temperature, and humidity, precipitation anomalies and nitrogenous (N) fertiliser are all things to think about. Plant breeding is a practical technological solution to environmental problems. constraints as a supplement to changes in agricultural management (e.g., Yadav et al., 2011), However achieving changes in cultivar yield or quality takes a long time and a lot of money. Environmental adaptation has a complicated genetic underpinning, making it difficult to determine which gene sets are necessary for maximum outcomes and explanations how genes controlling different features interact. Plant breeding capital needs determining when and where this is the most cost-effective approach for an industry in addressing CC. In a hotter environment, the reaction of distinctive plants to warming decides community-level efficiency. By contrast, warming-induced water push reduces graminoids and forbs cover in an elevated steppe (Ganjurjav et al., 2016) and grasses biomass in a semi-arid grassland (Yang et al., 2011). Plant community composition can be intervened by warming-induced changes in soil dampness conditions, which may be due in portion to the shifting root properties of diverse plant useful bunches (Xu et al., 2020). Deeper-rooted plants can utilize precipitation from the upper and more profound layers of the soil, and they are more dryspell safe (Xu et al., 2018). In a mesic snow-capped prairie on the northern Tibetan plains, under the effect of warming temperature and drought, plant community composition changes from sedges to graminoids and forbs medicines, referring to a four-year experiment (Liu et al., 2018a, 2018b). The roots of graminoids and forbs are regularly more profound than those of sedges (Li et al., 2018). A slew of bio-security threats has rendered plant species helpless. Environmental change will have an impact on canny relationships between trouble and trim species, affecting item effectiveness and quality, despite the expected repercussions for the rate of assault of interesting pests (Legreve & Duveiller, 2010) and the propagation of endemic microorganism strains (Chakraborty & Datta, 2003). Fundamental (i.e. cell dividers), biochemical, and physiological mechanisms mediate plant resistance to pests and diseases. These systems are being impacted by CC, and a seemingly safe assortment may become more vulnerable as a result (Chakraborty et al., 2008).

Both direct impacts of environment (e.g., filtering and volatilisation by precipitation) and circuitous necessities to match development prerequisites lower input prerequisites. Dwyer et al. (2007) investigated how CC affects nutritional status, notably nitrogen. They discovered a drop in N concentration in C_3 species when CO_2 levels were high. Erbs et al. (2010) recently established that when flour is produced under increased CO_2 with low to moderate doses of N fertiliser, the quality of the flour is reduced. There are currently reproduction drives with various field harvests to improve the efficiency with which these assets are utilised, such as more yield per unit of contribution for both nutrients and water (especially N and P). Several high-value plants, including the tomato, are known to exhibit hereditary diversity for responsiveness to increased CO_2 (Lindhout et al., 1991). Major vegetables (tomato, pepper, and cucumber) are frequently cultivated in growth chambers under artificially raised CO_2 circumstances in temperate regions of Europe, the United States, and Asia. As a result, vegetable breeding has long been focused on enhancing under these conditions (to a significant extent inside commercial firms).

11.5.2.4 Improved Cultivars

For risk reduction, soil and water conservation, and efficient water management, the Food and Agriculture Organization of the United Nations (FAO) advises adopting improvised harvesting techniques and assortments methods (containing both herbaceous and tree crops). The promotions of climate-resilient crops and cultivars (either yearly or perennial) makes a difference in minimizing the ill impacts of CC on agrarian systems while also ensuring stable agrarian generation. Presenting modern crops or assortments, or bringing back legacy crops, expands agrarian generation, which has positive effects on biodiversity and biological system administrations, particularly if developed in conjunction with preservation agribusiness techniques (for example, least soil unsettling influence, long-term soil natural cover).Matching phenology to developing season duration thru adjustments in duration of the cultivar and thermal sensitivity (Kumudini et al., 2014), changes in root structure allowing higher infiltration of soil water during the Green Revolution, as shown by the guidance of using the transition from photoperiod-sensitive landraces to photoperiod-unfeeling semi-bantam wheat and rice varieties (Lynch & Wojciechowski, 2015). The genetic structure of those capabilities is polygenic in nature. Many genes with minor outcomes had been proven to steer flowering time regulation (e.g. Buckler et al., 2009). Studies conducted by Tesfaye et al. (2017) revealed regional and temporal variations in CC consequences within South Asia, highlighting the need for carefully focused adaptation actions, such as crop breeding for improved heat stress tolerance. Heat tolerance characteristics were included into maize cultivars, which showed tremendous promise for adapting maize to changing climates. Punjab observed to have similarly excessive tiers of resilience. The truth that warmness tolerance has now no longer been extensively embraced became striking. Furthermore, new populations of overdue blight have arisen, posing a risk to (destiny) potato manufacturing in some of the regions (Chowdappa et al., 2015). Growers in numerous states, on the alternative hand, aren't making an investment in new substances which are proof against this virus. Investing in breeding for excessive warmness tolerance and overdue blight resistant varieties (Rana et al., 2013), in addition to powerful dissemination of these sorts to the states in which they may be maximum needed, want unique attention.

11.5.2.5 Cropping Patterns

There are various potential administration enhancements for farming frameworks to manage anticipated climatic and environmental changes. Among the adaptation strategies some are:

(a) Altering inputs such varieties and species to those with superior thermal time and vernalization requirements, as well as increased resistance to thermal shock and drought.
(b) Depending on the weather, adjusting fertilizer dosages to retain fruit or grain quality.
(c) Water management, including adjusting irrigation levels and timing and frequencies.
(d) Expanding the use of water-saving and soil-moisture-preserving technologies (e.g., crop residue retention).
(e) Improve water use and transportation in areas with declining rainfall. Regulating water to minimize nutrient leaching, erosion, and water logging as rainfall intensifies.

Simple and attainable changes modified major unfavourable consequences on sorghum (48–58%) to neutral to possibly beneficial ones (0–12%) in a modelling research at Modena, Italy (Tubiello et al., 2002). The adjustments for this situation were to change assortment and establishing times to forestall dry season and warmed spells during the more blazing and drier late spring months expected under environmental change. When a large number of variation studies are combined, most of the advantages of changing current structures will be recognised during direct warming (2 °C) and then level out as temperature changes increase. The benefits for rice and maize are less significant than for wheat, with a yield benefit of 10%, when compared to yields obtained without transformation. These improvements in results could result in temperature reductions of up to 12 °C in mild districts and 1.5–3 °C in tropical areas, thereby reducing the delay in implications for many years (Easterling et al., 2007) and providing more times for mitigation efforts.

11.5.3 Shelterbelts/ Windbreaks

The Intergovernmental Panel on CC claim, "there is evidence of long-term changes in the large-scale atmospheric circulation, such as the pole shift and the intensification of the westerly wind," and these changes are irreversible and will continue in the near future. A windbreak is a column or clump of trees, bushes, or other major elements that steer and obstruct the breeze (e.g. walls). Windbreaks protect fields from winning breeze patterns by reducing breeze speeds before they reach the crops. Shelterbelts are widely utilised to mitigate the effects of wind, flood, and desertification everywhere (Fig. 11.6). The viability of windbreaks is significantly influenced by the abundance of trees, the shelterbelt's permeability, its width, and its direction (Tsognamsrai & Dugarjay, 2016; Pablo & Mark, 2002). Easterling et al. (1997) claimed that in shelter, ET is reduced, lessening crop moisture stress. The good influence of haven on dryland maize is most notable for the most dramatic changes: highest precipitation setback and most notable expansion in windspeed, at all degrees of temperature increase. Trees in the rural environment reduced N_2O outflows from the soil and higher level of CH_4 oxidation (Evers et al., 2010). Deep-rooted trees can prevent denitrification by collecting residual NO_3 and excess soil water, which are vulnerable to N_2O emissions and NO_3 leaching. Despite the fact that a small portion of this N is stored in tree biomass, a larger portion is replenished by soil microbial fusion through dissipation. The function of the tree root safety net is understood (Allen et al., 2004), and it results in more effective N cycling, decreased interest in N manure by neighbouring soils, and fewer N_2O emissions from N treatment (Amadi et al., 2016). Shelterbelts also enhance C sequestration in stable SOC pools. When compared to the adjacent developed field, Sauer et al. (2007) found that SOC fixations in the 0–7.5 cm soil layer beneath a shelterbelt of red cedar (*Juniperus virginiana*) and Scots pine (*Pinus sylvestris*) were 55% higher (3.04%) and 12% higher (7.5–15 cm soil profundity). Perera (1981) found that even with an extremely thick shelterbelt (under 0.25 porosity), the breeze speed on the shelterbelt's leeward side was still exceptionally high. As per past wind disintegration research, the ideal level of shelterbelt porosity is somewhere in the range of 0.25 and 0.50 (Jiang & Zhu, 2002). Zhang et al. showed that rice dwelling decreased the creation loss of farmland safeguarded by shelterbelts by two thirds contrasted with unprotected farmland in Jiangsu Province (North of Shanghai); the most elevated insurance happened where the essential windbreak had more than 90% tree inclusion. Scattering can be diminished by up to 35% on the leeward side of windbreak borders, and evaporative drops are related with decreases in wind scrambling, as per Lomas and Schlesinger (1971). Miller et al. (1973) found a 20% drop in evaporation rate on the leeward side of a shelter belt over the course of six days of quantifications. Regional evaporation in China was truncated by 14% thanks to an astronomically immense-scale shelterbelt network (Wang et al., 2001).

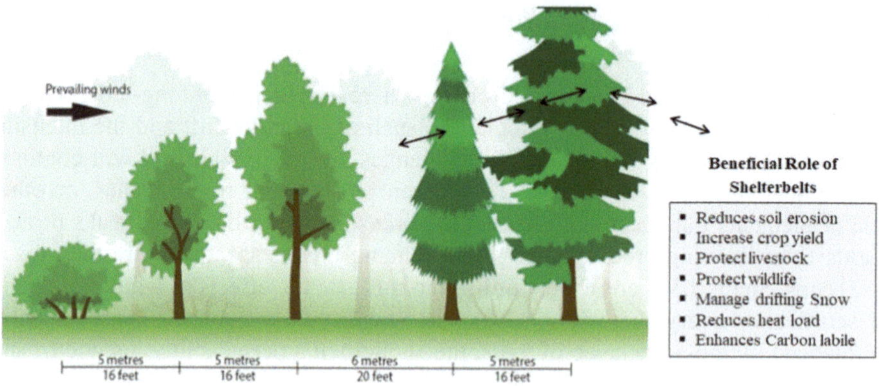

Fig. 11.6 Beneficial Role of Shelterbelts

11.6 Conclusion

The future climatic projections are predicting increase in ambient temperature, extreme precipitations, atmospheric carbon dioxide concentration, and outrageous climate occasions. Therefore, agriculture and its allied sectors will mostly be impacted negatively causing reduction in its performance. Therefore, one of the options is to adjust/modify the microclimate inside the crop canopy, both in field and horticultural crops to counteract the extremes and abrupt variability of climatic activities. These adjustments in microclimate, including radiation, soil temperature, soil moisture, among other variables, have proven to be highly beneficial for crop growth and development.

References

Abid, M., Ali, S., Qi, L. K., Zahoor, R., Tian, Z., Jiang, D, & Dai, T. (2018). Physiological and biochemical changes during drought and recovery periods at tillering and jointing stages in wheat (*Triticum aestivum* L.). *Scientific Reports, 8*(1), 1–15. https://doi.org/10.1038/s41598-018-21441-7

Abrahao, G. M., & Costa, M. H. (2018). Evolution of rain and photoperiod limitations on the soybean growing season in Brazil: The rise (and possible fall) of double-cropping systems. *Agricultural and Forest Meteorology, 256*, 32–45. https://doi.org/10.1016/j.agrformet.2018.02.031

Abrol, Y. P., & Ingram, K. T. (1996). Effects of higher day and night temperatures on growth and yields of some crop plants. In *Global climate change and agricultural production. Direct and indirect effects of changing hydrological, pedological and plant physiological processes* (pp. 124–140). Chichester: Wiley.

Adams, R. M., Hurd, B. H., Lenhart, S., & Leary, N. (1998). Effects of global CC on agriculture: An interpretative review. *Climate Research, 11*(1), 19–30.

Ahmed, M., Hassan, F. U., Asim, M., Aslam, M. A., & Akram, M. N. (2010). Correlation of photothermal quotient with spring wheat yield. *African Journal of Biotechnology, 9*, 7869–7852.

Allen, S. C., Jose, S., Nair, P. K. R., Brecke, B. J., Nkedi-Kizza, P., & Ramsey, C. L. (2004). Safety-net role of tree roots: Evidence from a pecan (Caryaillinoensis K. Koch)–cotton (Gossypium hirsutum L.) alley cropping system in the southern United States. *Forest Ecology and Management, 192*, 395–407. https://doi.org/10.1016/j.foreco.2004.02.009

Amadi, C. C., Van Rees, K. C. J., & Farrell, R. E. (2016). Soil-atmosphere exchange of carbon dioxide, methane and nitrous oxide in shelterbelts compared with adjacent cropped fields. *Agriculture, Ecosystems & Environment, 223*, 123–134. https://doi.org/10.1016/j.agee.2016.02.026

Anderson, G. C., Fillery, I. R. P., Dunin, F. X., Dolling, P. J., & Asseng, S. (1998). Nitrogen and water flows for pasture-ley and lupin-wheat rotations in deep sands. II. Water drainage and nitrate leaching. *Australian Journal of Agricultural Research, 49*, 345–361.

Andrea, M. C. D., Dallacort, R., Tieppo, R. C., et al. (2020). Assessment of climate change impact on double-cropping systems. *SN Appl Sci, 2*, 544. https://doi.org/10.1007/s42452-020-2325-z

Apel, K., & Hirt, H. (2004). Reactive oxygen species: Metabolism, oxidative stress, and signal transduction. *Annual Review of Plant Biology, 55*, 373–399.

Ashofteh, P. S., Bozorg-Haddad, O., & Loáiciga, H. A. (2017). Development of adaptive strategies for irrigation water demand management under CC. *Journal of Irrigation and Drainage Engineering, 143*(2), 04016077. https://doi.org/10.1061/(ASCE)IR.1943-4774.0001123

Ashraf, M., Saeed, M. M., & Qureshi, M. J. (1994). Tolerance to high temperature in cotton (Gossypium hirsutum L.) at initial growth stages. *Environmental and Experimental Botany, 34*(3), 275–283. https://doi.org/10.1016/0098-8472(94)90048-5

Atkin, O. K., & Tjoelker, M. G. (2003). Thermal acclimation and the dynamic response of plant respiration to temperature. *Trends in Plant Science, 8*, 343–351.

Aydinalp, C., & Cresser, M. S. (2008). The effects of global CC on agriculture. *American-Eurasian Journal of Agricultural & Environmental Sciences, 3*(5), 672–676.

Babé, A., Lavigne, T., Séverin, J. P., Nagel, K., Walter, A., Chaumont, F., Batoko, H., Beeckman, T., & Draye, X. (2012). Repression of early lateral root initiation events by transient water deficit in barley and maize. *Philosophical Transactions of the Royal Society London B Biological Sciences, 367*, 1534–1541. https://doi.org/10.1098/rstb.2011.0240

Bal, S. K., Dhakar, R., Kumar, P. V., Mishra, A., Pramod, V. P., Chandran, M. S., Sandeep, V. M., Rao, A. S., Gill, K. K., & Prasad, R. (2021). Temporal trends in frost occurrence and their prediction models using multivariate statistical techniques for two diverse locations of Northern India. *Theoretical and Applied Climatology, 146*, 1097–1110.

Bal, S. K., Manikandan, N., Sandeep, V. M., Vijaya Kumar, P., Lunagaria, M. M., Subba Rao, A. V. M., Pramod, V. P., & Singh, V. K. (2022a). Criteria based decisions for determining agroclimatic onset of the crop growing season. *Agricultural and Forest Meteorology, 317*, 108903. https://doi.org/10.1016/j.agrformet.2022.108903

Bal, S. K., Prasad, J. V. N. S., & Singh, V. K. (2022b). Heat Wave 2022—Causes, impacts and way forward for indian agriculture. ICAR-CRIDA Technical Bulletin 01–2022, p. 50. https://doi.org/10.13140/RG.2.2.15040.20482

Bal, S. K., Sandeep, V. M., Kumar, P. V., Rao, A. S., Pramod, V. P., Manikandan, N., & Bhaskar, S. (2022c). Assessing impact of dry spells on the principal rainfed crops in major dryland regions of India. *Agricultural and Forest Meteorology, 313*, 108768. https://doi.org/10.1016/j.agrformet.2021.108768

Balal, R. M., Shahid, M. A., Javaid, M. M., Iqbal, Z., Anjum, M. A., Garcia-Sanchez, F., & Mattson, N. S. (2016). The role of selenium in amelioration of heat-induced oxidative damage in cucumber under high temperature stress. *Acta Physiologiae Plantarum, 38*(6), 1–14. https://doi.org/10.1007/s11738-016-2174-y

Balasubramanian, S., Sureshkumar, S., Lempe, J., & Weigel, D. (2006). Potent induction of Arabidopsis thaliana flowering by elevated growth temperature. *PLoS Genetics, 2*, 980–989. https://doi.org/10.1371/journal.pgen.0020106

Barnabás, B., Jäger, K., & Fehér, A. (2008). The effect of drought and heat stress on reproductive processes in cereals. *Plant, Cell & Environment, 31*(1), 11–38. https://doi.org/10.1111/j.1365-3040.2007.01727.x

Baroowa, B., & Gogoi, N. (2014). Biochemical changes in black gram and green gram genotypes after imposition of drought stress. *Journal of Food Legumes, 27*, 350–353.

Bassu, S., Asseng, S., Motzo, R., & Giunta, F. (2009). Optimising sowing date of durum wheat in a variable Mediterranean environment. *Field Crops Research, 111*, 109–118.

Bhagat, K. P., Bal, S. K., Singh, Y., Potekar, S. V., Saha, S., Ratnakumar, P., Wakchaure, G. C., & Minhas, P. S. (2017). Effect of reduced PAR on growth and photosynthetic efficiency of soybean genotypes. *Journal of Agrometeorology, 19*(1), 1–9.

Bonan, G. (2016). *Ecological climatology; concepts and applications* (3rd ed.). University of Colorado.

Borst, H. L., & Woodburn, R. (1942). The effect of mulching and methods of cultivation on runoff and erosion from Muskingham silt loam. *Journal of Agricultural Engineering, 23*, 19–22.

Bot, A., & Benites, J. (2005). The importance of soil organic matter: Key to drought-resistant soil and sustained food production (No. 80). Food & Agriculture Organization.

Bramer, I., Anderson, B. J., Bennie, J., Bladon, A. J., De Frenne, P., Hemming, D., Hill, R. A., Kearney, M. R., Körner, C., Korstjens, A. H., & Lenoir, J. (2018). Advances in monitoring and modelling climate at ecologically relevant scales. *Advances in Ecological Research, 58*, 101–161.

Branquinho, S., Rolim, J., & Teixeira, J. (2021). CC adaptation measures in the irrigation of a super-intensive olive orchard in the south of Portugal. *Agronomy, 11*(8), 1658. https://doi.org/10.3390/agronomy11081658

Britannica, T. (2020). Editors of Encyclopaedia. Argon. Encyclopedia Britannica.

Buckler, E. S., Holland, J. B., Bradbury, P. J., Acharya, C. B., Brown, P. J., Browne, C., & McMullen, M. D. (2009). The genetic architecture of maize flowering time. *Science, 325*(5941), 714–718.

Busari, M. A., Kukal, S. S., Kaur, A., Bhatt, R., & Dulazi, A. A. (2015). Conservation tillage impacts on soil, crop and the environment. *International Soil and Water Conservation Research, 3*(2), 119–129.

Chakraborty, S., & Datta, S. (2003). How will plant pathogens adapt to host plant 45 resistance at elevated CO_2 under a changing climate? *New Phytologist, 159*, 733–742. https://doi.org/10.1046/j.1469-8137.2003.00842.x

Chakraborty, S., Luck, J., Hollaway, G., Freeman, A., Norton, R., & Garrett, K. A. (2008). Impacts of global change on diseases of agricultural crops and forest trees. *CABI Reviews: Perspectives in Agriculture, Veterinary Science, Nutrition and Natural Resources, 3*, 1–15. https://doi.org/10.1111/j.1365-3059.2010.02411.x

Challinor, A. J., Watson, J., Lobell, D. B., Howden, S. M., Smith, D. R., & Chhetri, N. (2014). A meta-analysis of crop yield under CC and adaptation. *Nature Climate Change, 4*(4), 287–291.

Chaves, M. M., Flexas, J., & Pinheiro, C. (2009). Photosynthesis under drought and salt stress: Regulation mechanisms from whole plant to cell. *Annals of Botany, 103*, 551–560. https://doi.org/10.1093/aob/mcn125

Cheesman, A. W., & Winter, K. (2013). Growth response and acclimation of CO_2 exchange characteristics to elevated temperatures in tropical tree seedlings. *Journal of Experimental Botany, 64*, 3817–3828.

Chen, J., Saunders, S. C., Crow, T. R., Naiman, R. J., Brosofske, K. D., Mroz, G. D., Brookshire, B. L., & Franklin, J. F. (1999). Microclimate in forest ecosystem and landscape ecology: Variations in local climate can be used to monitor and compare the effects of different management regimes. *BioScience, 49*(4), 288–297.

Chen, C. Q., Qian, C. R., Deng, A. X., & Zhang, W. J. (2012). Progressive and active adaptations of cropping system to climate change in Northeast China. *European Journal of Agronomy, 38*, 94–103.

Chowdappa, P., Nirmal Kumar, B. J., Madhura, S., Mohan Kumar, S. P., Myers, K. L., Fry, W. E., & Cooke, D. E. L. (2015). Severe outbreaks of late blight on potato and tomato in South

India caused by recent changes in the Phytophthora infestans population. *Plant Pathology, 64*, 191–199.
Cline, W. R. (2007). *Global warming and agriculture: Impact estimates by country*. Peterson Institute.
Cornic, G. (2000). Drought stress inhibits photosynthesis by decreasing stomatal aperture not by affecting ATP synthesis. *Trends in Plant Science, 5*(5), 187–188.
Corsi, S., Friedrich, T., Kassam, A., Pisante, M., & Sà, J. D. M. (2012). Soil organic carbon accumulation and greenhouse gas emission reductions from conservation agriculture: a literature review. Food and Agriculture Organization of the United Nations (FAO).
Crohn, D. M., & Bishop, M. L. (1999). Proximate carbon analysis for compost production and mulch use. *Transactions of ASAE, 42*, 791–797.
Crous, K. Y., Quentin, A. G., Lin, Y. S., Medlyn, B. E., Williams, D. G., Barton, C. V. M., & Ellsworth, D. S. (2013). Photosynthesis of temperate Eucalyptus globulus trees outside their native range has limited adjustment to elevated CO_2 and climate warming. *Global Change Biology, 19*, 3790–3807.
Cunha, R. L. D. F., Silva, B., & Avegliano, P. B. (2023). A comprehensive modeling approach for crop yield forecasts using AI-based methods and crop simulation models. arXiv preprint arXiv: 2306.10121.
da Cunha, E. R., Santos, C. A. G., da Silva, R. M., Bacani, V. M., & Pott, A. (2021). Future scenarios based on a CA-Markov land use and land cover simulation model for a tropical humid basin in the Cerrado/Atlantic forest ecotone of Brazil. *Land Use Policy, 101*, 105141.
Dai, A. (2013). Increasing drought under global warming in observations and models. *Nature CC, 3*, 52–58.
Dela, G., Or, E., Ovadia, R., Nissim-Levi, A., Weiss, D., & Oren-Shamir, M. (2003). Changes in anthocyanin concentration and composition in 'Jaguar'rose flowers due to transient high-temperature conditions. *Plant Science, 164*(3), 333–340.
Deligios, P. A., Chergia, A. P., Sanna, G., Solinas, S., Todde, G., Narvarte, L., & Ledda, L. (2019). Climate change adaptation and water saving by innovative irrigation management applied on open field globe artichoke. *Science of the Total Environment, 649*, 461–472. https://doi.org/10.1016/j.scitotenv.2018.08.349
Deng, L., Meng, X., Yu, R., & Wang, Q. (2019). Assessment of the effect of mulch film on crops in the arid agricultural region of China under future climate scenarios. *Water, 11*(9), 1819. https://doi.org/10.3390/w11091819
Deressa, T., Hassan, R. M., & Ringler, C. (2008). Measuring Ethiopian Farmers' Vulnerability to CC Across Regional States. IFPRI Discussion Paper 806. Washington, DC: International Food Policy Research Institute (IFPRI).
Dermody, O., Long, S. P., & DeLucia, E. H. (2006). How does elevated CO_2 or ozone affect the leaf-area index of soybean when applied independently? *New Phytologist, 169*, 145–155. https://doi.org/10.1111/j.1469-8137.2005.01565.x
Deutsch, C. A., Tewksbury, J. J., Tigchelaar, M., Battisti, D. S., Merrill, S. C., Huey, R. B., & Naylor, R. L. (2018). Increase in crop losses to insect pests in a warming climate. *Science, 361*(6405), 916–919.
Dey, M. M., Gosh, K., Valmonte-Santos, R., Rosegrant, M. W., & Chen, O. L. (2016). Economic impact of CC and CC adaptation strategies for fisheries sector in Solomon Islands: Implication for food security. *Marine Policy, 67*, 171–178.
Dhankher, O. P., & Foyer, C. H. (2018). Climate resilient crops for improving global food security and safety. *Plant, Cell and Environment, 41*, 877–884.
Ding, D. Y., Feng, H., Zhao, Y., He, J. Q., Zou, Y. F., & Jin, J. M. (2016). Modifying winter wheat sowing date as an adaptation to climate change on the Loess Plateau. *Agronomy Journal, 108*(1), 53–63. https://doi.org/10.2134/agronj15.0262
Downer, J., Faber, B., & Menge, J. (2002). Factors affecting root rot control in mulched avocado orchards. *Hort Tech, 12*, 601–605.

Duku, C., Zwart, S. J., & Hein, L. (2018). Impacts of climate change on cropping patterns in a tropical, sub-humid watershed. *PLoS ONE, 13*(3), e0192642. https://doi.org/10.1371/journal.pone.0192642

Dwyer, S. A., Ghannoum, O., Nicotra, A., & Von Caemmerer, S. (2007). High temperature acclimation of C4 photosynthesis is linked to changes in photosynthetic biochemistry. *Plant, Cell and Environment, 30*(1), 53–66. https://doi.org/10.1111/j.1365-3040.2006.01605.x

Eastburn, D. M., Degennaro, M. M., Delucia, E. H., Dermody, O., & McElrone, A. J. (2010). Elevated atmospheric carbon dioxide and ozone alter soybean diseases at SoyFACE. *Global Change Biology, 16*, 320–330.

Easterling, W. E., Hays, C. J., Easterling, M. M., & Brandle, J. R. (1997). Modelling the effect of shelterbelts on maize productivity under CC: An application of the EPIC model. *Agriculture, Ecosystems & Environment, 61*(2–3), 163–176. https://doi.org/10.1016/S0167-8809(96)01098-5

Easterling, W., Aggarwal, P., Batima, P., Brander, K., Erda, L., Howden, M., Kirilenko, A., Morton, J., Soussana, J. -F., Schmidhuber, J., & Tubiello, F. (2007) In M. L. Parry, O. F. Canziani, J. P. Palutikof, P. J. van der Linden & C. E. Hanson (Eds.), *CC 2007: Impacts, adaptation and vulnerability* (pp. 273–313). Cambridge University Press.

Entry, J. A., Strausbaugh, C. A., & Sojka, R. E. (2005). Compost amendments decrease Verticilliumdahliae infection on potato. *Compost Science & Utilization, 13*, 43–49.

Erbs, M., Manderscheid, R., Jansen, G., Seddig, S., Pacholski, A., & Weigel, H. J. (2010). Effects of free-air CO_2 enrichment and nitrogen supply on grain quality parameters and elemental composition of wheat and barley grown in a crop rotation. *Agriculture, Ecosystems & Environment, 136*, 59–68. https://doi.org/10.1016/j.agee.2009.11.009

Evers, A. K., Bambrick, A., Lacombe, S., Dougherty, M. C., Peichl, M., Gordon, A. M., Thevathasan, N. V., Whalen, J., Bradley, R. L. (2010). Potential greenhouse gas mitigation through temperate tree-based intercropping systems. *Open Agriculture Journal, 4*, 49–57. https://doi.org/10.2174/1874331501004010049

FAO. (2007). Modernising irrigation management—The MASSCOTE approach. FAO irrigation and drainage paper No. 63. Rome.

FAO. (2012). Advancement and impact of conservation agriculture/no-till technology adoption in Kazakhstan [Internet]. FAO Information note. http://www.fao.org/ag/ca/doc/Importance_Zero_Tillage_Northern_Kazakhstan.pdf. Accessed 01 Feb 2022.

FAO. (2016). Physical factors affecting soil organisms. Cited from the World Wide Web. http://www.fao.org/agriculture/crops/thematic-sitemap/theme/spi/soil-biodiversity/soil-organisms/physical-factorsaffecting-soil-organisms/en/

Farooq, M., Wahid, A., Kobayashi, N., Fujita, D., & Basra, S. M. A. (2009). Plant drought stress: Effects, mechanisms and management. *Agronomy for Sustainable Development, 29*, 185–212.

Feeley, K. J., Wright, S. J., Supardi, M. N. N., Kassim, A. R., & Davies, S. J. (2007). Decelerating growth in tropical forest trees. *Ecology Letters, 10*, 461–469.

Feng, Y. L., & Cao, K. F. (2005). Photosynthesis and photoinhibition after night chilling in seedlings of two tropical tree species grown under three irradiances. *Photosynthetica, 43*, 567–574. https://doi.org/10.1007/s11099-005-0089-8

Foken, T., & Napo, C. J. (2008). *Micrometeorology* (Vol. 308). Springer.

Folland, C. K., Rayner, N. A., Brown, S. J., Smith, T. M., Shen, S. S. P., Parker, D. E., Macadam, I., Jones, P. D., Jones, R. N., Nicholls, N., & Sexton, D. M. H. (2001). Global temperature change and its uncertainties since 1861. *Geophysical Research Letters, 28*, 2621–2624.

Francis, C. A. (1986) *Multiple cropping systems*, No. 631.58 M961. Macmillan.

Ganjurjav, H., Gao, Q., Gornish, E., Schwartz, M., Liang, Y., & Cao, X., et al. (2016). Differential response of alpine steppe and alpine meadow to climate warming in the Central Qinghai–Tibetan plateau. *Agricultural and Forest Meteorology, 223*, 233–240. https://doi.org/10.1016/j.agrformet.2016.03.017

Garrity, D. P., & O'Toole, J. C. (1995). Selection for reproductive stage drought avoidance in rice, using infrared thermometry. *Agronomy Journal, 87*(4), 773–779.

Garruna-Hernandez, R., Orellana, R., Larque-Saavedra, A., & Canto, A. (2014). Understanding the physiological responses of a tropical crop (*Capsicum chinense* Jacq.) at high temperature. *PLoS one, 9*(11), e111402.

Gay, C., Estrada, F., Conde, C., Eakin, H., & Villers, L. (2006). Potential impacts of CC on agriculture: A case of study of coffee production in Veracruz, Mexico. *Climate Change, 79*(3), 259–288.

Georgiou, P. E., & Karpouzos, D. K. (2017). Optimal irrigation water management for adaptation to CC. *International Journal of Sustainable Agricultural Management and Informatics, 3*(4), 271–285. https://doi.org/10.1504/IJSAMI.2017.090608

Gholipoor, M., & Shahsavani, S. (2008). Simulation study of past CC effect on Chickpea phenology at different sowing dates in Gorgan, Iran. In International Meeting on Soil Fertility Land Management and Agroclimatology, Turkey, pp. 13–36.

Gliessman, S. R. (2015). *Agroecology: The ecology of sustainable food systems* (3rd ed.). Taylor & Francis Group.

Gunderson, C. A., O'Hara, K. H., Campion, C. M., Walker, A. V., & Edwards, N. T. (2010). Thermal plasticity of photosynthesis: The role of acclimation in forest responses to a warming climate. *Global Change Biology, 16*, 2272–2286.

Hadid, A., & Toknok, B. (2020). Vegetable and food crop production with microclimate modification. *Journal of Physics: Conference Series* (1434, No. 1, p. 012027). IOP Publishing.

Haggblade, S., & Tembo, G. (2003). *Development, diffusion and impact of conservation farming in Zambia*. Lusaka: IFPRI and FSRP - MSU.

Hakala, K., Jauhiainen, L., Himanen, S. J., Ro¨tter, R., Salo, T., & Kahiluoto, H. (2012). Sensitivity of barley varieties to weather in Finland. *Journal of Agricultural Science, 150*, 145–160.

Ham, J. M., Kluitenberg, G. J., & Lamont, W. J. (1993). Optical properties of plastic mulches affect the field temperature regime. *Journal of the American Society for Horticultural Science, 118*, 88–193.

Hare, P. D., & Cress, W. A. (1997). Metabolic implications of stress-induced proline accumulation in plants. *Plant Growth Regulation, 21*(2), 79–102.

Hatfield, J. L., & Prueger, J. H. (2011). *Agroecology: Implications for plant response to CC* (pp. 27–43). Wiley-Blackwell.

Hayat, R., & Ahmad, S. (2014). Biochemical, physiological and agronomic response of wheat to changing climate of rainfed Pakistan. *Pakistan Journal of Botany, 51*, 2.

He, L., Asseng, S., Zhao, G., Wu, D., Yang, X., Zhuang, W., & Yu, Q. (2015). Impacts of recent climate warming, cultivar changes, and crop management on winter wheat phenology across the Loess Plateau of China. *Agricultural and Forest Meteorology, 200*, 135–143. https://doi.org/10.1016/j.agrformet.2014.09.011

Higashi, Y., Okazaki, Y., Myouga, F., Shinozaki, K., & Saito, K. (2015). Landscape of the lipidome and transcriptome under heat stress in Arabidopsis thaliana. *Scientific Reports, 5*(1), 10533.

Hikosaka, K., Kazumasa, I., Almaz, B., Onno, M., & Yusuke, O. (2006). Temperature acclimation of photosynthesis: Mechanisms involved in the changes in temperature dependence of photosynthetic rate. *Journal of Experimental Botany, 57*, 291–302.

Hillel, D. (1982). *Introduction to soil physics*. Academic Press.

Hu, Q., Yang, N., Pan, F., Pan, X., Wang, X., & Yang, P. (2017). Adjusting sowing dates improved potato adaptation to CC in semiarid region, China. *Sustainability, 9*(4), 615.

Huang, M., Wang, J., Wang, B., Li Liu, D., Yu, Q., He, D., & Pan, X. (2020). Optimizing sowing window and cultivar choice can boost China's maize yield under 1.5 °C and 2 °C global warming. *Environmental Research Letters, 15*(2), 024015. https://doi.org/10.1088/1748-9326/ab66ca

Huang, F., Liu, Z., Zhang, P., & Jia, Z. (2021). Hydrothermal effects on maize productivity with different planting patterns in a rainfed farmland area. *Soil and Tillage Research, 205*, 104794. https://doi.org/10.1016/j.still.2020.104794

Hussain, M., Malik, M. A., Farooq, M., Ashraf, M. Y., & Cheema, M. A. (2008). Improving drought tolerance by exogenous application of glycinebetaine and salicylic acid in sunflower. *Journal of Agronomy and Crop Science, 194*, 193–199.

Hyles, J., Bloomfeld, M. T., Hunt, J. R., Trethowan, R. M., & Trevaskis, B. (2020). Phenology and related traits for wheat adaptation. *Heredity*. https://doi.org/10.1038/s41437-020-0320-1

Iqbal, R., Raza, M. A., Saleem, M. F., Khan, I. H., Ahmad, S., Zaheer, M. S., & Haider, I. (2019). Physiological and biochemical appraisal for mulching and partial rhizosphere drying of cotton. *Journal of Arid Land, 11*, 785–794.

Jackson, R. B., Randerson, J. T., Canadell, J. G., Anderson, R. G., Avissar, R., Baldocchi, D. D., Bonan, G. B., Caldeira, K., Diffenbaugh, N. S., Field, C. B., & Hungate, B. A. (2008). Protecting climate with forests. *Environmental Research Letters, 3*(4), 044006.

Jiang, F. Q., & Zhu, J. J. (2002). Phase-directional management of protective plantations I. Fundamnetals. *Chinese Journal of Applied Ecology, 13*, 1352–1355.

Jiang, D., Fan, X., Dai, T., & Cao, W. (2008). Nitrogen fertiliser rate and post-anthesis waterlogging effects on carbohydrate and nitrogen dynamics in wheat. *Plant and Soil, 304*, 301–314.

Kader, M. A., Singha, A., Begum, M. A., Jewel, A., Khan, F. H., & Khan, N. I. (2019). Mulching as water-saving technique in dry land agriculture. *Bulletin of the National Research Centre, 43*, 1–6.

Kalra, S. K., Sidhu, P. J. S., Dhaliwal, G. S., & Singh, R. (1994). Effect of different spacings on yield of guava cv. Allahabad Sufeda. *Indian Journal of Horticulture, 51*(3), 272–274.

Kamal, S., & Singh, A. K. (2011). Effect of black plastic mulch on soil temperature and tomato yield. *Progressive Horticulture, 43*(2), 337–339.

Kang, Y., Khan, S., & Ma, X. (2009). CC impacts on crop yield, crop water productivity and food security-A review. *Progress in Natural Science, 19*, 1665–1674.

Karimi, V., Karami, E., & Keshavarz, M. (2018). CC and agriculture: Impacts and adaptive responses in Iran. *Journal of Integrative Agriculture, 17*(1), 1–15.

Klepeckas, M., Januškaitienė, I., Vagusevičienė, I., & Juknys, R. (2020). Effects of different sowing time to phenology and yield of winter wheat. Agricultural and food science. *Helsinki: Scientific Agricultural Society of Finland, 29*(4).

Kotak, S., Larkindale, J., Lee, U., Von, P., Vierling, E., & Scharf, K. D. (2007). Complexity of the heat stress response in plants. *Current Opinion in Plant Biology, 10*, 310–316.

Kumudini, S., Andrade, F. H., Boote, K. J., Brown, G. A., Dzotsi, K. A., Edmeades, G. O., Gocken, T., Goodwin, M., Halter, A. L., Hammer, G. L., Hatfield, J. L., & Tollenaar, M. (2014). Predicting maize phenology: Intercomparison of functions for developmental response to temperature. *Agronomy Journal, 106*(6), 2087–2097.

Kuntashula, E., Chabala, L. M., & Mulenga, B. P. (2014). Impact of minimum tillage and crop rotation as climate change adaptation strategies on farmer welfare in smallholder farming systems of Zambia. *Journal of Sustainable Development, 7*(4), 95. https://doi.org/10.5539/jsd.v7n4p95

Lal, M., Singh, K. K., Rathore, L. S., Srinivasan, G., & Saseendran, S. A. (1998). Vulnerability of rice and wheat yields in NW—India to future changes in climate. *Agricultural and Forest Meteorology, 89*, 101–114.

Lalljee, B. (2013). Mulching as a mitigation agricultural technology against land degradation in the wake of CC. *International Soil and Water Conservation Research, 1*(3), 68–74.

Legreve, A., & Duveiller, E. (2010). Preventing potential disease and pest 30 epidemics under a changing climate. In M. P. Reynolds (Ed.), *'CC and crop production, vol. 1'* (pp. 50–70). CABI.

Li, Y., Ye, W., Wang, M., & Yan, X. (2009). Climate change and drought: A risk assessment of crop-yield impacts. *Climate research, 39*(1), 31–46.

Li, S., Wheelerbm, T., Challinorc, A., Lind, E., Jud, H., & Xud, Y. (2010). The observed relationships between wheat and climate in China. *Agricultural and Forest Meteorology, 150*, 1412–1419.

Li, C., Peng, F., Xue, X., You, Q., Lai, C., & Zhang, W., et al. (2018). Productivity and quality of alpine grassland vary with soil water availability under experimental warming. *Frontiers in Plant Science, 9*, 1790. https://doi.org/10.3389/fpls.2018.01790

Lindhout, P., Pet, G., Jansen, R., & Jansen, H. (1991). Genetic differences in growth within and between Lycopersicon species. *Euphytica, 57*, 259–265.

Liu, H., Mi, Z., Lin, L. I., Wang, Y., Zhang, Z., Zhang, F., & He, J. S. (2018a). Shifting plant species composition in response to CC stabilizes grassland primary production. *Proceedings of the National Academy of Sciences, 115*(16), 4051–4056.

Liu, H., Mi, Z., Lin, L., Wang, Y., Zhang, Z., & Zhang, F., et al. (2018b). Shifting plant species composition in response to CC stabilizes grassland primary production. *Proceedings of the National Academy of Sciences USA, 115*, 4051–4056. https://doi.org/10.1073/pnas.1713895115

Lobell, D. B., & Asner, G. P. (2003). Climate and management contributions to recent trends in U.S. agricultural yields. *Science, 299*, 1032.

Lobell, D., & Field, C. (2007). Global scale climate–crop yield relationships and the impacts of recent warming. *Environmental Research Letters, 2*, 014002.

Lomas, J., & Schlesinger, E. (1971). The influence of a windbreak on evaporation. *Agricultural Meteorology, 8*, 107–115.

Long, S. P. (1991). Modification of the response of photosynthetic productivity to rising temperature by atmospheric CO_2 concentrations: Has its importance been underestimated? *Plant, Cell and Environment, 14*, 729–739.

Loomis, R. S., & Amthor, J. S. (1999). Yield potential, plant assimilatory capacity, and metabolic efficiencies. *Crop Science, 39*(6), 1584–1596.

Lynch, J. P., & Wojciechowski, T. (2015). Opportunities and challenges in the subsoil: Pathways to deeper rooted crops. *Journal of Experimental Botany, 66*(8), 2199–2210. https://doi.org/10.1093/jxb/eru508

Madhu, M., & Hatfield, J. L. (2013). Dynamics of plant root growth under increased atmospheric carbon dioxide. *Agronomy Journal, 105*, 657–669. https://doi.org/10.2134/agronj2013.0018

Mahato, A. (2014). CC and its impact on agriculture. *International Journal of Scientific Research Publications, 4*(4), 1–6.

Mahato, M., & Adhikari, B. B. (2017). Effect of planting geometry on growth of rice varieties. *International Journal of Applied Science and Biotechnology, 5*, 423–429.

Majumder, D., Kingra, P. K., & Kukal, S. S. (2016). Water productivity of spring maize under modified soil microenvironment. *Journal of Agrometeorology, 18*(1), 134.

Maleki, A., Naderi, A., Naseri, R., Fathi, A., Bahamin, S., & Maleki, R. (2013). Physiological performance of soybean cultivars under drought stress. *Bulletin of Environment Pharmacology and Life Sciences, 2*, 38–44.

Marie-Noel, M. F., Joseph, B. M., Liliane, T., Felix, F., Dorothy, M., & Salvador, D. P. (2021). Effect of plant spacing on the growth and yield of rainfed rice (Oryza Sativa) in the bimodal rain forest zone of Cameroon. *Journal of Agriculture and Crops, 7*(2), 48–59.

Matthew, G., Pierre, M., & Ariel, O. B. (2017). Negative impacts of CC on cereal yields: Statistical evidence from France. *Environmental Research Letters, 12*(5), 054007.

Mendelsohn, R. (2009). The impact of CC on agriculture in developing countries. *Journal of Natural Resources Policy Research, 1*(1), 5–19.

Meziane, D., & Shipley, B. (2001). Direct and indirect relationships between specific leaf area, leaf nitrogen and gas exchange. Effects of irradiance and nutrient supply. *Annals of Botany, 88*, 915–927.

Mhlanga, B., Ercoli, L., Pellegrino, E., Onofri, A., & Thierfelder, C. (2021). The crucial role of mulch to enhance the stability and resilience of cropping systems in southern Africa. *Agronomy for Sustainable Development, 41*(2), 1–14. https://doi.org/10.1007/s13593-021-00687-y

Miller, D. R., Rosenberg, N. J., & Bagley, W. T. (1973). Soybean water use in the shelter of a slat fence windbreak. *Agricultural Meteorology, 11*, 405–418.

Mislan, K. A. S., & Helmuth, B. (2008). Microclimate. *General Ecology*, 2389–2393.

Mittler, R. (2002). Oxidative stress, antioxidants and stress tolerance. *Trends in Plant Science, 7*, 405–410.

Mollo, L., Martins, M. C. M., Oliveira, V. F., Nievola, C. C. L., & Figueiredo-Ribeiro, R. D. C. (2011). Effects of low temperature on growth and non-structural carbohydrates of the imperial bromeliad Alcantareaimperialis cultured in vitro. *Plant Cell, Tissue and Organ Culture, 107*(1), 141–149. https://doi.org/10.1007/s11240-011-9966-y

Munne-Bosch, S., Jubany-Mari, T., & Alegre, L. (2003). Enhanced photo-and antioxidative protection, and hydrogen peroxide accumulation in drought-stressed Cistus clusii and Cistus albidus plants. *Tree Physiology, 23*(1), 1–12. https://doi.org/10.1093/treephys/23.1.1
Naiman, R. J., Decamps, H., McClain, M. E., & Lickens, G. E. (Eds.). (2005). *Riparia: Ecology, conservation, and management of streamside communities*. Academic Press.
Nam, N. H., Subbaroa, G. V., Chauhan, Y. S., & Johansen, C. (1998). Importance of canopy attributes in determining dry matter accumulation of pigeon pea under contrasting moisture regimes. *Crop Science, 38*, 955–961.
Nicholson, S. E. (2012). *Dryland climatology*. Florida State University.
Nicotra, A. B., Atkin, O. K., Bonser, S. P., Davidson, A. M., Finnegan, E. J., Mathesius, U., Poot, P., Purugganan, M. D., Richards, C. L., Valladares, F., & van Kleunen, M. (2010). Plant phenotypic plasticity in a changing climate. *Trends in Plant Science, 15*, 684–692.
Nonami, H. (1998). Plant water relations and control of cell elongation at low water potentials. *Journal of Plant Research, 111*, 373–382.
Nyanga, P. H., Johnsen, F. H., Aune, J. B., & Kalinda, T. H. (2011). Smallholder farmers perceptions of CC and conservation agriculture: evidence from Zambia. *Journal of Sustainable Development, 4*(4), 73–85. https://doi.org/10.5539/jsd.v4n4p73
Olen, B., Wu, J., & Langpap, C. (2016). Irrigation decisions for major west coast crops: Water scarcity and climatic determinants. *American Journal of Agricultural Economics, 98*(1), 254–275.
Olesen, J. E., Børgesen, C. D., Elsgaard, L., Palosuo, T., Rötter, R. P., Skjelvåg, A. O., Peltonen-Sainio, P., Börjesson, T., Trnka, M., Ewert, F., Siebert, S., & Van Der Fels-Klerx, H. J. (2012). Changes in time of sowing, flowering and maturity of cereals in Europe under CC. *Food Additives & Contaminants: Part A, 29*(10), 1527–1542.https://doi.org/10.1080/19440049.2012.712060
Ong, C. K., Black, C., & Wilson, J. (Eds.). (2015). *Tree-crop interactions: Agroforestry in a changing climate*. CABI.
Ottman, M. J., Kimball, B. A., White, J. W., & Wall, G. W. (2012). Wheat growth response to increased temperature from varied planting dates and supplemental infrared heating. *Agronomy Journal, 104*(1), 7–16.
Pablo, L. P., & Mark, B. (2002). Windbreaks in southern Patagonia, Argentina: A review of research on growth models, windspeed reduction, and effects on corps. *Agroforestry Systems, 56*, 129–144.
Pathak, H., Ladha, J. K., Aggarwal, P. K., Peng, S., Das, S., Singh, Y., Singh, B., Kamra, S. K., Mishra, B., Sastri, A. S. R. A. S., Aggarwal, H. P., Das, D. K., & Gupta, R. K. (2003). Trends of climatic potential and on-farm yields of rice and wheat in the Indo-Gangetic Plains. *Field Crops Research, 80*, 223–234.
Peng, S. B., Huang, J. L., Sheehy, J. E., Laza, R. C., Visperas, R. M., Zhong, X. H., Centeno, G. S., Khush, G. S., & Cassman, K. G. (2004). Rice yields decline with higher night temperature from global warming. *Proceedings of the National Academy of Sciences, 101*, 9971–9975.
Perera, M. D. A. E. S. (1981). Shelter behind two-dimensional solid and porous fences. *Journal of Wind Engineering and Industrial Aerodynamics, 8*(1–2), 93–104.
Pires, G. F., Abrahão, G. M., Brumatti, L. M., Oliveira, L. J. C., Costa, M. H., Liddicoat, S., Kato, E., & Ladle, R. J. (2016). Increased climate risk in Brazilian double cropping agriculture systems: Implications for land use in Northern Brazil. *Agricultural and Forest Meteorology, 228–229*, 286–298. https://doi.org/10.1016/j.agrformet.2016.07.005
Pirttioja, N., Carter, T. R., Fronzek, S., Bindi, M., Hoffmann, H., Palosuo, T., & Asseng, S. (2015). Temperature and precipitation effects on wheat yield across a European transect: A crop model ensemble analysis using impact response surfaces. *Climate Research, 65*, 87–105. https://doi.org/10.3354/cr01322
Polley, H. W. (2002). Implications of atmospheric and CC for crop yield. *Crop Science, 42*, 131–140.
Poorter, H., Niinemets, U., Poorter, L., Wright, I. J., & Villar, R. (2009). Causes and consequences of variation in leaf mass per area (LMA): A meta-analysis. *New Phytologist, 182*, 565–588.

Porter, J. R., & Semenov, M. A. (2005). Crop responses to climatic variation. *Philosophical Transactions of the Royal Society b: Biological Sciences, 360*, 2021–2035.

Posmyk, M. M., Bailly, C., Szafranska, K., Janas, K. M., & Corbineau, F. (2005). Antioxidant enzymes and isoflavonoids in chilled soybean (Glycine max (L.) Merr.) seedlings. *Journal of Plant Physiology, 162*, 403–412. https://doi.org/10.1016/j.jplph.2004.08.004

Powell, J. P., & Reinhard, S. (2016). Measuring the effects of extreme weather events on yields. *Weather and Climate Extremes, 12*, 69–79.

Rana, R. K., Sharma, N., Arya, S., Singh, B. P., Kadian, M. S., Chaturvedi, R., & Pandey, S. K. (2013). Tackling moisture stress with drought-tolerant potato (Solanum tuberosum) varieties: Perception of Karnataka farmers. *Indian Journal of Agricultural Sciences, 83*(2), 216–222.

Raza, M. A., et al. (2019). Effect of planting patterns on yield, nutrient accumulation and distribution in maize and soybean under relay intercropping systems. *Scientific Reports, 9*, 4947. https://doi.org/10.1038/s41598-019-41364-1

Reckling, M., Döring, T. F., Bergkvist, G., Chmielewski, F., Stoddard, F., Watson, C., Seddig, S., & Bachinger, J. (2018). Grain legume yield instability has increased over 60 years in long-term field experiments as measured by a scale-adjusted coefficient of variation. *Aspects of Applied Biology, 138*, 15–20.

Reich, P. B., Hobbie, S. E., & Lee, T. D. (2014). Plant growth enhancement by elevated CO_2 eliminated by joint water and nitrogen limitation. *Nature Geoscience, 7*, 920–924.

Richards, K. (2004). Observation and simulation of dew in rural and urban environments. *Progress Physical Geography, 28*(1), 76–94.

Richards, R. A. (2006). Physiological traits used in the breeding of new cultivars for water-scarce environments. *Agricultural Water Management, 80*, 197–211.

Richards, R., Condon, A., & Rebetzke, G. (2011). Traits to improve yield in dry environments. Application of Physiology in Wheat Breeding. CIMMYT, pp. 88–100.

Rivero, R. M., Ruiz, J. M., Garcıa, P. C., Lopez-Lefebre, L. R., Sánchez, E., & Romero, L. (2001). Resistance to cold and heat stress: Accumulation of phenolic compounds in tomato and watermelon plants. *Plant Science, 160*(2), 315–321.

Robert, J. L., Baumhardt, R. L., Hicks, S. K., & Heilman, J. L. (2000). Soil and plant water evaporation from strip-tilled cotton: Measurement and simulation. *Agronomy Journal, 86*, 987–994.

Rockstrom, J., Williams, J., Daily, G., Noble, A., Matthews, N., Gordon, L., Wetterstrand, H., DeClerck, F., Shah, M., Steduto, P., de Fraiture, C., & Smith, J. (2017). Sustainable intensification of agriculture for human prosperity and global sustainability. *Ambio, 46*(1), 4–17.

Roderick, M. L., & Farquhar, G. D. (2002). The cause of decreased pan evaporation over the past 50 years. *Science, 299*, 1997–1998.

Rosa, L., Chiarelli, D. D., Sangiorgio, M., Beltran-Peña, A. A., Rulli, M. C., D'Odorico, P., & Fung, I. (2020). Potential for sustainable irrigation expansion in a 3 °C warmer climate. *Proceedings of the National Academy of Sciences, 117*(47), 29526–29534.

Rosenzweig, C., & Parry, M. L. (1994). Potential impact of climate change on world food supply. *Nature, 367*(6459), 133–138. https://doi.org/10.1038/367133a0

Rosenzweig, C., Elliott, J., Deryng, D., Ruane, A. C., Müller, C., Arneth, A., Boote, K. J., Folberth, C., Glotter, M., Khabarov, N., Neumann, K., Piontek, F., Pugh, T. A. M., Schmid, E., Stehfest, E., Yang, H., & Jones, J. W. (2014). Assessing agricultural risks of CC in the 21st century in a global gridded crop model intercomparison. *Proceedings of the National Academy of Sciences USA, 111*(9), 3268–3273.

Rucker, K. S., Kvien, C. K., Holbrook, C. C., & Hook, J. E. (1995). Identification of peanut genotypes with improved drought avoidance traits. *Peanut Science, 24*, 14–18.

Ruf, F., Schroth, G., & Doffangui, K. C. C. (2015). Cocoa migrations and deforestation in West Africa: What does the past tell us about the future? *Sustainability Science, 10*, 101–111.

Sadras, V. O., & Connor, D. J. (1991). Physiological basis of the response of harvest index to the fraction of water transpired after anthesis—A simple model to estimate harvest index for determinate species. *Field Crops Research, 26*, 227–239.

Sage, R. F., Way, D. A., & Kubien, D. S. (2008). Rubisco, Rubisco activase, and global CC. *Journal of Experimental Botany, 59*, 1581–1595.

Sandhu, S. K., & Dhaliwal, L. K. (2016). Crop geometry effects on relative humidity variation within wheat crop. *Asian Journal of Environmental Science, 11*(1), 94–101. https://doi.org/10.15740/HAS/AJES/11.1/94-101

Sandhu, S. S., Kaur, P., Gill, K. K., & Vashisth, B. B. (2020). The effect of recent climate shifts on optimal sowing windows for wheat in Punjab, India. *Journal of Water and Climate Change, 11*(4), 1177–1190.

Sankaranarayanan, K., Prakash, A. H., & Rajendran, K. (2020). Effect of sowing time on productivity of Bt and non Bt cotton under climate change. *Bulletin of the National Research Centre, 44*, 1–12.

Sardans, J., & Peñuelas, J. (2005). Drought decreases soil enzyme activity in a Mediterranean Quercus ilex L. forest. *Soil Biology and Biochemistry, 37*(3), 455–461.

Sarkar, S., & Singh, S. R. (2007). Interactive eff ect of tillage depth and mulch on soil temperature, productivity and water use pattern of rainfed barley. *Soil Tillage and Research, 92*, 79–86. https://doi.org/10.1016/j.still.2006.01.014

Sauer, T., Cambardella, C., & Brandle, J. (2007). Soil carbon and tree litter dynamics in a red cedar–scotch pine shelterbelt. *Agroforestry Systems, 71*, 163–174. https://doi.org/10.1007/s10457-007-9072-7

Schlenker, W., & Roberts, M. J. (2009). Nonlinear temperature effects indicate severe damages to US crop yields under CC. *Proceedings of the National Academy of Sciences, 106*(37), 15594–15598.

Seneviratne, S. I., Corti, T., Davin, E. L., Hirschi, M., Jaeger, E. B., Lehner, I., Orlowsky, B., & Teuling, A. J. (2010). Investigating soil moisture-climate interactions in a changing climate: A review. *Earth Science Reviews, 99*(3–4), 125–161.

Sharkey, T. D., & Schrader, S. M. (2006). High temperature stress. In R. K. V. Madhava, A. S. Raghavendra, & K. Janardhan (Eds.), *Physiology and molecular biology of stress tolerance in plants* (pp. 101–129). Springer.

Singh, S., Sandhu, S., Dhaliwal, L. K., & Singh, I. (2012). Effect of planting geometry on microclimate, growth and yield of mung-bean (*Vigna radiata* L.). *Journal of Agricultural Physics, 12*, 70–73.

Singh, A., Dhaliwal, G. S., & Hundal, S. S. (2005). Effect of planting distance on radiation interception behaviour of guava (*Psidium guajava* L.). *Journal of Agrometeorology, 7*, 220–224.

Slama, I., Ghnaya, T., Hessini, K., Messedi, D., Savouré, A., & Abdelly, C. (2007). Comparative study of the effects of mannitol and PEG osmotic stress on growth and solute accumulation in Sesuvium portulacastrum. *Environmental and Experimental Botany, 61*, 10–17. https://doi.org/10.1016/j.envexpbot.2007.02.004

Smith, M. W. (2000). Cultivar and mulch affect cold injury of young pecan trees. *Journal of the American Pomological Society, 54*, 29–33.

Souza, T. T., Antolin, L. A., Bianchini, V. D., Pereira, R. A., Silva, E. H., & Marin, F. R. (2019). Longer crop cycle lengths could offset the negative effects of CC on Brazilian maize. *Bragantia, 78*, 622–631. https://doi.org/10.1590/1678-4499.20190085

Stigter, C. J. (1994a). Management and manipulation of microclimate. In F. Griffiths (Ed.), *Handbook of agricultural meteorology* (pp. 273–284). Oxford University Press.

Stigter, C. J. (1994b). Micrometeorology. In J.F. Griffiths (Ed.), *Handbook of agricultural meteorology* (pp. 59–72). Oxford University Press.

Subba Rao, A. V. M., Sarath Chandran, M. A., Bal, S. K., Pramod, V. P., Sandeep, V. M., Manikandan, N., Raju, B. M. K., Prabhakar, M., Islam, A., Naresh Kumar, S., & Singh, V. K. (2022). Evaluating area-specific adaptation strategies for maize under future climates of India. *Science of the Total Environment, 836*, 155511. https://doi.org/10.1016/j.scitotenv.2022.155511

Tamm, L., Smit, A. B., Hospers, M., Janssens, S. R. M., Buurma, J. S., Molgaard, J. P., & Leifert, C. (2004). Assessment of the socio-economic impact of late blight and state-of-the-art management in European organic potato production systems. http://www.orgprints.org/2936

Teasdale, J. R., & Mohler, C. L. (1993). Light transmittance, soil temperature, and soil moisture under residue of hairy vetch and rye. *Agronomy Journal, 85*, 673–680.

Tesfaye, K., Zaidi, P. H., Gbegbelegbe, S., Boeber, C., Rahut, D. B., Getaneh, F., Seetharam, K., Erenstein, O., & Stirling, C. (2017). CC impacts and potential benefits of heat-tolerant maize in South Asia. *Theoretical and Applied Climatology, 130*(3), 959–970. https://doi.org/10.1007/s00704-016-1931-6

Tsognamsrai, D., & Dugarjav, C. (2016). Effectiveness of some methods for restoration of degraded pastures in the desert–steppe zone of Mongolia. *Arid Ecosystems, 6*, 189194.

Tubiello, F. N., Rosenzweig, C., Goldberg, R. A., Jagtap, S., & Jones, J. (2002). *Climate Research, 20*, 259–270.

Turner, N. C. (2004). Agronomic options for improving rainfall-use efficiency of crops in dryland farming systems. *Journal of Experimental Botany, 55*, 2413–2425.

Unanimous. (2021). https://gml.noaa.gov/ccgg/trends/

Unwin, D. M., & Corbet, S. A. (1991). *Insects, plants and microclimate* (Naturalists' handbooks Series, Band 15). Richmond Publishing Company Ltd.

Uprety, D. C., Dhar, S., Hongmin, D., Kimball, B. A., Garg, A., & Upadhyay, J. (2012). Technologies for climate change mitigation: Agriculture sector. Department of Management Engineering, Technical University of Denmark. https://backend.orbit.dtu.dk/ws/portalfiles/portal/10267141/Technologies_for_Climate_Change_Mitigation_Agriculture_for_upload.pdf

Vijaya Kumar, P., Bal, S. K., Dhakar, R., Sarath Chandran, M. A., Subba Rao, A. V., Sandeep, V. M., Pramod, V. P., Malleswari, S. N., Sudhakar, G., Solanki, N. S., & Shivaramu, H. S. (2021). Algorithms for weather-based management decisions in major rainfed crops of India: Validation using data from multi-location field experiments. *Agronomy Journal, 113*(2), 1816–1830.

Vose, R. S., Easterling, D. R., & Gleason, B. (2005). Maximum and minimum temperature trends for the globe: An update through 2004. *Geophysical Research Letters, 32*, 23–30.

Waha, K., Müller, C., Bondeau, A., Dietrich, J. P., Kurukulasuriya, P., Heinke, J., & Lotze-Campen, H. (2013). Adaptation to CC through the choice of cropping system and sowing date in sub-Saharan Africa. *Global Environ Change, 23*(1), 130–143.

Wakchaure, G. C., Choudhary, R. L., Kaledhonkar, M. J., & Bal, S. K. (2015). Effect of precision land leveling on microenvironment and sorghum productivity in water scarce Deccan region. *Journal of Agrometeorology, 17*(2), 149–157.

Wang, Z., & Huang, B. (2004). Physiological recovery of Kentucky bluegrass from simultaneous drought and heat stress. *Crop Science, 44*, 1729–1736.

Wang, H., Takle, E. S., & Shen, J. (2001). Shelterbelts and windbreaks: Mathematical modelling and computer simulations of turbulent flows. *Annual Review of Fluid Mechanics, 33*(1), 549–586.

Wardlaw, I. F., Dawson, I. A., & Munibi, P. (1989). The tolerance of wheat to high temperatures during reproductive growth. II. Grain development. *Australian Journal of Agricultural Research, 40*, 15–24.

Way, D. A., & Oren, R. (2010). Differential responses to changes in growth temperature between trees from different functional groups and biomes: A review and synthesis of data. *Tree Physiology, 30*, 669–688.

Way, D. A., & Sage, R. F. (2008). Thermal acclimation of photosynthesis in black spruce Piceamariana (Mill.) BSP. *Plant, Cell and Environment, 31*, 1250–1262.

Wheeler, T., & von Braun, J. (2013). CC impacts on global food security. *Science, 341*(6145), 508–513.

Xiao, D. P., Tao, F. L., Liu, Y. J., Shi, W. J., Wang, M., Liu, F. S., Zhang, S., & Zhu, Z. (2013). Observed changes in winter wheat phenology in the North China Plain for 1981–2009. *International Journal of Biometeorology, 57*, 275–285.

Xu, W., Zhu, M., Zhang, Z., Ma, Z., Liu, H., & Chen, L., et al. (2018). Experimentally simulating warmer and wetter climate additively improves rangeland quality on the Tibetan plateau. *Journal of Applied Ecology, 55*, 1486–1497. https://doi.org/10.1111/1365-2664.13066

Xu, M., Li, X., Liu, M., Shi, Y., Zhou, H., & Zhang, B., et al. (2020). Spatial variation patterns of plant herbaceous community response to warming along latitudinal and altitudinal gradients in mountainous forests of the loess plateau. *China Environmental and Experimental Botany, 172*, 103983. https://doi.org/10.1016/j.envexpbot.2020.103983

Yadav, S. S., Redden, R. J., Hatfield, J. L., Lotze-Campen, H., & Hall, A. E. (Eds.). (2011). *Crop adaptation to Climate Change*. Chichester, UK.

Yang, J., Zhang, J., Wang, Z., Zhu, Q., & Liu, L. (2001). Water deficit–induced senescence and its relationship to the remobilization of pre-stored carbon in wheat during grain filling. *Agronomy Journal, 93*(1), 196–206.

Yang, H., Wu, M., Liu, W., Zhang, Z., Zhang, N., & Wan, S. (2011). Community structure and composition in response to CC in a temperate steppe. *Global Change and Biology, 17*, 452–465. https://doi.org/10.1111/j.1365-2486.2010.02253.x

Yin, H., Chen, Q., & Yi, M. (2008). Effects of short-term heat stress on oxidative damage and responses of antioxidant system in Lilium longiflorum. *Plant Growth Regulation, 54*(1), 45–54. https://doi.org/10.1007/s10725-007-9227-6

Yoshino, M. (1987). Local climatology. In J. E. Oliver & R. W. Fairbridge (Eds.), *The Encyclopedia of climatology*. Springer.

You, L., Rosegrant, M. W., Fang, C., & Wood, S. (2005). The impact of global warming on Chinese wheat productivity. *China's Dilemma, 265*.

Yu, K., Zhu, K., Ye, M., Zhao, Y., Chen, W., & Guo, W. (2016). Heat tolerance of highbush blueberry is related to the antioxidative enzymes and oxidative protein-repairing enzymes. *Scientia Horticulturae (Amsterdam, Netherland), 198*, 36–43. https://doi.org/10.1016/j.scienta.2015.11.018

Zhang, S., Jiang, H., Peng, S., Korpelainen, H., & Li, C. (2011). Sex related differences in morphological, physiological, and ultrastructural responses of Populuscathayana to chilling. *Journal of Experimental Botany, 62*(2), 675–686. https://doi.org/10.1093/jxb/erq306

Zhang, T. Y., Huang, Y., & Yang, X. G. (2013). Climate warming over the past three decades has shortened rice growth duration in China and cultivar shifts have further accelerated the process for late rice. *Global Change Biology, 19*, 563–570.

Zhang, P., Zhang, J., & Chen, M. (2017). Economic impacts of CC on agriculture: The importance of additional climatic variables other than temperature and precipitation. *Journal of Environmental Economics and Management, 83*, 8–31.

Zhang, N., Tian, L., Feng, L., Xu, W., Li, Y., Xing, F., & Wang, F. (2021). Boll characteristics and yield of cotton in relation to the canopy microclimate under varying plant densities in an arid area. *Peer J, 9*, e12111.

Zhao, T. J., Sun, S., Liu, Y., Liu, J. M., Liu, Q., Yan, Y. B., & Zhou, H. M. (2006). Regulating the drought-responsive element (DRE)-mediated signaling pathway by synergic functions of trans-active and trans-inactive DRE binding factors in Brassica napus. *Journal of Biological Chemistry, 281*, 10752–10759.

Zilberman, D., Liu, X., Roland-Holst, D., & Sunding, D. (2004). The economics of CC in agriculture. *Mitigation Adaptation Strategies Global Change, 9*(4), 365–382.

Zou, M., Yuan, L., Zhu, S., Liu, S., Ge, J., & Wang, C. (2017). Effects of heat stress on photosynthetic characteristics and chloroplast ultrastructure of a heat-sensitive and heat-tolerant cultivar of wucai (*Brassica campestris* L.). *Acta Physiologiae Plantarum, 39*, 1–10. https://doi.org/10.1007/s11738-016-2319-z

Chapter 12
Dynamics of Plant Water Uptake Under Modified Environment

Saurav Saha, B. U. Chaudhury, Bappa Das, Prashant Pandey, and Dibyendu Chatterjee

Abstract The environmental control of crop physiology increases relative sensitivity of crop water movement within the soil plant atmosphere continuum (SPAC), so as the dynamics of crop water usage under modified climate. The variable environmental conditions determines the relative direction of change. Elevated CO_2 exposure depressed the water movement of transpiration flux through reduced stomatal conductance and facilitate partial stomatal closure. However, the net impact may not be ensued the significant reduction in crop water usage at the end of crop season due to modified leaf area dynamics, but with obvious rise in crop water use efficiency (WUE). Thermal stress are often combined with drought events depressed crop WUE beyond a threshold sourced from adverse impact on total dry matter production under elevated temperature condition. The pathogenic role O3 exposure induced stomatal sluggishness and reduction in dry matter accumulation (or yield) are sourced from reduced photosynthetic assimilation and aberration in stomtal conductance thereby reduction in crop WUE under well-watered condition. However, the protective role of other co-existing abiotic stress factors are subjected to more explanatory research. The genetic resilience towards such climate change induced abiotic stress factors and supportive field management options will provide our future facets of sustainable crop production towards higher crop WUE under variable environmental conditions.

Keywords Stomatal conductance · Transpiration · Crop water use · Water use efficiency · Abiotic stress · Climate change

---

S. Saha (✉) · P. Pandey
ICAR-Research Complex NEH Region, Sikkim Center, Tadong, Sikkim 737102, India
e-mail: sauravs.saha@gmail.com

B. U. Chaudhury
ICAR Research Complex for NEH Region, Umiam, Meghalaya 793122, India

B. Das
ICAR-Central Coastal Agricultural Research Institute, Ela, Old Goa 403402, India

D. Chatterjee
ICAR-National Rice Research Institute, Cuttack 753006, India

H. Pathak et al. (eds.), *Climate Change Impacts on Soil-Plant-Atmosphere Continuum*, Advances in Global Change Research 78,
https://doi.org/10.1007/978-981-99-7935-6_12

12.1 Introduction

Efficient crop water use is a vital physiological strategy to combat water insufficiency in surrounding environment and restoring sufficient water to endure ecosystem services towards sustainable crop production (Morison et al., 2008). With the expansion of modern civilization, increased radiative forcing from increased anthropogenic greenhouse gas (GHG) emission poses an additional threat to our existing crop production system and food security with increased frequency of different weather aberrations from the recorded changes in atmospheric composition and warming, frequent exposure of biotic—abiotic stress factors and environmental pollution (Bhatia et al., 2013; Asseng et al., 2015; Chen et al., 2021). Globally, 1.5–4.0 °C rise in atmospheric temperatures is predicted for the next century (IPCC, 2013); are coupled with rising atmospheric CO_2 concentration @ 2.4 μmol mol^{-1} year^{-1} (https://research.noaa.gov/article/ArtMID/587/ArticleID/2636/Rise-of-carbon-dioxide-unabated). Multimodal projections predicted the rise to continue by 794–1142 μmol mol^{-1} by 2100 without any desist scenarios against 270 μmol mol^{-1} in preindustrial era. Water is one of the most precious resources and its utilization pattern became the crucial issue for attending both environmental and agricultural sustainability under projected climate change scenarios in near future. The rise in water holding capacity of air (7% °C^{-1}) and surface evaporation could result the intensification of earth hydrological cycle (Koutsoyiannis, 2020; Williams et al., 2007). Climate change research studies often accentuated the obvious climate change impact that could exacerbate the net impact on seasonal crop water use and regional crop production (Saha et al., 2021; Upadhyaya et al., 2011). However, reduction of global radiation from global dimming often interacts with the unprecedented ongoing atmospheric warming (Yang et al., 2013). Seasonal crop water use and final yield components of different agricultural crops are often significantly and differently impacted by such climate variations (Lovelli et al., 2010; Stanhill & Cohen, 2001). Over past few decades a noticeable warming of the climate were evident, which has been consistently linked to changes in a number of hydrological cycle and hydrological system components, including: altered precipitation patterns, intensities, and extremes; rising atmospheric water vapor; rising evaporation; and altered soil moisture and runoff (Huntington, 2006; Saha et al., 2017, 2022). The observed pattern of such environmental changes are not been uniform across the globe. Crop, climate, water, and soil connections are intricate, involving several biological, physiological, physical, and chemical processes (Fig. 12.1).

Projected increase in dry spells will make the issue of crop yields constrained by water worse (Saha et al., 2017). Water shortages are caused by limited water accessibility in addition to poor water quantity and availability. Realistic estimation of crop water requirement determines the extent of successful irrigation scheduling, planning and design of water management under any particular set of climatological circumstances of a region (English et al., 2002; Koutsoyiannis, 2020). Under limited resource availability, the availability and effectiveness of resource usage became the dominant co factors of maintaining crop productivity. Therefore, a clear understanding on the

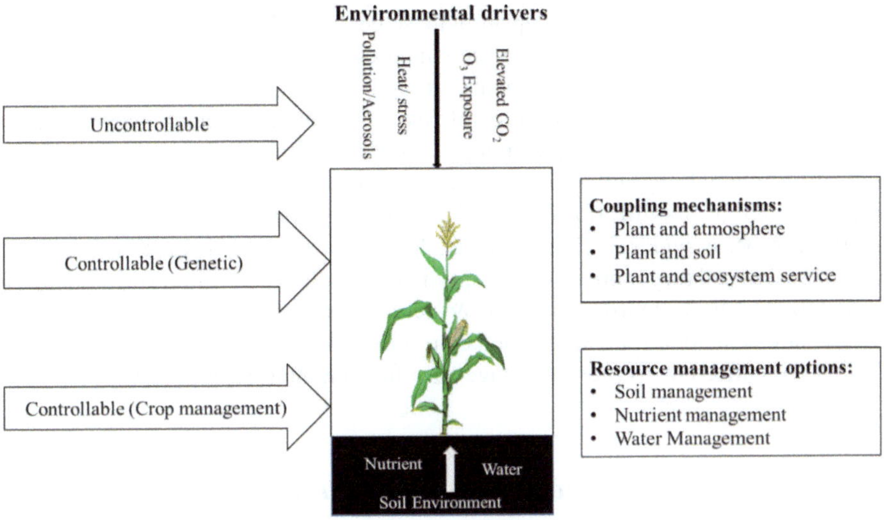

Fig. 12.1 Drivers of plant water uptake under modified environmental conditions

eco-physiological basis for improved water resource management became increasingly important, in order to achieve higher system productivity and modelling crop growth under modified environmental conditions. In this section, we outline the paradigm of crop water dynamics under modified environment affecting the effective resource capture through both aerial (solar radiation, atmospheric CO_2) and soil (moisture and nutrients) medium, in order to sustain crop growth and agricultural production system worldwide.

12.2 Physiology of Crop Water Use Under Modified Environment

Root water uptake regulates the plant growth and development and have indirect control on soil moisture availability and other biophysical processes, including nutrient cycling and transport within soil environment. Mass-flow and diffusion porometry contributed to the combined nutrient–water flow and supportive plant uptake phenomenon in viable textured soils, depending on differences in water potential, availability of transpiring leaf surface and differential relative permeability of water within soil–plant–atmosphere continuum (SPAC) system (Koch & Flu¨hler, 1993; Brusseau et al., 1997). Due to the shared mechanism of transpiration and photosynthetic assimilation process controlling water flow through SPAC, plant growth process is defined as mass- energy transport and energy transformation via photosynthesis into biological entity. Seasonal water availability often determined degree of alteration across different crop growth stages. The C4 plants have more efficient water utilization system from their inherent photosynthetic pathway (higher carboxylation efficiency) and C4 pump strength over C3 crop species (Way et al., 2014). Potassium also plays an integral role in maintaining plant–water relations through modification of plant physiological and biochemical processes viz. enzyme activation, maintaining cell osmotic potential, neutralization of soluble and insoluble molecular anions, stabilization of cell pH and modification of metabolic/ biomolecule synthesis process synthesis (Mengel et al., 2001). External application of K fertilization increases cell turgor pressure and optimize cell solute concentration, particularly under abiotic stress environment (Taiz & Zeman, 2010). In absence of any environmental stress factors, root water uptake are often eloquent with evapotranspiration (ET) across different crop growth stage specific computation of 'consumptive use'.

12.2.1 Rising Atmospheric CO_2 Concentration

Reduction in crop water use is quite evident under rising atmospheric CO_2 levels (Saha et al., 2021). Changes in stomatal density for elevated CO_2 grown plants (Kim et al., 2006; Rowland-Bamford et al., 1990), are often argued against the decreased stomatal conductance through partially closed stomata under short term exposure towards elevated CO_2 environment. Under sufficient irradiation availability, absence of water deficit & temperature stress, the response of stomatal density are inversely related to atmospheric CO_2 concentration (Kim et al., 2006; Royer, 2001). Structural plasticity of plant traits often enable plants to respond favorability towards rising CO_2 levels, particularly under suboptimal range (Beerling & Chaloner, 1993). However, several reviews on short term elevated CO_2 exposure impact studies on differential crop species confirmed that stomatal density barely decreased by an average of 5% (Ainsworth & Rogers, 2007) or did not alter significantly (Kumar et al., 2014). Limited photosynthetic activities under suboptimal level atmospheric CO_2 exposure compel plant to expand the stomatal pore area for enhanced CO_2 uptake, not evident for exposure towards elevated atmospheric CO_2 concentrations (KuÈrschner et al., 1998). In contrast, experimental outcomes confirmed the variable sensitivity (C3 plants) to respond to respond elevated CO_2 level (up to 700 μmol mol^{-1}) with down-regulation of leaf photosynthesis, partially mediated by sufficient substrate availability and reduced photorespiration without any prominent 'CO_2 ceiling effect' for current CO_2 levels; so as for stomatal density and stomatal index (Royer, 2001). Moreover, the plant leaves' altered stomatal densities are sought to be genetic adaptation, having gene regulated impact on stomatal distribution, development and functioning for differential expression towards long-term increased atmospheric CO_2 concentration (Haworth et al., 2015).

The regulation of stomata by guard cells often resulted widespread decline in leaf level stomatal conductance under elevated atmospheric CO_2 exposure that contributed the most for lowering seasonal crop water use to a greater extent, irrespective of changes in photosynthetic assimilation for both C3 and C4 plants (Bunce,

2004). The effective threshold for such reduction are crop specific and may vary between 10–20 s cm^{-1}. Nevertheless, the C3 crop species actually had a greater drop in transpiration due to increasing CO_2 than the C4 crop species (Dugas et al., 1997). Stimulated ABA biosynthesis (Chater et al., 2015), hydrogen peroxide and nitric oxide signaling (Shi et al., 2015), modulation of anion channel and decrease in K + /Ca2 + transport, organic solute (sucrose, malate etc.) concentration in guard cells ensued partial closure of stomatal aperture reducing leaf level transpiration under elevated CO_2 environment (Allen et al., 2011; Lawson et al., 2014). The reduction in leaf stomatal conductance is typically regulated by soil water potential and vapour pressure deficit (Saha et al., 2021; Wullschleger et al., 2002) and generally contributed to reduced bimolecular availability and nutrient concentration (e.g. protein, iron, zinc) under elevated CO_2 environment (Ainsworth et al., 2019); but not always resulted in reduced transpiration flux (Lenka et al., 2020). Reduced stomatal conductance enhanced leaf temperature build up that facilitated more transpiration from leaf to air vapour pressure deficit under elevated CO_2 environment (Gray et al., 2016; Saha et al., 2011). The canopy temperature of chickpea varied between 0.86∘C–1.3∘C in semiarid environment of north India, sourced from reduced leaf level transpiration, but with no significant changes in cumulative crop water use (Saha et al., 2015). Kim et al. (2006) reported rise in 1 °C daytime leaf temperature from reduced leaf-level transpiration that ameliorated the negative impact of water deficit stress and conserve soil moisture under elevated CO_2 environment (Leakey et al., 2009).

The net reduction in crop water use under elevated CO_2 exposure were mostly attributed to reduced canopy conductance followed by plant canopy transpiration (Ainsworth & Rogers, 2007; Kirschbaum & McMillan, 2018). However, elevated CO_2 induced rise in leaf area index (LAI) often determines the extent of reduction in leaf-level transpiration rate towards the determination of net elevated CO_2 impact on cumulative crop water use at plant canopy level. Lowering of canopy conductance with expanded leaf area under elevated CO_2 exposure offset each other impact depending on canopy size, with no significant change in crop water use (Jin et al., 2018). The observed reduction in leaf-level transpiration increases the leaf-level intrinsic water use efficiency under elevated CO_2 exposure (Ainsworth & Long, 2005; Hussain et al., 2013). However, the enhanced leaf internal resistance to CO_2 diffusion induced by the substantial reduction in stomatal conductance under elevated CO_2, gradually countered the stimulation of rubisco carboxylation rates in C3/C4 grassland species (Maherali et al., 2002); thus indirectly hampering photosynthetic assimilation process (Streit et al., 2014). Enhanced belowground allocation (particularly towards) root anchored the crop with finer root proliferation increases plants capacity to extract soil water under elevated CO_2 environment (Ma et al., 2020; Saha et al., 2016). The cumulative impact of amplified LAI counteracted the leaf level transpiration, often failed to ensure soil water savings for larger plant canopies with no significant change in cumulative crop water use under elevated CO_2 environment (Li et al., 2004; Saha et al., 2021). The reported observations on changing magnitude in crop water use efficiency and plant water dynamics under elevated CO_2 exposure are summarized in Table 12.1.

Table 12.1 Dynamics of crop water use under elevated CO_2 environment

Sl. no	Elevated/ ambient CO_2 (ppm)	Facility used	Crop (variety)	Dynamics of crop water use (%)	Changes in water use efficiency (%)	Reference citations
1	700/350	Open top chambers (OTC)	winter wheat (Triticum aestivum L. cv. MV16)	−32 to -44 (canopy conductance)	+10 to +37.4	Tuba et al. (1994)
2	717/363	Glass house	Maize (Zea mays)	−29 (transpiration)	+27	Samarakoon and Gifford (1996)
3	700/350	Open-top chambers	Cherry (Prunus avium)	No change in crop water use	+56 to +103	Centritto et al. (1999)
4	570/370	Free-air CO_2 enrichment	Spring wheat (Triticum aestivum L., cv. Yecora Rojo)	−1.2 to −4 (evapotranspiration)	+20	Hunsaker et al. (2000)
5	570/370	Free-air CO_2 enrichment	Sorghum (Sorghum bicolor)	−10 (evapotranspiration)	+9 to +16	Conley et al. (2001)
6	700/350	Soil–plant–atmosphere research	Soybean [Glycine max (L.) Merr.]	−9 (evapotranspiration)	+50 to +60	Allen et al. (2003)
7	700/350	Controlled growth chambers	Wheat (Triticum aestivum L.)	+16.7 (root water uptake)	+62 to +128	Wu et al. (2004)
8	712/358	Open top chambers	Winter wheat (Triticum aestivum L. cv. Kenong9204)	+52.9 (root water uptake)	+9.9	Qiao et al. (2010)
9	720/360	Sunlit controlled-environment chambers	Maize (Zea mays cv. Saturn Yellow)	−28.6 (transpiration)	+37.84	Allen et al. (2011)
			Grain sorghum (Sorghum bicolor cv. DeKalb 28E)	−27.38 (transpiration)	+40.27	
10	528–615/ 368–454	Free air CO_2 enrichment	poplar plantation (Populus alba L., Populus nigra L., Populus x euramericana Dode Guiner)	-	+73 to +77	Battipaglia et al. (2013)

(continued)

Table 12.1 (continued)

Sl. no	Elevated/ ambient CO_2 (ppm)	Facility used	Crop (variety)	Dynamics of crop water use (%)	Changes in water use efficiency (%)	Reference citations
11	700/380	Controlled environmental growth chambers	Rye (Secale cereale L. Winter -cv Musketeer, Spring- cv SR 4A)	−25 to −35 (transpiration)	+40 to +85	Dahal et al. (2014)
12	1000/700/380	Environmental chambers	Potato (Solanum tuberosum L.)	−19 to −40 (transpiration)	+43 to +62	Kaminski et al. (2014)
13	741/380	Open-top chambers	Soybean (Glycine max, cv. JiHuang13)	−18.1 (canopy conductance)	+26.2 to + 55.4	Li et al. (2013)
14	590/380	Greenhouse	Tomato (Solanum lycopersicum L.; ST 22 and ST 52)	−15 (stomatal conductance)	+14 to +22	Pazzagli et al. (2016)
15	580/380	Open-top chambers	(Glycine max (L.) Merr.; Zhonghuang 35)	No change in canopy conductance	+115.6	Wang et al. (2018)
16	700/550	Open top chambers	Cocoa (Theobroma cacao L.)	+25 to +27 (leaf potential)/ −15.9 to −23.8 (transpiration)	+10.3 to + 33.3	Hebbar et al. (2019)
17	867/421	Gas exchange chambers	Pinus halepensis (Aleppo pine)	−39.1 (canopy conductance)/ −22.3 midday leaf water potential	+80	Birami et al. (2020)
18	550/400	Free-air CO_2 enrichment	Field pea (Pisum sativum L. cv BPBA Twilight)	+28 (Relative leaf water content)	+37	Parvin et al. (2019)
19	700/400	Climate chamber and weighing lysimeters	Maize (Zea mays cv. Zheng Dan 958)	+15.4 to +22.7 (root water uptake)/ -28.1 (transpiration)	+30.5	Ma et al. (2020)
20	580/384	Open top chambers	kabuli chickpea (Cicer arietinum L.; Pusa 1,105)	+21 to +46 (leaf water potential), no change in consumptive use	+30	Saha et al. (2021)

12.2.2 Thermal Stress and Drought Events

Exposure to high temperature (thermal/ heat stress) are often coupled with drought (water deficit stress) that triggered significant reduction of crop productivity/grain yield (Wang & Huang, 2004; Ordó˜nez et al., 2015). Modelling studies estimated global wheat production was projected to decline by 6% °C-1 rise in temperature with scattered beneficial effect across cooler regions (Asseng et al., 2015; Challinor et al., 2014). The negative impact of heat stress are more pronounced for below-ground allocation in roots, nutrient-uptake (particularly Zn; Raj et al., 2015) and cell metabolism (enzyme inactivation; Hungria & Kaschuk, 2014); rather than other normal eco-physiological processes like photosynthesis and stomatal conductance (Giri et al., 2017). Heat stress reduces plant water content, significantly impairing the structure and function of membranes (Tiwari et al., 2018). Hence, three DHNs with apparent molecular masses of 21, 23, and 27 kDa played the protective role under thermal stress exposure and their expression was unaffected by the changes in leaf water relations (Wahid & Close, 2007). Reduction of root water content (Liu & Huang, 2005), root water uptake (Lal, 1974), root hydraulic conductivity (Dodd et al., 2000; Morales et al., 2004) are likely lead to stomatal closure, under extended heat stress or severe elevated soil temperature stress exposure only. With respect to the efficiency of photosynthetic pathway, plant N status, and pre heat-stress growth temperature exposure, elevated CO_2 influences photosynthetic resistance to acute heat stress (Heckathorn et al., 2013). Nonetheless, decrease in leaf level transpiration and root water uptake were apparent from narrowing down of vapour pressure difference between leaf surface and surrounding atmosphere under low temperature exposure (~ 0 °C–15 °C; Aroca et al., 2012). Rise in viscosity and decline in water absorption rate hinders root water uptake under low temperature environment (Toselli et. al., 1999).

Periodic moisture availability in SPAC system determines the extent and variability of crop response towards warming process in soil environment (Streit et al., 2014). Hence, alteration of gene expression through rapid investiture of protein molecules sensitive to dehydration process are effective to alleviate the negative impact thermal stress in standing crop species (Sastry & Barua, 2017; Takahashi & Badger, 2011). Plant hormones like abscisic acid (ABA), indole-3-acetic acid (IAA) and ethylene are involved in the responses of crops to elevated temperature by balancing transpiration through affecting stomatal conductance (Dodd et al., 2004). The association between crop ET and CO_2 levels is subjective by the rising temperature range (Allen et al., 2003). Under sufficient water availability, the positive effect of restricted stomatal aperture under elevated CO_2 environment towards limited crop water use, got nullified with higher evaporation and transpiration loss under elevated air temperature exposure (Luo et al., 2015).

Typically, higher temperatures and increased atmospheric evaporative demand co-occur naturally towards the advent of seasonal drought occurrences (De Boeck & Verbeeck, 2011). When compared to high temperature exposure, drought had a greater impact on plant water relations viz. relative water content and relative leaf area (Zhang et al., 2010). Opening of leaf stomata facilitates transpiration cooling as plants' feedback activated under lower soil moisture tensions (3 to 10 bars; Wanjura and Buxton, 1971) under heat stress exposure (Haldimann & Feller, 2005); if coupled with water deficit stress stomatal closure is inevitable to reduce water loss with subsequent rise in leaf temperature (Rizhsky et al., 2002), increased electrolyte leakage, leaf wilting and altered oxidative metabolism with the symptoms of oxidative injury i.e. increased lipid peroxidation and reduction in antioxidant enzymes activity (Anjum et al., 2011; Liu et al., 2008) but increased phosphorus uptake and accumulation (Jin et al., 2006). Plants experienced water deficit stress either sourced from limited water supply to the roots or excess water loss through transpiration cooling (Anjum et al., 2011); thus affecting crop growth, nutrient uptake (Da Silva et al., 2011), water relations, photosynthetic assimilation and its relative partitioning towards final yield (Praba et al., 2009). Under combined heat-drought stress exposure, water saving acclimation became evident either by increase in water use efficiency or reduction in available leaf area (Ruehr et al., 2016). C4 plants often exhibited more tolerance to combined drought and heat stress than C3 crop species (Killi et al., 2017). Moreover, ABA signaling under limited moisture availability may also facilitate stomatal closure while coupled with heat stress and often hinder transpiration cooling towards attaining higher leaf temperature, often lethal for plant survival (Blum, 2011).

12.2.3 Elevated Ozone Exposure

The majority of crop species are sensitive to elevated ozone (O3) exposure. The net magnitude of crop response is cultivar specific and often determined by internal anti-oxidative defense systems (Herbinger et al., 2002; Xu et al., 2021). Atmospheric O3 enters in to the plant system via stomatal pores as powerful oxidant and triggers cellular reaction with biomolecules to generate reactive oxygen species (ROS); adversely affect cell functioning and reduction in net productivity/ crop yield (Ainsworth, 2017; Bhatia et al., 2012). Stomatal closure is a common adaptive strategy for plants to prevent O3 uptake (Wittig et al., 2007). The reduction in hydraulic conductance in root and leaflet were contrasted with unchanged shoot hydraulic conductance (Zhang et al., 2018) with reduced stomatal conductance (Morgan et al., 2003) and increased canopy temperature (VanLoocke et al., 2012). Moreover, atmospheric O3 exposure depressed leaf cuticle development (Hoshika et al., 2014; Mills et al., 2016) that have dissension with plant's internal leaf water conservation strategy for plant survival under intensive drought stress exposure (Burghardt & Riederer, 2003; Hao et al., 2010). Elevated O3 exposure also disrupts ion of K + channels in guard cells inhibited stomatal opening in leaf epidermis, in order to reduce the capacity of photosynthetic assimilation in plant system (Torsethaugen et al., 1999). Under ample soil water supply, the O3 led disruption on stomatal control may resulted in to reduced relative leaf water content and leaf water potential, so as water use efficiency both at leaf and canopy level during early morning time, but reversed under mid-day sun due to stomatal sluggishness (VanLoocke et al., 2012; Zhang et al., 2018). Elevated O3 induced stomatal sluggishness is effective physiognomic trait for assessing plant susceptibility towards O3 pollution with unrestrained crop water use; often combined concurrent drought stress under variable light environment (Paoletti & Grulke, 2010). Such aberrant stomatal responses determines the inherent capacity of plants to endure drought stress (Pollastrini et al. 2014). Stomatal sluggishness, as modified after elevated O3 exposure endures modifies plant signaling system that extend the time for stomatal opening (reducing atmospheric CO_2 capture) and slower rate of partial stomatal closure (facilitating water loss through transpiration) even during night time (Grulke et al., 2007; Hoshika et al., 2015). Lower stomatal conductance and early leaf shedding of damaged leaves counterbalance for net canopy level water loss, even though stomatal sluggishness is expected to exacerbate water loss (Hoshika et al., 2012). Moreover, elevated temperature exposure restricts plant O3 uptake (Klingberg et al., 2011). Simultaneous exposure of water deficit/ drought stress played protective role against similar negative impact ozone on stomatal conductance (Hoshika et al., 2013).

12.2.4 Modified Radiation Environment

Increased air pollution reduces light quality of solar radiation with proportionate increase in diffuse light fraction. The impact of higher diffuse is still very less studied. Increased diffuse radiation enhances primary production often counterweighed by enhanced leaf transpiration loss (Urban et al., 2012). Under diffuse light, canopy photosynthesis surpassed the rise in leaf transpiration i.e. increased water use efficiency and often modify the community structure in long run (Knohl & Baldocchi, 2008). Under increased UV exposure, gene regulated response of stomatal control is regulated by the salicylic and abscisic acid hormones with observed delay in leaf senescence process (Mannucci et al., 2022).

12.3 Crop Water Use Efficiency Under Modified Climate

Plant growth and water loss has opposing role in crop production. In order to maintain higher dry matter production, plant facilitates rapid gas exchange in the expense of water loss through transpiration (Stanhill, 1986). Physiologically, water use efficiency (WUE) is commonly defined as amount of biomass (total dry matter or yield) produced per unit of crop water usage (mostly evapotranspiration). Increased CO_2 and temperature as a result of climate change has an impact on evapotranspiration (ET) and crop water use efficiency (WUE) (Thomas, 2008). In contrast, the hydrological definition of WUE refers to proportion of water volume used productively includes total contribution from rainfall, irrigation and soil moisture storage that represents the coupling of water and carbon cycles which could act as a useful trait for analysing the impact of climate change on vegetation (Mi et al., 2012) The hydrological concept of WUE follows the interest of soil water conservation engineering. Our interest on assessing WUE under modified climate is confined to physiological definition of WUE restricted for agronomy or crop physiological studies rather than hydrological definition. Rise in WUE may refers to significant increase dry matter production per unit water usage or dry matter production with significant decline in crop growth specific net water usage integrated over any specific time period. Changes in photosynthetic assimilation rate plays the pivotal role in determining the degree of changes in crop WUE under modified climate (Ainsworth & Long, 2005).

Statistical trend analysis studies confirmed WUE has increased over time because the grain yields have increased while water use has remained relatively constant (Basso & Ritchie, 2018). It is anticipated that the crop WUE would rise as a result of the minor decrease in water consumption and significantly greater yields under rising CO_2 levels. There are enormous number of reports showing that higher CO_2 levels lowers stomatal conductance which is sustained over time. The increase in above ground biomass and grain yield was found larger than the total consumption of water in growth-chamber conditions leading to enhanced WUE under elevated CO_2 (Wu et al., 2004). In their experiment, wheat crop exposed to higher CO_2 under high soil moisture conditions consumed more water throughout the growth season, with the exception of the dough stage. Under high soil N and well-watered circumstances, a small drop in ET of wheat was noticed (Hunsaker et al., 2000). Under conditions of high CO_2, an increase in soil moisture can momentarily relieve plants of their water stress (Saha et al., 2021; Sarker & Hara, 2009). Though reports have also shown no statistically significant trend of soil moisture since the commencement of the CO_2 enrichment experiment (Kammann et al., 2005). The impacts of high CO_2 on soil water availability may be mitigated by elevated temperature (Gray et al., 2016) and increased transpiration due to higher leaf area (Saha et al., 2021). In the same line, Saha et al., (2012) reported increased water use in pigeon pea under elevated CO_2 due to higher leaf area index.

Increase in temperature might increase evaporation from soil leading to rapid depletion of soil water, while increases in plant growth under increased CO_2 may hasten the exhaustion of other finite soil resources (Qiao et al., 2010). The lowest soil moisture levels were found when seasonal temperatures were at their highest, and warming had the largest influence on soil moisture (Dermody et al., 2008). With increasing CO_2 to a level of 550 μmol mol^{-1}, cotton ET assessed using three different methods-soil water balance (Hunsaker et al., 1994), sap flow gauges (Dugas et al., 1994), and energy balance (Kimball et al., 1994) did not alter appreciably. In a well-watered treatment, seasonal ET fell by 4.5 and 5.8%, but increased under water-deficit stress by an average of 4.8 and 0.9% over the course of the two years (Hunsaker et al., 1996). It has been discovered that eggplants growing in high CO_2 environments may utilise water more effectively when there is a soil water scarcity (Sarker & Hara, 2009). According to (Schapendonk et al., 1997), total transpiration on a canopy basis was the same for both ambient and elevated CO_2 concentrations, even though the average WUE increased at elevated (doubled) CO_2. This higher WUE was fully offset by a higher leaf area index. Overall, temperature and soil moisture had a greater impact (both positive and negative) on species and community-level responses than did high CO_2 levels (Engel et al., 2009). Crop simulation studies reported that wheat and maize yields would rise 38% and 12%, respectively, if atmospheric CO_2 concentrations reach approximately 600 ppm, and WUE will improve 40% and 25%, respectively, compared to crops grown without CO_2 fertilisation (Guo et al., 2010).

However, C3 plants respond differently to rising CO_2 levels than C4 plants. There is a positive impact due to the functioning of the carboxylation pathway in C3 plants under elevated CO_2 while the effect of elevated CO_2 is negligible in C4 plants under ideal soil water conditions (Hatfield & Dold, 2019). The benefits of elevated CO_2 in C4 plants is only observed under water-deficit stress conditions due to partial closure of stomata which reduces transpiration and their capacity to assimilate carbon even when stomata remain closed (Lopes et al., 2011). Analysis of free-air CO_2 enrichment (FACE) experimentations over last 27 years reported a 10% decrease in ET both in C3 and C4 plants with 19% increase in yields for C3 crops while no change in yield was observed in C4 crops under optimal water supply (Kimball, 2016). Under water-deficit stress, the WUE was found to increase at a greater degree compared to well-watered conditions in C3 crops like wheat (Qiao et al., 2010), barley (Robredo et al., 2007) and even in C4 crops with elevated CO_2 (Kimball, 2016). Therefore, elevated CO_2 may help to overcome the impacts of water-deficit stress.

Under elevated temperature, rise in leaf stomatal conductance below a particular temperature threshold increases leaf level WUE increases with greater due to greater photosynthetic assimilation than in transpiration (Mi et al., 2012). Increase in temperature beyond optimum will cause reduction in rate of photosynthesis, biomass accumulation leading to reduced WUE (Ortiz et al., 2008). Further temperature rise decreased WUE once the threshold was reached because domination of higher rates of ET. The temperature threshold depends on plant species (Wang et al., 2010). Crop ET was significantly lower under combined elevated CO_2 and temperature exposure for soybean crop, with significantly higher WUE (Lenka et al., 2020). Similarly doubling CO_2 caused a 9% drop in ET at a temperature combination of 28/18 °C, while there was minimal effect at 40/30 °C and 44/34 °C under controlled environmental chambers (Allen et al., 2003). The combined impact of rise in temperature and CO_2 under future climate change scenarios may be beneficial for growing crops until the temperature crosses the optimum thresholds under diverse agro-climatic conditions.

Elevated O3 exposure modifies leaf microenvironment through modification of canopy energy fluxes (VanLoocke et al., 2012; Wittig et al., 2007). Reduction in latent heat dissipation from stomatal sluggishness often modulated canopy thermal environment (Paoletti & Grulke, 2010). The the combined contribution reduced ecosystem services from decline in leaf level transpiration/ canopy evapotranspiration (Grulke et al., 2007; Zhang et al., 2018) and reduction in dry matter accumulation (Morgan et al., 2003) are often attributed to the linear reduction in WUE even up to 50% under elevated O3 exposure (VanLoocke et al., 2012). Environmental pollution enforced aerosol forcing that resulted decrease of global solar radiation is reported to be 1.3% per decade during 1960 to 2000 (IPCC, 2007). It has been also reported that diffuse fraction solar radiation is increasing worldwide (Yang et al., 2013). The decrease in global solar radiation may be due to rise in atmospheric aerosols caused by urbanization, industrialization, fossil fuel burning and other anthropogenic activities. Aerosols absorb and/or scatter incoming solar radiation. Reduction in solar radiation due to global dimming with increased diffuse radiation in places with high radiation will either have no impact on yield or will increase the yield leading to increased WUE (Stanhill & Cohen, 2001). Another reason for increased WUE under reduced solar radiation may be due to reduced evaporative demand resulting in lower crop ET (Supit et al., 2010). In the places with low solar radiation, global dimming may cause significant yield reductions (Yang et al., 2013). Moreover, iincreased Ultraviolet-B radiation exposure is reported to decrease stomatal density and stomatal conductance, thereby increasing leaf level water use efficiency (Gitz III et al., 2005).

12.4 Conclusion

Relative dynamics of photosynthetic assimilations engulf its active control on plant water uptake under modified environmental conditions. The nature and degree of exposure with co-existing abiotic stress factors often modifies the degree of response. It is quite evident that most of the plant species will have more efficient crop water utilization system in future elevated CO_2 world. The sole or concurrent exposure of other abiotic stress factors (heat, drought, ozone exposure etc.) need to be studied more intensively that will determine the net magnitude of local, region and global feedback system within the planetary boundary layer. However, future research initiatives for achieving higher crop WUE may source from either selection of efficient genotypes with higher assimilation rates exposed under abiotic stress environment or shifting management practices towards facilitating root zone soil moisture conservation, facilitating plants adaptive capacity for attending higher yield whenever exposed under modified surrounding environmental condition.

References

Ainsworth, E. A. (2017). Understanding and improving global crop response to ozone pollution. *The Plant Journal, 90*, 886–897. https://doi.org/10.1111/tpj.13298.

Ainsworth, E. A., & Long, S. P. (2005). What have we learned from 15 years of free-air CO_2 enrichment (FACE)? A meta-analytic review of the responses of photosynthesis, canopy properties and plant production to rising CO_2. *New Phytologist, 165*, 351–372. https://doi.org/10.1111/j.1469-8137.2004.01224.x.

Ainsworth, E. A., & Rogers, A. (2007). The response of photosynthesis and stomatal conductance to rising [CO_2]: Mechanisms and environmental interactions. *Plant, Cell and Environment, 30*, 258–270. https://doi.org/10.1111/j.1365-3040.2007.01641.x.

Ainsworth, E. A., Lemonnier, P., & Wedow, J. M. (2019). The influence of rising tropospheric carbon dioxide and ozone on plant productivity. *Plant Biology, 22*(51), 5–11. https://doi.org/10.1111/plb.12973.

Allen, L. H., Pan, D., Boote, K. J., Pickering, N. B., & Jones, J. W. (2003). Carbon dioxide and temperature effects on evapotranspiration and water use efficiency of soybean. *Agronomy Journal, 95*, 1071–1081. https://doi.org/10.2134/agronj2003.1071.

Allen, L. H., Kakani, V. G., Vu, J. C. V., & Boote, K. J. (2011). Elevated CO_2 increases water use efficiency by sustaining photosynthesis of water-limited maize and sorghum. *Journal of Plant Physiology, 168*, 1909–1918. https://doi.org/10.1016/j.jplph.2011.05.005.

Anjum, S. A., Wang, L. C., Farooq, M., Hussain, M., Xue, L. L., & Zou, C. M. (2011). Brassinolide application improves the drought tolerance in maize through modulation of enzymatic antioxidants and leaf gas exchange. *Journal of Agronomy and Crop Science, 197*, 177–185. https://doi.org/10.1111/j.1439-037X.2010.00459.x.

Aroca, R., Porcel, R., Ruiz-Lozano, J. M. (2012). Regulation of root water uptake under abiotic stress conditions. *Journal of Experimental Botany, 63*(1), 43–57. https://doi.org/10.1093/jxb/err266.

Asseng, S., Ewert, F., Martre, P., Rötter, R. P., Lobell, D. B., Cammarano, D., Kimball, B. A., Ottman, M. J., Wall, G. W., White, J. W., Reynolds, M. P., Alderman, P. D., Prasad, P. V. V., Aggarwal, P. K., Anothai, J., Basso, B., Biernath, C., Challinor, A. J., de Sanctis, G., Doltra, J., Fereres, E., Garcia-Vila, M., Gayler, S., Hoogenboom, G., Hunt, L. A., Izaurralde, R. C., Jabloun, M., Jones, C. D., Kersebaum, K. C., Koehler, A. K., Müller, C., Naresh Kumar, S., Nendel, C., O'leary, G., Olesen, J. E., Palosuo, T., Priesack, E., Eyshi Rezaei, E., Ruane, A. C., Semenov, M. A., Shcherbak, I., Stöckle, C., Stratonovitch, P., Streck, T., Supit, I., Tao, F., Thorburn, P. J., Waha, K., Wang, E., Wallach, D., Wolf, J., Zhao, Z., Zhu, Y. (2015). Rising temperatures reduce global wheat production. *Nature Climate Chang Flong, 5,* 143–147.https://doi.org/10.1038/nclimate2470.

Basso, B., & Ritchie, J. T. (2018). Evapotranspiration in high-yielding maize and under increased vapor pressure deicit in the US Midwest. *Agricultural and Environmental Letters, 3*(1), 170039. https://doi.org/10.2134/ael2017.11.0039.

Battipaglia, G., Saurer, M., Cherubini, P., Calfapietra, C., McCarthy, H. R., Norby, R. J., & Cotrufo, M. F. (2013). Elevated CO_2 increases tree-level intrinsic water use efficiency: Insights from carbon and oxygen isotope analyses in tree rings across three forest FACE sites. *New Phytologist, 197*(2), 544–554. https://doi.org/10.1111/nph.12044.

Beerling, D. J., & Chaloner, W. G. (1993). The impact of atmospheric CO_2 and temperature changes on stomatal density: Observation from Quercus Robur Lammas leaves. *Annals of Botany, 71*, 231–235. https://doi.org/10.1006/ANBO.1993.1029.

Bhatia, A., Tomer, R., Kumar, V., Singh, S. D., & Pathak, H. (2012). Impact of tropospheric ozone on crop growth and productivity-a review. *Journal of Scientific & Industrial Research, 71*, 97–112.

Bhatia, A., Kumar, V., Kumar, A., Tomer, R., Singh, B., & Singh, S. D. (2013). Effect of elevated ozone and carbon dioxide interaction on growth and yield of maize. *Maydica, 58*, 291–229.

Birami, B., Nägele, T., Gattmann, M., Preisler, Y., Gast, A., Arneth, A., & Ruehr, N. K. (2020). Hot drought reduces the effects of elevated CO_2 on tree water use efficiency and carbon metabolism. *New Phytologist*. https://doi.org/10.1111/nph.16471.

Blum, A. (2011). Plant water relations, plant stress and plant production. In: *Plant breeding for water-limited environments.* Springer New York, pp. 11–52. https://doi.org/10.1007/978-1-4419-7491-4_2.

Brusseau, M. L., Hu, Q., & Srivastava, R. (1997). Hydrology using flow interruption to identify factors causing nonideal contaminant transport. *Journal of Contaminant Hydrology, 24*, 205–219.

Bunce, J. A. (2004). Carbon dioxide effects on stomatal responses to the environment and water use by crops under field conditions. *Oecologia, 140*, 1–10. https://doi.org/10.1007/s00442-003-1401-6.

Burghardt, M., & Riederer, M. (2003). Ecophysiological relevance of cuticular transpiration of deciduous and evergreen plants in relation to stomatal closure and leaf water potential. *Journal of Experimental Botany, 54*, 1941–1949. https://doi.org/10.1093/jxb/erg195.

Centritto, M., Magnani, F., Lee, H. S. J., Jarvis, P. G. (1999). Interactive effects of elevated [CO2] and drought on cherry (Prunus avium) seedlings II. Photosynthetic capacity and water relations. *New Phytologst, 141*(1), 141–153. https://doi.org/10.1046/j.1469-8137.1999.00327.x.

Challinor, A. J., Watson, J., Lobell, D. B., Howden, S. M., Smith, D. R., & Chhetri, N. (2014). A meta-analysis of crop yield under climate change and adaptation. *Nature Clinical Practice Endocrinology & Metabolism, 4*, 287–291. https://doi.org/10.1038/nclimate2153.

Chater, C., Peng, K., Movahedi, M., Dunn, J. A., Walker, H. J., Liang, Y. K., McLachlan, D. H., Casson, S., Isner, J. C., Wilson, I., Neill, S. J., Hedrich, R., Gray, J. E., & Hetherington, A. M. (2015). Elevated CO_2-induced responses in stomata require ABA and ABA signaling. *Current Biology, 25*, 2709–2716. https://doi.org/10.1016/j.cub.2015.09.013.

Chen, L., Xie, H., Wang, G., Qian, X., Wang, W., Xu, Y., Zhang, W., Zhang, H., Liu, L., Wang, Z., Gu, J., & Yang, J. (2021). Reducing environmental risk by improving crop management practices at high crop yield levels. *Field Crops Research, 265*, 108123. https://doi.org/10.1016/j.fcr.2021.108123.

Conley, M. M., Kimball, B. A., Brooks, T. J., Pinter, P. J., Hunsaker, D. J., Wall, G. W., Adam, N. R., LaMorte, R. L., Matthias, A. D., Thompson, T. L., Leavitt, S. W., Ottman, M. J., Cousins, A. B., & Triggs, J. M. (2001). CO_2 enrichment increases water-use efficiency in sorghum. *New Phytologist, 151*(2), 407–412. https://doi.org/10.1046/j.1469-8137.2001.00184.x.

Da Silva, E., Nogueira, R. J. M. C., Da Silva, M. A., & De Albuquerque, M. B. (2011). Drought stress and plant nutrition. *Plant Stress, 5*(1), 32–41.

Dahal, K., Knowles, V. L., Plaxton, W. C., & Hüner, N. P. A. (2014). Enhancement of photosynthetic performance, water use efficiency and grain yield during long-term growth under elevated CO_2 in wheat and rye is growth temperature and cultivar dependent. *Environmental and Experimental Botany, 106*, 207–220. https://doi.org/10.1016/j.envexpbot.2013.11.015.

De Boeck, H. J., & Verbeeck, H. (2011). Drought-associated changes in climate and their relevance for ecosystem experiments and models. *Biogeosciences, 8*, 1121–1130. https://doi.org/10.5194/bg-8-1121-2011.

Dermody, O., Long, S. P., Mcconnaughay, K., & Delucia, E. H. (2008). How do elevated CO_2 and O3 affect the interception and utilization of radiation by a soybean canopy? *Global Change Biology, 14*, 556–564. https://doi.org/10.1111/j.1365-2486.2007.01502.x.

Dodd, I. C., & Davies, W. J. (2004). Hormones and the regulation of water balance. In P. J. Davies (Ed.), *Plant hormones: Biosynthesis, signal transduction, action* (3rd ed., pp. 519–548). Kluwer.

Dodd, I. C., He, J., Turnbull, C. G. N., Lee, S. K., & Critchley, C. (2000). The influence of supra-optimal root-zone temperatures on growth and stomatal conductance in Capsicum annuum L. *Journal of Experimental Botany, 51*, 239–248. https://doi.org/10.1093/jexbot/51.343.239.

Dugas, W. A., Heuer, M. L., Hunsaker, D., Kimball, B. A., Lewin, K. F., Nagy, J., & Johnson, M. (1994). Sap flow measurements of transpiration from cotton grown under ambient and enriched CO_2 concentrations. *Agricultural and Forest Meteorology, 70*, 231–245. https://doi.org/10.1016/0168-1923(94)90060-4.

Dugas, W. A., Prior ' S. A., Rogers, H. H. (1997). Transpiration from sorghum and soybean growing under ambient and elevated CO2 concentrations. *Agricultural and Forest Meteorology, 83*, 37–48.https://doi.org/10.1016/S0168-1923(96)02353-2.

Engel, E. C., Weltzin, J. F., Norby, R. J., & Classen, A. T. (2009). Responses of an old-field plant community to interacting factors of elevated [CO_2], warming, and soil moisture. *Journal of Plant Ecology, 2*, 1–11. https://doi.org/10.1093/jpe/rtn026.

English, M. J., Solomon, K. H., & Hoffman, G. J. (2002). A paradigm shift in irrigation management. *Journal of Irrigation and Drainage Engineering, 128*, 267–277. https://doi.org/10.1061/(ASCE)0733-9437(2002)128:5(267).

Giri, A., Heckathorn, S., Mishra, S., & Krause, C. (2017). Heat stress decreases levels of nutrient-uptake and -assimilation proteins in tomato roots. *Plants, 6*(1), 443–448. https://doi.org/10.3390/plants6010006.

GitzIII, D. C., Liu-Gitz, L., Britz, S. J., & Sullivan, J. H. (2005). Ultraviolet-B effects on stomatal density, water-use efficiency, and stable carbon isotope discrimination in four glasshouse-grown soybean (Glyicine max) cultivars. *Environmental and Experimental Botany, 53*, 343–355. https://doi.org/10.1016/j.envexpbot.2004.04.005.

Gray, S. B., Dermody, O., Klein, S. P., Locke, A. M., McGrath, J. M., Paul, R. E., Rosenthal, D. M., Ruiz-Vera, U. M., Siebers, M. H., Strellner, R., Ainsworth, E. A., Bernacchi, C. J., Long, S. P., Ort, D. R., & Leakey, A. D. B. (2016). Intensifying drought eliminates the expected benefits of elevated carbon dioxide for soybean. *Nature Plants, 2*, 16132. https://doi.org/10.1038/nplants.2016.132.

Grulke, N. E., Paoletti, E., Heath, R. L. (2007). Chronic vs. short-term acute O3 exposure effects on nocturnal transpiration in two Californian oaks. *The Scientific World Journal, 7*(S1), 134–140. https://doi.org/10.1100/tsw.2007.33.

Guo, R., Lin, Z., Mo, X., & Yang, C. (2010). Responses of crop yield and water use efficiency to climate change in the North China Plain. *Agricultural Water Management, 97*, 1185–1194. https://doi.org/10.1016/j.agwat.2009.07.006.

Haldimann, P., & Feller, U. (2005). Growth at moderately elevated temperature alters the physiological response of the photosynthetic apparatus to heat stress in pea (Pisum sativum L.) leaves. *Plant, Cell and Environment, 28*, 302–317. https://doi.org/10.1111/j.1365-3040.2005.01289.x.

Hao, G. Y., Sack, L., Wang, A. Y., Cao, K. F., & Goldstein, G. (2010). Differentiation of leaf water flux and drought tolerance traits in hemiepiphytic and non-hemiepiphytic Ficus tree species. *Functional Ecology, 24*, 731–740. https://doi.org/10.1111/j.1365-2435.2010.01724.x.

Hatfield, J. L., & Dold, C. (2019). Water-use efficiency: Advances and challenges in a changing climate. *Frontiers in Plant Science, 10*, 1–14. https://doi.org/10.3389/fpls.2019.00103.

Haworth, M., Killi, D., Materassi, A., & Raschi, A. (2015). Coordination of stomatal physiological behavior and morphology with carbon dioxide determines stomatal control. *American Journal of Botany, 102*, 677–688. https://doi.org/10.3732/ajb.1400508.

Hebbar, K. B., Apshara, E., Chandran, K. P., Prasad, P. V. V. (2019). Effect of elevated CO_2, high temperature, and water deficit on growth, photosynthesis, and whole plant water use efficiency of cocoa (Theobroma cacao L.). International Journal of Biometeorology. https://doi.org/10.1007/s00484-019-01792-0.

Heckathorn, S. A., Giri, A., Mishra, S., & Bista, D. (2013). Heat stress and roots. In N. Tuteja & S. S. Gill (Eds.), *Climate change and plant abiotic stress tolerance* (pp. 109–136). Wiley-VCH Verlag.

Herbinger, K., Tausz, M., Wonisch, A., Soja, G., Sorger, A., & Grill, D. (2002). Complex interactive effects of drought and ozone stress on the antioxidant defence systems of two wheat cultivars. *Plant Physiology and Biochemistry, 40*(6–8), 691–696. https://doi.org/10.1016/s0981-428(02)01410-9.

Hoshika, Y., Omasa, K., & Paoletti, E. (2012). Whole-tree water use efficiency is decreased by ambient ozone and not affected by O3-induced stomatal sluggishness. *PLoS ONE, 7*(6), e39270. https://doi.org/10.1371/journal.pone.0039270.

Hoshika, Y., Omasa, K., & Paoletti, E. (2013). Both ozone exposure and soil water stress are able to induce stomatal sluggishness. *Environmental and Experimental Botany, 88*, 19–23. https://doi.org/10.1016/j.envexpbot.2011.12.004.

Hoshika, Y., Carriero, G., Feng, Z., Zhang, Y., & Paoletti, E. (2014). Determinants of stomatal sluggishness in ozone-exposed deciduous tree species. *Science of the Total Environment, 481*, 453–458. https://doi.org/10.1016/j.scitotenv.2014.02.080.

Hoshika, Y., Katata, G., Deushi, M., Watanabe, M., Koike, T., & Paoletti, E. (2015). Ozone-induced stomatal sluggishness changes carbon and water balance of temperate deciduous forests. *Science and Reports, 5*, 9871. https://doi.org/10.1038/srep09871.

Huang, T., Joshi, V., & Jander, G. (2014). The catabolic enzyme methionine gamma-lyase limits methionine accumulation in potato tubers. *Plant Biotechnology Journal, 12*, 883–893. https://doi.org/10.1111/pbi.12191.

Hungria, M., & Kaschuk, G. (2014). Regulation of N2 fixation and NO3−/NH4+ assimilation in nodulated and N-fertilized Phaseolus vulgaris L. exposed to high temperature stress.

Environmental and Experimental Botany, 98, 32–39. https://doi.org/10.1016/j.envexpbot.2013.10.010.
Hunsaker, D. J., Hendrey, G. R., Kimball, B. A., Lewin, K. F., Mauney, J. R., & Nagy, J. (1994). Cotton evapotranspiration under field conditions with CO_2 enrichment and variable soil moisture regimes. *Agricultural and Forest Meteorology, 70*, 247–258. https://doi.org/10.1016/0168-1923(94)90061-2.
Hunsaker, D. J., Kimball, B. A., Pinter, P. J., Wall, G. W., Lamorte, R. L., Adamsen, F. J., Leavitt, S. W., Thompson, T. L., Matthias, A. D., & Brooks, T. J. (2000). CO_2 enrichment and soil nitrogen effects on wheat evapotranspiration and water use efficiency. *Agricultural and Forest Meteorology, 104*, 85–105. https://doi.org/10.1016/S0168-1923(00)00157-X.
Hunsaker, D. J., Kimball, B. A., Pinter, P. J., LaMorte, R. L., Wall, G. W. (1996). Carbon dioxide enrichment and irrigation effects on wheat evapotranspiration and water use efficiency. *Transactions of the ASAE, 39*, 1345–1355. https://doi.org/10.13031/2013.27626.
Huntington, T. G. (2006). Evidence for intensification of the global water cycle: Review and synthesis. *Journal of Hydrology (amst), 319*, 83–95. https://doi.org/10.1016/j.jhydrol.2005.07.003.
Hussain, M. Z., Vanloocke, A., Siebers, M. H., Ruiz-Vera, U. M., Cody Markelz, R. J., Leakey, A. D. B., Ort, D. R., & Bernacchi, C. J. (2013). Future carbon dioxide concentration decreases canopy evapotranspiration and soil water depletion by field-grown maize. *Global Change Biology, 19*, 1572–1584. https://doi.org/10.1111/gcb.12155.
IPCC. (2007). Climate Change 2007, the physical science basis, contribution of working group I to the fourth assessment report of the intergovernmental panel on climate change. Cambridge: Cambridge University Press.
IPCC. (2013). IPCC, 2013: summary for policymakers. Climate change 2013: The physical science basis. Contribution of working group I to the fifth assessment report of the intergovernmental panel on climate change, pp. 1–28.
Jin, J., Wang, G., Liu, X., Pan, X., Herbert, S. J., & Tang, C. (2006). Interaction between phosphorus nutrition and drought on grain yield, and assimilation of phosphorus and nitrogen in two soybean cultivars differing in protein concentration in grains. *Journal of Plant Nutrition, 29*(8), 1433–1449. https://doi.org/10.1080/01904160600837089.
Jin, Z., Ainsworth, E. A., Leakey, A. D. B., & Lobell, D. B. (2018). Increasing drought and diminishing benefits of elevated carbon dioxide for soybean yields across the US Midwest. *Global Change Biology, 24*, e522–e533. https://doi.org/10.1111/gcb.13946.
Kaminski, K. P., Kørup, K., Nielsen, K. L., Liu, F., Topbjerg, H. B., Kirk, H. G., & Andersen, M. N. (2014). Gas-exchange, water use efficiency and yield responses of elite potato (Solanum tuberosum L.) cultivars to changes in atmospheric carbon dioxide concentration, temperature and relative humidity. *Agricultural and Forest Meteorology, 187*, 36–45. https://doi.org/10.1016/j.agrformet.2013.12.001.
Kammann, C., Grünhage, L., Grüters, U., Janze, S., & Jäger, H.-J. (2005). Response of aboveground grassland biomass and soil moisture to moderate long-term CO_2 enrichment. *Basic and Applied Ecology, 6*, 351–365. https://doi.org/10.1016/j.baae.2005.01.011.
Killi, D., Bussotti, F., Raschi, A., & Haworth, M. (2017). Adaptation to high temperature mitigates the impact of water deficit during combined heat and drought stress in C3 sunflower and C4 maize varieties with contrasting drought tolerance. *Physiologia Plantarum, 159*, 130–147. https://doi.org/10.1111/ppl.12490.
Kim, S. H., Sicher, R. C., Bae, H., Gitz, D. C., Baker, J. T., Timlin, D. J., & Reddy, V. R. (2006). Canopy photosynthesis, evapotranspiration, leaf nitrogen, and transcription profiles of maize in response to CO_2 enrichment. *Global Change Biology, 12*, 588–600. https://doi.org/10.1111/j.1365-2486.2006.01110.x.
Kimball, B. A. (2016). Crop responses to elevated CO_2 and interactions with H2O, N, and temperature. *Current Opinion in Plant Biology, 31*, 36–43. https://doi.org/10.1016/j.pbi.2016.03.006.

Kimball, B. A., LaMorte, R. L., Seay, R. S., Pinter, P. J., Rokey, R. R., Hunsaker, D. J., Dugas, W. A., Heuer, M. L., Mauney, J. R., Hendrey, G. R., Lewin, K. F., & Nagy, J. (1994). Effects of free-air CO_2 enrichment on energy balance and evapotranspiration of cotton. *Agricultural and Forest Meteorology, 70*, 259–278. https://doi.org/10.1016/0168-1923(94)90062-0.

Kirschbaum, M. U. F., & McMillan, A. M. S. (2018). Warming and elevated CO_2 have opposing influences on transpiration. Which is more important? *Current Forestry Reports, 4*, 51–71. https://doi.org/10.1007/s40725-018-0073-8.

Klingberg, J., Engardt, M., Uddling, J., Karlsson, P. E., & Pleijel, H. (2011). Ozone risk for vegetation in the future climate of Europe based on stomatal ozone uptake calculations. *Tellus, Series a: Dynamic Meteorology and Oceanography, 63*(1), 174–187. https://doi.org/10.1111/j.1600-0870.2010.00465.x.

Knohl, A., & Baldocchi, D. D. (2008). Effects of diffuse radiation on canopy gas exchange processes in a forest ecosystem. *Journal of Geophysical Research, 113*, G02023. https://doi.org/10.1029/2007JG000663.

Koch, S., & Flühler, H. (1993). Non-reactive solute transport with micropore diffusion in aggregated porous media determined by a flow-interruption method. *Journal of Contaminant Hydrology, 14*, 39–54. https://doi.org/10.1016/0169-7722(93)90040-Y.

Koutsoyiannis, D. (2020). Revisiting the global hydrological cycle: Is it intensifying? *Hydrology and Earth System Sciences, 24*, 3899–3932. https://doi.org/10.5194/hess-24-3899-2020.

KuÈrschner, W. M., Stulen, I., Wagner, F., & Kuiper, P. J. C. (1998). Comparison of palaeobotanical observations with experimental data on the leaf anatomy of durmast oak [Quercus petraea (Fagaceae)] in response to environmental change. *Annals of Botany, 81*, 657–664.

Kumar, S., Chaitanya, B. S. K., Ghatty, S., & Reddy, A. R. (2014). Growth, reproductive phenology and yield responses of a potential biofuel plant, Jatropha curcas grown under projected 2050 levels of elevated CO_2. *Plant Physiology, 152*, 501–519. https://doi.org/10.1111/ppl.12195.

Lal, R. (1974). Effect of constant and fluctuating soil temperature on growth, development and nutrient uptake of maize seedlings. *Plant and Soil, 40*, 589–606.

Lawson, T., Simkin, A. J., Kelly, G., & Granot, D. (2014). Mesophyll photosynthesis and guard cell metabolism impacts on stomatal behaviour. *New Phytologist, 203*, 1064–1081.https://doi.org/10.1111/nph.12945.

Leakey, A. D. B., Ainsworth, E. A., Bernacchi, C. J., Rogers, A., Long, S. P., & Ort, D. R. (2009). Elevated CO_2 effects on plant carbon, nitrogen and water relations: Six important lessons from FACE. *Journal of Experimental Botany, 60*, 2859–2876. https://doi.org/10.1093/jxb/erp096.

Lenka, N. K., Lenka, S., Thakur, J. K., Yashona, D. S., Shukla, A. K., Elanchezhian, R., Singh, K. K., Biswas, A. K., & Patra, A. K. (2020). Carbon dioxide and temperature elevation effects on crop evapotranspiration and water use efficiency in soybean as affected by different nitrogen levels. *Agricultural Water Management, 230*, 105936. https://doi.org/10.1016/j.agwat.2019.105936.

Li, F., Kang, S., & Zhang, J. (2004). Interactive effects of elevated CO_2, nitrogen and drought on leaf area, stomatal conductance, and evapotranspiration of wheat. *Agricultural Water Management, 67*, 221–233. https://doi.org/10.1016/j.agwat.2004.01.005.

Li, D., Liu, H., Qiao, Y., Wang, Y., Cai, Z., Dong, B., Shi, C., Liu, Y., Li, X., & Liu, M. (2013). Effects of elevated CO_2 on the growth, seed yield, and water use efficiency of soybean (Glycine max (L.) Merr.) under drought stress. *Agricultural Water Management, 129*, 105–112. https://doi.org/10.1016/j.agwat.2013.07.014.

Liu, X., & Huang, B. (2005). Root physiological factors involved in cool-season grass response to high soil temperature. *Environmental and Experimental Botany, 53*, 233–245. https://doi.org/10.1016/J.ENVEXPBOT.2004.03.016.

Liu, J., Xie, X., Du, J., Sun, J., & Bai, X. (2008). Effects of simultaneous drought and heat stress on Kentucky bluegrass. *Scientia Horticulturae, 115*(2), 190–195. https://doi.org/10.1016/j.scienta.2007.08.003.

Lopes, M. S., Araus, J. L., Van Heerden, P. D. R., & Foyer, C. H. (2011). Enhancing drought tolerance in C4 crops. *Journal of Experimental Botany, 62*, 3135–3153. https://doi.org/10.1093/jxb/err105.

Lovelli, S., Perniola, M., di Tommaso, T., Ventrella, D., Moriondo, M., & Amato, M. (2010). Effects of rising atmospheric CO_2 on crop evapotranspiration in a Mediterranean area. *Agricultural Water Management, 97*, 1287–1292. https://doi.org/10.1016/j.agwat.2010.03.005.

Luo, Q., Bange, M., Johnston, D., & Braunack, M. (2015). Cotton crop water use and water use efficiency in a changing climate. *Agriculture, Ecosystems & Environment, 202*, 126–134. https://doi.org/10.1016/j.agee.2015.01.006.

Ma, Y., Wu, Y., & Song, X. (2020). How elevated CO_2 shifts root water uptake pattern of crop? Lessons from climate chamber experiments and isotopic tracing technique. *Water (switzerland), 12*, 1–15. https://doi.org/10.3390/w12113194.

Maherali H. H., Reid, C. D., Polley, H. W., Johnson, H. B., Jackson, R. B. (2002). Stomatal acclimation over a sub ambient to elevated CO2 gradient in a C3/C4 grassland. *Plant, Cell and Environment, 25*, 557–566.https://doi.org/10.1046/j.1365-3040.2002.00832.x.

Mannucci, A., Scartazza, A., Santaniello, A., Castagna, A., Santin, M., Quartacci, M. F., & Ranieri, A. (2022). Short daily ultraviolet exposure enhances intrinsic water-use efficiency and delays senescence in Micro-Tom tomato plants. *Functional Plant Biology, 49*(9), 810–821. https://doi.org/10.1071/FP22013.

Mengel, K., Kirkby, E. A., Kosegarten, H., Appel, T. (2001). Principles of plant nutrition. Dordrecht, the Netherlands: Kluwer Academic Publishers, pp. 481–511.

Mi, N., Zhang, Y. S., Ji, R. P., Cai, F., Zhang, S. J., & Zhao, X. L. (2012). Effects of climate change on water use efficiency in rain-fed plants. *International Journal of Plant Production, 6*, 513–534.

Mills, G., Harmens, H., Wagg, S., Sharps, K., Hayes, F., Fowler, D., Sutton, M., & Davies, B. (2016). Ozone impacts on vegetation in a nitrogen enriched and changing climate. *Environmental Pollution, 208*, 898–908. https://doi.org/10.1016/j.envpol.2015.09.038.

Morales, D., Rodriguez, P., Amico dell, J., Nicloas, E., Torrecillas, A., Sanchez-blanco, M. J. (2004). High-temperature preconditioning and thermal shock imposition affects water relations, gas exchange and root hydraulic conductivity in tomato. *Biologia Plantarum, 47*, 203–228.

Morgan, P. B., Ainsworth, E. A., Long, S. P. (2003). How does elevated ozone impact soybean? A meta-analysis of photosynthesis, growth and yield. *Plant Cell and Environment, 26*(8), 1317–1328. https://doi.org/10.1046/j.0016-8025.2003.01056.x.

Morison, J. I. L., Baker, N. R., Mullineaux, P. M., Davies, W. J. (2008). Improving water use in crop production. *Philosophical Transactions of the Royal Society of London. Series B, Biological Sciences, 363*, 639–658.https://doi.org/10.1098/rstb.2007.2175.

Ordóñez, R. A., Savin, R., Cossani, C. M., & Slafer, G. A. (2015). Yield response to heat stress as affected by nitrogen availability in maize. *Field Crops Research, 183*, 184–203. https://doi.org/10.1016/j.fcr.2015.07.010.

Ortiz, R., Sayre, K. D., Govaerts, B., Gupta, R., Subbarao, G. V., Ban, T., Hodson, D., Dixon, J. M., Iván Ortiz-Monasterio, J., & Reynolds, M. (2008). Climate change: Can wheat beat the heat? *Agriculture, Ecosystems & Environment, 126*, 46–58. https://doi.org/10.1016/j.agee.2008.01.019.

Paoletti, E., & Grulke, N. E. (2010). Ozone exposure and stomatal sluggishness in different plant physiognomic classes. *Environmental Pollution, 158*(8), 2664–2671. https://doi.org/10.1016/j.envpol.2010.04.024.

Parvin, S., Uddin, S., Fitzgerald, G. J., Tausz-Posch, S., Armstrong, R., & Tausz, M. (2019). Free air CO_2 enrichment (FACE) improves water use efficiency and moderates drought effect on N2 fixation of Pisum sativum L. *Plant and Soil, 436*(1–2), 587–606. https://doi.org/10.1007/s11104-019-03949-7.

Pazzagli, P. T., Weiner, J., & Liu, F. (2016). Effects of CO_2 elevation and irrigation regimes on leaf gas exchange, plant water relations, and water use efficiency of two tomato cultivars. *Agricultural Water Management, 169*, 26–33.

Pollastrini, M., Desotgiu, R., Camin, F., Ziller, L., Gerosa, G., Marzuoli, R., & Bussotti, F. (2014). Severe drought events increase the sensitivity to ozone on poplar clones. *Environmental and Experimental Botany, 100*, 94–104.

Polley, H. W. (2002). Implications of atmospheric and climatic change for crop yield and water use efficiency. *Crop Science, 42*, 131–140. https://doi.org/10.2135/cropsci2002.1310.

Praba, M. L., Cairns, J. E., Babu, R. C., & Lafitte, H. R. (2009). Identification of physiological traits underlying cultivar differences in drought tolerance in rice and wheat. *Journal of Agronomy and Crop Science, 195*, 30–46. https://doi.org/10.1111/j.1439-037X.2008.00341.x.

Qiao, Y., Zhang, H., Dong, B., Shi, C., Li, Y., Zhai, H., & Liu, M. (2010). Effects of elevated CO_2 concentration on growth and water use efficiency of winter wheat under two soil water regimes. *Agricultural Water Management, 97*, 1742–1748. https://doi.org/10.1016/j.agwat.2010.06.007.

Raj, A., Chakrabarti, B., Pathak, H., Singh, S. D., & Pratap, D. (2015). Impact of elevated temperature on iron and zinc uptake in rice crop. *International Journal of Agriculture Environment and Biotechnology, 8*, 691. https://doi.org/10.5958/2230-732x.2015.00077.7.

Rizhsky, L., Liang, H., & Mittler, R. (2002). The combined effect of drought stress and heat shock on gene expression in tobacco. *Plant Physiology, 130*, 1143–1151. https://doi.org/10.1104/pp.006858.

Robredo, A., Pérez-López, U., de la Maza, H. S., González-Moro, B., Lacuesta, M., Mena-Petite, A., & Muñoz-Rueda, A. (2007). Elevated CO_2 alleviates the impact of drought on barley improving water status by lowering stomatal conductance and delaying its effects on photosynthesis. *Environmental and Experimental Botany, 59*, 252–263. https://doi.org/10.1016/j.envexpbot.2006.01.001.

Rowland-Bamford, A. J., Nordenbrock, C., Baker, J. T., Bowes, G., & Hartwell Allen, L. (1990). Changes in stomatal density in rice grown under various CO_2 regimes with natural solar irradiance. *Environmental and Experimental Botany, 30*, 175–180. https://doi.org/10.1016/0098-8472(90)90062-9.

Royer, D. L. (2001). Stomatal density and stomatal index as indicators of paleo-atmospheric CO_2 concentration. *Review of Palaeobotany and Palynology, 114*, 1–28.

Ruehr, N. K., Gast, A., Weber, C., Daub, B., & Arneth, A. (2016). Water availability as dominant control of heat stress responses in two contrasting tree species. *Tree Physiology, 36*, 164–178. https://doi.org/10.1093/treephys/tpv102.

Saha, S., Sehgal, V. K., Nagarajan, S., & Pal, M. (2012). Impact of elevated atmospheric CO_2 on radiation utilization and related plant biophysical properties in pigeon pea (Cajanus cajan L.). *Agricultural and Forest Meteorology, 158–159*, 63–70. https://doi.org/10.1016/j.agrformet.2012.02.003.

Saha, S., Sehgal, V. K., Chakraborty, D., & Pal, M. (2015). Atmospheric carbon dioxide enrichment induced modifications in canopy radiation utilization, growth and yield of chickpea [Cicer arietinum L.)]. *Agricultural and Forest Meteorology, 202*, 102–111. https://doi.org/10.1016/j.agrformet.2014.12.004.

Saha, S., Chakraborty, D., Sehgal, V. K., Nain, L., & Pal, M. (2016). Long-term atmospheric CO_2 enrichment impact on soil biophysical properties and root nodule biophysics in chickpea (Cicer arietinum L.) Eur. *Journal of Agronomy, 75*, 1–11. https://doi.org/10.1016/j.eja.2015.12.005.

Saha, S., Chakraborty, D., Paul, R. K., Samanta, S., & Singh, S. B. (2017). Disparity in rainfall trend and patterns among different regions: Analysis of 158 years' time series of rainfall dataset across India. *Theoretical and Applied Climatology*. https://doi.org/10.1007/s00704-017-2280-9.

Saha, S., Chakraborty, D., Hazarika, S., et al. (2022). Spatiotemporal variability of weather extremes over eastern India: Evidences of ascertained long-term trend persistence and effective global climate controls. *Theoretical and Applied Climatology, 148*, 643–659. https://doi.org/10.1007/s00704-022-03949-1.

Saha, S., Chakraborty, D., Lata, Pal, M., Nagarajan, S. (2011). Impact of elevated CO_2 on utilization of soil moisture and associated soil biophysical parameters in pigeon pea (Cajanas cajan L.). *Agriculture, Ecosystems & Environment, 142* (3–4), 213–221. https://doi.org/10.1016/j.agee.2011.05.008.

Saha, S., Das, B., Chatterjee, D., Sehgal, V. K., Chakraborty, D., Pal, M. (2020). Crop growth responses towards elevated atmospheric CO_2. In Plant ecophysiology and adaptation under climate change: mechanisms and perspectives I. Singapore; Springer Singapore, pp. 147–198.

Saha, S., Chakraborty, D., Sehgal, V. K., Pal, M. (2021). Atmospheric CO2 enrichment effect on water use efficiency in chickpea (Cicer arietinum L.). *Journal of Agronomy and Crop Science, 207*(4), 733–744. https://doi.org/10.111/jac.12505.

Samarakoon, A., & Gifford, R. (1996). Elevated CO_2 effects on water use and growth of maize in wet and drying soil. *Australian Journal of Plant Physiology, 23*(1), 53. https://doi.org/10.1071/pp9960053.

Sarker, B. C., & Hara, M. (2009). Effects of elevated CO_2 and water stress on root structure and hydraulic conductance of Solanum melongena L. *Bangladesh Journal of Botany, 38*, 55–63.

Sastry, A., & Barua, D. (2017). Leaf thermotolerance in tropical trees from a seasonally dry climate varies along the slow-fast resource acquisition spectrum. *Science and Reports, 7*, 11246. https://doi.org/10.1038/s41598-017-11343-5.

Schapendonk, A. H. C. M., Dijkstra, P., Groenwold, J., Pot, C. S., & Van De Geijn, S. C. (1997). Carbon balance and water use efficiency of frequently cut Lolium perenne L. swards at elevated carbon dioxide. *Global Change Biology, 3*, 207–216. https://doi.org/10.1046/j.1365-2486.1997.00099.x.

Shi, K., Li, X., Zhang, H., Zhang, G., Liu, Y., Zhou, Y., Xia, X., Chen, Z., & Yu, J. (2015). Guard cell hydrogen peroxide and nitric oxide mediate elevated CO_2-induced stomatal movement in tomato. *New Phytologist, 208*, 342–353. https://doi.org/10.1111/nph.13621.

Stanhill, G. (1986). Water use efficiency. *Advances in Agronomy,* 53–85.https://doi.org/10.1016/s0065-2113(08)60465-4.

Stanhill, G., & Cohen, S. (2001). Global dimming: A review of the evidence for a widespread and significant reduction in global radiation with discussion of its probable causes and possible agricultural consequences. *Agricultural and Forest Meteorology, 107*(4), 255–278. https://doi.org/10.1016/s0168-1923(00)00241-0.

Streit, K., Siegwolf, R. T. W., Hagedorn, F., Schaub, M., & Buchmann, N. (2014). Lack of photosynthetic or stomatal regulation after 9 years of elevated [CO_2] and 4 years of soil warming in two conifer species at the alpine treeline. *Plant, Cell and Environment, 37*, 315–326.

Supit, I., van Diepen, C. A., Boogaard, H. L., Ludwig, F., & Baruth, B. (2010). Trend analysis of the water requirements, consumption and deficit of field crops in Europe. *Agricultural and Forest Meteorology, 150*, 77–88. https://doi.org/10.1016/j.agrformet.2009.09.002.

Suzuki, N., Rivero, R. M., Shulaev, V., Blumwald, E., & Mittler, R. (2014). Abiotic and biotic stress combinations. *New Phytologist, 203*, 32–43. https://doi.org/10.1111/nph.12797.

Taiz, L., & Zeman, F. (2010). Photosynthesis: The light reaction. *Plant Physiology* (pp. 163–198). Sinauer Associates Inc.

Takahashi, S., Badger, M. R. (2011). Photoprotection in plants: A new light on photosystem II damage. *Trends in Plant Science, 16*(1), 53–60. https://doi.org/10.1016/j.tplants.2010.10.001.

Thomas, A. (2008). Development and properties of 0.25-degree gridded evapotranspiration data fields of China for hydrological studies. *Journal of Hydrology, 358*, 145–158. https://doi.org/10.1016/j.jhydrol.2008.05.034.

Tiwari, B., Kalim, S., Bangar, P., Kumari, R., Kumar, S., Gaikwad, A., & Baht, K. V. (2018). Physiological, biochemical, and molecular responses of thermotolerance in moth bean (Vigna aconitifolia (Jacq.) Marechal). *Turkish Journal of Agriculture and Forestry, 42*, 176–184.

Torsethaugen, G., Pell, E. J., & Assmann, S. M. (1999). Ozone inhibits guard cell K+ channels implicated in stomatal opening. *Proceedings of the National Academy of Sciences, 96*(23), 13577–13582. https://doi.org/10.1073/pnas.96.23.13577.

Toselli, M., Flore, J. A., Marogoni, B., & Masia, A. (1999). Effects of root-zone temperature on nitrogen accumulation by non-breeding apple trees. *Journal of Horticultural Science and Biotechnology, 74*, 118–124. https://doi.org/10.1080/14620316.1999.11511083.

Tuba, Z., Szente, K., & Koch, J. (1994). Response of photosynthesis, stomatal conductance, water use efficiency and production to long-term elevated CO_2 in winter wheat. *Journal of Plant Physiology, 144*(6), 661–668. https://doi.org/10.1016/s0176-1617(11)80657-7.

Upadhyaya, H. D., Dronavalli, N., Gowda, C. L. L., & Singh, S. (2011). Identification and evaluation of chickpea germplasm for tolerance to heat stress. *Crop Science, 51*, 2079–2094. https://doi.org/10.2135/cropsci2011.01.0018.

Urban, O., Klem, K., Aˇc, A., Havra´nkova´, K., Holiˇsova´, P., Navra´til, M., Zitová, M., Kozlová, K., Pokorný, R., Šprtová, M., Tomášková, I., Špunda, V., & Grace, J. (2012). Impact of clear and cloudy sky conditions on the vertical distribution of photosynthetic CO2 uptake within a spruce canopy. *Functional Ecology, 16*, 46–55.https://doi.org/10.1111/j.1365-2435.2011.01934.x.

VanLoocke, A., Betzelberger, A. M., Ainsworth, E. A., & Bernacchi, C. J. (2012). Rising ozone concentrations decrease soybean evapotranspiration and water use efficiency whilst increasing canopy temperature. *New Phytologist, 195*, 164–171. https://doi.org/10.1111/j.1469-8137.2012.04152.x.

Wahid, A., & Close, T. J. (2007). Expression of dehydrins under heat stress and their relationship with water relations of sugarcane leaves. *Biologia Plantarum, 51*, 104–109.

Wang, Z., & Huang, B. (2004). Physiological recovery of Kentucky bluegrass from simultaneous drought and heat stress. *Crop Science, 44*, 1729–1736.

Wang, Q., Yu, D., Dai, L., Zhou, L., Zhou, W., Qi, G., Qi, L., & Ye, Y. (2010). Research progress in water use efficiency of plants under global climate change. *Yingyong Shengtai Xuebao, 21*(12), 3255–3265.

Wang, A., Lam, S. K., Hao, X., Li, F. Y., Zong, Y., Wang, H., & Li, P. (2018). Elevated CO_2 reduces the adverse effects of drought stress on a high-yielding soybean (Glycine max (L.) Merr.) cultivar by increasing water use efficiency. *Plant Physiology and Biochemistry, 132*, 660–665. https://doi.org/10.1016/j.plaphy.2018.10.016.

Wanjura, D. F., Buxton, D. R., & Stapleton, H. N. (1970). A temperature model for predicting initial cotton emergence. *Agronomy Journal, 62*, 741–743.

Way, D. A., Katul, G. G., Manzoni, S., & Vico, G. (2014). Increasing water use efficiency along the C3 to C4 evolutionary pathway: A stomatal optimization perspective. *Journal of Experimental Botany, 65*, 3683–3693. https://doi.org/10.1093/jxb/eru205.

Williams, P. D., Guilyardi, E., Sutton, R., Gregory, J., & Madec, G. (2007). A new feedback on climate change from the hydrological cycle. *Geophysical Research Letters, 34*(8). https://doi.org/10.1029/2007GL029275.

Wittig, V. E., Ainsworth, E. A., & Long, S. P. (2007). To what extent do current and projected increases in surface ozone affect photosynthesis and stomatal conductance of trees? A meta-analytic review of the last 3 decades of experiments. *Plant, Cell and Environment, 30*, 1150–1162. https://doi.org/10.1111/j.1365-3040.2007.01717.x.

Wu, D. X., Wang, G. X., Bai, Y. F., & Liao, J. X. (2004). Effects of elevated CO_2 concentration on growth, water use, yield and grain quality of wheat under two soil water levels. *Agriculture, Ecosystems & Environment, 104*, 493–507. https://doi.org/10.1016/j.agee.2004.01.018.

Wullschleger, S. D., Gunderson, C. A., Hanson, P. J., Wilson, K. B., & Norby, R. J. (2002). Sensitivity of stomatal and canopy conductance to elevated CO_2 concentration—interacting variables and perspectives of scale. *New Phytologist, 153*, 485–496. https://doi.org/10.1046/j.0028-646X.2001.00333.x.

Xu, Y., Shang, B., Peng, J., Feng, Z., & Tarvainen, L. (2021). Stomatal response drives between-species difference in predicted leaf water-use efficiency under elevated ozone. *Environmental Pollution, 269*, 116137. https://doi.org/10.1016/j.envpol.2020.116137.

Yang, X., Asseng, S., Wong, M. T. F., Yu, Q., Li, J., & Liu, E. (2013). Quantifying the interactive impacts of global dimming and warming on wheat yield and water use in China. *Agricultural and Forest Meteorology, 182–183*, 342–351. https://doi.org/10.1016/j.agrformet.2013.07.006.

Zhang, B., Liu, W., Chang, S. X., & Anyia, A. O. (2010). Water-deficit and high temperature affected water use efficiency and arabinoxylan concentration in spring wheat. *Journal of Cereal Science, 52*, 263–269. https://doi.org/10.1016/j.jcs.2010.05.014.

Zhang, W. W., Wang, M., Wang, A. Y., Yin, X. H., Feng, Z. Z., & Hao, G. Y. (2018). Elevated ozone concentration decreases whole-plant hydraulic conductance and disturbs water use regulation in soybean plants. *Physiologia Plantarum, 163*(2), 183–195. https://doi.org/10.1111/ppl.12673.

Chapter 13
Climate Change on Seeds Physiology

Debasish Roy, Niraj Biswakarma, Tridiv Ghosh, Koushik Bag, Ayan Sarkar, Krishnayan Paul, Bappa Das, Saikat Chowdhury, and B. Hari Krishna

Abstract Climate change leads to take modifications of both biotic and abiotic constituents. There is evidence that some species have genetic variation that can respond to such selection, such as important traits that influence distribution and abundance. In this chapter we have revise the literature information about an impact of the elevated CO_2, and temperature or heat stress on seeds's physiology, quality traits and yield determining factors. The physiological responses of the crops indicate that they will grow more quickly during the following few years, with minor alterations to their development like fruiting and flowering depending on the species. Moreover, the high temperature may offset the positive effect of elevated CO_2 on crop yield attributes or elevated CO_2 would compensates for the negative effects of high temperature. It is pointed out that studies related to changes in seed yields and their attributes due to consequence of global climate changes, need priority areas to further research as the they will highly associate with food security.

Keywords Food security · Elevated CO_2 · Seed quality · Seed physiology · Seed purity · Yield

13.1 Introduction

Climate change is a significant and enduring change in average weather conditions that spans decades to millions of years. It is considered the biggest threat to food and nutritional security (FAO, 2015; Raj et al., 2022). As it alters the temperature, rainfall pattern, and CO_2 levels, etc. which results in both positive and negative influences on agricultural production (Malhi et al., 2021). Seed is the most fundamental

D. Roy (✉) · N. Biswakarma · T. Ghosh · K. Bag · A. Sarkar · K. Paul · S. Chowdhury · B. Hari Krishna
ICAR-Indian Agricultural Research Institute, New Delhi 110012, India
e-mail: roydeba93@gmail.com

B. Das
ICAR-Central Coastal Agricultural Research Institute, 403402, Old Goa, India

H. Pathak et al. (eds.), *Climate Change Impacts on Soil-Plant-Atmosphere Continuum*, Advances in Global Change Research 78,
https://doi.org/10.1007/978-981-99-7935-6_13

agricultural input, and a source of the genetic material of crop species and their varieties (Delouche & Potts, 1983). The prerequisite for achieving higher agricultural productivity and profitability quality seeds is essential (Louwaars & Manicad, 2022). It has been estimated that the usage of high-quality seeds expressively influences crop yield by ~15–20%. The seeds are a critical resource to the global food and feed supply, which accounts for ~70% calories consumption by individuals and animals on a regular basis (Gupta et al., 2022). The seed contains tiny plants, and as it grows and emerges, a series of processes, including imbibition, cell multiplication, cellular metabolism, and radicle protrusion, takes place (Arun et al., 2022). Under natural conditions, it detects environmental signals and germinates only when there occurs a favourable environment. However, abnormalities in any growth factors, including salinity, alkalinity stressors, heat, moisture, air, and radiation, are anticipated to impair the quality of seed production and plant reproduction (Hedhly et al., 2008). Furthermore, Savvides et al. (2016) reported that extreme climatic events can have a substantial negative impact on seed germination due to delayed seed germination, inhibiting radicle emergence resulting in poor seedling establishment, and lowering final economic yields (Cotrina et al., 2023). Thus, our prime focus must be on assuring a quality seed to the stakeholders in order to acclimatize to the effect of climate change.

Singh et al. (2013) and Abebe et al. (2016) estimated the influence of climate change on seed production of rice, soybean, cotton, and peanut and reported that the yields of C4 crops are projected to remain stable, while the yield of C3 increases by 11–30% due to atmospheric CO_2 enrichment. A corresponding, report has also been stated by Hatfield et al. (2011). However, temperature stress associated with global climate change can impede various physiological aspects of seeds such as reducing seed number, and seed weight, accelerating seed growth rate, and shortening grain filling time, which might conceal the beneficial impact of increasing CO_2 on seed production (Singh et al., 2013). Furthermore, high temperature stress can reduce the dry matter partitioning during the primary stages of kernel advances as it affects the cell division, sugar, and starch metabolism in the endosperm, leading to reduced seed quality and yield (Essemine et al., 2010). Similarly, changing rainfall patterns results in prolonged dry spell stress at any stage of crop growth and among them, seed germination is largely exaggerated due to reduced hydraulic reduction resulting from poor water entry into the seed, thus affecting overall physiological and metabolic processes (Bareke, 2018; Fahad et al., 2017). Seed dormancy has also been reported to be altered by changing environment as the intensity and timing of environmental stimuli changes over time. It has been also reported that kharif crops are more affected due to increased incidence of extreme weather like early/late monsoon, flood or drought (Krishnan et al., 2009).

Climate change has contrary effects on both the physiological and biochemical configuration of seeds and thus has a detrimental impact on seed quality. Reactive oxygen species (ROS), which are constantly produced as metabolic by-products, are necessary for seed physiology. At low concentrations of ROS can function as signalling molecules that control gene expression and plant growth (Huang et al., 2019). Yet, at large concentrations, ROS can be deadly and harm biological elements like DNA, lipids, and proteins (Schieber & Chandel, 2014). Photosynthetic metabolism and reactions to environmental stress are significantly impacted by the effects of temperature stress, drought stress, and atmospheric CO_2 perturb redox imbalances (Munné-Bosch et al., 2013). Further, it also accelerates the seed ageing in mitochondria and peroxisomes and reduces seed viability, in turn affecting post-harvest storage and seed processing (Kijowska-Oberc et al., 2021). Thus, it is evident from many instances that adverse climate during the crop growth and post-harvesting period leading to poor seed quality and reduce agricultural sustainability. Thus, we have critically examined the consequences of climate change on seed yield and quality, with a focus on increasing CO_2, higher temperature, drought, and their interactions.

13.2 Climate Change and Agriculture

Climate change is one of the deadliest disasters of the present century for mother earth, its environment, ecosystem, and human habitation mainly intensified due to human-induced activities. There is a gamet of adverse consequences of climate change, and temperature rise is the most prominent and visible result among them. IPCC's 6th assessment report. This suggests that the global mean temperature has risen about 1 °C between 1901–2020 (IPCC, 2021). The primary cause of global warming is quicker accumulation of greenhouse gases (GHGs) in the atmosphere through anthropogenic activities. Principally, they act as a potential barrier to extra-terrestrial long-wave radiations that leave the earth and thus trap the heat in the atmosphere. As the projection of future emissions that influence climate change is extremely difficult, several climate change scenarios are assumed considering several socio-economic, technological, and environmental conditions (Fig. 13.1). In all scenarios, the average global temperature is expected to exceed 1.5 °C by 2100 (IPCC, 2014), whereas the greenhouse gas concentrations are assumed to increase continuously throughout the century under every emission pathway. Thus, increasing GHGs will aggravate the change in precipitation patterns, and increased regularity and intensity of extreme events will threaten agricultural production in a long run (Mendelsohn, 2009).

Further, the negative impacts of global warming, sea level rise, desertification, droughts, GHGs (except CO_2), and increased frequencies of weather extremes can be offset by adopting several cultural, crop management, and the CO_2 fertilization effect in particular. Though the extent of positive results depends on several factors like soil, latitude, agro-climate, etc. the adaptive capabilities to climate risk, and magnitude of the positive impact of these factors is still debatable. So, we summarize the factors that may increase or decrease crop production under changing climate scenarios, rather than the integrated effects of these factors on crop production (Table 13.1).

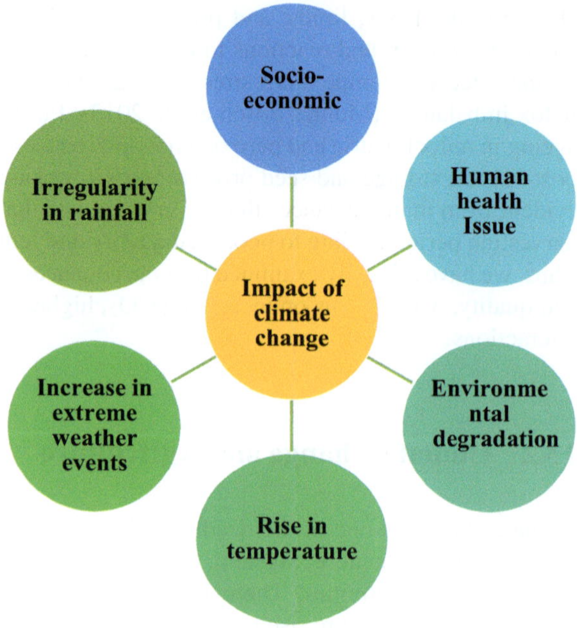

Fig. 13.1 A summary of the effects of climate change

Table 13.1 Climate change and crop yield determining factors

Reasons for yield increase	Reasons for yield decrease
• Longer growth period for temperate crops • Increase in spring soil temperatures • Modest to unaffected maximum summer temperature • Increased nicto temperatures • Probability of frost-free spring • More precipitation • More soil moisture • Increased carbon uptake by crops • Higher CO_2 increases the water-use efficiency of crops	• Accelerated precipitation extremes • Heavy rainfall events • Frequent droughts • High flood incidents • Higher incidence of over-wintering pests • Susceptibility to pathogens due to over humidity • Vigorous weed growth etc

13.3 Physiological Aspects of Seeds

In case of angiosperms, development of seed is a very crucial event in its life cycle. A seed is developed by the occurrence of following events (Fig. 13.2).

A seed has three parts viz. seed coat, endosperm, and an embryo. During the germination of a seed, at first, the imbibition of water occurs. As water enters a seed, the breaking of the seed coat occurs due to imbibition pressure. Subsequently, gibberellin is secreted by the embryo due to the activation of the respiratory mechanism. This gibberellin induces hydrolytic enzyme secretion from the aleurone layer.

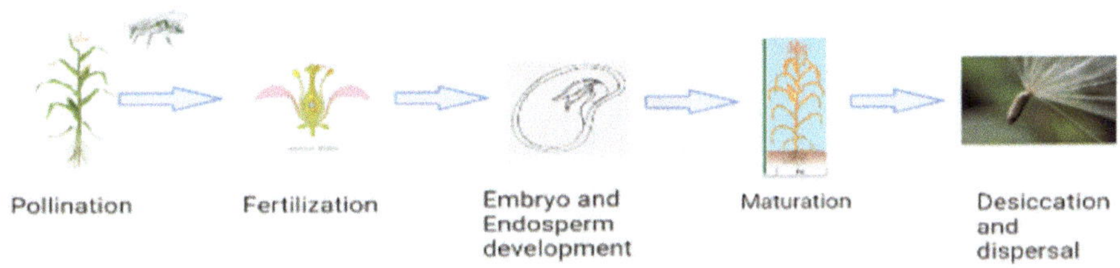

Fig. 13.2 Seed development phage

Hydrolytic enzymes break down complex materials from seed reserves and simple sugar and amino acid are formed which helps the plant to germinate and emerge from the soil. Then radicle protrusion occurs first, followed by plumule emergence. Seed physiology indicates the process of germination followed by seedling emergence and understanding this by nexus of several genetic as well as external environmental factors like water, air, temperature, light, etc. (Reddy & Gupta et al., 2022; Reddy, 2019).

13.4 Climate change's Impact on Seed Quality

Seed yield and quality are affected by climatic conditions throughout the crop growth cycle, notably during the reproductive stage (Alan et al., 2022). Seed faces the difficult task of generating more food every day to feed the world's constantly growing population. Seed yield and quality are reduced as a result of climate anomalies that occur during blooming, seed set, and seed development due to lower nutrient uptake, partitioning, and remobilization of nutrients in the seeds (Parihar et al., 2022). Seed quality is a broad term. In this context, the phrase "seed quality" refers to a seed's entire planting value, as measured by seed germination, vigour, viability%, longevity, seed nutrition, seed water content, diameter, morphology, mass, colour, and other factors. The three prime aspects of quality seeds are the size of the individual seed, and the ability of a seed to germinate and grow. Thus, seed quality has a pronounced influence on the superiority of planting stock.

13.4.1 Elevated CO_2 Effects on Seed Quality

The experiment on how rising CO_2 impacts on crop development in greenhouse environments was initiated by (Enoch et al., 1971). Over the past few decades, anthropogenic activities have caused a steady rise in the amount of CO_2 in the atmosphere. The annual CO_2 increase, according to current projections, is 2.2 ppm (Le Quéré et al., 2018). A rise in atmospheric CO_2 is forecast by climate models, with 538–936 ppm by 2100 (Meinshausen et al., 2011; Van Vuuren et al., 2011). The consequences

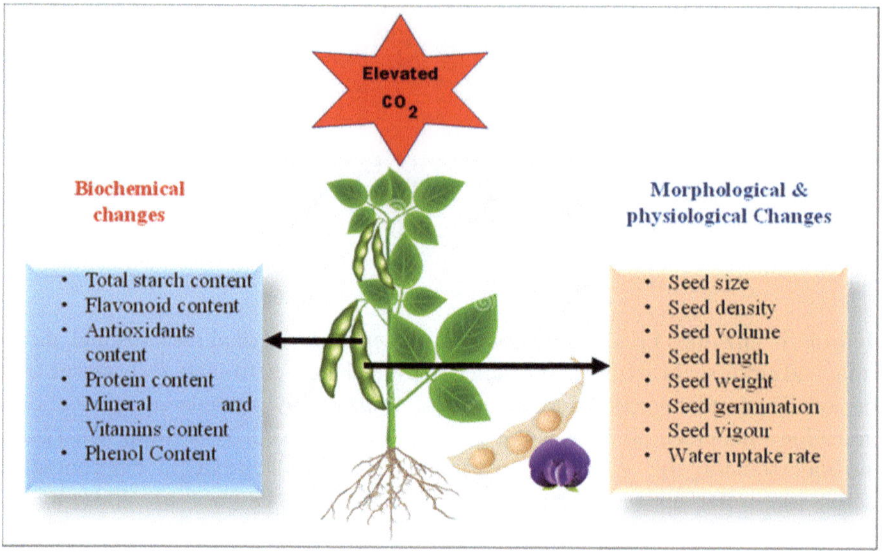

Fig. 13.3 Under elevated CO_2, potential morphological, physiological, and biochemical alterations in seed quality

on agriculture are unknown, and crop performance and seed quality under a new climatic scenario are important issues to be concerned about (Korres et al., 2016). In numerous crops, including rice, it has been demonstrated that a rise in atmospheric CO_2 has an impact on the rate of photosynthesis, carbon assimilation, seed set, and grain quality (Ainsworth & Long, 2021). Various factors such as type of plant species, developmental stages, duration and intensity of exposure to elevated CO_2 decide the influence of elevated CO_2 on seed quality. There are enormous reports, particularly on the impact of high CO_2 levels on the biochemical makeup of seeds (Almuhayawi et al., 2021; Lamichaney et al., 2021a, b; Li et al., 2018; Zheng et al., 2020). As a result of rising atmospheric CO_2, the potential morphological, physiological, and biochemical alterations in seed quality are depicted in (Fig. 13.3).

13.4.2 *Impact of Elevated CO_2 on Seeds' Viability*

Increasing CO_2 concentration in the atmosphere can alter the C:N ratio of seeds due to the CO_2 fertilization effect, and this increased carbon content reduces the C: N in seeds. This reduced nitrogen and affects the amino acid supply for de novo protein synthesis resulting in the inhibition of embryo growth. Which ultimately can reduce seed viability %, particularly for herbaceous plants (Andalo et al., 1996; Ward & Strain, 1997; Jablonski et al., 2002; Hampton et al., 2013). But in the case of trees like Pinus taeda L., increased CO_2 content did not affect seed viability % (La Deau &

Clark, 2001, 2006) as trees might have preferential nitrogen allocation to seeds or they can increase nitrogen uptake (Finzi et al., 2007). Temperature also has a role in deciding the degree of affecting seed viability as the increased temperature can reduce the time required for reducing seed viability through elevated CO_2 concentration as evident from Amaranthus retroflexus L. (Weller et al., 2021).

13.4.3 *Impact of Increased CO_2 on Seed Genetic Purity*

An increase in CO_2 can affect the genetic purity of the seed, which is a very important trait with respect to uniformity, adaptability, and disease resistance traits of the seed which have an influence on seed physiology. Elevated CO_2 along with increasing temperature can cause an interruption in self-incompatibility or male sterility of many cross-pollinated crops like clover or perennial rye grass. This will promote self-pollination in those species and reduce the genetic purity of the seeds of those crops. These seeds will show inbreeding depression resulting in poor performance of plants (Hampton et al., 2016). Elevated CO_2 can affect the pollination behaviours of insects which also can indirectly affect the genetic purity of seeds (Crowley et al., 2021).

13.4.4 *Impact of Increased CO_2 on Seed Longevity*

Seed longevity can be defined as the viability of the seed after storing it in dry conditions (Rajjou & Debeaujon, 2008). The longevity of a seed is decided by genetic factors along with environmental conditions like temperature, relative humidity, moisture content, oxygen pressure, etc. (Groot et al., 2012; Sano et al., 2016). The survival of seeds in soil is largely determined by seed mass (Fenner & Thompson, 2005; Poschlod et al., 2005). In temperate grassland, it was found that only CO_2 enrichment, affect the seed mass vis a vis seed longevity. Only CO_2 enrichment increased seed longevity in grass Elymus scaber (C3) but decreased in Austrodanthonia caespitosa and Hypochaeris radicata and did not influence on 13 other species taken into that experiment (Hovenden et al., 2007). Reative oxygen species (ROS) and Reative nitrogen species (RNS) are two types of compounds that impart oxidative and nitrosative stress on seeds. Under elevated CO_2 conditions, the production of these compounds increases and affects the longevity of seeds by means of two mechanisms. The first is that as ROS is formed, it is subsequently scavenged by glutathione-s hydrolase, causing the EGSSG/2GSH ratio to rise (oxidizing condition). The thioredoxin-apoptosis signal-regulating kinase 1 (Trx-ASK 1) complex splits under these circumstances, and one of its components, ASK-1, causes the degradation of DNA and nucleic acids (Chen et al., 2014). In another mechanism programmed cell death (PCD) occurs due to calcium release from the endoplasmic reticulum (ER) (Kumar et al., 2021a, b). So, an increase in CO_2 can significantly affect the soil seed bank of various plants by affecting longevity of them.

13.4.5 Effect of Elevated CO_2 on Physical Purity of Seeds

Physical purity of seed refers to the percent (%) of pure seed from a seed lot, free inert mater, undersized seeds, and broken seeds, other crop seeds, weed seed, and soil and dust particles, etc. There isn't much evidence suggesting how high CO_2 levels impair seed physical purity directly, although they can have an indirect impact. As already discussed, elevated CO_2 can affect the seed developmental process in some cases, so undersized seeds or broken seed proportion may increase due to climate change, hampering the physical purity of seeds. Similarly increase in CO_2 may also increase weed seed production, resulting in more chance of contamination as well as a reduction of physical purity (Singh et al., 2011).

13.4.6 Effect of Elevated CO_2 on Seed Weight

Seed weight or mass is the least variable trait in terms of yield, because of the breeding for uniformity of seed in normal conditions (Manan et al., 2022). But in changing environmental conditions these characteristics may change significantly. The availability of assimilates is supposed to increase as atmospheric CO_2 rises, resulting in an increase in grain weight however, previous studies indicated that crops' responses to elevated CO_2 are diverse, like seed weight was found to increase (Quaderi & Reid, 2005; Soba et al., 2020; Usui et al., 2014), decrease (Mao et al., 2021; Smith et al., 2000; Wagner et al., 2001), or exhibit no change (Lamichaney et al., 2021a, b). Increased CO_2 generated an average increase in seed weight of 4%, according to Jablonski et al. (2002), with legume crops responding most strongly (+8%), followed by non-legume crops (+3%), and C4 plants (+3%).

Non-legumes showed a significant loss in seed nitrogen content because of elevated CO_2 due to the dilution effect, however, there was no decrease in the nitrogen content of legume seeds. (Jablonski et al., 2002). This suggested that if nitrogen was not a limiting factor, seed weight might grow with high CO_2. Excess carbon in legumes may be transferred to nodules as an energy source, allowing them to get the most out of elevated CO_2 and hence increase seed weight without changing seed nitrogen. While, increased seed weight may be achieved in non-legumes at the cost of decreased seed nitrogen content. According to Hikosaka et al. (2011), the seed weight per plant increased from 0.75 to 4.45 g in rice and from 0.88 to 2.07 in wheat (Triticum aestivum L.). However, Lamichaney et al. (2019) found no increase in seed weight under elevated CO_2. Liu et al. (2019) also stated no change in the test weight of wheat when subjected to elevated CO_2. Given that the response to elevated CO_2 is genotype-specific, differences in cultivars utilized by the same research group could explain such discrepancies in results (Maphosa et al., 2019).

13.4.7 The Influence of Increased CO_2 on Seed Germination

Germination of seed is a very important stage in plants' life cycle as establishment depends on the success of germination (Gorai & Neffati, 2007). The percentage of seeds that sprout normally is determined by testing germination in a lab under ideal conditions (ISTA, 2020). When seeds germinate, a succession of processes that start with seed hydration followed by radicle protruding through the seed coat. The procedures in which seeds are harvested, dried, cleaned, and stored might affect germination, although unfavourable weather conditions, in particular high or low temperatures, precipitation, and relative humidity in the field during seed development, can also hinder germination (Dornbos, 2020; Egli et al., 2005). Several studies have been done to look into the impact of elevated CO_2 levels on seed germination rates. According to studies, increased CO_2 can cause seed germination to decrease (Lamichaney et al., 2019; Saha et al., 2015), remain the same (Chen et al., 2015; Thinh et al., 2017), or even increase (Edwards et al., 2001). Debouza et al. (2021) reported the reduction in seed germination % due to higher CO_2 concentration. The result was found decline in the germination of winter beans where soil CO_2 level was increased from 20 to 50% (Al-Traboulsi et al., 2013). The elevated CO_2 in rice did not change the germination rate, and energy (Chen et al., 2015). The reduction in protease enzymes, and seed N content reduced the chickpea germination by 45–47% due to exposure of a higher CO_2 levels (580 ppm) (Saha et al., 2015), a). Reduced source-to-sink N translocation under elevated CO_2 levels may decrease ~13% N content (Jena et al., 2018; Yang et al., 2007). Research on Pinus taeda, wheat, and yam indicated that introducing the mother plants to elevated CO_2 resulted in non-significant effect on the seed germination % (Sanhewe et al., 1996; Thinh et al., 2017; Way et al., 2010). The responses vary with species to species (Ziska & Bunce, 1993), and there is significant genotypic variance (Andalo et al., 1996). This effect is more pronounced by means of maternal parent CO_2 enrichment than paternal parent CO_2 enrichment. According to Ziska and Bunce (1993), Esashi et al. (1986), Leubner-Metzger (2006) elevated CO_2 can stimulates ethylene production that results in enhanced germination in M. sativa, A. hybridus and C. album. Moreover, seed germination in monocotyledonous plants may believe to be more resilient to higher CO_2 concentrations than the dicotyledonous plants (West et al., 2009).

13.4.8 Effect of Elevated CO_2 on Seed Vigour

Seed vigour is determined by the seed size, texture, appearance, storability, rapidity, and uniformity in germination. Further synthesis in research is needed to determine the effect of an elevated CO_2 on the seed vigour (Lamichaney & Maity, 2021a). The rice plants grown at 610 ppm levels of CO_2, reduces the seed vigour, due to lower nitrogen content, substrate availability, and substrate translocation (Lamichaney et al., 2019). Further, Lamichaney and Maity (2021a) reported that elevated CO_2 had no significant effect on the germination and vigour of the chickpea. Chen et al. (2015) also reported no difference in seed vigour of six rice cultivars. So, it emphasizes the importance of exploring seed quality in the future circumstances. Seed storability is another major predictor of seed vigor, although there are no studies on the lifetime of seeds generated under elevated CO_2 conditions. An increase in seed vigor, as well as germination % in rice, was recorded due to elevated CO_2 by Abzar et al. (2017). Yet, chickpeas seed vigour decreased in the presence of high CO_2 levels (Saha et al., 2015).

13.4.9 Elevated CO_2 on Seed Biochemical Composition

The biochemical makeup directly determines the seed performance, and crop growth (Maity & Pramanik, 2013). Researchers have found that elevated CO_2 can affect most of the signalling pathways involving primary and secondary metabolites (Foyer & Noctor, 2020; Misra & Chen, 2015). In this regard, Buchner and co-workers (2015) stated that elevated CO_2 can up-regulate the genes involved in carbohydrate, nitrogen assimilation, and amino acid transport affecting the biochemical seed properties. Increased CO_2, lowers the amount of grain protein in edible portion of the crops (Medek et al., 2017). The meta-analysis by Ben Mariem et al. (2021) found that increasing CO_2 significantly decreased grain protein contents, increased photosynthesis, and grain carbohydrate accumulation, which lead to reduction in grain protein content. The CO_2 enrichment decreased the seed N content leading to a greater C:N ratio in wheat, and barley seeds (Jablonski et al., 2002). In grains and tubers, higher CO_2 increased (6–47%) the ratio of non-structural carbohydrates (TNC) to protein. while lowering the mineral concentrations in wheat, rice, and barley grains by 7.6%, 6.9%, and 7.6%, respectively (Loladze, 2014). However, Högy et al. (2010a) had shown non-significant effect on the protein content and grain yield in Brassica napus (cv. Campino). Saha et al. (2015) found that, higher CO_2 reduced the chickpea seed's protease activity and nitrogen content. It breaks off the equilibrium between the nitrogen and carbon contents of tissues, which lowers the amount of protein in seeds. Seed protein concentration is decreased, which implies there are fewer amino acids available for appropriate embryo growth and development in germinating seeds, which might affect seed germination and viability. The concentrations of the majority of minerals were significantly reduced, but B and Na had little impact and K increased by 2% (Broberg et al., 2017). The outcomes of this meta-analysis concur with those of the individual wheat FACE experiment (Fernando et al., 2014a, 2014b). Balkhyour et al. (2021) reported that elevated has induced biological and nutritive values of aniseeds on their maturity condition by increasing antioxidant and essential oil precursors (cinnamic and shikimic acids). The influence of higher CO_2 on biochemical compositions of the different crops are listed in Table 13.2.

13.5 Impact of Temperature on Seed Quality

Currently, increasing the temperatures in many places of the world is big concern among the researchers (Schuurmans, 2021; Yadav et al., 2015). High-temperature had a significant influence on the reproduction of a different crops, resulting in poor final economic yields and deteriorate seed quality. The grain-filling stage is crucial for determining average seed weight, seed composition, and quantitative and qualitative yield, even though these pressures can affect crop performance at any developmental stage (Prasad et al., 2017; Sehgal et al., 2018). Jumrani and Bhatia (2018) found that increased temperature and water-deficit had major influences on the seed quality of soybean in combination. Furthermore, more seed degradation was brought on by water deficit during the reproductive stage. High temperature stress effect on seed setting, seed fullness, and quality in wheat (Djanaguiraman et al., 2020; Fahad et al., 2019; Talukder et al., 2014) and rice (Lin et al., 2010a, b; Pansrithong et al., 2019).

13.5.1 Effect of Heat Stress on Seed Germination and Vigour

Temperature is a crucial factor which affecting germination of seed. Higher temperature had a considerable harmful effect on germination potential, leading to compact viability of seed and germination (Fahad et al., 2017; Javaid et al., 2022). Wheat embryonic cells are negatively affected by temperatures of about 45 °C, which suppresses seed germination and emergence and thus reduces crop stands (Essemine et al., 2010). Suriyasak et al. (2020) found that rice seeds from plants associated with high temperature stress at gain-filling, germinated significantly faster than those from non-stressed plants. Rice seed germination and seedling growth are aided by high temperatures during early endosperm development (Begcy et al., 2018). Many studies have suggested selecting quality seeds for planting, seed priming and breeding variants with superior seed performance improves the seed germination processes. Lamichaney et al. (2021b) report that climate extremes, mostly high temperatures, affect cool-season grain legumes. Moreover, they investigated how high-temperature stress impacts the phenology, seed establishment, and seed quality parameters of Pisum sativum L., and found a significant reduction in seed yields (24–60%) and seeds germination (4–8%) with a pronounced effect on long-duration genotypes. Iloh et al. (2014) reported that an increase in temperature from 37 to 50 °C for 96 h

Table 13.2 The impact of elevated CO_2 on different seed biochemical compositions

Common name	Botanical name	Family	Biochemical property	Impact of elevated CO_2	References
Rice	Oryza sativa	Poaceae	Protein	No impact	(Wang et al., 2011)
Wheat	Triticum aestivum	Poaceae	Protein	No impact	(Högy & Fangmeier, 2008); (Piikki et al., 2008); (Lam et al., 2012)
Peanut	Arachis hypogaea	Fabaceae	Protein	No impact	(Burkey et al., 2007)
Rapeseed	Brassica napus (cv. Campino)	Brassicaceae	Protein	Positive	(Högy et al., 2010a)
Wheat	Triticum aestivum	Poaceae	N content	Negative	(Jablonski et al., 2002)
Soybean	Glycine max	Fabaceae	N content	Negative	(Rosenthal et al., 2014)
Wheat	Triticum aestivum	Poaceae	Starch Content	Positive	(Högy & Fangmeier, 2008)
Wheat	Triticum aestivum	Poaceae	Sugars and Sugar Derivatives	Positive	(Högy et al., 2010b)
Rice	Oryza sativa	Poaceae	Seed Soluble Sugar	Negative	(Lamichaney et al., 2019)
Potato	Solanum tuberosum	Solanaceae	Carbohydrate	Positive	(Högy & Fangmeier, 2009)
Sugarbeet	Beta vulgaris	Amaranthaceae	Carbohydrate	Positive	(Manderscheid et al., 2010)
Soybean	Glycine max	Fabaceae	P, Ca, S and The Natural Antioxidant-Isoflavones and δ-Tocopherol	Negative	(Zheng et al., 2020)
Soybean	Glycine max	Fabaceae	Fatty Acid Peroxidation and Soluble Sugars	No impact	(Li et al., 2018)
Chickpea	Cicer arietinum	Fabaceae	Protease Activity & N Content	Negative	(Saha et al., 2015)
Soybean	Glycine max	Fabaceae	Isoflavone content	Positive	(Caldwell et al., 2005)

(continued)

Table 13.2 (continued)

Common name	Botanical name	Family	Biochemical property	Impact of elevated CO_2	References
Soybean	Glycine max	Fabaceae	seed Protein, total oil, and fatty acid	No impact	(Thomas et al., 2003)
Rape	Brassica napus L	Brassicaceae	Oil quality	No impact	(Franzaring et al., 2008)
Rape	Brassica napus L	Brassicaceae	N content	Decrease	(Franzaring et al., 2012)
Foxtail millet	Setaria italica L	Poaceae	Protein, Fe	Decrease	(Gong et al., 2021)
Foxtail millet	Setaria italica L	Poaceae	Zn, Mn, Ca, Mg, Cu, K, or B	No significant effect	(Gong et al., 2021)
Foxtail millet	Setaria italica L	Poaceae	P, K, and Mn	Increase	(Gong et al., 2021)
Rice	Oryza sativa	Poaceae	Vitamin complexes (B1, B2, B5, and B9), Zn, Fe	Decreases	(Zhu et al., 2018)
Ajwain	Trachyspermum ammi	Apiaceae	Soluble sugars, essential amino acids, organic acids, phenolics, and favonoids	Increase	(Hassan et al., 2021)
Soybean	Glycine max	Fabaceae	Oleic acid	Increase	(Li et al., 2018)

exposure had reduced 60–100% and 40% rice and sorghum/maize seed germination, respectively. In case of Arabidopsis, a comparable outcome was reported by Silva-Correia et al. (2014) and Tamura et al. (2006). Climate change, however, can also affect the rate of germination and trigger germination in plants like Corylopsis gotoanavar. coreana and Prunus padus possess hard seed coats that can be activated by high temperatures to facilitate rapid and effective germination (Balkaya, 2004; Leubner-Metzger, 2006). Establishment success and duration are determined by seed vigor, a complex agronomic attribute that includes seed lifetime, germination rate, seedling development, and early stress tolerance. Early seed development in cereals, legumes, and vegetable crops can be adversely affected by increased temperature (Reed et al., 2022). When seeds are subjected to high temperatures and moisture content during the storage loses its viability. Physiological and biochemical properties of oat seeds were studied by Chen et al. (2016) after being exposed to high temperatures for 24 h and 2 days (35, 45, and 50 °C), respectively. It has been reported that increase in temperature from 35 to 50 °C, can reduce seed vigour in lettuce, oats, and chickpeas. The crops grown in low temperature can be negatively affected by higher temperature due to seed dormancy and lower germination rates. (Reed et al., 2022).

13.5.2 Elevated Temperature on Seed Longevity

Seed moisture content and storage temperature are the important factors that determines a lifespan of the seeds (Ellis, 2022). Reactive oxygen species (ROS), increases with increasing temperature and thus increases cell membrane permeability due to oxidative stress. For an example, Seed longevity of Wahlenbergia tumidifructa was reduced in warm wet compared to cool wet condition as evident from accelerated aging test (Kochanek et al., 2010). In the same way, Li et al., (2017a, b) also found that mean seed longevity was lower due to elevated temperature in case of three cultivars of Vicia sativa, a leguminous forage species. However, in this aspect also different result was found in case of four alpine species were climate warming resulted in increase in seed longevity (Bernareggi et al., 2015). The type of cultivar is key in determining how high temperatures affect a crop as well on seed longevity as Li et al., (2017a, b) found that, increased temperature significantly reduced the longevity of two cultivar of Vicia sativa, "Lanjian 1" and "Lanjian 2" but not in "Lanjian 3". So, there is a scope for selecting suitable cultivar in this changing climate scenario of increasing temperature.

13.5.3 Heat Stress on Weight and Number of Seeds

Sehgal et al. (2018) has been noted that increased respiration, accelerated seed growth, and shortened seed filling periods all have a significant impact on seed mass. There is scientific evidence that seed mass may decrease, increase or remain unchanged due to elevated temperature (Baskin & Baskin, 2014). Generally, it has been observed that, if the mother plant is subjected to temperature stress at seed development stage either higher or lower temperature, the seed mass is declined (Bewley et al., 2013). Seed weight and seed number are the critical attributes that affects the reproductive success and final economic yield. Heat stress caused a reduction in both seed weight and number in temperate cereals (Suryavanshi & Buttar, 2016). During the reproductive stage, heat stress (>30 °C) significantly reduced seed number, and seed weight in wheat crops (Prasad & Djanaguiraman, 2014). The vegetative is considered as a most heat resistant than the reproductive stage (Aiqing et al., 2018; Prasad et al., 2017). The stages 8 to 6 days before anthesis and anthesis in wheat have been found as the most vulnerable to heat stress, leading to the greatest possible reduction in floret fertility and ultimately seed set (Prasad & Djanaguiraman, 2014). During flowering stage, heat stress for 10 days reduced the number of seeds, and seed weight by 15.4 to 23% and 32.3 to 34.6%, respectively. Bheemanahalli et al. (2019) suggested that increase in temperatures during the flowering and grain filling stages in wheat can decreased seed set and seed weight. Heat stress reduced average grain filling duration in wheat by 8.6 days under late sowing and 12.6 days under very late sowing conditions, resulting in a 17 and 39% drop in test weight, respectively (Dwivedi et al., 2017).

Rice productivity in the tropical, sub-tropical and temperate regions is endangered by frequent occurrences of heat and cold shocks (Arshad et al., 2017). Compared to a high daytime temperature, the night temperature can expressively reduce the rice grain weight (Morita et al., 2005) with each 1 °C increase in ambient temperature (25 °C), the duration of grain development in rice reduces by up to three days, lowering the average grain weight and proportion of mature grain (Tashiro & Wardlaw, 1991) (Tashiro & Wardlaw, 1991). Arshad et al. (2017) reported that heat stress increases grain growth during ripening, but negatively affects the grain development, weight and yield. Shi et al. (2013) used artificial temperature chambers to imitate field conditions to study the impact of persistently high night temperatures on rice grain yield (from panicle start to physiological maturity). The outcomes show that a reduction in nitrogen and non-structural carbohydrate translocation following anthesis, which resulted in a decrease in maximum and mean grain-filling as well as grain weight, was mostly responsible for a decrease in grain yield at the high night temperature (28 °C).

Maize is also a high-yielding crop when grown under ideal environmental and crop management conditions. However, maize plants are sensitive to drought and heat, and each year, 15–20% of the world's potential maize yield is lost owing to these conditions (Yadav et al., 2016). The reproductive stage of maize is more sensitive to high temperatures than the vegetative stage, and tassel blast is the most visible phenotype, with reduced pollen shedding, viability, and pollination efficiency, affecting kernel size, weight, and development, and resulting a decrease in seed number. Plants exposed to high temperatures during grain filling intensify senescence, lower seed set, seed weight, and reduce yield. Temperatures above 30 °C slowed results in lower grain weight of maize and yield due to reduced cell division and amyloplast replication (Yadav et al., 2016).

The quality and productivity of legumes are severely impaired by heat stress during grain filling stage and thus induced poor pod development, and reduced grain yield (Farooq et al., 2017). Wang et al. (2006) reported reduced chickpea yield due to a high senescence rate, reduced grain set, and grain weight. Li et al., (2017a, b) found that, increase in temperature during seed developmental stage have decreased seed mass in Vicia sativa. Which was in line with the findings of (Lambert & Linck, 1958) in Pisum sp. (Thomas et al., 2009) in Phaseolus vulgaris and (Wulff, 1986) in Desmodium paniculatum. Seeds of some plants like soybean developed thicker seed coat in higher temperature than lower temperature (Keigley & Mullen, 1986). Opposite trend was found by Lacey et al. (1997). Lowering seed dry matter may not reduce seed germination or vigour (Hampton et al., 2013). Increase in temperature has increased seed mass in alpine herbs (Sandvik & Eide, 2009; Totland, 1999; Totland & Alatalo, 2002).

13.5.4 Effect of Elevated Temperature on Seed Biochemical Composition

A faster rate of grain filling under heat stress may be due to increased enzymatic and metabolic processes that speed up the process of seed development. Sehgal et al. (2018) reported that reduced dry matter accumulation length may be balance by an upsurge in the amount of seed dry matter accumulation. Moreover, increasing CO_2 concentration with temperature increase the changes of total nitrogen accumulation in barley crops (Rezaei et al., 2022), the storage protein composition was also altered during heat stress or drought. Albumins and globulins, in particular, were damaged by heating and drying in pea seeds (Yang et al., 2022). According to Mession et al. (2013), high temperatures causes aggregation and denaturation of the pea seed proteins such as vicilin, legumin, and globulins. Which may be due to covalent and non-covalent connections were lost (Sun & Arntfield, 2012). Heat stress can reduced the ADP glucose pyrophosphorylase, glucokinase, sucrose synthase, and soluble starch synthase, and thus results in decrease starch, protein, and oil content of a wheat grains (Wilhelm et al., 1999; Sita et al., 2018; Liu et al., 2011). Heat stress has variable impacts on the fatty acid content and composition. Temperature directly affects the total protein and oil content in the soybean crops (Jumrani & Bhatia, 2018). The effect of prolonged high temperatures has also been examined in B. napus cultivar (N99-508), and it has a negative impact on the grain yield, and oil content (Elferjani & Soolanayakanahally, 2018). The elevated temperature or heat stress response encourages the formation of secondary metabolites. Brassicaceae family secondary metabolites known as indole glucosinolates are produced from tryptophan and are associated with the stress response as well as auxin production (Eom et al., 2018) as well as secondary seed dormancy (Liu et al., 2019). Climate change is threatening soybean cultivation, due to rising temperatures and diminishing water resource availability. Alsajri et al. (2019) has also examined the effect high temperatures on total biomass, yield, and seed quality attributes of the soybean cultivars. High temperature of 39 °C, just after one day of fertilization affects the cellularization of rice endosperm, and alteration in gene expression leading to reduction in starch content (Begcy et al., 2018). High temperature stress affect plants at cellular level, causing degradation of mitochondrial degeneration, ATP accumulation and oxygen uptake in wheat embryo, affecting the seed quality in wheat (Grass & Burris, 1995). Mitochondrial damage further allow release of peroxidative enzymes, which could do subsequent damage of seeds (McDonald, 1999). Krishnan et al. (2020) examined the impact of heat stress on protein composition and found that the temperature stress has negatively affected the seed lipoxygenase activity, sucrose binding protein β-subunit and Bowman-Birk protease inhibitor. Additionally heat shock proteins (HSSSSP 70 and HSP 17.6) were accumulated in seeds along with structural change in protein storage vacuole. Thus, climate change mitigation requires a systematic research under a broad temperature continuum.

13.6 Climate Change on Seed Dormancy

Hoyle et al. (2008a) reported that seeds lose its dormancy faster under warmer environments. The term dormancy and germination two are dissimilar progressions (Fenner & Thompson, 2005), where the climate system affects them independently. The dormancy is of foremost importance in preserving species strengthen and sustaining biodiversity (Donohue et al., 2010), besides colonization and distribution of a species (Chen et al., 2020; Donohue et al., 2010). In the present context, climate change affects the environmental variables which are likely to change the reproduction of seeds, alter the phenology, and also cause fatal to diversity in plant communities (Walck et al., 2011). It has a significant effect on long-term and short-term of seed dormancy. The tenacious change in climate may have a long-term consequences and can results in inheritable dormancy differences due to species, ecotypes, and clonal variation (Montague et al., 2008). While seeds mature, various meteorological conditions have short-term consequences (Donohue, 2009; Fenner, 1991; Huang et al., 2015). In seasonal climates, where seedling success is only feasible once the soil is moist and relatively warm, seed dormancy is anticipated to be most adaptable. Taken an example, at fire prone environments, the optimal conditions were favoured by post-fire environments with large bare patches, and have levels of light, moisture and nutrients. Therefore, if smoke from a fire that precedes ideal recruiting conditions breaks a natural dormancy, seeds stored in the soil will benefit from adaptation (Pausas & Lamont, 2022). The global meta-analysis by Lamont, (2023), has shown the smoke released seed dormancy in 40% of C3 and C4 grasses.

13.7 Climate Change Affects Seed Ageing

It has evidenced that the climate change will effect plant reproduction and warming will speed up reproduction process across a different agro-climatic regions (De Frenne et al., 2011, 2013). While population dynamics to seed characteristics influence all stages of plant ecology (Huang et al., 2016) to community formation (Larson & Funk, 2016). Rapid climate change has increased environmental stress, which has caused the seed ageing process in mitochondria and peroxisomes to begin. It is a natural process which tends to decrease the rate of germination of seed. Ageing consists of an accumulation of ROS (Huang et al., 2019; Kurek et al., 2019), and thus affects lipid membrane, proteins, carbohydrates and genetic materials (Sharma et al., 2012). High temperatures diminishing seed's viability and hasten its ageing. The "free-radical theory" of ageing presumptions that ageing is an advanced process brought on by oxidative stressors of diverse origins that progressively raise the probability of mortality in all living things. Where, reduced species dispersal and the acquisition of new habitats brought on by climate change could have long-term effects on tree reproduction (Thurman et al., 2020). The global warming effect on seeds follows the path with their long-term storage. There are fewer seeds produced when seed yields for plants with cyclic yields are synchronised. St Clair and Lynch, (2005) reported climate change occurs at a rate that is too rapid for such species to adapt genetically, and thus these plants may have significantly reduced seed germination ability. Tolerant to desiccation, orthodox seeds may endure any unfavourable environmental circumstances, while the recalcitrant seeds loses its viability during the storage (Walters, 2015). Pukacka et al. (2003) reported that, European beech seeds lose its viability after 10 years resulting from an increased lipid peroxidation.

The hydroxyl radical, is most violent form of ROS that cause structural damage in purines, pyrimidines and deoxyribose (Garcia-Caparros et al., 2021). Furthermore, ROS and reactive nitrogen species (RNS) control the seed quality traits such as germination, vigour, seed dormancy and longevity under various environment (Kumar et al., 2016). The long-term accumulation of ROS with antioxidant decline leads to deterioration of stored seeds (Sano et al., 2016).

In addition, farmers in developing countries traditionally store their farm produce for next season showing under unfavourable conditions resulting in seed deteriorate at increasing rate especially for soybean, peanut and groundnut seeds (Rai et al., 2020). Several aging techniques were explored predominantly by increasing relative humidity and accelerated ageing-controlled deterioration treatment (CDT). Further, ROS homeostasis and DNA/RNA/protein repair mechanisms help in sustaining the viability of seeds (He & Kermode, 2010; Kibinza et al., 2011; Mahakham et al., 2017). Chen et al. (2014) transcriptomics study on the pea (Pisum sativum), ageing circumstances start with the accumulation of ROS, and a corresponding decline in the cell's antioxidant potential results in the degeneration of seeds.

13.8 Drought on Seed Physiology

Drought frequency and intensity are both increasing due to change in global climate. Many global organisations, including Climatic Research Unit (CRU) and the Global Precipitation Climatology Center (GPCC), have demonstrated a positive trend of rising frequency, intensity, and duration of drought episodes worldwide across all continents (Chiang et al., 2021). The prolonged dry spell during the crop growth period expressively distresses the plant physiology (Chaves et al., 2011), leads to reduced seed setting (Al-Ghzawi et al., 2009; Westgate & Boyer, 1986) mainly due to several factors like reduction in viability of pollen grain (Al-Ghzawi et al., 2009), abortion of ovary (Andersen et al., 2002) etc. The seed germination is vulnerable to drought stress due to altered physiology in wheat crops (Qayyum et al., 2011), Maize (Queiroz et al., 2019) and sorghum (Patanè et al., 2013). As drought affects water availability, so imbibition process disturbed, resulting in poor seed germination (Murillo-Amador et al., 2002). Although there are some evidences that, this effect of drought on seed germination may be species specific, like there are some species which can germinate easily up to a certain degree of drought (Li et al., 2013). These species can be considered during structuring the species community during ecological succession. Seed size is also an important consideration with respect to drought. As drought leads to decrease in photosynthates production and transport to seeds, so seed mass also generally gets reduced (Brunner et al., 2015). There was approximately 9% reduction in seed germination in drought stress compared to irrigated condition (Smiciklas et al., 1992). Sometimes, instead of source, sink may also get limited due to decrease in cell expansion in sink, resulting decreased seed size (Keipp et al., 2020). Seed carbohydrate, protein, lipid, other various biochemical compounds get affected by drought stress at various life stages of plants (Sehgal et al., 2018). Drought stress also affects the soluble starch synthase (SSS) in wheat grains compared to other enzymes (Hurkman et al., 2003). Decrease in digestible carbohydrate content like glucose, sucrose and fructose indirectly affects lipid content as well (Bellaloui et al., 2013). Other various biochemical changes due to drought stress are listed below in Table 13.3.

13.9 Altered Pattern of Precipitation Affecting Seed Physiology

Altered pattern of precipitation is another abnormality caused by climate change phenomena. This alteration is uncertain and subjected location to location (Orlowsky & Seneviratne, 2011) with more chance of decreasing precipitation frequency, but higher amount of precipitation event (Allan & Soden, 2008). Increasing or decreasing water availability, change in relative humidity, pollination pattern is some of the factors which is going to affect seeds due to altered precipitation pattern (Scaven & Rafferty, 2013). Seed germination and emergence is severely

Table 13.3 Bio-chemical properties changes in seed under drought conditions

Common name	Botanical name	Family	Biochemical property affected by drought	References
Sesame	Sesamum indicum	Pedaliaceae	Decrease in Sesamin, Sesamolin and linoleic acid; increase in lignin glycosides and oleic acid content	Kim et al. (2006)
Wheat	Triticum aestivum L	Poaceae	Increase in albumin and gliadin content, but unchanged amylose, globulin and glutenin content	Zhang et al. (2014)
Rapeseed	Bassica napus	Brassicaceae	Increase in seed protein and total glucosinolate content, decrease in sinapin along with some other phenolic compound	Bouchereau et al. (1996)
Sunflower	Helianthus annus L	Compositae	Increases the oleic acid and decreases the linoleic acid	Flagella et al. (2000)
Common bean	Phaseolus vulgaris	Fabaceae	Fe, Zn, P and N content decrease	Ghanbari et al. (2013)
Maize	Zea mays	Poaceae	Decrease in seed oil, total tocopherol, flavonoid but increase in linolenic acid and oleic acid	Ali (2011)
Groundnut	Arachis hypogaea	Fabaceae	Decrease in linoleic, palmitic and linolenic acid; increase in stearic and oleic acid content	Gao et al. (2009)
Rice	Oryza sativa	Poaceae	Increase in storage protein with decrease in prolamin contant	Lin et al. (2010a, b), Ohdaira et al. (2010)

affected with changing precipitation pattern. Similarly effect of altered precipitation pattern on seed germination was found by other workers (Gutterman, 1993, 2002; Gutterman & Evenari, 1994). So, altered precipitation pattern might have direct impact on future climate change scenario as well, influencing the soil seed bank of plants on various location (White et al., 2012).

13.10 Adaption of Breeding Programme to Combat Climate Change

Fujita (2013) reported that, the deployment of a new varieties is required to sustain the system productivity. While, Spielman and Smale (2017) suggested the placement of a new varieties across agro-climatic regions were positively related to higher crop productivity, whereas varietal age in farmers' fields would contribute negatively to the final yields. In order to increase the varietal replacement rate (VRR) in emerging nations, it is essential that these countries have access to the best germplasm, quick breeding cycles, multi-locational testing, and efficient seed systems (Atlin et al., 2017). Any cropping system's ability to adapt to climate change depends primarily on its ability to produce sustainable yields under biotic and abiotic pressures (Shahzad et al., 2021). In African and South Asian countries, varieties are bred for the past several years that leads to incompatibility to cope with recent climate change. According to Challinor et al. (2016) it takes ~30 years for a maize variety to develop and spread throughout Sub-Saharan Africa, indicating that significant climate change has already occurred. In Indian scenario, aged varieties of most crops are still foremost and some of them has achieved the position of luminary varieties (Singh et al., 2017). Thus, rapid breeding cycles along with competitive seed markets are essential to combat the effects of climate change. Additionally, adaption to the rapidly changing climate, different nations must collaborate internationally and share elite germplasm.

13.11 Seed Priming for Climate Change Adaptation

Climate change alters the crop phenology, pollination, seed performance, pollen viability, and final economic output (Bhadouria et al., 2019). Numerous alternative strategies are employed now to breed plants that can endure these circumstances. Currently, seed priming is considered as a crucial method for climate change adaptation and mitigation (Chakraborti et al., 2022). Seed priming is defined as treating seeds with natural or synthetic substances prior to germination in order to induce a particular physiological condition (Lutts et al., 2016). Priming is a physiological mechanism used in plant defence that gets a plant ready to react to approaching abiotic stress more swiftly or vehemently (Mittal et al., 2020). A strong and quick cellular defence response against biotic stressors was also demonstrated by plants grown from primed seeds. Definitely, there are various metabolic pathways involved in distinct metabolic processes during priming for increased resistance to biotic stress. Early and consistent germination was visible in the seedlings from prepared seeds.

Significant results were obtained with seed priming using bio-agents, which is known as bio-priming (Reddy, 2013; Singh et al., 2020). Similarly, other biotechnological approaches like halo priming (Lemmens et al., 2019), anti-oxidant priming (Hossain et al., 2015), hormonal and rhizo-bacteria priming (Sneideris et al., 2015; Xu et al., 2020) had most significant impact. In fragile ecosystems, seed priming is a potential approach for improving crop production. A deeper understanding of the metabolic processes involved in priming and subsequent germination should make it simpler to use this low-cost technology to maximise seed performance in inventive ways across a broad spectrum of environments.

13.12 Policy Options

The climate change adaptation and mitigation measures on the seed physiological aspect might be of general tactics, rather than the specific programmes. The policy implemented for each vulnerability sector should be redesigned in such a manner that, it would supplement the specific policies for addressing the issues within different sectors. Thus, long-term planning must be taken into an account, which may lead to adjustments and improving capacity of the future generations to adapt the impact of a climate change. List the current actions and choices made towards climate adaptation. An inventory can concentrate on actual social and economic choices made in the context of varying climate regimes across time or between different places. It is necessary to have a comprehensive policy option for seed longevity as well as maintain quality. A cost-effective method to find the adaptable options is perhaps needed to compile data on various practices to identify, and analyse adaptation strategies for making a long-term decision (Joda, 1989; IPCC, 1990). This option is of the utmost importance due to long-term decisions, catastrophic or irreversible effects, and unfavourable trends. For farmers to succeed, it's essential to provide them with appropriate training, early warning systems, credits, and market access options (Anzum et al., 2023) and other informal industries (Debela & Beyene, 2023) resilience to climate change.

13.13 Conclusion

The ever-increasing global population, anticipated climate change, and agroecological transitions, demands a resilient technology to sustain the food and nutritional security. The availability of a quality seed is equally important for feeding growing population, and improving the socio-economic status of the poor and marginal farmers. However, the impact of climate change on seed physiology and quality is often neglected. Thus, in this chapter we have critically examine the impact of climate change on different seed physiological aspects like seed germination, vigour, weight, number, biochemical composition, and dormancy. In addition to the aforementioned tactics, local landraces should be assessed and screened, conserved, and multiplied for their innate resistance to climate change. Further, future probabilistic research should be oriented towards adoption of a biotechnological approaches on the seed physiological quality regulation to cope up the adverse impact of climate change. A smooth path towards a successful response to climate change has been made possible by innovative breeding for better germinable genotypes, photo- and thermosensitive cultivars, along with greater seed sugars, protein, and lipids concentration. Thus, in nutshell private and public extension systems, and government interventions must be strengthened to improve the livelihood of the farmers.

References

Abebe, A., Pathak, H., Singh, S. D., Bhatia, A., Harit, R. C., & Kumar, V. (2016). Growth, yield and quality of maize with elevated atmospheric carbon dioxide and temperature in north–west India. *Agriculture, Ecosystems & Environment, 218*, 66–72.

Abzar, A., Said, M. N. M., Ahmad, W. J. W., Yusoff, W. M. W., Doni, F., Fathurrahman, F., & Zainet, C. R. M. (2017). Elevated CO_2 concentration enhance germination, seedling growth and vigor of rice. *Ecology Environment Conservation, 23*, 2017–2058.

Ainsworth, E. A., & Long, S. P. (2021). 30 years of free-air carbon dioxide enrichment (FACE): What have we learned about future crop productivity and its potential for adaptation? *Global Change Biology, 27*(1), 27–49.

Aiqing, S., Somayanda, I., Sebastian, S. V., Singh, K., Gill, K., Prasad, P. V. V., & Jagadish, S. K. (2018). Heat stress during flowering affects time of day of flowering, seed set, and grain quality in spring wheat. *Crop Science, 58*(1), 380–392.

Alan, O., Kanturer, D., Powell, A. A., & Ilbi, H. (2022). Growing season climate affects phenological development, seed yield and seed quality of dill (Anethum graveolens). *Seed Science and Technology, 50*(1–2), 11–20.

Al-Ghzawi, A. A., Zaitoun, S., Gosheh, H. Z., & Alqudah, A. M. (2009). The impacts of drought stress on bee attractively and flower pollination of Trigonella moabitica (fabaceae). *Archives of Agronomy and Soil Science., 55*(6), 683–692.

Ali, Q. (2011). Exogenous use of some potential organic osmolytes in enhancing drought tolerance in maize (Zae mays L.).–A thesis submitted in partial fulfillment of the requirements for the degree of doctor of philosophy. Botany Dep., Faculty of Science, University of Agriculture, Faisalabad, Pakistan, pp. 312

Allan, R. P., & Soden, B. J. (2008). Atmospheric warming and the amplification of precipitation extremes. *Science, 321*(5895), 1481–1484.

Almuhayawi, M. S., Hassan, A. H., Al Jaouni, S. K., Alkhalifah, D. H. M., Hozzein, W. N., Selim, S., & Khamis, G. (2021). Influence of elevated CO_2 on nutritive value and health-promoting prospective of three genotypes of Alfalfa sprouts (Medicago Sativa). *Food Chemistry, 340*, 128147.

Alsajri, F. A., Singh, B., Wijewardana, C., Irby, J. T., Gao, W., & Reddy, K. R. (2019). Evaluating soybean cultivars for low-and high-temperature tolerance during the seedling growth stage. *Agronomy, 9*(1), 13.

Al-Traboulsi, M., Sjögersten, S., Colls, J., Steven, M., & Black, C. (2013). Potential impact of CO_2 leakage from carbon capture and storage (CCS) systems on growth and yield in maize. *Plant and Soil, 365*(1), 267–281. https://doi.org/10.1007/s11104-012-1390-5

Andalo, C., Godelle, B., Lefranc, M., Mousseau, M., & TILL-BOTTRAUD, I. R. É. N. E. (1996). Elevated CO_2 decreases seed germination in Arabidopsis thaliana. *Global Change Biology, 2*(2), 129–135.

Andersen, M. N., Ash, F., Jensen, C. R., Naested, H., Mogensen, V. O., & Koch, K. E. (2002). Soluble invertase expression is an early target of drought stress during the critical, abortion-sensitive phase of young ovary development in maize. *Plant Physiology, 130*, 591–604.

Anzum, H. M. N., Chakraborty, T. K., & Bosu, H. (2023). Farmer's perception and factors influencing adoption of adaptation measures to cope with climate change: An evidence from coastal Bangladesh. *Environment and Natural Resources Journal, 21*(2), 113–126. https://doi.org/10.32526/ennrj/21/202200186

Arshad, M. S., Farooq, M., Asch, F., Krishna, J. S., Prasad, P. V., & Siddique, K. H. (2017). Thermal stress impacts reproductive development and grain yield in rice. *Plant Physiology and Biochemistry, 115*, 57–72.

Arun, M. N., Hebbar, S. S., Senthivel, T., Nair, A. K., Padmavathi, G., Pandey, P., & Singh, A. (2022). *Seed priming: The way forward to mitigate abiotic stress in crops* (Vol. 11, p. 173). London, UK: IntechOpen.

Atlin, G. N., Cairns, J. E., & Das, B. (2017). Rapid breeding and varietal replacement are critical to adaption of cropping systems in the developing world to climate change. *Global Food Security, 12*, 31–37.

Balkaya, A. (2004). Modelling the effect of temperature on the germination speed in some legume crops. *Journal of Agronomy, 3*, 179–183. https://doi.org/10.3923/ja.2004.179.183

Balkhyour, M. A., Hassan, A. H. A., Halawani, R. F., Summan, A. S. & Elgawad, H. (2021). Effect of elevated CO_2 on seed yield, essential oil metabolism, nutritive value, and biological activity of Pimpinella Anisum L. Accessions at different seed maturity stages. *Biology, 10*(10), 979. https://doi.org/10.3390/biology1010097

Bareke, T. (2018). Biology of seed development and germination physiology. *Advances in Plants & Agriculture Research, 8*(4), 336–346.

Baskin, C. C., & Baskin, J. M. (2014). *Seeds: Ecology, biogeography and evolution of dormancy and germination* (2nd ed.). Elsevier.

Begcy, K., Sandhu, J., & Walia, H. (2018). Transient heat stress during early seed development primes germination and seedling establishment in rice. *Frontiers in Plant Science, 9*, 1768.

Bellaloui, N., Hu, Y., Mengistu, A., Kassem, M. A., & Abel, C. A. (2013). Effects of foliar boron application on seed composition, cell wall boron, and seed δ15N and δ13C isotopes in water-stressed soybean plants. *Frontiers in Plant Science, 4*, 270. https://doi.org/10.3389/fpls.2013.00270

Ben Mariem, S., Soba, D., Zhou, B., Loladze, I., Morales, F., & Aranjuelo, I. (2021). Climate change, crop yields, and grain quality of C3 cereals: A meta-analysis of [CO_2], temperature, and drought effects. *Plants, 10*(6), 1052.

Bernareggi, G., Carbognani, M., Petraglia, A., & Mondoni, A. (2015). Climate warming could increase seed longevity of alpine snowbed plants. *Alpine Botany, 125*, 69–78.

Bewley, J. D., Bradford, K. J., Hilhorst, H. W. M., & Nonogaki, H. (2013). *Seeds: Physiology of development, germination and dormancy* (3rd ed.). Springer. https://doi.org/10.1007/978-1-4614-4693-4

Bhadouria, R., Singh, R., Singh, V. K., Borthakur, A., Ahamad, A., Kumar, G., & Singh, P. (2019). Agriculture in the era of climate change: Consequences and effects. In *Climate change and agricultural ecosystems* (pp. 1–23). Woodhead Publishing.

Bheemanahalli, R., Sunoj, V. J., Saripalli, G., Prasad, P. V., Balyan, H. S., Gupta, P. K., Grant, N., Gill, K. S., & Jagadish, S. K. (2019). Quantifying the impact of heat stress on pollen germination, seed set, and grain filling in spring wheat. *Crop Science, 59*(2), 684–696.

Bouchereau, A., Clossais-Besnard, N., Bensaoud, A., Leport, L., & Renard, M. (1996). Water stress effects on rapeseed quality. *European Journal of Agronomy, 5*, 19–30. https://doi.org/10.1016/s11610301(96)02005-9

Broberg, M. C., Högy, P., & Pleijel, H. (2017). CO_2-induced changes in wheat grain composition: Meta-analysis and response functions. *Agronomy, 7*(2), 32.

Brunner, I., Herzog, C., Dawes, M. A., Arend, M., & Sperisen, C. (2015). How tree roots respond to drought. *Frontiers in Plant Science, 6*, 547. https://doi.org/10.3389/fpls.2015.00547

Buchner, P., Tausz, M., Ford, R., Leo, A., Fitzgerald, G. J., Hawkesford, M. J., & Tausz-Posch, S. (2015). Expression patterns of C- and N-metabolism related genes in wheat are changed during senescence under elevated CO_2 in dry-land agriculture. *Plant Science, 236*, 239–249. https://doi.org/10.1016/j.plantsci.2015.04.006

Burkey, K. O., Booker, F. L., Pursley, W. A., & Heagle, A. S. (2007). Elevated carbon dioxide and ozone effects on peanut: II. Seed yield and quality. *Crop Science, 47*(4), 1488–1497.

Caldwell, C. R., Britz, S. J., & Mirecki, R. M. (2005). Effect of temperature, elevated carbon dioxide, and drought during seed development on the isoflavone content of dwarf soybean [Glycine max (L.) Merrill] grown in controlled environments. *Journal of Agricultural and Food Chemistry, 53*(4), 1125–1129.

Carón, M. M., De Frenne, P., Brunet, J., Chabrerie, O., Cousins, S. A., Decocq, G., & Verheyen, K. (2015). Divergent regeneration responses of two closely related tree species to direct abiotic and indirect biotic effects of climate change. *Forest Ecology and Management, 342*, 21–29.

Chakraborti, S., Bera, K., Sadhukhan, S., & Dutta, P. (2022). Bio-priming of seeds: Plant stress management and its underlying cellular, biochemical and molecular mechanisms. *Plant Stress, 3*, 100052.

Challinor, A. J., Koehler, A. K., Ramirez-Villegas, J., Whitfield, S., & Das, B. (2016). Current warming will reduce yields unless maize breeding and seed systems adapt immediately. *Nature Climate Change, 6*(10), 954–958.

Change, I. P. O. C. (2021). Climate change widespread, rapid, and intensifying–IPCC. Climate change.

Chaves, M. M., Costa, J. M., & Saibo, N. J. M. (2011). Recent advances in photosynthesis under drought and salinity. *Advances in Botanical Research, 57*, 49–104.

Chen, C., Jiang, Q., Ziska, L. H., Zhu, J., Liu, G., Zhang, J., & Zhu, C. (2015). Seed vigor of contrasting rice cultivars in response to elevated carbon dioxide. *Field Crops Research, 178*, 63–68.

Chen, L., Chen, Q., Kong, L., Xia, F., Yan, H., Zhu, Y., & Mao, P. (2016). Proteomic and physiological analysis of the response of oat (Avena sativa) seeds to heat stress under different moisture conditions. *Frontiers in Plant Science, 7*, 896.

Chen, S. C., Poschlod, P., Antonelli, A., Liu, U., & Dickie, J. B. (2020). Trade-off between seed dispersal in space and time. *Ecology Letters, 23*(11), 1635–1642.

Chen, H., Osuna, D., Colville, L., Lorenzo, O., Graeber, K., Küster, H., & Kranner, I. (2014). Correction: Transcriptome-wide mapping of pea seed ageing reveals a pivotal role for genes related to oxidative stress and programmed cell death. *Plos One*, 9(1).

Chiang, F., Mazdiyasni, O., & AghaKouchak, A. (2021). Evidence of anthropogenic impacts on global drought frequency, duration, and intensity. *Nature Communication, 12*, 2754. https://doi.org/10.1038/s41467-021-22314-w

Cotrina Cabello, G. G., Ruiz Rodriguez, A., Husnain Gondal, A., Areche, F. O., Flores, D. D. C., Astete, J. A. Q., ... & Cruz Nieto, D. D. (2023). Plant adaptability to climate change and drought stress for crop growth and production. CABI Reviews.

Crowley, L. M., Sadler, J. P., Pritchard, J., & Hayward, S. A. (2021). Elevated CO_2 impacts on plant-pollinator interactions: A systematic review and free air carbon enrichment field study. *Insects, 12*(6), 512.

De Frenne, P., Brunet, J., Shevtsova, A., Kolb, A., Graae, B. J., Chabrerie, O., & Verheyen, K. (2011). Temperature effects on forest herbs assessed by warming and transplant experiments along a latitudinal gradient. *Global Change Biology, 17*(10), 3240–3253.

De Frenne, P., Graae, B. J., Rodríguez-Sánchez, F., Kolb, A., Chabrerie, O., Decocq, G., & Verheyen, K. (2013). Latitudinal gradients as natural laboratories to infer species' responses to temperature. *Journal of Ecology, 101*(3), 784–795.

Debela, M. P., & Beyene, F. (2023). *Social Network in pluralistic seed system for climate change adaptation: Eastern Oromia, Hararghe, Ethiopia* (Doctoral Dissertation).

Debouza, N. E., Babu, T. S., Gopi, K., Zain, S., & Ksiksi, T. (2021). Plant and seed germination responses to global change, with a focus on CO: A review. *One Ecosystem, 6*, e74260.

Delouche, J. C., & Potts, H. C. (1983). The importance of seed in agriculture and the need for a seed program.

Djanaguiraman, M., Narayanan, S., Erdayani, E., & Prasad, P. V. (2020). Effects of high temperature stress during anthesis and grain filling periods on photosynthesis, lipids and grain yield in wheat. *BMC Plant Biology, 20*, 1–12.

Donohue, K. (2009). Completing the cycle: Maternal effects as the missing link in plant life histories. *Philosophical Transactions of the Royal Society b: Biological Sciences, 364*(1520), 1059–1074.

Donohue, K., Rubio de Casas, R., Burghardt, L., Kovach, K., & Willis, C. G. (2010). Germination, postgermination adaptation, and species ecological ranges. *Annual Review of Ecology, Evolution, and Systematics, 41*, 293–319.

Dornbos, D. L. (2020). Production environment and seed quality. In *Seed quality* (pp. 119–152). CRC Press.

Dwivedi, S. K., Basu, S., Kumar, S., Kumar, G., Prakash, V., Kumar, S., & Arora, A. (2017). Heat stress induced impairment of starch mobilisation regulates pollen viability and grain yield in wheat: Study in Eastern Indo-gangetic plains. *Field Crops Research, 206*, 106–114.

Edwards, G., Clark, H., & Newton, P. (2001). The effects of elevated CO_2 on seed production and seedling recruitment in a sheep-grazed pasture. *Oecologia, 127*(3), 383–394.

Egli, D. B., TeKrony, D. M., Heitholt, J. J., & Rupe, J. (2005). Air temperature during seed filling and soybean seed germination and vigor. *Crop Science, 45*(4), 1329–1335.

Elferjani, R., & Soolanayakanahally, R. (2018). Canola responses to drought, heat, and combined stress: shared and specific effects on carbon assimilation, seed yield, and oil composition. *Frontiers in Plant Science*, 1224.

Ellis, R. H. (2022). Seed ageing, survival, and the improved seed viability equation; forty years on. *Seed Science and Technology.*

Enoch, H. Z., Zieslin, N., Biran, Y. (1971). Principles of CO_2 research: Symposium on greenhouse climate: Evaluation of research methods. *ISHS Acta Horticulturae* https://doi.org/10.17660/ActaHortic.1973.32

Eom, S. H., Baek, S. A., Kim, J. K., & Hyun, T. K. (2018). Transcriptome analysis in Chinese cabbage (Brassica rapa ssp. pekinensis) provides the role of glucosinolate metabolism in response to drought stress. *Molecules, 23*(5), 1186.

Esashi, Y., Ooshima, Y., Abe, M., Kurota, A., & Satoh, S. (1986). CO_2-enhanced C2H4 production in tissues of imbibed cocklebur seeds. *Functional Plant Biology, 13*(3), 417–429.

Essemine, J., Ammar, S., & Bouzid, S. (2010). Impact of heat stress on germination and growth in higher plants: Physiological, biochemical and molecular repercussions and mechanisms of defence. *Journal of Biological Sciences, 10*(6), 565–572.

Fahad, S., Bajwa, A. A., Nazir, U., Anjum, S. A., Farooq, A., Zohaib, A., et al. (2017). Crop production under drought and heat stress: Plant responses and management options. *Frontiers in Plant Science, 8*, 1147. https://doi.org/10.3389/fpls.2017.01147

Fahad, S., Adnan, M., Hassan, S., Saud, S., Hussain, S., Wu, C., Wang, D., Hakeem, K. R., Alharby, H. F., Turan, V., & Khan, M. A. (2019). Rice responses and tolerance to high temperature. *Advances in rice research for abiotic stress tolerance* (pp. 201–224). Woodhead Publishing.

FAO; International Fund for Agricultural Development; World Food Program. The state of food insecurity in the world 2015. (2015). Strengthening the enabling environment for food security and nutrition [Internet]. Rome (Italy): FAO. [cited 2017 Jul 5]. http://www.fao.org/hunger/en/.

Farooq, M., Nadeem, F., Gogoi, N., Ullah, A., Alghamdi, S. S., Nayyar, H., & Siddique, K. H. (2017). Heat stress in grain legumes during reproductive and grain-filling phases. *Crop and Pasture Science, 68*(11), 985–1005.

Fenner, M. (1991). The effects of the parent environment on seed germinability. *Seed Science Research, 1*(2), 75–84.

Fenner, M., & Thompson, K. (2005). *The ecology of seeds.* Cambridge University Press.

Fernando, N., Panozzo, J., Tausz, M., Norton, R. M., Fitzgerald, G. J., Myers, S., & Seneweera, S. (2014a). Intra-specific variation of wheat grain quality in response to elevated [CO_2] at two sowing times under rain-fed and irrigation treatments. *Journal of Cereal Science, 59*(2), 137–144.

Fernando, N., Panozzo, J., Tausz, M., Norton, R. M., Neumann, N., Fitzgerald, G. J., & Seneweera, S. (2014b). Elevated CO_2 alters grain quality of two bread wheat cultivars grown under different environmental conditions. *Agriculture, Ecosystems & Environment, 185*, 24–33.

Finzi, A. C., Norby, R. J., Calfapietra, C., Gallet-Budynek, A., Gielen, B., Holmes, W. E., & Ceulemans, R. (2007). Increases in nitrogen uptake rather than nitrogen-use efficiency support higher rates of temperate forest productivity under elevated CO_2. *Proceedings of the National Academy of Sciences, 104*(35), 14014–14019.

Flagella, Z., Rotunno, T., Di Caterina, R., De Simone, G., Ciciretti, L. & De Caro, A. (2000). Effect of supplementary irrigation on seed yield and oil quality of sunflower (Helianthus annus L.) grown in a sub-arid environment. In 15th International Sunflower Conference. (Vol. 1, pp. 139–144).

Foyer, C. H. & Noctor, G. (2020). Redox homeostasis and signaling in a higher-CO_2 world. *Annual Review of Plant Biology*, 71, 157–182. https://doi.org/10.1146/annurev-arplant-050718-095955
Franzaring, J., Gensheimer, G., Weller, S., Schmid, I., & Fangmeier, A. (2012). Allocation and remobilisation of nitrogen in spring oilseed rape (*Brassica napus L. cv. Mozart*) as affected by N supply and elevated CO_2. *Environmental and Experimental Botany, 83*, 12–22. https://doi.org/10.1016/j.envexpbot.2012.03.015
Franzaring, J., Högy, P., Fangmeier, A. (2008). Effects of free-air CO_2 enrichment on the growth of summer oilseed rape (*Brassica napus cv. Campino*). *Agriculture, Ecosystems and Environment*, 128(1–2), 127–134. https://doi.org/10.1016/j.agee.2008.05.011
Fujita, K. (2013). Growth and stagnation of agriculture in Eastern India and Bangladesh. *The Brown Journal of World Affairs, 20*, 179.
Gao, J., Hao, X., Thelen, K. D., & Robertson, G. P. (2009). Agronomic management system and precipitation effects on soybean oil and fatty acid profiles. *Crop Science, 49*, 1049–1057.
Garcia-Caparros, P., De Filippis, L., Gul, A., Hasanuzzaman, M., Ozturk, M., Altay, V., & Lao, M. T. (2021). Oxidative stress and antioxidant metabolism under adverse environmental conditions: A review. *The Botanical Review, 87*(4), 421–466.
Ghanbari, A. A., Shakiba, M. R., Toorchi, M., & Choukan, R. (2013). Nitrogen changes in the leaves and accumulation of some minerals in the seeds of red, white and chitti beans (Phaseolus vulgaris) under water deficit conditions. *Australian Journal of Crop Science, 7*, 706–712.
Gong, Z., Dong, L., Lam, S., Zhang, D., Zong, Y., Hao, X., & Li, P. (2021). Nutritional quality in response to elevated CO_2 concentration in foxtail millet (Setaria italica). *Journal of Cereal Science, 103318.*https://doi.org/10.1016/j.jcs.2021.103318
Gorai, M. & Neffati, M. (2007) Germination responses of Reaumuria vermiculata to salinity and temperature. *Annals of Applied Biology, 151* (1): 53–59. https://doi.org/10.1111/j.1744-7348.2007.00151.x
Grass, L., & Burris, J. S. (1995). Effect of heat stress during seed development and maturation on wheat (*Triticum durum*) seed quality. II. Mitochondrial respiration and nucleotide pools during early germination. *Canadian Journal of Plant Science, 75*, 831–839.
Groot, S. P. C., Surki, A. A., De Vos, R. C. H., & Kodde, J. (2012). Seed storage at elevated partial pressure of oxygen, a fast method for analysing seed ageing under dry conditions. *Annals of Botany, 110*(6), 1149–1159.
Gupta, S., Van Staden, J., & Doležal, K. (2022). An understanding of the role of seed physiology for better crop productivity and food security. *Plant Growth Regulation, 97*(2), 171–173.
Gutterman, Y. (1993). *Seed germination in desert plants.* Springer.
Gutterman, Y. (2002). *Survival strategies of annual desert plants: Adaptations of desert organisms.* Springer.
Gutterman, Y., & Evenari, M. (1994). The influences of amounts and distribution of irrigation during the hot and dry season on emergence and survival of some desert winter annual plants in the Negev Desert. *Israel Journal of Plant Sciences, 42*(1), 1–14.
Hampton, J. G., Boelt, B., Rolston, M. P., & Chastain, T. G. (2013). Effects of elevated CO_2 and temperature on seed quality. *The Journal of Agricultural Science., 151*, 154–162. https://doi.org/10.1017/S0021859612000026
Hampton, J., Conner, A., Boelt, B., Chastain, T., & Rolston, P. (2016). Climate change: Seed production and options for adaptation. *Agriculture, 6*(3), 33. https://doi.org/10.3390/agriculture6030033
Harman, D. (1994). Free-radical theory of aging: Increasing the functional life span. *Annals of the New York Academy of Sciences, 717*(1), 1–15.
Hassan, A. H. A., Okla, M. K., Al-amri, S. A., El-Tayeb, M. A., Moussa, I. M., Elbadawi, Y., Adel-Maksoud, M. A., Ali, F. H. M., S. Almaary, K., Selim, S., AbdElgawad, H. (2021). Exploratory assessment to evaluate seed sprouting under elevated CO_2 revealed improved biomass, physiology, and nutritional value of Trachyspermum ammi. *Agronomy* :11, 830. https://doi.org/10.3390/agronomy11050830

Hatfield, J. L., Boote, K. J., Kimball, B. A., Ziska, L. H., Izaurralde, R. C., Ort, D., & Wolfe, D. (2011). Climate impacts on agriculture: Implications for crop production. *Agronomy Journal, 103*(2), 351–370.

He, X., & Kermode, A. R. (2010). Programmed cell death of the megagametophyte during post-germinative growth of white spruce (Picea glauca) seeds is regulated by reactive oxygen species and the ubiquitin-mediated proteolytic system. *Plant and Cell Physiology, 51*(10), 1707–1720.

Hedhly, A., Hormaza, J. I., & Herrero, M. (2008). Global warming and plant sexual reproduction. *Trends in Plant Science, 14*(1), 30–36.

Hikosaka, K., Kinugasa, T., Oikawa, S., Onoda, Y., & Hirose, T. (2011). Effects of elevated CO_2 concentration on seed production in C_3 annual plants. *Journal of Experimental Botany, 62*(4), 1523–1530.

Högy, P., & Fangmeier, A. (2008). Effects of elevated atmospheric CO_2 on grain quality of wheat. *Journal of Cereal Science, 48*(3), 580–591.

Högy, P., Wieser, H., Köhler, P., Schwadorf, K., Breuer, J., Erbs, M., & Fangmeier, A. (2009). Does elevated atmospheric CO_2 allow for sufficient wheat grain quality in the future? *Journal of Applied Botany and Food Quality, 82*(2), 114–121.

Högy, P., Franzaring, J., Schwadorf, K., Breuer, J., Schütze, W., & Fangmeier, A. (2010a). Effects of free-air CO_2 enrichment on energy traits and seed quality of oilseed rape. *Agriculture, Ecosystems and Environment, 139*(1–2), 239–244.

Högy, P., Keck, M., Niehaus, K., Franzaring, J., & Fangmeier, A. (2010b). Effects of atmospheric CO_2 enrichment on biomass, yield and low molecular weight metabolites in wheat grain. *Journal of Cereal Science, 52*(2), 215–220.

Högy, P., & Fangmeier, A. (2009a). Atmospheric CO_2 enrichment affects potatoes: 2. Tuber quality traits. *European Journal of Agronomy, 30*(2), 85–94.

Hossain, M. A., Bhattacharjee, S., Armin, S. M., Qian, P., Xin, W., Li, H. Y., & Tran, L. S. P. (2015). Hydrogen peroxide priming modulates abiotic oxidative stress tolerance: Insights from ROS detoxification and scavenging. *Frontiers in Plant Science, 6*, 420.

Hovenden, M. J., Wills, K. E., Vander, J. K., S., Chaplin, R. E., Williams, A. L., Nolan, M. J., Newton, P. C. D. (2007). Flowering, seed production and seed mass in a species-rich temperate grassland exposed to FACE and warming. *Australian Journal of Botany, 55*, 780–794.

Hoyle, G. L., Daws, M. I., Steadman, K. J., & Adkin, S. W. (2008). Pre- and post-harvest influences on physiological dormancy alleviation of an Australian Asteraceae species: *Actinobole uliginosum* (A. Gray) H. Eichler. *Seed Science Research, 18*, 191–199.

Huang, Z., Ölçer-Footitt, H., Footitt, S., & Finch-Savage, W. E. (2015). Seed dormancy is a dynamic state: Variable responses to pre-and post-shedding environmental signals in seeds of contrasting Arabidopsis ecotypes. *Seed Science Research, 25*(2), 159–169.

Huang, Z., Liu, S., Bradford, K. J., Huxman, T. E., & Venable, D. L. (2016). The contribution of germination functional traits to population dynamics of a desert plant community. *Ecology, 97*(1), 250–261.

Huang, H., Ullah, F., Zhou, D. X., Yi, M., & Zhao, Y. (2019). Mechanisms of ROS regulation of plant development and stress responses. *Frontiers in Plant Science, 10*, 800.

Hurkman, W. J., McCue, K. F., Altenbach, S. B., Korn, A., Tanaka, C. K., Kothari, K. M., Johnson, E. L., Bechtel, D. B., Wilson, J. D., Anderson, O. D., & DuPont, F. M. (2003). Effect of temperature on expression of genes encoding enzymes for starch biosynthesis in developing wheat endosperm. *Plant Science, 164*, 873–881. https://doi.org/10.1016/S0168-9452(03)00076-1

Iloh, A. C., Omatta, G., Ogbadu, G. H., & Onyenekwe, P. C. (2014). Effects of elevated temperature on seed germination and seedling growth on three cereal crops in Nigeria. *Scientific Research and Essays, 9*(18), 806–813. https://doi.org/10.5897/SRE2014.5968

IPCC (Intergovernmental Panel on Climate Change). (1990). Policy makers summary of the formulation of response strategies: Report of working group Ill, WMO/UNEP intergovernmental panel on climate change. IPCC, Geneva

IPCC. (2014). Climate change impacts, adaptation, and vulnerability. Part A: Global and sectoral aspects. Contribution of working group II to the Fifth assessment report of the intergovernmental panel on climate change. Cambridge: Cambridge university press

ISTA. (2020). *International rules for seed testing*. Bassersdorf, Switzerland: International Seed Testing Association

Jablonski, L. M., Wang, X., & Curtis, P. S. (2002). Plant reproduction under elevated CO_2 conditions: A meta-analysis of reports on 79 crop and wild species. *New Phytologist, 156*(1), 9–26.

Javaid, M. M., Mahmood, A., Alshaya, D. S., AlKahtani, M. D., Waheed, H., Wasaya, A., ... & Fiaz, S. (2022). Influence of environmental factors on seed germination and seedling characteristics of perennial ryegrass (Lolium perenne L.). *Scientific Reports, 12*(1), 1–11.

Jeevan Kumar, S. P., Rajendra Prasad, S., Banerjee, R., & Thammineni, C. (2015). Seed birth to death: Dual functions of reactive oxygen species in seed physiology. *Annals of Botany, 116*(4), 663–668.

Jena, U. R., Swain, D. K., Hazra, K. K., & Maiti, M. K. (2018). Effect of elevated [CO_2] on yield, intra-plant nutrient dynamics, and grain quality of rice cultivars in eastern India. *Journal of the Science of Food and Agriculture, 98*(15), 5841–5852.

Joda, N. S. (1989). Potential strategies for adapting to greenhouse warming: perspectives from the develop~ng world. In: N. Rosenberg et al. (Eds.), *Greenhouse warming: Abatement and adaptation. Resources for the future* (pp. 147–158). Washington. DC.

Jumrani, K., & Bhatia, V. S. (2018). Impact of combined stress of high temperature and water deficit on growth and seed yield of soybean. *Physiology and Molecular Biology of Plants, 24*(1), 37–50.

Keigley, P. J., & Mullen, R. E. (1986). Changes in soybean seed quality from high-temperature during seed fill and maturation. *Crop Science, 26*, 1212–1216. https://doi.org/10.2135/cropsci1986.0011183X002600060028x

Keipp, K., Hutsch, B. W., Ehlers, K., & Schubert, S. (2020). Drought stress in sunflower causes inhibition of seed filling due to reduced cell-extension growth. *Journal of Agronomy and Crop Science, 00*, 1–12. https://doi.org/10.1111/jac.12400

Kibinza, S., Bazin, J., Bailly, C., Farrant, J. M., Corbineau, F., & El-Maarouf-Bouteau, H. (2011). Catalase is a key enzyme in seed recovery from ageing during priming. *Plant Science, 181*(3), 309–315.

Kijowska-Oberc, J., Staszak, A. M., & Ratajczak, E. (2021). Climate change affects seed aging? Initiation mechanism and consequences of loss of forest tree seed viability. *Trees*, 1–10.

Kim, K. S., Ryu, S. N., & Chung, H. G. (2006). Influence of drought stress on chemical composition of sesame seed. *Korean Journal of Crop Science., 5*(1), 73–80.

Kochanek, J., Buckley, Y. M., Probert, R. J., Adkins, S. W., & Steadman, K. J. (2010). Pre-zygotic parental environment modulates seed longevity. *Austral Ecology, 35*(7), 837–848.

Korres, N. E., Norsworthy, J. K., Tehranchian, P., Gitsopoulos, T. K., Loka, D. A., Oosterhuis, D. M., & Palhano, M. (2016). Cultivars to face climate change effects on crops and weeds: A review. *Agronomy for Sustainable Development, 36*(1), 1–22.

Krishnan, H. B., Kim, W., Oehrle, N. W., Smith, J. R., & Gillman, J. D. (2020). Effect of heat stress on seed protein composition and ultrastructure of protein storage vacuoles in the cotyledonary parenchyma cells of soybean genotypes that are either tolerant or sensitive to elevated temperatures. *International Journal of Molecular Sciences, 21*(13), 4775. https://doi.org/10.3390/ijms21134775

Krishnan, P., Ramakrishnan, B., Rao, K. S., & Dash, R. N. (2009). Simulation studies to characterize the impact of climate change on crop production and to identify strategies for adaptation and mitigation. *In Climate Change and Crops* (pp. 39–61). Berlin, Heidelberg: Springer Berlin Heidelberg.

Kumar, S. P. J., Prasad, S. P., Kumar, M., Singh, C., Sinha, A. K., & Pathak, A. (2016). Seed quality markers: A review. Research & Reviews. *Journal of Botanical Sciences, 5*, 13–17.

Kumar, S. J., Susmita, C., Agarwal, D. K., Pal, G., Rai, A. K., & Simal-Gandara, J. (2021a). Assessment of genetic purity in rice using polymorphic SSR markers and its economic analysis with grow-out-test. *Food Analytical Methods, 14*, 856–864.

Kumar, S. P. J., Chintagunta, A. D., Reddy, Y. M., Rajjou, L., Garlapati, V. K., Agarwal, D. K., Prasad, S. R. & Simal-Gandara, J. (2021). Implications of reactive oxygen and nitrogen species in seed physiology for sustainable crop productivity under changing climate conditions. *Current Plant Biology, 26* (100197). https://doi.org/10.1016/j.cpb.2021.10019

Kurek, K., Plitta-Michalak, B., & Ratajczak, E. (2019). Reactive oxygen species as potential drivers of the seed aging process. *Plants, 8*(6), 174.

Lacey, E., Smith, S., & Case, A. (1997). Parental effects on seed mass: Seed coat but not embryo/endosperm effects. *American Journal of Botany, 84*, 1617–1620. https://doi.org/10.2307/2446624

LaDeau, S. L., & Clark, J. S. (2001). Rising CO_2 levels and the fecundity of forest trees. *Science, 292*, 95–98.

LaDeau, S. L., & Clark, J. S. (2006). Elevated CO_2 and tree fecundity: The role of tree size, inter-. annual variability and population heterogeneity. *Global Change Biology, 12*, 822–833.

Lam, S. K., Chen, D., Norton, R., & Armstrong, R. (2012). Nitrogen demand and the recovery of 15N-labelled fertilizer in wheat grown under elevated carbon dioxide in southern Australia. *Nutrient Cycling in Agroecosystems, 92*(2), 133–144.

Lambert, R. G., & Linck, A. J. (1958). Effects of high temperature on yield of peas. *Plant Physiology, 33*, 347–350. https://doi.org/10.1104/pp.33.5.347

Lamichaney, A., & Maity, A. (2021). Implications of rising atmospheric carbon dioxide concentration on seed quality. *International Journal of Biometeorology, 65*(6), 805–812.

Lamichaney, A., Swain, D. K., Biswal, P., Kumar, V., Singh, N. P., & Hazra, K. K. (2019). Elevated atmospheric carbon–dioxide affects seed vigour of rice (Oryza sativa L.). *Environmental and Experimental Botany, 157*, 171–176.

Lamichaney, A., Parihar, A. K., Hazra, K. K., Dixit, G. P., Katiyar, P. K., Singh, D., & Singh, N. P. (2021b). Untangling the influence of heat stress on crop phenology, seed set, seed weight, and germination in field pea (Pisum sativum L.). *Frontiers in Plant Science, 12*, 437.

Lamichaney, A., Tewari, K., Basu, P. S., Katiyar, P. K., & Singh, N. P. (2021). Effect of elevated carbon-dioxide on plant growth, physiology, yield and seed quality of chickpea (Cicer arietinum L.) in Indo-Gangetic plains. *Physiology and Molecular Biology of Plants, 27*(2), 251–263.

Lamont, B. B. (2023). Grass species with smoke-released seed dormancy: A response to climate and fire regime but not photosynthetic pathway. *Plant Biology, 25*(1), 24–31.

Larson, J. E., & Funk, J. L. (2016). Regeneration: An overlooked aspect of trait-based plant community assembly models. *Journal of Ecology, 104*(5), 1284–1298.

Le Quéré, C., Andrew, R. M., Friedlingstein, P., Sitch, S., Hauck, J., Pongratz, J., & Zheng, B. (2018). Global carbon budget 2018. *Earth System Science Data, 10*(4), 2141–2194.

Lemmens, E., Deleu, L. J., De Brier, N., De Man, W. L., De Proft, M., Prinsen, E., & Delcour, J. A. (2019). The impact of hydro-priming and osmo-priming on seedling characteristics, plant hormone concentrations, activity of selected hydrolytic enzymes, and cell wall and phytate hydrolysis in sprouted wheat (Triticum aestivum L.). *ACS Omega, 4*(26), 22089–22100.

Leubner-Metzger, G. (2006). Seed dormancy and the control of germination. *New Phytologist, 171*, 501–523.

Li, H., Li, X., Zhang, D., Liu, H., & Guan, K. (2013). Effects of drought stress on the seed germination and early seedling growth of the endemic desert plant Eremosparton songoricum (Fabaceae). *EXCLI Journal, 12*, 89.

Li, R., Chen, L., Wu, Y., Zhang, R., Baskin, C. C., Baskin, J. M., & Hu, X. (2017a). Effects of cultivar and maternal environment on seed quality in Vicia sativa. *Frontiers in Plant Science, 8*, 1411.

Li, R., Chen, L., Wu, Y., Zhang, R., Baskin, C. C., Baskin, J. M., & Hu, X. (2017b). Effects of cultivar and maternal environment on seed quality in Vicia sativa. *Frontiers of Plant Science, 8*, 1411. https://doi.org/10.3389/fpls.2017.01411

Li, Y., Yu, Z., Jin, J., Zhang, Q., Wang, G., Liu, C., & Liu, X. (2018). Impact of elevated CO_2 on seed quality of soybean at the fresh edible and mature stages. *Frontiers in Plant Science, 9*, 1413.

Lin, Y. L., Chao, Y. Y., & Kao, C. H. (2010b). Exposure of rice seedlings to heat shock protects against subsequent Cd-induced decrease in glutamine synthetase activity and increase in specific protease activity in leaves. *Journal of Plant Physiology, 167*(13), 1061–1065.

Lin, C. J., Li, C. Y., Lin, S. K., Yang, F. H., Huang, J. J., Liu, Y. H., & Lur, H. S. (2010). Influence of high temperature during grain filling on the accumulation of storage proteins and grain quality in rice (Oryza sativa L.). *Journal of Agriculture and Food Chemistry, 58*, 10545–10552. https://doi.org/10.1021/jf101575

Liu, J., Hasanuzzaman, M., Wen, H., Zhang, J., Peng, T., Sun, H., & Zhao, Q. (2019). High temperature and drought stress cause abscisic acid and reactive oxygen species accumulation and suppress seed germination growth in rice. *Protoplasma, 256*(5), 1217–1227.

Liu, P., Guo, W., Jiang, Z., Pu, H., Feng, C., Zhu, X., & Little, C. R. (2011). Effects of high temperature after anthesis on starch granules in grains of wheat (Triticum aestivum L.). *The Journal of Agricultural Science, 149*(2), 159–169.

Loladze, I. (2014). Hidden shift of the ionome of plants exposed to elevated CO_2 depletes minerals at the base of human nutrition. *eLife, 3*, p. e02245.

Louwaars, N. P., & Manicad, G. (2022). Seed systems resilience—an overview. *Seeds, 1*, 340–356. https://doi.org/10.3390/seeds1040028

Lutts, S., Benincasa, P., Wojtyla, L., Kubala, S., Pace, R., Lechowska, K., ... & Garnczarska, M. (2016). Seed priming: new comprehensive approaches for an old empirical technique. *New Challenges in Seed Biology-Basic and Translational Research Driving Seed Technology, 46*.

Mahakham, W., Sarmah, A. K., Maensiri, S., & Theerakulpisut, P. (2017). Nanopriming technology for enhancing germination and starch metabolism of aged rice seeds using phytosynthesized silver nanoparticles. *Scientific Reports, 7*, 1–21.

Maity, A., & Pramanik, P. (2013). Climate change and seed quality: An alarming issue in crop husbandry. *Current Science, 105*(10), 1336–1338.

Malhi, G. S., Kaur, M., & Kaushik, P. (2021). Impact of climate change on agriculture and its mitigation strategies: A review. *Sustainability, 13*(3), 1318.

Manan, A., Zafar, M. M., Ren, M., Khurshid, M., Sahar, A., Rehman, A., & Shakeel, A. (2022). Genetic analysis of biochemical, fiber yield and quality traits of upland cotton under high-temperature. *Plant Production Science, 25*(1), 105–119.

Manderscheid, R., Pacholski, A., & Weigel, H. J. (2010). Effect of free air carbon dioxide enrichment combined with two nitrogen levels on growth, yield and yield quality of sugar beet: Evidence for a sink limitation of beet growth under elevated CO_2. *European Journal of Agronomy, 32*(3), 228–239.

Mao, Q., Tang, L., Ji, W., Rennenberg, H., Hu, B., & Ma, M. (2021). Elevated CO_2 and soil mercury stress affect photosynthetic characteristics and mercury accumulation of rice. *Ecotoxicology and Environmental Safety, 208*, 111605.

Maphosa, L., Fitzgerald, G. J., Panozzo, J., Partington, D., Walker, C., & Kant, S. (2019). Genotypic response of wheat under semi-arid conditions showed no specific responsive traits when grown under elevated CO_2. *Plant Production Science, 22*(3), 333–344.

Marty, C., Bassiri, R., Hormoz. (2014). Seed germination and rising atmospheric CO_2 concentration: a meta-analysis of parental and direct effects. *New Phytologist, 202*(2), 401 414. https://doi.org/10.1111/nph.12691

McDonald, M. B. (1999). Seed deterioration: Physiology, repair and assessment. *Seed Science and Technology., 27*, 177–237.

Medek, D. E., Schwartz, J., & Myers, S. S. (2017). Estimated effects of future atmospheric CO 2 concentrations on protein intake and the risk of protein deficiency by country and region. *Environmental Health Perspectives, 125*(8), 087002.

Meinshausen, M., Smith, S. J., Calvin, K., Daniel, J. S., Kainuma, M. L., Lamarque, J. F., & van Vuuren, D. P. (2011). The RCP greenhouse gas concentrations and their extensions from 1765 to 2300. *Climatic Change, 109*, 213–241.

Mendelsohn, R. (2009). The impact of climate change on agriculture in developing countries. *Journal of Natural Resources Policy Research, 1*(1), 5–19.

Mession, J. L., Sok, N., Assifaoui, A., & Saurel, R. (2013). Thermal denaturation of pea globulins (Pisum sativum L.) molecular interactions leading to heat-induced protein aggregation. *Journal of Agricultural and Food Chemistry, 61*(6), 1196–1204.

Michał, B., Kelly, D., Thomas, P. A., GA, L. J., & Hacket-Pain, A. (2020). Climate warming disrupts mast seeding and its fitness benefits in European beech. *Nature Plants, 6*(2), 88–94.

Misra, B. B., & Chen, S. (2015). Advances in understanding CO_2 responsive plant metabolomes in the era of climate change. *Metabolomics, 11*, 1478–1491. https://doi.org/10.1007/s11306-015-0825-4

Mittal, D., Kaur, G., Singh, P., Yadav, K., & Ali, S. A. (2020). Nanoparticle-based sustainable agriculture and food science: Recent advances and future outlook. *Frontiers in Nanotechnology, 2*, 579954.

Montague, J. L., Barrett, S. C. H., & Eckert, C. G. (2008). Re-establishment of clinal variation in flowering time among introduced populations of purple loosestrife (Lythrum salicaria, Lythraceae). *Journal of Evolutionary Biology, 21*(1), 234–245.

Morita, S., Yonemaru, J. I., & Takanashi, J. I. (2005). Grain growth and endosperm cell size under high night temperatures in rice (Oryza sativa L.). *Annals of Botany, 95*(4), 695–701.

Munné-Bosch, S., Queval, G., & Foyer, C. H. (2013). The impact of global change factors on redox signaling underpinning stress tolerance. *Plant Physiology, 161*(1), 5–19.

Murillo-Amador, B., Lopez-Aguilar, R., Kaya, C., Larrinaga-Mayoral, J., & Flores-Hernandez, A. (2002). Comparative effects of NaCl and polyethylene glycol on germination, emergence and seedling growth of cowpea. *Journal of Agronomy and Crop Science, 188*, 235–247.

Ohdaira, Y., Takeda, H., & Sasaki, R. (2010). Effects of temperature on the digestible protein content of grains during ripening in a seed-protein mutant rice cultivar LGC soft. *Plant Production Science, 13*(2), 132–140. https://doi.org/10.1626/pps.13.132

Orlowsky, B., & Seneviratne, S. (2011). Investigating spatial climate relations using CARTs: An application to persistent hot days in a multimodel ensemble. *Journal of Geophysical Research, 116*, D14106.

Pansrithong, M., Romkaew, J., Malumpong, C., Thongket, P., & Thongjo, C. (2019). Effect of high temperature at reproductive stage on seed set and seed quality of rice. *Kaen Kaset= Khon Kaen Agriculture Journal, 47*(3), 445–458.

Parihar, A. K., Hazra, K. K., Lamichaney, A., Dixit, G. P., Singh, D., Singh, A. K., & Singh, N. P. (2022). Characterizing plant trait (s) for improved heat tolerance in field pea (Pisum sativum L.) under subtropical climate. *International Journal of Biometeorology, 66*(6), 1267–1281.

Patanè, C., Saita, A., & Sortino, O. (2013). Comparative effects of salt and water stress on seed germination and early embryo growth in two cultivars of sweet sorghum. *Journal of Agronomy and Crop Science, 199*, 30–37.

Pausas, J. G., & Lamont, B. B. (2022). Fire-released seed dormancy-a global synthesis. *Biological Reviews, 97*(4), 1612–1639.

Piikki, K., De Temmerman, L., Ojanperä, K., Danielsson, H., & Pleijel, H. (2008). The grain quality of spring wheat (Triticum aestivum L.) in relation to elevated ozone uptake and carbon dioxide exposure. *European Journal of Agronomy, 28*(3), 245–254.

Poschlod, P., Tackenberg, O., & Bonn, S. (2005). Plant dispersal potential and its relation to species frequency and coexistence. *Vegetation Ecology*, 147–171.

Prasad, P. V., & Djanaguiraman, M. (2014). Response of floret fertility and individual grain weight of wheat to high temperature stress: Sensitive stages and thresholds for temperature and duration. *Functional Plant Biology, 41*(12), 1261–1269.

Prasad, P. V., Bheemanahalli, R., & Jagadish, S. K. (2017). Field crops and the fear of heat stress—opportunities, challenges and future directions. *Field Crops Research, 200*, 114–121.

Pukacka, S., Hoffmann, S. K., Goslar, J., et al. (2003). Water and lipid relations in beech (*Fagus sylvatica* L.) seeds and its effect on storage behaviour. *BBA General Subjects, 1621*, 48–56. https://doi.org/10.1016/S0304-4165(03)00046-1

Qayyum, A., Razzaq, A., Ahmad, M., Jenks, M. A. (2011). Water stress causes differential effects on germination indices, total soluble sugar and proline content in wheat (Triticum aestivum L.) genotypes. *African Journal of Biotechnology*, 10, 14038–14045.

Quaderi, M. M., & Reid, D. M. (2005). Growth and physiological responses of canola (*Brassica napus*) to UV-B and CO_2 under controlled environment conditions. *Physiologia Plantarum, 125*, 247–259.

Queiroz, M. S., Oliveira, C. E. S., Steiner, F., Zuffo, A. M., Zoz, T., Vendruscolo, E. P., Menis, V. S., Mello, B. F. F. R., Cabral, R. C., & Menis, T. F. (2019). Drought stresses on seed germination and early growth of maize and sorghum. *The Journal of Agricultural Science*, 11, 310–318.

Rai, A. K., Sahni, P., Agarwal, D. K., Pal, G., Susmita, C., Chintagunta, A. D., & Kumar, S. P. J. (2020). Assessment of seed deterioration mechanisms in groundnut (Arachis hypogaea L.). *Indian Journal of Ecology*, *47*(11), 181–184.

Raj, S., Roodbar, S., Brinkley, C., & Wolfe, D. W. (2022). Food Security and climate change: Differences in impacts and adaptation strategies for rural communities in the Global South and North. *Frontiers in Sustainable Food Systems, 5*.

Rajjou, L., & Debeaujon, I. (2008). Seed longevity: Survival and maintenance of high germination ability of dry seeds. *Comptes Rendus Biologies, 331*(10), 796–805.

Reddy, P. P., & Reddy, P. P. (2013). Bio-priming of seeds. *Recent Advances in Crop Protection*, 83–90.

Reddy, T. Y., & Reddy, G. H. (2019). *Principles of agronomy*. Kalyani publishers.

Reed, R. C., Bradford, K. J., & Khanday, I. (2022). Seed germination and vigor: Ensuring crop sustainability in a changing climate. *Heredity*, 1–10.

Rezaei, E. E., Rojas, L. V., Zhu, W., & Cammarano, D. (2022). The potential of crop models in simulation of barley quality traits under changing climates: A review. *Field Crops Research, 286*, 108624.

Rosenthal, D. M., Ruiz-Vera, U. M., Siebers, M. H., Gray, S. B., Bernacchi, C. J., & Ort, D. R. (2014). Biochemical acclimation, stomatal limitation and precipitation patterns underlie decreases in photosynthetic stimulation of soybean (Glycine max) at elevated [CO_2] and temperatures under fully open-air field conditions. *Plant Science, 226*, 136–146.

Saed-Moucheshi, A., Pakniyat, H., Pirasteh-Anosheh, H., Azooz, M. M., & Ahmad, P. (2014). Role of ROS as signaling molecules in plants in oxidative damage to plants antioxidant networks and signaling.

Saha, S., Chakraborty, D., Sehgal, V. K., & Pal, M. (2015). Rising atmospheric CO_2: Potential impacts on chickpea seed quality. *Agriculture, Ecosystems & Environment, 203*, 140–146.

Sandvik, S. M., & Eide, W. (2009). Costs of reproduction in circumpolar *Parnassia palustris* L. in light of global warming. *Plant Ecology, 205*, 1–11.

Sanhewe, A. J., Ellis, R. H., Hong, T. D., Wheeler, T. R., Batts, G. R., Hadley, P., & Morison, J. I. L. (1996). The effect of temperature and CO_2 on seed quality development in wheat (Triticum aestivum L.). *Journal of Experimental Botany*, *47*(5), 631–637.

Sano, N., Rajjou, L., North, H. M., Debeaujon, I., Marion-Poll, A., & Seo, M. (2016). Staying alive: Molecular aspects of seed longevity. *Plant and Cell Physiology, 57*(4), 660–674.

Sarkar, A. K., & Sadhukhan, S. (2022). Proteomics—a powerful tool for understanding saline stress response in germinating seed. In *Plant stress: Challenges and management in the new decade* (pp. 375–399). Springer, Cham.

Savvides, A., Ali, S., Tester, M., & Fotopoulos, V. (2016). Chemical priming of plants against multiple abiotic stresses: Mission possible? *Trends in Plant Science, 21*(4), 329–340.

Scaven, V. L., & Rafferty, N. E. (2013). Physiological effects of climate warming on flowering plants and insect pollinators and potential consequences for their interactions. *Current Zoology, 59*(3), 418–426.

Schieber, M., & Chandel, N. S. (2014). ROS function in redox signaling and oxidative stress. *Current Biology, 24*(10), R453–R462.

Schuurmans, C. J. E. (2021). The world heat budget: Expected changes. *Climate Change Impact on Coastal Habitation*, 1.

Sehgal, A., Sita, K., Siddique, K. H., Kumar, R., Bhogireddy, S., Varshney, R. K., & Nayyar, H. (2018). Drought or/and heat-stress effects on seed filling in food crops: Impacts on functional biochemistry, seed yields, and nutritional quality. *Frontiers in Plant Science, 9*, 1705.

Shahzad, A., Ullah, S., Dar, A. A., Sardar, M. F., Mehmood, T., Tufail, M. A., & Haris, M. (2021). Nexus on climate change: Agriculture and possible solution to cope future climate change stresses. *Environmental Science and Pollution Research, 28*(12), 14211–14232.

Sharma, P., Jha, A. B., Dubey, R. S., & Pessarakli, M. (2012). Reactive oxygen species, oxidative damage, and antioxidative defense mechanism in plants under stressful conditions. *Journal of botany, 2012.*

Shi, W., Muthurajan, R., Rahman, H., Selvam, J., Peng, S., Zou, Y., & Jagadish, K. S. (2013). Source–sink dynamics and proteomic reprogramming under elevated night temperature and their impact on rice yield and grain quality. *New Phytologist, 197*(3), 825–837.

Silva-Correia, J., Freitas, S., Tavares, R. M., Lino-Neto, T., & Azevedo, H. (2014). Phenotypic analysis of the Arabidopsis heat stress response during germination and early seedling development. *Plant Methods, 10*(7), 1–11. https://doi.org/10.1186/1746-4811-10-7

Singh, R. P., Singh, R. K., & Singh, M. K. (2011). Impact of climate and carbon dioxide change on weeds and their management–a review. *Indian Journal of Weed Science, 43*(1 & 2), 1–11.

Singh, R. P., Prasad, P. V., & Reddy, K. R. (2013). Impacts of changing climate and climate variability on seed production and seed industry. *Advances in Agronomy, 118*, 49–110.

Singh, R. P., Chintagunta, A. D., Agarwal, D. K., Kureel, R. S., & Kumar, S. J. (2020). Varietal replacement rate: Prospects and challenges for global food security. *Global Food Security, 25*, 100324.

Singh, R. P., Agarwal, D. K., Rajendra Prasad, S., Sripathy, K. V., Jeevan Kumar, S. P. (2017). Varietal and Seed Replacement in the Era of Climate Change. pp. 156 ISBN: 978-81- 925128 2-6 published by Indian Council of Agricultural Research -Indian Institute of Seed Science (ICAR-IISS), Mau, U.P, India.

Sita, K., Sehgal, A., Bhandari, K., Kumar, J., Kumar, S., Singh, S., & Nayyar, H. (2018). Impact of heat stress during seed filling on seed quality and seed yield in lentil (Lens culinaris Medikus) genotypes. *Journal of the Science of Food and Agriculture, 98*(13), 5134–5141.

Smiciklas, K. D., Mullen, R. E., Carlson, R. E., & Knapp, A. D. (1992). Soybean seed quality response to drought stress and pod position. *Agronomy Journal, 84*(2), 166–170.

Smith, S. D., Huxman, T. E., Zitzer, S. F., Charlet, T. N., Housman, D. C., Coleman, J. S., & Nowak, R. S. (2000). Elevated CO_2 increases productivity and invasive species success in an arid ecosystem. *Nature, 408*(6808), 79–82.

Sneideris, L. C., Gavassi, M. A., Campos, M. L., D'Amico-Damiao, V., & Carvalho, R. F. (2015). Effects of hormonal priming on seed germination of pigeon pea under cadmium stress. *Anais Da Academia Brasileira De Ciências, 87*, 1847–1852.

Soba, D., Shu, T., Runion, G. B., Prior, S. A., Fritschi, F. B., Aranjuelo, I., & Sanz-Saez, A. (2020). Effects of elevated [CO_2] on photosynthesis and seed yield parameters in two soybean genotypes with contrasting water use efficiency. *Environmental and Experimental Botany, 178*, 104154.

Spielman, D. J., Smale, M. (2017). Policy options to accelerate variety change among smallholder farmers in South Asia and Africa South of the Sahara. Environment and Production Technology Division, pp. 68 IFPRI Discussion Paper 01666.

St Clair, S. B., & Lynch, J. P. (2005). Base cation stimulation of mycorrhization and photosynthesis of sugar maple on acid soils are coupled by foliar nutrient dynamics. *New Phytologist*, 581–590.

Sun, X. D., & Arntfield, S. D. (2012). Molecular forces involved in heat-induced pea protein gelation: Effects of various reagents on the rheological properties of salt-extracted pea protein gels. *Food Hydrocolloids, 28*(2), 325–332.

Suriyasak, C., Oyama, Y., Ishida, T., Mashiguchi, K., Yamaguchi, S., Hamaoka, N., & Ishibashi, Y. (2020). Mechanism of delayed seed germination caused by high temperature during grain filling in rice (Oryza sativa L.). *Scientific Reports, 10*(1), 1–11.

Suryavanshi, P., & Buttar, G. S. (2016). Mitigating terminal heat stress in wheat. *International Journal of Bio-Resource and Stress Management, 7*(1), 142–150.

Talukder, A. S. M. H. M., McDonald, G. K., & Gill, G. S. (2014). Effect of short-term heat stress prior to flowering and early grain set on the grain yield of wheat. *Field Crops Research, 160*, 54–63.

Tamura, N., Yoshida, T., Tanaka, A., Sasaki, R., Bando, A., Toh, S., Lepiniec, L., & Kawakami, N. (2006). Isolation and characterization of high temperature-resistant germination mutants of *Arabidopsis thaliana. Plant Cell Physiology, 47*, 1081–1094. https://doi.org/10.1093/pcp/pcj078

Tashiro, T. W. I. F., & Wardlaw, I. F. (1991). The effect of high temperature on kernel dimensions and the type and occurrence of kernel damage in rice. *Australian Journal of Agricultural Research, 42*(3), 485–496.

Thinh, N. C., Kumagai, E., Shimono, H., & Kawasaki, M. (2017). Effects of elevated CO_2 concentration on bulbil germination and early seedling growth in Chinese yam under different air temperatures. *Plant Production Science, 20*(3), 313–322.

Thomas, J. M. G., Boote, K. J., Allen, L. H., Gallo-Meagher, M., & Davis, J. M. (2003). Elevated temperature and carbon dioxide effects on soybean seed composition and transcript abundance. *Crop Science, 43*(4), 1548–1557.

Thomas, J. M. G., Prasad, P. V. V., Boote, K. J., & Allen, L. H. J. R. (2009). Seed composition, seedling emergence and early seedling vigour of red kidney bean seed produced at elevated temperature and carbon dioxide. *Journal of Agronomy and Crop Science, 195*, 148–156. https://doi.org/10.1111/j.1439-037X.2008.00348.x

Thurman, L. L., Stein, B. A., Beever, E. A., Foden, W., Geange, S. R., Green, N., & Young, B. E. (2020). Persist in place or shift in space? Evaluating the adaptive capacity of species to climate change. *Frontiers in Ecology and the Environment, 18*(9), 520–528.

Totland, Ø. (1999). Effects of temperature on performance and phenotypic selection on plant traits in alpine *Ranunculus acris. Oecologia, 120*, 242–251.

Totland, Ø., & Alatalo, J. M. (2002). Effects of temperature and date of snowmelt on growth, reproduction, and flowering phenology in the arctic/alpine herb, *Ranunculus glacialis. Oecologia, 133*, 168–175.

USGCRP. (2017). Climate science special report: Fourth national climate assessment, Vol. I. In [D. J. Wuebbles, D. W. Fahey, K. A. Hibbard, D. J. Dokken, B. C. Stewart, & T. K. Maycock (Eds.) (p. 470). Washington, DC, USA: U.S. Global Change Research Program. https://doi.org/10.7930/J0J964J6.

Usui, Y., Sakai, H., Tokida, T., Nakamura, H., Nakagawa, H., & Hasegawa, T. (2014). Heat-tolerant rice cultivars retain grain appearance quality under free-air CO_2 enrichment. *Rice, 7*(1), 1–9.

Van Vuuren, D. P., Edmonds, J., Kainuma, M., Riahi, K., Thomson, A., Hibbard, K., & Rose, S. K. (2011). The representative concentration pathways: An overview. *Climatic Change, 109*, 5–31.

Wagner, J., Lüscher, A., Hillebrand, C., Kobald, B., Spitaler, N., & Larcher, W. (2001). Sexual reproduction of Lolium perenne L. and Trifolium repens L. under free air CO_2 enrichment (FACE) at two levels of nitrogen application. *Plant, Cell & Environment, 24*(9), 957–966.

Walck, J., Hidayati, S., Dixon, K., Thompson, K., & Poschlod, P. (2011). Climate change and plant regeneration from seed. *Global Change Biology, 17*, 2145–2161.

Walters, C. (2015). Orthodoxy, recalcitrance and in-between: Describing variation in seed storage characteristics using threshold responses to water loss. *Planta, 242*, 397–406.

Wang, J., Gan, Y. T., Clarke, F., & McDonald, C. L. (2006). Response of chickpea yield to high temperature stress during reproductive development. *Crop Science, 46*(5), 2171–2178.

Wang, Y., Frei, M., Song, Q., & Yang, L. (2011). The impact of atmospheric CO_2 concentration enrichment on rice quality—a research review. *Acta Ecologica Sinica, 31*(6), 277–282.

Ward, J., & Strain, B. (1997). Effects of low and elevated CO_2 partial pressure on growth and reproduction of *Arabidopsis thaliana* from different elevations. *Plant, Cell and Environment, 20*, 254–260.

Way, D. A., Ladeau, S. L., McCarthy, H. R., Clark, J. S., Oren, R. A. M., Finzi, A. C., & Jackson, R. B. (2010). Greater seed production in elevated CO_2 is not accompanied by reduced seed quality in Pinus taeda L. *Global Change Biology, 16*(3), 1046–1056.

Weller, S., Florentine, S., Javaid, M. M., Welgama, A., Chadha, A., Chauhan, B. S., & Turville, C. (2021). Amaranthus retroflexus L. (Redroot Pigweed): Effects of elevated CO_2 and soil moisture on growth and biomass and the effect of radiant heat on seed germination. *Agronomy, 11*(4), 728.

West, J. M., Pearce, J. M., Coombs, P., Ford, J. R., Scheib, C., Colls, J. J., Smith, K. L., & Steven, M. D. (2009). The impact of controlled injection of CO_2 on the soil ecosystem and chemistry of an English lowland pasture. *Energy Procedia, 1*(1), 1863–1870. https://doi.org/10.1016/j.egypro.2009.01.243

Westgate, M. E., & Boyer, J. S. (1986). Reproduction at low silk and pollen water potentials in maize. *Crop Science, 26*, 951–956.

White, S., Bork, E., Karst, J., & Cahill, J. (2012). Similarity between grassland vegetation and seed bank shifts with altered precipitation and clipping, but not warming. *Community Ecology, 13*(2), 129–136. https://doi.org/10.1556/comec.13.2012.2.1

Wilhelm, E. P., Mullen, R. E., Keeling, P. L., & Singletary, G. W. (1999). Heat stress during grain filling in maize: Effects on kernel growth and metabolism. *Crop Science, 39*(6), 1733–1741.

Wulff, R. D. (1986). Seed size variation in *Desmodium paniculatum*. I. Factors affecting seed size. *Journal of Ecology, 74*, 87–97. https://doi.org/10.2307/2260350

Xu, L. X., Xin, X., Yin, G. K., Zhou, J., Zhou, Y. C., & Lu, X. X. (2020). Timing for antioxidant-priming against rice seed ageing: Optimal only in non-resistant stage. *Scientific Reports, 10*(1), 13294.

Yadav, M. K., Singh, R. S., Singh, K. K., Mall, R. K., Patel, C. B., Yadav, S. K., & Singh, M. K. (2015). Assessment of climate change impact on productivity of different cereal crops in Varanasi, India. *Journal of Agrometeorology, 17*(2), 179–184.

Yadav, S. K., Tiwari, Y. K., Pavan Kumar, D., Shanker, A. K., Jyothi Lakshmi, N., Vanaja, M., & Maheswari, M. (2016). Genotypic variation in physiological traits under high temperature stress in maize. *Agricultural Research, 5*(2), 119–126.

Yang, L., Wang, Y., Dong, G., Gu, H., Huang, J., Zhu, J., & Han, Y. (2007). The impact of free-air CO_2 enrichment (FACE) and nitrogen supply on grain quality of rice. *Field Crops Research, 102*(2), 128–140.

Yang, J., Mocking-Bode, H. C., van den Hoek, I. A., Theunissen, M., Voudouris, P., Meinders, M. B., & Sagis, L. M. (2022). The impact of heating and freeze or spray drying on the interface and foam stabilising properties of pea protein extracts: Explained by aggregation and protein composition. *Food Hydrocolloids*, 107913.

Zhang, Y., Huang, X., Wang, L., Wei, L., Wu, Z., You, M., & Li, B. (2014). Proteomic analysis of wheat seed in response to drought stress. *Journal of Integrative Agriculture, 13*(5), 919–925. https://doi.org/10.1016/S2095-3119(13)60601-2

Zheng, G., Chen, J., & Li, W. (2020). Impacts of CO_2 elevation on the physiology and seed quality of soybean. *Plant Diversity, 42*(1), 44–51.

Zhu, C., Kobayashi, K., Loladze, I., Zhu, J., Jiang, Q., Xu, X., Liu, G., Seneweera, S., Ebi, K. L., Drewnowski, A., Fukagawa, N. K., Ziska, L. H. (2018). Carbon dioxide (CO_2) levels this century will alter the protein, micronutrients, and vitamin content of rice grains with potential health consequences for the poorest rice-dependent countries. *Science Advances. 4*(5), eaaq1012. https://doi.org/10.1126/sciadv.aaq1012

Ziska, L. H., & Bunce, J. A. (1993). The influence of elevated CO_2 and temperature on seed germination and emergence from soil. *Field Crops Research, 34*(2), 147–157.

Chapter 14
Climate Change Impacts and Adaptation Strategies for Agronomic Crops

Gopal Dutta, Bappa Paramanik, Basabduttaa Bhabai, Jayanta Layek, Ashok Choudhury, Susanta Dutta, Sayantan Bhattacharjee, and Feroze Hasan Rahman

Abstract Agriculture and food security are seriously threatened by climate change. Changes in precipitation patterns and numerous weather fluctuations pose a danger to agricultural output. IPCC has predicted that temperature increases for the Indian subcontinent will be 0.85–3.15 °C by 2050 and 1.55–5.40 °C by 2080, depending on the future growth scenario, and India could experience a 40% decline in agricultural productivity by the 2080s. Different predictive models inferred that a rise in temperature and changes in rainfall distribution would reduce the production of major agronomic crops. 60% of the net cultivated area relies on rainwater for irrigation and is therefore more vulnerable to biotic and abiotic stresses caused by climatic aberrations. As a result, adaptation measures are needed to reduce climatic variability and

G. Dutta (✉)
Bidhan Chandra Krishi Viswavidyalaya, Nadia, West Bengal 741252, India
e-mail: gopaldutta411@gmail.com

B. Paramanik
Dakshin Dinajpur KVK, Uttar Banga Krishi Viswavidyalaya, Dakshin Dinajpur, West Bengal 733133, India

B. Bhabai
Central Seed Research Station for Jute and Allied Fibre, ICAR-CRIJAF, Budbud, West Bengal 713403, India

J. Layek
ICAR-Indian Institute of Agricultural Biotechnology, Ranchi, Jharkhand 834003, India

A. Choudhury
Uttar Banga Krishi Viswavidyalaya, Cooch behar, West Bengal 736165, India

S. Dutta
Birsa Agricultural University, West Singhbhum, Jharkhand 833204, India

S. Bhattacharjee
West Bengal University of Animal and Fishery Sciences, Kolkata, West Bengal 700037, India

F. H. Rahman
ICAR-National Bureau of Soil Survey and Land Use Planning, Regional Center, Kolkata, West Bengal 700091, India

H. Pathak et al. (eds.), *Climate Change Impacts on Soil-Plant-Atmosphere Continuum*, Advances in Global Change Research 78,
https://doi.org/10.1007/978-981-99-7935-6_14

sustain crop production. Climate change's negative consequences can be minimised by developing different stress-tolerant varieties and adopting innovative agricultural practices such as resource conservation technologies. The assessment of the effects of global climatic changes on agronomic crops and possible adaptation approaches provides researchers and farmers with useful information for adopting farming and improving agricultural production.

Keywords Climate change · Agronomic crops · Adaptation strategies

14.1 Introduction

Changes in climate will persist to have a major consequence on the worldwide thermal atmosphere (Field & Barros, 2014), especially enhances in local mean temperatures and the incidence of heatwaves (Buckley & Huey, 2016). These changes pose a significant threat to biodiversity and to have serious effects for the distribution and plenty of native vegetation (Walsh et al., 2019). Devastating changes in the environment have been proven to be detrimental for natural systems, crop production and peoples' health (Arunanondchai et al., 2018). Since the world's population already reached 8 billion and increasing further, the demand for food will correspondingly rise and for that more food to be produced from less area without harming the environment. Agricultural productivity is significantly influenced by various factors such as air pollution, soil fertility and water availability (Noya et al., 2018). With sudden alters in environmental states, the severe effects on plant productivity are intensifying due to both indirect and direct impacts of abiotic stresses. The amount of CO_2 in the atmosphere has enhanced from 280 μmol^{-1} since pre-industrial era to 400 μmol^{-1} due to constant destruction of forest and excess use of fossil fuels (Raza et al., 2019). It is predicted to be increase further. The greenhouse effect and warmer average earth temperature are mostly caused by the release of hazardous gases, particularly CO_2 (Vaughan et al., 2018). The consequences of climate change and fluctuation in environment are measured primarily based on the stress spells number, related effect on everyday living, and agricultural loss (FAO, 2018). Excessive rainfall may cause destructive floods while lack or the deficiency of precipitation for an extended time results in drought (Khan et al., 2016). The frequency of global warming is predicted to increase because of extreme weather, which will eventually disturb the ecosystems of globe (Kanojia & Dijkwel, 2018).

The world's major cereal crops, including paddy, wheat and maize, are vital for meeting daily dietary requirements. Among these, one of the most widely grown staples is wheat (Tack et al., 2015). As global temperature rises, the climate experiences severe variations and becomes abiotically stressed. Changes in the environment are extremely detrimental and pose numerous hazards to naturally occurring species of plant (Espeland & Kettenring, 2018). Heat and drought are the two common challenges in field condition that considerably affect plants (Pereira, 2016). Variations of temperature have a major impact on plant physiology as crops need an optimal temperature for proper growth and flowering (Hatfield & prueger, 2015). Heat stress reduces yield, cold stress induces sterility, and drought stress has a negative impact on plants morpho-physiology (Salehi-Lisar & Bakhshayeshan-Agdam, 2016). These climatic issues severely disrupt plant growth and productivity, generating enormous negative responses within plants, involving molecular, biochemical, physiological, and morphological alterations (Zandalinas et al., 2018). In this situation, understanding the plants stress-resistance mechanisms has appeared as a highly challenging task for plant researchers to generate stress-resistant plants (Singh et al., 2018). Hence, maintaining crop productivity is a crucial challenge in current agriculture, and producing stress-tolerant plants is needed of the hour (Abhinandan et al., 2018). As a result, adaptation measures are needed to reduce climatic variability and maintain crop production. The assessment of the effects of global climatic changes on agronomic crops and possible adaptation approaches provides researchers and farmers with useful information for adopting farming and improving agricultural production (India: The Impact of Climate Change to 2030 a Commissioned Research Report- Sambal Member Upload, 2017).

14.2 Future Projection

This is predicted that the temperature will enhance significantly in dry regions of India, Pakistan and Western part of China (Change, 2013). The world is facing alarming challenges because of climate change which is showed by climatic models that temperature of the earth surface may surpass 1.5 °C compared to 1850–1900 for each representative concentration pathways (RCP) scenarios for the end of this century (Ahmed et al., 2019). Predictions of models pointed out that there would be more erratic precipitation with higher intensity over some areas especially for, summer monsoon season. Because of reduction in rainfall in winter, in South and Southeast Asia enhanced aridity anticipated. Climate change threatens millions of poor rural people of Asian countries. Various extreme occurrence including heat waves, irregularity and uncertainty of precipitation and increasing number of hot days, dust storms, sustained dry spells and tropical cyclones will be more frequent in the region (Ahmed et al., 2019). Agricultural systems are being disturbed by unusual climate changes that affect the physical, chemical and biological processes of the systems. From 1961 to 2003, there is an increase in the cunt of hot days and warm nights in Asia and decrease in cold days and nights was found particularly in the years following the onset of El Nino (Cinco et al., 2014). The intensity and frequency of tropical cyclones in the Pacific Ocean have increased over the past few decades (Walsh et al., 2019).

According to the report of World Food Programme (WFP), 2018, enhance in crop productivity per hectare is notably slower than the rates of population growth. According to Food and Agriculture Organization (FAO) data, 2016, if the present situation of greenhouse gas emissions and climate change continues, the productivity of main cereal crops will decrease (by 20–45% for maize, 5–50% in wheat and 20–30% in rice) by the year 2100 (Officer, 2016). Globally, South Asia which is the major food insecure area and approx. 262 million people suffers from malnutrition (Field et al., 2012). Hence if the current trends continue, crop losses could enhance in the near future an aberrant rate which would significantly lead to diminished production, rising cost of food, and will make it difficult to meet the increasing demand of a growing population (Arora, 2019).

The Indian Institute of Tropical Meteorology (IITM) recently developed a regional model that predicts an average temperature increase of 4 °C across India by the year 2100 (India: The Impact of Climate Change to 2030 a Commissioned Research Report- Sambal Member Upload, 2017). However, the model indicates that, a delay in the rise in monsoon precipitation would occur leading to drought in some regions due to the longer time scales for system-wide changes in atmosphere–ocean interactions (Parry et al., 2007).It is expected that India will see a warming of 0.5 °C by the year 2030 (roughly equivalent to the warming over the twentieth century) and a warming of 2–4 °C in the end of the century, with the highest rise across northern part of India (Shukla, 2003).

14.3 Current Scenario of Climate Change in India

India is one of the world's most vulnerable countries to anticipated climate change has already been facing climate changes and its effects, including water scarcity, heat waves and dry spells, dreadful storms and floods, as well as negative effects on health and livelihoods that come with them (India: The Impact of Climate Change to 2030 a Commissioned Research Report- Sambal Member Upload 2017). Over the past few decades, the burning of fossil fuels and massive deforestation have caused increased concentrations of atmospheric Greenhouse Gases (GHG), and these changes in GHG lead to significant climate changes around the world (Shukla et al., 2019). As a result of these, the temperature of the earth has risen by 1.41 °C from the preindustrial time (1850–1900) to the present day (1998–2018) (Shukla et al., 2019). After declining of global fossil carbon di-oxide releases by 5.4% in 2020 as a result of extensive lockdowns, global fossil carbon di-oxide releases in 2021 came back to the pre-pandemic levels of 2019. As per preliminary data global CO_2 emissions in 2022 are 1.2% greater than the levels documented over the same timeframe in 2019, driven by enhances in the India, United States and majority of the European countries. The last seven years, from 2015 to 2021, were the warmest on history. Global mean temperature for 2018–2022 (basis of information up to May or June 2022) is projected to be 1.17 ± 0.13 °C over 1850–1900 mean. Temperatures in 2021–22 are slightly cool due to La Nina event however this will be temporary (United in Science: We are Heading in the Wrong Direction, 2022). In the earth system, the ocean holds about 90% of the accumulated heat and the oceanic heat during 2018–2022 was greater than for any previous five year, with rates of warming of sea exhibiting a especially sharp rise over the last two decades (Eight warmest years on record witness upsurge in climate change impacts, 2022). Over the 30 years of the satellite altimeter record (1993–2022), the average annual increase in global sea level is estimated to be 3.4 ± 0.3 mm. Between 1993–2002 and 2013–2022 the rate has doubled and sea level enhanced by around 5 mm between January 2021 and August 2022. The acceleration is because of rising ice melt (Eight Warmest Years on Record Witness Upsurge in Climate Change Impacts, 2022). The number and magnitude of climatic calamities have enhanced globally. We cannot deny the danger caused by the change in climate. There is a high chance that hazardous climate change will destroy the stability of earth's system. Around the world, number and magnitude of climatic calamities has enhanced. There is a substantial chance that dangerous climate change could undermine the stability of the earth's system. The numbers related to the effects of climate change are alarming and of course, quite dangerous (Climate Change: Current Scenario & Challenges for India 2022).

- Between 1970 and 2019, the world experienced over 11,000 disasters resulting 2 million deaths
- By 2050, over 40% of India's population would face water scarcity
- It is anticipated that the frequency of heat waves will increase significantly as global average temperature increases
- Mean temperature had already been increased by 0.7 °C during 1901–2018 and by the end of the twenty-first century, average temperatures might rise 4.4 °C.

14.4 Impact of Climate Change on Agronomic Crops

Agronomic crops, such as rice, wheat, maize, millets, pulses, and oilseeds, are called primary crops because the majority of the global population relies on them for food. Considering forecasted patterns such as changes in dietary composition, demand, and income development, it is necessary to increase the production of cereals by 70–100% to feed the 9.8 billion of the estimated global population by 2050 (Godfray et al., 2010). Rising demand is commonly thought to be solved by increasing production rates; however, historical data shows that current production rates fall far short of what is required to achieve the objectives. The IPCC has predicted temperature increases for the Indian subcontinent of 0.5–1.20 °C by 2020, 0.85–3.15 °C by 2050, and 1.55–5.40 °C by 2080, depending on the future growth scenario (Mackay, 2008). Extreme temperatures, heat waves, and heavy and non-uniform precipitation are predicted to increase in frequency, which affects cereal production. Researchers concur that global climate change can affect crop yields and must be handled in order to achieve food security (Tripathi et al., 2016). In light of the above issues, we have to address the detrimental effects of climate change on agronomic crop production.

14.4.1 Impact of Climate Change on Rice

Rice (*Oryza sativa* L.) is one of the primary staple foods in developing countries; it provides a key source of daily energy and nutritional carbohydrates. Asian countries produce 90% of the world's rice, with China and India alone accounting for 50% of worldwide rice output (Muthayya et al., 2012). Demand for rice is increasing day by day, but climate change and global warming affect global rice production in different ways. Among the consequences of climate change, high temperatures and altered patterns of precipitation are the primary reasons for fluctuations in the global output and price of rice. (Sheffield et al., 2012). Temperature and precipitation are key climatic elements that have both good and bad effects on the growth and development of rice. It is cultivated in tropical and subtropical climates and requires somewhat warmer temperatures, ideally between 20 and 40 °C (Khanal et al., 2018). At different development phases, rice requires varying critical temperatures, as a range of 16–20 °C is ideal for flowering and fertilisation stages, whereas 18–32 °C is ideal for ripening. However, grain filling is affected by temperatures above 35 °C (Khanal et al., 2018). Effects of climate change, such as warmer temperatures and water stress, directly impact rice farming and production. Recent research demonstrated that an increase in the global average temperature will result in a considerable decrease in worldwide rice production due to an increase in evapotranspiration triggered by the higher temperature. Changes in the precipitation pattern and the late start of the monsoon also cause poor germination, less tillering, delayed panicle initiation, and reduced grain and panicle numbers (Khanal et al., 2018). Another study found that rice production in Nigeria increased by 22% for every 1% increase in precipitation but decreased by 4% for every 1% increase in temperature (Nwalieji & Uzuegbunam, 2013). Drought also negatively affects rice's growth and development, decreasing spikelet fertility, panicle initiation, and ultimately yield. Frequent drought not only decreases water supply but also raises plant transpiration water requirements (Felkner et al., 2009). Interactions between crops and pests may shift when various life stages occur for both hosts and pests. Large quantities of chemical pesticides used to combat this infestation are a potential source of pollution and food contamination (Khanal et al., 2018). Paddy fields are responsible for 17.9% of the total methane (CH_4) emissions that are causing global warming (Yan et al., 2009). It has also been found that rice plants release the most CH_4 when they are in the tillering and heading stage (Sriphirom et al., 2019). Due to more GHGs being released into the atmosphere, the average global temperature has also gone up which may reduce rice spikelets and make them less fertile, which is detrimental for crop production. However, another study reported that an increase in atmospheric CO_2 levels could improve rice output (Dabi & Khanna, 2018). Although the higher concentration of CO_2 may increase rice yield slightly, higher temperatures and rainfall anomalies will significantly reduce the rice yield. Overall, natural and anthropogenic climate change may have direct or indirect consequences on rice growth and output.

14.4.2 Impact of Climate Change on Wheat

Wheat (*Triticum aestivum* L.) is the second most significant cereal crop in southern Asia, behind rice. In the next two decades, wheat demand is anticipated to expand at a rate of 2.5% per year (Rosegrant et al., 2002). But it is thought that climate change will have a big effect on South Asia, and most crops could see big drops in their yields. In a lot of places, both rainfed and irrigated wheat production will go down (Nelson et al., 2009). In order to estimate how climate change would affect wheat production, numerous studies have used a wide variety of simulation models (Kersebaum & Nendel, 2014). According to one model, climate change will reduce wheat yield by 3.5–12.9% between 2037 and 2065, while winter wheat output will decrease by 14.6–17.2% by the end of the century (Gammans et al., 2017). Weather events like drought and heat waves affect soil and groundwater storage, which reduced winter wheat production as soil water reserve is a crucial factor for winter wheat cultivation (Elliott et al., 2013). Fluctuation of precipitation and temperature as a result of global warming can affect wheat production by altering soil qualities and processes (Brevik, 2012). Warming of the air temperature also slows the rate of grain filling in wheat, and extreme canopy temperatures can damage the leaf's photosynthesis machinery, which speeds up senescence (Zhao et al., 2007). Crop phenology is influenced by temperature, as each species has a unique base temperature, optimal temperature, and maximum temperature range (Hatfield, 2011). Wheat yield was reduced by 5.3% for every 1 °C increase in temperature (Innes et al., 2015). Significant damage to wheat output can be caused by cold or hot temperatures during pollination (Rezaei et al., 2015). Another study observed that wheat yield was negatively correlated with temperature while a positive correlation was found between precipitation and wheat yield (Lobell & Burke, 2008). It has been shown that doubling the amount of CO_2 in the air decreases seasonal evaporation by 8% in wheat grown at 28 °C during the day and 18 °C at night (Hatfield et al., 2011). Higher CO_2 levels in the air, called the "CO_2 fertilisation effect," can cause plants to grow bigger and stronger, have significantly greater total dry matter and produce more output (Lobell & Field, 2007). Because of higher CO_2 levels directly affect plant growth by speeding up photosynthesis and slowing down evaporation (Ortiz et al., 2008). Experiments with CO_2 enrichment showed that wheat lost nitrogen, calcium, magnesium, and sulphur, although results varied for phosphorus and potassium (Högy & Fangmeier, 2008). Sucrose and stachyose were the most notable changes in carbohydrate content in response to a rise in temperature, while glucose, fructose and raffinose observed little or no change (DaMatta et al., 2010).

14.4.3 Impact of Climate Change on Maize

Maize (*Zea mays* L.) is one of the most common cereal crops, and it grows well in a wide range of climates. In terms of maize output, the United States came in first with a 52.9% share, followed by Asia (28.5%), Europe (11.4%), and Africa (7.1%) (Hatfield et al., 2017). By 2050, maize demand is expected to double, and by 2025, it is anticipated to have the highest worldwide production of any crop (Rosegrant et al., 2001). Production area under maize is still expanding as consumption increased, but at the same time, changing climatic conditions also have an impact on maize output. Several biotic and abiotic factors like temperature, water stress, drought, pest and diseases have an impact on the world's maize production (Oseni & Masarirambi, 2011). Study reported that global maize yield is predicted to reduce by 3% to 10% by 2050 due to climate change (Rosegrant et al., 2001). A 2 °C increase in temperature and a 20% decrease in precipitation reduced maize yield by around 10% and grain output drops by 1–1.7% for every degree above 30 °C (Lobell et al., 2011). The flowering stage of maize is the most susceptible to drought stress, although all stages of maize development and production may be negatively impacted by rising temperatures and reducing precipitation (Cairns et al., 2012). Maize is susceptible to a variety of detrimental consequences when exposed to high temperatures over a short period of time, including a shorter life cycle, less photosynthesis, and more male sterility. However, yields are predicted to fall in certain areas; they may rise in places where maize is not the primary crop (Wang et al., 2011). Reproductive procedures in maize are also on the verge of being threatened. High temperatures produce asynchrony in the tasselling and silking of maize because they reduce the development and responsiveness of the style and increase the number of antipodal cells in the embryo sac (Ribaut et al., 2009). Among male and female flowers of maize, female flowers are more susceptible to damage from high temperatures and water stress (Jablonski et al., 2002). Heat stress has detrimental impacts on maize grain yield and alters the starch's quality by lowering its amylose concentration in grain (Wang et al., 2011). Heat stress increases the risk of pollen damage by lowering sink capacity, which in turn alters cellular division and carbohydrate production (Cicchino et al., 2013). Agricultural pests and diseases in maize crops, in particular, will alter in both diversity and receptiveness as climates shift. Climate change has the capacity to affect the stages of pathogen development, and its most prominent effects on maize are alterations to the host's physiology and resistance (Rosenzweig et al., 2001). Increased maize cultivation in changing climatic scenarios will open up previously inaccessible regions to the introduction, establishment, and spread of diseases like *Cercospora zeae-maydis* and *Cercospora zeina*, which cause grey leaf spot and corn rootworm, respectively, and are extremely sensitive to surrounding climatic factors (Crous et al., 2006).

14.4.4 Impact of Climate Change on Millets

Millets are globally important cereal crops with an outstanding nutritional profile and non-glutenous composition. Millets are found in numerous places around the world, each of which requires a unique type of soil for optimal growth. The most commonly cultivated millet types include Pearl (*Pennisetum glaucum*), Foxtail (*Setaria italica*),

Japanese/Barnyard (*Echinochloa esculenta*), Finger (*Eleusine coracana*), and Kodo (*Paspalum scrobiculatum), which* are regarded as essential crops for ensuring food security in India and Africa (Passot et al., 2016). Pearl millet makes up 50% of the world's total millet production (Singh et al., 2017). In India, millets are cultivated on 15.48 million hectares with a production of 1111 kg ha^{-1}, yielding approximately 17.2 million t (Malathi et al., 2016). Between 1950–1951 and 2011–2012, this crop's contribution to global food grain output decreased considerably from around 22 to 7% (Michaelraj & Shanmugam, 2013). Millet is more drought-resistant than other cereal crops like rice, wheat, and maize, requires less water for growth, and provides more nutrients like phosphorous, potassium, iron, and magnesium (Awika, 2011). Millets can be cultivated under stress conditions, and it is not a water-intensive crop, so the absence of irrigation facilities in Asia and Africa's low economies is not a barrier to adopting its production. Millet has a strong root structure that enables it to flourish in locations that frequently suffer from dry weather (Hadebe et al., 2016). Another study that used a modified CERES-millet simulation model to test the impact of temperature changes on millet production reported that millet production increased by 8% during simulations of heat and by 6% during simulations of drought (Singh et al., 2017). According to another study, drought stress causes poor and unpredictable yields from blooming to grain filling. Post-flowering drought stress is one of the most important environmental factors limiting pearl millet grain yield by as much as 70%, while crop failure might also result from too much rain before blossoming (Chitalu, 2014). From all the studies, it can be reported that millets have the highest resistance to global warming while also being resistant to its impacts, such as a greater frequency of droughts and higher average temperatures. This crop has the ability to significantly minimise the carbon footprint.

14.4.5 Impact of Climate Change on Pulses

Pulses are an excellent source of vegetable protein, vitamins, and minerals that our bodies need. They are very important for keeping the soil fertile because they fix nitrogen biologically from the air. India is the largest producer of legumes in the world, accounting for 25% of total production (Gull et al., 2020). Since pulses are a big part of rain-fed agriculture, climate change will have a bigger effect on their yields. This is especially true in the Indo-Gangetic plains, where water is scarce and there are no ways to deal with changes in rainfall. Predicted temperature shifts and their effects on water availability, pests, and diseases are all potential threats to pulse production. (Gull et al., 2020). When grown in rainfed conditions, major pulses like chickpea, pigeon pea, and lentil are exposed to a number of stresses, such as water stress, high and low temperature stress, increased solar radiation, salinity stress, and waterlogging conditions. Temperatures below 5 °C and above 45 °C are considered to be lethal for most pulses because they cause irreversible changes at the cellular level, the clumping together of macromolecules, and the massive destruction of the photosynthesis process, which stops it completely and at these extreme/low temperatures, no fertilisation takes place (Basu et al., 2016). However, the duration and intensity of heat exposure varies amongst different pulse crops like the heat tolerance of chickpeas and lentils is greater than that of field pea (Egli et al. 2005). Another study reported that with every 0.1 °C increased in maximum temperature, chickpea, lentil, and pigeon pea yields decreased by 38.49, 40.70, and 26.86 kg ha^{-1} respectively although, every 0.1 °C increased in minimum temperature reduced pulse crop yields by 13.47, 14.23 and 9.38 kg ha^{-1} (Dubey et al., 2011). Temperatures above 21 °C (26/16 °C day/night) diminish chickpea (*Cicer arietinum*) yields in greenhouse tests (Jumrani & Bhatia, 2014). Changes in temperature trigger flower drop, drying, and bending of apical leaves in pigeon pea (Basu et al., 2009). Another study reported that the seed yield of garden pea is negatively co-related with canopy temperature and stress degree days (Dutta et al., 2020). Rainfall also has significant effect on pulse production for example, chickpea, lentil and pigeon pea average yields drop by 12.35, 13.05, and 8.05 kg ha^{-1} respectively for every 10 mm drop in annual precipitation (Dubey et al., 2011). Lentil production reduced by 33% under high temperatures and increased CO_2 levels (Bourgault et al., 2018). But the harvest index and the weight of each grain went down by 11% and 4%, respectively, when CO_2 levels were high (Delahunty et al., 2018). Another study found that pigeon pea and chickpea yields went up by 20.0% and 33.0%, respectively, when CO_2 levels were expected to rise to 470 ppm and 760 ppm, respectively, from normal levels (Yadav et al., 2016). Grain yield was higher in the double CO_2 scenario than in normal simulations because radiation use efficiency was better in higher CO_2 concentration (Yadav et al., 2016). Different crops do better when there is more CO_2 in the air.

14.4.6 Impact of Climate Change on Oilseeds

Land, production, and value-wise, oilseeds are second only to food grains in importance to the agricultural economy. Many different plants are grown for their oils, but some of the most notable are groundnut (*Arachis hypogaea*), mustard (*Brassica sp.*), soybeans (*Glycine max*), sunflowers (*Helianthus annuus*) and sesame (*Sesamum indicum*). India produced 10.1 MT of edible oils from all sources in 2017, a growth of 84% from 5.5 MT in 2001. This difference between demand and production is only filled out through imports, which have climbed by a factor of 267% from 4.2 MT in 2001 to 15.3 MT in 2017 (Jat et al., 2019). Drought and other abiotic stressors can affect the oil content and composition of a plant. This is because the manufacturing process and chemical composition of fatty acids are typically altered by a lack of water, leading to lower oil output and a different chemical profile (Baldini et al., 2000). Nonetheless, another study reported that oilseed crop seeds improved under water-deficit stress, particularly in terms of protein content (Amini et al., 2014). Climate change significantly affect the yield of mustard because it is extremely susceptible to environmental variables. Eastern India's irrigated mustard grain yield fell sharply as temperatures rose. Temperature increases of 5 °C reduced yield by 86.6% in this location and CO_2 increases of 450 and 550 ppm reduced irrigated mustard yield by 82.4 and 79.4%. But, rainfed mustard in eastern India would lose 78.4% production with a 5 °C temperature rise and central India would lose 40.2% yield whereas in northern India, a 5 °C temperature increase led a 20.9% yield reduction, which fell to 8.5% and 7.3% with higher CO_2 levels (Boomiraj et al. 2010). Increased level of CO_2 helps plants grow and produce more but temperature rises reduced leaf area index, grain number, and grain weight, reducing crop production. It also reported that significant impact of environmental conditions such as drought on the fatty acid profile of canola seed oil (Dawood & Sadak, 2014). Sunflower may live under stress circumstances better than other oilseed crops, even though it is very susceptible to drought stress from blooming to grain development in arid and semi-arid regions because there isn't enough soil moisture and the temperature is high (Hussain et al., 2018). Another study found that the oil quality of sunflower was reduced due to the combination of low rainfall in arid and semi-arid areas and high atmospheric temperature, which sped up the period of sprouting and flowering periods and stopping the growth of steam and leaf area (Fulda et al., 2011). Future climate change scenarios could reduce rainfed mustard yields. Future rainfall variations affect the simulated production of rainfed mustard and in all sites except eastern India, rainfed mustard had a higher predicted yield loss (Boomiraj et al. 2010). Regardless of climate change scenarios, northern and central India's rainfall will drop in 2020, 2050, and 2080, which reduced the region's rainfed mustard output, it was predicted that in 2080, northern India could lose 29.8–40.4% of its production, central India 31–42.5%, and eastern India 57% (Boomiraj et al. 2010). Another study reported that changing climatic condition reduced leaf area index of mustard which decreases the crop's radiation use efficiency and ultimately reduced net photosynthesis and total dry matter production (Pidgeon et al., 2001).

14.5 Adaptation Strategies

The primary goal of adaptation planning is to strengthen societies' capacities to deal with climate change (Smit & Wandel, 2006). The agricultural industry is affected by climate change and decline in the yield of agricultural crops is a major challenge to food availability, especially in the rapid growth in the global population (Rogelj et al., 2016). It is anticipated that the enhanced incidence of drought, excessive precipitation, rising temperature, pest infestation and salinity will reduce crop productivity, thereby increasing the risk of starvation (Dhankher & Foyer, 2018). Crop yields are mainly affected severely because of unfavorable environments in developing countries, so increasing temperature, changing pattern of precipitation and excessive accumulations of CO_2 have forced researchers to develop new approaches to deal with fewer predictable issues (Raza et al., 2019). Adaptation is a crucial component that has the potential to decrease the negative consequences of climate change by reducing the vulnerabilities in agriculture caused by it. Adaptation tactics offer a chance to overcome the problems caused by climate change and ensure crop output (Fischer et al., 2002).

14.5.1 Adaptive Measures to Reduce Climate Change's Effects Include

- Cultivating cultivars that are resistant to environmental stresses like as flooding and drought as well as heat and salinity
- Restoration of damaged land and mitigation of soil erosion
- Modifying crop management practices
- Drought and flood resilience and Improving water management and water use efficiency
- Adopting innovative agricultural practices such as resource-conserving technologies (RCTs), climate smart agriculture (CSA), climate resilient agriculture, organic Farming
- Diversification of crops based on climatic situation
- Integrated pest and diseases management
- Enhanced predictions of weather and insurance for agricultural risks
- Utilizing farmers' indigenous technical expertise.

A number of agronomic practices, such as RCTs, mulching, and the sequestration of carbon by agroforestry and cropping systems, have the potential to reduce greenhouse gas emissions by increasing agricultural systems' capacity to consume carbon and store it in biomass, wood, and soil (Koundinya et al., 2018). Changing the planting or sowing date is essential for maximizing crop potential in the context of climate change, as high temperatures during the grain-filling stage shorten the period for grain-filling, resulting in lower yields (Ahmed et al., 2019). Already, farmers have investigated numerous strategies for adjusting to the effects of climate change. Farmers in Pakistan and Brazil, for instance, have responded to the variability brought on by climate change by shifting planting times and optimising plant population (Ahmad et al., 2015).

Organic farming has the potential to slow climate change since it eliminates the use of synthetic fertilisers, pesticides, and herbicides while increasing soil carbon and incorporating legumes into crop rotations. By not disturbing the soil structure, reducing soil erosion, increasing plant cover, and using minimum fertilisers and pesticides, organic farming is demonstrably more kind to the environment and results in lower CO_2 emissions than conventional agriculture. (Sartaj et al., 2013). Therefore, organic farming practices reduced the impact of climate change by avoiding soil erosion, improving soil health, increasing water use efficiency, conserving soil water, increasing drought and flood resilience, reducing greenhouse gas emissions and agrobiodiversity (IFOAM, 2009).

Climate change adaptation decisions can be aided by the carbon sequestration potential of integrated agricultural systems and associated cropping practices. According to Wang et al. (2010) crop rotation using legumes aids in fixing atmospheric N, which in turn reduces the need for artificial fertilisers, the manufacturing of which requires the combustion of fossil fuels. The cultivation of cover crops is an efficient method that can boost carbon capture and the amount of organic carbon stored in soil (Chatterjee, 2011).

Increasing the genetic diversity of agricultural plants is primarily an adaptation technique since it serves as a type of preparation for crop plants to adjust to the expected climate of the future. The modification of crop plant's genes in order to provide them the ability to survive the negative effects of climate change is a key means for ensuring both the long-term sustainability of agricultural production and the safety of food supplies (Koundinya et al., 2018). To maintain productivity in a changing climate, early-maturing, drought- and heat-tolerant cultivars are needed. New cultivars would boost per-area production under moisture stress and severe temperatures (Deressa et al., 2009).

Increased crop resilience can be achieved by increased fertiliser productivity through the application of precision nutrient management. The precise control of fertilizer in crops, in particular maize, led to a reduction in the use of fertilizer, which led to an increase in production and improved soil health, which in turn led to a reduction in the emission of greenhouse gases (GHGs) (Srinivasan, 2006). Another study reported that, when oilseed crops were subjected to heat stresses, foliar application of Zinc (Zn) regulated the plant physiological properties, leading to increases plant biomass, seed number, seed weight, oil and linoleic acid content (Ahmad et al., 2021).

The cultivation of crops is severely hindered by weeds. Weeds limit crop growth and productivity by competing with plants for water and nutrients. That's why integrating weed management into your farming practices is so important for increasing productivity. The decreased usage of synthetic weedicides also contributes to a reduction in greenhouse gas emissions (Pedde et al., 2013). The ability of transgenic crops to manage insect pests will be diminished due to global warming and climate variability, which will in turn affect the expression of host plant resistance to insects (Sharma et al., 2015). As a result, it is important to breed insect-resistant agricultural cultivars that can withstand changes in climate and soil. Spinosads and avermectins are two examples of the fungal, nuclear polyhedrosis virus, plant, and Bacillus thuringiensis derived bio-pesticides that are currently in widespread use because of their low environmental impact. (Sharma et al., 2015).

Irrigation water management under climate change relies heavily on timely irrigation and preserving soil moisture. Mulching and cover cropping are both effective ways to prevent soil erosion and weed growth by keeping the soil moist and limiting weed growth. Among all irrigation technologies farmers can use subsurface irrigation, sprinklers, center pivot irrigation which will cooled the plants during heat stress also reduced water losses from conventional irrigation methods (Ahmad et al., 2021). In order to reduce methane emission and maximize water use efficiency, modern rice-planting techniques such as direct seeded rice and the system of rice intensification with alternate wetting and draying have been developed (Latif et al., 2005).

In addition, heat stress is being detected before symptoms are visible by using remote sensing technologies such as multispectral and hyperspectral imaging, thermal imaging (Hernández-Clemente et al., 2019). An unmanned aerial vehicles (UAVs) equipped with an infrared imaging spectrometer were utilised to measure heat stress (Shivers et al., 2019). Whereas, multispectral, hyperspectral or visible sensor mounted UAVs or drones can be used to monitor water stress, heat stress, nutrient disorder, diseases in agricultural field, which will be helpful for early decision making and ultimately reduces crop losses (Dutta & Goswami, 2020).

14.5.2 Government of India Initiatives for Climate Change Adaptations in Agriculture

In order to effectively combat the threat posed by climate change, we need to concurrently act in a focused manner across various levels. The National Action Plan on Climate Change (NAPCC) is focused on the creation and implementation of innovative technologies. Many National Missions implemented by NAPCC and government of India which reflect different but complementary approaches to achieve major objective in the face of climate change. Some important policies are detailed below.

14.6 Conclusion

Global warming and changing climate have adverse consequences on agronomic crop production. The increased number and severity of extreme weather events like floods, droughts, and heat waves significantly reduced crop yield. The average temperature has already risen by 0.7–1.41 °C, and future projections show that it will rise by 4 °C across India by the year 2100. The severity and frequency of tropical cyclones in the Pacific Ocean have increased over the last few decades, and greenhouse gas emissions have increased year after year. If current levels of greenhouse gas emissions and climate change continue, the yield of primary cereal crops, such as maize, wheat, and rice, will decrease by 20–45%, 5–50%, and 20–30%, respectively, by the end of this century. The adverse and negative consequences of climate change can be reduced by the implementation of adaptation strategies. The introduction of stress-tolerant varieties, alteration of crop management practices, and the implementation of innovative agricultural technologies can ameliorate the adverse impact of climate change. Such simulation studies can help us see the detrimental effect of climate change on agronomic crop production and possible adaptation techniques that, if strategically employed, not just minimise the impact of climate change but it also increases sustainable production and food security.

Table 14.1 Climate change adaptations for agronomic crops

Agronomic crops	Adaptation strategies
General	• Water-saving techniques such as laser land levelling, direct-seeded rice, no-till rice, alternate wetting and drying method, System of rice intensification (SRI) practices • Heat, drought, salt and alkali tolerant cultivar • Adjusting sowing date • Balance nutrient application • Organic farming • Residue management
Rice	• Integrated pest management • Weather forecast and crop insurance • Direct seeded rice • SRI
Wheat	• No tilled wheat • Furrow bed cultivation • Heat tolerant variety, Drought tolerant variety • Adjustment of planting date • Integrated pest management • Crop diversification • Weather forecast and crop insurance
Maize	• Precision nutrient management • Raise bed panting • Early maturing and drought tolerant cultivar • Mixed cropping • High yielding variety • Land rotation • Weather forecast and crop insurance
Pulses	• Early sowing with pre-sowing irrigation • Photo and thermo sensitive variety • Disease tolerant variety and Improve pest management practices • Integrated weed control • Crop rotation, inter or multiple cropping • Fallow or conservation tillage • No-till cultivation of winter pulse in rice and maize fallow • Water saving technology (Drip or sprinkler irrigation)
Oil seed	• Biotic and abiotic stress tolerant varieties • Application of Zinc • Proper spacing maintain • Crop rotation • Water saving technology • No-till cultivation of oilseed in rice fallow • Weather forecast and crop insurance

Table 14.2 Policies of Govt. of India to combat climate change with their main objectives

Policies	Objectives
1. Agro-Advisory Services	To provide real time crop-specific agro-met services in order to reduce the adverse effect of weather
2. National Solar Mission	To enhance renewable energy and reduced fossil fuel
3. National Water Mission (NWM)	For preserving water resources, reducing water waste, and increasing water use efficiency by 20% across all sectors, including agriculture
4. Green India Mission (GIM)	To improve forest, grassland and wetland cover and improve carbon sequestration
5. National Mission on Sustainable Habitat (NMSH)	To make habitat sustainable improvement in energy efficiency and waste management
6. Pradhan Mantri Krishi Sinchayee Yojana (PMSKY)	To expand cultivable area under assured irrigation with sustainable water conservation practices
7. National Adaptation Fund for Climate Change (NAFCC)	To minimize the consequences of socioeconomic and environmental losses caused by climate change
8. Pradhan Mantri Fasal Bima Yojana (PMFBY)	To support sustainable agricultural production by providing financial assistance to farmers who have lost crops due to environmental hazards
9. National Mission on Strategic Knowledge for Climate Change	To identify the challenges of, and the responses to climate change
10. National Mission for Sustainable Agriculture (NMSA)	To make more resilient agriculture under changed climatic scenario

References

Abhinandan, K., Skori, L., Stanic, M., Hickerson, N. M., Jamshed, M., & Samuel, M. A. (2018). Abiotic stress signaling in wheat–an inclusive overview of hormonal interactions during abiotic stress responses in wheat. *Frontiers in Plant Science, 9*, 734.

Ahmad, A., Ashfaq, M., Rasul, G., Wajid, S. A., Khaliq, T., Rasul, F., & Valdivia, R. O. (2015). Impact of climate change on the rice–wheat cropping system of Pakistan. In *Handbook of climate change and agroecosystems: The agricultural model intercomparison and improvement project integrated crop and economic assessments*, Part 2 (pp. 219–258).

Ahmad, M., Waraich, E. A., Skalicky, M., Hussain, S., Zulfiqar, U., Anjum, M. Z., & Sabagh, A. E. (2021). Adaptation strategies to improve the resistance of oilseed crops to heat stress under a changing climate: An overview. *Frontiers in Plant Science 12.*

Ahmed, I., Ullah, A., Rahman, M. H., Ahmad, B., Wajid, S. A., Ahmad, A., & Hassain, S. (2019). Climate change impacts and adaptation strategies for agronomic crops. In *Climate change and agriculture* (pp. 1–14). London, UK: IntechOpen.

Amini, H., Arzani, A., & Karami, M. (2014). Effect of water deficiency on seed quality and physiological traits of different safflower genotypes. *Turkish Journal of Biology, 38*, 271–282. https://doi.org/10.3906/biy-1308-22.

Arora, N. K. (2019). Impact of climate change on agriculture production and its sustainable solutions. *Environmental Sustainability, 2*(2), 95–96. https://doi.org/10.1007/s42398-019-00078-w.

Arunanondchai, P., Fei, C., Fisher, A., McCarl, B. A., Wang, W., Yang, Y. (2018). *How does climate change affect agriculture? In The Routledge handbook of agricultural economics* (pp. 191–210). Routledge.

Awika, J. M. (2011). Major cereal grains production and use around the world. In *Advances in cereal science: Implications to food processing and health promotion* (pp. 1–13). American Chemical Society.

Baldini, M., Giovanardi, R., & Vannozzi, G. (2000). Effect of different water availability on fatty acid composition of the oil in standard and high oleic sunflower hybrids. In Proceedings of 15th International Sunflower Conference, A.

Basu, P. S., Singh, U. M., Kumar, A. N., Praharaj, C. S., & Shivran, R. K. (2016). Climate change and its mitigation strategies in pulses production. *Indian Journal of Agronomy, 61*, S71–S82.

Basu, P. S., Ali, M., & Chaturvedi, S. K. (2009). Terminal heat stress adversely affects chickpea productivity in northern India—Strategies to improve thermo tolerance in the crop under climate change. In *W3 Workshop Proceedings: Impact of Climate Change on Agriculture* (Vol. 23, pp. 189–193). New Delhi: International Society for Photogrammetry and Remote Sensing.

Boomiraj, K., Chakrabarti, B., Aggarwal, P., Choudhary, R., & Chander, S. (2010). Assessing the vulnerability of Indian mustard to climate change. *Agriculture, Ecosystems & Environment, 138*(3–4), 265–273. https://doi.org/10.1016/j.agee.2010.05.010.

Bourgault, M., Löw, M., Tausz-Posch, S., Nuttall, J. G., Delahunty, A. J., Brand, J., Panozzo, J. F., & Tausz, M. (2018). Effect of a heat wave on lentil grown under free-air CO2 enrichment (FACE) in a semi-arid environment. *Crop Science, 58*(2), 803–812. https://doi.org/10.2135/cropsci2017.09.0565.

Brevik, E. C. (2012). Soils and climate change: gas fluxes and soil processes. *Soil Horizons, 53*(4), 12. https://doi.org/10.2136/sh12-04-0012.

Buckley, L. B., & Huey, R. B. (2016). Temperature extremes: Geographic patterns, recent changes, and implications for organismal vulnerabilities. *Global Change Biology, 22*(12), 3829–3842. https://doi.org/10.1111/gcb.13313.

Cairns, J. E., Sonder, K., Zaidi, P. H., Verhulst, N., Mahuku, G., Babu, R., & Prasanna, B. M. (2012). Maize production in a changing climate: Impacts, adaptation, and mitigation strategies. *Advances in Agronomy, 114*, 1–58. https://doi.org/10.1016/B978-0-12-394275-3.00006-7.

Change, I. C. (2013). The physical science basis. Contribution of working group I to the fifth assessment report of the intergovernmental panel on climate change 1535.

Chatterjee, C. S. (2011). Mitigating climate change through agriculture: An untapped potential. deutsche bank research. Working Papers id:4474, eSocialSciences Frankfurt, Germany.

Chitalu, G. (2014). Genetic characterization of pearl millet (Pennisetum glaucum (L.) R. Br) genotypes in Zambia (Doctoral dissertation).

Cicchino, M. A., Rattalino Edreira, J. I., & Otegui, M. E. (2013). Maize physiological responses to heat stress and hormonal plant growth regulators related to ethylene metabolism. *Crop Science, 53*(5), 2135–2146.https://doi.org/10.2135/cropsci2013.03.0136.

Cinco, T. A., de Guzman, R. G., Hilario, F. D., & Wilson, D. M. (2014). Long-term trends and extremes in observed daily precipitation and near surface air temperature in the Philippines for the period 1951–2010. *Atmospheric Research, 145*, 12–26. https://doi.org/10.1016/j.atmosres.2014.03.025.

Field, C. B., & Barros, V. R. (Eds). (2014). *Climate change 2014–impacts, adaptation and vulnerability: Regional aspects.* Cambridge University Press.

Climate Change: Current Scenario & Challenges for India. (2022). LeadsConnect. Retrieved November 14, 2022, from https://www.leadsconnect.in/climate-change-current-scenario-challenges-for-india/.

Crous, P. W., Groenewald, J. Z., Groenewald, M., Caldwell, P., Braun, U., & Harrington, T. C. (2006). Species of Cercospora associated with grey leaf spot of maize. *Studies in Mycology, 55*, 189–197. https://doi.org/10.3114/sim.55.1.189.

Dabi, T., & Khanna, V. K. (2018). Effect of climate change on rice. *Agrotechnology, 7*(2), 2–7. https://doi.org/10.4172/2168-9881.1000181.

DaMatta, F. M., Grandis, A., Arenque, B. C., & Buckeridge, M. S. (2010). Impacts of climate changes on crop physiology and food quality. *Food Research International, 43*(7), 1814–1823. https://doi.org/10.1016/j.foodres.2009.11.001.

Dawood, M. G., & Sadak, M. S. (2014). Physiological role of glycinebetaine in alleviating the deleterious effects of drought stress on canola plants (Brassica napus L.). *Middle East Journal of Agriculture Research, 3*(4), 943–954.

Delahunty, A., Nuttall, J., Nicolas, M., & Brand, J. (2018). Response of lentil to high temperature under variable water supply and carbon dioxide enrichment. *Crop and Pasture Science, 69*(11), 1103. https://doi.org/10.1071/cp18004.

Deressa, T. T., Hassan, R. M., Ringler, C., Alemu, T., & Yesuf, M. (2009). Determinants of farmers' choice of adaptation methods to climate change in the Nile Basin of Ethiopia. *Global Environmental Change, 19*(2), 248–255. https://doi.org/10.1016/j.gloenvcha.2009.01.002.

Dhankher, O. P., & Foyer, C. H. (2018). Climate resilient crops for improving global food security and safety. *Plant, Cell & Environment, 41*(5), 877–884. https://doi.org/10.1111/pce.13207.

Dubey, S. K., Sah, U., & Singh, S. K. (2011). Impact of climate change on pulse productivity and adaptation strategies as practiced by the pulse growers of Bundelkhand region of Uttar Pradesh. *Journal of Food Legumes, 24*(3), 230–234.

Dutta, G., & Goswami, P. (2020). Application of drone in agriculture: a review. *IJCS, 8*, 181–187. https://doi.org/10.22271/chemi.2020.v8.i5d.10529.

Dutta, G., Saha, T., Dey, S., Mandal, B., & Saha, G. (2020). Productivity of Garden Pea (Pisum sativum L.) Under Abiotic Stress and Modified Bio-climatic condition. *International Journal of Current Microbiology and Applied Sciences, 9*(7), 1197–1206. https://doi.org/10.20546/ijcmas.2020.907.138.

Egli, D. B., TeKrony, D. M., & Spears, J. F. (2005). Effect of high temperature stress during different stages of seed development in soybean [Glycine max (L.) Merrill]. *Seed Technology, 27*(2), 177–189. http://www.jstor.org/stable/23433336.

Eight warmest years on record witness upsurge in climate change impacts. (2022). World Meteorological Organization. Retrieved November 14, 2022, from https://public.wmo.int/en/media/press-release/eight-warmest-years-record-witness-upsurge-climate-change-impacts.

Elliott, J., Deryng, D., Müller, C., Frieler, K., Konzmann, M., Gerten, D., Glotter, M., & Wisser, D. (2013). Constraints and potentials of future irrigation water availability on agricultural production under climate change. *Proceedings of the National Academy of Sciences, 111*(9), 3239–3244. https://doi.org/10.1073/pnas.1222474110.

Espeland, E. K., & Kettenring, K. M. (2018). Strategic plant choices can alleviate climate change impacts: A review. *Journal of Environmental Management, 222*, 316–324. https://doi.org/10.1016/j.jenvman.2018.05.042.

FAO; Unicef; WFP; WHO. (2018). The state of food security and nutrition in the world: Building resilience for peace and food security. Rome, Italy: Food and Agriculture Organization of the United Nations (FAO). https://doi.org/10.1016/j.tree.2018.12.002.

Felkner, J., Tazhibayeva, K., & Townsend, R. (2009). Impact of climate change on rice production in Thailand. *American Economic Review, 99*(2), 205–210. https://doi.org/10.1257/aer.99.2.205.

Field, C. B., Barros, V., Stocker, T. F., Dahe, Q. (Eds). (2012). Managing the risks of extreme events and disasters to advance climate change adaptation: special report of the intergovernmental panel on climate change. Cambridge University Press.

Fischer, G., Shah, M. M., & Van Velthuizen, H. T. (2002). *Climate change and agricultural vulnerability*. IIASA, Laxenburg, Austria. https://pure.iiasa.ac.at/6670.

Fulda, S., Mikkat, S., Stegmann, H., & Horn, R. (2011). Physiology and proteomics of drought stress acclimation in sunflower (Helianthus annuus L.). *Plant Biology, 13*(4), 632–642. https://doi.org/10.1111/j.1438-8677.2010.00426.x.

Gammans, M., Mérel, P., & Ortiz-Bobea, A. (2017). Negative impacts of climate change on cereal yields: Statistical evidence from France. *Environmental Research Letters, 12*(5), 054007. https://doi.org/10.1088/1748-9326/aa6b0c.

Godfray, H. C. J., Beddington, J. R., Crute, I. R., Haddad, L., Lawrence, D., Muir, J. F., Pretty, J., Robinson, S., Thomas, S. M., & Toulmin, C. (2010). Food security: The challenge of feeding 9 billion people. *Science, 327*(5967), 812–818. https://doi.org/10.1126/science.1185383.

Gull, R., Bhat, T. A., Sheikh, T. A., Wani, O. A., Fayaz, S., Nazir, A., & Nazir, I. (2020). Climate change impact on pulse in India-A review. *Journal of Pharmacognosy and Phytochemistry, 9*(4), 3159–3166.

Hadebe, S. T., Modi, A. T., & Mabhaudhi, T. (2016). Drought tolerance and water use of cereal crops: A focus on sorghum as a food security crop in Sub-Saharan Africa. *Journal of Agronomy and Crop Science, 203*(3), 177–191. https://doi.org/10.1111/jac.12191.

Srinivasan, A. (Ed.). (2006). Handbook of precision agriculture: Principles and applications. CRC press.

Hatfield, J. L., & Prueger, J. H. (2015). Temperature extremes: Effect on plant growth and development. *Weather and Climate Extremes, 10*, 4–10. https://doi.org/10.1016/j.wace.2015.08.001.

Hatfield, J. L., Boote, K. J., Kimball, B. A., Ziska, L. H., Izaurralde, R. C., Ort, D., Thomson, A. M., & Wolfe, D. (2011). Climate impacts on agriculture: Implications for crop production. *Agronomy Journal, 103*(2), 351–370. https://doi.org/10.2134/agronj2010.0303.

Hatfield, J. L., Wright-Morton, L., & Hall, B. (2017). Vulnerability of grain crops and croplands in the Midwest to climatic variability and adaptation strategies. *Climatic Change, 146*(1–2), 263–275. https://doi.org/10.1007/s10584-017-1997-x.

Hatfield, J. L. (2011). Soil management for increasing water use efficiency in field crops under changing climates. *Soil Management: Building a Stable Base for Agriculture* 161–173. https://doi.org/10.2136/2011.soilmanagement.c10.

Hernández-Clemente, R., Hornero, A., Mottus, M., Peñuelas, J., González-Dugo, V., Jiménez, J. C., & Zarco-Tejada, P. J. (2019). Early diagnosis of vegetation health from high-resolution hyperspectral and thermal imagery: Lessons learned from empirical relationships and radiative transfer modelling. *Current Forestry Reports, 5*(3), 169–183. https://doi.org/10.1007/s40725-019-00096-1.

Högy, P., & Fangmeier, A. (2008). Effects of elevated atmospheric CO2 on grain quality of wheat. *Journal of Cereal Science, 48*(3), 580–591. https://doi.org/10.1016/j.jcs.2008.01.006.

Hussain, M., Farooq, S., Hasan, W., Ul-Allah, S., Tanveer, M., Farooq, M., & Nawaz, A. (2018). Drought stress in sunflower: Physiological effects and its management through breeding and agronomic alternatives. *Agricultural Water Management, 201*, 152–166. https://doi.org/10.1016/j.agwat.2018.01.028.

IFOAM. (2009). Organic agriculture–A guide to climate change and food security. International Federation of Organic Agriculture Movements EU.

India: The Impact of Climate Change to 2030 A Commissioned Research Report-Sambal member upload. (2017). India: The impact of climate change to 2030 a commissioned research report- sambal member upload. Retrieved November 21, 2022, from http://sambal.bihar.gov.in/document/view/327/India--The-Impact-of-Climate-Change-to-2030-A-Commissioned-Research-Report.

Innes, P., Tan, D., Van Ogtrop, F., & Amthor, J. (2015). Effects of high-temperature episodes on wheat yields in New South Wales, Australia. *Agricultural and Forest Meteorology, 208*, 95–107. https://doi.org/10.1016/j.agrformet.2015.03.018.

Jablonski, L. M., Wang, X., & Curtis, P. S. (2002). Plant reproduction under elevated CO2 conditions: A meta-analysis of reports on 79 crop and wild species. *New Phytologist, 156*(1), 9–26. https://doi.org/10.1046/j.1469-8137.2002.00494.x.

Jat, R. S., Singh, V. V., Sharma, P., & Rai, P. K. (2019). Oilseed brassica in India: Demand, supply, policy perspective and future potential. *OCL, 26*, 8. https://doi.org/10.1051/ocl/2019005.

Jumrani, K., & Bhatia, V. S. (2014). Impact of elevated temperatures on growth and yield of chickpea (Cicer arietinum L.). *Field Crops Research, 164*, 90–97. https://doi.org/10.1016/j.fcr.2014.06.003.

Kanojia, A., & Dijkwel, P. P. (2018). Abiotic stress responses are governed by reactive oxygen species and age. *Annual Plant Reviews Online* 295–326.https://doi.org/10.1002/9781119312994.apr0611.

Kersebaum, K., & Nendel, C. (2014). Site-specific impacts of climate change on wheat production across regions of Germany using different CO2 response functions. *European Journal of Agronomy, 52*, 22–32. https://doi.org/10.1016/j.eja.2013.04.005.

Khan, A. A., Ijaz, M., Muhammad, J., Goheer, A. R., Akbar, G., & Adnan, M. (2016). Climate change implications for wheat crop in Dera Ismail Khan district of Khyber Pakhtunkhwa. *Pakistan Journal of Meteorology, 13*(25).

Khanal, U., Wilson, C., Hoang, V. N., & Lee, B. (2018). Farmers' adaptation to climate change, its determinants and impacts on rice yield in Nepal. *Ecological Economics, 144*, 139–147. https://doi.org/10.1016/j.ecolecon.2017.08.006.

Koundinya, A. V. V., Kumar, P. P., Ashadevi, R. K., Hegde, V., & Kumar, P. A. (2018). Adaptation and mitigation of climate change in vegetable cultivation: A review. *Journal of Water and Climate Change, 9*(1), 17–36. https://doi.org/10.2166/wcc.2017.045.

Latif, M. A., Islam, M. R., Ali, M. Y., & Saleque, M. A. (2005). Validation of the system of rice intensification (SRI) in Bangladesh. *Field Crops Research, 93*(2–3), 281–292. https://doi.org/10.1016/j.fcr.2004.10.005.

Lobell, D. B., & Burke, M. B. (2008). Why are agricultural impacts of climate change so uncertain? The importance of temperature relative to precipitation. *Environmental Research Letters, 3*(3), 034007. https://doi.org/10.1088/1748-9326/3/3/034007.

Lobell, D. B., & Field, C. B. (2007). Global scale climate–crop yield relationships and the impacts of recent warming. *Environmental Research Letters, 2*(1), 014002. https://doi.org/10.1088/1748-9326/2/1/014002.

Lobell, D. B., Schlenker, W., & Costa-Roberts, J. (2011). Climate trends and global crop production since 1980. *Science, 333*(6042), 616–620. https://doi.org/10.1126/science.1204531.

Mackay, A. (2008). Climate change 2007: Impacts, adaptation and vulnerability. contribution of working group II to the fourth assessment report of the intergovernmental panel on climate change. *Journal of Environmental Quality, 37*(6), 2407–2407. https://doi.org/10.2134/jeq2008.0015br.

Malathi, B., Appaji, C., Reddy, G. R., Dattatri, K., & Sudhakar, N. (2016). Growth pattern of millets in India. *Indian Journal of Agricultural Research, 50*(4), 382–386.

Michaelraj, P. S. J., & Shanmugam, A. (2013). A study on millets-based cultivation and consumption in India. *International Journal of Marketing, Financial Services & Management Research, 2*(4), 49–58.

Muthayya, S., Hall, J., Bagriansky, J., Sugimoto, J., Gundry, D., Matthias, D., & Maberly, G. (2012). Rice fortification: An emerging opportunity to contribute to the elimination of vitamin and mineral deficiency worldwide. *Food and Nutrition Bulletin, 33*(4), 296–307. https://doi.org/10.1111/nyas.12540.

Nelson, G. C., Rosegrant, M. W., Koo, J., Robertson, R., Sulser, T., Zhu, T., ... & Lee, D. (2009). Climate change: Impact on agriculture and costs of adaptation (Vol. 21). International Food Policy Research Institute.

Noya, I., González-García, S., Bacenetti, J., Fiala, M., & Moreira, M. T. (2018). Environmental impacts of the cultivation-phase associated with agricultural crops for feed production. *Journal of Cleaner Production, 172*, 3721–3733. https://doi.org/10.1016/j.jclepro.2017.07.132.

Nwalieji, H., & Uzuegbunam, C. (2013). Effect of climate change on rice production in Anambra State, Nigeria. *Journal of Agricultural Extension, 16*(2). https://doi.org/10.4314/jae.v16i2.7.

Officer, P. (2016). Food and agriculture organization of the United Nations. FAO, Italy.

Ortiz, R., Sayre, K. D., Govaerts, B., Gupta, R., Subbarao, G., Ban, T., Hodson, D., Dixon, J. M., Iván Ortiz-Monasterio, J., & Reynolds, M. (2008). Climate change: Can wheat beat the heat?

Agriculture, Ecosystems & Environment, 126(1–2), 46–58. https://doi.org/10.1016/j.agee.2008.01.019.
Oseni, T. O., & Masarirambi, M. T. (2011). Effect of climate change on maize (Zea mays) production and food security in Swaziland. *Change, 2*(3), 386–391.
Parry, M. L., Canziani, O., Palutikof, J., Van der Linden, P., & Hanson, C. (Eds). (2007). Climate change 2007-impacts, adaptation and vulnerability: Working group II contribution to the fourth assessment report of the IPCC (Vol. 4). Cambridge University Press.
Passot, S., Gnacko, F., Moukouanga, D., Lucas, M., Guyomarc'h, S., Ortega, B. M., Laplaze, L. (2016). Characterization of pearl millet root architecture and anatomy reveals three types of lateral roots. *Frontiers in Plant Science 7*.https://doi.org/10.3389/fpls.2016.00829.
Pedde, K. C., Gore, A. K., & Chavan, A. S. (2013). Integrated weed management in chickpea. *Indian Journal of Weed Science, 45*(4), 299.
Pereira, A. (2016). Plant abiotic stress challenges from the changing environment. *Frontiers in Plant Science 7*.https://doi.org/10.3389/fpls.2016.01123.
Pidgeon, J., Werker, A., Jaggard, K., Richter, G., Lister, D., & Jones, P. (2001). Climatic impact on the productivity of sugar beet in Europe, 1961–1995. *Agricultural and Forest Meteorology, 109*(1), 27–37. https://doi.org/10.1016/s0168-1923(01)00254-4.
Raza, A., Razzaq, A., Mehmood, S. S., Zou, X., Zhang, X., Lv, Y., & Xu, J. (2019). Impact of climate change on crops adaptation and strategies to tackle its outcome: A review. *Plants, 8*(2), 34. https://doi.org/10.3390/plants8020034.
Rezaei, E. E., Siebert, S., & Ewert, F. (2015). Intensity of heat stress in winter wheat—phenology compensates for the adverse effect of global warming. *Environmental Research Letters, 10*(2), 024012. https://doi.org/10.1088/1748-9326/10/2/024012.
Ribaut, J. M., Betran, J., Monneveux, P., & Setter, T. (2009). Drought tolerance in maize. In *Handbook of maize: its biology* (pp. 311–344). New York, NY: Springer.
Rogelj, J., Den Elzen, M., Höhne, N., Fransen, T., Fekete, H., Winkler, H., & Meinshausen, M. (2016). Paris agreement climate proposals need a boost to keep warming well below 2 C. *Nature, 534*(7609), 631–639. https://doi.org/10.1038/nature18307.
Rosegrant, M. W., Paisner, M. S., Meijer, S., & Witcover, J. (2001), 2020 Global food outlook: Trends, alternatives, and choices (Vol. 11). International Food Policy Research Institute.
Rosegrant, M. W., Cai, X., & Cline, S. A. (2002). World water and food to 2025: dealing with scarcity. International Food Policy Research Institute.
Rosenzweig, C., Iglesius, A., Yang, X. B., Epstein, P. R, & Chivian, E. (2001). Climate change and extreme weather events-Implications for food production, plant diseases, and pests. https://digitalcommons.unl.edu/nasapub/24.
Salehi-Lisar, S. Y., & Bakhshayeshan-Agdam, H. (2016). Drought stress in plants: Causes, consequences, and tolerance. *Drought Stress Tolerance in Plants, 1*, 1–16. https://doi.org/10.1007/978-3-319-28899-4_1.
Sartaj, A. W., Chand, S., Najar, G. R., & Teli, M. A. (2013). Organic farming: As a climate change adaptation and mitigation strategy. *Current Agriculture Research Journal, 1*(1), 45.
Sharma, H. C., Pathania, M., War, A. R., Pavani, T., & Vashisth, S. (2015) Climate change effects on pest spectrum and incidence in grain legumes. In: *Pulses: Challenges & opportunities: Under changing climatic scenario.* Kanpur: Indian Society of Pulses Research and Development (ISPRD) (pp. 124–137). http://oar.icrisat.org/id/eprint/9406.
Sheffield, J., Wood, E. F., & Roderick, M. L. (2012). Little change in global drought over the past 60 years. *Nature, 491*(7424), 435–438. https://doi.org/10.1038/nature11575.
Shivers, S. W., Roberts, D. A., & McFadden, J. P. (2019). Using paired thermal and hyperspectral aerial imagery to quantify land surface temperature variability and assess crop stress within California orchards. *Remote Sensing of Environment, 222*, 215–231. https://doi.org/10.1016/j.rse.2018.12.030.
Shukla, P. R., Skeg, J., Buendia, E. C., Masson-Delmotte, V., Pörtner, H. O., Roberts, D. C., & Malley, J. (2019). Climate change and land: An IPCC special report on climate change, desertification,

land degradation, sustainable land management, food security, and greenhouse gas fluxes in terrestrial ecosystems.
Shukla, P. R. (2003). Climate change and India: vulnerability assessment and adaptation. Universities Press. ISBN 8173714711, 9788173714719.
Singh, P., Boote, K., Kadiyala, M., Nedumaran, S., Gupta, S., Srinivas, K., & Bantilan, M. (2017). An assessment of yield gains under climate change due to genetic modification of pearl millet. *Science of the Total Environment, 601–602*, 1226–1237. https://doi.org/10.1016/j.scitotenv.2017.06.002.
Singh, P., Basu, S., & Kumar, G. (2018). Polyamines metabolism: A way ahead for abiotic stress tolerance in crop plants. In *Biochemical, physiological and molecular avenues for combating abiotic stress tolerance in plants* (pp. 39–55). Academic Press. https://doi.org/10.1016/B978-0-12-813066-7.00003-6.
Smit, B., & Wandel, J. (2006). Adaptation, adaptive capacity and vulnerability. *Global Environmental Change, 16*(3), 282–292. https://doi.org/10.1016/j.gloenvcha.2006.03.008.
Sriphirom, P., Chidthaisong, A., & Towprayoon, S. (2019). Effect of alternate wetting and drying water management on rice cultivation with low emissions and low water used during wet and dry season. *Journal of Cleaner Production, 223*, 980–988. https://doi.org/10.1016/j.jclepro.2019.03.212.
Tack, J., Barkley, A., & Nalley, L. L. (2015). Effect of warming temperatures on US wheat yields. *Proceedings of the National Academy of Sciences, 112*(22), 6931–6936. https://doi.org/10.1073/pnas.1415181112.
Tripathi, A., Tripathi, D. K., Chauhan, D., Kumar, N., & Singh, G. (2016). Paradigms of climate change impacts on some major food sources of the world: A review on current knowledge and future prospects. *Agriculture, Ecosystems & Environment, 216*, 356–373. https://doi.org/10.1016/j.agee.2015.09.034.
United in Science: We are Heading in the Wrong Direction. (2022). *United Nations Climate Change*. Retrieved November 15, 2022, from https://unfccc.int/news/united-in-science-we-are-heading-in-the-wrong-direction.
Vaughan, M. M., Block, A., Christensen, S. A., Allen, L. H., & Schmelz, E. A. (2018). The effects of climate change associated abiotic stresses on maize phytochemical defenses. *Phytochemistry Reviews, 17*(1), 37–49. https://doi.org/10.1007/s11101-017-9508-2.
Walsh, B. S., Parratt, S. R., Hoffmann, A. A., Atkinson, D., Snook, R. R., Bretman, A., & Price, T. A. (2019). The impact of climate change on fertility. *Trends in Ecology & Evolution, 34*(3), 249–259. https://doi.org/10.1016/j.tree.2018.12.002.
Wang, Q., Li, Y., & Alva, A. (2010). Cropping systems to improve carbon sequestration for mitigation of climate change. *Journal of Environmental Protection, 1*(03), 207–215. https://doi.org/10.4236/jep.2010.13025.
Wang, M., Li, Y., Ye, W., Bornman, J., & Yan, X. (2011). Effects of climate change on maize production, and potential adaptation measures: A case study in Jilin Province, China. *Climate Research, 46*(3), 223–242. https://doi.org/10.3354/cr00986.
Yadav, M. K., Singh, R. S., Singh, K. K., Mall, R. K., Patel, C., Yadav, S. K., Singh, M. K. (2016). Assessment of climate change impact on pulse, oilseed and vegetable crops at Varanasi, India. *Journal of Agrometeorology, 18*(1), 13–21. https://doi.org/10.54386/jam.v18i1.881.
Yan, X., Akiyama, H., Yagi, K., & Akimoto, H. (2009). Global estimations of the inventory and mitigation potential of methane emissions from rice cultivation conducted using the 2006 Intergovernmental Panel on Climate Change Guidelines. *Global Biogeochemical Cycles, 23*(2) https://doi.org/10.1029/2008gb003299.
Zandalinas, S. I., Mittler, R., Balfagón, D., Arbona, V., & Gómez-Cadenas, A. (2018). Plant adaptations to the combination of drought and high temperatures. *Physiologia Plantarum, 162*(1), 2–12. https://doi.org/10.1111/ppl.12540.
Zhao, H., Dai, T., Jing, Q., Jiang, D., & Cao, W. (2007). Leaf senescence and grain filling affected by post-anthesis high temperatures in two different wheat cultivars. *Plant Growth Regulation, 51*(2), 149–158. https://doi.org/10.1007/s10725-006-9157-8.

Chapter 15
Understanding the Effects of Changing Climate on Weeds and Their Management

Sushmita Munda, Bitish Kumar Nayak, Saikat Ranjan Das, Souvik Dey, Abhijit Pradhan, Chinmaya Kumar Swain, and Baishnab Charan Muduli

Abstract Global temperature is shifting, and in addition to monitoring variations in carbon dioxide (CO_2) and temperature levels, which are thought to contribute most to global warming, more attention is being paid to the implications of this change for agricultural production systems, including weeds. Future weed management and agricultural production are seriously threatened by rising CO_2 levels and resulting changes in global precipitation and temperature. Most plant pests are largely influenced by climatic conditions in terms of their distribution, population structure, life cycle length, infestation stress, and general occurrence. The distribution of Competition among weed species in a weed population and among crops is influenced by changes in atmospheric CO_2, precipitation, temperature, and other growth factors. In addition, interactions among these environmental elements can have unanticipated consequences for weed development. According to recent studies, these weeds as a group may have an effective reaction to the recent increase in atmospheric carbon dioxide. Many of these weeds reproduce by vegetative propagation. One of the most important features of the changing climate that can affect existing plants and weeds, allowing the displacement of certain native weeds by others and their spread into previously uninhabited regions, is the change and increase in temperature. Even some weeds develop allelic compounds that enable them to coexist and compete with crops in arid environments. The main aspects of this study are the effects of climate change variables on weed development and the likely variations in weed species. Comprehensive research of the climate change effects components and their interactions on all common weed dynamics are required in order to understand the significance for management of weeds under future climatic forecasts.

S. Munda (✉) · B. K. Nayak · S. R. Das · A. Pradhan · C. K. Swain · B. C. Muduli
ICAR National Rice Research Institute, Cuttack, Odisha 753 006, India
e-mail: sustot@gmail.com

S. Dey
Bidhan Chandra Krishi Viswavidyalaya, Mohanpur, Nadia, West Bengal 741 252, India

H. Pathak et al. (eds.), *Climate Change Impacts on Soil-Plant-Atmosphere Continuum*, Advances in Global Change Research 78,
https://doi.org/10.1007/978-981-99-7935-6_15

Keywords Climate change · Weed · Temperature · Carbon dioxide · Precipitation · Weed management

15.1 Introduction

Climate plays an important role in Earth's ecology, and even small climate changes can have a significant impact on the natural world. Although historical climate changes occurred over hundreds of years, modern changes have attracted more attention because they caused massive changes in a very short period of time (Treut et al., 2007a, b). Numerous aspects of life on our planet have already begun to change as a result of regional and global climate adaptations, but the impact on agriculture and food availability could pose the greatest threat to the continuation of life (Chatterjee et al., 2020). Changes in climate could have significant impacts on crop yield quality and quantity, as well as on agronomic practises such as fertilisation, irrigation, and weed and insect control (Wing et al., 2021). To predict and adapt cropping practises that optimise crop production under future climate scenarios, it is critical to evaluate the effects of climate change worldwide variables on agriculture and cropping practises, particularly the consequences of increasing CO_2 concentrations and temperature rise (Raza et al., 2019). Achieving sustainable crop production under uncertain conditions viz., droughts, floods etc. requires a comprehensive strategy that not only improves crop yields but also effectively manages crop pests such as weeds (Nayak et al., 2020).

Weeds are the oldest and most pervasive issue in agricultural production, which are considered by people as undesirable plants that affect natural resources and agriculture (Chatterjee et al., 2016; Rao et al., 2020). Weeds are inherently complex and have detrimental effects on many aspects of human activities, including agriculture, grasslands, forestry, and public health. Weeds pose a constant threat to agricultural production (>90% damage if not controlled), unlike insects (up to 40% yield loss) and diseases (>40% damage), whose outbreaks can be unpredictable and irregular (Baite et al., 2020; Bhagat et al., 2017). The amount of yield loss done by weeds varies by species, crop, and management techniques (Kumar et al., 2016). Crop losses caused by competition are greatest when weeds occur in high densities. They also degrade product quality and infect agricultural seed (Satpathy et al., 2021). Weeds also have an impact on harvest operations. They produce allelochemicals and serve as hosts for pests, diseases, and insects (Wato, 2020). Therefore, weed management is a critical and essential element of good crop production. Cultural, mechanical, biological, and chemical methods can be used to control weeds. Chemical weed control with herbicides has gained popularity because it is inexpensive, easy to use, more effective, and saves time (Kraehmer et al., 2014). Environmental factors, including CO_2 enrichment, have been found to alter the dynamics of plant-weed competition (Ramesh et al., 2017). Competitive plant interactions between C3 and C4 may change due to differences in how they respond to elevated temperatures and CO_2.

India is the 4th largest economy and the 5th largest producer of greenhouse gases (GHG), contributing 5% of global emissions (Friedrich et al., 2020). Climate is the primary factor influencing plant dispersal. The floral composition of many ecosystems will change as a result of climate change, as variations in temperature and humidity will affect seed dormancy, fruiting, and flowering. In general, competitive relationships between plants and weeds will change as a result of climate change, either direct or indirect effects, which will differentially affect the development or health of weeds and plants (Peters et al., 2014). The result may be beneficial to crops or weeds. It is predicted that new weed species will spread and that the composition of the weed flora will change as a result of climate change. Compared to C4 plants, CO_2 enrichment is expected to be more favourable to the growth of C3 plants. Weeds will become more competitive with crops due to their increased tolerance to a changing environment. For example, *Amaranthus viridis* (C4) and *Parthenium hysterophorus* (C3) showed improved growth and biomass production under increasing CO_2, which in turn improved flower production (Anwar et al., 2021). Many crops, including rice and maize, face the dual problem of lower productivity in a challenging environment and increasing weed competition. Because most crops belong to the C3 group and most weeds belong to the C4 category, the differential effects of CO_2 enrichment on C3 and C4 plants may have significant implications for crop-weed interactions Climate change will alter weed populations, and the risk of invasion may increase. Weeds are very active and adapt quickly to new circumstances, so management strategies must take this into account. Chemical, mechanical, and biological methods of weed control could be affected by climate change (Satpathy et al., 2022).

Few research studies have examined the interactions among many sources of change, whereas most studies have examined the influence of a single component (increased CO_2). Research is required to evaluate the simultaneous interacting impacts of various climate change components in order to create adaptive integrated weed management practises based on a knowledge of the biology and ecology of weeds. This makes it feasible to forecast how weed issues may alter as a result of climate change. Therefore, this study's goal is to give a broad overview of the possible impacts of climate change on the emergence and spread of different weed species. The study examines the effects of climate forcing on the distribution and abundance of weed species in weed and crop ecosystems, specifically increases in CO_2 levels, increases in air surface temperatures, and variations in precipitation events. Climate change not only affects weed species composition and distribution, but can also influence weed management. In addition, optimising herbicide treatments for future weed management is critical to understanding how climate change will affect weed development.

15.2 Threats Due to Climate Change on Weeds

Significant advancement has been made in weed research regarding the effect of climate change and some threats have been identified:

i. Changes to weed dynamics and the spread of invasive weed species in croplands
ii. Response of weeds to various climatic factors
iii. Climatic change factors' effects on crop-weed competition.

i. **Changes to weed dynamics and the spread of invasive weed species in croplands**

Plants will have the opportunity to invade other favourable habitats as a result of climate change. The interplay of various environmental conditions will be critical in transforming currently harmless species (both native and non-native) into invasive species (Finch et al., 2021). The vast weed population and the fact that weed seeds often spread more rapidly through the field than native plant seeds make native plants particularly vulnerable to displacement by weeds (Travlos et al., 2020). Some weed species may benefit from climate change, as studies of potential impacts on invasive species have shown (Marambe & Wijesundara, 2021). Invasive species can often replace native species due to climate change (Rai & Singh, 2020). It is also expected that dormant weeds could become more active weeds. Plants known as "dormant weeds" may spread unexpectedly rapidly following certain natural disasters such as floods, cyclones, droughts, flooding, or poor drainage that are probably going to happen more often in the future (Chauhan et al., 2017). Following cyclones, floods, or wildfires that create optimal conditions for their growth, many of Australia's worst weeds have thrived. The spread of certain hydrophytic plants, such as *Marsilea quadrifolia* in India, is significantly affected by flooding. Non-native weed species that reproduce well and are competitive can easily establish and spread to other regions. For example, during the 2004 tsunami, Lantana (*Lantana camara*), Mesquite (*Prosopis juliflora*), Siam weed (*Chromolaena odorata*), and prickly pear (*Opuntia stricta*) were found to have increased significantly in Sri Lanka's southern coastal regions (Iqbal et al., 2014). Because invasive weeds can successfully reproduce in large numbers and at great distances from their parents, they are capable of rapidly colonising large areas (Clements & Jones, 2021). Recent research indicates that many invasive weeds grow vegetatively (via stolons, roots, etc.) and that these weeds are very sensitive to changes in atmospheric CO_2 (Amare, 2016). Increased air temperatures may favour the invasion of certain weeds. The biomass and leaf area of itch grass (*Rottboellia cochinchinensis*) increased 88 and 68%, respectively, with a 3 ºC increase in average temperature (Meksawat & Pornprom, 2010). This invasive plant significantly reduced sugarcane yields and was also a competitive pest in maize, cotton, sorghum, soybean, and rice crops (Shahzad et al., 2021). Populations of this weed in the Midwest and its spread have been attributed to an increase in low winter temperatures as this specific plant is restricted by low winter temperatures of −15 °C. (Ziska et al., 2019).

The absence of natural enemies in the imported range is a significant factor in the proliferation of invasive weeds, as is the case with *Eichhornia crassipes* in many parts of the world, the invasion of *Acacia longifolia* in South Africa, and *Chrysanthemoides monilifera* in Australia. (Grice, 2006). Reportedly, the invasion of *Ficus microcarpa* in the United States and New Zealand, originally planted as an ornamental, was triggered by the start of a pollinator called *Parapristina verticillata* (a fig wasp) (Wang, 2014). In Australia, the introduction of cattle instead of sheep and regular, above-average rainfall promoted the spread of *Acacia nilotica* species (Kriticos et al., 2003).

ii. **Response of weeds to various climatic factors**

Climate change inevitably impacts a variety of mutualistic and antagonistic relationships among terrestrial species, as well as extinctions of plant and animal species and changes in their distributions (Tylianakis et al., 2008). The widely held belief that CO_2 enrichment favours C3 plant species over C4 plant species by enhancing net photosynthesis is modified by other associated climate conditions affecting this response (Hikosaka et al., 2005). As opposed to just having an impact on the photosynthetic process, the interaction effect of CO_2 enrichment with moisture and temperature would alter competition between plants and weeds simultaneously or sequentially in complicated ways. Previous climate change studies have focused more on controlling the response of plants to CO_2 concentrations than on the associated increase in heat or drought (Fuhrer, 2003). Undoubtedly, projected changes in temperature and moisture due to climate change will have an impact on the geographical and temporal development of weed seeds and seedlings (IPCC, 2007), which requires more thorough investigation. For instance, increasing humidity and rising temperatures are predicted to cause dormancy, one of the primary barriers to weed germination, to be disrupted earlier or even earlier (Jaganathan & Liu, 2015). It is well established that soil temperature, rather than a species' CO_2 response, primarily controls the dormancy observed in some temperate species where water is not periodically restricted (Batlla & Benech-Arnold, 2004). The sensitivity of C3 plants to elevated CO_2 is limited by the cumulative effects of soil nutrient and water stress, but not that of C4 species (Ramesh et al., 2017). A critical element affecting how species and communities respond to increasing CO_2 is water supply (Belote et al., 2003). The interactions between higher CO_2 levels and lower temperatures in C4 weeds or plants are poorly understood. There is limited research that has examined the development and reproduction of C3 weeds in response to the interaction of temperature and CO_2. While some weeds (*Abutilon theophrasti* and *Triticum repen* L.) did not respond in the same way to high temperatures and CO_2, weeds such as *Senna obtusifolia* and *Anoda cristata* did (Patterson et al., 1988). In addition, it is important to note how high CO_2 at unfavourable temperatures promotes C3 crops and weeds differently. According to Alberto et al. (1996), higher CO_2 alone can improve the ability of a C3 plant species to compete with a C4 weed species (*Echinochloa glabrescens*), but a combined increase in temperature and CO_2 remained beneficial to the plant. Surprisingly, the relationship between CO_2 and water availability has only been explored in crop plants (Bunce, 2004), with little attention paid to differences between crop plants and weeds with the same photosynthetic pathway. *Digitaria ciliaris* (Retz.), *Eleusine indica* (L.), and *Echinochloa crusgalli* L., three C4 weeds, all grew more favourably as CO_2 levels increased, although the C3 plant was expected to provide greater growth stimulation (Ramesh et al., 2017). There are not many studies examining the interactions between drought, weed competition, and increasing CO_2 levels. It is possible to hypothesise that C4 weeds might still outcompete C3 crops in an environment of elevated CO_2 and drought if their water requirements are reduced by CO_2 relative to C3 crops (Ramesh et al., 2017). Surprisingly, little research has examined how other environmental parameters such as water availability and/or temperature may affect CO_2-induced changes in morphological and physiological development (Springer & Ward, 2007).

Both the prediction made ten years later by Wolfe et al. (2008) that weeds might increase more than crops and the assessment by Rosenzweig and Hillel (1998) that rising temperatures and CO_2 concentrations might reduce crop competition with weeds have come to pass. Despite lower weed and barley growth in southern Finland, in the barley crop, *Amaranthus retroflexus* produced more seeds (Hyvonen, 2011).

The increase in CO_2 in the Hoagland solution to 675 ppm stimulated the development of soybean and two related weeds, *Crotalaria spectabilis* and *Senna obtusifolia* (L.) (Ramesh et al., 2017). Using a system model with a C3 crop (rice) and a C4 weed (*E. crusgalli*), Zhu et al. (2008) examined the effects of nutrients and CO_2 on crop-weed competition and found that under optimal nitrogen fertilization, the biomass of rice increased proportionally compared to *E. crusgalli.* In contrast, the ability of rice to compete with *E. crusgalli* was lower when elevated CO_2 was present along with inadequate nitrogen levels. Therefore, an increase in atmospheric carbon dioxide will result in significant rice yield losses caused by competition with C4 weeds when soil nitrogen levels are low. Systematic studies are needed to evaluate the interactions between key environmental factors on different weed species and populations in different habitats.

iii. **Climatic change factors' effects on crop-weed competition**

Because of their broad genetic pool and greater physiological flexibility, weeds are thought to be more resilient and adaptable to changes in CO_2 concentration and rising temperatures than crops. Depending on how C3-and C4-photosynthetic weeds respond to elevated CO_2 concentrations and temperatures, the mechanisms of conflict between weeds and crops may change. In addition to the beneficial effects on weed production, climate change may reduce the efficacy of many herbicides, making weed control a serious obstacle to sustainable agricultural production. Crop losses caused by weed infestations peak when weeds emerge in significant densities at the same time as or before crop emergence. Ziska (2000) discovered that soybean seed production decreased by 39% while the biomass of a C3 weed called common lambsquarters (*C. album* L.) increased by 65% at high CO_2 concentrations. In contrast, redroot weed (*A. retrofexus* L.), a C4 weed, showed no change in biomass at high CO_2 concentrations, and soybean yield losses decreased from 45 to 30% due to antagonism of this weed. Different nutrient balances between crops and weeds may be affected by CO_2 increase. Due to the CO_2 increase in the atmosphere, the competition between rice and switchgrass shifted in favour of rice (Zeng et al., 2011). Rice exhibited higher absolute C, N, P, and K nutrient uptake and tissue concentration under barnyard grass at high CO_2, which increased the net assimilation rate, leaf area index, tillering, and biomass of the rice plant. Weedy rice (red rice) has better physiological adaptability and genetic diversity than cultivated rice, which may affect weed-plant competition when atmospheric CO_2 increases (Ziska & McClung, 2008). According to Asch et al. (2005), drought-resistant rice varieties are critical to reducing weed competition for water under rainfed conditions and preventing yield losses caused by water stress. The shift from rice transplanting to no-till rice in South Asia, weed competition has already risen and weed ecology has evolved (Matloob et al., 2015). Due to increasing rainfall, the competition of C. *arvense* with wheat growth and yield increased (Donald & Khan, 1992). According to Patterson (1995), weed competition has little impact on crops during periods of water deficit because water deficit has already reduced crop yields. Rice was unable to withstand water stress. However, *Leptochloa chinensis* (L) and *Amaranthus spinosus* produced a considerable number of leaves and shoots even at the lowest soil water content, so they were able to withstand water stress. One of the key barriers to weed emergence, dormancy, is anticipated to be disrupted early as a result of increasing moisture availability and higher temperatures (Ooi et al., 2014). According to researchers, the growth cycles of certain species in temperate regions where water supply is not seasonally restricted are predominantly affected by soil temperature, regardless of their CO_2 response (Batlla & Benech-Arnold, 2004). The combined effects of soil water stress and nutrient stress reduced the sensitivity of C3 plants to increasing CO_2 concentrations, but not C4 species (Ward et al., 1999). Increases in temperature and CO2 may make it more difficult for crop plants to suppress weeds (Rosenzweig & Hillel, 1998).

15.3 Response of Weed to Interactions Between Climatic Factors

Increasing CO_2 concentrations, rising temperatures, changing rainfall and solar radiation, and generating extreme weather events including severe storms, summer droughts, and extreme cold spells all have an effect on agriculture (Diaz et al., 1999). Seasonal changes in local climate are therefore likely to occur more frequently and with greater amplitude (Walther et al., 2002). Range expansion, including the introduction or migration to new areas, changes to species life cycles, and adjustments to the population dynamics of weed vegetation are all possible effects of climate change. In controlled and natural environments, weed community structure and composition will change as a result of weed migration. In response to climate change, changes in weed ecology, biology, and disturbance potential will lead to complex interactions between crops and weeds that will require different adaptation strategies. The response of weeds to different climatic variables is discussed below:

15.3.1 Weed Response to Increasing CO_2 Levels

The greenhouse gas that has the greatest global impact on climate is CO_2 because of its ability to absorb heat and its additional atmospheric lifespan (Andreae & Merlet,

2001). At current emission rates, by the end of the twenty-first century, CO_2 concentrations are predicted to reach about 1000 ppm, which will result in an increase in the Earth's annual surface temperature of 2 to 4 °C (IPCC, 2007). It is well known that C3 plants will benefit from net photosynthesis as a result of the anticipated increase in atmospheric CO_2 concentration since it will reduce CO_2 loss through photorespiration and raise the gradient of CO_2 concentration from the environment to the interior of leaves (Ziska, 2000). As a result of an internal mechanism for concentrating CO_2 at the site of CO_2 carboxylation, plants having a C4 photosynthetic pathway are less sensitive to high CO_2 concentrations. According to this theory, increasing atmospheric CO_2 will have a major impact on competition between crops and weeds and on reducing crop yields. Numerous studies have examined the relationships between weeds and crops by comparing the development and physiology of C3 crops and C4 weeds. They concluded that higher CO_2 concentrations promote the vegetative growth of C3 plant species relative to C4 plant species (Patterson, 1995). However, not all crops and weeds rely on C4 metabolic pathways (Ziska et al., 2010). Although C3 crops such as rice generally struggle with C4 grasses and broadleaf weeds, the above concept is relevant to these crops as well. Many commercially important C4 crops, including maize, sugarcane, and sorghum, compete with important C3 weeds such as *Chenopodium album* L. (Ziska, 2000). Because elevated CO_2 is required to reap the potential benefits of CO_2 fertilisation, weed-induced yield losses are typically expected to increase for C4 crops under elevated CO_2, whereas this is not the case for C3 crops.

The increase in CO_2 significantly decreased relative seed yields of one rice cultivar (a C3 plant species), but had the opposite effect on a weedy rice biotype (Ziska et al., 2010). This is thought to be due to greater genetic diversity and physiological adaptation between wild and domesticated strains (Treharne, 1989). An increase in atmospheric carbon dioxide concentration may affect the biology of invasive weeds (Ziska et al., 2011). Increases in CO_2 concentration are thought to favour the establishment and growth of many invasive plant species. One invasive C3 perennial weed species, *Cirsium arvense* L., exhibited a 70 percent increase in growth at high CO_2 concentrations (Ziska et al., 2011).

15.3.2 Weed Response to Increased Temperature

Atmospheric temperature is considered a key indicator of the spread of weed species in a geographic region (Patterson et al., 1999). Their growth can influence the spread and competitive behaviour of weeds in both weed vegetation and agricultural stands. Climate change could favour C4 weeds over C3 weeds because photosynthesis of C3 weeds would be inhibited by increasing CO_2, decreasing CO_2 solubility and CO_2 affinity of RUBISCO enzyme (Tubiello et al., 2007). Due to the shift of weeds to higher latitudes, changes in weed distribution will affect major agricultural systems around the world, including rice-based cropping systems. Arable weeds will increase as a result of climate change (Patterson, 1995). Under the effects of climate change, plants such as datura (*Datura stramonium* L.), which require high temperatures for lush development (Cavero et al., 1999), may become more competitive adversaries. By accelerating growth and reproduction, warm temperatures have promoted the spread of *Hieracium aurantiacum* L. in Australia (Brinkley & Bomford, 2002).

15.3.3 Weed Reaction to Changing Rainfall and Drought Periods

The distribution of weeds and their impact on agricultural production could change due to altered precipitation patterns and greater drought resulting from a warmer environment. Rising temperatures increase evaporation, and global patterns of precipitation variability indicate that monsoon regions will be drier (Giannini et al., 2008), increasing the likelihood of droughts by 5–8% (Rodenburg et al., 2011). In such a case, the spread and prevalence of weeds in crop ecosystems will be a problem, and summer drought will have a particular impact on weed control in spring crops (Peters & Gerowitt, 2014). According to Rodenburg et al. (2011), C4 and parasitic weeds such as *S. hermonthica* thrive better during prolonged dry periods. Weeds such as *Rhamphicarpa fistulosa.* thrive in areas with an abundance of water. Alternating rainfall patterns favour hydromorphic weeds, but prolonged dry periods favour C4 weeds over C3 weeds. It is challenging to adequately prepare the soil for wet rice in dry or wet places because there is little water available for flooding, especially in the early part of the season when weed competition threatens rice the greatest. As a result, conventional weed control is ineffective in flooded rice, requiring the use of pesticides. Under these conditions, rice yield losses are probably to be greater.

According to climate change models, rainfall variability will increase worldwide, with a decrease in the average frequency of occurrence and an increase in the frequency of extreme rainfall events (IPCC, 2007). Studies of annual plant species have found that precipitation variability has implications for plant community dynamics. However, the effects of precipitation variability on commercially important agricultural systems are still poorly understood. The amount and distribution of precipitation affect the behaviour of different plant species. Changes in precipitation can be anticipated to alter the composition and relative abundance of weeds important to agriculture (Robinson & Gross, 2010). Understanding the potential consequences of changes in precipitation's temporal variability on plants in various habitats is crucial.

Drought response under agricultural conditions is then influenced by species and cultural factors. Unpredictable precipitation caused by climate change can also be expected to alter the species composition of weeds. A crop may be exposed to increased environmental stress due to sudden climatic changes, making it more susceptible to pests and diseases and less competitive for weeds. Under warm and dry conditions, the C4 photosynthetic pathway is most beneficial. Compared to C3 plants, C4 plants have a higher WUE. While C4 plants predominate in dry soils,

C3 plants are more common in flooded soils (Rudov et al., 2020). C4 weeds, which compete strongly with rice plants, are protected by flooding. On the other hand, C4 weeds pose a serious threat to upland rice and rainfed lowland rice in areas with little rainfall. *Echinochloa crusgalli, Digitaria ciliaris*, and *Eleusine indica*, three C4 grasses, exhibited significantly higher leaf area and total dry weight when exposed to CO_2 enrichment during drought, according to Patterson (1986). Both C3 and C4 plants can grow more when exposed to water stress when CO2 enrichment is present, but it seems reasonable that C3 plants are stimulated to grow more. However, *Phalaris minor* outperformed wheat when moisture stress and CO_2 enrichment were present (Naidu & Varshney, 2011). Germination of invasive weeds in pastures such as cheatgrass (*Bromus tectorum*) and yellowstar thistle (*Centaurea solstitialis*) is highly dependent on available soil moisture. Prolonged or cold weather that dramatically increases soil moisture promotes the production of more seeds in both weed species (Patterson, 1995).

15.3.4 Weed Response to Wind

Intergovernmental Panel on Climate Change forecasts slowing winds for the coming decades due to climate change. By 2100, average annual wind speeds could drop by up to 10 percent (Abell et al., 2021). Wind affects plant growth, reproduction, distribution, death and ultimately plant evolution (Ennos, 1997). Research suggest that changes resulting from the effect of wind may have a greater effect on ecology than the more acute effects of destructive storms. Weeds that depend on vegetative dispersal transport pollen or seeds more slowly than those that rely on efficient methods of seed dispersal, such as wind, which allows them to spread to new sites (Singh et al., 2011). Increased height or development of the weed plant may contribute to wind dispersal of weed seeds. Examples of invasive weed species that are windborne include *Carduus nutans* L., *Sonchus arvensis* L., *Cirsium arvense* L., and *Sonchus oleraceus* L (Keighery & Longman, 2004).

15.3.5 Weed Response to Light

Weeds compete with plants both below and above ground. Crop type and cultivation techniques, such as seeding direction, seeding density, and variety (Gaba et al., 2010), affect aboveground competition for light. Emergence of many weed species from dormancy and subsequent germination are controlled by light. Under field conditions, seed light exposure, which is primarily recognised by photoreceptors of the phytochrome family, provides essential knowledge for initiating germination in the proper environmental context (Batlla and Benech-Arnold, 2014). Weed seeds can determine their location in the soil profile, the existence of a leaf canopy that blocks light and other resources, and the presence of tillage by using the spectral composition and irradiance of the light environment. From an agronomic perspective, the regulation of weed germination by light is an aspect of the weed life cycle that can be critical to developing efficient management strategies. Besides the direct affect, comprehensive studies published by the Intergovernmental Panel on Climate Change suggest that, averaged over the solar cycle, the sun's brightness has changed between the pre-industrial period and the present (2019) which is quantified closely to 0.06 Watts per square meter. This means an increase of temperature of about 0.01 degrees Celsius i.e. around 1 percent of the warming the planet has experienced over the industrial era (0.95–1.2 degrees Celsius in 2011–2020 versus 1850–1900).

15.4 Weed Management Under Climate Change

Environmental factors are also thought to influence the effectiveness of weed control. Weeds may thicken their leaf cuticles, halt their vegetative growth, and produce more blooms more quickly in response to moisture stress (Gupta, 2013). It is harder to treat drought-stricken weeds with post-emergent herbicides than active plants because systemic herbicides that are anchored in the weed, for example, need an active growth stage of the plant to be effective. The soil must be moist and plants must be actively rooting for pre-emergent herbicides or herbicides absorbed by plant roots to reach their target locations. Drought conditions can reduce the effectiveness of pre-emergent herbicides (Davies et al., 2022). Overall, there are valid reasons to believe that the changing environment and/or increasing CO_2 levels will affect weed control. Adaptation measures are available, but it is uncertain how much their use will cost (e.g., new herbicides, higher chemical dosages, and new biological control agents).

15.4.1 Mechanical Weed Control

Mechanical removal of weeds is a common method of weed control, especially popular in developing countries. In agronomic systems, tillage (by animals or mechanical methods) is recognized as a universal method of weed control (Rana & Rana, 2016). Due to soil contamination caused by heavy, repeated rainfall, it is very difficult to implement tillage and other mechanical weed control measures. Under these conditions, tillage can help reduce soil moisture, which can slow crop growth and reduce weed competitiveness. Increased CO_2 storage in the soil can lead to increased root or rhizome growth, especially in perennial weeds (Al Souki et al., 2021). Therefore, mechanical tillage may result in increased asexual reproduction from belowground structures, increased plant growth in a higher CO_2 environment, and adverse effects on weed control.

15.4.2 Chemical Weed Control

Chemical management is undoubtedly significantly affected by direct or indirect effects of a changing environment. Variations in temperature, wind speed, soil moisture, and humidity can have an impact on how well treatments work. Drought, for example, can result in a thicker cuticle or more pubescence on leaves, resulting in less penetration of herbicides into the leaf. After pesticide application, these same variables can also hinder plant development and recovery. In general, pesticides work best when applied to plants that are not under environmental stress and that develop and metabolize quickly. Environmental factors including as temperature, precipitation, wind, and relative humidity have an impact on the efficacy of herbicides (Varanasi et al., 2016). Most herbicides require the ideal temperature to be effective. However, efficacy and selectivity can be affected by sustained high temperatures after application (of pre-emergence and post-emergence herbicides). Therefore, selective herbicides may become unselective at high temperatures. A growing number of studies showed that chemical efficacy decreases as CO_2 levels increased. However, some researchers believe that when CO_2 levels increase, certain weeds (C3) may grow and develop more vigorously and plants may develop defences against infection or detoxification mechanisms (high tissue volume). Recent research on *Cirsium arvense* grown in monocultures under field conditions and elevated CO_2 showed a greater ratio of roots to shoots and thus a dilution effect of glyphosate. The expert emphasized that CO_2 reduces pesticide efficacy. But more investigation is required to identify herbicide concentration, specificity, and application rates as potential adaptation mechanisms.

In addition, it is possible that rising CO_2 and heat reduce herbicide efficacy, so treatment frequency and herbicide concentration must be increased to achieve adequate control. Although the effects of rising atmospheric CO_2 concentrations on interactions between weeds, crops, and herbicides are not fully known, a recent study indicated that higher CO_2 levels diminish the efficacy of the widely used herbicide glyphosate. If these impacts are frequent and widespread, agriculture will be significantly impacted. Weeds may also thicken their leaf cuticles in response to moisture stress, which can slow their vegetative growth and cause them to flower more quickly (Amare, 2016). It is more challenging to treat drought-stricken weeds with post-emergent herbicides than it is to treat plants that are actively growing because systemic herbicides, such as those that translocate inside the weed, need the plant to be in an active growth stage in order to be effective. It is necessary for the soil to be moist and for plants to be actively rooting for pre-emergent herbicides or herbicides that are absorbed by plant roots to reach their target locations. Pre-emergent herbicides may perform less effectively in dry environments. In general, different scientists at different times have come up with their own theories about climate change's effects on chemical weed-control techniques. Which are:

1. When plants are grown in a high CO_2 environment (like Canada thistle), they have a larger root to shoot ratio, resulting in dilution of glyphosate.

2. Reduced stomatal opening or number, thickening of the cuticle, formation or increase in leaf hairiness, all of which result in less herbicide reaching the leaf.
3. Changes in transpiration resulting in less herbicide uptake in the soil.
4. Herbicide degradation and microbial activity are impacted by changes in the wetting and drying of soil profiles.
5. Herbicide efficacy and selectivity are affected by increasing temperatures.
6. Increased growth rate results in smaller application windows for herbicides that must be applied before weeds emerge and before production is lost due to weed competition.
7. The window of opportunity for spraying is reduced by low relative humidity, high temperatures, and unfavourable wind conditions.

15.4.3 Biological Control

To control a weed species, biological control introduces host-specific agents, usually insects or fungi. Over time, biological control agents have changed to only consume the target plant (Aguilera et al., 2011). A stable environment is best for most biological control systems. But in addition to the anticipated rise in average temperatures, CO_2 and a change in the pattern of precipitation are anticipated to cause an increase in climate unpredictability. It is anticipated that extreme weather occurrences including droughts, floods, and even seasonal frosts will increase in frequency. Despite the fact that many species have protection against extremes, it takes time for them to adapt or become resistant. Biological control agents' resistance to these conditions and the susceptibility of host plants to temperature, drought, and flood extremes determine whether a drought or flood is followed by a weed or pest epidemic (Gerard et al., 2010). Plant phenology, growth, and distribution change as a result of fluctuations in water availability, temperature, and CO_2 levels, which also affect herbivores and those that feed on them. Aspects of the temperature-dependent life cycle of plants, such as the timing and length of seed germination, bud burst, and blooming, might be anticipated to change with rising temperatures depending on photoperiod and water availability. The degree to which life cycle events are coordinated to increase survival will also affect the benefits that each species receives.

Changes in the fitness of herbivores and natural enemies can also result from changes in host plant quality as a consequence of elevated CO_2, drought, or flooding (War et al., 2016). The nutritional content of plant and plant produce generally declines as CO_2 and temperature levels rise because it causes reduction in absorption and assimilation of nutrients in plants (Hedhly, 2009). Foliar predators typically injure more plants by consuming more of the plant's tissue, but they also grow and develop more slowly, have lower fecundity, and have poorer survival rates. Climate change has a direct impact on the biology of the biological control agent and/or the host plant's ability to deal with or adapt to the presence of the herbivore or plant pathogen. Both biological control agents and weed life cycles are likely to accelerate with increasing temperatures. They may be less effective under drier conditions, as increased water stress affects host plant development and thus biological control agent production. Increases in greenhouse gases may affect both the spatial and temporal dimensions of the link between herbivorous plants and their prey. The formation of plant defences against herbivores may be affected by changes in temperature. Drought is known to increase the concentration of many allelo-chemicals that are resistant to insects (Narwal & Haouala, 2013). Variations in the ideal distribution of the active components and the target herb are likely to be evident at the geographic level as a result of the impacts of humidity and temperature. For instance, a rise in temperature would make it possible for subtropical bio control agents to endure on temperate weeds. For instance, prickly pear (*Opuntia* spp.), which may be favoured in southern regions of its distribution, and salvinia (*Salvinia molesta*) and alligator weed, both of which are aquatic weeds, are examples of terrestrial weeds (*Alternanther aphiloxeroides*). To determine whether a weed would migrate and "escape" its control agents under climate change, one must know the physiology of the agent and the weed, as well as their evolutionary potential and ability to disperse. CO_2can affect the growth, morphology, and reproduction of the target pest, which can alter the effectiveness of weed control agents. The ratio of Carbon to Nitrogen and changes in herbivore dietary patterns and growth rates would all be directly affected by CO_2 (Bale et al., 2002). As mentioned earlier, warming could lead to an increase in overwintering insect populations and a change in their projected range. This would improve the ability of some weeds to be bio controlled, but could also lead to an increase in certain plant pests, which would indirectly affect competition between plants and weeds. In general, it is unlikely that biological control pests will continue to grow and reproduce at the same rates as their designated targets during times of rapid climatic change or climatic extremes. Describe the range of possible effects that have been linked to changes in plant chemistry and herbivore performance caused by CO2 emissions (based primarily on polyphagous invertebrate herbivores). Effects include:

1. The number of herbivores decreased significantly (−21.6%) (Stiling & Cornelissen, 2007).
2. Increases in development time (+3.87%), overall consumption (+9.2%), and herbivores' relative consumption rate (+16.5%) (Stiling & Cornelissen, 2007).
3. Significant declines were seen in relative growth rate (−8.3%), conversion efficiency (−19.9%), and pupal weight (−5.03%) (Berliner et al., 2023).
4. There are no variations in herbivore connections.
5. Plant biomass increased sharply (+34.4%), as did C: N ratio (+26.57%), N concentration (−16.4%), tannin content (+29.9%) and other phenolic compounds (Amare, 2016).

15.4.4 Cultural Weed Management

Crop management methods (fire, competing crops, cultivation, etc.) can be used to control weeds in both environmental and agricultural settings. By burning or cultivating, these weed management techniques may change successional patterns, lessening plant competition for some species, or they may result in direct confrontation between crops and weeds that are engaged in concurrent germination competition. Root-to-stem ratios can vary, showing how changing climate affects plant competition (Duboscq-Carra et al., 2018). If the root-to-stem ratio of some weeds that recover from root pieces left behind after mechanical ploughing increases as roots or rhizomes emerge, such as Canada thistle, it may become more challenging to control *Cirsium arvense* (Canada thistle). Weeds such as *Solanum elaeagnifolium* (silver leaf nightshade) and *Chondrilla juncea* (skeleton weed) are examples of weeds that may respond similarly (*Solanumelae agnifolium*).

15.4.5 Competitive Crop Species

Competitive varieties, in contrast, might be more alluring because they don't have additional expenses. Such varieties may be better able to compete with weed species for scarce resources, form chemical exudates that inhibit growth, and reduce the financial costs of weeds by avoiding yield losses (Abbas et al., 2018). Competitive varieties could help with medium- to long-term weed control, reduce herbicide use, and improve agricultural sustainability by reducing seed yield of a weed species (Andrew et al., 2015).

15.5 Conclusion

Factors related to global changing climate have significant implications for crop growth and production, as well as weed species diversity, competitiveness, and weed control effectiveness. Effective weed management and crop development may be strongly impacted by the progressive rise in atmospheric carbon dioxide and its possible effects on other climate parameters like temperature, relative humidity, precipitation, and light. The spread of weeds, their population dynamics, the length of their life cycles, infestation pressure, and their general abundance are all significantly affected by climate. The findings of our study indicate that any beneficial effects of climate change on crop output may be cancelled out by an increased weed response. Weeds typically have more effective survival strategies during climate change due to their higher interspecific genetic diversity and physiological resilience. Future climatic conditions may alter the consequences of current weed control methods, which rely primarily on the use of herbicides. This necessitates a careful examination of the current and future effects of climate conditions on weed populations. Experiments must be carried out with a variety of climatic parameters in order to examine the interplay between the effects of climate change and weed management. The majority of research has concentrated on single factor studies, which have limited real-world applicability because plant responses to interacting environmental circumstances differ dramatically from responses of individual components. Furthermore, results from earlier studies sometimes pertain to controlled environmental circumstances and might not be appropriate to responses in the field. Long-term field research under experimental conditions that mimic future climate is crucial to better predict the implications of global climate change. This is because there hasn't been much research done to foresee how global climate change will affect weed management in the field.

References

Abbas, T., Zahir, Z. A., Naveed, M., & Kremer, R. J. (2018). Limitations of existing weed control practices necessitate development of alternative techniques based on biological approaches. *Advances in Agronomy, 147*, 239–280.

Abell, J. T., Winckler, G., Anderson, R. F., & Herbert, T. D. (2021). Poleward and weakened westerlies during Pliocene warmth. *Nature, 589*(7840), 70–75.

Aguilera, J., Gomes, A. R., & Nielsen, K. M. (2011). Genetically modified microbial symbionts as arthropod pest controllers: Risk assessment through the European legislations. *Journal of Applied Entomology, 7*, 494–502.

Al Souki, K. S., Burdová, H., Trubač, J., Štojdl, J., Kuráň, P., Kříženecká, S., Machová, I., Kubát, K., Popelka, J., Malinská, H. A., & Nebeská, D. (2021). Enhanced carbon sequestration in marginal land upon shift towards perennial C4 miscanthus× giganteus: A case study in North-Western Czechia. *Agronomy, 11*(2), 293.

Alberto, A. M., Ziska, L. H., Cervancia, C. R., & Manalo, P. A. (1996). The influence of increasing carbon dioxide and temperature on competitive interactions between a C3 crop, rice (Oryza sativa) and a C4 weed (*Echinochloa glabrescens*). *Functional Plant Biology, 23*(6), 795–802.

Amare, T. (2016). Review on impact of climate change on weed and their management. *American Journal of Biological and Environmental Statistics, 2*(3), 21–27.

Andreae, M. O., & Merlet, P. (2001). Emission of trace gases and aerosols from biomass burning. *Global Biogeochemical Cycles, 15*(4), 955–966.

Andrew, I. K., Storkey, J., & Sparkes, D. L. (2015). A review of the potential for competitive cereal cultivars as a tool in integrated weed management. *Weed Research, 55*(3), 239–248.

Anwar, M. P., Islam, A. M., Yeasmin, S., Rashid, M. H., Juraimi, A. S., Ahmed, S., & Shrestha, A. (2021). Weeds and their responses to management efforts in a changing climate. *Agronomy, 11*(10), 1921.

Asch, F., Dingkuhn, M., Sow, A., & Audebert, A. (2005). Drought-induced changes in rooting patterns and assimilate partitioning between root and shoot in upland rice. *Field Crops Research, 93*(2–3), 223–236.

Baite, M. S., Raghu, S., Prabhukarthikeyan, S. R., Keerthana, U., Jambhulkar, N. N., & Rath, P. C. (2020). Disease incidence and yield loss in rice due to grain discolouration. *Journal of Plant Diseases and Protection, 127*(1), 9–13.

Bale, J. S., Masters, G. J., Hodkinson, I. D., Awmack, C., Bezemer, T. M., Brown, V. K., Butterfield, J., Buse, A., Coulson, J. C., Farrar, J., & Good, J. E. (2002). Herbivory in global climate change research: Direct effects of rising temperature on insect herbivores. *Global Change Biology, 8*(1), 1–6.

Batlla D, Benech-Arnold RL (2004) A predictive model for dormancy loss in Polygonum aviculare L. seeds based on changes in population hydro time parameters. Seed Science Research. 14(3):277–86.

Batlla, D., & Benech-Arnold, R. L. (2014). Weed seed germination and the light environment: Implications for weed management. *Weed Biology and Management, 14*(2), 77–87.

Belote, R. T., Weltzin, J. F., & Norby, R. J. (2003). Response of an understory plant community to elevated [CO_2] depends on differential responses of dominant invasive species and is mediated by soil water availability. *New Phytologist, 161*(3), 827–835. https://doi.org/10.1111/j.1469-8137.2004.00977.x

Berliner, J., Ganguly, A. K., Kamra, A., & Sirohi, A. (2023). Effect of elevated carbon dioxide on population growth of root-knot nematode, Meloidogyne incognita in tomato. Indian Phytopathology 1–7.

Bhagat, S., Kumar, A., Punniya, R., Banotra, M., Sharma, A., Gupta, V., & Sharma, A. (2017). Impact of climate change on crop weed association and their eminent threat and perspective in crop culture. *International Journal of Current Microbiology and Applied Sciences, 6*(12), 3761–3770.

Brinkley, T. R., & Bomford, M. (2002). Agricultural sleeper weeds: What is the potential threat? Kingston, ACT: Bureau of Rural Sciences.

Bunce, J. A. (2004). Carbon dioxide effects on stomatal responses to the environment and water use by crops under field conditions. *Oecologia, 140*(1), 1.

Cavero, J., Zaragoza, C., Suso, M. L., & Pardo, A. (1999). Competition between maize and Datura stramonium in an irrigated field under semi-arid conditions. *Weed Research-Oxford, 39*, 225–240.

Chamberlain, S. A., Bronstein, J. L., & Rudgers, J. A. (2014). How context dependent are species interactions? *Ecology Letters, 17*(7), 881–890.

Chatterjee, D., Kumar, R., Kuotsu, R., & Deka, B. C. (2016). Validation of traditional weed control method through common salt application in the hill region of Nagaland. *Current Science, 25*, 1459–1467.

Chatterjee, D., Saha, S., Swain, B., Chakraborty, D., Nayak, A. K., Pathak, H., & Singh, M. P. (2020). Monitoring and impact assessment of climate change on agriculture using advanced research techniques. pp. 33–53. In: A. Rakshit, S. Ghosh, S. Chakraborty, V. Philip, & A. Datta (Eds.) *Soil analysis: Recent trends and applications* Springer, Singapore. p. 338+xv.

Chauhan, B. S., Matloob, A., Mahajan, G., Aslam, F., Florentine, S. K., & Jha, P. (2017). Emerging challenges and opportunities for education and research in weed science. *Frontiers in Plant Science, 8*, 1537.

Clements, D. R., & Jones, V. L. (2021). Rapid evolution of invasive weeds under climate change: Present evidence and future research needs. *Frontiers in Agronomy, 3*, 664034.

Davies, K. W., Bates, J. D., & Svejcar, L. (2022). Reducing exotic annual grass competition did not improve shrub restoration success during a drought. *Rangeland Ecology & Management, 85*, 9–14.

Díaz, S., Cabido, M., Zak, M., Martínez Carretero, E., & Aranı́bar, J. (1999). Plant functional traits, ecosystem structure and land-use history along a climatic gradient in central-western Argentina. *Journal of Vegetation Science, 10*(5), 651–660.

Donald, W. W., & Khan, M. (1992). Yield loss assessment for spring wheat (Triticum aestivum) infested with Canada thistle (Cirsium arvense). *Weed Science, 40*(4), 590–598.

Duboscq-Carra, V. G., Letourneau, F. J., & Pastorino, M. J. (2018). Looking at the forest from below: The role of seedling root traits in the adaptation to climate change of two Nothofagus species in Argentina. *New Forests, 49*(5), 613–635.

Ennos, A. R. (1997). Wind as an ecological factor. *Trends in Ecology & Evolution, 12*(3), 108–111.

Finch, D. M., Butler, J. L., Runyon, J. B., Fettig, C. J., Kilkenny, F. F., Jose, S., Frankel, S. J., Cushman, S. A., Cobb, R. C., Dukes, J. S., & Hicke, J. A. (2021). Effects of climate change on invasive species.

Friedrich, J., Ge, M., & Pickens, A. (2020). This interactive chart shows changes in the World's Top 10 Emitters.

Fuhrer, J. (2003). Agroecosystem responses to combination of elevated CO_2, ozone and global climate change. *Agriculture, Ecosystems & Environment, 97*, 1–20.

Gaba, S., Chauvel, B., Dessaint, F., Bretagnolle, V., & Petit, S. (2010). Weed species richness in winter wheat increases with landscape heterogeneity. *Agriculture, Ecosystems & Environment, 138*(3–4), 318–323.

Gerard, P. J., Kean, J. M., Phillips, C. B., Fowler, S. V., Withers, T. M., Walker, G. P., & Charles, J. G. (2010). *Possible impacts of climate change on biocontrol systems in New Zealand*. Report for MAF Pol Project 0910–11689.

Giannini, A., Biasutti, M., Held, I. M., & Sobel, A. H. (2008). A global perspective on African climate. *Climatic Change, 90*(4), 359–383.

Grice, A. C. (2006). The impacts of invasive plant species on the biodiversity of Australian rangelands. *The Rangeland Journal, 28*(1), 27–35.

Gupta, R. L. (2013). Impact of climate change on pesticide usage. Climate Change and Environment, p. 128.

Hedhly, A., Hormaza, J. I., & Herrero, M. (2009). Global warming and plant sexual reproduction. *Trends in Plant Science, 14*(1), 30–36. https://doi.org/10.1016/j.tplants.2008.11.001

Hikosaka, K., Onoda, Y., Kinugasa, T., Nagashima, H., Anten, N. P., & Hirose, T. (2005). Plant responses to elevated CO_2 concentration at different scales: leaf, whole plant, canopy, and population. In *Forest ecosystems and environments* (pp. 3–13). Springer, Tokyo.

Hyvönen, T. (2011). Impact of temperature and germination time on the success of a C4 weed in a C3 crop: Amaranthus retroflexus and spring barley. *Agricultural and Food Science, 20*(2), 183–189.

IPCC. (2007). *Climate change: Impacts, adaptation and vulnerability* (p. 986). IPCC Secretariat.

Iqbal, M. C., Wijesundera, D. S., & Ranwala, S. M. (2014). Climate change, invasive alien flora and concerns for their management in Sri Lanka. *Ceylon Journal of Science (Bio. Sci.), 43*(2), 1–5.

Jaganathan, G. K., & Liu, B. (2015). Role of seed sowing time and microclimate on germination and seedling establishment of *Dodonaea viscosa* (Sapindaceae) in a seasonal dry tropical environment—an insight into restoration efforts. *Botany, 93*(1), 23–29.

Keighery, G., & Longman, V. (2004). The naturalized vascular plants of Western Australia 1: Checklist, environmental weeds and distribution in IBRA regions. *Plant Protection Quarterly, 19*(1), 12–32.

Kraehmer, H., Laber, B., Rosinger, C., & Schulz, A. (2014). Herbicides as weed control agents: state of the art: I. Weed control research and safener technology: the path to modern agriculture. *Plant Physiology, 166*(3), 1119–31.

Kriticos, D. J., Sutherst, R. W., Brown, J. R., Adkins, S. W., & Maywald, G. F. (2003). Climate change and the potential distribution of an invasive alien plant: Acacia nilotica ssp. indica in Australia. *Journal of Applied Ecology, 40*(1), 111–24.

Kumar, R., Chatterjee, D., Deka, B. C., Kumar, M., Kuotsu, R., Merasenla, A. O., & Ngachan, S. V. (2016). Weed management practices in upland direct seeded jhum rice under the Eastern Himalaya. *Research on Crops, 17*(2), 199–204.

Marambe, B., & Wijesundara, S. (2021). Effects of climate change on weeds and invasive alien plants in Sri Lankan agro-ecosystems: Policy and management implications. *Frontiers in Agronomy, 3*, 641006. https://doi.org/10.3389/fagro.2021.641006

Matloob, A., Khaliq, A., & Chauhan, B. S. (2015). Weeds of rice in Asia: Problems and opportunities. *Advances of Agronomy, 130*, 291–336.

Meksawat, S., & Pornprom, T. (2010). Allelopathic effect of itchgrass (Rottboellia cochinchinensis) on seed germination and plant growth. *Weed Biology and Management, 10*(1), 16–24.

Mishra, J. S., Kumar, R., Kumar, R., Rao, K. K., & Bhatt, B. P. (2019). Weed density and species composition in rice-based cropping systems as affected by tillage and crop rotation. *Indian Journal of Weed Science, 51*(2), 116.

Naidu, V. S., & Varshney, J. A. (2011). Interactive effect of elevated CO. *Indian Journal of Agricultural Sciences, 81*(11), 1026–1029.

Narwal, S. S., & Haouala, R. (2013). Role of allelopathy in weed management for sustainable agriculture. InAllelopathy (pp. 217–249). Springer, Berlin, Heidelberg.

Nayak, A, K., Chatterjee, D., Tripathi, R., Shahid, M., Vijayakumar, S., Satapathy, B. S., Kumar, A., Mohanty, S., Bhattacharyya, P., Mishra, P., Kumar, U., Mohapatra, S. D., Panda, B. B., Rajak, M., Bhaduri, D., Munda, S., Chakraborty, K., Priyadarsani, S., Swain, C. K., Moharana, K. C., Nayak, P. K., Kumar, G. A. K., Swain, P., Tesfai, M., Nagaothu, U. S., & Pathak, H. (2020). *Climate smart agricultural technologies for rice production system in Odisha*. ICAR-National Rice Research Institute, Cuttack, Odisha, 753006, India, pp. 366.

Ooi, M. K., Denham, A. J., Santana, V. M., & Auld, T. D. (2014). Temperature thresholds of physically dormant seeds and plant functional response to fire: Variation among species and relative impact of climate change. *Ecology and Evolution, 4*(5), 656–671.

Patterson, D. T. (1986). Responses of soybean (Glycine max) and three C4 grass weeds to CO2 enrichment during drought. *Weed Science, 34*(2), 203–210.

Patterson, D. T. (1995). Weeds in a changing climate. *Weed Science, 43*, 685–701.

Patterson, D. T., Highsmith, M. T., & Flint, E. P. (1988). Effects of temperature and CO_2 concentration on the growth of cotton (*Gossypium hirsutum*), spurred anoda (*Anoda cristata*), and velvetleaf (*Abutilon theophrasti*). *Weed Science, 36*(6), 751–757.

Patterson, D. T., Westbrook, J. K., Lingren, P. D., & Rogasik, J. (1999). Weeds, insects, and diseases. *Climatic Change, 43*(4), 711–727.

Peters, K., & Gerowitt, B. (2014). Important maize weeds profit in growth and reproduction from climate change conditions represented by higher temperatures and reduced humidity. *Journal of Applied Botany and Food Quality, 21*, 87.

Peters, K., Breitsameter, L., & Gerowitt, B. (2014). Impact of climate change on weeds in agriculture: A review. *Agronomy for Sustainable Development, 34*(4), 707–721.

Rai, P. K., & Singh, J. S. (2020). Invasive alien plant species: Their impact on environment, ecosystem services and human health. *Ecological Indicators, 111*, 106020.

Ramesh, K., Matloob, A., Aslam, F., Florentine, S. K., & Chauhan, B. S. (2017). Weeds in a changing climate: Vulnerabilities, consequences, and implications for future weed management. *Frontiers in Plant Science, 8*, 95.

Rana, S. S., & Rana, M. C. (2016). *Principles and practices of weed management*. Department of Agronomy, College of Agriculture, CSK Himachal Pradesh Krishi Vishvavidyalaya, Palampur 138. https://doi.org/10.13140/RG.2.2.33785.47207

Rao, A. N., Singh, R. G., Mahajan, G., & Wani, S. P. (2020). Weed research issues, challenges, and opportunities in India. *Crop Protection, 134*, 104451.

Raza, A., Razzaq, A., Mehmood, S. S., Zou, X., Zhang, X., Lv, Y., & Xu, J. (2019). Impact of climate change on crops adaptation and strategies to tackle its outcome: A review. *Plants, 8*(2), 34.

Robinson, T. M., & Gross, K. L. (2010). The impact of altered precipitation variability on annual weed species. *American Journal of Botany, 97*(10), 1625–1629.

Rodenburg, J., Meinke, H., & Johnson, D. E. (2011). Challenges for weed management in African rice systems in a changing climate. *The Journal of Agricultural Science, 149*(4), 427–435.

Rosenzweig, C., & Hillel, D. (1998). *Climate change and the global harvest* (pp. 1–324). Oxford University Press.

Rudov, A., Mashkour, M., Djamali, M., & Akhani, H. (2020). A review of C4 plants in southwest Asia: An ecological, geographical and taxonomical analysis of a region with high diversity of C4 eudicots. *Frontiers in Plant Science, 11*, 546518.

Satapathy, B. S., Duary, B., Saha, S., Munda, S., & Chatterjee, D. (2021). Impact of sowing methods and weed control practices on yield and economics of wet direct seeded rice. *Oryza, 58*(3), 375–383.

Satapathy, B. S., Chatterjee, D., Saha, S., Duary, B., & Singh, T. (2022). Weed management in a direct-seeded rice-ratoon rice cropping system. *The Journal of Agricultural Science, 7*, 1.

Shahzad, A., Ullah, S., Dar, A. A., Sardar, M. F., Mehmood, T., Tufail, M. A., Shakoor, A., & Haris, M. (2021). Nexus on climate change: Agriculture and possible solution to cope future climate change stresses. *Environmental Science and Pollution Research, 28*(12), 14211–14232.

Singh, R. P., Singh, R. K., & Singh, M. K. (2011). Impact of climate and carbon dioxide change on weeds and their management-a review. *Indian Journal of Weed Science, 43*(1&2), 1–1.

Springer, C. J., & Ward, J. K. (2007). Flowering time and elevated atmospheric CO2. *New Phytologist, 176*(2), 243–255.

Stiling, P., & Cornelissen, T. (2007). How does elevated carbon dioxide (CO_2) affect plant–herbivore interactions? A field experiment and meta-analysis of CO_2-mediated changes on plant chemistry and herbivore performance. *Global Change Biology, 13*(9), 1823–1842.

Travlos, I., Gazoulis, I., Kanatas, P., Tsekoura, A., Zannopoulos, S., & Papastylianou, P. (2020). Key factors affecting weed seeds' germination, weed emergence, and their possible role for the efficacy of false seedbed technique as weed management practice. *Frontiers in Agronomy, 2*, 1.

Treharne, K. (1989). The implications of the 'greenhouse effect' for fertilizers and agrochemicals, in The Greenhouse Effect and UK Agriculture, CAS Paper 19, ed. R. C. Bennet (Reading: University. of Reading Press), pp. 67–78.

Treut, Le., Somerville, H. R., Cubasch, U., Ding, Y., Mauritzen, C., Mokssit, A., Peterson, T., & Prather, M. (2007a), Historical overview of climate change. In: Solomon, S., Qin, D., Manning, M., Chen, Z., Marquis, M., Averyt, K. B., Tignor, M., & Miller, H. L. (Eds.) *Climate change 2007: The physical science basis. Contribution of working group I to the fourth assessment report of the intergovernmental panel on climate change.* Cambridge University Press, Cambridge, United Kingdom and New York, NY, USA.

Treut, L., Somerville, R., Cubasch, U., Ding, Y., Mauritzen, C., Mokssit, A., Peterson, T., Prather, M., Qin, D., Manning, M., Chen, Z., Marquis, M., Averyt, K., Tignor, M., Miller, H.L., & Solomon, S. (2007b). Historical overview of climate change science. *Earth, 1*, 93–127.

Tubiello, F. N., Soussana, J. F., & Howden, S. M. (2007). Crop and pasture response to climate change. *Proceedings of the National Academy of Sciences, 104*(50), 19686–19690.

Tylianakis, J., Didham, R., Bascompte, J., & Wardle, D. (2008). Global change and species interactions in terrestrial ecosystems. *Ecology Letters, 11*, 1351–1363. https://doi.org/10.1111/j.1461-0248.2008.01250.x

Varanasi, A., Prasad, P. V., & Jugulam, M. (2016). Impact of climate change factors on weeds and herbicide efficacy. *Advances in Agronomy, 135*, 107–146.

Walther, G. R., Post, E., Convey, P., Menzel, A., Parmesan, C., Beebee, T. J., Fromentin, J. M., Hoegh-Guldberg, O., & Bairlein, F. (2002). Ecological responses to recent climate change. *Nature, 416*(6879), 389–395.

Wang, R, (2014), The fig wasps associated with Ficus microcarpa, an invasive fig tree. University of Leeds.

War, A. R., Taggar, G. K., War, M. Y., & Hussain, B. (2016). Impact of climate change on insect pests, plant chemical ecology, tritrophic interactions and food production. *International Journal of Clinical and Biological Sciences, 1*(02), 16–29.

Ward, J. K., Tissue, D. T., Thomas, R. B., & Strain, B. R. (1999). Comparative responses of model C3 and C4 plants to drought in low and elevated CO_2. *Global Change Biology, 5*(8), 857–867.

Wato, T. (2020). The role of allelopathy in pest management and crop production-A review. *Food Science and Quality Management, 93*, 13–21.

Wing, I. S., De Cian, E., & Mistry, M. N. (2021). Global vulnerability of crop yields to climate change. *Journal of Environmental Economics and Management, 109*, 102462.

Wolfe, D. W., Ziska, L., Petzoldt, C., Seaman, A., Chase, L., & Hayhoe, K. (2008). Projected change in climate thresholds in the Northeastern US: Implications for crops, pests, livestock, and farmers. *Mitigation and Adaptation Strategies for Global Change, 13*(5), 555–575.

Zeng, Q., Liu, B., Gilna, B., Zhang, Y., Zhu, C., Ma, H., Pang, J., Chen, G., & Zhu, J. (2011). Elevated CO2 effects on nutrient competition between a C3 crop (Oryza sativa L.) and a C4 weed (Echinochloa crusgalli L.). *Nutrient Cycling in Agroecosystems, 89*(1), 93–104.

Zhu, C., Zeng, Q., Ziska, L. H., Zhu, J., Xie, Z., & Liu, G. (2008). Effect of nitrogen supply on carbon dioxide–induced changes in competition between rice and barnyardgrass (Echinochloa crus-galli). *Weed Science, 56*(1), 66–71.

Ziska, L. H. (2000). The impact of elevated CO_2 on yield loss from a C3 and C4 weed in field-grown soybean. *Global Change Biology, 6*(8), 899–905.

Ziska, L. H., & McClung, A. (2008). Differential response of cultivated and weedy (red) rice to recent and projected increases in atmospheric carbon dioxide. *Agronomy Journal, 100*, 1259–1263.

Ziska, L. H., Tomecek, M. B., & Gealy, D. R. (2010). Competitive interactions between cultivated and red rice as a function of recent and projected increases in atmospheric carbon dioxide. *Agronomy Journal, 102*(1), 118–123.

Ziska, L. H., Blumenthal, D. M., & Franks, S. J. (2019). Understanding the nexus of rising CO2, climate change, and evolution in weed biology. *Invasive Plant Science and Management, 12*(2), 79–88.

Ziska, L. H., Blumenthal, D. M., Runion, G. B., Hunt, ERJr., & Diaz-Soltero, H. (2011). Invasive species and climate change: an agronomic perspective. *Climate Change, 105*, 13–42.

Chapter 16
Impacts of Abiotic Stresses on Eco-Physiology of Crop in Changing Climate

Jagadish Rane and Krishna Kumar Jangid

Abstract Eco-physiology describes the inherent mechanisms used by plants to respond to external stimuli that contribute to adaptations to particular environment. In nature, plants are often simultaneously challenged by individual and interactive effects of several biotic and abiotic stresses such as drought, supra- or sub-optimal ambient temperatures, waterlogging and salinity. These factors directly or indirectly influence plant growth and development process. Internal cell osmoregulation, modulated metabolic activity, hormonal superintendence and supportive alteration of antioxidant activities at cellular level are the crucial controlling factors for maintaining plant growth plasticity under variable abiotic stress environment. Plants have inbuilt mechanisms for stress tolerance that thrive upon many intrinsic factors such as genotypic and phenotypic constitutions and developmental circumstances. Many of these traits involved in tolerance to abiotic stress were helpful during the conventional selection procedures of crop improvement. However, modern omics technologies can help to accelerate the incremental accumulation of favorable alleles for physiological traits for climate adaptation. Challenges of ensuring global food security in the current situation of predicted adverse effects of climate change can be addressed by managing abiotic stresses for sustainable agriculture by deep insights into the eco-physiology of crops.

Keywords Abiotic stress · Eco-physiology · Physiological traits · Sustainable agriculture · Climate change

J. Rane (✉) · K. K. Jangid
ICAR-National Institute of Abiotic Stress Management, Baramati, Maharashtra 413115, India
e-mail: jagadish.rane@icar.gov.in

H. Pathak et al. (eds.), *Climate Change Impacts on Soil-Plant-Atmosphere Continuum*, Advances in Global Change Research 78,
https://doi.org/10.1007/978-981-99-7935-6_16

16.1 Introduction

Eco-physiology establishes a connection between the physical and biological components of a plant's environment and interacts with its physiology (Flexas & Gago, 2018). While organisms, which coexist with crop plants can be beneficial or harmful, soil and atmospheric factors that influence the growth and ultimately the productivity of the crops make essential components of crop eco-physiology (Fig. 16.1). When at the optimum level, these factors provide scope for the best performance of crops in terms of economic yield, which is critical for the maximisation of profit for farmers. However, if in excess or deficit at critical stages of crop growth, soil and atmospheric factors create stress, primarily referred to as abiotic stresses (Gong et al., 2020). Crop ecophysiology gains immense importance as crops during the domestication process appears to have lost many traits that continue to exist in natural vegetation, which is resilient to harsh environmental conditions and display an incredible potential to recover from biotic and abiotic stresses (Mammadov et al., 2018; van Zonneveld et al., 2020). It can be attributed to the crop improvement programme's preference for crop productivity over resilience to the eco-system until recently. It is increasingly realised that the challenge of ensuring global food security in the context of predicted adverse effects of climate change can be addressed by managing abiotic stresses for sustainable agriculture informed by deep insights into the ecophysiology of crops. The objectives of the chapter are to (i) to assess the impact of abiotic stress factors on plant species from cellular to canopy level, (ii) to explore the possible plant response mechanism towards the induced abiotic stress environment and (iii) to study the influence of soil environment on crop-ecology under variable abiotic stress environment exposure.

16.2 Abiotic Stresses in the Changing Climate

Stresses caused by non-living factors include vital components of the atmosphere such as ambient temperature and solar radiation, water and soil (Onwuka & Mang, 2018). Extreme levels of all these factors affect the physiology of crop plants and trigger protective mechanisms as illustrated in Fig. 16.2. Supra-optimal or suboptimal temperatures may cause heat and cold stress, respectively. Solar radiation is the only source for photosynthesis in plants, which drives the process of biomass accumulation and growth of plants. However, high and low solar radiation can adversely influence crop performance (Zakurin et al., 2020). The soil moisture deficit and excess lead to drought and waterlogging, respectively, affecting the crop if they occur during critical crop growth stages (Qian et al., 2020). Soil provides essential nutrients for crop growth; however, these mineral nutrients can cause stress for plants if their soil levels cross the minimum and maximum limits. These stresses are manifested in mineral deficiencies or toxicities in crop plants (Gong et al., 2020). Accumulation of salts in soil is a natural process but enhanced by anthropogenic interference such as

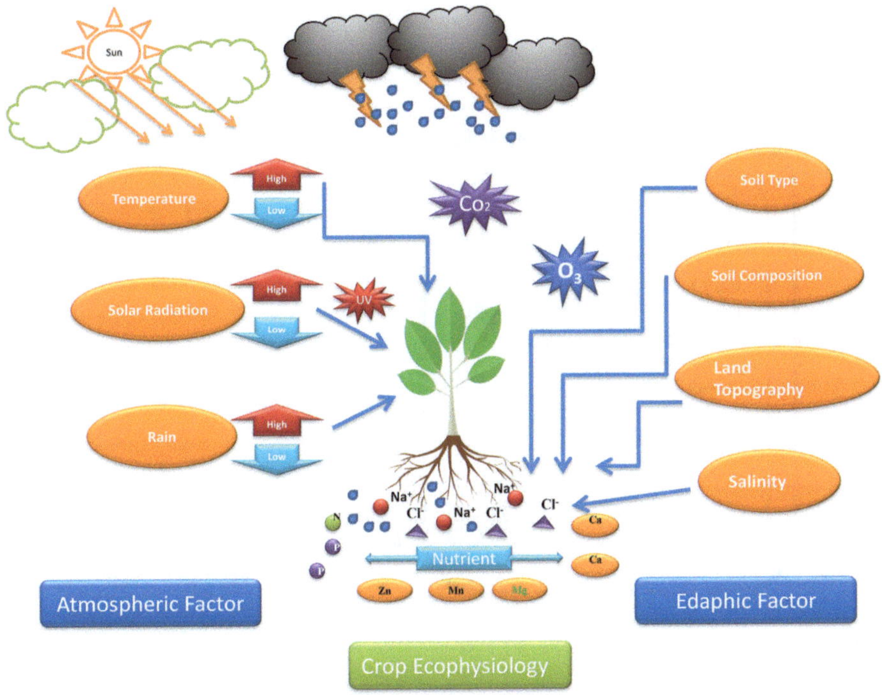

Fig. 16.1 Factors affecting crop eco-physiology

indiscriminate use of saline water for irrigation, which has tendered the fertile lands into unproductive lands as salinity and alkalinity substantially reduce crop growth (Kamran et al., 2020; Kumar & Sharma, 2020). These factors adversely affect the functions and processes leading to plant growth and development and hence play crucial roles in crop ecophysiology.

Changes in weather are driven by atmospheric factors, which can also influence edaphic factors and hence predicted climate change has great implications for interactions among atmospheric, edaphic and crop plants (Raza et al., 2019). Environmental challenges for plants in general and crops, in particular, have been the focus of investigations in many of the studies (Arunanondchai et al., 2018; Noya et al., 2018). However, it is essential to understand the mode and intensity of influence of abiotic stresses on crop-ecophysiology to address these challenges in the context of climate change. The following session deals with mechanisms, which influence the impact of abiotic stresses on plants.

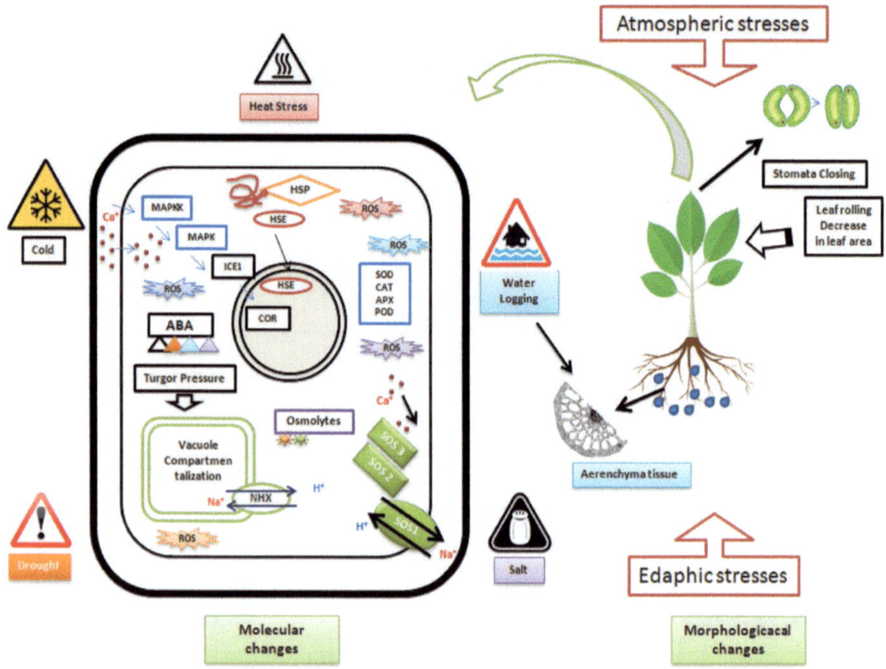

Fig. 16.2 Plant mechanisms triggered by abiotic stresses in crop plants

16.3 Impacts of Abiotic Stresses and Response Mechanisms of Plants

16.3.1 High Temperature

It is well accepted that enhanced ambient temperatures are the driving force for many of the adverse impact of global climate change (Teshome et al., 2020). Moreover, the prevailing attention is an increase in mean global temperature while the increase in land surface temperature is higher than that of overall global temperature, including land and sea ambient temperature (Cox et al., 2020). Hence, this difference can influence the intensity and duration of high-temperature stress on crop physiology directly or indirectly.

At whole plant level, the direct effect of temperature on crop results from its influence on phenology, can reduce the crop growth cycle as each phenological phase is usually determined by specific growing degree days. Enhanced temperatures can directly affect physiological processes such as photosynthesis, respiration, water relations, and plants' reproductive physiology (Kumari et al., 2020). Temperatures can also indirectly influence nutrient uptake by reducing soil moisture and plants transpiration process (Soares et al., 2019; Pregitzer & King, 2005).

An important mechanism that can influence crops' temperature effect includes the plasticity of the crop growth cycle, which can vary across the crop cultivars. Plants that can trigger the developmental process after sensing the increase in ambient temperatures can provide better scope for adaptation of agriculture to climate change, and this mechanism is being increasingly explored in the form of short duration crops (Aryal et al., 2019). However, the reduction in the duration comes at the cost of yield penalty, which necessitates some level of tolerance to high temperatures in crops at the whole plant, tissue, organ and cellular level.

Mechanisms to avoid the coincidence of flowering with temperature peaks during the day allow regular fertilization and grain formation (Hatfield & Prueger, 2015). Hence, crop species or cultivars having this diurnal escape capacity can produce more grains than those lacking them. It has been first reported in rice, and the genotypes, which flower in the early hours of the day, are being increasingly used in crop breeding programmes (Bheemanahalli et al., 2017). Leaf canopy cooling mechanism is pivotal in maintaining the optimal temperature inside the plants in a way to favour all the processes and functions at the tissue and cellular level (Deva et al., 2020; Nelson & Bugbee, 2015). This feature of the plant is determined by the plants capacity to take up water from the soil and maintain transpiration, which keeps the leaves cool (Zhang et al., 2019). Plant's root and shoot system architectures that vary across the crops and the crop cultivars greatly influence the transpiration process, which is effective, mainly when soil moisture is not a limiting factor (Wasaya et al., 2018).

Shoot system architecture of crop plants provides other mechanisms to reduce the impact of high temperatures on crop growth and development at the whole plant level. Vertical leaf orientation over horizontal one can reduce the direct effect of radiation and heat load, and hence the annual crops with erect leaves are common in regions that experience high temperatures during crop growth (Pérez-Llorca et al., 2019). Further, the protection of ear heads by leaf canopy at early stages of growth is also a mechanism that can reduce the impact of high temperature on grain formation and productivity of cereal crops (Lu et al., 2018).

At the tissue level, stomatal distribution and mechanism play a crucial role in determining the canopy cooling process through transpiration (Pirasteh-Anosheh et al., 2016; Schymanski et al., 2013). Mechanisms to retain the viability of pollens and stigma critical for reproductive success can reduce the impact of high temperature on grain formation and productivity (Zulkarnain et al., 2019). At the cellular level, mechanisms that ensure cell membrane stability contribute to crucial components such as thylakoid membranes, which are essential components of the photosynthesis system. The stability of crucial enzymes such as high-temperature sensitive Rubisco activase involved in photosynthesis can also reduce the adverse effect of global warming on crop productivity (Perdomo et al., 2017; Raza et al., 2019). Heat shock proteins which act as a protective cover for many of the key enzymes in plant processes have been extensively investigated and suggested as an avenue for improvement of crops capacity to tolerate high temperatures (Khan et al., 2019; Lu et al., 2020).

16.3.2 Solar Radiation

Solar radiation is the key input for photosynthesis's primary process and hence crop productivity (Alvarado et al., 2020); however, light can also be harmful at high intensities (Bourbousse et al., 2020) primarily when it occurs along with water stress and high temperature. Plants with erect leaves receive less proportion of incident light intercepted by the crop canopy, but they contribute to improved crop productivity per unit of light interception (Perez et al., 2019). Thus, leaf and shoot architecture reduces stress due to high solar radiation (Zhu et al., 2020).

Reproduction can be affected by high solar radiation, including sterility, reduction in reproduction rate, and developmental abnormalities or reduction in viability of offspring. Plant cells contain chromosomes, i.e., the genetic material responsible for plant reproduction. If the cell is much damaged by radiation, then reproduction is hindered. With the virtue of its capacity to generate energy in leaf epidermal tissues and contributing to vapour pressure deficit, solar radiation can influence stomatal mechanisms and water relations of plants (Pieruschka et al., 2010). If so, energy balance mechanisms, if any at epidermal tissues of plants, will influence extreme solar radiation in crop plants. The radiation composition can influence plant growth and development as solar radiation has an electromagnetic spectrum spanning different wavelengths. For example, the ultraviolet range of the spectrum, when in excess, can affect plant growth and sprout, and the amount of damage is proportional to the radiation received. UV radiation harms the plants by generating superoxide anion radicals that affect plant processes, as reported in soybean (Kataria et al., 2020).

Since chlorophyll molecules trap the solar energy essential for photosynthesis and excess energy absorption is harmful, protective mechanisms exist at the cellular level. One of the mechanisms to protect chlorophyll molecules includes the increased concentration of carotenoids, which act as antioxidants and shield chlorophyll against photo-damage due to excess light intensity (Shen et al., 2017). Plant hormones are involved in mechanisms that protect the plants from high-intensity light (Cortleven & Schmülling, 2015; Cortleven et al., 2019). This hormone protects the plants from superoxide molecules (Procházková et al., 2008). Hence, existence and variation among the crop species concerning these mechanisms can also reduce climate change-induced impact of abiotic stresses on crop plants.

Overcast skies during the rainy season may cause incidences of suboptimal solar radiation for plant growth and development. The low intensity of solar radiation during the grain filling phase of crops negatively influences grain yield, as evident from maize experiments (Ren et al., 2016; Wang et al., 2020). It can be attributed to the reduction in photosynthesis to the extent of 48%, mainly if solar radiation reduction occurs at the crop's reproductive stage. The cause of the reduction in low radiation-induced stress was damage to ultra-structures of photosynthetic apparatus such as chloroplast in mesophyll cells (Rai & Agrawal, 2017). Besides, photosynthate transport can also get disrupted by the reduction in solar radiation and reduction in receipt of solar radiation can affect leaf nitrogen metabolism; as a result, 50% light deprivation after silking can lead to yield reduction to the extent of 70% in maize

(Wang et al., 2020). These reports indicate that intrinsic mechanisms to protect the ultrastructure of photosynthetic apparatus can reduce the effect of low radiation stress in crop plants. Increase in UV absorbing compounds such as flavonoids and derivative of hydroxycinnamic acid also help in UV Protection (Neugart et al., 2012).

16.3.3 Drought

Stress caused by soil moisture deficit is a common constraint for crop growth and development during short or long dry spells that occur in the arid and semiarid region and common in the humid region during the dry spells that lead to agricultural drought. Plant survive drought by escape, dehydration avoidance and dehydration tolerance mechanisms (Fang & Xiong, 2015).

Most of the mechanisms that regulate plants water balance revolve around stomata, tiny pores on the surface of plants, which serve as gateways for the exchange of gas between the atmosphere and the leaf tissues. Loss of water through transpiration is regulated by stomatal mechanisms, which contribute to the productive use of water for the plant's growth and development as the latter depends on biomass accumulated through photosynthesis, which also depends on CO_2 exchange. Environmental factors, including radiation, influence mechanisms that allow a trade-off between transpiration and photosynthesis (Harrison et al., 2020) and influence plants mechanisms to tolerate stress. Though there are many reports about metabolites contributing to osmotic adjustment for the survival of water deficit in soil, abscisic acid (ABA) metabolism is one of the most cited internal metabolites that play a crucial role in sensing the drought and regulate stomatal mechanisms. Besides, ABA, the plant stress hormone, can contribute to root system architectural changes in plants. Deep rooting in the absence of sufficient moisture at the soil's surface is another mechanism by which plant try to survive drought, and the introduction of this trait can enhance the productivity of crops such as rice (Uga et al., 2013). Architectural changes in root in response to drought also include the emergence of roots only in the rhizosphere's moist region, avoiding the dry region. This mechanism, which involves hydro-patterning response, is driven by auxin, the plant hormone reported in Arabidopsis (Orosa-Puente et al., 2018).

16.3.4 Submergence

Plants survive submergence through inbuilt mechanisms, enabling them to extend their shoots above the water surface to access oxygen from the air and avoid hypoxia. Such mechanisms are found in wild relatives of rice (Stein et al., 2018). In this mechanism, plants escape the submergence by extending their inter nodal length through the influence of plant growth hormones such as gibberellins and ethylene, as reported in diverse species of plants, which survive in the wetland (Voesenek &

Bailey-Serres, 2015). Some mechanisms check the assimilate degradation during submergence and make the plant survive prolonged submergence (Xu et al., 2006). Such regulations have been attributed to enzymatic components that regulate protein degradation through the N-end rule pathway (Graciet & Wellmer, 2010). The benefits of introducing genes such as SUB1A, SK1, SK2 and SD1 contributing to crop ecophysiology of flooding tolerant plants have been suggested for crop improvement (Bailey-Serres & Voesenek, 2020).

Modification in root traits is another mechanism that enables plants to survival under excess soil moisture conditions (Voesenek & Bailey-Serres, 2015). It involves oxygen driven regulation of genes contributing to submergence tolerance. The formation of air sacs in response to hypoxia or anoxia is a common mechanism to survive transient situations arising out of waterlogging and submergence caused by floods. With this feature, crops other than rice also can survive waterlogging conditions.

16.3.5 Salinity

Out of the 227 million hectares of irrigated land worldwide, 45 million hectares (20%) are impacted by salinity-related issues (Shahid et al., 2018). This section aims to highlight plant systems that can affect crop resistance to excessive amounts of in-situ soil salts or those that are sporadically introduced into coastal soils by sea incursions, which are anticipated to intensify owing to climate change. Any mechanism that might lessen osmotic stress and ion accumulation can immediately make it possible for plants to survive longer in soil environments that are salt-enriched. Ion toxicity and osmotic stress are the two main causes of the detrimental effects of salts on plant growth and development, respectively (Arif et al., 2020). For plants that are sensitive to salts inside the plant tissues, salt stress exposure triggers plant mechanisms that can lessen the impact of accumulated salts on vital functions like water relations, photosynthesis, and reproductive physiology (Hessini et al., 2019; Soori et al., 2019; Wani et al., 2019). Climate-resilient agriculture can benefit from crop species and cultivars that successfully implement the mechanisms to shield plants' ultrastructure from the direct and indirect effects of salt.

Although the plant's growth functions are protected by mechanisms, the vulnerability of the pollens and stigma to abiotic conditions such salinity, waterlogging, drought, and high temperatures could cause the plant to falter (Fábián et al., 2019), which are highly critical for plant's reproductive success. The impact of climate change on abiotic pressures can be influenced through the utilisation of genetic variation in systems to protect pollens and stigma from abiotic stresses. Highly tolerant plants can reduce the concentration of incompatible solutes in the cytoplasm, avoid salt ion uptake, and sequester salt ions within leaf vacuoles. (Munns & Tester, 2008) and maintain a balanced cytosolic Na^+/K^+ ratio (Flowers & Colmer, 2015). For sensitive species, exposure to salinity changes the ion balance, which inhibits growth by reducing photosynthesis, protein synthesis, and metabolic activities. (Parida & Das, 2005). However, by maintaining water balance and being trait flexible to osmotic stress, plants can typically recover from brief exposure to high salinity (Munns, 2002). The plants with moderate tolerance to soil salinity often exhibits higher water use efficiency (WUE; Parida & Das, 2005), more leaf mass per area (LMA) (Munns, 2005), regulated leaf transpiration rates (Harris et al., 2010) and consistent or enhanced growth (Munns & Tester, 2008) under high salinity exposure (Lum & Barton, 2020).

16.3.6 Mineral Nutrients Deficiency

One of the events of concern in the predicted climate change is heavy rains, which can cause soil erosion. The value of agricultural land is substantially reduced by soil erosion, which leads to physical, chemical degradations. It is reported that nutrient losses due to erosion can be more in maize than in the legumes like cowpea (Bashagaluke et al., 2018). The legume-based cropping system's advantage is also evident in the most resource use an intensive combination of crops like rice and wheat (Bhardwaj et al., 2019). It indicates that the choice of plant types can help in preventing soil nutrient losses due to differential mechanisms that operate for nutrient uptake and possible soil retention in the event of erosion.

The negative effect of drought on the growth and productivity of crop can be partially attributed to an adverse effect on mobility and extractability of soil nutrients, the two processes impeded by depletion of soil moisture (Moreno-Jiménez et al., 2019). Organic manures and exogenous application of micronutrients can modulate the effects of drought on plants' nutrient status (Dimkpa et al., 2020). Such approaches may be critical for the arid region vulnerable to desertification, which affects carbon, nitrogen and phosphorus cycles (An et al., 2019). Hence, plant nutrient uptake mechanisms can influence the impact of climate change on plant stress experienced due to climate change impact.

Since the root is the entry point for nutrients inside the plants, mechanisms operating at the root level play an important role in coping with plants' nutrient stress. It is observed that root impedance due to soil compactness caused by drought rather than moisture stress can be responsible for poor nutrient uptake by plants (Hawkesford & Griffiths, 2019). Plant mechanisms that enable the roots to penetrate through hard soil can also influence drought on crop growth and productivity. Besides, the mechanisms triggered by specific genes involved in root system architecture (Uga et al., 2013) and root growth (Ramireddy et al., 2018; Zhang et al., 2020) can also influence nutrient uptake mechanisms under drought.

Remarkable changes in soil nutrient status, which occur during waterlogging, is due to a significant decrease in soil redox potential (Setter et al., 2009; Khabaz-Saberi et al., 2010; Shabala, 2011). Also, an increase in the ratio of CO_2/O_2 level in the root can impede root growth and its capacity to take up nutrients (Shabala, 2011). Built mechanisms in plants to cope with this situation through ultrastructural changes like aerenchyma or modulated metabolism can reduce the impact of waterlogging on crop plants. For example, compounds like 24-epibrassinolide activate the antioxidant system to protect soybean plants from waterlogging induced damage at the shoot and root level (Pereira et al., 2020).

Potassium, a macronutrient of immense importance for maintaining electrical balance and osmotic potential in plant cells, has a prominent role in abiotic stress alleviation in plants, as comprehensively reviewed by Nordstedt et al. (2020). Plant mechanisms that involve the transport of potassium and nitrate can reduce the impact of abiotic stresses. Demonstration of the possibility of engaging microorganisms like pseudomonas species, as shown in ornamental crops, suggests that plants mechanisms to interact with soil microbiome can help alleviate drought and nutrient stress in plants (Nordstedt et al. 2020). Further, the nutrient availability for crop plants is determined by weed infestation and this is conspicuous during the soil moisture deficit conditions, particularly in crops like rice under aerobic cultivation. Mechanisms that enable plants to compete with weeds can also contribute to alleviating nutrient deficiency in crop plants.

16.3.7 Unseasonal Rains and Wind

In the events of unseasonal rains, the crops yield loss occurs due to shattering of pods as in oilseeds or precocious germination of seeds as in wheat or pulses. Plants mechanisms to accumulate germination inhibitors are an essential component of plants' ecophysiology, exhibiting extraordinary adaptation to torrential rains and hailstorm. Mechanisms that control pod shattering can also influence crop plants' performance exposed to unseasonal rains and heavy wind (Liljegren et al., 2000).

16.4 Common Mechanisms that Influence Different Abiotic Stresses

It is observed that some common mechanisms get activated within the plant in responses to environmental stimuli in the form of extreme temperature, drought, salinity, high level of radiation (Fig. 16.3, Table 16.1). Osmotic adjustment and antioxidant mechanisms have been reported on several occasions. These mechanisms involve the accumulation of stress-induced metabolites (Alhaithloul, 2019). The plant requires a threshold level of reactive oxygen species (ROS) to accomplish routine physiological functions. Supra optimal concentration of ROS can be deleterious as it causes oxidative stress. Due to impaired photosynthesis, which utilises trapped solar energy, ROS concentration increases due to drought, high temperature, and salinity. Excess ROS can severely affect cell membranes and hence the ultrastructure of organelles and leakage of valuable electrolytes. Similar to drought-imposed stress situations, plants face difficulty in water uptake despite abundant water in salt-affected soils. It results in reduced water potential and turgor in plant shoots (Betzen

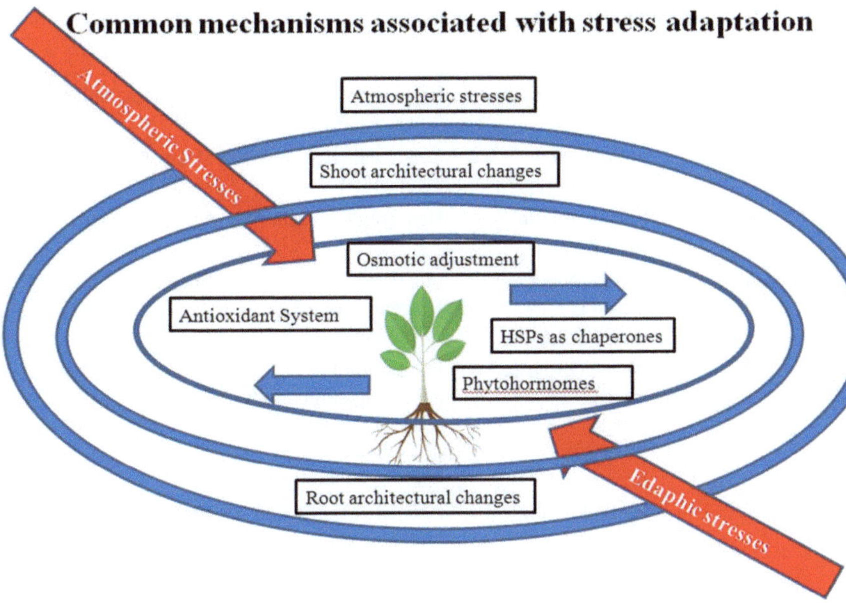

Fig. 16.3 Common mechanisms that influence different abiotic stress effects on crop plants

et al., 2019). Plant mechanisms that produce osmolytes, stress hormones, organic acids as protective mechanisms against stresses like drought and salinity (Benjamin et al., 2019; D'Amelia et al., 2018) can strengthen the ecophysiology of crop plants. The impact of abiotic stress on photosynthesis has been attributed to damage to electron transmission components (Arif et al., 2020; Zhou et al., 2019). Plants exhibit differential responses in preserving their photosystem efficiency as reflected by chlorophyll fluorescence kinetics in resilient crops like karonda and dragon fruits (Rane et al. unpublished). It indicates that protective mechanisms operating inside the plants can drive efforts to develop abiotic resilient crops.

16.5 Soil Processes Influencing Effects of Abiotic Stress on Crop-Ecology

This topic is widely investigated and discussed to enhance crop productivity in both underwater sufficient and insufficient agro-ecosystems. Predicted global climate change adds a new dimension to the approaches essential for managing soil moisture as it can extend the duration and intensity of dry and wet spells with unequal spatial and temporal distribution. Mechanisms that influence plant water relations make an essential component of eco-physiological studies.

Table 16.1 Mechanisms influencing impact of abiotic stresses on crop eco-physiology

Environmental stress	Crop	Mechanisms	Reference
High temperatures	Rice (Oryza sativa)	Early morning flowering (EMF) trait	Bheemanahalli et al. (2017)
	Common bean (Phaseolus vulgaris)	Leaf canopy cooling mechanism	Deva et al. (2020)
	Wheat (Triticum aestivum L.)	Heat shock proteins	Lu et al. (2020)
Solar radiation	Tobacco (Nicotiana tabacum L.)	Increased concentration of carotenoids	Shen et al. (2017)
	Cabbage (Brassica oleracea)	Increase in UV absorbing compounds (Flavonoids and derivative of hydroxycinnamic acid)	Neugart et al. (2012)
Drought stress	Wheat (Triticum aestivum)	Abscisic acid (ABA) metabolism	Mega et al. (2019)
	Wheat (Triticum aestivum)	Transpiration regulation by stomatal mechanisms	Kulkarni et al. (2017)
	Rice (Oryza sativa L.)	Root architectural Traits	Canales et al. (2019)
Waterlogging stress	Arabidopsis (Arabidopsis thaliana) and Rice (Oryza sativa L.)	Formation of adventitious roots	Bellini et al. (2014)
	Sunflower *(Helianthus annuus)* and Maize (Zea nicaraguensis)	Aerenchyma formation	Ni et al. (2019) Gong et al. (2020)
Salinity stress	Rice (Oryza sativa L.)	Salt overly sensitive (SOS) Pathway	Martínez-Atienza et al. (2007)
	Wheat (Triticum aestivum L.)	High sodium affinity transporter (HKT)	Munns et al. (2012)
	Tomato (Solanum pennellii)	$Na+, K+/H+$ exchangers (NHXs)	Albaladejo et al. (2017)
Nutrients deficiency	Wheat (Triticum aestivum L.)	Root system architecture	Stangoulis (2019)

Under water deficit conditions, plants adaption to their ecosystem has been attributed to three fundamental mechanisms viz. escape dehydration avoidance and dehydration tolerance that act at the whole plant and cellular levels. As in high temperature, water deficit accelerated development allows the plant to escape the terminal drought stress but at the cost of yield penalty (Rani et al., 2020). Crop plants can maintain the tissue water level by reducing the transpiration or maintaining the water uptake from deeper soil layers. Dehydration tolerance has been attributed to plants internal mechanisms to keep their functions and processes less influenced by dehydration at the cellular level. Crop plants having dehydration tolerance can recover from prolonged drought if soil moisture is replenished by natural precipitation or lifesaving irrigation.

When solar radiation is intercepted by the crop canopy, the difference between the solar incident radiation and reflected radiation by the soil surface (Villalobos et al., 2002) often determines crop development as solar energy is highly essential for fundamental physiological processes such as photosynthesis (Campillo et al., 2012). There are reports that due to radiation exposure, soil can become compact and lose the nutrients needed for plants to grow. Reduction in seed germination also reported by high light intensity (Campillo et al., 2012).

16.6 Conclusion

Abiotic stresses such as temperature, precipitation, radiation extremes, wind can be traced to the atmosphere, while salinity, alkalinity, anoxia or hypoxia due to excess moisture, suboptimum level of ions/salts are the soil factors that limit plant growth and development. Individual and interactive effects of these factors influence ambient relative humidity and soil moisture available for plant growth. Various processes inside the plant and the soil can influence the impact of predicted climate change on crop growth, development and productivity. Plants have inbuilt mechanisms to cope with abiotic stresses, and they are more conspicuous in naturally growing plants on this earth than in crop plants. Many of the traits involved in tolerance to abiotic stress were ignored during the selection and improvement of crop plants as the significant focus was on the productivity of crops. Accelerated attempts to recover the lost traits relevant to abiotic stresses in plants can add to the agriculture resilient to climate change. This task which necessarily involves knowledge of advances in sciences for genetic improvement should be complemented by parallel efforts to make the production practices climate-smart. The later will focus on soil management for climate impact management. The massive set of information generated through systematic experiments on plant responses to soil factors and agronomic practices can inform modern agriculture to mitigate greenhouse gas emissions from agriculture. Sensor-based studies can help improve resource use efficiency and precision agriculture, which can significantly contribute to coping with climate change. Many of these approaches will be agro-ecosystem specific and guided by fare knowledge of target environments for which technologies are destined. Thus, the framework for employing crop ecophysiology knowledge for climate resilience should necessarily include target delineation, genetic improvement of crop and resource management options customised for each agro-ecosystem.

References

Albaladejo, I., Meco, V., Plasencia, F., Flores, F. B., Bolarin, M. C., & Egea, I. (2017). Unravelling the strategies used by the wild tomato species Solanum pennellii to confront salt stress: From leaf anatomical adaptations to molecular responses. *Environmental and Experimental Botany, 135*, 1–2.

Alhaithloul, H. A. (2019). Impact of combined heat and drought stress on the potential growth responses of the desert grass Artemisia sieberi alba: Relation to biochemical and molecular adaptation. *Plants, 8*(10), 416.

Alvarado, K. A., Mill, A., Pearce, J. M., Vocaet, A., & Denkenberger, D. (2020). Scaling of greenhouse crop production in low sunlight scenarios. *Science of the Total Environment, 707*, 136012.

An, H., Tang, Z., Keesstra, S., & Shangguan, Z. (2019). Impact of desertification on soil and plant nutrient stoichiometry in a desert grassland. *Scientific Reports, 9*(1), 1–8.

Arif, Y., Singh, P., Siddiqui, H., Bajguz, A., & Hayat, S. (2020). Salinity induced physiological and biochemical changes in plants: An omic approach towards salt stress tolerance. *Plant Physiology and Biochemistry, 156*, 64–77.

Arunanondchai, P., Fei, C., Fisher, A., McCarl, B. A., Wang, W., & Yang, Y. (2018). How does climate change affect agriculture? *The Routledge Handbook of Agricultural Economics, 17*, 191.

Aryal, J. P., Sapkota, T. B., Khurana, R., Khatri-Chhetri, A., Jat, M. L. (2019). Climate change and agriculture in South Asia: Adaptation options in smallholder production systems. *Environment, Development and Sustainability*, 1–31.

Bailey-Serres, J., Voesenek, L. A. (2020). Genetic drivers of high-rise rice that survives deep floods. *Nature, 584*(7819).

Bashagaluke, J. B., Logah, V., Opoku, A., Sarkodie-Addo, J., & Quansah, C. (2018). Soil nutrient loss through erosion: Impact of different cropping systems and soil amendments in Ghana. *PLoS ONE, 13*(12), e0208250.

Bellini, C., Pacurar, D. I., & Perrone, I. (2014). Adventitious roots and lateral roots: Similarities and differences. *Annual Review of Plant Biology, 65*, 639–666.

Benjamin, J. J., Lucini, L., Jothiramshekar, S., & Parida, A. (2019). Metabolomic insights into the mechanisms underlying tolerance to salinity in different halophytes. *Plant Physiology and Biochemistry, 135*, 528–545.

Betzen, B. M., Smart, C. M., Maricle, K. L., & MariCle, B. R. (2019). Effects of increasing salinity on photosynthesis and plant water potential in kansas salt marsh species. *Transactions of the Kansas Academy of Science, 122*(1–2), 49–58.

Bhardwaj, A. K., Rajwar, D., Mandal, U. K., Ahamad, S., Kaphaliya, B., Minhas, P. S., Prabhakar, M., Banyal, R., Singh, R., Chaudhari, S. K., & Sharma, P. C. (2019). Impact of carbon inputs on soil carbon fractionation, sequestration and biological responses under major nutrient management practices for rice-wheat cropping systems. *Scientific Reports, 24*, 9114.

Bheemanahalli, R., Sathishraj, R., Manoharan, M., Sumanth, H. N., Muthurajan, R., Ishimaru, T., & Krishna, J. S. (2017). Is early morning flowering an effective trait to minimize heat stress damage during flowering in rice? *Field Crops Research, 203*, 238–242.

Bourbousse, C., Barneche, F., & Laloi, C. (2020). Plant chromatin catches the sun. *Frontiers in Plant Science, 10*, 1–11. https://doi.org/10.3389/fpls.2019.01728

Campillo, C., Fortes, R., Prieto, M. D. H., & Babatunde, E. B. (2012). Solar radiation effect on crop production. *Solar Radiation, 1*, 494.

Canales, F. J., Nagel, K. A., Müller, C., Rispail, N., & Prats, E. (2019). Deciphering root architectural traits involved to cope with water deficit in Oat. *Frontiers in Plant Science, 10*, 1558.

Cortleven, A., & Schmülling, T. (2015). Regulation of chloroplast development and function by cytokinin. *Journal of Experimental Botany, 66*(16), 4999–5013.

Cortleven, A., Leuendorf, J. E., Frank, M., Pezzetta, D., Bolt, S., & Schmülling, T. (2019). Cytokinin action in response to abiotic and biotic stresses in plants. *Plant, Cell & Environment, 42*(3), 998–1018.

Cox, D. T., Maclean, I. M., Gardner, A. S., & Gaston, K. J. (2020). Global variation in diurnal asymmetry in temperature, cloud cover, specific humidity and precipitation and its association with leaf area index. *Global Change Biology, 26*(12), 7099–7111.

D'Amelia, L., Dell'Aversana, E., Woodrow, P., Ciarmiello, L. F., Carillo, P. (2018). Metabolomics for crop improvement against salinity stress. In: *Salinity responses and tolerance in plants* (vol 2, pp. 267–287). Springer, Cham.

Deva, C. R., Urban, M. O., Challinor, A. J., Falloon, P., & Svitákova, L. (2020). Enhanced leaf cooling is a pathway to heat tolerance in common bean. *Frontiers in Plant Science, 11*, 19.

Dimkpa, C. O., Andrews, J., Sanabria, J., Bindraban, P. S., Singh, U., Elmer, W. H., Gardea-Torresdey, J. L., & White, J. C. (2020). Interactive effects of drought, organic fertilizer, and zinc oxide nanoscale and bulk particles on wheat performance and grain nutrient accumulation. *Science of the Total Environment, 722*, 137808.

Fábián, A., Sáfrán, E., Szabó-Eitel, G., Barnabás, B., & Jäger, K. (2019). Stigma functionality and fertility are reduced by heat and drought co-stress in wheat. *Frontiers in Plant Science, 10*, 244.

Fang, Y., & Xiong, L. (2015). General mechanisms of drought response and their application in drought resistance improvement in plants. *Cellular and Molecular Life Sciences, 72*(4), 673–689.

Flexas, J., & Gago, J. (2018). A role for ecophysiology in the'omics' era. *The Plant Journal, 96*(2), 251–259.

Flowers, T. J., & Colmer, T. D. (2015). Plant salt tolerance: Adaptations in halophytes. *Annals of Botany, 115*(3), 327–331.

Gong, Z., Xiong, L., Shi, H., Yang, S., Herrera-Estrella, L. R., Xu, G., Chao, D. Y., Li, J., Wang, P. Y., Qin, F., & Li, J. (2020). Plant abiotic stress response and nutrient use efficiency. *Science China Life Sciences., 63*(5), 635–674.

Graciet, E., & Wellmer, F. (2010). The plant N-end rule pathway: Structure and functions. *Trends in Plant Science, 15*(8), 447–453.

Harris, B. N., Sadras, V. O., & Tester, M. (2010). A water-centred framework to assess the effects of salinity on the growth and yield of wheat and barley. *Plant and Soil, 336*, 377–389.

Harrison, E. L., ArceCubas, L., Gray, J. E., & Hepworth, C. (2020). The influence of stomatal morphology and distribution on photosynthetic gas exchange. *The Plant Journal, 101*(4), 768–779.

Hatfield, J. L., & Prueger, J. H. (2015). Temperature extremes: Effect on plant growth and development. *Weather and Climate Extremes, 10*, 4–10.

Hawkesford, M. J., & Griffiths, S. (2019). Exploiting genetic variation in nitrogen use efficiency for cereal crop improvement. *Current Opinion in Plant Biology, 49*, 35–42.

Hessini, K., Issaoui, K., Ferchichi, S., Saif, T., Abdelly, C., Siddique, K. H., & Cruz, C. (2019). Interactive effects of salinity and nitrogen forms on plant growth, photosynthesis and osmotic adjustment in maize. *Plant Physiology and Biochemistry, 139*, 171–178.

Kamran, M., Parveen, A., Ahmar, S., Malik, Z., Hussain, S., Chattha, M. S., Saleem, M. H., Adil, M., Heidari, P., & Chen, J. T. (2020). An overview of hazardous impacts of soil salinity in crops, tolerance mechanisms, and amelioration through selenium supplementation. *International Journal of Molecular Sciences, 21*(1), 148.

Kataria, S., Rastogi, A., Bele, A., & Jain, M. (2020). Role of nitric oxide and reactive oxygen species in static magnetic field pre-treatment induced tolerance to ambient UV-B stress in soybean. *Physiology and Molecular Biology of Plants, 26*(5), 931–945.

Khabaz-Saberi, H., Rengel, Z., Wilson, R., & Setter, T. L. (2010). Variation of tolerance to manganese toxicity in Australian hexaploid wheat. *Journal of Plant Nutrition and Soil Science, 173*(1), 103–112.

Khan, A., Ali, M., Khattak, A. M., Gai, W. X., Zhang, H. X., Wei, A. M., & Gong, Z. H. (2019). Heat shock proteins: Dynamic biomolecules to counter plant biotic and abiotic stresses. *International Journal of Molecular Sciences, 20*(21), 5321.

Kulkarni, M., Soolanayakanahally, R., Ogawa, S., Uga, Y., Selvaraj, M. G., & Kagale, S. (2017). Drought response in wheat: Key genes and regulatory mechanisms controlling root system architecture and transpiration efficiency. *Frontiers in Chemistry, 5*, 106.

Kumar, P., & Sharma, P. K. (2020). Soil salinity and food security in India. *Frontiers in Sustainable Food Systems, 4*, 533781. https://doi.org/10.3389/fsufs

Kumari, P., Rastogi, A., & Yadav, S. (2020). Effects of Heat stress and molecular mitigation approaches in orphan legume, Chickpea. *Molecular Biology Reports, 47*(6), 4659–4670.

Liljegren, S. J., Ditta, G. S., Eshed, Y., Savidge, B., Bowman, J. L., & Yanofsky, M. F. (2000). SHATTERPROOF MADS-box genes control seed dispersal in Arabidopsis. *Nature, 404*(6779), 766–770.

Lu, S., Zhang, M., Zhang, Z., Wang, Z., Wu, N., Song, Y., & Wang, P. (2018). Screening and verification of genes associated with leaf angle and leaf orientation value in inbred maize lines. *PLoS ONE, 13*(12), e0208386.

Lu, Y., Zhao, P., Zhang, A., Ma, L., Xu, S., Wang, X. (2020). Alternative splicing diversified the heat response and evolutionary strategy of conserved heat shock protein 90s in Hexaploid wheat (Triticum aestivum L.). *Frontiers in Genetics, 11*.

Lum, T. D., & Barton, K. E. (2020). Ontogenetic variation in salinity tolerance and ecophysiology of coastal dune plants. *Annals of Botany, 125*(2), 301–314.

Mammadov, J., Buyyarapu, R., Guttikonda, S. K., Parliament, K., Abdurakhmonov, I. Y., & Kumpatla, S. P. (2018). Wild relatives of maize, rice, cotton, and soybean: Treasure troves for tolerance to biotic and abiotic stresses. *Frontiers in Plant Science, 9*, 886.

Martínez-Atienza, J., Jiang, X., Garciadeblas, B., Mendoza, I., Zhu, J. K., Pardo, J. M., & Quintero, F. J. (2007). Conservation of the salt overly sensitive pathway in rice. *Plant Physiology, 143*(2), 1001–1012.

Mega, R., Abe, F., Kim, J. S., Tsuboi, Y., Tanaka, K., Kobayashi, H., Sakata, Y., Hanada, K., Tsujimoto, H., Kikuchi, J., & Cutler, S. R. (2019). Tuning water-use efficiency and drought tolerance in wheat using abscisic acid receptors. *Nature Plants, 5*(2), 153–159.

Moreno-Jiménez, E., Plaza, C., Saiz, H., Manzano, R., Flagmeier, M., & Maestre, F. T. (2019). Aridity and reduced soil micronutrient availability in global drylands. *Nature Sustainability, 2*(5), 371–377.

Munns, R. (2002). Comparative physiology of salt and water stress. *Plant, Cell & Environment, 25*(2), 239–250.

Munns, R. (2005). Genes and salt tolerance: Bringing them together. *New Phytologist, 167*(3), 645–663.

Munns, R., & Tester, M. (2008). Mechanisms of salinity tolerance. *Annual Review of Plant Biology, 59*, 651–681.

Munns, R., James, R. A., Xu, B., Athman, A., Conn, S. J., Jordans, C., Byrt, C. S., Hare, R. A., Tyerman, S. D., Tester, M., & Plett, D. (2012). Wheat grain yield on saline soils is improved by an ancestral Na+ transporter gene. *Nature Biotechnology., 30*(4), 360–364.

Nelson, J. A., & Bugbee, B. (2015). Analysis of environmental effects on leaf temperature under sunlight, high pressure sodium and light emitting diodes. *PLoS ONE, 10*(10), e0138930.

Neugart, S., Zietz, M., Schreiner, M., Rohn, S., Kroh, L. W., & Krumbein, A. (2012). Structurally different flavonol glycosides and hydroxycinnamic acid derivatives respond differently to moderate UV-B radiation exposure. *Physiologia Plantarum, 145*(4), 582–593.

Ni, X. L., Gui, M. Y., Tan, L. L., Zhu, Q., Liu, W. Z., & Li, C. X. (2019). Programmed cell death and aerenchyma formation in water-logged sunflower stems and its promotion by ethylene and ROS. *Frontiers in Plant Science, 9*, 1928.

Norstedt, G., Axelsson, A. L., Laudon, H., & Östlund, L. (2020). Detecting cultural remains in boreal forests in Sweden using airborne laser scanning data of different resolutions. *Journal of Field Archaeology, 45*(1), 16–28.

Noya, I., González-García, S., Bacenetti, J., Fiala, M., & Moreira, M. T. (2018). Environmental impacts of the cultivation-phase associated with agricultural crops for feed production. *Journal of Cleaner Production, 172*, 3721–3733.

Onwuka, B., & Mang, B. (2018). Effects of soil temperature on some soil properties and plant growth. *Advances in Plants & Agriculture Research, 8*, 34–37.

Orosa-Puente, B., Leftley, N., von Wangenheim, D., Banda, J., Srivastava, A. K., Hill, K., Truskina, J., Bhosale, R., Morris, E., Srivastava, M., & Kümpers, B. (2018). Root branching toward water involves posttranslational modification of transcription factor ARF7. *Science, 362*(6421), 1407–1410.

Parida, A. K., & Das, A. B. (2005). Salt tolerance and salinity effects on plants: A review. *Ecotoxicology and Environmental Safety, 60*(3), 324–349.

Perdomo, J. A., Capó-Bauçà, S., Carmo-Silva, E., & Galmés, J. (2017). Rubisco and rubisco activase play an important role in the biochemical limitations of photosynthesis in rice, wheat, and maize under high temperature and water deficit. *Frontiers in Plant Science, 8*, 490.

Pereira, Y. C., Silva, F. R. D., Silva, B. R. S. D., Cruz, F. J. R., Marques, D. J., & Lobato, A. K. D. S. (2020). 24-epibrassinolide induces protection against waterlogging and alleviates impacts on the root structures, photosynthetic machinery and biomass in soybean. *Plant Signaling & Behavior, 15*(11), 1805885.

Perez, R. P., Fournier, C., Cabrera-Bosquet, L., Artzet, S., Pradal, C., Brichet, N., Chen, T. W., Chapuis, R., Welcker, C., & Tardieu, F. (2019). Changes in the vertical distribution of leaf area enhanced light interception efficiency in maize over generations of selection. *Plant, Cell & Environment, 42*(7), 2105–2119.

Pérez-Llorca, M., Casadesús, A., Müller, M., Munné-Bosch, S. (2019). Leaf orientation as part of the leaf developmental program in the semi-deciduous shrub, Cistus albidus L.: Diurnal, positional, and photoprotective effects during winter. *Frontiers in plant science, 10*, 767.

Pieruschka, R., Huber, G., & Berry, J. A. (2010). Control of transpiration by radiation. *Proceedings of the National Academy of Sciences, 107*(30), 13372–13377.

Pirasteh-Anosheh, H., Saed-Moucheshi, A., Pakniyat, H., Pessarakli, M. (2016). Stomatal responses to drought stress. *Water Stress and Crop Plants*, 24–40.

Pregitzer, K.S., King, J. S. (2005). Effects of soil temperature on nutrient uptake. In: *Nutrient acquisition by plants* (pp. 277–310).Springer, Berlin, Heidelberg.

Procházková, D., Haisel, D., & Wilhelmová, N. A. (2008). Antioxidant protection during ageing and senescence in chloroplasts of tobacco with modulated life span. *Cell Biochemistry and Function: Cellular Biochemistry and Its Modulation by Active Agents or Disease, 26*(5), 582–590.

Qian, L., Chen, X., Wang, X., Huang, S., & Luo, Y. (2020). The effects of flood, drought, and flood followed by drought on yield in cotton. *Agronomy, 10*(4), 555.

Rai, K., & Agrawal, S. B. (2017). Effects of UV-B radiation on morphological, physiological and biochemical aspects of plants: An overview. *Journal of Scientific Research, 61*, 87–113.

Ramireddy, E., Hosseini, S. A., Eggert, K., Gillandt, S., Gnad, H., von Wirén, N., & Schmülling, T. (2018). Root engineering in barley: Increasing cytokinin degradation produces a larger root system, mineral enrichment in the shoot and improved drought tolerance. *Plant Physiology, 177*(3), 1078–1095.

Rani, A., Devi, P., Jha, U. C., Sharma, K. D., Siddique, K. H., & Nayyar, H. (2020). Developing climate-resilient chickpea involving physiological and molecular approaches with a focus on temperature and drought stresses. *Frontiers in Plant Science, 10*, 1759.

Raza, A., Razzaq, A., Mehmood, S. S., Zou, X., Zhang, X., Lv, Y., & Xu, J. (2019). Impact of climate change on crops adaptation and strategies to tackle its outcome: A review. *Plants, 8*(2), 34.

Ren, X., Sun, D., & Wang, Q. (2016). Modeling the effects of plant density on maize productivity and water balance in the Loess Plateau of China. *Agricultural Water Management, 171*, 40–48.

Schymanski, S. J., Or, D., & Zwieniecki, M. (2013). Stomatal control and leaf thermal and hydraulic capacitances under rapid environmental fluctuations. *PLoS ONE, 8*(1), e54231.

Setter, T. L., Waters, I., Sharma, S. K., Singh, K. N., Kulshreshtha, N., Yaduvanshi, N. P. S., & Cakir, M. (2009). Review of wheat improvement for waterlogging tolerance in Australia and India: The importance of anaerobiosis and element toxicities associated with different soils. *Annals of Botany, 103*(2), 221–235.

Shabala, S. (2011). Physiological and cellular aspects of phytotoxicity tolerance in plants: The role of membrane transporters and implications for crop breeding for waterlogging tolerance. *New Phytologist, 190*(2), 289–298.

Shahid, S. A., Zaman, M., Heng, L. (2018). Soil salinity: historical perspectives and a world overview of the problem. In: *Guideline for salinity assessment, mitigation and adaptation using nuclear and related techniques* (pp. 43–53). Springer, Cham.

Shen, J., Jiang, C., Yan, Y., Liu, B., Zu, C. (2017). Effect of increased UV-B radiation on carotenoid accumulation and total antioxidant capacity in tobacco (Nicotiana tabacum L.) leaves. *Genetics and Molecular Research, 16*(1). https://doi.org/10.4238/gmr16018438

Soares, J. C., Santos, C. S., Carvalho, S. M., Pintado, M. M., & Vasconcelos, M. W. (2019). Preserving the nutritional quality of crop plants under a changing climate: Importance and strategies. *Plant and Soil, 443*(1), 1–26.

Soori, N., Bakhshi, D., RezaeiNejad, A., & Faizian, M. (2019). Effect of salinity stress on some physiological characteristics and photosynthetic parameters of several Iranian commercial pomegranate genotypes. *Journal of Plant Process and Function, 8*(30), 155–170.

Stangoulis, J. (2019). Variation in root system architecture and morphology of two wheat genotypes is a predictor of their tolerance to phosphorus deficiency. *Acta Physiologiae Plantarum, 41*(7), 1–13.

Stein, J. C., Yu, Y., Copetti, D., Zwick, D. J., Zhang, L., Zhang, C., Chougule, K., Gao, D., Iwata, A., Goicoechea, J. L., & Wei, S. (2018). Genomes of 13 domesticated and wild rice relatives highlight genetic conservation, turnover and innovation across the genus Oryza. *Nature Genetics, 50*(2), 285–296.

Teshome, D. T., Zharare, G. E., & Naidoo, S. (2020). The threat of the combined effect of biotic and abiotic stress factors in forestry under a changing climate. *Frontiers in Plant Science, 11*, 1874.

Uga, Y., Sugimoto, K., Ogawa, S., Rane, J., Ishitani, M., Hara, N., Kitomi, Y., Inukai, Y., Ono, K., Kanno, N., Inoue, H., Takehisa, H., Motoyama, R., Nagamura, Y., Wu, J., Matsumoto, T., Takai, T., Okuno, K., & Yano, M. (2013). Control of root system architecture by DEEPER ROOTING 1 increases rice yield under drought conditions. *Nature Genetics, 45*(9), 1097–1102.

Villalobos, F. J., Mateos, L., Orgaz, F., Fereres, E. (2002). Fitotecnia: Bases y tecnologías de la producciónagrícola. 1ª. Edición. Ediciones Mundi-Prensa. España.

Voesenek, L. A., & Bailey-Serres, J. (2015). Flood adaptive traits and processes: An overview. *New Phytologist, 206*(1), 57–73.

Wang, J., Zhang, J., Zhou, Y., Liu, H., Xue, Q., Li, X., & Jen, A. K. (2020). Highly efficient all-inorganic perovskite solar cells with suppressed non-radiative recombination by a Lewis base. *Nature Communications, 11*(1), 1–9.

Wani, A. S., Ahmad, A., Hayat, S., & Tahir, I. (2019). Epibrassinolide and proline alleviate the photosynthetic and yield inhibition under salt stress by acting on antioxidant system in mustard. *Plant Physiology and Biochemistry, 135*, 385–394.

Wasaya, A., Zhang, X., Fang, Q., & Yan, Z. (2018). Root phenotyping for drought tolerance: A review. *Agronomy, 8*(11), 241.

Xu, K., Xu, X., Fukao, T., Canlas, P., Maghirang-Rodriguez, R., Heuer, S., Ismail, A. M., Bailey-Serres, J., Ronald, P. C., & Mackill, D. J. (2006). Sub1A is an ethylene-response-factor-like gene that confers submergence tolerance to rice. *Nature, 442*(7103), 705–708.

Zakurin, A. O., Shchennikova, A. V., & Kamionskaya, A. M. (2020). Artificial-light culture in protected ground plant growing: Photosynthesis, photomorphogenesis, and prospects of LED application. *Russian Journal of Plant Physiology, 67*, 413–424.

Zhang, R., Zhou, Y., Yue, Z., Chen, X., Cao, X., Ai, X., Jiang, B., & Xing, Y. (2019). The leaf-air temperature difference reflects the variation in water status and photosynthesis of sorghum under waterlogged conditions. *PLoS ONE, 14*(7), e0219209.

Zhang, X., Mi, Y., Mao, H., Liu, S., Chen, L., & Qin, F. (2020). Genetic variation in ZmTIP1 contributes to root hair elongation and drought tolerance in maize. *Plant Biotechnology Journal, 18*(5), 1271–1283. https://doi.org/10.1111/pbi.13290

Zhou, R., Kan, X., Chen, J., Hua, H., Li, Y., Ren, J., & Yin, Z. (2019). Drought-induced changes in photosynthetic electron transport in maize probed by prompt fluorescence, delayed fluorescence, P700 and cyclic electron flow signals. *Environmental and Experimental Botany, 158*, 51–62.

Zhu, L., Bloomfield, K. J., Asao, S., Tjoelker, M. G., Egerton, J. J., Hayes, L., Lasantha, K., Weerasinghe, D. C., & Atkin, O. K. (2020). Acclimation of leaf respiration temperature responses across thermally contrasting biomes. *New Phytologist*. https://doi.org/10.1111/nph.16929

van Zonneveld, M., Rakha, M., Tan, Y. S., Chou, Y. Y., Chang, C. H., Yen, J. Y., Schafleitner, R., Nair, R., Naito, K., Solberg, S. Ø. (2020). Mapping patterns of abiotic and biotic stress resilience uncovers conservation gaps and breeding potential of Vigna wild relatives. *Scientific Reports, 10*(1), 1.

Zulkarnain, Z., Eliyanti, E., Swari, E. I. (2019). Pollen viability and stigma receptivity in Swainsona formosa (G. Don) J. Thompson (Fabaceae), an ornamental legume native to Australia. *Ornamental Horticulture, 25*(2):158–167.

Chapter 17
Impact of Climate Change on Agriculturally Important Insects and Nematodes

Victor Phani, Tushar K. Dutta, Achintya Pramanik, and Jaydeep Halder

Abstract Extreme weather conditions accompanying the climate change potentially impact the global food production system including its pest profile scenario. Insect and nematode life strategies are largely governed by several climatic factors like temperature, precipitation, humidity, and atmospheric carbon dioxide level. Being the prominent environmental drivers of organismal biology, any fluctuation in these parameters affect the agriculturally important pest species as well. Temperature is one of the most pivotal environmental components, and global warming by rise in average temperature of the earth impacts the biogeographical range, distribution, abundance, survival, fitness, reproduction and growth rate of the agriculturally important insect and nematode pests. Further, changes in the macro- and micro-environment also affect the insect and nematode relied tritrophic interactions. However, the climate change driven consequences exert both positive and negative effects on these organisms. On a generalized account, a noticeable shift can be expected in the plant protection scenario in near future due to shifted pest problems. In this line, adequate monitoring of the pest population dynamics in relation to global climatic fluctuations will lead us to development of modeling based prediction tools and also help to modify the integrated pest management tactics to combat the disaster in a species and location specific manner.

V. Phani (✉)
Department of Agricultural Entomology, College of Agriculture, Uttar Banga Krishi Viswavidyalaya, Dakshin Dinajpur, West Bengal 733133, India
e-mail: victorphani@gmail.com

T. K. Dutta
Division of Nematology, ICAR-Indian Agricultural Research Institute, New Delhi 110012, India

A. Pramanik
Discipline of Agricultural Entomology, Quality and Productivity Improvement Division, ICAR-Indian Institute of Natural Resins and Gums, Ranchi, Jharkhand 834010, India

J. Halder
Division of Crop Protection, ICAR-Indian Institute of Vegetable Research, Varanasi, Uttar Pradesh 221305, India

H. Pathak et al. (eds.), *Climate Change Impacts on Soil-Plant-Atmosphere Continuum*, Advances in Global Change Research 78,
https://doi.org/10.1007/978-981-99-7935-6_17

Keywords Agriculture · Climate change · Insect · Nematode · Pests

17.1 Introduction

Climate change refers to the paradigm long-term changes in earth's meteorological conditions like temperature, precipitation and wind pattern by natural and anthropogenic causes. One of the most pervasive outcomes of climate change is reflected through global warming, which has resulted in rise of earth's average temperature. According to United Nation's report, the last decade (2011–2020) was 1.1 °C warmer as compared to the pre-industrialization era, and the concentration of greenhouse gases measured to be the highest in previous two million years. Recently, the UNEP Emissions Gap Report (2022) predicted that the current track for temperature rise may reach up to 2.4–2.6 °C increase by the end of this century (www.unep.org/resources/emissions-gap-report-2022). Although, the most prevalent drivers of this catastrophe are reckoned to be the greenhouse gases (GHGs; methane, carbon dioxide and nitrous oxide) enormously emitted by human activities, fossil-fuel burning and changes in land-use patterns; the role of external forces like solar luminosity, volcanic eruptions, radiation budget, and earth's orbital variations round the sun cannot also be overruled. Amongst all the contributing factors, carbon dioxide is the most abundant driver that has dramatically increased from a pre-industrial average of 280 to 416 ppm at present; and is likely to be doubled by 2100 (https://gml.noaa.gov/ccgg/trends/). Many people link the consequences of climate change to be abiotically restricted to warmer temperature only; but it has far-reaching effects on the earth's biome pertaining to agriculture, wildlife, ecosystems, and biological extinction rates.

With exceptional population hike in the present era, the global food production needs to be doubled by 2050 for ensuring everyday meal to all (Tilman et al., 2011). Several recommendations have been made to maintain the food security in a sustainable way, where the environment needs protection with its components. Globally climate change exerts simultaneous positive and negative effects on agriculture bringing various uncertainties to the production systems (Tirado et al., 2013). The rising atmospheric temperature, elevated level of carbon dioxide, and other natural extremes like heat wave, altered precipitation, flood and drought have drawn considerable attention as potential factors to affect the agricultural production. When studied in relation to agriculture, a vast majority of the insects and nematodes offer beneficial services as inseparable partners and drivers of the agroecosystem. Besides, they also act as potential biotic stressors. The change in climatic factors directly affects the distribution, reproduction, population dynamics and survival of the pest species along with their interactions with natural enemies and environment (Skendžic et al., 2021). The patterned shift of pest biology and ecology eventually pose an alarming threat to the crop growers, where advocating a seminal solution to mitigate the damage becomes problematic.

17.2 Insects and Nematodes in Agroecosystem—Climate Change Continuum

A dominant form of earth's land management involves agriculture, and agroecosystem constitutes nearly 40% of the terrestrial land surfaces. Although agroecosystem is primarily provisioned by the humans, it strongly depends on natural unmanaged services (Power, 2010). For example, the two main services flowing in an agroecosystem, i.e., biological control of pests and pollination services are supported by the natural ecosystem. Besides, nutrient cycling, structure and fertility of soil, and landscape structures also contribute to this (Power, 2010). Greatly influenced by anthropogenic practices, these natural processes though provide foundation services to agroecosystems; they also favor few disservices like loss of biodiversity conservation, nutrient loss, and pesticide poisoning to the non-targets (Zhang et al., 2007). Worldwide the agroecosystems reveal enormous variation in their structure and function, and such variability results in large miscellany of the ecosystem services. Thus, conversion of undisturbed natural system into agriculture exerts strong pressure on the ability of existing systems for ecosystem service generation. However, an agroecosystem offers several beneficial deeds providing the opportunity to consider it as an intermediate phase of urban and wild ecosystems in the human impact continuum (Swinton et al., 2007).

The insects and nematodes form an integral part as ecosystem drivers playing major roles in the natural and agricultural ecosystems. Natural biodiversity aid in agricultural sustenance, as the agroecosystems cannot survive extensive disturbances for lack of diversity. In land ecosystems, the insects play key roles like bioturbation, seed dispersal, pollination, nutrient cycling and pest control (De Groot et al., 2002; Gabriel & Tscharntke, 2006; Jankielsohn, 2018; Nichols et al., 2008; Slade et al., 2016). Although in conventional agriculture, emphasis is given on few preferred insect species, the collective role of many species remains ignored or un-acknowledged (Hoehn et al., 2008). Insects constitute about 66% of all the known living species (Zhang, 2011), and show great variation in respect to abundance and species richness. They occupy all the trophic niches ranging their functions as herbivores, carnivores and detritus feeders. For ecosystem recycling, the herbivorous insects spatiotemporally modify the quality and quantity of plant detrital inputs that potentially exert large effects by transforming living vegetal biomass into throughfall, greedfall and frass (Belovsky & Slade, 2000; Hunter, 2001). Apart from nutrient recycling, insects also participate in other ecological processes like pollination, predation, parasitism and decomposition (Jankielsohn, 2018). About two-third of the flowering plant species, including the agricultural crops, rely on insects like butterflies, bees, beetles and flies for pollination (Schoonhoven et al., 2005). Further, many insects of higher trophic levels regulate the density and abundance of the primary consumers keeping their numbers under a definite threshold level (van Lenteren, 2012). The nematodes on the other hand are soil-dwelling organisms, and play pivotal roles in essential soil processes contributing directly or indirectly. The bacterivorous, fungivorous and predatory nematodes feed upon the soil microbes and other nematode species contributing to nitrogen and carbon mineralization process and their recycling (Ferris & Bongers, 2006). They also help in distribution of the microbes, regulating their turnover in uninfected soils indirectly helping in mineralization. Furthermore, the feeding of plant-parasitic nematodes (PPNs) on the plants affect host physiology that in turn change the root exudation chemistry and help to modify the microbial population in the plant rhizosphere (Neher, 2001; Tu et al., 2003). The herbivory changes the biomass allocation pattern and shapes the above-ground vegetation community (Brinkman et al., 2015; Franco et al., 2020; Hedenec et al., 2022). The plant parasites also aid in nutrient cycling and soil carbon increase (Yeates et al., 1993).

While enumerating the effects of climate change in different trophic levels of an ecosystem, not only the temperature variability poses the greatest threat; changes in soil and atmospheric chemistry, precipitation and water balance also alter the ecosystem stability. As a result, the ecosystems differ in their sensitivity and response because of remodeling in the underlying interactions among its organisms (Malhi et al., 2020). Changes in natural and agricultural ecosystems primarily put biodiversity at risk by bringing abrupt and irreversible changes. Nonetheless, these changes are difficult to observe and stochastically seldom predictable (Malhi et al., 2020; Turner et al., 2020). For example, climate change driven drought, fires and floods lead to abrupt changes in the population dynamics and structure of the land species, and can make the population sensitive to an introduced pathogen that affect multiple ecosystem drivers. The resilience and vulnerability of the natural species population, in response to the biophysical feedbacks, can also lead to complete shift of ecosystem composition. The frequency and magnitude of disturbances exert strong pressure on the below-ground faunal community making them vulnerable (Bardgett & Caruso, 2020). As a result, the growth rate, resource use efficiency and energy flow pattern changes within the trophic levels. The complementary attributes of the slow and fast energy channels in the soil may destabilize the ecosystem when intrinsically altered (Malhi et al., 2020). But, the effects on soil macro- and microbial communities are well recognized under chronic stress than stress posed by abrupt changes. Further, climate change drives the terrestrial biodiversity loss via change in land use pattern. Molotoks et al. (2020) found the expansion of crop lands to negatively impact the biodiversity and carbon storage especially in the biodiversity hotspots changing the beneficial and harmful species diversity. Another serious phenomenon linked to climate change is the extinction of specific community in an ecosystem. As the biologists agree that the sixth major extinction event is rapidly proceeding behind our eyesight, we can only estimate the extinction rates from continuous observation of the fossil records (Raven & Wagner, 2021). Climate change acts as a major driver of the biological extinctions, aggravating its rates. Additionally, the elevated rate of disease and habitat degradation also indirectly aid to this catastrophe along with physiological and genetic change-driven extinctions ultimately affecting the species richness and associated services in the natural and agricultural ecosystems.

17.3 Insects and Nematodes as Agricultural Threat Under Climate Change

The Food and Agricultural Organization (FAO) estimates around 20–40% annual loss in global food production by the agricultural pests and diseases, where invasive insects alone account for approximately US$ 70 billion each year (https://www.fao.org/). Insect pests have always been recognized as a major competitor of humans for sharing the agricultural resources. These macroscopic flying organisms are largely favored by the intensive use of agricultural inputs and continuous monocultures in large areas practiced to meet the demand of the increasing population (Oerke & Dehne, 2004). They incur severe crop losses under open-field (pre-harvest) and storage (post-harvest) conditions (Oerke, 2006). Precise estimation of insect pest damage in the agri-horticultural systems is difficult as a number of factors like cultivated plant genotypes, environmental conditions, agronomic practices and socioeconomic conditions of the growers altogether contribute to the damage occurrence. In this line, certain governmental approaches in few countries were able to systematically evaluate the crop losses by insects in a scattered way (Yudelman et al., 1998). Additionally, the direct damage of insect pests also predisposes the affected plant parts to other microorganisms that exacerbate the damage intensity.

The PPNs, on the other hand, are microscopic soil-dwelling obligate parasites of the below-ground plant parts and occasionally damage the above-ground parts. Around 4100 species of PPNs exert an estimated yearly loss of US$ 173 billion to the agricultural and horticultural crops under open-field and protected conditions (Elling, 2013; Jones et al., 2013; Phani et al., 2021). They either induce differentiated feeding cells on vascular tissues (e.g., syncytia by cyst nematodes, giant cells by root-knot nematodes) and pericycle (nurse cells by reniform nematode), or feed on the cortical tissues from inside or outside the roots affecting the nutrient availability to the plants. Few of the PPN species remarkably maneuver the plant host physiology, thus developing intimate host–pathogen relationship (Siddique et al., 2022). Additionally, the PPNs also predispose the damaged plants to fungal, bacterial and viral pathogens synergizing the damage toll. Unlike insect pests, in spite of the heavy damage caused by the PPNs, they often remain underrated which eventually leads to development of severe population build-up beyond satisfactory management.

It is difficult to generalize the effects of climate change on the insect and nematode pests. The diversity, abundance and pathogenic potential of these parasites and pathogens are largely determined by climatic factors. Development, survival and behavior of the host seeking stages of insects and nematodes are directly governed by temperature, humidity, precipitation pattern and climatic extremes. It can be apprehended that the warmer and humid climate will favor the parasites and pathogens in terms of increased and longer period of transmission, higher reproduction, quick development, and spatiotemporal shift in their diversity (Lafferty, 2009; Pascual & Bouma, 2009; Randolph, 2009). As the macro- and micro-parasitic abundance in an agroecosystem is directly correlated to the degree of host damage and crop loss, the increased pest burden will definitely bring more sufferings to the hosts. But, under certain geo-spatial isolations the effects are also observed to be neutral or negative (to the biological and ecological facts) as climate change effects are not equally distributed allover the earth.

17.4 Effect of Climate Change on Insect Pests

Global climate change exerts both direct and indirect effects on the agriculturally important insects. Direct impacts are likely to be on the insect feeding, development, reproduction, population dynamics, distribution, summer and winter survival, dispersal, abundance, voltinism; whereas indirectly it will affect the relationships between pest species, environment and other insects including vectors, competitors, mutualists and natural enemies (Prakash et al., 2014).

17.4.1 Impact on Insect Distribution and Abundance

On a general note, insect distribution and abundance is fundamentally governed by natural biogeography, distribution of crop pattern, agricultural practices and climatic factors (Skendžic et al., 2021). Climate change exerts pronounced effects on geographical distribution of the insect pests, and the temperate areas are often more affected than the tropical or sub-tropical regimes (Meynard et al., 2013). Farmers face severe problem due to spread and invasion of insects in newer areas along with alteration in their biological and physiological traits (Fand et al., 2012). The insects tend to shift towards the earth poles (Bebber et al., 2013). For example, European corn borer *Ostrinia nubilalis* has shifted about 1000 km northward from its native region of infestation (Porter et al., 1991). Similar shifts were also noticed for various Lepidopteran species. Increase in global average temperature has resulted in expansion of range and better hibernation survival of the cotton bollworm *Helicoverpa armigera* and corn earworm *H. zeae* in the American states (Fand et al., 2012). Coulson et al. (2002) pointed that higher fluctuations of warm air masses towards the Arctic Ocean drove the diamond back moth *Plutella xylostella* around 800 km north away from its native range in Russia widening the area of infestation. Similarly, the temperature driven shifts are also being observed for the cotton pink boll worm *Pectinophora gossypiella* and olive fly *Bactrocera oleae* in several regions (Gutierrez et al., 2006, 2009). Besides, elevated temperature and humidity levels positively correlates with the population build-up of many sucking insects in newer areas (Halder et al., 2019; Pathania et al., 2020). It was observed that the adult females of rice leaf folder *Cnaphalocrocis medinalis* cannot lay eggs if the pupae are reared at higher temperature than 35 °C (Heong et al., 1995). These shifts suggest that the formerly unfavorable regions are becoming favorable for the insects due to change in major and minor climatic factors, and vice versa.

Climate change alters the flight phenology of the insects, as seen in the sucking pests, thereby changing their time of migration and establishment. For example, the potato aphid *Myzus persicae* begins spring flight about two weeks earlier for every 1 °C increase in the mean temperature (Shrestha, 2019). The shift in migration timing of aphids eventually leads to an abundance of the concerned species by providing the population better chance to proliferate the damaging population (Zhou et al., 1995). Decrease in the frequency of spring frosting due to global warming results in extended periods and intensity of insect epizootics (Raza et al., 2014). Furthermore, the rise in average temperature also increases the chance of overwinter survival of the insects at higher altitudes, thereby expanding their abundance and geographic range (Pareek et al., 2017).

The increase in temperature adversely affects the insect pests and their natural enemies relationships. It was observed that with 2 °C increase in the atmospheric temperature rice brown plant hopper *Nilaparvata lugens* adds additional life cycles per season (Chakravarthy et al., 2013). This insect is almost 17 times more tolerant to a temperature hike up to 40 °C than its natural predator *Cyrtorrhynus lividipennis.* Certainly, the rapid temperature increase in summer months of tropical and subtropical regimes leads to huge proliferation of the pest in rice fields. But, Dabhi et al. (2011) showed a negative correlation between the increasing temperature and activity of the Hymenopteran parasitoid *Bracon hebetor*. A decrease in efficiency of the natural enemies thus affects the abundance of the existing insect species.

It was found that an increase in the atmospheric temperature by 10 °C influences the insect physiology by doubling their metabolic rates (Clarke & Fraser, 2004; Lactin et al., 1995). Even a modest change in the temperature, humidity or other intricate factors immediately affect the dispersal and abundance of the insects through their fecundity and reproduction rate. In this line, the flour beetle species *Tribolium confusum* surpasses *T. castaneum* when reared at 30% relative humidity and 34 °C temperature; whereas at same temperature the outcome become reversed if the RH is increased to 70%. Organismal reproduction and growth also sustain better at higher temperatures than in low temperatures. Higher temperatures result in extra generations in different insect species and alter voltinism characters (Pollard & Yates, 1993; Martín-Vertedor et al., 2010; Chen et al., 2011). For example, Morgan (1996) found 2 °C increase in temperature to shorten the generation period of the bird cherry aphid *Rhopalosiphum padi*. Elevated temperatures also hasten the development of onion maggot, cabbage maggot, Colorado potato beetle, European corn borer by adding more generations resulting in extended crop damage per year. For example, the cabbage maggot *Delia radicum* completes three generations when emerge early, however only two generations are completed for the late emerging population (Hance et al., 2007; Turnock & Boivin, 1997). The lately emerged insects also suffer from higher rate of mortality as seen in North America during long summers. An increase in number of generations subsequently lead to greater number of insects in a season, which are forced to disperse to newer areas in search of food and better access to resources. Such effects on distribution and abundance alteration are not only restricted to the temperate parts of the world but also reach to the tropical and sub-tropical parts.

It has been predicted that the number of generations for coffee berry borer *Hypothenemus hampei* will increase from five to ten in the *Coffea arabica* producing areas thereby affecting the production in East Africa (Jaramillo et al, 2011). However, it is difficult to enumerate and describe all the precise effects of climate change on each species of the insect pests for prevalence of huge diversity in global agriculture with respect to its ecosystem and environmental frameworks. Moreover, the shifts in insect migration and abundance also correspond to several geo-physiological phenomena that alter the gross damage profile in the host plants.

17.4.2 Impact on Insect Herbivory and Direct Injury to Crops

If all other factors remain constant, at a higher temperature, insect pests are expected to feed faster and complete the life cycle within a shorter span of time as per the theory of thermal constant and growing degree days model. Higher ambient temperature will be accompanied by elevated CO_2 and water stress leading to altered host plant philology. Hence, impact on insect pest feeding and direct injury to crops will mainly be imparted through combination of all the major climate change drivers. Elevated CO_2 levels are likely to change the C:N ratio of plants, which negatively affect the foliage quality with nutrients determining the palatability to leaf-eating insects (Lincoln, 1993). Nitrogen is an essential element required for development and reproduction of the insects, and therefore elevated atmospheric CO_2 concentration is expected to cause increased foliage feeding rate in most insects (Bezemer et al., 1998). It may result in increased damage to plant vegetation, as pests must consume more tissue with compensatory feeding to obtain an equivalent level of essential nutrients like nitrogen (DeLucia et al., 2008; Hamilton et al., 2005). Hence increased consumption rates are more obvious in the defoliaters like caterpillars, miners, nibblers and chewers. Several studies have been conducted in recent years where crop plants were grown under elevated and ambient levels of atmospheric CO_2 concentrations to assess the impact on insect pests. For example, Hamilton et al. (2005) demonstrated about 57% more damage in soybean due to major pests like Mexican bean beetle (*Epilachna varivestis*), Western corn rootworm (*Diabrotica virgifera virgifera*), Potato leafhopper (*Empoasca fabae*), and Japanese beetle (*Popillia japonica*) when grown under elevated CO_2 conditions. The study further concluded that increase in C:N ratio in soybean foliage might have triggered the compensatory insect pest feeding causing more plant damage. However, increased feeding rates and consumption of plant tissues do not always compensate for poor diet quality under elevated CO_2 conditions, but may also reduce the efficiency of the feeding insect species (Zvereva & Kozlov, 2005). Moreover, insect responses to elevated CO_2 fertilization may vary depending on the type of insect feeding. Whole-cell foliage feeders like thrips showed a surge in population size (Bezemer et al., 1998), whereas phloem-feeding sap suckers (e.g., whiteflies, jassids, mealy bugs, aphids) showed contrasting responses with decreased population density and increased growth rates (Sutherst et al., 2011).

The effect of elevated CO_2 on pest feeding and direct injury is not only explicit to the insect species, but also to individual insect pest–host plant system (Coviella & Trumble, 1999). Elevated levels of CO_2 would likely to have higher impact on C_3 crops (rice, wheat, cotton, pulses etc.) than on the C_4 crops (sorghum, corn, sugarcane etc.). These differential effects of on C_3 and C_4 plants may result in asymmetric effects on insect herbivory, since their feeding response may vary for C_3 and C_4 plants (Skendžic et al., 2021). Hence, elevated CO_2 is expected to positively affect the C_3 plants with negative impact by increased insect pest damage; whereas the C_4 plants are less responsive to elevated CO_2 and insect pest feeding are less likely to be affected by the level of CO_2 rise (Lincoln et al., 1984).

17.4.3 Impact on Insect Development, Reproduction and Population Dynamics

Climate change will impact the insect development, reproduction and population dynamics directly through temperature hike (DeLucia et al., 2008; Deutsch et al., 2018), and indirectly in terms of elevated CO_2-mediated alteration in host physiology. Insect physiological processes are highly temperature-sensitive where metabolic rates become almost doubled with an increase of 10 °C (Dukes et al., 2009). In this context, it is inevitable that elevated temperature will accelerate the development, food consumption and movement of the insects. As ambient temperatures move toward optimum for growth and development reducing the most critical thermal constraints on insect population dynamics, pest infestation severity is likely to escalate (Deutsch et al., 2008). Nevertheless, given the physiological tolerance limit, narrow ecological niche requirements, and differential effects on individual species phenology and life history due to variable temperatures, global warming is not likely to uniformly impact economic crop losses in all the places (Lehmann et al., 2020; Skendžic et al., 2021). Certainly, there will be spatial difference in the type of response received from insect pests. In the tropical regions where ambient temperature is already on the higher side of the optimum range for survival, insect population may decline with further increase in temperature. Whereas in the temperate regions where insect growth is limited due to short span available growing temperature are most likely to experience gradual surge in existing insect pest populations as well as introduction of newer pests from tropical regions (Deutsch et al., 2018). For example Deutsch et al. (2018) estimated changes in the pest population growth in relation to world's top three staple food crops (maize, rice and wheat) under variable climate change scenarios. The study demonstrated that pest population growth would accelerate for wheat due to global warming, which is predominantly grown in temperate areas. For rice grown in the tropical zones, a decrease in the pest population growth was predicted, whereas mixed responses were predicted for maize grown in temperate and tropical regions. In another study, Lehmann et al. (2020) showed increased pest severity in response to temperature rise. Around 59% of all species analysed in the study showed reduced harmful impact mostly via range contraction and jeopardized physiological performance (Skendžic et al., 2021). Another study involving about 1100 species from different insect genera found global warming to reduce the chance of survival for many species, and drive about 15–37% of these species to extinction by 2050 (Hance et al., 2007; Thomas et al., 2004).

The effects of increased temperatures can be different for above-ground (e.g., aphids, whiteflies, stem borers, defoliators etc.) and soil-dwelling insects (e.g., termites, root grubs etc.). The impact is expected to be greater for aboveground insects than for soil-dwelling insects that spend most of the time in their life under the soil. This opposite effect can be expected for soil being a thermally insulating medium that can efficiently buffer the changes in atmospheric temperature and thus impact less on the soil-dwelling insects (Bale et al., 2002).

In comparison, scant reports are available on the effects of elevated CO_2 on insect pest development and reproduction. It is anticipated that abundance and fecundity may increase for most insect pests (Hance et al., 2007). Further, the effect may be different for chewing and sucking pests. Stiling and Cornelissen (2007) conducted a meta-analytic study to interpret the indirect effects of elevated CO_2 on life history parameters of herbivores. The study found an average of (I) 22% decrease in pest abundance; (II) 17% increase in consumption rates; (III) 9% decrease in relative growth rate; and (IV) 4% increase in development time. More substantial effects of elevated CO_2 were recorded for chewing pests in contrast to sap suckers. Overall the results showed robust responses of insect pests to elevated CO_2 compared to ambient CO_2 (Skendžic et al., 2021).

Frequent occurrences of extreme climate events like drought, storms, heat waves, floods and wildfires are expected due to climate change scenarios. Such events might have a significant effect on insect pest population dynamics. The occurrence of drought may affect the herbivorous insects through different ways, like (1) it may provide a congenial ambience for the growth and developments of certain pest species (e.g., whiteflies, thrips etc.); (2) drought-stressed plants may turn more attractive to certain insect pests (e.g., some bark beetles are more attracted to drought-stressed plants); (3) plants stressed by drought become resource constraint leading to decrease in the production of defensive secondary metabolites. This may lead to breakdown of host plant resistance and cause more insect pest damage (Skendžic et al., 2021). In contrast, frequent tropical storms may wipe out local insect population and cause temporal extinction over a region. Wildfires and untimely heavy rain and flooding is also detrimental for soil-dwelling insects and their frequent occurrence may cause population decline of certain species.

17.4.4 Impact on Insect Voltinism

Amongst the major drivers of climate change, elevated temperature is expected to positively impact the insects while the others like drought stress and extreme climatic events might affect them negatively. According to the ambient energy hypothesis rate of growth and reproduction are higher at elevated temperatures. Hence, global warming is most likely to cause greater population sizes of insect pests that in turn may lead to higher species numbers in dynamic equilibrium (Menéndez, 2007; Menéndez et al., 2007). Within a certain range of preferred temperature, the number of generations of pest species will upsurge, leading to more crop damage (Yamamura & Kiritani, 1998). Further, to predict the insect pest and plant phenology in agriculture growing degree days (GDD) model can also be followed where the daily total sum of degrees between a maximum and minimum temperature threshold is measured and annual heat accumulation is calculated (Cayton et al., 2015).

Analyses of long-term phonological data showed that the timing of insect pest emergence would advance under climate change (Pathak et al., 2012). The relative abundance of spring populations of aphids ranges from very high to very low depending on the duration of exposure to the prevalent temperature of the previous winter (Harrington et al., 2007). Advancement in the first occurrence after winter has also been reported for butterflies in UK, where 26 insect species have been found to advance their first appearance (Menéndez, 2007; Roy & Sparks, 2000). In Spain, the first appearance of 17 species of butterflies has advanced earlier by 1–7 weeks in a short span of 15 years (Menéndez, 2007; Stefanescu et al., 2003). Global warming has most likely resulted in early emergence of European grapevine moth (*Lobesia botrana*) in Spain. This trivoltine pest was able to add one additional generation leading to increased damage potential in grapevine orchards (Martín-Vertedor et al., 2010). On the other hand, impact of elevated temperature will be different in extent and magnitude for univoltine and multivoltine species. If all other factors remain constant, then multivoltine species like aphids, whiteflies and most caterpillars will develop rapidly and will likely complete additional generations within the growing season (Bale et al., 2002; Pollard & Yates, 1993). Using several models, Yamamura and Kiritani (1998) inferred up to five annual additional life cycles of insect pests due to 2 °C increase in temperature. Similar observations were also observed for aphids, where a low developmental threshold led to shorter generation time with four to five additional generations per year (Menéndez, 2007). Moreover, higher temperatures during the early developmental phase will shorten the time for larval and nymphal stages and might favor rapid population build-up due to escape from predators and parasitoids during the susceptible life stages (Bernays, 1997; Menéndez, 2007).

17.4.5 Impact on Insect-Vectored Disease Transmission

Along with direct injury through feeding insect pests also cause indirect damage as the vector of plant viruses with varying transmission efficiency. The prevalence, persistence and spread of viruses depend on the type of insect vectors, their host plants and the surrounding climate where they thrive (Hull, 2014; Skendžic et al., 2021; Trębicki, 2020). Since insect physiology and behaviour is directly affected by climate, it is supposed to indirectly affect the virus transmission. Thus climate change is most likely to play a pivotal role on the epidemiology of plant viral diseases (Trębicki et al., 2016). Such impact can bear positive, negative or neutral consequences for viral diseases in crop plants (Trębicki, 2020).

Most of the plant viruses are transmitted by sap-sucking Hemipterans like aphids, whiteflies and leafhoppers. While the aphid transmitted diseases are more prevalent in temperate zones, whiteflies are serious vectors in warmer regions and also thrive well in greenhouse conditions under temperate climate (Canto et al., 2009). High reproductive potential coupled with short generation time makes all such insect vectors very sensitive to climate change (Hance et al., 2007). Being small in size they can be dispersed through passive flight assisted by atmospheric air movements. Aphids also have long-distance migration potential, which is most likely to make the vector-transmitted disease problem more severe (Fereres et al., 2017). Such long-distance passive flight by low-pressure North American winds has been linked with viral epidemics of corn diseases vectored by aphids in northern corn-growing areas in Minnesota (Zeyen et al., 1987). Moreover, aphid survival rate will increase in warmer winters, and rise in early spring temperatures will also increase their rate of reproduction and development. The severity of viral diseases increases with early inoculation and a higher amount of initial viral load which is enhanced by the over-wintering success of its insect vectors (Irwin et al., 2000). All these factors lead to higher incidence of viral disease in crop plants (Alonso-Prados et al., 2003). This is evident from the report that viral diseases of potato have increased during recent years in Northern Europe due to earlier (at the onset of growing season) infestation of aphids that are the main vectors of the viruses (Robert et al., 2000). Amongst the white fly species, the silver leaf whitefly (*Bemisia tabaci*) and greenhouse whitefly (*Trialeurodes vaporariorum*) are considered to be highly notorious virus vectors. Irrigated crop plants grown under dry climate with high temperature are the favorable host for *B. tabaci* population build-up (Sseruwagi et al., 2004). Thus climate change may lead to development of large breeding populations in early spring or summer. Moreover, such scenario may lead to faster evolution of the plant viruses due to longer available proliferation period, resulting in the development of virulent viral strains with greater transmission efficiency and broader host range.

While global warming may favor greater incidence of insect-transmitted viral diseases due to increased abundance and expansion of geographic range of the vectors (Sharma, 2014; Sharma et al., 2005), the outcome may not always be positive. For example, barley yellow dwarf virus (BYDV) is a serious disease which is transmitted by different aphid species. In Central Europe, it has been found that 8 °C is minimally required for spring migration of the bird cherry aphid *R. padi*, the primary vector of BYDV. Thus global warming should favor early aphid population build-up and subsequently greater incidence of BYDB. But that is less likely to be happen as when temperatures reach above 36 °C in summer months, aphid survival decreases leading to lower spread of BYDV in subsequent months (Parry et al., 2012). Moreover, prolonged drought may negatively affect the sucking pests by decreasing survival rate, disrupting development and dispersal (Sutherst et al., 2011).

Another important effect can be observed through change in geographical distribution of prevalent vector mediated diseases. Flavescence dorée is an alarming phytoplasma disease of the grape vine in Europe (Boudon-Padieu, 2003), vectored by

American grapevine leafhopper *Scaphoideus titanus* (Falzoi et al., 2014). Currently, the disease is most prevalent in the middle of Europe while the northern limit of this species is Ukraine (Mirutenko et al., 2018). Due to global warming, *S. titanus* is expanding its range towards the north (Boudon-Padieu & Maixner, 2007). Earlier short summer growing season was a barrier to the northern spread of this species due to the insect's inability to complete life cycle (Boudon-Padieu, 2000; Mirutenko et al., 2018). But, global warming may favor its spread even in the northern vineyards of Germany due to availability of favorable longer development period (Boudon-Padieu & Maixner, 2007). Moreover, its current distribution range may be contracted in areas like southern Italy due to warming, leading to small-range local extinctions or population declines of *S. titanus* (Boudon-Padieu, 2000). Thus global climate change can add new areas for certain vector borne diseases while it is also likely that some of the endemic areas may become disease free due to local extinction of vector populations.

17.4.6 Impact on Insect Survival

Insects are one of the most successful animals due to their ability to control metabolic rates in response to adverse environments. Such a hormonally regulated state of suppressed metabolic activity is called diapause which adaptively helps in the seasonal regulation of insect life cycles. Diapause can be obligate (mostly in univoltine species) or facultative (mostly in multivoltine species) and is influenced by temperature, photoperiod and humidity (Gill et al., 2017; Taylor et al., 1986). Diapause is of two type, i.e., hibernation (for winter survival) and aestivation (for summer survival). Insects being cold-blooded animals, winter survival is critical for many insect pests since freezing during winter months can become major mortality factor and consequently, lower breeding population is available for following crop season (Hill, 1987). Hence, insects that need to hibernate are most likely to experience the greatest changes in their thermal environment (Bale & Hayward, 2010).

Climate change will make normal weather events more unpredictable significantly impacting diapause and its induction, maintenance, termination, and post-diapause quiescence (Bale & Hayward, 2010). For example, if late autumn temperature increases in future, hibernation initiation is expected to be delayed for many temperate insect species. In such a scenario, sudden onset of cold wave during early winter will create the risk of cold stress for the hibernation-destined species prior to establishing their bodily cold tolerance mechanisms leading to significant mortality (Bradshaw & Holzapfel, 2010). Diapause duration is also going to change in future as hibernation duration is shorter at higher temperatures since metabolic rate is higher under higher temperature and energy reserves are depleted quickly leading to early diapause termination. For example, the diapausing stage lasts for 118, 70 and 57 days at 17, 25 and 28 °C, respectively after hibernation initiation in flesh fly *Sarcophaga crassipalpis* (Bale & Hayward, 2010; Hahn & Denlinger, 2011).

Early diapause termination will lead to phonological asynchrony with the host plant during early summer (Bale & Hayward, 2010; Parmesan, 2007). For example, many insect pests of plantation crops synchronize oviposition and the emergence of neonates with the timing of a new spring flash. Such synchrony between different trophic levels could become uncoupled if climate change affects plant phenology and insect physiology differently (Harrington et al., 1999). For example, neonate larval emergence of winter moth (*Operophtera brumata*) is markedly advanced compared to budburst timing on its host plant pedunculate oak (*Quercus robur*). A mere 2 °C increase in temperature is least likely to advance bud breaking in the host plant, but the same can advance the egg hatching of *O. brumata* significantly. The early larva will be starved to death and can lead to population reduction and extinction of this species if it fails to adapt to climate change (Harrington et al., 2001). Such impact of asynchrony will be more detrimental for the subsequent trophic level, i.e., natural enemies of the insect pests. Another impact of early diapause termination is greater aphid and whitefly outbreaks and increased viral disease incidence. Many studies have shown that higher winter temperature favors overwintering aphid survival in the United Kingdom, and even their first flight is reported to be advanced by one month (Bale, 1991; Cannon, 1998; Harrington et al., 1995; Zhou et al., 1995). Moreover, greenhouse pests like *Liriomyza huidobrensis* will cause more damage due to higher overwintering survival outside greenhouses (Cannon, 1998; Baker et al., 1996).

Climate change is going to change the amount, frequency, and intensity of precipitation with more incidences of extreme events. This can prove detrimental to insect pests hibernating in soil since untimely heavy rainfall during the winter months can result in flooding and prolonged water stagnation leading to high mortality of the diapausing stage. For insect pests of tropical and sub-tropical regions, aestivation is critical for survival during extremely hot summer months. Further increase in temperature along with moisture stress may render the existing aestivation mechanism inefficient and force the insect pest species to change their distribution or may reduce their local population.

17.4.7 Impact on Insect Invasion and Host Range Expansion

How most native and invasive insect species respond to climate change is still unknown, but it can be apprehended that the new normal climatic conditions would not certainly benefit all (Masters & Norgrove, 2010). Climate change is expected to increase the problems of invasive alien species. Such insect pests already possess several adaptive traits enabling them to survive under newer geography and host plants (Pareek et al., 2017; Patterson et al., 1999). Under climate change, they can outsmart the native species. Global warming will allow them to start feeding earlier resulting in greater damage with extra generations completed during the growing season (Raza et al., 2014). With increased survival due to warmer winter for non-diapausing, the frost-sensitive species can overwinter in their active stages. Such insect pests can build up their populations rapidly during early spring and expand their geographic ranges to higher altitudes and latitudes (Bale et al., 2002). For example, increased overwinter survival with expanded range has been reported for *Helicoverpa zea* and *H. armigera* in USA (Diffenbaugh et al., 2008; Fand et al., 2012).

Another indirect negative effect of climate change, like frequent cyclonic storms, is the rapid spread of invasive alien species. For example, many coconut growers of coastal states of eastern India (West Bengal, Odisha, Andhra Pradesh) have experienced a sudden and massive increase in the spiraling whitefly (*Aleurodicus disperses*) population after the frequent occurrence of cyclonic storms during recent years. Such range expansion along with greater pest survival and damage is going to be a major threat to global food security.

17.4.8 Impact on Insect Relied Tritrophic Interaction

Crop plants, insect pests and their natural enemies (predators and parasitoids) are all trophically interconnected and their tritrophic interactions will be seriously impacted due to change in climate factors. Insect herbivore populations are naturally controlled in two ways, i.e., bottom-up (quality and host availability) and top-down (natural enemies) (Jamieson et al., 2012). Simultaneously, climate change may affect the natural enemies by altering behaviour or maneuvering the physiology and behaviour of host plants as well as physiological traits in host plants. Elevated temperature can impact each component of the tritrophic interactions differently, causing spatial and temporal asynchrony (Hance et al., 2007). Natural enemies are at higher trophic levels and will be affected by the climate change-mediated asynchrony in phenological processes (Furlong & Zalucki, 2017).

Consequently, climate change is also likely to negatively impact the degree of success of biological control programs (Thomson et al., 2010). For example, if aphid predators start developing at lower temperatures faster than prey (likely due to elevated temperatures); a warm and early spring will lead to the early emergence of the natural enemies. As a result, those natural enemies may die due to the lack of availability of sufficient prey (Hance et al., 2007). Repetition of such phenomenon over time will result in failing adaptation of the natural enemy species to changing climate and may cause local extinctions. In another case study, rising temperature was found to disrupt the biocontrol potential of cereal leaf beetle *Oulema melanopus* due to temporal phenological asynchrony between natural enemy species and prey (Evans et al., 2013).

Climate change will alter the distribution ranges of crop plants and associated insect pests. Consequently, insect pests may migrate to newer areas with or without being followed by their natural parasitoids and predators, resulting in spatial asynchrony (Hullé et al., 2010). However, the final outcome of the tritrophic interaction depends on the ability of the natural enemy species to expand their geographic range or on the efficiency of native natural enemy populations in the expanded geography that could suppress the pest population in its new habitat (Thomson et al., 2010). In the absence of these conditions, the invasive pest population may increase unabated due to the absence of natural biological control in their new habitat (Crowder and Harwood, 2014). The potential for natural enemies to respond under such changing scenarios primarily depends on their ecological tolerance relative to the hosts and rate of movement as well (Thomson et al., 2010). In such cases, climate change will affect the specialist natural enemies more than the generalists due to their limited ability to adapt to spatial asynchrony with host communities (Gilman et al., 2010).

While the impact of elevated temperature is largely negative on natural enemies, elevated CO_2 levels are likely to have a positive, negative or neutral impact on the third trophic level contingent upon the life history traits (Chen et al., 2007). Thomson et al. (2010) showed that the plants grown under elevated CO_2, extreme temperature and scant precipitation can provide varied nutritional resources for herbivores, thereby indirectly impacting the fitness of natural enemies. For example, parasitism efficiency of aphid parasitoid *Aphidius picipes* surviving on the grain aphid *Sitobion avenae* increased under elevated CO_2, but reduced fecundity was reported for the wasps (Chen et al., 2007). In another study, Dyer et al. (2013) reported that nutritional properties of alfalfa plants were reduced under elevated temperature and CO_2, which consequently shortened the development time of the larvae of beet armyworm *Spodoptera exigua*. So, its larval parasitoid *Cotesia marginiventris* was unable to fully develop due to nutritionally poor host, leading to local extinction of *C. marginiventris*. Other studies have revealed how elevated CO_2 concentration affects predator performance. Chen et al. (2005, 2007) found that the Asian ladybird beetle (*Harmonia axyridis*) preferred aphids as prey under elevated CO_2 concentration than under ambient CO_2 concentration. Although overall predation performance was unaffected by elevated CO_2 levels, the required time for larval development of *H. axydiris* was notably reduced or remained unchanged. Thus, it is difficult to generalize the impact of climate change on tritrophic interaction; hence, it is vital to assess the interconnected tritrophic system as a whole. It can only be done by developing models incorporating long-term data on the phenological processes of host plants, herbivores and their associated natural enemies as well as their collective response to existing climate and future climate change (Thomson et al., 2010).

17.4.9 Climate Change Adaptation and Mitigation Strategies for Insect Pests

It is evident from the above discussion that climate change significantly affects the population dynamics and the extent of crop losses by insects. Therefore, crop growers can expect to face intensified unforeseen pest problems in the forthcoming years. Such an increase in crop pest problems across the geography is most likely to threaten global food security irrespective of countries and their agroecosystems. In this context, planning and executing climate change adaptation and mitigation tactics for new and existing insect problems is the need of the hour. Such strategies are briefly discussed below.

17.4.9.1 Adoption of Newer Methods of Crop Production

Climate change will add significant pest problems, and hence the cost of plant protection will increase in traditional farming practices. Moreover, due to the rapid depletion of ground water resources and unpredictable rainfall patterns, crop plants are expected to experience more moisture stress in the future. Recent innovations in protected crop production methods like aeroponics, hydroponics and vertical farming have immense potential to address this issue since water use efficiency is greatly improved in such systems. Additionally, such protected cultivation methodologies can efficiently manage insect pest problems and vector-borne viral diseases. However, the broader uses of such technologies are currently limited due to the requirement of high initial investment. Nevertheless, if crop protection costs increase further in the future, more crop growers are expected to adopt these technologies for crop production, particularly for high-value horticultural crops.

17.4.9.2 Use of Improved Genetic Resources

The introduction of CRISPR/Cas9-assisted genome editing has opened many avenues to encounter the emerging challenges of crop production under climate change. With introduction of this technology, researchers are able to rewrite the genetic code of different crop plants to remove yield limiting factors, improve different quality traits along with improvement in biotic and abiotic stress tolerance. Such gene edited crops will expectedly be more tolerant to insect pest attack, with reduction in yield losses. Further, the genome of insect gut microbes can also be edited to potentially decrease insecticide resistance and virus transmission. In the same way, heat tolerant and more adaptive natural enemies can be developed, mass-reared and released to enhance the efficiency of biological control under climate change scenarios.

17.4.9.3 Coordinated Pest Monitoring and Surveillance

Pest monitoring and surveillance are the backbones of successful IPM practices. Under a climate change scenario, monitoring, surveillance and risk assessment must be effectively executed to develop any unified approach. Firstly, a system to share the critical information on pest population dynamics, behaviour of invasive species, disease outbreak and epidemiology including weather data is urgently needed (Skendžic et al., 2021). Hence, enhanced cooperation is required to ensure data sharing between countries and regions at global scale (Perrings et al., 2010). Moreover, quarantine measures and early detection for invasive alien species, including entry point monitoring and rapid eradication, as exemplified by USDA Early Warning and Rapid Response Program and the EPPO Early Warning and Information System for IAS, need to be strengthened in other countries (Joyce et al., 2013; Ziska et al., 2011). Thus, growers can adopt pest prevention tools beforehand to reduce crop damage due to increased insect pest problems (Heeb et al., 2019). Additionally, the increase in new plant diseases vectored by the insects can also be anticipated. Hence, research and development efforts must be directed to developing cheap and affordable plant virus diagnostic tools and trained personnel (Skendžic et al., 2021).

17.4.9.4 Enhanced Pest Modeling and Forecasting

Under the climate change scenario, reliable pest forecasting models can aid crop growers with the necessary tools to adopt crucial preventive measures to minimize the impact of increased insect pest attacks. Climate models combining the pest species with their environmental requirements can be developed as an effective device to project the possible range expansion on a global scale under a climate change regime. Modeling the pest risk together with host plant response to climate change, such as ecological niche models (ENMs), can be used to predict the outcome of particular host plant–insect interaction (Raza et al., 2014). Two types of ENMs are in use, i.e., correlative models (e.g., MaxEnt, Bioclim, Random Forest, etc.) and mechanistic models (e.g., CLIMEX) (Kumar et al., 2014). In addition to model development, a comprehensive analysis of available long-term climate data and historical weather records of different regions should also be used for pest risk prediction. This should aid in proactive mitigation strategy development under the climate change scenario (Yonow et al., 2018).

17.4.9.5 Dynamic Modification in IPM

Many recent studies have enumerated the problems of insect pest management under climate change conditions. There is an urgent need to reconsider modifying the existing Integrated Pest Management (IPM) tools in agroecosystems that are resilient to withstand the weather variabilities (Barzman et al., 2015; Skendžic et al., 2021). For this, many of the existing IPM tactics in practice need provisioning to respond to the significant global climate change-driven impacts (Barzman et al., 2015). Some of the important aspects can be highlighted as follows:

(i) Economic threshold level (ETL) and economic injury level (EIL) are the backbone for decision making in IPM. Due to climate change and heat shock, crop plants are expected to be under drought stress frequently. Such condition will significantly reduce their ability to tolerate additional injury caused by insect pests, lowering the economic threshold (Lamichhane et al., 2015). Further, the elevated temperature will cause faster development of insects and crop damage will also be accelerated. Therefore, ETL must be lowered to prevent avoidable yield losses (Trumble & Butler, 2009).

(ii) Under elevated temperature, there might be breakdown of resistance in insect-resistant crop varieties, which are the first line of defence in IPM. Hence climate-resilient and insect-resistant crop varieties need to be developed using genetic engineering and genome editing.
(iii) Pheromones are widely used now for monitoring, mating disruption and mass trapping. Such tools will become less viable due to the rapid degradation and short life of pheromone molecules under elevated temperatures (Heuskin et al., 2011). They may require additional synergists or other adjuvants to reduce their volatility and maintain the normal efficacy under high-temperature conditions (Andrew & Hill, 2017).
(iv) Most bio-pesticides based on entomopathogenic bacteria, fungi, nematodes and viruses are also likely to be less effective under higher temperature, low relative humidity and altered UV radiation. Therefore, development of climate-resilient strains along with discovery of suitable protecting adjuvants requires urgent research and development effort.
(v) Synthetic insecticides are considered as the last resort in IPM when the pest population reaches EIL. But their efficacy is also expected to decrease due to rapid metabolism inside insect body as well as microbe mediated degradation in open environment, such as in soil, under elevated temperature (Nihal, 2020). Moreover, the chance of development of insecticide resistant pest population will also increase due to rapid multiplication of insect pests and more number of generations per year (Vadez et al., 2012). In this context, researchers should focus on developing new climate smart insecticide molecules with novel mode of action and formulations with attractants and/or repellents so that such tactics remain viable for long.

Additionally, to modify the cropping practices and adaptive management approaches for curbing pest damage in a changing climate, planting different crop varieties, altering the time of planting, and increasing biodiversity at field borders (non-crop refuge) to increase the number of natural enemies should be practiced (Andrew & Hill, 2017; Skendžic et al., 2021; Thomson et al., 2010).

17.4.9.6 Technological Interventions and Improved Communication with Farmers

In recent years, we have seen a technological revolution in the digital world. The wide use of smartphones has made it easy for new technological interventions in traditional plant protection practices. Some interesting areas are automated sensor-based pest alert system; artificial intelligence (AI) based pest monitoring; areal drones for monitoring and spraying etc. Timely and effective communication with the farmer is essential for the success of any pest management strategies. Digitization has added many new avenues for communication with crop growers. With such developments, crop protection advisories can be delivered to large number of crop growers through social media. Many new developments are expected in these areas, which will revolutionize the concept of pest management in farmers' fields and add additional tools to our arsenal to cope with increased insect pest challenges. Hence, there is a great need for a proactive and scientific approach for planning and formulating suitable mitigation strategies in the form of revised monitoring, pest forecasting and IPM tactics to reduce the risk of economic losses challenging our food security (Skendžic et al., 2021).

17.5 Effect of Climate Change on Plant-Parasitic Nematodes

PPNs are obligate parasites that mostly feed on the plant roots (as ecto, endo or semi-endoparasites), while few species feed on the above-ground plant parts (e.g. *Anguina* spp., *Aphelenchoides* spp., *Bursaphelenchus* spp.). The economically important genera of the PPNs include root-knot (*Meloidogyne* spp.), cyst (*Heterodera* spp., *Globodera* spp.), lesion (*Pratylenchus* spp.), burrowing (*Radopholus* spp.), reniform (*Rotylenchulus* spp.), false root-knot (*Nacobbus* spp.), stem and bulb (*Ditylenchus* spp.), foliar (*Aphelenchoides* spp.), dagger (*Xiphinema* spp.), needle (*Longidorus* spp.), stubby-root (*Trichodorus* spp.), ring (the criconematids), awl (*Belonolaimus* spp.), sting (*Belonolaimus* spp.), spiral (*Helicotylenchus* spp.), stunt (*Tylenchorhynchus* spp.) and lance (*Hoplolaimus* spp.) nematode (Jones et al., 2013; Phani et al., 2021). By contrast to the negative impacts brought by the PPNs on plant health, the other functional groups of nematodes like bacterivores, fungivores, omnivores, predators, and entomopathogens play a positive role in the plant and soil health improvement (Ferris et al., 2001). The bacterial and fungal feeders regulate the soil organic matter decomposition, alter C:N ratio, aid in carbon sequestration and nitrogen mineralization (Neher et al., 2010). Note that the nematode biomass (0.3 gigatonnes) in the global topsoil corresponds to approximately 80% of collective human biomass on planet Earth. Besides, the quantity of carbon respired (0.11 gigatonnes C per month) by the soil inhabiting nematodes is equivalent to approximately 15% of carbon emissions from fossil fuel use (van den Hoogen et al., 2019). Since all the trophic groups of Phylum Nematoda are interconnected via complex ecological networks of plant, soil and environment; understanding the impact of changing global climate is necessary to achieve our ultimate goal of sustainable crop production. Amongst the multitude of climate change factors, hike in atmospheric CO_2, increased temperature, altered precipitation and the climatic extremes predominantly drive the nematode community composition and biology (Franco et al., 2022). Alike the insects, the climate change effects on nematodes are perceived in both positive and negative directions as follows.

17.5.1 Impact on Nematode Distribution and Abundance

As evident from several field trials and simulated experiments, changes in climatic factors profoundly influence the PPN distribution, abundance and population dynamics at different parts of the world. Edaphic factors including soil structure and texture along with environmental factors such as temperature and moisture determine the spatial distribution of PPNs in a given area. As the PPNs lack capacity of active migration to long distances by their own accord it is relatively easier to forecast their geographical distribution in response to long term climate change. However the effects are perceived to be highly species specific. For example, Jones et al. (2017) found elevated soil temperature to reduce the population of potato cyst nematode *G. pallida* in southern United Kingdom, whereas an increase in the population of *G. rostochiensis* was observed. Previously also, it has been predicted using simulated models that global warming would expand the distribution of *G. rostochiensis* in Finland (Carter et al., 1996). The soybean cyst nematode *H. glycines*, after being identified in Canada in 1987, shifted in different northern countries (Boland et al., 2004). In a more recent study, Bebber et al. (2013) found different species of *Meloidogyne*, *Pratylenchus*, *Helicotylenchus* and *Ditylenchus* to be latitudinally shifted in the northern hemisphere. The scenario even gets worsened for the vector nematode species that transmit viruses. For example, Neilson and Boag (1996) predicted 1 °C in the mean temperature to expand the Longidorids and Trichodorids in Great Britain by around 200 km northward. Thus several species experience a northward shift due to climate change that threatens the crop production in the new areas of introduction. Recently, Pridannikov et al. (2022) recorded *G. rostochiensis* to become a serious invasive pest in Russia due to global warming. Apart from range shifting the distribution pattern of the nematode species can also change. A strong correlation between 1 °C rise in temperature and increased incidence of *L. caespiticola* was evidenced in the northern territories of United Kingdom (Boag & Neilson, 1996). Higher PPN abundance with elevated temperature was noticed in the Kashmir valley of India (Nisa et al., 2021). Similarly, the rising temperature has expanded the distribution of the above-ground peril pine wilt nematode *B. xylophilus* and its vector beetle *Monochamus alternatus* in several parts of the Asia, Europe and America (Hirata et al., 2017; Li et al., 2022). However, the climate change effects may also result in decreasing the nematode abundance in certain parts of the world. For example, Liu et al. (2022) found the elevated temperature in the Tibetan plateau to reduce the PPN abundance. Similar consequences were also observed in the semiarid grasslands (Mueller et al., 2016).

PPN abundance under elevated and ambient CO_2 plots in pasture land of New Zealand has been studied extensively, and findings suggest the PPNs to become more abundant under elevated CO_2 conditions (Yeates et al., 2003). The authors assumed that the elevated CO_2 level to induce greater root biomass that consequently escalated the PPN incidence. Another study encountered higher abundance of lesion nematodes of *Pratylenchus* spp. in the soil samples from wetland or hydric soil collected from elevated CO_2 plots (Yeates et al., 1999). In Germany, the PPN numbers associated with wheat and sugar beet were found to be significantly higher at the physiologically mature stage in elevated CO_2 (Sticht et al., 2009). The virus transmitting Trichodorids also increase in abundance under elevated CO_2 (Hoeksema et al., 2000). However, elevated level of CO_2 does not always positively affect the PPN abundance and distribution. For example, Ayres et al. (2008) reported no change in nematode abundance, species richness and species diversity under CO_2 enrichment at the pastures of Colorado and California in the USA and Montpellier in France. Li et al., (2007, 2009) also found no significant change in the PPN abundance under rice–wheat cropping system in China following elevation of CO_2. The increased abundance also affects the corresponding soil food web structure (Eisenhauer et al., 2012). However, the study of Neher and Weicht (2013) concluded no change in abundance for the aphelenchids in pine forest, whereas the dorylaimids were increased. Hence, the effects can contribute in both positive and negative directions.

Climate change also mediates the spatial and temporal variation in global precipitation, ultimately affecting the functional composition of nematode communities in soil ecosystems. For example, Fleming et al. (2016) recorded significant increase in prevalence of *M. minor*, *Heterodera* spp. and *Pratylenchus* spp. under increased precipitation conditions at grasslands and cereal cropping systems of Northern Ireland. But such results are largely variable across nematode species. Torode et al. (2016) found negligible changes in PPN abundance under altered precipitation. Landesman et al. (2011) observed no change for the aphelenchids. However, the droughts exert more pronounced effects on the PPNs. Being essentially aquatic in nature nematode abundance generally decreases with drought conditions, and severe drought for prolonged condition may alter the community structure. The projected increase in the frequency of extreme dry years due to global warming lessenes the predation pressure on PPNs, thereby increasing the PPN population pressure in the mesic grasslands of the USA (Franco et al., 2019). In continuation of that study, authors demonstrated under extreme drought in sub-humid grasslands to alter the below-ground trophic interactions that affect plant growth (Franco et al., 2020). Further, it can be predicted that the climatic extremes like heat wave, storms and wildfires will severely affect the community composition and abundance of the soil nematodes. Taken together all the records it is evident that climate change effects are not similar on every species and locations to drive the nematode abundance and distribution uniformly. In this line, it can be predicted that some species will face positive effects, whereas some will face negative or neutral effects.

17.5.2 Impact on Nematode Reproduction and Survival

For majority of the PPNs low temperature favors slower development, whereas faster development is observed under warmer condition (Tyler, 1933; Tzortzakakis & Trudgill, 2005). Global warming due to climate change affects the nematode survival and reproduction potential. Using simulated models, Carter et al. (1996) predicted that global warming would increase the number of *G. rostochiensis* generations per year by 2050 in Finland. Similar study with *M. incognita* in coffee plantations of Brazil indicates increase in number of generations per month in the future decades of 2020, 2050 and 2080 (Ghini et al., 2008). Increase in number of generations will surely aggravate the PPN problems in respective areas. Rise in the temperature owing to the global warming results in the multiple nematode generations per season that facilitate the PPN spread to newer geographical localities (Trudgill et al., 2005). In parallel, an increase in CO_2 level and surface air temperature lead to greater plant photosynthetic rate in summer, enhanced respiration in winter, and escalated amplitude of seasonal cycle. All of these contribute to increased terrestrial plant growth, which ultimately will support the newly introduced nematode pests. But contrarily, phenomena like sex reversal (as seen in the root-knot nematodes) will decrease the damaging population levels (Papadopoulou & Triantaphyllou, 1982). Further, temperature increase will also predictably escalate the metabolic activities and O_2 consumption of the soil inhabiting nematode stages to reduce their open-environment survival (Robinson et al., 2023). Similar contrasting assumptions can also be made for the altered precipitation effects, where precipitation might support the population increase of the nematodes but drought will act in counter direction. In this line, future research should be directed on the economically important PPN species to assess how the climate change effects grossly intervene in re-shaping the reproduction and survival potential of the nematodes in precise geographical areas.

17.5.3 Impact on Nematode Pathogenicity and Plant-Nematode Interaction

Susceptible host, virulent pest or pathogen and conducive environment develop a complex interaction to manifest the plant diseases (Agrios, 2005). As an important component of disease triangle, the environmental factors are pliable target of climate change determining the result of plant-PPN interaction. Although rising CO_2 level exerts neutral or positive effect on the incidence of PPN species, the effect on plant-nematode interaction is complex. In this line, Wilsey (2001) found the PPN damage in grasses to be comparatively higher under elevated CO_2 than that of normal CO_2 level due to greater root development. Similarly, Prasad and Somasekhar (2009) found *M. graminicola* to cause more damage to rice roots at higher CO_2 level (700 ppm) than ambient level (400 ppm). On the contrary, Ayres et al. (2008) found the PPNs to respond neutrally to CO_2 enrichment under grassland ecosystems despite having greater root biomass. The authors hypothesized that under elevated CO_2 conditions the feeding of PPNs probably increases leading to greater plant damage (Ayres et al., 2008). However, the ectoparasitic PPNs of family Longidoridae do not derive much benefit from higher root biomass that occurs at elevated CO_2 (Cesarz et al., 2015). This may occur due to the activation of plant defence provoked by robust root system or least availability of soil nitrogen under enriched CO_2. However, generalizing such events to get a group-specific trend for the PPNs is difficult with the currently available studies.

Higher temperature condition is likely to worsen the water stress symptoms in the PPN infected plants thereby affecting their nutrition. Global warming was found to accredit pine tree mortality in Switzerland due to abundance of pine wilt nematodes *B. mucronatus* and *B. xylophilus* and their insect vectors at elevated temperature. But, drought stress negatively affected the host resistance to the pathogens in the respective areas (Rebetez & Dobbertin, 2004). Ankrom et al. (2020) reported the changes in precipitation to differentially affect PPN feeding thereby affecting their pathogenicity. They found that the abundance of endoparasitic PPNs remains unaltered with increase in precipitation at arid grassland site; but rises and drops at semiarid and mesic sites, respectively. Besides, the ectoparasitic PPNs remain unaltered in the semiarid and arid sites, but decreases at mesic sites. Subsequently, Ankrom et al. (2022) found the fungal feeder-dominated decomposition pathways to be higher at the arid ecosystem; whereas nematode maturity indices declined at the semiarid sites with increase in PPN species.

Climate change also interferes the plant defense mechanism. It was found that higher CO_2 level represses jasmonic acid-mediated defense in tomato when infected with the root-knot nematode *M. incognita* (Sun et al., 2011). But, in the defense-regressive wild tomato cultivars showed neutral response. Climate change may also affect the *R*-gene mediated host resistance especially active against the *Meloidogyne* spp. It is an established fact that the *Mi* gene mediated resistance response dwindles when the temperature goes beyond 28 °C (Williamson & Kumar, 2006). A few heat stable *Mi* genes were identified (El-Sappah et al., 2019), but their efficacy is yet to be evaluated at high heat stress. Besides, Jones et al. (2018) indicated higher expression of thermotolerant genes in response to heat stress while working with *Globodera* spp. Similar effects may also be observed for the other PPN species where climate change may eventually aid in deveoping climate-adapted virulent biotypes.

17.5.4 Impact on Nematode Relied Tritrophic Interaction

The climate change also alters the intricate tritrophic interactions between the hosts, PPNs and their natural enemies. In this line, extensive studies are scant on how the natural enemies of PPNs are specifically affected with the changing climates. The alteration in precipitation has been found to disrupt the balance between the PPN communities and their natural enemies in soil (Chiew et al., 2011). The projected increase in the frequency of extreme dry years due to global warming lessens the predation pressure on PPNs thereby increasing their population pressure and plant infestation in the mesic grasslands of USA (Franco et al., 2019). The increased plant damage then worsens the drought effects by altering carbon cycling and ecosystem production. Further, it has been predicted the altered soil moisture regime will affect the PPN biocontrol agents by altering their densities, emergence, development and spread (Colagiero & Ciancio, 2012). The authors expected an increase in the spread of some moisture dependent biocontrol agents like *Catenaria* under high soil moisture content. The increase in temperature may also aggravate the pathogenicity of the nematode antagonistic fungi and bacteria thereby aiding in escalated natural biocontrol. The increase natural control will definitely maintain the PPN population and the related host infection dynamics; but how the economically important nematode species in a particular location being simultaneously influenced by various environmental factors will concomitantly interact with crops and natural enemies is yet to be understood in detail.

17.5.5 Climate Change Adaptation and Mitigation Strategies for Nematode Pests

Under the changing climate scenario we may no longer rely on the current management tactics advocated for the PPNs, and accordingly the integrated nematode management (INM) practices need to be refined to suppress the PPN population level below economic threshold. In this line, we may approach to the 'climate-smart' management practices where the crop production and nematode management are equally prioritized with environmental protection (Heeb et al., 2019). It is not the case that climate change will aggravate the PPN abundance and pathogenicity in all the crop growing areas uniformly. The effects are duly species and location specific. Hence management options should be optimized accordingly.

The widely adopted cultural and physical practices for PPN control includes flooding, fallowing and solarization (Sikora et al., 2005). However, studies have found these practices to contribute to global warming by GHG emission (Lee et al., 2021; Lemke et al., 1999; Mosier et al., 1991; Sánchez-Rodríguez et al., 2019; Toma et al., 2019). Hence, introduction of antagonistic cover crops like grain sorghum, millets, oil radish, Mulato grass can benefit the environmental quality. Sustainable management tactics like biofumigation and crop rotation also offer significant promise under changing climate scenario and improving the soil carbon sequestration. Use of FYM, than animal manure, will be relatively safe (Mori, 2018).

Climate change may also affect the PPN biocontrol by the fungal and bacterial antagonists. It has been found that increased temperature reduces the pathogenic ability of the root-knot nematode antagonist *Pasteuria penetrans* (Freitas et al., 1997). Elevated soil temperature and increased CO_2 level increase the parasitism ability of *Pochonia chlamydosporia* and *Metarrhizium anisopliae* (Karabörklü et al., 2022; Luambano et al., 2015). Thus a dualism can be predicted among the PPN antagonists, where they will affect the nematode communities both positively and negatively. Hence, it can be apprehended that the natural soil suppressiveness will be altered. But how the changes will affect the PPN biocontrol is difficult to predict now.

Climate change effect is also correlated with host physiology changes especially with *R*-gene mediated resistance (to PPNs) deployed in crops either via conventional or molecular breeding (Sun et al., 2011). Nevertheless, the most considerable challenge to genetic resistance mechanism posed by PPNs is that under changing climate, more selection pressure may lead to resistance-breaking or virulent subspecies.

Temperature fluctuations and erratic precipitation can potentially affect the persistence and availability of the chemical nematicides in soil (Delcour et al., 2015). This may alter their effectiveness to control the PPNs when applied. It may be assumed that the demand for nematicides among the farming community is likely to be higher because of the greater incidence of PPNs under changing climate scenario, despite the fact that only a handful of nematicides are currently available. Although the fate of pesticides in soil upon the change in global climate is fairly understood (Gatto et al., 2016), information must be generated on the effect of climate change on the fate of nematicides in the soil in order to gain a better understanding of climate change implications on PPN management tactics. We should also look for alternatives to reduce the nematicide usage following the global aim to curb pesticide usage in the coming decades. This will be definitely challenging for the growers to cope with the impending situation. Thus a comprehensive scenario planning is necessary with all the available data from different locations to find the best suited management option.

17.6 Conclusion and Future Prospect

Although the specific effects of climate change on insect and nematode pests are difficult to ascertain with the available knowledge to date, some generalizations for the farmers can be made. For example, the elevated temperature in the winter season and temperate areas is likely to lead to higher abundance of the pests and pathogens. It will eventually impact the economic production in the respective regions. The expansion of range and invasion to newer areas may cause havoc due to the availability of 'no-hindrance' for the insects and nematodes in those areas. The situation may be worsened with the increased reproduction rate and completion of above-normal generations per season. Besides, frequent changes in humidity and precipitation profile will also favor some pathogens over the others. As a result, minor, occasional and seasonal pests are likely to achieve major pest status. However, certain climate change effects may also act in counter direction to jeopardize the pathogenic potential of these pests. This seems to lead to an contraposition effect for the pests where a seminal solution will be difficult to find. In this context we must go for species and location specific management options that can be best suited to the farmers. Newer mitigation systems should be created or existing methods should be improvised to monitor the pest population dynamics with wider inter- and intra-center collaboration support. Thus a win–win state can be achieved for sustainable pest management and climate protection with prior emphasis towards primary production.

References

Agrios, G. N. (2005). *Plant pathology*. Elsevier.

Alonso-Prados, J. L., Luis-Arteaga, M., Alvarez, J. M., Moriones, E., Batlle, A., Laviña, A., García-Arenal, F., & Fraile, A. (2003). Epidemics of aphid-transmitted viruses in melon crops in Spain. *European Journal of Plant Pathology, 109*, 129–138.

Andrew, N. R., Hill, S. J. (2017). Effect of climate change on insect pest management. In: M. Coll, E. Wajnberg (Eds.), *Environmental pest management: challenges for agronomists, ecologists, economists and policymakers* (pp. 197–215). John Wiley and Sons Ltd, NJ, USA.

Ankrom, K. E., Franco, A. L., Fonte, S. J., Gherardi, L. A., de Tomasel, C. M., Andriuzzi, W. S., Shaw, E. A., Sala, O. E., & Wall, D. H. (2020). Ecto-and endoparasitic nematodes respond differently across sites to changes in precipitation. *Oecologia, 193*, 761–771.

Ankrom, K. E., Franco, A. L., Fonte, S. J., Gherardi, L. A., de Tomasel, C. M., Wepking, C., Guan, P., Cui, S., Sala, O. E., & Hall, D. H. (2022). Ecological maturity and stability of nematode communities in response to precipitation manipulations in grasslands. *Applied Soil Ecology, 170*, 104263.

Ayres, E., Wall, D. H., Simmons, B. L., Field, C. B., Milchunas, D. G., Morgan, J. A., & Roy, J. (2008). Below-ground nematode herbivores are resistant to elevated atmospheric CO 2 concentrations in grassland ecosystems. *Soil Biology & Biochemistry, 40*, 978–985.

Baker, R. A., Cannon, R. C., & Walters, K. A. (1996). An assessment of the risks posed by selected non-indigenous pests to UK crops under climate change. *Aspects of Applied Biology, 45*, 323–330.

Bale, J. S., & Hayward, S. A. L. (2010). Insect overwintering in a changing climate. *Journal of Experimental Biology, 213*, 980–994.

Bale, J. S., Masters, G. J., Hodkinson, I. D., Awmack, C., Bezemer, T. M., Brown, V. K., Butterfield, J., Buse, A., Coulson, J. C., Farrar, J., Good, J. E. G., Harrington, R., Hartley, S., Jones, T. H., Lindroth, R. L., Press, M. C., Symrnioudis, I., Watt, A. D., & Whittaker, J. B. (2002). Herbivory in global climate change research: Direct effects of rising temperature on insect herbivores. *Global Change Biology, 8*, 1–16.

Bale, J. S. (1991). Implications of cold hardiness for pest management. In: R. E. Lee (ed.), *Insects at low temperature* (pp. 461–498). Springer, NY, USA.

Bardgett, R. D., & Caruso, T. (2020). Soil microbial community responses to climate extremes: Resistance, resilience and transitions to alternative states. *Philosophical Transactions of the Royal Society B, 375*, 20190112.

Barzman, M., Bàrberi, P., Birch, A. N. E., Boonekamp, P., Dachbrodt-Saaydeh, S., Graf, B., Hommel, B., Jensen, J. E., Kiss, J., Kudsk, P., & Lamichhane, J. R. (2015). Eight principles of integrated pest management. *Agronomy for Sustainable Development, 35*, 1199–1215.

Bebber, D. P., Ramotowski, M. A. T., & Gurr, S. J. (2013). Crop pests and pathogens move polewards in a warming world. *Nature Clinical Practice Endocrinology & Metabolism, 3*, 985–988.

Belovsky, G. E., & Slade, J. B. (2000). Insect herbivory accelerates nutrient cycling and increases plant production. *Proceedings of the National Academy of Sciences of the United States of America, 97*, 14412–14417.

Bernays, E. A. (1997). Feeding by lepidopteran larvae is dangerous. *Ecological Entomology, 22*, 121–123.

Bezemer, T. M., Jones, T. H., & Knight, K. J. (1998). Long-term effects of elevated CO_2 and temperature on populations of the peach potato aphid *Myzus persicae* and its parasitoid *Aphidius matricariae*. *Oecologia, 116*, 128–135.

Boag, B., & Neilson, R. (1996). Effects of potential climate changes on plant-parasitic nematodes. *Aspects of Applied Biology, 45*, 331–334.

Boland, G. J., Melzer, M. S., Hopkin, A., Higgins, V., & Nassuth, A. (2004). Climate change and plant diseases in Ontario. *Canadian Journal of Plant Pathology, 26*, 335–350.

Boudon-Padieu, É., Maixner, M. (2007, March 28–30). Potential effects of climate change on distribution and activity of insect vectors of grapevine pathogens. In *Proceedings of the international and multi-disciplinary "global warming, which potential impacts on the vineyards?* (p. 23). Beaune, France.

Boudon-Padieu, E. (2000). Cicadelle vectrice de la flavescence dorée, *Scaphoideus titanus* Ball, 1932. In: J. Stockel (ed.), *Ravageurs de la Vigne, Féret* (pp. 110–120). Bordeaux, France.

Boudon-Padieu, E. (2003, September 12–17). The situation of grapevine yellows and current research directions: Distribution, diversity, vectors, diffusion and control. In *Proceedings of the Extended Abstracts, 14th Meeting of the ICVG* (pp. 47–53). Locorotondo, Italy.

Bradshaw, W. E., & Holzapfel, C. M. (2010). Light, time, and the physiology of biotic response to rapid climate change in animals. *Annual Review of Plant Biology, 72*, 147–166.

Brinkman, E. P., Duyts, H., Karssen, G., Van der Stoel, C. D., & van der Putten, W. H. (2015). Plant-feeding nematodes in coastal sand dunes: Occurrence, host specificity and effects on plant growth. *Plant and Soil, 397*, 17–30.

Cannon, R. J. (1998). The implications of predicted climate change for insect pests in the UK, with emphasis on non-indigenous species. *Global Change Biology, 4*, 785–796.

Canto, T., Aranda, M. A., & Fereres, A. (2009). Climate change effects on physiology and population processes of hosts and vectors that influence the spread of hemipteran-borne plant viruses. *Global Change Biology, 15*, 1884–1894.

Carter, T. R., Saarikko, R. A., & Niemi, K. J. (1996). Assessing the risks and uncertainties of regional crop potential under a changing climate in Finland. *Agric Food Sci Finl, 5*, 329–350.

Cayton, H. L., Haddad, N. M., Gross, K., Diamond, S. E., & Ries, L. (2015). Do growing degree days predict phenology across butterfly species? *Ecology, 96*, 1473–1479.

Cesarz, S., Reich, P. B., Scheu, S., Ruess, L., Schaefer, M., & Eisenhauer, N. (2015). Nematode functional guilds, not trophic groups, reflect shifts in soilfood webs and processes in response to interacting global change factors. *Pedologia, 58*, 23–32.

Chakravarthy, A. K., Doddabasappa, B., & Shashank, P. R. (2013). The potential impacts of climate change on insect pests in cultivated ecosystems: An Indian perspective. In S. Nautiyal, K. S. Rao, H. Kaechele, K. V. Raju, & R. Schaldach (Eds.), *Knowledge systems of societies for adaptation and mitigation of impacts of climate change, environmental science and engineering* (pp. 143–162). Springer.

Chen, F., Ge, F., & Parajulee, M. N. (2005). Impact of elevated CO_2 on tri-trophic interaction of *Gossypium hirsutum, Aphis gossypii*, and *Leis axyridis. Environmental Entomology, 34*, 37–46.

Chen, F., Wu, G., Parajulee, M. N., & Ge, F. (2007). Impact of elevated CO_2 on the third trophic level: A predator *Harmonia axyridis* and a parasitoid *Aphidius picipes. Biocontrol Science and Technology, 17*, 313–324.

Chen, S., Shelby Fleischer, J., Tobin, P. C., & Saunders, M. C. (2011). Projecting insect voltinism under high and low greenhouse gas emission conditions. *Environmental Entomology, 40*, 505–515.

Chiew, F. H., Young, W. J., Cai, W., & Teng, J. (2011). Current drought and future hydroclimate projections in southeast Australia and implications for water resources management. *Stochastic Environmental Research and Risk Assessment, 25*, 601–612.

Clarke, A., & Fraser, K. P. P. (2004). Why does metabolism scale with temperature? *Functional Ecology, 18*, 243–251.

Colagiero, M., & Ciancio, A. (2012). Climate changes and nematodes: Expected effects and perspectives for plant protection. *Redia, 94*, 113–118.

Coulson, S. J., Hodkinson, I. D., Webb, N. R., Mikkola, K., Harrison, J. A., & Pedgley, D. E. (2002). Aerial colonization of high Arctic islands by invertebrates: The diamondback moth *Plutella xylostella* (Lepidoptera: Yponomeutidae) as a potential indicator species. *Diversity and Distributions, 8*, 327–334.

Coviella, C. E., & Trumble, J. T. (1999). Effects of elevated atmospheric carbon dioxide on insect-plant interactions. *Conservation Biology, 13*, 700–712.

Crowder, D. W., & Harwood, J. D. (2014). Promoting biological control in a rapidly changing world. *Biological Control, 75*, 1–7.

Dabhi, M. R., Korat, D. M., & Borad, P. K. (2011). Seasonal abundance of *Bracon hebetor* Say in relation to temperature and relative humidity. *Insect Environ, 16*, 149–150.

De Groot, R. S., Wilson, M. A., & Boumans, R. M. J. (2002). A typology for the classification, description and valuation of ecosystem functions, goods and services. *Ecological Economics, 41*, 393–408.

Delcour, I., Spanoghe, P., & Uyttendaele, M. (2015). Impact of climate change on pesticide use. *Food Research International, 68*, 7–15.

DeLucia, E. H., Casteel, C. L., Nabity, P. D., & O'Neill, B. F. (2008). Insects take a bigger bite out of plants in a warmer, higher carbon dioxide world. *Proceedings of the National Academy of Sciences of the United States of America, 105*, 1781–1782.

Deutsch, C. A., Tewksbury, J. J., Huey, R. B., Sheldon, K. S., Ghalambor, C. K., Haak, D. C., & Martin, P. R. (2008). Impacts of climate warming on terrestrial ectotherms across latitude. *Proceedings of the National Academy of Sciences of the United States of America, 105*, 6668–6672.

Deutsch, C. A., Tewksbury, J. J., Tigchelaar, M., Battisti, D. S., Merrill, S. C., Huey, R. B., & Naylor, R. L. (2018). Increase in crop losses to insect pests in a warming climate. *Science, 361*, 916–919.

Diffenbaugh, N. S., Krupke, C. H., White, M. A., & Alexander, C. E. (2008). Global warming presents new challenges for maize pest management. *Environmental Research Letters, 3*, 044007.

Dukes, J. S. D. S., Pontius, J., Orwig, D., Garnas, J. R. G. R., Rodgers, V. L., Brazee, N., Cooke, B., Theoharides, K. A. T. A., Stange, E. E. S. E., Harrington, R., Ehrenfeld, J., Gurevitch, J., Lerdau, M., Stinson, K., Wick, R., & Ayres, M. (2009). Responses of insect pests, pathogens, and invasive plant species to climate change in the forests of northeastern North America: What can we predict? *Canadian Journal of Forest Research, 39*, 231–248.

Dyer, L. A., Richards, L. A., Short, S. A., & Dodson, C. D. (2013). Effects of CO_2 and temperature on tritrophic interactions. *PLoS ONE, 8*, e62528.

Eisenhauer, N., Cesarz, S., Koller, R., Worm, K., & Reich, P. B. (2012). Global change belowground: Impacts of elevated CO_2, nitrogen, and summer drought on soil food webs and biodiversity. *Global Change Biology, 18*, 435–447.

Elling, A. A. (2013). Major emerging problems with minor *Meloidogyne* species. *Phytopathology, 103*, 1092–1102.

El-Sappah, A. H., Islam, M. M., El-awady, H., Yan, S., Qi, S., Liu, J., Cheng, G., & Liang, Y. (2019). Tomato natural resistance genes in controlling the root-knot nematode. *Genes, 10*, 925.

Evans, E. W., Carlile, N. R., Innes, M. B., & Pitigala, N. (2013). Warm springs reduce parasitism of the cereal leaf beetle through phonological mismatch. *Journal of Applied Entomology, 137*, 383–391.

Falzoi, S., Lessio, F., Spanna, F., & Alma, A. (2014). Influence of temperature on the embryonic and post-embryonic development of *Scaphoideus titanus* (Hemiptera: Cicadellidae), vector of grapevine Flavescence dorée. *International Journal of Pest Management, 60*, 246–257.

Fand, B. B., Kamble, A. L., & Kumar, M. (2012). Will climate change pose serious threat to crop pest management: A critical review. *International journal of scientific and Research publications, 2*, 1–14.

Fereres, A., Irwin, M. E., & Kampmeier, G. E. (2017). Aphid movement: Process and consequences. In H. F. van Emden & R. Harrington (Eds.), *Aphids as crop pests* (pp. 196–224). CABI Publishing, Wallingford.

Ferris, H., & Bongers, T. (2006). Nematode indicators of organic enrichment. *Journal of Nematology, 38*, 3–12.

Ferris, H., Bongers, T., & de Goede, R. G. M. (2001). A framework for soil food web diagnostics: Extension of the nematode faunal analysis concept. *Applied Soil Ecology, 18*, 13–29.

Fleming, T. R., McGowan, N. E., Maule, A. G., & Fleming, C. C. (2016). Prevalence and diversity of plant parasitic nematodes in Northern Ireland grassland and cereals, and the influence of soils and rainfall. *Plant Pathology, 65*, 1539–1550.

Franco, A. L., Gherardi, L. A., de Tomasel, C. M., Andriuzzi, W. S., Ankrom, K. E., Shaw, E. A., Bach, E. M., Sala, O. E., & Wall, D. H. (2019). Drought suppresses soil predators and promotes root herbivores in mesic, but not in xeric grasslands. *Proceedings of the National Academy of Sciences, 116*, 12883–12888.

Franco, A. L., Gherardi, L. A., de Tomasel, C. M., Andriuzzi, W. S., Ankrom, K. E., Bach, E. M., Guan, P., Sala, O. E., & Hall, D. H. (2020). Root herbivory controls the effects of water availability on the partitioning between above-and below-ground grass biomass. *Functional Ecology, 34*, 2403–2410.

Franco, A. L., Guan, P., Cui, S., de Tomasel, C. M., Gherardi, L. A., Sala, O. E., & Wall, D. H. (2022). Precipitation effects on nematode diversity and carbon footprint across grasslands. *Global Change Biology, 28*, 2124–2132.

Freitas, L. G., Mitchell, D. J., & Dickson, D. W. (1997). Temperature effects on the attachment of *Pasteuria penetrans* endospores to *Meloidogyne arenaria* race 1. *Journal of Nematology, 29*, 547–555.

Furlong, M. J., & Zalucki, M. P. (2017). Climate change and biological control: The consequences of increasing temperatures on host– parasitoid interactions. *Current Opinion in Insect Science, 20*, 39–44.

Gabriel, D., & Tscharntke, T. (2006). Insect pollinated plants benefit from organic farming. *Agriculture, Ecosystems & Environment, 118*, 43–48.

Gatto, M. P., Cabella, R., & Gherardi, M. (2016). Climate change: The potential impact on occupational exposure to pesticides. *Annali dell'Istituto Superiore di Sanita, 52*, 374–385.

Ghini, R., Hamada, E., Pedro Júnior, M. J., Marengo, J. A., & Gonçalves, R. R. V. (2008). Risk analysis of climate change on coffee nematodes and leaf miner in Brazil. *Pesquisa agropecuária brasileira, 43*, 187–194.

Gill, H. K., Goyal, G., & Chahil, G. (2017). Insect diapause: A review. *Journal of Agricultural Science and Technology, 7*, 454–473.

Gilman, S. E., Urban, M. C., Tewksbury, J., Gilchrist, G. W., & Holt, R. D. (2010). A framework for community interactions under climate change. *Trends in Ecology & Evolution, 25*, 325–331.

Gutierrez, A. P., D'Oultremont, T., Ellis, C., & Ponti, L. (2006). Climatic limits of pink bollworm in Arizona and California: Effects of climate warming. *Acta Oecologica, 30*, 353–364.

Gutierrez, A. P., Ponti, L., & Cossu, Q. A. (2009). Prospective comparative analysis of global warming effects on olive and olive fly. (*Bactrocera oleae* (Gmelin)) in Arizona-California and Italy. *Climate Change, 95*, 195–217.

Hahn, D. A., & Denlinger, D. L. (2011). Energetics of insect diapause. *Annual Review of Entomology, 56*, 103–121.

Halder, J., Kushwaha, D., Deb, D., & Rai, A. B. (2019). Abundance and distribution of sucking pests complex of okra in relation to meteorological parameters. *Journal of Agrometeorology, 21*, 227–229.

Hamilton, J. G., Dermody, O., Aldea, M., Zangerl, A. R., Rogers, A., Berenbaum, M. R., & DeLucia, E. H. (2005). Anthropogenic changes in tropospheric composition increase susceptibility of soybean to insect herbivory. *Environmental Entomology, 34*, 479–485.

Hance, T., Baaren, J. V., Vernon, P., & Boivin, G. (2007). Impact of extreme temperatures on parasitoids in a climate change perspective. *Annual Review of Entomology, 52*, 107–126.

Harrington, R., Bale, J. S., & Tatchell, G. M. (1995). Aphids in a changing climate. In R. Harrington & N. E. Stork (Eds.), *Insects in a changing environment* (pp. 125–155). Academic Press.

Harrington, R., Woiwod, I., & Sparks, T. (1999). Climate change and trophic interactions. *Trends in Ecology & Evolution, 14*, 146–150.

Harrington, R., Fleming, R. A., & Woiwod, I. P. (2001). Climate change impacts on insect management and conservation in temperate regions: Can they be predicted? *Agricultural and Forest Entomology, 3*, 233–240.

Harrington, R., Clark, S. J., Welham, S. J., Verrier, P. J., Denholm, C. H., & Hulle, M. (2007). Environmental change and the phenology of European aphids. *Global Change Biology, 13*, 1550–1564.

Hedenec, P., Jiménez, J. J., Moradi, J., Domene, X., Hackenberger, D., Barot, S., Frossard, A., Oktaba, L., Filser, J., Kindlmann, P., & Frouz, J. (2022). Global distribution of soil fauna functional groups and their estimated litter consumption across biomes. *Science and Reports, 12*, 17362.

Heeb, L., Jenner, E., & Cock, M. J. (2019). Climate-smart pest management: Building resilience of farms and landscapes to changing pest threats. *Journal of Pest Science, 92*, 951–969.

Heong, K. L., Song, Y. H., Pimsamarn, S., Zhang, R., & Bae, S. D. (1995). Global warming and rice arthropod communities. In S. Peng, K. T. Ingram, H. U. Neue, & L. H. Ziska (Eds.), *Climate Change and Rice* (pp. 327–335). Springer.

Heuskin, S., Verheggen, F. J., Haubruge, E., Wathelet, J. P., & Lognay, G. (2011). The use of semiochemical slow-release devices in integrated pest management strategies. *Biotechnologie, Agronomie, Societe et Environnement, 15*, 459–470.

Hill, D. S. (1987). *Agricultural insect pests of temperate regions and their control*. Cambridge University Press.

Hirata, A., Nakamura, K., Nakao, K., Kominami, Y., Tanaka, N., Ohashi, H., Takano, K. T., Takeuchi, W., & Matsui, T. (2017). Potential distribution of pine wilt disease under future climate change scenarios. *PLoS ONE, 12*, e0182837.

Hoehn, P., Tscharntke, T., Tylianakis, J. M., & Steffan-Dewenter, I. (2008). Functional group diversity of bee pollinators increases crop yield. *Proceedings of the Royal Society B: Biological Sciences, 275*, 2283–2291.

Hoeksema, J. D., Lussenhop, J., & Teeri, J. A. (2000). Soil nematodes indicate food web responses to elevated atmospheric CO_2. *Pedobiologia, 44*, 725–735.

Hull, R. (2014). *Plant virology*. Elsevier Academic Press.

Hullé, M., D'Acier, A. C., Bankhead-Dronnet, S., & Harrington, R. (2010). Aphids in the face of global changes. *Critical Review Biology, 333*, 497–503.

Hunter, M. D. (2001). Insect population dynamics meets ecosystem ecology: Effects of herbivory on soil nutrient dynamics. *Agricultural and Forest Entomology, 3*, 77–84.

Irwin, M. E., Ruesink, W. G., Isard, S. A., & Kampmeier, G. E. (2000). Mitigating epidemics caused by non-persistently transmitted aphidborne viruses: The role of the pliant environment. *Virus Research, 71*, 185–211.

Jamieson, M. A., Trowbridge, A. M., Raffa, K. F., & Lindroth, R. L. (2012). Consequences of climate warming and altered precipitation patterns for plant-insect and multitrophic interactions. *Plant Physiology, 160*, 1719–1727.

Jankielsohn, A. (2018). The importance of insects in agricultural ecosystems. *Advances in Entomology, 6*, 62–73.

Jaramillo, J., Muchugu, E., Vega, F. E., Davis, A., Borgemeister, C., & Chabi-Olaye, A. (2011). Some like it hot: The influence and implications of climate change on coffee berry borer (*Hypothenemus hampei*) and coffee production in east Africa. *PLoS ONE, 6*, e24528.

Jones, J. T., Haegeman, A., Danchin, E. G., Gaur, H. S., Helder, J., Jones, M. G., Kikuchi, T., Manzanilla-López, R., Palomares-Rius, J. E., & Wesemael, W. M. (2013). Top 10 plant parasitic nematodes in molecular plant pathology. *Molecular Plant Pathology, 14*, 946–961.

Jones, L. M., Koehler, A. K., Trnka, M., Balek, J., Challinor, A. J., Atkinson, H. J., Urwin, P. E. (2017). Climate change is predicted to alter the current pest status of *Globodera pallida* and *G. rostochiensis* in the United Kingdom. *Global Change Biology, 23*, 4497–4507.

Jones, L. M., Eves-van den Akker, S., van-Oosten Hawle, P., Atkinson, H. J., Urwin, P. E. (2018). Duplication of hsp-110 is implicated in differential success of Globodera species under climate change. *Molecular Biology and Evolution, 35*, 2401–2413

Joyce, L. A., Briske, D. D., Brown, J. R., Polley, H. W., McCarl, B. A., & Bailey, D. W. (2013). Climate change and North American rangelands: Assessment of mitigation and adaptation strategies. *Rangeland Ecology Management, 66*, 512–528.

Karabörklü, S., Aydınlı, V., & Dura, O. (2022). The potential of *Beauveria bassiana* and *Metarhizium anisopliae* in controlling the root-knot nematode *Meloidogyne incognita* in tomato and cucumber. *Journal of Asia-Pacific Entomology, 25*, 101846.

Kumar, S., Neven, L. G., & Yee, W. L. (2014). Evaluating correlative and mechanistic niche models for assessing the risk of pest establishment. *Ecosphere, 5*, 1–23.

Lactin, D. J., Holliday, N. J., Johnson, D. L., & Craigen, R. (1995). Improved rate model of temperature-dependent development by arthropods. *Environmental Entomology, 24*, 68–75.

Lafferty, K. D. (2009). The ecology of climate change and infectious diseases. *Ecology, 90*, 888–900.

Lamichhane, J. R., Barzman, M., Booij, K., Boonekamp, P., Desneux, N., Huber, L., Kudsk, P., Langrell, S. R. H., Ratnadass, A., Ricci, P., Sarah, J.-L., & Messean, A. (2015). Robust cropping systems to tackle pests under climate change. A review. *Agronomy for Sustainable Development, 35*, 443–459.

Landesman, W. J., Treonis, A. M., & Dighton, J. (2011). Effects of a one-year rainfall manipulation on soil nematode abundances and community composition. *Pedobiologia, 54*, 87–91.

Lee, H. H., Kim, S. U., Han, H. R., Owens, V. N., Kumar, S., & Hong, C. O. (2021). Mitigation of global warming potential and greenhouse gas intensity in arable soil with green manure as source of nitrogen. *Environmental Pollution, 288*, 117724.

Lehmann, P., Ammunét, T., Barton, M., Battisti, A., Eigenbrode, S. D., Jepsen, J. U., Kalinkat, G., Neuvonen, S., Niemelä, P., Terblanche, J. S., Okland, B., & Bjorkman, C. (2020). Complex responses of global insect pests to climate warming. *Frontiers in Ecology and the Environment, 18*, 141–150.

Lemke, R. L., Izaurralde, R. C., Nyborg, M., & Solberg, E. D. (1999). Tillage and N source influence soil-emitted nitrous oxide in the Alberta Parkland region. *Canadian Journal of Soil Science, 79*, 15–24.

Li, Q., Liang, W. J., Jiang, Y., & Neher, D. A. (2007). Effect of elevated CO_2 and N fertilization on soil nematode abundance and diversity in wheat field. *Applied Soil Ecology, 36*, 63–69.

Li, Q., Chonggang, X., Wenju, L., Shuang, Z., Xunhua, Z., & Jianguo, Z. (2009). Residue incorporation and N fertilization affect the response of soil nematodes to the elevated CO_2 in a Chinese wheat field. *Soil Biology & Biochemistry, 41*, 1497–1503.

Li, H., Xing, L., Liu, X., Pu, Y., Yang, Y., & Fu, Y. (2022). Potential impact of climate change on the distribution of the pinewood nematode *Bursaphelenchus xylophilus* in Chongqing, China. *Pakistan Journal of Zoology, 54*, 809–816.

Lincoln, D. E. (1993). The influence of plant carbon dioxide and nutrient supply on susceptibility to insect herbivores. *Vegetatio, 104*, 273–280.

Lincoln, D. E., Sionit, N., & Strain, B. R. (1984). Growth and feeding response of *Pseudoplusia includens* (Lepidoptera: Noctuidae) to host plants grown in controlled carbon dioxide atmospheres. *Environmental Entomology, 13*, 1527–1530.

Liu, Y., Wang, W., Liu, P., Zhou, H., Chen, Z., & Suonan, J. (2022). Plant-soil mediated effects of long-term warming on soil nematodes of Alpine Meadows on the Qinghai-Tibetan plateau. *Biology, 11*, 1596.

Luambano, N. D., Manzanilla-López, R. H., Kimenju, J. W., Powers, S. J., Narla, R. D., Wanjohi, W. J., & Kerry, B. R. (2015). Effect of temperature, pH, carbon and nitrogen ratios on the parasitic activity of *Pochonia chlamydosporia* on *Meloidogyne incognita. Biological Control, 80*, 23–29.

Malhi, Y., Franklin, J., Seddon, N., Solan, M., Turner, M. G., Field, C. B., & Knowlton, N. (2020). Climate change and ecosystems: Threats, opportunities and solutions. *Philosophical Transactions of the Royal Society B, 375*, 20190104.

Martín-Vertedor, D., Ferrero-García, J. J., & Torres-Vila, L. M. (2010). Global warming affects phenology and voltinism of *Lobesia botrana* in Spain. *Agricultural and Forest Entomology, 12*, 169–176.

Masters, G., Norgrove, L. (2010). *Climate change and invasive alien species* [CABI Working Paper 1]. CABI, Wallingford, UK, p. 30.

Menéndez, R. (2007). How are insects responding to global warming? *Tijdschrift voor Entomologie, 150*, 355–365.

Menéndez, R., González-Megías, A., Collingham, Y., Fox, R., Roy, D. B., Ohlemüller, R., & Thomas, C. D. (2007). Direct and indirect effects of climate and habitat factors on butterfly diversity. *Ecology, 88*, 605–611.

Meynard, C. N., Migeon, A., & Navajas, M. (2013). Uncertainties in predicting species distributions under climate change: A case study using *Tetranychus evansi* (Acari: Tetranychidae), a widespread agricultural pest. *PLoS ONE, 8*, e66445.

Mirutenko, V., Janský, V., & Margitay, V. (2018). First records of *Scaphoideus titanus* (Hemiptera, Cicadellidae) in Ukraine. *EPPO Bulletin, 48*, 167–168.

Molotoks, A., Henry, R., Stehfest, E., Doelman, J., Havlik, P., Krisztin, T., Alexander, P., Dawson, T. P., & Smith, P. (2020). Comparing the impact of future cropland expansion on global biodiversity and carbon storage across models and scenarios. *Philosophical Transactions of the Royal Society B, 375*, 20190189.

Morgan, D. (1996). Temperature changes and insect pests: A simulation study. *Aspects of Applied Biology, 45*, 277–283.

Mori, A. (2018). Application of farmyard manure rather than manure slurry mitigates the net greenhouse gas emissions from herbage production system in Nasu, Japan. *Atmosphere, 9*, 261.

Mosier, A., Schimel, D., Valentine, D., Bronson, K., & Parton, W. (1991). Methane and nitrous oxide fluxes in native, fertilized and cultivated grasslands. *Nature, 350*, 330–332.

Mueller, K. E., Blumenthal, D. M., Carrillo, Y., Cesarz, S., Ciobanu, M., Hines, J., Pabst, S., Pendall, E., de Tomasel, C. M., Wall, D. H., & Eisenhauer, N. (2016). Elevated CO_2 and warming shift the functional composition of soil nematode communities in a semiarid grassland. *Soil Biology & Biochemistry, 103*, 46–51.

Neher, D. A. (2001). Role of nematodes in soil health and their use as indicators. *Journal of Nematology, 33*, 161–168.

Neher, D. A., & Weicht, T. R. (2013). Nematode genera in forest soil respond differentially to elevated CO_2. *Journal of Nematology, 45*, 214–222.

Neilson, R., & Boag, B. (1996). The predicted impact of possible climatic change on virus-vector nematodes in great Britain. *European Journal of Plant Pathology, 102*, 193–199.

Nichols, E., Spector, S., Louzada, J., Larsen, T., Amezquita, S., & Favila, M. E. (2008). Ecological functions and ecosystem services provided by Scarabaeinae dung beetles. *Biological Conservation, 141*, 1461–1474.

Nihal, R. (2020). Global Climate change and its impact on integrated pest management. *Agro Economic International Journal, 7*, 133–137.

Nisa, R. U., Tantray, A. Y., Kouser, N., Allie, K. A., Wani, S. M., Alamri, S. A., Alyemeni, M. N., Wijaya, L., & Shah, A. A. (2021). Influence of ecological and edaphic factors on biodiversity of soil nematodes. *Saudi Journal of Biological Science, 28*, 3049–3059.

Oerke, E. C. (2006). Crop losses to pests. *Journal of Agricultural Science, 144*, 31–43.

Oerke, E. C., & Dehne, H. W. (2004). Safeguarding production-losses in major crops and the role of crop protection. *Crop Protection, 23*, 275–285.

Papadopoulou, J., & Triantaphyllou, A. C. (1982). Sex differentiation in *Meloidogyne incognita* and anatomical evidence of sex reversal. *Journal of Nematology, 14*, 549–566.

Pareek, A., Meena, B. M., Sharma, S., Tetarwal, M. L., Kalyan, R. K., & Meena, B. L. (2017). Impact of climate change on insect pests and their management strategies. In P. S. Kumar, M. Kanwat, P. D. Meena, V. Kumar, & R. A. Alone (Eds.), *Climate change and sustainable agriculture* (pp. 253–286). New India Publishing Agency.

Parmesan, C. (2007). Influences of species, latitudes and methodologies on estimates of phenological response to global warming. *Global Change Biology, 13*, 1860–1872.

Parry, H. R., Macfadyen, S., & Kriticos, D. J. (2012). The geographical distribution of Yellow dwarf viruses and their aphid vectors in Australian grasslands and wheat. *Australasian Plant Pathology, 41*, 375–387.

Pascual, M., & Bouma, M. J. (2009). Do rising temperatures matter? *Ecology, 90*, 906–912.

Pathak, H., Aggarwal, P. K., Singh, S. D. (2012). *Climate change impact, adaptation and mitigation in agriculture: Methodology for assessment and applications*. Indian Agricultural Research Institute.

Pathania, M., Verma, A., Singh, M., Arora, P. K., & Kaur, N. (2020). Influence of abiotic factors on the infestation dynamics of whitefly, *Bemisia tabaci* (Gennadius 1889) in cotton and its management strategies in North-Western India. *International Journal of Tropical Insect Science, 40*, 969–981.

Patterson, D. T., Westbrook, J. K., Joyce, R., Lingren, P. D., & Rogasik, J. (1999). Weeds, insects, and diseases. *Climate Change, 43*, 711–727.

Perrings, C., Burgiel, S., Lonsdale, M., Mooney, H., & Williamson, M. (2010). International cooperation in the solution to trade-related invasive species risks. *Annals New York Academy of Sciences, 1195*, 198–212.

Phani, V., Khan, M. R., & Dutta, T. K. (2021). Plant-parasitic nematodes as a potential threat to protected agriculture: Current status and management options. *Crop Protection, 144*, 105573.

Pollard, E., & Yates, T. J. (1993). *Monitoring butterflies for ecology and conservation: The British butterfly monitoring scheme*. Springer.

Porter, J. H., Parry, M. L., & Carter, T. R. (1991). The potential effects of climatic change on agricultural insect pests. *Agricultural and Forest Meteorology, 57*, 221–240.

Power, A. G. (2010). Ecosystem services and agriculture: Tradeoffs and synergies. *Philosophical transactions of the royal society B: biological sciences, 365*, 2959–2971.

Prakash, A., Rao, J., Mukherjee, A. K., Berliner, J., Pokhare, S. S., Adak, T., Munda, S., Shashank, P. R. (2014). *Climate change: Impact on crop pests*. Applied Zoologists Research Association (AZRA), Central Rice Research Institute.

Prasad, J. S., Somasekhar, N. (2009). *Nematode pest of rice: Diagnosis and management* (Technical Bulletin No. 38). Directorate of Rice Research (ICAR), p. 29.

Pridannikov, M. V., Zinovieva, S. V., Khudyakova, E. A., Limantseva, L. A., Osipov, F. A., Dergunova, N. N., & Petrosyan, V. G. (2022). Range dynamics of potato cyst nematode *Globodera rostochiensis* (Wollenweber, 1923) (Nematoda, Heteroderidae) under conditions of global climate change in Russia. *Russian Journal of Biological Invasions, 13*, 510–529.

Randolph, S. E. (2009). Perspectives on climate change impacts on infectious diseases. *Ecology, 90*, 927–931.

Raven, P. H., & Wagner, D. L. (2021). Agricultural intensification and climate change are rapidly decreasing insect biodiversity. *Proceedings of the National Academy of Sciences, 118*, e2002548117.

Raza, M. M., Khan, M. A., Arshad, M., Sagheer, M., Sattar, Z., Shafi, J., Haq, E. U., Ali, A., Aslam, U., Mushtaq, A., & Ishfaq, I. (2014). Impact of global warming on insects. *Archives of Phytopathology and Plant Protection, 48*, 84–94.

Rebetez, M., & Dobbertin, M. (2004). Climate change may already threaten scots pine stands in the Swiss Alps. *Theoretical and Applied Climatology, 79*, 1–9.

Robert, Y., Woodford, J. T., & Ducray-Bourdin, D. G. (2000). Some epidemiological approaches to the control of aphid-borne virus diseases in seed potato crops in northern Europe. *Virus Research, 71*, 33–47.

Robinson, C. M., Hansen, L. D., Xue, X., & Adams, B. J. (2023). Temperature response of metabolic activity of an Antarctic nematode. *Biology, 12*, 109.

Roy, D. B., & Sparks, T. H. (2000). Phenology of British butterflies and climate change. *Global Change Biology, 6*, 407–416.

Sánchez-Rodríguez, A. R., Nie, C., Hill, P. W., Chadwick, D. R., & Jones, D. L. (2019). Extreme flood events at higher temperatures exacerbate the loss of soil functionality and trace gas emissions in grassland. *Soil Biology & Biochemistry, 130*, 227–236.

Schoonhoven, L. M., Van Loon, J. J. A., & Dicke, M. (2005). *Insect-plant biology*. Oxford University Press.

Sharma, H. C. (2014). Climate change effects on insects: Implications for crop protection and food security. *Journal of Crop Improvement, 28*, 229–259.

Sharma, H. C., Dhillon, M. K., Kibuka, J., & Mukuru, S. Z. (2005). Plant defense responses to sorghum spotted stem borer, Chilo partellus under irrigated and drought conditions. *International Sorghum and Millets Newsletter, 46*, 49–52.

Shrestha, S. (2019). Effects of climate change in agricultural insect pest. *Acta Scientific Agriculture, 3*, 74–80.

Siddique, S., Coomer, A., Baum, T., & Williamson, V. M. (2022). Recognition and response in plant–nematode interactions. *Annual Review of Phytopathology, 60*, 143–162.

Sikora, R. A., Bridge, J., Starr, J. L. (2005). Management practices: an overview of integrated nematode management technologies. In R. A. Sikora, D. Coyne, J. Hallmann, P. Timper (Eds.), *Plant parasitic nematodes in subtropical and tropical agriculture* (pp. 793–825). CABI.

Skendžic, S., Zovko, M., Živkovic, I. P., Lešic, V., & Lemic, D. (2021). The impact of climate change on agricultural insect pests. *Insects, 12*, 440.

Slade, E. M., Riutta, T., Roslin, T., & Tuomisto, H. L. (2016). The role of dung beetles in reducing greenhouse gas emissions from cattle farming. *Scientific Reports, 6*, 8140.

Sseruwagi, P., Sserubombwe, W. S., Legg, J. P., Ndunguru, J., & Thresh, J. M. (2004). Methods of surveying the incidence and severity of cassava mosaic disease and whitefly vector populations on cassava in Africa: A review. *Virus Research, 100*, 129–142.

Stefanescu, C., Penuelas, J., & Filella, I. (2003). Effects of climatic change on the phenology of butterflies in the northwest Mediterranean basin. *Global Change Biology, 9*, 1494–1506.

Sticht, C., Schrader, S., Giesemann, A., & Weigel, H. J. (2009). Sensitivity of nematode feeding types in arable soil to free air CO_2 enrichment (FACE) is crop specific. *Pedobiologia, 52*, 337–349.

Stiling, P., & Cornelissen, T. (2007). How does elevated carbon dioxide (CO_2) affect plant–herbivore interactions? A field experiment and meta-analysis of CO_2-mediated changes on plant chemistry and herbivore performance. *Global Change Biology, 13*, 1823–1842.

Sun, Y., Yin, J., Cao, H., Li, C., & Ge, F. (2011). Elevated CO_2 influences nematode-induced defense responses of tomato genotypes differing in the JA pathway. *PLoS ONE, 6*, e19751.

Sutherst, R. W., Constable, F., Finlay, K. J., Harrington, R., Luck, J., & Zalucki, M. P. (2011). Adapting to crop pest and pathogen risks under a changing climate. *Wiley Interdisciplinary Reviews: Climate Change, 2*, 220–237.

Swinton, S. M., Lupi, F., Robertson, G. P., & Hamilton, S. K. (2007). Ecosystem services and agriculture: Cultivating agricultural ecosystems for diverse benefits. *Ecological Economics, 64*, 245–252.

Taylor, L. R., Tauber, M. J., Tauber, C. A., & Masaki, S. (1986). *Seasonal adaptations of insects*. Oxford University Press.

Thomas, C. D., Cameron, A., Green, R. E., Bakkenes, M., Beaumont, L. J., Collingham, Y. C., Erasmus, B. F. N., De Siqueira, M. F., Grainger, A., Hannah, L., et al. (2004). Extinction risk from climate change. *Nature, 427*, 145–148.

Thomson, L. J., Macfadyen, S., & Hoffmann, A. A. (2010). Predicting the effects of climate change on natural enemies of agricultural pests. *Biological Control, 52*, 296–306.

Tilman, D., Balzer, C., Hill, J., & Befort, B. L. (2011). Global food demand and the sustainable intensification of agriculture. *Proceedings of the National Academy of Sciences, 108*, 20260–20264.

Tirado, M. C., Crahay, P., Mahy, L., Zanev, C., Neira, M., Msangi, S., Brown, R., Scaramella, C., Coitinho, D. C., & Müller, A. (2013). Climate change and nutrition: Creating a climate for nutrition security. *Food and Nutrition Bulletin, 34*, 533–547.

Toma, Y., Nufita Sari, N., Akamatsu, K., Oomori, S., Nagata, O., Nishimura, S., Purwanto, B. H., & Ueno, H. (2019). Effects of green manure application and prolonging mid-season drainage on greenhouse gas emission from paddy fields in Ehime, Southwestern Japan. *Agriculture, 9*, 29.

Torode, M. D., Barnett, K. L., Facey, S. L., Nielsen, U. N., Power, S. A., & Johnson, S. N. (2016). Altered precipitation impacts on above- and below-ground grassland invertebrates: Summer drought leads to outbreaks in spring. *Frontiers in Plant Science, 7*, 1468.

Trębicki, P. (2020). Climate change and plant virus epidemiology. *Virus Research, 286*, 198059.

Trębicki, P., Vandegeer, R. K., Bosque-Pérez, N. A., Powell, K. S., Dader, B., Freeman, A. J., Yen, A. L., Fitzgerald, G. J., & Luck, J. E. (2016). Virus infection mediates the effects of elevated CO_2 on plants and vectors. *Science and Reports, 6*, 22785.

Trudgill, D. L., Honek, A., Li, D., & van Straalen, N. M. (2005). Thermal time–concepts and utility. *The Annals of Applied Biology, 146*, 1–14.

Trumble, J., & Butler, C. (2009). Climate change will exacerbate California's insect pest problems. *California Agriculture, 63*, 73–78.

Tu, C., Koenning, S. R., & Hu, S. (2003). Root-parasitic nematodes enhance soil microbial activities and nitrogen mineralization. *Microbial Ecology, 46*, 134–144.

Turner, M. G., Calder, W. J., Cumming, G. S., Hughes, T. P., Jentsch, A., LaDeau, S. L., Lenton, T. M., Shuman, B. N., Turetsky, M. R., Ratajczak, Z., & Williams, J. W. (2020). Climate change, ecosystems and abrupt change: Science priorities. *Philosophical Transactions of the Royal Society B, 375*, 20190105.

Turnock, W. J., & Boivin, G. (1997). Inter- and intrapopulation differences in the effects of temperature on post diapause development of *Delia radicum*. *Entomologia Experimentalis Et Applicata, 84*, 255–265.

Tyler, J. (1933). Development of the root-knot nematode as affected by temperature. *Hilgardia, 7*, 389–415.

Tzortzakakis, E. A., & Trudgill, D. L. (2005). A comparative study of the thermal time requirements for embryogenesis in *Meloidogyne javanica* and *M. incognita*. *Nematology, 7*, 313–315.

Vadez, V., Berger, J. D., Warkentin, T., Asseng, S., Ratnakumar, P., Poorna Chandra Rao, K., Gaur, P. M., Munier-Jolain, N., Larmure, A., Voisin, A.-S., et al. (2012). Adaptation of grain legumes to climate change: A review. *Agronomy for Sustainable Development, 32*, 31–44.

van den Hoogen, J., Geisen, S., Routh, D., Ferris, H., Traunspurger, W., Wardle, D. A., et al. (2019). Soil nematode abundance and functional group composition at a global scale. *Nature, 572*, 194–198.

van Lenteren, J. C. (2012). *Internet book of biological control*. International Organization for Biological Control.

Williamson, V. M., & Kumar, A. (2006). Nematode resistance in plants: The battle underground. *Trends in Genetics, 22*, 396–403.

Wilsey, B. J. (2001). Effects of elevated CO_2 on the response of *Phleum pratense* and *Poa pratensis* to aboveground defoliation and root-feeding nematodes. *International Journal of Plant Sciences, 162*, 1275–1282.

Yamamura, K., & Kiritani, K. (1998). A simple method to estimate the potential increase in the number of generations under global warming in temperate zones. *Applied Entomology and Zoology, 33*, 289–298.

Yeates, G. W., Bongers, T., de Goede, R. G. M., Freckman, D. W., & Georgieva, S. S. (1993). Feeding habits in soil nematode families and genera-an outline for soil ecologists. *Journal of Nematology, 25*, 315–331.

Yeates, G. W., Newton, P. C. D., & Ross, D. J. (1999). Response of soil nematode fauna to naturally elevated CO_2 levels influenced by soil pattern. *Nematology, 1*, 285–293.

Yeates, G. W., Newton, P. C. D., & Ross, D. J. (2003). Significant changes in soil microfauna in grazed pasture under elevated carbon dioxide. *Biology and Fertility of Soils, 38*, 319–326.

Yonow, T., Kriticos, D. J., Kirichenko, N., & Ota, N. (2018). Considering biology when inferring range-limiting stress mechanisms for agricultural pests: A case study of the beet armyworm. *Journal of Pest Science, 91*, 523–538.

Yudelman, M., Ratta, A., Nygaard, D. F. (1998). Pest management and food production: looking to the future. International Food Policy Research Institute.

Zeyen, R. J., Stromberg, E. L., & Kuehnast, E. L. (1987). Long-range aphid transport hypothesis for maize dwarf mosaic virus: History and distribution in Minnesota, USA. *The Annals of Applied Biology, 111*, 325–336.

Zhang, Z. Q. (2011). Animal biodiversity: An introduction to higher-level classification and taxonomic richness. *Zootaxa, 3148*, 7–12.

Zhang, W., Ricketts, T. H., Kremen, C., Carney, K., & Swinton, S. M. (2007). Ecosystem services and dis-services to agriculture. *Ecological Economics, 64*, 253–260.
Zhou, X.-L., Harrington, R., Woiwod, I. P., Perry, J. N., Bale, J. S., & Clark, S. J. (1995). Effects of temperature on aphid phenology. *Global Change Biology, 1*, 303–313.
Ziska, L. H., Blumenthal, D. M., Runion, G. B., Raymond Hunt Jr, E., & Diaz-Soltero, H. (2011). Invasive species and climate change: An agronomic perspective. *Climate Change, 105*, 13–42.
Zvereva, E. L., & Kozlov, M. V. (2005). Consequences of simultaneous elevation of carbon dioxide and temperature for plant-herbivore interactions: A metaanalysis. *Global Change Biology, 12*, 27–41.

Chapter 18
Plant Virus Diseases Dynamics Under Modified Environments and Their Impacts on Host Virus-Vector Landscape

Naorem Loya Mangang, Konjengbam Sarda Devi, Raghuveer Singh, Saurav Saha, Nitika Gupta, and Susheel Kumar Sharma

Abstract Global climate changes are instigated by natural and human activities. It has fast-tracked in the past few decades. In the twenty-first century, rising greenhouse gas levels are expected to continue to boost global temperatures and alter water availability. Plant viral diseases are profoundly influenced by the environment; if the environment is not favourable to disease development, a susceptible host will still not be infected by a virulent virus. Plant hosts, vectors and viral pathogens are affected by the direct repercussions of climate changes viz., changing rainfall patterns, drought, higher wind speeds, increased temperature, greenhouse gas emissions and indirectly through factors such as regional fluctuations in cultivated regions, crop ranges, farming practices, vector distribution, and its abundance, and also reservoir, potential weed hosts. These climatic factors can have positive, negative or neutral effects on disease development as each disease may respond differently to these variations. This actively demonstrates that to fully understand the multivariate nature of plant-pathogen and vector interactions and to produce disease-resistant crop plants that are resilient to climate change, future research must increasingly exploit dynamic environmental factors.

Keywords Plant diseases · Environment · Plant virus · Host · Vector

---

N. L. Mangang · K. S. Devi · S. K. Sharma (✉)
ICAR Research Complex for NEH Region, Manipur Centre, Imphal, Manipur 795004, India
e-mail: susheelsharma19@gmail.com

R. Singh
ICAR Research Complex for NEH Region, Arunachal Pradesh Centre, Basar, Arunachal Pradesh 791101, India

S. Saha
ICAR Research Complex for NEH Region, Sikkim Centre, Tadong, Sikkim 737102, India

N. Gupta · S. K. Sharma
Advanced Centre for Plant Virology, Division of Plant Pathology, ICAR-Indian Agricultural Research Institute, New Delhi 110012, India

H. Pathak et al. (eds.), *Climate Change Impacts on Soil-Plant-Atmosphere Continuum*, Advances in Global Change Research 78,
https://doi.org/10.1007/978-981-99-7935-6_18

18.1 Introduction

A disease in plants may be defined by the occurrence of abnormalities in the appearance, physiology and the normal productivity of the plants as compared to healthy plants. There are various agents that cause such disturbances, one of which is a virus that is a pathogenic, intracellular, sub-microscopic entity made up primarily of nucleic acid (either RNA or DNA) and protein that drives its own replication in living host cells (Schumann & D'Arcy, 2006). A nucleotide-based genome and a protective protein coat are found in most plant viruses. The virions of different shapes and sizes are formed viz., icosahedral/spheres, rigid rods twinned, bacilliform and flexuous filaments outside the host. These forms and sizes aid in the identification of distinct virus groups. Each virus is made up of strains/genotypes/variants, which emerge mostly through mutation or recombination (Jones, 2012). The 10th assessment of the International Committee for the Taxonomy of Viruses (ICTV) gives the notion that plant viruses are developing rapidly, and presently, 9110 species listed in the report. A significant surge in the known viruses has recently been aggrandized with the advancement of new tools in virology. This is evident from the confirmed number of 2285 virus species recorded in the 9th report of ICTV which increased three-folds (9110) in the 10th report (https://talk.ictvonline.org). Nonetheless, the overwhelming amount of them is from crop plants, and also recent metagenomic analyses of plant viral biodiversity have revealed the richness and diversity of plant viruses (Roossinck, 2015; Sidharthan et al., 2021). Viruses are either generalist, with a wide host range but poor adaptation to specific hosts, or specialists, with a confined host range but excellent adaptation to specific hosts. Considering climatic variability, generalists are better suited to changes in host range than specialists. Viruses adjust quickly to climatic factors due to their small genomes, which indicate why; according to Anderson et al. (2004) they contribute nearly half of new pathogens in plants (Jones, 2016). Plants get infected with viruses only and only if they have an injury or some vectors introduce the virus to them as it has natural physical barriers such as cuticles and cell walls. Propagation to the next generation of host plants known as vertical transmission has been seen in some circumstances as a result of seed infection (King et al., 2012). Viruses of plants transmitted by insects are a huge concern to agricultural production systems around the world, and these relationships interact with climate in certain ways to influence the food supply and also in the emergence of new diseases. Despite the availability of significant advances in understanding the climatic factors influencing infectious diseases in the case of animals, the impact of various environmental variables in relation to the plant virus interaction system that is known to affect major crops still remains as a difficulty with much to understand and reveal. It depends on the biotic and abiotic conditions whether such interspecific interactions are inimical, interdependent, or impartial. Explaining, identifying environmental conditions that drive the interactions has been used for elucidating patterns of species plethora and plant productiveness in agro ecosystems. Most weather incidents result in amplified abiotic stress and hence productivity of the plants is greatly affected. Arthropod vector's natural movement patterns and their populations are mostly affected due to the changing climate, which in turn influence the viral epiphytotics (Sharma et al., 2021). A few plant viruses increase adaptation to their hosts as the increasing environmental stresses such as thermal stress and drought thereby extend their role in potential ecological effects. Effects may be mutualistic, meaning plants and viruses help in each other's survival. The present review provides a comprehensive account of dynamics of disease caused by plant viruses under the modified environmental conditions and influences on host-virus-vector landscapes.

18.1.1 Economic Losses Caused by Plant Viral Diseases

Virus infections in various crops are well-known to cause massive failures of product quantity and quality over the entire globe. They are responsible for some of the most cataclysmic plant epiphytotics in the world, resulting in considerable annual productivity losses in agricultural crops (Jones, 2020; Rybicki, 2015). Numerous causes have led to the existing global plant viral disease situation, which is progressively developing. Three significant most devastating plant virus groups viz., begomoviruses, potyviruses and tospoviruses that are transmitted by whiteflies, aphids and thrips respectively, threaten the security of food in different countries by inflicting catastrophic diseases in tropical and sub-tropical food crops. In South and Southeast Asian countries, viruses such as groundnut bud necrosis orthotospovirus in peanut and tomato are destroying crops extensively (Oliver & Whitfield, 2016; Pappu et al., 2009) and begomoviruses causes cassava mosaic disease (CMD) in sub-Saharan Africa and the Indian subcontinent which causes huge economic losses to the production and productivity of these crop plants (Jacobson et al., 2018; Legg et al., 2011; Rey & Vanderschuren, 2017; Rojas et al., 2018). Plant virus pandemics and significant epidemics pose a hazard to cultivated plants grown for a variety of purposes, including feeding humans and cattle, as well as producing ornamental plants, and therapeutic items. When several plants get persistently infected by virus and causes severe disease symptoms, reductions in total agricultural yields or food quality might be disastrous. Furthermore, comparable losses occur in cultivars that are both seasonal and perennial. Also, when a viral pandemic or a significant epidemic of plants strike a key staple food crop that are critical to food security, they have the potential to drastically reduce food supplies, resulting in catastrophic food shortages. Viruses are a primary source of plant disease, with an annual economic effect of more than $30 billion (Sastry & Zitter, 2014). They account for over half of the pathogens that cause new and resurging plant diseases around the globe, and they harm both wild and domesticated crops (Anderson et al., 2004; Bos, 1981; Cooper & Jones, 2006; Jones, 2009; Sharma et al., 2021; Thresh, 2004, 2006a). Viruses that infect rice crops cause output losses worth more than $1.5 billion in Southeast Asia alone (Abo & Sy, 1997; Hull, 2013; Sasaya et al., 2014). In India, Sri Lanka and Africa cassava mosaic disease (CMD) caused by a group of begomoviruses inflict annual losses of 25 billion tonnes of cassava making them one of the most dangerous virus-related concerns (Nicaise, 2014). Cassava being a staple food for over 500 million people worldwide, disease is frequently linked to famine episodes (Calvert & Thresh, 2002; Legg, 1999; Legg & Thresh, 2000). In the United States, the potato leafroll virus (PLRV) costs rupees 7,483.35 million each year, while in the United Kingdom; it costs £30–50 million per year (Wale et al., 2008; Sastry & Zitter, 2014). Citrus tristeza virus (CTV) was responsible for the deaths of over 100 million trees in citrus tree cultures around the world (Harper, 2013; Moreno et al., 2008). The barley yellow dwarf virus costs UK growers almost £10 million in lost yield each year (Ordon et al., 2009; Sastry & Zitter, 2014). Viruses are thus important crop concerns, causing significant agronomic losses and a global food crisis (Nicaise, 2014).

18.1.2 Some Historical Point of View on Plant Viral Diseases

There is a history of citrus tristeza virus causing a devastating epidemic in the world. In addition to this there is a record of plum pox disease and wheat streak mosaic virus targeting different crops in the world. Pandemics and outbreaks of plant virus diseases are becoming harder to control as a result of climate uncertainty linked to climate change. "Historical crops introduced from another continent were reviewed and the principal virus disease examples described were rice yellow mottle, cassava brown streak, groundnut rosette and cocoa swollen shoot, which are limited to sub-Saharan Africa (SSA); cassava mosaic in SSA, the Indian subcontinent and Southeast Asia; and tomato yellow leaf curl, which to all extents and purposes now has a global distribution" (Jones, 2020).

18.1.3 Plant Virus Disease Triangle/Pentagon

Certain parameters contribute to the formation of a well-defined disease triangle or pentagon, most of them were contributed by the development of a pathogen and also by the degree of harm caused by the pathogen to the host, and in fact many events took place in the development of a disease which is described explicitly in the disease triangle or pentagon. The three major factors defined as virulent pathogens, susceptible host, and favourable environmental conditions, leads to the development and establishment of disease (Fig. 18.1). Time is another addition to the component of the conventional disease triangle, which plays a major role in disease development.

The disease pentagon is substantiated through various epidemics of plant diseases. To elucidate, instances of thrips led tospoviruses infection may be well mentioned as to the high propensity of thrips to act as vectors, availability of wide susceptible host plants stand under favourable environmental space and time fulfilling the parameters of a disease pentagon. Riley et al. (2011) correlated temporal tomato spotted wilt virus incidence, a tospovirus in tomato plant with the seasonal fluctuation of its thrips vectors. They further evaluated the relationship between incidence of tomato spotted wilt virus (TSWV) and susceptible tomato under field conditions in Georgia

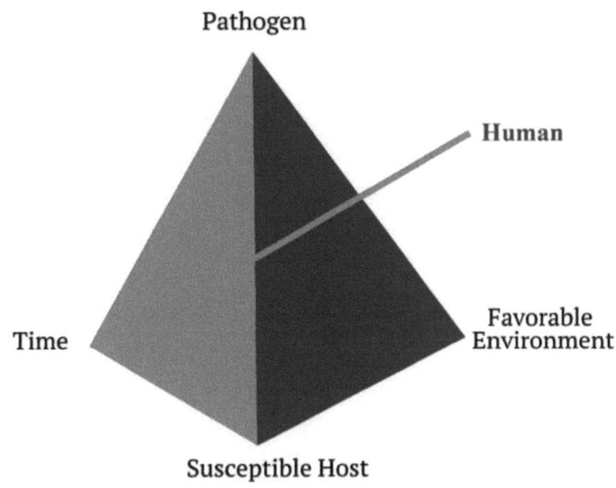

Fig. 18.1 An illustration of disease development parameters

to proportionate positively and also, in case of whitefly led epidemic of tomato yellow leaf curl virus (TYLCV) the wide abundance of host crop, as tomato is cultivated throughout the world as well as various weeds in the form of other alternate host crops; becomes available to whitefly (*Bemisia tabaci*), the vector. Under favourable environmental conditions and time, *B. tabaci* arises as a vector whose population size substantially gets boosted, thereby giving a natural conditioning of the required parameters which leads to devastating viral leaf curl disease of tomato crop and thus the ultimate epidemics. According to Legarrea et al. (2020) the role of presence of the hosts (whether tomato crop species or weeds, alternate host) by acting as green bridges allow for the persistence and yearly recurrence of TYLCV. The positive correlation of TYLCV incidence with the increasing densities of whiteflies was reported (Anco et al., 2020). Therefore, it is only logical to understand that the presence of host, alternate host as vector reservoir and vector population under apt environmental space and time, highlights the significance of role of the parameters in a disease pentagon.

18.1.4 Plant Virus Infection Cycle

Plant viruses have a typical infection cycle which is very different from the infections caused by fungi, bacteria, phytoplasma and other pathogens. Unlike the animal viruses, plant viruses are not able to make an active entry to the host cells and require a means to take the virions inside the cell. This entry is facilitated by the vectors (arthropods, fungi etc.) and in some cases by mechanical means (Fig. 1.2). Once the virions take entry to the healthy plant cell, genome uncoating happens so as to release the genomic nucleic acid. The viruses can have either of (i) positive sense single started RNA (+ss RNA), (ii) negative sense ssRNA (-ss RNA), (iii) single stranded DNA (ssDNA), (iv) double stranded DNA (dsDNA) as genome. The

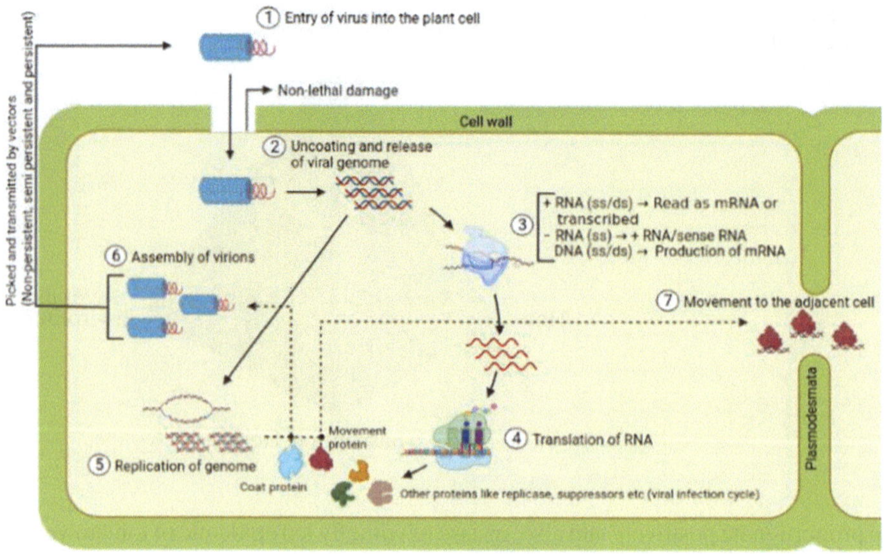

Fig. 18.2 A schematic illustration of the plant virus infection cycle indicating the different steps in establishment of successful infection. Entry of virus into the plant cell through non-lethal damage of cell wall as plant cells does not have any receptor for viral entry, uncoating and release of viral genome occurs only when the virus particles enter into the cell, viruses with genome dsRNA or sense ssRNA are transcribed into mRNA, they have the viral RNA-dependent RNA polymerase (RdRp), for dsDNA they are first transcribed into mRNA by the host RNA-dependent RNA polymerase inside the host cell nucleus, whereas viruses which has ssDNA genome are first replicated and then transcribed into mRNA inside the nucleus

mRNA are transcribed from the viral genome and translated to viral proteins by the host ribosome. Viral genome information is translated into functional proteins either during early stages of infection cycle or late stages of infection cycle depending on the requirement. The viral genome is then replicated and assembled to infectious virions which are picked up by the vectors for horizontal transmission in the host population. The virus movement from initially infected cells to the adjacent cells happens through plasmodesmata channels thus leading to the systemic infection in the host (Fig. 18.2).

18.1.5 Host-Virus-Vector Landscape

Plant viruses as such cannot take active entry to the host plant cells due to the presence of cellular barriers like cell walls. Hence, they need a route to carry them inside the live cells and make successful infection. Plant viruses thus have evolved to make it easier for insect vectors to acquire, retain, and inoculate them to the plant cell. To establish a successful viral infection in the plants, many perspectives have to be covered and to a great extent the environmental factors as well as the host susceptibility plays an important role (Fig. 18.3). It was previously thought that the virus's ability to adapt was restricted to some proteins that allow them to adhere to the mouth parts of vectors or to the internal tissues of the vectors. However, recent study implies that viruses were able to change host plant morphological characters and vector behaviour in different ways that help them spread. In a particular case of thrips (*Frankliniella occidentalis*) that is infected with TSWV, modification in their feeding pattern can be seen which helps in increasing the spread of the TSWV (Dietzgen et al., 2016). Viruses manipulate or change their host so as to affect their vectors in different ways; it changes their actions, their orientation of the vectors, and consequently changing the feeding patterns and dispersal behaviour of the vectors. The phenotypic effects on hosts vary for different viral pathosystems, yet there are a considerable number of convergences among different types of viruses that are spread by the action of the same characteristic of a vector. This convergence on transmission of viruses supports the concept of adaptive nature of the virus effects and not simply a by-product of its infection. Depending on this we can suggest that manipulation on hosts by viruses occurs in the host proteins that are multifunctional and help them exploit the resources of hosts and somehow alter the phenotypes of the host. However, this idea is rarely considered in light of the multiple constraints on virus development imposed by molecular and environmental variables, which are important in virus-host interactions study that does not involve host modification.

Insect vectors are the primary carriers of plant viruses under field conditions. The insect vectors such as thrips play an important role in the development and spread of tomato spotted wilt disease which is evolving as one of the potentials economically damaging plant viral diseases in different crops. Temperature, photoperiod, humidity, rainfall, and the availability of suitable hosts for feeding and oviposition are all seasonal elements that have been proven to affect thrips population sizes (Chaisuekul & Riley, 2005). The degree of development in the population size of thrips is most likely influenced by temperature. In *Frankliniella fusca* and *Frankliniella occidentalis*, the number of degree days necessary to finish one cycle of life to adulthood is recorded to be 234 (lower threshold 10.5 °C) and 253 (6.5 °C), respectively and the western flower thrips needed almost fourteen days at 25 °C to mature to adulthood which is shorter in case of tobacco thrips (Morsello & Kennedy, 2009). Gamarra et al., 2020, have studied the effect of temperature on the efficiency of virus transmission by the whitefly under controlled laboratory conditions and developed a temperature-dependent phenology model for the whitefly vector using the Insect Life Cycle Modeling (ILCYM) software (Gamarra et al., 2020). Plant-virus and insect-vector management strategies can be benefitted by using new technology like metagenomics which can improve our understanding of complex plant-virus-vector interactions. Infected plants show physiological changes that might include alteration in cellulose deposition, amino acid levels, down-regulation of plant defense pathways and many more and these changes could attract massive vector populations, resulting in an epidemic breakout of viral diseases (Islam et al., 2020). The first step to have a profound knowledge on effective management techniques of the virus vectors such as whitefly, aphids, thrips, etc. is to have an immense knowledge on how the host

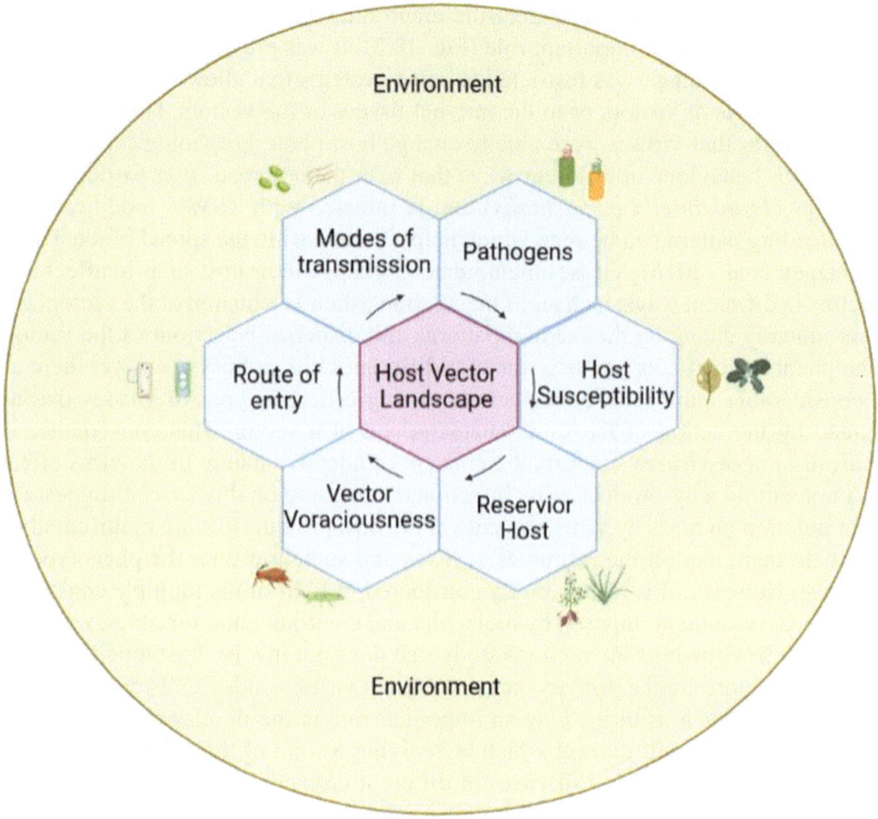

Fig. 18.3 Diagrammatic representation of host-vector landscape as influenced by the environment

vector as well as the virus ecology is implementing on the development of the plant viral pandemics.

18.1.6 Epidemiology of Plant Virus Diseases

Epidemics are frequently caused by novel viruses or new strains/variants of old viruses that have spread from reservoir species to cultivated plants (Elena et al., 2011). Epidemics and pandemics of virus infection endanger all sorts of cultivars, such as those used as foods for human as well as for domesticated animals, and also for those planted for ornate, fiber, or therapeutic purposes (Anderson et al., 2004; Hull, 2014; Jones, 2009, 2014; Jones & Naidu, 2019; Thresh, 1980a, 2006a). Virus outbreaks also pose a hazard to wild plant groups in natural habitats. Strategies implemented in controlling viral epidemics in plants were done of intuition until the twentieth century, without any knowledge of the nature of the virus and how it was transmitted. Fortunately, a great deal has been learned since then about many plant viruses and their basic nature such as what variables control their spread and destruction, how they cause epidemics and why the yield of cultivars is greatly reduced by them (Bos, 1982; Cooper & Jones, 2006; Jones, 1981, 2004, 2006; Thresh, 1982, 1983, 2006a). Numerous causes are there for the current global plant viral disease scenario, which is progressively developing. Few reasons for this are multinational corporations' actions of worldwide distribution of various plant products bringing harmful virus epiphytotics to places of the world where they were previously unknown. For example, multinational seed firms' proclivity for producing seed crops in far-flung underdeveloped countries that have hotter-regions at seasons of the year in times that they can't be cultivated in warm climes facilitate virus expansion crisis to the greatest extent. Empowerment of the supply chain, which includes international agreements on free trade and tariff reductions, has paved the door for large-scale crop transfers from one region or country to another. Exports of agricultural products from developing countries have increased because developed countries' subsidies have declined. This commerce has been facilitated by more efficient air and sea transportation techniques, as well as the relaxation of plant quarantine requirements to comply with new WTO guidelines (Constable et al., 2018; Dombrovsky & Smith, 2017; IMF, 2000; James & Steger, 2014; Jones, 2009, 2014; Jones & Naidu, 2019; Rodoni, 2009). Because of the displacement of crops from their native domesticated land to foreign land, new diseases of viruses are appearing at an increasing pace (Jones, 2009, 2018, 2020; Thresh, 1980a, 1980b, 1981, 1982, 2006a, 2006c). Pandemics and outbreaks of plant virus diseases are becoming harder to control as a result of climate uncertainty linked to climate change (Canto et al., 2009; Jones, 2009, 2012, 2014, 2016; Trebicki, 2020). The arrival of novel crops to parts of the region where they are unknown, and also the adoption of extremely intensified agricultural approaches, resulted in the fresh interactions with viruses to crops which help in the development of destructive viral epidemics. They are also helped by the widespread cultivation of susceptible improved varieties engineered for higher yields (Jones, 2009; Loebenstein & Thottappilly, 2013; Thresh, 1980b, 1982, 2004, 2006a, 2006b, 2006c). Whenever plant virus pandemics or significant epidemics strike key food crops that are critical to subsistence, they have the potential to reduce food supplies to the point where extreme food shortages result in famine (Bos, 1992; Brunt et al., 1990; Jones, 2009, 2014; Jones & Naidu, 2019; Loebenstein & Thottappilly, 2013; Thresh, 1980a, 2004, 2006a, 2006c). Analysing viruses in wild natural vegetation and introduced crops has the ability to give pertinent data not only on the biodiversity threats and cultured species, but also on the development and its adaptation and over the past millions of years (Jeger et al., 2006; Lovisolo et al., 2003; Webster et al., 2007). The development of novel viruses has frequently occurred as a result of host introductions. Cacao swollen shoot virus (CSSV) became a major cacao infecting virus in the twentieth century in West Africa which was a result of introduced crops from the USA, soon after the crop was introduced from America (Posnette, 1981; Thresh, 1980b). Similar case happens with the Maize streak virus, another American invasion (Fargette et al., 2006; Martin et al., 2001).

Cassava, a Latin American crop, is an important source of nutrition for more than 700 million people living in tropical countries. The African continent contributes half of the overall production of 238 million tonnes (FAOSTAT, 2008). Cassava is susceptible to at least 20 distinct viral infections as a consequence of its vegetative proliferation, which makes it sensitive to a variety of pests and diseases, of all this, cassava mosaic disease (CMD) possesses the greatest threat in cassava cultivation throughout Africa and Indian regions (Legg & Fauquet, 2004; Legg et al., 2006). It is caused by a group of begomoviruses. The CMD outbreak has spread to at least nine countries in East and Central Africa, encompassing 2.6 million square kilometres and costing 1.9–2.7 billion USD in annual financial losses and is described as the world's most dangerous plant virus disease, causing starvation and the deaths of thousands of people (Legg et al., 2006). A very virulent strain of East African cassava mosaic virus (EACMV-Ug; resulting from EACMV and African cassava mosaic virus: ACMV interspecific recombination) caused an epidemic in 1988 (Zhou et al., 1997) that started from Uganda and expanded across Central and East Africa, creating massive losses that necessitated international assistance to avoid widespread starvation (Anderson et al., 2004).

Sharka, the most common viral infection of stone fruit crops, is caused by the plum pox virus (PPV) (Cambra et al., 2006; Garcia & Cambra, 2007). For most of the twentieth century, PPV was restricted to Europe only but it was discovered in Africa, South and North America, and Asia in the last 20 years (Candresse & Cambra, 2006). As a result, practically all significant producing locations are now affected in some way. Several nations have expensive eradication or control measures. These attempts have largely delayed, but not prevented, the infection and spread of PPV, although successful in a few situations. In the last 30 years, the global cost of Sharka disease and control measures has been estimated at US$10 billion (Cambra et al., 2006). In banana plantations, banana bunchy top disease (BBTD) caused by banana bunchy top virus (BBTV) creates a disastrous epidemic. The first occurrence of BBTD was in Fiji, Oceania, in 1889. It is the most destructive viral disease that attacks banana plantations on a global scale, posing a severe threat to developing countries' food security (Dale, 1987; Kumar et al., 2011, 2015). The plant is vegetatively propagated, and the virus has spread widely due to anthropogenic activities that unintentionally carried contaminated planting material between countries and continents (Dale, 1987; Qazi, 2016; Rybicki & Pietersen, 1999).

Peanut Rx, a tool, is available to growers to inform peanut management decisions based on the assessment of the risk factors associated with tomato spotted wilt (TSW) disease, caused by tomato spotted wilt virus and transmitted by thrips. Chappell et al., 2020, studied the relationships between weather and TSW risk mediated by thrips vectors, and developed an addition to Peanut Rx that proved informative and easy to implement (Chappell et al., 2020).

18.1.7 Role of Climate Variables on Plant Viral Diseases Development

The temperature is one among the various factors that influence the interactions between plant pathogens, and is also an important factor that governs the development of healthy plants (Obrępalska-Stęplowska et al., 2015). As with relative humidity, temperature is an important climatic variable for plant virus infection (Huber & Gillespie, 1992). The comparison of performance of plants at various temperature conditions are of particular significance when facing climate change, both when the plants are made to grow normally and the other is provoked by various biotic stimuli such as viruses that strictly depend on the host cell life-cycle (Obrępalska-Stęplowska et al., 2015). According to the Intergovernmental Panel on Climate Change (IPCC, 2014), the global temperature is predicted to rise from 1.5 to 4.8 °C, by the end of the twenty-first century based on average global surface temperature. Although pathogens have already begun expanding pole wards in terms of geographic expansion (Bebber et al., 2013), it is predicted that temperature elevation will modify the fitness of pathogens that evolve at higher latitudes as well as increase their epidemic risk since they have adapted to temperatures below their physiological optimum (Bebber, 2015a, 2015b; Elad & Pertot, 2014; McDonald & Stukenbrock, 2016). When temperature rises, the plant defenses are weakened and the spread of tobacco mosaic virus or turnip mosaic virus is encouraged (Wang et al., 2009). Arabidopsis plants replicate the turnip crinkle virus (TCV) more rigorously at higher temperatures thereby favouring the development of the virus (Király et al., 2008; Prasch & Sonnewald, 2013; Zhang et al., 2012). Diverse processes can be involved regarding the temperature sensitive mechanisms that help in the virus spread and its development. In addition to increasing membrane fluidity and cytoskeletal dynamics that likely accelerate the membrane- and cytoskeletal-associated processes involved in virus replication and transport, the regulation of intercellular communication channels in plant cell-walls may also be affected by temperature. The plasmodesmata (PD) act as conduits through which viruses move their genomes in order to spread infection between cells, into the phloem, and finally throughout the plant. A shift in temperature can significantly alter the composition and regulation of plasmodesmata, which affects virus spread from cell to cell (Amari et al., 2021). A temperature gradient strongly impacts virus dissemination, replication, translocation and susceptibility in plants, as well as symptom expression in the host (Jensen, 1973; Matthews, 1991). Plants that are heated to slightly higher temperatures than normal are more prone to virus inoculation with infective sap (Kassanis, 1952; Matthews, 1970). Increasing the temperature of previously infected plants enhances the rate of virus multiplication and general movement within cells (Matthews, 1970). It was observed that at 18–22 °C, the tobacco rattle virus multiplied the fastest, but at 26 °C, it multiplied the least (Frost & Harrison, 1967). In tomato plants, Goodman et al. (1986) observed that tomato spotted wilt virus (TSWV) replication was higher at 20 °C, but symptoms were more severe at 36 °C. Chrysanthemums may display TSWV symptoms over a temperature range from 18 to 27 °C, but infected plants were more common at a temperature regime of 21/16 °C (day/night) than at 24/18 °C (Llamas-Llamas et al., 1998). Ochoa et al. (1996) reported that the symptoms caused by tospovirus in chrysanthemums were most severe when the average temperature was over 28 °C. As the temperature rises, new pathogens are likely to emerge thus resulting in to the more prevalent and severe epidemics (Jones, 2009). By altering selection pressures and virus evolution rates, fluctuations in average temperature can change the magnitude of virus outbreaks. Warmer temperatures can speed up viral development, resulting in more virulent strains with greater natural host ranges, faster virus multiplication rates in reservoir animals, and higher vector transmission efficiency (Picó et al., 1996) e.g. when CMV-infected tobacco plants were kept at 25 °C or 33 °C, the greater temperature increased spontaneous mutations in the virus's coat protein gene, altering pathogenicity in terms of systemic mobility (Yoon et al., 2011). When temperatures exceed a threshold, the effectiveness of single dominant gene resistances, which are temperature-dependent, is reduced (Canto & Palukaitis, 2002; Fraser & Loughlin, 1982; Pfitzner, 2006). As a result, when previously virus-resistant plants are cultivated in warmer temperatures, temperature-sensitive resistance genes that are now efficient may become ineffective, resulting in the occurrence of epidemics in these plants (Fraser, 1990; Fraser & Loughlin, 1982; Jones, 2009).

Using climatic and biological frameworks, Jones and Barbetti (2012a, 2012b) assessed the probabilities of direct and indirect effects of climate change on the vectors, viruses, and hosts that constitute plant virus pathosystems as a whole. With respect to viruses, this review highlighted many of the composite and geographically diverse effects of climate variations that could impact epidemics of vector and non-vector-mediated plant virus pathogens around the world. Furthermore, it was concluded that many critical epidemic components of viruses will change due to climate change in different ways, often enhancing them or the other way around varying upon the host and circumstances. Furthermore, since vector-borne pathogens and new encounter scenarios are complicated by climate change parameters, emergence of previously unknown virus pathogens and implications of climate change on vector types added additional factors to be considered. Plant virus epidemics are predicted to become increasingly difficult to control as a result of future climate instability, requiring extensive research to ensure food security and biodiversity throughout the world. Plant host, vectors and pathogens are affected by climate changes direct repercussions, viz., changing rainfall patterns, drought, higher wind speeds, and increased temperature and greenhouse gas emissions indirectly via things like regional fluctuations in cultivated regions, crop ranges, farming practices, vector distribution and abundance, and reservoir and weed potential hosts (Jones, 2016). If rainfall patterns shift as a consequence of global warming, outbreaks of infections spread by insects are expected to change. The likelihood of detrimental epidemics of viruses conveyed by *B. tabaci,* which includes many begomoviruses, occurring in mid-latitude regions of Asia, Australia, Latin America, and Africa increases significantly when they establish the prolonged dry seasons that favour *B. tabaci* population expansion (Morales & Jones, 2004).

Moisture variations have a major effect on cropping patterns, which has implications in agriculture. Heavy precipitation and inundation, as well as severe drought, are the result of severe weather events. Massive virus epidemics in agricultural and wild vegetation can be exacerbated by such weather events (Jones, 2012). Because of enhanced vector zoospore activity spurred by an increase in soil dampness, outbreaks of fungus-transmitted viruses are anticipated to spread throughout larger areas in temperate locations where soil-borne viruses have received more rainfall than previously. As a result, Furovirus and Bymovirus epidemics in crops such as cereals, potatoes, and sugar beets would become more common as illustrated in Fig. 18.4. The decrease in soil wetness of non-irrigated fields reduces the activity of the vector zoospores; epidemics of warm temperature-adapted viruses like peanut clump virus (PCV) and Indian peanut clump virus (IPCV) are predicted to become less common as rainfall drops in drier sub-tropical and mid-latitude regions. In drier mid-latitude locations with etesian climate, falling moisture content is expected to reduce the prevalence of nematode-transmitted viruses in rain fed crops (Jones, 2016).

The primary rains in Kenya's coastal areas, where rainfall is occurring twice, the main rain falls between April and July, while the short rains fall between September and November. Around May and July, a single large period of African cassava mosaic virus (ACMV) dissemination occurs in these areas, followed by a significant drop in rates of infection from August to October (Bock, 1994). Later in the season, a following, less discernible surge of spread follows, which appears to be linked to the periodic rains. ACMV distribution was larger in the wettest areas (>1200 mm annual rainfall), however it was low or non-existent in the drier zone (800–1000 mm annual rainfall) (Bock, 1994). A parallel finding was recorded in Côte d'Ivoire, where, in the Savanna, ACMV spread slowly but more quickly in the damp forest zone (Fauquet et al., 1988). It should also be emphasized that in certain Savanna regions, as in Uganda, expansion can be swift (Otim-Nape et al., 1994).

The extent of leaf wetness is regulated by relative humidity, which is directly linked to temperature. The luxuriant plant growth and soft delicate leaves that form under situations of high relative humidity promote the dissemination of contact-transmitted viruses in crops and pastures. Soft-leaved plants are more receptive to infection than hard-leafed plants found in low-humidity environments. It is because when growth is soft, the wounds through which such viruses must enter a plant's protective cuticle prior to actually invading damaged cells develop much more rapidly (Bawden, 1964; Jones, 2004). As a result, higher humidity climate projections are likely to favour epidemics, while low relative humidity climate change scenarios are expected to have the opposite impact (Fig. 18.5). The speed of replication and dispersion of air-borne virus vectors can be affected significantly by relative humidity (Matthews, 1970). When the flight efficiency of *Aphis glycines*, the vector of soybean mosaic virus, was evaluated, the ideal relative humidity for flight was 75 percent at the optimum flight temperature (16–28 °C) (Zhang et al., 2008).

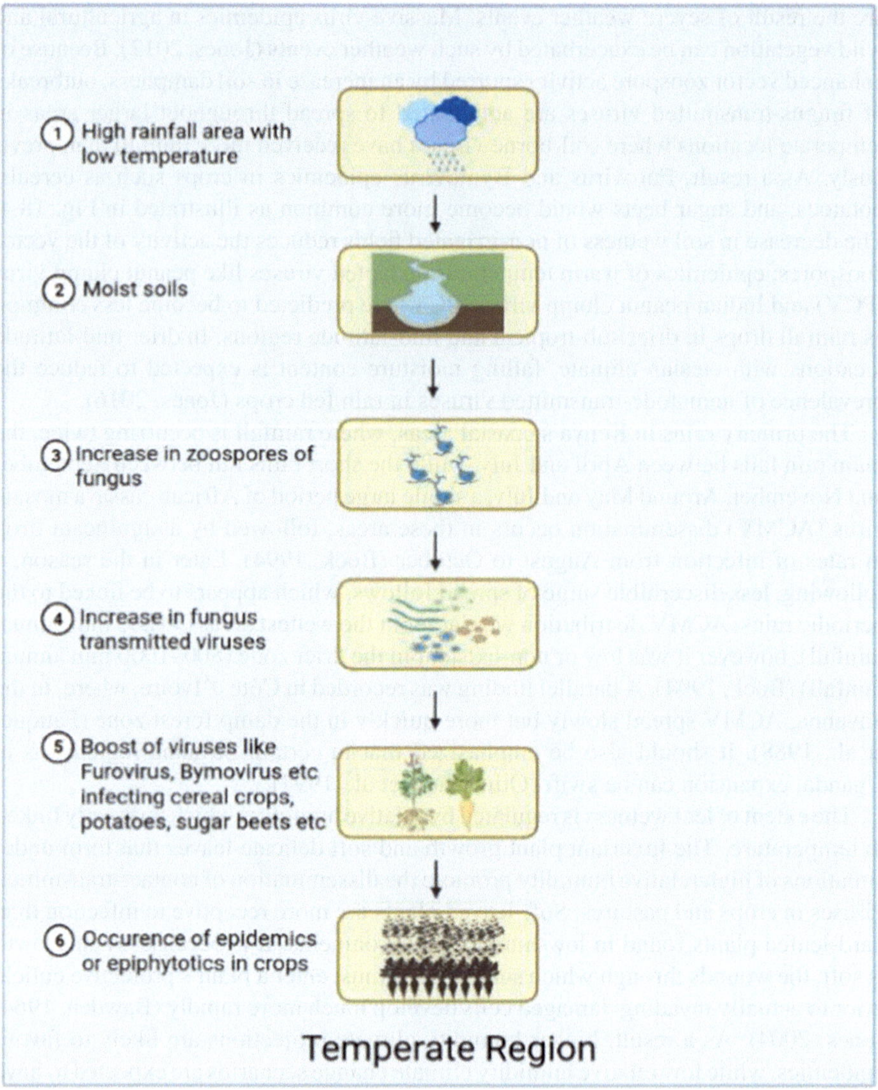

Fig. 18.4 Schematic illustration of effects of rainfall pattern shift in the development of viral epiphytotics with respect to temperate regions

18.1.8 Conclusion and Future Prospects

Deciphering the best possible treatment for plant viral epidemics along with the management measures in the host vector landscape needs a convergence of knowledge from geneticists, agronomists and climate experts. New scientific leads can be found to help manage the vectors of the viruses as we go further to integrated

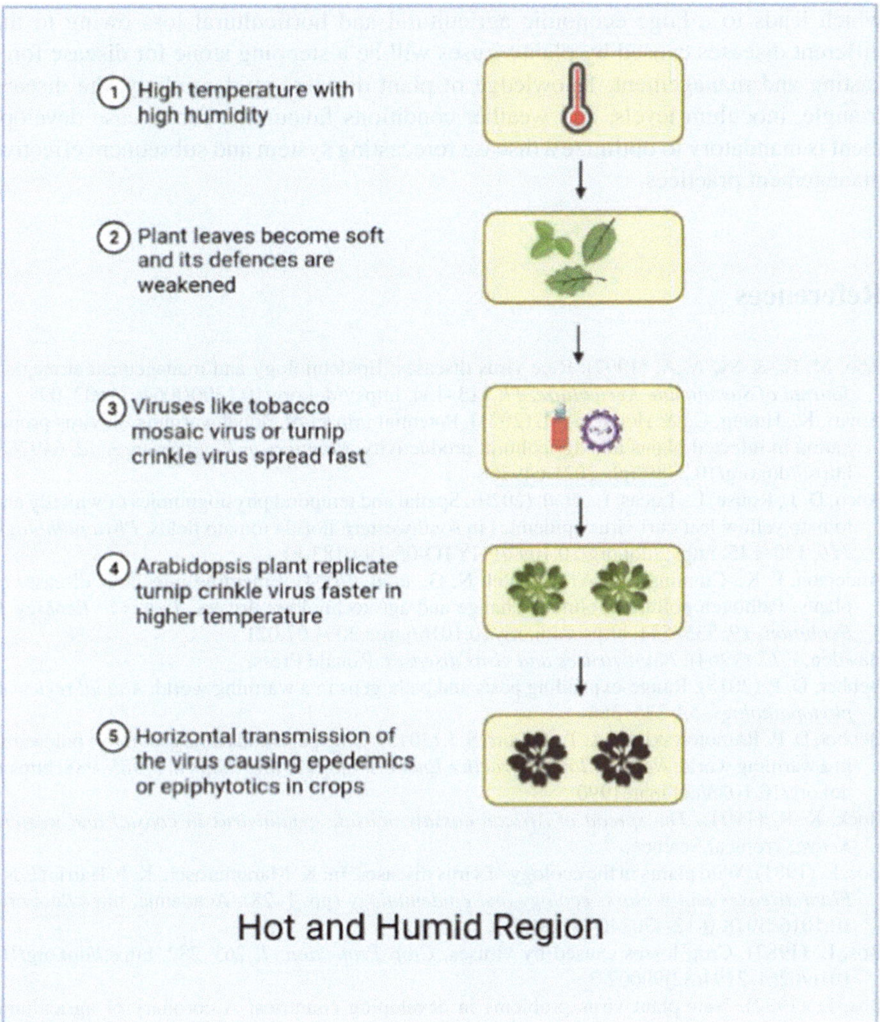

Fig. 18.5 Schematic illustration of effects of temperature and humidity in the development of viral epiphytotics

research of climatic variations on the spread and development of plant viruses. The understanding of plant virus vector disease dynamics under a modified environment provides a foundation for deployment of host resistance under favourable circumstances. Further, with this information in hand we can perhaps start and venture effective establishment of modelling abiotic and biotic relationships with disease dynamics and manage crop losses for feeding the present generation without compromising the needs of the future generations. Understanding a vivid scenario of the various processes of plant virus vector disease dynamics occurring in time and space which leads to a huge economic agricultural and horticultural loss owing to the different diseases caused by plant viruses will be a stepping stone for disease forecasting and management. Knowledge of plant diseases epidemiology, the disease triangle, inoculum levels, and weather conditions favourable for disease development is mandatory to optimize a disease forecasting system and subsequent effective management practices.

References

Abo, M. E., & Sy, A. A. (1997). Rice virus diseases: Epidemiology and management strategies. *Journal of Sustainable Agriculture, 11*, 113–134. https://doi.org/10.1300/J064v11n02_09

Amari, K., Huang, C., & Heinlein, M. (2021). Potential impact of global warming on virus propagation in infected plants and agricultural productivity. *Frontiers in Plant Science, 12*, 649768. https://doi.org/10.3389/fpls.2021.649768

Anco, D. J., Rouse, L., Lucas, L., et al. (2020). Spatial and temporal physiognomies of whitefly and tomato yellow leaf curl virus epidemics in southwestern florida tomato fields. *Phytopathology, 110*, 130–145. https://doi.org/10.1094/PHYTO-05-19-0183-FI

Anderson, P. K., Cunningham, A. A., Patel, N. G., et al. (2004). Emerging infectious diseases of plants: Pathogen pollution, climate change and agrotechnology drivers. *Trends in Ecology & Evolution, 19*, 535–544. https://doi.org/10.1016/j.tree.2004.07.021

Bawden, F. C. (1964). *Plant viruses and virus diseases.* Ronald Press.

Bebber, D. P. (2015). Range-expanding pests and pathogens in a warming world. *Annual review of phytopathology, 53*, 335–356

Bebber, D. P., Ramotowski, M. A. T., & Gurr, S. J. (2013). Crop pests and pathogens move polewards in a warming world. *Nature Clinical Practice Endocrinology & Metabolism, 3*, 985–988. https://doi.org/10.1038/nclimate1990

Bock, K. R. (1994). *The spread of African cassava mosaic geminivirus in coastal and western Kenya.* Tropical Science.

Bos, L. (1981). Wild plants in the ecology of virus diseases. In: K. Maramorosch, K. F. Harris (Eds.) *Plant diseases and vectors: ecology and epidemiology* (pp. 1–28). Academic. https://doi.org/10.1016/B978-0-12-470240-0.X5001-X

Bos, L. (1982). Crop losses caused by viruses. *Crop Protection, 1*, 263–282. https://doi.org/10.1016/0261-2194(82)90002-3

Bos, L. (1992). New plant virus problems in developing countries: A corollary of agricultural modernization. *Advances in Virus Research, 41*, 349–407. https://doi.org/10.1016/s0065-3527(08)60040-8

Brunt, A., Crabtree, K., Gibbs, A. J. (1990). *Viruses of tropical plants: Descriptions and lists from the VIDE database.* CAB Int

Calvert, L. A., Thresh, J. M. (2002). The viruses and virus diseases of cassava. In: R. J. Hillocks, & J. M. Thresh (Eds.), *Cassava: Biology, production and utilization* (pp. 237–260). CABI Publishing. https://doi.org/10.1079/9780851995243.0237

Cambra, M., Capote, N., Myrta, A., & Llácer, G. (2006). Plum pox virus and the estimated costs associated with sharka disease. *EPPO Bulletin, 36*, 202–204. https://doi.org/10.1111/j.1365-2338.2006.01027.x

Candresse, T., & Cambra, M. (2006). Causal agent of sharka disease: Historical perspective and current status of Plum pox virus strains. *EPPO Bulletin, 36*, 239–246. https://doi.org/10.1111/j.1365-2338.2006.00980.x

Canto, T., Aranda, M. A., & Fereres, A. (2009). Climate change effects on physiology and population processes of hosts and vectors that influence the spread of hemipteran-borne plant viruses. *Global Change Biology, 15*, 1884–1894. https://doi.org/10.1111/j.1365-2486.2008.01820.x

Canto, T., & Palukaitis, P. (2002). Novel N gene-associated, temperature-independent resistance to the movement of tobacco mosaic virus vectors neutralized by a cucumber mosaic virus RNA1 transgene. *Journal of Virology, 76*, 12908–12916. https://doi.org/10.1128/jvi.76.24.12908-12916.2002

Chaisuekul, C., & Riley, D. G. (2005). Host plant, temperature, and photoperiod effects on ovipositional preference of Frankliniella occidentalis and Frankliniella fusca (Thysanoptera: Thripidae). *Journal of Economic Entomology, 98*, 2107–2113. https://doi.org/10.1093/jee/98.6.2107

Chappell, T. M., Codod, C. B., Williams, B. W., Kemerait, R. C., Culbreath, A. K., & Kennedy, G. G. (2020). Adding epidemiologically important meteorological data to peanut Rx, the risk assessment framework for spotted wilt of peanut. *Phytopathology, 110*, 1199–1207.

Constable, F., Daly, A., Terras, M. A., et al. (2018). Detection in Australia of Cucumber green mottle mosaic virus in seed lots of cucurbit crops. *Australasian Plant Disease Notes, 13*, 18. https://doi.org/10.1007/s13314-018-0302-9

Cooper, I., Jones, R. A. C. (2006). Wild plants and viruses: under-investigated ecosystems. In: *Plant virus epidemiology* (pp. 1–47). Elsevier. https://doi.org/10.1016/S0065-3527(06)67001-2

Dale, J. L. (1987). Banana bunchy top: An economically important tropical plant virus disease. *Advances in Virus Research, 33*, 301–325. https://doi.org/10.1016/s0065-3527(08)60321-8

Dietzgen, R. G., Mann, K. S., Johnson, K. N. (2016). Plant virus-insect vector interactions: current and potential future research directions. *Viruses, 8*. https://doi.org/10.3390/v8110303

Dombrovsky, A., Smith, E. (2017). Seed transmission of tobamoviruses: aspects of global disease distribution. In: J. C. Jimenez-Lopez (ed.), *Advances in seed biology* (pp. 233–260). InTechOpen. https://doi.org/10.5772/intechopen.70244

Elad, Y., & Pertot, I. (2014). Climate change impacts on plant pathogens and plant diseases. *Journal of Crop Improvement, 28*, 99–139. https://doi.org/10.1080/15427528.2014.865412

Elena, S. F., Bedhomme, S., Carrasco, P., et al. (2011). The evolutionary genetics of emerging plant RNA viruses. *Molecular Plant-Microbe Interactions, 24*, 287–293. https://doi.org/10.1094/MPMI-09-10-0214

FAOSTAT. (2008). *FAO database*. Food and Agriculture Organization of the United Nations. http://faostat.fao.org/site/567/default.aspx

Fargette, D., Konaté, G., Fauquet, C., et al. (2006). Molecular ecology and emergence of tropical plant viruses. *Annual Review of Phytopathology, 44*, 235–260. https://doi.org/10.1146/annurev.phyto.44.120705.104644

Fauquet, C., Fargette, D., & Thouvenel, J. C. (1988). Some aspects of the epidemiology of African cassava mosaic virus in Ivory Coast. *Tropical Pest Management, 34*, 92–96. https://doi.org/10.1080/09670878809371217

Fraser, R. S. S. (1990). The genetics of resistance to plant viruses. *Annual Review of Phytopathology, 28*, 179–200. https://doi.org/10.1146/annurev.py.28.090190.001143

Fraser, R. S. S., & Loughlin, S. A. R. (1982). Effects of temperature on the Tm-1 gene for resistance to tobacco mosaic virus in tomato. *Physiological Plant Pathology, 20*, 109–117. https://doi.org/10.1016/0048-4059(82)90029-7

Frost, R. R., & Harrison, B. D. (1967). Comparative effects of temperature on the multiplication in tobacco leaves of two tobacco rattle viruses. *Journal of General Virology, 1*, 455–464. https://doi.org/10.1099/0022-1317-1-4-455

Gamarra, H., Carhuapoma, P., Cumapa, L., Gonzalez, G., Munoz, J., Sporleder, M., & Kreuze, J. (2020). A temperature-driven model for potato yellow vein virus transmission efficacy by *Trialeurodes vaporariorum* (Hemiptera: Aleyrodidae). *Virus Research, 289*, 198109.

Garcia, J. A., & Cambra, M. (2007). Plum pox virus and sharka disease. In: *Plant viruses* (pp. 69–79).

Goodman, N. R., Kirali, Z., & Wood, K. R. (1986). *The biochemistry and physiology of plant disease*. University of Missouri Press.

Harper, S. J. (2013). Citrus tristeza virus: Evolution of complex and varied genotypic groups. *Frontiers in Microbiology, 4*, 93. https://doi.org/10.3389/fmicb.2013.00093
Huber, L., & Gillespie, T. J. (1992). Modeling leaf wetness in relation to plant disease epidemiology. *Annual Review of Phytopathology, 30*, 553–577. https://doi.org/10.1146/annurev.py.30.090192.003005
Hull, R. (2013). *Plant virology*, 5th edn. Academic Press. https://doi.org/10.1016/C2010-0-64974-1
Hull, R. (2014). *Matthews' plant virology*. Academic Press.
IMF. (2000). *International monetary fund. Globalization: Threat or Opportunity?*
IPPC. (2014). *Intergovernmental panel on climate change, fifth assessment report.*
Islam, W., Noman, A., NaveedH, A. S. A., Hashem, M., Huang, Z., Han, Y. H., & Chen, H. Y. (2020). Plant-insect vector-virus interactions under environmental change. *Science of the Total Environment, 701*, 135044.
Jacobson, A. L., Duffy, S., & Sseruwagi, P. (2018). Whitefly-transmitted viruses threatening cassava production in Africa. *Current Opinion in Virology, 33*, 167–176. https://doi.org/10.1016/j.coviro.2018.08.016
James, P., & Steger, M. B. (2014). A genealogy of 'globalization': The career of a concept. *Globalizations, 11*, 417–434. https://doi.org/10.1080/14747731.2014.951186
Jeger, M. J., Seal, S. E., & Van den Bosch, F. (2006). Evolutionary epidemiology of plant virus disease. In: *Plant virus epidemiology* (pp. 163–203). Elsevier. https://doi.org/10.1016/S0065-3527(06)67005-X
Jensen, S. G. (1973). Systemic movement of barley yellow dwarf virus in small grains. *Phytopathology, 63*, 854. https://doi.org/10.1094/Phyto-63-854
Jones, R. (2012). Influence of climate change on plant disease infections and epidemics caused by viruses and bacteria. *CAB Reviews, 7*. https://doi.org/10.1079/PAVSNNR20127022
Jones, R., & Barbetti, M. (2012a). Influence of climate change on plant disease infections and epidemics caused by viruses and bacteria. *CAB Reviews: Perspectives in Agriculture, Veterinary Science, Nutrition and Natural Resources, 7*(22), 1–31. https://doi.org/10.1079/PAVSNNR20127022
Jones, R. A. C. (1981). The ecology of viruses infecting wild and cultivated potatoes in the Andean Region of South America. In J. M. Thresh (Ed.), *Pests, Pathogens and Vegetation* (pp. 89–107). Pitman.
Jones, R. A. C. (2004). Using epidemiological information to develop effective integrated virus disease management strategies. *Virus Research, 100*, 5–30. https://doi.org/10.1016/j.virusres.2003.12.011
Jones, R. A. C. (2006). Control of plant virus diseases. In: *Plant virus epidemiology* (pp. 205–244). Elsevier. https://doi.org/10.1016/S0065-3527(06)67006-1
Jones, R. A. C. (2009). Plant virus emergence and evolution: Origins, new encounter scenarios, factors driving emergence, effects of changing world conditions, and prospects for control. *Virus Research, 141*, 113–130. https://doi.org/10.1016/j.virusres.2008.07.028
Jones, R. A. C. (2014). Plant virus ecology and epidemiology: Historical perspectives, recent progress and future prospects. *The Annals of Applied Biology, 164*, 320–347. https://doi.org/10.1111/aab.12123
Jones, R. A. C. (2016). Future scenarios for plant virus pathogens as climate change progresses. *Advances in Virus Research, 95*, 87–147. https://doi.org/10.1016/bs.aivir.2016.02.004
Jones, R. A. C. (2018). Plant and insect viruses in managed and natural environments: Novel and neglected transmission pathways. *Advances in Virus Research, 101*, 149–187. https://doi.org/10.1016/bs.aivir.2018.02.006
Jones, R. A. C. (2020). Disease pandemics and major epidemics arising from new encounters between indigenous viruses and introduced crops. *Viruses, 12*. https://doi.org/10.3390/v12121388
Jones, R. A. C., & Barbetti, M. J. (2012b). Influence of climate change on plant disease infections and epidemics caused by viruses and bacteria. *CABI Review*. https://doi.org/10.1079/PAVSNNR20127022

Jones, R. A. C., & Naidu, R. A. (2019). Global dimensions of plant virus diseases: Current status and future perspectives. *Annual Review of Virology, 6*, 387–409. https://doi.org/10.1146/annurev-virology-092818-015606

Kassanis, B. (1952). Some effects of high temperature on the susceptibility of plants to infection with viruses. *Annals of Applied Biology, 39*, 358–369. https://doi.org/10.1111/j.1744-7348.1952.tb01018.x

King, A. M. Q., Adams, M. J., Carstens, E. B., & Lefkowitz, E. (2012). *Virus taxonomy: Classification and nomenclature of viruses: Ninth report of the international committee on taxonomy of viruses*. Elsevier Academic Press.

Király, L., Hafez, Y. M., Fodor, J., & Király, Z. (2008). Suppression of tobacco mosaic virus-induced hypersensitive-type necrotization in tobacco at high temperature is associated with downregulation of NADPH oxidase and superoxide and stimulation of dehydroascorbate reductase. *Journal of General Virology, 89*, 799–808. https://doi.org/10.1099/vir.0.83328-0

Kumar, P. L., Hanna, R., Alabi, O. J., et al. (2011). Banana bunchy top virus in sub-Saharan Africa: Investigations on virus distribution and diversity. *Virus Research, 159*, 171–182. https://doi.org/10.1016/j.virusres.2011.021

Kumar, P. L., Selvarajan, R., Iskra-Caruana, M.-L., et al. (2015). Biology, etiology, and control of virus diseases of banana and plantain. *Advances in Virus Research, 91*, 229–269. https://doi.org/10.1016/bs.aivir.2014.10.006

Legarrea, S., Barman, A., Diffie, S., & Srinivasan, R. (2020). Virus accumulation and whitefly performance modulate the role of alternate host species as inoculum sources of tomato yellow leaf curl virus. *Plant Disease, 104*, 2958–2966. https://doi.org/10.1094/PDIS-09-19-1853-RE

Legg, J. P. (1999). Emergence, spread and strategies for controlling the pandemic of cassava mosaic virus disease in east and central Africa. *Crop Protection, 18*, 627–637. https://doi.org/10.1016/S0261-2194(99)00062-9

Legg, J. P., & Fauquet, C. M. (2004). Cassava mosaic geminiviruses in Africa. *Plant Molecular Biology, 56*, 585–599. https://doi.org/10.1007/s11103-004-1651-7

Legg, J. P., Jeremiah, S. C., Obiero, H. M., et al. (2011). Comparing the regional epidemiology of the cassava mosaic and cassava brown streak virus pandemics in Africa. *Virus Research, 159*, 161–170. https://doi.org/10.1016/j.virusres.2011.04.018

Legg, J. P., Owor, B., Sseruwagi, P., & Ndunguru, J. (2006). Cassava mosaic virus disease in East and Central Africa: epidemiology and management of a regional pandemic. In: *Plant virus epidemiology* (pp. 355–418). Elsevier. https://doi.org/10.1016/S0065-3527(06)67010-3

Legg, J. P., & Thresh, J. M. (2000). Cassava mosaic virus disease in East Africa: A dynamic disease in a changing environment. *Virus Research, 71*, 135–149. https://doi.org/10.1016/s0168-1702(00)00194-5

Llamas-Llamas, M. E., Zavaleta-Mejia, E., Gonzalez-Hernandez, V. A., et al. (1998). Effect of temperature on symptom expression and accumulation of tomato spotted wilt virus in different host species. *Plant Pathology, 47*, 341–347. https://doi.org/10.1046/j.1365-3059.1998.00249.x

Loebenstein, G., Thottappilly, G. (2013). *Virus and virus-like diseases of major crops in developing countries*. Springer. https://doi.org/10.1007/978-94-007-0791-7

Lovisolo, O., Hull, R., & Rösler, O. (2003). Coevolution of viruses with hosts and vectors and possible paleontology. *Advances in Virus Research, 62*, 325–379. https://doi.org/10.1016/s0065-3527(03)62006-3

Martin, D. P., Willment, J. A., Billharz, R., et al. (2001). Sequence diversity and virulence in Zea mays of Maize streak virus isolates. *Virology, 288*, 247–255. https://doi.org/10.1006/viro.2001.1075

Matthews, R. E. F. (1970). *Plant virology*. Academic Press.

Matthews, R. E. F. (1991). *Plant virology*. Academic Press Inc.

McDonald, B. A., Stukenbrock, E. H. (2016). Rapid emergence of pathogens in agro-ecosystems: Global threats to agricultural sustainability and food security. *Philosophical Transactions of the Royal Society of London. Series B, Biological Sciences, 371*. https://doi.org/10.1098/rstb.2016.0026

Morales, F. J., & Jones, P. G. (2004). The ecology and epidemiology of whitefly-transmitted viruses in Latin America. *Virus Research, 100*, 57–65. https://doi.org/10.1016/j.virusres.2003.12.014

Moreno, P., Ambrós, S., Albiach-Martí, M. R., et al. (2008). Citrus tristeza virus: A pathogen that changed the course of the citrus industry. *Molecular Plant Pathology, 9*, 251–268. https://doi.org/10.1111/j.1364-3703.2007.00455.x

Morsello, S. C., & Kennedy, G. G. (2009). Spring temperature and precipitation affect tobacco thrips, *Frankliniella fusca*, population growth and tomato spotted wilt virus spread within patches of the winter annual weed *Stellaria media. Entomologia Experimentalis Et Applicata, 130*, 138–148.

Nicaise, V. (2014). Crop immunity against viruses: Outcomes and future challenges. *Frontiers in Plant Science, 5*, 660. https://doi.org/10.3389/fpls.2014.00660

Obrępalska-Stęplowska, A., Renaut, J., Planchon, S., et al. (2015). Effect of temperature on the pathogenesis, accumulation of viral and satellite RNAs and on plant proteome in peanut stunt virus and satellite RNA-infected plants. *Frontiers in Plant Science, 6*, 903. https://doi.org/10.3389/fpls.2015.00903

Ochoa, M. D. L., Zavaleta-Mejía, E., Johansen, N. R. M., et al. (1996). Tospoviruses, weeds and thrips associated with chrysanthemum (*Dendranthema grandiflora* Tzvelev cv. Polaris). *International Journal of Pest Management, 42*, 157–159.

Oliver, J. E., & Whitfield, A. E. (2016). The genus tospovirus: Emerging bunyaviruses that threaten food security. *Annual Review of Virology, 3*, 101–124. https://doi.org/10.1146/annurev-virology-100114-055036

Ordon, F., Habekuss, A., Kastirr, U., et al. (2009). Virus resistance in cereals: Sources of resistance, genetics and breeding. *Journal of Phytopathology, 157*, 535–545. https://doi.org/10.1111/j.1439-0434.2009.01540.x

Otim-Nape, G. W., Shaw, M., & Thresh, J. M. (1994). *The effects of African cassava mosaic geminivirus on the growth and yield of cassava in Uganda*. Tropical Science.

Pappu, H. R., Jones, R. A. C., & Jain, R. K. (2009). Global status of tospovirus epidemics in diverse cropping systems: Successes achieved and challenges ahead. *Virus Research, 141*, 219–236. https://doi.org/10.1016/j.virusres.2009.01.009

Pfitzner, A. J. P. (2006). Resistance to *Tobacco Mosaic Virus* and *Tomato Mosaic Virus* in Tomato. In: G. Loebenstein, J. P. Carr (Eds.), *Natural resistance mechanisms of plants to viruses*. Springer. https://doi.org/10.1007/1-4020-3780-5_18

Bebber, D. P. (2015b). Range-expanding pests and pathogens in a warming world. *Annual Review of Phytopathology, 53*, 335–356. https://doi.org/10.1146/annurev-phyto-080614-120207

Picó, B., Díez, M. J., & Nuez, F. (1996). Viral diseases causing the greatest economic losses to the tomato crop. II. The Tomato yellow leaf curl virus—a review. *Scientia Horticulturae, 67*, 151–196. https://doi.org/10.1016/S0304-4238(96)00945-4

Posnette, A. F. (1981). The role of wild hosts in cocoa swollen shoot disease. In: J. M. Thresh (ed.), Pests, pathogens and vegetation. Pitman.

Prasch, C. M., & Sonnewald, U. (2013). Simultaneous application of heat, drought, and virus to Arabidopsis plants reveals significant shifts in signaling networks. *Plant Physiology, 162*, 1849–1866. https://doi.org/10.1104/pp.113.221044

Qazi, J. (2016). Banana bunchy top virus and the bunchy top disease. *Journal of General Plant Pathology, 82*, 2–11. https://doi.org/10.1007/s10327-015-0642-7

Rey, C., & Vanderschuren, H. (2017). Cassava mosaic and brown streak diseases: Current perspectives and beyond. *Annual Review of Virology, 4*, 429–452. https://doi.org/10.1146/annurev-virology-101416-041913

Riley, D. G., Joseph, S. V., Srinivasan, R., Diffie, S. (2011). Thrips vectors of tospoviruses. *Journal of Integrated Pest Management, 2*, I1–I10. https://doi.org/10.1603/IPM10020

Rodoni, B. (2009). The role of plant biosecurity in preventing and controlling emerging plant virus disease epidemics. *Virus Research, 141*, 150–157. https://doi.org/10.1016/j.virusres.2008.11.019

Rojas, M. R., Macedo, M. A., Maliano, M. R., et al. (2018). World management of geminiviruses. *Annual Review of Phytopathology, 56*, 637–677. https://doi.org/10.1146/annurev-phyto-080615-100327

Roossinck, M. J. (2015). Plants, viruses and the environment: Ecology and mutualism. *Virology, 479–480*, 271–277. https://doi.org/10.1016/j.virol.2015.03.041

Rybicki, E. P. (2015). A top ten list for economically important plant viruses. *Archives of Virology, 160*, 17–20. https://doi.org/10.1007/s00705-014-2295-9

Rybicki, E. P., & Pietersen, G. (1999). Plant virus disease problems in the developing world. *Advances in Virus Research, 53*, 127–175. https://doi.org/10.1016/s0065-3527(08)60346-2

Sasaya, T., Nakazono-Nagaoka, E., Saika, H., et al. (2014). Transgenic strategies to confer resistance against viruses in rice plants. *Frontiers in Microbiology, 4*, 409. https://doi.org/10.3389/fmicb.2013.00409

Sastry, S. K., Zitter, T. A. (Eds.) (2014). Management of virus and viroid diseases of crops in the tropics. In: Plant virus and viroid diseases in the tropics, epidemiology and management (vol. 2, 1st ed., pp. 149–480). Springer. https://doi.org/10.1007/978-94-007-7820-7

Schumann, G. L., & D'Arcy, C. J. (2006). *Essential plant pathology*. American Phytopathological Society.

Sharma, S. K., Gupta, O. P., Pathaw, N., et al. (2021). CRISPR-cas-led revolution in diagnosis and management of emerging plant viruses: new avenues toward food and nutritional security. *Frontiers in Nutrition, 8*, 751512. https://doi.org/10.3389/fnut.2021.751512

Sidharthan, V. K., Sharma, S. K., & Baranwal, V. K. (2021). The first near-complete genome sequence of large cardamom chirke virus mined from the transcriptome dataset of large cardamom. *Plant Gene, 28*, 100324. https://doi.org/10.1016/j.plgene.2021.100324

Thresh, J. M. (1980a). An ecological approach to the epidemiology of plant virus diseases. In: J. Palti, J. Kranz (Eds.), *Comparative epidemiology: A tool for better disease management* (pp 57–70). Centre for Agricultural Publishing and Documentation.

Thresh, J. M. (1980b). The origins and epidemiology of some important plant virus diseases. *Applied Biology, 5*, 1–65.

Thresh, J. M. (1981). *Pests, pathogens and vegetation*. Pitman.

Thresh, J. M. (1982). Cropping practices and virus spread. *Annual Review of Phytopathology, 20*, 193–216. https://doi.org/10.1146/annurev.py.20.090182.001205

Thresh, J. M. (1983). Plant virus epidemiology and control: Current trends and future prospects. In R. T. Plumb & J. M. Thresh (Eds.), *Plant virus epidemiology—the spread and control of insect-borne pathogens* (pp. 349–360). Blackwell.

Thresh, J. M. (2004). The impact of plant virus diseases in developing countries. In: G. Loebenstein, G. Thottappilly (Eds.), *Virus and virus-like diseases of major crops in developing countries* (pp. 1–30). Springer. https://doi.org/10.1007/978-94-007-0791-7

Thresh, J. M. (2006a). Viruses and virus diseases: a global perspective. In: J. I. Cooper, T. Kuehne, V. P. Polischuk (Eds.), *Virus diseases and crop biosecurity* (1st ed., pp. 9–32). Springer. https://doi.org/10.1007/978-1-4020-5298-9

Thresh, J. M. (2006b). Control of tropical plant virus diseases. In *Plant virus epidemiology* (pp. 245–295). Elsevier. https://doi.org/10.1016/S0065-3527(06)67007-3

Thresh, J. M. (2006c). Plant virus epidemiology: the concept of host genetic vulnerability. In *Plant virus epidemiology* (pp. 89–125). Elsevier. https://doi.org/10.1016/S0065-3527(06)67003-6

Trebicki, P. (2020). Climate change and plant virus epidemiology. *Virus Research, 286*, 198059. https://doi.org/10.1016/j.virusres.2020.198059

Wale, S., Platt, B., & Cattlin, N. D. (2008). *Diseases, pests and disorders of potatoes: A colour handbook*. Manson Publishing Ltd.

Wang, Y., Bao, Z., Zhu, Y., & Hua, J. (2009). Analysis of temperature modulation of plant defense against biotrophic microbes. *Molecular Plant-Microbe Interactions, 22*, 498–506. https://doi.org/10.1094/MPMI-22-5-0498

Webster, C. G., Coutts, B. A., Jones, R. A. C., et al. (2007). Virus impact at the interface of an ancient ecosystem and a recent agroecosystem: Studies on three legume-infecting potyviruses in the southwest Australian floristic region. *Plant Pathology, 56*, 729–742. https://doi.org/10.1111/j.1365-3059.2007.01653.x

Yoon, J. Y., Chung, B. N., & Choi, S. C. (2011). High-temperature mediated spontaneous mutations in the coat protein of cucumber mosaic virus in Nicotiana tabacum. *Archives of Virology, 156*, 2173–2180.

Zhang, X., Singh, J., et al. (2012). Temperature-dependent survival of Turnip crinkle virus-infected arabidopsis plants relies on an RNA silencing-based defense that requires dcl2, AGO2, and HEN1. *Journal of Virology, 86*, 6847–6854. https://doi.org/10.1128/JVI.00497-12

Zhang, Y., Wang, L., Wu, K., et al. (2008). Flight performance of the soybean aphid, Aphis glycines (Hemiptera: Aphididae) under different temperature and humidity regimens. *Environmental Entomology, 37*, 301–306.

Zhou, X., Liu, Y., Calvert, L., et al. (1997). Evidence that DNA-A of a geminivirus associated with severe cassava mosaic disease in Uganda has arisen by interspecific recombination. *Journal of General Virology, 78*(Pt 8), 2101–2111. https://doi.org/10.1099/0022-1317-78-8-2101

Chapter 19
Breeding Rice Variety with Suitable Plant Ideotype for Next-Generation Indian Agriculture in the Changing Climatic Conditions

Annamalai Anandan, Siddharth Panda, Anumalla Mahender, and Varunseelan Murugaiyan

Abstract The latter half of the twentieth century saw a tremendous increase in food grain production, especially rice, a major component of dietary intake in Asia. The Green revolution ensured food security but this was at the cost of deteriorating soil health and groundwater reserves. The rate of environmental degradation and the increased susceptibility of rice varieties to various stress conditions has compelled researchers to work towards tailor-made specifically bred varieties that cater to the growing climatic aberrations. The majority of breeding efforts to date focused on increasing the yield compromising their hardiness, and tolerance to biotic and abiotic stress. This chapter highlights the perils of climate change complemented by the faulty management practices in rice cultivation and suggests a plant ideotype that can complement the changing climatic conditions, and react to different biotic and abiotic stress with minimum impact on its yield.

Keywords Climate · Nutrient use efficiency · Aerobic · Abiotic stress · Biotic stress

---

A. Anandan (✉)
Indian Council of Agricultural Research, Indian Institute of Seed Science, Regional Station, Bangalore, India
e-mail: anandanau@yahoo.com

S. Panda
Institute of Agricultural Sciences, SOA (DU), Bhubaneswar, Odisha, India

A. Mahender · V. Murugaiyan
Rice Breeding Innovation Platform, International Rice Research Institute (IRRI), Los Baños, Philippines

H. Pathak et al. (eds.), *Climate Change Impacts on Soil-Plant-Atmosphere Continuum*, Advances in Global Change Research 78,
https://doi.org/10.1007/978-981-99-7935-6_19

19.1 Introduction

The Indian subcontinent has a variety of ecologies defined by its topography, rainfall, edaphic characteristics, and ecological diversity. Crops growing in such diversified regions offer a huge variability unique to their places of adaptation. Continuous selection pressure through human selection has made these crops more adaptive to different ecological niche enabling their cultivation throughout the country. With the advances in varietal breeding techniques and improved cultural practices, India is self-sufficient in its food grain production, especially rice. The rice production in India is 127.93 million tons, and the consumption is 103.5 million tons as estimated in the year 2021–2022 (www.statista.com). This shows the crop's high demand, which has compelled the farmer towards intensive cultivation with high chemical fertilizers and water resources inputs. The fertilizer use data shows that in 1969, during the green revolution, 12.4 kg/ha of fertilizer was consumed in India, which increased to 175 kg/ha in 2018 (www.knoema.com/atlas/India/Fertilizer-consumption). This high use of fertilizers doesn't reflect well in the nutrient recovery efficiency which remains below 50%. It is observed that the excess application of fertilizers, which is an ongoing trend also weakens the system. It is reported that 80–90% of phosphorus applied to the plants is not utilized but lost in some fixated form in the soil, sometimes leading to eutrophication (Mahender et al., 2018). These nutrient losses in the soil may be attributed to drainage, volatilization, leaching, etc., or fixation in the ground making them unavailable to the plant. This highlights the need to focus on improving the Nutrient Use Efficiency of crop plants (NUE).

Additionally, groundwater mining and freshwater diverted to agriculture have also increased. According to Bouman et al., 2002, it is predicted that approximately 2Mha of irrigated dry season rice and 13Mha of irrigated wet season rice in Asia may soon experience physical water scarcity, and some form of economic water scarcity. Besides, there will be a 10% to 15% reduction in the available water for agriculture by 2025 due to the ever-increasing demand from non-agricultural sectors such as industry, domestic uses, urban settlements, etc. (Peng et al., 2009). This amount of groundwater and freshwater exploitation is greater than the volume of water replenished by rainfall (Fig. 19.1). When we examine this incessant use of water resources for agriculture, it is learned that groundwater exploitation in India for irrigation purposes has increased 100 folds since 1950, causing a depletion in the water table (Kukal, 2004). In Asia, 50% of the freshwater is utilized for rice cultivation alone (Tuong & Bouman, 2003). The average agricultural water use for low-income, middle-income, and high-income countries is 90, 79, and 41% respectively by the year of 2017 (Ritchie & Roser, 2017). An average of 2500 L of water is consumed in a rice field to produce 1 kg of rice, sometimes reaching up to 5000 L in the transplanted cultivation system (Surendran et al., 2021).

The transplanted system of rice is one of the most commonly followed systems of cultivation in the present times. While this system has led the country towards self-sufficiency, we cannot overlook the ecological damages. In a transplanted system, the field is continuously maintained in flooded with 5–10 cm of water almost throughout

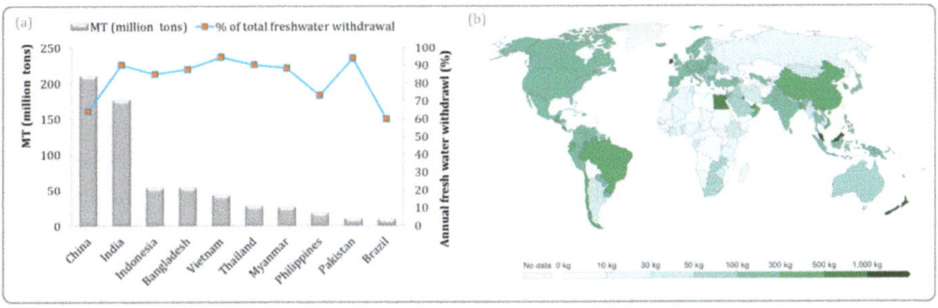

Fig. 19.1 **a** Global rice production of the top 10 countries in 2019 (based on UN's FAOSTAT and the USDA). It is clearly shown that ~84% of rice is being harvested in these countries, and globally, 70% percent of freshwater withdrawals for agriculture. **b** Consumption of fertilizer (nitrogen, potash, and phosphate) inputs per hectare of cropland in 2018 (based on the FAO database (https://www.fao.org; https://ourworldindata.org/; https://www.visualcapitalist.com/worlds-biggest-rice-producers/)

the growing period. Methane and Nitrous oxide (N_2O) are simultaneously emitted as irrigated rice fields offer favourable conditions for their production and emission (Majumdar, 2003). It is estimated that methane emissions from paddy fields can range from 31 to $112Tgy^{-1}$ which is, almost 19% of the total anthropogenic emissions (Win et al., 2020). Apart from methane, nitrous oxide released from the paddy fields also contributes to global warming, which is again about 11% of the global share (Win et al., 2020). India's self-sufficiency in rice production largely depends on the irrigated rice fields. But the incessant fertilizer application, depleting water table, and increasing water scarcity pose a serious threat to the sustainability of the transplanted rice cultivation system. Adding to this, high labor demand is increasingly becoming a menace in many parts of the country.

Global food security is one of the biggest challenges in feeding the rapidly growing worldwide population and to meet this demand, the global food production needs to increase by over 40% by 2030 and 70% by 2050 (Islam & Karim, 2019). Climate change has become a worldwide phenomenon that has ramifications for agriculture (Fig. 19.2). The increasing temperature hamper fertility, the insect life cycle pattern and reduces yield. High temperatures also alter the moisture availability for the plants, i.e., both from the soil and leaves through evaporation and transpiration, respectively (collectively termed as evapotranspiration). The elitist minds of the world are now concerned about the changing climate scenario and its effect on the biological entities living on earth, as the consequences are becoming hazardous. It is obvious that we look at agriculture under the lenses of climate change effects and sustainability in the long run. Thus, it is imperative to scout for alternate methods of cultivation practices that have a sustainable approach with a lower ecological footprint and at the same time compensate for our food grain needs. So far, the breeding activities have focused on developing high-yielding genotypes when all the required inputs are supplied in ample amounts externally (fertilizers, pesticides, and irrigation water). The focus should be on developing and/or improving genotypes that can survive under limited external inputs by utilizing the available resources to the optimum, maintaining a balance between profitability margins and essential resource input costs. Researchers have suggested and modified the existing practices to maintain productivity with less water or efficient water utilization in rice cultivation, keeping food security in mind. At the same time, new varieties are now being bred to adapt to the fluctuating climatic scenario and utilize nutrient resources available to the maximum. For instance, drought-tolerant plants with higher root proliferation, plants with higher nutrient uptake capacity or tolerance towards a new race of pathogen/ insect, etc., have become the focus of research. This chapter is an effort to highlight the alternates of transplanted paddy fields and identify the developments in improving the genetics of the contributing plant traits towards the development of climate-smart rice genotype with the minimum ecological footprint. Figure 19.2 represents the overall aim of the chapter highlighting the road map ahead utilising the available plant genetic resources assisted with advanced phenotypic and genotypic platforms. The combined study of phenotypic and genotypic variabilities can aid in discovering genetic associations with the target traits exploiting interdisciplinary methods such as omics technologies, genome-wide association studies, and genomic selection, etc. Such explorations and achievements will ultimately help in identifying the donors for GCVs related traits.

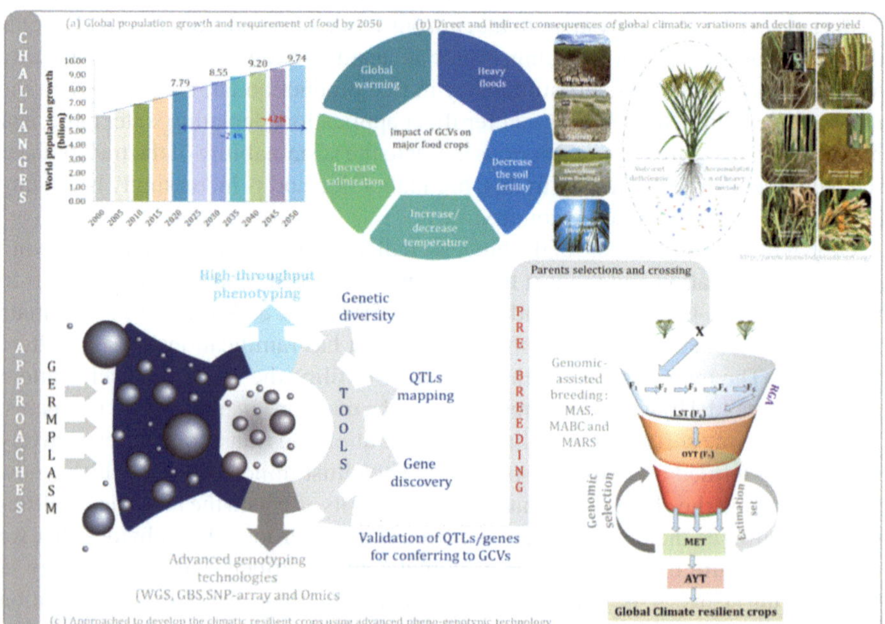

Fig. 19.2 A schematic representation of the multi-disciplinary approach for identifying and developing the climate-resilient rice varieties to increase crop yield to sustain the global population

19.2 Adaptive Approaches for Sustainable Rice Cultivation

The nearest kin to the transplanted rice cultivation system is the alternate wetting and drying (AWD) method, whereby the modification is focused on reducing water use and increasing water use efficiency. However, this method is not new to the rice scenario (Lampayan et al., 2015). AWD and other such systems, such as single or multiple drying systems, have been in practice for decades. However, a scientific approach to its importance has been a recent development. Such kind of non-flooded systems is common in north-western India and other places where irrigation facilities are not in ample quantity. This practice also addresses the issue of greenhouse gas emissions and takes care of rice productivity. Depending on weather, soil type, degree of dryness, crop duration, and crop growth stage, AWD irrigation technology can maintain same or greater paddy production (10–20%) than the conventional farming while reducing total water inputs (25–70%), CH4 emission (11–95%), As (13–90%), and Hg (5–90%) in rice grains. (Ishfaq et al., 2020). The anaerobic condition in the flooded rice fields decomposes organic material and produces methane but in the case of AWD, intermittent drying decelerates the decomposition process and limits the methane emissions. PCC 2006 Guidelines for National Greenhouse Gas Inventories estimate a 48% reduction in methane emissions from AWD. A study on the economics of the AWD system showed that it has increased "economic efficiency in social terms" by reducing the water used without affecting the profits from the rice cultivation (Moya et al., 2001). The lowland areas accessible to be drained in five-day intervals are best suitable for AWD. This is because high rainfall may destroy the sole cause of AWD, where the precipitation may exceed the loss due to evapotranspiration. Before applying irrigation in AWD, the ponded water on the field (also known as "perched water") is permitted to sink to a depth of 10–15 cm below the soil surface. A perforated or punctured water tube buried in the soil is used to measure the depth of perched water. Roots can still draw water from the perched water table at a threshold of 10 to 15 cm, and plants are not stressed (Mote et al., 2021). AWD irrigation significantly increased rice water production and water savings. Compared to typical submersion, irrigation water was saved by 20–50% (Kumar & Rajitha, 2019). However, this practice is not advised in rainfed areas where the rainfall is unpredictable and flooding is the only reliable irrigation method.

The system of rice intensification (SRI) is proposed more recently and has given some promising results with reduced water use and high production potential. In this system, rice plants are comparatively widely spaced, with one seedling per hill, without continuous flooding, weeding, and disturbing the soil, enhancing soil biotic activities by aerating the soil. As a result, this system increased grain yield by 10–25%, and water use decreased by 29% (https://www.icar-iirr.org). Aerobic rice cultivation is another unique system advocated for upland or rainfed areas where water shortage is a regular phenomenon. The crop is established by direct dry seeding in a non-puddle field under unflooded conditions (Anandan et al., 2021a). The peculiarity here is the planting method, which is similar to that of other cereals such as wheat or maize. By now, it is the most promising method of rice cultivation in resource-poor areas with abiotic stress factors to limit water and labour costs with zero to a minimum penalty on yield. The term aerobic justifies the growth condition of the crop, i.e., the soil is unflooded and maintains an aerobic environment. The aerobic system has its own set of practices devised for seed treatment, stringent agronomic practices, weed, water, and nutrient management. Growing rice under aerobic conditions requires 36–41% less water than under puddled system of rice cultivation (PTR) (Belder et al., 2005). Compared with flooded rice, aerobic rice had lower production costs, higher water productivity, and an equivalent outcome (Jabran & Chauhan, 2015). In China, the water use for aerobic rice production was 55–56% lower than the flooded rice with 1.6–1.9 times higher water productivity. It indicates that aerobic rice may be a viable option where the shortage of water does not allow the growing of lowland or transplanted systems of rice (Priyanka et al., 2012). Around 90% of all freshwater is diverted to agriculture; aerobic rice would be an incentive to save this water quantity. The grain yield potential of 6–7 t ha^{-1} has been reported in direct-seeded rice varieties (Huaqi et al., 2002). The technique of direct seeding followed are of three types; wet DSR, where sprouted rice seeds are broadcasted or sown in lines on wet puddled soil, dry DSR where dry rice seeds are either drilled or broadcasted under unpuddled dry conditions and water DSR wherein the seeds are broadcasted in standing water (Mahender et al., 2015a) (Fig. 19.3).The standing water can also lead to salinity or salt stress as a consequence of frequent irrigation, summer fallowing, excess use of muriate of potash, and ammonium sulphate. The ill-effects of salt stress start as early as from seed germination, lower water uptake, decreased nutrient uptake, reduced rate of photosynthesis, and oxidative stress ultimately reducing plant growth and yield (Singh et al., 2021). However, in some cases, low moisture can also be a stress for the plants leading to drought stress, mainly in the rainfed areas.

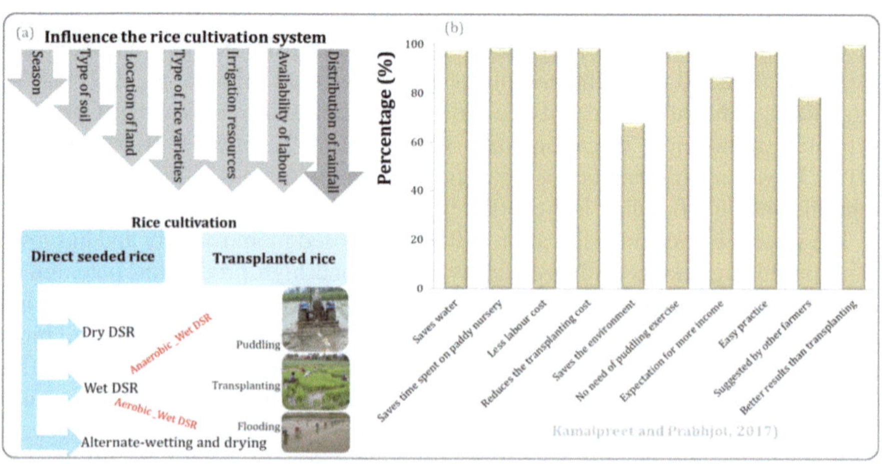

Fig. 19.3 **a** Different factors are involved in the rice ecosystem and follow the major rice cultivations; **b** Among the various methods direct-seed rice technology has many advantages and benefits

Drought stress reduces rice output by lowering cell development and disrupting plant antioxidant activity by boosting the formation of reactive oxygen species. It stress-induced gene and protein functions, altering morphological, biochemical, physiological, and molecular responses. Thus, to achieve the yield targets in these areas, there have to be target-specific trait introgression or trait improvement schemes. The above-mentioned systems' practical benefits can only be realized when they are complemented with ideal genotypes that withstand these systems' management and cultural practices. Thus, it is of utmost importance to follow an ideotype for an ideal aerobic variety that forms a blueprint for breeding the next generation rice varieties that are climate resilient and also take care of food security. The following sections would focus on the genetic and morphological needs of such a plant ideotype and the traits it would require while growing till grains, the constraints, and mitigation strategies.

19.3 Adaptive Traits for Next-Generation Rice Plant

Developing a next-generation plant would require specific traits that can stand under biotic and abiotic stress-prone environments. In addition, some important technological approaches (phenotypic and genotypic) are needed to develop climate-resilient crops. Characterization of genetic resources and phenotypic evaluation of the crops under targeted environments are most important for selecting the germplasm, allowing greater precision in identifying sources of heritable traits used in the breeding programs. The tremendous development in the availability of substantial genomic resources and modern breeding approaches, including the genotype and phenotypic technologies, provided a strong platform for precise target trait mapping, and gene identification and developed the markers for use in the breeding program. With these, trait-based markers and targeted phenotype-trait markers in rice have already demonstrated the applications of marker-assisted breeding approaches by developing several new rice varieties for tolerance to biotic and abiotic stresses. As Fig. 19.4 shows, the trait-based markers for the various environment have been represented in each chromosome-wise. These markers are highly utilized and important to breed in next-generation plants for developing stress-tolerance rice varieties.

19.3.1 Adaptive Traits for Seedling Establishment

The varieties to be grown under aerobic conditions or that are to be direct-seeded need to face a different set of abiotic stresses while growing. The earliest challenge that it can encounter is during germination. Sometimes after seeding, there is a chance of rain's onset, making the seedbed saturated and creating an anaerobic condition for the germinating seed. The seed must then be able to germinate, sprouting its coleoptile (before the coleorhiza) even without any oxygen (Itoh et al., 2005). Such

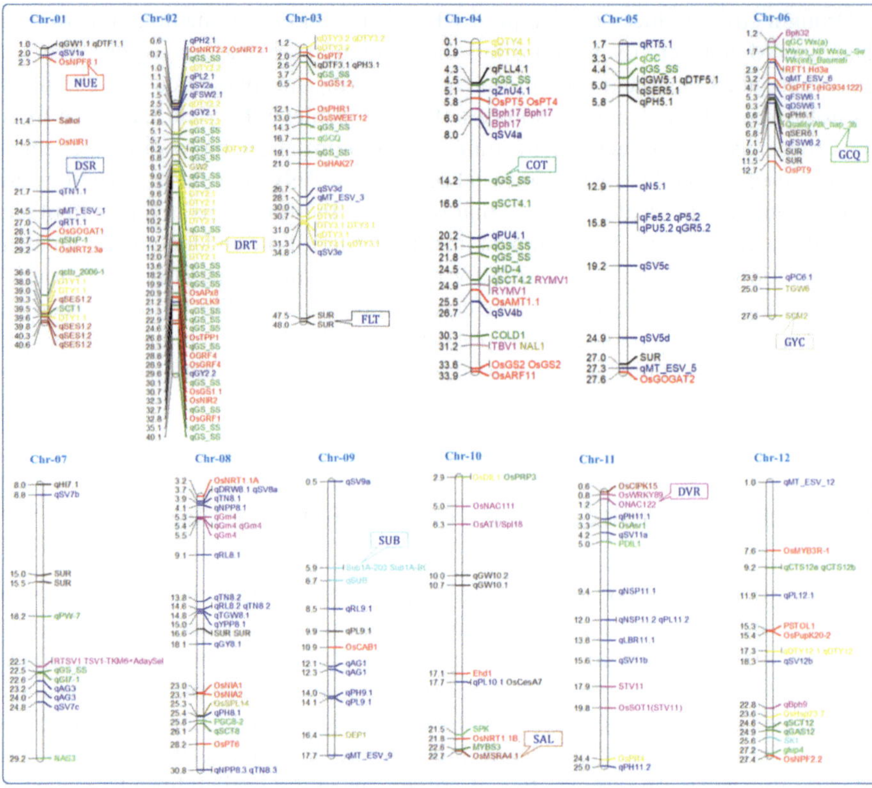

Fig. 19.4 A schematic representation of Quantitative trait loci (QTLs) and genes position on 12 chromosomes. The number on the left of each chromosome is the QTLs/genes location (Mb) and associated with target traits. The different color pattern indicates each QTLs/genes for traits related to rice biotic and abiotic stress tolerance. The total of 10 box outline color represent the various traits associated with the different parameters., like NUE_Nutrient use efficiency; DSR-Direct seeded rice; DRT_Drought tolerance; SAL_salinity tolerance; COL_Cold tolerance; SUB_Submergence tolerance; DVR_Disease insect and virus resistance; GCQ_Grain cooking and quality; GYC_Grain yield component traits; FLT_Flood tolerance. (*Source* Arbelaez et al., 2019; Ghosal et al., 2019; Mahender et al., 2015b; Sagare et al., 2020; Sandhu et al., 2015, 2021; Yadav et al., 2017; Yang et al., 2021), http://qtaro.abr.affrc.go.jp/; https://isl.irri.org/services/genotyping/trait-based-genotyping)

conditions can lead to hypoxia. Thus, varieties need to be bred for the next generation ideotype must possess the quality of anaerobic germination, i.e., germinate and grow to survive under anaerobic conditions. In this regard, many genes and QTLs have been reported to provide tolerance to anaerobic germination effectively. The major one is the AG1 gene in the QTL *qAG-9–2*, which encodes the trehalose-6-phosphate phosphatase family gene enabling coleoptile to elongate and rise above the anaerobic environment (Kretzschmar et al., 2015). Another gene *AG2* has also been reported to help in anaerobic germination, and both the genes have displayed surprisingly higher survivability under anaerobic conditions in the introgressed lines than in the recurrent parent, Dongan, a japonica cultivar (Kim et al., 2019). A novel QTL, *qSUR6-1* has been identified as potential for anaerobic condition survivability with high germination percentage in a mapping population using a common parent (Kalarata) with the anaerobic germination trait (Ghosal et al., 2019). QTLs (*qAG7.1* and q*AG7.2*) are located on the short arm of chromosome 7, (Septiningsih et al., 2013) and was found to promote efficient germination under anaerobic conditions. Apart from these, there are expansin genes that help in elongating the epidermal cells and promoting the growth of the coleorhiza hairs, such as *Os-EXP1, OsEXP2, Os-EXP4, Os-EXP5, Os-EXP6, Os-EXP10, OsEXPB3, Os-EXPB4, Os-EXPB6, Os-EXPB11, Os-EXPL1, Os-EXPL2 and Os-EXPL3* (Huang et al., 2000). Once the seed elongates its hypocotyl to the surface, it becomes easier to acquire oxygen. By this stage, the seed has almost exhausted all its reserve starch to elongate its hypocotyl and thus would require rapid nutrient acquisition from the rhizosphere to suffice the energy requirements of the cells. Therefore, early seedling vigor is a highly essential trait in the DSR system. This includes uniform germination and emergence, establishment, and nutrient uptake by the young seedling. Early Uniform Emergence (EUE) is contributed by traits like early and high germination rate, rapid root-shoot development, and enhanced seedling vigor (Fig. 19.4). QTLs for early uniform emergence (*qEUE11.1*) (Dixit et al., 2015) and early vigor (*qEVV9.1*) (Sandhu et al., 2015) have been reported that can be used for the improvement of dry-DSR varieties. Two significant major QTLs (*qEMM1.1* on chromosome 1 and *qEMM11.1* on chromosome 11) for EUE derived from Moroberekan (BC_2F_3; Moroberekan × Swarna) were reported by Dixit et al. (2015). The other aspect is the early and vigorous uptake of nutrients by the roots, this is facilitated by the early development of crown roots, a major trait in rice that controls nutrient and water acquisition in the initial days of seedling growth. Thus, their number and distribution are deciding factors of early seedling vigor, and they also help in P uptake due to their shallow rooting. Therefore, rapid nodal root growth, longer root, and high density of roots are imperative for EUE under DSR (Sandhu et al., 2019). For their timely normal emergence from the shoot base, seedlings with high vigor have been reported to utilize the WUSCHEL-related homeobox (WOX) gene, *WOX11* (Zhao et al., 2009).

Seedling vigor or the early vegetative vigor is a function of the assimilate source (light intensity and photosynthetic rate) and the sink (structural growth: leaf appearance rate, potential size, and tiller outgrowth). The photosynthetic rate, SPAD, leaf canopy spread, and canopy closure time is important determinants of the shoot vigour traits that can help to select ideal genotypes (Rebetzke et al., 2004). The varieties bred specifically for low moisture abiotic stress must have a vigorous growth at the seedling stage, an enhanced foliar growth to combat weeds at the vegetative stage, moderate tillering (10–12 tillers) and enhanced assimilation export from leaves to stems during the late vegetative and reproductive phases (Dingkuhn et al., 1999), also in reducing the competitive effects of weeds and increase yield stability in dry direct-seeded environments (Okami et al., 2011). The increased canopy delimits the sunlight and other resources that inhibit weed growth, a major menace in the aerobic rice cultivation system (Anandan et al., 2016). A back cross program using the donor Weed Tolerant Rice 1, has identified four promising lines; G-6-Y7-WL-3, G-6-Y6-WU-3, G-6-Y3-WL-3, and G-6-Y8-WU-1 that performed well under upland weedy conditions with high yields (Dimaano et al., 2017). Several QTLs have been reported related to weed competitiveness; nine QTL hotspots for competitive weed traits such as *qWCA2a, qWCA2b, qWCA2c, qWCA3, qWCA5, qWCA7, qWCA8, qWCA9,* and *qWCA10* have been reported in backcross populations of *O. sativa L. and O. glaberrima* Steud. (Bharamappanavara et al., 2020).

19.3.2 Nutrient and Moisture Acquisition and Utilization Through the Growth Period

Phosphorus (P) availability is a limitation in low moisture systems. It is highly mobile in plants but becomes unavailable to plants in the soil system, this is because phosphorus forms calcium salts in the soil which is inaccessible to the plants (Abel et al., 2002). The two-week-old seedling required at least 6 ppm of P (Anandan et al., 2021b) and the uptake of P start as early as 2–3 days after germination. In relation to uptake, two transporters have been identified; *OsPHT1* and *OsPHT8*, play a major role in the reallocation of P from source to sink organs and help in the nutrient homeostasis in the growing embryo (Jia et al., 2011). Many other such transporters molecules of the *pht* have been identified but are yet to be fully understood. Phosphorus starvation tolerance 1 (*PSTOL1*), a significant gene identified in the phosphorus-deficiency tolerance QTL, *phosphorus uptake 1* (*Pup1*) QTL aids in the uptake of P and enhancer of early root growth (Gamuyao et al., 2012). Besides, nitrogen and phosphorus, iron and zinc can also be the limiting factors. The roots take up the iron with the help of *YELLOW STRIPE-LIKE PROTEINS* (*YSLs*) (Nozoye et al., 2011) through the root exudates; phytosiderophores. Many Zn transporters of the ZIP (*Zrt*, *Irt*-like Protein) family have been reported that sense and get activated in response to Zn deficiency; *OsZIP1, OsZIP3, OsZIP4*, and *OsZIP5*(Ishimaru et al., 2006; Lee et al., 2010; Ramesh et al., 2003). These *ZIP* transporters are also involved in the transport of iron and manganese (Lou, 2000).

Mycorrhizal fungi also aid in nutrient uptake, i.e., zinc and phosphorus, they colonize cortical cells of the roots and extend their hyphae into the rhizosphere for nutrient mining (Johnson & Gehring, 2007). This helps to supplement the existing surface area of the roots, thereby exploring larger parts of the soil and facilitating increased uptake. Such a symbiotic relationship can be more beneficial, especially when the nutrient in question is immobile, such as zinc or phosphorus (Smith & Read, 2010). According to a study, an aerobic environment enhanced root colonization by 28–57% as compared to the anaerobic flooded conditions (Anandan et al., 2022; Gao et al., 2007). Inadequate water supply is the primary constraint to yield under upland conditions, followed by different nutrient deprivation. The most limiting in an aerobic system after moisture is nitrogen which is available in the oxidized environment, usually in the form of nitrate ions (NO_3^-). The plant needs to be loaded with required genes and transporter molecules to facilitate nutrient uptake. Genes like Two different transporter families; NPF and NRT2 (NAR2) are involved in nitrate uptake. The NPF family of transporters displays a wide range of activities with nitrogen that includes root-shoot transport (OsNPF2.4, OsNPF2.2), lateral root promotion (OsNPF8.20), panicle elongation (OsNPF4.1), and increasing grain yield (OsNPF8.20) (Huang et al., 2018). Out of the five NRT genes discovered so far, OsNRT2.3b under high expression levels enhanced N, Fe, and P uptake. A gene; *Tolerance of Nitrogen Deficiency 1* (*TOND1*), has been identified in relation to N deficiency tolerance (Zhang et al., 2015). Sometimes flooding due to rains can change the soil dynamics and lead to a reduced rhizosphere, in such cases ammonium transporters can come to the rescue (Panda et al., 2021). At least ten putative AMT genes have been operating in rice (three members in each group of *OsAMT1, OsAMT2*, and *OsAMT3* and one member of *OsAMT4*). They are involved mostly in root uptake and root-shoot transport (Sonoda et al., 2003).

19.3.3 Adaptive Traits for Early and Late Vegetative Growth Stages

A variety bred to sustain abiotic stresses must predominantly survive a system of low soil moisture. The plant may be frequently exposed to soil dryness, thus conserving moisture and restricting transpirational losses and leaf area index (LAI) are more likely to be delimited compared to its counterparts in transplanted systems. The leaf area and specific leaf area index need to be enhanced in such a situation, this can be facilitated by a higher allotment of assimilates in the stem (de Groot et al., 2002). Availability of both water and nutrients throughout the growth period would help in higher LAI. Stomatal pore regulation is another important mechanism. Stomatal pores close swiftly as soon as the plant senses internal water deficit (Boyer, 1996). Leaves with lower stomatal density can be exploited while breeding aerobic rice as they withstand drought with their more conservative water use and suit an elevated carbon dioxide environment (a trait likely to be ideal in the near future). Stomatal response and its closure timing decide the fate of the tolerance reaction. Studies on the rice epidermal patterning factor (*OsEPF1*) gene revealed its role in producing leaves with lower stomatal density (Caine et al., 2019). Another study involving the overexpression of the *OsASR5* gene (isolated from an upland japonica variety IRAT109) has increased sensitivity to drought stress that led to rapid closure of stomata, there conserving the internal moisture level (Li et al., 2017). A study at IRRI, suggested that top-yielding aerobic rice genotypes have erect stands (Zhao et al., 2006), but erect plants do not always ensure high yield. Plants taller than 110 cm, increase the chance of lodging. To have durable culm strength, a brittle culm would be advantageous over a thick culm, which requires more nutrients and moisture to increase the circumference. In selection, a height of less than 120 cm under favorable aerobic soil conditions is acceptable as they provide lodging resistance as well as weed suppressiveness (Zhao & Kumar, 2016). Keeping the plant erect would require strengthening of the culm and lower positioning of panicles that can aid in reducing the risks of lodging. The QTL STRONG CULM3 (*SCM3*) develops culm strength in rice (Yano et al., 2015). Similarly, other QTLs contributing to erect plant traits have also been reported; culm diameter (*qCD1.1*) and culm strength (*qCS1.1*) (Yadav et al., 2017). Similar to plant erectness, there is also a relative decrease in the tiller number when grown in dry or low moisture conditions as compared to the transplanted ecology. A constant increase in the number of tillers per m^2 up to 90 DAS and thereafter mortality of some tillers caused due to water deficit and iron deficiency under aerobic conditions has been observed (Chandrika et al., 2017). Thus, the selection for higher tiller numbers must focus on aiding in the yield stability under water stress conditions.

The planting density in a field often determines the yield indirectly through many different traits such as the number of productive tillers, panicle length fertility, etc. High-density planting leads to earlier panicle initiation and higher grain yield when compared to its counterpart. It is seen that a density of 50 plants per meter square produces higher yields (Clerget et al., 2016). Adding to this is the panicle size, the yield is often compromised due to the panicle size. Several QTLs have been identified that influence the panicle size on chromosomes 1,2,3,5 and 11 (Kato et al., 2009). A difference in gene expression was seen in irrigation limited and full irrigation situations, on the short arm of chromosome 1, QTLs for florets per primary branches were detected while on chromosome 2 (Chr. 2–1), QTLs for florets per panicle were detected under limited irrigation and full irrigation respectively. *LONG PANICLE 1*(*LP1*), a major locus on Chromosome 9 encoding Remorin_C-containing protein of unknown function and its various alleles, has been associated with panicle length (Deng et al., 2020). Several QTLs related to grain yield under dry direct-seeded conditions, $qGY_{1.1}$, $qGY_{2.1}$, $qGY_{2.2}$, $qGY_{6.1}$, $qGY_{8.1}$, and $qGY_{10.1}$, have been identified (Sandhu et al., 2013, 2014) and can be incorporated into future QTL introgression work to improve yield under dry-DSR conditions.

19.3.4 Lodging Resistance

Lodging is a complex trait affected by a plethora of traits such as root spread, culm strength, soil type, wind, and rainfall. This becomes a bigger problem when the seeds are broadcasted and in cases of cyclone-prone areas. Heading and early grain development stages are relatively more prone to lodging. There are two types of lodging recognized in direct-seeded rice: i. stem lodging and ii. root lodging.

(a) ***Stem lodging:***

This is more frequent in the case of wet-DSR than transplanted in rice. This is due to the surface placement of the seeds in the wet bed method as a result the roots don't spread much laterally but rather grow deeper into the soil. This leads to weak anchorage in the soil and leaves it susceptible to heavy wind or rain. Stem lodging is again described as stem bending type and stem breaking type (Kashiwagi et al., 2005). Bending type lodging when the middle internode bends without breaking the culm, while in the breaking type the internodes below the third internode break. The latter is considered more fatal than the former because there is an obstruction in the transport of nutrients, moisture and assimilates caused by buckling of one of the bottom internodes and resulting in the breaking down of the upper stem and bringing the ear horizontal to the ground (Rani et al., 2017).

(b) ***Root lodging:***

Root lodging is a common problem, especially when the seeds are surface sown under a dry-DSR system (Setter et al., 1997). Leaning of the intact and unbroken culm from the crown due to weakening of the root system is the reason for root lodging. This occurs when the force exerted by all of the shoots belonging to a plant exceeds the strength of its root anchorage system. The root anchorage strength and root plate spread of a plant are the determinants of root lodging. Intermediate plant height, brittle culm, high lignin content on the stem, lower positioning of panicles in the plant's canopy, increased stem strength of the lower internodes and wider root plate with strong anchorage are traits of lodging tolerance (Fu et al., 2013; Setter et al., 1997; Zhang et al., 2013).

Zhu et al. (2008) have identified 17 QTL regions associated with basal internodal traits in rice which contribute to lodging resistance. QTL for breaking strength of the top three internodes on chromosome 11 was detected and this confers resistance to bending type of lodging in typhoon-affected areas (Kashiwagi, 2014). Additionally, the accumulation of starch enhances the culm's flexibility (Kashiwagi et al., 2006) and higher silicon deposition provides strength against lodging (Ma & Yamaji, 2006). Kashiwagi and Ishimaru reported a locus (*prl5*) that enhanced the weight of the lower stem, thus displaying resistance against lodging (Kashiwagi & Ishimaru, 2004). A NIL with QTLs *SCM2* + *SCM3* showed a much stronger culm than a plant with solitary QTL *SCM2* and *SCM3*. In addition, these QTLs are also reported to increase spikelet number due to the additive effect (Yano et al., 2015). QTLs for higher nutrient uptake with lodging resistance ($qLDG_{3.1}$, $qLDG_{4.1}$) have already been reported (Dixit et al., 2015), and a multi-QTL stacking program is being undertaken at NRRI, Cuttack to incorporate them in breeding DSR varieties (Anandan et al., 2018).

19.3.5 Biotic Stress Resistance

The dry-conditioned rice-growing system has a different set of diseases and pests when compared to the lowland conditions. Nematode infestations are one of the major roadblocks in the growth and yield of aerobic varieties accounting for up to 70% loss owing to the damage by *Pratylenchus spp.* and *Meloidogyne spp.* (Jain et al., 2007). Many genotypes have been identified as tolerant to nematodes; Three *O. glaberrima* genotypes (TOG 5674, TOG 7235, and CG 14) were found to be highly resistant to *M. graminicola*, while five *O. sativa* genotypes (IR78910-23-1-3-4, WAB 638-1, IRAT 216, IR78877-208-B-1-2, and IR72) were identified as partially resistant to *M. graminicola.* QTLs linked to *M. graminicola* tolerance have been identified on the chromosomes 3, 5, and 10 from a cross between *O. glaberrima* and *O. sativa* (Sandhu & Kumar, 2016).

The losses incurred due to diseases can sometimes reach up to 80% in severe epidemics. Few of the major diseases that pose a threat to rice cultivation have been highlighted here. The bacterial leaf blight (BB) is caused by *Xanthomonas oryzae pv. oryzae* is one such disease, it can damage up to 80% of the crop depending on its stage and environmental conditions. There have been multiple reports and identification of genes about BB resistance. 14 dominant (*Xa1, Xa2, Xa3, Xa4, Xa7, Xa10, Xa11, Xa12, Xa14, Xa16, Xa17, Xa18, Xa21*, and *Xa22(t)*) and six recessive (*xa5, xa8, xa13, xa15, xa19*, and *xa20*) resistance genes for BB have been identified. Among these genes, the combinations of three dominant resistance genes (*Xa4* + *Xa7* + *Xa21*) in IRBB 62, followed by*xa5* + *xa13* + *Xa21* (Pradhan et al., 2015) and*Xa4* + *xa5* + *xa13* + *Xa21* (Loan et al., 2006) were reported to be conferring high resistance. Blast is another major disease that has resulted in yield stability in many rice-growing areas caused by the fungus *Magnaporthe grisea.* More than 40 major genes as well as QTLs for blast resistance have been identified. Monogenic resistance to blast is less stable but varieties with pyramided monogenes or QTLs are durably resistant. The *Pita* locus in rice is a key player against blast disease, the genes *Pita* and *Pita2* near the centromeric region on chromosome 12 reportedly have broader resistance (Shikari et al., 2013). Rice brown spot caused by *Cochliobolus miyabeanus* is a chronic disease, that affects millions of hectares of rice every growing season, more often in South and Southeast Asian countries. It's a peculiar disease, rather considered a secondary problem associated with rice crops facing abiotic stresses like low moisture, poor fertility and most importantly facing nitrogen deficiency. Multiple QTLs have been discovered across chromosomes conferring resistance to brown spots; *qBS2, qBS9*, and *qBS11* on chromosomes 2, 9, and 11, respectively, *qBSfR1, qBSfR4,* and *qBSfR1* on chromosomes 1, 4, and 11, respectively (Sato et al., 2015). IET 27,539, a breeding line is reported to be resistant to brown spot, this line can pave the way for developing brown spot resistant varieties (ICAR-NRRI Annual Report, 2019). Areas with alternate wetting and drying face sheath blight are more commonly caused by *Rhizoctonia solani.* However, due to the lack of effective resistant donors for this trait, there have not been many resistant varieties released. A recent report has studied the variability of different tolerant lines for sheath blight; Tetep, Jasmine85, and Teqing. Such studies can be more helpful in developing tolerant varieties or in the introgression of tolerant genes (Naveenkumar et al., 2021).

Insects can create havoc, especially with the modern-day cultivars that more or less share a genetic background derived from a few handfuls of good yield cultivars as their ancestors. Thus the infestation rate becomes very high with huge crop losses. Brown planthoppers is one of the most destructive pests in Asia. The BPH increasingly exhibits resistance to insecticides and adaptation to resistant varieties, so they threaten rice production thus food security (Gurr et al., 2011). A relatively warm and high humidity environment favors the outbreak of this pest. By now more than 37 BPH-resistant genes have been identified across wild species and indica varieties (Sani Haliru et al., 2020). With the advent of molecular approaches, it is now easier to exploit these genes in different combinations and study their efficacy in long run. Gall midge (*Orseolia oryzae*) is prevalent in China, India, Sri Lanka, Indonesia, Bangladesh, Myanmar, and Vietnam; and causes significant yield loss mostly during the *Kharif* season (Sagare et al., 2020). Breeding rice varieties with pest resistance genes such as W1263 (with *Gm1*), Phalguna (*Gm2*), RP2068 (*gm3*), Abhaya (*Gm4*), and ARC5984 (*Gm5*), and with other resistance sources with unknown genes such as Bhumansan, NHTA8, Benglei, Aganni, and ARC6605, has been a viable, ecologically acceptable approach for the management of gall midge (Sandhu & Kumar, 2017).

19.4 Constraints Associated with Low Moisture Input Rice Cultivation

Despite several benefits, few constraints associated with the adoption of dry-DSR have restricted its popularity and adoption. When the system is changed from predominantly flooded to aerobic DSR, the plants experience a certain degree of water stress, also a change in soil chemistry and the availability of nutrients (Buresh & Haefele, 2010). There is no standing water to suppress the weed infestation in this system and thus, crop and competing weeds appear simultaneously in the direct seeded systems. They are subjected to much higher weed pressure than the conventional puddled transplanted system (Singh et al., 2008) in which weeds are suppressed by standing water and by transplanted rice seedlings, which have a 'head start' over germinating weeds. The average yield of aerobic rice varieties in farmers' fields is between 3.0 and 5.0 t ha^{-1} (Shashidhar, 2012) and little success has been achieved in enhancing this. This is due to the delayed or non-uniform emergence, pre-emergence death of seedlings, uneven plant population in the field, higher weed problem in the early crop growth stage, and lower nutrient availability to the crop under aerobic conditions. The roots structure (root length, volume, distribution of shallow and deep roots, number of later roots, etc.) adapted especially to such conditions are yet to be explored and incorporated into the breeding programs. While the SRI method of cultivation is giving good results, its adoption is limited to areas with high labour availability and irrigation facilities. This limits their wide-scale adoption of rice cultivation. The continuous practice of aerobic cultivation has shown a gradual yield decline in rice production fields as reported in Brazil (Pinheiro et al., 2006) and in the Philippines (Bouman et al., 2007) suggesting that it's not sustainable as a monocrop but can be circumvented by crop rotation. The yield decline in Brazil and the Philippines were caused by an auto-toxicity effect and nematode build-up, respectively, which were reminiscent of the soil sickness syndrome observed in other crops (Grodzinskii, 2016). Reduced nutrient uptake especially immobilization of P, increased N leaching, and volatilization under aerobic conditions compared to continuously flooded conditions has been the key factor for yield reduction under aerobic cultivation (Kumar & Ladha, 2011). Several biotic (brown spot, blast, BB, seedling blight, nematode, weeds, etc.) and abiotic (anaerobic germination, moisture deficiency, nutrient deficiency, etc.) stresses reduced the yield status. Another important factor is that the varieties developed for the transplanted condition are found unsuitable for aerobic-DSR and have shown a yield decline of 8–28.5% under aerobic conditions in India and Pakistan (Sandhu et al., 2019).

19.5 Conclusion

Breeding activities must progress in parallel to the change in environment and human needs. Climate-smart crops can shift the ongoing intensive and unstable system of cultivation towards a more stable, efficient, and resilient system. Thus, varieties must now be bred with the requisite genes/ QTLs as highlighted in this chapter catering to the needs of the rice-growing areas. For instance, a plant with genes like PSTOL or TOND1 would better utilize the available nutrients compared to their counterparts. Similarly, plants with biotic stress tolerant genes would stand a better chance of giving good yield while facing a disease or pest incidence. The hindrance in this is the maintenance and improvement of yield after introgressing multiple genes. Such a program of multiple QTL/ gene introgression would always have a linkage drag manifesting some undesirable effects reflected in the yield. Modern genomics and biotechnological tools like gene editing, and omics platforms are now more efficient in the selection and controlled expression of genomic segments. Therefore, continuous research efforts and agricultural investments in this direction are essential to develop climate-smart rice ideotypes designed to be high yielding as well as climate-resilient.

References

Abel, S., Ticconi, C. A., & Delatorre, C. A. (2002). Minireview phosphate sensing in higher plants. *Physiologia Plantarum, 115*, 1–8.

Anandan, A., Anumalla, M., Pradhan, S. K., Ali, J. (2016). Population structure, diversity and trait association analysis in rice (Oryza sativa L.) germplasm for early seedling vigor (ESV) using trait linked SSR markers. *PLoS One, 11*. https://doi.org/10.1371/journal.pone.0152406

Anandan, A., Nagireddy, R., Sabarinathan, S., Bhatta, B. B., Mahender, A., Vinothkumar, M., Parameswaran, C., Panneerselvam, P., Subudhi, H., & Meher, J. (2022). Multi-trait association study identifies loci associated with tolerance of low phosphorus in Oryza sativa and its wild relatives. *Science and Reports, 12*, 1–24.

Anandan, A., Panda, S., Pradhan, S. K., Dash, S. K., Sarkar, S., Patra, B. C. (2021a). Aerobic dry direct-seeded rice A sustainable approach in rice cultivation

Anandan, A., Parameswaran, C., Azharudheen, T. M., & Singh, O. (2018). Towards the development of dry direct seeded rice varieties by stacking of multiple QTLs/genes. *ORYZA-An International Journal on Rice, 55*, 51. https://doi.org/10.5958/2249-5266.2018.00006.1

Anandan, A., Parameswaran, C., Mahender, A., Nayak, A. K., Vellaikumar, S., Balasubramaniasai, C., & Ali, J. (2021b). Trait variations and expression profiling of OsPHT1 gene family at the early growth-stages under phosphorus-limited conditions. *Science and Reports, 11*, 1–19.

Arbelaez, J. D., Dwiyanti, M. S., Tandayu, E., Llantada, K., Jarana, A., Ignacio, J. C., Platten, J. D., Cobb, J., Rutkoski, J. E., & Thomson, M. J. (2019). 1k-RiCA (1K-Rice Custom Amplicon) a novel genotyping amplicon-based SNP assay for genetics and breeding applications in rice. *Rice, 12*, 1–15.

Belder, P., Bouman, B. A. M., Spiertz, J. H. J., Peng, S., Castaneda, A. R., & Visperas, R. M. (2005). Crop performance, nitrogen and water use in flooded and aerobic rice. *Plant and Soil, 273*, 167–182.

Bharamappanavara, M., Siddaiah, A. M., Ponnuvel, S., Ramappa, L., Patil, B., Appaiah, M., Maganti, S. M., Sundaram, R. M., Shankarappa, S. K., Tuti, M. D., Banugu, S., Parmar, B., Rathod, S., Barbadikar, K. M., Kota, S., Subbarao, L. V., Mondal, T. K., & Channappa, G. (2020). Mapping QTL hotspots associated with weed competitive traits in backcross population derived from Oryza sativa L. and O. glaberrima Steud. *Science and Reports, 10*, 1–13. https://doi.org/10.1038/s41598-020-78675-7

Bouman, B. A. M., Xiaoguang, Y., Huaqi, W., Zhiming, W., Junfang, Z., Channggui, W., Bin, C. (2002). Aerobic rice (Han Dao): a new way of growing rice in water-short areas. In *Proceedings of the 12th international soil conservation organization conference*, pp. 175–181.

Bouman, B. A. M., Humphreys, E., Tuong, T. P., Barker, R. (2007). Rice and Water. *Advances in Agronomy, 92*, 187–237.

Boyer, J. S. (1996). Advances in drought tolerance in plants. *Advances in Agronomy, 56*, 187–218.

Buresh, R. J., Haefele, S. M. (2010). Changes in paddy soils under transition to water-saving and diversified cropping systems. In: *World congress of soil science*, Brisbane, Australia, pp. 1–6.

Caine, R. S., Yin, X., Sloan, J., Harrison, E. L., Mohammed, U., Fulton, T., Biswal, A. K., Dionora, J., Chater, C. C., & Coe, R. A. (2019). Rice with reduced stomatal density conserves water and has improved drought tolerance under future climate conditions. *New Phytologist, 221*, 371–384.

Chandrika, M., Devi, M. U., Ramulu, V., & Ramana, M. V. (2017). Evaluation of different varieties of aerobic rice (Oryza sativa L.) under different fertigation levels on growth and yield parameters. *International Journal of Current Microbiology and Applied Sciences, 6*(5), 2793–2801.

Clerget, B., Bueno, C., Domingo, A. J., Layaoen, H. L., & Vial, L. (2016). Leaf emergence, tillering, plant growth, and yield in response to plant density in a high-yielding aerobic rice crop. *Fields Crop Research, 199*, 52–64. https://doi.org/10.1016/j.fcr.2016.09.018

De Groot, C. C., Marcelis, L. F. M., van den Boogaard, R., & Lambers, H. (2002). Interactive effects of nitrogen and irradiance on growth and partitioning of dry mass and nitrogen in young tomato plants. *Functional Plant Biology, 29*, 1319–1328.

Deng, J., Feng, X., Wang, D., Lu, J., Chong, H., Shang, C., Liu, K., Huang, L., Tian, X., & Zhang, Y. (2020). Root morphological traits and distribution in direct-seeded rice under dense planting with reduced nitrogen. *PLoS ONE, 15*, 1–13. https://doi.org/10.1371/journal.pone.0238362

Dimaano, N. G. B., Ali, J., Cruz, P. C. S., Baltazar, A. M., Diaz, M. G. Q. Acero, B. L., Li, Z. (2017). Performance of newly developed weed-competitive rice cultivars under lowland and upland weedy conditions. *Weed Science, 65*, 798–817. https://doi.org/10.1017/wsc.2017.57

Dingkuhn, M., Johnson, D. E., Sow, A., & Audebert, A. Y. (1999). Relationships between upland rice canopy characteristics and weed competitiveness. *Fields Crop Research*. https://doi.org/10.1016/S0378-4290(98)00152-X

Dixit, S., Grondin, A., Lee, C. R., Henry, A., Olds, T. M., & Kumar, A. (2015). Understanding rice adaptation to varying agro-ecosystems: Trait interactions and quantitative trait loci. *BMC Genetics*. https://doi.org/10.1186/s12863-015-0249-1

Fu, X., Jing, F., Bin, Y., Gao, Y., Zheng, Y., Yue, B. (2013). Morphological, biochemical and genetic analysis of a brittle stalk mutant of maize inserted by mutator. *Journal of Integrative Agriculture, 12*, 12–18. https://doi.org/10.1016/S2095-3119(13)60200-2

Gamuyao, R., Chin, J. H., Pariasca-Tanaka, J., Pesaresi, P., Catausan, S., Dalid, C., Slamet-Loedin, I., Tecson-Mendoza, E. M., Wissuwa, M., & Heuer, S. (2012). The protein kinase Pstol1 from traditional rice confers tolerance of phosphorus deficiency. *Nature, 488*, 535–539. https://doi.org/10.1038/nature11346

Gao, X., Kuyper, T. W., Zou, C., Zhang, F., & Hoffland, E. (2007). Mycorrhizal responsiveness of aerobic rice genotypes is negatively correlated with their zinc uptake when nonmycorrhizal. *Plant and Soil, 290*, 283–291. https://doi.org/10.1007/s11104-006-9160-x

Ghosal, S., Casal, C., Quilloy, F. A., Septiningsih, E. M., Mendioro, M. S., Dixit, S. (2019). Deciphering genetics underlying stable anaerobic germination in rice: Phenotyping, QTL identification, and interaction analysis. *Rice, 12*. https://doi.org/10.1186/s12284-019-0305-y

Grodzinskii, A. M. (2016). *Allelopathy in the life of plants and their communities*. Scientific Publishers

Guerinot, M. L. (2000). The ZIP family of metal transporters. *Biochimica et Biophysica Acta (BBA)-Biomembranes*.

Gurr, G. M., Liu, J., Read, D. M. Y., Catindig, J. L. A., Cheng, J. A., Lan, L. P., & Heong, K. L. (2011). Parasitoids of Asian rice planthopper (Hemiptera: Delphacidae) pests and prospects for enhancing biological control by ecological engineering. *The Annals of Applied Biology, 158*, 149–176.

Huang, J., Takano, T., Akita, S. (2000). Expression of α-expansin genes in young seedlings of rice (Oryza sativa L.). *Planta*. https://doi.org/10.1007/s004250000311

Huang, S., Zhao, C., Zhang, Y., & Wang, C. (2018). Nitrogen use efficiency in rice. *Nitrogen in Agriculture-Updates*. https://doi.org/10.5772/intechopen.69052

Huaqi, W., Bouman, B. A. M., Zhao, D., Changgui, W., Moya, P. F. (2002). Aerobic rice in northern China: opportunities and challenges. In *Water-wise rice production* (pp. 143–154). International Rice Research Institute.

ICAR-NRRI ANNUAL REPORT. (2019). *NRRI Annual Report 2019*. Cuttack.

Ishfaq, M., Farooq, M., Zulfiqar, U., Hussain, S., Akbar, N., Nawaz, A., & Anjum, S. A. (2020). Alternate wetting and drying: A water-saving and ecofriendly rice production system. *Agricultural Water Management, 241*, 106363.

Ishimaru, Y., Suzuki, M., Tsukamoto, T., Suzuki, K., Nakazono, M., Kobayashi, T., Wada, Y., Watanabe, S., Matsuhashi, S., Takahashi, M., Nakanishi, H., Mori, S., & Nishizawa, N. K. (2006). Rice plants take up iron as an Fe3+-phytosiderophore and as Fe2+. *The Plant Journal*. https://doi.org/10.1111/j.1365-313X.2005.02624.x

Islam, S. M. F., Karim, Z. (2019). World's demand for food and water: The consequences of climate change. *Desalination-Challenges Opportunities*, 1–27.

Itoh, J. I., Nonomura, K. I., Ikeda, K., Yamaki, S., Inukai, Y., Yamagishi, H., Kitano, H., Nagato, Y. (2005). Rice plant development: From zygote to spikelet. *Plant and Cell Physiology*.

Jabran, K., & Chauhan, B. S. (2015). Weed management in aerobic rice systems. *Crop Protection, 78*, 151–163. https://doi.org/10.1016/j.cropro.2015.09.005

Jain, R. K., Mathur, K. N., & Singh, R. V. (2007). Estimation of losses due to plant parasitic nematodes on different crops in India. *Indian Journal of Nematology, 37*(2), 219–221.

Jia, H., Ren, H., Gu, M., Zhao, J., Sun, S., Zhang, X., Chen, J., Wu, P., & Xu, G. (2011). The phosphate transporter gene ospht1;8 is involved in phosphate homeostasis in rice. *Plant Physiology*. https://doi.org/10.1104/pp.111.175240

Johnson, N. C., Gehring, C. A. (2007). Mycorrhizas: Symbiotic mediators of rhizosphere and ecosystem processes. In The Rhizosphere.

Kashiwagi, T. (2014). Identification of quantitative trait loci for resistance to bending-type lodging in rice (Oryza sativa L.). *Euphytica*. https://doi.org/10.1007/s10681-014-1111-7

Kashiwagi, T., & Ishimaru, K. (2004). Identification and functional analysis of a locus for improvement of lodging resistance in rice. *Plant Physiology*. https://doi.org/10.1104/pp.103.029355

Kashiwagi, T., Madoka, Y., Hirotsu, N., & Ishimaru, K. (2006). Locus prl5 improves lodging resistance of rice by delaying senescence and increasing carbohydrate reaccumulation. *Plant Physiology and Biochemistry*. https://doi.org/10.1016/j.plaphy.2006.02.004

Kashiwagi, T., Sasaki, H., Ishimaru, K. (2005). Factors responsible for decreasing sturdiness of the lower part in lodging of rice (Oryza sativa L.). *Plant Production Science*. https://doi.org/10.1626/pps.8.166

Kato, Y., Nemoto, K., & Yamagishi, J. (2009). QTL analysis of panicle morphology response to irrigation regime in aerobic rice culture. *Fields Crop Research, 114*, 295–303. https://doi.org/10.1016/j.fcr.2009.08.014

Kim, S. M., Kim, C. S., Jeong, J. U., Reinke, R. F., & Jeong, J. M. (2019). Marker-assisted breeding for improvement of anaerobic germination in japonica rice (Oryza sativa). *Plant Breeding*. https://doi.org/10.1111/pbr.12719

Kretzschmar, T., Pelayo, M. A. F., Trijatmiko, K. R., Gabunada, L. F. M., Alam, R., Jimenez, R., Mendioro, M. S., Slamet-Loedin, I. H., Sreenivasulu, N., Bailey-Serres, J., Ismail, A. M., Mackill, D. J., & Septiningsih, E. M. (2015). A trehalose-6-phosphate phosphatase enhances anaerobic germination tolerance in rice. *Nat Plants*. https://doi.org/10.1038/nplants.2015.124

Kukal, S. S. (2004). Water-saving irrigation scheduling to rice (Oryza sativa) in indo-gangetic plains of India. In G. Huang, & L. S. Pereira (Eds.), *Land and water management: decision tools and practices* (vol. 1, pp. 83–87).

Kumar, K. A., & Rajitha, G. (2019). Alternate wetting and drying (AWD) irrigation-a smart water saving technology for rice: A review. *International Journal of Current Microbiology and Applied Sciences, 8*, 2561–2571.

Kumar, V., Ladha, J. K. (2011). *Direct seeding of rice. Recent developments and future research needs*.

Lampayan, R. M., Rejesus, R. M., Singleton, G. R., & Bouman, B. A. M. (2015). Adoption and economics of alternate wetting and drying water management for irrigated lowland rice. *Fields Crop Research, 170*, 95–108.

Lee, S., Jeong, H. J., Kim, S. A., Lee, J., Lou, G. M., & An, G. (2010). OsZIP5 is a plasma membrane zinc transporter in rice. *Plant Molecular Biology*. https://doi.org/10.1007/s11103-010-9637-0

Li, J., Li, Y., Yin, Z., Jiang, J., Zhang, M., Guo, X., Ye, Z., Zhao, Y., Xiong, H., & Zhang, Z. (2017). Os ASR 5 enhances drought tolerance through a stomatal closure pathway associated with ABA and H2O2 signalling in rice. *Plant Biotechnology Journal, 15*, 183–196.

Loan, L. C., Ngan, V. T. T., & Du, P. V. (2006). Preliminary evaluation on resistance genes against rice bacterial leaf blight in Can Tho Province-Vietnam. *Omonrice, 14*, 44–47.

Ma, J. F., Yamaji, N. (2006). Silicon uptake and accumulation in higher plants. *Trends in Plant Science*.

Mahender, A., Anandan, A., & Pradhan, S. K. (2015a). Early seedling vigour, an imperative trait for direct-seeded rice: An overview on physio-morphological parameters and molecular markers. *Planta*. https://doi.org/10.1007/s00425-015-2273-9

Mahender, A., Anandan, A., & Pradhan, S. K. (2015b). Early seedling vigour, an imperative trait for direct-seeded rice: An overview on physio-morphological parameters and molecular markers. *Planta, 241*, 1027–1050. https://doi.org/10.1007/s00425-015-2273-9

Mahender, A., Anandan, A., Pradhan, S. K., & Singh, O. N. (2018). Traits-related QTLs and genes and their potential applications in rice improvement under low phosphorus condition. *Archives of Agronomy and Soil Science, 64*, 449–464. https://doi.org/10.1080/03650340.2017.1373764

Majumdar, D. (2003). Methane and nitrous oxide emission from irrigated rice fields: Proposed mitigation strategies. *Current Science*, 1317–1326.

Mote, K., Rao, V. P., Ramulu, V., Kumar, K. A., Devi, M. U. (2021). Performance of rice (Oryza sativa (L.)) under AWD irrigation practice—A brief review. *Paddy and Water Environment*, 1–21.

Moya, P., Hong, L., Dawe, D., Chen, C. D. (2001). Comparative assessment of on-farm water saving irrigation techniques in the Zhanghe Irrigation System. In: *Water saving irrigation for rice: Proceedings of an International Workshop* (pp. 81–96). Wuhan, China.

Naveenkumar, R., Singh, V., Singh, P. K., Anandan, A. (2021). Pattern of genetic variation in rice (Oryza sativa L .) population for sheath blight resistance over the seasons Pattern of genetic variation in rice (Oryza sativa L .) population for sheath blight resistance over the seasons. 7–10. https://doi.org/10.31742/IJGPB.81.1.15

Nozoye, T., Nagasaka, S., Kobayashi, T., Takahashi, M., Sato, Y., Sato, Y., Uozumi, N., Nakanishi, H., & Nishizawa, N. K. (2011). Phytosiderophore efflux transporters are crucial for iron acquisition in graminaceous plants. *Journal of Biological Chemistry.* https://doi.org/10.1074/jbc.M110.180026

Okami, M., Kato, Y., & Yamagishi, J. (2011). Role of early vigor in adaptation of rice to water-saving aerobic culture: Effects of nitrogen utilization and leaf growth. *Fields Crop Research.* https://doi.org/10.1016/j.fcr.2011.06.013

Panda, S., Majhi, P. K., Anandan, A., Mahender, A., Veludandi, S., Bastia, D., Guttala, S. B., Singh, S. K., Saha, S., & Ali, J. (2021). Proofing direct-seeded rice with better root plasticity and architecture. *International Journal of Molecular Sciences, 22*, 6058.

Peng, S., Tang, Q., & Zou, Y. (2009). Current status and challenges of rice production in China. *Plant Production Sciences, 12*, 3–8.

Pinheiro, B. D. S., Castro, E. D. M. D., & Guimarães, C. M. (2006). Sustainability and profitability of aerobic rice production in Brazil. *Fields Crop Research, 97*, 34–42. https://doi.org/10.1016/j.fcr.2005.08.013

Pradhan, S. K., Nayak, D. K., Mohanty, S., Behera, L., Barik, S. R., Pandit, E., Lenka, S., & Anandan, A. (2015). Pyramiding of three bacterial blight resistance genes for broad-spectrum resistance in deepwater rice variety, Jalmagna. *Rice, 8*, 1–14.

Priyanka, S., Jitesh, B., & Babu, S. (2012). Aerobic rice, a new approach of rice cultivation. *International Journal of Research in BioSciences, 1*, 1–6.

Ramesh, S. A., Shin, R., Eide, D. J., & Schachtman, D. P. (2003). Differential metal selectivity and gene expression of two zinc transporters from rice. *Plant Physiology.* https://doi.org/10.1104/pp.103.026815

Rani, M., Satyanarayana, P. V., Jhansirani, P. (2017). Breeding strategies for lodging resistance in rice. *International Journal of Bio-resource and Stress Management, 8*, 895–903. https://doi.org/10.23910/ijbsm/2017.8.6.1793a

Rebetzke, G. J., Botwright, T. L., Moore, C. S., Richards, R. A., & Condon, A. G. (2004). Genotypic variation in specific leaf area for genetic improvement of early vigour in wheat. *Fields Crop Research, 88*, 179–189.

Ritchie, H., Roser, M. (2017). Water use and stress. *Our World in Data.*

Sagare, D. B., Abbai, R., Jain, A., Jayadevappa, P. K., Dixit, S., Singh, A. K., Challa, V., Alam, S., Singh, U. M., Yadav, S., Sandhu, N., Kabade, P. G., Singh, V. K., & Kumar, A. (2020). More and more of less and less: Is genomics-based breeding of dry direct-seeded rice (DDSR) varieties the need of hour? *Plant Biotechnology Journal, 18*, 2173–2186. https://doi.org/10.1111/pbi.13454

Sandhu, N., Jain, S., Kumar, A., Mehla, B. S., Jain, R. (2013). Genetic variation, linkage mapping of QTL and correlation studies for yield, root, and agronomic traits for aerobic adaptation. *BMC Genetics, 14.* https://doi.org/10.1186/1471-2156-14-104

Sandhu, N., & Kumar, A. (2016). Paper 16 traits for dry direct-seeded rice. *Regional: Development and dissemination of climate-resilient rice varieties for water-short areas of South Asia and Southeast Asia*, 378.

Sandhu, N., & Kumar, A. (2017). Bridging the rice yield gaps under drought: QTLs, genes, and their use in breeding programs. *Agronomy, 7*, 27.

Sandhu, N., Singh, A., Dixit, S., Sta Cruz, M. T., Maturan, P. C., Jain, R. K., & Kumar, A. (2014). Identification and mapping of stable QTL with main and epistasis effect on rice grain yield under upland drought stress. *BMC Genetics.* https://doi.org/10.1186/1471-2156-15-63

Sandhu, N., Subedi, S. R., Singh, V. K., Sinha, P., Kumar, S., Singh, S. P., Ghimire, S. K., Pandey, M., Yadaw, R. B., Varshney, R. K., & Kumar, A. (2019). Deciphering the genetic basis of root morphology, nutrient uptake, yield, and yield-related traits in rice under dry direct-seeded cultivation systems. *Science and Reports, 9*, 1–16. https://doi.org/10.1038/s41598-019-45770-3

Sandhu, N., Torres, R. O., Sta Cruz, M. T., Maturan, P. C., Jain, R., Kumar, A., & Henry, A. (2015). Traits and QTLs for development of dry direct-seeded rainfed rice varieties. *Journal of Experimental Botany*. https://doi.org/10.1093/jxb/eru413

Sandhu, N., Yadav, S., Catolos, M., Cruz, M. T. S., Kumar, A. (2021). Developing climate-resilient, direct-seeded, adapted multiple-stress-tolerant rice applying genomics-assisted breeding. *Frontiers in Plant Science, 12*.

Sani Haliru, B., Rafii, M. Y., Mazlan, N., Ramlee, S. I., Muhammad, I., Silas Akos, I., Halidu, J., Swaray, S., & Rini Bashir, Y. (2020). Recent strategies for detection and improvement of brown planthopper resistance genes in rice: A review. *Plants, 9*, 1202.

Sato, H., Matsumoto, K., Ota, C., Yamakawa, T., Kihara, J., & Mizobuchi, R. (2015). Confirming a major QTL and finding additional loci responsible for field resistance to brown spot (Bipolaris oryzae) in rice. *Breeding Science, 65*, 170–175.

Septiningsih, E. M., Ignacio, J. C. I., Sendon, P. M. D., Sanchez, D. L., Ismail, A. M., & Mackill, D. J. (2013). QTL mapping and confirmation for tolerance of anaerobic conditions during germination derived from the rice landrace Ma-Zhan Red. *Theoretical and Applied Genetics*. https://doi.org/10.1007/s00122-013-2057-1

Setter, T. L., Laureles, E. V., & Mazaredo, A. M. (1997). Lodging reduces yield of rice by self-shading and reductions in canopy photosynthesis. *Fields Crop Research*. https://doi.org/10.1016/S0378-4290(96)01058-1

Shashidhar, H. E. (2012). An eco-friendly aerobic rice BI 33 (Anagha). www Aerob rice

Shikari, A. B., Khanna, A., Krishnan, S. G., Singh, U. D., Rathour, R., Tonapi, V., Sharma, T. R., Nagarajan, M., Prabhu, K. V., & Singh, A. K. (2013). Molecular analysis and phenotypic validation of blast resistance genes Pita and Pita2 in landraces of rice (Oryza sativa L.). *Indian Journal of Genetics and Plant Breeding, 73*, 131–141.

Singh, R. K., Kota, S., & Flowers, T. J. (2021). Salt tolerance in rice: Seedling and reproductive stage QTL mapping come of age. *Theoretical and Applied Genetics, 134*, 3495–3533.

Singh, S., Ladha, J. K., Gupta, R. K., Bhushan, L., & Rao, A. N. (2008). Weed management in aerobic rice systems under varying establishment methods. *Crop Protection, 27*, 660–671. https://doi.org/10.1016/j.cropro.2007.09.012

Smith, S. E., Read, D. J. (2010). *Mycorrhizal symbiosis*. Academic press.

Sonoda, Y., Ikeda, A., Saiki, S., Von Wirén, N., Yamaya, T., & Yamaguchi, J. (2003). Distinct expression and function of three ammonium transporter genes (OsAMT1;1–1;3) in rice. *Plant and Cell Physiology*. https://doi.org/10.1093/pcp/pcg083

Surendran, U., Raja, P., Jayakumar, M., & Subramoniam, S. R. (2021). Use of efficient water saving techniques for production of rice in India under climate change scenario: A critical review. *Journal of Cleaner Production, 309*, 127272.

Tuong, T. P., & Bouman, B. A. M. (2003). Rice production in water scarce environments. In *Water productivity in agriculture: Limits and opportunities for improvement*. CABI Publications.

Win, E. P., Win, K. K., Bellingrath-Kimura, S. D., & Oo, A. Z. (2020). Greenhouse gas emissions, grain yield and water productivity: A paddy rice field case study based in Myanmar. *Greenhouse Gases: Science and Technology, 10*, 884–897.

Yadav, S., Singh, U. M., Naik, S. M., Venkateshwarlu, C., Ramayya, P. J., Raman, K. A., Sandhu, N., & Kumar, A. (2017). Molecular mapping of QTLs associated with lodging resistance in dry direct-seeded rice (Oryza sativa L.). *Frontiers in Plant Science, 8*, 1–10. https://doi.org/10.3389/fpls.2017.01431

Yang, J., Guo, Z., Luo, L., Gao, Q., Xiao, W., Wang, J., Wang, H., Chen, Z., & Guo, T. (2021). Identification of QTL and candidate genes involved in early seedling growth in rice via high-density genetic mapping and RNA-seq. *The Crop Journal, 9*, 360–371.

Yano, K., Ookawa, T., Aya, K., Ochiai, Y., Hirasawa, T., Ebitani, T., Takarada, T., Yano, M., Yamamoto, T., Fukuoka, S., Wu, J., Ando, T., Ordonio, R. L., Hirano, K., & Matsuoka, M. (2015). Isolation of a novel lodging resistance QTL gene involved in strigolactone signaling and its pyramiding with a QTL gene involved in another mechanism. *Molecular Plant*. https://doi.org/10.1016/j.molp.2014.10.009

Zhang, J., Li, G., Song, Y., Zhang, W., Yang, C., Wang, S., & Ding, Y. (2013). Lodging resistance of super-hybrid rice Y Liangyou 2 in two ecological regions. *Acta Agronomica Sinica, 39*, 682–692.

Zhang, Y., Tan, L., Zhu, Z., Yuan, L., Xie, D., & Sun, C. (2015). TOND1 confers tolerance to nitrogen deficiency in rice. *The Plant Journal, 81*, 367–376. https://doi.org/10.1111/tpj.12736

Zhao, D. L., Atlin, G. N., Bastiaans, L., & Spiertz, J. H. J. (2006). Comparing rice germplasm groups for growth, grain yield and weed-suppressive ability under aerobic soil conditions. *Weed Research, 46*(6), 444–452.

Zhao, Y., Hu, Y., Dai, M., Huang, L., & Zhou, D. X. (2009). The WUSCHEL-Related homeobox gene WOX11 is required to activate shoot-borne crown root development in rice. *The Plant Cell*. https://doi.org/10.1105/tpc.108.061655

Zhao, D., & Kumar, A. (2016). Paper 2 developing aerobic rice varieties at IRRI. *Regional: Development and dissemination of climate-resilient rice varieties for water-short areas of South Asia and Southeast Asia*, 29.

Zhu, C., Gore, M., Buckler, E. S., Yu, J. (2008). Status and prospects of association mapping in plants. *The Plant Genome, 1*.

Chapter 20
Crop Modelling for Climate Change Studies

Shivani Kothiyal, Sonam Sah, Aaradhana Chilwal, and R. N. Singh

Abstract Global climate change is likely to affect world's food production and poses a serious threat to the agricultural production system. With the increasing pressure of the growing world population agricultural systems needs to take adaptive measures to cater the need of the increasing food demand. Crop models are major tools available to assess the impacts of climate change on the agricultural production system and formulate adaptation measures for the same. This chapter covers the recent trends in greenhouse gases, rainfall, temperature, sea level and extreme weather events. The associated impacts of these changes on food crops were also explained briefly. We also discussed the basics of crop models in brief followed by their applications in climate change study including adaptation strategies. At last, we discussed the key limitations, challenges and uncertainties in the use of crop models in the agricultural production system.

Keywords Adaptation · Climate change · Crop model · Greenhouse gases

20.1 Introduction

Global food production growth superseded population growth over the second half of the twentieth century. However, the growth in food production is reaching a plateau and further increase in global food production with the existing technologies will be a difficult task. Further, the global demand for food has also risen over the past decades due to population escalation raising a question about future global food

S. Kothiyal (✉) · A. Chilwal
Punjab Agricultural University, Ludhiana, Punjab 141004, India
e-mail: shivani.kothiyal41418@gmail.com

S. Sah
G.B. Pant, University of Agriculture and Technology, Pantnagar, Uttarakhand 263145, India

R. N. Singh
ICAR-National Institute of Abiotic Stress Management, Baramati, Maharashtra 413115, India

H. Pathak et al. (eds.), *Climate Change Impacts on Soil-Plant-Atmosphere Continuum*,
Advances in Global Change Research 78,
https://doi.org/10.1007/978-981-99-7935-6_20

security. According to the recent report by the UN, the projected global population will be 9.7 billion by 2050 and this will cause the global food demand to rise by 59–98% by 2050. The challenges in meeting the forecasted demand for food are further aggravated by the constraints like climate change, urbanization, soil degradation and groundwater depletion. Climate change is one of the biggest threats to world's food security. It has both direct and indirect impacts on global food security. Direct impacts include changes in weather and environmental condition along with the occurrence of extreme weather events. Indirect impacts include the changes in demography and income of people, thereby bringing changes in food demands. In the past few decades, human-induced climate change has enhanced significantly which is evident from the increasing concentrations of greenhouse gases like Carbon dioxide (CO_2), Methane (CH_4) and Nitrous oxide (N_2O). This rise in GHGs leads to rising atmospheric temperature due to the greenhouse effect. Many weather parameters like rainfall, and temperature are changing drastically due to climate change (Sah et al., 2021; Singh et al., 2021c; Kaur et al., 2022).

Agriculture is the most vulnerable sector to climate change, especially in developing countries like India. Due to its high sensitivity to weather parameters and the dependency of the huge amount of population on agriculture, climate change can cause huge economic impacts. The rising temperatures, rainfall vagaries, and increasing pests and diseases impose serious threats to crop production (Bhatia et al., 2021; Singh et al., 2021a, 2021b, 2021c, 2021d, 2021e). Though the impact of climate change is positive in some low latitude areas, it is limited to a very small scale and cannot offset the negative impacts on the larger part of the world. Crop models provide a handy tool for the assessment of crop response to the various key drivers like weather variables, agricultural inputs, management practices etc. This allows the assessment of climate change impact and planning adaptation and mitigation strategies accordingly for the same (Sah et al., 2020). The development of the crop model dates back to the 1960s. With the passage of almost six decades, currently, there is a vast range of crop models depending on the crop type, research area, complexity, target questions, and type of study. Earlier, crop models were designed mainly for a small spatial scale and were crop specific with fewer basic input and management options. Therefore, climate change studies using crop models were limited to small areas. However, lately, the emphasis has shifted to make models more diverse in terms of crops and their management and for large area applications. Crop modelling has become an increasingly important and integral part of climate change studies.

20.2 Climate Change Trend

Not a single region in the world remains untouched by the vagaries of climate change which can be in any form like floods, wildfires, earthquakes, cyclones and many more. Increased frequency of droughts, heat waves and floods have already threatened the livelihood and food security of billions of people. Since 1961, the growth of crop productivity in Africa has shrunk by one-third. Wildfires are spreading without control leading to irreparable losses to the landscape and turning people homeless. Threats to water and vector-borne diseases have also increased as the increasing temperatures form the congenial environment for them. Various species are getting extinct and many are on the verge of getting extinct like the golden toad and Bramble Cays Melomys, which now don't exist and mass die-off is a common incident under climate change. Species are shifting and habitats are changing to avoid the climate change havoc but the projected trends give no relief until humans understand their responsibility to save the planet. Not only animals but many plant species are getting extinct and major shifts in their habitats have also been observed. Following are the current trends of changing climate.

20.2.1 Greenhouse Gases

The concentration of greenhouse gases has reached new levels which are 413.2 parts per million (ppm) CO_2 (carbon dioxide), 1889 parts per billion (ppb) CH_4 (methane) and 333.2 ppb N_2O (nitrous oxide) in 2020 with respect to the pre-industrial (1750) levels. This increase continued in 2021 and is predicted to further rise in 2022.

20.2.2 Temperatures

By 2021, the global mean temperature was above the 1850–1900 average by 1.09 °C (NOAA, 2021). Recently, an analysis of six datasets by WMO placed 2021 as the sixth warmest year globally and 2015–2021 were indicated as the seven warmest years on record (WMO, 2022). However, the influence of moderate La Niña during the start of the year 2021 resulted it to be less warm in comparison to recent years as its cooling phenomenon influenced the weather and climate at the regional level leading to lesser temperature rise globally (Matt MacGrath, 2020).

20.2.3 Ocean

The oceans store around 90% of the heat accumulation occurring in the Earth system which is measured using Ocean Heat Content. A study by Schuckmann et al. (2020) reported that the ocean depth up to 2000 m reached a new record in 2019 as it continued to warm and an analysis of seven global datasets indicated 2020 to exceed these warm records. The past two decades indicate a strong upsurge in the rate of ocean warming which will continue in the future as well. During the period from January to April 2021, the Laptev and Beaufort Sea faced "extreme" marine heatwaves. One strong' Marine Heatwave was experienced by most of the oceans during 2021 except the eastern equatorial Pacific Ocean (due to La Niña) and much of the Southern Ocean. In the past 40 years, open ocean surface pH has declined globally and is now at its lowest in the past 26,000 years which is due to the absorption of 23% of the annual anthropogenic CO_2 emissions that occur in the atmosphere (UNCC & WMO, 2021). Reduction in the pH also reduces the capacity of oceans to absorb CO_2 from the atmosphere.

20.2.4 Sea Level

The global mean sea level rise occurs due to ocean warming and ice melting which has been indicated as 2.1 mm per year between 1993 and 2002 (Beckley et al., 2017) and 4.4 mm per year between 2013 and 2021 (NASA, 2021). These measurements have been made since the early 1990s using altimeter satellites with high precision.

20.2.5 Glaciers and Ice Sheets

The loss of glaciers in North America has accelerated over the last two decades and it has nearly doubled for the period 2015–2019 in comparison to 2000–2004 (Pelto & WGMS Network, 2020). Western North America suffered an unusually warm, dry summer in 2021 which further led to extensive loss of mountain glaciers (EPA, 2022). The warm humid air conditions in the Greenland Ice Sheet in mid-August led to high temperatures and runoff above normal. Also, the Summit station (highest point in Greenland Ice Sheet 3,216 m) experienced rainfall for several hours and the air temperatures in the region remained above freezing for nine hours. Such continuous rainfall has never been reported at this point and such melting conditions have also occurred the third time (NSIDC, 2021). The twentieth century recorded a similar melting event once.

20.2.6 Precipitation

Extreme rainfall has been recorded in various parts of the world which have led to extreme events. The Henan province in China experienced extreme rainfall from 17 to 21 July while the city of Zhengzhou gave a Chinese national record with a recorded rainfall of 201.9 mm in one hour and 720 mm for the entire occurrence on 20 July which was more than the annual average (Wei et al., 2022). Flash floods are a common result of extreme rainfall events and these contribute to fatalities as well as economic losses. In mid-July, western Germany and eastern Belgium recorded their most severe flooding event with 100–150 mm rainfall over the saturated ground leading to floods, landslides and 200 demises (World Weather Attribution, 2021).

The second outcome of extreme rainfall events is drought which was observed significantly in most of the regions of subtropical South America (southern Brazil, Paraguay, Uruguay and northern Argentina) where the rainfall was below average. Brazil faced ruthless weather impact on agriculture where the drought accompanied by a cold outbreak at end of July damaged several coffee farms and the low levels of rivers resulted in poor hydroelectricity generation and disruption in river transportation (UNCC & WMO, 2021). The southwestern United States experienced below-normal rainfall indicating January 2020 to August 2021 to be the driest months on record (Allison Chinchar, 2021).

20.2.7 Extreme Weather Events as Per Reports by UNCC and WMO (2021)

1. Many heatwave events affected the western region of North America where station records were broken with temperatures rising by 4–6 °C during June and July months and Lytton city in south-central British Columbia experienced devastating fires due to temperature rise up to 49.6 °C.
2. The hottest summer was recorded in the United States when California reached an extremely high temperature (54.4 °C) on the 9th of July which is the highest since the 1930s.
3. One of the major wildfires recorded was in northern California, The Dixie fire which destroyed 390,000 hectares of land.
4. Many parts of Algeria, southern Turkey and Greece experienced the worst wildfires due to rising temperatures.
5. Mid-February in northern Mexico and the central United States faced unusual cold outbreaks with Texas experiencing extremely low temperatures since 1989.

20.3 Response of Crops to Climate Change

The future scenarios indicate and support the current climate changes across the globe and their continuation in the future as well. Changing trends of solar radiation, temperature, precipitation, ozone and carbon dioxide content for the next 30 years under the continued scenario holds a challenge for crop production. These challenges need to be critically studied and understood for developing a climate resilient cropping system which can stand the consequences of climate change impact. Depending upon crop type, response to CO_2, temperature and precipitation is highly variable. In addition to this, local/ regional differences in predicted climate are also expected which will further deteriorate the situation in which the crop responses will be more complicated to understand and manage. The various interactions of climate change with biotic and abiotic stresses are already known but the challenge lies in studying and understanding these responses under the climate change scenario, predicted for the future.

20.3.1 Temperature

Throughout the life cycle, crop responds differently to temperature during different growth stages. Crop plants have a specific range of temperatures within which they perform well and these include maximum, minimum and optimum temperatures. Within the range of maximum and minimum temperatures growth of the crops takes place while at optimum temperature plant growth takes place at its fastest pace. The growth rate of crops is largely affected by temperature which slows down the growth of plants exposed to temperatures above optimum and ceases growth in plants exposed to maximum temperatures and higher. Vegetative development (rate of formation of nodes and appearance of leaves) accelerates as temperatures reach the optimum level of a particular species and these may be higher than that during reproductive development. However, understanding the difference between the plant and air temperatures is important as plant temperatures could be warmer under water stress and cooler underwater adequacy in comparison to air temperatures. Annual plants respond variably to increased temperatures and their exposure to higher temperatures results in faster growth but this is not equivalent to maximum production in these crops. Faster development in annual plants leads to a shorter life-cycle of the plants which reduces the duration of the reproductive stage and reduce cumulative light interception during the growth of these crops causing declination in yield potential. When the temperatures fall below or rise above the threshold for a particular crop species, then the yield might be affected thus the life cycle of a crop is determined largely by the location and temperature of that cultivar. The reproductive stage is the most sensitive stage to temperature stress as a high-temperature impact can be observed on the viability of pollen, fertilization and fruit/grain formation. An 8.3% yield reduction in maize, 1.3% in soybean and 5.4% in wheat were observed by Lobell and Field (2007) when temperatures were risen by 1 °C during the reproductive stage. Welch et al. (2010) revealed the significant impact of solar radiation and temperature on the rice yield where the assessment showed that increasing minimum temperature decreases the yield and increasing maximum temperature increases the yield only when it reaches the optimum threshold of maximum temperature. Prasad et al. (2006) found the grain filling duration to reduce with rising temperatures. Many other researchers reported also reported a decline in crop yield due to rising temperature (Singh et al., 2021b; Kothiyal et al., 2022a). The projected temperature changes and rainfall variability in the Punjab state indicates decline in yield of major cereal crops i.e. wheat, maize and rice under different emission scenarios as reported by Prabhjyot-Kaur et al. (2023).

20.3.2 Ozone

Maximum attention has been directed towards an increase in CO_2 content under climate change scenarios and therefore, less attention has been given to an increase in O_3 level in the troposphere. However, the past 50 years have observed a significant rise in O_3 level and is predicted to rise continuously in the future years. These increasing O_3 levels do impact crop production in different ways as a consequence of which efforts in direction of increasing ozone tolerance among plants will soon become important in order to maintain the competitiveness of crops with climate change. How O_3 concentrations will vary in coming years is now linked with IPCC scenarios, so the impact of varying O_3 levels on agricultural crop production is given due importance and can be considered as important as the other components in climate change like temperature, CO_2, precipitation etc. The foremost agricultural areas in the Northern Hemisphere as per model predictions done based on economic development and emission control planning show a significant rise in the annual mean surface concentrations of O_3 (Dentener et al., 2005). The last 100 years indicate a steady rise in the daytime ozone levels, in response to which photosynthesis, dry matter, and yield of crops have decreased (Morgan et al., 2004). Such crop response is attributed to the fact that unremitting levels of O_3 cause a reduction in photosynthetic capacity and leaf area of plants as these levels hamper the carbon uptake capacity in plants (Singh et al., 2017). For example, soybean plants exposed to persistently high levels of O_3 caused the plants to be shorter with the reduction in the yield contributing characters like dry mass, pods and seeds and these damages increased with the age of the seedling which shows the cumulative effect of O_3 damage over time. The exposure to current ambient levels of O_3 shows small changes in the stomatal conductance of field crops till 2030 but its variation throughout the globe and its interaction with other climate variables need to be studied for the development of better adaptive strategies to avoid potential yield losses. Thus, these interactions need critical study for a better fate of our crops.

20.3.3 Carbon Dioxide

A steady increase has been observed in the carbon dioxide levels over the past 50 years and is expected to increase over the next 30 years too. The current level of CO_2 which is around 387 μmol mol^{-1} is expected to reach up to 450 μmol mol^{-1} by 2050 and this rise is taking place at a rapid rate (Karl et al., 2009). The impact of changing CO_2 concentrations can be studied with the combination of enclosed CO_2 chambers and free-air carbon dioxide enrichment (FACE) where the CO_2 can be maintained above ambient concentrations (Singh et al., 2021a, 2021b, 2021c, 2021d, 2021e). The impact of CO_2 varies among different crop species where the plant growth may increase or decrease with the increasing CO_2 levels. The crops consist of two pathways, C3 and C4 which respond accordingly to changing levels of CO_2 and the

C4 plants like maize and sorghum have smaller responses than the C3 plants. This is due to the higher efficiency of C4 plants in capturing the CO_2 in the atmosphere in comparison to C3 plants. However, an increase in CO_2 levels has a positive impact on all the crop plants either C3 or C4. Several studies have shown that weeds amongst the various plant species show a strong response to increasing CO_2 (Ziska, 2004). The rice or wheat crop individually might show a positive response to increasing CO_2 but the positive response of weed species can be alarming as this would lead to competition and crop yield losses. If the rising of CO_2 continues, then the agricultural systems where weed control is not practised will suffer yield reductions. The crop species respond differently to changing CO_2 concentrations as these interactions with other soil and atmospheric components also impact the crop yield. The future projections show that variability in precipitation and drought conditions along with rising temperatures would offset the positive impact of increasing CO_2 levels on the growth of plants.

20.3.4 Solar Radiation (Light)

The most important environmental factor influencing the growth and development of plants is light. As the light intensity varies, variations in the anatomy, morphology and growth of plants take place which indirectly affects the occurrence of weeds, pests and diseases. The translocation of assimilates takes place through phloem and the rate of this translocation is determined by the light intensity which influences the photosynthesis rate. The increase in light intensity increases the net photosynthesis which results in increased movement of foliar-applied systemic insecticides, pesticides and herbicides that play an important role in plant health maintenance. Along with maintaining plant health, light plays a major role in the development of seeds and leaves. Higher light intensities keep the stomata open, thus, improving CO_2 penetration through tissue and loss of water through transpiration. Excessive light intensity becomes harmful for the plants under which plants acquire adaptation measures such as the increase in the leaf thickness while at low intensity, plants produce thinner leaves with greater height and leaf area so that enough light is captured to meet the photosynthetic demand of the plants.

20.3.5 Rainfall

Precipitation is the major climatic factor influencing the growth and yield of crops as the future predictions on variations in rainfall patterns hold great uncertainty. The future projections show a rise in temperatures and CO_2 levels but the distribution and intensity of rainfall are unpredictable and these variations are a major concern for crop production (Singh et al., 2021a). The frequency of occurrence of high precipitation affects the infiltration rate and availability of water as these cause flash floods and high surface runoff. The growing season of a crop requires an adequate and consistent supply of moisture for fruitful completion of the crop life cycle as moisture plays a key role in the seedling establishment which is the primary step for crop growth. Drought is a primary limiting factor in the growth and productivity of crop plants for regions where the projected summer temperatures are expected to increase and rainfall to decrease (Mishra & Singh, 2010). The frequency and intensity of rainfall vary under climate change leading to the occurrence of extreme weather events like floods and droughts, and crops adaptive to such events will have a competitive edge in the survival of the fittest. Vagaries in patterns of rainfall and availability of water impact the plant size as well as the production and dispersal of seeds (Rodenburg et al., 2011). Vegetation shifts are a major occurrence under drought conditions where the composition of crop and weed species changes like cheatgrass and yellow thistle which outgrows other crop species as cheatgrass completes its life cycle in available soil moisture and yellow thistle star competes for moisture with the deeper root system (Hatfield et al., 2011). C3 and C4 crops' response varies with the water stress as the increase in CO_2 concentrations decreases the stomatal conductance leading to the lower rate of transpiration and 70–100% higher water use efficiency in both species. The higher water use efficiency of C4 plants helps them to perform better under arid conditions as they can produce higher biomass, more root growth and seeds than the C3 plants.

The altering crop characteristics and growing season under climate change can impact the agricultural pest and weed population, thus, making their management difficult. The perennial weeds will flourish under global warming conditions which will cause warmer nights and winters and these conditions will also cause expansion of the habitable temperature ranges for agricultural pests. Climate change impact varies over the crop and weeds as any adaptation for better crop management will also influence the weed flora and its distribution. Therefore, the response of crops to climate change is a multifaceted subject and needs a thorough study of every aspect which then only can support better planning and mitigation for future climate change trends.

20.4 Crop Models and Components

In simplest terms, crop simulation models are the mathematical representation of the real-world system. It is a tool to assist researchers and scientists to find solutions to complex farming problems through an integrated approach. The outputs simulated by crop models are based on soil-plant-atmosphere dynamics. Based on the scale of data and information available, models are designed for different levels of complexity of a system. The models may range from a simple multivariate regression equation (empirical model) to process-based mechanistic models developed for understanding crop physiological interaction. Traditional multi-year field trials are conducted on relatively small plots or fields in one location or multi-locations. The recommendations from these trials are communicated to the farmers for their on-field applications. These recommendations may work well for developed countries where farmers have sufficient resources and inputs to manage their farms and the farms are relatively large and homogeneous. However, for most developing countries where farmers lack resources and have small or marginal farmland holdings with non-uniform growing conditions, these recommendations fail to produce desired results. Therefore, crop modelling plays a very important role in addressing these problems as it gives outputs which are very specific to the farmland, crop genotype, crop environment, applied inputs and management conditions.

Risks and uncertainties of varying degrees are inevitable components of agriculture. They play a very important role in agriculture management, decision making and policy planning. The large-scale studies involving risk assessment associated with weather aberrations on crop production can be modelled using the historical weather data records. Similar assessment, if done with traditional field experimentation, would require a huge amount of time and money making it virtually impossible. Therefore, crop models are considered an important tool for policy planners at national and international levels.

20.4.1 Important Applications of Crop Models in Agriculture:

Crop simulation models have been used in agriculture for various purposes like.

- Identifying sensitive parameters for crop growth or response to climate change.
- Evaluation of the effect of varying rates of natural inputs like water, light, temperature, CO_2, O_3 etc. on crop growth, development and yield.
- Study of the effect of different biotic and abiotic stressors on the crop.
- Determination of best variety/cultivar for specific conditions.
- Irrigation scheduling.
- Optimization of agricultural inputs (irrigation, fertilizer, ameliorants) and management practices (tillage, date of sowing/planting, plant density) for a particular environment.
- Yield gap analysis: Potential yields can be estimated using crop models. Comparison of potential with attainable and actual yield can also be done to find out the yield limiting and yield-reducing factors, respectively in the region.
- Evaluation of the effect of variable weather conditions on the crop.
- Genetic improvement of the crop
  - Identification of desirable genetic traits in plants.
  - Partitioning gene × environment interactions for better sensitivity analysis of genetic traits on crop yield.
  - Improvement of genotypes for a predefined target environment and Characterization of the target environment for different germplasms
  - Assessing the response of a novel genotype to a particular environment.

- Crop modelling along with remote sensing can also be used for crop yield prediction.
- Supporting policy-makers and investment decisions.

20.5 Types of Crop Models

20.5.1 On the Basis of Purpose, the Models Can Be Classified into the Following Groups

1. **Empirical-Statistical model:** These models express relationships between dependent (crop yield and yield components) and independent variable (weather parameters) using statistical techniques like Multivariate regression equations, and correlation.
2. **Mechanistic model:** These models along with predicting the relationship between the variables, also explain the mechanism underlying the processes.
3. **Stochastic model:** In these models, different outputs are given for each set of inputs and with each output, a probability element is attached.
4. **Deterministic model:** These models have distinct genotypic coefficients which are used to predict the precise value of the dependent variable.
5. **Static models:** Both the variables, dependent and independent remain constant over a given period because these models do not consider time as a variable.
6. **Dynamic models:** Time is considered as a variable.
7. **Simulation model:** Crop growth, development and yield are considered a function of crop, soil, weather and management conditions. Differential equations are used to calculate rate and state variables over a period of time.

20.5.2 Steps in Crop Modelling

There are 6 major steps in crop modelling:

1. Defining goals
2. Defining system and boundaries
3. Define key variables: State, rate and auxiliary variables
4. Calibration: It is the most important step in developing a crop model
5. Validation
6. Sensitivity analysis.

20.6 Data for Modelling

The crop models run on certain input data which is known as Minimum Data Set (MDS). MDS is the minimum amount of data required to run a crop simulation model and evaluate simulation outcomes. It is also used for model development and improvement. The concept of MDS was introduced during the workshop held at International Crops Research Institute for the Semi-Arid Tropics (ICRISAT), India. The following data are considered MDS in order to run a crop model.

- **Soil data:**
  - Soil classification, surface slope, colour, permeability, and drainage class.
  - Soil profile data- upper and lower horizon depths (cm), percentage sand, silt, and clay content, 1/3 bar bulk density, organic carbon, pH in water, aluminium saturation, and root abundance information.
- **Weather Data:**
  - Latitude and longitude of the weather station,
  - Daily incoming solar radiation (MJ/m^2-day),
  - Maximum and minimum daily air temperature (°C), and
  - Daily rainfall (mm).
- **Genotype data—**crop variety
- **Crop management data**—planting date, planting density, row spacing, planting depth, irrigation, and fertilizer practices

20.6.1 Sensitivity Analysis

The sensitivity analysis is a procedure to determine the sensitivity of the model output towards the changes done in the input parameters which helps in understanding the model behaviour. The procedure involves making small alterations in the input parameters and if these alterations cause a relatively larger change in the output of the model, then the output is sensitive to that particular input parameter. This infers that this parameter needs to be determined with accuracy as it affects the output.

20.6.2 Calibration of the Model

Calibration of the model is the process of adjusting the model parameters according to the local conditions of the region and it is necessary for determining the genetic coefficients for new cultivars being used in the study. The model consists of cultivar-specific coefficients (CSPs), also known as genetic coefficients which depict the different stages of crop growth. These CSPs are specific to the crop cultivar as well as the local climate and they need to be calibrated individually. The calibration process in the model determines its truthfulness and it requires some key information including planting dates, anthesis, maturity dates and grain yield of that particular cultivar for which calibration is being done. Calibration of the time for crop phenological stages is possible by altering the coefficients, characterizing the juvenile phase duration (P1), the sensitivity of the photoperiod (P2) and reproductive phase duration (P5). The cultivar coefficients dealing with vegetative growth and phenology of the crop (P1, P2, P5 and PHINT) and those with the grain number per cob and grain filling (G1 and G2) are determined using several iterations till they give close simulations to that observed on the field.

20.6.3 Validation of the Model

The validation of the model helps in determining the usefulness and its applicability for further studies. The validation procedure is evaluated using various statistical measures that include the coefficient of determination (R^2), Root Mean Square Error (RMSE), Normalized Root Mean Square Error (NRMSE), index of agreement (d-stat) and Nash–Sutcliffe model efficiency (EF). The validation of the model is very important when it is being applied for further climate change studies where the crop is going to perform under different environments as accurate results can only be obtained using a well-validated model.

The simulated results are assured to be reliable with the accurate calibration and validation process held for the crop models as reported by many researchers (Kothiyal et al., 2021; Prabhjyot-Kaur et al., 2022). Calibration of crop models is an adjustment of the genetic coefficients to reduce the difference between the observed and simulated crop parameters. The validation process is an assessment of the model ability to simulate accurate results by comparing the simulated and observed crop parameters.

20.7 Application of Crop Simulation Models in Climate Change Studies

Crop simulation models (CSMs) are widely used as strategic research and decision support tools in the area of crop production and resource management. These are largely involved in studying the climate change impact on crops under rising greenhouse gases (Rosenzweig & Liverman, 1992; White et al., 2011). Models play a significant role in resource optimization and managing farming practices under future climate scenarios. Earlier the impact of the environment on crop growth and yield was determined using special structures or chambers but these are expensive and are not affordable by developing countries which are the major victim of climate change, so, CSMs were developed to study these impacts. These models help in developing the appropriate conditions for monitoring the phenology, growth and yield of various crops. Several models have been developed for the crop modelling studies as different crops require different models which include Decision Support System for Agrotechnology Transfer (DSSAT (Jones et al., 2003)) covering 42 crops, Agricultural Production Systems Simulator (APSIM (Keating et al., 2003)) including a diverse range of crops, pasture and trees, AquaCrop (Vanuytrecht et al., 2014) is used for simulating crop yield response to water and is suitable for conditions where water is the crucial limiting factor and STICS (Brisson et al., 2003) is a soil–plant simulation model efficient in simulating agricultural and environmental variables using daily climatic data. These are some of the widely used models which help in discovering the yield gaps in various crops and how these can be resolved using certain agronomic practices. Climate change is visible with the increasing temperatures and high variability in rainfall, so, all these events indicate a major blow on crop production. Thus, optimization of farm resources for better crop production in future is necessary. Crop models help in creating such environments where one can analyse the growth and yield of various crops.

The early CSMs were developed in the 1980s which included ARCWHEAT1 (Porter, 1984) which doesn't take into account the water or nutrient limitations, Dutch models Simple and Universal Crop growth Simulator: SUCROS (Goudriaan & van Laar, 1994), a mechanistic CSM taking in account the environmental conditions while simulating the potential as well as water stressed growth in crops, and five models from the ARS Wheat Yield Project where CERES-Wheat (Ritchie et al. 1985) and Winter Wheat (Baker et al., 1985) were the prominent ones. Following these in the 1990s, various crop models were merged into a common crop modelling platform which was DSSAT (Jones et al., 2003), APSIM (Keating et al., 2003), CropSyst (Stöckle et al., 2003), STICS (Brisson et al., 2003) and EPIC (Kiniry et al., 1995). The advent of the 2000s introduced new crop simulation models including the SIRIUS (Martre et al., 2006), Wheat Grow (Pan et al., 2006) and Rice Grow model (Zhu et al., 2009). A vast number of crop models have been developed over the last 10 years and these are variable in their approach and complexity. Some important crop simulation models have been given in Table 20.1.

Till now a significant number of crop models with wide applications have been developed and all of these operate on a daily time step (Jamieson et al., 2010).

Model, in general, is an imitation of the real system which improves our understanding of the system, its behaviour, functioning and how it would react to the changing conditions. With time, these models have evolved and they act as a practical tool in variable fields for improvement of their systems by scientists, engineers, planners, managers, and growers (Hoogenboom et al., 1992). Testing of model simulations has been done under varied climatic conditions by a number of scientists like under monsoonal (Liu et al., 2013), semiarid tropical (Carberry et al., 1989), subhumid, subtropical (Behera and Panda, 2009), Mediterranean (Hasegawa et al., 2000), cold winter, dry summer (Thaler et al., 2012) and humid temperate (Otegui et al., 1996). Extensive studies on crop models under varied irrigation and fertilizer

Table 20.1 Some important crop simulation models

S. No	Crop simulation model	Purpose	Scientist involved	Year
1	CROPWAT	Irrigation scheduling through calculation of water requirements	Martin Smitt	1992
2	CANEGRO	Sugarcane potential and water stress conditions	Inman-Bamber	1994
3	CropSyst	Wheat and other crops	Stockle et al.	1994
4	WOFOST	Wheat and maize, water and nutrient	Supit et al.	1994
5	EPIC	Erosion productivity impact calculator	Williams et al.	1984
6	COMAX	Cotton	Lemmon	1986
7	GWN	General weed model in row crop	Wiles et al.	1996
8	SALUS	Crop residue management strategies	Basso et al.	2001
9	HERMES	Nitrogen dynamics in agricultural plant-soil-systems	Kersebaum	2007
10	LPJml	Vegetation composition and land–atmosphere exchange flows of carbon and water	Bondeau et al.	2007
11	AquaCrop	Assess the effect of environment and management on crop production	Steduto et al.	2009
12	IRRIGATE	Irrigation scheduling model	Merot and Bergez	2010
13	MONICA	Nitrogen and carbon dynamics in agro-ecosystems	Nendel et al.	2011
14	SIMDualKc	Computation of water balance and irrigation scheduling	Rosa et al.	2011
15	SIMPLE	Generic model covering simulation of growth and yield of 14 crops	Zhao et al.	2019
16	GreenLight	Greenhouse model with supplemental lighting	Katzin et al.	2020

amounts, timing and method of application, sowing dates and other altered management practices has been conducted in different parts of the world (Basso et al., 2016).

20.8 Adaptation Strategies Under Climate Change Scenario

The trends and predictions for the future depict their effect on crop production, thus for the social and economic stability of the world, there is a requirement for proper adaptation and mitigation strategies so that efficient management of crops is possible. A large number of studies have been performed to help farmers recognize the impact of climate change and how management strategies can be planned for better crop production. The current farming conditions and practices won't support the crops with the wavering temperatures and rainfall as till then the soil would be degraded and soil reclamation techniques would be required. Similarly, the phenology of the crops would also be affected which can be maintained with shifts in sowing time, adjustment of fertilizer and irrigation application and selection of the suitable cultivar.

The works of scientists over the world though show an increase with shifts in current farm management practices but this is not true for all crops. Some species are predicted to get extinct under the scorching Earth conditions which is an alarm for the people to save their planet and the human race. Proper planning of actions and policies can help in mitigating the climate change impact globally. At various places, scientists are experimenting with plants that can survive climate change and can bear extreme weather conditions. As reported by Jenny Morber (2019), the stifling heat of southern Israel supported the rows of Salicornia which are commonly known as sea beans and these plants performed well in more sandy and salty conditions which are not supportive of normal plants. These are turning out to be a great alternative to those species which might get extinct in near future and are also being exported by the local farmers for their living. These sea beans might become climate saviours and they are being extensively experimented with extreme conditions as well as what methods can increase their production in the dry coastal regions of Turkey, India, Israel and the United Arab Emirates. Dan Charles (2018) reported the performance of various crops under varied temperature and rainfall conditions as a warning for the human race and Californian farmers in future. The extreme temperatures are a major drawback for horticultural crops requiring chilling temperatures which have already seen a major shift but may not be able to survive the extreme temperatures in future. Similarly, coffee too might not be able to stand against the high temperature and variable rainfall. The maize crop also is predicted to suffer a high night temperature and wavering rainfall with storms which will affect the corn yield badly thus, replacing them with other suitable crops.

20.8.1 Local Adaptation Measures

The climate change scenario has led the farmers to adapt to local measures for better performance of their crops and yield like a study by Kaboré and Reij (2004) and Sawadogo (2011) in Burkino Faso showed farmers using low-cost crop management strategies like stone bunds and following water harvesting techniques like Zaï at micro-level and Demi-Lune. So, local adaptation measures are a great alternative to changing climate conditions as these vary from region to region but there exist some regions and countries where the capacity to combat climate change events is not sufficient like sub-Saharan African regions. These regions require attention as these are already drought-hit countries and the researchers require to select crop management strategies that can be supported by these environmental conditions. These measures for soil conservation and water management might help them in their food security as well as adaptation to climate change. Local adaptation measures are useful as well as environment friendly but may not be applicable for many parts of the world because the climatic and local conditions vary from country to country, thus, these can't be applicable on a large scale. Also, many countries like India have a diverse climatic condition where one region might be affected by severe drought while the other by extreme rains, thus, these require strong adaptation measures.

This is where the crop simulation models come into action as the environmental modifications, as well as the modification of farm management practices, helps in determining the suitable conditions as well as the alternatives under the changing scenario.

The impact of climate change has been recently investigated on a large scale and many strategies have been analysed by scientists for plant adaptation to climate variations. The farmers have started adopting approaches such as shifts in the planting and harvesting time of crops, practising crop rotation, selection of crops with shorter life cycles and those resistant to thermal stress or droughts and variation in irrigation techniques along with suitable cropping schemes. These approaches under climate stress conditions can help in better adaptability and food security. The importance of fertilizer doses in crop plants is unavoidable as they act as a substantial energy source for the plants and also help in maintaining the soil fertility and increasing productivity (Raza et al., 2019).

20.8.2 Shifts in Sowing Time

This is a major adaptation strategy suggested by the researchers according to the various studies done using crop models. The assessed rise in crop yield observed under this management strategy against the current was high. However, the combined approach of sowing dates and other management practices can help combat the climate change impact worldwide. These other management strategies can be the optimization of planting dates according to the climate forecasts for the region. These predictions can also help in selecting the suitable cultivar as well as planning the labour for field preparation. Almost all the research studies concluded that delayed sowing is the best measure for maize planting as this could help in avoiding the thermal stress during the growth period of the crop. However, the optimized sowing dates responded varyingly in different countries around the world according to the predicted temperature and rainfall. Haris and Chhabra (2015) indicated advanced sowing (last week of April for the hybrids, the first week of May for early composite variety and the first week of May for local variety) in combination with single irrigation to be helpful in certain regions of Bihar during all the future periods and all varieties were found to perform well at the split application of 90 kg N ha^{-1} (60 kg N ha^{-1} at sowing and 30 kg N ha^{-1} at knee high stage). The adverse impact of climate variation in Central India could be evaded with the delayed sowing which would cause the crop water requirement to coincide with the peak of seasonal rainfall, thus, preventing drought during the most sensitive stage of the crops (Patidar et al., 2020). A study on two important maize hybrids of Punjab through iterative combinations of sowing period revealed that shift in sowing dates i.e. early June in Ballowal Saunkhri, mid- to end-June in Ludhiana and Amritsar and end-May to mid-June in Patiala was an efficient method to nullify the negative impact of climate change under two extreme climate scenarios. The study also reported suitable agroclimatic zones for maize cultivation in the state (Kothiyal et al., 2023a).

20.8.3 Variation in Fertilizer and Irrigation Application

DSSAT CERES-Maize model was used in sandy soils of Florida to analyse the best management practices in sweet corn for nitrogen and irrigation application and found that increasing the nitrogen application did not improve the yield because nitrogen leaching was the most prevalent event in sandy soils. However, the model evaluated the nitrogen application and found that a small application of nitrogen could help in fixing the leaching problem (He et al., 2012). The model provides useful interactions between the crop, soil, weather and field management strategies but the results obtained depend on many model assumptions so the application of results sometimes becomes difficult. Fertilizer application though can help in increasing the socio-economic stability of the world by providing sufficient food but still, the energy required to produce it and the waste generated by it is unavoidable. So, farmers can go for replacements like legume crops, not completely but to some extent for a better future. A study revealed the reduced application of farm yard manure could be beneficial in enhancing the yield in northeastern region of India (Deb et al., 2015). A multi-model approach (APSIM and DSSAT CERES-Maize model) in the rainfed region of Tamil Nadu indicated earlier sowing date, supplemental irrigation and no additional fertilizer helped perform maize well with a median increase of 20% (McDermid et al., 2016). Shift in sowing date to second week of June coupled with nitrogen @ 165 kg ha^{-1} and 185 kg ha^{-1} was found to be an adaptable viable option under future scenarios for sustainable maize cultivation in Punjab state (Kothiyal et al., 2023b).

20.8.4 Selection of Cultivars and Breeding for New Cultivars Adaptable to Climate Change

The entire process for development and release of variety for the breeders is a complex process which may take up to 5–15 years but the modelling approach can use a combination of different environments to understand the performance of a particular variety. Crop modelling integrates different components of agroecosystems and can be used for conducting multi-location field experiments which help in understanding the G × E combination (genotype by environment) (Oteng-Darko et al., 2013). The selection of a suitable cultivar for a region is very important under current conditions as well as future projections. Amongst the many adaptation measures, Lin et al. (2017) used the growing of late-maturing cultivar with a long growth period for improvement of maize yield for the farmers. Some cultivars are heat tolerant and have the minimum effect of climate change, selection of such varieties can help combat climate change. Plant breeding is a dynamic technique which helps in the betterment of crop cultivars and guarantees the food security and safety of crop plants under adverse weather conditions. It also helps in developing resistant cultivars which can escape stress during crucial stages of their life cycle (Blum, 2018).

20.8.5 Crop Modelling for Climate Change Impact Assessment

Crop modelling is a very easy approach to understanding and analysing the climate change impact on various crops. The models help us know the crop yield trend and its performance over the past years and future years with the available weather data. A study in central India showed the grain yield reduction under 1 °C rise in temperature by 8% as simulated by the DSSAT CERES-Maize model and a similar trend was simulated with a 2, 3 and 4 °C rise in temperature (Patidar et al., 2020). Many international studies (Bannayan et al., 2016; Wang et al., 2011) used the DSSAT model to determine maize yield decline in China (by 10–30%) and Iran (2.6–82%), respectively under future IPCC scenarios. Modelling is easily applicable for future projection studies and is widely used in studying the IPCC scenarios: their trend and what impact would these have on the future yields of various crops. Scientists have performed these experiments in many parts of the world using DSSAT, APSIM, AQUACROP, etc. The impact assessment can be easily performed by using the weather files for future scenarios and these can then be compared with the baseline for understanding the quantitative changes in phenology as well as the yield of the plant. Even when the future scenarios were not available, many studies were done by environmental modifications in the crop model itself in Ludhiana, Punjab an advancement in maturity days by 1, 3, 3, 1, 2, and 2 days and grain yield decrease of 4.26, 0.02, 12.15, 10.99, 15.92 and 16.85% for a temperature increase from normal by 0.5, 1.0, 1.5, 2.5 and 3.0 °C, respectively was observed (Prabhjyot-Kaur & Hundal,

2008). A climate change assessment under stabilization scenarios showed that the maize hybrids grown in the Punjab state for the current dates of sowing indicated a strong negative correlation between the yield and the weather parameters. The study also showed that the rising temperatures and rainfall variability in the state led to lowering of the maize yield from high yield category (>5000 kg/ha) to low (<3000 kg/ha) yield category in different agroclimatic zones (Kothiyal et al., 2022b).

Tackling the climate change problem is not an easy task and deciding the appropriate strategies to mitigate or adapt to it is more difficult as the studies show ambiguous results for their applicability. The focus currently is required on the management of farm practices and optimization of plant growth against abiotic stresses. In future, many novel schemes like breeding of stress-resistant cultivars, implementation of proper cropping schemes, and many other conventional methods can help combat climate change and save agriculture during future times.

20.9 Challenges, Limitations and Uncertainties

The crop models are used in the prediction of crop performance in various environmental conditions, however, many conditions are not provided by the models and these include pest and disease attacks on the crops, frost and heat damage, extreme weather conditions like floods, droughts, fire, etc. and lodging which are some major issues not covered by the simulation models. A diverse model is necessary which takes well account of farmer's practices (Jeuffroy et al., 2014). The usefulness of the model depends on its testing and most of the models are poorly tested which indicates the easy formulation of models but poor validation. The validation of models for agronomic purposes is required at a good scale for their further application but most of the untrained agronomists use the crop models indiscriminately leading to poor results. Crop models significantly contribute to save money and time if they have been intensely calibrated and validated. They can also help in fulfilling the world food demand (Rauff & Bello, 2015).

Misuse of models is a common practice as they are often unable to provide accurate projections due to computer power limitations and inadequate understanding of various natural processes occurring in the environment leading to assessing possible effects of climate change to be based on estimations. Real-world systems are complex. Therefore, it is very difficult to comprehensively simulate a real-world system representing all its essential components and mechanisms operating within it. Crop models are usually regional or local for a region, thus, the selection of appropriate models for a region is important for reliable results. The complications in the validation process of the model lie because the available field data (plant, soil and meteorological data) is never accurate. These measured parameters may vary at small distances with the heterogeneity of the soil and variations in environmental factors. The impact of projected climate on the crops and their response depends on the representation of the processes within the crop models. Various studies confirm the upscaling of crop simulation models for their large area applications and this upscaling can be done by including the spatial variability of soil and management for fertilization, sowing and varieties; adequate structure of the model and use of ensemble means (Ewert et al., 2014). Modelling approaches consist of a wide number of challenges. For accuracy and precision of the model, its complexity needs to be upgraded but this increase in complexity comes with constraints which mainly include the accuracy of the input files (predicted weather data, soil data and crop management data) used in the simulation models. Though the increase in simulations and the spatial resolution seems to be profitable but these create larger ensembles leading to more uncertainty (Challinor et al., 2009). The crop models for a region are developed based on the aim of the study but regional differences result in different crop responses which makes their reproducibility at different sites difficult. Thus, the climate change studies and understanding of crop response to future predictions have to be balanced between the region specificity and scale which require synergistic approaches (quantification of uncertainty using crop simulation ensembles; techniques including a combination of diverse modelling approaches and observations; judicious choice of models and their usage) for good results.

Climate change impact assessment and use of adaptation measures to combat the same using crop simulation models lies with lots of uncertainties which is due to improper input data, technical uncertainty in the model or use of a single model, but improvements are possible with the following measures (Kumar, 2016):

(i) Climate models should use high spatial resolution data so that their input in crop models gives accurate results.
(ii) Uncertainties in predicted data can be reduced using an ensemble approach through multi-model intercomparison.
(iii) The field data generated and compiled should be high quality so that when accessed gives true results.
(iv) To improve models for better representation of reality, there is a need for an integrated multidisciplinary approach involving crop modellers, agronomists, crop physiologists, plant breeders, soil scientists, agro-meteorologists, etc.
(v) An increase in the predictive capacity of existing simulation models or new models can be created by combining crop modelling with physiology, biochemistry, molecular genetics and gene-based modelling.
(vi) The indirect impacts on the crops caused due to pests, diseases, changes in nutrient availability, soil moisture, weeds and other climate-induced adversities should also be analysed by combining the soil and pest models with the crop models.

20.10 Conclusion

The increasing world population along with climate change has set immense pressure on the agricultural production system to ensure the food security of the world. Even though future climatic scenarios may have some uncertainty, most of the researchers reported a decrease in agricultural productivity. The adverse impact of climate change can be reduced by changing agricultural practices in response to climate change. Crop models play an important role in the evaluation of possible impacts of climate change like increasing temperature, reduced rainfall, increasing CO_2, etc. It also helps in optimising resource use and the selection of robust and sustainable management practices to reduce the impact of climate change on agricultural productivity. Though there are some limitations and uncertainties in crop model estimates, the use of a multi-model approach, high-quality field data and weather data, and integrating multiple disciplines like physiology, biochemistry, and gene-based modelling can improve the model performances to a fairly acceptable limit.

References

Baker, D. N., Whisler, F. D., Parton, W.J., et al. (1985). *The development of winter wheat: A physical physiological process model* (vol. 38, pp. 176–187). ARS United States Department of Agriculture, Agricultural Research Service.

Bannayan, M., Paymard, P., & Ashraf, B. (2016). Vulnerability of maize production under future climate change: Possible adaptation strategies. *Journal of the Science of Food and Agriculture, 96*(13), 4465–4474.

Basso, B., Liu, L., & Ritchie, J. T. (2016). A comprehensive review of the CERES-Wheat, -Maize and –Rice models' performances. *Advances in Agronomy, 136*, 27–132.

Basso, B., Ritchie, J. T., Pierce, F. J., Braga, R. P., & Jones, J. W. (2001). Spatial validation of crop models for precision agriculture. *Agriculture System, 68*, 97–112.

Beckley, B. D., Callahan, P. S., Hancock, D. W., Mitchum, G. T., & Ray, R. D. (2017). On the "cal-mode" correction to TOPEX satellite altimetry and its effect on the global mean sea level time series. *Journal of Geophysical Research: Oceans, 122*, 8371–8384. https://doi.org/10.1002/2017JC013090

Behera, S. K., & Panda, R. K. (2009). Integrated management of irrigation water and fertilizers for wheat crop using field experiments and simulation modelling. *Agricultural Water Management, 96,* 1532–1540.

Bhatia, A., Mina, U., Kumar, V., et al. (2021). Effect of elevated ozone and carbon dioxide interaction on growth, yield, nutrient content and wilt disease severity in chickpea grown in Northern India. *Heliyon, 7*, e06049. https://doi.org/10.1016/j.heliyon.2021.e06049

Blum, A. (2018). *Plant breeding for stress environments.* CRC Press.

Bondeau, A., Smith, P., Zaehle, S., et al. (2007). Modelling the role of agriculture for the 20th century global terrestrial carbon balance. *Global Change Biology, 13*, 679–706.

Brisson, N., Gary, C., Justes, E., Roche, R., Mary, B., Ripoche, D., Zimmer, D., Sierra, J., Bertuzzi, P., Burger, P., Bussiere, F., Cabidoche, Y. M., Cellier, P., Debaeke, P., Gaudillere, J. P., Henault, C., Maraus, F., Seguin, B., & Sinoquet, H. (2003). An overview of the crop model STICS. *European Journal of Agronomy, 18*(3), 309–332.

Carberry, P. S., Muchow, R. C., & Mc Cown, R. L. (1989). Testing the CERES-Maize simulation model in semi-arid tropical environment. *Field Crops Research, 20*, 297–315.

Challinor, A. J., Ewert, F., Arnold, S., Simelton, E., & Fraser, E. (2009). Crops and climate change: Progress, trends and challenges in simulating impacts and informing adaptation. *Journal of Experimental Botany, 60*(10), 2775–2789. https://doi.org/10.1093/jxb/erp062

Chinchar, A. (2021). *After one of the driest years on record, it will take multiple years for some states to recover.* https://edition.cnn.com/2021/10/02/weather/weather-water-year-drought-forecast/

Charles, D. (2018). *5 major crops in the crosshairs of climate change*. https://www.npr.org/sections/thesalt/2018/10/25/658588158/5-major-crops-in-the-crosshairs-of-climate-change

Deb, P., Kiem, A. S., Babel, M. S., Chu, S. T., & Chakma, B. (2015). Evaluation of climate change impacts and adaptation strategies for maize cultivation in the Himalayan foothills of India. *Journal of Water and Climate Change, 6*(3), 596–612.

Dentener, F., Stevenson, D., Cofala, J., Mechler, R., Amann, M., Bergamaschi, P., Raes, F., & Derwent, R. (2005). The impact of air pollutant and methane emission controls on tropospheric ozone and radiative forcing: CTM calculations for the period 1990–2030. *Atmospheric Chemistry and Physics, 5*, 1731–1755.

EPA: United States Environmental Protection Agency. (2022). Climate Change Indicators: Glaciers. https://www.epa.gov/climate-indicators/climate-change-indicators-glaciers

Ewert, F., van Bussel Lenny, G. J., Zhao, G., Hoffmann, H., et al. (2014). *Uncertainties in scaling up crop models for large area climate change impact assessments* (pp. 261–277). Chapter 10.

Goudriaan, J., & van Laar, H. H. (1994). *Modelling potential growth processes: Textbook with exercises*. Kluwer Academic Publishers. ISBN: 9780792332190-238

Haris, A. A., & Chhabra, V. (2015). Studies on maize potential in Bihar under climate change scenario: impact and adaptation. *Journal of AgriSearch, 2*(1), 40–43.

Hasegawa, H., Bryant, D. C., & Denison, R. F. (2000). Testing CERES model predictions of crop growth and N dynamics, in cropping systems with leguminous green manures in a Mediterranean climate. *Field Crops Research, 67*, 239–255.

Hatfield, J. L., Boote, K. J., Kimball, B. A., Ziska, L. H., Izaurralde, R. C., Ort, D., Thomson, A. M., & Wolfe, D. (2011). Climate impacts on agriculture: Implications for crop production. *Agronomy Journal, 103*, 351–370.

He, J. Q., Dukes, M. D., Hochmuth, G. J., Jones, J. W., & Graham, W. D. (2012). Identifying irrigation and nitrogen best management practices for sweet corn production on sandy soils using CERES-Maize model. *Agricultural Water Management, 109*, 61–70.

Hoogenboom, G., Jones, J. W., & Boote, K. J. (1992). Modeling growth, development, and yield of grain legumes using Soygro, Pnutgro, and Beangro: A review. *Transactions of ASAE, 35*(6), 2043–2056.

Inman-Bamber, N. G. (1994). Effect of age and season on components of yield of sugarcane in South Africa. *Proc S Afr Sug Technol Ass, 68*, 23–27.

Jamieson, P. D., Asseng, S., Chapman, S.C., et al. (2010). Modelling wheat production. In: M. van Ginkel, A. Bonjean, W. Angus (Eds.), *The world wheat book* (p. 40). Lavoisier Publishing.

Jeuffroy, M.-H., Casadebaig, P., Debaeke, P., Loyce, C. & Meynard, J.-M. (2014). Agronomic model uses to predict cultivar performance in various environments and cropping systems. A review. *Agronomy for Sustainable Development, 34*, 121–137.

Jones, J. W., Hoogenboom, G., Porter, C. H., Boote, K. J., Batchelor, W. D., Hunt, L. A., Wilkens, P. W., Singh, U., Gijsman, A. J., & Ritchie, J. T. (2003). The DSSAT cropping system model. *European Journal of Agronomy, 18*, 235–265.

Kaboré, P. D., & Reij, C. (2004). *The emergence and spreading of an improved traditional soil and water conservation practice in Burkina Faso*. Discussion paper 114. Technical Report. Environment and Production Technology Division, IFPRI, Washington, DC.

Karl, T. R., Melillo, J. M., & Peterson, T. C. (Eds.). (2009). *Global climate change impacts in the United States*. Cambridge University Press.

Katzin, D., van Mourik, S., Kempkes, F., van Henten, J. E. (2020). GreenLight-An open source model for greenhouses with supplemental lighting: Evaluation of heat requirements under LED and HPS lamps. *Biosystems Engineering, 194*, 61–81. https://doi.org/10.1016/j.biosystemseng.2020.03.010

Kaur, P., & Hundal, S. S. (2008). Climatic trends in different agroclimatic zones of Punjab: II rainfall features and its variability. *Journal of Research (Punjab Agriculture University), 45*, 41–50.

Kaur, J., Prabhjyot-Kaur, & Kothiyal, S. (2022). Futuristic changes in monthly meteorological parameters as simulated by four GCMs under four emission based scenarios at Ludhiana, Punjab. *Arabian Journal of Geosciences, 15*, 906. https://doi.org/10.1007/s12517-022-10199-y

Keating, B. A., Carberry, P. S., Hammer, G. L., Probert, M. E., Robertson, M. J., Holzworth, D., et al. (2003). An overview of APSIM, a model designed for farming systems simulation. *European Journal of Agronomy, 18*(3–4), 267–288.

Kersebaum, K. (2007). Modelling nitrogen dynamics in soil crop systems with HERMES. *Nutrient Cycling in Agroecosystems, 77*, 39–52.

Kiniry, J., Major, D., Izaurralde, R., et al. (1995). EPIC model parameters for cereal, oilseed and forage crops in the northern great plains region. *Canadian Journal of Plant Science, 75*, 679–688.

Kothiyal, S., Prabhjyot-Kaur, Sandhu, S. S. (2021). Determination of sowing window for kharif maize in Punjab, India using sensitized, calibrated and validated CERES-Maize model. *Maydica, 66*(3), M29

Kothiyal, S., Prabhjyot-Kaur, & Kaur, J. (2022a). Simulation and analysis of maize (Zea mays L.) yield trend under different emission scenarios in Ludhiana district of Punjab. *Agricultural Research Journal, 59*(4), 639–652. https://doi.org/10.5958/2395-146X.2022.00093.X

Kothiyal, S., Prabhjyot-Kaur, & Kaur, J. (2022b). A critical analysis of the effect of projected temperature and rainfall for differential sowing of maize cultivars under RCP 4.5 and RCP 6.0 scenarios for Punjab. *Theoretical and Applied Climatology, 151*, 329–354. https://doi.org/10.1007/s00704-022-04291-2

Kothiyal, S., Prabhjyot-Kaur, Sandhu, S. S., & Kaur, J. (2023a). Modelling the climate change impact of mitigation (RCP 2.6) and high emission (RCP 8.5) scenarios on maize yield and possible adaptation measures in different agroclimatic zones of Punjab, India. *Journal of the Science of Food and Agriculture, 103*, 6984–6994.

Kothiyal, S., Prabhjyot-Kaur, Sandhu, S. S., & Kaur, J. (2023b). Modelling viable adaptive options under climate change scenarios to increase maize productivity in Indian Punjab. *Arabian Journal of Geosciences, 16*, 403. https://doi.org/10.1007/s12517-023-11516-9

Kumar. (2016). Impact of climate change on crop yield and role of model for achieving food security. *Environmental Monitoring Assessment, 188*, 465. https://doi.org/10.1007/s10661-016-5472-3

Lemmon, H. (1986). COMAX: An expert system for cotton crop management. *American Association for the Advancement of Science, 233*(4759), 29–33.

Lin, Y., Feng, Z., Wu, W., Yang, Y., Zhou, Y., & Xu, C. (2017). Potential impacts of climate change and adaptation on maize in Northeast China. *Agronomy Journal, 109*, 1476–1490.

Liu, J. P., Song, M., Horton, R. M., & Hu, Y. (2013). Reducing spread in climate model projections of a September ice-free Arctic. *Proceedings of the National Academy of Sciences, 110*, 12571–12576.

Lobell, D. B., & Field, C. B. (2007). Global scale climate-crop yield relationships and the impact of recent warming. *Environmental Research Letters, 2*, 1–7.

Martre, P., Jamieson, P. D., Semenov, M. A., Zyskowski, R. F., Porter, J. R., & Triboi, E. (2006). Modelling protein content and composition in relation to crop nitrogen dynamics for wheat. *European Journal of Agronomy, 25*, 138–154.

MacGrath, M. (2020). *Climate Change: 2021 will be cooler but still in top six warmest*. https://www.bbc.com/news/science-environment-55365414

McDermid, S., Gowtham, R., Bhvaneswari, K., & Vellingiri, G. (2016). The impacts of climate change on Tamil Nadu rainfed maize production: A multi-model approach to identify sensitivities and uncertainities. *Current Science, 110*(7), 1257–1271.

Merot, A., & Bergez, J. E. (2010). IRRIGATE: A dynamic integrated model combining a knowledge-based model and mechanistic biophysical models for border irrigation management. *Environmental Modelling and Software, 25*(4), 421–432.

Mishra, A. K., & Singh, V. P. (2010). A review of drought concepts. *Journal of Hydrology, 391*, 202–216.

Morber, J. (2019). *As the planet warms, unusual crops could become climate saviors-if we're willing to eat them*. https://ensia.com/features/climate-change-resilient-crops-plants-meat/

Morgan, P. B., Bernacchi, C. J., Ort, D. R., & Long, S. P. (2004). An in vivo analysis of the effect of season-long open-air elevation of ozone to anticipated 2050 levels on photosynthesis in soybean. *Plant Physiology, 135*, 2348–2357.

NASA: National Aeronautics and Space Administration (2021). *Understanding sea level: Global mean sea level*. https://sealevel.nasa.gov/understanding-sea-level/key-indicators/global-mean-sea-level

Nendel, C., Berg, M., Kersebaum, K. C., Mirschel, W., Specka, X., Wegehenkel, M., Wenkel, K. O., & Wieland, R. (2011). The MONICA model: testing predictability for cropgrowth, soil moisture and nitrogen dynamics. *Ecological Modelling, 222,* 1614–1625. https://doi.org/10.1016/j.ecolmodel.2011.02.018

NSIDC: National Snow & Ice Data Center (2021). *Greenland ice sheet today: Rain at the summit of Greenland*. https://nsidc.org/greenland-today/2021/08/rain-at-the-summit-of-greenland/

NOAA (2021). National Centers for Environmental Information, monthly global climate report for annual 2021, published online January 2022. https://www.ncei.noaa.gov/news/global-climate-202112

Otegui, M., Riuz, R. A., & Petruzzi, D. (1996). Modelling hybrid and sowing date effects on potential grain yield of maize in a humid temperate region. *Field Crops Research, 47*, 167–174.

Oteng-Darko, P., Yeboah, S., Addy, S. N. T., Amponsah, S., & Owusu Danquah, E. (2013). Crop modelling: A tool for agricultural research—A review. E3 *Journal of Agricultural Research and Development, 2*(1), 1–6.

Pan, J., Zhu, Y., Jiang, D., Dai, T. B., Li, Y. X., & Cao, W. X. (2006). Modeling plant nitrogen uptake and grain nitrogen accumulation in wheat. *Field Crops Research, 97*, 322–336.

Patidar, R., Mohanty, M., Sinha, N. K., Gupta, S. C., Somasundaram, J., Chaudhary, R. S., Soliya, R., Hati, K. M., Prabhakar, M., Reddy, K. S., Patra, A. K., & Rao, S. (2020). Potential impact of future climate change on maize under rainfed condition in central India. *Journal of Agrometeorology, 22*(1), 18–23.

Pelto, M., & WGMS Network. (2020). Alpine glaciers. *Bulletin of the American Meteorological Society, 101*(8), S37–S38. https://doi.org/10.1175/BAMS-D-20-0104.1

Porter, J. R. (1984). A model of canopy development in winter wheat. *Journal of Agricultural Science, 102*, 383–392.

Prabhjyot-Kaur, Sandhu, S. S., Kothiyal, S., Aryal, A., Kaur, J., Singh, J., Kaur, J., & Singh, G. (2023). Productivity and management of cereals under projected climatein Punjab. Department of Climate Change and Agricultural Meteorology, Punjab Agricultural University, Ludhiana. pp. 120.

Prabhjyot-Kaur, Sandhu, S. S., Kothiyal, S., & Kaur, J. (2022). Determination of sowing window for wheat in Punjab, India using sensitized, calibrated and validated CERES-Wheat model. *Journal of Agricultural Physics, 22*(2), 215–227.

Prasad, P. V. V., Boote, K. J., & Allen, L. H., Jr. (2006). Adverse high temperature effects on pollen viability, seed-set, seed yield and harvest index of grain sorghum *Sorghum bicolor* (L.) Moench are more severe at elevated carbon dioxide due to high tissue temperature. *Agricultural and Forest Meteorology, 139*, 237–251.

Rauff, K. O, & Bello, R. (2015). A review of crop growth simulation models as tools for agricultural meteorology. *Agricultural Sciences, 6,* 1098–1105. https://doi.org/10.4236/as.2015.69105

Raza, A., Razzaq, A., Mehmood, S. S., Zou, X., Zhang, X., Lv, Y., & Xu, J. (2019). Impact of climate change on crop adaptation and strategies to tackle its outcome: A review. *Plants, 8*(34). https://doi.org/10.3390/plants8020034

Ritchie, J., Godwin, D., & Otter-Nacke, S. (1985). *CERES-wheat: A user-oriented wheat yield model* (p. 252). Michigan State University, Michigan.

Rodenburg, J., Meinke, H., & Johnson, D. E. (2011). Challenges for weed management in African rice systems in a changing climate. *Journal of Agricultural Science, 149*, 427–435.

Rosa, R. D., Paredes, P., Rodrigues, G. C., Alves, I., Fernando, R. M., Pereira, L. S., & Allen, R. G. (2011). Implementing the dual crop coefficient approach in interactive software. 1. Background and computational strategy. *Agricultural Water Management*. https://doi.org/10.1016/j.agwat.2011.10.013

Rosenzweig, C., & Liverman, D. (1992). Predicted effects of climate change on agriculture: A comparison of temperate and tropical regions. In: S.K. Majumdar (Ed.), *Global climate change:*

Implications, challenges, and mitigation measures (pp. 342–361). The Pennsylvania Academy of Sciences.
Sah, S., Singh, R. N., Chaturvedi, G., & Das, B. (2021). Trends, variability, and teleconnections of long-term rainfall in the Terai region of India. *Theoretical and Applied Climatology, 143*, 291–307. https://doi.org/10.1007/s00704-020-03421-y
Sah, S., Singh, R. N., & Nain, A. S. (2020). Impact of weather parameters on simulated biophysical parameters of impact of weather parameters on simulated biophysical parameters of chickpea. *Journal of Agricultural Physics, 19*, 58–66.
Sawadogo, H. (2011). Using soil and water conservation techniques to rehabilitate degraded lands in northwestern Burkina Faso. *International Journal of Agricultural Sustainability, 9*, 120–128.
von Schuckmann, K., Cheng, L., Palmer Mathew, D., et al. (2020). Heat stored in the earth system: Where does the energy go? *Earth System Science Data, 12*, 2013–2041. https://doi.org/10.5194/essd-12-2013-2020
Singh, R. N., Mukherjee, J., Sehgal, V. K., et al. (2017). Effect of elevated ozone, carbon dioxide and their interaction on growth, biomass and water use efficiency of chickpea (Cicer arietinum L.). *Journal of Agrometeorology, 19*, 301–305.
Singh, R. N., Mukherjee, J., Sehgal, V. K., et al. (2021). Interactive effect of elevated tropospheric ozone and carbon dioxide on radiation utilisation, growth and yield of chickpea (Cicer arietinum L.). *International Journal of Biometeorology*. https://doi.org/10.1007/s00484-021-02150-9
Singh, R. N., Sah, S., Chaturvedi, G., et al. (2021b). *Innovative Trend Analysis of Rainfall in Relation to Soybean Productivity over Western Maharashtra., 23*, 228–235.
Singh, R. N., Sah, S., Das, B., et al. (2021c). Innovative trend analysis of spatio-temporal variations of rainfall in India during 1901–2019. *Theoretical and Applied Climatology, 145*, 821–838. https://doi.org/10.1007/s00704-021-03657-2
Singh, R. N., Sah, S., Das, B., et al. (2021d). Long-term spatiotemporal trends of temperature associated with sugarcane in west India. *Arabian Journal of Geosciences, 14*, 1–17.
Singh, R. N., Sah, S., Das, B., et al. (2021e). Spatio-temporal trends and variability of rainfall in Maharashtra, India: Analysis of 118 years. *Theoretical and Applied Climatology, 143*, 883–900. https://doi.org/10.1007/s00704-020-03452-5
Steduto, P., Hsiao, T. C., Raes, D., & Fereres, E. (2009). AquaCrop—The FAO crop model to simulate yield response to water: I concepts and underlying principles. *Agronomy Journal, 101*, 426–437.
Stockle, C. O., Martin, S., & Campbell, G. S. (1994). CropSyst, a cropping systems model: Water/nitrogen budgets and crop yield. *Agricultural Systems, 46*, 335–359.
Stöckle, C. O., Donatelli, M., & Nelson, R. (2003). CropSyst, a cropping systems simulation model. *European Journal of Agronomy, 18*, 289–307.
Supit, I., Hooijer, A. A., & van Diepen, C. A. (Eds.). (1994). *System description of the WOFOST 6.0 crop simulation model implemented in CGMS*. Publication EUR 15956 of the Office for Official Publications of the EU.
Thaler, S., Eitzinger, J., Trnka, M., & Dubrovsky, M. (2012). Impacts of climate change and alternative adaptation options on winter wheat yield and water productivity in a dry climate in Central Europe. *The Journal of Agricultural Science, 150*(5), 537–555.
UNCC and WMO: United Nations Climate Change and World Meteorological Organisation (2021). *State of climate in 2021: Extreme events and major impacts*. https://unfccc.int/news/state-of-climate-in-2021-extreme-events-and-major-impacts
Vanuytrecht, E., Raes, D., Steduto, P., Hsiao, T. C., Fereres, E., Heng, L. K., Vila, M. G., & Moreno, P. M. (2014). AquaCrop: FAO's crop water productivity and yield response model. *Environmental Modelling and Software, 62*, 351–360.
Wang, M., Li, Y., Ye, W., Bornman, J. F., & Xiaodong, Y. (2011). Effects of climate change on maize production, and potential adaptation measures: A case study in Jilin Province, China. *Climatic Research, 46*(3), 223–246.

Wei, P., Xu, X., Xue, M., et al. (2022). On the key dynamical processes supporting the 21.7 Zhengzhou record-breaking hourly rainfall in China. *Advances in Atmospheric Sciences*. https://doi.org/10.1007/s00376-022-2061-y

Welch, J. A., Vincent, J. R., Auffh ammer, M., Moya, P. F., Dobermann, A., & Dawe, D. (2010). Rice yields in tropical/subtropical Asia exhibit large but opposing sensitivities to minimum and maximum temperatures. *Proceedings of the National Academy of Sciences USA 107*, 14562–14567.

White, J. W., Hoogenboom, G., Kimball, B. A., & Wall, G. W. (2011). Methodologies for simulating impacts of climate change on crop production. *Field Crop Research, 124*, 357–368.

Wiles, L. J., King, R. P., Schweizer, E. E., Lybecker, D. W., & Swinton, S. M. (1996). GWN: General Weed Management Model *50*(4), 355–376. https://doi.org/10.1016/0308-521X(95)00016-X

WMO: World Meteorological Organization (2022). *More bad news for the planet: Greenhouse gas levels hit new highs*. https://public.wmo.int/en/resources/united_in_science

World weather attribution (2021). *Heavy rainfall which led to severe flooding in Western Europe made more likely by climate change*. https://www.worldweatherattribution.org/heavy-rainfall-which-led-to-severe-flooding-in-western-europe-made-more-likely-by-climate-change/

Zhao, C., Liu, B., Xiao, L., Hoogenboom, G., et al. (2019). A SIMPLE crop model. *European Journal of Agronomy, 104*, 97–106. https://doi.org/10.1016/j.eja.2019.01.009

Zhu, Y., Chang, L., Tang, L., Jiang, H., Zhang, W., & Cao, W. (2009). Modelling leaf shape dynamics in rice. *Journal of Life Sciences, 57*, 73–81.

Ziska, L. H. (2004). Rising carbon dioxide and weed ecology. In Inderjit (Ed.), *Weed biology and management* (pp. 159–176). Kluwer Academic Publ.

Part III
Advanced Climate Change Research on Atmospheric Phenomenon

Chapter 21
Advanced Facilities for Climate Change Research and Greenhouse Gas Mitigation

Sandeep Kumar, Bidisha Chakrabarti, Niveta Jain, Sudha Kanojia, Disha Sandilaya, Abhiliasha Sharma, and Arti Bhatia

Abstract Advanced facilities for climate change research and greenhouse gas mitigation are essential in our efforts to understand and address the challenges posed by climate change. These facilities include eddy covariance system, Free Air CO_2 Enrichment (FACE), Temperature gradient tunnel (TGT), Carbon dioxide and Temperature Gradient Chamber (CTGC), Free air ozone and carbon dioxide enrichment (FAOCE), Open Top Chambers (OTC), Free Air Temperature Elevation (FATE) and Temperature FACE (T-FACE) which helps to monitor greenhouse gas and also assist to develop new technologies to reduce greenhouse gas emissions and mitigate the impacts of climate change. These facilities help us understand the complex interactions between the Earth's atmosphere and land, and how anthropogenic activities are contributing to global warming and climate change. Through continued research and development of new technologies and strategies, we can better understand and address the impacts of climate change and work towards a more sustainable future.

Keywords Open top chambers · Climate change · Greenhouse gas · Free air temperature elevation · Free air CO_2 enrichment · Temperature gradient tunnel

21.1 Introduction to Climate Change Research

Accelerated anthropogenic greenhouse Gas viz. Carbon dioxide (CO_2), methane (CH_4) and nitrous oxide (N_2O) emission attained a record high of 55.3 Gt CO_2 eq yr^{-1} in 2018 (UNEP, 2019) that has been emerged as the most prominent environmental issue worldwide. Agriculture and allied sector contributes about 24% of the net GHG emissions (IPCC, 2014). Increasing atmospheric GHG concentrations facilitates positive radiative forcing towards unequivocal warming of our climatic system. IPCC (2018) concluded that to limit global warming to 1.5 °C until

S. Kumar · B. Chakrabarti · N. Jain · S. Kanojia · D. Sandilaya · A. Sharma · A. Bhatia (✉)
Division of Environment Science, ICAR-Indian Agricultural Research Institute,
New Delhi 110012, India
e-mail: arti.bhatia@icar.gov.in

H. Pathak et al. (eds.), *Climate Change Impacts on Soil-Plant-Atmosphere Continuum*,
Advances in Global Change Research 78,
https://doi.org/10.1007/978-981-99-7935-6_21

2030 relative to 2010, global decline in methane and nitrous oxide emissions up to 48% and 26%, respectively from agriculture are required. Atmospheric warming accelerated ocean acidification, rapid snow melt of glaciers and sea level rise with its considerable impact on agro ecosystem functioning including crop, soil, livestock, fishery and pest. Therefore, climate change research aims to develop understanding on ecosystem response or feedback from human society.

21.2 Greenhouse Gas Emissions

The agriculture sector is both a victim of and a contributor to climate change. The "Agriculture, Forestry and Other Land Use (AFOLU)" sector emits 18.4% of global GHGs, wherein 5.8% contribution is from enteric fermentation and manure management followed by 4.1% from managed agricultural soils, thus contributing to climate change. Agriculture activities produce a considerable amount of GHGs (CO_2, CH_4 and N_2O), contributing significantly to global warming and climate change (Holmberg et al., 2021). Globally, agricultural actions contribute around 30% of total GHGs. In the agricultural sector, GHGs emissions are mainly due to chemical fertilizers, pesticides, residue burning, and animal waste. This rate is tied to further increase due to an upsurge in demand for food by a growing population, the more robust need for dairy and meat products, and the intensification of modern agricultural practices. Soils function as sources and sink for GHG such as CO_2, CH_4 and N_2O. Since storage and emission capacities may be significant, specific quantifications are required to get trustworthy global budgets for land-use management (agriculture, forestry), global change, and climate research (Oertel et al., 2016). Soils are the largest pool/sink for carbon (C) and nitrogen (N), containing almost 1500 Pg of total C and 136 Pg of total N accumulated in the uppermost layer of soil (Oertel et al., 2016; Schaufler et al., 2010). Structural changes in soil aggregation may affect its source-sink relationship and net storage capacity. Schulze et al. (2009) revealed GHGs fluxes, i.e. of CO_2, CH_4 and N_2O, (2000–2005) were near neutral as grasslands and forests acted as CO_2 sink of agricultural CH_4 and N_2O emissions. Bahn et al. (2010) observed that soils emit around 98 Pg C a^{-1} (roughly 15-year turnover time for the reservoir) across the vegetation types and biomes worldwide.

The soil GHGs emission depends on several biotic and abiotic factors. Source and sink chemistry for GHGs is variable under different land use systems. Carbon can enter the soil through litter, crops, plant roots, crop residues, and animal manure on agricultural lands. This carbon can be accumulated mainly as soil organic matter (CRS, 2021) (Fig. 21.1). CO_2 emission from the soil is mainly contributed by microbial respiration, including roots. Almost $\sim$50% (10–95%) of the total soil respiration alone through root respiration. CH_4 emission from soil under anaerobic conditions by methanogens and consumed by methanotrophs. The oxides of nitrogen (N_2O and NO) are released by nitrification and denitrification processes (Ussiri & Lal, 2013). Land-use changes may greatly change the related source or sink of GHGs

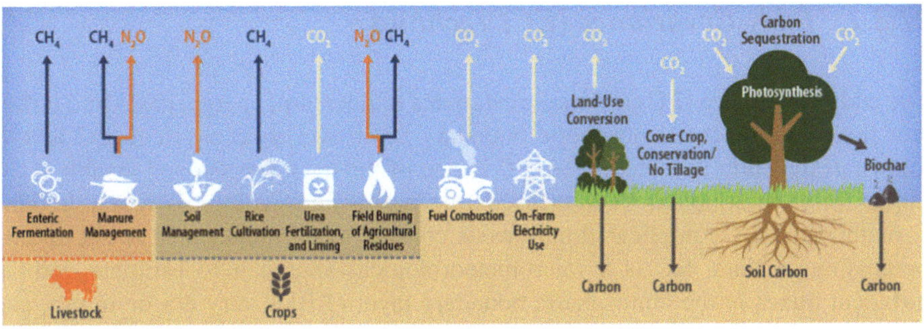

Fig. 21.1 Sources and sinks of GHG emission from agricultural activities. *Source* CRS (2021)

(Poeplau & Don, 2013). According to Heikkinen et al. (2013) assessment of soil inventories showed that croplands had lost C in 1974–2009.

21.3 Monitoring of Greenhouse Gas Emissions

In agricultural research, several methods, i.e. closed-chamber method, open-chamber method, micrometeorological method and eddy covariance (EC) technique, are used to estimate GHGs emissions from soil. There are two measurement systems for gases emitted from soil, i.e. large scale and small-scale measurement methods. The traditional means of assessing an ecosystem's net carbon exchange over several years concerns quantifying temporal differences of biomass and soil carbon (Lal et al., 2001). First time CO_2 emission measurements from soils were conducted in laboratories in the 19th Century. Agricultural soil gas emission measurements with chamber methods were introduced at the beginning of the 20th Century. Measurement of other trace gas emissions was initiated later when the gas analyzers became available as N_2O in the 1950s by Arnold (1954), and CH_4 in the 1980s by Holzapfel-Pschorn et al. (1985).The large-scale measurement methods include the nocturnal box method, daytime boundary layer budget method, and small-scale methods include static chambers and dynamic chambers (Cape & Fowler, 2014). Among these methods, eddy covariance (EC) is the most advanced and accurate technique to estimate flux since it is the only direct flux determination method (Kumar et al., 2021). Eddy covariance (EC) technique is used to estimate the vertical fluxes of the GHGs (CO_2, CH_4, N_2O) from various land cover systems (agricultural fields, forests, aquatic bodies, and built environment) and is crucial to quantify the source sink strength of an ecosystem and for identifying likely climate change mitigation strategies and adaptation options.

21.3.1 Eddy Covariance (EC) Technique

The eddy covariance (EC) technique estimates momentum, energy, gases and turbulent fluxes within the atmospheric boundary layer over an ecosystem. EC is also known as the eddy flux technique. EC is one of the most precise and tenable techniques in environmental monitoring, hydrological and ecological research. This is mathematically complicated and needs considerable maintenance to set up and processing of data. EC is a micrometeorological approach to estimate vertical turbulent fluxes in the atmospheric boundary layer (ABL), very elastic in use, and broadly employed in measuring CO_2, CH_4, N_2O, H_2O, heat, momentum, VOCs and isotopes. This technique has tremendous application in the area of carbon storage and sequestration, industrial monitoring, landfill and environmental management.

The technique is also used to verify and test climate-related models. Combined ecological and physiological measurements help partition carbon fluxes into plant and soil segments and indicate mechanisms accountable for these fluxes. There are many significant applications of EC technique in agricultural research such as Gross Primary Production (GPP), Ecosystem Respiration (RE), Net Primary Production (NPP), trace gas emissions, hydrologic and energy partitioning, Net Ecosystem Production (NEP), biogeochemistry and GHGs measurement (Pathak et al., 2013). The major challenge of the technique for a non-expert is the complexity of system design, execution, and processing a large amount of data.

The basic EC system mainly constitute the fast-response trace gas analyzer, fast-response 3-dimensional sonic anemometer, data logger, and uninterrupted power supply (Fig. 21.2). Besides these, there are additional sensors to support meteorological data, is air temperature and relative humidity sensor, net radiation sensors, canopy temperature sensor, photosynthetic photon flux density (PPFD) sensor, precipitation sensors, quantum sensor, snow depth sensor, soil heat flux plates, TDR moisture probes, surface/soil thermocouple array, barometric pressure, etc. Fast-response datalogger is required for data acquisition, transmission/storage, and multiplexers power source and mobile or fixed tripod or tower for mounting the sensors. The EC process depends on the integrated measurements of meteorological data to compute flux rates. In EC, there are open and closed path fast response sensors and both systems have specific benefits and drawbacks (Burba and Anderson, 2007). Depending of the objective of the study, the particular sensor type is chosen.

Principle and methodology of Eddy covariance flux measurement

The EC method is authentic when the atmospheric conditions are constant, the underlying vegetation is homogenous, and situated on a flat landscape for an extended distance upwind. Deviation from these assumptions can induce systematic errors in the performance of the eddy covariance measurements. The working process of the Eddy Covariance technique is based on high frequency (10–20 Hz) measurements of wind speed and wind direction and CH_4, N_2O, CO_2 and H_2O level at a point over the canopy utilising a three-axis sonic anemometer and rapid reaction infrared (IR) gas analyser. Airflow can be supposed as a horizontal flow of numerous rotating eddies.

Fig. 21.2 Eddy covariance flux tower and its components

Each eddy has 3-D elements, including the vertical direction of the air. The condition examines messy at first, but these elements can be readily measured from the EC tower. The GHG sensors, sonic anemometer and IR gas analyzer, are generally established at 2–3 m above crop canopy and 30–40 m height with a sensor separation of 15–20 cm in case of forests. The data obtained from sensors is sampled at 10 and or 20 Hz using a fast-response data logger. The basic principle for flux measurement is to measure how many molecules are pushing up and down over time, and how fast they cross. This vertical flux of molecules can be described as the covariance between vertical velocity, up and down movements, and concentration of gas. Gap-filling of eddy flux data is an important step, which are rejected by quality control tests or due to instrument error. Nowadays, EdiRe, TSA, ECpack, Alteddy, TK2, EddySoft and EddyPro software are used to generate mission data or gap-filling.

The CO_2 NEE is the difference between carbon assimilation during photosynthesis and CO_2 release via respiration. A negative flux denotes CO_2 uptake by the ecosystem, and a positive flux represents CO_2 loss. Forest trees play a vital role in sequestering atmospheric CO_2. EC method is installed specially to study the ecosystem physiology and flux measurement (Table 21.1).

Errors associated with flux calculation

EC data processing is performed operating any of the multiple flux computation applications. Several methods have been applied to remove uncertainty due to instrumental noise, and the "random shuffle method" proposed by Billesbach (2011) has

Table 21.1 Measurement of GHGs flux using EC system

Study site	Type of ecosystem	Studied parameters	Fluxand system variations	References
Siikaneva Fen, Southern Finland	Wetland	CH_4 emission	Average wintertime emissions <1 mg m^{-2} h^{-1}, and summertime emissions about 3.5 mg m^{-2} h^{-1}	Rinne et al. (2007)
Helsinki, Finland	Semi-urban area	N_2O fluxes	1.7 ± 2.0 μmol m^{-2} h^{-1}	Jarvi et al. (2014)
Edinburgh, UK	Urban	N_2O fluxes	11.7 μmol m^{-2} h^{-1}	Famulariet al. (2010)
Harford, NY, USA	Agricultural field	N_2O fluxes	2.8 μmol m^{-2} h^{-1}	Molodovskayaet al. (2011)
Lägeren, Switzerland	Mixed forests	N_2O fluxes	0.8 μmol m^{-2} h^{-1}	Eugsteret al., (2007)
Easter Bush, UK	Grasslands	N_2O fluxes	0.2–0.5 μmol m^{-2} h^{-1}	Fletchard et al. (2007)
Ssland of Sealand, Denmark	Agricultural field	N_2O emission	149–495 μg N_2O-N m^{-2} h^{-1}	Christensen et al. (2004)
Ontario, Canada	Cropland	N_2O fluxes	−10 and 10 ng N_2O–N m^{-2} s^{-1}	Brown et al. (2018)
Maaninka, Eastern Finland	Agricultural field	N_2O exchange	Cumulative sum of 2.8 kg N_2O ha^{-1}	Shurpali et al. (2016)
Nolensville, Tennessee	Corn field	N_2O fluxes	Cumulative 6.87 kg N_2O-N ha^{-1}	Huang et al. (2014)
Western Quebe, Canada	Agricultural field	N_2O fluxes	Cumulative 0.7 kg N_2O-N ha^{-1}	Pattey et al. (2008)
New Delhi, India	Rice-wheat system	CH_4 flux	Cumulative 39.45 CH_4 kg ha^{-1}	Bhatia et al.(2023)
Cuttack, India	Rice-Rice system	CH_4 flux	+2.80 and +4.34 mg m^{-2} h^{-1} in wet and dry cropping seasons respectively	Bhattacharyya et al. (2014)
New Delhi, India	Rice-wheat system	CO_2 fluxes	−576 g Cm^{-2} (wheat), −368 g Cm^{-2} (rice)	Kumar et al. (2021)

been used by several authors. These errors required minimization to perform the process (Pathak et al., 2013).

Advantages of the Eddy covariance system

Using non-invasive sampling, EC can be utilised for continuous in-situ measurements over a vast area. The principles of gas exchange are quite reliable, justifiable, and established. The system's reaction time is quick. With quick and accurate sampling, it also provides information on the short-term volatility of fluxes. With little assistance, the automatic measurement method assures uniform coverage (Pathak et al., 2013).

Limitations of the Eddy covariance system

Despite many advantages, the EC system has several limitations (Pathak et al., 2013).

- The research region ought to be level, uniform, and represent a comparable environment.
- The EC system supplies specific information for a single ecosystem.
- The process is mathematically complex.

- The noise in the EC Flux data is primarily the result of random measurement errors.
- The EC approach is not appropriate for small areas with weak winds, point flux sources with complicated topography, etc.
- It requires a number of presumptions and corrections, careful design, execution, and processing that suits the individual experimental site's detailed purpose.

21.3.2 Chamber Technique for GHG Monitoring

Closed-chamber methods are most commonly used to measure GHGs flux from the soil. In these methods, gas samples are periodically collected from installed chambers (on surface soil) to measure the changes in gases flux. Chamber measurements are relatively easy to use and flexible to a wide variety of analyses, and they are essential devices in circumstances where the EC technique cannot be applied/installed. Likewise, they are helpful to measure the spatial heterogeneity of GHG fluxes, NEE, and also offer to support data for the gap-filling of the EC data. On a shorter scale (area of 0.1–10 m^2), gas fluxes can be measured using chambers positioned on the ground surface, in which change in gas concentrations is calculated. The gas chambers are generally fabricated of inflexible plastic, perspex or acrylic sheets. These chambers have variable shapes and sizes, but typically 50 cm × 30 cm × 100 cm chambers made of 6-mm acrylic sheets are used for sampling in the crop field. A mixing fan, thermometer, and sampling vent are installed on the top of the chamber to mix the gases, the inner ambient temperature measurement, and sampling, respectively. The headspace volume of the chamber is recorded to calculate the flux of the gas. First sampling should be done immediately after placing and sealing the chamber on the stand from the headspace. Other samples at equal time intervals from headspace not exceeding 2 h. To confirm the linearity of the gas concentration increase in the chamber, three measurements must be taken. During the collection of gas samples from the field, the air inside the chamber should be homogenized using a battery-operated fan or air circulation pump with an air displacement of 1.5 L min^{-1}. With the aid of a 24-gauge hypodermic needle and a 20–50 mL syringe, gas samples can be taken every 0, 1/4, 1/2, and 1 h. After drawing the sample, the syringe should be made airtight with a three-way stop cock. Dynamic close chamber is shown in.

A rectangular-shaped closed chamber with no openings for air to enter or depart and no potential for CO_2 diffusion interaction. The dynamic chamber approach allows air to enter and exit the chamber space at a volume basis flow rate; however, no diffusion through the chamber walls is possible during this operation. If there is a divergence from linearity, there is either a leak in the chamber seal or the gradient of the gas concentration between the production zone in the soil and chamber atmosphere has decreased. The gas samples are collected by syringes, stored in airtight vials and gas samples are examined in a gas chromatograph (GC). In order to reduce the impact on the environment inside the enclosure created by the chamber wall, the chamber cover should be removed as soon as the last sample is taken. The samples that were taken should be measured right away. Samples should be collected into evacuated vials with rubber or silicone septa if sample analysis is not practicable (e.g. vacutainers/exetainers). An alternative approach is using glass serum bottles fitted with butyl rubber stoppers. Contamination problems are controlled by injecting adequate samples to achieve an over-pressure, e.g. 10 mL, into a 9-mL vial.

Gas Chromatograph (GC) analyzes the concentration of methane and nitrous oxide in the gas samples fitted with a flame ionization detector (FID) and electron capture detector (ECD), respectively. The CH_4 standards (1, 5 and 10 ppm) and N_2O standard (300, 500 and 1000 ppbV) are used as a primary standard.

Calculation of methane and nitrous oxide flux

The following equations are used for the calculation of methane and nitrous oxide flux of the field soil.

$$CH_4\text{flux} = [(C_t - C_o)/t] \times H \times 16/22.4 \times 10000 \times 24 \text{ mg ha}^{-1}\text{d}^{-1}$$

$$N_2O \text{ flux} = [(C_t - C_o)/t] \times H \times 44/22.4 \times 10000 \times 24 \text{ mg ha}^{-1}\text{d}^{-1}$$

Here, A is cross-sectional area of the chamber (m^2); H represents Headspace (m); C_o and C_t denote gas concentration at 0 time (μL L^{-1}) and after 0-time t (μL L^{-1}), respectively; (Ct-Co) denotes change in concentration in time t (μL L^{-1}). The static chamber method is the most common technique to calculate CH_4 fluxes from soils. The errors quantification due to chamber structure, function and flux calculation techniques though are not calculated correctly. The impact of chamber dimensions and geometry on gases fluxes has not been as widely examined or tested, although they are critical subjects in evaluating how well the chamber can detect the gases fluxes. Poor sealing of the chamber may create error during measurement.

The estimation of N_2O and CH_4 losses from field soil using chamber method is very common practice. Pathak et al. (2003) studied the effect of irrigation, farmyard manure (FYM) and dicyandiamide (DCD) application using closed chamber technique in rice–wheat cropping system on CH_4 emission. CH_4 fluxes were observed up to -0.1 kg ha^{-1} day^{-1}, no difference in CH_4 flux among irrigation treatments in wheat. Bhatia et al. (2012a) also observed that the application of 120 kg N/ha at LCC ≤ 4 reduced N_2O emissions by 16% and CH_4 by 11% over the conventional split application of urea in rice.

Advantages of closed chamber technique

Closed chambers are inexpensive and straightforward to build from various materials in various designs and dimensions.

- Measurement of gases and operation of chambers are simple, and thus this method provides a possibility to measure GHG at diverse sites with the same equipment.
- In a short period of time, closed chambers can measure extremely low rates of GHG fluxes from agricultural fields, pastures, marsh, drains, and ditches (from 30 min to 1 h).

- Mostly static chamber methods do not require an electrical supply.
- Easy to transport from one location to another.
- Easily detection of leakage problems.
- Minimum disturbance to farm operations.

Disadvantages of closed chamber technique

Gas accumulated in chamber headspace leads to a decline in the concentration gradient and, therefore, a drop-in gas diffusion, causing nonlinear fluxes between the soil and the air. The suppression of fluxes can be minimized by reducing the enclosure period.

- Closed chambers modify the changes of atmospheric pressure; hence a special vent can maintain air pressure inside and outside of the chamber.
- Soil and atmospheric temperature modifications within the chamber can occur. The coating of reflective and insulating material on the chamber can reduce the temperature effects.
- Soil and atmospheric moisture modifications.

Precautions during the use of closed-chamber technique

The following precautions should be taken care while using the closed-chamber technique for GHG flux measurement.

- For the sampling from the rice field, approximately 4–6 plants should be inside the chamber. For upland crops, the chamber should be positioned between the rows/plants to capture the gas samples.
- The standard height of the chamber needs to be higher than 30 cm.
- The chamber headspace gases concentration at zero time demands highly accurate measurement.
- It is best to obtain as few air samples as possible in order to detect a discernible rise in headspace gas concentration. A longer chamber setup could have detrimental effects.
- To reduce the uncertainty caused by spatiotemporal fluctuation, more chambers should be used.
- To avoid diurnal fluctuation, gas samples from the field should be gathered at specific times of day throughout the cropping season, ideally between 09:00–11:00 and 15:00–17:00 h.

Comparison of flux measurements between closed chamber and EC techniques

Several investigations comparing GHG fluxes as measured by the CC and EC techniques revealed wide variances and discrepancies (Table 21.2). Many studies showed that the fluxes estimated by the CC approach were greater than those estimated by the EC method. The probable causes may be the non-homogeneous soil factors and the different spatial and temporal scales. The CO_2 fluxes from soil were estimated with the application of chambers covering a small soil area. But, such structures are occasionally criticized because they are not well suited to sample the variability of soil CO_2 efflux, called chamber effects. Due to surface cooling and persistent stratification in the late afternoon, which reduces turbulent exchange, chamber methods measure fluxes that are larger than EC fluxes. A rather consistent chamber flow and an EC flux with a lot more variability were both seen during the night. Studies have demonstrated that while EC does not maintain flux uniformity, the nighttime stable stratification does, along with low wind speeds and low outgoing longwave radiation. Yu et al. (2013) compared three methods, i.e. manual static chamber (MSC), continuous automated chamber (CAC) and EC, to monitor the CH_4 fluxes in an alpine wetland, China. They saw a satisfactory agreement among these techniques in the seasonal CH4 flux, but the diurnal patterns differed. The chamber measurement methods have made advancements in recent years, still due to its insensitivity to diverse atmospheric conditions, micrometeorological tools such as EC are preferable for determining ecosystem fluxes. Comparison of GHGs flux measurement of the ecosystem between CC and EC technique has been carried out by many researchers. Wang et al. (2013) carried out a comparison study to NEE observations in a cotton field and wheat field using transparent automatic chamber (AC) and EC techniques. They observed that half-hourly NEE fluxes estimated with the AC fluxes were 0.78 (cotton field) and 1.06 (wheat field) times the size of the EC fluxes. During the crop periods with similar field conditions, the cumulative NEE fluxes indicated slight differences between both techniques (~7%). The surface N_2O emissions were measured from the agricultural field by closed and open-flow chamber methods, using gas chromatographic analysis and over bigger areas by EC. The average N_2O emission was 162–202 $\mu gN_2O\text{-}N\ m^{-2}\ h^{-1}$ from a wheat stubble and 328–467 $\mu gN_2O\text{-}N\ m^{-2}\ h^{-1}$ from a carrot field with chamber procedures and using the EC method the N_2O emissions were 149–495 $\mu gN_2O\text{-}N\ m^{-2}\ h^{-1}$. Pihlatie et al. (2012) experimented with fifteen static chambers (height 0.06 to 1.8 m, volume 0.02–0.195 m^3) against four flux levels (FL) varying from 200 to 2300 $\mu g\ CH_4\ m^{-2}\ h^{-1}$.

21.4 Quantifying Climate Change Impact and Adaptation

21.4.1 Open Top Chambers (OTC)

Open top chambers are widely used to study the effects of high CO_2 and other atmospheric gases on crops and soils (OTCs). OTCs are enclosures with transparent vertical sidewalls made up of plexi glass, plastic sheet or polyvinylchloride (PVC) sheet built around metal frames (Singh et al., 2012). The glass and plastic walls of the enclosed space work as a greenhouse, warming the air within. The top of the OTCs is left open to promote natural air exchange, which aids in keeping relative humidity and temperature levels close to ambient values (Vanaja et al., 2006). Carbon dioxide (CO_2) enriched air is supplied inside with circular tubes placed near the bottom of the chamber. CO_2 is supplied from cylinders containing commercial grade CO_2. The

Table 21.2 Comparison of flux measurement of ecosystem between closed chamber and EC technique

Location	Ecosystem	Closed chamber (CC) technique	EC technique	References
Yuncheng city, Shanxi, China	Agricultural field	• Cumulative NEE of -42 gCm^{-2} yr^{-1} • Cumulative seasonal NEE of -205 gCm^{-2}	• Cumulative NEE $-$ 40 gCm^{-2} yr^{-1} • Cumulative seasonal NEE of -251 gCm^{-2}	Wang et al. (2013)
Fuchu, Tokyo, Japan	Flooded paddy field	• CH_4 flux averages 58, 81, 94 and 57% higher than EC at 30, 60 and 90 DAT and after harvest, respectively	• Measure low CH_4 flux than close chamber	Chaichana et al. (2018)
Glencar, Ireland	Patterned peatland	• Annual NEE 242 gCO_2 m^{-2}	• Annual NEE 206 gCO_2 m^{-2}	Laine et al. (2006)
Florida[a]	Scrub-oak ecosystem	• 8–26% higher NEE flux than EC	• Measure low NEE flux than close chamber	Dore et al. (2003)
Bavaria, Germany	Submontane grassland	• Mean CO_2 value-5.6 $\mu molm^{-2}$ s^{-1} • Chamber sink flux was 40% larger than EC	• Mean CO_2 -4.0 $\mu molm^{-2}$ s^{-1}	Riederer et al. (2013)
Tibetan Plateau, China	Alpine wetland	• 7.6% greater CH_4 fluxes than EC	• Measure low CH_4 flux	Yu et al. (2013)
Laramie, US	Glacier lake ecosystem	• Higher ecosystem respiration rate (ER) than EC	• Lower ER value for the same time periods 34% (2011) to 58% (2005), and	Speckman, (2013)
Changbai Mountain Natural Reserve, China	Mature temperate mixed forest	• CC estimates of NEE and ER were 146, 1240 and 1220 gC m^{-2} $year^{-1}$, respectively • NEE was 9.8% of actual GPP	• EC estimates of NEE, ER and GPP were 188, 1030 and 1220 gC m^{-2} $year^{-1}$, respectively • NEE was 22.5%higher than chamber-based measurements	Wang et al. (2019)

[a] Dore et al. (2003), used open top chamber and EC technique to measure NEE in scrub-oak ecosystem of Florida

basic structure of OTCs present in the farm of Indian Agricultural Research Institute (IARI), New Delhi is built of galvanised iron (GI) pipes with side walls of PVC sheets that allow transparency of more than 90% (Fig. 21.3). Central Research Institute for Dryland Agriculture (CRIDA) has developed completely automated OTC facility which can maintain the desired atmospheric CO_2 concentration using manifold gas regulators, solenoid valves, and supervisory control and data acquisition (SCADA). It also continuously records the temperature and relative humidity inside with the help of temperature and relative humidity sensors (Srinivasa Rao et al., 2018). Air compressor is used to maintain the homogeneity of CO_2 concentration. Air inside the OTCs is sampled from the centre through height adjustable copper tubes and further analyzed for the CO_2 concentrationusing InfraRed Gas Analyzer (IRGA).

Fig. 21.3 Open top chamber at ICAR-IARI, New Delhi, India

Several experiments have been conducted using OTCs to study the effect of elevated CO_2 and ozone (O_3) on crops and soil, the influence of moisture deficit and its interaction at different CO_2 concentrations on crop yield. Bhatia et al. (2012b) studied the effect of elevated tropospheric ozone (O_3) and CO_2 on growth and productivity of high quality protein maize in open top chambers in IARI. The yield decreased by 13 and 20% under elevated O_3 while the presence of elevated CO_2 along with elevated O_3 increased the yield by 9–10% over elevated O_3 alone. Severity of Maydis leaf blight disease was found to be highest in maize varieties at 3 °C elevated temperature and ambient CO_2 level (Kumar et al., 2016). In pigeon pea crop, profile moisture content was found to be lower under elevated CO_2 indicating that the crop was using more water. However, increasing biomass resulted to higher crop water use efficiency. In case of chickpea, seed germination, and seedling vigour were reduced under elevated CO_2 levels. CO_2 enrichment resulted a significant increase in leaf area index in chickpea but no change in fractional intercepted photosynthetically active radiation (PAR). Cumulative carbon emitted was more than the carbon fixed under elevated CO_2 as compared to ambient CO_2 levels (Daripa et al., 2014). The effects of elevated CO_2 and temperature on soil organic N fractions and enzyme activity in the rice rhizosphere was quantified in OTCs (Maity et al., 2020). OTCs are also used to study the effect of air pollutant sulphur dioxide (SO_2) on maize crop in South

Africa (Heyneke et al., 2012). Guru-Pirasanna-Pandi et al. (2018) studied the effect of elevated CO_2 on brown planthopper (BPH) in rice grown in OTCs. Elevated CO_2 showed a positive effect on BPH population through increase in fecundity. Rice crops suffered yield loss under elevated CO_2 level due to increased BPH infestation.

21.4.2 *Free Air CO_2 Enrichment (FACE)*

Free air carbon dioxide enrichment (FACE) allows studying the impact of elevated CO_2 on crops under natural conditions. It does not require enclosures unlike growth chambers and greenhouses. Hence, crops are exposed to elevated CO_2 concentration under ambient environment without causing any changes in other weather variables like temperature, precipitation, wind, relative humidity and sunlight. Besides this, the confined enclosures do not cause any problems related to crop growth and development. In India, South Asian Mid FACE facility was developed at IARI, New Delhi to study responses of crops to elevated CO_2 under natural field condition (Uprety et al., 2000). Another fully automated FACE facility was later on developed at IARI which consisted of a ring of eight horizontal PVC pipes arranged in an octagonal shape (Fig. 21.4) (Chakrabarti et al., 2012a, 2012b). The FACE ring consisted of a circular array of horizontal or vertical pipes which release CO_2 enriched air at the crop canopy level. The supply and monitoring of CO_2 concentration inside the FACE was regulated by the computer-based Supervisory Control Data Acquisition (SCADA) system. The CO_2 enriched air was supplied from CO_2 cylinders passed through air mixing compressor and regulated by flow meter, andsolenoid valves. CO_2 enriched air was released inside the FACE rings by the solenoid valves and opening and closing of these valves were regulated by wind direction and wind speed. Air samples were three points inside the FACE rings and analysed using IRGA. Data on CO_2 concentration, temperature, wind speed, relative humidity were logged automatically at every 5 min interval.

A meta-analysis was carried out to synthesise the findings of previously published studies carried out in FACE analysed data from 12 FACE sites covering more than 40 plant species (Ainsworth & Long, 2005). The response of trees to increased CO_2 was found to be greater than that of herbaceous plants. While C4 plant species did not respond much to high CO_2, crop yields increased less than in earlier enclosure tests. Throughout the course of the FACE trials, elevated CO_2 accelerated leaf photosynthetic rate while reducing stomatal conductance.

A study conducted in FACE at IARI, showed that elevated CO_2 level (550 ppm) increased yield by 15.1–16.7% in wheat and 21.1–21.9% in chickpea over ambient (Chakrabarti et al., 2020). Raj et al. (2019) reported that grain yield of rice increased by 25.5% under elevated CO_2 (550 ppm) level than ambient (400 ppm) using FACE facility at IARI. Grain N concentration in rice decreased in elevated CO_2 treatment. Other studies conducted in FACE facility with leguminous crops like mungbean and cowpea showed that elevated CO_2 enhanced photosynthesis rate, leaf area, seed yield,

Fig. 21.4 FACE facility at ICAR-IARI, New Delhi, India

seed P uptake and also improved nodulation, nitrogen fixation and availability of P in soil (Dey et al., 2017a, 2017b).

A study was conducted with five different Japanese rice cultivars inside the FACE facility in farmers' field located at Tsukubamirai City, Japan to examine the response of those cultivars bred in different time periods elevated CO_2 (Sakai et al., 2019). Results suggested that breeding had not improved the response of those cultivars to elevated CO_2 level. Selection of traits like sink capacity, nitrogen uptake can help in improving rice yield under elevated CO_2 condition. A study conducted by growing soybean crop at the SoyFACE research facility in University of Illinois showed that elevated CO_2 concentration delayed reproductive development by up to 3 days in soybean. Soybean grown under elevated CO_2 required more growing degree days (GDD) to complete full bloom stage. Table 21.3 shows the list of different experiments carried out in FACE facilities located in various countries.

Comparison between OTC and FACE facility

Both OTC as well as FACE facilities are being used to study the impact of elevated CO_2 on crop plants. However, there are certain advantages and limitations associated with both the systems (Table 21.4).

21.4.3 Temperature Gradient Tunnel (TGT)

Different technologies have been adopted worldwide to study the impact of elevated temperature on plants. Enclosed chambers, temperature tunnels, walk-in tunnels etc. have been used to expose the plants to high temperature condition. The temperature gradient tunnels (TGT) allow researchers to study the response of plants to a range of temperatures within the same tunnel under field condition. The advantage of TGT is that it allows exposing the crop plants to a wide range of temperature combinations.

Table 21.3 Yield enhancement in different crops from experiments conducted in different free air CO_2 enrichment (FACE) facilities in different countries

Site	Crop	Year	CO_2 concentration (ppm)	Percentage yield change	References
Tsukubamirai City, Japan	Rice	2010–2011	560–585	3.4–30.3%	Sakai et al., (2019)
FACE experiments, China	Rice	2001–2016	+200	13.5–32.8%	Lv et al., (2020)
FACE experiment Shizukuishi, Iwate, Japan	Rice	1998–2000	+200	7–15%	Kim et al., (2003)
SoyFACE, USA	Soybean	2001–2003	550	15%	Morgan et al., (2005)
SoyFACE, USA	Soybean	2004–2008	550	9%	Bishop et al., (2015)
FACE facility Stuttgart Germany	Oilseed rape	2007	500	18.4%	Hogy et al., (2010)
Mini-FACE, Stuttgart	Wheat	2004–2006	+150	10.4%	Hogy et al., (2009)
AGFACE, Australia	Field pea	2010–2012	550	26%	Bourgault et al., (2016)
AGFACE, Australia	Wheat	2007–2009	550	24–53%	Fitzgerald et al., (2016)
FACE facility, Tsukuba, Japan	Rice	2010	+200	3–36%	Hasegawa et al., (2013)
FACE, IARI	Wheat	2011–2013	550	15.1–16.7%	Chakrabarti et al., (2020)
FACE, IARI	Chickpea	2011–2013, 2017	550	21.1–21.9%	Chakrabarti et al., (2020), Bhatia et al., (2021)
FACE, IARI	Rice	2013	550	25.5%	Raj et al., (2019)
FACE, IARI	Cowpea	2014	550	27.3%	Dey et al., (2019)
FAOCE, IARI	Wheat	2016–2018	550	8.2–20.9%	Yadav et al., (2019)

Table 21.4 Advantages and limitations of OTC and FACE system

Facility	Advantages	Limitations
OTC	• CO_2 concentrations can be regulated more easily as it is partially enclosed • More appropriate for examining the physiological reactions of plants • Provide the crop with near to reality environment • Easy to move and can be left in the field for several seasons • Structurally sturdy enough to survive heat, cold and winds • OTCs are simple, low-cost systems. Capital as well as operational cost is quite low than FACE system	• It is difficult to conduct the study under natural condition • The experimental area is less as size is a constraint • The side walls of the OTCs have a screening effect which lowers the amount of photosynthetically active radiation (PAR) reaching the plants • The microclimate inside gets modified. Hence the results are not as realistic as FACE studies • Due to the chamber effect temperatures are slightly higher inside as compared to the ambient condition
FACE	• FACE systems are in open air and have benefits over controlled environment experiments • Allows the study to be conducted in an undisturbed ecosystem • The system does not modify the microclimate (light, temperature, precipitation, wind direction, wind speed, pests, pollinators etc.) as experienced by the plants • Due to their open design and huge diameter, these facilities are more suited for examining how increased CO_2 affects forest ecosystems • Avoids problems of edge effects prevalent in enclosed chambers • More amount of plant material can be harvested due to the larger area • Lot of data generated helps in better interpretation of the findings and making realistic projections of plant response to elevated atmospheric CO_2 condition • In FACE the absolute responses are obtained while in enclosed chambers or OTCs relative responses are obtained	• The major limitation is the huge cost involved in constructing and operating the system • In most of the FACE studies CO_2 concentration is not elevated during the night time due to the difficulties associated with controlling $CO_2$and the high expenditure involved • Trees might get damaged by tiny herbivores which remain protected in enclosed chambers

The TGTs constructed may be covered with UV stabilized polythene sheets which have 90% transparency (Fig. 21.5). This creates a greenhouse like atmosphere inside as the air gets naturally heated by the incoming sunlight. Ambient air is sucked from one end of the tunnel and released to the other end leading to creation of the temperature gradient. The regulation of fan speed generates desired temperature gradient. Temperature sensors installed inside record temperature at different points of the tunnel and data is logged at 1 min interval. The area inside is divided into plots equal

Fig. 21.5 TGT facility at ICAR-IARI, New Delhi, India

to the number of temperature treatments. Plants in each plots are exposed to different temperatures which allows to study the crop response to varying temperatures under similar condition.

The sheets should be washed frequently to maintain high transparency level. Temperature sensors also need to be calibrated to get real values of temperature. Construction of double door ensures maintenance of desired temperature levels inside the TGT.

A study conducted inside TGT at IARI for two consecutive years showed that crop duration reduced in wheat and chickpea crop under elevated temperature condition. In wheat, reduction in yield was around 10.5% per degree rise in temperature (Chakrabarti et al., 2013). But temperature rise was found to be beneficial for chickpea crop to a certain extent. Another study in wheat showed that temperature rise by 3.5 °C increased accumulated growing degree days (GDD) and photo-thermal unit (PTU), thereby hastening maturity of the crop. Grain yield reduced by 17.1%, with 3.5 °C rise in temperature (Chakrabarti et al., 2021). Kannojiya et al. (2019) conducted experiment inside the TGT to study the effect of elevated CO_2 and temperature on growth, yield and weed infestation in wheat crop. The results indicated that elevated CO_2 concentration of 550 ppm was able to compensate the yield loss in wheat upto 1.5 °C rise in temperature. The yield loss due to weed infestation accelerated due to enhanced crop-weed competition. As weeds respond more to elevated CO_2 level than crops, new weeds could also become problematic in the future. Hence, better weed management practices will be required under future climatic condition.

Carbon dioxide and Temperature Gradient Chamber (CTGC)

CTGC is designed to study the impact of both elevated CO_2 and temperature on crop plants. A CTGC facility was developed at ICAR-CRIDA in field with both elevated CO_2 and high temperature conditions (Fig. 21.6). The heating inside the CTGC system is done with infrared heaters while CO_2 is supplied from liquid CO_2 tank. CO_2 concentration is measured by CO_2analyzer while temperature and humidity sensors

Fig. 21.6 CTGC facility at ICAR-CRIDA, Hyderabad, India

measure air temperature and relative humidity. Temperature gradient is maintained through the air inlet and outlet and cooling/heating system. Data is recorded through computer system with SCADA software. The polycarbonate sheets are cleaned regularly to remove the dust, soil and other deposits. This facility helps in studying the interactive effects of elevated CO_2 and temperature on crops and insect's-pests.

21.4.4 Free Air Temperature Elevation (FATE) and Temperature FACE (T-FACE)

Free Air Temperature Elevation (FATE) facility helps in increasing the canopy temperature of crop plants under open field condition. The FATE system developed at ICAR-CRIDA consists of nine rings where there is provision of exposing the crops to both elevated CO_2, and temperature condition (Srinivasa Rao et al., 2018). For effective temperature increase it is connected to an automated temperature and CO_2 control and monitoring system. Different weather parameters like air temperature, CO_2 concentration, relative humidity, wind speed, wind direction etc. are measured at regular intervals and logged by the data logger.

The Temperature FACE (T-FACE) facility developed at ICAR-IARI also helps to study the interactive effect of both elevated CO_2 and temperature of crop plants under open field condition (Fig. 21.7). The system is installed with infrared heaters which increases the canopy temperature inside the rings. This new technology is a fully automated system where temperature and CO_2 concentrations are regulated and monitored by the computer-based Supervisory Control Data Acquisition (SCADA)

Fig. 21.7 T-FACE facility at ICAR-IARI, New Delhi, India

system. Different weather parameters are measured by sensors at regular intervals and logged by the data logger. Both FATE and T-FACE facilities allow us to conduct experiments with elevated CO_2 and temperature under open field condition without using enclosures.

A FATE facility located at the Yangtze River valley China was developed to study the impact of global warming on rice-wheat cropping system. The system was able to create uniform warming in the plots with 67% of the temperature readings were within ±0.5 °C of the target value. However, the effectiveness of heating was lowered at high wind condition.

The interactive effect of elevated CO_2 and temperature was studied on groundnut crop under FATE facility at CRIDA (Vanaja et al., 2019). Elevated temperature significantly decreased photosynthetic rate in groundnut genotypes while elevated CO_2 plus elevated temperature was able to mitigate the ill effects of high temperature. A study conducted by growing rice crop in a free-air temperature increase (FATI) facility in subtropical China showed that temperature increase by 1.4–2.1 °C did not significantly affect grain yield in rice crop. But nitrogen loss significantly increased in high temperature treatments in the late rice crop. Hence, global warming might increase N losses and reduce N use efficiency in rice (Yang et al., 2019).

21.4.5 Free Air Ozone and Carbon Dioxide Enrichment (FAOCE)

Tropospheric ozone (O_3) is an air pollutant harmful for both plants and animals. The effects of ozone on plants begin in the leaves when the gas enters through the stomata and affects cellular processes, resulting in a reduction in photosynthesis and yield.

Fig. 21.8 FAOCE facility at ICAR-IARI, New Delhi, India

Ozone enrichment in the open air allows studying the impact of elevated O_3 on plants without causing any change to other environmental factors. The free air ozone and carbon dioxide enrichment (FAOCE) facility allows studying the impact of elevated ozone (O_3) and CO_2 on crops under natural field conditions. The FAOCE facility located at ICAR-IARI are rings with diameter of 5 m (Fig. 21.8). CO_2 is released at the canopy level inside the rings through PVC tubing controlled by the solenoid valves. The wind vane and anemometer installed inside monitor the wind direction and wind speed and CO_2 is released as per the wind direction and speed inside the rings. Ozone generators convert atmospheric oxygen (O_2) into O_3 by ultraviolet (UV) lamps. Blowers mix O_3 with air and it is controlled by the flappers using solenoid valves. CO_2 and O_3 concentrations inside the rings are measured by CO_2 analyzer and O_3 analyzer respectively. The microprocessor based system helps in its operation, as well as data logging at regular interval.

21.5 Green House Gases Mitigation in Agriculture

Agriculture sector has also been identified as one of the potential sectors for GHG mitigation after Parris agreement by adopting of low carbon technologies. According to UNEP report on Roadmap to carbon free future, it has been estimated that agriculture sector has the potential for reducing 6.7 Gt of CO_2 eq. Out of this approximately 2 Gt of emissions can be mitigated by reducing food wastage and shifting to sustainable plant based diets. However, there is no quick, simple and shortcut method for mitigation of GHG emission from agriculture sector. Several technological interventions/agricultural practices have been reported in literature which has the potential to reduce the ill-effect of climate change by reducing/avoiding GHG emissions (Table 21.5) (Smith et al., 2014). These agricultural management practices can improve crop yield and nutrient use efficiency and reduce the losses of different forms of carbon and nitrogen from soil. Energy crops and surplus crop residues could be the potential source of biofuel after conversion to ethanol/diesel or electricity. The main target of these technologies is to avoid the fossil fuel-based emissions.

Indian rice fields emit 3.4 Tg of CH4 annually. The conventional practice of rice cultivation leads to emission of methane due to prevailing anoxic conditions after flooding the fields. Efficient water management in rice crop such as direct seeded rice (DSR), system of rice intensification (SRI), aerobic rice (AR), alternate wetting and drying (AWD), mid-season drainage (MSD) of the paddy fields etc. has the potential to reduce the CH4 emission. It has been evaluated that practices such as modified rice drainage and straw incorporation could diminish worldwide CH4 outflows from rice cultivation by up to 30%. However, there will be increase in emission of nitrous oxide. Using the slow release/coated fertilizers, nano fertilisers, site specific nitrogen application using green seeker/leaf colour chart can help in reducing emission of

Table 21.5 Possible mitigation techniques for agriculture sector

Mitigation technique	Measures
1. Management of cropland	Agronomic and nutrient management to enhance nutrient efficiency and reduce losses, tillage, residue management, water management in rice, crop diversification, intercropping, agroforestry and land cover use
2. Grazing land management and pasture improvement	This measure includes techniques like Grazing intensity, Increased productivity (including fertilization), Nutrient management, Fire management and Species introduction
3. Management of organic soils	Avoiding deep ploughing, maintaining a shallower water table, avoiding the drainage of these soils in the first place and re-establishing a high water table where GHG emissions are still high
4. Restoration of degraded lands	Application of organic amendments, and nutrient
5. Livestock management	Improved feeding practices with dietary additives/supplements
6. Manure management	Improved storage and handling of manure during anaerobic digestion, efficient use as a source of nutrients
7. Bioenergy	Cultivation of energy crops, biogas from manure, use of surplus agriresidues for energy

Source Adapted from Smith et al. (2014)

Table 21.6 Mitigation Potential and possible constraints of different GHG mitigation options

Mitigation strategies	Mitigation potential (%)	Yield (%)	Constraints
1. Methane emission from rice field			
• Intermittent drying/AWD	25–35	(−5)–(−8)	Assured irrigation
• Direct-seeded rice	30–40	(−5)–(−8)	Machine, herbicide
• SRI	20–25	−5–10	Labour, assured irrigation
• Midseason drainage	15–20	0–5	Assured irrigation
• Crop diversification	10–15	–	MSP for produce, Inadequate market
2. Nitrous oxide emission from soil			
• Site-specific N use	10–15	10–15	Awareness, fertiliser policy
• Nitrification inhibitor	10–15	10–20	Additional cost, No incentive
• Nano fertilizers	10–12	5–10	Additional cost, No incentive
• Coated/slow release fertilisers	15–20	10–15	Additional cost, No incentive
3. Residue management			
• Conservation agriculture	15–20	5–8	Continuity, small farm holding
• Zero tillage	15–20	5–8	Continuity

Source Adapted from Pathak et al. (2014)

N_2O from crop land. The GHGs mitigation potential of the technologies are given in Table 21.6.

21.5.1 *Water Management for Mitigating Methane Emission from Rice Fields*

Alternate wetting and drying (intermittent irrigation)

Alternate Wetting and Drying (AWD)/intermittent flooding (IF) reduces water use by up to 30% and can save cost incurred on pumping irrigation water by farmers. It has been used widely across the globe as a water saving practice for several decades. AWD is a practice wherein the periodic drying of rice fields is done either naturally through percolation and evapotranspiration or intentionally and re-irrigating utilizing a 'field water tube to save water instead of keeping the field continuously flooded AWD is generally acknowledged as the most encouraging practice for decreasing GHG outflows from flooded rice and can reduce methane (CH_4) emissions by an average of 40–50% compared to continuous flooding, but this comes at a cost of increased N_2O emission and reduced yield.

Several studies have demonstrated the impact of AWD on GHG emission in Asian region. In the region of in Hue, Jakenan, PrachinBuri, and Muñoz, AWD reduced CH_4 emissions relative to CF by 151 (25%), 166 (37%), 9 (31%), and 22 (32%) kg CH_4 ha^{-1} $season^{-1}$, respectively. However, in PrachinBuri and Muñoz, reduction in CH_4 emissions was observed only during the dry season. The change in soil redox potential due to shift from anaerobic condition during continuous flooding to aerobic soil conditions for short period by AWD is the main cause for this significant reduction. During this time CH_4 production is suppressed while CH_4 oxidation is accelerated thereby resulting in CH_4 mitigation from rice. AWD at 15 and 30 cm water level reduced CH_4 emission by 57.06 and 84.29% compared with continuous flooding (CF) water regime in rice. However, under AWD practices increases in N_2O emission ranging from 14–50% has been observed by several researchers. Cowan et al. (2021) observed that converting from CF to IF irrigation does lower CH_4 emissions (by approximately 18%); however, this comes at a cost of significant decrease in grain yield (6.1%) and NUE (22.5%) when compared to CF. Bhatia et al. (2023) have shown that by practicing intermittent flooding in irrigated rice, CH4 flux could be reduced by 40%. However, this will lead to increase in N_2O-N fluxes by approximately 6%. However, the total GWP from the irrigated rice areas of the country could be reduced from 41.1 Tg C to 36.2 Tg C per annum with AWD practice in rice. The increased N_2O emission from rice field with AWD/intermittent flooding can be mitigated with efficient use of nitrogenous fertilisers.

Midseason drainage

Practice of short-term drainage of water from rice fields for one to two weeks just before maximum tillering stage in rice is known as midseason drainage (MSD). It has been reported in literature that MSD has a potential of reducing CH4 emission by 36- 65% due to availability of plenty of oxygen.

Prolonged MSD was able to mitigate total CH_4 flux by 69% in Japan as compared to conventional irrigation practice (Itoh et al., 2011). Total CH_4 emanations under constant flooding and MSD were in the range of of 6.4–12.0 and 1.7–7.8 Tg C $year^{-1}$, respectively with the potential abatement of around 5 Tg C $year^{-1}$ with MSD (Li et al., 2002). Mid-season drainage makes unsaturated soils conditions, which advances N2O production. Therefore, management practice like MSD may be an effective tool to reduce CH4 emission in irrigated rice ecosystem. However, this practice may be specific for locations with assured irrigation facility.

System of Rice Intensification (SRI)

The System of Rice Intensification (SRI) was introduced for lowland irrigated rice to save irrigation water by Uphoff, in which young rice seedlings of 8 to 12 days old and in some places 14–15 days old seedlings (2–3 leaf stage) are transplanted singly in square pattern with moist field condition. SRI leads to better plant growth, economizes use water, chemicals and fertilizer use, and increases productivity in sustainable manner. Similar to AWD/IF, SRI has the potential to reduce the emission of methane as this technique does not require continuous soil submergence. Jain et al. (2014) observed 61–64% decrease in methane emission with SRI compared to conventional method of growing rice with flood irrigation. However, the study shows the increase in N_2O-N emission by 22.5% under the SRI. The GWP however, was reduced by 28–30% in SRI over the conventional method. Another study shows the reduction in GWP up to 37.5% in Indonesia (Hidayah et al., 2009). Adoption of SRI also saved total energy inputs by 4350 MJ/ ha along with lowering the emission of 0.280 kg CO_2 e/kg rice grain (Nirmala et al., 2021). Higher yields with SRI than conventional paddy cultivation (10–40%) have been reported.

21.5.2 Direct Seeding of Rice (DSR)

Water scarcity and mitigation of GHG are the drivers to explore different rice cultivation method. Due to its high working potential under water and labour scarce conditions, this method has gained acceptance by farmers. Dry seeding, wet seeding and water seeding are 3 prime methods of DSR. In dry-seeding, rice seeds are broadcasted or drilled into non-puddled and unsaturated soil and then covered till germination. Upland rice and aerobic rice are the examples of dry DSR. Pre-germinated seeds (radicle length 1–3 mm) are sown on or into puddled soil in wet direct seeding. In this, it is necessary to ensure that landscape is well levelled, well puddled and weed free. The process can be done both in aerobic and anaerobic condition. During the process, seeds can either be spread or planted in-line by utilizing drum seeder. Moreover, an anaerobic seeder along with a furrow opener and closer can also be utilized. Water seeding of rice involves broadcasting of pre germinated seeds in standing water on puddled or un-puddled soil. It has become popular among the landscapes facing issues of weedy rice and red rice. Similar to water management practices in rice, the methane emission is drastically reduced in both the DSR method i.e. dry and wet seeded rice, however due to aerobic soil conditions emission of N_2O is increased (Bhatia et al., 2013).

A study conducted at three villages of Jalandhar district, Punjab documented that total GWP in the DSR ranged from 1.3 to 2.9 t CO_2 eq. ha^{-1} which is quite lower than TPR (2.0–4.6 t CO_2 eq. ha^{-1} (Pathak et al., 2013), however, N_2O emission increased. The usage of UDP (urea deep placement) and DCD (dicyandiamide) along with DDSR shows the reduction in N_2O emission by 95% and 73% respectively, along with yield increase in the range of 16–19% (Gaihre et al., 2020). Global warming potential has been estimated to be 76.2% lower for dry direct-seeded rice in China (Tao et al., 2016).

21.5.3 Crop Diversification

Rice cultivation occupies a large area especially in Asian countries. Rice-based cropping system has complex effect on GHG emissions. Rice is conventionally grown under submerged soil conditions resulting in higher global warming potential than other crops due to more methane emission. Diversified cropping systems along with intercrops or mixed crops will not only increase the systems yield but also reduce GHG emissions and lower the foot prints of cereal crops. Crop diversification improves productivity and lowers the carbon footprints and thus can be a viable option to mitigate the risk of climate change (Pathak et al., 2014). Diversification of rice crop with other cereal crop such as maize, pulses and oil seed crops can reduce the GHG emission. In several places land remains fallow after wet season rice cultivation emitting more N_2O due to root biomass decomposition, thereby reducing the C:N ratio of soil. However, the GHG intensity per unit of land may vary significantly under different cropping systems. It has been well recognised that diversification has the potential for developing resilience in agricultural systems to climate change and GHG emission can be drastically reduced by 70–80% with crop diversification from rice–wheat or rice-rice to maize-wheat system.

21.5.4 Mitigation of Nitrous Oxide Emission

Nitrification (aerobic soil conditions resulting in oxidation of ammonium to nitrate to N_2O by auto and heterotrophic bacteria) and denitrification (anaerobic soil condition leading to reduction of nitrate to N_2O and N_2 by heterotrophic bacteria and fungi) are the two important pathways responsible for N_2O production and emission from agricultural soils. Soil N_2O emissions are sensitive to different management, environmental, and soil factors that can affect N_2O-production processes. Soil nitrogen and organic carbon, pH, temperature, texture, moisture and pore structure play a substantial role in N_2O emissions (Bhatia et al., 2010b). Different agri-management practices can affect N_2O production and emissions from soils.

Mitigation of N_2O emissions from managed agroecosystems is quiet challenging due to complex processes of N_2O production in the soil. However, reduction in N_2O emission can be done through increasing the N use efficiency and decreasing N losses by efficient management of N applied through manure and synthetic fertiliser using 4R principal i.e. use of Right source, Right rate, Right time, and Right place). Use of nitrification/urease inhibitors, slow release fertilisers, nanofertilisers, site specific and demand driven N application, management of crop residues can mitigate N_2O emission from soil (Pathak et al., 2016). However, mitigation measures have to be implemented by farmers. Therefore, no matter how extensive is the list of mitigation option, it has to be simple, cost effective and easy to implement at field/farm level. In the following section different N_2O mitigation options are discussed in detail.

Site specific Nutrient Management

Site-specific nutrient management (SSNM) is the management of agricultural crops in which nutrients are supplied to plants based on their requirement and the soil's nutrient supplying capacity to account real-time variation in crop N demand at major growth stages. Chlorophyll Meter, Leaf Color Chart (LCC), Green seeker and decision support systems like Nutrient Expert (NE) (http://software.ipni.net) are different tools available to apply nutrients especially nitrogen based on crop demand. SSNM generally maintains or increases crop yields. In a study conducted at 13 sites in Southeast Asia, SSNM led to grain yield increases of 13% over a three-year period, although yields declined slightly in the first year (Pasuquin et al., 2014). LCC card is the cheapest method for synchronization of crop N demand with supply. The LCC was developed by International Rice Research Institute, and Philippines Rice Research Institute, Philippines of measuring the required quantity of Nitrogen to be applied in rice fields. LCC has four green strips, with shading going from yellow green to dull green. The greenness of the rice leaf is related to the N content of leaf. Benefits of using LCC or other SSNM tools is that the crop growth is better resulting in higher or similar yields with less amount of nutrient, cost effective and helps in reducing GHG emission.

SPAD and LCC based N application in wheat increased N-use efficiency over 3 split of N application with saving of 60 and 40 kg N ha, respectively without any yield penalty. Bhatia et al. (2012a) observed that LCC based application of N in rice wheat system has the potential to reduce the GWP by 10.5%. Sapkota et al. (2019) reported that nutrient expert based (SSNM) fertilizer recommendation application in all rice and wheat acreage in India would increase the productivity of rice–wheat system by 13.92 million tonnes (Mt) with 1.44 Mt less N fertilizer and 5.34 Mt CO_2 e per year of GHG reduction as compared to current farmers practice. Besides reduction in GHG emission application of N on the basis of LCC also reduced NO_3-N leaching by 39.8% (Mohanty et al., 2017). Xia et al., (2016) observed significant increase in grain yield (1.3–10%), N use efficiency (8–48%) and reduction in N_2O emission (5.4–39.8%) by use of different SSNM practices.

Nitrification/urease inhibitors

Nitrification inhibitors slow the nitrification of N, when applied to soil along with fertilisers. Whereas, application of urease inhibitors is known to slow down the process of urea hydrolysis in soil. These inhibitors not only reduce the emission of nitrous oxide but are also reported to increase the economic yield. Several chemical (e.g. dicyandiamide (DCD), nitrapyrin (2-chloro-6-(trichloromethyl) pyridine; NP) and N-(n-butyl) thiophosphorictriamide (NBPT) thiosulphateetc) sodium chlorate and sodium azideand plant based inhibitor (neem oil, neem cake, karange oil, karange cake, etc.) are reported in literature for their potential to reduce the emission of GHGs mainly N_2O due to their capability of inhibition of urea hydrolysis (Pathak et al., 2002). Globally, several field investigations were conducted to study the mitigation potential of different chemical and plant based nitrification/urease inhibitors alone or in combination for reduction in emission of N_2O. However, many of these NIs have limitations to their usefulness, and have been restricted to the experimental stage only because of the cost effectiveness, limiting availability and having adverse impacts on beneficial soil microorganisms.

Impact of hydroquinone along with DCD application at basal, tillering and panicle initiation stage on N_2O emission and observed of 23.59%, 55.64% and 17.18% mitigation of seasonal N2O emission respectively over control. Malla et al (2005) have reported reduction in N_2O emission ranging from 5 to 10% with different NI from wheat soils in north western part of India. The use of dicyandiamide (DCD), nitrapyrin (2-chloro-6-(trichloromethyl) pyridine; NP) and N-(n-butyl) thiophosphorictriamide (NBPT) nitrification inhibitors improved rice growth mainly due to the increase in urea N use efficiency and reduced the GWP (Meng et al., 2020). Dawar et al. (2021) have observed that application of nitrification inhibitor with or without biochar Significantly reduced the N_2O emission over control. Among the natural NI, neem oil and neem oil coated urea is widely used in India in agricultural crops. Application of neem oil coated urea is able to reduce N_2O emissions by 10–20% over urea along with 5–7% increase in economic yield. However, use of NI although reduced N_2O emission but enhances the NH_3 volatilization losses by 3–38% (Cowan et al., 2021). Combination of Urease and Nitrification Inhibitors used in maize—wheat in China was able to reduce 48.3, 25.1 and 38.2% of N_2O emission with the application of NI, UI and NIUI with urea respectively compared to urea alone.

Slow release fertilizers and Nano fertilisers

Nitrogenous fertilizers are prone to losses via different pathways like nitrification, denitrification, leaching, volatilization, and immobilization contributing approximately 60% or more of applied N losses. In order to reduce these losses of N the best option is to enhance the N use efficiency where slow release fertilisers and nano fertilisers may hold a promise (Kirti et al., 2021). A slow-release fertilizer is composed of a coating the water soluble fertiliser with a semipermeable membrane which allows the sustained release of small amount of nutrient over the period of time enabling the plant to uptake the nutrients as and when required and reducing the losses. Several coating materials have been used across the world to prepare these fertilisers such as sulphur, neem oil, attapulgite, starch, chitosan, zeolite, ca alginate to name a few. Nano fertilisers are the materials containing macro and micronutrients in the form of nanoparticles delivering the nutrients to the crops in a controlled mode gradually and slowly.

Recently, a study conducted by Jumadi et al. (2019), in Indonesia found that application of urea coated with zeolite, neem and DCD significantly reduced N_2O emission by 49.15% and 45.76% respectively over urea. Bordoloi et al. (2020) conducted field experiment with starch coated and neem coated urea and observed 20.88% and 11.65% N_2O emission mitigation respectively, over urea. The application of urea crosslinked with formaldehyde reduced N_2O emissions by ~42% and 15–26% during maize and wheat growing season, respectively (Jiang et al., 2010). Kirti et al. (2021) have reported that nano clay polymer coated composites when applied with 75% of recommended dose of N fertiliser reduced N_2O emission by 16.1% in rice and 12.4% in wheat as compared to urea. In India, nano urea has been developed by a fertiliser industry and is reported to reduce the quantity of N fertiliser by 25–40%. However in-depth studies are required to evaluate the impact of Nano fertilisers on environmental losses.

Crop residue management and tillage

Conservation agriculture, which involves minimal or no mechanical soil disturbance and permanent soil cover with crop residues, offers prospects for input savings, increasing resource efficiency, and GHG emissions reduction. Farmers are becoming more interested in zero-tillage farming. Using this method, rice–wheat farmers may drill wheat directly after harvesting rice without any prior tillage, ensuring that grain filling in wheat crop happen before the commencement of pre-monsoon hot weather. These techniques are becoming more popular across the world. When compared to conventional tillage (CT), zero-tillage (ZT) systems provide higher soil quality, carbon sequestration, and maintain a profitable and sustainable agricultural system owing to the build-up of soil organic carbon. The CO_2 emission has been found to be significantly lower in the ZT plots than CT plots (Bhattacharya et al. 2018). ZT may potentially have an influence on N_2O emissions, however the overall effects are variable and poorly documented internationally. Decreased N_2O emissions due to reduced tillage may vary depending on soil and climatic conditions. In some regions, reduced tillage enhances N_2O emissions and in other areas, it may lower emissions or have no observable effect. However, higher emission of N_2O under ZT compared to CT practice has been reported by several researchers (Bhatia et al., 2010a; Bhattacharya et al., 2018), which has been attributed to enhanced denitrification with reduced tillage due to the production of micro-aggregates inside macro-aggregates, resulting in anaerobic micro sites (Bhattacharya et al., 2018).

Under CT systems net global warming potential is significantly higher (26–31%) than ZT systems (Mangalassery et al., 2015). Retention or incorporation of crop residues also tend to increase soil C as the crop residues are the precursors for soil organic matter in the soil. However, soil respiration also increases by the residue incorporation and the effect is dependent on amount of residue incorporated or retained. Shifting from CT to zero-tillage can decrease soil erosion and increment organic carbon by 7.9% of arable land around the world.

21.6 Conclusion

Advanced facilities for climate change research and greenhouse gas mitigation provide the technology and resources necessary to better understand the complexities of climate change and develop effective strategies for mitigating its impacts. Various advanced facilities, such as, eddy covariance system, free air CO_2 Enrichment (FACE), temperature gradient tunnel (TGT), carbon dioxide and temperature gradient chamber (CTGC), free air ozone and carbon dioxide enrichment (FAOCE), open top chambers (OTC), free air temperature elevation (FATE) and temperature FACE (T-FACE) play a critical role in our understanding of climate change and in developing strategies. These facilities use cutting-edge technology and methods to measure atmospheric levels of greenhouse gases, model climate scenarios, and test new technologies and strategies for reducing emissions. With continued investment in these facilities, we can improve our understanding of climate change and work towards a more sustainable future.

References

Ainsworth, E. A., & Long, S. P. (2005). What have we learned from 15 years of free air CO_2 enrichment (FACE)? A meta-analytic review of the responses of photosynthesis, canopy properties and plant production to rising CO_2. *New Phytologist, 165*, 351–372.

Arnold, P. W. (1954). Losses of nitrous oxide from soil. *Journal of Soil Science, 5*(1), 116–128.

Bahn, M., Reichstein, M., Davidson, E. A., Grünzweig, J., Jung, M., Carbone, M. S., Epron, D., Misson, L., Nouvellon, Y., Roupsard, O., Savage, K., Trumbore, S. E., Gimeno, C., CurielYuste, J., Tang, J., Vargas, R., & Janssens, I. A. (2010). Soilrespiration at mean annual temperature predicts annual total acrossvegetation types and biomes. *Biogeoscience, 7*(7), 2147–2157.

Bhatia, A., Sasmal, S., Jain, N., Pathak, H., Kumar, R., & Singh, A. (2010a). Mitigating nitrous oxide emission from soil under conventional and no-tillage in wheat using nitrification inhibitors. *Agriculture, Ecosystems and Environment., 136*, 247–253. https://doi.org/10.1016/j.agee.2010.01.004

Bhatia, A., Pathak, H., Aggarwal, P. K., & Jain, N. (2010b). Trade-off between productivity enhancement and global warming potential of rice and wheat in India. *Nutrient Cycling in Agroecosystems, 86*, 413–424.

Bhatia, A., Pathak, H., Jain, N., Singh, P. K., & Tomer, R. (2012a). Greenhouse gas mitigation in rice–wheat system with leaf color chart-based urea application. *Environmental Monitoring and Assessment, 184*(5), 3095–3107.

Bhatia, A., Tomer, R., Kumar, V., Singh, S. D., & Pathak, H. (2012b). Impact of tropospheric ozone on crop growth and productivity–a review. *Journal of Scientific and Industrial Research, 71*, 97-112.

Bhatia, A., Kumar, A., Das, T.K., Singh, J., Jain, N. and Pathak, H. (2013). Methane and nitrous oxide emissions from soils under direct seeded rice. International *Journal of Agricultural and Statistical Sciences , 9*, 729–736.

Bhatia, A., Mina, U., Kumar, V., Tomer, R., Kumar, A., Chakrabarti, B., Singh, R. N., & Singh, B. (2021). Effect of elevated ozone and carbon dioxide interaction on growth, yield, nutrient content and wilt disease severity in chickpea grown in Northern India. *Heliyon, 7*(1), e06049. ISSN 2405-8440, https://doi.org/10.1016/j.heliyon.2021.e06049

Bhatia, A., Cowan, N.J., Drewer, J., Tomer, R., Kumar, V., Sharma, S., Paul, A., Jain, N., Kumar, S., Jha, G., & Singh, R. (2023). The impact of different fertiliser management options and cultivars on nitrogen use efficiency and yield for rice cropping in the Indo-Gangetic Plain: Two seasons of methane, nitrous oxide and ammonia emissions. *Agriculture, Ecosystems and Environment,* 355, 108593.

Bhattacharyya, P., Neogi, S., Roy, K. S., Dash, P. K., Nayak, A. K., & Mohapatra, T. (2014). Tropical low land rice ecosystem is a net carbon sink. *Agriculture, Ecosystems and Environment, 189*, 127–135.

Bhattacharyya, R., Bhatia, A., Das, T. K., Lata, S., Kumar, A., Tomer, R., Singh, G., Kumar, S., & Biswas, A. K. (2018). Aggregate-associated N and global warming potential of conservation agriculture-based cropping of maize-wheat system in the north-western Indo-Gangetic Plains. *Soil and Tillage Research, 182*, 66–77.

Billesbach, D. P. (2011). Estimating uncertainties in individual eddy covariance flux measurements: A comparison of methods and a proposed new method. *Agricultural and Forest Meteorology, 151*(3), 394–405.

Bishop, K. A., Betzelberger, A. M., Long, S. P., & Ainsworth, E. A. (2015). Is there potential to adapt soybean (Glycine max Merr.) to future [CO2]? An analysis of the yield response of 18 genotypes in free-air CO2 enrichment. *Plant, Cell and Environment, 38*, 1765–1774.

Bordoloi, N., Baruah, K. K., & Hazarika, B. (2020). Fertilizer management through coated urea to mitigate greenhouse gas (N2O) emission and improve soil quality in agroclimatic zone of Northeast India. *Environmental Science and Pollution Research, 27*, 11919–11931.

Bourgault, M., Brand, J., Tausz, M., & Fitzgerald, G. J. (2016). Yield, growth and grain nitrogen response to elevated CO_2 of five fieldpea (Pisumsativum L.) cultivars in a low rainfall environment. *Field Crops Research, 196*, 1–9.

Brown, S. E., Sargent, S., & Wagner-Riddle, C. (2018). Evaluation of a lower-powered analyzer and sampling system for eddy-covariance measurements of nitrous oxide fluxes. *Atmospheric Measurement Techniques, 11*(3), 1583–1597.

Burba, G. G., & Anderson, D. J. (2007). Introduction to the Eddy covariance method: General guidelines and conventional workflow. *LI-COR Biosciences,* 141 http://www.licor.com

Cape, J. N., & Fowler, D. (2014). Trace gas exchange. 256–261.

Chaichana, N., Bellingrath-Kimura, S. D., Komiya, S., Fujii, Y., Noborio, K., Dietrich, O., & Pakoktom, T. (2018). Comparison of closed chamber and eddy covariance methods to improve the understanding of methane fluxes from rice paddy fields in Japan. *Atmosphere, 9*(9), 356.

Chakrabarti, B., Singh, S. D., Kumar, S. N., Aggarwal, P. K., Pathak, H., & Nagarajan, S. (2012b). Low-cost facility for assessing impact of carbon dioxide on crops. *Current Science, 102*(7), 1035–1040.

Chakrabarti, B., Singh, S. D., Bhatia, A., Kumar, V., & Harit, R. C. (2020). Yield and nitrogen uptake in wheat and chickpea grown under elevated carbon dioxide level. *National Academy Science Letters, 43*, 109–113. https://doi.org/10.1007/s40009-019-00816-y

Chakrabarti, B., Bhatia, A., Pramanik, P., Singh, S. D., Jatav, R. S., Das Saha, N., Raj, A., Joshi, R., & Kumar, V. (2021). Changes in thermal requirements, growth and yield of wheat under the elevated temperature. *Indian Journal of Agricultural Sciences, 91*(3), 435–439.

Chakrabarti, B., Singh, S. D., Anand, A., Singh, M.P., Pathak, H., Nagarajan, S. (2012b). Impact of high temperature on crop and soil. In H. Pathak, P. K. Aggarwal, & S. D. Singh (Eds.), *Climate change impact, adaptation and mitigation in agriculture: Methodology for assessment and application* (pp. 88–95).

Chakrabarti, B., Singh, S. D., Kumar, V., Harit, R. C. & Misra, S. (2013). Growth and yield response of wheat and chickpea crops under high temperature. *Plant Physiology Reports (Indian Jouranl of Plant. Physiol), 18*(1), 7–14.

Christensen, T. R., Johansson, T., Akerman, H. J., Mastepanov, M., Malmer, N., Friborg, T., Crill, P., & Svensson, B. H. (2004). Thawing sub-arctic permafrost: effects on vegetation and methane emissions. *Geophysical Research Letters* 31, L04501. https://doi.org/10.1029/2003GL018680

Cowan, N., Bhatia, A., Jain, N., Singh, R., Tomer, R., et al. (2021). Experimental comparison of continuousand intermittent flooding of rice in relation to methane, nitrous oxide and ammonia emissions and the implications for nitrogen use efficiency and yield. *Agriculture, Ecosystems and Environment, 319*, 107571. https://doi.org/10.1016/j.agee.2021.107571

CRS (Congressional Research Service). (2021). Greenhouse Gas Emissions and Sinks in U.S. Agriculture. In focus, updated September 2, 2021. https://crsreports.congress.gov

Daripa, A., Bhatia, A., Tomer, R., Singh, S. D., Jain, N., Pathak, H. (2014). Nitrous oxide and carbon dioxide emission from maize (*Zea mays* L.) under fertiliser application and elevated carbon dioxide in northwest India. *Expl Agric.* https://doi.org/10.1017/S0014479714000118

Dawar, K., Sardar, A., Muhammad, K., Zaman, M., Muller, C., Sanz-Cobena, A., Khan, A., Borzouei, A., & Pérez-Castillo, A. (2021). Effects of the nitrification inhibitor nitrapyrin and the plant growth regulator gibberellic acid on yield-scale nitrous oxide emission in maize fields under hot climatic conditions. *Pedosphere, 31*, 323–331. https://doi.org/10.1016/S1002-0160(20)60076-5

Dey, S. K., Chakrabarti, B., Prasanna, R., Pratap, D., Singh, S. D., Purakayastha, T. J., & Pathak, H. (2017a). Elevated carbon dioxide level along with phosphorus application and cyanobacterial inoculation enhances nitrogen fixation and uptake in cowpea crop. *Archives of Agronomy and Soil Science, 63*(13), 1927–1937. https://doi.org/10.1080/03650340.2017.1315105

Dey, S. K., Chakrabarti, B., Prasanna, R., Singh, S. D., Purakayastha, T. J., Datta, A., & Pathak, H. (2017b). Productivity of mungbean (Vignaradiata) with elevated carbon dioxide at various phosphorus levels and cyanobacteria inoculation. *Legume Research, 40*(3), 497–505.

Dey, S. K., Chakrabarti, B., Purakayastha, T. J., Prasanna, R., Mittal, R., Singh, S. D., & Pathak, H. (2019). Interplay of phosphorus doses, cyanobacterial inoculation, and elevated carbon dioxide on yield and phosphorus dynamics in cowpea. *Environmental Monitoring and Assessment, 191*(4), 223. https://doi.org/10.1007/s10661-019-7378-3

Dore, S., Hymus, G. J., Johnson, D. P., Hinkle, C. R., Valentini, R., & Drake, B. G. (2003). Cross validation of open-top chamber and eddy covariance measurements of ecosystem CO2 exchange in a Florida scrub-oak ecosystem. *Global Change Biology, 9*(1), 84–95.

Eugster, W., Zeyer, K., Zeeman, M., Michna, P., Zingg, A., Buchmann, N., & Emmenegger, L. (2007). Methodical study of nitrous oxide eddy covariance measurements using quantum cascade laser spectrometery over a Swiss forest. *Biogeosciences, 4*, 927–939.

Famulari, D., Nemitz, E., Di Marco, C., Phillips, G. J., Thomas, R., House, E., & Fowler, D. (2010). Eddy-covariance measurements of nitrous oxide fluxes above a city. *Agricultural and Forest Meteorology, 150*, 786–793.

Fitzgerald, G. J., Tausz, M., O'Leary, G., Mollah, M., Tausz-Posch, S., Seneweera, S., Mock, I., Lowe, M., Partington, D., Argall, R., McNeil, D., & Norton, R. M. (2016). Elevated atmospheric [CO_2] can dramatically increase wheat yields in semi-arid environments and buffer against heat waves. *Global Change Biology, 22*, 2269–2284.

Flechard, C. R., Ambus, P., Skiba, U., Rees, R. M., Hensen, A., Van Amstel, A., & Grosz, B. (2007). Effects of climate and management intensity on nitrous oxide emissions in grassland systems across Europe. *Agriculture, Ecosystems and Environment, 121*(1–2), 135–152.

Gaihre, Y., Singh, U., Bible, W., Fugice, J., & Sanabria, J. (2020). Mitigating N2O and NO emissions from direct-seeded rice with nitrification inhibitor and urea deep placement. *Rice Science, 27*, 434–444. https://doi.org/10.1016/j.rsci.2020.03.005

Guru-Pirasanna-Pandi, G., Chander, S., Pal, M., & Soumia, P. S. (2018). Impact of elevated CO_2 on *Oryza sativa* phenology and brown planthopper, Nilaparvatalugens (Hemiptera: Delphacidae) population. *Current Science, 114*(8), 1767–1777.

Hasegawa, T., Sakai, H., Tokida, T., Nakamura, H., Zhu, C., Usui, Y., Yoshimoto, M., Fukuoka, M., Wakatsuki, H., Katayanagi, N., Matsunami, T., Kaneta, Y., Sato, T., Takakai, F., Sameshima, R., Okada, M., Mae, T., & Makino, A. (2013). Rice cultivar responses to elevated CO_2 at two free-air CO_2 enrichment (FACE) sites in Japan. *Functional Plant Biology, 40*(2), 148–159.

Heikkinen, J., Ketoja, E., Nuutinen, V., & Regina, K. (2013). Declining trend of carbon in Finnish cropland soils in 1974–2009. *Global Change Biology, 19*(5), 1456–1469.

Heyneke, E., Strauss, A. J., Strasser, R. J., Krüger, G. H. J. (2012). Open-top chamber facility to study air pollution impacts in South Africa. Part II: SO_2–drought interactions on yield, photosynthesis and symbiotic nitrogen fixation in soybean. *South African Journal of Plant and Soil, 29*(1), 9–23.

Hidayah, S., Agustina, D. A., Joubert, M. D., Soekrasno (2009) Intermittent irrigation in System of Rice intensification potential as an adaptation and mitigation option of negative impacts of rice cultivation in irrigated paddy field. Royal Irrigation Department (Thailand).

Hogy, P., Wieser, H., Kohler, P., Kohler, P., Schwadorf, K., Breuer, J., Franzaring, J., Muntifering, R., & Fangmeier, A. (2009). Effects of elevated CO_2 on grain yield and quality of wheat: Results from a 3-year free-air CO_2 enrichment experiment. *Plant Biology, 11*, 60–69.

Hogy, P., Franzaring, J., Schwadorf, K., Breuer, J., Schütze, W., & Fangmeier, A. (2010). Effects of free-air CO2 enrichment on energy traits and seed quality of oilseed rape. *Agriculture, Ecosystems and Environment, 139*, 239–244.

Holmberg, M., Akujärvi, A., Anttila, S., Autio, I., Haakana, M., Junttila, V., Karvosenoja, N., Kortelainen, P., Mäkelä, A., Minkkinen, K., & Forsius, M. (2021). Sources and sinks of greenhouse gases in the landscape: Approach for spatially explicit estimates. *Science of the Total Environment, 781*, 146668.

Holzapfel-Pschorn, A., Conrad, R., & Seiler, W. (1985). Production, oxidation and emission of methane in rice paddies. *FEMS Microbiology Letters, 31*, 343–351. https://doi.org/10.1016/j.agee.2014.03.013

Huang, H., Wang, J., Hui, D., Miller, D. R., Bhattarai, S., Dennis, S., & Reddy, K. C. (2014). Nitrous oxide emissions from a commercial cornfield (Zea mays) measured using the eddy covariance technique. *Atmospheric Chemistry and Physics, 14*(23), 12839–12854.

IPCC. (2014). Climate change 2014: Synthesis report, in contribution of working groups I, II and III to the fifth assessment report of the intergovernmental panel on climate change. In R. K. Pachauri, & L. A. Meyer (Eds.), *Core writing team.* IPCC

Itoh, M., Sudo, S; Mori, S., Saito, H., Yoshida, T., Shiratori, Y., Suga, S., Yoshikawa, N., Suzue, Y., Mizukami, H., Mochida, T., & Yagi, K. (2011). Mitigation of methane emissions from paddy fields by prolonging midseason drainage. *Agriculture Ecosystems and Environment, 141*, 359–372. https://doi.org/10.1016/j.agee.2011.03.019

Jain, N., Dubey, R., Dubey, D. S., Singh, J., Khanna, M., Pathak, H., & Bhatia, A. (2014). Mitigation of greenhouse gas emission with system of rice intensification in the Indo-Gangetic Plains. *Paddy and Water Environment, 12,* 355–363

Järvi, L., Nordbo, A., Rannik, Ü., Haapanala, S., Riikonen, A., Mammarella, I., Pihlatie, M., & Vesala, T. (2014). Urban nitrous-oxide fluxes measured using the eddy-covariance technique in Helsinki, Finland

Jiang, J., Hu, Z., Sun, W., & Huang, Y. (2010). Nitrous oxide emissions from Chinese cropland fertilized with a range of slow-release nitrogen compounds. *Agriculture, Ecosystems and Environment, 135*, 216–225. https://doi.org/10.1016/j.agee.2009.09.014

Jumadi, O., Hartono, H., Masniawati, A., Iriany, R. N., Makkulawu, Andi T., & Inubushi, K. (2019) Emissions of nitrous oxide and methane from rice field after granulated urea application with nitrification inhibitors and zeolite under different water managements. *Paddy and Water Environment.* https://doi.org/10.1007/s10333-019-00724-3

Kannojiya, S., Singh, S. D., Prasad, S., Kumar, S., Malav, L. C., & Kumar, V. (2019). Effect of elevated temperature and carbon dioxide on wheat (*Triticumaestivum*) productivity with and without weed interaction. *Indian Journal of Agricultural Sciences, 89*(5), 751–756.

Kim, H. Y., Lieffering, M., Kobayashi, K., Okada, M., & Miura, S. H. U. (2003). Seasonal changes in the effects of elevated CO_2 on rice at three levels of nitrogen supply: A free air CO_2 enrichment (FACE) experiment. *Global Change Biology, 9*, 826–837. https://doi.org/10.1046/j.1365-2486.2003.00641.x

Kirti, S., Manjaiah, K. M., Datta, S. C., Biswas, D. R., Bhatia, A., Bandyopadhyay, K. K., & Tomer, R. (2021). Mitigating nitrous oxide emission using nanoclay-polymer composites in rice-wheat cropping system. *Archives of Agronomy and Soil Science,* 1–15.

Kumar, R., Mina, U., Gogoi, R., Bhatia, A., & Harit, R. C. (2016). Effect of elevated temperature and carbon dioxide levels on maydis leaf blight disease tolerance attributes in maize. *Agriculture, Ecosystems and Environment, 231*, 98–104.

Kumar, A., Bhatia, A., Sehgal, V., Tomer, R., Jain, N., & Pathak, H. (2021). Net ecosystem exchange of carbon dioxide in rice-spring wheat system of northwestern Indo-Gangetic plains. *Land, 10*, https://doi.org/10.3390/land10070701

Laine, A., Sottocornola, M., Kiely, G., Byrne, K. A., Wilson, D., & Tuittila, E. S. (2006). Estimating net ecosystem exchange in a patterned ecosystem: Example from blanket bog. *Agricultural and Forest Meteorology, 138*(1–4), 231–243.

Lal, R., Kimble, J. M., Follett, R. F., & Stewart, B. A. (2001). Assessment methods for soil carbon (advances in soil science). 696.

Li, C., Qui, J., Frolking, S., Xiao, X., Salas, W., & Moore, B. E. T., III. (2002). Reduced methane emissions from large-scale changes in water management of China's rice paddies during 1980–2000. *Geophysical Research Letters, 29*(20), 1972. https://doi.org/10.1029/2002gl015370

Lv, C., Huang, Y., Sun, W., Yu, L., & Zhu, J. (2020). Response of rice yield and yield components to elevated [CO_2]: A synthesis of updated data from FACE experiments. *European Journal of Agronomy, 112*, 125961. https://doi.org/10.1016/j.eja.2019.125961

Maity, P. P., Chakrabarti, B., Purakayastha, T. J., Bhatia, A., Saha Das, N. D., Jatav, R. S., Sharma, A., Bhowmik, A., Kumar, V., & Chakraborty, D. (2020). Do elevated CO_2 and temperature affect organic nitrogen fractions and enzyme activities in soil under rice crop? *Soil Research, 58*, 400–410. https://doi.org/10.1071/SR19270

Malla, G., Bhatia, A., Pathak, H., Prasad, S., Jain, N., & Singh, J. (2005). Mitigating nitrous oxide and methane emissions from soil in rice–wheat system of the Indo-Gangetic plain with nitrification and urease inhibitors. *Chemosphere, 58*, 141–147.

Mangalassery, S., Sjögersten, S., Sparkes, D., & Mooney, S. (2015). Examining the potential for climate change mitigation from zero tillage. *The Journal of Agricultural Science, 153*. https://doi.org/10.1017/S0021859614001002

IPCC. (2018). Masson-Delmotte, P. V., Zhai, H. O., Pörtner, D., Roberts, J., Skea, P. R., Shukla, A., Pirani, W., Moufouma-Okia, C., Péan, R., Pidcock, S. (2018). Summary for policymakers. In *Global warming of 1.5°C. An IPCC special report on the impacts of global warming of 1.5°C above pre-industrial levels and related global greenhouse gas emission pathways, in the context of strengthening the global response to the threat of climate change, sustainable development, and efforts to eradicate poverty* (pp. 3–24.). Eds World Meteorological Organization, United Nations Environmental Programme.

Meng, X., Li, Y., Yao, H., Wang, J., Dai, F., Wu, Y., & Chapman, S. (2020). Nitrification and urease inhibitors improve rice nitrogen uptake and prevent denitrification in alkaline paddy soil. *Applied Soil Ecology, 154*, 103665. https://doi.org/10.1016/j.apsoil.2020.103665

Mohanty, S., Swain, C. K., Tripathi, R., Sethi, S. K., Bhattacharyya, P., Kumar, A., Raja, R., Shahid M., Panda, B.B., Lal, B., Gautam, P., Munda, S., Nayak, A. K. (2017). Nitrate leaching, nitrous oxide emission and N use efficiency of aerobic rice under different N application strategy. *Archives of Agronomy and Soil Science*. https://doi.org/10.1080/03650340.2017.1359414

Molodovskaya, M., Warland, J., Richards, B. K., Öberg, G., & Steenhuis, T. S. (2011). Nitrous oxide from heterogeneous agricultural landscapes: Source contribution analysis by eddy covariance and chambers. *Soil Science Society of America Journal* 1829–1838

Morgan, P. B., Bollero, G., Nelson, R. L., Dohleman, F. G., & Long, S. P. (2005). Smaller than predicted increase in above-ground net primary production and yield of field-grown soybean was found when [CO2] is elevated in fully open-air. *Global Change Biology, 11*, 1856–1865.

Nirmala, B., Tuti, M. D., Mahender Kumar, R., Waris, A., Muthuraman, P., Parmar, B., & Vidhan Singh, T. (2021). Integrated assessment of system of rice intensification vs. conventional method of transplanting for economic benefit, energy efficiency and lower global warming potential in India. *Agroecology and Sustainable Food Systems, 45*(5), 745–766. https://doi.org/10.1080/21683565.2020.1868648.

Oertel, C., Matschullat, J., Zurba, K., Zimmermann, F., & Erasmi, S. (2016). Greenhouse gas emissions from soils-A review. *Geochemistry, 76*(3), 327–352.

Pasuquin, J. M., Pampolino, M. F., Witt, C., Dobermann, A., Oberthür, T., Fisher, M. J., Inubushi, K. (2014). Closing yield gaps in maize production in Southeast Asia through site-specific nutrient management. *Field Crops Research, 156*, 219–230. https://doi.org/10.1016/j.fcr.2013.11.016

Pathak, H., Bhatia, A., Prasad, S., Singh, S., Kumar, S., Jain, M. C., & Kumar, U. (2002). Emission of nitrous oxide from rice-wheat systems of Indo-Gangetic plains of India. *Environmental Monitoring and Assessment, 77*(2), 163–178.

Pathak, H., Prasad, S., Bhatia, A., Singh, S., Kumar, S., Singh, J., & Jain, M. C. (2003). Methane emission from rice–wheat cropping system in the Indo-Gangetic plain in relation to irrigation, farmyard manure and dicyandiamide application. *Agriculture, Ecosystems and Environment, 97*(1–3), 309–316.

Pathak, H., Bhatia, A., & Jain, N. (2014). Greenhouse gas emission from Indian agriculture: Trends, mitigation and policy needs. *Indian Agricultural Research Institute, New Delhi, 110012*, p39.

Pathak, H., Jain, N., Bhatia, A., Kumar, A., & Chatterjee, D. (2016). Improved nitrogen management: A key to climate change adaptation and mitigation. *Indian Journal of Fertilisers, 12*(11), 151–162.

Pathak, H., Upadhyay, R. C., Muralidhar, M., Bhattacharyya, P., Venkateswarlu, B. (2013). *Measurement of greenhouse gas emission from crop, livestock and aquaculture* (p. 101). Indian Agricultural Research Institute, New Delhi.

Pattey, E., Blackburn, L. G., Strachan, I. B., Desjardins, R., & Dow, D. (2008). Spring thaw and growing season N2O emissions from a field planted with edible peas and a cover crop. *Canadian Journal of Soil Science, 88*(2), 241–249.

Pihlatie, M. K., Christiansen, J. R., Aaltonen, H., Korhonen, J. F., Nordbo, A., Rasilo, T., & Pumpanen, J. (2012). Comparison of static chambers to measure CH4 emissions from soils. *Agricultural and Forest Meteorology, 171*, 124–136.

Poeplau, C., & Don, A. (2013). Sensitivity of soil organic carbon stocks and fractions todifferent land-use changes across Europe. *Geoderma, 192*, 189–201.

Raj, A., Chakrabarti, B., Pathak, H., Singh, S. D., Mina, U., & Purakayastha, T. J. (2019). Growth, yield and nitrogen uptake in rice crop grown under elevated carbon dioxide and different doses of nitrogen fertilizer. *Indian Journal of Experimental Biology, 57*, 181–187.

Rawson, H. M., Gifford, R. M., & Condon, B. N. (1995). Temperature gradient chambers for research on global environment change. I. Portable chambers for research on short-stature vegetation. *Plant, Cell and Environment, 18*(9), 1048–1054.

Riederer, M., Serafimovich, A., & Foken, T. (2013). Net ecosystem CO_2 exchange measurements by the closed chamber method and the eddy covariance technique and their dependence on atmospheric conditions–a case study. *Atmospheric Measurement Techniques Discussions, 6*(5), 8783–8805.

Rinne, J., Riutta, T., Pihlatie, M., Aurela, M., Haapanala, S., Tuovinen, J. P., & Vesala, T. (2007). Annual cycle of methane emission from a boreal fen measured by the eddy covariance technique. *Tellus b: Chemical and Physical Meteorology, 59*(3), 449–457.

Sakai, H., Tokida, T., Usui, Y., Nakamura, H., & Hasegawa, T. (2019). Yield responses to elevated CO_2 concentration among Japanese rice cultivars released since 1882. *Plant Production Science, 22*(3), 352–366. https://doi.org/10.1080/1343943X.2019.1626255 (Taylor & Francis Online, Web of Science)

Sapkota, T. B., Vetter, S. H., Jat, M. L., Sirohi, S., Shirsath, P. B., Singh, R., Jat, H. S., Smith, P., Hillier, J., & Stirling, C. M. (2019). Cost-effective opportunities for climate change mitigation in Indian agriculture. *Science of the Total Environment, 655*, 1342–1354.

Schaufler, G., Kitzler, B., Schindlbacher, A., Skiba, U., Sutton, M. A., & Zechmeister-Boltenstern, S. (2010). Greenhouse gas emissions from Europeansoils under different land use: Effects of soil moisture and temperature. *European Journal of Soil Science, 61*, 683–696.

Schulze, E. D., Luyssaert, S., Ciais, P., Freibauer, A., Janssens, I. A., Soussana, J. F.,Smith, P., Grace, J., Levin, I., Thiruchittampalam, B., Heimann, M., Dolman, A. J.,Valentini, R., Bousquet, P., Peylin, P., Peters, W., R¨odenbeck, C., Etiope, G.,Vuichard, N., Wattenbach, M., Nabuurs, G. J., Poussi, Z., Nieschulze, J., & Gash, J. H. (2009). CarboEurope Team. Importance of methane and nitrous oxide forEurop¨es terrestrial greenhouse-gas balance. *Nature Geoscience, 2*, 842–850.

Shurpali, N. J., Rannik, Ü., Jokinen, S., Lind, S., Biasi, C., Mammarella, I., & Martikainen, P. J. (2016). Neglecting diurnal variations leads to uncertainties in terrestrial nitrous oxide emissions. *Scientific Reports, 6*(1), 1–9.

Singh, S. D., Chakrabarti, B., Kumar, S. N., & Pathak, H. (2012). Impact of elevated carbon dioxide concentration on crop and soil. In H. Pathak, P. K. Aggarwal, S. D. Singh (Eds.), *Climate change impact, adaptation and mitigation in agriculture: Methodology for assessment and application* (pp. 79–87).

Smith, P., Bustamante, M., Ahammad, H., Clark, H., Dong, H., Elsiddig, E. A., Haberl, H., Harper, R., House, J., Jafari, M., & Masera, O. (2014). Agriculture, forestry and other land use (AFOLU).

In Climate change 2014: mitigation of climate change. *Contribution of working group III to the fifth assessment report of the intergovernmental panel on climate change* (pp. 811–922). Cambridge University Press.

Speckman, H. (2013). Forest respiration from Eddy covariance and chamber measurements under high turbulence and a bark beetle epidemic (Doctoral dissertation, Colorado State University).

Srinivasa Rao, M., Vanaja M., Srinivas I., Nageshwara Rao C. V. K., Srinivas, K., Maheswari, M., Prabhakar, M., Sreelakshmi, P., Bhaskar, S., & Sammi Reddy, K. (2018). CTGC: A facility to study the interactive effects of CO_2 and temperature (p. 44). Bulletin No.01/2018, ICAR-Central Research Institute for Dryland Agriculture, Santoshnagar, Hyderabad, India.

Tao, Y., Chen, Q., Peng, S., Wang, W., Nie, L. (2016). Lower global warming potential and higher yield of wet direct-seeded rice in Central China. *Agronomy for Sustainable Development, 36*(2), 24. https://doi.org/10.1007/s13593-016-0361-2

UNEP. (2019). Emissions Gap Report, 2019. Nairobi: United Nations Environment Programme (UNEP)

Uprety, D. C., Garg, S. C., Tiwari, M. K., & Mitra, A. P. (2000). Crop responses to elevated CO_2: Technology and Research (Indian studies). *Global Environmental Research Japan, 3*, 155–167.

Ussiri, D., & Lal, R. (2013). *Soil emission of nitrous oxide and its mitigation* (p. 378). Springer.

Vanaja, M., Maheswari, M., Ratna Kumar, P., & Ramakrishna, Y. S. (2006). Monitoring and controlling of CO_2 concentration in open top chambers for better understanding of plants response to elevated CO_2 levels. *Indian Journal of Radio and Space Physics, 35*, 193–197.

Vanaja, M., Satish, P., Jyothi Lakshmi, N., Vijay Kumar, G., Vagheera, P., Mohan, C. H., Yadav, S. K., Sarkar, B., Maheswari, M., & Sammi Reddy, K. (2019). Elevated CO_2 ameliorate the negative effects of high temperature on groundnut (*Arachishypogaea*)-Studies under free-air temperature elevation. *Journal of Agrometeorology, 21*(4), 411–419.

Wang, K., Liu, C., Zheng, X., Pihlatie, M., Li, B., Haapanala, S., & Hu, F. (2013). Comparison between eddy covariance and automatic chamber techniques for measuring net ecosystem exchange of carbon dioxide in cotton and wheat fields. *Biogeosciences, 10*(11), 6865–6877.

Wang, Q., Zhou, F., Shang, Z., Ciais, P., Winiwarter,W., Jackson,R. B., & Canadell, J. G. (2019). Data-driven estimates of global nitrous oxide emissions from croplands. *National Science Review, 7*, 441–452.

Xia, L., Lam, S. K., Chen, D., Wang, J., Tang, Q., & Yan, X. (2016). Can knowledge-based N management produce more staple grain with lower greenhouse gas emission and reactive nitrogen pollution? A Meta-Analysis. *Global Change Biology, 23*(5), 1917–1925. https://doi.org/10.1111/gcb.13455

Yadav, A., Bhatia, A., Yadav, S., Kumar, V., & Singh, B. (2019). The effects of elevated CO_2 and elevated O_3 exposure on plant growth, yield and quality of grains of two wheat cultivars grown in north India. *Heliyon, 5*, e02317.

Yang, T., Zeng, Y., Sun, Y., Zhang, J., Tan, X., Zeng, Y., Huang, S., & Pan, X. (2019). Experimental warming reduces fertilizer nitrogen use efficiency in a double rice cropping system. *Plant, Soil and Environment, 65*(10), 483–489.

Yu, L., Wang, H., Wang, G., Song, W., Huang, Y., Li, S. G., & He, J. S. (2013). A comparison of methane emission measurements using eddy covariance and manual and automated chamber-based techniques in Tibetan Plateau alpine wetland. *Environmental Pollution, 181*, 81–90.

In: Climate change 2014: mitigation of climate change. Contribution of Working group III to the fifth assessment report of the intergovernmental panel on climate change, pp [illegible] Cambridge University Press

[illegible]

Chapter 22
Monitoring and Impact Assessment of Climate Change: Eddy Covariance Technique

Sunayan Saha, Nilimesh Mridha, Debasish Chakraborty, Dibyendu Chatterjee, Manoj Kumar Nanda, M. S. Dhanya, Chinmaya Kumar Swain, Saikat Ranjan Das, and Koushik Banerjee

Abstract The human population is witnessing discernible changes in earth's climate and its negative impacts during the recent decades. Positive radiative forcing from accumulation of greenhouse gases (GHGs) is driving this change. Climate is simulated using regional or global circulation models which are 3-D models of earth-atmosphere system. Proper data driven monitoring of model parameters and input variables is important for greater confidence in future projections. Quantification of GHG fluxes has been carried out by chamber-based, micrometeorological and remote sensing approaches. This chapter discusses the measurements of GHGs and other surface fluxes with the help of a micrometeorological eddy covariance (EC) technique to understand the two-way feedback between climate and ecosystems. The prediction and impact assessment of climate change involving EC technique has gained prominence due to specific advantages like direct, continuous and real time

---

S. Saha (✉)
ICAR-Central Potato Research Institute, Regional Station Jalandhar, Punjab 144026, India
e-mail: sunayan.iari@gmail.com

N. Mridha
ICAR-National Institute of Natural Fibre Engineering and Technology, Kolkata, West Bengal 700040, India

D. Chakraborty
ICAR-Research Complex for NEH Region, Umiam, Meghalaya 793103, India

D. Chatterjee · C. K. Swain · S. R. Das
ICAR-National Rice Research Institute, Cuttack, Odisha 753006, India

M. K. Nanda
Bidhan Chandra Krishi Vishwavidyalaya, Mohanpur, West Bengal 741252, India

M. S. Dhanya
Central University of Punjab, Bhatinda, Punjab 151401, India

K. Banerjee
ICAR - Mahatma Gandhi Integrated Farming Research Institute, Piprakothi, Motihari, Bihar 845429, India

H. Pathak et al. (eds.), *Climate Change Impacts on Soil-Plant-Atmosphere Continuum*, Advances in Global Change Research 78,
https://doi.org/10.1007/978-981-99-7935-6_22

monitoring of CO_2, CH_4, H_2O, heat and momentum fluxes. EC- fluxes are useful for validation of carbon, water and energy budget models. Activities under various flux-networks and net carbon exchanges of various ecosystems based on long-term monitoring have been presented. EC-tower site fluxes can be upscaled in conjunction with remote sensing measurements for climate change studies.

Keywords Greenhouse gases · Fluxes · Eddy covariance · Models · Climate change

22.1 Introduction

Climate change is already being witnessed, strongly so after 1980s, in every corner of the earth with anthropogenic factors as the driver of this change beginning with industrial revolutions in the late eighteenth century. Of the observed changes and future projections, overall warming of the globe is most crucial as it sets in other changes too in climate. Average surface temperature of the globe during the first 20 years of the twenty-first century was about 1 °C higher than the baseline period of 1850–1900 with landmass experiencing greater increase compared to the ocean. Decades were progressively warmer after 1980s and each of those was warmer than any decade during 1850 and 1980. In most land areas, extreme hot days and heat waves are nowadays more frequent and also more intense. On the contrary, extreme cold days and cold waves have become less frequent as well as the severity of such events has reduced. Heavy precipitation events, both in terms of frequency and intensity, have increased over most land regions since 1950s. It can be stated that the share of major category of tropical cyclones out of total number of cyclonic events have risen since 1980s, the likelihood of which is 66–100% probability. In the West-North Pacific Ocean region, there has been a shift in the location of peak intensity tropical cyclones towards north (IPCC, 2021). Heating of the climate has caused a rise in the global mean sea level (0.20 m increase between 1901 and 2018), through melting of ice cover on land and thermal expansion due to ocean warming. Evapo-transpiration over land region has increased which also resulted in higher number of drought events in some ecological regions and agro-ecosystems. With global warming, the response of land biosphere has been consistent since 1970s: in both north and south hemispheres, climate zones have shifted poleward. Since 1950s, in the extratropical regions of N-Hemisphere, the length of the growing season has increased, on an average, by up to 2-days per decade.

Rise in atmospheric concentrations of various GHGs, particularly the well-mixed GHGs are the principal cause of global warming and climate change and CO_2 is the biggest contributor in this saga. Continued emissions would cause further warming and the total warming shall be determined by the quantum of CO_2 emissions, both in the past and future. It is anticipated that the earth's average surface temperature by the end of twenty-first century (2081–2100) would be higher by atleast 1.5 °C compared to 1850–1900 and the magnitude of such increase could well reach anywhere between

3.3 °C and 5.7 °C under SSP5-8.5 i.e., very high greenhouse gas emissions scenario (IPCC, 2021).

If we see the global carbon budget in the last about 50 years, about 9 billion tones (BT) of C has been emitted due to anthropogenic activities, of which 2.3 BT has been absorbed by ocean and another 2.5 BT by undefined sinks. The rest 4.2 BT i.e., 46% of total emission has been retained in the atmosphere. This has caused a phenomenal rise in CO_2 concentration by about 100 ppm i.e., increasing at a rate of ~2 ppm per year. Increasing GHGs has been causing a net positive radiative forcing i.e., net long wave radiation flux towards the earth surface is increasing, resulting in accrual of additional heat and warming of the climate system. It has been estimated that compared to the pre-industrial time of 1750, human-induced radiative forcing in 2019 has been 2.7 W m^{-2}. Apart from global warming, this radiative forcing has also impacted carbon, water and other important biogeochemical cycles on earth, the overall result of which we know now as climate change.

When we talk of carbon cycle, the role of vast ocean is very clear as a sink. Also, ocean is less prone to anthropogenic disturbances and show less dynamism. On the other hand, terrestrial vegetation although smaller in size than ocean, is more prone to anthropogenic disturbances and effect of climate change. Also, the soil over which the vegetation lies is the largest terrestrial repository for carbon on the surface of the earth. Remote sensing-based studies have indicated that during two-and half decades after 1980, the net (vegetation) primary production (NPP), an important component of C-cycle, has increased at a rate of 3–8.5% per decade in India and at global level, albeit with large interannual variability primarily driven by precipitation and temperature anomaly (Bala et al., 2013; Singh et al., 2011). During the first decade of twenty-first century (2000–2009), there has been a decrease in global NPP by about 0.55 PgC which was likely due to the occurrence of droughts and drying trends in large regions of the S-hemisphere (Zhao & Running, 2010). Notwithstanding the above trends, NPP estimates vary quite widely depending upon the carbon cycle model chosen for simulation and the type of observational dataset used as model inputs (Bala et al., 2013; Chhabra & Dadhwal, 2004). Models that do not include vegetation feedback to climate predict gain of Carbon would continue in future due to enhanced growing period in temperate climate and CO_2 fertilization effect. However, if the carbon cycle feedback on climate is considered such trend may get reversed with land may become huge source of Carbon to the atmosphere. Such reversal would have great implication on the quantum of warming and changes in other parameters in future climate. Hence, it is very important to improve our understanding of the exchange processes of energy, momentum and greenhouse gases (GHGs) at biosphere–atmosphere interface and clear knowledge about the role of various terrestrial ecosystems as sources or sinks of atmospheric CO_2. Eddy covariance is a state-of-the-art micrometeorological technique for generating such data on important land surface fluxes (net ecosystem exchange or NEE of CO_2, CH_4, H_2O/latent heat, sensible heat and momentum). This system of data generation has some typical advantages over other approaches in the entire gamut of climate modeling schemes. This chapter has focused on eddy covariance methodology from principle to potential related to the science of climate change.

22.2 Land Surface Fluxes in Climate Change Research

Fluxes of mass and energy at the earth surface-atmosphere interface demands special consideration while simulating weather and climate using numerical models (Patil et al., 2015). Climate is predicted using different three-dimensional, regional or global climate models (RCMs and GCMs), which have generally horizontal grid resolutions of the order of 50–300 km and typical time resolutions of about 30 min. These models work with the present status of the ocean surface temperatures, circulation pattern and at least some basic processes occurring over the land surface. Coupling of the atmosphere to the land and the ocean takes place at their interfaces via exchanges of latent heat or moisture, sensible heat and momentum.

The output from RCMs or GCMs gives only a mean value of variables or parameters for the associated grid boxes whereas there could be huge intra-grid variations in terms of actual observations. Some physical processes act at a much smaller scale than the characteristic grid interval of the climate models. Further, there are processes that are too complex to clearly understand and hence difficult to physically represent them in the model. Such sub-grid scale and or complex processes like turbulent exchanges of heat, moisture and momentum as well as their transformation in atmosphere e.g., cloud formation, precipitation, development of weather systems, etc. are often parameterized. However, in parameterization, those processes are simplified -or their known properties are averaged over the larger scale of model-grid. This simplification exercise adds to uncertainties in the projected future climate obtained through simulation runs of any RCM or GCM.

Considerable research efforts are on to develop further efficient ways towards incorporation of various sub-grid processes into climatic models. Among the various inputs, climate models require spatio-temporal data fields of land surface parameters such as leaf area index (LAI), fPAR (fraction of photosynthetically active radiation absorbed by vegetation), albedo, net fluxes of heat energy, water vapour, GHGs and atmospheric concentrations of GHGs (Fig. 22.1). Hence, reliable estimation of surface fluxes is crucial for climate change monitoring and impact assessment.

Carbon dioxide, water vapor and methane are the three most important GHGs, determining the trajectory of global climate change. Scientists needed to know how the biome specific sources and sinks of important GHGs across different regions of the world adds up to the total atmospheric pool of GHGs over a reasonable span of time. Measurement of fluxes of CO_2 are needed to know the rates of net carbon emission and to construct carbon budgets of various natural and managed ecosystems such as forests, wetlands, agricultural or urban ecosystems, industrial zones like areas demarcated for C-sequestration, landfill sites etc. Long-term observations are needed to confirm the periods in which vegetation primary productivity, soil or ecosystem respiration and net ecosystem productivity are most sensitive to climate change. Agriculture, being part of managed ecosystems, have great flexibility in manipulating the global carbon budget. As there are prospects of expansion of agriculture driven by temperature increase in temperate regions and through targeted land use change to cope up with ever increasing population pressure, the contribution of

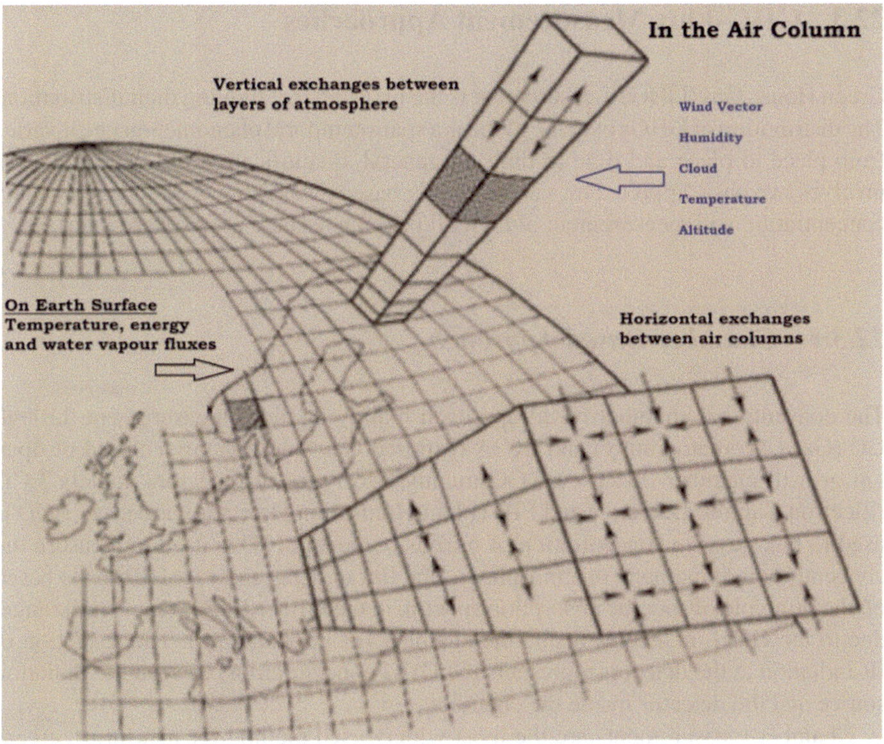

Fig. 22.1 A simplified representation for climate modelling scheme

agro-ecosystems in bringing climate change as well as impact of climate change on it should be of our interest. Further, the carbon flux data is also used to estimate carbon footprints, verifying compliance with carbon emissions regulations and for settling carbon credit claims. Water vapor fluxes (latent heat fluxes) are linked to hydrological cycle, therefore regulates evapo-transpiration, formation of cloud and rains and occurrence of flood or drought. Methane has much higher global warming potential than CO_2 or H_2O, hence methane load of the atmosphere has far reaching consequences.

So, these land surface flux data could be used to evaluate carbon cycle, water budget and energy budget models, validating remote sensing data products and as input for climate modelling. This also helps in real time monitoring of the impact of climate change on agriculture and serves as critical input for formulating mitigation strategies through appropriate nutrient and manure management, irrigation scheduling and precision agriculture among other things.

22.3 GHG-Flux Measurement Approaches

Green House Gas (GHG) measurement is the first step for studying their distribution. The distribution of GHGs over the globe is a spatiotemporal phenomenon i.e., it varies from place to place and time to time. In general, quantification of GHG exchange involves two basic approaches, viz., Chamber based measurement of change in GHG concentration and measurement of GHG flux at the surface-atmosphere interface.

22.3.1 Chamber Based Approach

The conventional chamber-based approach mainly includes measurement through GC (Gas Chromatography) and IRGA (Infrared Gas Analyzer) in a closed or open top growth chamber. In case of GC mainly two types of detectors namely ECD (Electron Capture Detector) and FID (Flame Ionisation Detectors) are used. ECD is used to measure the concentration of nitrous oxide and FID is used to measure the concentration of methane in atmosphere. The Infrared gas analyzer (IRGA) is based on the principle of specific absorption spectra, in the infrared (IR) wavelength range, due to the GHG molecules inside the chamber (or in open path). The reduction in IR radiation at the detector represents the GHG concentration between the radiation source and the detector inside the chamber.

Chamber measurements are the most widely used method for measuring greenhouse gases because they allow storage of gas samples for later analysis and do not require on-site power supply. Using the chambers, the treatment effects (e.g., fertilizer, irrigation, crop or variety) can be studied more conveniently in small experimental plots. Also, the effects of topography, land use and land cover on GHG exchange can be studied by it using less resources. However, because the placement of the chamber affects environmental conditions and modifies the natural turbulence (eddies), the observed fluxes may not always represent true fluxes. Hence, care must be taken to correct acquired data.

Dynamic chambers and static chambers are the two types of chambers normally used for the purpose. The headspace air exchange rate in dynamic chambers is high (>1–2 times the rate in static chambers) (Butterbach-Bahl et al., 1997a, 1997b). The static chambers are made gas-tight with provision for ventilation so that the pressure within the chamber headspace remains equal with that of outside ambient air pressure. The volume of gas measured at each sampling interval should be greater than the capacity of the vent tube (Xu et al., 2007). Dynamic chambers are used in two situations: first, to measure reactive gas fluxes, such as NO_2 emissions from the ground, and second, to reduce the bias caused by variations in airspace concentration in the headspace on the flow (Butterbach-Bahl et al., 1997a, 1997b). Because the flow at the soil-atmosphere interface, is the result of simultaneous production and consumption processes, significant differences in gas concentrations in the chamber and ambient air have an effect on the soil-atmosphere exchange process. However, there are solutions, such as using nonlinear rather than linear models to calculate the fluxes captured by the static chamber approach (Kroon et al., 2008), using quantum cascade lasers (QCLs) on field (Hansen et al. 2006) and reducing the chamber closure time in general. The magnitude of gas flow and chamber height have an impact on chamber closure time. As a result, chambers that can be increased in sections depending on plant development are recommended in agricultural systems. Including plants within the chambers make the measurements representative of the system under study (Barton et al., 2008). It is recommended that during the sampling interval which is generally 30 min or so, four or more gas samples be drawn from the chamber headspace so that gas flux calculation is accurate (Rochette, 2011). The sampling protocol is an important issue in chamber based measurement to represent the agricultural system. The sampling days generally begin with the crop's early growth stage and progress through the vegetative, reproductive, and physiological maturation stages. In case of rice, gas samples are taken shortly after sowing/ transplanting and at regular intervals throughout the cropping season. This can also be used in rice nurseries or on bare soil during fallow periods. In general, six hills are covered with a Perspex chamber [51 cm (h) × 53 cm (l) × 37 cm (w)] from seedling to tillering stage and maximum tillering stage onwards the height (h) is raised up to 71 cm, leaving other dimensions intact. These are placed on a battery-powered air circulating pump having air displacement rate of 1.5 l min^{-1} for sampling and estimation of GHG fluxes, and the air samples are analyzed by gas chromatography for CO_2, N_2O and CH_4.

The most commonly used technique for assessing soil respiration rates is IRGA-chamber based field measurement. The method calculates, increase of CO_2 concentration in a closed chamber over a given period. The differences in IRGA-based measurements of soil respiration are due to differences in IRGA, measurement parameters, and CO_2-flux algorithms. The type of soil and vegetation in which measurements are made affects these impacts. Greenhouse gas fluxes from soil are measured and analyzed in a monitoring chamber equipped with an infrared gas analyzer connected to a monitor. A marker is placed at the specific point within the plot from which a greenhouse gas flux is monitored during the study period. The IRGA chamber is placed on the soil surface at each site for two minutes until the GHGs flux measurements are recorded in the data collection device (Mohanty et al., 2017).

These conventional chamber-based methods have some limitations.

- These are point measurements to represent very small footprint (one or few plants only depending on the size of chamber). Hence, a large number of samples are required to represent a typical crop surface.
- These are taken at discrete time interval which is required to be scaled up in diurnal and seasonal time scale
- Both in closed and open chambers (more in the closed chamber) the microclimate is modified because the natural eddies are restricted by the physical barriers as well as radiative forcing of the walls of the chamber.

While upscaling, the small errors in the chamber-based measurements are generally magnified to yield wide deviations from the expected results at larger spatio-temporal scale.

22.3.2 Micrometeorological Approach

There are various micrometeorological methods available for area-integrated field scale measurement of surface fluxes in a non-destructive way. These techniques are viz.,eddy accumulation, relaxed eddy accumulation, Bowen ratio, aerodynamic resistance and eddy covariance (also called as eddy correlation) approach. Of these, eddy covariance (EC) is the most sophisticated, state-of-the-art technique that is capable to provide most direct, continuous measurement of energy, momentum and GHG fluxes in real time. The EC-technique and its usefulness for climate change related research are given in the following sections.

22.4 Eddy Covariance (EC) Technique

22.4.1 Principle of Measurements and Components

The eddy covariance (EC) method is based on the assumption that the exchange of heat (sensible heat), vapour (latent heat) and gases (CO_2 and other GHGs) takes place through upward moving eddies. Thus, the exchange process is quantified by simultaneous measurement of vertical wind speed in the eddies along with scalar quantities e.g., temperature, water vapour and GHGs concentration at high temporal frequency to match with the rapid fluctuation associated with the turbulent eddies at the height of interception. A schematic pictorial diagram of measurements at an eddy covariance station is presented in Fig. 22.2.

The mean vertical flux densities (Fz) are computed as covariance between deviations in instantaneous vertical wind speed from mean (ω') and that of the scalar quantity (c'), e.g., air temperature (for sensible heat flux), or water vapour density (for latent heat flux) or CO_2 mixing ratio (for CO_2 flux) or CH_4 mixing ratio (for CH_4 flux) (Baldocchi, 2003; Burba, 2013; Saha et al., 2016). Information such as specific heat of air and latent heat for vaporization are also required to convert the appropriate fluxes into suitable energy units. A simplified general form of the flux computing equation is given below:

$$\mathrm{Fz} = \overline{\rho \mathrm{a}} * \overline{\omega' \mathrm{c}'}. \tag{22.1}$$

In Eq. 22.1, ρa is used for air density, the overbars stand for time averaging, and primes represent fluctuations of instantaneous data from its mean value. As per the

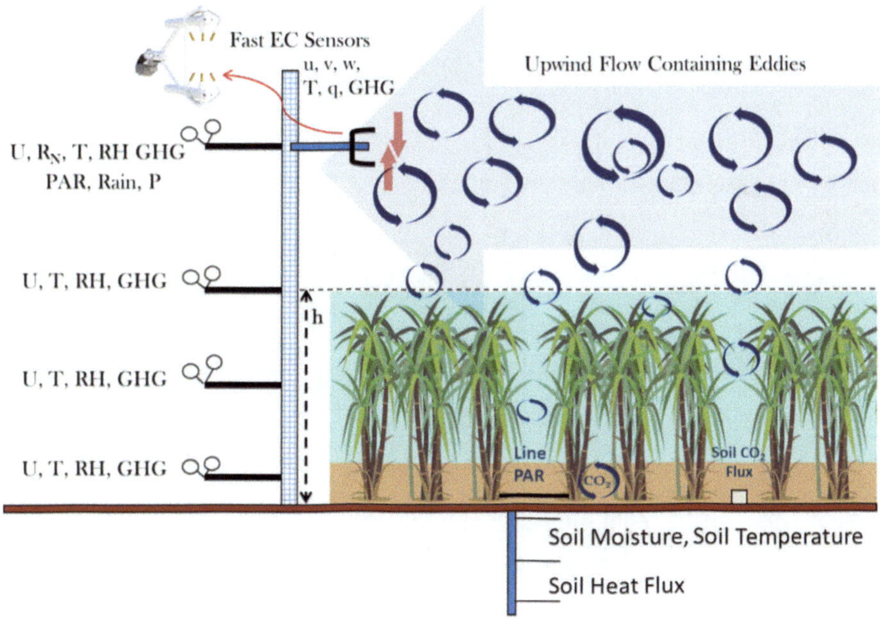

Fig. 22.2 An extended eddy covariance station for measurement of upwind vertical turbulent fluxes and auxiliary ecosystem parameters. The letters represent as follows: U—wind speed, T—air temperature, RH—Relative Humidity, R_N—Net Radiation, PAR—Photosynthetically active radiation, P—Pressure, u and v—lateral components of wind speed, w—vertical component of windspeed, q—water vapour density, GHG—any greenhouse gas concn

sign-convention followed by atmospheric physicists or micrometeorologists, positive covariances between ω' and c' point to net transfers into atmosphere whereas negative values signify net flow from atmosphere towards surface which in case of CO_2 is absorption by the vegetation. For tall crops or tree ecosystems, often the carbon stored within the canopy height is significant. So, for these cases, net ecosystem exchange (NEE) of CO_2 is calculated as the sum of the eddy CO_2 flux (Fz) and the change in CO_2 storage (Fs) within the air volume below EC-sensor height.

A typical eddy covariance system includes an infrared gas analyzer (IRGA), a sonic anemometer, a data logger and an incessant source of power supply (solar PV cell or power-grid connection). Modern EC technique requires high frequency (generally, 10–20 Hz) measurements of windspeed in three directions [i.e. horizontal (x and y) and vertical (z)] and concomitant measurements of trace gas concentrations (e.g. CO_2, H_2O, CH_4) and air temperature at some height over the surface under study using a 3-D sonic anemometer and an infrared gas analyzer (open path or close path type CO_2-H_2O analyzer; open path CH_4-analyzer) (Fig. 22.3). The ideal height of sensor placement is ever dynamic, depending on crop height, available fetch area and wind factors. As it is assumed the eddies carry the flux, the frequency response of the sensors has to be synchronized with the size of the eddy. The eddy size normally increases with height of measurement, wind speed and roughness of

Fig. 22.3 An open path eddy covariance system installed at ICAR-NIASM farm in Baramati, Pune, India (crop: wheat, 2014–2015)

the surface. For smaller eddy size close to the surface the frequency response of the instrument should to be high. On the other hand, the height of the EC tower is also related to the fetch length. As a thumb rule 1:100 instrument height to fetch ratio is required for EC measurement. Ideally, the sensors are placed in the constant flux layer (inertial sublayer) which generally starts at about 1.5 to 2 times the canopy height. Nevertheless, more the height of the sensors, the wider will be the foot print area. In various experimental sites, the height of EC-sensors varies somewhere between ~2 m (short crops) to 40 m or even more (tall trees). Accordingly, a tripod or a tower is selected as the mounting hardware.

To ensure that the covariance measurements (wind vis-a-vis scalar quantities) at each sampling are done in the same volume of air, the aforesaid fast response sensors should be placed minimally apart with no vertical separation and the mid-axis horizontal separation distance not exceeding 15–20 cm. The system is generally installed in the middle of flat, homogeneous landscape of the target vegetation or crop having sufficient fetch area in all prevalent directions which can be determined from the wind rose pattern of the site. A fast response data logger is used to synchronize among the typical fast response EC-sensors as well as a number of slow response meteorological sensors that, in a way, is integral part of all modern EC-systems used in the micrometeorological research. The fast data acquired by the logger is processed either using an in-built data logger program or by separate software installed in the computer to obtain fluxes of interest (sensible heat, momentum and GHGs) at desired time scale. The preferred frequency of final flux output is generally 30 min or

1 h in most cases. Alteddy, BARFlux, EdiRe, EddyPro, Eddysol, ECPack, ECO_2S, LundFlux and TK3.0 are examples of some flux data processing softwares.

The main advantage of the EC method over conventional chamber method lies in its scalability which can range from a few square meters to a hectare and beyond in spatial scale and from fractions of a second to a typical averaging interval of 30 min and longer (Swain et al., 2016, 2018) depending on the height of measurement and turbulent characteristic of the boundary layer.

A list of sensors and instruments that are either integral or relevant in research using eddy covariance system is provided in Table 22.1.

22.4.2 Eddy Covariance with Open Path and Close Path System

When selecting instrumentation for eddy covariance research, the decision whether to use an open path or closed path IRGA is thus, very critical. The open path eddy covariance (OPEC) system measures scalar concentrations in situ, close to the point of interception of the sonic anemometer. The sensors in the closed path (CPEC) systems, on the other hand, are placed in a weather-proof enclosure (cell) at some distance from the sonic anemometer. In CPEC the air from near the anemometer sucked down to the cell containing the sensors through a tube for analysis of scalar concentrations. The relative advantages and disadvantages of the open and closed path systems are as follows.

- The CPEC system consumes more power for operation of its suction pump than the OPEC system. Hence, an OPEC system is better suited for remote locations where power supply is limited.
- The OPEC requires less maintenance than the CPEC because in case of the later air filters in the tube and the suction pump needs regular monitoring.
- The CPEC is least affected by aerosols, rain, fog, dew as well as snowfall while the IRGA sensor in OPEC is exposed to all these atmospheric interruptions leading to large uncertainties and data gaps.
- During the data post-processing in CPEC system, appropriate frequency correction is required against the tube-related dampening and phase shift of concentration fluctuations (Massman, 2000) which may not be required in the OPEC systems
- The OPEC system requires density correction against density fluctuation caused by the rapid variations in air heat and water vapour content at the IRGA sensor interface (Webb et al., 1980) as well as the heating of IRGA sensor itself (Burba et al., 2008). The correction is popularly known as WPL correction i.e. Webb-Pearman-Leuning correction. The CPEC does not require WPL correction.

The open path system has gained worldwide acceptability, due to its low power consumption and less maintenance. Nowadays, a hybrid of the above two type EC

Table 22.1 Sensors and instruments of a typical, modern eddy covariance system

Sensors	Parameters and purpose
Fast response devices	
Combined CO_2-H_2O analyzer	CO_2 and Water vapour concentration
CH_4-analyzer	Methane concentration
Sonic anemometer	Wind speed (high frequency) in three-dimensions
Fine Wire Thermocouple	Atmospheric temperature fluctuations with very high accuracy, required for flux correction
Slow response type auxilliary sensors for meteorological/environmental monitoring	
Net Radiometer (preferably four-component)	Incoming and outgoing short-and long- wave radiations; energy balance—for validating turbulent fluxes using energy balance closure (EBC) principle
Soil heat flux plates (in atleast two depths of soil, preferably three to four depths)	Ground heat flux; energy balance—for validation of turbulent fluxes from EBC principle
Soil temperature sensors (in atleast two depths of soil, preferably three to four depths)	Determining the environmental control by developing empirical/functional relationship with fluxes
Soil moisture sensors (in atleast two depths of soil, preferably three to four depths)	Determining the environmental control on fluxes; soil water balance
Raingauge/precipitation sensor	Soil water balance computation
AT-RH (Air temperature and relative humidity sensor)	Determining the environmental control on fluxes
Barometric pressure	For measuring atmospheric pressure—for correction of different parameters
Vertical (multi-height) CO_2-concentration profiler	For measuring CO_2 storage change within the vegetation canopy
Other auxiliary equipments for co-located measurements	
Photosynthetic photon flux density (PPFD) sensor [Point or line quantum sensors] (preferably two)	Photosynthetically active radiation (PAR); Absorbed/Intercepted PAR for determining radiation use efficiency
Infra red thermometer/canopy temperature sensor	Stress detection in vegetation canopy
Spectroradiometer	Portable instrument for sampling within flux footprint area—for computing vegetation index and stress characterization
TDR-probe	Portable instrument for sampling within flux footprint area for computing soil water balance

systems has come up which is known as EPEC i.e. enclosed path eddy covariance system. EPEC system could overcome to some extent the limitations of a closed path system by shortening the intake tube, power requirement etc.

22.4.3 Challenges with Eddy Covariance Technique

Unlike experiments under laboratory or controlled environmental conditions, measurements using eddy covariance system are done under natural conditions. Practically, it covers a variety of terrain and atmospheric conditions, many of which violate the basic theoretical assumptions. This poses great challenges in obtaining robust data. Some of the challenges are briefly mentioned below.

Designing experimental setup: Reliability of flux measurements using the eddy covariance methodology lies on the fact that how good the underlying theoretical assumptions hold true at the measurement sites. The EC system is applicable to a minimum footprint area as governed by a permissible fetch length. Hence, this technique is not applicable to agronomic experiments where a number treatments or interventions are evaluated in a cluster of experimental plots. Thus, it is essential to identify the goal of flux measurement at each site before finalizing the design of experimental setup which includes site selection, positioning of flux tower, choice of sensors, height and orientation of sensors placement among other things. For details in this aspect the readers are further referred to the book written by George Burba (2013). Adequate steps need to be taken in order to reduce systematic and random errors to the extent possible, thereby increasing the accuracy and precision in NEE estimates (Loescher et al., 2006).

Flux convergence or divergence: Due to a variety of reasons like slow wind speed condition, temperature inversion, undulating topography or mountainous landscape wind flow may become non-stationary. Under such conditions, advection, drainage, flow convergence or divergence may emerge dominant. These types of phenomena lead to non-uniformity in the fetch in terms of sources and sinks of fluxes. These violations of basic assumptions of the eddy covariance principle needs to addressed.

Dynamic flux footprints: Depending on the stability conditions of the atmosphere as well as wind speed and direction, the area under or surrounding the flux tower that contributes to the total flux gets either shrank or stretched or the foot print area may be shifted from the intended sampling area. It is important to know the contribution of source areas to the total flux registered. Single footprint model cannot be applied under all conditions.

Resolving cross scale phenomena: The flux contribution from very small size eddies (i.e. when frequency of an eddy is greater than the temporal resolution of the sensor) may not be detected. On the other hand, there are large eddies or planetary scale circulations that bring gases and energy, not related to local sources. Resolving the contributions of such cross-scale phenomena is a challenge.

Non-turbulent transport: The EC systems cannot take into account the contribution from non-turbulent transport. Vertical movement of molecules due to diffusion gradient i.e. non-turbulent transport cannot be detected.

Non-ideal study sites: Reliable measurement of fluxes in mixed vegetation, fragmented farm-holdings and undulating landscapes is challenging. The very basic boundary layer assumptions of eddy covariance theory are violated in these cases. Hence, there is a need to develop appropriate protocol.

Gap-filling of missing data: Invariably, there would be considerable (say, about 20% or more) missing data (either due to electronic noise, power failure or data screening for quality) at any sites even after taking all precautions. Appropriate gap-filling techniques must be applied to the time series flux data for a valid interpretation.

22.4.4 Data Quality Control and Corrections

Uncertainties in flux measurement by EC-method, even after all possible steps taken during system installation, are quite inevitable either due to limitations in experimental set up, complex terrain or atmospheric stability factors. To remove these uncertainties and ensure that these fluxes are true to the target site, proper data screening protocol must be followed for identification and removal of these errors. This aspect of EC-based measurement is so important that researchers, often in lighter vein, say EC (Eddy Covariance) is for "endless corrections". Depending on the characteristics of the site, prevailing weather (e.g. occurrence of rainfall or precipitation, direction of wind flow, atmospheric stability factors), sensor malfunctioning and also the ultimate purpose of the data varied type of corrections are required in the EC flux data.

Table 22.2 provides a generalized estimates of some commonly occurred errors in EC data and procedures applied for rectification. After screening out the bad quality data, there would be gaps in the time series flux data, which is often quite large. These gaps must be filled using appropriate missing data techniques before finally arriving at the ecosystem's behavior as a source or sink. Till now there is no single EC methodology that can be uniformly applied to all sites like flat terrain, hilly areas, agriculture, forest etc. under varying atmospheric conditions. This is also a challenge for different flux networks across the world from flux inter-comparison point of view.

22.4.5 EC-Measurements in Hilly Terrain

Under hilly terrain conditions, observation of NEE or any type of fluxes using EC system is complex research challenge. A single standalone EC-system, whether placed at any upslope or downslope location, cannot appropriately represent the terrain as the lateral advective fluxes along the slope (also referred as drainage flow)

Table 22.2 Various types of errors in EC-measured fluxes, its reasons and some preventive or corrective measures

Error type	Reason/source of errors	Flux-affected	Corrective measures
Frequency response	• Slow time response of fast sensors • Tube attenuation (for close path EC system) • Averaging effect of measurements over sensor-path length • Sensor separation • Sensor response mismatch • Low or high pass filtering • Digital sampling	All	• Select faster sensors to detect very rapid changes in eddy transport • Use shorter tube length in CPEC • The mid-axis separation distance (both horizontal and vertical) between sonic anemometer and gas analyzer should be as minimum as possible
Time delay	• Sensor time delay (especially important in closed-path analyzers with long intake tubes), sonic heat flux errors	All	• Adjust for time delay • Minimize the separation distance between the sonic anemometer and intake point for the gas analyzer
Wind angle of attack	• Wind flow from the back of the sonic head	All	• During installation sonic head should face the direction from which maximum wind blows into it • Angle of attack correction
Spike and noise	• Faults in sensor electronics, power issues	All	• Spike removal (based on certain threshold standard deviations)
Unlevelled instruments or flow	• Undulating topography/complex terrain • Non-steady tower • Unlevelled sonic anemometer	All	• Level the sonic anemometer • Coordinate rotation during processing
Density fluctuation	• Differences in temperature of ambient from and gas volume in the analyzer path (applicable for open path sensor)	CO_2, CH_4 and H_2O	• Webb-Pearman-Leuning (WPL) density correction with respect to dilution term, Thermal expansion term and pressure term

(continued)

Table 22.2 (continued)

Error type	Reason/source of errors	Flux-affected	Corrective measures
Sonic heat error	• When temperature derived from sonic anemometer is different than the gas analyzer path temperature	Sensible heat	• Use fine wire thermocouple • Apply sonic temperature correction
Band broadening	• Observed in NDIR-based measurements	Mostly CO_2 and CH_4	• Band broadening correction
Oxygen in flux measurement path	• Observed in krypton hygrometer based measurements	H_2O	• Oxygen correction
Spectroscopic error	• Observed in recent laser based technologies; studies are limited	Any gas	• No standard protocol; may require instrument-specific corrections
Missing data	• Power failure • Filtering/screening for quality during data processing	All	• Gap-filling using appropriate methods (moving averages, look-up table/LUT etc.)

will have a significant contribution (Fig. 22.4). This often results in underestimation of respiratory fluxes at the top portion of the slope whereas there is overestimation downhill. Use of two or multiple EC-towers along the slope or additional measurements like CO_2 concentrations at multiple heights from the ground to above the canopy using vertical profiler may be required. On the software part for flux data processing, site-specific quality control filter must be developed to resolve or reduce any inconsistency between the observations and general theoretical considerations. Kang et al. (2015) has reported to develop one such filter, referred as dynamical process network (DPN) to exclude CO_2 data that is strongly affected by advection. Such advection can be identified from strong downslope information flow in the above type software.

22.5 Disaggregating EC-Fluxes for Carbon Budget Study

Gross primary production (GPP) and ecosystem respiration (RE) are two important constituents in carbon budget studies. GPP is the amount of atmospheric carbon-dioxide that gets captured photosynthetically by the vegetation during daytime hours whereas RE represents the diurnal CO_2emission from the ecosystem due to respiration by soil microbial population as well as above and below ground plant parts as part of maintenance respiration. Net exchange of CO_2 (NEE) that is measured by an EC-system can be separated into the two aforesaid components, viz., GPP and RE

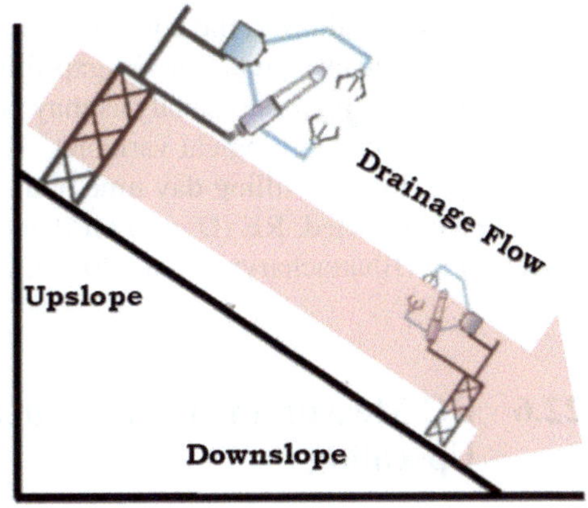

Fig. 22.4 Drainage flow and effect of EC-Tower position on flux measurement

by means of mathematical models. These models estimate GPP and RE as empirical functions of certain environmental variables.

During night hours, there is no photosynthesis, hence GPP equals zero. Thus, the nighttime NEE measured by an EC system, hereafter abbreviated as NEE (N), actually represents the night time respiration of the ecosystem [(RE (N)]. NEE(N) has been found to be exponentially related with prevailing air or soil temperature (T) (Eq. 22.2) and assuming the same response for the other part of the diurnal cycle, the function in eq. I could well be used to estimate daytime respiration, RE (D) too (Falge et al., 2001).

$$\mathrm{RE\ (N)} = \mathrm{R_0 Q_{10}[(T - T_0)/10]}. \quad (22.2)$$

Here, R_0 and Q_{10} are empirical regression constants and T_0 is the reference temperature. Once, hourly or sub-hourly NEE flux (F_{NEE}) is directly obtained from EC system and the corresponding period RE (D) is estimated from an empirical temperature dependent model, fluxes of GPP (F_{GPP}) can be calculated using Eq. 22.3 (Davis et al., 2003; Falge et al., 2002):

$$\mathrm{F_{GPP}} = -\mathrm{F_{NEE}} + \mathrm{RE\ (D)}. \quad (22.3)$$

Like, RE, F_{GPP} is also a function of environmental factors. Gu et al., 2002 has given a rectangular hyperbolic function to estimate it from incident photosynthetically active radiation (PAR) as shown in Eq. 22.4.

$$\mathrm{F_{GPP}} = \mathrm{P_{max}}\alpha\mathrm{PAR}/(\mathrm{P_{max}} + \alpha\mathrm{PAR}). \quad (22.4)$$

where, P_{max} and α are empirical constants that can be determined by doing regression analysis between GPP and PAR. P_{max} indicates the hypothetical maximum of GPP or the closeness to the linear response coefficient, and α denotes the ecosystem quantum yield which is the initial slope of the function.

The missing values in the eddy flux time series dataset must be filled up before final analysis and interpretation. For gap-filling of missing NEE fluxes of few hours or more, non-linear regression analysis may be performed so as to estimate those from environmental/meteorological variables. To gap-fill NEE (N), the form of Eq. 22.2 can be used. For gap-filling day time hours NEE, first RE (D) and GPP time series should be completed. RE (D) and GPP can be gap-filled using Eqs. 22.3 and 22.4, respectively (Bhattacharyya et al., 2013a, 2013b).

22.6 EC Measurements in Model Verification and Flux Upscaling

Many types of models are available that provide output on different type of fluxes and carbon or water balance components at varied temporal and spatial resolution. In one end, it may be a canopy scale model and on the other, a regional or global scale model. An eddy covariance system provides most direct measurements of fluxes at field scale, and also at a very high temporal frequency, thus fits well in the improvement and validation scheme for a wide range of models. Calibrated and validated models reduce the uncertainties in simulated fluxes.

Satellite remote sensing (RS) can provide dynamic data and synoptic coverage of large areas with pixel size (spatial resolution) of the order of meters to few kilometers. To validate flux related RS data products micrometeorological EC method is one of the best technological options available as of now. Remote sensing is useful for flux upscaling and such upscaled products are appropriate for climate models to run efficiently.

22.6.1 Flux Estimation Through Remote Sensing

Estimation of gas exchange through remote sensing is basically a model-based approach. It utilizes the information contained in the upwelling radiation (from atmosphere towards space) in order to derive the amounts and distributions of atmospheric constituents, parameters and selected surface phenomena. Determination of atmospheric GHG concentrations requires high quality measurements, accurate understanding of the absorption spectroscopy and the scattering of electromagnetic radiation in the atmosphere and at the Earth's surface. Nowadays, there are several remote sensing satellites like ENVISAT (SCIAMACHY), GOSAT, Metop-SG, Sentinel-5P, Aqua (AIRS) are available that offer GHG products (one or more of CO_2, CH_4 and N_2O). The general methodology for GHG-product development through remote sensing consists of four basic steps: (a) Selection of GHG absorption bands; (b) Retrieval of information from satellite data; (c) Data processing through Radiative Transfer (RT) models; and (d) Validation.

Carbon dioxide can be detected through absorption bands near 1.6, 2.0 or 15. Likewise, Methane has absorption bands at 1.7, 2.2 and 7 μm. O_2 has a absorption band at 0.75 μm which is used for removing interference from the GHG absorption bands. Suitable absorption bands are selected on the basis of Signal to Noise ratio of that band. By deducting various interferences in a given GHG absorption band, filtered spectrum can be derived. RT models are then used to simulate the spectrum of the atmosphere in that particular band. By varying the concentrations of input constituents of the atmosphere in the RT models, different possible spectra can be synthesized. By comparing the amplitude of filtered spectra (from satellite) with that of the synthesized spectra (from RT models), the amount of desired GHG present in the field-of-view of the satellite can be known (Fig. 22.5). The input constituents of the synthesized spectrum that matches with the filtered satellite spectrum, represent the atmospheric constituents like GHG concentrations at that particular condition.

Recently, Sun et al. (2021) undertook validation of a remote sensing model based output of GHG concentrations at different altitudes in the atmosphere using an EC-system mounted on an unmanned aerial vehicle (UAV).

The remote sensing approach has the following limitations: (a) Absolute measurement of physical parameters is not possible through remote sensing; (b) Satellite measurements are influenced by several disturbances like clouds, aerosols, sensor

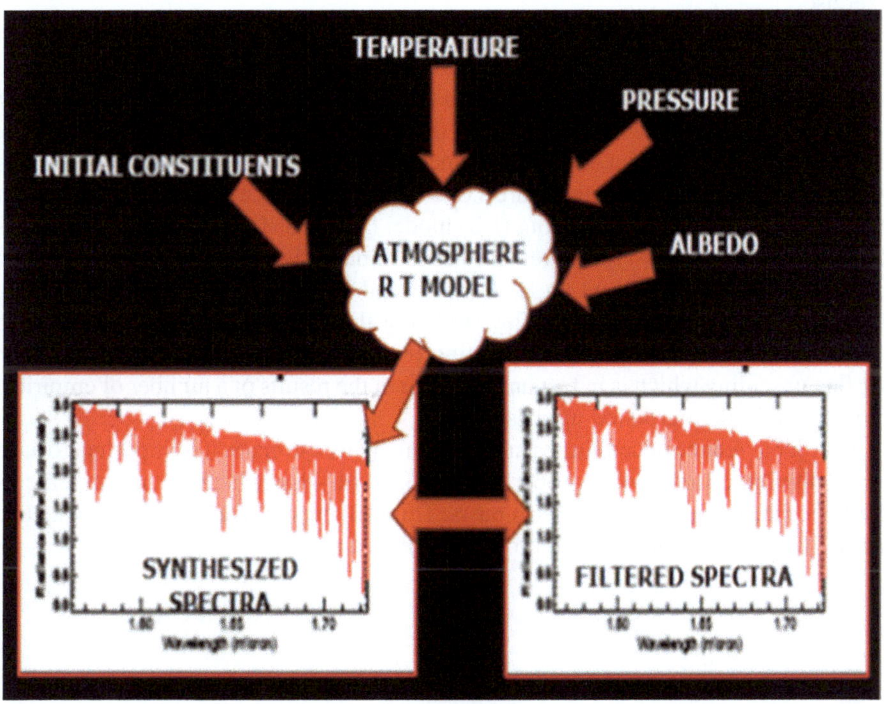

Fig. 22.5 Schematics of remote sensing data-driven radiative transfer model for GHG estimation

calibration and validation etc.; (c) Retrieval of GHGs through satellite data needs thorough understanding about interaction of radiation with GHGs and the interfering factors, thus the retrieval algorithms are often quite complex; (d) the measurement is temporally discrete.

22.6.2 Upscaling EC Fluxes

An EC-tower has a relatively small footprint, generally of the order of one or few hectares. To upscale means to develop, from EC-site points, a spatially distributed flux field/map of biosphere-atmosphere exchanges over a larger area. Upscaling would provide potent data flow to evaluate the regional or global variation in terrestrial Carbon and H_2O cycle. In principle, to start with the upscaling process, usually the first step is training of a machine learning algorithm for predicting or estimating fluxes of an entity based upon measured values of meteorological variables and remote sensing data on vegetation (type and properties). In this training exercise, high quality eddy covariance flux data of different biome type in multiple years should be used. The trained model is then applied spatially using grids of the input data. Both, interpolation and extrapolation are generally involved in this upscaling exercise.

To mention a few example, Chakraborty et al. (2015) used remote sensing based temperature-greenness (TG) model to upscale the daily mean GPP of flooded rice field that were derived from EC-tower flux data. They estimated carbon fluxes over low land tropical rice ecosystem of India during the *rabi* season of 2014. TG model estimates GPP as a product of scaled enhanced vegetation index i.e. canopy greenness and scaled land surface temperature i.e. environment down-regulation factors. Yu et al. (2015) used a remote sensing (RS) model (NEERSM) to generate NEE-fluxes of different vegetation types over northern China and the Tibetan plateau region using MODIS data products on EVI, LSWI and LST as input to the model. Data from ChinaFlux sites were used for validating the above RS model outputs. Jung et al. (2009) employed ERROR (Evolving tRees with RandOm gRowth) approach for flux upscaling which is in fact an ensemble of the results of a number of empirical models (multiple regression models).

In order to reduce uncertainties in the upscaled flux estimates it is important to have a well distributed network of properly maintained EC-sites where flux measurements could be carried out on a long-term basis. This will ensure temporally adequate and spatially distributed datasets for proper training of models.

22.7 Eddy Covariance Flux Monitoring Networks

Long-term EC flux data of typical ecosystems in various climatic zones helps improve the land-surface schemes of climate models, monsoon and climate prediction. For last few decades, considerable efforts has been made to estimate land-atmosphere fluxes of GHGs, energy or momentum using the eddy covariance technique (Aubinet et al., 2012) at different sites, to study ecosystem functions and determining the trajectory of earth's climate system. Many permanent study sites have come up under various regional or continental networks. The magnitude of land-atmosphere exchange processes in diverse ecologies and climatic conditions in different regions of the globe are being monitored through these networks. Such datasets offer insight for climate modeling and also provide indications of future behavior of land-atmosphere processes.

Broadly, the objectives of different eddy covariance study site networks can be stated as follows:

- Quantifying the bio-geographical variation in CO_2 and H_2O flux rates through direct measurement
- Quantifying the temporal dynamics of flux rates in order to assess the effect of ecological dynamics due to changes in vegetation phenology, growing season length, drought or wet spell, heat spell, El Niño or La Nina, snow cover on canopy-level fluxes
- Quantifying the CO_2 and H_2O flux variations due to changes in environment, soil and vegetation attributes

22.7.1 *Regional Flux Networks*

Some EC-networks catering to specific regions of the world are briefly introduced here.

AmeriFlux: More than 300 sites are distributed throughout the Western Hemisphere under AmeriFlux network (science.energy.gov/ber/). It covers the full spectrum of ecosystems and climates, from Amazonian rainforests to the North Slope of Alaska.

OzFlux: This is part of the Australian Terrestrial Ecosystem Research Network (TERN), consists of nearly 30 flux towers in Australia and New Zealand. Many of these sites are also members of the Australian Supersite Network (ASN).

ChinaFLUX: The Chinese Terrestrial Ecosystem Flux Research Network (ChinaFLUX) is basically a national network for measuring long-term micrometeorological flux covering 10 ecosystem types in a wide range of Latitudes (http://www.chinaflux.org/). In ChinaFLUX, the open-path eddy covariance system is the standard flux measurement technique instead of closed path system due some logistical reasons and its performance has been evaluated against a closed-path eddy covariance system by Yu et al. (2006).

AsiaFlux: It is a network for Asian region that has united scientists from different universities and research institutions to investigate the carbon-, water- and energy exchanges between ecosystems and the atmosphere from daily to interannual time scales. As of June, 2022, there are 115 sites registered under AsiaFlux. The sites represent forest/plantation, horticultural farms, rice paddy, pasture, agricultural crops, grassland, wetland, shrubland, steppe, tundra, desert and urban areas (http://www.asiaflux.net).

There are many other countries specific or region specific or purpose specific flux network projects exist. To name a few are EuroFlux and JapanFlux, KoFlux, Urban Flux etc. India is pre-dominantly a tropical monsoon climatology, a vast landmass and one of the least explored and least understood regions from land surface studies point of view. Due to significant dependence of vegetation/ecosystem and agricultural productivity on monsoon rainfall, climate in India exerts significant control, particularly on the interannual variability of net CO_2, H_2O and sensible heat influx into the atmosphere (Nayak et al., 2015). A conscious effort for establishing an EC-network in India i.e., INDOFLUX on the line of Fluxnet has been underway during the recent past decade (Sundareshwar et al., 2007). INDOFLUX will serve as an important national network covering Indian subcontinent and adjoining oceans to know better this typical monsoon tropical/sub-tropical environment. It will further help in developing a baseline for evaluation of future environmental changes at regional and global scale.

22.7.2 Fluxnet

Fluxnet is an umbrella project which was initiated with the aim to unite regional and continental networks into an integrated global network of eddy covariance based micrometeorological flux measurement sites. This project is a novel approach as it includes both micro-meteorological method as well as remote sensing for measurement of regional to global scale fluxes. As of February, 2017, worldwide there are approximately 900 sites that are registered as members of the flux research community under such EC-network initiative (Pastorello et al., 2021). There are about 550 sites which are more than 5 years old and of which about 100 sites are working for more than 15 years now. As far as distribution is concerned, maximum EC-sites are located within natural ecosystems, viz., forest landscapes followed by grassland. This is followed by managed ecosystems i.e. croplands. Other vegetation classes (as per International Geosphere–Biosphere Programme, IGBP nomenclature) like wetlands, shrublands, savanna have also been covered under the EC-networks. Geographically, maximum EC-sites are distributed in Europe and N-America followed by Asia whereas South America, Oceania and Africa are scantly represented compared to its vast landmasses (Fig. 22.6). High frequency raw data are processed using standardized data processing routines to compute flux covariances and flux reporting is done at every 30 min to 1-h interval (Aubinet et al., 2000; Moncrieff et al., 1996). Readers are encouraged to go through published research papers such as Pastorello

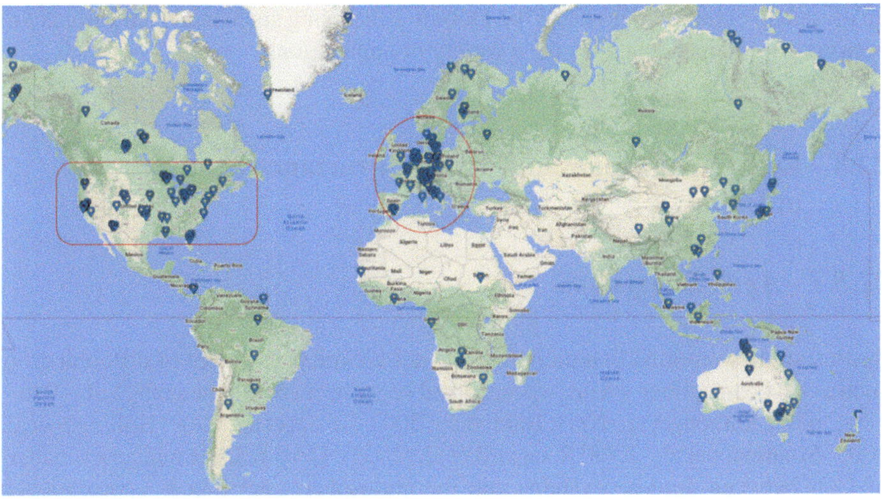

Fig. 22.6 Distribution of FLUXNET field sites on the global map

et al. (2021) to get a good idea of the annual rates of net ecosystem exchange (NEE) of CO_2, gross primary production (GPP) and ecosystem respiration (RECO) across different vegetation classes as derived from of the multi-year FUXNET site datasets.

22.7.3 Data Policy

One of the biggest flux data networking communities in the world is Ameriflux. It has evolved certain standard practices for generating, quality testing (QA/QC), archiving and sharing flux data. In general, the contributor of the data should maintain the relevant metadata of the site and maintain the instruments to ensure quality data. There are two data sharing policies, viz., AmeriFlux Legacy Data Policy and the AmeriFlux CC-BY-4.0 Licensing system, the later one was introduced lately, in August, 2021. The data contributor needs to indicate under which policy they wish to share it with the global AmeriFlux user community. As of now, most of the sites/project investigators are preferring the latest policy. The site from which data is shared under the historical AmeriFlux Legacy Data Policy is accorded lower priority for some services like site visits and such contributors are excluded in product generation which will be shared under a CC-BY-4.0 license. AmeriFlux data can be openly accessed through its data portal (https://ameriflux.lbl.gov). With each data downloaded from AmeriFlux website (ameriflux.lbl.gov), required citation along with DOI, email contacts of project investigators and information on source of funding are also provided to the user. With the courtesy of acknowledgement, citation and/or by letting know the data provider about the intended use of the data or co-authorships whenever appropriate, flux research communities across the globe are accessing such data from various flux networks or independent sites for a variety of applications.

22.8 Ecosystem Fluxes and Eddy Covariance Studies in India

22.8.1 *Agro-Ecosystems*

Several studies reported that daily average net assimilation fluxes of different crops such as soybean, winter wheat, rapeseed and sugar beet are within a range of −9 to −13 gC m^{-2} (Anthoni et al., 2004; Baldocchi, 2003; Béziat et al., 2009; Hollinger et al., 2005; Moureaux et al., 2008; Soegaard et al., 2003). Despite large annual exchanges of CO_2 between agro-ecosystems and the atmosphere, net flux is approximately balanced with soil contributing less than 1% of global anthropogenic CO_2 emissions and estimates for agricultural lands hovering around 0.05 Gt of net CO_2 emission per annum (US-EPA, 2007). Salient findings from various eddy covariance based flux measurement sites in agriculture and other ecosystems including those from Indian region are presented here.

22.8.1.1 Rice (*Oryza Sativa* L.)

Among the agro-ecosystems, rice paddies, particularly low land rice arouse special interest among researchers as far as GHG-flux studies are concerned. As the crop is largely grown by keeping a shallow layer of standing water, it releases significant amount of methane into the atmosphere, beside the crop is also an important source of N_2O (Bhaduri et al., 2017; Chatterjee et al., 2021a, 2021b; IPCC, 2007). Currently, the global acreage under rice crop is as much as about 164 m ha with south or south-east Asian region countries including China and India having the major share (Statista, 2022). Rice has 467 and 169% higher global warming potential (GWP) than wheat and maize, respectively (Linquist et al., 2012). Bhattacharyya et al., (2013a, 2013b) compared fluxes from rice paddy under both flooded and aerobic field conditions and from studies conducted across Asia (countries included were India, Bangladesh, Philippines, Taiwan and Japan). They reported NEE values ranging between −300 gC m^{-2} and −85 gC m^{-2} per growing season which when converted to daily scale would mean a maximum uptake rate of about −2.5 $gCm^{-2}d^{-1}$. Submerged rice with continuous waterlogging (34%) had the highest GHG emissions, followed by rainfed rice with flood-prone (18%) and irrigated single aeration rice (17%) (Bhatia et al., 2013).

Methane fluxes from rice: It is estimated that irrigated rice fields contribute 6–8% of the total 410–660 million tonnes of CH_4 produced each year globally (Tseng et al., 2010). In most rice-growing locations across the world, CH_4 emissions ranged from

14 to 375 mg $m^{-2}d^{-1}$ (Oo et al., 2015). Thailand's deep water rice fields are expected to emit 99 kg of CO_2 per season. (Wassmann et al., 2000). Average CH_4 emission rates in Beijing's rice fields ranged from 11 to 364 mg m^{-2} d^{-1} (Wang et al., 2000). In India, average rice-field CH_4 emissions ranged from 3.5–4.2 Tg yr^{-1} (Parashar et al., 1996). According to Bhatia et al. (2004), Indian paddy fields release 4.7 Tg yr^{-1} CH_4, with irrigated rice fields emitting 1.4 Tg yr^{-1}. Alberto et al. (2014) calculated seasonal CH_4 emissions in IRRI irrigated rice fields, observed the ecosystem emitted 3.03 g C m^2 of CH_4-C into the atmosphere throughout the growing season. Drought-prone rice varieties, deep-water rice ecosystems and irrigated multiple aeration rice ecosystems contributed 16%, 8% and 6% of methane, respectively (Pathak et al., 2014). Bhattacharyya et al. (2014) reported net methane emission (NME) in two consecutive rice growing seasons ranging between 84 and 129 kg ha^{-1} and annual value was 419 kg ha^{-1}.

Carbon dioxide fluxes from rice: Carbon dioxide fluxes from the soil are influenced by many factors such as soil texture, moisture content, pH and salinity. Elevated temperature affects labile pools of C which may act as a source for GHG emission (Chatterjee et al., 2019). SOM degradation and soil-plant respiration (heterotrophic + autotrophic) are primarily responsible for CO_2 emissions to atmosphere during rice cultivation tillage processes. However, in lowland flooded rice ecology, total CO_2 emissions from respiration are often negated by gross primary production (photosynthesis) and the net CO_2 flux remains negative NEE indicating that the ecosystem acts as a net CO_2-C sink. (Alberto et al., 2009). A different scenario is encountered in aerobic rice cultivation in the highlands. Daily NEE of rice in Taiwan ranged from −17.0 to 12.9 mol CO_2 m^{-2} s^{-1} during harvest season (Alberto et al., 2012). Chakraborty et al. (2015) reported half-hourly NEE fluxes over low land tropical rice ecosystem which ranged between +4.3 and –12.6 μmol CO_2 m^{-2} s^{-1} during the rabi season of 2014.

Energy fluxes or evapotranspiration

Chatterjee et al., (2021a, 2021b) measured the actual evapotranspiration (ETa) using EC-method in lowland rice crop in the dry (Jan–May) and wet seasons (Jul–Nov) of two years. They reported seasonal average ETa in the range of 2.3 to 3.32 mm d^{-1} with dry season having higher values.

22.8.1.2 Upland Cereals

In irrigated wheat (*Triticum aestivum*) grown on edaphically stressed shallow basaltic soil, Saha et al. (2016) observed the diurnal average rates during various phenophases ranged between 2.2 to 16.2 Wm^{-2} (sensible heat flux, H) and between 44.2 to 56.8 Wm^{-2} (latent heat flux, LE). The respective seasonal means were 11.0 Wm^{-2} and 49.0 Wm^{-2} and the maximum estimated value of total actual evapo-transpiration was 248 mm. The net daytime CO_2 exchange rates during various phenophases, viz., Sowing to CRI, CRI to Max. Tillering, Max. Tillering to Anthesis, Flowering to Dough and Dough to Maturity were 3.3, −2.0, −8.0, −7.5 and −3.1 μmol m^{-2} s^{-1}, respectively. From sowing to harvesting it acted as a sink during daytime with average NEE rate of −4.6 μmol m^{-2} s^{-1}. Among the environmental variables, soil and air temperature exerted significant control on daytime NEE rates.

Kumar et al. (2021) studied the carbon exchange in a rice-wheat cropping system of North-West Indo-Gangetic plains using EC method. They observed similar rates of GPP during the growing duration of both crops. However, RE was higher in rice compared to spring wheat with RE/GPP ratio of 0.51 and 0.34, respectively. The averages of daily NEE over the crop durations were −3.7 g C m^{-2} d^{-1} in rice and −5.0 g C m^{-2} d^{-1} in and wheat.

22.8.1.3 Pulse Crops

Chakraborty et al., (2021) used an OPEC system to study fluxes over a rainfed chickpea crop (*Cicer arietinum*). The fluxes of CO_2 and H_2O and all meteorological variables were averaged over 30 min. They observed that the crop was net sink on daily basis during 17–69 DAS. Over the entire season, it was a weak carbon sink with cumulative NEE was found to be −32.4 ± 4.2 gC m^{-2}, GPP, 163 ± 15 gC m^{-2} and ecosystem respiration, 127.5 ± 12.9 gC m^{-2}. Actual Evapotranspiration (AET) was 177 ± 18.1 mm. Peak daily values of NEE (CO_2), GPP and AET were −2.5 gC m^{-2} day^{-1}, 4.2 gC m^{-2} d^{-1} and 2.4 mm d^{-1}, respectively. In another study on chickpea, grown on soil with edaphic constraints in western Maharashtra, India seasonal mean (November–February, 2015–2016) of the diurnal sensible heat flux rates was observed to be 30.7 Wm^{-2} (ICAR-NIASM Annual Report, 2016). Sensible heat flux rate during the initial (LAI < 0.3), crop development (LAI: 0.3–3), mid season (LAI: 3 to start of senescence) and late season stages of crop growth were 33.4, 23.0, 23.2 and 52.3 Wm^{-2}, respectively. Vegetation parameter, EVI and an abiotic stress index (CATD) could explain about 61% ($p < 0.01$) and 63% ($p < 0.01$) of variability in Sensible heat flux, respectively.

22.8.1.4 Green Manure Crops

Saha et al. (2016) monitored energy and mass fluxes using an OPEC system over dhaincha (*Sesbania aculeata)* crop raised on a shallow basaltic soil of limited depth under rainfed condition in West-Central region of India. Daily average NEE during 45–126 DAS of the crop ranged between +3.1 and −7.2 μmol m^{-2} s^{-1}. It was a weak but net sink with average NEE of −1.6 gC m^{-2} d^{-1}. Daytime NEE was strongly controlled by soil moisture, soil temperature, net radiation and insolation.

22.8.1.5 Oilseed Crops

Saha et al. (2015) monitored fluxes over soybean (*Glycine max*) crop. They reported seasonal (Jul–Oct, 2014) average NEE rates under daytime condition (6 AM to 6 PM) as -5.2 μmol m^{-2} s^{-1}, under nighttime as 3.2 μmol m^{-2} s^{-1} and the diurnal rate was -1.0 μmol m^{-2} s^{-1}.

Deb Burman et al. (2020) carried out carbon flux observations in a wet season (kharif/monsoon) oilseed crop, sesame (Sesamum indicum) in an eastern Uttar Pradesh site of Barkachha (25°N, 82.6°E, ~170 m AMSL), India. Based on observations over a two crop-year period, they reported half-hourly NEE values, diurnally, ranging between two extremes i.e. –25 and 8 μmol m^{-2} s^{-1}. The above ranges during the non-growing season (winter) were much lower with the NEE amplitudes moved between -5.0 and 5.0 μmol m^{-2} s^{-1}.

22.8.1.6 Fibre Crops

Barman et al. (2022) observed fluxes over jute crop (*Corchorus olitorius*) in eastern India. Combining both daytime and night time fluxes, jute ecosystem was a net CO_2 sink on a daily basis. They reported seasonal total NEE value as -268.5 gC m^{-2} (i.e. 10.3 t CO_2 ha^{-1}).

22.8.2 Other Ecosystems

22.8.2.1 Wetland Ecosystem

Wetland ecosystems are somewhat intermediate between ecosystems with mainly land and mainly water (Van der Valk Arnold, 2012). Some typical characteristics are presence of shallow water, consistently saturated or anaerobic soils and distinct flora and fauna that can be exclusively found in these defined areas. Tropical mangroves, arctic or sub-arctic peatlands, coastal salt marsh, freshwater inland marsh, coastal lowlands, coral reefs are some such examples. Wetlands are storehouses of carbon, however, given its characteristic GHG fluxes, particularly CH_4 emissions, under changing climate these ecosystems may turn to net GHG sources (Mitsch & Gosselink, 2007).

Gnanamoorthy et al. (2022) observed methane fluxes over Cauvery Vellar-Coleroon estuarine (wetland) using an Open path methane analyser (frequency: 10 Hz) installed at 10-m height. EC-fluxes of CH_4 and all meteorological variables were averaged over a 30 min period. They found that the wet season flux rate was 12–26 nmol CH_4 m^{-2} $s^{-1}d^{-1}$ i.e. 0.8–1.8 g CH_4 m^{-2} $month^{-1}$ whereas the dry Season rate was 6–20 nmol CH_4 m^{-2} s^{-1} i.e. 0.4–0.6 g CH_4 m^{-2} $month^{-1}$. Li et al. (2020) reported EC-measured CO_2 and CH_4 fluxes from urban wetland of boreal region. The annual NEE was 8.0 g C-CO_2 m^{-2} yr^{-1} and for methane it was 3.9 g C-CH_4 m^{-2} yr^{-1}. Barring the summer season, the ecosystem acted as a source of CO_2. CH_4 displayed strong seasonal dynamics and sporadic emission episodes, higher in summer and lower in winter. During summer months, compared to open water part in the vegetative areas of the wetland stronger diurnal dynamics was observed in both CH_4 and CO_2 fluxes with peaks around noontime. Overall, the radiative forcing factor of the open-water areas was found to be higher than that of the vegetation covered areas.

22.8.2.2 Grassland Ecosystem

Grassland ecosystems, depending on how it is defined, are estimated to occupy as much as about 31–69% of the Earth's land area (Conant, 2010). Thus, grasslands have an important say, in determining the response of earth to climate change and its variability, in terms of carbon sequestration potential. Accuracy in flux measurement from such ecosystems is very crucial due to the potential magnification of even small errors at field scale measurements during upscaling. Balogh et al. (2007) reported flux measurements from four different grassland sites located under varied environments. Half-hourly average rates of CO_2 fluxes varied between –0.04 and –11.8 $\mu mol\ m^{-2}\ s^{-1}$. These EC-based values were within a close range when compared with the open chamber based measurements in those sites, however, the scatter was large in EC technique, especially during the peak of vegetative growth when CO_2 uptake rates were higher. In two sites with two contrasting years (rainfall) EC based H_2O fluxes (0.33 and 6.84 mmol $m^{-2}\ s^{-1}$) were also in the comparable ranges as that of open chamber based flux data (0.98 and 5.98 mmol $m^{-2}\ s^{-1}$). Rogger et al. (2022) carried out long-term studies on temperate mountainous grassland ecosystem carbon budget using EC-system. The effect of management (fertilization, cutting and grazing) and inter annual climate variability on carbon fluxes was studied. With average annual NEE of about -350 g Cm^{-2} and a NBP of about -150 g Cm^{-2}, the grassland behaved as a net C sink. However, there have been exceptional years of warming when net C loss was observed.

22.8.2.3 Lakes and Other Water Bodies

Accurate measurement of evaporation from small or large water bodies is important for formulating appropriate strategies of water allocation to agriculture, industries or other sectors after analysing the demand-supply situation. This is crucial keeping in view the sustainability of future water resources and especially useful during times of prolonged and severe drought. Eddy covariance method has been used to quantify evaporation from lakes and water bodies and for validation of various evaporation models (McGowan et al., 2011; Pérez et al., 2020; Tanny et al., 2008). Tanny et al. (2008) computed the ratio of daily pan (Class-A)—and EC-evaporation rates that varied between 0.96 to 1.94. Water heat fluxes predicted by Penman–Monteith–Unsworth and Penman–Brutsaert models were within -3% of that EC measured values. McJannet et al. (2011) reported very good agreement of EC-latent heat fluxes over open water with that of scintillometer based estimation under a wide range of weather conditions and flux rates.

22.9 Conclusion

Land surface process models, used as part of climate models, are required to reflect the current environmental conditions and also the situations that would be experienced in future. EC-Fluxes are the most direct measures of the sensible heat energy, moisture, momentum and GHG fluxes in real time at field or landscape scale on a continuous basis. These flux datasets help scientists to examine crucial linkages between ecosystem processes and climate responses. Changes in vegetation primary productivity is the most important indicator of climate change as it is directly linked to human or societal wellbeing through food production, ecosystem sustainability and resilience. Combined physiological, ecological and meteorological measurements at the EC-flux sites enable partitioning of carbon fluxes (NEE) into carbon cycle components of primary production (GPP, NPP) and ecosystem respiration (RE) and its environmental regulation. EC-Tower based fluxes can be upscaled to the regional/ national/global estimates by utilizing remote sensing data products such as vegetation indices (EVI, NDVI etc.) and Land Surface Temperature (LST) in data-driven models of vegetation primary productivity. Eddy covariance data are used widely for improvement and validation of complex physiological, ecological and biogeochemical models, weather and climatic models, and estimates obtained through aircraft or satellite based remote sensing. Although, eddy covariance is a state-of-the art technique, there is scope for research and improvements with regard to instrumentations, footprint models, flux correction theory, gap-filling methods or standardization of measurement protocol in complex sites. Knowledge gained from various flux networks through long-term observations including inter annual flux variations lead to improved understanding of important ecosystem processes and climate prediction, as well as advanced our foundational understanding of the Earth system to formulate future policy decisions. However, the present distribution of the EC-study sites in the world is quite asymmetric with evident bias towards the mid-latitude belt of the N-hemisphere (Jung et al., 2009, 2020). As there is no benchmark algorithm available currently for upscaling and as the effect of extrapolation on upscaling has not been investigated well the issue deserves attention of the researchers. There is a need to intensify as well as diversify EC-based flux measurements under coordinated networks in relatively less explored regions and or soil-vegetation-ecosystem classes like Africa, tropical monsoon Asia, tundra and boreal regions, transition zones like coastal lands, mangroves and wetlands, organic peat soils etc. In developing countries like India where there is prevalence of mixed cropping and very small unit landholdings, characterizing flux dynamics with eddy covariance technique offers new challenges. Measurement under these conditions may often lead to large mismatch of fluxes with that of available energy i.e. possibilities of large gaps in the energy balance closure. Moreover, frequent calibrations of sensitive fast response sensors, changing height or orientations of sensor in synchrony with the dynamics of vegetation cover or the subjects under study etc. warrants adequate trainings, technical manpower, backup sensors and other logistics support for proper maintenance of such high-end systems. That means to properly operate modern, sophisticated EC-systems in field conditions and to get reliable flux data consistently one requires considerable theoretical knowledge on micrometeorology and technical knowhow. Research programs involving EC-technology truly demand collaborative, multi-disciplinary and well-coordinated network effort for unraveling various facets of climate change.

References

Alberto, M. C. R., Hirano, T., Miyata, A., Wassmann, R., Kumar, A., Padre, A., & Amante, M. (2012). Influence of climate variability on seasonal and interannual variations of ecosystem CO_2 exchange in flooded and non-flooded rice fields in the Philippines. *Field Crops Research, 134*, 80–94.

Alberto, M. C. R., Quilty, J. R., Buresh, R. J., Wassmann, R., Haidar, S., Correa, T. Q., Jr., & Sandro, J. M. (2014). Actual evapotranspiration and dual crop coefficients for dry-seeded rice and hybrid maize grown with overhead sprinkler irrigation. *Agricultural Water Management, 136*, 1–12.

Alberto, M. C. R., Wassmann, R., Hirano, T., Miyata, A., Kumar, A., Padre, A., & Amante, M. (2009). CO_2/heat fluxes in rice fields: Comparative assessment of flooded and non-flooded fields in the Philippines. *Agricultural and Forest Management, 149*(10), 1737–1750.

Anthoni, P. M., Freibauer, A., Kolle, O., & Schulze, E. D. (2004). Winter wheat carbon exchange in Thuringia, Germany. *Agricultural and Forest Management, 121*, 55–67.

Aubinet, M., Vesala, T., & Papale, D. (2012). *Eddy covariance: A practical guide to measurement and data analysis*. Springer.

Aubinet, M., Grelle, A., Ibrom, A., Rannik, S., Moncrieff, J., Foken, T., Kowalski, A. S., Martin, P. H., Berbigier, P., Bernhofer, C., Clement, R., Elbers, J. A., Granier, A., Grünwald, T., Morgenstern, K., Pilegaard, K., Rebmann, C., Snijders, W., Valentini, R., & Vesa, T. (2000). Estimates of the annual net carbon and water exchange of forests: The EUROFLUX methodology. *Advances in Ecological Research, 30*(1), 113–175.

Bala, G., Joshi, J., Chaturvedi, R. K., Gangamani, H. V., Hashimoto, H., & Nemani, R. (2013). Trends and variability of AVHRR-derived NPP in India. *Remote Sensing, 5*, 810–829. https://doi.org/10.3390/rs5020810

Baldocchi, D. D. (2003). Assessing the eddy covariance technique for evaluating carbon dioxide exchange rates of ecosystem: Past, present and future. *Global Change Biology, 9*, 479–492.

Baldocchi, D. D. (2020). How eddy covariance flux measurements have contributed to our understanding of global change biology. *Global Change Biology, 26*, 242–260.

Ballav, S., Naja, M., Patra, P., Machida, T., & Mukai, H. (2015). High resolution CO_2 transport simulation by the WRF-CO_2 model: South Asia. In *Proceedings of the joint conference of the Asiaflux workshop 2015 and ISPRS* (p. 107), IITM, Pune.

Balogh, J., Nagy, Z., Sz, F., Pintér, K., Sz, C., Péli, E. R., Acosta, M., Marek, M. V., Zs, C., & Tuba, Z. (2007). Comparison of CO_2 and H_2O fluxes over grassland vegetations measured by the eddy-covariance technique and by open system chamber. *Photosynthetica, 45*(2), 288–292.

Barman, D., Chakraborty, A., Das, P., Roy, S., Saha, R., Mazumdar, S., Bandyopadhyay, S., Singh, A. K., Sabyasachi, M., Kundu, D., Bagui, A., Murthy, C., Rao, P., & Choudhury, S., & Kar, G. (2022). Net ecosystem CO_2 exchange from jute crop (Corchorus olitorius L.) and its environmental drivers in tropical Indo-Gangetic plain using open-path eddy covariance technique. *Environmental Monitoring and Assessment, 194*, 251. https://doi.org/10.1007/s10661-022-09872-2

Barton, J. R., Issaias, I., & Stentiford, E. I. (2008). Carbon–making the right choice for waste management in developing countries. *Waste Management, 28*(4), 690–698.

Béziat, P., Ceschia, E., & Dedieu, G. (2009). Carbon balance of a three crop succession over two cropland sites in South West France. *Agricultural and Forest Management, 149*, 1628–1645.

Bhaduri, D., Mandal, A., Chakraborty, K., Chatterjee, D., & Dey, R. (2017). Interlinked chemical-biological processes in anoxic waterlogged soil-a review. *Indian Journal of Agricultural Sciences, 87*(12), 1587–1599.

Bhatia, A., Jain, N., & Pathak, H. (2013). Methane and nitrous oxide emissions from Indian rice paddies, agricultural soils and crop residue burning. *Greenhouse Gases: Science and Technology, 3*(3), 196–211.

Bhatia, A., Pathak, H., & Aggarwal, P. K. (2004). Inventory of methane and nitrous oxide emissions from agriculture soils of India and their global warming potential. *Current Science, 87*(3), 317–324.

Bhattacharyya, P., Neogi, S., Roy, K. S., & Rao, K. S. (2013a). Gross primary production, ecosystem respiration and net ecosystem exchange in Asian rice paddy: An eddy covariance based approach. *Current Science, 104*, 67–75.

Bhattacharyya, P., Neogi, S., Roy, K. S., Dash, P. K., Nayak, A. K., & Mohapatra, T. (2014). Tropical low land rice ecosystem is a net carbon sink. *Agriculture, Ecosystems & Environment, 189*, 127–135.

Bhattacharyya, P., Neogi, S., Roy, K. S., Dash, P. K., Tripathi, R., & Rao, K. S. (2013b). Net ecosystem CO_2 exchange and carbon cycling in tropical lowland flooded rice ecosystem. *Nutrient Cycling in Agroecosystems, 95*, 133–144.

Blanken, P. D., Rouse, W. R., Culf, A. D., Spence, C., Boudreau, L. D., Jasper, J. N., Kochtubajda, B., SchertzerWM, M. P., & Verseghy, D. (2000). Eddy covariance measurements of evaporation from Great Slave Lake, Northwest Territories, Canada. *Surface Water and Climate, 36*(4), 1069–1077.

Burba, G. (2013). Eddy covariance method for scientific, industrial, agricultural, and regulatory applications—A field book on measuring ecosystem gas exchange and areal emission rates. LI-COR Biosciences, Lincoln, Nebraska.

Burba, G.G., Mcdermitt, D.K., Grelle, A., Anderson, D.J., & Xu, L. (2008). Addressing the influence of instrument surface heat exchange on the measurements of CO_2 flux from open-path gas analyzers. *Global Change Biology, 14*, 1854–1876. https://doi.org/10.1111/j.1365-2486.2008.01606.x.

Butterbach-Bahl, K., Gasche, R., Breuer, L., & Papen, H. (1997a). Fluxes of NO and N_2O from temperate forest soils: Impact of forest type, N deposition and of liming on the NO and N_2O emissions. *Nutrient Cycling in Agroecosystems, 48*(1), 79–90.

Butterbach-Bahl, K., Papen, H., & Rennenberg, H. (1997b). Impact of gas transport through rice cultivars on methane emission from rice paddy fields. *Plant, Cell and Environment, 20*(9), 1175–1183.

Chakraborty, A., Choudhary, K. K., Ramana, K. V., Srikant, P., Seshasai, M. V. R, Dadhwal, V. K., Rao, S. V.C.K., & Srinivas, C. (2015). Upscaling carbon fluxes over low land tropical rice ecosystem using flux tower and remote sensing data. In *Joint conference of the Asia flux workshop 2015 and ISPRS* (International Society for Photogrammetry and Remote Sensing) TC WG VIII/3 (p. 14), IITM, Pune.

Chakraborty, A., Choudhary, K. K., Srikanth, P., Ramana, K. V., Seshasai, M. V. R., Narayanarao, K., Amaregouda, A., Yadav, Y., & Annapurna, G. (2021). CO_2, H_2O and energy fluxes from chickpea crop grown under residual soil moisture condition in rainfed peninsular India using eddy covariance techniques. *Field Crops Research, 273*, 108307. https://doi.org/10.1016/j.fcr.2021.108307

Chakraborty, A., Venugopalan, M. V., Mani, J. K., Bagadkar, A. J., & Manikandan, A. (2022). Rainfed cotton crop in Central India is a strong net CO_2 sink: An eddy covariance-based analysis of ecosystem fluxes. *Field Crops Research, 286*, 108595. https://doi.org/10.1016/j.fcr.2022.108595

Chatterjee, D., Kuotsu, R., Ao, M., Saha, S., Ray, S. K., & Ngachan, S. V. (2019). Does rise in temperature effect adversely on soil fertility, carbon fractions, microbial biomass and enzymatic activities under different land use? *Current Science, 116*(12), 2044–2054.

Chatterjee, D., Saha, S., Pradhan, A., Swain, C. K., Venkatramaiah, E., Nayak, A. K., & Pathak, H. (2021) Reducing methane emission from lowland rice ecosystem. In *Soil science: Fundamentals to recent advances* (pp. 493–511). Springer.

Chatterjee, S., Stoy Paul, C., Debnath, M., Nayak, A. K., Kumar, S. C., Tripathi, R., Chatterjee, D., Mahapatra, S. S., Talib, A., & Pathak, H. (2021b). Actual evapotranspiration and crop coefficients for tropical lowland rice (Oryza sativa L.) in eastern India. *Theoretical and Applied Climatology, 146*, 155–171. https://doi.org/10.1007/s00704-021-03710-0

Chhabra, A., & Dadhwal, V. K. (2004). Assessment of major pools and fluxes of carbon in indian forests. *Climatic Change, 64*, 341–360. https://doi.org/10.1023/B:CLIM.0000025740.50082.e7

Conant, R. T. (2010). Challenges and opportunities for carbon sequestration in grassland systems: A technical report on grassland management and climate change mitigation. Integrated Crop Management (vol. 9, pp. 57). Plant Production and Protection Division, Food and Agriculture Organization of the United Nations (FAO), Italy.

Cui, J. I., Li, C., & Trettin, C. (2005). Analyzing the ecosystem carbon and hydrologic characteristics of forested wetland using a biogeochemical process model. *Global Change Biology, 11*, 278–289. https://doi.org/10.1111/j.1365-2486.2005.00900.x

Dadhwal, V. K. (2015). Vegetation and soil carbon pools and fluxes in India and results from flux tower networks. *In Proceedings of the joint conference of the Asiaflux workshop 2015 and ISPRS* (p. 2) IITM, Pune, November.

Davis, K. J., Bakwin, P. S., Yi, C., Berger, B. W., Zhaos, C., Teclaw, R. M., & Isebrands, J. G. (2003). The annual cycles of CO_2 and H_2O exchange over a northern mixed forest as observed from a very tall tower. *Global Change Biology, 9*, 1278–1293.

Deb Burman, P. K., Shurpali, N. J., Chowdhuri, S., Karipot, A., Chakraborty, S., Lind, S. E., Martikainen, P. J., Chellappan, S., Arola, A., Tiwari, Y. K., Pmurugavel, G. D., Todekar, K., & Prabha, T. V. (2020). Eddy covariance measurements of CO_2 exchange from agro-ecosystems located in subtropical (India) and boreal (Finland) climatic conditions. *Journal of Earth System Science, 129*(43), 1–18. https://doi.org/10.1007/s12040-019-1305-4

Falge, E., Baldocchi, D., Olson, R., Anthoni, P., Aubinet, M., Bernhofer, C., Burba, G., Ceulemans, R., Clement, R., Dolman, H., Granier, A., Gross, P., Grünwald, T., Hollinger, D., Jensen, N. O., Katul, G., Keronen, P., Kowalski, A., Lai, C. T., … Moncrieff, J. (2001). Gap filling strategies for defensible annual sums of net ecosystem exchange. *Agricultural and Forest Management, 107*, 43–69. https://doi.org/10.1016/S0168-1923(00)00225-2

Falge, E., Baldocchi, D., Tenhunen, J., Aubinet, M., Bakwin, P., Berbigier, P., Bernhofer, C., Burba, G., Clement, R., Davis, K. J., Elbers, J. A., Goldstein, A. H., Grelle, A., Granier, A., Guðmundsson, J., Hollinger, D., Kowalski, A. S., Katul, G., Law, B. E., Malhi, Y., Meyers, T., Monson, R. K., Munger, J. W., Oechel, W., & Paw, U. K. T., et al. (2002). Seasonality of ecosystem respiration and gross primary production as derived from FLUXNET measurements. *Agricultural and Forest Management, 78*, 83–105.

Gnanamoorthy, P., Chakraborty, S., Nagarajan, R., Ramasubramanian, R., Selvam, V., Burman, P. K. D., Sarathy, P. P., Zeeshan, M., Song, Q., & Zhang, Y. (2022). Seasonal variation of methane fluxes in a mangrove ecosystem in South India: An eddy covariance-based approach. *Estuaries and Coasts, 45*, 551–566. https://doi.org/10.1007/s12237-021-00988-1

Greco, S., & Baldocchi, D. D. (1996). Seasonal variations of CO_2 and water vapour exchange rates over a temperate deciduous forest. *Global Change Biology, 2*, 183–198.

Gu, L., Baldocchi, D., Verma, S. B., Black, T. A., Vesala, T., Falge, E. M., & Dowty, P. R. (2002). Advantages of diffuse radiation for terrestrial ecosystem productivity. *Journal of Geophysical Research, 107*(D6), 4050. https://doi.org/10.1029/2001JD001242

Guo, Q., Li, W. W., Liu, D. D., Wu, W., Liu, Y., Wen, X. X., & Liao, Y. C. (2013). Seasonal characteristics of CO_2 fluxes in a rain-fed wheat field ecosystem at the loess plateau. *Spanish Journal of Agricultural Research, 11*(4), 980–988.

Hansen, M. N., Henriksen, K., & Sommer, S. G. (2006). Observations of production and emission of greenhouse gases and ammonia during storage of solids separated from pig slurry: Effects of covering. *Atmospheric Environment, 40*(22), 4172–4181.

Hernandez-Ramirez, G., Hatfield, J. L., Parkin, T. B., Sauer, T. J., & Prueger, J. H. (2011). Carbon dioxide fluxes in corn-soybean rotation in the midwestern U.S.: Inter- and intra-annual variations, and biophysical controls. *Agricultural and Forest Management, 151*, 1831–1842.

Hollinger, S. E., Bernacchi, C. J., & Meyers, T. P. (2005). Carbon budget of mature no-till ecosystem in north central region of the United States. *Agricultural and Forest Management, 130*, 59–69.

Retrieved August 17, 2022, from https://wrrc.arizona.edu/sites/wrrc.arizona.edu/files/pdfs/Shamir_RioRicoTransferability13Nov2014Methodology.pdf

Retrieved Augurst 6, 2022, from https://www.statista.com/statistics/271969/world-rice-acreage-since-2008/

ICAR-NIASM. (2016). Monitoring energy and mass fluxes in stressed crops. In J. Rane, Y. Singh, K. K. Meena, B. B. Fand, B. Sajjanar, & N. Kumar (Eds.), *Annual report* 2015–16 (pp. 10). Director, ICAR-National Institute of Abiotic Stress Management, Malegaon, Baramati, 413 115, Pune, Maharashtra, India.

IPCC. (2021). Summary for policymakers. In V. Masson-Delmotte, P. Zhai, A. Pirani, S. L. Connors, C. Péan, S. Berger, N. Caud, Y. Chen, L. Goldfarb, M. Gomis, K. Huang, E. Leitzell, J. B. R. Lonnoy, T. K. Matthews, T. Maycock, M. I. Waterfield, O. Yelekçi, R. Yu, & B. Zhou (Eds.), *Climate change 2021: The physical science basis. Contribution of working group I to the sixth assessment report (AR6) of the intergovernmental panel on climate change.* Cambridge University Press.

IPCC. (2007). Agriculture. In B. Metz, O. R. Davidson, & P.R. Bosch (Eds.), *Climate change 2007: Mitigation. Contribution of working group III to the fourth assessment report of the intergovernmental panel on climate change.* Cambridge University Press.

Joyson, A., Patel, N. R., Kushwha, S. P. S., & Dadhwal, V. K. (2015). Ecosystem respiration over a mixed forest plantation of Northern India using MODIS-derived vegetation indices and flux-tower observations. In *Proceedings of joint conference AsiaFlux 2015 and ISPRS* (p. 104). IITM, Pune.

Jung, M., Reichstein, M., & Bondeau, A. (2009). Towards global empirical upscaling of FLUXNET eddy covariance observations: Validation of a model tree ensemble approach using a biosphere model. *Biogeosciences, 6*, 2001–2013. https://doi.org/10.5194/bg-6-2001-2009

Jung, M., Schwalm, C., Migliavacca, M., Walther, S., Camps-Valls, G., Koirala, S., Anthoni, P., Besnard, S., Bodesheim, P., Carvalhais, N., Chevallier, F., Gans, F., Goll, D. S., Haverd, V., Köhler, P., Ichii, K., Jain, A. K., Liu, J., Lombardozzi, D., … Reichstein, M. (2020). Scaling carbon fluxes from eddy covariance sites to globe: Synthesis and evaluation of the FLUXCOM approach. *Biogeosciences, 17*, 1343–1365. https://doi.org/10.5194/bg-17-1343-2020

Kang, M., Ruddell, B. L., Cho, C., Chun, J. H., & Kim, J. (2015). Filtering drainage-affected eddy covariance observations on a forest hill slope using dynamical process networks. In *Proceedings of joint conference AsiaFlux 2015 and ISPRS* (p. 97). IITM, Pune.

Kroon, P. S., Hensen, A., Van den Bulk, W. C. M., Jongejan, P. A. C., & Vermeulen, A. T. (2008). The importance of reducing the systematic error due to non-linearity in N_2O flux measurements by static chambers. *Nutrient Cycling in Agroecosystems, 82*(2), 175–186.

Kumar, A., Bhatia, A., Sehgal, V. K., Tomer, R., Jain, N., & Pathak, H. (2021). Net ecosystem exchange of carbon dioxide in rice-spring wheat system of Northwestern Indo-Gangetic Plains. *Land, 10*(701), 19. https://doi.org/10.3390/land10070701

Law, B. E., Arkebauer, T., Campbell, J. L., Chen, J., Sun, O., Schwartz, M., van Ingen, C., & Verma, S. (2008). Terrestrial carbon observations: Protocols for vegetation sampling and data submission. Report No.55, Global Terrestrial Observing System, FAO, Rome.

Li, X., Wahlroos, O., Haapanala, S., Pumpanen, J., Vasander, H., Ojala, A., Vesala, T., & Mammarella, I. (2020). Carbon dioxide and methane fluxes from different surface types in a created urban wetland. *Biogeosciences, 17*, 3409–3425.

Linquist, B. A., Adviento-Borbe, M. A., Pittelkow, C. M., Van Kessel, C., & Van Groenigen, K. J. (2012). Fertilizer management practices and greenhouse gas emissions from rice systems: A quantitative review and analysis. *Field Crops Research, 135*, 10–21.

Loescher, H. W., Law, B. E., Mahrt, L., Hollinger, D. Y., Campbell, J., & Wofsy, S. C. (2006). Uncertainties in, and interpretation of, carbon flux estimates using the eddy covariance technique. *Journal of Geophysical Research, 111*, D21S90. https://doi.org/10.1029/2005JD006932

Massman, W. J. (2000). A simple method for estimating frequency response corrections for eddy covariance systems. *Agricultural and Forest Meteorology, 104,* 185–198. https://doi.org/10.1016/S0168-1923(00)00164-7.

McGowan, H. A., McGloin, R., McJannet, D., & Burn, S. (2011). Long term measurement of lake evaporation using a pontoon mounted Eddy Covariance system. *Conference paper. AGU fall meeting abstracts-Id*. H33A-1266. American Geophysical Union.

McJannet, D. L., Cook, F. J., McGloin, R. P., McGowan, H. A., & Burn, S. (2011). Estimation of evaporation and sensible heat flux from open water using a large-aperture scintillometer. *Water Resources Research, 47*, W05545. https://doi.org/10.1029/2010WR010155,2011

Mitsch, W. J., & Gosselink, J. G. (2007). *Wetlands* (4th ed.). John Wiley & Sons Inc.

Mohanty, S., Verma, S. K., & Nayak, S. K. (2006). Dynamic mechanical and thermal properties of MAPE treated jute/HDPE composites. *Composites Science and Technology, 66*(3–4), 538–547.

Moncrieff, J. B., Malhi, Y., & Leuning, R. (1996). The propagation of errors in long-term measurements of land-atmosphere fluxes of carbon and water. *Global Change Biology, 2*(3), 231–240. https://doi.org/10.1111/j.1365-2486.1996.tb00075.x

Moureaux, C., Debacq, A., Hoyaux, J., Suleau, M., Tourneur, D., Vancutsem, F., Bodson, B., & Aubinet, M. (2008). Carbon balance assessment of a Belgian winter wheat crop (Triticum aestivum L). *Global Change Biology, 14*, 1353–1366.

Nayak, R. K., Mishra, N., Patel, N. R., & Dadhwal, V. K. (2015). Net CO_2 sink potential of Indian ecosystem in present decade. In *Proceedings of the joint conference of the AsiaFlux Workshop 2015 and ISPRS* (p. 13). IITM, Pune.

Oo, A. Z., Win, K. T.,& Bellingrath-Kimura, S. D. (2015). Within field spatial variation in methane emissions from lowland rice in Myanmar. *Springer Plus, 4*, 145. https://doi.org/10.1186/s40064-015-0901-2

Parashar, D. C., Mitra, A. P., Gupta, P. K., Rai, J., Sharma, R. C., Singh, N., Koul, S., Ray, H. S., Das, S. N., Parida, K. M., Rao, S. B., Kanungo, S. P., Ramasami, T., Nair, B. U., Swamy, M., Singh, G., Gupta, S. K., Singh, A. R., Saikia, B. K., … Sinha, S. K. (1996). Methane budget from paddy fields in India. *Chemosphere, 33*, 737–757.

Pastorello, G., Trotta, C., Canfora, E., Chu, H., Christianson, D., Cheah, Y. W., Poindexter, C., Chen, J., et al. (2021). The FLUXNET2015 dataset and the ONEFlux processing pipeline for eddy covariance data. *Scientific Data, 8*, 72. https://doi.org/10.1038/s41597-021-00851-9

Patel, N. R., Kushwha, S. P. S., Tripathi, P., & Dadhwal, V. K. (2015). Evaluation of biome-Bgc model for simulating biophysical attributes and carbon fluxes over mixed forest plantation in foot hills of Lower Shiwalik. In *Proceedings of the joint conference of AsiaFlux workshop 2015 and ISPRS* (p. 105). IITM, Pune.

Pathak, H., Bhatia, A., & Jain, N. (2014*). Greenhouse gas emission from Indian agriculture: Trends, mitigation and policy needs*. Indian Agricultural Research Institute.

Pathak, H., Upadhyay, R. C., Muralidhar, M., Bhattacharyya, P., & Venkateswarlu, B. (Eds.). (2013). *Measurement of greenhouse gas emission from crop, livestock and aquaculture*. Indian Agricultural Research Institute.

Patil, M. N., Dharmaraj, T., Waghmare, R. T., Chinthalu, G. R., Singh, D. K., Prabha, T. (2015). Evaluating relationship between fluxes of heat and CO_2 using direct eddy covariance measurements. In *Proceedings of the joint conference of the AsiaFlux Workshop 2015 and ISPRS* (p. 102). IITM, Pune.

Pérez, A., Lagos, O., Lillo-Saavedra, M., Souto, C., Paredes, J., & Arumí, J. L. (2020). Mountain lake evaporation: a comparative study between hourly estimations models and in situ measurements. *Water, 12*, 2648. https://doi.org/10.3390/w12092648

Posse, G., Richter, K., Corin, J. M., Lewczuk, N. A., Achkar, A., & Rebella, C. (2010). Carbon dioxide fluxes on a soybean field in Argentina: Influence of crop growth stages. *Open Agriculture Journal, 4*, 8–63.

Rochette, P. (2011). Towards a standard non-steady-state chamber methodology for measuring soil N_2O emissions. *Animal Feed Science and Technology, 166*, 141–146.

Rogger, J., Hörtnagl, L., Buchmann, N., & Eugster, W. (2022). Carbon dioxide fluxes of a mountain grassland: Drivers, anomalies and annual budgets. *Agricultural and Forest Management, 314*, 108801. https://doi.org/10.1016/j.agrformet.2021.108801

Saha, S., Minhas, P. S., Bal, S.K., & Singh, Y. (2015). Trends in climatic features and greenhouse gas exchange of crops in scarcity zone (Baramati) of western Maharashtra. Technical Bulletin No. 6, ICAR-National Institute of Abiotic Stress Management, Malegaon, Baramati- 413115, Pune, Maharashtra, India, p. 42.

Saha, S., Bal, S. K., & Bhagat, K. (2016). Fluxes and production efficiency of irrigated wheat ecosystem under edaphic constraints of Western Maharashtra plateau: A micrometeorological investigation. *Journal of Agrometeorology, 18*(2), 175–183.

Schneising, O., Buchwitz, M., Burrows, J. P., Bovensmann, H., Reuter, M., Notholt, J., Macatangay, R., & Warneke, T. (2008). Three years of greenhouse gas column-averaged dry air mole fractions retrieved from satellite–Part 1: Carbon dioxide. *Atmospheric Chemistry and Physics, 8*(14), 3827–3853.

Singh, R. P., Rovshan, S., Goroshi, S., Panigrahy, S., Parihar, J. (2011). Spatial and temporal variability of Net Primary Productivity (NPP) over terrestrial biosphere of India using NOAA-A VHRR based GloPEM Model. *Journal of the Indian Society of Remote Sensing, 39*(3). https://doi.org/10.1007/s12524-011-0123-1

Soegaard, H., Jensen, N. O., Boegh, E., Hasager, C. B., Schelde, K., & Thomsen, A. (2003). Carbon dioxide exchange over agricultural landscape using eddy correlation and footprint modelling. *Agricultural and Forest Management, 114*, 153–173.

Sun, Y., Ma, J., Sude, B., Lin, X., Shang, H., Geng, B., Diao, Z., Du, J., & Quan, Z. (2021). A UAV-based eddy covariance system for measurement of mass and energy exchange of the ecosystem: Preliminary results. *Sensors, 21*(2), 403. https://doi.org/10.3390/s21020403

Sundareshwar, P. V., Murtugudde, R., Srinivasan, G., Singh, J. S., Ramesh, K. J., Ramachandran, R., Verma, S. B., Agarwal, D. A., Baldocchi, D., Baru, C., Baruah, K. K., Chowdhury, G. R., Dadhwal, V. K., Dutt, C. B. S., et al. (2007). Environment: Environmental monitoring network for India. *Science, 316*(5822), 204–205. https://doi.org/10.1126/science.1137417

Swain, C. K., Bhattacharyya, P., Nayak, A. K., Singh, N. R., Chatterjee, D., Dash, P. K., Neogi, S., & Pathak, H. (2018). Temporal variation of energy fluxes during dry season in tropical lowland rice. *MAPAN- Journal of Metrology Society of India, 33*(3), 241–251. https://doi.org/10.1007/s12647-018-0260-x

Swain, C. K., Bhattacharyya, P., Singh, N. R., Neogi, S., Sahoo, R. K., Nayak, A. K., Zhang, G., & Leclerc, M. Y. (2016). Net ecosystem methane and carbon dioxide exchange in relation to heat and carbon balance in lowland tropical rice. *Ecological Engineering, 95*, 364–374.

Tanny, J., Cohen, S., Assouline, S., Lange, F., Grava, A., Berger, D., Teltch, B., & Parlange, M. B. (2008). Evaporation from a small water reservoir: Direct measurements and estimates. *Journal of Hydrology, 351*(1–2), 218–229.

Tseng, H.K., Tsai, L.J., Alagesan, A., Tsuang, J.B., Yao, H.M., & Kuo, H.P. (2010). Determination of methane and carbon dioxide fluxes during the rice maturity period in Taiwan by combining profile and eddy covariance measurements. *Agricultural and Forest Meteorology, 150*, 852–859.

Van der Valk, A. G. (2012). The biology of freshwater wetlands. Oxford University Press.

Wang, Z. Y., Xu, Y. C., Li, Z., Guo, Y. X., Wassmann, R., Neue, H. U., Lantin, R. S., Buendia, L. V., Ding, Y. P., & Wang, Z. Z. (2000). A four-year record of methane emissions from irrigated rice fields in the Beijing region of China. *Nutrient Cycling in Agroecosystem, 58*, 55–63.

Wassmann, R., Buendia, L. V., Lantin, R. S., Bueno, C. S., Lubigan, L. A., Umali, A., Nocon, N. N., Javellana, A. M., & Neue, H. U. (2000). Mechanisms of crop management impact on methane emissions from rice fields in Los Baños, Philippines. *Nutrient Cycling in Agroecosystems, 58*, 107–119.

Webb, E. K., Pearman, G. I., & Leuning, I. (1980). Corrections of flux measurements for density effects due to heat and water vapour transfer. *Quarterly Journal of the Royal Meteorological Society, 106,* 85–100.

Xu, S., Hao, X., Stanford, K., McAllister, T., Larney, F. J., & Wang, J. (2007). Greenhouse gas emissions during co-composting of cattle mortalities with manure. *Nutrient Cycling in Agroecosystems, 78,* 177–187.

Yu, G., Gao, Y., Yan, H., Zhu, X., Wang, Q., Zhang, J., Wang, Y., Li, Y., & Shi, P. (2015). Modeling net ecosystem CO_2 exchange in Northern China and the Tibetan plateau based solely on MODIS imagery. In *Proceedings of the Joint conference of the AsiaFlux Workshop 2015 and ISPRS* (p. 100) IITM, Pune.

Yu, G. R., Wen, X. F., Sun, X. M., Tanner, B. D., Lee, X., & Chen, J. Y. (2006). Overview of China FLUX and evaluation of its eddy covariance measurement. *Agricultural and Forest Management, 137*(3–4), 125–137. https://doi.org/10.1016/j.agrformet.2006.02.011

Zhao, M., & Running, S. W. (2010). Drought-induced reduction in global terrestrial net primary production from 2000 through 2009. *Science, 329*, 940–943. https://doi.org/10.1126/science.1192666

Chapter 23
Greenhouse Gas and Energy Flux Measurements with Eddy Covariance Technique Under Lowland Rice Ecology

Chinmaya Kumar Swain, Dibyendu Chatterjee, Amaresh Kumar Nayak, Kiran Kumar Mohapatra, Rajalaxmi Sahoo, Abhijit Pradhan, Khitish Chandra Moharana, and Nihar Ranjan Singh

Abstract In the present climate change scenario, understanding the response of terrestrial ecosystems to environmental changes on temporal scale is a key issue. The changes in weather and climate are closely inter-linked with land surface processes, therefore, interest has been given for establishing global networks for monitoring surface exchanges of matters (trace gases) and energy between ecosystems and atmosphere. Agriculture is responsible for emission of all three major GHGs, carbon dioxide (CO_2), methane (CH_4) and nitrous oxide (N_2O). Rice paddy contributes toward the emissions of the most important GHGs responsible for global warming: CO_2, CH_4 and N_2O. Sometimes agriculture has potential role in the global carbon budget of greenhouse gases such as CO_2 and CH_4 as they sequester the former but release the latter and their relative contribution changes with agricultural management practices. To monitor the GHGs emission, net ecosystem carbon dioxide (NEE), methane exchange (NEME), energy exchange components and net ecosystem carbon balance in lowland tropical rice, the eddy covariance (EC) techniques was employed to linking mass exchange to energy components, linking relationship with soil microbial activities and labile carbon pools with net ecosystem exchange and to quantify carbon budget in lowland rice. Net ecosystem carbon balance was made by computing all the input components and output components. Net ecosystem exchange displayed a distinct daily and seasonal pattern throughout the cropping season and it has shown that the low land rice ecology has the capacity to sequester carbon from the atmosphere in long run. Long-term trails incorporating agricultural practices and the physico-chemical-biological factors in ecosystem are worthwhile research topics to

C. K. Swain (✉) · D. Chatterjee · A. K. Nayak · K. K. Mohapatra · R. Sahoo · A. Pradhan · K. C. Moharana
ICAR-National Rice Research Institute, Cuttack, India
e-mail: chinu.swain@gmail.com

N. R. Singh
Ravenshaw University, Cuttack, Odisha, India

H. Pathak et al. (eds.), *Climate Change Impacts on Soil-Plant-Atmosphere Continuum*, Advances in Global Change Research 78,
https://doi.org/10.1007/978-981-99-7935-6_23

understand the mechanism of C flow in the soil–plant-atmosphere in the lowland rice ecosystem.

Keywords Greenhouse · Energy flux · Eddy Covariance · Lowland · Rice ecology

23.1 Introduction

Agriculture is significantly impacted by changing climate, with higher temperatures diminishing optimal agricultural production and increasing weed and pest spread. Understanding how terrestrial ecosystems respond to environmental changes over time is critical in a changing climate scenario (IPCC, 2014). Because weather and climate change are inseparably tied to land surface processes, worldwide networks for monitoring surface transfers of matter and energy between ground and the atmosphere have been established (Gao et al., 2016). Agriculture is one of the most significant anthropogenic contributors of greenhouse gas in the atmosphere (Lal, 2000). Annual greenhouse gas emissions from agricultural output in terrestrial ecosystems were projected to be 5.0–5.8 $GtCO_2$ $eqyr^1$ (one billion tonnes of carbon dioxide or gigatonnes) between 2000- 2010, consisting of approximately 10 ~ 12% of global anthropogenic emissions (IPCC, 2014) and contribute to regional carbon and heat budgets. Both natural and anthropogenic activities significantly impact greenhouse gas (GHGs) emissions (Bouwman, 1990). During the industrial revolution, atmospheric greenhouse gases such as carbon dioxide (CO_2), methane (CH_4), and nitrous oxide (N_2O) increased dramatically. Fossil fuel consumption, deforestation, land-use change, and intensive agriculture are the causes of rising GHG concentrations (Bhattacharyya et al., 2013). The most significant greenhouse gas, CO_2 has a concentration of 413.2 parts per million as of 2020, 149% above pre-industrial levels. Methane and nitrous N_2O levels are now 262% and 123% lower than in 1750, when human activities first began to disturb the planet's natural equilibrium (WMO, 2022). The global mean temperature has risen by 0.4–0.76 °C since the end of the nineteenth century due to GHG accumulation in the atmosphere. Biotic and abiotic soil activities are a significant source of GHGs. Rice is grown on approximately 44 million hectares in India, with three significant ecosystems: rainfed uplands (16% area, 7.04 million hectares), irrigated medium lands (45% area, 19.8 million hectares), and rainfed lowlands (39% area, 17.16 m ha). Rain-fed lowland rice is farmed in roughly 13 million hectares of land in Eastern India, primarily in Odisha, West Bengal, Bihar, Jharkhand, and Chhattisgarh. Rice paddy contributes to the emissions of the three leading greenhouse gases that cause global warming: CO_2, CH_4, and N_2O. Soil is one of the significant sources and sinks of GHGs that contribute to global warming and climate change (Janseens et al., 2006). Soil makes up around 20% of the total CO_2 emissions through soil and root respiration, CH_4 emissions (12%), and anthropogenic N_2O emissions (60%) (IPCC, 2007). Soil health refers to how well soil currently performs all of its activities and how those functions are being conserved for future use. Crop production, water quality, or any other single outcome cannot be used to define soil health. Because soil health cannot be tested directly, we use indicators to assess it. As a result, measuring GHGs from soil is essential to climate change research. GHGs emitted into the atmosphere from soils, such as CO_2, CH_4, and N_2O, are largely biogenic. As a result, the closed-chamber method, open-chamber method, and micrometeorological method i.e., eddy covariance technique, are all commonly used to measure GHG emissions from soil, whereas Eddy covariance (EC) is the primary method for measuring flux since it is the only method that allows for direct flux determination.

23.2 Soil Flux Measurement Methods

Close chamber-based and micrometeorological methods are the two primary methods for measuring greenhouse gas fluxes between agro-ecosystems and the atmosphere. However, combining the two methodologies should be used for a more comprehensive investigation.

23.2.1 Manual Closed Chamber Methods

Gas samples from closed chambers are obtained manually or automatically in these methods. The samples are then examined in laboratory using a process called gas chromatography. The chamber approach has a substantial limitation in spatial variability in estimated GHG fluxes, especially for very different smallholding farm systems like those present in India. Another disadvantage is that imposing artificial chambers may disrupt the microclimate, perhaps causing severe damage, differences in temperature and obstruction in natural wind flow with the ambient environment introducing errors in the measured flows. Furthermore, because more laboratory employees, tools, and chemicals are required to reflect temporal variability appropriately or to track the day-night trend, the chamber approach becomes onerous and often impracticable.

23.2.2 Micrometeorological Methods

Vertical flux-profile or flux-gradient techniques and EC techniques are two micrometeorological methodologies. Micrometeorological approaches, in contrast to chamber techniques, enable high-frequency temporal measurements integrated across a vast region while preserving the natural microclimate of the location. With the development of cutting-edge sensor technology, automation, and very efficient computing tools, eddy covariance techniques have become increasingly popular and have emerged as the most popular choice among today's researchers for GHG monitoring at the ecological scale (Aubinet et al., 2000; Baldocchi et al., 2001). However, numerous flux networks worldwide, such as FLUXNET, ICOS, and NEON, are leading the charge in this direction with this single methodology for the EC approach. After 1980s, the EC approach was widely used (Miyata et al., 2000). On the other hand, the majority of research used short-term measures ranging from a few days to a few weeks. The relaxed eddy accumulation approach estimated CO_2 flux over two consecutive growing seasons in a rice field in Texas, USA. However, as previously said, there are differences in the carbon (C) budgets of paddy fields that are affected by changes in cultivation practices and field management.

In contrast to chamber methods, micrometeorological procedures enable high-frequency temporal data integrated across a broad territory while conserving the locality's native microclimate. Eddy covariance techniques have become more and more widespread as a result of the advancement of cutting-edge sensor technology, automation, and highly effective computational tools, and are currently the most widely used method for GHG monitoring at the ecological scale (Aubinet et al., 2000; Baldocchi et al., 2001). However, the EC approach lacks standardized language and suitable methods, with many flux networks around the world, such as FLUXNET, ICOS, and NEON, leading the way in this direction. The EC technique became commonly employed after the 1980s (Miyata et al., 2000). On the other hand, most of those studies utilized short-term measures lasting a few days to a few weeks. In a rice field in Texas, USA, the relaxed eddy accumulation method was employed to calculate CO_2 flux across two successive growing seasons. However, variations in the carbon (C) budgets of paddy fields are impacted by adjustments to farming methods and field maintenance.

Area-integrated gas exchange measurements can be made using micrometeorological techniques while maintaining the natural canopy microclimate. Due to recent advancements in technology, it is now possible to monitor fluxes at very high temporal frequencies, ranging from hours to minutes. On the other hand, the instrumentation is expensive, and a complete GHG budgeting typically requires a power supply and a uniform crop cover covering a specific minimum area. Agricultural fields' net exchange fluxes of GHGs are tracked using chamber-based measurement techniques or micrometeorological methods. In the landscape and regional scales and the vertical scale at the profile (Lobsey and Viscarra Rossel, 2016; Ramifehiarivo et al., 2017), measurement standards for estimating soil carbon content and carbon sequestration on a wide scale are currently evolving. A recent attempt was made to assess carbon sequestration at the ecosystem level using eddy-covariance flux towers and agricultural life cycle analysis. Integrating past and present SOC sequestration strategies is crucial, while keeping in mind the value of measuring and quantifying soil carbon at point-field, regional, and ecosystem scales. This method is used to validate and tune regional weather models, global climate models, complex biochemical and ecological models, and remote sensing forecasts from satellites and aircraft.

23.2.3 *Eddy Covariance with Different Sensors and Its Function*

The EC methodology is a sophisticated meteorological method for calculating vertical turbulent fluxes of energy components and GHG concentrations within atmospheric boundary layers. It comprises several sensors (Fig. 23.1) that track GHG concentrations, energy, and momentum. Using vertical wind speed covariance, these data are then converted to fluxes. A sonic anemometer that measures three-dimensional wind speed, sonic temperature, and wind direction is part of the eddy covariance system. The speed of sound in air sent through a transducer serves as the basis for the sonic anemometer's measurement. Another transducer captures sound reflections (i) The interval between transmission and reception determines the wind speed. (ii) an LI-7500A open-path infrared gas analyzer for monitoring CO_2 and water vapor variations densities, (iii) an open-path infrared methane gas analyzer (LI7700) (iv) This four-part radiation sensor measures net radiation (NR), shortwave upwell (SU), shortwave down well (SD), longwave upwell (LU), and longwave down well (LD) radiations. The soil temperature probe measures the temperature (Tg) at three depths (5, 15, and 30 cm). Various sensors, such as their measurements and manufacturer. The sensors are all fixed to a tripod-shaped metal tower that faces away from the predominant southeast wind. The head of the LI-7500A is tilted 15 degrees to lessen raindrop interference. A 2 GB memory card attached to a data logger stores all of the data (10 Hz) (CR3000).

Fig. 23.1 Open path eddy covariance system in lowland rice ecosystem

23.3 Purpose and Application of Eddy Covariance System

The validation of physiological models of C exchange and the connection between fluxes and remote sensing data depend on data from flux sites. Physiological and ecological data can be used to partition carbon fluxes into plant and soil components, revealing the mechanisms underlying these fluxes. At some places, C budgets obtained from direct flow measurements have been validated by biomass-based estimates of C storage and vice versa. In particular, sites with numerous years of data have been used to quantify inter-annual flux fluctuations. The information from the flux sites has also been used in ecology, weather forecasting, and climate studies. The existence of anthropogenic climate change has now been established beyond a doubt. Making predictions and preparing for the future depends on understanding current emissions levels and terrestrial sink dynamics. Gross Primary Production (GPP) and Ecosystem Respiration (RE) are some of the significant applications of the EC flux tower in agricultural studies. The eddy covariance flux tower is the best method for observing greenhouse gas flux from agricultural areas. Eddy covariance is a possible tool for comprehending the mechanisms and processes governing the carbon and energy balance of the agricultural ecosystem.

23.4 Principle of the Eddy Covariance Measurements

Airflow refers to a horizontal wind flow that contains numerous whirling eddies. There are three types of velocity components for each eddy: vertical (v), lateral (u), and horizontal (w). The eddy covariance method measures the covariance of variations in vertical wind velocity (w) and the physical attribute to be examined to quantify turbulent flux data (Burba, 2013): The vertical flux in a turbulent flow is expressed as

$$\mathrm{F} = \overline{\rho_a \mathrm{ws}} \tag{23.1}$$

"ρ_a" is the air density, and "s" is the measured gas's dry mole percentage. Using Reynolds decomposition, each term may be separated into its mean and variance.:

$$\mathrm{F} = \overline{\left(\overline{\rho}_a + \rho_a\right)'(\overline{w} + w')(\overline{s} + s')} \tag{23.2}$$

Equation 23.2 may be expanded as:

$$\mathrm{F} = \overline{\left(\overline{\rho}_a \overline{w}\,\overline{s} + \overline{\rho}_a \overline{ws}' + \overline{\rho}_a w'\overline{s} + \overline{\rho}_a w's' + \rho_a' \overline{w}\,\overline{s} + \overline{\rho}_a' \overline{ws}' + \overline{\rho}_a' w'\overline{s} + \overline{\rho}_a' w's'\right)} \tag{23.3}$$

where the averaged variation from the average is zero (for horizontal homogeneous terrain, air density is assumed to be negligible with mean vertical velocity to be negligible), the Eq. (23.3) may be rewritten as:

$$F = \overline{\overline{\rho}_a w' s'} \tag{23.4}$$

mass exchange and energy balance parameters are measured using the above principle. On a half-hourly basis, the mean vertical CO_2 flux density (FCO_2) is calculated as covariance between variations in w (w') and the CO_2 mixing ratio (cCO_2'):

$$F_{CO2} = \overline{\rho_a w' c'_{CO2}} \tag{23.5}$$

Here, the air density is represented by ρ_a, the overbars represent the average over time, and the primes represent the oscillations around the mean in the equation.

The 30-min covariance between w' and the CH_4 mixing ratio (cCH_4´) is used to compute the mean vertical flux density of CH_4(F_{CH4}). by the following formula:

$$F_{CH4} = \overline{\rho_a w' c'_{CH4}} \tag{23.6}$$

Water vapor flux is calculated as covariance of w´ and water vapor concentration:

$$F_{H2O} = \overline{\rho_a w' q'} \tag{23.7}$$

q' is the fluctuation in specific humidity, ρa is the dry air density, and the overbar denotes the mean value.

Simultaneously, based on the EC principle, the sensible heat flux (H) is calculated as the sum of the vertical covariance between wind speed, temperature deviations, and the mean air density. The specific heat flux is then multiplied by the quantity of particular heat flux units to obtain the energy units.

$$H = \rho_a C_p w' T' \tag{23.8}$$

EC system also measures the Latent heat flux (LE) using the measurement of water vapor and converted to energy units.

$$LE = \pi \tag{23.9}$$

In Eqs. (23.8 and 23.9), a denotes air density, Cp denotes the specific heat capacity of air at constant pressure, w denotes vertical wind speed, T denotes air temperature, denotes the psychrometric constant, Mw/Ma denotes the ratio of wet to dry air's molecular weights, e denotes water vapor pressure, and P indicates air pressure.

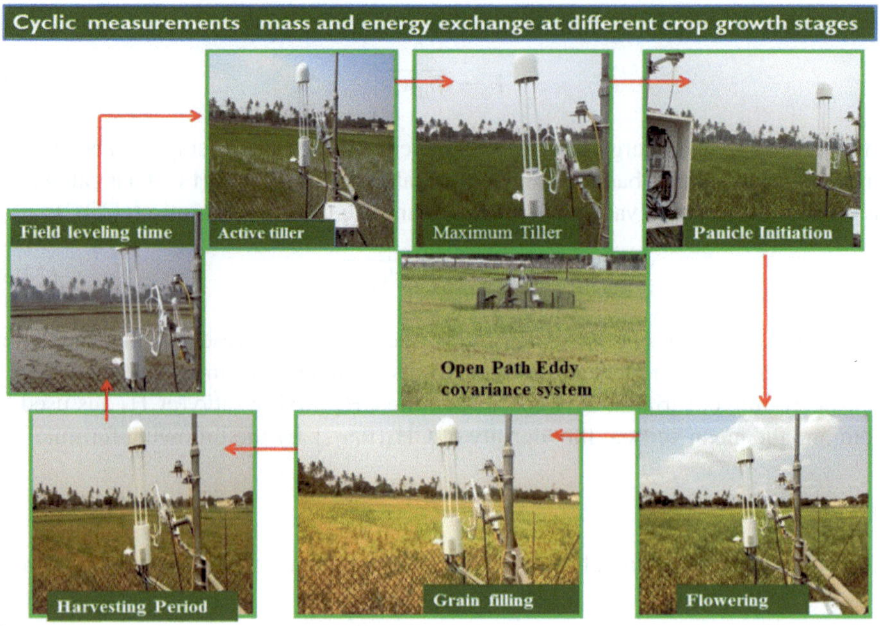

Fig. 23.2 Cyclic process of GHG measurement at different crop growth stages of rice at lowland ecology

23.5 Connection of EC Technique

The ICAR-National Rice Research Institute (NRRI) installed an EC system in the experimental field, which is 2.25 hectares in size and situated at 20°26′60.0" N latitude and 85°56′10.9" E longitude on a tripod. All other sensors are installed 1.5 m above ground (Swain et al., 2016, 2018; Chatterjee et al., 2019a, b). An adequate fetch (150 m) is maintained in the dominant wind direction, the south and southeast. Standard spacing is used for transplanting seedlings, i.e., 21–25 days old. The crop is harvested in May and November during the dry and wet seasons. Water management techniques are kept up throughout the growing seasons to maintain an average of 5–8 cm of standing water drained during the last two weeks. The usual nutrition management practices were used. The EC system captured data on fallows and all crop growth stages (Fig. 23.2).

23.6 Analysis of Real-Time Binary Data of EC

To process EC data, several flux calculation algorithms are employed. The process of computing fluxes involves examining the data for mistakes or gaps, syncing the data to account for temporal lags, and estimating fluxes based on average intervals

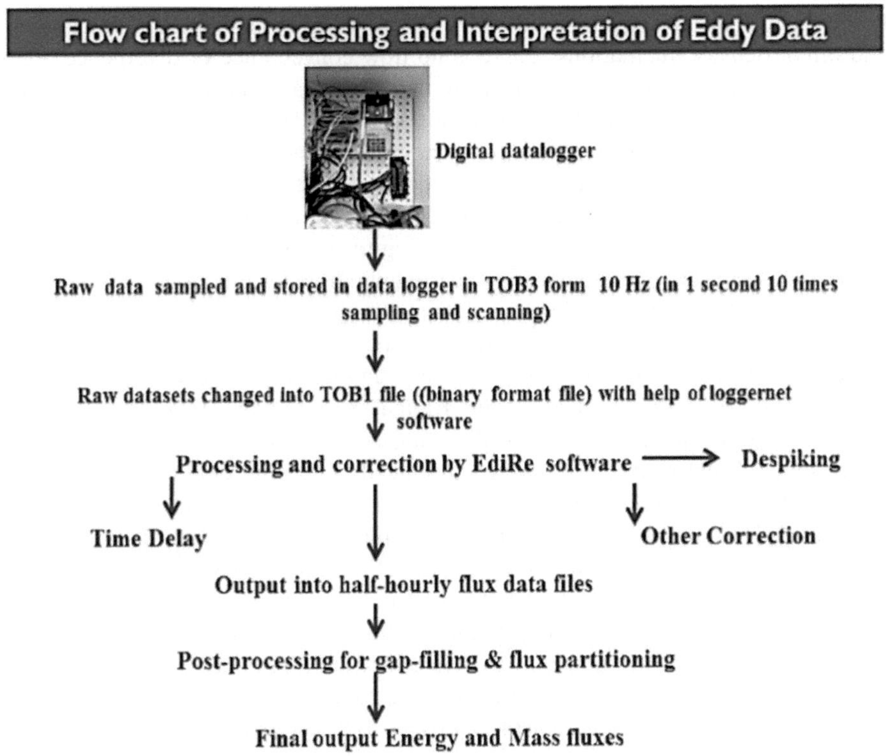

Fig. 23.3 Processing of real-time raw data from the EC data logger

of half an hour or an hour. Software is typically used to construct the EC system. The program typically offers a graphical presentation option for real-time data. A complete set of tools for using Campbell Scientific data loggers is included in the LoggerNet program. Data on the Eddy covariance is recorded at a frequency of 10 Hz. The information is immediately saved in TOA3 format (i.e., ASCII text form size nearly about 1.94 GB). This raw data is then transmitted via a Campbell Scientific data logger to record data from Shape Arrays. The flow chart for EC raw data processing data is shown in Fig. 23.3.

23.7 Flux Correction

Flux computation programs can be used to process EC data. Checking data for mistakes or gaps, aligning data to account for temporal delays, and estimating fluxes based on half-hour or one-hour averaging intervals are all part of computing fluxes. The following are some key EC data analysis assumptions. Only the fluxes in the area of interest are measured. Measurements should be taken within the interest layer's border layer. The terrain should be horizontal, uniform with no average fluctuations and minor density fluctuations. There is no flow convergence or divergence. Only an upwind area is represented by measurements taken at a specific location. Flux is thoroughly turbulent, and eddies handle most of the net vertical transfer. Instruments can detect minute changes in a high-frequency environment ranging from 5 to 40 Hz.

Optimizing the link between vertical wind speed and concentrations was necessary to determine the amount of time needed to carry air from the sampling location to the analyzer (Goulden et al., 1996). According to Aubinet et al. (2000), frequency response losses were corrected using experimental transfer functions. A data logger program sampled the time series' raw data at a frequency of 10 Hz. EdiRe software was used to process all of the data in real-time. Off-line coordinate rotation was subjected to cross products (second moments) and the flux correction term (Tanner and Thurtell, 1969; Webb et al., 1980; Kaimal & Finnigan, 1994; Wilczak et al. 2001). By doing quality checks on the data, large spikes in data brought on by unfavorable climatic conditions were eliminated. By doing quality checks on the EC flux data, large spikes in data due to adverse meteorological conditions were eliminated. The data sets were submitted to U* filtering for spike detection (Vickers & Mahrt, 1996; Reichstein et al., 2005; Papale et al., 2006). It was calculated using the lowest nocturnal air velocity threshold limit, below which ecosystem gaseous-C fluxes were ignored. Based on the hypotheses, the friction velocity (U*) threshold for nighttime data in this environment was determined at 0.1 m s^{-1} (Bhattacharyya et al., 2014). To calculate the seasonal NEE, missing data gaps were filled using a "look-up" table method (Falge et al., 2001). Data may contain flux measurement errors due to any of these elements, assumptions, equipment constraints, physical processes, and topographical peculiarities. The most common flux measurement errors are the mismatch between time and sensor response, sensor separation, scalar path averaging, tube attenuation, high and low pass filtering, density fluctuation, sonic heat flux error, oxygen in the "krypton" path, and data filling. Before processing and examining the data, numerous changes are often performed to reduce these inaccuracies. Data modifications for quality checks are presented separately in the part that follows.

23.7.1 Eddy Covariance Data Corrections for Quality Assurance

23.7.1.1 Delay

The delay between the two-time series (three-axis sonic anemometer and CO_2 / H_2O / CH_4 analyzer) is caused by differences in the spatial separation of the wind and scalar sensors, electronic signal processing, and airflow via tubes in closed-path eddy covariance systems. Time delays associated with electrical signal treatment are well-known, small, and constant and may be taken into immediate consideration. Additionally, important is the temporal delay brought on by sensor separation. The wind speed, direction, and space between the sensors affect how long it takes an air parcel to pass through both devices. Air passing through tubes in closed-path eddy covariance systems created a substantial delay. The interior volume of the air-conducting components controls this delay. This delay is affected by the measured gas, the mass flow through the system, and the interior volume of the eddy covariance system's air-conducting elements (filters, tubes, valves, and detection cell). To combine chemical and velocity measurements in a single sampling container, new sensor research is being conducted. Study between the scalar of interest and the vertical wind component can be used to calculate the lag times for each averaging interval. The selected time lags are those that offer the best correlation. However, this method could result in perplexing lag times, especially if the connection needs to be improved. Thus, when fluxes are large enough, an efficient automatic method for calculating lag periods could use a defined search window based on mass flow, tube diameters, and typical wall interactions.

23.7.1.2 Despiking

Instrumental issues or any modification to the measuring volume may result in spikes in high-frequency raw data (bird droppings, cobwebs, precipitation, etc.). Despiking solutions use algorithms to locate and eliminate spikes like disproportionately high variances, skewnesses, kurtosis, and discontinuities. In brief spikes, the approach eliminates them and fills the resulting space. Other times, the spikes might be shown by highlighting them, giving the user choice of whether or not to take them out of the data set. After removal, there may still be some uncertainty regarding flagged data, which primarily depends on the volume and data gap-filling technique.

23.7.1.3 Planar Fit and Coordinate Rotation

The sonic anemometer level is in accordance with the EC measurement's fundamental tenet. The surface is not consistently flat, or even frequently, it is a slight incline. That data and analytical theories, which are best cast in the stream-wise coordinate system, can be compared with ease; we must streamline the coordinate system. To achieve results that are comparable to those produced on a flat region, parameterization that lessens the effect of the sloping terrain must also be created. Implementing the turbulent parameterization in numerical models for this system would be simple. The apparent stress will increase if the anemometer is positioned correctly vertically over sloping terrain because changes in the stream-wise velocity will do so. Comparisons to turbulence measurement over steep terrain would be tricky since turbulence measurement over flat terrain depends on the slope angle about the wind direction. Instead, the surface layer turbulence data is examined in a coordinate system parallel to the mean streamline, preventing the apparent stress caused by the deviation along the slope in the windward direction. The focus would then be independent of terrain slope and direction to first order in the stream-wise coordinate system, enabling comparisons with data from level terrain. Reduce inaccuracies brought on by an improperly positioned acoustic anemometer with coordinate rotation. Coordinate rotation results in high-pass filtering of the scalar covariance. To avoid losing the low-frequency component, it is advised to utilize a longer duration for averaging and coordinating rotation. If the vertical axis of the sonic is not parallel to the local normal to the surface, there will be cross-contamination between the flux divergence's parts. "Tilt mistakes" are used to describe this. It has been shown that tilt errors significantly affect the momentum flux. Since anemometers are often set at a fixed position at a tower, the anemometer coordinate system cannot be adjusted to the changing flow field. The operator should place the sonic near the required vertical axis alignment as is practical and reasonable (Wilczak et al., 2001).

23.7.1.4 Detrending

It is the process of removing non-fluctuating portions of time series data that do not exhibit turbulence. High-pass filtering and detrending are two methods for getting rid of trends. Detrending and high-pass filtering remove random or systematic noise from flux calculations brought on by low-frequency bias in turbulent time series. The scalar concentrations, wind speed, direction changes that occur during the day or sporadically, as well as measurement artifacts such as sudden or brief instrument drifts, can all lead to bias. High-pass filtering cannot be avoided when estimating covariance from a finite measurement time, therefore, corrections are always necessary. During detrending, mean values are removed from instantaneous readings to calculate flow. The three traditional methods are block averaging, linear detrending, and non-linear filtering, each of which has a unique set of procedures. Detrending and high-pass filtering remove random noise from flux calculations brought on by low-frequency bias in turbulent time series. The scalar concentrations, wind speed, direction changes that occur during the day or sporadically, as well as measurement artifacts such as sudden or brief instrument drifts, can all lead to bias. There are three conventional methods: block averaging, linear detrending, and non-linear filtering. Each has a unique set of algorithms. Each approach might be appropriate in a particular situation, even though block averaging is the most common technique. When isolating the absolute turbulent flux from the potentially skewed measured signal, detrending is still widely used in trending.

23.7.1.5 WPL Correction

The Webb-Pearman-Leuning term is used to account for changes in the volume of CO_2, H_2O, CH_4, and other gases brought on by variations in the atmosphere's temperature and water vapor content. This correction, formerly known as the Webb-correction to regulate water vapor fluxes, is now known as the WPL-correction in honor of the three scientists Webb, Pearman, and Leuning (Webb et al., 1980). They increased its scope of use by adding the measurement of eddy fluxes of trace gases. The correction is necessary because changes in temperature and humidity lead to volume/concentration changes in trace gases unrelated to the flow of the trace gas we want to measure. The modification can considerably impact the measured flux; for instance, the additive correction decreases the CO_2 emissions flux calculated using vertical velocity and density covariance. This adjustment must be used with extreme caution for any trace gases.

23.7.1.6 Gaps Filling

Missing data may be gap-filled using a "look-up" table method based on empirical relationships between meteorological factors and flow data. This method can estimate seasonal NEE or NEME (Falge et al., 2001). Other methods, like the preceding table, mean, model-based, or other logistic methods, can also be used to fill gaps. Other scientists occasionally publish these methods.

23.8 A Case Study of Lowland Rice Employing Mass Exchange Analysis Based on Eddy Covariance

23.8.1 Mass Exchange Calculation

The EC methodology is a common technique for tracking CO_2, CH_4, and N_2O fluxes in local and global networks (Aubinet et al., 2000; Baldocchi, 2003; Dabberdt et al., 1993; Peltola et al., 2013). Woods (Aubinet et al., 2012), grasslands (Wohlfahrt & Gu, 2015), crops (Moureaux et al., 2006), grazed pasture, marshes (Laurila et al., 2012), and lakes have all been the subject of EC observations (Wohlfahrt & Gu, 2015; Vesala et al., 2012). When nutrients are abundant (fertilization), and irrigation is used in dry conditions, cropland ecosystems proliferate throughout the crop growing season (Moureaux et al., 2006). Another characteristic of agricultural ecosystems is the range of management techniques used, such as tillage, planting, fertilizer applications, herbicides, fungicides, insecticides, irrigation, harvest, and so on. Crop rotation, pedoclimatic conditions, and cultivated crops all affect management activities. In addition to impacting net ecosystem exchange, the machinery generates CO2, which could skew CO2 concentration and flow measurements. To evaluate management strategies and gauge their effect on carbon fluxes and budgets to follow changes, flux measurements can be performed in subplots (Eugster & Plüss, 2010). To measure the fluxes, and change over different management systems, a moveable measuring EC system can be used as opposed to a permanent installation.

After planting, the EC system and additional meteorological sensors may be put on tripod platforms with an aluminum pole and positioned in the agricultural field. On the other hand, deep soil sensors might be set up before planting without harming the rice crop. However, the EC systems with numerous sensors must be removed and reinstalled before harvesting or doing other agricultural tasks (Chen & Hu, 2003). Agricultural practices are anticipated to impact carbon fluxes and the carbon budget. As part of the system for mitigating carbon emissions, these activities must be evaluated in terms of carbon fluxes and budgets. As a result, by splitting a crop area into subplots managed in various ways and using several EC masts, the approach may be utilized to compare agricultural techniques (Pattey et al., 2006; Davis et al., 2010).

23.8.1.1 Net Carbon Dioxide Exchange in the Environment

Eddy covariance is used to assess the NEE of mass and energy between ecosystems and the atmosphere. The NEE comprises two components: CO_2 emission from ecosystem respiration and CO_2 uptake via photosynthesis (i.e., GPP). NEE partitioning is necessary to better comprehend carbon fluxes' inter-annual and geographical variability (Reichstein et al., 2005; Shi et al., 2012). The carbon budget must consider GPP, NPP, and net ecosystem production (NEP). The CO_2 exchange between a terrestrial ecosystem and the atmosphere, which takes place in all other ecosystems, involves plants absorbing CO_2 through photosynthesis and releasing it to the atmosphere via RE (Heimann & Reichstein, 2008). An imbalance would impact the amount of CO_2 in the atmosphere, which affects the world's climate (Taneva & Gonzalez-Meler, 2011). To determine if a particular ecosystem is a net CO_2 source or sink over time, CO_2 flux measurements are used. Any direction can cause the NEE to break. Insolation duration, temperature, nutrient availability, and hydrology are only a few variables that control vegetation community type, which is related to plant primary production (Frolking et al., 1998; Waddington et al., 1998; Griffis et al., 2000). Contrarily, vegetation communities are connected to ecosystem respiration, which includes autotrophic and heterotrophic respiration, as well as the decomposition of organic matter substrates and different vegetation types (Moore & Basiliko, 2006). Conversely, ecosystem respiration can be influenced by temperature, substrate composition, and hydrology. As a result, environmental changes could cause the crop to act as a source or sink of CO_2 during the growing season (Chimner & Cooper, 2003). Eddy covariance measurements have found great use in ecological problems, and their knowledge has advanced significantly. The most trustworthy EC observations for annual carbon budgets come from locations with excellent micro meteorological conditions. Such information can characterize how stand-scale carbon fluxes react to environmental changes, year-to-year variability in carbon flow, and the relationship between carbon fluxes and phenology (Baldocchi et al., 2001).

23.8.1.2 Net Methane Exchange in the Environment

Data is continuously generated, and fluxes are reported using the EC technique at the ecosystem scale. These characteristics are beneficial for determining ecosystem-scale CH_4 exchange because CH_4 has a complex source/sink distribution that is challenging to account for using manual/static chamber measurements (Sachs et al., 2010). The global CH_4 budget and the consequences of CH_4 on the climate system are both unclear at this time (Kirschke et al., 2013). Several studies have revealed that global wetlands' responses to recent climate variability and changes in rice farming techniques had a substantial influence on the stability of CH_4 levels in the atmosphere, even though the causes are unknown (Heimann, 2011; Kai et al., 2011). The ratios between the rates of CH_4 production, oxidation, and soil redox potential affect the net ecosystem CH_4 exchange (Bhattacharyya et al., 2014). The NEME was controlled by plant-mediated methane transport that passed through the aerenchyma pore space of rice plants (Butterbach-Bahl et al., 1997), as well as soil emission to the atmosphere by ebullition and diffusion through the water column. This NEME balance method thus fills the gap between static chambers and sophisticated micrometeorology methods. While advanced micrometeorology is preferable for larger areas, the manual chamber is excellent for measuring gas fluxes over a small area with clearly defined source sites. The NEME balance methods don't need a unique form for the wind profile to maintain atmospheric stability because they are based on a straightforward hypothetical premise. However, to ascertain wind vectors and fetch area in closed and open systems, wind directions must be continuously monitored. Mass balancing (NEME) techniques work best when the surface flow of the gas outside the emission area which are suitable for identifying ammonia (NH_3) and N_2O emissions after fertilizer applications, pesticide volatilization, and CH_4 and NH_3 emissions from small-scale grazing activities, but they are not suitable for evaporation studies (Denmead, 2008). Shortly, atmospheric CH_4 is anticipated to have a significant influence on future climates (Whalen, 2005). The ability of methanogens to make and methanotrophs to consume CH_4 in the soil profile, as well as the transmission of CH_4 to the surface via diffusion, ebullition, and plant-mediated transport, all have an impact on methane emissions to the atmosphere. Higher temperatures in the peat profile should increase CH_4 emission because methanogenesis is more sensitive to temperature than CH_4 consumption (Dunfield et al., 1993). Given that there is a significant association between the CH_4 flux and the presence of waterlogged conditions in rice fields, creating the anaerobic conditions necessary for methanogenesis is significant (Moore & Dalva, 1993; Bubier et al., 1995; Liblik et al., 1997). Although there is significant regional and temporal heterogeneity in CH_4 flow in the field, a strong correlation between average summertime CH_4 flux and typical levels of waterlogging has been found for several places in Canada. The intercepts of the regression, such as the CH_4 flux when the water table is at the surface, are noticeably different between sites, with the northernmost regions having the largest CH_4 fregression slopes regression are similar across locations (Liblik et al., 1997). Additional research suggests a connection between CH_4 flux and NEE or sedge biomass where the water level is at or near the ground surface because vascular plant mobility promotes CH_4 flux by avoiding the zone of CH_4 oxidation (Bellisario et al., 1999; Waddington & Roulet, 1996; Whiting & Chanton, 1992). Greater plant growth in response to higher atmospheric CO_2 concentrations may result in increased CH_4 emission, as shown in salt marsh habitats (Dacey & Lee, 1994). Rice paddy's physiological adaptation to growing wet soil makes it ideal for CH_4 emissions. The plant's vascular system is responsible for passively transferring up to 90% of the CH_4 generated in rice paddies into the atmosphere (Nouchi et al., 1990). The majority of the growing season for rice occurs in wet soils, which makes them perfect for the production of CH_4. Large amounts of water are required for flooding and puddling in the field to stabilize the rice ecosystem and improve soil fertility by balancing pH, enhancing nitrogen fixation, increasing carbon supply, and increasing diffusion rates. Organic matter from exudation, the death of rice plants, and organic additions are the main sources of CH_4 in wetland rice soils (Watanabe & Roger, 1985; Yagi & Minami, 1990). The synthesis of CH_4 can still be constrained by soil characteristics like high redox potential (Conrad, 1989; Ponnamperuna, 1972), low cation exchange capacity, and soil texture (Neue, 1990; Sass et al., 1990).

23.8.2 NEE and NEME Seasonal, Diurnal Fluctuations

Over the crop growing period, which included the crop seasons and fallow period, average NEE varied from -6.54 to 2.87 g C $m^{-2}d^{-1}$ (Fig. 23.4). The average NEE after harvest is 1.58 g C $m^{-2}d^{-1}$ for the dry season and 0.23 g C $m^{-2}d^{-1}$ for the wet season, whereas the total NEE is 98.04 g C m^{-2} and 14.11 g C m^{-2} during the dry and wet seasons, respectively. While the cumulative NEE was 383.35, and 251.1 g C m^{-2} during the dry and wet seasons, the average during the dry season was 2.71 g C $m^{-2}d^{-1}$, and the wet season was 1.69 g C $m^{-2}d^{-1}$. NEE in the rice canopy is influenced by several ecological and environmental factors, including latent heat, vapor pressure deficit, canopy irradiance, heat stress, stomatal response, high evaporative demand, circadian rhythm, rice crop growth phases, leaf area index, biomass, and others (Patel et al., 2011). During the day, plants use respiration to remove CO_2 from the soil and standing water, as well as photosynthesis to remove CO_2 from the atmosphere (Bhattacharyya et al., 2013). Plant respiration released a net amount of CO2 into the atmosphere at night. Water management, leaf area index (LAI), and specific developmental stages all affected the NEE during the growing season (Saito et al., 2005). By the study's findings, LAI increased at various growth phases, peaked around the time of heading to blooming, and then gradually decreased until maturity as a result of leaf senescence or a loss of leaf greenness over the course of maturity (Pakoktom et al., 2009; Patel et al., 2010). The flooded rice field continuously acted as a net sink of CO_2 from the vegetative stage until shortly before maturity. The NEE started to fall in the flowering (FL) and Panicle initiation (PI)-FL stages and peaked at the reproductive stage as a result of increased physiological activity and higher photosynthetic CO_2-C absorption (Campbell et al., 2001; Alberto et al., 2012; Bhattacharyya et al., 2013). Due to the buried conditions in which rice

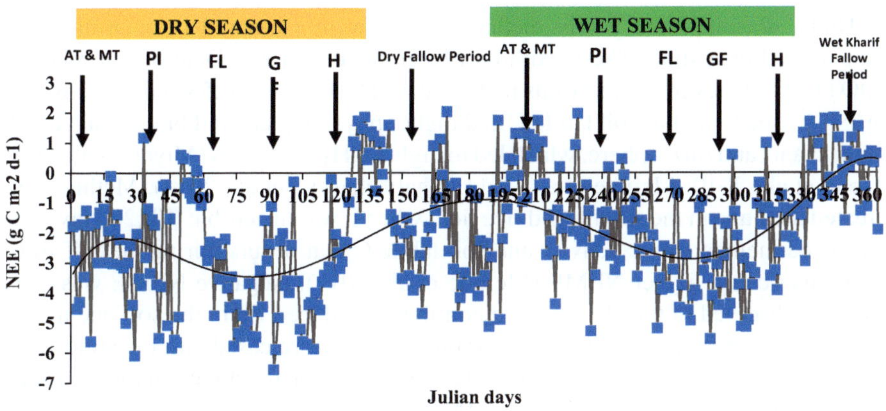

Fig. 23.4 Daily mean of NEE during different crop growth periods in lowland tropical rice (*Courtesy*: Swain et al., 2016)

was grown, various factors affected NEE between rice paddies and the atmosphere. Changes in NEE brought on by floods impact the root activity, photosynthesis, and respiration of rice plants. NEE between rice and the atmosphere may have been influenced by the actions of aquatic plants in floodwaters, like algae (Bhattacharyya et al., 2013; Miyata et al., 2000). The presence of sunlight in the form of Ig, NR, and leaf gas exchange between the canopy's pattern of light interception affected the seasonal way of NEE and C absorption (Swain et al., 2016).

The mean NEME throughout the fallow period after harvest of the dry and wet season crops was 2.93 and 0.73 mg CH_4 m^{-2} h^{-1}, respectively. Seasonal fluctuations in daily NEME are shown (Fig. 23.5).

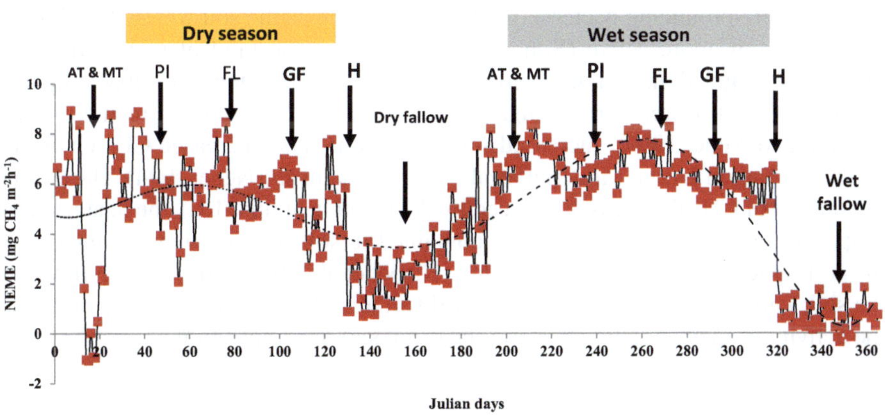

Fig. 23.5 Daily mean of NEME during different crop growth periods in lowland tropical rice (*Courtesy*: Swain et al., 2016)

Daily, seasonal, and yearly methane fluxes were measured to determine how they altered over time. The maximum tillering (MT)-PI and Date of transplanting (DOT)-MT stages of both seasons had increased CH_4 fluxes, 6.58 and 6.98 mg CH_4 m^{-2} h^{-1}, due to higher soil labile C and higher physiological and biological activity in the plants and rhizosphere, which led to higher CH_4 emissions (Miyata et al., 2000; Tseng et al., 2010; Meijide et al., 2011; Bhattacharyya et al., 2014). Methanogens created methane in the submerged rice field when it was anaerobic (Eh:200 mv). The enhanced methanogenic activity and soil labile C that occurs during the MT to FL stage may cause higher NEME (Meijide et al., 2011). Negative NEME was occasionally observed in the absence of a continuous reduced Eh in the soil due to CH_4 oxidation (Frenzel et al., 1992). Soil-trapped CH_4 was released into the atmosphere through ebullition, diffusion, and transfer through the aerenchyma in the stems and shoots of rice plants (Nouchi et al., 1994). The NEME was the result of concurrent production and consumption activities, which were fueled by plant-mediated CH_4 transport through the aerenchyma system of the rice plant and CH_4 emission from the soil to the atmosphere via bubbles and diffusion through the water column (Butterbach-Bahl et al., 1997). For most of the year, the soil in our study was saturated with water, creating the perfect environment for methanogenesis. Furthermore, significant CH4 fluxes occurred under waterlogged conditions at specified periods when the temperature was favorable, and the rice rhizosphere was most active beneath a mature rice canopy. In addition to limiting CH_4 oxidation, soil submersion created the anaerobic conditions needed for methanogenesis (Neue, 1993). Additionally, it promoted the development of rice plants, which were in charge of more than 90% of the transfer of CH_4 from the ground to the atmosphere (Inubushi et al., 1992).

Mechanisms by which biophysical factors affected CH_4 fluxes, daily patterns of some critical environmental variables). In addition to being detected in rice fields in Japan, Italy, and India, our research discovered a strong correlation between CH_4 emissions and temperature (Bhattacharyya et al., 2014). Ideal temperatures encouraged the CH_4-producing bacteria (methanogens), and as the soil warmed during the day, they grew more active (Simpson et al., 1995). The kind of rice cultivar, air temperature, nutrition, and aerenchyma orientation in the stem all impacted plant-mediated CH_4 transfer (Armstrong, 1971; Denier van der Gon & Breemen, 1993; Minami and Neue, 1994). Increased soil temperature also increased CH_4 ebullition and molecular diffusion (Moncrieff et al., 1997; Minami and Neue, 1994). The diurnal fluctuation of CH_4 flow was influenced by the latent heat flux (LE), the heat used for evapotranspiration, because the CH_4 contained in soil solution may be released to the atmosphere via the transpiration flux (Nisbet et al., 2009a, b). The heat flow into the soil, known as soil heat flux, increased soil temperature and probably increased the activity of methanogenic bacteria, favoring soil decomposition.

23.8.3 Dividing NEE into GPP and RE

The advancement of climate change on various scales is greatly aided by the growing number of eddy covariance towers across the world (Baldocchi, 2008; Baldocchi et al., 2001). Around the world, it is routine practice to use EC measurements of NEE to estimate the role of vegetation in estimating the size and variability of the terrestrial carbon sink (Baldocchi et al., 2001; Sarmiento & Wofsy, 1999). Eddy covariance data are usually utilized with models that break NEE into two components to investigate the factors that cause NEE: RE and GPP (Beer et al., 2010; Jung et al., 2011). A variety of measuring and modeling techniques are used for NEE partitioning, each with its own set of benefits and drawbacks (Desai et al., 2008). These "flux partitioning" models (FPMs) can be beneficial for estimating CO_2 flow over long timescales and filling in missing data gaps (Falge Baldocchi et al., 2001) since few studies have thoroughly examined how the selection of a particular FPM affects the magnitude and variability of carbon flux estimates in a specific ecosystem, even though FPMs are valuable tools for inferring EC data. To better understand how NEE is managed at the process level, there is much interest in partitioning CO_2 NEE measurements. Because RE equals NEE, partitioning is accessible at night; however, partitioning during the day depends on the model selected. As a result, there is much uncertainty in the GPP and RE estimates (Hagen et al., 2006; Richardson et al., 2006). According to Desai et al. (2008), more variability was observed when additional gaps were added to the data. Most partitioning algorithms varied in terms of yearly integrals by less than 10%. When only one partitioning technique was applied, GPP patterns tended to be similar across sites. This suggests that algorithm preference leads to an ordered bias of an unknown extent because the GPP is unknowable. When splitting RE at shorter time intervals, there was more significant variability among algorithms in terms of diurnal cycles (Lasslop et al., 2010).

The inter-annual variability of GPP and RE is more closely related to temperature in cold, cool, and humid places. Although RE inter-annual variability is more closely correlated with precipitation than with temperature. Still, GPP is more strongly associated with temperature, and RE inter-annual variability appears to dominate NEE inter-annual variability in some tropical regions. In the tropical areas, soil respiration is likely to be far less restricted by temperature than by litter moisture content. We can see that the climatic domain contains the hotspot regions of inter-annual variability of land–atmosphere fluxes, where moisture supply appears to be the primary driver of variability (Jung et al., 2011). Gross primary production becomes a dominant control if its variation is 20% higher than its variation in annual RE. According to specific regional to continental studies for Europe (Ciais et al., 2005; Luyssaert et al., 2007; Vetter et al., 2008) and Africa (Ciais et al., 2005; Luyssaert et al., 2007; Vetter et al., 2008), the bulk of the land surface determines the inter-annual variability of NEE (Weber, 2009). According to global simulations by dynamical global vegetation models, GPP variability encourages NEE variability throughout most of the planet's regions. Simulations, models, and up-scaling products indicate that GPP is the primary driver of NEE inter-annual changes in the hot spots of inter-annual variability that have been found. In East Asia and South America's tropics, RE seems to impact inter-annual variations in NEE. Since climatic factors do not considerably restrict photosynthesis, RE exhibits a higher inter-annual variability than GPP.

23.8.4 Net Ecosystem Nitrous Oxide Exchange

Agricultural soils account for over 42% of human N2O emissions and are vital to the global nitrous oxide budget (IPCC, 2007; Wang et al., 2013). The vast bulk of these emissions are extensively related to using nitrogen fertilizers. However, there are still many unknowns when calculating anthropogenic influence (Crutzen et al., 2008; IPCC, 1997). To further comprehend such mistakes in manual chambers and to enhance regional and global N_2O emission inventories, studies on long-term assessments of the comparability of N_2O fluxes observed by eddy covariance were done (Stella et al., 2013; Wang et al., 2013). The EC technique is the direct micrometeorological method for monitoring N_2O flows. It does larger-scale calculations of the mean flows through an ecosystem without impacting the surrounding area. Advanced fast-response detectors, including the tunable diode laser and quantum cascade laser, have made it possible in recent years to conduct in-depth EC studies of N_2O emissions from a variety of terrestrial environments (Jones et al., 2011; Eugster et al., 2007; Pihlatie et al., 2005; Neftel et al., 2007; Wang et al., 2013). However, obstacles to adopting the EC technique to monitor N_2O fluxes include signal offset caused by fringe effects (Mammarella et al., 2010), time lag between vertical wind and gas concentration (Pihlatie et al., 2005), and water vapor interference (Neftel et al., 2010). While the EC technique based on a tunable diode laser operated wonderfully and with remarkable precision, the manual chamber technique based on gas chromatography analysis was also exceedingly dependable in determining N_2O fluxes throughout the year (Wang et al., 2013).

23.8.5 EC Water Vapor Flux Measurement

One of the most prevalent greenhouse gases is water vapor, although little is known about how it affects the greenhouse effect. The management of water resources, irrigation, agriculture, hydrological modeling, and remote sensing benefit from accurate water vapor flow observations. Additionally, determining the water vapor flux is essential for the density correction of other atmospheric gas fluxes (CO_2 and CH_4). The amount of water vapor flux seen throughout the dry season increases as crop phenology develops, with maturity having the highest value and vegetative stage having the lowest. Net radiation and maximum and minimum temperature variations affect the water vapor flux (Schneider et al., 2010). The daily water vapor flux variation was found to have a bell-shaped curve, peaking between 13:30 and 14:00 IST. The leaf stomata close for a few hours, about 14:00, to decrease moisture loss

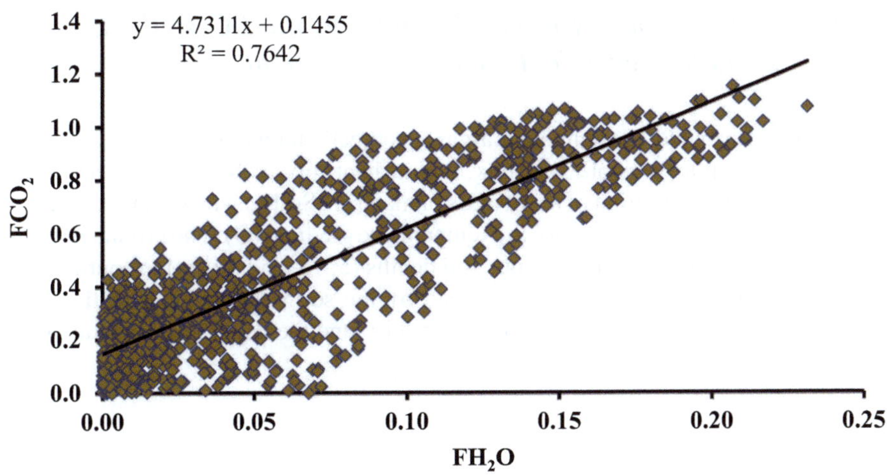

Fig. 23.6 FCO_2 and FH_2O correlation in a paddy field using half-hourly average data (*Courtesy*: Chatterjee et al., 2019a, b)

in an environment with arable crops and limited water supply (Huang et al., 2002; Chatterjee et al., 2019a). After reaching its peak, the water vapor flux progressively decreased until it reached its lowest point at midnight or early the following day. Latent heat of vaporization, NR, PAR, air, soil, and water temperatures, shortwave radiation, evaporation, and RH are some of the micrometeorological factors that affect water vapor flux. In an experiment at ICAR-NRRI, Cuttack, favorable correlations between carbon dioxide flow (FCO_2) and water vapor flux (FH_2O) were found when data were aggregated by growing stage, ignoring direction (sign) (Fig. 23.6). The air and leaf stomata's diffusion of CO_2 and water vapor, respectively, are thought to be the causes of the positive correlation between FCO_2 and FH_2O (Chatterjee et al., 2019a, b).

23.8.6 Energy Exchange Measurement by EC Technique

The quantity of energy flux in a given direction per unit of time passing through a unit area normal to that direction is called flux (W m^{-1}or J m^{-2} s^{-1}). The main determinants of surface energy exchange are four forms of energy fluxes that enter or exit the surface, namely Rn, H, LE, and G. The rice crop is flooded with a continuous water layer of 5–8 cm in the field, making it work differently in energy exchange with the environment. A variety of environmental factors such as rainfall, temperature, sun radiation, humidity, and PAR influences rice productivity. Furthermore, the nature and type of vegetation cover impact heat fluxes. As a result, measuring energy exchange in rice is critical for understanding and modeling the rice ecosystem (Chatterjee et al., 2019b).

23.8.7 Surface Energy Flux, Measured by the EC Method, Includes Sensible, Latent, and Soil Heat

The energy attributed to changes in a gas or object's temperature without a phase change is known as sensible heat flux. Latent heat flux is the term for the energy needed to alter the condition of a unit mass of material. Soil heat flux is the conduction of energy per unit area in response to a temperature gradient (G). The ground heat flux sensor plates are positioned at various soil depths (5, 15, and 30 cm) to continuously monitor the flux (G), which is determined by the soil's thermal conductivity, the temperature gradient (dT/dZ) at a given depth (z) and the depth's temperature (T).

$$G = k.\frac{dT}{dZ} \tag{23.10}$$

As rice farming develops, the three land surface energy fluxes vary. In contrast to LE, which predominates from active tillering until grain filling and drops at harvest, H is more significant in the early phases of rice growth. H is negative during the flowering and grain-filling phases. H peaks between 11:00 a.m. and 2:30 p.m., with maximal values of LE and G reached between 12:30–1:30 PM and 9:00–10:00 AM, respectively. The H is constantly close to zero or negative before sunrise and after nightfall due to the absence of insolation and the low turbulence during those times. Positive upward H during the day was caused mainly by rising turbulence and sunlight (Montazar et al., 2016). After the sunrises, daily G starts to change. This process is delayed by a few hours in the morning (Roxy et al., 2014). Diurnal fluctuations in different seasons, including fallows, can be caused by various factors, such as rainfall, soil moisture, NR, skin temperature, vegetation, etc. An ICAR-NRRI experiment showed that LE supplied 23–66 percent of the energy, while H contributed 1–23%. LE exceeded H with a magnitude 1.2–63.9 times higher than H in all cropping seasons and fallows. Higher evapotranspiration in rice fields was caused by increased canopy cover, and higher evapotranspiration, which increased LE relative to H. 5–10% of the total energy on hand came from the G. However, at night, G is the dominant supplier of energy, and NR, as well as LE, are the key receivers. During the day, NR is the primary source of energy flux to the surface, while LE is the primary receiver from the surface (Swain et al. 2018).

23.8.8 Net Radiation Measurement

Net radiation (NR) flow is the equilibrium between entering and outgoing energy that affects the climate. It is essential to comprehend net radiation because it powers the hydrologic cycle. Energy surpluses are produced if incoming radiation exceeds exiting radiation , whereas net radiation deficits are made if the opposite is true. In an eddy covariance system, a 4-component net radiometer that measures net radiation consists of a downward-facing CM3 pyranometer for reflected solar radiation and an upward-facing CM3 pyranometer for incoming solar radiation (i.e., ISR: Incoming shortwave radiation, ILR: inbound longwave radiation, OSR: outgoing shortwave radiation, and OLR: outgoing longwave radiation are all measured using a four-component net radiometer. The resultant of these four radiations is utilized to compute NR.

$$\text{NR} = (ISR + OLR) - (OSR + OLR) \tag{23.11}$$

23.8.9 Bowen Ratio Estimation (B)

The Bowen ratio will calculate the relative dominance of sensible and latent heat fluxes or the H/LE ratio.

$$\text{B} = \frac{H}{LE} \tag{23.12}$$

A greater Bowen ratio demonstrates significant Rn partitioning into LE (Alberto et al. 2009). Bowen ratio can be used as a water stress indicator and is important for understanding a region's hydrological cycle, boundary layer characteristics, and weather variability (Alves & Pereira, 2000). The range of rice's average B is 0.21 to 0.64. There is a considerable change in B between agricultural seasons and fallow periods. Due to its increased photosynthetic capacity and decreased stomatal resistance, rice has a lower B than other arable crops worldwide (Wilson et al., 2002).

23.8.10 Closure of Energy Balance with the EC Approach

Since the soil surface of flooded rice has a persistent water layer, it has different energy budgeting and energy balance closure (EBC) requirements than other upland crops (Alberto et al., 2011). The surface energy budget in rice is the total surface LE and H fluxes matching all other energy sources and sinks (Tsai et al., 2007). Energy conservation is the basis.

$$= LE + H \tag{23.13}$$

The EBC is investigated using ordinary least squares, energy balance ratio, and residual heat flow. These techniques advise underestimating the available heat flow because full closure is impossible. To evaluate how well the curve matches the data, one can contrast these three EBC methods.

a. **Ordinary least squares (alpha)**

In the first technique, the slope and intercept of linear regression are generated using the ordinary least squares (OLS) method to examine the relationship between turbulent heat flux (H + LE) and available heat flux (V) (Wohlfahrt & Widmoser, 2013). This strategy is successful if the independent variable contains no random errors. This can be stated as follows:

$$LE + H = a(Rn - G) + b \tag{23.14}$$

where the slope and intercept of the linear regression are denoted as "a" and "b," respectively. The perfect closure is achieved when the slope is one, and the intercept is zero. One kind of intercept is "a." On both the X (Rn-G) and Y (H + LE) axes, a = 0 and a perfect curve crossing at 45 degrees (b = 1) are achieved when the energy balance closure is completely closed. However, it is essentially not conceivable. The EBC can be improved if storage terms are supplied. In actuality, overvaluation and mistake are two options that depend on seasonal variations. Overvaluation is present when b1 and a > 0 and Rn-G = 0, H + LE have a positive value. When Rn-G = 0 and H + LE have a negative value, a computation error occurs if both b > 1 and a0 are true.

b. **Energy balance ratio (EBR)**

Additionally, the EBR is employed to evaluate the closure (Wilson et al., 2002). This is defined as the ratio of the available heat flow to the sum of the turbulent heat fluxes over time (LE + H)/ (V).

$$\text{EBR} = \frac{\sum(LE + H)}{\sum(V)} \tag{23.15}$$

Improvement of several storage terms (S, W, F, and C) helps to obtain more accurate EBR values when used with the standard EBR values (Eq. 23.15). The coefficients of the EBC measured by EBR did not vary between the cropping seasons and fallows. However, EBR in the rainy season was low compared to the dry season and fallows. The dry season and dry fallow were practically rain-free, but for a few cloudy days in these seasons. As a result, the energy components were almost balanced during the days without rain. Applying the correction to H and LE before adding the storage term could create a new and enhanced EBC. These new restrictions result in a minor improvement in the energy balance during the wet season and the wet fallow.

c. **Residual heat flux (RHF)**

The third method, known as RHF, determines if the LE + H recorded by the EC is inflated by measuring the difference between V and turbulent heat flux (H + LE), which evaluates the level of EBC (Wohlfahrt and Widmoser, 2013).

$$\mathrm{RHF} = \mathrm{V} - \mathrm{H} - \mathrm{LE} \tag{23.16}$$

The RHF should be zero after the surface energy budget is closed. When RHF is more than zero, energy supply outweighs energy loss; the opposite is true when it is less than zero. The average RHF in Cuttack grew favorably from dry to wet fallow, with a value of 10.3–20% greater in the wet season than the dry season and 8.3–20% higher in the dry season fallow than the wet fallow. The surface energy budget should be zero when it is completely closed. Because the average value of RHF was higher than zero in all seasons and years, the available heat flux was more significant than the cumulative turbulent flux (Liu et al., 2017). To get a better closure when measuring EBR, standing water heat storage in the canopy and photosynthetic energy flow are considered for better results. The EBC computed using RHF is the best way for calculating closure because it can detect different crop growth seasons, including both the fallow periods, followed by OLS when three approaches for calculating closure in lowland paddy are tested.

When the surface energy balance is successfully closed, there is a high degree of confidence in the flux observation approach. Improper closure suggests issues with the eddy covariance system measurement or a lack of heat storage measurement. On the other hand, recent research indicates that scale is a problem with imperfect energy balance closure (Masseroni et al., 2014). Data quality can be assessed using the EBR test (Aubinet et al., 2000). Errors in CO_2 flux measurements have occasionally been connected to an incomplete EBR (Liu et al., 2006). Closing the energy budget is challenging during the rainy season due to increased water depth. Freshwater from rain enters the system, which allows energy to be advected into it. It can also be challenging to maintain the regular water depth during the rainy season after a downpour event. The measurement of energy balance closure was more erratic during the wet season.

23.8.11 Measurement of Water Balance Through Eddy Covariance

Estimating the water balance is necessary to analyze the current state and trends in water resource availability and support water management decisions. The water balance equation (Falalakis and Gemitzi, 2020) summarizes the movement of water in various forms into and out of a hydrological system:

$$\mathrm{P} + \mathrm{L} = \mathrm{ET} + \mathrm{SR} + \mathrm{I} \tag{23.17}$$

"P" represents precipitation, ET represents evapotranspiration, SR represents surface runoff, I represent infiltration to the vadose zone and groundwater. L represents moisture loss from soil by the capillary rise or transfer through plants. ET can be calculated using eddy covariance data, and the remaining parameters can be tested at the field level to ensure the correct water balance.

23.9 Carbon Budgeting

Eddy Covariance measures the net exchange of materials and energy between ecosystems and atmosphere. The EC method can determine the carbon flux, balance, and budget in the terrestrial ecosystem. The difference between two larger fluxes, net photosynthetic intake and respiratory outflow at night, determines a terrestrial ecosystem's net carbon exchange. Even a slight exclusion of nighttime flow could cause a long-term, significant overestimation of carbon sequestration. The yearly NEP is frequently used to determine the rate of carbon sequestration. Taking into account a very minimal carbon exchange in a form other than CO_2 and by processes other than turbulent transport, the NEP is believed to be equivalent to the annual net exchange of CO_2 with the atmosphere (a positive value indicates carbon sink, a negative value suggests carbon source). The CO_2 uptake via photosynthesis (i.e., GPP) and the CO_2 release by RE result in two larger fluxes of the same sign, which combine to form the NEE.

Direct measurements of canopy CO_2 exchange on daily and annual timeframes are crucial to understanding ecosystem carbon budgets. They offer data sets for parameterizing and testing carbon balance models. Successful measurements require a multidisciplinary protocol combining micrometeorologists, ecophysiologists, biogeochemists, modelers, and experimenters (Baldocchi et al., 1996). to determine the NEE between the ground and the atmosphere. The NEE between the atmosphere and any terrestrial system is calculated as the equilibrium between uptake by photosynthesis and release from both dead and live organic matter. Net primary productivity is the amount of organic matter stored in plant tissues over the measurement period that exceeds the amount used for respiration by the plants. The calculations of the NEE and NEP measurements do not take non-CO_2 carbon exchanges into account. Even though the latter is lost in minute amounts from cropland, these carbons can be measured using the EC technique employing specialized gas analyzers, such as those used for methane and volatile organic compounds (VOCs) (Smith et al., 2010). Furthermore, ecosystem carbon budgets usually exclude any carbon gains or losses unrelated to turbulent fluxes, such as advection losses, harvesting removals, or manure applications. However, over time, some of the NEP may be released into the environment as microbially produced methane, dissolved organic carbon (DOC), particulate organic carbon (POC), dissolved inorganic carbon (DIC), pollen, and seeds, or it may be leached into groundwater as DOC, particulate organic carbon (POC) (Eugster, 2007). In an agricultural system, organic carbon can be moved using fertilization, manure application, re-deposition from somewhere else in the form of sediments due to erosion, and DOC deposition in water. As a result, the carbon budget should be examined and adjusted to account for non-CO_2 and non-respiratory carbon losses and gains (Fig. 23.7). For accurate carbon sequestration estimation, a detailed

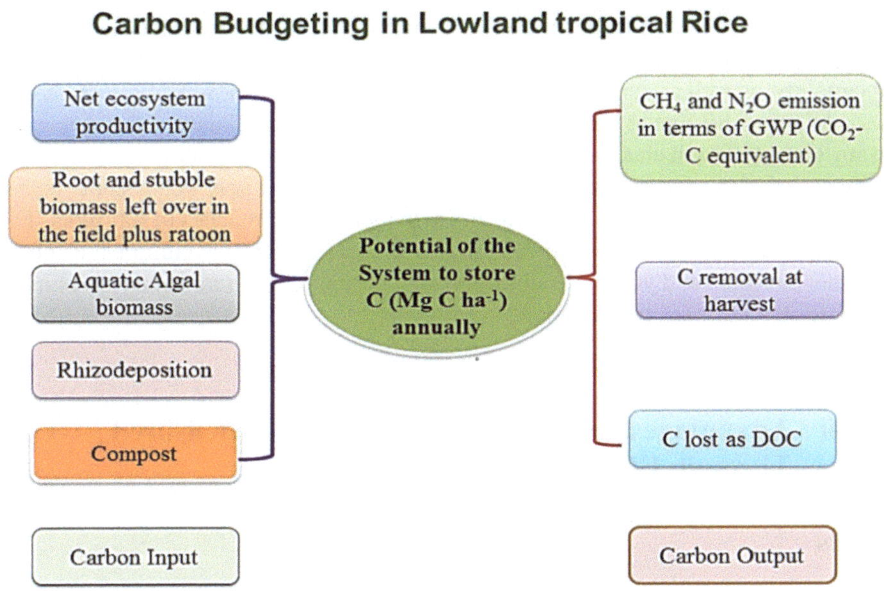

Fig. 23.7 Schematic diagram of annual carbon budget in lowland tropical rice

spatiotemporal measurement of net biome productivity (NBP) is required (Davis et al., 2010; Tao & Wang, 2019). Ascertaining if an ecological system is stable, where inputs of material and energy can balance outflows, is one of the trickiest parts of measuring productivity (Odum, 1971). Agricultural systems, where external fertilizer and irrigation inputs are frequent, among other things, exacerbate the issue. Although repeated harvest data make it relatively simple to estimate aboveground net primary production (ANPP), root harvests are almost always imprecise, making it more challenging to estimate belowground output. The root: shoot ratio is typically used to convert ANPP to NPP, but conflicting connections between shoot and root production have frequently been noted (Mitchell et al., 1990).

In the form of litter and soil carbon pools, a portion of the biomass produced each year is kept in the soil, where it decomposes and aids in heterotopic respiration (Rh), a multi-year process (Smith et al., 2010). Net biome production is equivalent to the integrated net ecosystem carbon balance (NECB) over time and space (NBP; Schulze & Heimann, 1998; Buchmann & Schulze, 1999; Chapin et al., 2006). Numerous studies have estimated NBP and used EC flow to determine changes in agricultural carbon stocks (Davis et al., 2010; Schmidt-Rohr & Spiess, 2012; Swain et al., 2018). While Alberto et al. (2018) discovered that flooded rice-rice cropping systems (862 kg C $ha^{-1}y^{-1}$) could sequester more SOC than rice-maize (293 kg C $ha^{-1}y^{-1}$) and rice-mung bean systems, it has been reported that tropical lowland rice fields act as sinks (Swain et al., 2018). Using this method, Schmidt-Rohr and Spiess (2012) identified temperate winter wheat as a carbon source (6844 kg C $ha^{-1}y^{-1}$). On the other hand, it's possible that short-term predictions need to calculate NBP when examining carbon sequestration accurately. Using the EC technique to calculate agricultural carbon sequestration could result in turbulent flux calculation errors and uncertainty, as well as carbon gain and loss via non-CO_2 and non-respiratory carbon loss. Additionally, due to inter-annual volatility, the carbon balance of the same crop in any given year will change. Because of this, we recommend employing a long-term eddy covariance approach for flux estimation, in particular agro-ecosystems with a variety of flux towers for various crop and management approaches in sub-plots. The distinction of random error will help lessen the uncertainty for comprehension of the inter-seasonal and yearly variability trends. It requires both monetary investment and inter-institutional coordination.

23.10 Advantages and Disadvantages of Flux Measured Through EC

Merits

- In-situ measurements over the area of interest could be achievable
- Non-invasive sampling, not disrupt the area where fluxes are recorded
- Reliable, verifiable, and defensible values of gas exchange or emission rates

Demerits

- Automated measurement system with minimal intervention that provides continuous coverage.
- It necessitates a continuous series of high-time-resolution measurements (e.g., 5–40 Hz).
- The method is mathematically challenging and necessitates great caution while setting up and processing data.
- Flux towers give data specific to a particular ecosystem type or state.
- Flux data is noisy, and this unpredictability is mainly attributable to measurement error.
- The EC method cannot be used to measure fluxes in many scenarios, or it is not the ideal way to utilize.
- Environmental conditions with a small research area, primarily low wind speed, complicated topography, point flux sources, and so on, are examples.
- In addition, some gases, such as ammonia and volatile chemicals, may require a more sensitive sensor system to detect minor changes at 10 or 20 Hz frequencies.
- It necessitates a variety of adjustments, as well as careful planning, execution, and processing tailored to the unique purpose at the experimental site.

23.11 Conclusion

The classic micrometeorological approach for observing CO_2, CH_4, and heat fluxes, which are used to determine the carbon and heat budget of terrestrial ecosystems, is known as the EC method. GHG emissions from any environment can be monitored in real-time, accurately, and precisely using open-path EC approaches. This helps in the accurate short- and long-term quantification of GHGs to compile a database inventory for a particular environment. The flow partitioning method can be used to calculate ecological respiration and gross primary output. In various ways, the EC system's information helps quantify net biome production and net ecosystem carbon balance. Long-term experiments incorporating agricultural practices, life cycles of crop components, and the physico-chemical-biological factors in ecosystem are worthwhile research topics to understand the mechanism of C flow in the soil–plant-atmosphere in the lowland rice ecosystem. To help predict emission scenarios, these data on high-frequency GHGs and heat fluxes can be up-scaled from any point size and extrapolated to larger scales for similar ecologies. A range of modeling techniques that will help in the simulation of future GHG emission levels will benefit from the data collected utilizing EC real-time data. Using the information from the GHG fluxes of the EC system, one may assess if a system is a source or a sink of atmospheric gaseous carbon.

References

Alberto, M. C. R., Hirano, T., Miyata, A., Wassmann, R., Kumar, A., Padre, A., & Amante, M. (2012). Influence of climate variability on seasonal and inter annual variations of ecosystem CO_2 exchange in flooded and non-flooded rice fields in the Philippines. *Agricultural and Forest Meteorology, 149*, 1737–1750.

Alberto, M. C. R., Wassmann, R., Hirano, T., Miyata, A., Kumar, A., Padre, A., & Amante, M. (2009). CO_2/heat fluxes in rice fields: Comparative assessment of flooded and non-flooded fields in the Philippines. *Agricultural and Forest Meteorology, 149*(10), 1737–1750.

Alberto, M. C. R., Wassmann, R., Hirano, T., Miyata, A., Hatano, R., Kumar, A., ... & Amante, M. (2011). Comparisons of energy balance and evapotranspiration between flooded and aerobic rice fields in the Philippines. *Agricultural Water Management, 98*(9), 1417–1430.

Alberto, N., Domingues, M. F., Marques, C., André, P., & Antunes, P. (2018). Optical fiber magnetic field sensors based on magnetic fluid: A review. *Sensors, 18*(12), 4325.

Alves, I., & Pereira, L. S. (2000). Modelling surface resistance from climatic variables? *Agricultural Water Management, 42*(3), 371–385.

Armstrong, W. (1971). Radial oxygen losses from intact rice roots as affected by distance from the apex respiration and waterlogging. *Physiologia Plantarum, 25*(2), 192–197.

Aubinet, M., Grelle, A., Ibrom, A., Rannik, Ü., Moncrieff, J., Foken, T., Kowalski, A. S., Martin, P. H., Berbigier, P., Bernhofer, Ch., Clement, R., Elbers, J., Granier, A., Grün-wald, T., Morgenstern, K., Pilegaard, K., Rebmann, C., Snijders, W., Valentini, R., & Vesala, T. (2000). Estimates of the annual net carbon and water exchange of forests: The EUROFLUX methodology. *Advance Ecology Research*, 30113–175.

Aubinet, M., Vesala, T., & Papale, D. (Eds.) (2012). Eddy covariance: A practical guide to measurement and data analysis. Springer Science & Business Media.

Baldocchi, D., Valentini, R., Running, S., Oechel, W., & Dahlman, R. (1996). Strategies for measuring and modelling carbon dioxide and water vapour fluxes over terrestrial ecosystems. *Global Change Biology, 2*(3), 159–168.
Baldocchi, D. (2008). 'Breathing'of the terrestrial biosphere: lessons learned from a global network of carbon dioxide flux measurement systems. *Australian Journal of Botany, 56*(1), 1–26.
Baldocchi, D., Falge, E., Gu, L., Olson, R., Hollinger, D., Running, S., Anthoni, P., Bernhofer, C., Davis, K., Evans, R., & Fuentes, J. (2001). FLUXNET: A new tool to study the temporal and spatial variability of ecosystem-scale carbon dioxide water vapor and energy flux densities. *Bulletin of the American Meteorological Society, 82*(11), 2415–2434.
Baldocchi, D. D. (2003). Assessing the eddy covariance technique for evaluating carbon dioxide exchange rates of ecosystems: Past present and future. *Global Change Biology, 9*(4), 479–492.
Beer, C., Reichstein, M., Tomelleri, E., Ciais, P., Jung, M., Carvalhais, N., Rödenbeck, C., Arain, M. A., Baldocchi, D., Bonan, G. B., & Bondeau, A. (2010). Terrestrial gross carbon dioxide uptake: Global distribution and covariation with climate. *Science, 329*(5993), 834–838.
Bellisario, L. M., Bubier, J. L., Moore, T. R., & Chanton, J. P. (1999). Controls on CH_4 emissions from a northern peatland. *Global Biogeochemical Cycles, 13*(1), 81–91.
Bhattacharyya, P., Neogi, S., Roy, K. S., Dash, P. K., Tripathi, R., & Rao, K. S. (2013). Netecosystem CO_2 exchange and carbon cycling in tropical low land flooded rice ecosystem. *Nutrient Cyclying in Agroecosystems, 95*, 133–144.
Bhattacharyya, P., Neogi, S., Roy, K. S., Dash, P. K., Nayak, A. K., & Mohapatra, T. (2014). Tropical low land rice ecosystem is a net carbon sink. *Agriculture, Ecosystems & Environment, 189*, 127–135.
Bouwman, A. F. (1990). Soils and the greenhouse effect fluxes of momentum sensible heat water vapour and carbon dioxide. *Journal of Plants Proceedings of Royal Society, 276*, 1347–1354.
Bubier, J. L., Moore, T. R., Bellisario, L., Comer, N. T., & Crill, P. M. (1995). Ecological controls on methane emissions from a northern peatland complex in the zone of discontinuous permafrost Manitoba Canada. *Global Biogeochemical Cycles, 9*(4), 455–470.
Buchmann, N., & Schulze, E. D. (1999). Net CO_2 and H_2O fluxes of terrestrial ecosystems. *Global Biogeochemical Cycles, 13*(3), 751–760.
Burba, G. (2013). Eddy covariance method for scientific industrial agricultural and regulatory applications: A field book on measuring ecosystem gas exchange and areal emission rates. *LI-Cor Biosciences.*
Butterbach-Bahl, K., Papen, H., & Rennenberg, H. (1997). Impact of gas transport through rice cultivars on methane emission from rice paddy fields. *Plant, Cell and Environment, 20*, 1175–1183.
Campbell, C. S., Heilman, J. L., Mclnnes, K. J., Wilson, L. T., Medley, J. C., Wu, G., & Cobos, D. R. (2001). Diel and seasonal variation in CO_2 flux of irrigated rice. *Agricultural and Forest Meteorology, 108*, 15–27.
Chapin, F. S., Woodwell, G. M., Randerson, J. T., Rastetter, E. B., Lovett, G. M., Baldocchi, D. D., Clark, D. A., Harmon, M. E., Schimel, D. S., Valentini, R., & Wirth, C. (2006). Reconciling carboncycle concepts terminology and methods. *Ecosystems, 9*(7), 1041–50. 101007/s10021-005-0105-7.
Chatterjee, D., Nayak, A. K., Vijayakumar, S., Debnath, M., Chatterjee, S., Swain, C. K., Bihari, P., Mohanty, S., Tripathi, R., Shahid, M., Kumar, A., & Pathak, H. (2019a). Water vapor flux in tropical lowland rice. *Environmental Monitoring and Assessment, 191*, 550. 101007/s10661-019-7709-4.
Chatterjee, D., Tripathi, R., Chatterjee, S., Debnath, M., Shahid, M., Bhattacharyya, P., Swain, C. K., Tripathy, R., Bhattacharya, B. K., & Nayak, A. K. (2019b) Characterization of land surface energy fluxes in a tropical lowland rice paddy *Theoretical and Applied Climatology, 136*(1–2), 157–168. 101007/s00704-018-2472-y.
Chen, J., & Hu, F. (2003). Coherent structures detected in atmospheric boundary-layer turbulence using wavelet transforms at Huaihe River Basin China. *Boundary-Layer Meteorology, 107*(2), 429–444.

Chimner, R. A., & Cooper, D. J. (2003). Influence of water table levels on CO_2 emissions in a Colorado subalpine fen: An in site microcosm study. *Soil Biology and Biochemistry, 35*, 245–351.

Ciais, P., Reichstein, M., Viovy, N., Granier, A., Ogée, J., Allard, V., Aubinet, M., Buchmann, N., Bernhofer, C., Carrara, A., & Chevallier, F. (2005). Europe-wide reduction in primary productivity caused by the heat and drought in 2003. *Nature, 437*(7058), 529.

Conrad, J. (1989). The price effect of option introduction The. *Journal of Finance, 44*(2), 487–498.

Crutzen, R., De Nooijer, J., Brouwer, W., Oenema, A., Brug, J., & De Vries, N. K. (2008). Internet-delivered interventions aimed at adolescents: A Delphi study on dissemination and exposure. *Health Education Research, 23*(3), 427–439.

Dabberdt, W. F., Lenschow, D. H., Horst, T. W., Zimmerman, P. R., Oncley, S. P., & Delany, A. C. (1993). Atmosphere-surface exchange measurements. *Science, 260*(5113), 1472–1481.

Dacey, D. M., & Lee, B. B. (1994). The 'blue-on' opponent pathway in primate retina originates from a distinct bistratified ganglion cell type. *Nature, 367*(6465), 731–735.

Davis, S. J., Caldeira, K., & Matthews, H. D. (2010). Future CO_2 emissions and climate change from existing energy infrastructure. *Science, 329*(5997), 1330–1333.

Den Van der Gon, H. A., & Van Breemen, N. (1993). Diffusion-controlled transport of methane from soil to atmosphere as mediated by rice plants. *Biogeochemistry, 21*(3), 177–190.

Denmead, O. T. (2008). Approaches to measuring fluxes of methane and nitrous oxide between landscapes and the atmosphere. *Plant and Soil, 309*(1–2), 5–24.

Desai, A. R., Richardson, A. D., Moffat, A. M., Kattge, J., Hollinger, D. Y., Barr, A., Falge, E., Noormets, A., Papale, D., Reichstein, M., & Stauch, V. J. (2008). Cross-site evaluation of eddy covariance GPP and RE decomposition techniques. *Agricultural and Forest Meteorology, 148*(6–7), 821–838.

Dunfield, P., Dumont, R., & Moore, T. R. (1993). Methane production and consumption in temperate and subarctic peat soils: Response to temperature and pH. *Soil Biology and Biochemistry, 25*(3), 321–326.

Eugster, W., & Plüss, P. (2010). A fault-tolerant eddy covariance system for measuring CH_4 fluxes. *Agricultural and Forest Meteorology, 150*(6), 841–851.

Eugster, W., Zeyer, K., Zeeman, M., Michna, P., Zingg, A., Buchmann, N., & Emmenegger, L. (2007). Methodical study of nitrous oxide eddy covariance measurements using quantum cascade laser spectrometry over a Swiss forest. *Biogeosciences, 4*, 927–939.

Falge Baldocchi, D., Olson, R., Anthoni, P., Aubinet, M., Bernhofer, C., Burba, G., Ceulemans, R., Clement, R., Dolman, H., & Granier, A. (2001). Gap filling strategies for defensible annual sums of net ecosystem exchange. *Agricultural and Forest Meteorology, 107*(1), 43–69.

Frenzel, P., Rothfuss, F., & Conrad, R. (1992). Oxygen profiles and methane turnover in a flooded rice microcosm. *Biology and Fertility of Soils, 14*(2), 84–89.

Frolking, S. E., Bubier, J. L., Moore, T. R., Ball, T., Bellisario, L. M., Bhardwaj, A., Carroll, P., Crill, P. M., Lafleur, P. M., McCaughey, J. H., & Roulet, N. T. (1998). Relationship between ecosystem productivity and photosynthetically active radiation for northern peatlands. *Global Biogeochemical Cycles, 12*(1), 115–126.

Gao, Y., Zhang, Z., Liu, X., Yi, N., Zhang, L., Song, W., Wang, Y., Mazumder, A., & Yan, S. (2016). Seasonal and diurnal dynamics of physicochemical parameters and gas production in vertical water column of a eutrophic pond. *Ecological Engineering, 87*, 313–323.

Goulden, M. L., Munger, J. W., Fan, S. M., Daube, B. C., & Wofsy, S. (1996). Measurements of carbon sequestration by long-term eddy covariance: Methods and a critical evaluation of accuracy. *Global Change Biology, 2*, 169–182.

Griffis, T. J., Rouse, W. R., & Waddington, J. M. (2000). Interannual variability of net ecosystem CO_2 exchange at a subarctic fen. *Global Biogeochemical Cycles, 14*(4), 1109–1121.

Hagen, S. C., Braswell, B. H., Linder, E., Frolking, S., Richardson, A. D., & Hollinger, D. Y. (2006). Statistical uncertainty of eddy flux–based estimates of gross ecosystem carbon exchange at Howland Forest, Maine. *Journal of Geophysical Research: Atmospheres, 111*(D8).

Heimann, M., & Reichstein, M. (2008). Terrestrial ecosystem carbon dynamics and climate feedbacks. *Nature, 451*(7176), 289.
Heimann, M. (2011). Atmospheric science: Enigma of the recent methane budget. *Nature, 476*(7359), 157.
Huang, C., Davis, L. S., & Townshend, J. R. G. (2002). An assessment of support vector machines for land cover classification. *International Journal of Remote Sensing, 23*(4), 725–749.
Nisbet, R. E., Fisher, R., Nimmo, R. H., Bendall, D. S., Crill, P. M., Gallego-Sala, A. V., Hornibrook, E. R., López-Juez, E., Lowry, D., Nisbet, P. B., & Shuckburgh, E. F. (2009a). Emission of methane from plants. *Proceedings of the Royal Society b: Biological Sciences, 276*(1660), 1347–1354.
Inubushi, K., Shibahara, F., Tsuji, T., & Umebayashi, M. (1992). Assay for microbial biomass nitrogen in unflooded paddy soil by chloroform fumigation-extraction method. *Journal of Plant Nutrition Soil Science, 63*, 342–344.
IPCC. (1997). *Revised 1996 IPCC guidelines for national greenhouse gas inventories: Greenhouse gas inventory reference manual IPCC/OECD/IGES.*
IPCC. (2007). Agriculture In B. Metz, O. R. Davidson, P. R. Bosch (Eds.), *Climate change 2007: Mitigation contribution of working group III to the fourth assessment report of the intergovernmental panel on climate change.* Cambridge University Press, Cambridge, United Kingdom/ New York NY, USA.
IPCC. (2014). In C. B. Field, V. R. Barros, D. J. Dokken, K. J. Mach, M. D. Mastrandrea, T. E. Bilir, M. Chatterjee, K. L. Ebi, Y. O. Estrada, R. C. Genova, B. Girma, E. S. Kissel, A. N. Levy, S. MacCracken, P. R. Mastrandrea, L. L. White (Eds.), *Climate change 2014: Impacts adaptation and vulnerability part A: Global and sectoral aspects contribution of working group II to the fifth assessment report of the intergovernmental panel on climate change.* Cambridge University Press, Cambridge, United Kingdom and New York NY, p. 1132.
Janssens, I. A., Freibauer, A., Schlamadinger, B., Ceulemans, R., Ciais, P., Dolman, A. J., Heimann, M., Nabuurs, G. J., Smith, P., Valentini, R., & Schulze, E. D., (2006). The carbon budget of terrestrial ecosystems at country-scale–a European case study.
Jones, S. K., Famulari, D., Di Marco, C. F., Nemitz, E., Skiba, U. M., Rees, R. M., & Sutton, M. A. (2011). Nitrous oxide emissions from managed grassland: A comparison of eddy covariance and static chamber measurements. *Atmospheric Measurement Technology, 4*, 2179–2194.
Jung, M., Reichstein, M., Margolis, H. A., Cescatti, A., Richardson, A. D., Arain, M. A., Arneth, A., Bernhofer, C., Bonal, D., Chen, J., & Gianelle, D. (2011). Global patterns of land-atmosphere fluxes of carbon dioxide latent heat and sensible heat derived from eddy covariance satellite and meteorological observations. *Journal of Geophysical Research: Biogeosciences, 116*(G3).
Kai, F. M., Tyler, S. C., Randerson, J. T., & Blake, D. R. (2011). Reduced methane growth rate explained by decreased Northern Hemisphere microbial sources. *Nature, 476*(7359), 194.
Kaimal, J. C., & Finnigan, J. J. (1994). *Atmospheric boundary layer flows.* Oxford University Press, p. 289
Kirschke, S., Bousquet, P., Ciais, P., Saunois, M., Canadell, J. G., Dlugokencky, E. J., Bergamaschi, P., Bergmann, D., Blake, D. R., Bruhwiler, L., & Cameron-Smith, P. (2013). Three decades of global methane sources and sinks. *Nature Geoscience, 6*, 813–823.
Lal, R. (2000). World cropland soils as a source or sink for atmospheric carbon. *Advances Agronomy, 71*, 145–191.
Lasslop, G., Reichstein, M., Papale, D., Richardson, A. D., Arneth, A., Barr, A., Stoy, P., & Wohlfahrt, G. (2010). Separation of net ecosystem exchange into assimilation and respiration using a light response curve approach: Critical issues and global evaluation. *Global Change Biology, 16*(1), 187–208.
Laurila, T., Aurela, M., & Tuovinen, J. P. (2012). Eddy covariance measurements over wetlands. In *Eddy Covariance*, pp. 345–364.
Liblik, L. K., Moore, T. R., Bubier, J. L., & Robinson, S. D. (1997). Methane emissions from wetlands in the zone of discontinuous permafrost: Fort Simpson Northwest Territories Canada. *Global Biogeochemical Cycles, 11*(4), 485–494.

Liu, Y., Yu, G., Wen, X., Wang, Y., Song, X., Li, J. U., ... & Liu, Q. (2006). Seasonal dynamics of CO_2 fluxes from subtropical plantation coniferous ecosystem. *Science in China Series D: Earth Sciences, 49,* 99–109.

Liu, Y., Hill, M. J., Zhang, X., Wang, Z., Richardson, A. D., Hufkens, K., ... & Schaaf, C. B. (2017). Using data from Landsat, MODIS, VIIRS and PhenoCams to monitor the phenology of California oak/grass savanna and open grassland across spatial scales. *Agricultural and Forest Meteorology, 237,* 311–325.

Lobsey, C. R., & Viscarra Rossel, R. A. (2016). Sensing of soil bulk density for more accurate carbon accounting. *European Journal of Soil Science, 67*(4), 504–513.

Luyssaert, S., Inglima, I., Jung, M., Richardson, A. D., Reichstein, M., Papale, D., Piao, S. L., Schulze, E. D., Wingate, L., Matteucci, G., & Aragao, L. E. O. C. (2007). CO_2 balance of boreal temperate and tropical forests derived from a global database. *Global Change Biology, 13*(12), 2509–2537.

Mammarella, I., Werle, P., Pihlatie, M., Eugster, W., Haapanala, S., Kiese, R., Markkanen, T., Rannik, Ü., Vesala, T. (2010). A case study of eddy covariance flux of N_2O measured within forest ecosystems: quality control and flux error analysis. *Biogeosciences, 7,* 427–440.

Masseroni, D., Corbari, C., & Mancini, M. (2014). Limitations and improvements of the energy balance closure with reference to experimental data measured over a maize field. *Atmósfera, 27*(4), 335–352.

Meijide, A., Manca, G., Goded, I., Magliulo, V., Di Tommasi, P., Seufert, G., & Cescatti, A. (2011). Seasonal trends and environmental controls of methane emissions in a rice paddy field in Northern Italy. *Biogeosciences, 8*(12), 3809–3821.

Minami, K., & Neue, H. U. (1994). Rice paddies as a methane source. *Climate Change, 27.*

Mitchell, J. F. B., Manabe, S., Meleshko, V., & Tokioka, T. (1990). Equilibrium climate change and its implications for the future climate change: The IPCC scientific assessment, vol. 131, p. 172.

Miyata, A., Leuning, R., Denmead, O. T., Kim, J., & Harazono, Y. (2000). Carbon dioxide and methane fluxes from an intermittently flooded paddy field. *Agricultural and Forest Meteorology, 102*(4), 287–303.

Moncrieff, J. B., Massheder, J. M., De Bruin, H., Elbers, J., Friborg, T., Heusinkveld, B., Kabat, P., Scott, S., Soegaard, H., & Verhoef, A. (1997). A system to measure surface fluxes of momentum sensible heat water vapour and carbon dioxide. *Journal of Hydrology, 188,* 589–611.

Montazar, A., Radawich, J., Zaccaria, D., Bali, K., & Putnam, D. (2016). Increasing water use efficiency in alfalfa production through deficit irrigation strategies under subsurface drip irrigation. *Water Shortage and Drought: From Challenges to Solution,* 195.

Moore, T. R., & Basiliko, N. (2006). Decomposition. In R. K. Wieder, D. H. Vitt (Eds.) *Boreal peatland ecosystems.* Ecological studies, Vol. 188. Springer-Verlag, Berlin.

Moore, T. R., & Dalva, M. (1993). The influence of temperature and water table position on carbon dioxide and methane emissions from laboratory columns of peatland soils. *European Journal of Soil Science, 44*(4), 651–664.

Moureaux, C., Debacq, A., Bodson, B., Heinesch, B., & Aubinet, M. (2006). Annual net ecosystem carbon exchange by a sugar beet crop. *Agricultural and Forest Meteorology, 139*(1–2), 25–39.

Neftel, A., Ammann, C., Fischer, C., Spirig, C., Conen, F., Emmenegger, L., Tuzson, B., & Wahlen, S. (2010). N_2O exchange over managed grassland: Application of a quantum cascade laser spectrometer for micrometeorological flux measurements. *Agricultural and Forest Meteorology, 150,* 775–785.

Neftel, A., Flechard, C., Ammann, C., Conen, F., Emmenegger, L., & Zeyer, K. (2007). Experimental assessment of N_2O background fluxes in grassland systems. *Tellus, 59B,* 470–482.

Neue, H. U. (1990). Organic matter dynamics soil properties and cultural practices in rice lands and their relationship to methane production. *Soils and the Greenhouse Effect, 1990,* 457–466.

Neue, H. U. (1993). Methane emission from rice fields. *BioScience, 43*(7), 466–474.

Nisbet, R. E. R., Fisher, R., Nimmo, R. H., Bendall, D. S., Crill, P. M., Gallego-Sala, A. V., Hornibrook, E. R. C., López-Juez, E., Lowry, D., Nisbet, P. B. R., & Shuckburgh, E. F. (2009b).

Emission of methane from plants. *Proceedings of the Royal Society b: Biological Sciences, 276*(1660), 1347–1354.
Nouchi, I., Hosono, T., Aoki, K., & Minami, K. (1994). Seasonal variation in methane flux from rice paddies associated with methane concentration in soil water, rice biomass and temperature, and its modelling. *Plant and Soil, 161*, 195–208.
Nouchi, I., Mariko, S., & Aki, K. (1990). Mechanism of methane transport from the rhizosphere to the atmosphere through the rice plants. *Plant Physiology, 94*, 59–66.
Odum, H. T. (1971). *Environment power and society*. Wiley-Interscience, New York, USA.
Pakoktom, T., Aoki, M., Kasemsap, P., Boonyawat, S., & Attarod, P. (2009). CO_2 and H_2O fluxes ratio in paddy fields of Thailand and Japan. *Hydrological Research Letters, 3*, 10–13.
Papale, D., Reichstein, M., Aubinet, M., et al. (2006). Towards a standardized processing of net ecosystem exchange measured with eddy covariance technique: Algorithms and uncertainty estimation. *Biogeosciences, 3*, 571–583.
Patel, N., Chittamuru, D., Jain, A., Dave, P., & Parikh, T. S. (2010). Avaaj otalo: A field study of an interactive voice forum for small farmers in rural india. In *Proceedings of the SIGCHI conference on human factors in computing systems* (pp. 733–742).
Patel, N. R., Dadhwal, V. K., & Saha, S. K. (2011). Measurement and scaling of carbon dioxide (CO_2) exchanges in wheat using flux-tower and remote sensing. *Journal of the Indian Society of Remote Sensing,* 39, 383–391.
Pattey, E., Strachan, I. B., Desjardins, R. L., Edwards, G. C., Dow, D., & MacPherson, J. I. (2006). Application of a tunable diode laser to the measurement of CH_4 and N_2O fluxes from field to landscape scale using several micrometeorological techniques. *Agricultural and Forest Meteorology, 136*, 222–236.
Peltola, O., Mammarella, I., Haapanala, S., Burba, G., & Vesala, T. (2013). Field intercomparison of four methane gas analyzers suitable for eddy covariance flux measurements. *Biogeosciences, 10*, 3749–3765.
Pihlatie, M., Rinne, J., Ambus, P., Pilegaard, K., Dorsey, J. R., Markkanen, Ü., Launiainen, S., & Vesala, T. (2005). Nitrous oxide emissions from a beech forest floor measured by eddy covariance and soil enclosure techniques. *Biogeosciences, 2*, 377–387.
Ponnamperuna, F. N. (1972). The chemistry of submerged soils. *Advances in Agronomy, 24*, 29–96.
Ramifehiarivo, N., Brossard, M., Grinand, C., Andriamananjara, A., Razafimbelo, T., Rasolohery, A., Razafimahatratra, H., Seyler, F., Ranaivoson, N., Rabenarivo, M., & Albrecht, A. (2017). Mapping soil organic carbon on a national scale: TOwards an improved and updated map of Madagascar. *Geoderma Regional, 9*, 29–38.
Reichstein, M., Falge, E., Baldocchi, D., Papale, D., Aubinet, M., Berbigier, P., Bernhofer, C., Buchmann, N., Gilmanov, T., Granier, A., & Grünwald, T. (2005). On the separation of net ecosystem exchange into assimilation and ecosystem respiration: Review and improved algorithm. *Global Change Biology, 11*(9), 1424–1439.
Richardson, A. D., Hollinger, D. Y., Burba, G. G., Davis, K. J., Flanagan, L. B., Katul, G. G., Munger, J. W., Ricciuto, D. M., Stoy, P. C., Suyker, A. E., & Verma, S. B. (2006). A multi-site analysis of random error in tower-based measurements of carbon and energy fluxes. *Agricultural and Forest Meteorology, 136*(1–2), 1–18.
Roxy, M. K., Ritika, K., Terray, P., & Masson, S. (2014). The curious case of Indian Ocean warming. *Journal of Climate, 27*(22), 8501–8509.
Sachs, J., Remans, R., Smukler, S., Winowiecki, L., Andelman, S. J., Cassman, K. G., Castle, D., DeFries, R., Denning, G., Fanzo, J., & Jackson, L. E. (2010). Monitoring the world's agriculture. *Nature, 466*(7306), 558.
Saito, M., Miyata, A., Nagai, H., & Yamada, T. (2005). Seasonal variation of carbon dioxide exchange in rice paddy field in Japan. *Agricultural and Forest Meteorology, 135*(1–4), 93–109.
Sarmiento, J. L., & Wofsy, S. C. (1999). *A US carbon cycle science plan.* US Global Change Research Program.
Sass, R. L., Fisher, F. M., Harcombe, P. A., & Turner, F. T. (1990). Methane production and emission in a Texas rice field. *Global Biogeochemical Cycles, 4*(1), 47–68.

Schmidt-Rohr, K., & Spiess, H. W. (2012). *Multidimensional solid-state NMR and polymers*. Elsevier

Schneider, A., Friedl, M. A., & Potere, D. (2010). Mapping global urban areas using MODIS 500-m data: New methods and datasets based on 'urban ecoregions'. *Remote Sensing of Environment, 114*(8), 1733–1746.

Schulze, E. D., & Heimann, M. (1998). *Carbon and water exchange of terrestrial Asian change in the context of global climate change: Impact of natural and anthropogenic changes in Asia on Global Biogeochemical Cycles*, vol 3, p. 145.

Shi, W. Y., Zhang, J. G., Yan, M. J., Yamanaka, N., & Du, S. (2012). Seasonal and diurnal dynamics of soil respiration fluxes in two typical forests on the semiarid Loess Plateau of China: Temperature sensitivities of autotrophs and heterotrophs and analyses of integrated driving factors. *Soil Biology and Biochemistry, 52*, 99–107.

Simpson, I. J., Thurtell, G. W., Kidd, G. E., Lin, M., Demetriades-Shah, T. H., Flitcroft, I. D., Kanemasu, E. T., Nie, D., Bronson, K. F., & Neue, H. U. (1995). Tunable diode laser measurements of methane fluxes from an irrigated rice paddy field in Philippines. *Journal of Geophysical Research, 100*, 7283–7290.

Smith, P., Lanigan, G., Kutsch, W. L., Buchmann, N., Eugster, W., Aubinet, M., Ceschia, E., Béziat, P., Yeluripati, J. B., Osborne, B., & Moors, E. J. (2010). Measurements necessary for assessing the net ecosystem carbon budget of croplands. *Agriculture Ecosystems & Environment, 139*(3), 302–315.

Stella, P., Kortner, M., Ammann, C., Foken, T., Meixner, F. X., & Trebs, I. (2013). Measurements of nitrogen oxides and ozone fluxes by eddy covariance at a meadow: evidence for an internal leaf resistance to NO_2. *Biogeosciences, 10*(9), 5997.

Swain, C. K., Bhattacharyya, P., Singh, N. R., Neogi, S., Sahoo, R. K., Nayak, A. K., Zhang, G., & Leclerc, M. Y. (2016). Net ecosystem methane and carbon dioxide exchange in relation to heat and carbon balance in lowland tropical rice. *Ecological Engineering, 95*, 364–374.

Swain, C. K., Nayak, A. K., Bhattacharyya, P., Chatterjee, D., Chatterjee, S., Tripathi, R., Singh, N. R., & Dhal, B. (2018). Greenhouse gas emissions and energy exchange in wet and dry season rice: Eddy covariance-based approach. *Environmental Monitoring and Assessment, 190*(7), 1–17.

Taneva, L., & Gonzalez-Meler, M. A. (2011). Distinct patterns in the diurnal and seasonal variability in four components of soil respiration in a temperate forest under free-air CO_2 enrichment. *Biogeosciences, 8*(10), 3077–3092.

Tanner, C. B., & Thurtell, G. W. (1969). Anemoclinometer measurements of reynolds stress and heat transport in the atmospheric surface layer. University of Wisconsin Tech Rep ECOM-66-G22-F, p 82

Tao, T., & Wang, H. (2019). Modelling of longitudinal evolutionary power spectral density of typhoon winds considering high-frequency subrange. *Journal of Wind Engineering and Industrial Aerodynamics, 193*, 103957.

Tsai, L. W., Hsieh, J. W., & Fan, K. C. (2007). Vehicle detection using normalized color and edge map. *IEEE Transactions on Image Processing, 16*(3), 850–864.

Tseng, K. H., Tsai, J. L., Alagesan, A., Tsuang, B. J., Yao, M. H., & Kuo, P. H. (2010). Determination of methane and carbon dioxide fluxes during the rice maturity period in Taiwan by combining profile and eddy covariance measurements. *Agricultural and Forest Meteorology, 150*(6), 852–859.

Vesala, T., Eugster, W., & Ojala, A. (2012). *Eddy covariance measurements over lakes In Eddy covariance* (pp. 365–376). Springer.

Vetter, M., Churkina, G., Jung, M., Reichstein, M., Zaehle, S., Bondeau, A., Chen, Y., Ciais, P., Feser, F., Freibauer, A., & Geyer, R. (2008). Analyzing the causes and spatial pattern of the European 2003 carbon flux anomaly using seven models. *Biogeosciences 5*(2), 561–583.

Vickers, D., & Mahrt, L. (1996). Quality control and flux sampling problems for tower and aircraft data. *Journal of Atmospheric and Oceanic Technology, 14*, 512–526.

Waddington, J. M., & Roulet, N. T. (1996). Atmosphere-wetland carbon exchanges: Scale dependency of CO_2 and CH_4 exchange on the developmental topography of a peatland. *Global Biogeochemical Cycles, 10*(2), 233–245.

Waddington, J. M., Griffis, T. J., & Rouse, W. R. (1998). Northern Canadian wetlands: Net ecosystem CO_2 exchange and climatic change. *Climatic Change, 40*(2), 267–275.

Wang, K., Zheng, X., Pihlatie, M., Vesala, T., Liu, C., Haapanala, S., Mammarella, I., Rannik, Ü., & Liu, H. (2013). Comparison between static chamber and tunable diode laser-based eddy covariance techniques for measuring nitrous oxide fluxes from a cotton field. *Agricultural and Forest Meteorology, 171*, 9–19.

Watanabe, I., & Roger, P. A. (1985). Ecology of flooded rice fields In: Wetland soils: Characterization classification and utilization. International Rice Research Institute Los Baiios Philippines, pp. 229–243.

Webb, E. K., Pearman, G. I., & Leuning, R. (1980). Correction of flux measurements for density effects due to heat and water vapour transfer. *Quarterly Journal of the Royal Meteorological Society, 106*, 85–100.

Weber, E. P. (2009). Explaining institutional change in tough cases of collaboration: "Ideas" in the Blackfoot Watershed. *Public Administration Review, 69*(2), 314–327.

Whalen, S. C. (2005). Biogeochemistry of methane exchange between natural wetlands and the atmosphere. *Environmental Engineering Science, 22*(1), 73–94.

Whiting, G. J., & Chanton, J. P. (1992). Plant-dependent CH_4 emission in a subarctic Canadian fen. *Global Biogeochemical Cycles, 6*(3), 225–231.

Wilczak, J. M., Oncley, S. P., & Stage, S. A. (2001) Sonic anemometers tilt correction algorithms. *Boundary Layer Meteorology, 99*, 127–150.

Wilson, K., Goldstein, A., Falge, E., Aubinet, M., Baldocchi, D., Berbigier, P., ... & Verma, S. (2002). Energy balance closure at FLUXNET sites. *Agricultural and Forest Meteorology, 113*(1–4), 223–243.

Wohlfahrt, G., & Widmoser, P. (2013). Can an energy balance model provide additional constraints on how to close the energy imbalance? *Agricultural and Forest Meteorology, 169*, 85–91.

Wohlfahrt, G., & Gu, L. (2015). The many meanings of gross photosynthesis and their implication for photosynthesis research from leaf to globe. *Plant Cell & Environment, 38*(12), 2500–2507.

World Meteorological Organization. (2022). 2021 One of the seven warmest years on record, WMO consolidated data shows (Geneva: WMO).

Yagi, K., & Minami, K. (1990). Effect of organic matter application on methane emission from some Japanese paddy fields. *Soil Science and Plant Nutrition, 36*(4), 599–610.

Chapter 24
Climate Change Scenarios of India with Special Emphasis on Sundarbans Delta and Western Himalayan Region

Lalu Das, Javed Akhter, Manish Kumar Naskar, Ratul Roy Choudhury, Ratul Chakraborty, and Jitendra Kumar Meher

Abstract In recent times, creation of climate change scenarios at country scale as well as different homogeneous zones, and some identified vulnerable zones are more demanding for the implementation of different adaptation and mitigation strategies by the multi-level stakeholders and administrators towards combating climate change issues. An attempt was made in the present study to generate climate change scenarios over different homogeneous zones of India with special focus over two most vulnerable regions namely Western Himalayan Region (WHR) in the north and the Indian Sundarbans in the east which are technically known as "climatic hot-spots" of India. To put more reliability of the climate change results and to minimise the inherent biases of the present generation General Circulation Models (GCMs) simulations, two widely accepted downscaling methods namely Model Output Statistics (MOS) and Perfect Prognosis (PP) is utilised to develop downscaled rainfall and temperature scenarios over above-mentioned homogeneous zones and two "climatic hot spots" regions of India. In general, future downscaled projection indicates an increment in annual rainfall and temperature in all India level with varying magnitudes over different homogeneous zones. Significant warming is observed for all seasons during 1969–2009 over WHR with winter warming of 1.41 °C is substantially higher than global average changing the hydrological system of this region. Annual rainfall is found to be declined over WHR which affects the agricultural pattern and crop yield. Annual rainfall of this region is projected to increase by both representative concentration pathway (RCP) scenarios which may significantly contribute towards

L. Das (✉) · M. K. Naskar · R. Chakraborty
Department of Agricultural Meteorology and Physics, Bidhan Chandra Krishi Viswavidyalaya, Mohanpur 741252, Nadia, West Bengal, India
e-mail: das.lalu@bckv.edu.in; daslalu@yahoo.co.in

J. Akhter · R. R. Choudhury
Department of Atmospheric Sciences, University of Calcutta, Kolkata 700019, India

J. K. Meher
Kreate Energy (I) Pvt. Ltd. Ansal Plaza, Sector-1, Vaishali, Ghaziabad 201010, India

H. Pathak et al. (eds.), *Climate Change Impacts on Soil-Plant-Atmosphere Continuum*, Advances in Global Change Research 78,
https://doi.org/10.1007/978-981-99-7935-6_24

Sea-level-rise. However, decreased winter rainfall under RCP 8.5 may cause higher salinity in the mangrove region hampering the ecosystem.

Keywords Sundarbans · Western Himalayan Region · Climate change · General circulation models · Representative concentration pathway · Statistical downscaling

24.1 Introduction

Scientists all over the world are deeply involved to quantify climate change information on the global, regional, sub-regional and local scales from the observational and different simulations of General Circulation Models (GCMs), Regional Climate Model (RCMs), and Empirical Statistical Downscaling models (ESDs). After analysing the observational and model outputs for historical data and its future projection, the climate scientists around the globe are able to identify some regions/domains which are most vulnerable under the context of climate change and global warming which are technically recognized as "climate hot spots" for each country or climatic region. Some of the climatic hot-spots are the Himalayan region, Antarctica region, Arctic region and some small Islands like Sundarbans in different parts of the globe. These regions are affected adversely due to climate change. So, increasing attention for generating historical climate change information and its future projection over these vulnerable regions is more demanding but at the same time it is more challenging due to remote location and non-availability of adequate observational data.

Climate of a particular region depends upon its latitude, altitude and typical geographical orientation. So, the climate change acts and behave different upon different climate regions such as mountainous climate, coastal climate, forest climate, island climate etc. In some places, the change is in alarming rate whereas some places experiences negligible changes. In case of India, areas like the Himalayan Region as well as the Sundarbans region are the most vulnerable zones under changing climate because slightly change of air temperature resulting a great alteration of ice-cap/ice cover and finally melt water can caused the rise of sea level and many small islands are in the seas over Sundarbans delta. Though the change in climate is a very slow and long term process a minute change in climate (such as decadal temperature increase or decadal rainfall decrease) can cause extreme weather condition (severity of cyclones, desertification, inundation of low-lying islands, muddy runoff in mountains etc.) causing enormous hazards for human existence. Rising global sea level is threatening low lying coastal areas and islands like Sundarbans as mentioned above. The changes in rainfall and temperature patterns can alter the earth's surface's green coverage and surface characteristics significantly resulting in changes in agricultural scenarios and economic status. Therefore, it is known that the rapid increase of global warming rate is a growing threat to biodiversity all over the world (Rull & Vegas-Vilarrúbia, 2006). All available literatures indicate the unavoidable earth surface warming in the coming century but pockets of areas would experience more rain and other pockets will receive deficit rainfall under the warming condition. As a result, the heat waves will occur more frequently and last longer. The extreme rainfall events will become more intense and frequent in several zones.

To identify the climate vulnerable places on earth surface especially in the Indian domain, it is essential to have a complete understanding of different climatic drivers and factors. The climate analysis of Indian homogeneous rainfall and temperature zones thus gained importance to identify the vulnerable places. Several studies (Sontakke et al., 2008; Kothawale & Rupakumar, 2005) highlighted the seven homogeneous rainfall and temperature zones in India that are having clearly distinguished climatic features. Names of seven homogeneous zones based on rainfall distribution are: North Mountainous India (NMI), North Central India (NCI), Northeast India (NEI), Northwest India (NWI), East Peninsular India (EPI), West Peninsular India (WPI) and South Peninsular India (SPI). Similarly based on temperature distribution the names of seven zones are as: Northwest India (NWI), Interior Peninsular India (IPI), Northeast India (NEI), North Central India (NCI), Western Himalayas of India (NMI), West coastal India (WCOI), East coastal India (ECOI).

24.2 What is Downscaling?

How the generation of climatic information from global to local scales evolved over time has been schematically represented by Fig. 24.1. The development of GCMs and their role in generating climate change scenarios in different horizontal scales pay huge scientific attention all over the world since the 1980s. Since then, the understanding of the complex earth-atmospheric-ocean system is increasing and the GCMs are showing their better performance generation to next. The information that is generated from the global scale needs to be calibrated in the local scale and then validated if any observational data sources are available. The linkage between the large scale GCMs outputs to the local scale surface parameters through observation is technically known as downscaling (Fig. 24.1).

Although models are improving performance from their predecessors, GCMs are still faced criticism for not adequately representing the local scale climatic features like clouds due to their coarser horizontal resolution (Meher et al., 2017). Now, this mismatch in scales of GCM data and local hydro climatic data is one of the main hurdles in the field of climate research. To get around this problem, downscaling is recommended as it acts as a bridge that connects large scale GCMs and local hydro climatic data. So, many downscaling methods are developed. Here in this chapter, two types of statistical downscaling methods are discussed namely Model Output Statistics (MOS) and Perfect Prognosis (PP).

The GCMs included in this study are as follows, bcc-csm1-1, bcc-csm1-1-m, BNU-ESM, CanESM2, CMCC-CM, CMCC-CMS, CNRM-CM5, FIO-ESM, MPI-ESM-LR, MPI-ESM-MR, NorESM1-ME, NorESM1-M. Detailed information of the GCMs can be found in the article of Das and Akhter (2019).

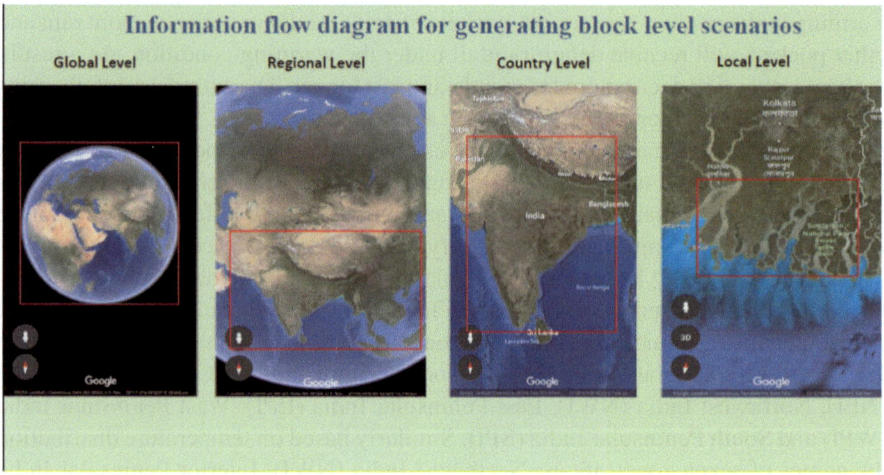

Fig. 24.1 Climate information flow

24.2.1 Downscaling by Model Output Statistics (MOS)

In general, the transfer functions make a connection between the Cumulative Density Functions (CDFs) of observation and model rainfall data within a general time frame. The use of transfer functions to projected rainfall or temperature is a very common procedure in the MOS technique (Block et al., 2009; Piani et al., 2010; Déqué, 2007). Normally the transfer function is indices, which can be achieved by mapping an empirical/theoretical distribution on the cumulative density functions of GCMs and observation rainfall/temperature. Techniques of establishing relationships between GCMs simulation and observed CDFs are also called quantile mapping (Wood et al., 2004).

24.2.1.1 Methods of MOS

Normally in the MOS methods GCMs outputs and observation are used for downscaling the model but no predictors from any reanalysis data are considered. As it was stated earlier that GCMs can perform well but most GCMs do not demonstrate excellent viability in simulating observation adequately indicating their inability. To overcome this difficulty to generate zone specific climate change scenarios, the present study attempted to create rainfall change scenarios over seven homogeneous zones (Sontakke et al., 2008) of India using following methods of MOS.

Scaling: Simplest bias correction method is the scaling method, which has wide application in climate scenarios (Wetterhall et al., 2012). It deals with the ratio between observation to the GCMs output in the reference period multiplied by GCMs outputs in the future period.

Standardized Reconstruction (SdRc): In this technique, mean and standard deviation of reference period are used for standardizing the future GCMs (Johnson & Sharma, 2012). It is basically the ratio of anomaly between the GCMs output in future period and GCMs output in reference period to the standard deviation of GCMs output in reference period.

Quantile Mapping: This technique tries to map some modeled variable in a way that the observed variable distribution is the new distribution. The Gamma Quantile Mapping is where the initial assumption is that the gamma distribution is used to estimate the intensity distribution including both modeled and observed data (Piani et al., 2010). The Empirical Quantile Mapping is where empirical percentile is used to evaluate modeled and observed rainfall and the linear interpolation technique is used for approximating the values that come within the percentiles (Gudmundsson et al., 2012).

The calibration period of the MOS model is 1961–1990 and validation period is 1991–2005. For the sake of simplicity, Simple Ensemble Mean (SEM) collected using every raw model has been considered while implementing MOS (MOS-SEM).

24.2.2 Method of Perfect Prognosis (PP)

According to Klein et al. (1959), in order to use the deterministic dynamical forecast from Numerical Weather Prediction models efficiently, the Perfect Prognosis (PP) method was established. The establishment of connection between local-scale predictand and large-scale predictors is executed by transfer functions such as, regression, Artificial Neural Network (ANN) and Support Vector Machine (SVM) etc. The most widely used downscaling method is the PP method, which is performed by using regression (Benestad, 2002; Mtongori et al., 2016). In addition to regression, the ANN and SVM techniques are also used as a transfer function for developing downscaling models by many researchers worldwide (Okkan & Inan, 2015; Tripathi et al., 2006).

In PP-based downscaling, it is very important to select suitable predictors from a wide range of predictors available in the reanalysis fields in addition to choosing suitable predictor domains. Normally a suitable predictor must have its great predictive power. In general, based on the values of correlation between predictors and predictands, the most informative predictors can be identified (Maraun et al., 2010). It is found that combined predictors are working better compared to single predictors (Huth, 1999, 2002). The main obstacle of using raw predictors is that it has a high dimensional field of gridded values and in addition to that information in nearby grid points is not independent (Maraun et al., 2010). So, to reduce dimensionality one of the most used techniques is Empirical orthogonal function (EOF)/Principal Component Analysis (PCA).

24.3 Future Temperature and Rainfall Scenario Over Indian Subcontinent

24.3.1 Future Temperature and Rainfall Scenario Using MOS

24.3.1.1 Future Temperature Scenario Using MOS Over Indian Subcontinent

Future warming rates in different time periods over different zones of India were shown in Fig. 24.2 for RCP 4.5 and in Fig. 24.3 for RCP 8.5. In general, higher warming is found in RCP 8.5 scenarios compared to the RCP 4.5 scenarios.

According to RCP 4.5 scenarios WHI zone experienced temperature increment of 0.72–0.99 °C and 2.32–2.90 °C during 2006–35 and 2066–95 respectively. Based on RCP 8.5 scenarios, WHI region of India showed a range of temperature increment by 0.80–1.08 °C, and 4.25–5.19 °C, for the periods of 2006–35, and 2066–95 respectively. From here, the values in bracket will stand for the RCP 8.5 scenario. Over NWI, the range of warming was 0.55–0.81 (0.63–0.90) °C in 2006–35, while for the year 2066–95, the warming was by 1.74–2.48 (3.16–4.59) °C for RCP4.5 (RCP8.5). Over the region of NCI, the warming found in 2006–35 year is 0.61–0.88 (0.63–0.94) °C, in 2066–95 year is 1.95–2.69 (3.56–4.88) °C according to RCP 4.5 (RCP 8.5). Greater temperature increment will be in northeastern NEI according to RCP 4.5 and 8.5. Over NEI, in 2006–35 year, 0.52–0.87 (0.52–0.89) °C and in 2066–95 year, 1.67–2.58 (2.93–4.50) °C warming were found according to RCP 4.5 (RCP 8.5). Over the region of IPI, the temperature is found to be increasing in 2006–35 year, by 0.49–0.66 (0.52–0.68) °C and in 2066–95 year, by 1.46–2.03 (2.69–3.74) °C according to RCP 4.5 (RCP 8.5). The temperature is found to increase over ECOI in 2006–35 year, by 0.49–0.59 (0.49–0.64) °C and in 2066–95 year, by 1.49–1.84 (2.72–3.40) °C according to RCP 4.5 (RCP 8.5). The temperature is set to increase over WCOI in 2006–35 year, by 0.49–0.63(0.52–0.71)°C and in 2066–95 year, by 1.46–2.01 (2.70–3.63) °C according to RCP 4.5 (RCP 8.5). All in all, over northern zone of India more warming is found than that of south of India.

24.3.1.2 Future Rainfall Scenario Using MOS Over Indian Subcontinent

Similar work was done over the homogeneous rainfall zones of India. Main results were described below.

In the dry northeast zone of NMI, rainfall increment is found to be 3 to 6% according to both RCP 4.5 and RCP 8.5 and 1 to 3% increment over the other areas in 2006–35. Both RCP 4.5 and RCP 8.5 projected rainfall increments (3 to 6% in 2036–65 and 6–12% in 2066–95) over the southern region. Almost 3–12% of rainfall

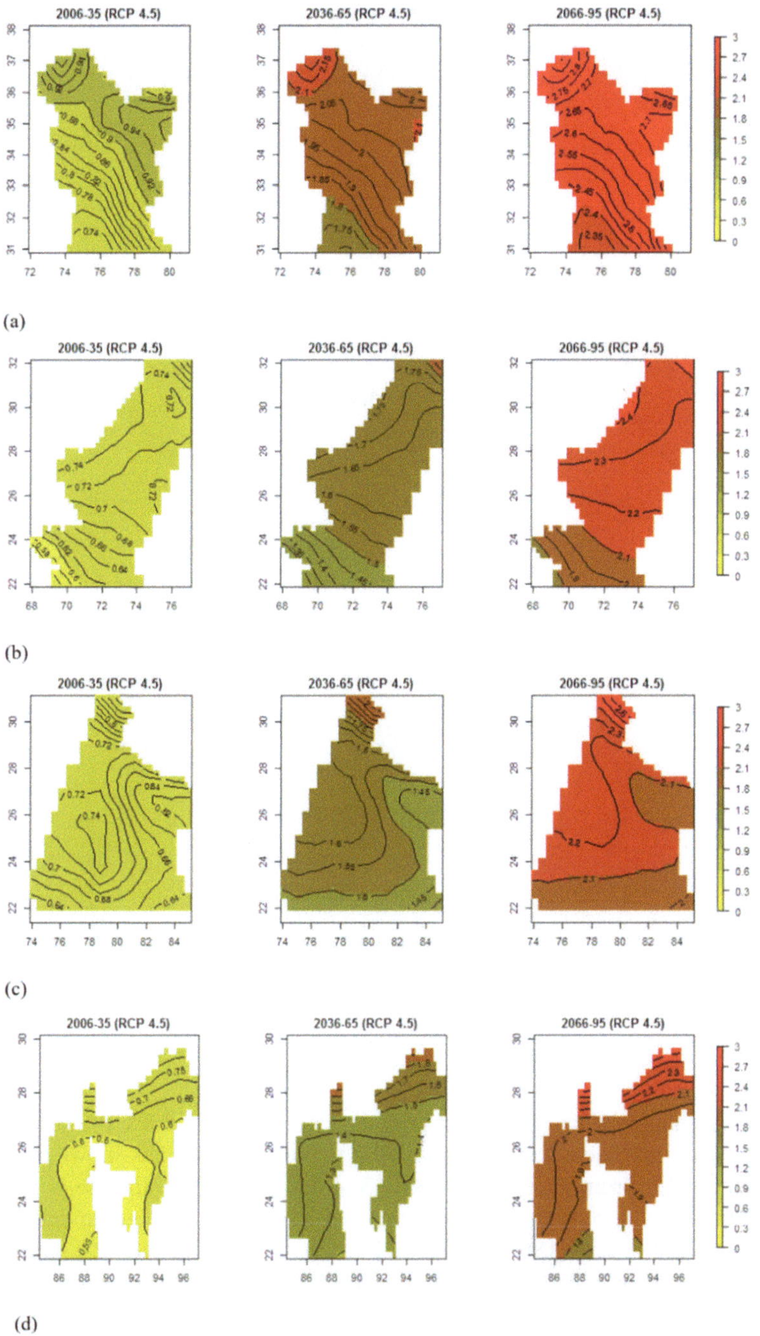

Fig. 24.2 Future temperature change (°C) based on RCP 4.5 for different homogeneous zones of India using MOS-SEM during the period of 2006–35, 2036–65 and 2066–95

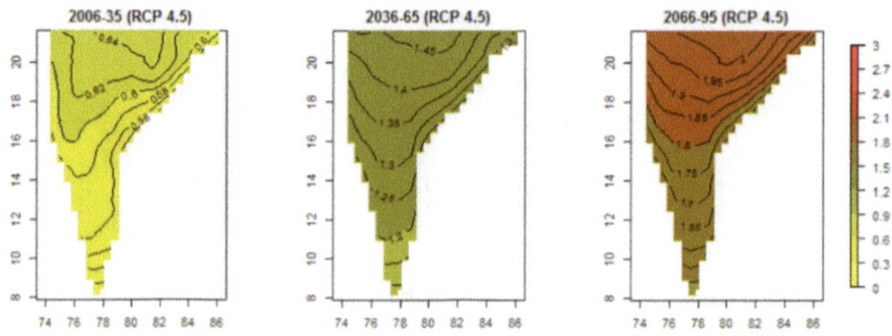

(e)

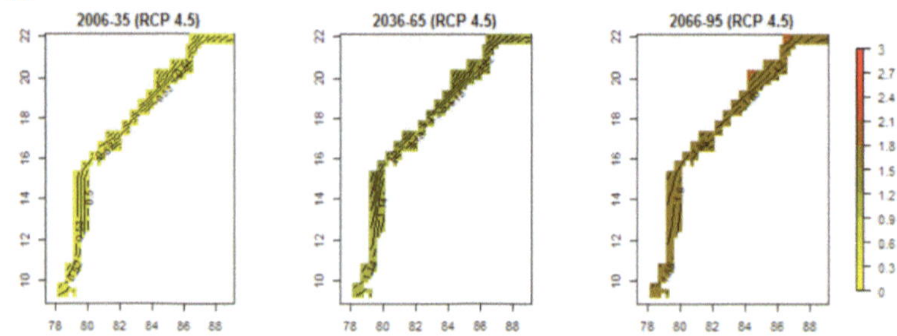

(f)

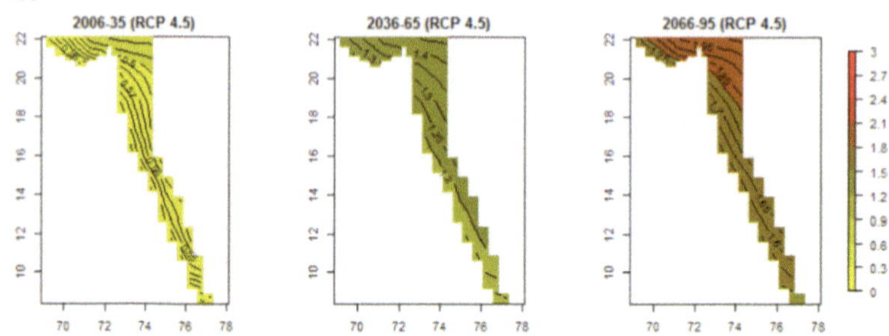

(g)

Fig. 24.2 (continued)

increase can happen during two later periods according to both RCPs. Similarly, 3 to 6% of rainfall increase is found in southwest to northeast NWI as shown by RCP 4.5 during 2006–35. From RCP 8.5 scenario 3 to 6% of rainfall increase is also found during 2006–35 but only in the western part of this region. An increment of rainfall based on RCP 4.5 is 12–18% and for RCP 8.5 is 15–30% have been noticed for the later two periods. Small decrease in annual rainfall projecting 3% decrease during the period of 2006–35 in southern NCI was found in both RCPs. Whereas 3–6% more rainfall will occur in southern NEI than other zones in 2006–35 period and for the other two periods rainfall amount will also be increased. During 2006–35, both RCP 4.5 and RCP 8.5 simulated a decrease of 3% in rainfall over the eastern EPI and for

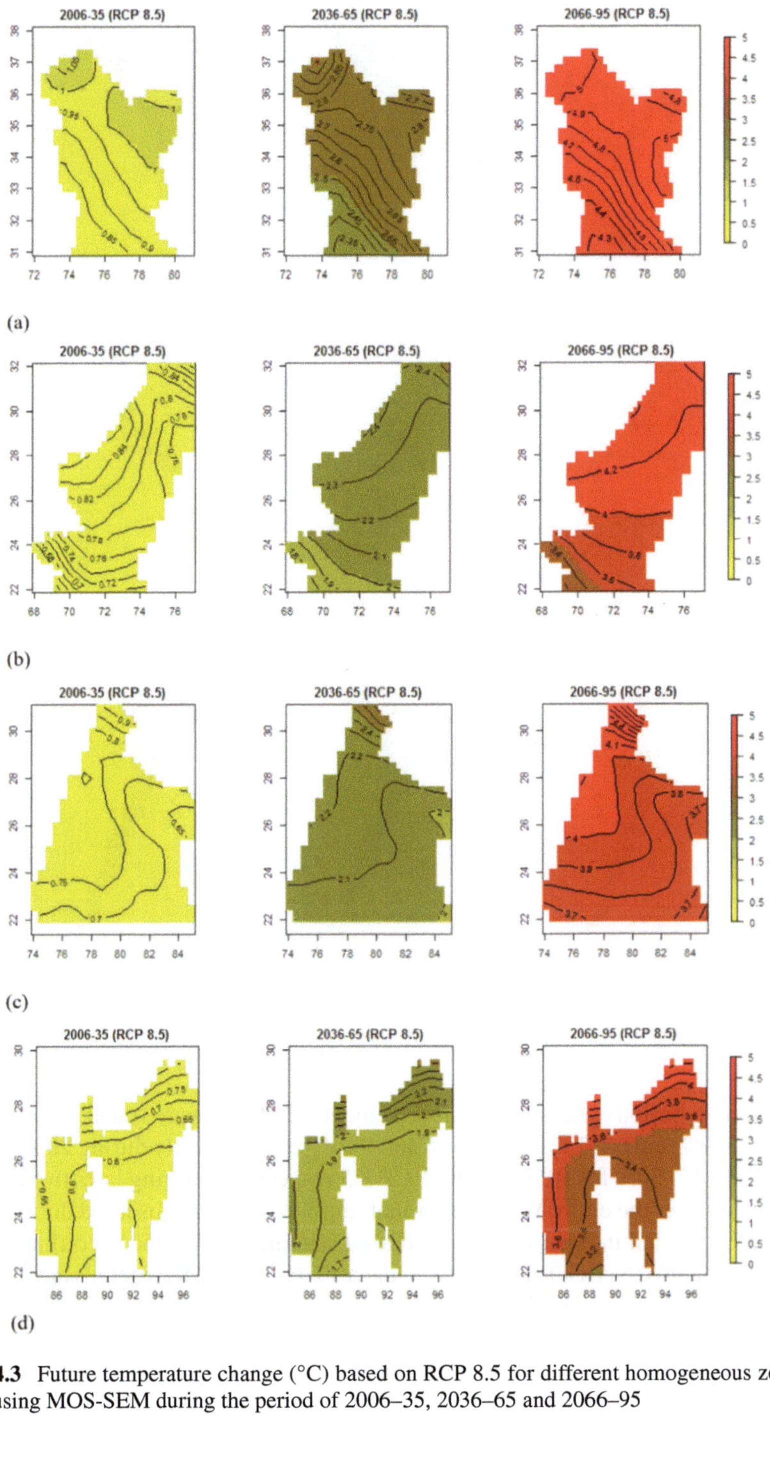

Fig. 24.3 Future temperature change (°C) based on RCP 8.5 for different homogeneous zones of India using MOS-SEM during the period of 2006–35, 2036–65 and 2066–95

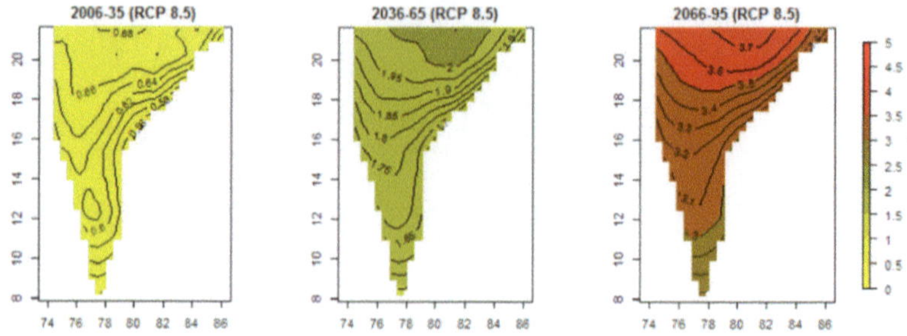

(e)

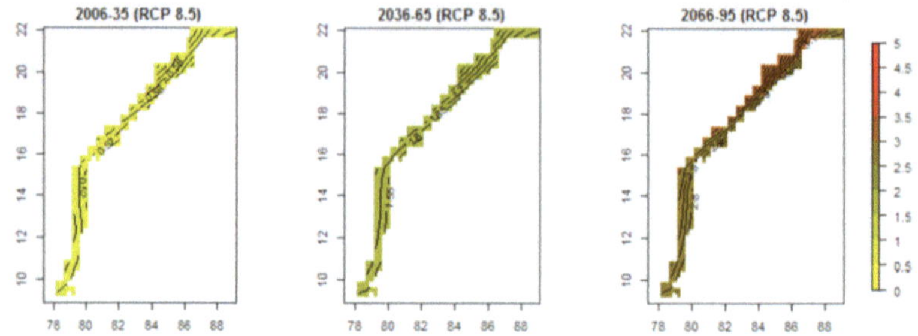

(f)

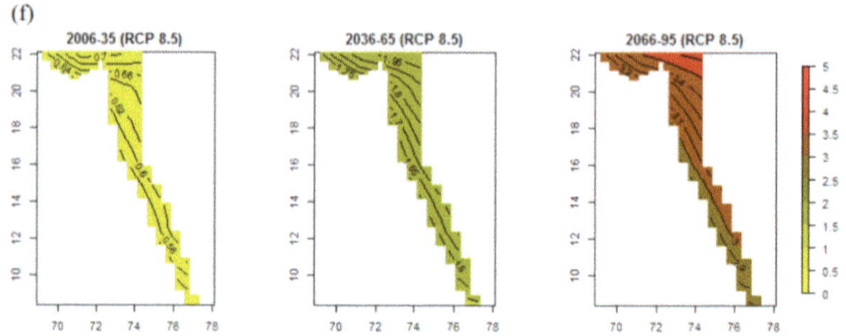

(g)

Fig. 24.3 (continued)

the other two periods gradual rainfall increments (3–15%) were found. Increment of rainfall (3–9% and 6–12% for RCP 4.5 and RCP 8.5 respectively during 2006–35 and 9–18% and 12–30% for RCP 4.5 and RCP 8.5 respectively during later two periods) is projected over the entire WPI. A nominal rainfall change is observed during 2006-35 in SPI zone and in other periods increased rainfall is projected according to both RCP 4.5 scenario (6–18%) and RCP 8.5 scenario (6–24%). More details are available in the article Akhter et al. (2017).

24.3.2 *Future Temperature and Rainfall Scenario Using Perfect Prognosis Method*

24.3.2.1 Temperature Projection Using Perfect Prognosis Over Indian Subcontinent

Season-wise selection of predictors for downscaling is performed before future projection of temperature. The detailed procedure can be found in Das and Akhter (2019). The selected predictors for maximum and minimum temperature are shown in Table 24.1 during four major Indian seasons, which are pre-monsoon (March, April, May—MAM), monsoon (June, July, August, September—JJAS), post-monsoon (October, November, December—OND), winter (January, February—JF).

The future projections using selected predictors are shown in Figs. 24.4 and 24.5 for maximum and minimum temperatures, respectively. It is found that the projection of both the maximum temperature and minimum temperature is increasing. In the case of RCP 4.5 the peak signals of temperature are found from mid to late-mid part of the century and then fall. Additionally, in case of RCP 8.5 the temperature signals are ever rising and as a result never reach their peak. Naturally the temperature increment rate is higher for RCP 8.5 compared to RCP 4.5. The future projection of minimum temperature demonstrated more increase than that of maximum temperature with the exception of region WHI and NCI.

According to RCP 4.5 (8.5) the future projected changes of maximum temperature is 0.39–0.80(0.78–1.21)°C for WHI, 0.50–1.05(0.91–1.56)°C for NWI, 0.47–0.92(0.99–1.38)°C for NCI, 0.41–0.62(0.71–0.89)°C for NEI, 0.48–0.72(0.81–1.08)°C for IPI, 0.26–0.43(0.48–0.69) °C for ECOI and 0.26–0.40(0.33–0.58) °C for WCOI during the period of 2066–95.

From the projection of RCP 4.5 (8.5) the future projected minimum temperature increment is 0.64–1.00(1.09–1.59)°C for WHI, 0.53–0.95(1.10–1.38)°C for NWI, 0.44–0.79(0.83–1.28)°C for NCI, 0.63–0.97(1.11–1.30)°C for NEI, 0.74–1.12(1.20-1.66)°C for IPI, 0.54–0.87(0.82–1.06)°C for ECOI and 0.72–1.04(1.00–1.30)°C for WCOI during the period of 2066–95.

24.3.2.2 Future Rainfall Scenario Over Indian Subcontinent Using Perfect Prognosis Method

Similar to previous section, suitable predictors ae chosen for downscaling of rainfall during four major seasons and provided in Table 24.2.

Downscaled precipitation scenarios created through PP method indicates an increasing trend of rainfall over all seven homogeneous rainfall zones of India. Multi Model Median (MMM) of downscaled projections of rainfall was prepared and results were displayed in Fig. 24.6. It is seen that in early years the rainfall increment rate is larger in case of RCP 4.5 compared to RCP 8.5 but in later years the rainfall increment for RCP 8.5 is found to be greater than that of RCP 4.5. So,

Table 24.1 Selected predictors for downscaling maximum and minimum temperature in 4 seasons

	JF		MAM		JJAS		OND	
	Maximum temperature	Minimum temperature	Maximum temperature	Minimum temperature	Maximum temperature	Minimum temperature	Maximum temperature	Minimum temperature
WHI	t2m + sh850	ta1000 + sh1000	ta1000 + slp	t2m + sh850	t2m + rh1000	t2m + u500	t2m + rh850	t2m + sh850
NWI	t2m + rh850	t2m + sh1000	t2m + rh1000 + v1000	t2m + rh1000	t2m + sh850	ta850 + sh850	t2m + rh1000	t2m + sh1000
NCI	t2m + rh850	t2m + rh850	t2m + rh1000	t2m + slp	t2m + sh850	t2m + sh850	t2m + u850	t2m + sh850
NEI	t2m + u500	t2m + sh1000	t2m + sh500	t2m + sh850	ta500 + sh500	ta1000 + sh1000	t2m + slp	t2m + sh1000
IPI	t2m + u1000	t2m + sh850	t2m + rh850	t2m + sh1000	t2m + sh1000	t2m + sh1000	ta1000 + slp	t2m + sh1000
ECOI	t2m + u1000	t2m + sh850	t2m + sh1000	t2m + sh1000	t2m + z500	t2m + sh1000	t2m + u1000	t2m + sh1000
WCOI	t2m + sh1000	t2m + sh1000	t2m + sh1000	t2m + sh1000	t2m + rh850	t2m + sh1000	t2m + u1000	t2m + sh1000

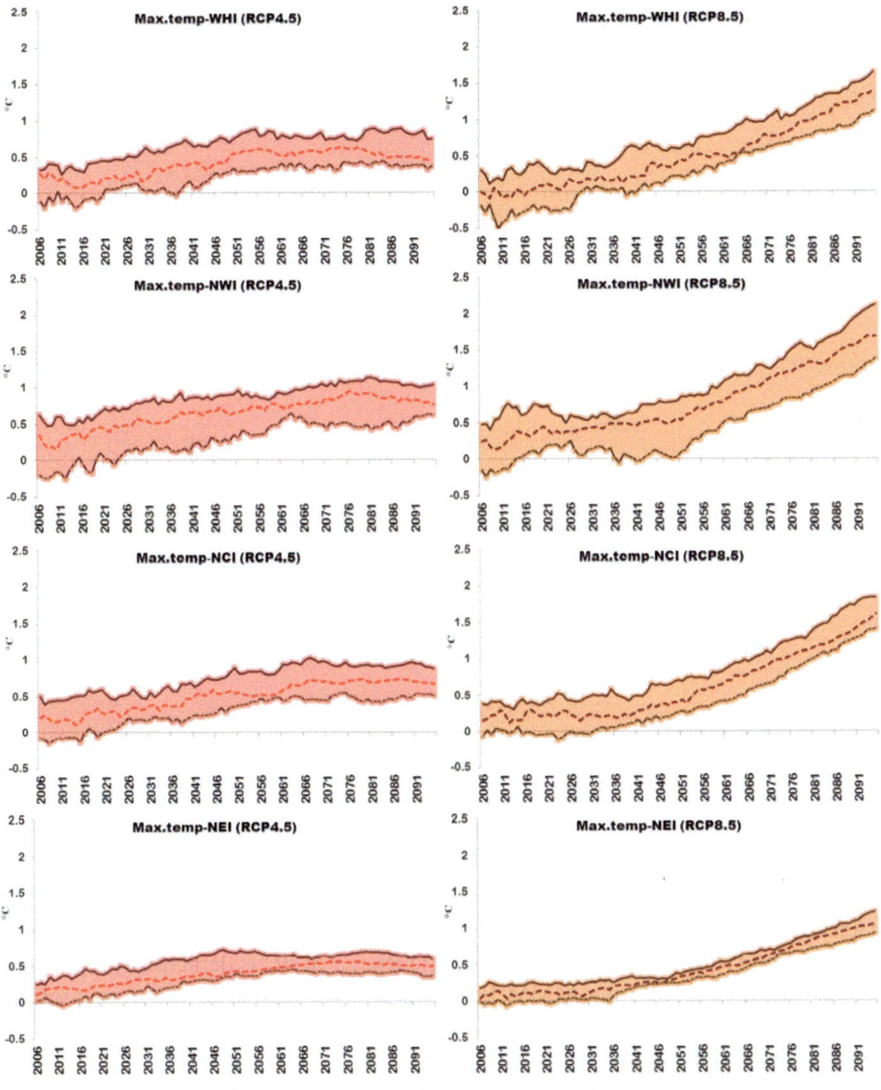

Fig. 24.4 Downscaled GCM changes in annual maximum temperature over all seven homogeneous zones

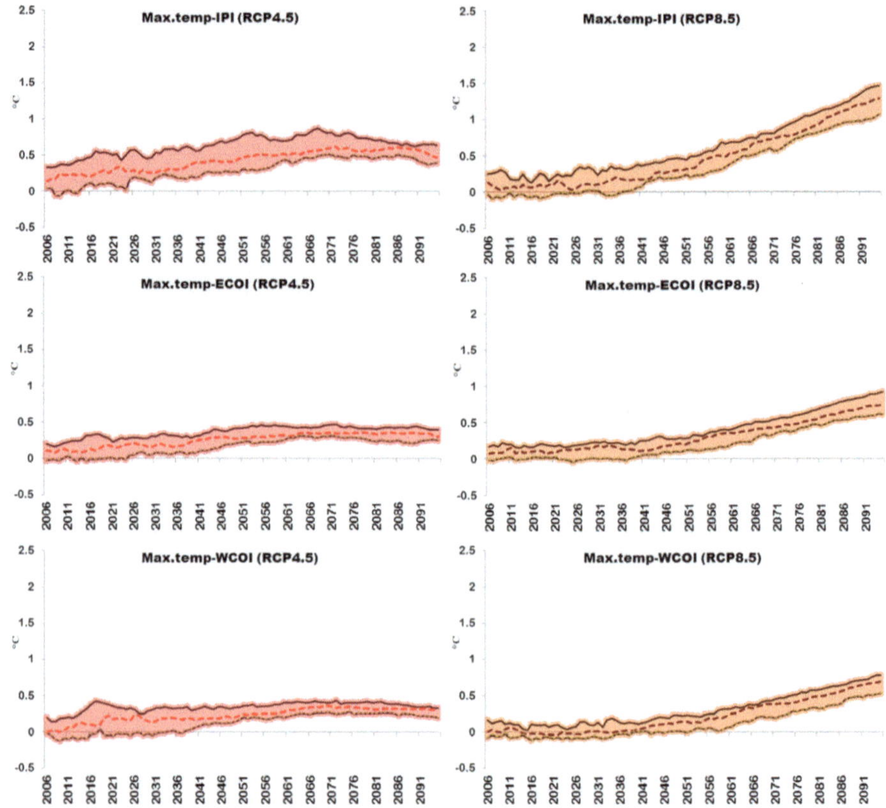

Fig. 24.4 (continued)

naturally the average trend of rainfall in the 2006–95-year span is greater for RCP 8.5 compared to RCP 4.5. The Inter Quartile Range (IQR) is depicted by the shaded region which is around the MMM line. It is a visualization of the most probable future projection as well as a projection of uncertainty. Based on IQR values, all regions denoted the increment of rainfall.

According to the projection based on RCP 8.5, the future zone-wise projected changes of rainfall are 10.7–23.0% for NMI, 9.2–24.9% for NWI, 13.1–24.8% for NCI, 11.8–18.8% for NEI, 6.1–22.8% for EPI, 12.1–26.9% for WPI and 9.4–18.8% for SPI during the period of 2066–95. Based on IQR values, the increment of rainfall was found only over regions of NMI (which is 5.7–17.4%), NCI (which is 1.5–18.6%) and NEI (which is 8.5–23.3%) according to RCP 4.5 and a combination of increment and decrement were observed for the other zones during the period of 2066–95. According to RCP 4.5, the future rainfall changes showed a decreasing trend over NWI, EPI, WPI and SPI by (−) 3.4–18.0%, (−) 3.2–13.8%, (−) 0.4–19.8% and (−) 3.7–12.5% respectively during 2066–95.

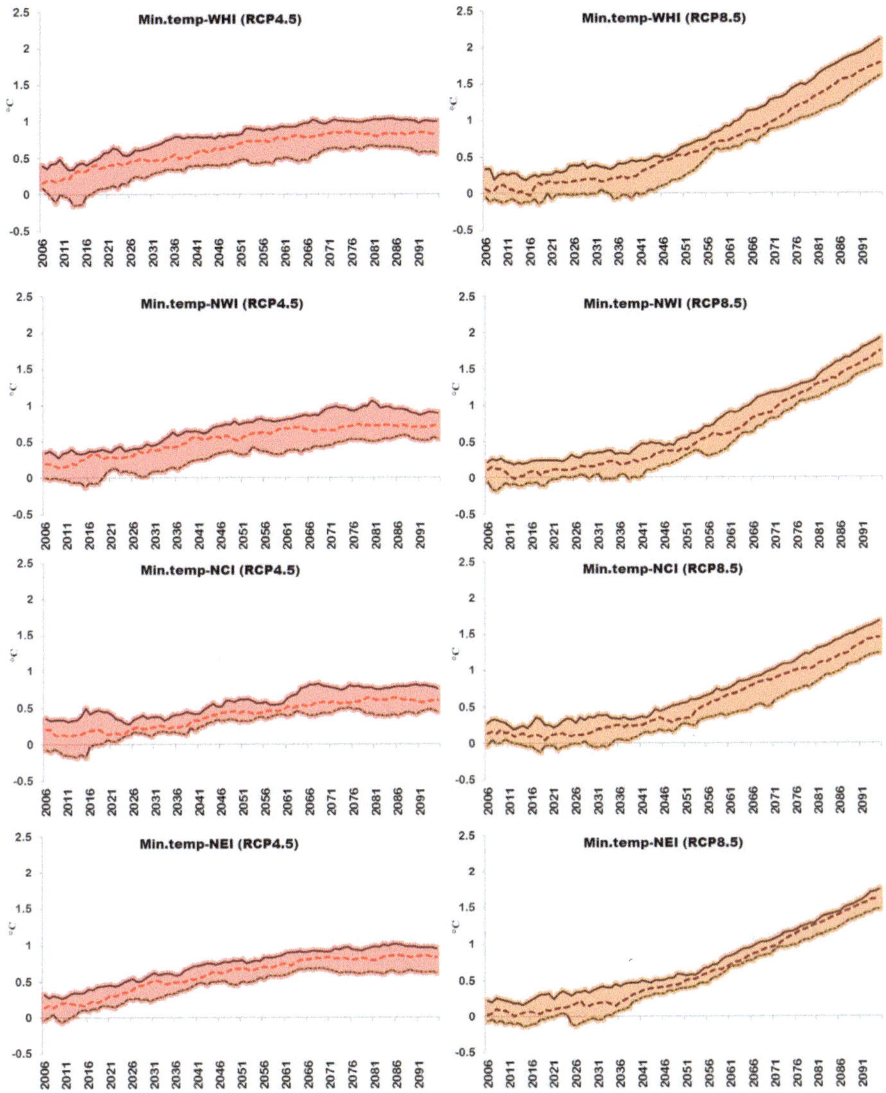

Fig. 24.5 Downscaled GCM changes in annual minimum temperature over all seven homogeneous zones

24.4 Climate Change Over Western Himalayan Region (WHR)

The Himalayas, are a mountain range in South-Asia separating the plains of the Indian subcontinent from the Tibetan Plateau. The width of the Himalayas is about 300 km, stretching through ~2500 km along the line of northern border of India.

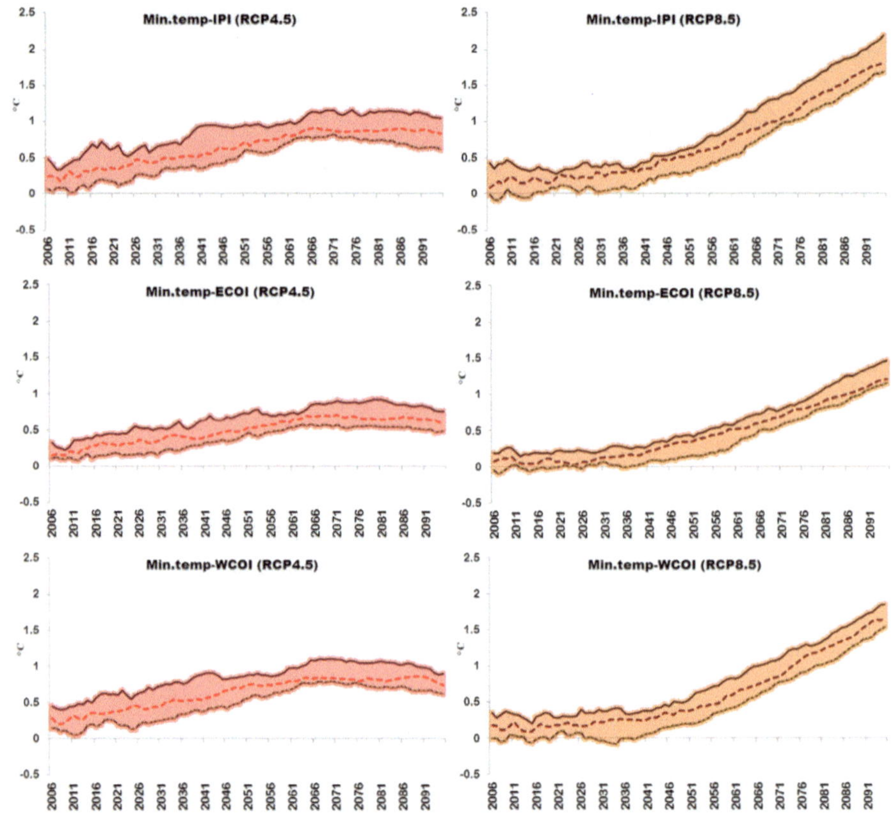

Fig. 24.5 (continued)

Table 24.2 Selected predictors used for downscaling rainfall for 4 seasons

	JF	MAM	JJAS	OND
NMI	v1000 + prw	v850 + rh850	sh850	v850 + prw
NWI	rh1000 + ta1000	rh1000 + slp	u850 + prw	u850 + prw
NCI	u1000 + rh1000	v1000 + sh1000	u850 + prw	rh500 + slp
NEI	v850 + sh850	v500 + prw	u850 + prw	sh500
EPI	z850 + prw	rh500 + ta500	sh850	v850 + ta850
WPI	v1000 + ta1000	v850 + slp	slp + prw	z850 + prw
SPI	z500 + prw	u850 + sh850	v1000 + prw	u850 + prw

The position and dimension of Himalayas makes it possible to intercept both the Arabian Sea and Bay of Bengal branch of the Indian summer monsoon bringing precipitation in the form of rainfall and snow over India (Ramisch et al., 2016). It acts as a persistent barrier to the cold continental air masses that come from the arid

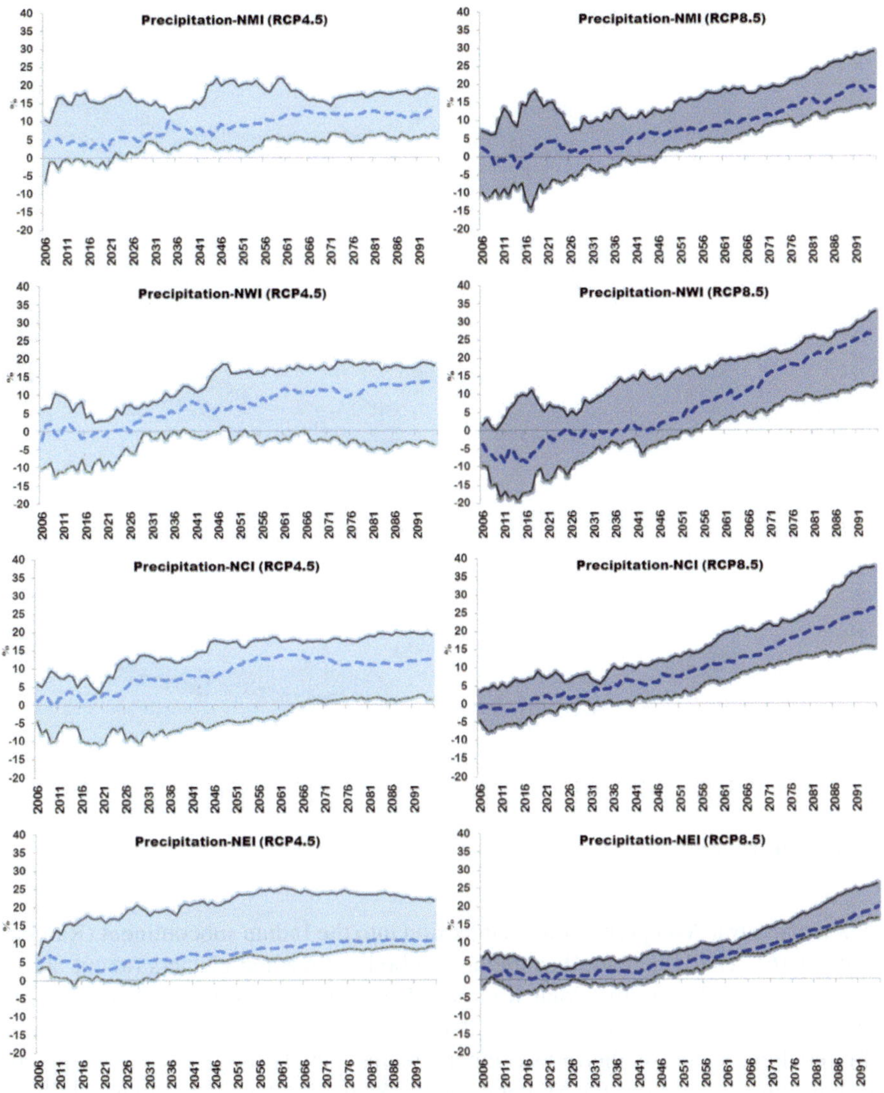

Fig. 24.6 Downscaled GCM projected annual rainfall change over all seven homogeneous rainfall zones of India

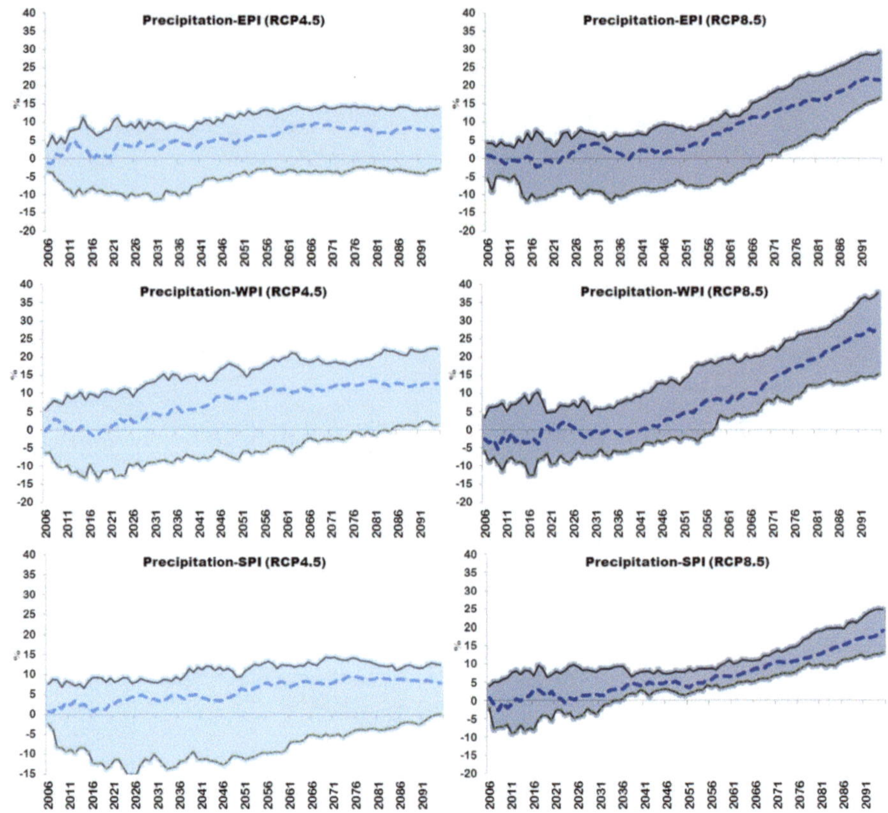

Fig. 24.6 (continued)

region of central Asia preventing its intrusion into the Indian subcontinent (Ramisch et al., 2016). The Himalayas has the largest body of ice and glaciers (about 30,000 km^2 in size) excluding the Polar Regions. That is why, it is also called the 'Third Pole'. The water present in the glaciers that exist in the Himalayas nourishes a number of rivers flowing across the Indo-Gangetic plains namely the Indus, the Ganga, the Brahmaputra, the Yellow, the Mekong and the Yang-Tze. Himalayas play significant role in the Indian subcontinent, on the other hand, tropical, Mediterranean and polar wind circulations influence the monsoon circulation over the Asia (Bhutiyani et al., 2009). The Himalayas are categorized under those mountain ecosystems which are very sensitive to the harmful effects of global climate change (Pant et al. 2018).

Himalayas are categorized into three sub-regions, viz. Central Himalayas, Western Himalayas, and Eastern Himalayas. A significant part of the WHR that outstretches over the Indian subcontinent wraps three states of India which are, Jammu and Kashmir (J & K), Uttarakhand, and Himachal Pradesh. In the current analysis, we have concentrated on the Western Himalaya Region (WHR). Before comprehending climate change in the WHR, it is essential to apprehend some conceivable indicators that can be treated as an indication of climate change.

24.4.1 Climate Change Indicators Over the Western Himalayan Region

The signs of climate change in the WHR can be pictured from climate change indicators such as, rainfall change, temperature change, retreating of glaciers, glacial lake outburst floods, and cloud burst. These are also the typical set of indicators usually documented by the Intergovernmental Panel on Climate Change (IPCC) to characterize the climate change information over the regional mountainous hotspots of the earth (IPCC, 2021). The foothills of WHR are rich in bio and geo-diversity. There are other climate change indicators connected to the ecosystem, flora, and fauna of the WHR with particular emphasis on livelihoods such as rise in invasive species, early flowering of wild as well as farmed plants, mainly horticulture crops, decline in crop yield, particularly in conventional crops such as potatoes, peas, beans, ginger and yam (Joshi & Joshi, 2011). A piece of detailed information on climate change over the WHR is portrayed in the subsequent section. Before comprehending the climate change over the WHR, the reader must understand the meteorology of the WHR.

24.4.2 Meteorology of Western Himalayan Region

Several natural hazards, localized weather events, large scale atmospheric circulation, geographical features, and feedback from different meteorological factors are impacting the weather and climate of WHR adversely. Notable meso and micro-scale circulation occurring in different seasons are created by the irregular characteristics of the Himalayan region. The southwest monsoon event brings sufficient rainfall to the Indian subcontinent from June to September. Additionally, Western Disturbances (WDs) develop a winter rainfall regime over the WHR during the months of December to March.

Several researches done over different south Asian subregions have revealed that Black Carbon (BC) (Meehl et al., 2008) and Eurasian snow-monsoon interactions (Peings & Douville, 2010) are the two components affecting the monsoon scenario of WHR. Darkening of the snow-covered area due to Black Carbone depositing is one of the main influencers contributing to the rapid melting of snow in the Himalayan region (Nair et al., 2013; Qian et al., 2011).

Warming in the winter season is a salient feature of the WHR. High elevation locations of WHR experiences more winter warming compared to the foothills. The 'Elevation-Dependent Warming' (EDW) which is the warming happens in the high elevation mountains is caused by phenomenon such as, aerosols chemistry, snow albedo and surface-based feedbacks, surface heat loss and temperature change, water vapor changes and latent heat release, surface water vapor and radiative flux changes (Pepin et al., 2015; Palazzi et al., 2017). Melting of the Himalayan glacier is accelerating because of the increased warming. As a result, biodiversity is suffering great loss (Mandal et al., 2014).

Weather over this region is quite uncertain due to unexpected occurrences of some of the major hazards—Glacial Lake Outburst Flood (GLOF), thunderstorm, cloudburst, and forest fire. A main cause of forest fire is anthropogenic and threatens biodiversity as well as, air pollution over WHR.

24.4.3 Status of Climate Change Over the WHR from Different Climate Change Indicators

24.4.3.1 Status of Observed Rainfall Over the WHR

How the observed rainfall has changed its characterestic over different local latitudinal areas of WHR have been investigated by several researchers recently (Meher et al. 2017; Gautam et al., 2013). In general, there are two principal seasons in which the WHR received rainfall.

First, is the monsoon season where the Western Himalyas Region received almost 75–80% of the entire annual rains during the months of June–September (Meher et al. 2017, 2018). Medina et al. (2010) affirmed that in the monsoon months, the low-level water vapor-rich air from the Arabian ocean crosses the Thar Desert, enriching the buoyancy of air transportation by adding sensible heat flux causing more rain. In the foothills of the western Himalayas, the air passage acquires extra moisture from the soil if there was a prior occurrence of precipitation. In the next phase, the elevated layers of warm and moisture-less continental flow at low-level caps the air passage and converts it into an unstable stream. This unstable passage is orographically raised to saturation and precipitates over the foothills of western Himalayas.

Second, the North-eastern wind system is commonly seen over the Himalayas during the winter months (Meher et al., 2018; Wang, 2006) causing some rain over the WHR. The North-east monsoon winds form due to the pressure fluctuations between the neighbouring warm oceans and cold Tibetan Plateau. As a result, the western Himalayas acquire ~10% of annual precipitation (Meher et al. 2017, 2018) during winter season. In addition to North-east monsoon circulation, the western Himalayas also encounters western disturbances (WD) in the winter seasons. WDs are cyclonic circulations in the mid and lower troposphere which originate over the Mediterranean and Caspian seas. They have a tendency to move eastwards in the direction of North-WHR with the westerly jet streams (Meher et al. 2017). The movement of western disturbances diminishes along its route from J & K to Uttarakhand and considerably declines further towards the far east (Meher et al., 2018). This decreasing tendency of movement of WDs resulting in less rainfall and number of rainy days is experienced over Uttarakhand than Himachal Pradesh and J & K. In the winter season, the rainy days and the rainfall increases with altitude up to ~1356 m and then decreases (Meher et al., 2018). Singh and Kumar (1997) also reported similar results.

The annual and monsoon season rainfall has shown significantly decreasing trend in the last century namely the period of 1901–2005 (Akhter et al. 2016; Singh & Mal, 2014; Meher et al. 2017) over the WHR of India. It is to be noted that monsoon and annual rainfall over the Western Himalayas has significantly (at 5% levels) decreased by 283.4 mm/100 years and 289.3 mm/100 years, respectively.

The frequency of western disturbances may has a direct impact on various rainfall parameters, viz., winter rainy days, total winter rainfall, cold days, and wet winter days. The change of winter rainfall in various sites of the WHR had shown both an inclining and declining signal (Meher et al., 2018; Das and Meher 2019). The likely reason for the decline of winter rainfall and frequency of cold days in the western Himalayas is the declining trend of frequency of western disturbances (Dash & Mamgain, 2011; Kumar et al., 2015a, b) and abnormal warming in the winter (Bhutiyani et al., 2007), respectively.

24.4.3.2 Status of Observed Temperature Over the WHR

Das et al. (2017) observed that the average monthly temperature over the WHR varies from 5.1 °C in January to 22.1 °C in July, based on temperature statistics from 16 meteorological sites from 1969 to 2009. The temperature range (variation of maximum temperature to minimum temperature) in the winter is higher compare to monsoon season. The snowpack over the WHR in elevated locations, as well as its impact on the local atmosphere, have a major effect on the temperature. The removal of snowpack reduces surface albedo and increases solar radiation absorption, resulting in increased warmth.

Numerous studies (Das et al. 2017; Shrestha et al., 1999; Stocker et al., 2013; Singh et al., 2010) revealed that winter temperature in the Himalayas was higher (1.41 °C) than the global mean (0.85 °C during the span of 1880 to 2012) temperature. During the period of 1969 to 2009, mean monsoon season temperature has shown a notable incline trend over the western Himalayas. During 1969–2009, in comparison to the warming in the monsoon (0.51 °C), warming in the winter (1.41 °C) was greater (Das et al., 2018). Warming rates were noticeably smaller in the other seasons than they were in the winter. For the 1969–2009 time frame, the heating rates were 0.31 °C, 0.35 °C, and 0.51 °C in post-monsoon, pre-monsoon, and monsoon seasons, respectively. In general, the WHR has warmed annually by 0.84 °C. As the warming rate is higher in winter season over WHR, the snows are unable to pile up and deposit over the mountains. On the other hand, relatively warm pre-monsoon accelerates glacial ice melting, resulting in a change in hydrological systems. The change in hydrological cycles of the WHR can alter the life and livelihood of the people residing in the IGP (Indo-Gangetic Plain) as well as the entire region of South Asia. It was discovered that the temperature change mutation point began after 1992 and continued further over WHR.

Rising and falling patterns of warm nights at several locations across the WHR during 1960 to 2000 were also reported by Panday et al. (2015). The seasonal and locational diversity of warming rates over WHR was estimated by Fowler and Archer (2006) where they reported a significant drop in average annual temperature of the upper Indus basin; however, substantial increase in winter average and maximum temperature was also reported. Due to higher warming over WHR, a clear phase change of water was noticed. As for example, the snow-dominated systems will shift to rainfall-dominated systems as winter temperatures rise in the highlands (Adam et al., 2009; Nolin & Daly, 2006).

24.4.4 Status of Natural and Anthropogenic Hazards Over the WHR

Normally, the WHR is prone to the occurrence of several natural and anthropogenic hazards like Glacial Lake Outburst Flood (GLOF), cloud burst, thunder storm, flood and forest fire. A brief description of the impact of each of these hazards over the WHR is given below.

The GLOF is a release of meltwater from an unstable natural dam formed from a glacial retreat causing catastrophic flooding downstream which has a huge short-term impact on buildings, environmental assets, and people's lives. Melted water flow from receding glaciers usually forms a reservoir or lake-like system near its surroundings, especially between the receding glacier and the frontal moraine (Ives et al., 2010), termed as a 'Glacial Lake.' The freshly formed glacial lakes are unstable and may be directly triggered by an earthquake or ice avalanche, resulting in a 'flash flood' type condition, called GLOF. The regular occurrences of such events over the WHR are indicated by an elevation of temperature trend driving glacier retreat and extreme hydro-meteorological conditions (Allen et al., 2015). Glacial lakes form whenever a section of the bed that has been over-deepened or significantly depressed as a result of the glacier's abrasive activity is revealed to the receding glacier. Allen et al. (2015) found that specific regions in Himachal Pradesh, such as the tehsils of Hangrang, Pangi, Dodra Kwar, Chamba, Multhan, Lahaul, and Aut are highly susceptible to the GLOF.

Now-a-days, the occurrence of cloudbursts is a regular phenomenon over the WHR. The magnitude and severity of casualty due to cloudbursts make it a serious natural hazard over the western Himalayas. According to meteorology research, there are almost 37 cloudburst incidences were observed in Himachal Pradesh, (Dimri et al., 2017; Prasad et al., 2016). Similarly, the research has described 25 incidences of cloudburst occurrences in Uttarakhand from 1970 to 2016 and 8 similar occurrences in J & K from 2010 to 2015 (Bharti, 2015). Surprisingly, there have been seven consecutive cloudbursts during the monsoon season in 2018 that caused severe damage to properties and loss of lives in Uttarakhand (SANDRP, 2018).

The full explanation of cloudburst and their associated anthropogenic and natural implications is highlighted in the following section.

Cloudburst accompanied with flash floods is among the world's most dangerous environmental disasters, wreaking havoc on climate, hydrology, local population, economy, and human life. The complicated mountain range over WHR provides a suitable environment for creating isolated cyclonic storms over the constricted valley, culminating in Cloudburst (Gupta et al., 2012). As per Dimri et al. (2017), the duration of cloudburst episode is very small but it has a wide range of almost 20–30 km^2 causing a highly intensified (>100 mm per hour) rainfall. The cloudburst incident that happened on the 5th and 6th of August 2010 throughout the Leh region was investigated by Kumar et al. (2014) using computer-based simulation model. It was revealed that over the Tibetan Plateau a storm was formed along the vertical edge due to a 500-hPa wind. The lower water vapor holding air comes from the Arabian Sea as well as the Bay of Bengal, gain enough strength to develop along the Himalayan border. Resulting a strong storm rotational winds known as mesoscale squall line with a convective line at the front, a stratiform zone at the back, and an inflow jet in the middle.

The WHR is a prone to exercise the Earth's most severe thunderstorms. Albrecht et al. (2016) revealed that large-scale systems related to the thunderstorm events could render the western disturbance (WDs) to run across J & K and north region of Pakistan.

The probability of occurrence of most frequent thunderstorm from June to August months during 1986–1990 period in North-western part of India was reported by Sahu (1996). It is revealed that some places like Bhadarwah, Batote, Jammu, Chamba, Kathua, and Udhampur on the Western Himalayas, the thunderstorm occur throughout the whole year while thunderstorms are rare in places like Kargil and Leh. Another study by Tyagi (2007) investigated that the states of Uttarakhand and Himachal Pradesh and J & K experiences a greater number of thunderstorms. As for example, places like—Qazigund, Bhuntar, Dehradun, and Mukteshwar acquired 60–70 thunderstorms/year while sites like Udhampur and Jammu acquired ~131 and ~95 thunderstorms/year during 1951–1980.

In recent years, many devastating floods like the Jammu & Kashmir floods (2014), the Uttarakhand floods (2013), the Western Himalayan flash floods (2012), and Leh-Ladakh floods (2010) had occurred over the WHR because of the unpredictability of local weather (Ghawana & Minoru, 2015). The water vapor transit from the neighbouring ocean into the densely raining clouds was credited to Uttarakhand's rapid and catastrophic inundation in 2013, resulting in a tremendous growth of buoyancy, significant vertical accelerations, and intense precipitation occurrences.

Forest fires are largely caused by humans (Dobriyal & Bijalwan, 2017). Approximately 28 percent of western Himalayas (Uttarakhand, Himachal Pradesh, and J & K) is covered with forest (ISFR, 2017). From 1910 to 2016, there have been 21 major forest fires in Uttarakhand, the majority of which were sparked by human intervention, such as shifting agriculture and deliberate ignition (Dobriyal & Bijalwan, 2017; Satendra et al. 2014). In the WHR, a forest fire is caused by a combination of factors. The sun, for example, begins to migrate towards the southern hemisphere on June

21. As a result, the solar output obtained by north-facing mountainside in Western Himalayas is low, as compared to the solar output obtained by the south-facing mountainsides. As a result, the vegetation cover in the 'southern' and the 'south-western' pine forests experiences harsh climate patterns and is vulnerable to windfall and forest fire (Kanga et al., 2017; Kumar et al., 2015a, 2015b). According to Dobriyal and Bijalwan (2016), potential time to trigger a forest fire in the WHR falls during the pre-monsoon seasons, when there is a shortage of moisture due to inadequate precipitation, and during the winter season, if there is a dry spell accompanied by leaf fall phase (from February to May). Forest fires have severe consequences in the WHR environment, including depletion of soil nutrition and fertility, soil degradation, water resource deterioration, and ecological threats. Forest fires in this region, according to Banerjee and Chand (1981), cause a rapid drop in total organic matter from 33 to 50 percent. Forest fires can also be attributed to the crop failure situation on a regional scale (ISFR, 2017; Mehta, 1996). Forest fires in India during the twentieth century, notably in 1921, 1930, 1942, 1970, and 1995, were linked to prolonged drought spells, according to Mehta (1996). Forest fires cause a rise in pH (potential of hydrogen) within the soil, as well as a brief loss in organic carbon and nitrogen, a decrease in biomass and yield, and excessive rainfall.

24.4.5 Status of Climate Change Indicators Linked to the Ecosystem, Flora and Fauna of the WHR

A quick checklist of indicators of climate change framed by Joshi and Joshi (2011) over the WHR, specifically in the Uttarakhand sub-division, documented the predominance of invasive exotic species in most forest zones in the area. The existence of Eupatorium sp. (locally called Kaali ghaas and Cancer ghaas) and *L. camara* have expanded since 2000–2005. Earlier, Eupatorium sp. was confined to a height of 1000 m above mean sea level. Now, it is seen in both open and degraded forests and some of the well-stocked forests up to 1800 m height above mean sea level. Increasing dry spell in winter makes favorable conditions for the *L. camara* species to invade the higher altitude. Their study also reported earlier flowering in Rhododendron by 1–2 months and ~3 inches reduction in its flower size. Berries of Kaphal tree (*Myrica sapinda*, *Myrica nagi*) and Hinsalu (*Rubus ellipticus*) ripen early but experiences reduction in total yield. Productivity of several other traditional tree fodder species on which the residents are heavily dependent is reducing.

The effect of climate change is considerably detectable in agriculture in the form of disruption in the crop cycle, lowering the yield and dimensions of fruits and vegetables. Inconsistent rainfall routines have led to a transition in the conventional crop patterns and the predicted yields of major crops and vegetables. For instance, some areas of Uttarakhand registered late sowing of non-irrigated varieties of wheat due to extended dry spells in the winter. Not only wheat but also several other conventional winter crops like chana, masoor, etc., are noted to have exhibited a

Fig. 24.7 Administrative headquarters of 19 blocks of Indian Sundarbans

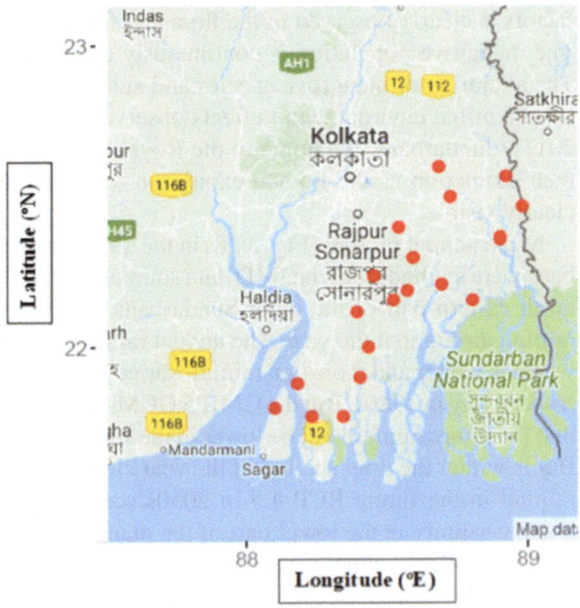

substantial decline in recent years due to sparse winter rainfall-induced infertility in the soil and snowless winter. Few villages like Bajoon and Adhora, which are at an altitude of 1550 m from MSL, have not witnessed snowfall since the 1990s. The rainfall deficit in the early Kharif season (June–July) and heavy rainfall in the late Kharif season (September–October) reduced standing crop yields in the fields.

24.5 Sundarbans in the Wink of Swipe

Sundarbans is one of the climatic hotspots in the present arena and is home to several mangrove species (Mandal & Naskar, 2008). It is a mangrove area located in between the river delta of Ganga, Brahmaputra, and Padma. This typical climatic hotspot spreads from the eastern bank of the Hoogly River in West Bengal to the Baleshwer river in Bangladesh. The Indian Sundarbans comprises 6 blocks of North and 13 blocks of South 24 Parganas along with a reserved forest area of a total of 4260 km^2 (Fig. 24.7).

The continuous hitting effect of tidal surges along the coastal shoreline is the reason for coastal area decay. The concentration of sediments is high in the region. The erosion process makes the area vulnerable to sea level rise. The high silt deposition in the mouth of the Karis creates a blockage of water flow within the inner premises of Sundarbans, resulting in a hike in salinity. The overall water quality of the mangrove region is typically of less BOD and high EC which hinders the aquatic ecosystem and biodiversity. The imbalance created by such natural and anthropogenic factors is clearly observed in the flora and fauna of Sundarbans (Raha et al., 2012). The mangrove population is continuously changing as influenced by such factors. The alteration of mangrove species and subsequent threat to terrestrial biodiversity are the prime environmental effects observed under changing climate (Mitra et al., 2011). Sundarbans, the home to the Royal Bengal Tiger is also facing the threat of their extinction as the human expansion to the northern part of the forest area is clearly visible.

Mean annual rainfall (Fig. 24.8) in the 19 administrative blocks of Indian Sundarbans were studied from the IMD data sources (1965–1995). The studies revealed that north eastern part of the inner Sundarbans area received relatively less amount of rainfall throughout the year. The annual rainfall varied between 1718 and 1800 mm, whereas winter and monsoon rainfall varied between 51–54 mm and 1290–1351 mm respectively in 1980s. But the CMIP5 GCMs ensembled forecast in RCP 4.5 revealed that the future rainfall will be lower in the south western block of Indian Sundarbans. But in winter the drier section of the year clearly indicates a higher decreased winter rainfall in the future RCP 4.5 in 2050s scenario of -26.5 to -40%. This indicates that the salinity in the lower part of the mangrove region will significantly increase. CMIP5 model simulations indicated an increase of 22.4 to 31.3% annual rainfall under RCP 8.5 scenario. Whereas winter rainfall is greatly reduced by 27.5 to 36.6% under extreme emission scenarios. Thus, ecological imbalances occur and many flora and fauna species come to extinction. In other words, the future winter rainfall indicates decreasing pattern which is a major concern in this region. Time series plots (1901–2095) for four seasons (pre-monsoon, monsoon, post-monsoon and winter) of two extreme points of Indian Sundarbans are presented in the Fig. 24.9. Pramanik (2016) mentioned that seasonal sea-level changes measured using tidal gauge observation of Diamond Harbour and Haldia that was recorded for 1952–2013 indicate an increase of 2.59 ± 1 mm/ year in sea-level rise. The health of the mangrove forest is studied through modern imageries obtained from satellites. The area under the wild habitat of mangroves is continuously changing under the context of climate change (Alongi, 2015).

In Sundarbans, a study revealed that the temperature is gradually increasing in the past and predicted to be increasing over the present century. In Sundarbans region the historical temperature increased about 0.33 to 0.36 °C (during 1969–2009). The next 90-year future projection of annual temperature indicated that, there will be an annual temperature increment of 1.8-2°C according to RCP 4.5 and 3.9–4.2 °C according to RCP 8.5 (Table 24.3). It clearly depicts that if the industrial emission contributing towards global warming is continued the temperature increase by 5 °C is evident in the region. Thus, the increase of temperature along with less winter rainfall causes the area more vulnerable to promote biodiversity and causes the significant losses in human livelihood.

16% (11–70 species) of the mangrove species are marked as the species under threat of extinction. Several studies suggested 30–40% of the low-lying area of the region along with the complete extinction of mangrove forest is inevitable within the next 100 years if the mangrove population declining rate prevails (Duke et al., 2007; IPCC 2001). The increased annual temperature along with lower winter rainfall

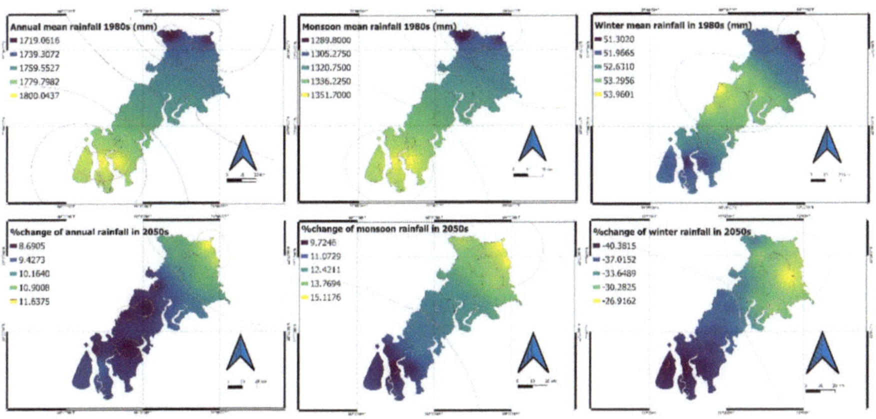

Fig. 24.8 Past and future rainfall scenario of Sundarbans region

gives rise to the salinity, turbidity, EC of the aquatic ecosystem. The high silt load thus prevents the further influx of oceanic water, which creates belts of blocked salty water basins along the southern part of the region. This gradual change in the physical environment is shaping the overall health of the biosphere. Moreover, a significant increase in sea level rise can completely swipe the entire Sundarbans along with its huge biodiversity, which is now only a matter of time.

24.6 Conclusions

Climate change is one of the most discussed topics in present days. The effect of global warming manipulates the dynamics of the atmosphere and changes rainfall and temperature scenarios. Identification of the areas which are disaster-prone and vulnerable to climate change and is the prime objective of the recent climate scientists. Hence, the present study analyzed different climate change scenarios which are possible to occur under different warming scenarios to assess the vulnerabilities of different climate zones as well as over two climate hotspots of India. Downscaled projections using MOS method indicate 3–18% (3–30%) increment for RCP 4.5 (8.5) in the later part of this century (2036–2095) over different homogeneous zones of India. PP method also shows up to 23.3% (26.9%) increase for RCP 4.5 (8.5) during 2066–95. RCP 8.5 projected greater rainfall change in future compared to its 4.5 counterpart in all seven regions. Temperature projection using MOS reveals 1.46 to 2.90 °C (1.46 to 2.90 °C) warming by RCP 4.5 (8.5) during 2066–95 over various zones. Maximum temperature is projected to increase by 0.26 to 1.05 °C (0.33 to 1.56 °C) while minimum temperature would rise by 0.44 to 1.12 °C (0.82 to 1.66 °C) as per RCP 4.5 (8.5) during this period.

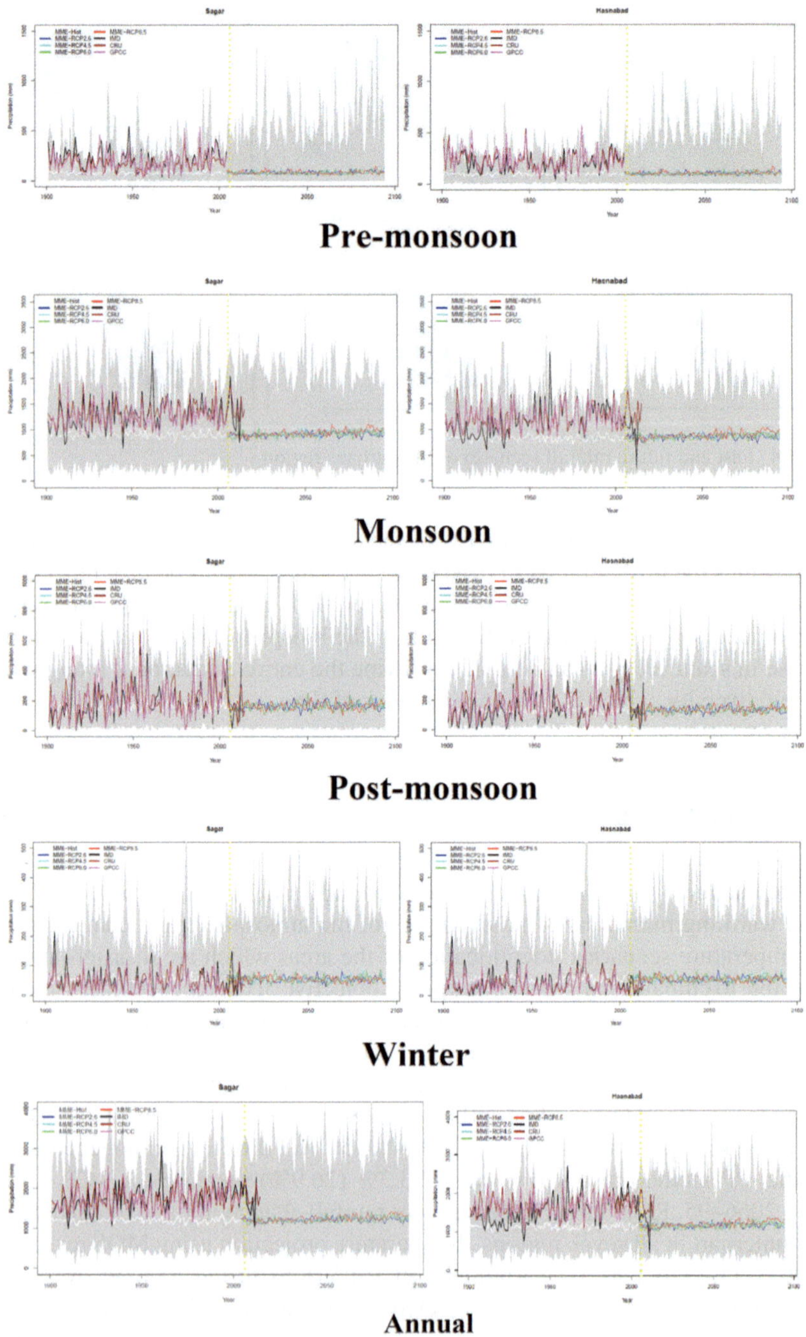

Fig. 24.9 Temporal variation of rainfall as simulated by 20 Models and IMD, CRU and GPCC data during 1901–2095 for Sagar and Hasnabad blocks

Table 24.3 Annual temperature trend of historical and future time frame

	1969–2009 (°C/41 year)	RCP 4.5 (°C/90 year)	RCP 8.5 (°C/90 year)
Sagar	0.33	1.8	3.9
Hasnabad	0.36	2	4.2

Mean temperature of WHR increased significantly for all four seasons during 1969–2009. The warming rates are found to be 0.31 °C, 0.35 °C, and 0.51 °C per 40 years for post-monsoon, pre-monsoon, and monsoon seasons respectively during this period. Comparatively higher warming (1.41 °C/ 40 years) is noticed over WHR during winter which also exceeds global average trend by a large margin. Snows are unable to pile up and deposit over the mountains due to increasing winter temperature. On the other hand, pre-monsoon warming accelerates glacial ice melting, resulting in a change of hydrological systems. Decreasing trends in monsoon (283.4 mm/100 years) and annual rainfall (289.3 mm/100 years) is found over WHR during 1901–2005. Winter rainfall is also found to be declining over many locations of WHR due to the declining trend of frequency of western disturbances and abnormal warming in the winter. During 1970–2016, there were 25 cloudburst occurrences in Uttarakhand as well as 8 events in J & K from 2010 to 2015. 21 major forest fires were reported in Uttarakhand during 1910–2016, the majority of which were sparked by human intervention, such as shifting agriculture and deliberate ignition. Forest fires have severe consequences in the WHR environment, including depletion of soil nutrition and fertility, soil degradation, water resource deterioration, and ecological threats, rapid drop (33–50%) in total organic matter. As a result of changing climate agriculture is affected through the disruption in the crop cycle, lowering the yield and dimensions of fruits and vegetables. A substantial decline in the yield of winter crops like chana, masoor, etc. is found in recent years due to sparse winter rainfall-induced infertility in the soil and snowless winter. Inconsistent rainfall routines have led to a transition in the conventional crop patterns. Late sowing of non-irrigated varieties of wheat is practiced in some areas of Uttarakhand due to extended dry spells in the winter.

The historical annual temperature during 1969–2009 increased by about 0.33 to 0.36 °C over the Sundarbans region. The future projection by RCP 4.5 (8.5) shows increment of 1.8–2 °C (3.9–4.2 °C) in annual temperature. Annual rainfall of this region is projected to increase by 8.5 to 11.5% (22.4 to 31.3%) under RCP 4.5 (8.5) scenario during 2065–2095 which may significantly contribute to sea-level-rise. However, decreased winter rainfall of 27.5 to 36.6% as shown by RCP 8.5 may cause higher salinity in the mangrove region hampering the ecosystem.

References

Adam, J. C., Hamlet, A. F., & Lettenmaier, D. P. (2009). Implications of global climate change for snowmelt hydrology in the twenty-first century. *Hydrological Processes, 23*, 962–972.

Akhter, J., Das, L., & Deb, A. (2017). CMIP5 ensemble-based spatial rainfall projection over homogeneous zones of India. *Climate Dynamics, 49*(5–6), 1885–1916. https://doi.org/10.1007/s00382-016-3409-8

Albrecht, R. I., Goodman, S. J., Buechler, D. E., Blakeslee, R. J., & Christian, H. J. (2016). Where are the lightning hotspots on Earth? *Bulletin of the American Meteorological Society, 97*(11), 2051–2068.

Allen, S. K., Rastner, P., Arora, M., Huggel, C., & Stoffel, M. (2015). Lake outburst and debris flow disaster at Kedarnath, June 2013: hydrometeorological triggering and topographic predisposition. *Landslides, 13*(6), 1479–1491. https://doi.org/10.1007/s10346-015-0584-3

Alongi, D. M. (2015). The impact of climate change on mangrove forests, pp. 32–35. https://doi.org/10.1007/s40641-015-0002-x

Banerjee, S. P., & Chand, S. (1981). Physico chemical properties and moisture characteristics of soils as influenced by forest fire. *Indian Forester, 107*(3), 178–182.

Benestad, R. E. (2002). Empirically downscaled temperature scenarios for northern Europe based on a multi-model ensemble. *Climate Research, 21*(2), 105–125.

Bharti, V. (2015). Investigation of extreme rainfall events over the northwest Himalaya Region using satellite data. MSc Thesis, Faculty of Geoinformation Science and Earth Observation, University of Twente, Enschede, The Netherlands.

Bhutiyani, M. R., Kale, V. S., & Pawar, N. J. (2007). Long-term trends in maximum, minimum and mean annual air temperatures across the northwestern Himalaya during the twentieth century. *Climate Change, 85*, 159–177.

Bhutiyani, M. R., Kale, V. S., & Pawar, N. J. (2009). Climate change and the precipitation variations in the northwestern Himalaya: 1866–2006. *International Journal of Climatology, 30*(4), 535–548.

Block, P. J., Souza Filho, F. A., Sun, L., & Kwon, H. H. (2009). A streamflow forecasting framework using multiple climate and hydrological models 1. *JAWRA Journal of the American Water Resources Association, 45*(4), 828–843.

Das, L., & Akhter, J. (2019). How well are the downscaled CMIP5 models able to reproduce the monsoon precipitation over seven homogeneous zones of India? *International Journal of Climatology, 39*(7), 3323–3333.

Das, L., Dutta, M., Mezghani, A., & Benestad, R. E. (2017). Use of observed temperature statistics in ranking CMIP5 model performance over the Western Himalayan Region of India. *International Journal of Climatology, 38*(2), 554–570. Portico. https://doi.org/10.1002/joc.5193

Das, L., Dutta, M., Mezghani, A., & Benestad, R. E. (2018). Use of observed temperature statistics in ranking CMIP5 model performance over the Western Himalayan Region of India. *International Journal of Climatology, 38*(2), 554–570.

Dash, S. K., & Mamgain, A. (2011). Changes in the frequency of different categories of temperature extremes in India. *Journal of Applied Meteorology and Climatology, 50*(9), 1842–1858.

Déqué, M. (2007). Frequency of precipitation and temperature extremes over France in an anthropogenic scenario: Model results and statistical correction according to observed values. *Global and Planetary Change, 57*(1–2), 16–26.

Dimri, A. P., Chevuturi, A., Niyogi, D., Thayyen, R. J., Ray, K., Tripathi, S. N., Pandey, A. K., & Mohanty, U. C. (2017). Cloudbursts in Indian Himalayas: A review. *Earth-Science Reviews, 168*, 1–23. https://doi.org/10.1016/j.earscirev.2017.03.006

Dobriyal, M. J. R., & Bijalwan, A. (2016). Why cutting down Chirpine is not a solution to Uttarakhand forest fires. Retrieved March 30, 2018 from http://www.downtoearth.org.in/blog/why-cutting-down-chirpine-is-not-a-solution-to-uttarakhand-forest-fires-51178

Dobriyal, M., & Bijalwan, A. (2017). Forest fire in western Himalayas of India: A Review. *New York Science Journal, 10*(6), 39–46.

Duke, N. C., Meynecke, J. O., Dittmann, S., Ellison, A. M., Anger, K., Berger, U., Cannicci, S., Diele, K., Ewel, K. C., & Field, C. D. (2007). A world without mangroves? *Science, 317*, 41–42.

Fowler, H. J., & Archer, D. R. (2006). Conflicting signals of climatic change in the Upper Indus Basin. *Journal of Climate, 19*, 4276–4293.

Gautam, M. R., Timilsina, G. R., & Acharya, K. (2013). *Climate change in the Himalayas: Current state of knowledge*. World Bank, Policy Research Working Paper 6516, p. 47.

Ghawana, T., & Minoru, K. (2015). Indian Flood management: A report on Institutional arrangements. Integrated Spatial Analytics Consultants, Delhi, India & International Centre for Water Hazard and Risk Management (ICHARM) under the auspices of UNESCO; Public Works Research Institute (PWRI). Retrieved May 30, 2018, from http://whrm-kamoto.com/assets/files/TARUN%20MINORU-ICHARM%20REOPRT%20ON%20INDIAN%20FLOOD%20MANAGEMENT%20INSTITUTIONS%20JANUARY%202015.pdf

Gudmundsson, L., Bremnes, J. B., Haugen, J. E., & Skaugen, T. E. (2012). Downscaling RCM precipitation to the station scale using quantile mapping–a comparison of methods. *Hydrology & Earth System Sciences Discussions, 9*(5), 6185–6201.

Gupta, V., Dobhal, D. P., & Vaideswaran, S. C. (2012). August (2012) Cloudburst and subsequent flash flood in the Asi Ganga, a tributary of the Bhagirathi river, Garhwal Himalaya India. *Current Science, 105*(2), 249–253.

Huth, R. (1999). Statistical downscaling in central Europe: Evaluation of methods and potential predictors. *Climate Research, 13*(2), 91–101.

Huth, R. (2002). Statistical downscaling of daily temperature in central Europe. *Journal of Climate, 15*(13), 1731–1742.

IPCC. (2001). Climate change 2001. The scientific basis. In J. T. Houghton, Y. Ding, D. J. Griggs, M. Noguer, P. J. Van der Linder, X. Dai, K. Maskell, C. A. Johnson (Eds.), *Contribution of working group I to the third assessment report of the IPCC*. Cambridge University Press, Cambridge, UK.

IPCC. (2021). Summary for policymakers. In V. Masson-Delmotte, P. Zhai, A. Pirani, S. L. Connors, C. Péan, S. Berger, N. Caud, Y. Chen, L. Goldfarb, M. I. Gomis, M. Huang, K. Leitzell, E. Lonnoy, J. B. R. Matthews, T. K. Maycock, T. Waterfield, O. Yelekçi, R. Yu, & B. Zhou (Eds.), *Climate change 2021: The physical science basis. Contribution of working group I to the sixth assessment report of the intergovernmental panel on climate change*. In Press.

ISFR. (2017). India State of Forest Report 2017. Forest survey of India, Dehradun, India. Retrieved June 30, 2018, from http://fsi.nic.in/isfr2017/isfr-forest-cover-2017.pdf

Ives, J., Shresta, R., & Mool, P. (2010). *Formation of glacial lakes in the Hindu Kush–Himalayas and GLOF risk assessment*. International Centre for Integrated Mountain Development, Kathmandu, Nepal.

Johnson, F., & Sharma, A. (2012). A nesting model for bias correction of variability at multiple time scales in general circulation model precipitation simulations. *Water Resources Research, 48*(1).

Joshi, A. K., & Joshi, P. K. (2011). A rapid inventory of indicators of climate change in the middle Himalaya. *Current Science, 100*(6), 831–833.

Kanga, S., Kumar, S., & Singh, S. K. (2017). Climate induced variation in forest fire using Remote Sensing and GIS in Bilaspur District of Himachal Pradesh. *International Journal of Engineering and Computer Science, 6*(6), 21695–21702.

Klein, W. H., Lewis, B. M., & Enger, I. (1959). Objective prediction of five-day mean temperatures during winter. *Journal of Atmospheric Sciences 16*(6):672–682

Kothawale, D. R., & Rupa Kumar, K. (2005). On the recent changes in surface temperature trends over India. *Geophysical Research Letters, 32*(18).

Kumar, A., Houze, R. A., Jr., Rasmussen, K. L., & Peters-Lidard, C. (2014). Simulation of a flash flooding storm at the steep edge of the Himalayas. *Journal of Hydrometeorology, 15*(1), 212–228.

Kumar, N., Yadav, B., Gahlot, S., & Singh, M. (2015a). Winter frequency of western disturbances and precipitation indices over Himachal Pradesh, India: 1977–2007. *Atmósfera, 28*, 63–70.

Kumar, S., Bairagi, G. D., & Kumar, A. (2015b). Identifying triggers for forest fire and assessing fire susceptibility of forests in Indian western Himalaya using geospatial techniques. *Natural Hazards, 78*(1), 203–217.

Mandal, R. N., & Naskar, K. R. (2008). Diversity and classification of Indian mangroves: A review. *Tropical Ecology, 49*(2), 131.

Mandal, A., Ramanathan, A. L., & Angchuk, T. (2014). Assessment of Lahaul-Spiti (western Himalaya, India) glaciers–an overview of mass balance and climate. *Journal of Earth Science Climate Change, 11*, 003.

Maraun, D., Wetterhall, F., Ireson, A. M., Chandler, R. E., Kendon, E. J., Widmann, M., Thiele-Eich, I., et al. (2010) Precipitation downscaling under climate change: Recent developments to bridge the gap between dynamical models and the end user. *Reviews of Geophysics, 48*(3).

Medina, S., Houze, R. A., Kumar, A., & Niyogi, D. (2010). Summer monsoon convection in the Himalayan region: Terrain and land cover effects. *Quarterly Journal of the Royal Meteorological Society, 136*(648), 593–616.

Meehl, G. A., Arblaster, J. M., & Collins, W. D. (2008). Effects of black carbon aerosols on the Indian Monsoon. *Journal of Climate, 21*, 2869–2882.

Meher, J. K., Das, L., Akhter, J., Benestad, R. E., & Mezghani, A. (2017). Performance of CMIP3 and CMIP5 GCMs to simulate observed rainfall characteristics over the Western Himalayan region. *Journal of Climate, 30*(19), 7777–7799.

Meher, J. K., Das, L., Benestad, R. E., & Mezghani, A. (2018). Analysis of winter rainfall change statistics over the Western Himalaya: The influence of internal variability and topography. *International Journal of Climatology, 38*, e475–e496.

Mehta, J. S. (1996). Forest fires and land degradation in Uttarakhand. In K. Kumar, P. P. Dhyani, & L. M. S. Palni (Eds.), *Land utilization in central Himalaya: Problem and management options* (pp. 125–133). Indus Publishing Company.

Mitra, A., Sengupta, K., & Banerjee, K. (2011). Standing biomass and carbon storage of above-ground structures in dominant mangrove trees in the Sundarbans. *Forest Ecology and Management, 261*(7), 1325–1335.

Mtongori, H. I., Stordal, F., & Benestad, R. E. (2016). Evaluation of empirical statistical downscaling models' skill in predicting Tanzanian rainfall and their application in providing future downscaled scenarios. *Journal of Climate, 29*(9), 3231–3252.

Nair, V. S., Babu, S. S., Moorthy, K. K., Sharma, A. K., Marinoni, A., & Ajai. (2013). Black carbon aerosols over the Himalayas: Direct and surface albedo forcing. *Tellus B, 65*(1), 19738.

Nolin, A. W., & Daly, C. (2006). Mapping "at risk" snow in the Pacific Northwest. *Journal of Hydrometeorology, 7*, 1164–1171.

Okkan, U., & Inan, G. (2015). Statistical downscaling of monthly reservoir inflows for Kemer watershed in Turkey: Use of machine learning methods, multiple GCMs and emission scenarios. *International Journal of Climatology, 35*(11), 3274–3295.

Palazzi, E., Luca, F., Jost. V. H., & Lindsey, H. (2017) Why are the Himalayas getting hotter? *Environment Science Journal for Teens*. Retrieved March 22, 2018, from http://www.sciencejournalforkids.org/uploads/5/4/2/8/54289603/himalaya-article.pdf

Panday, P. K., Thibeault, J., & Frey, K. E. (2015). Changing temperature and precipitation extremes in the Hindu Kush Himalayan region: An analysis of CMIP3 and CMIP5 simulations and projections. *International Journal of Climatology, 3*(10), 3058–3077.

Pant, G. B., Kumar, P. P., Revadekar, J. V., & Singh, N. (2018) *Climate change in the Himalayas*. Springer International Publishing, 145p.

Peings, Y., & Douville, H. (2010). Influence of the Eurasian snow cover on the Indian summer monsoon variability in observed climatologie and CMIP3 simulations. *Climate Dynamics, 34*, 643–660.

Pepin, N., Bradley, R. S., Diaz, H. F., Baraër, M., Caceres, E. B., Forsythe, N., Fowler, H., Greenwood, G., Hashmi, M. Z., Liu, X. D., & Miller, J. R. (2015). Elevation-dependent warming in mountain regions of the world. *Nature Climate Change, 5*(5), 424.

Piani, C., Haerter, J. O., & Coppola, E. (2010). Statistical bias correction for daily precipitation in regional climate models over Europe. *Theoretical and Applied Climatology, 99*(1), 187–192.

Pramanik, M. K. (2016). Assessment of the impacts of sea level rise on mangrove dynamics in the Indian part of Sundarbans using geospatial techniques. *Journal of Biodiversity, Bioprospecting and Development, 3*(155), 2376–3214.

Prasad, A. S., Pandey, B. W., Leimgruber, W., & Kunwar, R. M. (2016). Mountain hazard susceptibility and livelihood security in the upper catchment area of the river Beas, Kullu Valley, Himachal Pradesh, India. *Geoenvironmental Disasters, 3*(1), 3.

Qian, Y., Flanner, M. G., Leung, L. R., & Wang, W. (2011). Sensitivity studies on the impacts of Tibetan Plateau snowpack pollution on the Asian hydrological cycle and monsoon climate. *Atmospheric Chemistry and Physics, 11*, 1929–1948.

Raha, A., Das, S., Banerjee, K., & Mitra, A. (2012). Climate change impacts on Indian Sunderbans: A time series analysis (1924–2008). *Biodiversity and Conservation, 21*(5), 1289–1307.

Ramisch, A., Lockot, G., Haberzettl, T., Hartmann, K., Kuhn, G., Lehmkuhl, F., Schimpf, S., Schulte, P., Stauch, G., Wang, R., & Wünnemann, B. (2016). A persistent northern boundary of Indian Summer Monsoon precipitation over Central Asia during the Holocene. *Science and Reports, 6*, 25791.

Rull, V., & Vegas-Vilarrúbia, T. (2006). Unexpected biodiversity loss under global warming in the neotropical Guayana Highlands: A preliminary appraisal. *Global Change Biology, 12*(1), 1–9.

Sahu, J. K. (1996). Some thermodynamical and forecasting aspects of local severe storms over Northwest India. Doctoral dissertation, Department of Physics, Savitribai Phule Pune University, Pune, India. Retrieved March 27, 2018, from http://shodhganga.inflibnet.ac.in/handle/10603/133270

SANDRP. (2018). Uttrakhand cloudburst incidents 2018. South Asia network on dams, rivers and people. Retrieved November 12, 2018, from https://sandrp.in/2018/07/21/uttrakhand-cloudburst-incidents-2018/

Satendra, K., Gupta, A. K., Naik, V. K., Roy, T. K. S., Sharma, A. K., & Dwivedi, M. (2014) Uttarakhand disaster 2015. National Institute of Disaster Management, New Delhi, p. 184.

Shrestha, A. B., Wake, C. P., Mayewski, P. A., & Dibb, J. E. (1999). Maximum temperature trends in the Himalaya and its vicinity: An analysis based on temperature records from Nepal from period 1971–94. *Journal of Climate, 12*(9), 2775–2787.

Singh, P., & Kumar, N. (1997). Effect of orography on precipitation in the western Himalayan region. *Journal of Hydrology, 199*, 183–206.

Singh, R. B., & Mal, S. (2014). Trends and variability of monsoon and other rainfall seasons in western Himalaya, India. *Atmospheric Science Letters, 15*, 218–226.

Singh, S. P., Singh, V., & Skutsch, M. (2010). Rapid warming in the Himalayas: Ecosystem responses and development options. *Climate and Development, 2*(3), 221–232.

Sontakke, N. A., Singh, N., & Singh, H. N. (2008). Instrumental period rainfall series of the Indian region (AD 1813–2005): Revised reconstruction, update and analysis. *The Holocene, 18*(7), 1055–1066.

Stocker, T. F., Qin, D., Plattner, G. K., Tignor, M., Allen, S. K., Boschung, J., Nauels, A., Xia, Y., Bex, V., & Midgley, P. M. (2013). IPCC Climate Change 2013: The physical science basis. Contribution of working group I to the fifth assessment report of the intergovernmental panel on climate change. Cambridge University Press, Cambridge, UK/New York, NY, USA.

Tripathi, S., Srinivas, V. V., & Nanjundiah, R. S. (2006). Downscaling of precipitation for climate change scenarios: A support vector machine approach. *Journal of Hydrology, 330*(3–4), 621–640.

Tyagi, A. (2007). Thunderstorm climatology over Indian region. *Mausam, 58*(2), 189.

Wang, B. (2006). *The asian monsoon* (p. 787). Springer.

Wetterhall, F., Pappenberger, F., He, Y., Freer, J., & Cloke, H. L. (2012). Conditioning model output statistics of regional climate model precipitation on circulation patterns. *Nonlinear Processes in Geophysics, 19*(6), 623–633.

Wood, A. W., Leung, L. R., Sridhar, V., & Lettenmaier, D. P. (2004). Hydrologic implications of dynamical and statistical approaches to downscaling climate model outputs. *Climatic Change, 62*(1), 189–216.

Chapter 25
Drought Preparedness Under Changing Climate with Special Reference to Manipur

Aribam Priya Mahanta Sharma, Deepak Jhajharia, Ghanashyam Singh Yurembam, and Rajiv Kumar Srivastava

Abstract Drought is one of the biggest challenges limiting the food security and lives of people of the world. It deals with significant natural resource limitation caused by a shortage of water in many of the developing part of India. Managing drought and water scarcity is crucial for addressing key development goals in dry areas. These affected regions lacks contingency planning as a result of insufficient financial resources, an inadequate understanding of its impacts, and poor cooperation among government institutions. Although, drought is an unavoidable climatic phenomenon, however it can be intervened and its preparedness can lead to better preparation to construct more resilient ecosystems which will enable to recover from its impact. Drought preparedness involves geographical improvements of agricultural systems by adopting concept of seed, fodder, nutrient banks, and enhancing irrigation efficiency by encouraging water conservation and groundwater use. This chapter presents successful case study of accomplishment of drought preparedness involving Government and other stakeholders in Manipur.

Keywords Drought · Water resources use · Land use · Irrigation efficiency · Policy · Deforestation

---

A. P. M. Sharma (✉) · D. Jhajharia · G. S. Yurembam
Department of Soil and Water Conservation Engineering, College of Agricultural Engineering and Post Harvest Technology, Central Agricultural University (Imphal), Ranipool, Gangtok, Sikkim, India
e-mail: mahantasharma@gmail.com

R. K. Srivastava
Institute of Soil Science and Land Evaluation, Biogeophysics, University of Hohenheim, Stuttgart, Germany

H. Pathak et al. (eds.), *Climate Change Impacts on Soil-Plant-Atmosphere Continuum*, Advances in Global Change Research 78,
https://doi.org/10.1007/978-981-99-7935-6_25

25.1 Introduction

Drought is considered a natural hazard with a wide-ranging impact on water bodies. The definition of drought is an occurrence of low rain in the area leading to a shortage of water and increasing in temperature, high winds, and low relative humidity increase the severity of drought. It affects the wetlands, snowpacks, glaciers, rivers, and soil water content including groundwater content too. Based on occurrence, drought considers four types i.e. meteorological, hydrological, agriculture, and socioeconomic and it can develop anywhere within a month or slower over a season even in months to a decade (Sugg et al., 2020). In contrast, arid regions have more severe drought conditions due to heat waves and high temperatures (Mishra & Singh, 2010).

Climate change due to global warming increase the rising in temperature which causes heat waves, reduction in relative humidity, and changing rainfall pattern and these events increase the severity of drought. A recent study showed how climate change occurred throughout history and how its changes affect and destroyed the old civilization by increasing the severity of drought for a longer period (Sadhukhan et al., 2023). In addition, the impact of climate change extremes and adaptations on different crops is reported in Srivastava et al. (2019, 2023) and Halder et al. (2015). However, the success of adaptations to combat climate change depends on the local or regional environment and its suitability. Hence, effective drought management is required to reduce the impact of drought on water bodies and land.

Drought management has been attempted in the past through reactive crisis management, although this has been proven ineffective due to poor coordination and an untimely approach (Wilhite et al., 2005). In addition, global warming has been the factor believed for increasing the frequency of drought in the future coupled with the cost or impact associated with drought on sectors beyond agriculture, socio-economy, environment, and water resources have instilled the thought process towards improved drought preparedness and policy planning. Northeastern India has been more susceptible to drought than any other western region of India (Parida & Oinam, 2015). Amongst the state of northeastern India, Manipur possesses a distinctive character with its complicated physiographic of valleys and hills.

25.2 Material and Methodology

25.2.1 Study Region

Manipur, located in the extreme northeastern portion of India, is a small geographical area of about 22,327 km^2. Bounded by the neighboring States i.e., Nagaland, Mizoram, and Assam in the North, South, and West respectively, and also with Burma in the East demarcating an international boundary providing a feasible for trade between India and Myanmar and other Southeast Asian countries through the Moreh town. Characterized by two distinct physical regions, rugged hills surround

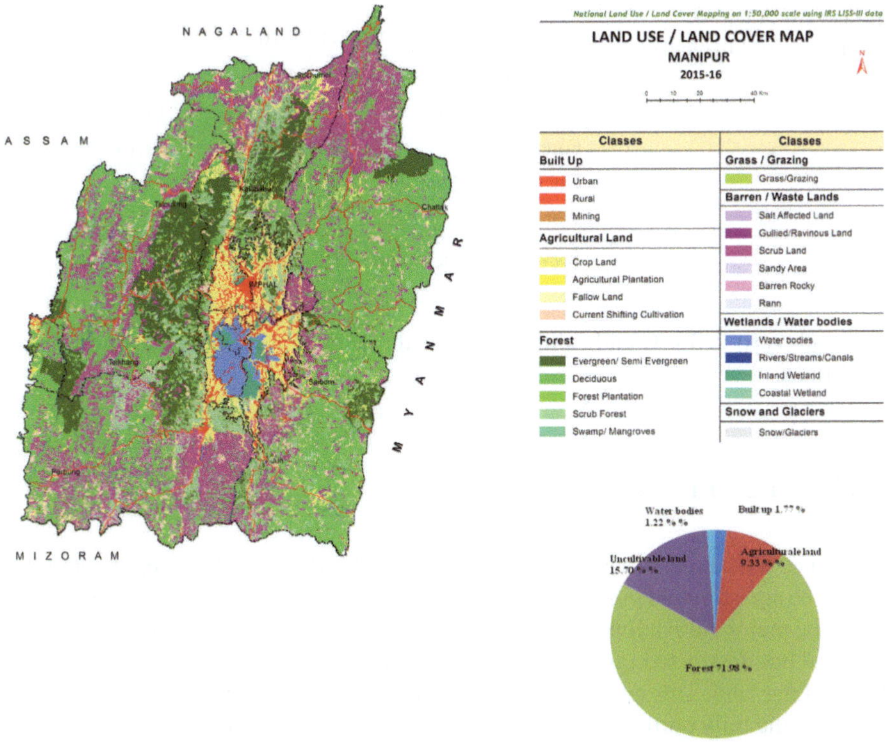

Fig. 25.1 Land use class of Manipur state in India. (*Source* Bhuvan, Indian Geo Platform Service of ISRO)

the inner plains of the valley. The valley oval-shaped valley region covers an area of 1,843 sq. km (10% of the state) surrounded by the Blue Mountains and comprises six districts viz., Imphal East, Imphal West, Thoubal, Lilong, Kakching and Bishnupur Districts. The Hilly region occupies a major portion of the state with an area of 20,571 sq. km. The state is demarcated into ten districts viz., Chandel, Senapati, Churachandpur, Tamenglong, Ukhrul, Kamjong, Noney, Kangpokpi, Jiribam, and Pherzawl districts (Fig. 25.1).

25.2.2 *Climate of Study Region*

Manipur bears a range of annual rainfall varying minimum of 1085 mm at Wangbal in Thoubal District to a maximum of 2871 mm in Jiribam. The average annual rainfall is about 1616.2 mm. The rainfall period ranges from mid-April and continues up to October. Being located near the Tropic of Cancer (23½ °N latitude) the state is under the influence of a subtropical monsoon climate receiving temperature ranges

Table 25.1 Areal distribution of different land use in Manipur

Land use	2015–16	Percentage
Built-up	545.56	2.44
Agriculture land	2592.93	11.61
Forest	14,987.26	67.13
Uncultivable land	3634.34	16.28
Water bodies	566.98	2.54

(*Source* Bhuvan, Indian Geo Platform Service of ISRO)

from sub-zero to 36 °C and exhibits a tropical to temperate climate depending upon elevations.

25.2.3 Land Use of Study Region

The state, of Manipur, is about 90% hill area which is dominantly covered with forest land. The agricultural lands are mostly confined to the valley region of the state. Over time, the major cultivation practice in the hill area has been shifting or Jhum Cultivation and has destructed immeasurable areas of the forests and hampered the productivity of the land. This cultivation practice has led to greater environmental degradation over the hilly portion of the state resulting in massive deforestation, nutrient loss by soil erosion, extinction of species, impede water holding capacity of soils, and persistence of more dry spells leading to adverse climate change. Figure 25.1 shows the changes in land use and water bodies in Manipur state, India and it indicates that forest areas share the most land use (71.98%) than uncultivated land (15.70%) and agricultural land (9.33%) while water bodies share the 1.22%. Further, the rapid growth and expansion of Imphal city have led to severe unregulated constructions of built-ups used for multiple purposes, disrupting the productivity of the limiting fertile agricultural lands (Abujam & Koijam, 2016).

Table 25.1 presented below justifies the changes in the land use of the State and it indicates that the changes in the forest were high 67.12% followed by uncultivable land (16.27%), agriculture land (11.61%), and water bodies (2.53%) while changes in the built-up area were low with 2.44%.

25.2.4 Data Collection

In this study, 51 years of meteorological data years 1979–1992 were collected for drought analysis and the impact on land use and water consumption was also evaluated for the state of Manipur. Moreover, spatial water consumption in the state of Manipur was also analyzed. Thereafter, the impact of drought on water bodies and land use was evaluated spatially. Further, groundwater data from a survey of the authority of the Central Ground Water Board (CGWB) were also considered to evaluate the impact of drought on groundwater.

25.2.5 Technical Suggestions for Policymakers

Further, some technical suggestions were evaluated and suggested for policymakers to combat the drought condition which is now severe due to increasing temperatures and changing rainfall patterns due to climate change. Table 25.1 shows that the changes in the forest were more which might be the possible impact of climate change which cause more soil erosion and soil degradation in the Manipur state.

25.3 Results and Discussions

25.3.1 Water Resources Use and Consumption

Figure 25.2 shows the drainage and water distribution map of the Manipur state and showed that the waterbodies in Manipur occupy about 0.75% of the total geographical area i.e., 166.77 km^2 with 15 major rivers/streams (Source: National Wetland Inventory and Assessment). Drinking water across the state is supplied predominantly by three major rivers namely the Imphal River, Nambul River, and Iril River. According to National Wetland Atlas, 2010, Manipur possesses 2.85% about 63,616 ha of the geographical area covering 700 wetlands, constituting lake/pond (61.5%), river/steam (26.2%), waterlogged (5.5%) and aquaculture pond (6.8%).

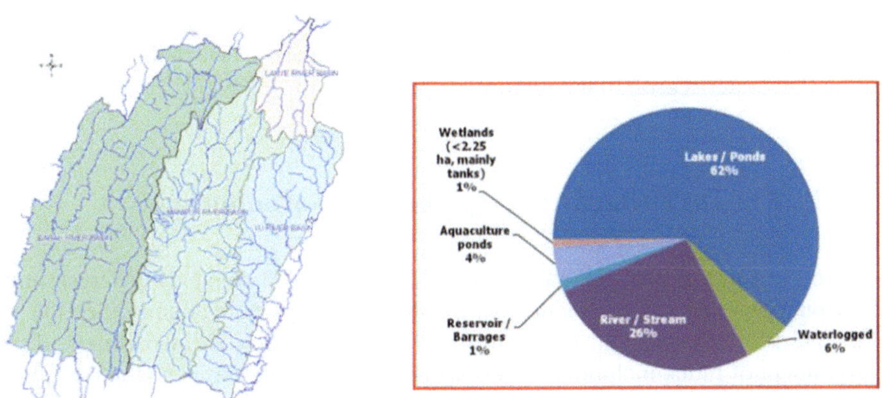

Fig. 25.2 Drainage map and water resources distribution of Manipur. *Source* National Wetland Atlas, 2010

Seasonal analysis of wetlands during pre-monsoon and post-monsoon reported an increase in open-water wetland areas from 62 to 71% respectively. Consequently, aquatic vegetation wetland areas showed decreasing trend i.e., 37 to 26% during pre-monsoon to post-monsoon.

Groundwater survey of Manipur by the authority of the CGWB, observed the existence of deeper aquifers under sub-artesian and artesian conditions due to the exploitation of groundwater through open wells, hand pumps, etc. The existence of granular zones in Imphal valley at depths of 150 m and 220 m in Jiribam valley was also reported by CGWB. Annually Manipur, reports a severe water shortage despite being in a high rainfall area, particularly during the dry/lean season, which lasts from January to May each year.

The state exhibits a unique river system that originates in the hills and distributes across the state during the annual monsoon. Deforestation in the state's hilly areas also causes heavy siltation and disrupts river flow. The problem of water scarcity has been exacerbated further by climate unevenness resulting in erratic monsoons extensively over the years. Moderate levels of pollution and contamination in the water bodies of Manipur have been reported by Environment R & D Laboratory, Directorate of Environment, Porompat. However, the Nambul river and Kongba river have been reported the highest degree of pollution due to increasing anthropogenic activity in the central valley districts of the state.

25.3.2 Drought and Its Impacts on Manipur

Six major drought events have been observed in the last five decades, (Fig. 25.3). For drought variable analysis the duration of drought events was identified as mostly 3 months. In the study of 51 years of meteorological data, the drought episode of 1979–1992 has been reported as the most severe. The maximum severity of drought events as recorded was 3.81 (1992) resulting in scarcity of domestic and industrial consumption, energy, and irrigation. The occurrence of two drought events was encountered within 63 months. The maximum number of dry spells is prevalent during the winter season in the region.

25.3.3 Present Drought Management and Monitoring

Section 11 of the Disaster Management (DM) Act of 2005 demands the creation of a National Disaster Management Plan (NDMP) for the whole country. Section 37 of the Disaster Management Act mandates that each ministry and department of the Government of India, including hazard-specific nodal ministries, develop DM plans stating their contributions to the country's efforts in disaster prevention, preparedness, response, and recovery. At present, India aims to voluntarily abide by the non-binding Sendai Framework for Disaster Risk Reduction 2015–2030 to create

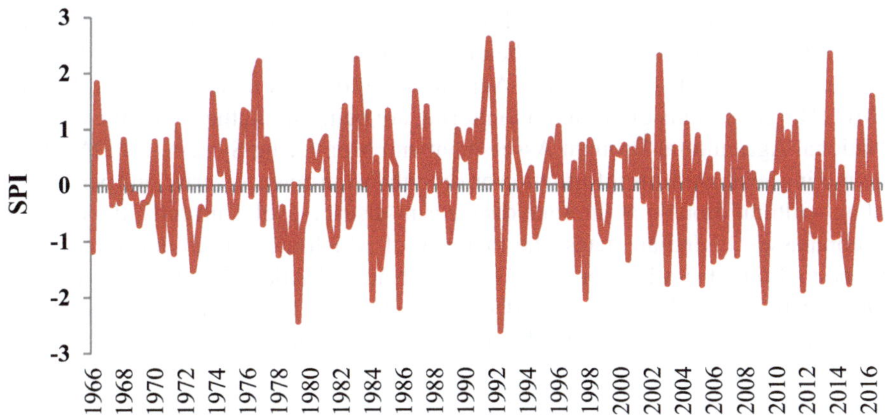

Fig. 25.3 Drought persistence during the period from 1966 to 2016

a Drought Management Plan (DMP) with the objectives and priorities set out for Disaster Risk Reduction (DRR). The plan will be adjusted as necessary during the periodic reviews and updates based on international best practices and national experiences. The task of coordinating drought-related relief efforts falls under the purview of the Drought Management Division, DAC & FW. Special power to initiate immediate relief measures to address the drought situation is entitled to the State Governments to transact from the available State Disaster Response Fund. Further, an annual Inter-Ministerial mechanism operates under the Crop Weather Watch Group for Drought Management (CWWGDM) of the DAC&FW to monitor the drought situation across the country once a week from June to September. The Additional Secretary & Central Drought Relief Commissioner led the Drought Crisis Management Group, constituted of representatives of Line Departments of States, and other stakeholders. The National Remote Sensing Centre, Department of Space (Government of India), associated center, Mahalanobis National Crop Forecast Centre (MNCFC), 2012 was established in the office of DAC&FW to serve agricultural assessment, drought assessment, and monitoring using space technology under the National Agricultural Drought Assessment and Monitoring System (NADAMS). Under the recommendation of the National Disaster Management Authority for Drought, in 2010 separate Drought Monitoring Cells (DMCs) with adequate staff under SDMA were established to prioritize drought monitoring effectively for the preparation of vulnerability maps for their respective states.

Indian Meteorological Department (IMD) superintends the spatial and temporal rainfall data across the country on multiple scales. Until 2012, these generated data are preferably utilized with percent normal deviation and Aridity Anomaly Index for drought monitoring. Percent normal deviation covers meteorological drought while Aridity Anomaly Index (AAI) is for agricultural drought. The Standardized Precipitation Index (SPI) has been ideally utilized over AAI, % normal as a drought index for monitoring since 2013, supplemented by the information of the Normalized

Difference Vegetation Index (NDVI) (Rathore et al., 2014). Medium-range weather forecast-based agro-advisories at the district level have been continuously delivered by the combined effort of ICAR's Agro Meteorological Field Units (AMFUs) and IMD. The constraints of drought monitoring are being strengthened by the initiative National Agricultural Drought Assessment and Monitoring System (NADAMS) by delivering near real-time information on drought occurrence, severity, and persistence of agricultural drought at the state/district/sub-district level. Currently, the initiative encompasses 13 Indian states, the majority of which are agriculturally based and prone to drought (Rathore et al., 2014). The National Atlas and Thematic Mapping Organization (NATMO) developed the drought atlas for India, which assisted in identifying and prioritizing specific risk management areas.

25.3.4 Boosting Drought Preparedness and Management

The process of water management involved the active participation of stakeholders only to provide consultation in various levels of management initiatives. However, the Government authorities are committed to gradually changing this scenario through private and public sectors and to intervene with technology and scheme to deliver and enhance these resources. In the situation when a drought is declared, active and continuous participation of the stalk holders is essential to limit its extent to control.

Stakeholders are also associated with the evaluation of drought impact and formulation of responses to it through identification and vulnerability assessment, and propose the best measures to cope with future occurrences (Kampragou et al., 2015). Instead of drought prediction, emphasis in the drought policy on preparedness should be given for a future drought of any intensity, frequency, and duration (Bandyopadhyay et al., 2020).

25.3.5 Policies and Measures to Improve Drought Preparedness and Management

In the present context, the crisis mitigation approach has been the only basis for drought management. So, there is a need to plan and develop a policy specifically for droughts. It necessities prompt measures and planning to minimize the impact of drought on the farmers and common people. Drought planning should be addressed with an exclusive policy laid for specific management tools. To carry out effective response and containment of drought careful preparation is very essential. Prior identification of drought-prone areas, and regular monitoring of seasonal forecasts of IMD. All the preparations have to be carried out in certain harmony amongst the stalk holders and the other agencies (Fig. 25.4).

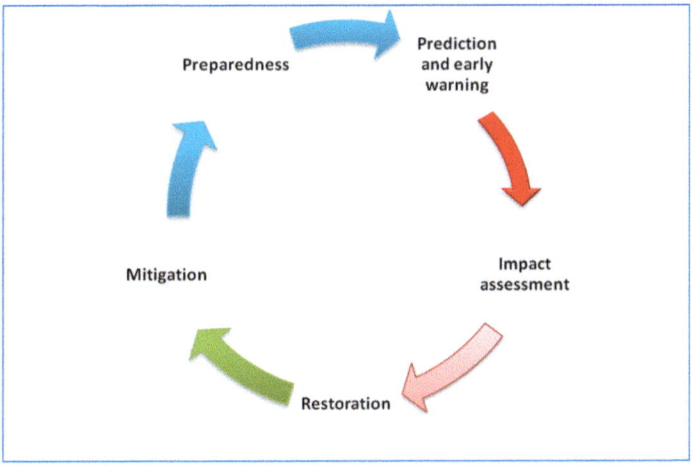

Fig. 25.4 Schematic pattern for drought preparedness

The action plan should be framed considering the various aspects of drought and improving operational drought management.

25.3.6 Suggested Strategies to Improve Drought Preparedness and Management

Based on the study, the suggested strategies to improve drought preparedness and management are given below.

1. Formulation of policies to ensure the availability of appropriate varieties of seed in sufficient quantity during drought for adequate production in advance.
2. Establishment of fodder bank at a strategic location for supply at deficit areas to maintain Livestock component in arid regions during drought years for ensuring livelihoods.
3. Introducing the concept of a Nutrient Bank for the local community to help with timely sowing by stockpiling essential manures and fertilizers, soil amendments, foliar spray chemicals, bio-fertilizers, and so on.
4. Prioritizing water conservation and evaluating the groundwater resources for sole use in the provision of drinking water. Management of operation of sluice gates on minor and large irrigation tanks to conserve water for use by people and animals.
5. Setting up a rapid task team at the district level to repair all of the hand pumps, and electric pumps installed on bore wells, and stock all accessories and spare parts inventories.

25.4 Conclusions

Drought has been considered a disaster and managed with measures adopted to overcome its impact on agricultural and rural and urban sectors. Long-term planning addressing the drought-like condition can be mitigated with adequate water supply, particularly in water-scarce regions. Specific environmentally-oriented measures have to be adapted to promote water-saving and efficient farming practices and technologies. Evaluation and exploration of the aspects of drought mitigation are essential, at the very least in a framework strategy for water conservation, to close policy gaps in the framework of water management.

References

Abujam, M. S., Koijam, R. D. (2016). Landuse and landcover change detection of fringe areas of Imphal City, Manipur, India. *IOSR Journal of Humanities and Social Science (IOSR-JHSS), 21*(2), 09–16.

Approach Paper on New Land Use Policy/Project, Manipur (2014).

Bandyopadhyay, N., Bhuiyan, C., & Saha, A. K. (2020). Drought mitigation: Critical analysis and proposal for new drought policies with special reference to Gujarat (India). *Progress in Disaster Science*. https://doi.org/10.1016/j.pdisas.2019.100049

Halder, D., Panda, R. K., & Srivastava, R. K. (2015). *Impact of elevated temperature and CO_2 on productivity of peanut in eastern India*. Paper presented at ASABE 1st Climate Change Symposium: Adaptation and Mitigation Conference Proceedings. American Society of Agricultural and Biological Engineers, pp. 1–3.

Kampragou, E., Assimacopoulos, D., Andreu, J., Bifulco, C., de Carli, A., Dias, S., González Tánago, I., Haro Monteagudo, D., Massarutto, A., Musolino, D., Paredes, J., Rego, F., Seidl, I., Solera, A., Urquijo Reguera, J., & Wolters, W. (2015). Systematic classification of drought vulnerability and relevant strategies—case study scale. Drought–R&SPI Technical Report No. 24, Athens, Greece, p. 43.

Mishra, A. K., & Singh, V. P. (2010). A review of drought concepts. *Journal of Hydrology, 391*(1–2), 202–216.

Parida, B. R., & Oinam, B. (2015). Unprecedented drought in North East India compared to Western India. *Current Science, 109*(11), 2121–2126.

Rathore, B. M. S., Sud, R., Saxena, V., Rathore, L. S., Rathore, T. S., Subrahmanyam, V. G., & Roy, M. M. (2014, May). Drought conditions and management strategies in India. In Country Workshop Report, Regional Workshop for Asia-Pacific, *UNWater Initiative on Capacity Development to Support National Drought Management Policies* (pp. 6–9).

Sadhukhan, B., Srivastava, R. K., Chakraborty, A., & Panda, R. K. (2023). Scientific evidence supporting the progression of climate change induced drought from the history. In *Integrated drought management* (Vol. 1). Taylor and Francis Group. ISBN 9781032231587.

Srivastava, R. K., Talla, A., Swain, D. K., & Panda, R. K. (2019). Quantitative approaches in adaptation strategies to cope with increased temperatures following climate change in potato crop. *Potato Research, 62*(2), 175–191.

Srivastava, R. K., Panda, R. K., Halder, D., & Chakraborty, A. (2023). Drought tolerance agronomic management strategies for rainfed and irrigated maize crop in Eastern India. In: *Integrated drought management* (Vol. 2). Taylor and Francis Group. ISBN 9781032231587.

Sugg, M., Runkle, J., Leeper, R., Bagli, H., Golden, A., Handwerger, L. H., Magee, T., Moreno, C., Reed-Kelly, R., Taylor, M., & Woolard, S. (2020). A scoping review of drought impacts on health and society in North America. *Climatic Change, 162*, 1177–1195.

Wilhite, D. A., Hayes, M. J., & Knutson, C. L. (2005). Drought preparedness planning: Building institutional capacity. *Drought and Water Crises: Science, Technology, and Management Issues, 107*, 93–136.

Chapter 26
Climate Change-Induced Spatiotemporal Monsoon Variability: Assessment of Its Implications on Global and Regional Production Systems

Lingaraj Huggi, M. N. Thimmegowda, Shankarappa Sridhara, M. H. Manjunatha, and Bappa Das

Abstract Monsoons are the major type of global climate named after characteristic seasonal reversal of winds and corresponding changes in rainfall because of pressure and temperature gradients created by the earth's tilted motion around its axis and differences in land-to-sea masses in the northern and southern hemispheres. Monsoons facilitate agricultural activities by their symphony with the crop water requirements, contributing to global food security as monsoons prevail over larger areas of the world. As evident from past studies, the monsoons are highly influenced by clime change due to alterations in regular pressure and thermal gradient systems. This enforces us to study climate change-induced monsoon variability and its influence on regional and global food security. Simulation studies have indicated impacts of variations in monsoonal precipitation besides increased temperature have resulted in decreased wheat and rice productivity in tropical Asia. The decreases in precipitations in the American monsoon regions like south Chile, south-west Argentina, south Peru, and west-central America have imparted yields of major crops like soybean, wheat, and maize due to abiotic stresses. In the US, a 0.01 to 0.05 tonnes/annum decrease in maize yield was reported due to internal variability in ENSO and the variability was more in rainfed maize than irrigated. The African region is comparatively more vulnerable to monsoon variability because of entirety of monsoon dependence. In Sahel region, a notable decrease in groundnut yields was reported (850 kg/

L. Huggi (✉)
CoEWM, UAS, GKVK, Bangalore 560065, India
e-mail: lingarajhuggi@gmail.com

M. N. Thimmegowda · M. H. Manjunatha
AICRP on Agrometeorology, UAS, GKVK, Bangalore 560065, India

S. Sridhara
Center for Climate Resilient Agriculture, KSNUAHS, Shivamoga 577412, India

B. Das
ICAR-CCARI, Ela, Old Goa 403402, India

H. Pathak et al. (eds.), *Climate Change Impacts on Soil-Plant-Atmosphere Continuum*, Advances in Global Change Research 78,
https://doi.org/10.1007/978-981-99-7935-6_26

ha in 1966–67 to 440 kg/ha in 1981) due to variations in rainfall distribution during crop growth period. On the other hand, seven African grasslands out of 31 world grassands under study showed vulnerability to monsoon variability leading to poor fodder availability to livestock. This highlights the socio-economic losses caused by the monsoon variability as an impact of climate change. In light of this information, the present chapter examines the impact of climate change on precipitation variability in monsoon regions of the world and assesses its implications on global and regional production systems.

Keywords Climate change · Monsoon variability · Spatio-temporal scale · Food security

26.1 Introduction

The climate of a region is the long-term weather pattern often averaged over a period of 30 years (Planton et al., 2008). Specifically, it is the depiction of mean and variability in weather variables across time, which can range from months to millions of years (Shepherd, 2009). In common, meteorological variables like temperature, humidity, atmospheric pressure, wind, and precipitation are measured for identification and classification of a region's climate. In broader terms, climate refers to the state of a region's climate system components encompassing ocean, land, and precipitation these are largely influenced by the region's geography, terrain, altitude, and proximity to water bodies. As these components vary spatially and temporally, garnered observations form a huge meteorological dataset, this calls attention to defining, and identification of similarities and distinctness in the observed patterns of climate in the world which is collectively called climate classification (Khlebnikova & Sall, 2009). Climate classification has scientific and practical use in both understanding physical mechanisms controlling climate and evaluating climatic potentialities and features of their use in diverse parts of the world, it primarily had been the domain of plant geographers, but due to contributions of statisticians, mathematicians and meteorologists made it a matter of discussion among the scientific community involving all scientific sectors (Virmani et al., 1980).

The climate classification can be split into two categories; genetic classification and descriptive classification. While genetic classification deals with the natural laws of climate formation and explain the reasons why climate patterns exist in different parts of the world. The descriptive classification distinguishes climatic types that are unrelated to climate genesis. The averages and range of various climatic variables, most often temperature and precipitation, are used to classify the world's climate. Even though, earlier the world climate was classified by various climatologists (Table 26.1) like Alisov's and Strahler's classification (Alisov & Poltaraus, 1974; Strahler, 1969), Berg climate classification based on the principle of distribution according to geographic zones (Berg, 1925). Holdridge life zones system based on an ecophysiological approach (Holdridge & Grenke, 1947), Lauer classification based on thermal and hydric thresholds for both natural vegetation and crops (Lauer, 1925), and Troll classification based on humid months in a year (Troll, 1965), all have not been accepted universally by scientific community for general applications. Only a few have proven their universal applicability. Despite being almost over a century old, Wladimir Koppen's climatic categorization, which has been modified by his partners and successors, is still widely used. The Köppen climate classification divides climates into five main classes (A: tropical, B: dry, C: temperate, D: continental, and E: polar) and thirty sub-types. Since the categorization is based on air temperature and precipitation threshold values and seasonality, makes use of existing vegetation of a location as a 'crystallized, visible climate' (Köppen, 1936). So, indirectly Koppen's classification aims at mapping biome distribution around the world which is taken as a proxy for climate of that particular region/location (Rubel & Kottek, 2011). Therefore, the Köppen-Geiger climate classification is an excellent choice for aggregating complicated climatic gradients into a straightforward yet ecologically important classification scheme (Webber, 2019). As a result, it is frequently utilized as input in studies of species distribution (Brugger & Rubel, 2013), growth behavior (Tarkan & Vilizzi, 2015), and the creation of dynamic global vegetation models (Poulter et al., 2011).

26.2 Monsoonal Climate

The 'monsoon' is the seasonal reversing of wind, characterized by corresponding changes in rainfall (Ramage, 1971). Most of us know that the seasonality i.e., periodicity of a meteorological event caused by skewness of the earth from its actual axis, unlike seasonality, the monsoon weather impacts are caused by land-sea temperature differences influenced by sunshine (Huffman, 1997). The earth's rotation and revolution around sun expose different masses of southern and northern hemispheres to the sun at different times leading to changing seasons. In the south Asian subcontinent, this seasonal reversing is caused by low-level summer winds of monsoon season and is explained by the powerful westerlies (850 hPa) over the Arabian Sea causing a low-level westerly jet (LLJ) and a large-scale cyclone vortex arising at the north Bay of Bengal to western India, called monsoon trough (Taraphdar et al., 2010). The easterly jet, which is located at about 5° N latitude, and the Tibetan anticyclone, which is located around 30° N latitude, are major characteristics of upper-level winds over the monsoon area throughout the summer. Summers with precipitation and winters with low humidity are linked to seasonal shifts in low-level winds, making them critical for agricultural productivity and the economies of many South Asian countries (Abrol & Gadgil, 1999). This seasonal reversal in south Asian region could be due to northward seasonal migration of the east–west oriented precipitation band (called as Tropical Convergence Zone; TCZ) from south hemisphere during winter to north hemisphere during summer (Gadgil, 2003). The largest northward journey of the rain band takes place over India where it moves from south (5°) during winter to north (20°) during northern summer (Waliser & Gautier, 1993). While equatorial easterlies are feeble

Table 26.1 Previous classifications and hypothesis behind classification of world's climate

Classification	Year	Hypothesis	Coverage	References
Alisov's classification	1954	Variability of air masses	Classification of world into seven climate belts according to the activity of the air masses: equatorial, subequatorial, tropical, subtropical, temperate, subpolar, and polar	Alisov
Bergs classification	1925	Principle of distribution according to geographic zones	Classified world climates into 2 groups; Lowland and Elevations	Berg (1925)
Holdridge life zone classification	1947	Mapping of soil and climax vegetation based on climate	Designed for tropical and subtropical areas, the system applies globally	Holdridge and Grenke (1947)
Lauer's classification	1996	Eco-physiological approach	Thermal and hydric thresholds for both natural vegetation and agricultural crops	Lauer (1925)
Strahler's classification	1969	Dependence of world climates on air masses that produce them	Divided world climate into 3 groups: Phenomenon of low-latitude climates influenced by Equatorial/ tropical air masses. Mid-latitude climates are influenced by Tropical and polar air masses. High-latitude climates are influenced by Polar and arctic air masses Subdivision of world climate into 14 groups based on variability in temperature and precipitation	Strahler (1969)
Thornthwaite's classification	1931 (revised 1948)	Effectiveness of precipitation on an index and Thermal efficiency, Inclusion of moisture index in 1948	World climate; 5 groups based on Effectiveness of precipitation, 6 groups based on thermal efficiency, 6 groups based on moisture index	Thornthwaite (1943, 1948)
Trewartha's classification	1968	Modifications in Koppen's criteria	Describes 6 principal climatic groups	Trewartha and Horn (1980)

(continued)

Table 26.1 (continued)

Classification	Year	Hypothesis	Coverage	References
Troll's classification	1965	Analysis of humid months in a year	Classification of tropical climates	Troll (1965)
Vahl's classification	1919	Modification of Köppen approach considering average annual temperatures	Classification of world climate into 4 zonal climates and subdivision of them based on temperature limits (warmest and coldest months.)	Reumert (1946)

and limited to 5° N and 10° S in the upper atmosphere, subtropical westerlies move to 10° N in northern winter. During the northern summer, the subtropical westerlies shift north (30°), and the equatorial upper atmosphere is characterized by a powerful easterly jet. Over the years, variations in long-term seasonal mean precipitation over the Indian subcontinent are substantially linked to regional food production (Knox et al., 2012). In Asian monsoon regions, occurrence of economic losses due to an estimated approximate decrement in rainfall of 8 mm/day has been estimated by many researchers, who highlighted the quantum of rainfall decrement per season or growing period will be high as compared to daily bases. On the other hand, the North American monsoon (aka Southwest US, Mexican, and Arizona) sowed a remarkable hike in precipitation from June (maximally dry) to July (more rain-bearing) aiding crop production (Froehlich, 1993). Cessation of these summer rains continues up to the middle of September which coincides with the crop growth period in these regions (Maddox et al., 1995).

26.2.1 Genesis

The earlier studies in the 1980s on atmospheric heat flux transfer and wind moments indicated distinct, large-scale, less repetitive changes in heat storage and transport in the Indian Ocean, related to wind forcing (Leetmaa, 1977). This concept paved the way to understand the causes behind this moment by noting the presence of significant wind-driven meridional oceanic heat flux. With an accurate understanding of the motion, the Northern Hemisphere heat flow is steered southward to the Winter Hemisphere in summer, while it reverses in the boreal winter. This reversal in heat flux in seasonal order maps the likely reaction between the ocean and the atmosphere that could regulate monsoon strength (Loschnigg et al., 2003).

As explained above, the seasonality is the product of tilted movement of Earth around its axis, and the monsoon weather systems are the results of differences in land-sea temperatures caused by solar radiation due to differences in the land masses of northern and southern hemispheres (Huffman, 1997). This difference also causes the antagonistic seasons in the middle of northern and southern hemispheres knowledgeable into two regimes, viz., Southeast Asian summer monsoon (10°–20° N) and the western North Pacific summer monsoon (10°–20° N, 130°–150° E) and are divided by a line over the South China Sea (Kripalani & Kulkarni, 1997). During cold season, the tilting of the Earth receives less energy from the sun in the northern hemisphere resulting in fast cooling by decreased air pressure. Anticyclones evolve over Siberia and the cold north-easterly air reaches the seaside waters of China before titling towards Southeast Asia (MMD, 2012). The East Asian winter monsoon (EAWM) is customarily dry in Southeast Asia. During the summer, the southwest monsoon precipitation is influenced by the warming of the northern hemisphere, as hot air rises and is carried to the southern hemisphere by the monsoon wind (Wolfson & Schneider, 2002).

On the other hand, the North American Monsoon, being centered on northwestern Mexico, its influence is observed more. The physical and relief features of south-western North America are the major factors behind the occurrence and movement of monsoonal winds in the region. The domain is surrounded by the Pacific Ocean in west, containing the California Gulf. During the summer, sea surface temperatures (SSTs) at central and subtropical latitudes in line with the Pacific Coast remain low (<25 °C), but SSTs in the Gulf of California and steamy eastward Pacific region remain high (>28 °C). As northwestern Mexico is covered by Gulf of Mexico having higher SSTs (>26 °C), dissimilarities in air and land temperature cause the pressure slope provoking wind movement from ocean to land (David et al., 1997). The interior of this cyclone domain can be explained by several large highland areas such as the Colorado Plateau which extends north and east to the Rocky Mountains of Arizona's Mogollon Rim.

In past studies, the evolution of the North American monsoon system has been divided into three steps, namely, development, mature, and decay phases (Douglas, 1995). In the development step (May–June), an ending period of cold season and starting up of a warm-season regime is observed. This was followed by a decrease in mid-latitude synoptic-scale transient action over the conterminous states of America and northern parts of Mexico because of weakening and migration of extratropical storm track towards pole to a point near Canadian border at the end of June (Hurford et al., 2012). Increases in the magnitude of the daily precipitation cycle (Higgins et al., 1996; Wallace, 1975) and the frequency of low-level jet occurrence (Bonner & Paegle, 1970; Higgins et al., 1996) influence the onset of monsoon in the region. This Mexican monsoon (Douglas et al., 1993; Stensrud et al., 1995) is heavy over southern Mexico, spreading suddenly towards north i.e. along west slopes of the Sierra Madre Occidental (hereafter SMO) and into Arizona and New Mexico during earlier periods of July. Enhanced vertical moisture transport via convection coincides with increased precipitation over northwestern Mexico (Douglas et al., 1993) and southerly winds flowing up the Gulf of California (Badan-Dangan et al. 1991).

Higher precipitation over southwest United States has been caused by many factors as explained by various climatologists. Such as anticyclones and high altitude winds (Okabe, 1995), depression in the desserts of the region due to thermal gradient (Rowson & Colucci, 1992), displacements of the Pacific and Bermuda highs towards

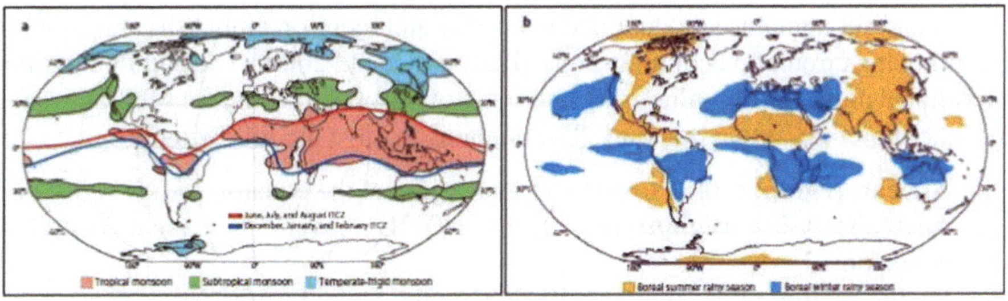

Fig. 26.1 Distribution of global monsoon indicated using dynamical normalized seasonality (DNS) a large-scale, unified monsoon index (Li & Zeng, 2002). *Source* Li and Zeng (2002)

north, formation of low-level jets over the Gulf of California inducing establishment of the Arizona cyclone horizon (Adang & Gall, 1989) and increases in sea surface temperature gradients on the east side of Pacific (Carleton et al., 1990). From June to July, still there an increase in oceanic surface pressure over the southwestern United States (Okabe, 1995) and a common crest in intervening latitudes guide the migratory warming of the tropospheric air. Greater increases in height happen over the western and southern United States and are linked to enhanced atmospheric heating over the higher terrains of the western states of America and Mexico (Fig. 26.1), this increased hidden warming is associated with the development of the monsoons in the Northern US. The resulting middle and superior tropospheric "monsoon high" doppelgangers the Tibetan high over Asia (Tang & Reiter, 1984) and the warm season Bolivian extreme over South America (Johnson, 1976).

26.2.2 Features and Global Spread

As the term 'monsoon' itself arises from seasonal variations in winds, more precisely the reversing of seasons in the atmospheric circulation of tropics and subtropics, and are associated with precipitation, the extremes of these events can often be best characterized as "wet" and "dry" seasons rather than summer and winter. As explained earlier, these reversals exist due to changes brought in by temperature gradients between continents and their adjacent oceans. Despite the lack of seasonal migration of the Pacific and Atlantic intertropical convergence zones (ITCZs) between the hemispheres, it is apparent that there is a global-scale cyclone, which means a global-scale persistent overturning of the atmosphere encompassing the Tropics and subtropics which changes with time of year. This provokes the emphasis on global aspect by virtues of coordination brought in by the annual repetitions of the heating of land and oceanic masses by solar energy and their visible connections in the global multidirectional circulation influenced by conservation of mass. Being enclosed within this system, global monsoons can be explained precisely, with the understanding of regional variations in monsoons.

Around six domains are identified to portray and focus on global structures of the over-moving circulations of monsoons (Trenberth et al., 2000). The author classifies meridional monsoon domains using outcomes of European Centre for Medium-range Weather Forecast, into 6 domains. The domains involve the following:

i. **African region:** (10° E to 40° E) Land region at the equator.
ii. **Australia-Asian region:** (60° E to 180° E) covers the Asian-Australian monsoon.
iii. **Pacific region:** (170° W to 90°W) Contain the South Pacific Convergence Zone (SPCZ) and the Inter Tropical Convergence Zone (ITCZ) towards north of the equator.
iv. **North American region:** (110° W to 80° W) Narrow strip of Central America roughly oriented in a north–south direction.
v. **South American region:** (80° W to 40° W) Area covering the main southern parts of South America.
vi. **Atlantic region:** (30° W to 10° E) Includes some western parts of South America and southeastern parts of North America.

The monsoon is a macro phenomenon and its climate characteristics fascinate meteorologists, oceanographers, and geographers. In addition, forecasting the monsoon is of great socio-economic importance. Li and Zeng (2002) identified and mapped global monsoons to geographic regions according to a broad-based monsoon index called the mean dynamic normalized seasonality (DNS) of a monsoon domain. As indicated in Fig. 26.1, the tropical monsoon lies basically between the seasonal migration perimeters of the ITCZ (Wang et al., 2009). Under the annual phase of insolation, the migratory journey of the ITCZ, which occurs through transtropic pressure gradients, forms a strong tropical monsoon (Fasullo & Webster, 2003). The global tropical monsoon is predominantly distributed over the following areas: tropical Asia, Australia, and Africa (Khromov, 1957; Ramage, 1971); South America (Zhou & Lau, 1998); and the tropical eastern Pacific adjacent to Central America (Lau et al., 2012; Li & Zeng, 2002). The tropical Asiatic-Australian monsoon is the typical system containing these main subsystems: the Indian/South Asian domain (Goswami & Xavier, 2003), South of China Sea domain (Ding et al., 2001; Feng et al., 2009) the Indochinese peninsula, and North-West Pacific domain (Wang et al., 2009), Australian tropical domain and Continental Maritime domain (Webster et al., 1998).

Monsoons in the subtropical regions of southern and northern hemispheres form individually because of differences in the land-sea distribution and migration of subtropical extremes which have a connection with relief, interaction with upper atmospheric winds, and Rossby deformity radius (Molnar et al., 2010). The Northern Hemisphere subtropical monsoon resides in the parts of orient Asia and North America, North African and Tibetan Plateau monsoons, in addition to subtropical North Atlantic and North Pacific domains. The subtropical Southern Hemisphere monsoon domain involves Southern Australia, South Africa, and subtropical South

Pacific regions (Li & Zeng, 2002). Globally, monsoon systems highlight a peculiar three-dimensional upright structure which is not mentioned widely in the bibliography. The formation of this upright climatic structure in the region is due to observable differences in pressure and temperature in the middle of the Northern and Southern Hemispheres accounting for differences in the land–ocean disposal. In the lower levels of troposphere, the global monsoons indicate a tropical monsoon in the South and Subtropical features in the northern hemisphere. But, in the central and higher troposphere, they account for subtropical monsoon systems in both the hemispheres with a characteristic seasonal shift in high ridges of subtropics, character sable with an increase in altitude. A stratospheric monsoon appears in both North and South Hemispheric extra-tropical regions (Li & Zeng, 2002).

Figure 26.1 demonstrates the spreading of partially temporal pressure arrangements and existing circulations in cold and hot seasons, in addition to lower tropospheric monsoonal and non-monsoonal regions. The south-west and south-east monsoon domains gather in the Asia-Australia monsoon domain (0°–20° north, 140° east to 20° west) inside the northern hemisphere during June, July, and August and enhance as north-easterly wind during December, January, and February, accompanying north-westerlies in the Southern Hemisphere (0°–10° south, 20°–160° east). During June July and August, the dominant south-easterlies from Africa, the South Indian Ocean, and Australia (20°-0° south, 20° west to 140° east) turn towards right over the equator later and will become south-westerlies in tropics of North Africa and Southern Asia. During December, January, and February, the dominant airs in the above said North and South Hemispheric monsoon regions will overturn. The ground-level winds in the subtropics and colder regions also go through seasonal migrations. This character indicates that, either, the semi-permanent pressure schemes associated with these monsoons will reverse (explained as turning off a system of high-pressure to a system of low-pressure, and vice versa), or significant shifts in their locations.

Furthermore, to aid in the monitoring, forecasting, and study of the variability and predictability of regional monsoons, various precipitation and distribution indices were developed and are used to describe the overall variations of each regional monsoon (ex: The All India Rainfall Index by Parthasarathy et al., 1992). In addition to precipitation indices, circulation variables were also used to measure the strength of the monsoon circulation system. For example, Webster and Yang (1992) suggested an Asian monsoon vertical shear index to characterize the inter-annual instability of the entire Asian summer monsoon circulation system. Subsequently, indices for the Indian monsoon circulation and the western North Pacific-East Asian monsoon were published by Wang et al. (2001), and Australian Monsoon Indices by Kajikawa et al. (2010) have been proposed and widely used to study inter-annual and decadal variability.

Table 26.2 compares three force-limit parameters averaged over each local monsoon region depicted by the corresponding rectangular domains (Yim et al., 2014): (a) local summer (Jun-Sept or Dec-Mar) mean precipitation rates, (b) annual range or AR (Dec-Mar rainfall is subtracted in Jun-Sept rainfall in the North Hemisphere and Jun-Sept rainfall subtracted in Dec-Mar rainfall in South Hemisphere of rainfall, and (c) Yearly mean precipitation rate. The local region's summer rainfall

Table 26.2 Comparison of three monsoon intensity parameters averaged over each monsoon region

Regional monsoon	Rectangular monsoon domain	Rainfall intensity during Summer	IP1	IP2
Indian region	10 to 30° north, 70° to 105° east	0.99	0.89	0.73
Northwest Pacific	12.5 to 22.5°north, 110 to 150°east	0.99	0.91	0.59
East America	22.5 to 45° north, 110 to135°east	0.68	0.84	0.52
North American	7.5 to 225°north, 110 to 80°west	0.96	0.95	0.74
North African	5 to15°north, 30°west to30°east	0.98	0.95	0.95
South American	5° south to 25°south, 70°west	0.98	0.88	0.73
South African	7.5 to 25°south, 25 to 70°east	0.95	0.95	0.86
Australian	5 to 20°south, 110 to 15°east	0.98	0.89	0.89

IP1: intensity parameter 1: Annual range of the precipitation rate in the rectangular monsoon region
IP2: intensity parameter2: Annual mean precipitation rate in the rectangular monsoon region
Source Yim et al. (2014), comparison of variations of regional monsoon using indices

strength is well comparable with the yearly range, displaying a very well representation of regional monsoon rainfall during summer as an amplitude of the region's yearly cycle of monsoon. Progressing further, indicating a big magnitude reflection of local rainfall intensity on the total yearly precipitation intensity in most monsoon regions. This reflection may be false for Western North Pacific and East American monsoons, as these have different peaks of rainfall (i.e. in the months from July to October and in May to August, respectively) and reversals of their rainy summer and dehydrating winter are inferior to remaining monsoon regions.

26.2.3 Influence on Food Production

The impact of climate change-induced monsoon variability may be varying according to the geographical region. In some monsoonal regions, the crop yield variability is due to changes in the distribution of rainfall, in some other regions, it may be coupled with increasing temperatures which hinder crop growth. Issues of present and future climate variability and changes in agricultural production in the arid and semi-arid tropics of the world are discussed and presented separately for Asia, Latin America, and Africa as follows.

In tropical Asia, changes in diurnal temperatures beside the changes in nature and characteristics of monsoon cause variations in agricultural productivity. Increment in canopy temperature has shown significant effects on metabolic systems of crops, thus, on yield production because of thermally induced moisture stress (Riha et al., 1996). The higher temperatures along with decreased radiation because of higher cloudiness have caused spikelet sterility in wheat and rice, thereby causing yield reductions to a level that increases in dry matter because CO_2 is nullified. An annual 1% decline in the rice yields of east Java as a result of higher temperatures has been reported by Amien et al. (1996). Even though wheat crop shows a sensitivity towards increments in maximum temperature, rice show vulnerability to increments in minimum temperature. There are reports of minimization of adverse impacts due to moisture shortage on the productivity of wheat in India under elevated CO_2 levels, but these impacts are not the same for rice, causing decreased rice productivity (Lal et al., 1998). Therefore, shortage of water along with thermal stress negatively affects major food crops like wheat and rice productivity in India, although, positive impacts of higher CO_2 levels in the future are been demonstrated.

Simulation studies on the effects of climate variability on wheat productivity at multiple sites in India, performed using crop models (WTGROWS), demonstrate the dependency of crop yield on the magnitude of temperature changes. A 1 °C rise in mean temperature had no significant influence on potential yields in northern India, while a 2 °C increase in mean temperature lowered potential wheat yields in most parts of India (Aggarwal & Sinha, 1993). Subsequent studies by Rao and Sinha (1994) using the CERES-Wheat model and climate scenarios from three global climate models and the GISS model to assess the physiological effects of increased CO_2 levels. The decrease in wheat yields was less than in the current climate, even with the positive influences of CO_2 on crop yields in all simulations. The yield decrease was presumably caused by a shorter growing season due to increased temperature. Such declines in yields of important food crops such as wheat and rice could have serious implications for regional food security, alongside a growing population and increasing demand for food grains. Mukherjee et al. (2019) have reported the vulnerability of wheat yields to climate change in Northwest India during the 2000s, the adverse productivity of dryland crops such as sorghum in India due to lack of soil moisture during crop growth period (Sandeep et al., 2018), and corn yields in the Philippines (Crost et al., 2018) to monsoon variabilities.

Leng and Hall (2019) highlighted that there was a >80% probability of a fall in wheat production below its long-term average when experiencing drought in major producing countries like USA and Canada, maize in India and rice in Vietnam and Thailand, soybean in USA, Russian and India. they compared to present growing conditions with future ones using an ensemble of 11 crop models, identified an increase in yield loss risk by 9–12%, 5.6–6.3%, 18.1–19.4%, and 15.1–16.1% in wheat, maize, rice, and soybeans by the end of twenty-first century, respectively. They compared the drought indices like standardized precipitation index (SPI) with standardized precipitation evapotranspiration index (SPEI) and checked for their reliability with the yield formation. In particular, the incorporation of temperature impacts results in a significant variation in the projected yield loss risk of crops such as wheat and maize. However, similar results were obtained in case of rice and soybean too.

Precipitation has increased in south-eastern South America, northwestern Peru, and Ecuador in recent decades, whereas it has decreased in southern Chile, south-western Argentina, southern Peru, and western Central America (Magrin et al., 2014). It is anticipated that by the end of the twenty-first century, mean warming for Latin America might reach 14 °C, with precipitation anomalies (positive or negative) in tropical Latin America. Future impacts of these regional climate anomalies include significant species extinctions (primarily in tropical Latin America), displacement of tropical forests by savannas and semiarid vegetation by dry vegetation, an increase in the number of people suffering from water stress, a reduction in rice yields and a possible increase in soybean yields in south-eastern South America and increases in crop pests and diseases, with sea-level rise, weather, and climate variability and extremes affecting some coastal areas.

Zhu et al. (2019) studied the impact of climate change-induced monsoon variability on US crop yields using the five different projected climate zones. The result of the study showed that irrigated and rain-fed corn shows a declining trend (0.01 to 0.005 anomaly/year) in the twenty-first century, also both irrigated and rain-fed spring wheat tend to decline slowly, and more than 80% of these trends are statistically significant at the 10% level. In contrast, winter wheat is projected to increase in the northwest, as all climate projections resulted in a positive trend except for the irrigated CCSM4 case. Comparing the irrigated and rain-fed cases, rain-fed crop yields tend to show greater variability than irrigated crop yields. Yield anomalies are more likely to fall below 0.5 negative standard deviations from 2010 to 2099, with an increased probability after 2040. This also results in greater concurrency of large yield declines in multiple crops under rainy conditions. This is possibly due to a mode of internal variability in ENSO and may require mitigation efforts to reduce adverse climate impacts. Irrigation is effective in mitigating the effects of climate variability, as agriculturally poor years are much less likely for irrigated farming projections than for rainfed agriculture (Kurukulasuriya & Mendelsohn, 2008).

Agriculture in the semi-arid tropics of Africa, which is predominantly rain-fed, has a close relationship with climate because it depends on the timely onset of rainfall and its even distribution during the rainy season. Therefore, even a slow, small change in climate and its major parameters can critically increase climate risks. These impacts of climate instability and change in the arid and semi-arid regions of Africa have been related to projected increases in temperature, which have the greatest potential impact on water regimes, thereby increasing the sensitivity of various crops/crop systems to projected changes in temperature and precipitation. Numerous cases of the devastating effects of instability in the interannual precipitation on crop yields and economies (national and international) have been studied in recent decades (Chen et al., 2022; Gilbert et al., 2020) but none have been more dramatic, as the Sahel droughts of the early 1970s explained. Groundnut yields in Niger fell from 850 kg/ha in 1966–67 to 440 kg/ha in 1981, mainly due to drought and disease. Using the CENTURY ecosystem process model of plant-soil interactions Ojima et al. (1996) analyzed the effects of climate and atmospheric CO_2 changes on grasslands around the world, including seven out of 31 sites in Africa. Results showed changes in overall crop productivity that correlated positively with changes in rainfall and nitrogen mineralization. The response to nitrogen mineralization compared to precipitation variability was consistent, with the general observation of a positive grassland response to the addition of nitrogenous fertilizers.

26.3 Climate Change: Scientific Background, Evidence of Aberrations

Though the term climate change is carefully associated with an increase in mean (regional) temperature, changes in mean temperatures often coincide with temperature changes and their extremes (Rummukainen, 2012). The number of warm days and nights is unmistakably increasing, while the opposite is true for cold days and nights (Zhang et al., 2011). There is an exceptionally strong increasing trend for the occurrence of heat waves (Kyseli & Beranova, 2009), while low-temperature extremes are expected to be limited, albeit necessarily less intense and prolonged (Parish et al., 2012). Climate change caused by global warming is also affecting precipitation patterns. According to the alleged Clausius-Clapeyron relationship, a one-degree increase in temperature leads to a 7% increase in humidity (Rummukainen, 2012). Higher humidity in combination with rising temperature leads to changes in the water cycle and thus to increased evaporation by plants. Similar to previously detailed temperature patterns, precipitation has grown completely heterogeneously in the spatial dimension. In the middle and higher latitudes of the northern hemisphere, precipitation has increased by 0.5 to 1% per decade since 1910 (especially in autumn and winter), in the subtropics, precipitation has decreased by about 0.3% per decade (Walther et al., 2002) have been reported.

Not only the mean precipitation is subject to changes, but also the precipitation extremes are influenced by climate change. While average precipitation is limited by evaporation and has a loose connection with atmospheric humidity, extremes of precipitation are closely related to the total water content of the atmosphere, particularly for convective precipitation (Berg et al., 2013). Lehmann et al. (2015) showed an increased number of record-breaking precipitation occurrences per year between 1980 and 2010 on a global scale. Precipitation extremes are also becoming more common over land areas of the northern hemisphere and in some regions closer to the equator (Orlowsky & Seneviratne, 2012). Notably, an increase in extreme precipitation events is expected for southeast Asia (Lehmann et al., 2015). There is a most likely increase in trends in the number of droughts besides extreme events like dry spells. Drought prevalence is mostly increasing in large parts of Africa, the Mediterranean region, parts of North- and South America, and Southeastern Asia (Rummukainen, 2012). On the other hand, there is an increase in occurrence of destructive storms like hurricanes (Keller & DeVecchio, 2016). Being defined as areas of low air pressure characterized by rotating winds, cyclones are classified as tropical or extratropical based on their region of origin. Tropical ones form between the 5th and 20th latitudes and thus over warm waters. In contrast, extratropical ones have cool cores because they form in combination with weather fronts between the 30th and 70th parallels. Both for tropic and extratropic winds, there is evidence that wind severity (speed, rate of precipitation, etc.) may be increasing. In addition, global warming will likely affect the regions where storms occur (Rummukainen, 2012).

26.4 Climate Change Impacts on Monsoons

The defining characteristics of Earth's climate are global monsoons, which are a forced impact of the coupled climate on the yearly solar radiation cycle. The monsoon domains are essentially defined by the rainfall contrast at the solstice. Western Mexico and Arizona are parts of the North American area, as are Central America and Venezuela. African regions near equator and Maritime continents also show an annual overturning of surface winds, former one being characterized by two peaks in the equinoctial seasons and the latter being heavily varied by complex terrain (Chang et al., 2005). In this section, an attempt has been made to conclude future global and regional variations in rainfall using novel climate model results. The level of confidence in the results varies due to inherited uncertainties and other model flaws, but they provide an approximate explanation of climate change's impact on monsoons.

26.4.1 Global Scale

Climate models (General Circulation Models; GCMs) are novel techniques to project future climate change and its impacts on crop production apart from the regular experimentations on increased CO_2. These latest simulations by the Intergovernmental Panel on Climate Change (IPCC) in the Shared Socio-economic Pathway (SSP) indicated an increase in global mean surface temperature, mean precipitation, and differences in land surface temperature increases over the oceans, high latitudes have higher temperatures than low latitudes, and northern hemisphere temperatures have higher temperatures than southern hemispheres. It has been hypothesized that tropical atmospheric moisture content, latent warming, and the overall hydrological cycle increase with increasing tropospheric temperature (Hlywiak and Nolan, 2021). Simulations studies with growing greenhouse gas concentrations show an increasing intensity of the Asian summer monsoon (Hulme et al., 1998). These outcomes represent a rigorous check for climate models used for assessment of future climate change (Roeckner et al., 1999).

Examination of monsoonal precipitation (1930–1970) shows an increasing trend over Bangladesh and northern China region (Kripalani & Kulkarni, 2001). Therefore, inclining trends in northern China and Bangladesh (Fig. 26.2), derived here from data representing a short period, should be viewed as accompanying caution. It is reasonable that the trend observed over the past 56 years indicates a transition from a powerful step to a weak chapter in multi-decadal instability. Previous studies indicated that there was a rapid change in atmospheric air movements and ENSO in the mid-1970s (Trenberth & Hurrell, 1994). Figures 26.3a–c imply that changes in Northern Hemisphere monsoon, reflect regime change, which may be part of the interdecadal Pacific Oscillations, a period of 50 to 70 years, or due to changes in the tropical/subtropical land and snow at higher latitudes (Chase, 1996). We still have much to learn about the causes of the observable trends in the global monsoon.

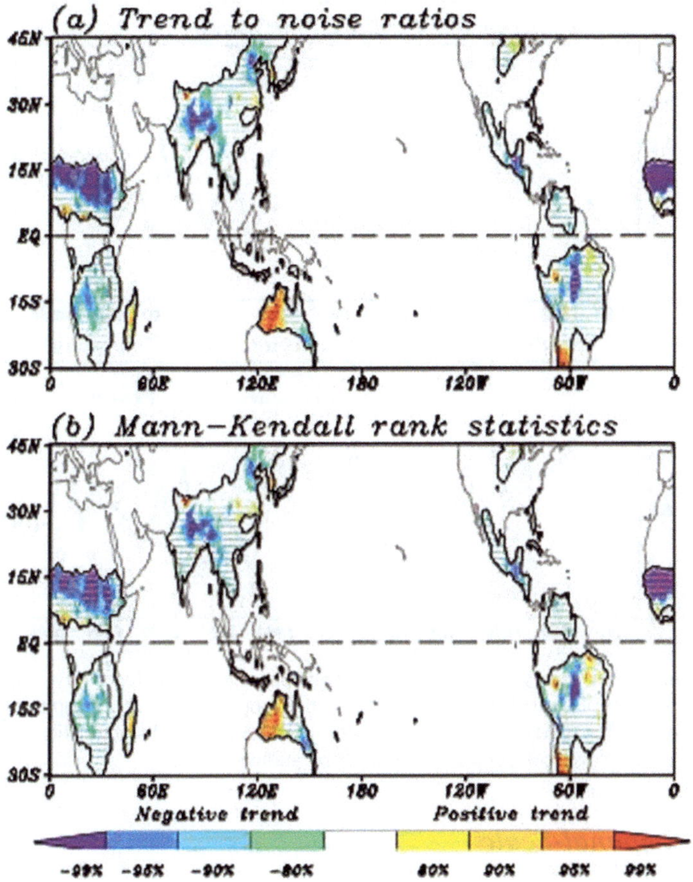

Fig. 26.2 Statistical significance of linear trends in summer monsoon precipitation. *Source* Kripalani and Kulkarni (2001), Monsoon rainfall variability and connections over Southern and Eastern Asia

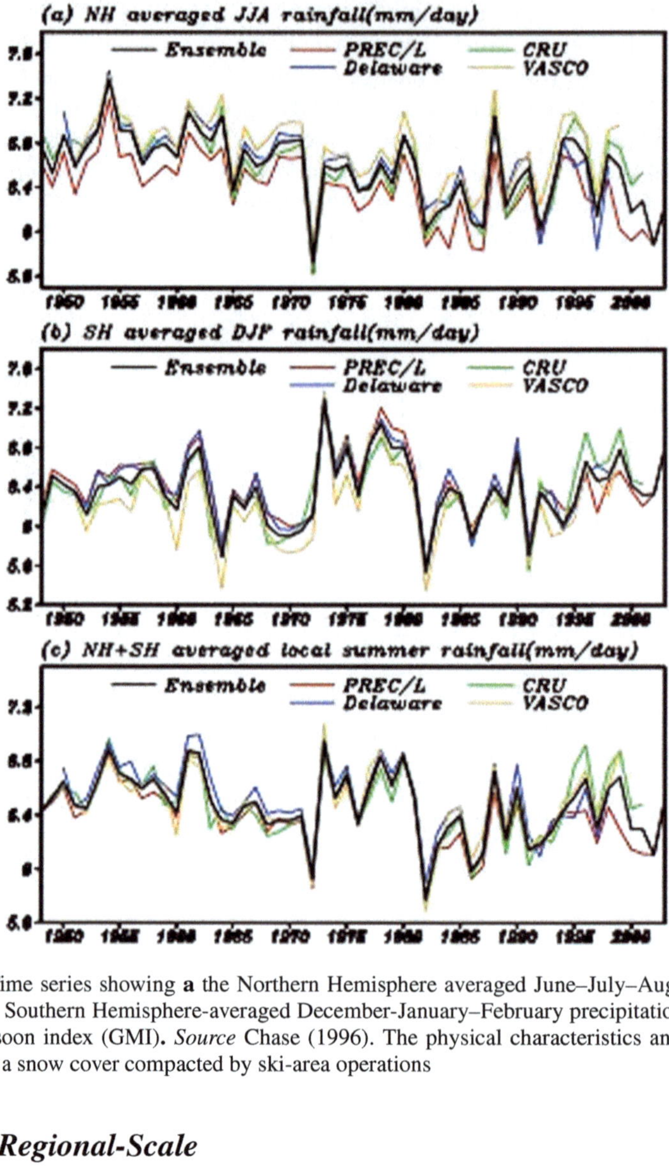

Fig. 26.3 Time series showing **a** the Northern Hemisphere averaged June–July–August precipitation, **b** the Southern Hemisphere-averaged December-January–February precipitation, and **c** the global monsoon index (GMI). *Source* Chase (1996). The physical characteristics and meltwater output from a snow cover compacted by ski-area operations

26.4.2 Regional-Scale

A. **Asia**

The two major regions of Asia were notified of the monsoon event, firstly the Indian subcontinent and secondly the East Asian region. The effects of climate change on the monsoons in these regions can be explained by the frequency and duration of droughts and floods. From a meteorological point of view, droughts, and floods mean an excess or deficit of precipitation compared to normal conditions, regardless of the severity of the disasters they cause. From the climatological perspective, several drought and flood indices have been developed. However, the deviation or anomaly of precipitation from the long-term mean is used to more broadly and practically describe droughts and floods. If the precipitation amount for the summer (June August) is in the range of −25 to −50% of the normal value, this period or season can be defined as a drought period or season, if the precipitation deviation is greater than −50% this corresponding period is defined as severe drought defined. In contrast, 25–50% and more than 50% of runoff are defined as floods and severe floods respectively (Ding et al., 2008).

The effects of climate change on monsoons are addressed by studying their onset and termination in a region. An early retreat or late end of the monsoon over a region affects the availability of water resources for various activities such as agriculture and hence food production. Wang and Ding (2006) studied the annual deviation of monsoon precipitation from normal over India (Fig. 26.4). It differed from year to year with a huge variation between years. They also found that there were significant negative deviations from monsoon rainfall in 1877, 1899, 1918, 1972, and 1987, suggesting major deficiencies in rainfall or drought. Recently, 2002 was a strong monsoon drought year, while 2003 was normal. In 2002 monsoon, daily precipitation remained low throughout the season, while in 2003 it was above normal. It should be noted that rainfall varies greatly from the seasonal mean in certain years, resulting in drought and flooding.

Another climatic anomaly that illustrates the impact of climate change on monsoons is flooding. In general, these are due to very heavy rainfall with monsoon disturbances such as tropical cyclones, depressions, and western disturbances. Single flood occurrences occur in all of the subcontinent's main rivers, except the Indus and Ganges river basins in Indo-flood Pakistan's plains, the Brahmaputra basin in northeastern India, and Bangladesh, and the Mahanadi, Tapi, and Narmada river basins in central India. The diversion of excessive monsoon runoff from Nepal's rivers exacerbates the flooding situation in India's Bihar state. Individual flooding occurrences in each of the above river basins occur even in drought-monsoon years, when a meteorological disturbance in a river basin results in unusually significant rainfall, even though total monsoon production may be below average for the time of year. A single flood event in a river basin is identified as exceeding the water discharge danger mark along with a gauging station, causing the river water to overflow its banks and devastate life, property, and crops in the surrounding areas (Kale, 2003).

It is a well-known fact that India's annual rainfall shows no particular trend in its annual amount and frequency. However, the seasonal precipitation series shows a decreasing trend (Malik et al., 2019; Praveen et al., 2020). This variability is due to the depressions formed in the neighboring areas of the subcontinent. Studies on repetitions of tropical depressions (TDs) moving across India and the monsoon precipitation over the 117 years from 1901 to 2017 were conducted by Nandargi and Mahto (2019), percentage anomaly of precipitation and the frequency of TDs showed that only on five occasions in country recorded 15% or more monsoon rains. While the country experienced ten times less than −15% of normal monsoon rainfall. The repetition of TDs was greater from 1921 to 1970, after which a sharply decreasing

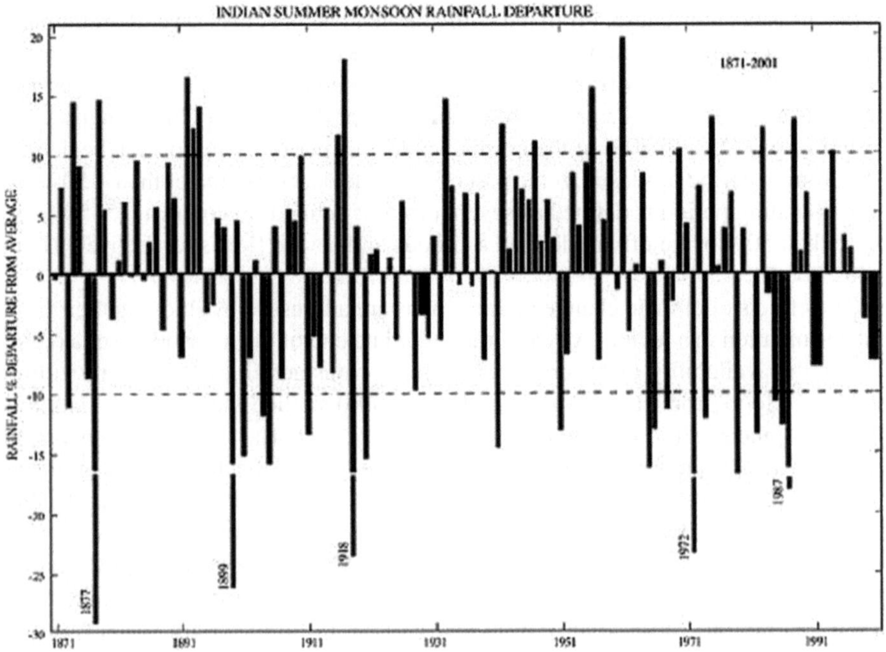

Fig. 26.4 The variation in monsoon rainfall from year to year across India (June–September). *Source* Wang and Ding (2006). Changes in Indian monsoon precipitation over the past 56 years

trend is observed. Although their frequency has increased over the past two years, there is no significant increase in rainfall. It can therefore be seen from above that over the past 50 years, the frequency and duration of formation of TDs in the Bay of Bengal, which is the main source of precipitation during the June–September monsoon season over the Indian region, has decreased. This led to a reduction in monsoon rainfall, which led to drought conditions 23 times (Table 26.3). This is mainly related to the increase in sea surface temperatures (SST), positive and intense El Nio phases between 1960 and 2017 (Akilan et al., 2018).

While differences in onset and finish cause droughts, at the other extreme, floods are caused by excessive rainfall in a specific location. Dhar and Nandargi (2000) provided a detailed overview of the severe flooding in India during summer monsoon, showing three bands of single-day flooding that produced severe east–west oriented rainstorms along the Indo-Gangetic Plains in the east-flowing Ganga Yamuna basins and along with central India, the eastern portion of which lies in the east-flowing Mahanadi and central and west portions in the west-flowing Tapi and Narbada basins.

B. **Africa**

The African monsoon (AM) is a weak stream of moisture that originates in the equatorial and southern tropical Atlantic basins and convects on the continent during the tropical rainy season, which lasts from July to September. For this reason, the

Table 26.3 Tropical depressions (TDs) in El Niño and La Niña years and monsoon rainfall as percentage of season normal rainfall

El Nino years	El Nino Category	Monsoonal rainfall (%)	TDs		La Nina Years	La Nina category	Monsoonal rainfall (%)	TDs	
			Sea	Land				Sea	Land
1902	Average	−9	7		1908	Average	7	5	
1904	Robust	−12	5	1	1909	Average	7	3	
1911	Robust	−15	4		1910	Average	8	3	
1913	Average	−10	5	1	1916	Robust	13	4	1
1918	Very robust	−25	4	1	1933	Robust	15	9	2
1925	Average	−3	8	2	1948	Average	2	9	
1930	Average	−5	4		1949	Average	1	6	
1935	Mild	−3	7	1	1954	Average	3	6	2
1944	Very mild	5	9	2	1955	Average	10	6	
1951	Robust	−19	7	1	1970	Average	12	6	1
1957	Average	−2	3		1973	Average	8	5	2
1965	Very robust	−18	5	1	1975	Robust	15	7	2
1968	Robust	−10	7	1	1988	Robust	18	3	
1972	Very robust	−24	5	2	1999	Mild	−4	3	1
1976	Neutral	2	3	2	2000	Mild	−7	1	1
1982	Robust	−15	6	2	2007	Average	6	5	
1987	Very robust	−18	1	3	2010	Average	2	–	–
1991	Average	−7	3		2011	Average	1	3	1
1997	Mild	2	6						
2002	Very robust	−19	–	–					
2006	Mild	0	5	1					
2009	Very robust	−23	3						
2015	Robust	−14	3	3					

Source Nandargi and Mahto (2019), Frequency and intensity of tropical disturbances over the Indian region and its neighboring seas with associated rainfall during the monsoon season

monsoon dynamics and the associated rainfall in the latitudes of the Sahel show strong intra-seasonal and inter-annual variability, which explains the high vulnerability of the region. Monerie et al. (2012) examined the possible future changes in the AM (from 2030 to 2070) using CMIP3 and CMIP5 models in a medium to low RCP scenario. They reported that the historical differences from RCP4.5 show significant increases over a regional window in temperature over the Sahara, southern Europe, Middle East (2.5 to 3°C), and the Mediterranean: 1.5 °C (~2°C) to the east. Precipitation amounts are expected to hike in the central Sahel and over the Atlantic region near equator, but a decrease in the western Sahel over Senegal and southern Mauritania, and especially over broad African regions.

Based on the multi-model, *Each Model Each Vote* concept by Santer et al., 2009, analysis of simulated monsoon dynamics and climate change for a current (1960–1999) and a future period (2031–2070), showed improvements in African precipitation. Positive trends in mean intensity (Lodoun et al., 2013), precipitation extremes (Sanogo et al., 2015), and the frequency of strong mesoscale convective systems all contributed to the rise in seasonal precipitation (Taylor et al., 2018). A wetter postseason and a later finish to rainfall have been seen in some West African nations (Lodoun et al., 2013). All of the alterations were in line with CMIP5's response to greenhouse gases (Sherwood et al., 2020). Preliminary results from CMIP6 confirm that the Sahel region is becoming wetter except for the west coast and that the rainy season is extending later. CMIP6 models predict that monsoon precipitation is anticipated to rise across the country, notwithstanding the vast variance. In East Africa, reported increases in boreal short fall rains are more robust than negative trends in long spring rains (Cattani et al., 2018; Maidment et al., 2015). There is significant regionality, as well as sensitivity to Indian Ocean SSTs and Pacific variability (Funk et al., 2014). Some CMIP6 models anticipate little consensus on how East Africa's precipitation would change, while others forecast greater precipitation during the short rainy season and a shorter long rainy season (Han et al., 2019). In the Congo Basin, observable trends in precipitation were explainable with doubts, but only one study reported an earlier onset of spring rains (Taylor et al., 2018). Analysis has revealed a complete improvement in CMIP6 models in overestimating Congo Basin precipitation, although projections of changes in the SSP24.5 scenario are inconsistent. The CMIP6 models predict that under the SSP24.5 scenario and in the second half of the twenty-first century, SAF land monsoon precipitation is likely to increase in summer but decrease significantly in winter, so the annual range will increase, but the annual mean precipitation will not change significantly (Cook & Vizy, 2019).

C. **Latin America**

Long-term precipitation patterns in the twentieth century were less than or equal to, however, trends within this region varied greatly (Pascale et al., 2019). Between the 1950s and the 2010s, the monsoon crest strengthened and modified the patterns of temporary reverse valleys, leading them to generate severe weather events less frequently (Lahmers et al., 2016). Recent research suggests that under anthropogenic global warming scenarios, severe events in North and Central America will become more powerful (Aguilar et al., 2005). Climate models suggest a redistribution of mean NAM precipitation from early to late with meager complete reduction (Cook & Seager, 2013) and a greater decrement in Central American rainfall (Colorado-Ruiz et al., 2018). However, outcomes are less confidently acceptable because of local biases (dynamics of land cover and land use, soil moisture, etc. representation by model) and distant biases (SST variability during current and upcoming years) (Prein et al., 2016). From the 1950s to the 1990s, a significant positive precipitation trend was observed in southeastern America, which, is associated with interdecadal change (Grimm & Saboia, 2015) due to ozone depletion, and rising greenhouse gases (Vera and Diaz, 2015).

Because of the tropical Atlantic's impact and potential to counteract spring-to-summer precipitation anomalies in central and eastern South America induced by land–atmosphere interactions, the tropical South American monsoon's patterns were less consistent (Grimm & Saboia, 2015). Recently, Fu et al., 2013 have identified the extension of dry season and winds turning into even drier (southern Amazonia) along with noticeable impacts on vegetation and moisture movement to the central SAM region. Future precipitation changes estimated by the CMIP6 models are identical to the variations guessed for El Nio: lesser change in annual mean precipitation, with more dry winter/spring and hiked monsoon precipitation peaks. This is being captured and identified as similar to El Nio impacts (Grimm & Saboia, 2015) and CMIP5 projections showing a late onset and early closure of seasons, higher number of high-intensity rainfall, and lengthy dry spells between simultaneous rain events (Christensen et al., 2013). On the other hand, the CMIP5 projections have an equal probability that they underestimate the climatic variations of South America and its sensitivity to climate forcing during monsoon periods (Fu et al., 2013). For overcoming this underestimation by projections Bias-corrected projections were employed, they too indicated a drier climate over eastern Amazonia (Duffy et al., 2005). Hence, the risks and uncertainties related to climatic drying of higher magnitudes and their ill impact on rainforests are expected to be happening for real in near future.

26.4.3 Spatial and Temporal Scale

The increased incidence of heavy rain on daily to multi-day time scales, as well as intensive rain on hourly ranges, is one of the most reliable signs of expected future changes. Heavy rainfall may rise significantly more than average precipitation, especially in Asia (Kitoh, 2017). Heavy and intense rainfall is more strongly regulated by ambient humidity, which is connected to the Clausius-Clapeyron relationship and changes in the convective circulation at scale, as opposed to mean variations in precipitation. Under RCP8.5 with a high signal-to-noise ratio, extreme 5-day GV precipitation reacts essentially linearly to global temperature change at a rate of 5.17 percent C1 (4.14 percent 5.75 percent C1) (Zhang et al., 2018). Extreme precipitation in the Asian monsoon area is the most sensitive to warming regionally, whereas changes in the North American and Australian monsoon regions are mild and have poor signal-to-noise ratios. Under the SSP24.5 scenario, the CMIP6 models predict changes in severe 1-day precipitation of +58 percent across South Asia and + 68 percent over East Asia in 20,652,100 compared to 19,792,014. With future forecasts of 1.52.5 °C global warming, modeling tests reveal a doubling of the frequency of rain extremes over the Indian subcontinent (Bhowmick et al., 2019). In meanwhile, mild to moderate rain showers may become less common (Sooraj et al., 2016).

Variations in monsoon precipitation can happen at a range of times. Brown et al. (2017) showed increases in precipitation variabilities (Scenario: RCP8.5) for individual timescale from each day rainfall to 10-year average rainfall over Australia, South Asia, and East Asia domains. The heaviest fractional increments in precipitation variability were observed on all sub-annual timescales for South Asian monsoon and on annual to decadal timescales for the East Asian monsoon. Forecasted variability in precipitation were having a high degree of relation with a variability of average wet season precipitation of all monsoon domains at all the time scales under study. Such chosen studies on CMIP5 projections on monsoons indicate more sequent floods, droughts, and other climatic aberrations over South Asia (Singh et al., 2019). Due to the more rapid increase in loss of moisture from the surface layers, 2015–2100 projections (CMIP6, SSP24.5, and SSP58.5) show that west–east Asia will experience more droughts and their related risks at a rate fast than East Asia.

Projections of future extreme changes in precipitation have special importance because of several reasons, especially in dense populations, and fast-growing coastal areas. The first reason is that there has been a lot of rain. The second argument is that anticipated changes in land-falling tropical cyclones will have an impact on changes in excessive precipitation in coastal locations. Simulated tropical cyclones (Knutson et al., 2019), for example, show a consistent (although with lesser confidence) northward tendency. Thinking that this signifies more occurrences, this would result in severe precipitation increases on the Korean peninsula and in Japan's coastal areas, as well as a decrease in China. The increment in monsoon rainfall and tropical cyclones, along with the rising sea levels, may cause exacerbated impact in several parts, coastal regions of the Indian subcontinent are not being an exception (Ray et al., 2019).

26.5 Recent Developments in Monsoon Variability Research

Assessing susceptibility to climatic aberrations is an important part of human perception with climate change studies. Assessments on vulnerabilities are conducted to find and characterize what factors are sensitive to climate risks at what levels, what are the causes staying behind, to formalize adaptation and factors for adaptations, and to identify opportunities for adaptation. Due to the recent developments in climate research and the development of climate models, everyone is aware of the changes that occur in various weather parameters such as temperature, precipitation, etc. due to climate change. Various climate models highlight continued and accelerating changes in future climate, with rising thermal regimes, changing amounts, and patterns of precipitation at rates not notified in recent human history. Superimposed on these, the variabilities are facing social, cultural, and economic pressures that can mitigate or hasten climate vulnerabilities. This impact on various human activities requires studies on the variability of monsoons caused by climate change.

26.5.1 Challenges for Assessment of Impacts

To fit the most appropriate method for managing drought and assessing adaptation, it is helpful to find a difference between simple, complex, or complex design interventions based on the degree of agreement. Those exact inputs, without a doubt, lead to specific results (Patton, 2011). In conclusion, simple interventions are required when there is a straightforward logical link between inputs, outputs, and results. Over a lengthy period, complicated interventions may include several components or parties. The ambiguities in the selection of input data and analytical techniques, and use of generated output to analyze the climate change impacts are the major complex challenges to be faced in the climate change impact assessment. A causal rain may become visible after a climatic event or may reach no end state where the problems are resolved (Shapiro et al., 2010). For minor interventions, being characterized by presence of consensus, and certainty, well-established monitoring and assessment methods are used in international development and other fields as there is an advantage of understanding short- to medium-term processes. Where there is consensus but not certainty, more complicated adaptation projects may necessitate a greater range of techniques (Shapiro et al., 2010). Many adaptation measures, however, are complicated, and the underlying uncertainties associated with climate change make assessment particularly difficult. Attributing observed changes to specific activities in complex contexts, setting baselines and targets with changing climatic hazards, and assessing the effectiveness of adaptation initiatives with short- and medium-term benefit-term evaluation cycles are all challenges to consider when monitoring and assessing these types of interventions (Lamhauge et al., 2012).

In recent years, theoretical frameworks for monitoring and assessing adaptation have arisen at the project, program, and fund levels (Bours et al., 2014). These frameworks give broad parameters within which to perform adaptation monitoring and evaluation. Adaptation planners and practitioners, on the other hand, have struggled to find methods within these guidelines to address the challenges of consistent donor accountability, measuring results within program or funding cycles, and the more specific challenges of assessing the effectiveness of climate change adaptation in the longterm. As a result, this paper focuses on the approaches that may be used within these wider frameworks, as well as an overall strategy to enhance CCAR monitoring and assessment.

26.5.2 *Methodologies of Impact Analysis and Their Feasibilities*

26.5.2.1 Social Science Methods

Surveys: Applicable to analyze both simple and complicated problems, address attribution and basic goals. Surveys by Heltberg et al. (2012) after changes in monsoon and global food and financial crises in 2008 provided a report from the perspective of affected people and a strong base for the generation of steps of relief programs.

Focus-group interview: a kind of interview in which respondents are questioned in a group setting and allowed to discuss among themselves is considered for analysis. Applicable to the analysis of both simple and complicated problems, addresses climate change attribution. Leeuw and Vaessen (2009) used this methodology for analysis of project beneficiaries using on-desk reviews, surveys, and focus group discussions.

Semi-structured interviews: One-to-one interviews with multiple questions on the matter to be discussed. A complicated method addresses complex challenges (Iturriza et al., 2020).

Most significant change analysis (MSC): involves asking participants or beneficiaries to express their success stories, and major changes either good or bad they experienced as a result of a particular relief program or action taken. (Davies & Dart, 2005). Applicable for complex problems addresses the challenge of attribution.

Outcome mapping: carried out to mark changes in the behavior of individuals, multiple people, and organizations on which program was inferred, instead of the physical variables that vary with the goals of the adoption program. Applicable to complicate and complex climate change problems. Complete the climate change characterization challenge (Chausson et al., 2020).

Limiting factor analysis: this technique is used for developing and sharing knowledge of the important factors that need to be analyzed and if necessary managed (and if possible) for a project or program to be sustainable over the long term. Particularly applicable to intricate and complex climate change problems addresses the long-term challenge of climate variability. Gullison and Hardner (2009) used the technique to identify constraining factors responsible for a wide range of projects, their types, and ecosystems.

Scenario building: involves generation of a range of alternative futures from modeled data and participatory scenarios to model data. Applicable to study the intricate and complex issues of climate change used to address the baselines/goals and long-term horizons.

26.5.2.2 Experimental Evaluation Methods

Experimental design with randomized controls: the method is applied to compare the treated individual or group (e.g., program participants) to a control individual or group (e.g., nonparticipants) using statistical techniques. Applicable to simple climate change studies, Addresses challenge of attribution and Baselines/Targets. Most of the climate change experiments employ this methodology for evaluating the impacts of climate change on plant growth (Piao et al., 2019).

Quasi-experimental design: identical to experimental design method, but has a different rationale for assigning control individuals or groups for comparison. However, this is not randomized. For example, can be done through a pipelined approach using groups at various stages of implementation. Analyzed using statistical techniques. Applicable to simple and complex climate change studies and addresses the challenge of attribution and baselines/targets. Kerr et al. (2018) used these techniques for climate change adaptation research on agroecology in Malawi.

26.5.2.3 Econometrics and Statistical Techniques

Modeling: Simulation of real-world, using physical relationships, expressed through mathematical equations. Applicable to Complicated and complex climate change studies, Addresses challenge of Long time horizons (Goodwin, 2007).

Stochastic baselines: Extraction of information on variables that are not captured by other techniques using probability distributions. The approach can be useful because of its ability to work on considerations of more than one alternative future scenario. Applicable complicated and complex climate change studies address the challenge of baselines/targets (Vesely et al., 2019).

Rolling and reconstructing baseline studies: Collecting baselines at various stages of the program rather than all at once. Reconstructing baselines utilizing secondary administrative data such as national surveys, as well as certain practical tools such as recall and mapping techniques. Applicable complicated and complex climate change studies address challenge of baselines/targets.

Normalization: Used to normalize data according to different trends. Carried out by fitting existing data according to existing trends and averages. Applicable complicated climate change studies address the challenge of baselines/targets.

Propensity score matching: Statistical procedure used to build a comparison group that has an identical propensity to get the intervention. Later this comparison group is used as the treatment group to compare adaptability. Applicable simple and complex climate change studies address the challenge of characterization, baselines/targets (Pradhan & Rawlings, 2002).

Difference in difference: Compares the effects between treatment and control (comparative) groups both before and after the delivery of an intervention. Applicable simple and complex climate change studies address the challenge of characterization, baselines/targets.

Regression analysis: a simple statistical methodology, that shows the level of variation of samples, linear or nonlinear relationship is developed after checking the significance, statistically. Applicable for simple and climate change studies only addresses the challenge of characterization (Moore et al., 2009).

26.5.2.4 Integrated Assessment Methodologies (IAMs)

To comprehend cause-and-effect links between human behavior and environmental and economic effects, integrated assessment (IA) entails the integration, interpretation, and sharing of knowledge from multiple scientific fields. It entails both the modeling and description of impacts, as well as the assessment of policy options. The goal of IA is to assist users, such as policymakers and stakeholders, in making better decisions by better understanding the repercussions of assumptions and policy choices. This link can be made in a variety of ways. One approach is to utilize decision-focused models, which incorporate formal decision analytic skills, such as trade-off displays, within IA models. Another type of impact assessment is policy-oriented impact assessment, which is aimed to offer effective data that may be utilized outside of the IA system (Bell et al., 2003).

26.5.3 Impacts of Monsoon Variability

Climate change is an important external factor causing internal variations in ecosystem vulnerability, which is an important measure of climate change-induced monsoon variability in any ecosystem and can be expressed as a function of the character, magnitude, and rate of climate change and the variation to which a system is exposed, its sensitivity and adaptability (IPCC, 2007). Such ecological impacts may influence nature's ability to generate timber, crops, livestock, etc. Concerns have also been raised concerning the influence on biodiversity in general, and specifically on species extinction.

26.5.3.1 Ecological Impacts

Assessing the impacts of climate change on the ecosystem involves several steps, particularly when assessing long-term impacts. Lack of care in this approach might result in arbitrary outcomes and underestimation of the uncertainty. A minimal set of criteria is proposed for several crucial phases in this process to prevent such unfavorable outcomes. The following decision criteria must be satisfied to get an accurate and reliable assessment.

i. Clear definition of emissions scenario,
ii. Selecting a suitable climate model,
iii. Evaluation of climate model skill and biases,
iv. Quantification of model variability of the General Circulation Models (GCMs).
v. Selecting an ecosystem model and assessing uncertainty,
vi. Proper consideration of responses,
vii. Inclusion of effects of elevated levels of pollutants like CO2 on plants,
viii. Evaluation of effects of simple assumptions,

ix. Considerations on relationships between animals and vegetation.

Liang et al., (2017) studied the Effects of monsoons on the distribution pattern of tropical Asian plants, revealing the influence of the monsoon on the migration and evolution of tropical plants is conducive to an in-depth understanding of the distribution of vegetation and the formation and maintenance of biodiversity in East Asia (Fig. 26.5). They also highlighted the effects of monsoons on the northward dispersal and species differentiation of tropical plants, indicating the tropical edge of Asia and the superimposed influence area of the three summer monsoons are precisely the areas of southwest China and the Indo-China Peninsula where high mountains, rivers, and limestone landforms are concentrated. Due to the common influence of summer monsoons in different directions and seasons, the tropical monsoon climate here is more obvious and longer, maintaining the basic hydrothermal conditions that can meet the growth and reproduction of tropical plants. The complex topography here and the ecotones formed by different vegetation types also provide a rich and diverse ecological niche for plants. Therefore, southwestern China and the Indo-China Peninsula are a hotspot for the adaptation and differentiation of tropical plants, resulting in a large number of endemic species, which may be one of the main reasons for the formation and maintenance of China-Myanmar biodiversity hotspots (Myers et al., 2000). Taking the species diversity and geographic distribution of endemic centers of four typical tropical plant groups as examples, this paper analyses how the monsoon affects the geographic distribution and species differentiation of tropical plants under the combined action of local topography.

East and Southeast Asia have been regarded as the origin and differentiation centers of modern angiosperms (Buerki et al., 2014), and the tropical monsoon may play an important role in it. The monsoon climate not only promotes the migration of the propagules of some tropical plants northward to the northern edge of the tropics but also intensifies the isolation of local habitats through intense rainfall and violent dry–wet alternation. Plants gradually adapted to the local special terrain and climate and differentiated into new groups, making the three superimposed areas affected by summer monsoon (the border of Yunnan, Guizhou, and Guangxi in southwest China and the Indo-China Peninsula) the center of species diversity and endemic species distribution of tropical plants in Asia one. With global warming, the effect of the Asian monsoon on the northward migration of tropical plants may be more pronounced, and it will have a more profound impact on plant diversity in East Asia and China. In future related research, further attention should be paid to the changing laws of the monsoon climate and its effects on the long-distance dispersal and species differentiation of tropical plants, the relationship between the flora of Southwest my country and Hainan Island and the Indo-China Peninsula, and a more accurate understanding of the flora of these regions.

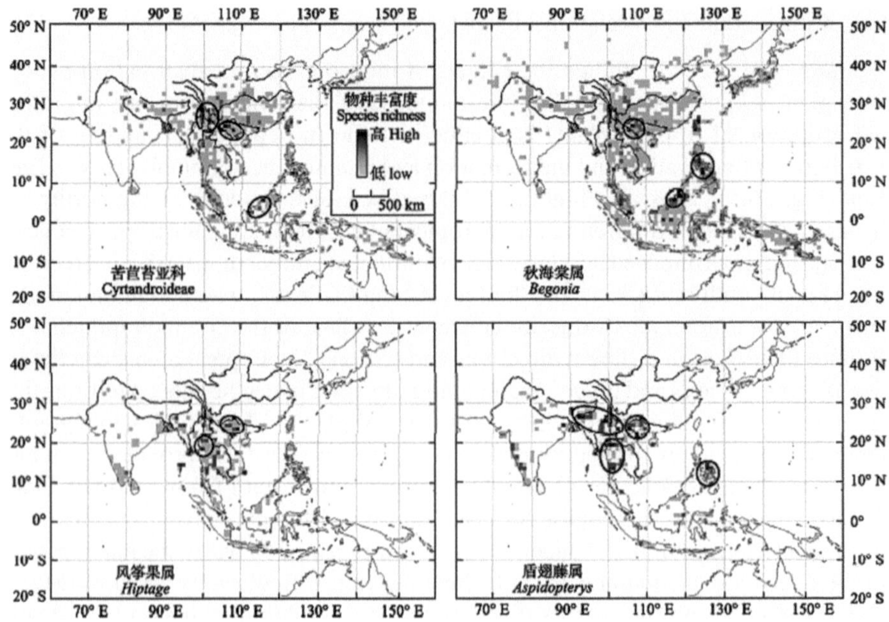

Fig. 26.5 Species distribution patterns of four representative tropical plant groups in Asia. *Source* Liang et al. (2017), application of vegetation condition index for estimation of long-term spatio-temporal drought affecting vegetation distribution. areas with the highest ratio between species richness and endemism are encircled

26.5.3.2 Economic Impacts

Monsoons, identified as one of the most energetic tropical climate systems, exert a great impact on social and economic state of billions of people around the world. Despite technological advances like release of improved crop varieties and usage of irrigation in crop production, agriculture productivity is still greatly influenced by the weather and climate. This can be illustrated by a situation of weak monsoon rains in 1987, resulting in a significant decrease in crop production in different developing countries like Bangladesh, Pakistan, and India which increased wheat importation by India and Pakistan (Rosenzweig et al., 1993). Similarly, in the previous two decades, there has also been a continued decline in production in major production regions like Africa, caused by persistent drought. Besides the decrements in food grain production, dramatic growth in agricultural trade provided significant food supplies for major importing nations and significant income for exporting nations. These instances highlight the interconnectedness of agriculture and the monsoon, the global aspect of food trade and security, and the necessity of taking into account the effects of climate change and monsoon variability in a larger context.

Parry et al. (2004), studied the effects of climate change on global food production under IPCC-Special Report on Emissions Scenarios (SRES) emissions and socio-economic scenarios. They reported the variations in regional crop yields under each

SRES scenario are due to interactions among temperature and precipitation, direct physiological effects of increased CO_2, and effectiveness and availability of adaptations for the 2020s, 2050s, and 2080s under the HadCM3 SRES scenarios with and without CO_2 effects as compared to 1990 food production. Selective reports on impact of observed climate change on monsoon variability during the twentieth century over India were gathered by Mall et al., (2021) which indicated the changes in monsoon for intra-season, inter-season, annual and decadal scales as indicated in Table 26.4.

The changes in rice productivity due to variations in rainfall have been recorded by different researchers in a typical monsoon country such as India (Table 26.5). Several researchers such as Aggarwal and Mall (2002) predicted rice yield using a

Table 26.4 Past studies and future analyses on observed rainfall variability over India

Region	Rainfall variability past trends	Source
All-India	Decreasing trend in monsoon season rainfall and is random if analyzed over a long period	Thapliyal et al. (1991)
Indo-Gangetic Plain Region (IGPR)	Monsoon precipitation during summer over the western Indo-Gangetic plain regions (IGPR) had an increasing trend (170 mm/100 years) from 1900 onwards, central IGPR had decreasing trend (5 mm/100 years), and over the eastern IGPR a decreasing trend (50 mm/100 years). 100 years)	Singh and Sontakke (2002)
Western Himalayas	No variability from last 100 years	Bhutiyani et al. (2010)
NE Peninsula, NE India, NW Peninsula, West Coast and Central Peninsula	The northeastern and western Peninsula, northeast India had a decreasing trend in rainfall (6 to 8% of normal/100 years), west coast and central peninsula had an increasing trend	Kumar et al. (1992)
Rajasthan desert	Slight increase in monsoon rainfall; despite large inter-annual variations	Pant and Higngne (1988)
Luni River Basin (Arid west Rajasthan)	Annual precipitation trending upwards at 19 stations (around Ajmer in the upper part of the Luni Basin)	Singh et al. (2005)
Year 2050s		
All India	Rainy days are decreasing in large parts of the country. Magnitude of decrement is stronger in western and central India	Ashrit et al. (2003)
Himalayan Foothills and northeastern India	5–10 days increments on rainy days	
Northwest India	Overall increase in the rainy day, rainfall intensity by 1–4 mm/day except for small areas in the northwest India where the rainfall intensities decrease by 1 mm/day	

↑: Increase, ↓: Decrease, ↔ : no Impact
Source Aggarwal and Mall (2002)

crop model. In the 2010 scenario, CERES rice responded to climate change with an increase of only 12% at 0 kg Nha^{-1}, while at 150 kg Nha^{-1} it increased by 25%. In the 2070 scenarios, the CERES rice-based simulation studies showed a large dependence of yields to the uncertainty of climate changes, i.e. rainfall at higher N values, especially under the optimistic climate change scenarios. In the pessimistic scenario, up to 6 percent, in the optimistic scenario, up to 9 percent yield improvements were observed, that too under lower N application rates. Singh et al., (2016) calibrated the CERES crop model for the Jaya rice variety, they showed a yield decrease in north-west India (1 to 1.5 mm/day decrease in rainfall) and a yield increase in the central region (up to 2 mm/day increase in rainfall) due to spatial variabilities in rainfall which are the impacts of changing climate.

During the crop life cycle, evapotranspiration was higher by 2.1–9.4% and 6.6–9.0% under A2 for rainfed and irrigated conditions, respectively, for normally sown cultivars, but it was lower by 3.2–7.8% and 1.0–5.1% under A1 for rainfed and irrigated conditions, respectively. In late-sown cultivars, however, the opposite pattern was observed, with higher evapotranspiration under A2 (1.1–6.0%) and 1.3–8.1% under rainfed and irrigated conditions, respectively, and lower evapotranspiration under A1 (0.8–5.7%) and 1.3–7.8% under rainfed and irrigated conditions, respectively, under the projected climate (Attri & Rathore, 2003). To meet increasing evapotranspiration needs, the crop grown in rainfed conditions needs to receive a well-distributed rainfall throughout the growth period, which is not possible under changing climatic scenarios, and may result in productivity decrements. Likewise, productivity decrements in crops like Soybean, Maize, Chickpea, Pigeonpea, Groundnut, Sorghum, and Brassica were reported by various researchers due to rainfall variability (Table 26.5).

El Nino Southern Oscillation has an impact on monsoon variability and indirectly, influences the yields of crops, globally. Variabilities in yields from the period 1984–2004 indicate significant negative impacts of El Nino (Fig. 26.6) and are evident in up to 22–24% of harvested areas worldwide (Toshichika & Toru, 2020). Maize yields in the south-eastern USA, parts of China, East and West Africa, Mexico, and Indonesia, oilseed crops like soybean yields in India and China, cereal food crops like rice yields in south China, Myanmar, and Tanzania, and wheat yield in major growing regions of China, US, Australia, Mexico and parts of Europe showed a significantly negative response due to moisture variability during crop growth period because of poor monsoon rains (Hammer et al., 2002).

The global demand for major crops like rice, wheat, soybean, and maize is expected to increase by 100–110% in 2050 from that in 2005 (Ray et al., 2013). To tackle this increasing demand, the rate of crop production must increase by 2.2–2.4% in the coming four decades. This can only be achieved by increasing crop productivity in areas that have low yields in the present day. These low yields may be due to a lack of adaption of technological advancements which nullify the effect of climatic aberrations like rainfall variability, necessitating research impetus to be put on the nullification of negative impacts of ENSO or maximizing the positive impacts of ENSO. This nullification of negative impact and exploitation of positive impacts are made possible by choice of planting date, crop type, and/or additional inputs

Table 26.5 Impacts of projected monsoon variability as an effect of changing climate on productivity of major crops (India)

Crop	Growing Region	Yield impact
Rice	All India	↑ (regional, based on scenarios)
	North West Central	↓ in North West, ↑ in central region
	South	↓/↑ (scenarios based)
	North-West	↓/↑ (scenarios based)
	Punjab	↓/↑ (scenarios based)
	All India	↑
	North	↑
	All India	↓
	Northern India	↓
Wheat	North	↔ / ↓ (scenarios based)
	Northwest	+ 16 to + 37%
	North West, Central India	No variability in north-west region, 10–15% ↓ in Central region
	Northwest	↓/↑ (varying according to scenarios)
	Punjab	↓/↑ (varying according to scenarios)
	All India	↓
	All India	↓/↑ (region wise and scenarios based)
	All India	↓/↑ (region wise and scenarios based)
	All India	↓
Soybean	All India	↓, ↑ (region wise and scenarios based)
	Central	↓, ↑ (scenarios based)
Maize	North	↓ ↑ (region wise and scenarios based)
	Punjab	↓
Chickpea	All India	↓
Pigeonpea	All India	↓
Groundnut	Punjab	↓
Sorghum	All India	↓
	All India	↓ ↑ (region wise and scenarios based)
Brassica	North	↑

of seeds, fertilizer, chemicals and irrigation, and other crop management strategies (Rosenzweig & Hillel, 2008). These crop management strategies along with skillful monsoon forecasts at the lead crop growth periods/seasons pave the way for rescuing the future generation from climate change-induced food scarcity.

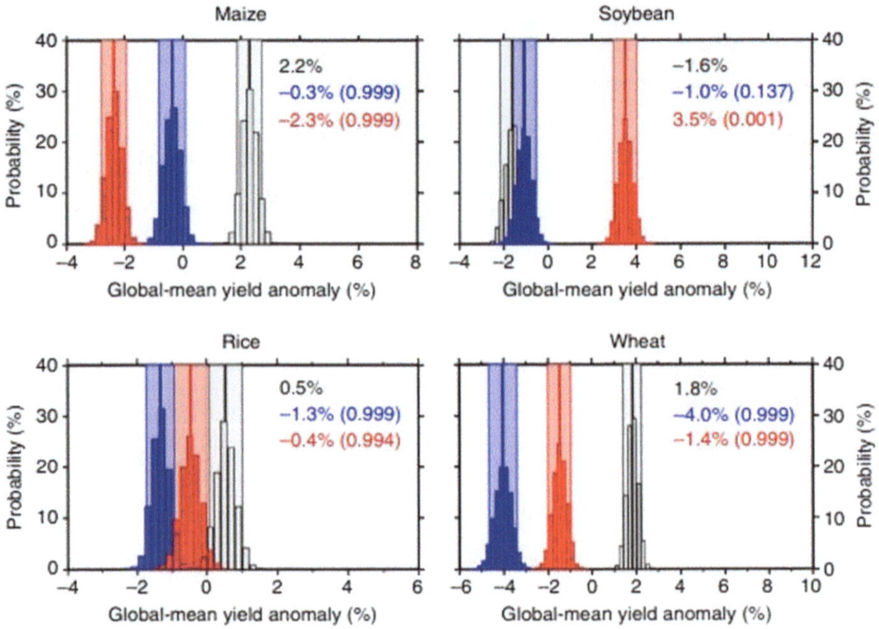

Fig. 26.6 global yield anomalies due to climate change (%). *Source* Toshichika and Toru (2020), The global dataset of historical yields for major crops 1981–2016

26.6 Consensus on Climate Change and Monsoonal Variability

Since the 1995 IPCC announcement on observable anthropogenic climate change, numerous independent studies have been conducted and confirmed that human populations are exposed to a climate change scenario. Most challenges to this claim arise from interested parties other than scientific community. Even if the communities admit that challenges faced by climate science motivated by politics or economic bodies, this does not affect the agreement of the scientific community is false. if we accept the fact of agreement on climate change on a scientific base, how can one prove that agreement is not wrong? Even though we cannot be assured about scientists' conclusions, studies based on the village surveys, farmers' experiences on occurrences of extreme events, etc. suggest alterations in past and present rainfall patterns along with other climatic parameters.

A study of 926 abstracts published in peer-reviewed climate change journals (searching for papers with keyword; 'global climate change') over the years from 1993 to 2003 provided a rough classification of these abstracts into the following categories: (a) those supporting the agreement, (b) those who disagree with the change, (c) those on studies for measuring, monitoring and predicting climate change, (d) those involve the discussion or documentation of the effects of climate change, (e) those that study ancient climate, and (f) those proposing different strategies to deal with climate change, pointed out that none of the publications dealt with the non-occurrence of climate change, people are not part of the reasons and the population does not experience the negative effects of climatic deviations (Oreskes, 2018). Most of the work dealt with the impacts of changing climate, 15 percent dealt with methods, and 10 percent with ancient climate change.

Recently, Lynas et al. (2021) evaluated over 80,000 studies published between 2012 and 2020 on climate change. Analysis of a subset of 3,000 studies, only four of which were sceptical about the human cause of climate change, representing 99.6299.96 percent agreement in the scientific community on changing climate and its impact on the monsoons. These types of studies demonstrate the global scientific community's acceptance of climate change and emphasize the need for action to address climate change.

26.6.1 Synthesis of Assessment Reports

To achieve a broad level of understanding or to provide a maximum level of knowledge on a specific topic or problem, it is necessary to conduct reviews of scientific literature, which compiles the results of a series of stand-alone studies. Compilations of many such assessment studies are called synthesis reports and are published by the IPCC in the field of climate change. The higher variations of precipitation for varying timescales are found in both constructed ancient climate and present-day measurements done instrumentally. This variability throws an uphill task for robust projection of future precipitation and changes in it at minor spatial and time scales. There is little confidence that internal variability contributes the greatest uncertainty to projected changes, at least in the short term (2021–2040) under global and regional projections.

Crumbled Atlantic Multidecadal Overturning Circulation (AMOC) have capacities to deteriorate the African and Asian monsoons but has a proven capacity to build up the Southern Hemisphere monsoons. All-inclusive, future changes in regional monsoons such as the South and Southeast Asian monsoons are generally consistent across global (including high-resolution) and regional climate models and supported by theoretical arguments if studied for the long term (2081–2100). Uncertainties in methodologies opted for simulating regional monsoons have a relation with various complexities of monsoon processes, outside factors, variability within, and difficulties in representing warm monsoon precipitation, and many others. Since from 5th assessment report, clarities in observation-based estimates and information by studying ancient climate data provide complete info on individual components of the climate and changes brought in it from ancient times to till date. Simulations by newer models, new analytical techniques, and methods encompassing multiple lines of evidence are leading to a better understanding of human interaction with the environment i.e. with climate, including extreme situations. In the IPCC 6th Assessment Report, they have provided comprehensive information on common regional changes in climate influence drivers (CIDs) related to temperature, precipitation, wind movement, snow, coastal and ocean variability.

Alterations in thermal CIDs

Thermal CIDs such as average thermal regimes, lengths of growing season, and extremes such as heatwaves and frost have already been reported worldwide, most of these alterations are linked to human activities. Already, in many of the places observed temperature changes have occurred well outside the range of internal variability as compared to 1850–1900 to date. Especially, positive shifts in early twentieth century temperatures have been noticed as compared to recent past temperature distributions in tropical regions. There are reports of land areas getting warmed more (0.1 °C per decade) since 1960, and the process is getting fast in recent decades. Studies on trends in increased occurrence of hot weather extremes and decreased occurrence of cold weather extremes indicate a consistent relationship with global mean temperature. On the other hand, the assessment of such trends and relationships themselves is difficult due to various limitations, unavailability of data being the major one.

Changes in precipitation related CIDs

The variabilities in precipitation at a regional scale are stronger than those variabilities on a global scale while, the uncertainties in observation, modeling, and forcing are all higher. However, greenhouse gas forcing has resulted in increments in contrasts in the amount of precipitation if wet and dry seasons of tropical land areas were considered. A high-magnitude precipitation increment at northern high latitudes was observed. Thus, indicating that, in most land regions, frequency and intensity of heavy precipitation events have increased. But also, there are reports of declined dry season availability of water. This can be attributed to ill distribution of rainfall, such as downpouring in a day or twice, leading to the loss of most of the rainfall due to flooding instead of infiltration, high-intensity rainfall causes loss of top layer of soil which better exposes the soil for solar radiation leading to increased evapotranspiration. Extreme rainfall events and flooding will increase in many regions of the world on nearly all continents, but the regional changes are more uncertain than pluvial ones because of the complexity in hydrological process, land cover, etc. over a global scale.

For getting a clear picture of regional precipitation variability due to climate change, one should know about observed changes in Africa, even though the changes in historical variability are not consistent, regions like North East Africa, east-South Africa, and Central Africa have instances of decreased rainfall. An exception to this, in parts of West Africa, increased precipitation has been noticed since about the 1980s. This can be supported by an increase in a number of flood events after 1980 in east and west parts of south Africa and western Africa.

Shifts in precipitation and droughts in African and Mediterranean parts have been associated with shifts in increased agricultural and environmental droughts. A shift towards an increased number of hydrological droughts was observable in the

Mediterranean and West African regions. These trends are consistent with simulated regional increases in drought and hot weather situations.

There will be an increment in observable mean wind speed in most areas, and it likely will go on increasing in future too. Over last 4 decades, the intensity of major tropical cyclones (TC) has shown a positive shift, i.e. the increased intensity of occurrence in most parts of the world.

26.6.2 Strengths of Present Day Policies

As early as 1976, the World Meteorological Organization (WMO) warned of the possible negative impact of an increased atmospheric build-up of greenhouse gases in future long and short-term weather. Together with the United Nations Environment Program, WMO initiated the IPCC in 1988 with a major task of informing governments scientifically about risks and uncertainties of climate change and changes brought about by it on humans and other natural systems. UNFCC's action along with the scientific information provided by IPCC and WMO help in mediating international negotiations thereby initializing international climate policies. For years, the mainstays of climatic policies have focused on emissions reductions, transfer of technology, equity, building up the capacity, and financing based on climate.

International conferences held on climate change such as UN Conference on Envt. and Devt.-Stockholm, 1972, Rio de Janeiro, 1992, World Summit on Sustainable Devt.–Johannesburg, 2002, and recently Conference of the UNs on Sustainable Devt., Rio de Janeiro, 2012 are remarkable with respect to environmental policymaking. The birth of agencies like United Nations Framework Convention on Climate Change (UNFCCC) is indeed a significant moment.

Processes and conditions affecting the adaptation of climate change policy by various systems of society:

Recently, studies using various approaches have been conducted to explain how demographic factors influence climate change policies. Then also, there is a need for integration of policy adoption by a social system since there is no thorough synthesis presently available, and the average impact of any associations between local conditions and different policy formulations. A study by Yeganeh et al. (Yeganeh et al., 2020) on determinants of climate change policy adoption indicated various processes and constraints in the US. They studied the processes and conditions enabling climate change policy integration at system, organization, and individual levels. They indicated major attributes of a system that are favoring or against the policy adoption, as follows.

At system level:

Political ideology and party positions: For example, the positions of political parties, driven by their political mandates, etc. influence the gathering of people towards tackling the climatic changes.

Institutional alignment: different social, political, and administrative systems of every country on earth has a unique way of functioning and traditions of administration. Some may have a positivity towards climate change and may show interest in nullifying the same, but some may show their interest in development, this difference in ideologies, considered as a special functioning way, may influence climate change policymaking and adoption to existing policies (Biesbroek et al., 2018).

Issue attention: the level of attention for a particular change brought in in weather parameters plays a key role in adaptation to climate change. For example, due to a change in rainfall in a region, more people will get affected, thereby that particular issue gets more attention from media and policymakers. This paves a path for setting up agenda on which the policy has to be made thereby its adoption by more people (Bromley-Trujilo & Poe, 2020).

Issue framing: policies with the least adjustment to them as compared to normal carryover of regular activities have proven more wins as compared to policies needing major adjustments. Even though the adoption of the policy itself is not the end, successful adaptation always depends on the results too (Runhaar et al., 2018).

At organization levels:

Capacity: The ability to integrate towards the common mindset and adjustment to changes brought in by the policy involves the level of trust each individual of an organization has.

Stickiness: Even though integration among individuals of an organization exists at a point in time, ensuring the same for longer periods has meager proves (Levin et al., 2012). Climate change adoption is a long-term process, and sticking to that policy for longer periods may not be possible at organizational levels due to the presence of people with varied mind-set.

Accountability: even though climate change has been identified and quantified for a long back, the identification of the individual/organization responsible for the change was not possible indeed it was a barrier to getting together for solving the vagaries. Recent developments in analytics have shown measuring the changes brought in and holding individuals and organizations accountable, making them work for goals set by policies (Molenveld et al., 2021).

Design: The design of political integration agreements pose a question of bureaucrats'/policymakers' intervention in the adaptation (Peters, 2015). Previously, studies have reported a remarkable negative impact on adoption of a policy due to varied mindsets of bureaucrats/policymakers themselves (Verhoest et al., 2016).

Timing: The point of time in which climate policy was made, and the socio-political condition of the particular group which has to adopt the policy influences the level of adaptation to a particular policy. It also has an influence on the speed of policy adaption, speed at which the changes are nullified (Time, 2021).

At the individual level:

Political commitment: Commitment of an individual to a political group ensures the integration of similar-minded individuals for effective action (Glover & Granberg, 2020).

MIX
Papier aus verantwortungsvollen Quellen
Paper from responsible sources
FSC® C105338

If you have any concerns about our products,
you can contact us on
ProductSafety@springernature.com

In case Publisher is established outside the EU,
the EU authorized representative is:
Springer Nature Customer Service Center GmbH
Europaplatz 3, 69115 Heidelberg, Germany

Printed by Libri Plureos GmbH
in Hamburg, Germany

Bonner, W. D., & Paegle, J. (1970). Diurnal variations in boundary layer winds over the south-central United States in summer. *Monthly Weather Review, 98*(10), 735–744.
Bours, D., McGinn, C., Pringle, P. (2014) Monitoring & evaluation for climate change adaptation and resilience: A synthesis of tools, frameworks and approaches. SEA Change Community of Practice, and UK Climate Impacts Programme.
Bromley-Trujillo, R., & Poe, J. (2020). The importance of salience: Public opinion and state policy action on climate change. *Journal of Public Policy, 40*(2), 280–304.
Brown, J. R., Moise, A. F., & Colman, R. A. (2017). Projected increases in daily to decadal variability of Asian-Australian monsoon rainfall. *Geophysical Research Letters, 44*(11), 5683–5690.
Brugger, K., & Rubel, F. (2013). Characterizing the species composition of European Culicoides vectors by means of the Köppen-Geiger climate classification. *Parasites & Vectors, 6*(1), 1–6.
Buerki, S., Forest, F., & Alvarez, N. (2014). Proto-South-East Asia as a trigger of early angiosperm diversification. *Botanical Journal of the Linnean Society, 174*(3), 326–333.
Carleton, A. M., Carpenter, D. A., & Weser, P. J. (1990). Mechanisms of interannual variability of the southwest United States summer rainfall maximum. *Journal of Climate, 3*(9), 999–1015.
Cattani, E., Merino, A., Guijarro, J. A., & Levizzani, V. (2018). East Africa rainfall trends and variability 1983–2015 using three long-term satellite products. *Remote Sensing, 10*(6), 931.
Chang, C. P., Wang, Z., McBride, J., & Liu, C. H. (2005). Annual cycle of Southeast Asia-Maritime Continent rainfall and the asymmetric monsoon transition. *Journal of Climate, 18*(2), 287–301.
Chase, J. E. (1996). *The physical characteristics and meltwater output from a snow cover compacted by ski-area operations.* Doctoral dissertation, University of New Hampshire.
Chausson, A., Turner, B., Seddon, D., Chabaneix, N., Girardin, C. A., Kapos, V., & Seddon, N. (2020). Mapping the effectiveness of nature-based solutions for climate change adaptation. *Global Change Biology, 26*(11), 6134–6155.
Chen, G., Wu, P., Wang, J., Zhang, P., & Jia, Z. (2022). Ridge–furrow rainfall harvesting system helps to improve stability, benefits and precipitation utilization efficiency of maize production in Loess Plateau region of China. *Agricultural Water Management, 261*, 107360.
Christensen, J. H., Kanikicharla, K. K., Aldrian, E., An, S. I., Cavalcanti, I. F., de Castro, M., & Zou, L. (2013). Climate phenomena and their relevance for future regional climate change. In *Climate change 2013 the physical science basis: Working group I contribution to the fifth assessment report of the intergovernmental panel on climate change* (pp. 1217–1308). Cambridge University Press.
Colorado-Ruiz, G., Cavazos, T., Salinas, J. A., De Grau, P., & Ayala, R. (2018). Climate change projections from coupled model intercomparison project phase 5 multi-model weighted ensembles for Mexico, the North American monsoon, and the mid-summer drought region. *International Journal of Climatology, 38*(15), 5699–5716.
Cook, B. I., & Seager, R. I. (2013). The response of the North American Monsoon to increased greenhouse gas forcing. *Journal of Geophysical Research: Atmospheres, 118*(4), 1690–1699.
Cook, K. H., & Vizy, E. K. (2019). Contemporary climate change of the African monsoon systems. *Current Climate Change Reports, 5*(3), 145–159.
Crost, B., Duquennois, C., Felter, J. H., & Rees, D. I. (2018). Climate change, agricultural production and civil conflict: Evidence from the Philippines. *Journal of Environmental Economics and Management, 88*, 379–395.
David, T. S., Ferreira, M. I., David, J. S., & Pereira, J. S. (1997). Transpiration from a mature Eucalyptus globulus plantation in Portugal during a spring-summer period of progressively higher water deficit. *Oecologia, 110*(2), 153–159.
Davies, R., & Dart. J. (2005) The 'Most Significant Change (MSC) technique'. A guide to its use. https://www.wikifplan.org/WIKIPLAN/1%201%20151%2020Most_significant_change_methodology_pa_abril%202005.pdf
Dhar, O. N., & Nandargi, S. (2000). A study of floods in the Brahmaputra basin in India. *International Journal of Climatology: A Journal of the Royal Meteorological Society, 20*(7), 771–781.
Ding, Y., Wang, Z., & Sun, Y. (2008). Inter-decadal variation of the summer precipitation in East China and its association with decreasing Asian summer monsoon. Part I: Observed

evidences. *International Journal of Climatology: A Journal of the Royal Meteorological Society*, *28*(9), 1139–1161.
Ding, Z. L., Yang, S. L., Sun, J., & Liu, T. S. (2001). Iron geochemistry of loess and red clay deposits in the Chinese Loess Plateau and implications for long-term Asian monsoon evolution in the last 7.0 Ma. *Earth and Planetary Science Letters*, *185*(1–2), 99–109.
Douglas, M. W. (1995). The summertime low-level jet over the Gulf of California. *Monthly Weather Review, 123*, 2334–2347.
Douglas, M. W., Maddox, R. A., Howard, K., & Reyes, S. (1993). The mexican monsoon. *Journal of Climate, 6*(8), 1665–1677.
Duffy, P. A., Walsh, J. E., Graham, J. M., Mann, D. H., & Rupp, T. S. (2005). Impacts of large-scale atmospheric–ocean variability on Alaskan fire season severity. *Ecological Applications, 15*(4), 1317–1330.
Fasullo, J., & Webster, P. J. (2003). A hydrological definition of Indian monsoon onset and withdrawal. *Journal of Climate, 16*(19), 3200–3211.
Feng, W., Zou, X., & Schaefer, D. (2009). Above-and belowground carbon inputs affect seasonal variations of soil microbial biomass in a subtropical monsoon forest of southwest China. *Soil Biology and Biochemistry, 41*(5), 978–983.
Froehlich, D. C. (1993). Short-duration-rainfall intensity equations for drainage design. *Journal of Irrigation and Drainage Engineering, 119*(5), 814–828.
Fu, Y., Argus, D. F., Freymueller, J. T., & Heflin, M. B. (2013). Horizontal motion in elastic response to seasonal loading of rain water in the Amazon Basin and monsoon water in Southeast Asia observed by GPS and inferred from GRACE. *Geophysical Research Letters, 40*(23), 6048–6053.
Funk, C., Hoell, A., Shukla, S., Blade, I., Liebmann, B., Roberts, J. B., & Husak, G. (2014). Predicting East African spring droughts using Pacific and Indian Ocean sea surface temperature indices. *Hydrology and Earth System Sciences, 18*(12), 4965–4978.
Gadgil, S. (2003). The Indian monsoon and its variability. *Annual Review of Earth and Planetary Sciences, 31*(1), 429–467.
Gilbert, B., MacDougall, A. S., Kadoya, T., Akasaka, M., Bennett, J. R., Lind, E. M., & Fay, P. A. (2020). Climate and local environment structure asynchrony and the stability of primary production in grasslands. *Global Ecology and Biogeography, 29*(7), 1177–1188.
Glover, L., & Granberg, M. (2020). The politics of adapting to climate change. In *The politics of adapting to climate change* (pp. 3–22). Palgrave Pivot, Cham.
Goodwin, C. (2007). Participation, stance and affect in the organization of activities. *Discourse and Society, 18*(1), 53–73.
Goswami, B. N., & Xavier, P. K. (2003). Potential predictability and extended range prediction of Indian summer monsoon breaks. *Geophysical Research Letters*, *30*(18).
Grimm, A. M., & Saboia, J. P. (2015). Interdecadal variability of the South American precipitation in the monsoon season. *Journal of Climate, 28*(2), 755–775.
Gullison, R., & Hardner, J. (2009). Using limiting factors analysis to overcome the problem of long time horizons. *New Directions for Evaluation, 122*, 19–29.
Hammer, G. L., Kropff, M. J., Sinclair, T. R., & Porter, J. R. (2002). Future contributions of crop modelling from heuristics and supporting decision making to understanding genetic regulation and aiding crop improvement. *European Journal of Agronomy, 18*(1–2), 15–31.
Han, F., Cook, K. H., & Vizy, E. K. (2019). Changes in intense rainfall events and dry periods across Africa in the twenty-first century. *Climate Dynamics, 53*(5), 2757–2777.
Heltberg, R., Reva, A., Turk, C., Hossain, N., & Heltberg, R. (2012) *Anatomy of coping: Evidence from people living through the crises of 2008–11* (Vol. 5957). World Bank.
Higgins, S. I., Richardson, D. M., & Cowling, R. M. (1996). Modeling invasive plant spread: The role of plant-environment interactions and model structure. *Ecology, 77*(7), 2043–2054.
Hlywiak, J., & Nolan, D. S. (2021). The response of the near-surface tropical cyclone wind field to inland surface roughness length and soil moisture content during and after landfall. *Journal of the Atmospheric Sciences, 78*(3), 983–1000.
Holdridge, L. R., & Grenke, W. C. (1947). Forest environments in tropical life zones: A pilot study.

Huffman, G. J. (1997). Estimates of root-mean-square random error for finite samples of estimated precipitation. *Journal of Applied Meteorology, 36*(9), 1191–1201.

Hulme, M., Osborn, T. J., & Johns, T. C. (1998). Precipitation sensitivity to global warming: Comparison of observations with HadCM2 simulations. *Geophysical Research Letters, 25*(17), 3379–3382.

Hurford, A. P., Parker, D. J., Priest, S. J., & Lumbroso, D. M. (2012). Validating the return period of rainfall thresholds used for extreme rainfall alerts by linking rainfall intensities with observed surface water flood events. *Journal of Flood Risk Management, 5*(2), 34–142.

Iturriza, M., Labaka, L., Ormazabal, M., & Borges, M. (2020). Awareness-development in the context of climate change resilience. *Urban Climate, 32*, 100613.

Johnson, R. H. (1976). The role of convective-scale precipitation downdrafts in cumulus and synoptic scale interactions. *Journal of Atmospheric Sciences, 33*(10), 1890–1910.

Kajikawa, Y., Wang, B., & Yang, J. (2010). A multi-time scale Australian monsoon index. *International Journal of Climatology, 30*(8), 1114–1120.

Kale, V. S. (2003) Geomorphic effects of monsoon floods on Indian rivers. In *Flood problem and management in South Asia* (pp. 65–84). Springer, Dordrecht.

Keller, E. A., & DeVecchio, D. E. (2016). *Natural hazards: Earth's processes as hazards, disasters, and catastrophes*. Routledge.

Kerr, R. B., Nyantakyi-Frimpong, H., Dakishoni, L., Lupafya, E., Shumba, L., Luginaah, I., & Snapp, S. S. (2018). Knowledge politics in participatory climate change adaptation research on agroecology in Malawi. *Renewable Agriculture and Food Systems, 33*(3), 238–251.

Khlebnikova, E. I., & Sall, I. A. (2009). Peculiarities of climatic changes in cloud cover over the Russian Federation. *Russian Meteorology and Hydrology, 34*(7), 411–417.

Khromov, S. P. (1957) Die Geographische Verbreitung der Monsune. *Petermanns Geogr. Mitteilung*, 234–7.

Kitoh, A. (2017) The Asian monsoon and its future change in climate models: A review. *Journal of the Meteorological Society of Japan. Ser. II*.

Knox, J., Hess, T., Daccache, A., & Wheeler, T. (2012). Climate change impacts on crop productivity in Africa and South Asia. *Environmental Research Letters, 7*(3), 034032.

Knutson, T., Camargo, S. J., Chan, J. C., Emanuel, K., Ho, C. H., Kossin, J., & Wu, L. (2019). Tropical cyclones and climate change assessment: Part I: Detection and attribution. *Bulletin of the American Meteorological Society, 100*(10), 1987–2007.

Köppen, W. (1936). Das geographische System der Klimate (Handbuch der Klimatologie, Bd. 1, Teil C).

Kripalani, R. H., & Kulkarni, A. (1997). Climatic impact of El Nino/La Nina on the Indian monsoon: A new perspective. *Weather, 52*(2), 39–46.

Kripalani, R. H., & Kulkarni, A. (2001). Monsoon rainfall variations and teleconnections over South and East Asia. *International Journal of Climatology: A Journal of the Royal Meteorological Society, 21*(5), 603–616.

Kumar, K. R., Pant, G. B., Parthasarathy, B., & Sontakke, N. A. (1992). Spatial and subseasonal patterns of the long-term trends of Indian summer monsoon rainfall. *International Journal of Climatology, 12*(3), 257–268.

Kurukulasuriy, P., & Mendelsohn, R. O. (2008) How will climate change shift agro-ecological zones and impact African agriculture? World Bank Policy Research Working Paper (4717).

Kyseli, J., & Beranova, R. (2009). Climate-change effects on extreme precipitation in central Europe: Uncertainties of scenarios based on regional climate models. *Theoretical and Applied Climatology, 95*(3), 361–374.

Lahmers, T. M., Castro, C. L., Adams, D. K., Serra, Y. L., Brost, J. J., & Luong, T. (2016). Long-term changes in the climatology of transient inverted troughs over the North American monsoon region and their effects on precipitation. *Journal of Climate, 29*(17), 6037–6064.

Lal, M., Singh, K. K., Rathore, L. S., Srinivas, G., & Saseendran, S. A. (1998). Vulnerability of rice and wheat yields in NW India to future changes in climate. *Agricultural and Forest Meteorology, 89*(2), 101–114.

Lamhauge, N., Lanzi, E., & Agrawala, S. (2012). Monitoring and evaluation for adaptation: lessons from development co-operation agencies.

Lau, W. K., Waliser, D. E., & Goswami, B. N. (2012). South asian monsoon. In *Intraseasonal variability in the atmosphere-ocean climate system* (pp. 21–72). Springer, Berlin, Heidelberg.

Lauer, C. N. (1925). Plant engineering as a service to production management. *The Annals of the American Academy of Political and Social Science, 119*(1), 97–102.

Leetmaa, A. (1977). Effects of the winter of 1976–1977 on the northwestern Sargasso Sea. *Science, 198*(4313), 188–189.

Leeuw, F. L., & Vaessen, J. (2009). Impact evaluations and development: NONIE guidance on impact evaluation.

Lehmann, J., Coumou, D., & Frieler, K. (2015). Increased record-breaking precipitation events under global warming. *Climatic Change, 132*(4), 501–515.

Leng, G., & Hall, J. (2019). Crop yield sensitivity of global major agricultural countries to droughts and the projected changes in the future. *Science of the Total Environment, 654*, 811–821.

Levin, K., Cashore, B., Bernstein, S., & Auld, G. (2012). Overcoming the tragedy of super wicked problems: Constraining our future selves to ameliorate global climate change. *Policy Sciences, 45*(2), 123–152.

Li, J., & Zeng, Q. (2002). Significance of the normalized seasonality of wind field and its rationality for characterizing the monsoon. *Science in China Series d: Earth Sciences, 43*(6), 646–653.

Liang, L., Sun, Q., Luo, X., Wang, J., Zhang, L., Deng, M., & Liu, Z. (2017). Long-term spatial and temporal variations of vegetative drought based on vegetation condition index in China. *Ecosphere, 8*(8), e01919.

Lodoun, T., Giannini, A., Traore, P. S., Some, L., Sanon, M., Vaksmann, M., & Rasolodimby, J. M. (2013). Changes in seasonal descriptors of precipitation in Burkina Faso associated with late 20th century drought and recovery in West Africa. *Environmental Development, 5*, 96–108.

Loschnigg, J., Meehl, G. A., Webster, P. J., Arblaster, J. M., & Compo, G. P. (2003). The Asian monsoon, the tropospheric biennial oscillation, and the Indian Ocean zonal mode in the NCAR CSM. *Journal of Climate, 16*(11), 1617–1642.

Lynas, M., Houlton, B. Z., & Perry, S. (2021). Greater than 99% consensus on human caused climate change in the peer-reviewed scientific literature. *Environmental Research Letters, 16*(11), 114005.

Maddox, R. A., McCollum, D. M., & Howard, K. W. (1995). Large-scale patterns associated with severe summertime thunderstorms over central Arizona. *Weather and Forecasting, 10*(4), 763–778.

Magrin, G. O., Marengo, J. A., Boulanger, J. P., Buckeridge, M. S., Castellanos, E., Poveda, G., Scarano, F. R., & Vicuña, S. (2014). *Central and South America. Climate change 2014: Impacts, Adaptation, and Vulnerability*.

Maidment, R. I., Allan, R. P., & Black, E. (2015). Recent observed and simulated changes in precipitation over Africa. *Geophysical Research Letters, 42*(19), 8155–8164.

Malik, A., Kumar, A., Guhathakurta, P., & Kisi, O. (2019). Spatial-temporal trend analysis of seasonal and annual rainfall (1966–2015) using innovative trend analysis method with significance test. *Arabian Journal of Geosciences, 12*(10), 1–23.

Mall, R. K., Chaturvedi, M., Singh, N., Bhatla, R., Singh, R. S., Gupta, A., & Niyogi, D. (2021). Evidence of asymmetric change in diurnal temperature range in recent decades over different agro-climatic zones of India. *International Journal of Climatology, 41*(4), 2597–2610.

Molenveld, A., Verhoest, K., & Wynen, J. (2021). Why public organizations contribute to cross-cutting policy programs: The role of structure, culture, and ministerial control. *Policy Sciences, 54*(1), 123–154.

Molnar, P., Boos, W. R., & Battisti, D. S. (2010). Orographic controls on climate and paleoclimate of Asia: Thermal and mechanical roles for the Tibetan Plateau. *Annual Review of Earth and Planetary Sciences, 38*, 77–102.

Monerie, P. A., Fontaine, B., & Roucou, P. (2012). Expected future changes in the African monsoon between 2030 and 2070 using some CMIP3 and CMIP5 models under a medium-low RCP scenario. *Journal of Geophysical Research: Atmospheres, 117*(D16).

Moore, R. E., Kearfott, R. B., & Cloud, M. J. (2009). *Introduction to interval analysis*. Society for Industrial and Applied Mathematics.

Mukherjee, A., Wang, S. Y. S., & Promchote, P. (2019). Examination of the climate factors that reduced wheat yield in Northwest India during the 2000s. *Water, 11*(2), 343.

Myers, N., Mittermeier, R. A., Mittermeier, C. G., Da Fonseca, G. A., & Kent, J. (2000). Biodiversity hotspots for conservation priorities. *Nature, 403*(6772), 853–858.

Nandargi, S. S., & Mahto, S. S. (2019). Frequency and intensity of tropical disturbances over the Indian region and its neighboring seas with associated rainfall during the monsoon season: A perspective. *Engineering Reports, 1*(5), e12069.

Ojima, D. S., Parton, W. J., Coughenour, M. B., Scurlock, J. M. O., Kirchner, T. B., Kittel, T. G. F., & Kinyamario, J. I. (1996). Impact of climate and atmospheric carbon dioxide changes on grasslands of the world.

Okabe, I. T. (1995). *The North American monsoon*. Doctoral dissertation, University of British Columbia.

Oreskes, N. (2018). The scientific consensus on climate change: How do we know we're not wrong? In *Climate modelling* (pp. 31–64). Palgrave Macmillan.

Orlowsky, B., & Seneviratne, S. I. (2012). Global changes in extreme events: Regional and seasonal dimension. *Climatic Change, 110*(3), 669–696.

Pant, G. B., & Hignane, L. S. (1988). Climate change in around Rajasthan during 29th century. *Journal of Climatology, 21*, 527–534.

Parish, E. S., Kodra, E., Steinhaeuser, K., & Ganguly, A. R. (2012). Estimating future global per capita water availability based on changes in climate and population. *Computers and Geosciences, 42*, 79–86.

Parry, M. L., Rosenzweig, C., Iglesias, A., Livermore, M., & Fischer, G. (2004). Effects of climate change on global food production under SRES emissions and socio-economic scenarios. *Global Environmental Change, 14*(1), 53–67.

Parthasarathy, B., Kumar, K. R., & Kothawale, D. R. (1992). Indian summer monsoon rainfall indices: 1871–1990. *Meteorological Magazine, 121*(1441), 174–186.

Pascale, S., Carvalho, L., Adams, D. K., Castro, C. L., & Cavalcanti, I. F. (2019). Current and future variations of the monsoons of the Americas in a warming climate. *Current Climate Change Reports, 5*(3), 125–144.

Patton, A. J. (2011). Volatility forecast comparison using imperfect volatility proxies. *Journal of Econometrics, 160*(1), 246–256.

Peters, B. G. (2015). *Pursuing horizontal management: The politics of public sector coordination* (pp. 1–214). University Press of Kansas.

Piao, S., Liu, Q., Chen, A., Janssens, I. A., Fu, Y., Dai, J., & Zhu, X. (2019). Plant phenology and global climate change: Current progresses and challenges. *Global Change Biology, 25*(6), 1922–1940.

Planton, S., Déqué, M., Chauvin, F., & Terray, L. (2008). Expected impacts of climate change on extreme climate events. *Comptes Rendus Geoscience, 340*(9–10), 564–574.

Poulter, B., Ciais, P., Hodson, E., Lischke, H., Maignan, F., Plummer, S., & Zimmermann, N. E. (2011). Plant functional type mapping for earth system models. *Geoscientific Model Development, 4*(4), 993–1010.

Pradhan, M., & Rawlings, L. B. (2002). The impact and targeting of social infrastructure investments: Lessons from the Nicaraguan Social Fund. *The World Bank Economic Review, 16*(2), 275–295.

Praveen, B., Talukdar, S., Mahato, S., Mondal, J., Sharma, P., Islam, A. R. M., & Rahman, A. (2020). Analyzing trend and forecasting of rainfall changes in India using non-parametrical and machine learning approaches. *Scientific Reports, 10*(1), 1–21.

Prein, A. F., Holland, G. J., Rasmussen, R. M., Clark, M. P., & Tye, M. R. (2016). Running dry: The US Southwest's drift into a drier climate state. *Geophysical Research Letters, 43*(3), 1272–1279.
Ramage, C. S. (1971). Monsoon meteorology (No. 551.518 R3).
Ray, D. K., Mueller, N. D., West, P. C., & Foley, J. A. (2013). Yield trends are insufficient to double global crop production by 2050. *PLoS ONE, 8*(6), e66428.
Ray, D. K., West, P. C., Clark, M., Gerber, J. S., Prishchepov, A. V., & Chatterjee, S. (2019). Climate change has likely already affected global food production. *PLoS ONE, 14*(5), e0217148.
Reumert, J. (1946). Vahls climatic divisions: An explanation. In: *Geografisk Tidsskrift*.
Riha, S. J., Wilks, D. S., & Simoens, P. (1996). Impact of temperature and precipitation variability on crop model predictions. *Climatic Change, 32*(3), 293–311.
Roeckner, E., Bengtsson, L., Feichter, J., Lelieveld, J., & Rodhe, H. (1999). Transient climate change simulations with a coupled atmosphere–ocean GCM including the tropospheric sulfur cycle. *Journal of Climate, 12*(10), 3004–3032.
Rosenzweig, C., & Hillel. D. (2008). *Climate variability and the global harvest: Impacts of El Niño and other oscillations on agro-ecosystems*. Oxford University Press.
Rosenzweig, C., Parry, M. L., Gischer, G., & Frohberg, K. (1993). *Climate change and world food supply*. Research Report-Environmental Change Unit, University of Oxford (United Kingdom).
Rowson, D. R., & Colucci, S. J. (1992). Synoptic climatology of thermal low-pressure systems over south-western north America. *International Journal of Climatology, 12*(6), 529–545.
Rubel, F., & Kottek, M. (2011). Comments on: "the thermal zones of the earth" by Wladimir Köppen (1884). *Meteorologische Zeitschrift, 20*(3), 361.
Rummukainen, M. (2012). Changes in climate and weather extremes in the 21st century. *Wiley Interdisciplinary Reviews: Climate Change, 3*(2), 115–129.
Runhaar, H., Wilk, B., Persson, A., Uittenbroek, C., & Wamsler, C. (2018). Mainstreaming climate adaptation: Taking stock about "what works" from empirical research worldwide. *Regional Environmental Change, 18*(4), 1201–1210.
Sandeep, V. M., Rao, V. U. M., Rao, B. B., Pramod, V., Chowdary, P. S., Kumar, P. V., & Mukesh, P. (2018). Impact of climate change on sorghum productivity in India and its adaptation strategies. *Journal of Agrometeorology, 20*(2), 89–96.
Sanogo, S., Fink, A. H., Omotosho, J. A., Ba, A., Redl, R., & Ermert, V. (2015). Spatio-temporal characteristics of the recent rainfall recovery in West Africa. *International Journal of Climatology, 35*(15), 4589–4605.
Santer, B. D., Taylor, K. E., Gleckler, P. J., Bonfils, C., Barnett, T. P., Pierce, D. W., & Wehner, M. F. (2009). Incorporating model quality information in climate change detection and attribution studies. *Proceedings of the National Academy of Sciences, 106*(35), 14778–14783.
Shapiro, M., Shukla, J., Brunet, G., Nobre, C., Béland, M., Dole, R., & Wallace, J. M. (2010). An earth-system prediction initiative for the twenty-first century. *Bulletin of the American Meteorological Society, 91*(10), 1377–1388.
Shepherd, J. G. (2009). Geoengineering the climate: Science, governance and uncertainty. Royal Society.
Sherwood, S. C., Webb, M. J., Annan, J. D., Armour, K. C., Forster, P. M., Hargreaves, J. C., & Zelinka, M. D. (2020). An assessment of Earth's climate sensitivity using multiple lines of evidence. *Reviews of Geophysics, 58*(4), e2019RG000678.
Singh, D., Ghosh, S., Roxy, M. K., & McDermid, S. (2019). Indian summer monsoon: Extreme events, historical changes, and role of anthropogenic forcings. *Wiley Interdisciplinary Reviews: Climate Change, 10*(2), e571.
Singh, K. P., Malik, A., Sinha, S., Singh, V. K., & Murthy, R. C. (2005). Estimation of source of heavy metal contamination in sediments of Gomti River (India) using principal component analysis. *Water, Air, and Soil Pollution, 166*(1), 321–341.
Singh, N., & Sontakke, N. A. (2002). On climatic fluctuations and environmental changes of the Indo-Gangetic plains, India. *Climatic Change, 52*(3), 287–313.

Singh, P. K., Singh, K. K., Rathore, L. S., Baxla, A. K., Bhan, S. C., Gupta, A., Mall, R. K. (2016). Rice (Oryza sativa L.) yield gap using the CERES-rice model of climate variability for different agroclimatic zones of India. *Current Science*, 405–413.

Sooraj, K. P., Terray, P., & Xavier, P. (2016). Sub-seasonal behaviour of Asian summer monsoon under a changing climate: Assessments using CMIP5 models. *Climate Dynamics, 46*(11), 4003–4025.

Stensrud, D. J., Gall, R. L., Mullen, S. L., & Howard, K. W. (1995). Model climatology of the Mexican monsoon. *Journal of Climate, 8*(7), 1775–1794.

Strahler, A. H. (1969). *Forest ecology of the Fairfax line*. The Johns Hopkins University.

Tang, M., & Reiter, E. R. (1984). Plateau monsoons of the Northern Hemisphere: A comparison between North America and Tibet. *Monthly Weather Review, 112*(4), 617–637.

Taraphdar, S., Mukhopadhyay, P., & Goswami, B. N. (2010). Predictability of Indian summer monsoon weather during active and break phases using a high-resolution regional model. *Geophysical Research Letters, 37*(21).

Tarkan, A. S., & Vilizzi, L. (2015). Patterns, latitudinal clines and countergradient variation in the growth of roach Rutilus rutilus (Cyprinidae) in its Eurasian area of distribution. *Reviews in Fish Biology and Fisheries, 25*(4), 587–602.

Taylor, C. M., Fink, A. H., Klein, C., Parker, D. J., Guichard, F., Harris, P. P., & Knapp, K. R. (2018). Earlier seasonal onset of intense mesoscale convective systems in the Congo Basin since 1999. *Geophysical Research Letters, 45*(24), 13–458.

Thapliyal, V., Kulshrestha, S. M., Mandal, G. S., Roy, N. S., & Sikka, D. R. (1991). A power regression model for long range forecast of southwest monsoon rainfall over India. *Mausam, 42*(2), 125–130.

Thornthwaite, C. W. (1943). Problems in the classification of climates. *Geographical Review, 33*(2), 233–255.

Thornthwaite, C. W. (1948). An approach toward a rational classification of climate. *Geographical Review, 38*(1), 55–94.

Time, M. S. (2021). How temporal discretion supports interagency coordination: Sweden's intersectoral fight against antimicrobial resistance. *Comparative European Politics, 19*(3), 360–379.

Toshichika, I., Toru, S. (2020). The global dataset of historical yields for major crops 1981–2016. *Scientific Data, 7*(1).

Trenberth, K. E., & Hurrell, J. W. (1994). Decadal atmosphere-ocean variations in the Pacific. *Climate Dynamics, 9*(6), 303–319.

Trenberth, K. E., Stepaniak, D. P., & Caron, J. M. (2000). The global monsoon as seen through the divergent atmospheric circulation. *Journal of Climate, 13*(22), 3969–3993.

Trewartha, G. T., & Horn, L. H. (1980). *An introduction to climate* (5th ed., p. 437). McGraw-Hill.

Troll, C. (1965). Seasonal climates of the earth. In *Weltkarten zur Klimakunde/World Maps of Climatology* (pp. 19–25). Springer, Berlin, Heidelberg.

Vera, C. S., & Díaz, L. (2015). Anthropogenic influence on summer precipitation trends over South America in CMIP5 models. *International Journal of Climatology, 35*(10), 3172–3177.

Verhoest, K., Van Thiel, S., Bouckaert, G., Lægreid, P.,Van Thiel, S. (Eds.). (2016). *Government agencies: Practices and lessons from 30 countries*. Springer.

Vesely, F. M., Paleari, L., Movedi, E., Bellocchi, G., & Confalonieri, R. (2019). Quantifying uncertainty due to stochastic weather generators in climate change impact studies. *Scientific Reports, 9*(1), 1–8.

Virmani, S. M., Sivakumar, M. V. K., & Reddy, S. J. (1980). Climatic classification of semi-arid tropics in relation to farming systems research.

Waliser, D. E., & Gautier, C. (1993). A satellite-derived climatology of the ITCZ. *Journal of Climate, 6*(11), 2162–2174.

Wallace, J. M. (1975). Diurnal variations in precipitation and thunderstorm frequency over the conterminous United States. *Monthly Weather Review, 103*(5), 406–419.

Walther, G. R., Post, E., Convey, P., Menzel, A., Parmesan, C., Beebee, T. J., & Bairlein, F. (2002). Ecological responses to recent climate change. *Nature, 416*(6879), 389–395.

Wang, B., & Ding, Q. (2006). Changes in global monsoon precipitation over the past 56 years. *Geophysical Research Letters, 33*(6).

Wang, B., Wu, R., & Lau, K. M. (2001). Interannual variability of the Asian summer monsoon: Contrasts between the Indian and the western North Pacific-East Asian monsoons. *Journal of Climate, 14*(20), 4073–4090.

Wang, Y., Xie, Z., Malhi, S. S., Vera, C. L., Zhang, Y., & Wang, J. (2009). Effects of rainfall harvesting and mulching technologies on water use efficiency and crop yield in the semi-arid Loess Plateau, China. *Agricultural Water Management, 96*(3), 374–382.

Webber, S. (2019). Putting climate services in contexts: Advancing multi-disciplinary understandings: Introduction to the special issue. *Climatic Change, 157*(1), 1–8.

Webster, P. J., & Yang, S. (1992). Monsoon and ENSO: Selectively interactive systems. *Quarterly Journal of the Royal Meteorological Society, 118*(507), 877–926.

Webster, P. J., Magana, V. O., Palmer, T. N., Shukla, J., Tomas, R. A., Yanai, M. U., & Yasunari, T. (1998). Monsoons: Processes, predictability, and the prospects for prediction. *Journal of Geophysical Research: Oceans, 103*(C7), 14451–14510.

Weible, C. M., Ingold, K., Nohrstedt, D., Henry, A. D., & Jenkins-Smith, H. C. (2020). Sharpening advocacy coalitions. *Policy Studies Journal, 48*(4), 1054–1081.

Wolfson, R., & Schneider, S. H. (2002). Understanding climate science. *Climate Change Policy: A Survey, 30.*

Yeganeh, A. J., McCoy, A. P., & Schenk, T. (2020). Determinants of climate change policy adoption: A meta-analysis. *Urban Climate, 31*, 100547.

Yim, S. Y., Wang, B., Liu, J., & Wu, Z. (2014). A comparison of regional monsoon variability using monsoon indices. *Climate Dynamics, 43*(5), 1423–1437.

Zhang, D. D., Lee, H. F., Wang, C., Li, B., Pei, Q., Zhang, J., & An, Y. (2011). The causality analysis of climate change and large-scale human crisis. *Proceedings of the National Academy of Sciences, 108*(42), 17296–17301.

Zhang, W., Zhou, T., Zou, L., Zhang, L., & Chen, X. (2018). Reduced exposure to extreme precipitation from 0.5 C less warming in global land monsoon regions. *Nature Communications*, *9*(1), 1–8.

Zhou, J., & Lau, K. M. (1998). Does a monsoon climate exist over South America? *Journal of Climate, 11*(5), 1020–1040.

Zhu, X., Troy, T. J., & Devineni, N. (2019). Stochastically modeling the projected impacts of climate change on rainfed and irrigated US crop yields. *Environmental Research Letters, 14*(7), 074021.

Chapter 27
Simple Holistic Method of Quantifying Local Versus Trans-boundary Air Pollution in NCR-Delhi

Sneha Aggarwal and Umesh Chandra Kulshrestha

Abstract This study reports a simple holistic approach to quantifying controllable and uncontrollable particulate load in Delhi-NCR. The concentration of carbonaceous aerosols was measured during a peculiar combination of days that included a major festival, a nationwide public holiday, a weekend, and a couple of working days during September–October 2017. The OC/EC ratio, which serves as an important diagnostic tool for source identification, varied between 1.35 and 1.60 with a mean value of 1.52 indicating fossil fuel combustion as the major source of both OC and EC in the region. The observed trend in AOD values was found to be similar to that in the concentrations of carbonaceous aerosols. An interesting observation was made during the study period wherein all the carbonaceous fractions exhibited a remarkable dip on the public holiday, which was also a non-working weekday. According to our method, around 31.40% of carbonaceous aerosols are emitted from local vehicles and industries (controllable), while the rest 68.60% are emitted from domestic activities and transboundary sources along with windblown resuspensions (uncontrollable). Based on our results, we propose a remedial 'Work from Home' policy over the 'Odd–Even Scheme' for an observable reduction in particulate load in Delhi-NCR.

Keywords Air pollution · Carbonaceous aerosols · NCR Delhi · Fine particles · Water-soluble organic carbon

S. Aggarwal · U. C. Kulshrestha (✉)
School of Environmental Sciences, Jawaharlal Nehru University, New Delhi 110067, India
e-mail: umeshkulshrestha@gmail.com

S. Aggarwal
Department of Environmental Science, Stockholm University, 11415 Stockholm, Sweden

H. Pathak et al. (eds.), *Climate Change Impacts on Soil-Plant-Atmosphere Continuum*, Advances in Global Change Research 78,
https://doi.org/10.1007/978-981-99-7935-6_27

27.1 Introduction

The air we breathe is polluted by several pollutants from diverse sources, such as vehicles, industries, thermal power plants, farmlands, construction sites, and also processes such as biomass burning, re-suspension of road dust, long-range transport (LRT) of gases and particles, etc. On a global scale, nearly 7 million deaths occur every year due to combined effects of ambient and indoor air pollution (WHO fact sheet, 2022). The number of deaths due to $PM_{2.5}$ alone is estimated at 1.64 million/ year (Institute for Health Metrics and Evaluation, 2017).

$PM_{2.5}$ which refers to the suspension of solid and liquid particles with aerodynamic diameter below 2.5 μm can be produced by both natural and anthropogenic processes. They are omnipresent in the boundary layer with their number concentration depending on factors like location, meteorological conditions, annual and diurnal cycles, and presence of local sources and sinks. The highest concentrations are usually found in urban areas, reaching up to 108 to 109 per cm^3 (Seinfeld & Pandis, 1998).

In this paper, we focus on carbonaceous aerosols that constitute a large fraction (~20% to 70%) of the particulate matter (Ramachandran et al., 2009). They are not only responsible for human morbidities but also for visibility degradation, haze formation, and atmospheric heating (Seinfeld & Pandis, 1998; Bond & Bergstrom, 2006; Ramanathan & Carmichael, 2008; Chung et al., 2012; Trompetter et al., 2013; Rao et al., 2013). Elemental carbon (EC) and organic carbon (OC) represent major fractions of carbonaceous aerosol loadings and are primarily responsible for health and climatic effects of aerosols. Anthropogenic carbonaceous emissions can even alter the chemical properties of urban dust thereby aggravating the risk to public health (Mishra & Kulshrestha, 2017, 2021). Most EC emissions are contributed by fossil fuel combustion. The COP-26 meeting held in Glasgow has provided some commitments from the member countries which India has committed to net zero by 2070. The coal phase-down approach instead of phase-out is a good option for India and China to switch over to clean energy (Kulshrestha & Katoch, 2022).

Although carbonaceous aerosols contribute largely to the atmospheric load and impact both human health and climate, uncertainties are associated with their budgets which pose a major challenge in atmospheric science. Thus, complete knowledge about their sources, sinks, physical and chemical properties, transport and secondary formation, etc., is needed to understand and simulate various atmospheric processes. Unfortunately, our understanding of different aspects of atmospheric aerosols is still poor in the south Asian region owing to the huge diversity in sources, climate, and land-use patterns (Kulshrestha, 2017). The situation in megacities, such as Delhi is further complex due to continuously developing sub-urban areas, the increasing number of industries and vehicles which serve as major local sources of air pollution. In addition to the local pollution, LRT and transboundary pollution from other states worsen the pollution scenario making it more complicated for source identification, impact prediction, and policy framing purposes (Kulshrestha, 2015; Kulshrestha & Mishra, 2020).

As per Environmental Performance Index 2020, India ranks 168 out of 180 countries over thirty-two metrics of environmental performance behind all South Asian countries except Afghanistan (Wending et al., 2020). It should be noted that India ranks second last, i.e., 179 out of 180 when evaluated based on air quality only. Particularly Delhi, the capital city of India, ranks 11th place in the global list for $PM_{2.5}$ concentration as per WHO Urban Air Quality Database, 2016. These rankings are a cause of alarm and draw our attention towards the pressing need of improving air quality owing to its negative impacts on both human health and climate.

In October–November 2016, Delhi faced one of the worst smog events when $PM_{2.5}$ concentrations increased up to about 13 times the permissible limit (Sawlani et al., 2019). As a result, the Delhi government implemented several adaptive measures, including the "Odd–Even Scheme" to reduce vehicular emissions, which was initially rolled out for a period of 15 days in January and April 2017. Additionally, the Supreme Court of India in November 2016 directed the concerned authorities and governments to formulate a plan to combat such episodes of pollution. The Ministry of Environment, Forest and Climate Change (MOEFCC) accordingly notified the Graded Response Action Plan (GRAP) in January 2017, which was implemented by the Environmental Pollution Prevention and Control Authority (EPCA). GRAP refers to a set of increasingly tougher actions in response to the air quality index of the city (Laura et al., 2018), which has been discussed in detail in the later section of this paper. Furthermore, the Central Pollution Control Board (CPCB) worked on expansion of the air quality monitoring network in Delhi-NCR to provide real-time ambient concentrations of major pollutants. This has facilitated identification of pollution hotspots and also reliable inputsfor policy decisions. However, despite the number of efforts taken by the government and the key authorities, Delhi remains to be in the top five most polluted cities in the world with an average $PM_{2.5}$ concentration of 98.6 $\mu g/m^3$ in 2019 (IQAir, 2019).

Thus, there is still a need to perform holistic and detailed studies to identify the major pollution sources and hotspots in Delhi to facilitate executive and policy decisions. Source apportionment can be done through a variety of techniques, such as Chemical Mass Balance (CMB) receptor model that uses chemical signatures to characterize source types or Principal Component Analysis (PCA) that uses variance of the data matrix to identify the possible sources (Begum et al., 2007). However, these techniques have their own sets of disadvantages, for instance, CMB cannot identify sources with emissions of similar chemical composition while PCA has limitations when dealing with heteroskedastic data, which is commonly observed in atmospheric measurements (Hopke & Jaffe, 2020). On the other hand, empirical calculations based on direct observations provide a more realistic overview, and can thus help frame policies to combat the high particulate load in Delhi.

Hence, the present study was undertaken as project work to determine EC-OC concentrations and their probable sources over the Southwest region of Delhi during the festive week of Dussehra, 2017. Our primary objective is to improve the current understanding of carbonaceous aerosols in Delhi during the festivities of Dussehra with severe vehicular traffic and burning of firecrackers besides identifying their major sources during the festive week. Based on the results of our study, we have also suggested a remedial policy to deal with air pollution in Delhi.

27.2 Methodology

27.2.1 Site Description

Sampling was carried out on the premises of Jawaharlal Nehru University (JNU) which is located in the Southwest part of New Delhi and the North most of the Aravali Hills (Fig. 27.1). The city shares borders with nearby states viz. Haryana in the East and Uttar Pradesh across the river Yamuna. The geography of Delhi has divided into three segments namely, the Yamuna flood plains, the Ridge, and the alluvial plains (Centre for Science and Environment). Most of the city area falls in the fertile land of plains. Delhi region has a semi-arid to semi-humid climate. It experiences four seasons with extreme temperatures in summers and winters. Wind directions vary with seasons predominantly being S and SW during summers, which bring particles via dust storms and N and NE during winters which are coupled with the phenomenon of temperature inversion. The major sources responsible for air pollution in Delhi include vehicular emissions, industries, and thermal power plants which operate all around the year as well as processes such as re-suspension of dust, construction activities, and open waste burning (Sharma & Dikshit, 2016). However, there are also some season-specific sources, namely dust storms in summer and biomass burning in winter.

The campus of JNU is a residential-cum-institutional area with a spread of 4 km^2 and a good vegetation cover. Sampling was carried out at a height of about 10 m from ground in the School of Environmental Sciences (SES). Traffic flow inside the campus is low, however, neighboring roads with moderate-heavy traffic are present. The nearest industrial areas (Mayapuri and Naraina) are located ~8–9 km NW of the site and house industrial units related to metallic-product manufacturing, scrap metal processing, plastic, and rubber pigments production, etc. We can consider JNU as a receptor site for atmospherically transported pollutants, as it lies in the downwind direction of the city for the major part of the year.

Therefore, we have segregated the probable sources of the carbonaceous aerosols in this study into three categories, namely, local (from local vehicles and industries), domestic (from commercial entities and food stalls in Delhi), and trans-boundary pollution (transported from nearby states) for ease of further segregation into controllable and non-controllable categories.

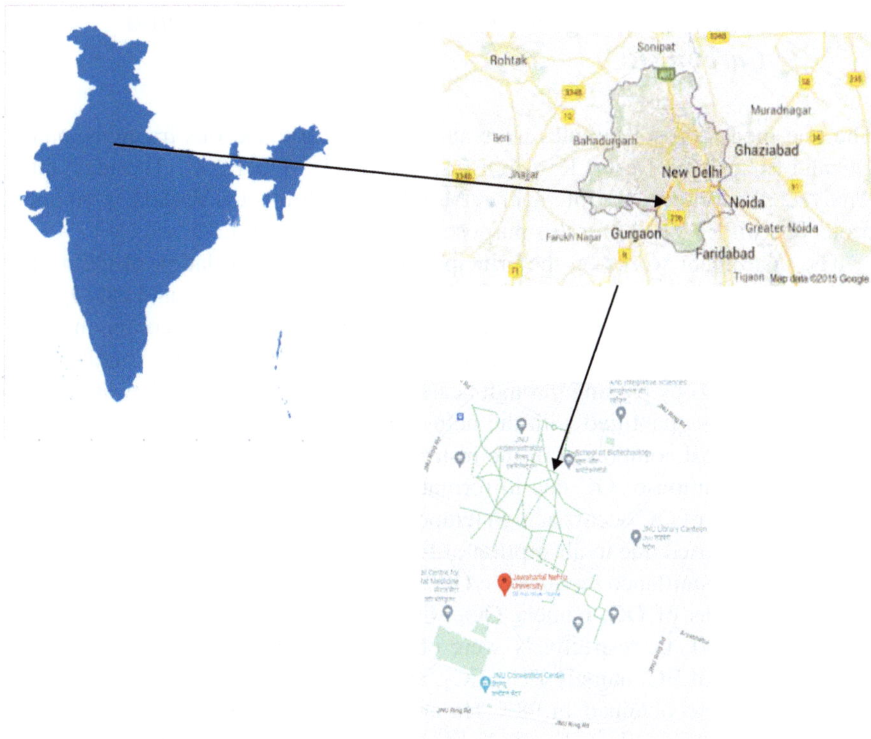

Fig. 27.1 Maps showing the location of the sampling site

27.2.2 Sampling Protocol

Samples of fine particulate matter (<$PM_{2.5}$) were collected on 47 mm Pallflex Tissue-quartz 2500QATUP microfiber filters with the help of Promivac air pump PHS SP01. The quartz filter was mounted on the polycarbonate filter holder which was placed at a height of about 10 m from the ground and was left exposed to air for a time interval of 8 h. The collection was carried out during the daytime from 9:00 am to 5:00 pm for a week during the festive season of Maha Navami, Dussehra, and Muharram followed by Gandhi Jayanti (29th September 2017–03rd October 2017). The collected samples were sealed using paraffin and stored at 4 °C for preservation until analysis.

27.2.3 *Analysis of Organic Carbon (OC) and Elemental Carbon (EC)*

The fine-mode aerosol samples were analyzed for carbonaceous fractions, namely OC and EC using a Desert Research Institute (DRI) Model 2001 Thermal/Optical Analyzer. We followed the Interagency Monitoring of Protected Visual Environments protocol (IMPROVE-A) for this purpose (Chow et al., 2007).

The instrument works on the principle of preferential oxidation of OC in pure Helium (He) atmosphere followed by oxidation of EC in 98% He and 2% O_2 atmosphere, in a stepwise fashion at different temperatures. All carbon compounds from the sample punch area of 0.5 cm^2 of quartz microfiber filter are first volatilized and converted to CO_2 by passing through heated MnO_2. Subsequently, they are reduced to CH_4 which is quantified with the help of a flame ionization detector. The main role of the optical component of the instrument is to prevent overestimation of EC and underestimation of OC due to formation of Pyrolyzed Organic Carbon (OP). When pyrolysis of OC occurs at high temperatures, reflectance or remittance of laser light gets decreased due to absorption of light by a filter. Continuous monitoring of reflectance/transmittance rectifies the OP from EC peak and assigns it to OC.

Four fractions of OC, namely OC_1, OC_2, OC_3, and OC_4 at 120 °C, 240 °C, 480 °C, and 580 °C, respectively were obtained in pure He atmosphere. Further, three fractions of EC, namely EC_1, EC_2, and EC_3 at 580 °C, 740 °C, and 840 °C, respectively were obtained in 98% He and 2% O_2 atmosphere. The OP fraction was obtained when reflected laser light reached its initial value after injection of oxygen into the analysis environment. Thus, the following formula was used for the determination of OC, EC, and TC concentrations:

$$\begin{aligned} \text{OC} &= \text{OC}_1 + \text{OC}_2 + \text{OC}_3 + \text{OC}_4 + \text{OP} \\ \text{EC} &= \text{EC}_1 + \text{EC}_2 + \text{EC}_3 - \text{OP} \\ \text{TotalCarbon(TC)} &= \text{OC} + \text{EC} \end{aligned}$$

27.2.4 *Analysis of Water-Soluble Organic Carbon (WSOC)*

The fine-mode aerosol samples were analyzed for WSOC with the help of Shimadzu Total Organic Carbon (TOC) Analyzer (Model 01,517; serial number H544354).

Sample Preparation

For the analysis of WSOC, sample solutions were prepared in MQ water. Two 0.5-cm^2 punches were taken from the microfiber filters deposited with fine particulate matter and added to a tube containing 30 ml MQ water. Subsequently, the tubes were sonicated for 30 min in an ice bath to prevent any thermal decomposition. The sonicated sample solutions were then filtered using a 0.2 μ syringe filter and were analyzed without any delay.

The extract was again dissolved and analyzed to assess the extraction efficiency, which turned out to be nearly 98% for TOC estimation.

27.2.4.1 Analysis

The instrument functions on the principle of catalytic oxidation coupled with non-dispersive infrared detection (NDIR). The sample was delivered to the combustion furnace and supplied with purified air. There, it was put to combustion at 680 °C with a Platinum catalyst. Consequently, the sample was decomposed and converted to carbon dioxide. The carbon dioxide generated was detected by NDIR after cooling and dehumidification. The concentration of TC (total carbon) in the sample was calculated through a calibration curve formula obtained from routine calibration of the instrument with standard solutions of sodium carbonate/sodium bicarbonate every 6 months. Furthermore, by subjecting the oxidized sample to the sparging process, the IC (inorganic carbon) in the sample was also converted to carbon dioxide, which was detected by NDIR. The WSOC concentration was finally calculated by subtracting the IC concentration from the obtained TC concentration (Fig. 27.2).

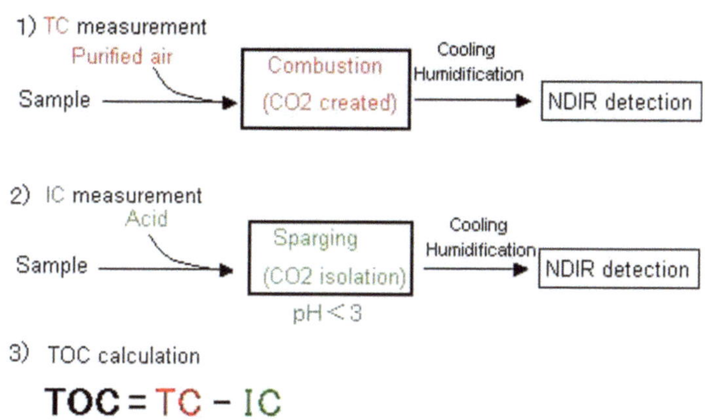

Fig. 27.2 WSOC Measurement (TC-IC) using 680 °C combustion catalytic oxidation and NDIR method (*Source* Shimadzu)

27.2.5 Quality Control in Sampling and Chemical Analysis

27.2.5.1 Sampling

The choice of sampling site plays a crucial role in representativeness and reliability of data. Hence, the sampling site for this study was selected after careful consideration. There was no obstacle or any emission source in the direct vicinity of the sampling site that could have introduced a bias in the observations. Moreover, the JNU campus lies in the downwind direction of the city and hence represents the typical air pollution scenario of Delhi. Further, we analyzed two filter blanks to reduce any background contamination. The blanks were sealed with parafilm and stored at 4 °C until analysis just as the samples. Finally, the average value of the blanks was used to correct data obtained after analyses of the filters loaded with particulate matter.

27.2.5.2 Chemical Analysis

Each sample was analyzed thrice to maintain precision and account for uniformity in PM deposition on filters. For this purpose, we used three 0.5-cm^2 punches, analyzed them using EC/OC analyzer and TOC analyzer, and took the average of the three values to obtain one reading. The mean value and standard deviation for all the samples and carbonaceous fractions have been presented in the Results section.

Further, to ensure accurate measurement of OC, EC, and WSOC concentrations, we calibrated the respective instruments with their certified reference materials. The EC/OC analyzer was calibrated with an injection of CH4 gas into the sample stream at the end and beginning of each work day. While the TOC analyzer was calibrated with standard solutions of Na_2CO_3 to obtain a calibration curve against which the samples were quantified.

27.3 Results and Discussion

27.3.1 Concentration of Carbonaceous Aerosols

The mean values of fine mode OC and EC during the 8-h daytime sampling were found to be 15 ± 2.12 μg/m^3 and 10 ± 1.02 μg/m^3, respectively. The samples were collected during the end of monsoon season in the festive week of Dussehra. The typical annual average concentrations of OC and EC in Delhi have been observed as 21.6 μg/m^3 and 10.7 μg/m^3 (Singh et al., 2016) and 23 μg/m^3 and 11 μg/m^3 (Srivastava et al, 2014). The observed lower than annual average values can be explained as sampling was performed at the end of the monsoon season, in 2017 (Sonwani & Kulshrestha, 2019).

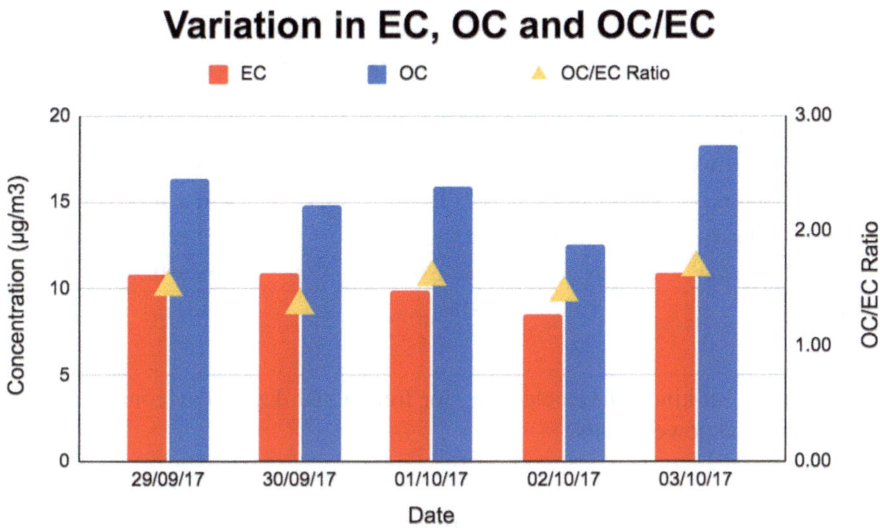

Fig. 27.3 Day-to-day variability in concentrations of OC and EC and OC/EC ratio during the sampling period

Figure 27.3 shows the day-to-day variation in concentrations of OC and EC as well as OC/EC ratio during the sampling period. It can be observed that a remarkable dip was exhibited by both the carbonaceous fractions on 2 October 2017, Gandhi Jayanti.

27.3.1.1 Variation in OC Concentration During the Sampling Period

During the entire study period, the maximum OC concentration was found on a regular working day (18.32 $\mu g/m^3$) while the lowest concentration was observed on a public holiday, i.e., Gandhi Jayanti (12.57 $\mu g/m^3$) (Table 27.1). If we compare OC concentration over festive days only, then the maximum concentration was observed on Maha Navmi (16.39 $\mu g/m^3$) followed by the days of Muharram (15.93 $\mu g/m^3$) and Dussehra (14.81 $\mu g/m^3$). These variations can be attributed to either changes in local emission sources or meteorological parameters or both, which have been discussed in the following sections.

OC is mainly released from biomass and fossil fuel burning. During the festive season, paper pulp industries and other agro-industries do not function leading to a decrease in OC emissions. However, additional sources like firecrackers, oil lamps, burning of cow dung cakes, holy rituals in temples, and community-wide functions exist along with vehicular traffic and thermal power plants. On the other hand, on the day of public holidays only thermal power plants and vehicles, mainly for freight, are expected to operate thereby giving the minimum OC concentration. Hence, maximum OC concentration was observed on a regular working day owing to the operation of

Table 27.1 The measured values of OC and EC during the sampling period

Date	Day	Occasion	OC ($\mu g/m^3$) (Mean ± Std Dev)	EC ($\mu g/m^3$) (Mean ± Std Dev)
29/09/17	Friday	Maha Navmi	16.39 ± 0.44	10.83 ± 0.74
30/09/17	Saturday	Dussehra	14.81 ± 0.40	10.93 ± 0.33
01/10/17	Sunday	Muharram	15.93 ± 0.12	9.94 ± 0.28
02/10/17	Monday	Gandhi Jayanti	2.57 ± 0.26	8.57 ± 0.42
03/10/17	Tuesday	No event	18.32 ± 0.27	10.89 ± 0.35
Range			**12–18**	**8–11**

all industries, all kinds of vehicles, i.e., for freight and day-to-day commutation as well as thermal power plants.

27.3.1.2 Variation in EC Concentration During the Sampling Period

The minimum EC concentration was observed on Gandhi Jayanti, a public holiday (8.57 $\mu g/m^3$) while the maximum concentration was observed on Dussehra (10.93 $\mu g/m^3$). This can be explained based on vehicular traffic. It was expected that almost 90% of vehicles used for day-to-day commutation did not operate on the public holiday owing to an official off as well as a holiday after a long tiring festive weekend. Hence, the observed EC concentration on Gandhi Jayanti could be attributed mainly to freight vehicles and thermal power plants.

Furthermore, no large-scale variations were observed in EC concentration over the festive and regular working days, unlike OC. This can be explained based on the fact that EC is mainly released from the vehicular exhaust and thermal power plants, and hence not affected by the festivities. Moreover, we expect that a similar number of vehicles operated on both festive and regular working days, though for different purposes.

27.3.2 Role of Meteorological Conditions

It can be seen from Table 27.2 that the minimum OC and EC concentrations coincide with the maximum average wind speed, thus highlighting the effect of meteorology along with the change in local emission sources. However, the average wind speed for festive days and the regular working day were comparable indicating emission source variability as the major reason for the observed disparity in OC and EC concentrations.

Table 27.3.2 The meteorological conditions during the sampling period. (*Source* Wunderground)

Date	Day	Average temperature (°C)	Average dew point (°C)	Average humidity (%)	Average wind (km/h)	Total precipitation (mm)
29/09/17	Friday	30	22	60	3	0
30/09/17	Saturday	29	21	60	3	0
01/10/17	Sunday	29	20	63	2	0
02/10/17	Monday	29	18	50	5	0
03/10/17	Tuesday	29	20	56	3	0

Table 27.3 OC/EC ratio during the sampling period

Date	Day	Occasion	OC/EC
29/09/17	Friday	Maha Navmi	1.51
30/09/17	Saturday	Dussehra	1.35
01/10/17	Sunday	Muharram	1.6
02/10/17	Monday	Gandhi Jayanti	1.46
03/10/17	Tuesday	No event	1.68
Range			1.3–1.6

27.3.3 OC/EC Ratio: A Diagnostic Tool

OC/EC ratio serves as an important diagnostic tool to trace the major sources of carbonaceous aerosols as well as has significant importance in determining the radiative properties (Carrico et al. 2003; Ram & Sarin, 2011). According to Chow et al (1996) and Safai et al (2014), a ratio greater than 2.0 indicates Secondary Organic Aerosol (SOA) formation. Schauer et al. (2002) suggested that OC/EC varies between 1–4.2 for diesel/gasoline vehicular exhaust and between 16.8–40.0 for wood combustion. Saarikoski et al. (2008) have also reported comparable results with OC/EC values of 6.6 for biomass burning and 0.71 for vehicular emissions.

The trend in OC/EC ratio for the present study is presented in Fig. 27.3. It was found to vary between 1.35 and 1.60 with a mean value of 1.52 (Table 27.3). These values indicate fossil fuel combustion as the major source for both OC and EC.

27.3.4 Concentration of Secondary Organic Carbon (SOC)

EC only has a primary origin, i.e., emitted directly from combustion of carbonaceous matter while OC may be emitted directly in the particulate phase or formed from gas-to-particle conversion processes in the atmosphere (Jones & Harrison, 2005; Saylor et al., 2006). EC is dominantly emitted from vehicular exhaust in urban areas and hence serves as a good tracer for fossil fuel combustion. Biomass burning also contributes to EC but at much lower rates and also such contributions are event-based. On the other hand, major sources of OC include fossil fuel combustion, biomass burning, and emissions from plants and microorganisms.

To quantify the OC formed from secondary processes, we used an indirect method of evaluation wherein the sample with a minimum value of OC/EC ratio is considered to exclusively contain primary carbonaceous aerosols from fossil fuel combustion (Turpin & Huntzicker, 1995; Castro-Tirado et al., 1999; Viidanoja et al., 2002; Pio et al., 2011). Using this approach, the concentration of secondary organic carbon (SOC) was calculated with the following formula:

$$\mathrm{SOC} = \mathrm{OCTotal} - (\mathrm{OC/EC})_{\min} \times \mathrm{EC}$$

Concentrations of primary organic carbon (POC) and SOC for all the samples are represented in Table 27.4. SOC concentration varied between 0.05 and 3.6 $\mu g/m^3$ with a mean value of 1.79 $\mu g/m^3$.

According to Castro-Tirado et al. (1999), estimation of SOC is correct if the sample with a minimum OC/EC ratio has a negligible concentration of SOC. This, in turn, verifies our estimation of SOC as our sample with minimum OC/EC ratio does have a negligible SOC concentration (0.33% only).

Consequently, it seems to be a reasonable hypothesis to consider low OC/EC ratios along with low SOC formation in an urban setup like Delhi, as an indication that vehicular exhaust is the major source for both OC and EC. Correlation and regression analyses were further performed to explore the relation between OC and EC and their sources, as described in one of the following sections.

Table 27.4 Concentrations of POC and SOC during the sampling period

Date	Day	Occasion	OC (ug/ m^3)	OC/EC	POC ($\mu g/m^3$)	SOC ($\mu g/m^3$)	%SOC of OC
29/09/17	Friday	Maha Navmi	16.39	1.51	14.62	1.77	10.79%
30/09/17	Saturday	Dussehra	14.81	1.35	14.76	0.05	0.33%
01/10/17	Sunday	Muharram	15.93	1.6	13.42	2.51	15.75%
02/10/17	Monday	Gandhi Jayanti	12.57	1.46	11.56	1	7.95%
03/10/17	Tuesday	No event	18.32	1.68	14.7	3.61	19.70%
Range			12–18	1.3–1.6	11–14	0.05–3.6	0.33–19.7%

Table 27.5 Concentration of WSOC during the sampling period

Date	Day	Occasion	WSOC ($\mu g/m^3$) (Mean ± Std Dev)	% WSOC of OC
29/09/17	Friday	Maha Navmi	15.85 ± 0.22	96.7
30/09/17	Saturday	Dussehra	12.01 ± 0.23	81.1
01/10/17	Sunday	Muharram	9.35 ± 0.22	58.7
02/10/17	Monday	Gandhi Jayanti	10.59 ± 0.21	84.2
03/10/17	Tuesday	No event	14.11 ± 0.24	77
Range			12–16	59–97

27.3.5 *Concentration of Water-Soluble Organic Carbon (WSOC)*

WSOC refers to the water-soluble fraction of organic carbon which is usually already oxygenated upon release or oxygenated later via secondary processes (Yang et al., 2005).

We measured the concentration of WSOC in our samples by preparing solutions in MQ water as discussed in Sect. 27.2.4. The concentration of WSOC was found to vary between 12–16 $\mu g/m^3$, i.e., about 59–97% of OC concentration in the samples as shown in Table 27.5. This indicates that the OC found in our samples was predominantly water-soluble. Moreover, we can infer that most of the WSOC were directly released into the atmosphere, i.e., of primary origin owing to the low levels of SOC formation (Table 27.4).

27.3.6 *Relationship Between OC and EC*

The correlation ($r = 0.78$) and regression ($R^2 = 0.597$) coefficients indicate that OC and EC are moderately correlated (Fig. 27.4). This suggests that although vehicular exhaust was the major source for both OC and EC as discussed in the previous section, some additional sources also operated throughout the sampling period for the two carbonaceous fractions.

We suggest that the major additional source for EC was coal combustion in thermal power plants which operate around the year irrespective of the holidays. While the additional source for OC on non-festive days was emissions from agro-based industries. The absence of such industrial emissions on festivals was compensated by activities like burning of firecrackers, oil lamps, cow dung cakes, etc., thereby leading to only a small decrease in OC emissions on the festive days than the normal working day.

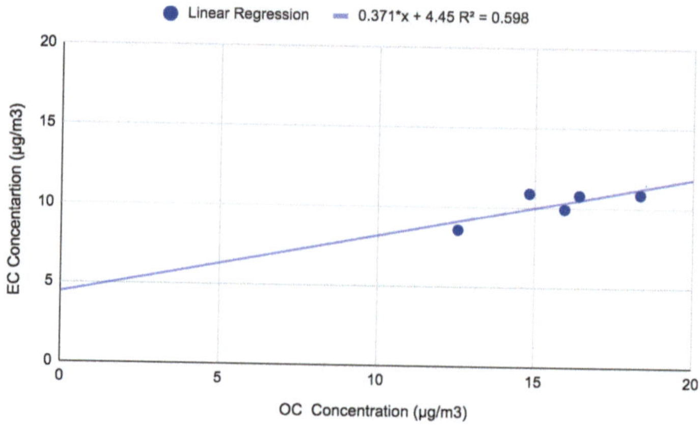

Fig. 27.4 Linear regression fit for EC-OC measured values during the sampling period (P value of 0.12522 with an alpha of 0.05)

27.3.7 Variation in OC and EC Fractions During Different days of the Study Period

To study the variation in OC and EC during the sampling period, we compared their concentrations on weekday versus weekend, festive day versus non-festive day, and working day versus non-working day bases (Fig. 27.5), which led us to the following observations.

a. **Weekday versus weekend:** Percentage changes in the concentrations of OC, EC, and TC on weekdays and weekends were found to be 2.47%, −3.37%, and 0.19%, respectively. This revealed that the concentration of carbonaceous aerosols did not get considerably affected by a day being a weekday or a weekend.
b. **Festive versus non-festive day:** Percentage changes in the concentrations of OC, EC, and TC on festive and non-festive days were found to be 1.72%, 7.86%, and 4.19%, respectively. This revealed that though the concentration of EC was reduced considerably on a non-festive day, concentrations of OC and TC were still not greatly affected.
c. **Working versus non-working day:** Percentage changes in the concentrations of OC, EC, and TC on working and non-working days were found to be 16.83%, 9.67%, and 14.04%, respectively. This revealed that the concentration of both fractions of carbonaceous aerosols was considerably decreased on a non-working day than on a working day.

In the above classification, the weekdays refer to Monday-Friday, and working days are weekdays which were not holidays.

These observations reinforce our previous inference that vehicular traffic is the major source of carbonaceous aerosols in Delhi's atmosphere as a significant decrease was observed in concentrations of both OC and EC on a non-working day rather

Fig. 27.5 Comparison of OC, EC, and TC average concentrations on **a** Weekday versus weekend **b** Festive versus non-festive **c** Working versus non-working bases

than on a working day. This can be attributed to the non-plying of vehicles meant for day-to-day office commutation on non-working days.

27.3.8 Aerosol Optical Depth (AOD) Analysis

To understand the effects of fine-mode carbonaceous aerosols on radiative forcing and hence climate, we compared the AOD values with concentrations of OC and EC during the sampling period. AOD provides a measure of the amount of sunlight that aerosols scatter or absorb, thereby revealing the concentration of optically active aerosols in the vertical column of the atmosphere (Van Donkelaar et al., 2006).

AOD values for this study were obtained from the MODIS instrument aboard the Terra satellite from Giovanni-MODIS Terra website. We used the Combined Dark Target and Deep Blue AOD product data from September 29 to October 03, 2017, for this purpose over regions 77E, 28N, 79E, and 29N (Delhi). Figure 27.6 shows the day-to-day variations in the AOD values taken at a wavelength of 550 nm during the sampling period. The observed trend in AOD values was found to be similar to the observed trends in OC and EC, which suggests the role of OC and EC in solar attenuation. However, the amplitude of variation in AOD values was more pronounced than that of variation in carbonaceous aerosol concentrations, thereby indicating the involvement of other optically active species. These species may include roadside re- suspended dust as well as transboundary emissions further highlighting the complexity of Delhi's air composition, which is not only determined by the local emissions but also by LRT and transboundary pollution sources (Kulshrestha & Mishra, 2020).

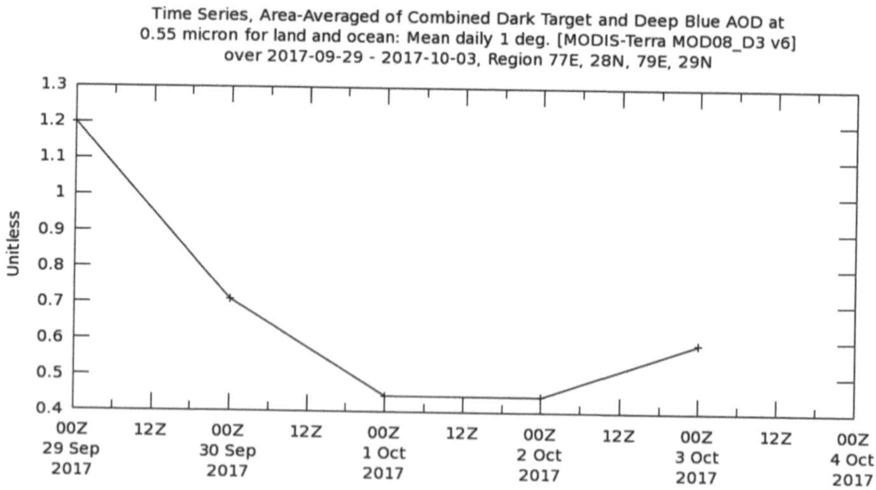

Fig. 27.6 AOD values as obtained from MODIS for the sampling period

27.3.9 Quantification of Controllable and Uncontrollable Carbonaceous Emissions in Delhi

We selected a peculiar combination of days for our study, which included festivals, public holidays, weekdays as well as weekends so that we can holistically assess variations in carbonaceous aerosol concentration, and hence identify their probable sources to help frame a policy for controlling Delhi's air pollution.

For this purpose, we divided the major sources of air pollution in Delhi into the following 3 categories:

A. Local: This includes emissions from local vehicles and industrial activities.
B. Domestic: This includes emissions from food stalls, restaurants, local shops, and other domestic commercial entities run by the inhabitants of Delhi.
C. Transboundary Pollution (TBP) and Wind Blown: This includes emissions from nearby states that cross the border and travel via air to enter Delhi's atmosphere as well as the local particulate matter re-suspended by the wind.

Subsequently, we analyzed OC concentrations on different sampling days to calculate the percentage contribution of these source categories. We selected OC for this purpose as it is contributed by all the above-mentioned sources, unlike EC which is majorly emitted from fossil fuel combustion only.

The analysis revealed that about 31.4% of fine-mode carbonaceous aerosols in Delhi are emitted from local transport and industrial activities, 18.3% from domestic activities while the rest 50.3% is contributed by transboundary sources and wind-blown resuspensions (Fig. 27.7). From a policymaking perspective, we believe that while it is possible to control local emission sources (31.4%) to some extent, it is difficult to control domestic activities, local windblown resuspensions and trans-boundary emissions (68.6%). Hence, we have suggested a remedial policy based on these results to control Delhi's air pollution in the following section to solve the pressing issue of degrading air quality in the capital.

27.4 Proposed Vehicular Policy for Significant Reductions in OC and EC Concentrations in Delhi-NCR

Low OC/EC ratio as well as low SOC formation, as discussed in previous sections indicate that vehicular emissions are the major source for both OC and EC in Delhi. Moreover, we observed a significant decrease of about 16.83% and 9.67% in concentrations of OC and EC, respectively on a non-working day than a working day, which can be attributed to non-plying of vehicles meant for day-to-day office commutation on the non-working day. This, in turn, suggests that an observable reduction in emissions can be achieved only if nearly all the vehicles (and not only half) shuttled for business or workplace commutation do not ply for an entire day.

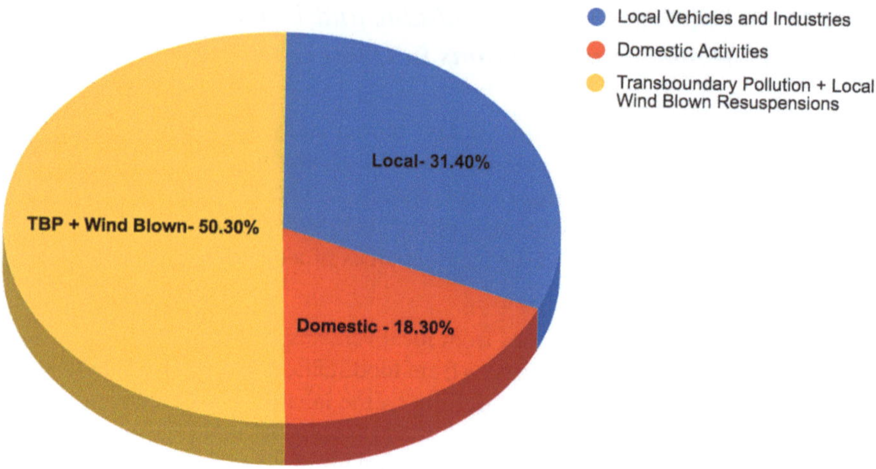

Fig. 27.7 Probable contribution of air pollution sources in Delhi (above) and their categorization from a policymaking perspective (below)

Hence, we suggest a "Work from Home" policy wherein personal vehicles shall be allowed to ply only once a week instead of an odd–even/50–50 scheme, wherein vehicles with registration numbers ending with an odd digit and those with an even digit are allowed to operate on the odd and even dates, respectively. Thus, on any given day, when the odd–even scheme is followed, half of the vehicles will be on the road whereas, when work from home policy is followed, only one-sixth of the vehicles will be on the road.

As a result, the work-from-home policy will produce an additional 33.3% reduction in emissions on any given day in comparison to the 50–50 scheme. The following table summarizes the expected differences in the efficiencies of both policies (Table 27.6).

Thus, the suggested policy should be followed to bring down the levels of fine-mode carbonaceous aerosols in Delhi, which in turn will contribute to improved air quality. Additionally, it is expected to minimize the health risks associated with traveling and driving stress, thereby decreasing psycho-nervous disorders and increasing productivity. Control of road accidents and noise pollution levels is also anticipated.

Table 27.6 Emission reductions expected from work-from-home and odd–even policies

Percentage reduction/day	Work from home policy (%)	Odd–even policy (%)
OC concentration	14.02	8.41
EC concentration	8.05	4.83
TC concentration	11.69	7.02

27.5 Comparison of the Present Micro-level Study with Meso-level and Macro-level Approaches

In this section, we compare the results of our micro-level study with those of meso-level and macro-level approaches, namely the GRAP by EPCA in Delhi-NCR region, and the nationwide lockdown due to COVID-19 pandemic.

27.5.1 *Meso-level Study-GRAP*

GRAP was formulated and approved by the Supreme Court in 2016 to combat the severe level of air pollution in Delhi (MoEF Notification, 2017). It is an incremental plan wherein specific sets of actions are taken in response to specific levels of pollution, which have been summarized in Table 27.7.

Although the implementation of GRAP has successfully solved the issue of accountability by getting several agencies on board, the improvement in air quality has been minor. As per a study by Delhi Integrated Multi-modal Transit System, vehicular traffic was reduced by only 2% during the odd–even scheme implemented under GRAP, thereby re-emphasizing the importance of 'Work from Home' policy to get significant results as suggested by the present study.

27.5.2 *Macro-level Study-COVID-19 Lockdown*

Because of COVID-19 pandemic, a nationwide lockdown was implemented in India from 24 March 2020 to 14 April 2020, which was later extended till 3 May 2020. This led to a halt on most of the sources of air pollutants, notably vehicular traffic, and non-essential industries. As a result, air quality improved significantly during this period. According to Mahato et al. (2020), the average concentrations of $PM_{2.5}$ and PM10 during the lockdown decreased by about 50% from the pre-lockdown levels. The fall in PM levels reached up to 60% in areas with usually high traffic and industrial activities indicating vehicular traffic and industrial emissions as major sources of air pollutants in Delhi.

Results from our micro-level study also suggest a decrease of about 22% in the concentration of fine-mode carbonaceous aerosols on a nationwide public holiday when compared to the sample average. It should be noted that the holiday occurred after a combination of busy festive and working days, which suggests minimal vehicular traffic. This was also confirmed by the lowest EC concentration on this day during the study period. Hence both the COVID-19 lockdown and our study suggest vehicular traffic as one of the major sources of air pollutants in Delhi-NCR.

Table 27.7 Graded Response Action Plan

Category	Action plan
Moderate to poor ($PM_{2.5}$ 61–120 μg/m^3 or PM_{10} 101–350 μg/m^3)	Heavy fines for burning garbage Enforce pollution control measures on industries and brick kilns Mechanized sweeping of roads Ensure watering of fly ash ponds Ban on firecrackers
Very Poor ($PM_{2.5}$ 121–250 μg/m^3 or PM_{10} 351–430 μg/m^3)	Discontinue the use of diesel generator sets Increase the parking fee by 3–4 times Improve bus and Metro services Direct housing associations to provide electric heaters to security staff to avoid bonfires Issue alerts to people with cardiac and respiratory conditions to control outdoor movement
Severe ($PM_{2.5}$ above 250 μg/m^3 or PM_{10} above 430 μg/m^3)	Shut brick kilns, hot mix plants, and stone crushers Shut down Badarpur power plant and maximize electricity generation from natural gas-based power plants Intensify public transport Mechanized cleaning of roads and sprinkling of water
Emergency ($PM_{2.5}$ above 300 μg/m^3 or PM_{10} above 500 μg/m^3 for 48 + hours)	Stop entry of trucks to Delhi except the ones carrying essential commodities Stop construction activities Implement the odd/even scheme with no exemptions Formation of the Task Force to decide on any additional steps

Source Central Pollution Control Board, Ministry of Environment, Forest & Climate Change

27.6 Advantages of a Holistic Approach

The results and inferences are drawn from our study coincide with those drawn from meso-level implementation of GRAP and macro-level COVID-19 lockdown, which indicates that a large-scale study may not always be required, and a micro-level study with a holistic approach can be as helpful. In the present study, we not only measured the concentrations of OC and EC in the samples but also quantified the fractions of secondary and water-soluble OC, assessed the OC/EC ratio, compared variations in carbonaceous aerosol concentrations on different combinations of days as well as with AOD during the study period. Additionally, we combined the knowledge from pre-existing literature and that gained from the experiments conducted in this study to understand the role of other pollutants and sources including long-range transport and transboundary pollution. However, additional studies with a larger dataset that involve analysis of chemical source markers are required to confirm our results.

27.7 Conclusion

The study suggests a "Work from Home" policy instead of the odd–even scheme to considerably reduce air pollution in Delhi. The suggested policy is expected to be 33.3% more efficient in reducing carbonaceous emissions than the 50–50 scheme. Finally, we compared the results of our study with those of the GRAP and nationwide lockdown due to COVID-19 pandemic. We found our results to match with the meso- and macro-level approaches, which highlights the importance of a holistic approach.

Additionally, we compared the concentrations of OC and EC on different combinations of days, namely weekday versus weekend, festive versus non-festive, and working versus non-working to assess emission source variability. The comparative analysis revealed an appreciable decrease in carbonaceous fractions only on a non-working day than a working day with the percentage reductions in OC and EC concentrations being 16.83% and 9.67%, respectively. These observations indicate that vehicles meant for commutation to the workplace contribute appreciably to carbonaceous emissions in Delhi. Furthermore, we analyzed variation in OC during the sampling period to investigate the probable sources. We found that 31.5% of the load was contributed by vehicles and industries, 18.3% from domestic activities, and 50.3% from transboundary and windblown sources.

The OC/EC ratio throughout the study varied between 1.35 and 1.60 suggesting vehicular emissions as the major source for both OC and EC. However, OC and EC were found to be only moderately correlated ($r = 0.78$) indicating the presence of some additional sources. We suggest that while thermal power plants contributed to the additional EC, festivities/agro-based industries contributed to the additional OC. We also calculated the fractions of SOC and WSOC, which revealed that OC present in Delhi's atmosphere is predominantly of primary origin and water-soluble.

Acknowledgements We sincerely thank Jawaharlal Nehru University, New Delhi for providing financial support to conduct this research work. We also gratefully acknowledge the Central Instrumentation Facility (CIF), SES, and Advanced Instrumentation Research Facility (AIRF), JNU for providing the analytical assistance. This work was carried out as part of DRSNet-India network.

References

Begum, B. A., Biswas, S. K., & Hopke, P. K. (2007). Source apportionment of air particulate matter by chemical mass balance (CMB) and comparison with positive matrix factorization (PMF) model. *Aerosol and Air Quality Research, 7*(4), 446–468.

Bond, T. C., & Bergstrom, R. W. (2006). Light absorption by carbonaceous particles: An investigative review. *Aerosol Science and Technology, 40*(1), 27–67.

Carrico, C. M., Bergin, M. H., Xu, J., Baumann, K., & Maring, H. (2003). Urban aerosol radiative properties: Measurements during the 1999 Atlanta Supersite Experiment. *Journal of Geophysical Research: Atmospheres*, 108(D7).

Castro-Tirado, A. J., Soldán, J., Bernas, M., Páta, P., Rezek, T., Hudec, R., Peña, A., et al. (1999). The burst observer and optical transient exploring system (BOOTES). *Astronomy and Astrophysics Supplement Series, 138*(3), 583–585.

Centre for Science and Environment http://www.rainwaterharvesting.org/index_files/geology.htm

Chow, J. C., Watson, J. G., Chen, L. W. A., Chang, M. O., Robinson, N. F., Trimble, D., & Kohl, S. (2007). The IMPROVE_A temperature protocol for thermal/optical carbon analysis: Maintaining consistency with a long-term database. *Journal of the Air & Waste Management Association, 57*(9), 1014–1023.

Chow, J. C., Watson, J. G., Lowenthal, D. H., & Countess, R. J. (1996). Sources and chemistry of PM10 aerosol in Santa Barbara County, CA. *Atmospheric Environment, 30*(9), 1489–1499.

Chung, C. E., Ramanathan, V., & Decremer, D. (2012). Observationally constrained estimates of carbonaceous aerosol radiative forcing. *Proceedings of the National Academy of Sciences, 109*(29), 11624–11629.

Hopke, P. K., & Jaffe, D. A. (2020). Letter to the editor: Ending the use of obsolete data analysis methods. *Aerosol and Air Quality Research, 20*(4), 688–689.

Institute for Health Metrics and Evaluation. (2017). http://www.healthdata.org/india

IQAir. (2019). World air quality report region and city $PM_{2.5}$ ranking. https://www.iqair.com/world-most-pollutedcities/world-air-quality-report-2019-en.pdf and https://www.iqair.com/world-most-polluted-cities

Jones, A. M., & Harrison, R. M. (2005). Interpretation of particulate elemental and organic carbon concentrations at rural, urban and kerbside sites. *Atmospheric Environment, 39*(37), 7114–7126.

Kulshrestha, U. (2015). Some facts about recent air pollution problem in Delhi. *Journal of Indian Geophysical Union, 19*(3), 351–352.

Kulshrestha, U. C. (2017). *Air pollution and climate change in South Asia: Issues, impact and initiatives.* Athena Academic, London, UK. ISBN 9781910390344

Kulshrestha, U. C., & Katoch, A. (2022). Coal phase down and phase out: Significance for India. *World Focus, 43*(505), 1–5.

Kulshrestha, U. C., & Mishra, M. (2020). A review on long range transport of air pollution in South Asia. *Vayumandal, 46*(2), 21–30.

Laura, D. V., Chatterton, T., Namdeo, A., Shivanagendra, S. G., Goyal, S., Bell, M., Sengupta, B., et al. (2018). Air pollution in Delhi: A review of past and current policy approaches. *Air Pollution XXVI, 230*, 441.

Mahato, S., Pal, S., & Ghosh, K. G. (2020). Effect of lockdown amid COVID-19 pandemic on air quality of the megacity Delhi, India. *Science of the Total Environment*, 139086.

Ministry of Environment, Forest and Climate Change Notification. (2017). https://cpcb.nic.in/uploads/GRAP_Notification.pdf

Mishra, M., & Kulshrestha, U. (2017). Chemical characteristics and deposition fluxes of dust-carbon mixed coarse aerosols at three sites of Delhi, NCR. *Journal of Atmospheric Chemistry, 74*(4), 399–421.

Mishra, M., & Kulshrestha, U. (2021). Source impact analysis using Char-EC/Soot-EC ratios in the central Indo-Gangetic Plain (IGP) of India. *Aerosol and Air Quality Research, 21*(9). https://doi.org/10.4209/aaqr.200628

Pio, C., Cerqueira, M., Harrison, R. M., Nunes, T., Mirante, F., Alves, C., & Matos, M. (2011). OC/EC ratio observations in Europe: Re-thinking the approach for apportionment between primary and secondary organic carbon. *Atmospheric Environment, 45*(34), 6121–6132.

Ramanathan, V., & Carmichael, G. (2008). Global and regional climate changes due to black carbon. *Nature Geoscience, 1*(4), 221–227.

Ramachandran, S., Rengarajan, R., & Sarin, M. M. (2009). Atmospheric carbonaceous aerosols: Issues, radiative forcing and climate impacts. *Current Science, 97*(1), 18–20.

Ram, K., & Sarin, M. M. (2011). Day–night variability of EC, OC, WSOC and inorganic ions in urban environment of Indo-Gangetic Plain: Implications to secondary aerosol formation. *Atmospheric Environment, 45*(2), 460–468.

Rao, S., Pachauri, S., Dentener, F., Kinney, P., Klimont, Z., Riahi, K., & Schoepp, W. (2013). Better air for better health: Forging synergies in policies for energy access, climate change and air pollution. *Global Environmental Change, 23*(5), 1122–1130.

Saarikoski, S., Timonen, H., Saarnio, K., Aurela, M., Järvi, L., Keronen, P., & Hillamo, R. (2008). Sources of organic carbon in fine particulate matter in northern European urban air. *Atmospheric Chemistry & Physics, 8*(20)

Safai, P. D., Raju, M. P., Rao, P. S. P., & Pandithurai, G. (2014). Characterization of carbonaceous aerosols over the urban tropical location and a new approach to evaluate their climatic importance. *Atmospheric Environment, 92*, 493–500.

Saylor, R. D., Edgerton, E. S., & Hartsell, B. E. (2006). Linear regression techniques for use in the EC tracer method of secondary organic aerosol estimation. *Atmospheric Environment, 40*(39), 7546–7556.

Sawlani, R., Agnihotri, R., Sharma, C., Patra, P. K., Dimri, A. P., Ram, K., & Verma, R. L. (2019). The severe Delhi SMOG of 2016: A case of delayed crop residue burning, coincident firecracker emissions, and atypical meteorology. *Atmospheric Pollution Research, 10*(3), 868–879.

Schauer, J. J. (2003). Evaluation of elemental carbon as a marker for diesel particulate matter. *Journal of Exposure Science & Environmental Epidemiology, 13*(6), 443–453.

Seinfeld, J. H., & Pandis, S. N. (1998). From air pollution to climate change. *Atmospheric Chemistry and Physics, 1326*.

Sharma, M., & Dikshit, O. (2016*). Comprehensive study on air pollution and green house gases (GHGs) in Delhi.* A report submitted to Government of NCT Delhi and DPCC Delhi, pp. 1–334.

Singh, R., Kulshrestha, M. J., Kumar, B., & Chandra, S. (2016). Impact of anthropogenic emissions and open biomass burning on carbonaceous aerosols in urban and rural environments of Indo-Gangetic Plain. *Air Quality, Atmosphere & Health, 9*(7), 809–822.

Sonwani, S., & Kulshrestha, U. C. (2019). PM 10 carbonaceous aerosols and their real-time wet scavenging during monsoon and non-monsoon seasons at Delhi, India. *Journal of Atmospheric Chemistry, 76*(3), 171–200.

Srivastava, A. K., Bisht, D. S., Ram, K., Tiwari, S., & Srivastava, M. K. (2014). Characterization of carbonaceous aerosols over Delhi in Ganga basin: Seasonal variability and possible sources. *Environmental Science and Pollution Research, 21*(14), 8610–8619.

Trompetter, W. J., Grange, S. K., Davy, P. K., & Ancelet, T. (2013). Vertical and temporal variations of black carbon in New Zealand urban areas during winter. *Atmospheric Environment, 75*, 179–187.

Turpin, B. J., & Huntzicker, J. J. (1995). Identification of secondary organic aerosol episodes and quantitation of primary and secondary organic aerosol concentrations during SCAQS. *Atmospheric Environment, 29*(23), 3527–3544.

Van Donkelaar, A., Martin, R. V., & Park, R. J. (2006). Estimating ground-level $PM_{2.5}$ using aerosol optical depth determined from satellite remote sensing. *Journal of Geophysical Research: Atmospheres, 111*(D21).

Viidanoja, J., Sillanpää, M., Laakia, J., Kerminen, V. M., Hillamo, R., Aarnio, P., Koskentalo, T. (2002). Organic and black carbon in $PM_{2.5}$ and PM_{10}: 1 year of data from an urban site in Helsinki, Finland. *Atmospheric Environment, 36*(19), 3183–3193.

Wendling, Z. A., Emerson, J. W., de Sherbinin, A., Esty, D.C., et al. (2020). *Environmental performance index.* New Haven, CT: Yale Center for Environmental Law & Policy (epi.yale.edu).

WHO Fact Sheet Ambient Air Pollution. (2022). https://www.who.int/news-room/fact-sheets/detail/ambient-(outdoor)-air-quality-and-health

Yang, H., Yu, J. Z., Ho, S. S. H., Xu, J., Wu, W. S., Wan, C. H., ... & Wang, L. (2005). The chemical composition of inorganic and carbonaceous materials in PM2. 5 in Nanjing, China. *Atmospheric Environment, 39*(20), 3735–3749.

Index

H. Pathak et al. (eds.), *Climate Change Impacts on Soil-Plant-Atmosphere Continuum*, Advances in Global Change Research 78, https://doi.org/10.1007/978-981-99-7935-6

U

V

W

X

Z